Lecture Notes in Computer Science   10991

Commenced Publication in 1973
Founding and Former Series Editors:
Gerhard Goos, Juris Hartmanis, and Jan van Leeuwen

More information about this series at http://www.springer.com/series/7410

Hovav Shacham · Alexandra Boldyreva (Eds.)

Advances in Cryptology – CRYPTO 2018

38th Annual International Cryptology Conference
Santa Barbara, CA, USA, August 19–23, 2018
Proceedings, Part I

 Springer

Editors
Hovav Shacham
The University of Texas at Austin
Austin, TX
USA

Alexandra Boldyreva
Georgia Institute of Technology
Atlanta, GA
USA

ISSN 0302-9743   ISSN 1611-3349 (electronic)
Lecture Notes in Computer Science
ISBN 978-3-319-96883-4   ISBN 978-3-319-96884-1 (eBook)
https://doi.org/10.1007/978-3-319-96884-1

Library of Congress Control Number: 2018949031

LNCS Sublibrary: SL4 – Security and Cryptology

This Springer imprint is published by the registered company Springer Nature Switzerland AG
The registered company address is: Gewerbestrasse 11, 6330 Cham, Switzerland

Preface

The 38th International Cryptology Conference (Crypto 2018) was held at the University of California, Santa Barbara, California, USA, during August 19–23, 2018. It was sponsored by the International Association for Cryptologic Research (IACR). For 2018, the conference was preceded by three days of workshops on various topics. And, of course, there was the awesome Beach BBQ at Goleta Beach.

Crypto continues to grow, year after year, and Crypto 2018 was no exception. The conference set new records for both submissions and publications, with a whopping 351 papers submitted for consideration. It took a Program Committee of 46 cryptography experts working with 272 external reviewers almost 2.5 months to select the 79 papers which were accepted for the conference. It also took one program chair about 30 minutes to dig up all those stats.

In order to minimize intentional and/or subconscious bias, papers were reviewed in the usual double-blind fashion. Program Committee members were limited to two submissions, and their submissions were scrutinized more closely and held to higher standards. The two program chairs were not allowed to submit papers. Of course, they were fine with that restriction since they were way too busy to actually write any papers.

The Program Committee recognized two papers and their authors for standing out among the rest. "Yes, There Is an Oblivious RAM Lower Bound!", by Kasper Green Larsen and Jesper Buus Nielsen, was voted best paper of the conference. Additionally, "Multi-Theorem Preprocessing NIZKs from Lattices," by Sam Kim and David J. Wu, was voted Best Paper Authored Exclusively By Young Researchers. There was no award for Best Paper Authored Exclusively by Old Researchers.

Crypto 2018 played host for the IACR Distinguished Lecture, delivered by Shafi Goldwasser. Crypto also welcomed Lea Kissner as an invited speaker from Google.

We would like to express our sincere gratitude to all the reviewers for volunteering their time and knowledge in order to select a great program for 2018. Additionally, we are very appreciative of the following individuals and organizations for helping make Crypto 2018 a success:

Tal Rabin - Crypto 2018 General Chair and Workshops Organizer
Elette Boyle - Workshops Chair
Fabrice Benhamouda - Workshops Organizer
Shafi Goldwasser - IACR Distinguished Lecturer
Lea Kissner - Invited Speaker from Google
Shai Halevi - Author of the IACR Web Submission and Review System
Anna Kramer and her colleagues at Springer
Sally Vito and UCSB Conference Services

We would also like to say thank you to our numerous sponsors, everyone who submitted papers, the session chairs, the rump session chair, and the presenters.

Lastly, a big thanks to everyone who attended the conference at UCSB. Without you, we would have had a lot of leftover potato salad at the Beach BBQ.

August 2018   Alexandra Boldyreva
                                                                                      Hovav Shacham

Crypto 2018

The 38th IACR International Cryptology Conference

University of California, Santa Barbara, CA, USA
August 19–23, 2018

Sponsored by the *International Association for Cryptologic Research*

General Chair

Tal Rabin   IBM T.J. Watson Research Center, USA

Program Chairs

Hovav Shacham   University of Texas at Austin, USA
Alexandra Boldyreva   Georgia Institute of Technology, USA

Program Committee

Shweta Agrawal	Indian Institute of Technology, Madras, India
Benny Applebaum	Tel Aviv University, Israel
Foteini Baldimtsi	George Mason University, USA
Gilles Barthe	IMDEA Software Institute, Spain
Fabrice Benhamouda	IBM Research, USA
Alex Biryukov	University of Luxembourg, Luxembourg
Jeremiah Blocki	Purdue University, USA
Anne Broadbent	University of Ottawa, Canada
Chris Brzuska	Aalto University, Finland
Chitchanok Chuengsatiansup	Inria and ENS de Lyon, France
Dana Dachman-Soled	University of Maryland, USA
Léo Ducas	Centrum Wiskunde & Informatica, The Netherlands
Pooya Farshim	CNRS and ENS, France
Dario Fiore	IMDEA Software Institute, Spain
Marc Fischlin	Darmstadt University of Technology, Germany
Georg Fuchsbauer	Inria and ENS, France
Steven D. Galbraith	University of Auckland, New Zealand
Christina Garman	Purdue University, USA
Daniel Genkin	University of Pennsylvania and University of Maryland, USA
Dov Gordon	George Mason University, USA
Viet Tung Hoang	Florida State University, USA

Tetsu Iwata	Nagoya University, Japan
Stanislaw Jarecki	University of California, Irvine, USA
Seny Kamara	Brown University, USA
Markulf Kohlweiss	University of Edinburgh, UK
Farinaz Koushanfar	University of California, San Diego, USA
Xuejia Lai	Shanghai Jiao Tong University, China
Tancrède Lepoint	SRI International, USA
Anna Lysyanskaya	Brown University, USA
Alex J. Malozemoff	Galois, USA
Sarah Meiklejohn	University College London, UK
Daniele Micciancio	University of California, San Diego, USA
María Naya-Plasencia	Inria, France
Kenneth G. Paterson	Royal Holloway, University of London, UK
Ananth Raghunathan	Google, USA
Mike Rosulek	Oregon State University, USA
Ron Rothblum	MIT and Northeastern University, USA
Alessandra Scafuro	North Carolina State University, USA
abhi shelat	Northeastern University, USA
Nigel P. Smart	Katholieke Universiteit Leuven, Belgium
Martijn Stam	University of Bristol, UK
Noah Stephens-Davidowitz	Princeton University, USA
Aishwarya Thiruvengadam	University of California, Santa Barbara, USA
Hoeteck Wee	CNRS and ENS, France
Daniel Wichs	Northeastern University, USA
Mark Zhandry	Princeton University, USA

Additional Reviewers

Aydin Abadi	Balthazar Bauer	Zvika Brakerski
Archita Agarwal	Carsten Baum	Jacqueline Brendel
Divesh Aggarwal	Amos Beimel	David Butler
Shashank Agrawal	Itay Berman	Matteo Campanelli
Adi Akavia	Marc Beunardeau	Brent Carmer
Navid Alamati	Sai Lakshmi Bhavana	Ignacio Cascudo
Martin Albrecht	Simon Blackburn	Wouter Castryck
Miguel Ambrona	Estuardo Alpirez Bock	Andrea Cerulli
Ghous Amjad	Andrej Bogdanov	André Chailloux
Megumi Ando	André Schrottenloher	Nishanth Chandran
Ralph Ankele	Xavier Bonnetain	Panagiotis Chatzigiannis
Gilad Asharov	Charlotte Bonte	Stephen Checkoway
Achiya Bar-On	Carl Bootland	Binyi Chen
Manuel Barbosa	Jonathan Bootle	Michele Ciampi
Paulo Barreto	Christina Boura	Benoit Cogliati
James Bartusek	Florian Bourse	Gil Cohen
Guy Barwell	Elette Boyle	Ran Cohen

Aisling Connolly
Sandro Coretti
Henry Corrigan-Gibbs
Geoffroy Couteau
Shujie Cui
Ting Cui
Joan Daemen
Wei Dai
Yuanxi Dai
Alex Davidson
Jean Paul Degabriele
Akshay Degwekar
Ioannis Demertzis
Itai Dinur
Jack Doerner
Nico Döttling
Benjamin Dowling
Tuyet Thi Anh Duong
Frédéric Dupuis
Betul Durak
Lior Eldar
Karim Eldefrawy
Lucas Enloe
Andre Esser
Antonio Faonio
Prastudy Fauzi
Daniel Feher
Serge Fehr
Nils Fleischhacker
Benjamin Fuller
Tommaso Gagliardoni
Martin Gagné
Adria Gascon
Pierrick Gaudry
Romain Gay
Nicholas Genise
Marilyn George
Ethan Gertler
Vlad Gheorghiu
Esha Ghosh
Brian Goncalves
Junqing Gong
Adam Groce
Johann Großschädl
Paul Grubbs
Jiaxin Guan

Jian Guo
Siyao Guo
Joanne Hall
Ariel Hamlin
Abida Haque
Patrick Harasser
Gottfried Herold
Naofumi Homma
Akinori Hosoyamada
Jialin Huang
Siam Umar Hussain
Chloé Hébant
Yuval Ishai
Ilia Iliashenko
Yuval Ishai
Håkon Jacobsen
Christian Janson
Ashwin Jha
Thomas Johansson
Chethan Kamath
Bhavana Kanukurthi
Marc Kaplan
Pierre Karpman
Sriram Keelveedhi
Dmitry Khovratovich
Franziskus Kiefer
Eike Kiltz
Sam Kim
Elena Kirshanova
Konrad Kohbrok
Lisa Maria Kohl
Ilan Komargodski
Yashvanth Kondi
Venkata Koppula
Lucas Kowalczyk
Hugo Krawczyk
Thijs Laarhoven
Marie-Sarah Lacharite
Virginie Lallemand
Esteban Landerreche
Phi Hung Le
Eysa Lee
Jooyoung Lee
Gaëtan Leurent
Baiyu Li
Benoit Libert

Fuchun Lin
Huijia Lin
Tingting Lin
Feng-Hao Liu
Qipeng Liu
Tianren Liu
Zhiqiang Liu
Alex Lombardi
Sébastien Lord
Steve Lu
Yiyuan Luo
Atul Luykx
Vadim Lyubashevsky
Fermi Ma
Varun Madathil
Mohammad Mahmoody
Mary Maller
Giorgia Azzurra Marson
Daniel P. Martin
Samiha Marwan
Christian Matt
Alexander May
Sogol Mazaheri
Bart Mennink
Carl Alexander Miller
Brice Minaud
Ilya Mironov
Tarik Moataz
Nicky Mouha
Fabrice Mouhartem
Pratyay Mukherjee
Mridul Nandi
Samuel Neves
Anca Nitulescu
Kaisa Nyberg
Adam O'Neill
Maciej Obremski
Olya Ohrimenko
Igor Carboni Oliveira
Claudio Orlandi
Michele Orrù
Emmanuela Orsini
Dag Arne Osvald
Elisabeth Oswald
Elena Pagnin
Chris Peikert

Léo Perrin
Edoardo Persichetti
Duong-Hieu Phan
Krzysztof Pietrzak
Bertram Poettering
David Pointcheval
Antigoni Polychroniadou
Eamonn Postlethwaite
Willy Quach
Elizabeth Quaglia
Samuel Ranellucci
Mariana Raykova
Christian Rechberger
Oded Regev
Nicolas Resch
Leo Reyzin
M. Sadegh Riazi
Silas Richelson
Peter Rindal
Phillip Rogaway
Miruna Rosca
Dragos Rotaru
Yann Rotella
Arnab Roy
Manuel Sabin
Sruthi Sekar
Amin Sakzad
Katerina Samari
Pedro Moreno Sanchez

Sven Schaege
Adam Sealfon
Yannick Seurin
Aria Shahverdi
Tom Shrimpton
Luisa Siniscalchi
Kit Smeets
Fang Song
Pratik Soni
Jessica Sorrell
Florian Speelman
Douglas Stebila
Marc Stevens
Bing Sun
Shifeng Sun
Siwei Sun
Qiang Tang
Seth Terashima
Tian Tian
Mehdi Tibouchi
Yosuke Todo
Aleksei Udovenko
Dominique Unruh
Bogdan Ursu
María Isabel González
  Vasco
Muthuramakrishnan
Venkitasubramaniam
Fre Vercauteren

Fernando Virdia
Alexandre Wallet
Michael Walter
Meiqin Wang
Qingju Wang
Boyang Wei
Mor Weiss
Jan Winkelmann
Tim Wood
David Wu
Hong Xu
Shota Yamada
Hailun Yan
LeCorre Yann
Kan Yasuda
Arkady Yerukhimovich
Eylon Yogev
Yang Yu
Yu Yu
Thomas Zacharias
Wentao Zhang
Hong-Sheng Zhou
Linfeng Zhou
Vassilis Zikas
Giorgos Zirdelis
Lukas Zobernig
Adi Ben Zvi

Sponsors

JUZIX

Contents – Part I

Searchable Encryption and Differential Privacy

Secret Sharing

Encryption

Symmetric Cryptography

Proofs of Work and Proofs of Stake

Contents – Part II

Trapdoor Functions

Round Optimal MPC

Foundations

Lattices

Lattice-Based ZK

Efficient MPC

Contents – Part III

Zero Knowledge

Obfuscation

Secure Messaging

Towards Bidirectional Ratcheted
Key Exchange

Bertram Poettering[1] and Paul Rösler[2(✉)]

[1] Information Security Group, Royal Holloway, University of London, Egham, UK
`bertram.poettering@rhul.ac.uk`
[2] Horst-Görtz Institute for IT Security, Chair for Network and Data Security,
Ruhr-University Bochum, Bochum, Germany
`paul.roesler@rub.de`

Abstract. Ratcheted key exchange (RKE) is a cryptographic technique used in instant messaging systems like Signal and the WhatsApp messenger for attaining strong security in the face of state exposure attacks. RKE received academic attention in the recent works of Cohn-Gordon et al. (EuroS&P 2017) and Bellare et al. (CRYPTO 2017). While the former is analytical in the sense that it aims primarily at assessing the security that one particular protocol *does* achieve (which might be weaker than the notion that it *should* achieve), the authors of the latter develop and instantiate a notion of security from scratch, independently of existing implementations. Unfortunately, however, their model is quite restricted, e.g. for considering only unidirectional communication and the exposure of only one of the two parties.

In this article we resolve the limitations of prior work by developing alternative security definitions, for unidirectional RKE as well as for RKE where both parties contribute. We follow a purist approach, aiming at finding strong yet convincing notions that cover a realistic communication model with fully concurrent operation of both participants. We further propose secure instantiations (as the protocols analyzed or proposed by Cohn-Gordon et al. and Bellare et al. turn out to be weak in our models). While our scheme for the unidirectional case builds on a generic KEM as the main building block (differently to prior work that requires explicitly Diffie–Hellman), our schemes for bidirectional RKE require a stronger, HIBE-like component.

1 Introduction

ASYNCHRONOUS TWO-PARTY COMMUNICATION. Assume an online chat situation where two parties, Alice and Bob, communicate by exchanging messages over the Internet (e.g., using a TCP/IP based protocol). Their communication shall follow the structure of a human conversation in the sense that participants send messages when they feel they want to contribute to the discussion,

---

The full version of this article is available in the IACR eprint archive as article 2018/296, at https://eprint.iacr.org/2018/296.

H. Shacham and A. Boldyreva (Eds.): CRYPTO 2018, LNCS 10991, pp. 3–32, 2018.
https://doi.org/10.1007/978-3-319-96884-1_1

as opposed to in lockstep, i.e., when it is 'their turn'. In particular, in the considered asynchronous setting, Alice and Bob may send messages concurrently, and they also may receive them concurrently after a small delay introduced by the network. With other words, their messages may 'cross' on the wire.

As Alice and Bob are concerned with adversaries attacking their conversation, they deploy cryptographic methods. Standard security goals in this setting are the preservation of confidentiality and integrity of exchanged messages. These can be achieved, for instance, by combining an encryption primitive, a message authentication code, and transmission counters, where the latter serve for identifying replay and reordering attacks. As the mentioned cryptographic primitives are based on symmetric keys, Alice and Bob typically engage in an interactive key agreement protocol prior to starting their conversation.

FORWARD SECRECY. In this classic first-key-agreement-then-symmetric-protocol setup for two-party chats, the advantage of investing in an interactive key agreement session goes beyond fulfilling the basic need of the symmetric protocol (the allocation of shared key material): If the key agreement involves a Diffie–Hellman key exchange (DHKE), and this is nowadays the default, then the communication between Alice and Bob may be protected with forward secrecy. The latter means that even if the adversary finds a way, at a point in time after Alice and Bob finish their conversation, to obtain a copy of the long-term secrets they used during key establishment (signature keys, passwords, etc.), then this cannot be exploited to reveal their communication contents. Most current designs of cryptographic chat protocols consider forward secrecy an indispensable design goal [18]. The reason is that inadvertently disclosing long-term secrets is often more likely to happen than expected: system intruders might steal the keys, thieves might extract them from stolen Smartphones, law enforcement agencies might lawfully coerce users to reveal their keys, backup software might unmindfully upload a copy onto network storage, and so on.

SECURITY WITH EXPOSED STATE. Modern chat protocols also aim at protecting users in case of a different kind of attack: the skimming of the session state of an ongoing conversation [18].[1] Note that the session state information is orthogonal to the long-term secrets discussed above and, intuitively, an artifact of exclusively the second (symmetric) phase of communication. The necessity of being able to recover from session state leakage is usually motivated with two observations: messaging sessions are in general long-lived, e.g., kept alive for weeks or months once established, so that state exposures are more damaging, more easily provoked, and more likely to happen by accident; and leaking state information is sometimes impossible to defend against (state information held in computer memory might eventually be swapped to disk and stolen from there, and in cloud computing it is standard to move virtual machine memory images around the world from one host to the other).

---

[1] In this article, we consider the terms state reveal, state compromise, state corruption, and state exposure synonyms.

RATCHETING. Modern messaging protocols are designed with the goal of providing security even in the face of adversaries that perform the two types of attack discussed above (compromise of long-term secrets and/or session states) [18]. One technique used towards achieving this is via 'hash chains' where the symmetric key material contained in the session state is replaced, after each use, by a new value derived from the old value by applying some one-way function. This method mainly targets forward security and has a long tradition in cryptography (e.g., is used in [17] in the context of secure logging). A second technique is to let participants routinely redo a DHKE and mix the newly established keys into the session state: As part of every outgoing message a fresh g^x value is combined with prior and later values g^y contributed by the peer, with the goal of refreshing the session state as often as possible. This was introduced with the off-the-record (OTR) messaging protocol from [3,13] and promises auto-healing after a state compromise, at least if the DHKE exponents are derived from fresh randomness gathered from an uncorrupted source after the state reveal took place. Of course the two methods are not mutually exclusive but can be combined. We say that a messaging protocol employs a 'key ratchet' (this name can be traced back to [9]) if it uses the described or similar techniques for achieving forward secrecy and security under state exposure attacks.

RATCHETING AS A PRIMITIVE. While many authors associate the word ratcheting with a set of techniques deployed with the aim of achieving certain (typically not formally defined) security goals, Bellare et al. recently pursued a different approach by proposing *ratcheted key exchange* (RKE) as a cryptographic primitive with clearly defined syntax, functionality, and security properties [1]. This primitive establishes a sequence of session keys that allows for the construction of higher-level protocols, where instant messaging is just one example.[2] Building a messaging protocol on top of RKE offers clear advantages over using ad-hoc designs (as all messaging apps we are aware of do): the modularity allows for easier cryptanalysis, the substitution of constructions by alternatives, etc. We note, however, that the RKE formalization considered in [1] is too limited to serve directly as a building block for secure messaging. In particular, the syntactical framework requires all communication to be unidirectional (in the Alice-to-Bob direction), and the security model counterintuitively assumes that exclusively Alice's state can be exposed.

We give more details on the results of [1]. In the proposed protocol, Alice's state has the form (i, K, Y), where integer i counts her send operations, K is a key for a PRF F, and $Y = g^y$ is a public key of Bob. Bob's state has the form (i, K, y). When Alice performs a send operation, she samples a fresh randomness x, computes $\mu \leftarrow F(K, g^x)$ and $(k, K') \leftarrow H(i, \mu, g^x, Y^x)$ where H is a random oracle, and outputs k as the established session key and (g^x, μ) as a ciphertext that is sent to Bob. (Value μ serves as a message authentication code for g^x.) The next round's PRF key is K', i.e., Alice's new state is $(i + 1, K', Y)$.

---

[2] Note that RKE, despite its name, is a tool to be used in the 'symmetric phase' that follows the preliminary key agreement. In [1], and also in this article, the latter is abstracted away into a dedicated state initialization algorithm (or: protocol).

In this protocol, observe that F and H together implement a 'hash chain' and lead to forward secrecy, while the g^x, Y^x inputs to the random oracle can be seen as implementing one DHKE per transmission (where one exponent is static). Turning to the proposed RKE security model, while the corresponding game offers an oracle for compromising Alice's state, there is no option for similarly exposing Bob. If the model had a corresponding oracle, the protocol would actually not be secure. Indeed, the following (fully passive) attack exploits that Alice 'encrypts' to always the same key Y of Bob: The adversary first reveals Alice's session state, learning (i, K, Y); it then makes Alice invoke her send routine a couple of times and delivers the respective ciphertexts to Bob's receive routine in unmodified form; in the final step the adversary exposes Bob and recovers his past session keys using the revealed exponent y. Note that in a pure RKE sense these session keys should remain unknown to the adversary: Alice should have recovered from the state exposure, and forward secrecy should have made revealing Bob's state useless.[3]

Contributions. We follow in the footsteps of [1] and study RKE as a general cryptographic primitive. However, we significantly improve on their results, in three independent directions:

Firstly, we extend the strictly unidirectional RKE concept of Bellare et al. towards bidirectional communication. In more detail, if we refer to the setting of [1] as URKE (unidirectional RKE), we introduce SRKE (sesquidirectional[4] RKE) and BRKE (bidirectional RKE; for space reasons only in the full version [14]). In SRKE, while both Alice and Bob can send ciphertexts to the respective peer, only the ciphertexts sent from Alice to Bob establish session keys. Those sent by Bob have no direct functionality but may help him healing from state exposure. Also in BRKE both parties send ciphertexts, but here the situation is symmetric in that all ciphertexts establish keys (plus allow for healing from state exposure).

Secondly, we propose an improved security model for URKE, and introduce security models for SRKE and BRKE (the latter only in [14]). Our SRKE and BRKE models assume the likely only practical communication setting for messaging protocols, namely the one in which the operations of both parties can happen concurrently (in contrast to, say, a ping-pong way). We develop our models following a purist approach: We start with giving the adversary the full set of options to undertake its attack (including state exposures of *both* parties), and then exclude, one by one, those configurations that unavoidably lead to a 'trivial win' (an example for the latter is if the adversary first compromises Bob's state and then correctly 'guesses' the next session key he recovers from an incoming ciphertext). This approach leads to strong and convincing security models (and it becomes quite challenging to actually meet them). We note that

---

[3] A protocol that achieves security in the described setting is developed in this paper; the central idea behind our construction is that Bob's key pair (y, Y) does not stay fixed but is updated each time a ciphertext is processed.

[4] Recall that 'sesqui' is Latin for one-and-a-half.

the (as we argued) insecure protocol from [1] is considered secure in the model of [1] because the latter was not designed with our strategy in mind, ultimately missing some attacks.

Thirdly, we give provably secure constructions of URKE and SRKE (and of BRKE in the full version [14]). While all prior RKE protocol proposals, including the one from [1], are explicitly based on DHKE as a low-level tool, our constructions use generic primitives like KEMs, MACs, one-time signatures, and random oracles. The increased level of abstraction not only clarifies on the role that these components play in the constructions, it also increases the freedom when picking acceptable hardness assumptions.

FURTHER DETAILS ON OUR URKE CONSTRUCTION. In brief, our (unidirectional) URKE scheme combines a hash chain and KEM encapsulations to achieve both forward secrecy and recoverability from state exposures. The crucial difference to the protocol from [1] is that in our scheme the public key of Bob is changed after each use. Concretely, but omitting many details, the state information of Alice is (i, K, Y) as in [1] (but where Y is the *current* public key of Bob), for sending Alice freshly encapsulates a key k^* to Y, then computes $(k, K', k') \leftarrow \text{H}(i, K, Y, k^*)$ using a random oracle H, and finally uses auxiliary key k' to update the old public key Y to a new public key Y that is to be used in her next sending operation. Bob does correspondingly, updating his secret key with each incoming ciphertext. Note that the attack against [1] that we sketched above does not work against this protocol (the adversary would obtain a useless decryption key when revealing Bob's state).

FURTHER DETAILS ON OUR SRKE CONSTRUCTION. Recall that, in SRKE, Bob can send update ciphertexts to Alice with the idea that this will help him recover from state exposures. Our protocol algorithms can handle fully concurrent operation of the two participants (in particular, ciphertexts may 'cross' on the wire). This unfortunately adds, as the algorithms need to handle multiple 'epochs' at the same time, considerably to their complexity. Interestingly, the more involved communication setting is also reflected in stronger primitives that we require for our construction: Our SRKE construction builds on a special KEM type that supports so-called key updates (also the latter primitive is constructed in this paper, from HIBE).

In a nutshell, in our SRKE construction, Bob heals from state exposures by generating a fresh (updatable) KEM key pair every now and then, and communicating the public key to Alice. Alice uses the key update functionality to 'fast-forward' these keys into a current state by making them aware of ciphertexts that were exchanged after the keys were sent (by Bob), but before they were received (by Alice). In her following sending operation, Alice encapsulates to a mix of old and new public keys.

OUTLOOK ON BRKE. We expose two BRKE constructions in the full version [14]. The first works via the amalgamation of two generic SRKE instances, deployed in reverse directions. To reach full security, the instances need to be carefully tied together (our solution does this with one-time signatures).

The second construction is less generic but slightly more efficient, namely by combining and interleaving the building blocks of our SRKE scheme in the right way.

Further related work. The idea of using 'hash chains' for achieving forward security of symmetric cryptographic primitives has been around for quite some time. For instance, [17] use this technique to protect the integrity of audit logs. The first formal treatment we are aware of is [2]. A messaging protocol that uses this technique is the (original) Silent Circle Instant Messaging Protocol [12].

The idea of mixing into the user state of messaging protocols additional key material that is continuously established with asymmetric techniques (in particular: DHKE) first appeared in the off-the-record (OTR) messaging protocol from [3,13]. Subsequently, the technique appeared in many communication protocols specifically designed to be privacy-friendly, including the ZRTP telephony protocol [19] and the messaging protocol *Double Ratchet Algorithm* [10] (formerly known as Axolotl). The latter, or close variants thereof, are used by WhatsApp, the Facebook Messenger, and Signal app. In the full version [14] we study these protocols more closely, proposing for each of them an attack that shows that it is not secure in our models.

Widely used messaging protocols were recently analyzed by Cohn-Gordon et al. [4] and Rösler et al. [16]. In particular, [4] contributes an analysis of the Signal messaging protocol [10] by developing a *"model with adversarial queries and freshness conditions that capture the security properties intended by Signal"*. While the work does propose a formal security model, for being geared towards confirming the security of one particular protocol, it may not necessarily serve as a reference notion for RKE.[5]

Academic work in a related field was conducted by [5] who study post-compromise security in (classic) key exchange. Here, security shall be achieved even for sessions established after a full compromise of user secrets. This necessarily requires mixing user state information with key material that is newly established via asymmetric techniques, and is thus related to RKE. However, we note the functionalities and models of (classic) key exchange and RKE are fundamentally different: The former generally considers multiple participants who have long-term keys and who can run multiple sessions, with the same or different peers, in parallel, while participants of the latter have no long-term keys at all, and thus any two sessions are completely independent.

Organization. In Sect. 2 we fix notation and describe the building blocks of our RKE constructions: MACs, KEMs (but with a non-standard syntax), one-time signatures. In Sect. 3 we develop the URKE syntax and a suitable security model, and present a corresponding construction in Sect. 4. In Sects. 5 and 6 we do the same for SRKE. In Sect. 7 we give an intuition of how SRKE can be extended to BRKE.

---

[5] In fact it defines weaker security than would be natural for RKE. We elaborate on this in the full version [14] where we explain why the Signal protocol is not secure in our model.

2   Preliminaries

2.1   Notation

If A is a (deterministic or randomized) algorithm we write $A(x)$ for an invocation of A on input x. If A is randomized, we write $A(x) \Rightarrow y$ for the event that an invocation results in value y being the output. We further write $[A(x)] := \{y : \Pr[A(x) \Rightarrow y] > 0\}$ for the effective range of $A(x)$.

If $a \leq b$ are integers, we write $[a \mathinner{..} b]$ for the set $\{a, \ldots, b\}$ and we write $[a, \ldots]$ for the set $\{x \in \mathbb{N} : a \leq x\}$. We also give symbolic names to intervals and their boundaries (smallest and largest elements): For an interval $I = [a \mathinner{..} b]$ we write $I^{\vdash}$ for a and $I^{\dashv}$ for b. We denote the Boolean constants True and False with $\mathtt{T}$ and $\mathtt{F}$, respectively. We use Iverson brackets to convert Boolean values into bit values: $[\mathtt{T}] = 1$ and $[\mathtt{F}] = 0$. To compactly write if-then-else expressions we use the ternary operator known from the C programming language: If C is a Boolean condition and e_1, e_2 are arbitrary expressions, the composed expression "$C\ ?\ e_1 : e_2$" evaluates to e_1 if $C = \mathtt{T}$ and to e_2 if $C = \mathtt{F}$.

When we refer to a *list* or *sequence* we mean a (row) vector that can hold arbitrary elements, where the empty list is denoted with ϵ. A list can be appended to another list with the concatenation operator $\|$, and we denote the is-prefix-of relation with $\preceq$. For instance, for lists $L_1 = \epsilon$ and $L_2 = a$ and $L_3 = b \| c$ we have $L_1 \| L_2 \| L_3 = a \| b \| c$ and $L_1 \preceq L_2 \not\preceq L_3$.

PROGRAM CODE. We describe algorithms and security experiments using (pseudo-)code. In such code we distinguish the following operators for assigning values to variables: We use symbol '$\leftarrow$' when the assigned value results from a constant expression (including the output of a deterministic algorithm), and we write '$\leftarrow_\$$' when the value is either sampled uniformly at random from a finite set or is the output of a randomized algorithm. If we assign a value that is a tuple but we are actually not interested in some of its components, we use symbol '$\_$' to mark positions that shall be ignored. For instance, $(\_, b, \_) \leftarrow (A, B, C)$ is equivalent to $b \leftarrow B$. If X, Y are sets we write $X \overset{\cup}{\leftarrow} Y$ shorthand for $X \leftarrow X \cup Y$, and if L_1, L_2 are lists we write $L_1 \overset{\|}{\leftarrow} L_2$ shorthand for $L_1 \leftarrow L_1 \| L_2$. We use bracket notation to denote associative arrays (a data structure that implements a dictionary). Associative arrays can be indexed with elements from arbitrary sets. For instance, for an associative array A the instruction $A[7] \leftarrow 3$ assigns value 3 to index 7, and the expression $A[\mathtt{abc}] = 5$ tests whether the value at index $\mathtt{abc}$ is equal to 5. We write $A[\cdot] \leftarrow x$ to initialize the associative array A by assigning the default value x to all possible indices. For an integer a we write $A[\ldots, a] \leftarrow x$ as a shortcut for 'For all $a' \leq a$: $A[a'] \leftarrow x$'.

GAMES. Our security definitions are based on games played between a challenger and an adversary. Such games are expressed using program code and terminate when the special 'Stop' instruction is executed; the argument of the latter is the outcome of the game. For instance, we write $\Pr[G \Rightarrow 1]$ for the probability that game G terminates by running into a 'Stop with 1' instruction. For a Boolean condition C, in games we write 'Require C' shorthand for 'If $\neg C$: Stop with 0' and we write 'Reward C' shorthand for 'If C: Stop with 1'. The two instructions are used for appraising the actions of the adversary: Intuitively, if the adversary behaves such that a required condition is violated then the adversary definitely 'loses' the game, and if it behaves such that a rewarded condition is met then it definitely 'wins'.

SCHEME SPECIFICATIONS. We also describe the algorithms of cryptographic schemes using program code. Some algorithms may abort or fail, indicating this by outputting the special symbol $\perp$. This is implicitly assumed to happen whenever an encoded data structure is to be parsed into components but the encoding turns out to be invalid. A more explicit way of aborting is via the 'Require C' shortcut which, in algorithm specifications, stands for 'If $\neg C$: Return $\perp$'. This instruction is typically used to assert that certain conditions hold for user-provided input.

2.2   Classic Cryptographic Building Blocks

Our RKE constructions use MACs, one-time signature schemes, and KEMs as building blocks. As the requirements on the MACs and one-time signatures are standard, we provide only very reduced definitions here and defer the full specifications to [14]. For KEMs, however, we assume a specific non-standard syntax, functionality, and notion of security; the details can be found below.

MACS AND ONE-TIME SIGNATURES. We denote the key space of a MAC M with $\mathcal{K}$, and assume that the tag and verification algorithms are called tag and $\mathrm{vfy}_{\mathsf{M}}$, respectively. Their syntax will always be clear from the context. As a security notion we define strong unforgeability, and the corresponding advantage of an adversary $\mathcal{A}$ we denote with $\mathrm{Adv}_{\mathsf{M}}^{\mathrm{suf}}(\mathcal{A})$. For a one-time signature scheme S we assume that the key generation algorithm, the signing algorithm, and the verification algorithm are called $\mathrm{gen}_{\mathsf{S}}$ and sgn and $\mathrm{vfy}_{\mathsf{S}}$, respectively. We assume that $\mathrm{vfy}_{\mathsf{S}}$ outputs values T or F to indicate its decision, and that the remaining syntax will again be clear from the context. As a security notion we define strong unforgeability, and the corresponding advantage of an adversary $\mathcal{A}$ we denote with $\mathrm{Adv}_{\mathsf{S}}^{\mathrm{suf}}(\mathcal{A})$.

KEY ENCAPSULATION MECHANISMS. We consider a type of KEM where key pairs are generated by first randomly sampling the secret key and then deterministically deriving the public key from it. While this syntax is non-standard, note that it can be assumed without loss of generality: One can always understand the coins used for (randomized) key generation of a classic KEM as the secret key in our sense.

A *key encapsulation mechanism* (KEM) for a finite session-key space $\mathcal{K}$ is a triple $\mathsf{K} = (\mathrm{gen}_{\mathsf{K}}, \mathrm{enc}, \mathrm{dec})$ of algorithms together with a samplable secret-key space $\mathcal{SK}$, a public-key space $\mathcal{PK}$, and a ciphertext space $\mathcal{C}$. In its regular form the public-key generation algorithm $\mathrm{gen}_{\mathsf{K}}$ is deterministic, takes a secret key $sk \in \mathcal{SK}$, and outputs a public key $pk \in \mathcal{PK}$. We also use a shorthand form, writing $\mathrm{gen}_{\mathsf{K}}$ for the randomized procedure of first picking $sk \leftarrow_{\$} \mathcal{SK}$, then deriving $pk \leftarrow \mathrm{gen}_K(sk)$, and finally outputting the pair (sk, pk). Two shortcut notations for key generation are thus

$$\mathcal{SK} \to \mathrm{gen}_K \to \mathcal{PK} \qquad \mathrm{gen}_K \to_{\$} \mathcal{SK} \times \mathcal{PK} \ .$$

The randomized encapsulation algorithm enc takes a public key $pk \in \mathcal{PK}$ and outputs a session key $k \in \mathcal{K}$ and a ciphertext $c \in \mathcal{C}$, and the deterministic decapsulation algorithm dec takes a secret key $sk \in \mathcal{SK}$ and a ciphertext $c \in \mathcal{C}$, and outputs either a session key $k \in \mathcal{K}$ or the special symbol $\bot \notin \mathcal{K}$ to indicate rejection. Shortcut notations for encapsulation and decapsulation are thus

$$\mathcal{PK} \to \mathrm{enc} \to_{\$} \mathcal{K} \times \mathcal{C} \qquad \mathcal{SK} \times \mathcal{C} \to \mathrm{dec} \to \mathcal{K} \ / \bot \ .$$

For correctness we require that for all $(sk, pk) \in [\mathrm{gen}_K]$ and $(k, c) \in [\mathrm{enc}(pk)]$ we have $\mathrm{dec}(sk, c) = k$.

We formalize a multi-receiver/multi-challenge version of one-way security as a security property for KEMs. In this notion, the adversary obtains challenge ciphertexts and has to recover any of the encapsulated keys. The adversary is supported by a key-checking oracle that, for a provided pair of ciphertext and (candidate) session key, tells whether the ciphertext decapsulates to the indicated key. The adversary is also allowed to expose receivers, learning their secret keys. The details of this notion are in game OW in the full version [14]. For a KEM K, we associate with any adversary $\mathcal{A}$ its one-way advantage $\mathrm{Adv}_K^{\mathrm{ow}}(\mathcal{A}) := \Pr[\mathrm{OW}(\mathcal{A}) \Rightarrow 1]$. Intuitively, the KEM is secure if all practical adversaries have a negligible advantage.

2.3   Key-Updatable Key Encapsulation Mechanisms

We introduce a type of KEM that we refer to as key-updatable. Like a regular KEM the new primitive establishes secure session keys, but in addition a dedicated key-update algorithm derives new ('updated') keys from old ones: Also taking an auxiliary input into account that we call the associated data, a secret key is updated to a new secret key, or a public key is updated to a new public key. A KEM key pair remains functional under such updates, meaning that session keys encapsulated for the public key can be recovered using the secret key if both keys are updated compatibly, i.e., with matching associated data. Concerning security we require a kind of forward secrecy: Session keys encapsulated to a (potentially updated) public key shall remain secure even if the adversary gets hold of any incompatibly updated version of the secret key.

A *key-updatable key encapsulation mechanism* (kuKEM) for a finite session-key space $\mathcal{K}$ is a quadruple $K = (\mathrm{gen}_K, \mathrm{enc}, \mathrm{dec}, \mathrm{up})$ of algorithms together with a samplable secret-key space $\mathcal{SK}$, a public-key space $\mathcal{PK}$, a ciphertext space $\mathcal{C}$, and an associated-data space $\mathcal{AD}$. Algorithms $\mathrm{gen}_K, \mathrm{enc}, \mathrm{dec}$ are as for regular KEMs. The key-update algorithm up is deterministic and comes in two shapes: either it takes a secret key $sk \in \mathcal{SK}$ and associated data $ad \in \mathcal{AD}$ and outputs an updated secret key $sk' \in \mathcal{SK}$, or it takes a public key $pk \in \mathcal{PK}$ and associated data $ad \in \mathcal{AD}$ and outputs an updated public key $pk' \in \mathcal{PK}$. Shortcut notations for the key update algorithm(s) are thus

$$\mathcal{SK} \times \mathcal{AD} \to \text{up} \to \mathcal{SK} \qquad \mathcal{PK} \times \mathcal{AD} \to \text{up} \to \mathcal{PK} \ .$$

For correctness we require that for all $(sk_0, pk_0) \in [\text{gen}_K]$ and $ad_1, \ldots, ad_n \in \mathcal{AD}$, if we let $sk_i = \text{up}(sk_{i-1}, ad_i)$ and $pk_i = \text{up}(pk_{i-1}, ad_i)$ for all i, then for all $(k, c) \in [\text{enc}(pk_n)]$ we have $\text{dec}(sk_n, c) = k$.

As a security property for kuKEMs we formalize a multi-receiver/multi-challenge version of one-way security that also reflects forward security in case of secret-key updates. It should be hard for an adversary to recover encapsulated keys even if it obtained secret keys that are further or differently updated than the challenge secret key(s). The details of the notion are in game KUOW in the full version [14]. For a key-updatable KEM K, we associate with any adversary $\mathcal{A}$ its one-way advantage $\text{Adv}_K^{\text{kuow}}(\mathcal{A}) := \Pr[\text{KUOW}(\mathcal{A}) \Rightarrow 1]$. Intuitively, the kuKEM is secure if all practical adversaries have a negligible advantage.

Observe that kuKEMs are related to hierarchical identity-based encryption (HIBE, [7]): Intuitively, updating a secret key using associated data ad in the kuKEM world corresponds in the HIBE world with extracting the decryption/ delegation key for the next-lower hierarchy level, using partial identity ad. Indeed, a kuKEM scheme is immediately constructed from a generic HIBE, with only cosmetic changes necessary when expressing the algorithms; we give the details and a specific construction in the full version [14].

3   Unidirectionally Ratcheted Key Exchange (URKE)

We give a definition of unidirectional RKE and its security. While, in principle, our syntactical definition is in line with the one from [1], our naming convention deviates significantly from the latter for the sake of a more clear distinction between (session) keys, (session) states, and ciphertexts[6]. Further, looking ahead, our security notion for URKE is stronger than the one of [1]. A speciality of our formalization is that we let the sending and receiving algorithms of Alice and Bob accept and process an associated data string [15] that, for functionality, has to match on both sides.

A *unidirectionally ratcheted key exchange* (URKE) for a finite key space $\mathcal{K}$ and an associated-data space $\mathcal{AD}$ is a triple $\mathsf{R} = (\text{init}, \text{snd}, \text{rcv})$ of algorithms together with a sender state space $\mathcal{S}_A$, a receiver state space $\mathcal{S}_B$, and a ciphertext space $\mathcal{C}$. The randomized initialization algorithm init returns a sender state $S_A \in \mathcal{S}_A$ and a receiver state $S_B \in \mathcal{S}_B$. The randomized sending algorithm snd takes a state $S_A \in \mathcal{S}_A$ and an associated-data string $ad \in \mathcal{AD}$, and produces an updated state $S'_A \in \mathcal{S}_A$, a key $k \in \mathcal{K}$, and a ciphertext $c \in \mathcal{C}$. Finally, the deterministic receiving algorithm rcv takes a state $S_B \in \mathcal{S}_B$, an associated-data

---

[6] The mapping between our names (on the left of the equality sign) and the ones of [1] (on the right) is as follows: '(session) key' = 'output key', '(session) state' = 'session key plus sender/receiver key', 'ciphertext' = 'update information'.

string $ad \in \mathcal{AD}$, and a ciphertext $c \in \mathcal{C}$, and either outputs an updated state $S'_B \in \mathcal{S}_B$ and a key $k \in \mathcal{K}$, or the special symbol $\perp$ to indicate rejection. A shortcut notation for these syntactical definitions and a visual illustration of the URKE communication setup is

$$
\begin{aligned}
\text{init} &\rightarrow \mathcal{S}_A \times \mathcal{S}_B \\
\mathcal{S}_A \times \mathcal{AD} &\rightarrow \text{snd} \rightarrow \mathcal{S}_A \times \mathcal{K} \times \mathcal{C} \\
\mathcal{S}_B \times \mathcal{AD} \times \mathcal{C} &\rightarrow \text{rcv} \rightarrow \mathcal{S}_B \times \mathcal{K} / \perp
\end{aligned}
$$

Correctness of URKE. Assume a sender and a receiver that were jointly initialized with init. Then, intuitively, the URKE scheme is correct if for all sequences (ad_i) of associated-data strings, if (k_i) and (c_i) are sequences of keys and ciphertexts successively produced by the sender on input the strings in (ad_i), and if (k'_i) is the sequence of keys output by the receiver on input the (same) strings in (ad_i) and the ciphertexts in (c_i), then the keys of the sender and the receiver match, i.e., it holds that $k_i = k'_i$ for all i.

We formalize this requirement via the FUNC game in Fig. 1.[7] Concretely, we say scheme R is *correct* if $\Pr[\text{FUNC}_R(\mathcal{A}) \Rightarrow 1] = 0$ for all adversaries $\mathcal{A}$. In the game, the adversary lets the sender and the receiver process associated-data strings and ciphertexts of its choosing, and its goal is to let the two parties compute keys that do not match when they should. Variables s_A and r_B count the send and receive operations, associative array adc_A jointly records the associated-data strings considered by and the ciphertexts produced by the sender, flag is_B is an indicator that tracks whether the receiver is still 'in-sync' (in contrast to: was exposed to non-matching associated-data strings or ciphertexts; note how the transition between in-sync and out-of-sync is detected and recorded in lines 13, 14), and associative array key_A records the keys established by the sender to allow for a comparison with the keys recovered (or not) by the receiver. The correctness requirement boils down to declaring the adversary successful (in line 17) if the sender and the receiver compute different keys while still being in-sync. Note finally that lines 12, 16 ensure that once the rcv algorithm rejects, the adversary is notified of this and further queries to the RcvB oracle are not accepted.

Security of URKE. We formalize a key indistinguishability notion for URKE. In a nutshell, from the point of view of the adversary, keys established by the sender and recovered by the receiver shall look uniformly distributed in the key space. In our model, the adversary, in addition to scheduling the regular URKE operations via the SndA and RcvB oracles, has to its disposal the four oracles ExposeA, ExposeB, Reveal, and Challenge, used for exposing users by obtaining copies of their current state, for learning established keys, and for requesting real-or-random challenges on established keys, respectively. For an URKE scheme R,

---

[7] Formalizing correctness of URKE via a game might at first seem overkill. However, for SRKE and BRKE, which allow for interleaved interaction in two directions, game-based definitions seem to be natural and notationally superior to any other approach. For consistency we use a game-based definition also for URKE.

Game $\text{FUNC}_R(\mathcal{A})$	Oracle $\text{SndA}(ad)$	Oracle $\text{RcvB}(ad, c)$
00 $s_A \leftarrow 0$; $r_B \leftarrow 0$	07 $(S_A, k, c) \leftarrow_\$ \text{snd}(S_A, ad)$	12 Require $S_B \neq \bot$
01 $adc_A[\cdot] \leftarrow \bot$	08 $adc_A[s_A] \leftarrow (ad, c)$	13 If $is_B \wedge adc_A[r_B] \neq (ad, c)$:
02 $is_B \leftarrow \mathbf{T}$	09 $key_A[s_A] \leftarrow k$	14 $\quad is_B \leftarrow \mathbf{F}$
03 $key_A[\cdot] \leftarrow \bot$	10 $s_A \leftarrow s_A + 1$	15 $(S_B, k) \leftarrow \text{rcv}(S_B, ad, c)$
04 $(S_A, S_B) \leftarrow_\$ \text{init}$	11 Return c	16 If $S_B = \bot$: Return $\bot$
05 Invoke $\mathcal{A}$		17 Reward $is_B \wedge k \neq key_A[r_B]$
06 Stop with 0		18 $r_B \leftarrow r_B + 1$
		19 Return

Fig. 1. Game FUNC for URKE scheme R.

in Fig. 2 we specify corresponding key indistinguishability games KIND_R^b, where $b \in \{0, 1\}$ is the challenge bit, and we associate with any adversary $\mathcal{A}$ its key distinguishing advantage $\text{Adv}_R^{\text{kind}}(\mathcal{A}) := |\Pr[\text{KIND}_R^1(\mathcal{A}) \Rightarrow 1] - \Pr[\text{KIND}_R^0(\mathcal{A}) \Rightarrow 1]|$. Intuitively, R offers key indistinguishability if all practical adversaries have a negligible key distinguishing advantage.

Most lines of code in the KIND^b games are tagged with a ' · ' right after the line number; to the subset of lines marked in this way we refer to as the games' *core*. Conceptually, the cores contain all relevant game logic (participant initialization, specifications of how queries are answered, etc.); the code lines available only in the full game, i.e., the untagged ones, introduce certain restrictions on the adversary that are necessary to exclude trivial attacks (see below). The games' cores should be self-explanatory, in particular when comparing them to the FUNC game, with the understanding that lines 18, 37 (in Fig. 2) ensure that only keys can be revealed or challenged that actually have been established before, and that line 38 assigns to variable k, depending on bit b, either the real key or a freshly sampled element from the key space.

Note that, in the pure core code, the adversary can use the four new oracles to bring itself into the position to distinguish real and random keys in a trivial way. In the following we discuss five different strategies to do so. We illustrate each strategy by specifying an example adversary in pseudocode and we explain what measures the full games take for disregarding the respective class of attack. (That is, the example adversaries would gain high advantage if the games consisted of just their cores, but in the full games their advantage is zero.)

The first two strategies leverage on the interplay of Reveal and Challenge queries; they do not involve exposing participants.

(a) The adversary requests a challenge on a key that it also reveals, it requests two challenges on the same key, or similar. Example: fix some ad; $c \leftarrow \text{SndA}(ad)$; $k \leftarrow \text{Reveal}(A, 0)$; $k' \leftarrow \text{Challenge}(A, 0)$; $b' \leftarrow [k = k']$; output b'. The full games, in lines 20, 39, overwrite keys that are revealed or challenged with the special symbol $\diamond \notin \mathcal{K}$. Because of lines 18, 37, this prevents any second Reveal or Challenge query involving the same key.

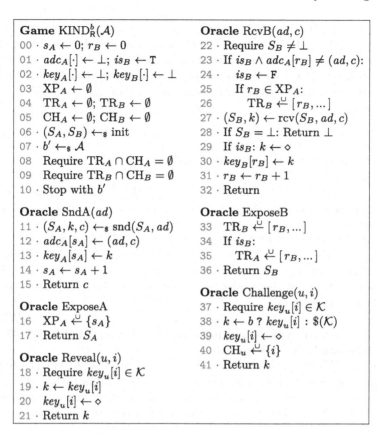

Fig. 2. Games KIND^b, $b \in \{0,1\}$, for URKE scheme R. We require $\diamond \notin \mathcal{K}$, and in Reveal and Challenge queries we require $u \in \{A, B\}$. If the notation in lines 26 or 38 is unclear, please consult Sect. 2.1.

(b) The adversary combines an attack from (a) with the correctness guarantee, i.e., that in-sync receivers recover the keys established by senders. For instance, the adversary reveals a sender key and requests a challenge on the corresponding receiver key. Example: fix some ad; $c \leftarrow \mathrm{SndA}(ad)$; $k \leftarrow \mathrm{Reveal}(A, 0)$; $\mathrm{RcvB}(ad, c)$; $k' \leftarrow \mathrm{Challenge}(B, 0)$; $b' \leftarrow [k = k']$; output b'. The full games, in line 29, overwrite in-sync receiver keys, as they are known (by correctness) to be the same on the sender side, with the special symbol $\diamond \notin \mathcal{K}$. By lines 18, 37, this rules out the attack.

The remaining three strategies involve exposing participants and using their state to either trace their computations or impersonate them to their peer. In the full games, the set variables $\mathrm{XP}_A, \mathrm{TR}_A, \mathrm{TR}_B, \mathrm{CH}_A, \mathrm{CH}_B$ (lines 03–05) help identifying when such attacks occur. Concretely, set XP_A tracks the points in time the sender is exposed (the unit of time being the number of past sending operations; see line 16), sets $\mathrm{TR}_A, \mathrm{TR}_B$ track the indices of keys that are 'traceable' (in particular: recoverable by the adversary) using an exposed state (see below), and sets $\mathrm{CH}_A, \mathrm{CH}_B$ record the indices of keys for which a challenge was requested (see line 40). Lines 08, 09 ensure that any adversary that requests to be challenged on a traceable key has advantage zero. Strategies (c) and (d) are state tracing attacks, while strategy (e) is based on impersonation.

(c) The adversary exposes the receiver and uses the obtained state to trace its computations: By iteratively applying the rcv algorithm to all later inputs of the receiver, and updating the exposed state correspondingly, the adversary implicitly obtains a copy of all later receiver keys. Example: fix some ad; $c \leftarrow \mathrm{SndA}(ad)$; $S_B^* \leftarrow \mathrm{ExposeB}()$; $(S_B^*, k) \leftarrow \mathrm{rcv}(S_B^*, ad, c)$; $\mathrm{RcvB}(ad, c)$; $k' \leftarrow \mathrm{Challenge}(B, 0)$; $b' \leftarrow [k = k']$; output b'. When an exposure of the receiver happens, the full games, in line 33, mark all future receiver keys as traceable.

(d) The adversary combines the attack from (c) with the correctness guarantee, i.e., that in-sync receivers recover the keys established by senders: After exposing an in-sync receiver, by iteratively applying the rcv algorithm to all later outputs of the sender, the adversary implicitly obtains a copy of all later sender keys. Example: fix some ad; $c \leftarrow \mathrm{SndA}(ad)$; $S_B^* \leftarrow \mathrm{ExposeB}()$; $(S_B^*, k) \leftarrow \mathrm{rcv}(S_B^*, ad, c)$; $k' \leftarrow \mathrm{Challenge}(A, 0)$; $b' \leftarrow [k = k']$; output b'. When an exposure of an in-sync receiver happens, the full games, in lines 34, 35, mark all future sender keys as traceable.

(e) Exposing the sender allows for impersonating it: The adversary obtains a copy of the sender's state and invokes the snd algorithm with it, obtaining a key and a ciphertext. The latter is provided to an in-sync receiver (rendering the latter out-of-sync), who recovers a key that is already known to the adversary. Example: fix some ad; $S_A^* \leftarrow \mathrm{ExposeA}()$; $(S_A^*, k, c) \leftarrow_\$ \mathrm{snd}(S_A^*, ad)$; $\mathrm{RcvB}(ad, c)$; $k' \leftarrow \mathrm{Challenge}(B, 0)$; $b' \leftarrow [k = k']$; output b'. The full games, in lines 25, 26, detect the described type of impersonation and mark all future receiver keys as traceable.

We conclude with some notes on our URKE model. First, the model excludes the (anyway unavoidable) trivial attack conditions we identified, but nothing else. This establishes confidence in the model, as no attacks can be missed. Further, observe that it is not possible to recover from an attack based on state exposure (i.e., of the (c)–(e) types): If *one* key of a participant becomes weak as a consequence of a state exposure, then necessarily *all* later keys of that participant become weak as well. On the other hand, exposing the sender and *not* bringing the receiver out-of-sync does not affect security at all.[8] Finally, exposing an out-of-sync receiver does not harm later sender keys. In later sections we consider ratcheting primitives (SRKE, BRKE) that resume safe operation after state exposure attacks.

4   Constructing URKE

We construct an URKE scheme that is provably secure in the model presented in the previous section. The ingredients are a KEM (with deterministic public-key

---

[8] This is precisely the distinguishing auto-recovery property of ratcheted key exchange.

Proc init	**Proc** snd(S_A, ad)	**Proc** rcv(S_B, ad, C)
00 $(sk, pk) \leftarrow_\$ \text{gen}_K$	06 $(pk, K, k.m, t) \leftarrow S_A$	15 $(sk, K, k.m, t) \leftarrow S_B$
01 $K \leftarrow_\$ \mathcal{K}; k.m \leftarrow_\$ \mathcal{K}$	07 $(k, c) \leftarrow_\$ \text{enc}(pk)$	16 $c \| \tau \leftarrow C$
02 $t \leftarrow \epsilon$	08 $\tau \leftarrow_\$ \text{tag}(k.m, ad \| c)$	17 Require $\text{vfy}_M(k.m, ad \| c, \tau)$
03 $S_A \leftarrow (pk, K, k.m, t)$	09 $C \leftarrow c \| \tau$	18 $k \leftarrow \text{dec}(sk, c)$
04 $S_B \leftarrow (sk, K, k.m, t)$	10 $t \overset{\shortmid\shortmid}{\leftarrow} ad \| C$	19 Require $k \neq \bot$
05 Return (S_A, S_B)	11 $k.o \| K \| k.m \| sk \leftarrow$	20 $t \overset{\shortmid\shortmid}{\leftarrow} ad \| C$
	$\qquad\qquad H(K, k, t)$	21 $k.o \| K \| k.m \| sk \leftarrow$
	12 $pk \leftarrow \text{gen}_K(sk)$	$\qquad\qquad H(K, k, t)$
	13 $S_A \leftarrow (pk, K, k.m, t)$	22 $S_B \leftarrow (sk, K, k.m, t)$
	14 Return $(S_A, k.o, C)$	23 Return $(S_B, k.o)$

Fig. 3. Construction of an URKE scheme from a key-encapsulation mechanism $K = (\text{gen}_K, \text{enc}, \text{dec})$, a message authentication code $M = (\text{tag}, \text{vfy}_M)$, and a random oracle H. For simplicity we denote the key space of the MAC and the space of chaining keys with the same symbol $\mathcal{K}$.

generation, see Sect. 2.2), a strongly unforgeable MAC, and a random oracle H. The algorithms of our scheme are specified in Fig. 3.

We describe protocol states and algorithms in more detail. The state of Alice consists of (Bob's) KEM public key pk, a chaining key K, a MAC key $k.m$, and a transcript variable t that accumulates the associated data strings and ciphertexts that Alice processed so far. The state of Bob is almost the same, but instead of the KEM public key he holds the corresponding secret key sk. Initially, sk and pk are freshly generated, random values are assigned to K and $k.m$, and the transcript accumulator t is set to the empty string. A sending operation of Alice consists of invoking the KEM encapsulation routine with Bob's current public key, computing a MAC tag over the ciphertext and the associated data, updating the transcript accumulator, and jointly processing the session key established by the KEM, the chaining key, and the current transcript with the random oracle H. The output of H is split into the URKE session key $k.o$, an updated chaining key, an updated MAC key, and, indirectly, the updated public key (of Bob) to which Alice encapsulates in the next round. The receiving operation of Bob is analogue to these instructions. While our scheme has some similarity with the one of [1], a considerable difference is that the public and secret keys held by Alice and Bob, respectively, are constantly changed. This rules out the attack described in the introduction.

Note that our scheme is specified such that participants accumulate in their state the full past communication history. While this eases the security analysis (random oracle evaluations of Alice and Bob are guaranteed to be on different inputs once the in-sync bit is cleared), it also seems to impose a severe implementation obstacle. However, as current hash functions like SHA2 and SHA3 process inputs in an online fashion (left-to-right with a small state overhead), they can process append-only inputs like transcripts such that computations are efficiently shared with prior invocations. In particular, with such a hash function our URKE scheme can be implemented with constant-size state. (This requires, though, rearranging the input of H such that t comes first).[9]

Theorem 1 (informal). *The URKE protocol R from Fig. 3 offers key indistinguishability if function H is modeled as a random oracle, the KEM provides OW security, the MAC provides SUF security, and the session-key space of the KEM is sufficiently large.*

The exact theorem statement and the respective proof are in the full version [14]. Briefly, the proof first shows that none of Alice's established session keys can be derived by the adversary without breaking the security of the KEM as long as no previous secret key of Alice's public keys was exposed. Then we show that Bob will only establish session keys out of sync if Alice was impersonated towards him, his state was exposed before, or a MAC forgery was conducted by the adversary. Consequently the adversary either breaks one of the employed primitives' security or has information-theoretically small advantage in winning the KIND game.

5   Sesquidirectionally Ratcheted Key Exchange (SRKE)

We introduce *sesquidirectionally ratcheted key exchange* (see Footnote 4) as a generalization of URKE. The basic functionality of the two primitives is the same: Sessions involve two parties, A and B, where A can establish keys and safely share them with B by providing the latter with ciphertexts. In contrast to the URKE case, in SRKE also party B can generate and send ciphertexts (to A); however, B's invocations of the sending routine do not establish keys. Rather, the idea behind B communicating ciphertexts to A is that this may increase the security of the keys established by A. Indeed, as we will see, in SRKE it is possible for B to recover from attacks involving state exposure. We proceed with formalizing syntax and correctness of SRKE.

Formally, a SRKE scheme for a finite key space $\mathcal{K}$ and an associated-data space $\mathcal{AD}$ is a tuple $R = (\text{init}, \text{snd}_A, \text{rcv}_B, \text{snd}_B, \text{rcv}_A)$ of algorithms together with a state space $\mathcal{S}_A$, a state space $\mathcal{S}_B$, and a ciphertext space $\mathcal{C}$. The randomized initialization algorithm init returns a state $S_A \in \mathcal{S}_A$ and a state $S_B \in \mathcal{S}_B$. The randomized sending algorithm snd_A takes a state $S_A \in \mathcal{S}_A$ and an associated-data string $ad \in \mathcal{AD}$, and produces an updated state $S'_A \in \mathcal{S}_A$, a key $k \in \mathcal{K}$, and a ciphertext $c \in \mathcal{C}$. The deterministic receiving algorithm rcv_B takes a state $S_B \in \mathcal{S}_B$, an associated-data string $ad \in \mathcal{AD}$, and a ciphertext $c \in \mathcal{C}$, and outputs either an updated state $S'_B \in \mathcal{S}_B$ and a key $k \in \mathcal{K}$, or the special symbol $\bot$ to indicate rejection. The randomized sending algorithm snd_B takes a state $S_B \in \mathcal{S}_B$ and an associated-data string $ad \in \mathcal{AD}$, and produces an

---

[9] A different approach to achieve a constant-size state is to replace lines 10 and 20 by the (non-accumulating) assignments $t \leftarrow (ad, C)$. We believe our scheme would also be secure in this case as, intuitively, chaining key K reflects the full past communication.

updated state $S'_B \in \mathcal{S}_B$ and a ciphertext $c \in \mathcal{C}$. Finally, the deterministic receiving algorithm rcv_A takes a state $S_A \in \mathcal{S}_A$, an associated-data string $ad \in \mathcal{AD}$, and a ciphertext $c \in \mathcal{C}$, and outputs either an updated state $S'_A \in \mathcal{S}_A$ or the special symbol $\bot$ to indicate rejection. A shortcut notation for these syntactical definitions is

$$
\begin{aligned}
\mathrm{init} &\to_{\$} \mathcal{S}_A \times \mathcal{S}_B \\
\mathcal{S}_A \times \mathcal{AD} &\to \mathrm{snd}_A \to_{\$} \mathcal{S}_A \times \mathcal{K} \times \mathcal{C} \\
\mathcal{S}_B \times \mathcal{AD} \times \mathcal{C} &\to \mathrm{rcv}_B \to \mathcal{S}_B \times \mathcal{K} \ / \bot \\
\mathcal{S}_B \times \mathcal{AD} &\to \mathrm{snd}_B \to_{\$} \mathcal{S}_B \times \mathcal{C} \\
\mathcal{S}_A \times \mathcal{AD} \times \mathcal{C} &\to \mathrm{rcv}_A \to \mathcal{S}_A \quad / \bot
\end{aligned}
$$

Correctness of SRKE. Our definition of SRKE functionality is via game FUNC in Fig. 4. We say scheme R is *correct* if $\Pr[\mathrm{FUNC}_R(\mathcal{A}) \Rightarrow 1] = 0$ for all adversaries $\mathcal{A}$. In the figure, the lines of code tagged with a '·' right after the line number also appear in the URKE FUNC game (Fig. 1). In comparison with that game, there are two more oracles, SndB and RcvA, and four new game variables, s_B, r_A, adc_B, is_A, that control and monitor the communication in the B-to-A direction akin to how SndA, RcvB, s_A, r_B, adc_A, is_B do (like in the URKE case) for the A-to-B direction. In particular, the is_A flag is the in-sync indicator of party A that tracks whether the latter was exposed to non-matching associated-data strings or ciphertexts (the transition between in-sync and out-of-sync is detected and recorded in lines 35, 36). Given that the specifications of oracles SndA and RcvB of Figs. 1 and 4 coincide (with one exception: lines 13, 21 are guarded by in-sync checks (in lines 12, 20) so that parties go out-of-sync not only when processing unauthentic associated data or ciphertexts, but also when they process ciphertexts that were generated by an out-of-sync peer[10]), and that also the specifications of oracles SndB and RcvA of Figs. 4 are quite similar to them (besides the reversion of the direction of communication, the difference is that all session-key related components were stripped off), the logics of the FUNC game in Fig. 4 should be clear. Overall, like in the URKE case, the correctness requirement boils down to declaring the adversary successful, in line 31, if A and B compute different keys while still being in-sync.

Epochs. The intuition behind having the B-to-A direction of communication in SRKE is that it allows B to refresh his state every now and then, and to inform A about this. The goal is to let B recover from state exposure.

Imagine, for example, a SRKE session where B has the following view on the communication: first he sends four refresh ciphertexts (to A) in a row; then he receives a key-establishing ciphertext (from A). As we assume a fully concurrent setting and do not impose timing constraints on the network delivery, the incoming ciphertext can have been crafted by A after her having received (from B) between zero and four ciphertexts. That is, even though B refreshed his state a

---

[10] This approach is borrowed from [6,11].

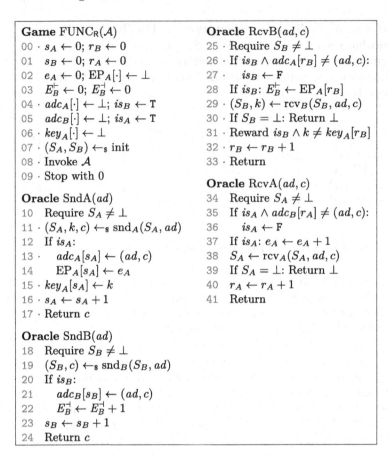

Fig. 4. Game FUNC for SRKE scheme R. The lines of code tagged with a '·' also appear in the URKE FUNC game. Note that the variables $e_A, \mathrm{EP}_A, E_B^{\vdash}, E_B^{\dashv}$ do not influence the the game outcome.

couple of times, to achieve correctness he has to remain prepared for recovering keys from ciphertexts that were generated by A before she recognized any of the refreshes. However, after processing A's ciphertext, if A created it after receiving some of B's ciphertexts (say, the first three), then the situation changes in that B is no longer required to process ciphertexts that refer to refreshes older than the one to which A's current answer is responding to (in the example: the first two).

These ideas turn out to be pivotal in the definition of SRKE security. We formalize them by introducing the notion of an *epoch*. Epochs start when the snd_B algorithm is invoked (each invocation starts one epoch), they are sequentially numbered, and the first epoch (with number zero) is implicitly started by the init algorithm. Each rcv_A invocation makes A aware of one new epoch, and subsequent snd_A invocations can be seen as occurring in its context. Finally, on

B's side multiple epochs may be active at the same time, reflecting that B has to be ready to process ciphertexts that were generated by A in the context of one of potentially many possible epochs. Intuitively, epochs end (on B's side) if a ciphertext is received (from A) that was sent in the context of a later epoch.

We represent the span of epochs supported by B with the interval variable E_B (see Sect. 2.1): its boundaries $E_B^\vdash$ and $E_B^\dashv$ reflect at any time the earliest and the latest such epoch. Further, we use variable e_A to track the latest epoch started by B that party A is aware of, and associative array EP_A to register for each of A's sending operations the context, i.e., the epoch number that A is (implicitly) referring to. In more detail, the invocation of init is accompanied by setting $E_B^\vdash, E_B^\dashv, e_A$ to zero (in lines 02, 03), each sending operation of B introduces one more supported epoch (line 22), each receiving operation of A increases the latter's awareness of epochs supported by B (line 37), the context of each sending operation of A is recorded in EP_A (line 14), and each receiving operation of B potentially reduces the number of supported epochs by dropping obsolete ones (line 28). Observe that tracking epochs is not meaningful after participants get out-of-sync; we thus guard lines 28, 37 with corresponding tests.

Security of SRKE. Our SRKE security model lifts the one for URKE to the bidirectional (more precisely: sesquidirectional) setting. The goal of the adversary is again to distinguish established keys from random. For a SRKE scheme R, the corresponding key indistinguishability games KIND_R^b, for challenge bit $b \in \{0, 1\}$, are specified in Fig. 5. With any adversary $\mathcal{A}$ we associate its key distinguishing advantage $\mathrm{Adv}_R^{\mathrm{kind}}(\mathcal{A}) := |\Pr[\mathrm{KIND}_R^1(\mathcal{A}) \Rightarrow 1] - \Pr[\mathrm{KIND}_R^0(\mathcal{A}) \Rightarrow 1]|$. Intuitively, R offers key indistinguishability if all practical adversaries have a negligible key distinguishing advantage.

The new KIND games are the natural amalgamation of the (URKE) KIND games of Fig. 2 with the (SRKE) FUNC game of Fig. 4 (with the exceptions discussed below). Concerning the trivial attack conditions on URKE that we identified in Sect. 3, we note that conditions (a) and (b) remain valid for SRKE without modification, conditions (c) and (d) (that consider attacks on participants by tracing their computations) need a slight adaptation to reflect that updating epochs repairs the damage of state exposures, and condition (e) (that considers impersonation of exposed A to B), besides needing a slight adaptation, requires to be complemented by a new condition that considers that exposing B allows for impersonating him to A.

When comparing the KIND games from Figs. 2 and 5, note that a crucial difference is that the key_A, key_B arrays in the URKE model are indexed with simple counters, while in the SRKE model they are indexed with pairs where the one element is the same counter as in the URKE case and the other element indicates the epoch for which the corresponding key was established[11]. The new indexing mechanism allows, when B is exposed, for marking as traceable only those future keys of A and B that belong to the epochs managed by B at the time of exposure (lines 54, 57). This already implements the necessary adaptation

---

[11] The adversary always knows the epoch numbers associated with keys, so it can pose meaningful Reveal and Challenge queries just as before.

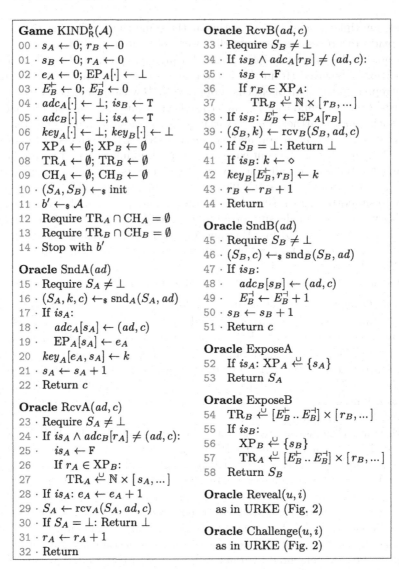

Game $\mathrm{KIND}_\mathsf{R}^b(\mathcal{A})$
00 · $s_A \leftarrow 0$; $r_B \leftarrow 0$
01 · $s_B \leftarrow 0$; $r_A \leftarrow 0$
02 · $e_A \leftarrow 0$; $\mathrm{EP}_A[\cdot] \leftarrow \bot$
03 · $E_B^\vdash \leftarrow 0$; $E_B^\dashv \leftarrow 0$
04 · $adc_A[\cdot] \leftarrow \bot$; $is_B \leftarrow \mathbf{T}$
05 · $adc_B[\cdot] \leftarrow \bot$; $is_A \leftarrow \mathbf{T}$
06 · $key_A[\cdot] \leftarrow \bot$; $key_B[\cdot] \leftarrow \bot$
07   $\mathrm{XP}_A \leftarrow \emptyset$; $\mathrm{XP}_B \leftarrow \emptyset$
08   $\mathrm{TR}_A \leftarrow \emptyset$; $\mathrm{TR}_B \leftarrow \emptyset$
09   $\mathrm{CH}_A \leftarrow \emptyset$; $\mathrm{CH}_B \leftarrow \emptyset$
10 · $(S_A, S_B) \leftarrow_\$ \mathrm{init}$
11 · $b' \leftarrow_\$ \mathcal{A}$
12   Require $\mathrm{TR}_A \cap \mathrm{CH}_A = \emptyset$
13   Require $\mathrm{TR}_B \cap \mathrm{CH}_B = \emptyset$
14 · Stop with b'

Oracle $\mathrm{SndA}(ad)$
15 · Require $S_A \neq \bot$
16 · $(S_A, k, c) \leftarrow_\$ \mathrm{snd}_A(S_A, ad)$
17 · If is_A:
18 · $adc_A[s_A] \leftarrow (ad, c)$
19 · $\mathrm{EP}_A[s_A] \leftarrow e_A$
20   $key_A[e_A, s_A] \leftarrow k$
21 · $s_A \leftarrow s_A + 1$
22 · Return c

Oracle $\mathrm{RcvA}(ad, c)$
23 · Require $S_A \neq \bot$
24 · If $is_A \wedge adc_B[r_A] \neq (ad, c)$:
25 · $is_A \leftarrow \mathbf{F}$
26   If $r_A \in \mathrm{XP}_B$:
27       $\mathrm{TR}_A \overset{\cup}{\leftarrow} \mathbb{N} \times [\, s_A, ... \,]$
28 · If is_A: $e_A \leftarrow e_A + 1$
29 · $S_A \leftarrow \mathrm{rcv}_A(S_A, ad, c)$
30 · If $S_A = \bot$: Return $\bot$
31 · $r_A \leftarrow r_A + 1$
32 · Return

Oracle $\mathrm{RcvB}(ad, c)$
33 · Require $S_B \neq \bot$
34 · If $is_B \wedge adc_A[r_B] \neq (ad, c)$:
35 · $is_B \leftarrow \mathbf{F}$
36   If $r_B \in \mathrm{XP}_A$:
37       $\mathrm{TR}_B \overset{\cup}{\leftarrow} \mathbb{N} \times [\, r_B, ... \,]$
38 · If is_B: $E_B^\dashv \leftarrow \mathrm{EP}_A[r_B]$
39 · $(S_B, k) \leftarrow \mathrm{rcv}_B(S_B, ad, c)$
40 · If $S_B = \bot$: Return $\bot$
41 · If is_B: $k \leftarrow \diamond$
42 · $key_B[E_B^\vdash, r_B] \leftarrow k$
43 · $r_B \leftarrow r_B + 1$
44 · Return

Oracle $\mathrm{SndB}(ad)$
45 · Require $S_B \neq \bot$
46 · $(S_B, c) \leftarrow_\$ \mathrm{snd}_B(S_B, ad)$
47 · If is_B:
48 · $adc_B[s_B] \leftarrow (ad, c)$
49 · $E_B^\dashv \leftarrow E_B^\dashv + 1$
50 · $s_B \leftarrow s_B + 1$
51 · Return c

Oracle $\mathrm{ExposeA}$
52   If is_A: $\mathrm{XP}_A \overset{\cup}{\leftarrow} \{s_A\}$
53   Return S_A

Oracle $\mathrm{ExposeB}$
54   $\mathrm{TR}_B \overset{\cup}{\leftarrow} [E_B^\vdash .. E_B^\dashv] \times [\, r_B, ... \,]$
55   If is_B:
56     $\mathrm{XP}_B \overset{\cup}{\leftarrow} \{s_B\}$
57     $\mathrm{TR}_A \overset{\cup}{\leftarrow} [E_B^\vdash .. E_B^\dashv] \times [\, r_B, ... \,]$
58   Return S_B

Oracle $\mathrm{Reveal}(u, i)$
  as in URKE (Fig. 2)

Oracle $\mathrm{Challenge}(u, i)$
  as in URKE (Fig. 2)

Fig. 5. Games KIND^b, $b \in \{0, 1\}$, for SRKE scheme R. Lines of code tagged with a '·' similarly appear in the SRKE FUNC game in Fig. 4.

of conditions (c) and (d) to the SRKE setting. The announced adaptation of condition (e) is executing line 52 only if $is_A = \mathbf{T}$; the change is due as the motivation given in Sect. 3 is valid only if A is in-sync (which is always the case in URKE, but not in SRKE). Finally, complementing condition (e), we identify the following new trivial attack condition:

(f) Exposing party B allows for impersonating it: Assume parties A and B are in-sync. The adversary obtains a copy of B's state and invokes the snd$_B$ algorithm with it, obtaining a ciphertext which it provides to party A (rendering the latter out-of-sync). If then A generates a new key using the snd$_A$ algorithm, the adversary can feed the resulting ciphertext into the rcv$_B$ algorithm, recovering the key. Example: fix some ad, ad'; $S_B^* \leftarrow \text{ExposeB}()$; $(S_B^*, c) \leftarrow_\$ \text{snd}_B(S_B^*, ad)$; $\text{RcvA}(ad, c)$; $c' \leftarrow \text{SndA}(ad')$; $(S_B^*, k) \leftarrow \text{rcv}_B(S_B^*, ad', c')$; $k' \leftarrow \text{Challenge}(A, 0)$; $b' \leftarrow [k = k']$; output b'. Lines 26, 27 (in conjunction with lines 07, 56) detect the described type of impersonation and mark all future keys of A as traceable.

This completes the description of our SRKE security model. As in URKE, it excludes the minimal set of attacks, indicating that it gives strong security guarantees.

6   Constructing SRKE

We present a SRKE construction that generalizes our URKE scheme to the sesquidirectional setting. The core intuition is as follows: Like in the URKE scheme, A-to-B ciphertexts correspond with KEM ciphertexts where the corresponding public and secret keys are held by A and B, respectively, and the two keys are evolved to new keys after each use. In addition to this, with the goal of letting B heal from state exposures, our SRKE construction gives him the option to sanitize his state by generating a fresh KEM key pair and communicating the corresponding public key to A (using the B-to-A link specific to SRKE). The algorithms of our protocol are specified in Fig. 6. Although the sketched approach might sound simple and natural, the algorithms, quite surprisingly, are involved and require strong cryptographic building blocks (a key-updatable KEM and a one-time signature scheme, see Sect. 2). Their complexity is a result of SRKE protocols having to simultaneously offer solutions to multiple inherent challenges. We discuss these in the following.

EPOCH MANAGEMENT. Recall that we assume a concurrent setting for SRKE and that, thus, if B refreshes his state via the snd$_B$ algorithm, then he still has to keep copies of the secret keys maintained for prior epochs (so that the rcv$_B$ algorithm can properly process incoming ciphertexts created for them). Our protocol algorithms implement this by including in B's state the array $SK[\cdot]$ in which snd$_B$ stores all keys it generates (line 27; obsolete keys of expired epochs are deleted by rcv$_B$ in line 47). Beyond that, both A and B maintain an interval variable E in their state: its boundaries $E^\vdash$ and $E^\dashv$ are used by B to reflect the earliest and latest supported epoch, and by A to keep track of epoch updates that occur in direct succession (i.e., that are still waiting for their 'activation' by snd$_A$). Note finally that the snd$_A$ algorithm communicates to rcv$_B$ in every outgoing ciphertext the epoch in which A is operating (line 12).

Proc init
00 $(sgk, vfk) \leftarrow_\$ \text{gen}_S$
01 · $(sk, pk) \leftarrow_\$ \text{gen}_K$
02 · $K \leftarrow_\$ \mathcal{K}; k.m \leftarrow_\$ \mathcal{K}; t \leftarrow \epsilon$
03 $E^\vdash \leftarrow 0; E^\dashv \leftarrow 0$
04 $s \leftarrow 0; r \leftarrow 0$
05 $PK[\cdot] \leftarrow \bot; PK[0] \leftarrow pk$
06 $SK[\cdot] \leftarrow \bot; SK[0] \leftarrow sk$
07 $L_A[\cdot] \leftarrow \bot; L_B[\cdot] \leftarrow \bot; L_A[0] \leftarrow \diamond$
08 $S_A \leftarrow (PK, E, s, L_A, vfk, K, k.m, t)$
09 $S_B \leftarrow (SK, E, r, L_B, sgk, K, k.m, t)$
10 Return (S_A, S_B)

Proc $\text{snd}_A(S_A, ad)$
11 $(PK, E, s, L, vfk, K, k.m, t) \leftarrow S_A$
12 $k^* \leftarrow \epsilon; C \leftarrow E^\dashv$
13 For $e' \leftarrow E^\vdash$ to $E^\dashv$:
14 · $(k, c) \leftarrow_\$ \text{enc}(PK[e'])$
15      $k^* \xleftarrow{"} k; C \xleftarrow{"} c$
16 · $\tau \leftarrow_\$ \text{tag}(k.m, ad \| C)$
17 · $C \xleftarrow{"} \tau; t \xleftarrow{"} \triangleright \| ad \| C$
18 · $k.o \| K \| k.m \| sk \leftarrow H(K, k^*, t)$
19 · $pk \leftarrow \text{gen}_K(sk)$
20 $PK[..., (E^\dashv - 1)] \leftarrow \bot; PK[E^\dashv] \leftarrow pk$
21 $E^\vdash \leftarrow E^\dashv; s \leftarrow s + 1; L[s] \leftarrow ad \| C$
22 $S_A \leftarrow (PK, E, s, L, vfk, K, k.m, t)$
23 Return $(S, k.o, C)$

Proc $\text{snd}_B(S_B, ad)$
24 $(SK, E, r, L, sgk, K, k.m, t) \leftarrow S_B$
25 $(sk^*, pk^*) \leftarrow_\$ \text{gen}_K$
26 $(sgk^*, vfk^*) \leftarrow_\$ \text{gen}_S$
27 $E^\dashv \leftarrow E^\dashv + 1; SK[E^\dashv] \leftarrow sk^*$
28 $C \leftarrow r \| pk^* \| vfk^*$
29 $\sigma \leftarrow_\$ \text{sgn}(sgk, ad \| C)$
30 $C \xleftarrow{"} \sigma; L[E^\dashv] \leftarrow \triangleleft \| ad \| C$
31 $S_B \leftarrow (SK, E, r, L, sgk^*, K, k.m, t)$
32 Return (S_B, C)

Proc $\text{rcv}_B(S_B, ad, C)$
33 $(SK, E, r, L, sgk, K, k.m, t) \leftarrow S_B$
34 $t^* \leftarrow ad \| C; C \| \tau \leftarrow C$
35 · Require $\text{vfy}_M(k.m, ad \| C, \tau)$
36 $k^* \leftarrow \epsilon; e \| C \leftarrow C$
37 Require $E^\vdash \leq e \leq E^\dashv$
38 $t \xleftarrow{"} L[E^\vdash + 1] \| \ldots \| L[e]$
39 $L[..., e] \leftarrow \bot$
40 For $e' \leftarrow E^\vdash$ to e:
41 · $c \| C \leftarrow C$
42 · $k \leftarrow \text{dec}(SK[e'], c)$
43 ·   Require $k \neq \bot$
44      $k^* \xleftarrow{"} k$
45 · $t \xleftarrow{"} \triangleright \| t^*$
46 · $k.o \| K \| k.m \| sk \leftarrow H(K, k^*, t)$
47 $SK[..., (e-1)] \leftarrow \bot; SK[e] \leftarrow sk$
48 For $e' \leftarrow e + 1$ to $E^\dashv$:
49      $SK[e'] \leftarrow \text{up}(SK[e'], t^*)$
50 $E^\vdash \leftarrow e; r \leftarrow r + 1$
51 $S_B \leftarrow (SK, E, r, L, sgk, K, k.m, t)$
52 Return $(S_B, k.o)$

Proc $\text{rcv}_A(S_A, ad, C)$
53 $(PK, E, s, L, vfk, K, k.m, t) \leftarrow S_A$
54 $t \xleftarrow{"} \triangleleft \| ad \| C; C \| \sigma \leftarrow C$
55 Require $\text{vfy}_S(vfk, ad \| C, \sigma)$
56 $r \| pk^* \| vfk \leftarrow C$
57 Require $L[r] \neq \bot$
58 $L[..., (r-1)] \leftarrow \bot; L[r] \leftarrow \diamond$
59 For $s' \leftarrow r + 1$ to s:
60      $pk^* \leftarrow \text{up}(pk^*, L[s'])$
61 $E^\dashv \leftarrow E^\dashv + 1; PK[E^\dashv] \leftarrow pk^*$
62 $S_A \leftarrow (PK, E, s, L, vfk, K, k.m, t)$
63 Return S_A

Fig. 6. Construction of a SRKE scheme from a key-updatable KEM $K = (\text{gen}_K, \text{enc}, \text{dec})$, a message authentication code $M = (\text{tag}, \text{vfy}_M)$, a one-time signature scheme $S = (\text{gen}_S, \text{sgn}, \text{vfy}_S)$, and a random oracle H. For simplicity we denote the key space of the MAC and the space of chaining keys with the same symbol $\mathcal{K}$. Notation: Lines 07, 58: If an entry of an array is expected to contain a ciphertext, but clearly the value of the ciphertext will not any more matter, we instead store the placeholder symbol $\diamond$. Line 38: If $E^\vdash = e$ then no value shall be concatenated to t. Line 41: The last iteration of the loop is meant to clear C; a more precise version of the line would say "If $e' < e$ then $c \| C \leftarrow C$ else $c \leftarrow C$". Lines 17, 45, 54, 30: We use labels $\triangleright$ and $\triangleleft$ in transcript fragments to distinguish whether they emerged in the A-to-B or B-to-A direction. Lines of code tagged with a '·' depict the URKE construction's core.

SECURE STATE UPDATE. Assume A executes once the snd_A algorithm, then twice the rcv_A algorithm, and then again once the snd_A algorithm. That is, following the above sketch of our protocol, as part of her first snd_A invocation she will encapsulate to a public key that she subsequently updates (akin to how she would do in our URKE solution, see lines 07, 12 of Fig. 3), then she will receive two fresh public keys from B, and finally she will again encapsulate to a public key that she subsequently updates. The question is: Which public key shall she use in the last step? The one resulting from the update during her first snd_A invocation, the one obtained in her first rcv_A invocation, or the one obtained in her second rcv_A invocation? We found that only one configuration is safe against key distinguishing attacks: Our SRKE protocol is such that she encapsulates to all three, combining the established session keys into one via concatenation.[12,13] The algorithms implement this by including in A's state the array $PK[\cdot]$ in which rcv_A stores incoming public keys (line 61) and which snd_A consults when establishing outgoing ciphertexts (lines 13–15; the counterpart on B's side is in lines 40–44). Once the switch to the new epoch is completed, the obsolete public keys are removed from A's state (line 20). If A executes snd_A many times in succession, then all but the first invocation will, akin to the URKE case, just encapsulate to the (one) evolved public key from the preceding invocation.

We discuss a second issue related to state updates. Assume B executes three times the snd_B algorithm and then once the rcv_B algorithm, the latter on input a well-formed but non-authentic ciphertext (e.g., the adversary could have created the ciphertext, after exposing A's state, using the snd_A algorithm). In the terms of our security model the latter action brings B out-of-sync, which means that if he is subsequently exposed then this should not affect the security of further session keys established by A. On the other hand, according to the description provided so far, exposing B's state means obtaining a copy of array $SK[\cdot]$, i.e., of the decapsulation keys of all epochs still supported by B. We found that this easily leads to key distinguishing attacks,[14] so in order to protect the elements of $SK[\cdot]$ they are evolved by the rcv_B algorithm whenever an incoming ciphertext is processed. We implement the latter via the dedicated update procedure up provided by the key-updatable KEM. The corresponding lines are 48–49 (note that t^* is the current transcript fragment, see line 34). Of course A has to synchronize on B's key updates, which she does in lines 59–60, where array $L[\cdot]$ is the state variable that keeps track of the corresponding past A-to-B transcript fragments. (Outgoing ciphertexts are stored in $L[\cdot]$ in line 21, and obsolete ones are removed from it in line 58.) Note that A, for staying synchronized with B, also needs to keep track of the ciphertexts that he received (from her) so far; for this reason, B indicates in every outgoing ciphertext the number r of incoming ciphertexts he has been exposed to (lines 56, 28).

---

[12] We discuss why it is unsafe to encapsulate to only a subset of the keys in Appendix A.3.

[13] The concatenation of keys of an OW secure KEM can be seen as the implementation of a secure combiner in the spirit of [8].

[14] We discuss this further in Appendix A.2.

TRANSCRIPT MANAGEMENT. Recall that one element of the participants' state in our URKE scheme (in Fig. 3) is the variable t that accumulates transcript information (associated data and ciphertexts) of prior communication so that it can be input to key derivation. This is a common technique to ensure that the keys established on the two sides start diverging in the moment an active attack occurs. Also our SRKE construction follows this approach, but accumulating transcripts is more involved if communication is concurrent: If both A and B would add outgoing ciphertexts to their transcript accumulator directly after creating them, then concurrent sending would immediately desynchronize the two parties. This issue is resolved in our construction as follows: In the B-to-A direction, while A appends incoming ciphertexts (from B) to her transcript variable in the moment she receives them (line 54), when creating the ciphertexts, B will just record them in his state variable $L[\cdot]$ (line 30), and postpone adding them to his transcript variable to the point when he is able to deduce (from A's ciphertexts) the position of when she did (line 38; obsolete entries are removed in line 39). The A-to-B direction is simpler[15] and handled as in our URKE protocol: A updates her transcript when sending a ciphertext (line 17), and B updates his transcript when receiving it (lines 34, 45). Note we tag transcript fragments with labels $\triangleright$ or $\triangleleft$ to indicate whether they emerged in the A-to-B or B-to-A direction of communication (e.g., in lines 17, 30).

AUTHENTICATION. To reach security against active adversaries we protect the SRKE ciphertexts against manipulation. Recall that in our URKE scheme a MAC was sufficient for this. In SRKE, a MAC is still sufficient for the A-to-B direction (lines 16, 35), but for the B-to-A direction, to defend against attacks where the adversary first exposes A's state and then uses the obtained MAC key to impersonate B to her,[16] we need to employ a one-time signature scheme: Each ciphertext created by B includes a freshly generated verification key that is used to authenticate the next B-to-A ciphertext (lines 26, 28, 29, 55, 56).

The only lines we did not comment on are 18, 19, 25, 46 — those that also form the core of our URKE protocol (which are discussed in Sect. 4).

Practicality of our construction. We remark that the number of updates per kuKEM key pair is bounded by the number of ciphertexts sent by A during one round-trip time (RTT) on the network between A and B (intuitively by the number of ciphertexts sent by A that *cross* the wire with one epoch update ciphertext from B). Ciphertexts that B did not know of when proposing an epoch (1/2 RTT) and ciphertexts A sent until she received the epoch proposal (1/2 RTT) are regarded for an update of a key pair. As a result, the hierarchy of an HIBE can be bounded by this number of ciphertexts when used for building a kuKEM for SRKE.

---

[15] Intuitively the disbalance comes from the fact that keys are only established by A-to-B ciphertexts and that transcripts are only used for key derivation.

[16] Note this is not an issue in the A-to-B direction: Exposing B and impersonating A to him leads to marking all future keys of B as traceable anyway, without any option to recover. We expand on this in Appendix A.1.

Theorem 2 (informal). *The SRKE protocol* R *from Fig. 6 offers key indistinguishability if function* H *is modeled as a random oracle, the kuKEM provides* KUOW *security, the one-time signature scheme provides* SUF *security, the MAC provides* SUF *security, and the session-key space of the kuKEM is sufficiently large.*

The exact theorem statement and the respective proof are in the full version [14]. The approach of the proof is the same as in our URKE proof but with small yet important differences: (1) the proof reduces signature forgeries to the SUF security of the signature scheme to show that communication from B to A is authentic, (2) the security of session keys established by A is reduced to the KUOW security of the kuKEM. The reduction to the KUOW game is split into three cases: (a) session keys established by A in sync, (b) the first session key established by A out of sync, and (c) all remaining session keys established by A out of sync. This distinction is made as in each of these cases a different encapsulated key—as part of the random oracle input—is assumed to be unknown to the adversary. Finally the SRKE proof—as in the URKE proof—makes use of the MAC's SUF security to show that B will never establish challengeable keys out of sync.

7   From URKE and SRKE to BRKE

In Sects. 3–6 we proposed security models and constructions for URKE and SRKE. For space reasons we defer the corresponding formalizations for BRKE (bidirectional RKE) to the full version [14]. Here we quickly sketch how one can obtain notions and constructions for the latter from the former.

The syntax, correctness, and security definitions for BRKE can be seen as an amalgamation of two copies of the corresponding definitions for SRKE, one in each direction of communication. Fortunately, several of the game variables can be unified so that the games remain relatively compact.

The same type of amalgamation can be applied to obtain a BRKE construction: While just running two generic SRKE instances side by side (in reverse directions) is not sufficient to obtain a secure solution, carefully binding them together, in our case with one-time signatures as an auxiliary tool, is. More precisely, each BRKE send operation results in (1) the creation of a fresh one-time signature key pair, (2) the invocation of the two SRKE send routines (the one in the A-to-B and the other in the B-to-A direction) where the signature verification key is provided as associated data, (3) encoding the verification key and the two SRKE ciphertexts into a single ciphertext and securing the latter with a signature. See [14] for the details.

Acknowledgments. We thank Fabian Weißberg for very inspiring discussions at the time we first explored the topic of ratcheted key exchange. We further thank Giorgia Azzurra Marson and anonymous reviewers for comments and feedback on the article. (This holds especially for a EUROCRYPT 2018 reviewer who identified an issue in a prior version of our URKE construction.) Bertram Poettering conducted part of the work at Ruhr University Bochum supported by ERC Project ERCC (FP7/615074). Paul Rösler received support by SyncEnc, funded by the German Federal Ministry of Education and Research (BMBF, FKZ: 16KIS0412K).

References

1. Bellare, M., Singh, A.C., Jaeger, J., Nyayapati, M., Stepanovs, I.: Ratcheted encryption and key exchange: the security of messaging. In: Katz, J., Shacham, H. (eds.) CRYPTO 2017, Part III. LNCS, vol. 10403, pp. 619–650. Springer, Cham (2017). https://doi.org/10.1007/978-3-319-63697-9_21
2. Bellare, M., Yee, B.: Forward-security in private-key cryptography. In: Joye, M. (ed.) CT-RSA 2003. LNCS, vol. 2612, pp. 1–18. Springer, Heidelberg (2003). https://doi.org/10.1007/3-540-36563-X_1
3. Borisov, N., Goldberg, I., Brewer, E.A.: Off-the-record communication, or, why not to use PGP. In: Atluri, V., Syverson, P.F., di Vimercati, S.D.C. (eds.) Proceedings of the 2004 ACM WPES 2004, Washington, DC, USA, 28 October 2004, pp. 77–84. ACM (2004)
4. Cohn-Gordon, K., Cremers, C.J.F., Dowling, B., Garratt, L., Stebila, D.: A formal security analysis of the signal messaging protocol. In: 2017 IEEE EuroS&P 2017, Paris, France, 26–28 April 2017, pp. 451–466. IEEE (2017)
5. Cohn-Gordon, K., Cremers, C.J.F., Garratt, L.: On post-compromise security. In: IEEE CSF 2016, Lisbon, Portugal, 27 June–1 July 2016, pp. 164–178. IEEE Computer Society (2016)
6. Eugster, P.T., Marson, G.A., Poettering, B.: A cryptographic look at multi-party channels. In: 31st IEEE Computer Security Foundations Symposium (2018, to appear)
7. Gentry, C., Silverberg, A.: Hierarchical ID-based cryptography. In: Zheng, Y. (ed.) ASIACRYPT 2002. LNCS, vol. 2501, pp. 548–566. Springer, Heidelberg (2002). https://doi.org/10.1007/3-540-36178-2_34
8. Giacon, F., Heuer, F., Poettering, B.: KEM combiners. In: Abdalla, M., Dahab, R. (eds.) PKC 2018, Part I. LNCS, vol. 10769, pp. 190–218. Springer, Cham (2018). https://doi.org/10.1007/978-3-319-76578-5_7
9. Langley, A.: Source code of Pond, May 2016. https://github.com/agl/pond
10. Marlinspike, M., Perrin, T.: The double Ratchet algorithm, November 2016. https://whispersystems.org/docs/specifications/doubleratchet/doubleratchet.pdf
11. Marson, G.A., Poettering, B.: Security notions for bidirectional channels. IACR Trans. Symm. Cryptol. **2017**(1), 405–426 (2017)
12. Moscaritolo, V., Belvin, G., Zimmermann, P.: Silent Circle Instant Messaging Protocol: Protocol specification (2012). https://silentcircle.com/sites/default/themes/silentcircle/assets/downloads/SCIMP_paper.pdf
13. Off-the-record messaging (2016). http://otr.cypherpunks.ca
14. Poettering, B., Rösler, P.: Asynchronous ratcheted key exchange. Cryptology ePrint Archive, Report 2018/296 (2018). https://eprint.iacr.org/2018/296

15. Rogaway, P.: Authenticated-encryption with associated-data. In: Atluri, V. (ed.) ACM CCS 2002, Washington D.C., USA, 18–22 November 2002, pp. 98–107. ACM Press (2002)
16. Rösler, P., Mainka, C., Schwenk, J.: More is less: on the end-to-end security of group chats in Signal, WhatsApp, and Threema. In: IEEE EuroS&P 2018 (2018)
17. Schneier, B., Kelsey, J.: Secure audit logs to support computer forensics. ACM Trans. Inf. Syst. Secur. **2**(2), 159–176 (1999)
18. Unger, N., Dechand, S., Bonneau, J., Fahl, S., Perl, H., Goldberg, I., Smith, M.: SoK: secure messaging. In: 2015 IEEE Symposium on Security and Privacy, San Jose, CA, USA, 17–21 May 2015, pp. 232–249. IEEE Computer Society Press (2015)
19. Zimmermann, P., Johnston, A., Callas, J.: ZRTP: media path key agreement for unicast secure RTP. RFC 6189, RFC Editor, April 2011. http://www.rfc-editor.org/rfc/rfc6189.txt

A   Rationale for SRKE Design

We sketched the reasons for employing sophisticated primitives as basic blocks for our design of SRKE in the main body. In this section we develop more detailed arguments for our design choices by providing attacks on constructions different from our design. At first it is described why SRKE requires signatures for protecting the communication from B to A—in contrast to employing a MAC from A to B. Then we will evaluate the requirements for the KEM key pair update in the setting of concurrent sending of A and B.

A.1   Signatures from A to B

While a MAC suffices to protect authenticity for ciphertexts sent from A to B it does not suffice to protect the authenticity in the counter direction. The reason for this lies within the conditions with which future session keys of A and B are marked traceable in the KIND_R game of SRKE. An impersonation of A towards B has the same effect on the traceability of B's future session keys as if the adversary exposes B's state and then brings B out of sync. Either way all future session keys of B are marked traceable (see Fig. 5 lines 37 and 54, 38). In the first scenario, the adversary can compute the same session keys as B because the adversary initiates the key establishment impersonating A. In the second scenario, the adversary can comprehend B's computations during the receipt of ciphertexts because it possesses the same state information as B.

For computations of A, however, only the former scenario is applicable: if the adversary impersonated B towards A, then again the adversary is in the position to trace the establishment of session keys of A because it can simulate the respective counterpart's receiver computations. In contrast to this, when exposing A and bringing her out of sync, according to the KIND_R game, the adversary must not obtain information on her future session keys (see Fig. 5 lines 52 et seqq.). As a result, the exposure of A's state should not enable the adversary to impersonate B towards A. Consequently the authentication of the communication from B to A cannot be reached by a primitive with a symmetric secret but rather the protocol needs to ensure that B needs to be exposed in order to impersonate him towards A.

The non-trivial attack that is defended by employing signatures consists of the following adversary behavior: $S_A \leftarrow$ ExposeA; *Extract authentication secrets from S_A to derive S_B';* $(C', S_B'') \leftarrow_\$ \mathrm{snd}_B(S_B', \epsilon)$; RcvA$(C', \epsilon)$; $C_{A1} \leftarrow_\$$ SndA(ϵ); $k_b \leftarrow_\$$ Challenge$(A, 1)$. Thereby the adversary must not be able to decide whether it obtained the real or random key for ciphertext C_{A1} from the challenge oracle. Please note that this is related to *key-compromise impersonation* resilience (while in this case ephemeral signing keys are compromised).

A.2   Key-Updatable KEM for Concurrent Sending

There exist two crucial properties that are required from the key pair update of the KEM in the setting in which A and B send concurrently. Firstly, the key update needs to be forward secure which means that an updated secret key does not reveal information on encapsulations to previous secret keys or to differently updated secret keys. Secondly, the update of the public key must not reveal information on keys that will be encapsulated to its respective secret key. We will explain the necessity of these requirements one after another.

The key pair update for concurrently sending only affects epochs that have been proposed by B, but that have not been processed by A yet. These updates have to consider ciphertexts that A sent during the transmission of the public key for a new epoch from B to A. Subsequently we describe an example scenario in which these updates are necessary for defending a non-trivial attack: In the worst case, all secrets among A and B have been exposed to the adversary before B proposes a new epoch ($S_A \leftarrow$ ExposeA; $S_B \leftarrow$ ExposeB). Thereby only a public key sent by B after the exposure will provide security for future session key establishments initiated by A. Now consider a scenario in which B proposes this new public key to A ($C_{B1} \leftarrow_\$$ SndB(ϵ); RcvA(C_{B1}, ϵ)) and A is simultaneously impersonated towards B (($S_A', k', C') \leftarrow_\$ \mathrm{snd}_A(S_A, \epsilon)$; RcvB$(C', \epsilon)$). Since B proposed the new public key within C_{B1} in sync and A received it in sync respectively—and B was not exposed under the new state—, future established session keys of A are considered to be indistinguishable from random key space elements again ($C_{A1} \leftarrow_\$$ SndA(ϵ); $k_b \leftarrow_\$$ Challenge$(A, 1)$). Due to the impersonation of A towards B, however, B became out of sync. Becoming out of sync cannot be detected by B because the adversary can send a valid ciphertext C' under the exposed state of A S_A. Exposing B out of sync afterwards ($S_B' \leftarrow$ ExposeB), by definition, must not have an impact on the security of session keys established by A (see Fig. 5 line 55). As a result, after the adversary performed these steps, the challenged session key is required to be indistinguishable from a random element from the key space. Consequently B must perform an update of the secret key for the newest epoch when receiving C' such that the public key transmitted in C_{B1} still provides its security guarantees when using it in A's final send operation (remember that all previous secrets among A and B were exposed before).

When accepting that an update of B's future epoch's secret keys is required at the receipt of ciphertexts, another condition arises for the respective update of A's public keys. For maintaining correctness, A of course needs to compute updates of a received new public key with respect to all previously sent ciphertexts that B was not aware of when sending the public key. Suppose A's and B's secrets have all been exposed towards the adversary again ($S_A \leftarrow$ ExposeA; $S_B \leftarrow$ ExposeB). Now A sends a new key establishing ciphertext and B proposes a new epoch public key ($C_{A1} \leftarrow_\$ \mathrm{SndA}(\epsilon)$; $C_{B1} \leftarrow_\$ \mathrm{SndB}(\epsilon)$). According to the previous paragraph, A needs to update the received public key in C_{B1} with respect to C_{A1} after receiving C_{B1} ($\mathrm{RcvA}(C_{B1}, \epsilon)$). Since C_{B1} introduces a new epoch, the next send operation of A needs to establish a secure session key again ($C_{A2} \leftarrow_\$ \mathrm{SndA}(\epsilon)$; $k_b \leftarrow_\$ \mathrm{Challenge}(A, 2)$). Now observe that in order to update the received public key, A can only use information from her state S_A—which is known by the adversary—, public information like the transmitted ciphertexts, and randomness. Essentially, the update can hence only depend on information that the adversary knows plus random coins which cannot be transmitted confidentially to B before performing the update (because there exist no secrets apart from the key pair that first needs to be updated). Since B probably received C_{A1} before A received C_{B1}, A cannot influence the update performed by B on his secret key. This means that the updates of A and B need to be conducted independently. As such, the adversary is able to perform the update on the same information that A has (only randomness of A and the adversary can differ). Nevertheless, both updates—the one performed by the adversary and the one performed by A—need to be compatible to the secret key that B derives from his update. As a result, the update of the public key must not reveal the respective secret key (or any other information that can be used to obtain information on keys encapsulated to this updated public key). Otherwise, the adversary would obtain this information as well (and thereby the security of key $(A, 2)$ would not be preserved).

Both requirements are reflected in the security game of the kuKEM (see full version [14]).

A.3   Encapsulation to All Public Keys

Subsequently we describe a scenario in which A only maintains one public key in her state to which she can securely encapsulate keys (while the state contains multiple *useless* public keys). This scenario is crucial because A does not know, which of her public keys provides security, and the SRKE protocol is required to output secure session keys in this scenario. Consequently only encapsulating to all public keys in A's state solves the underlying issue. The reasons for encapsulating to all public keys in A's state is closely related to the reasons for employing a kuKEM in SRKE (see the previous subsection).

Assume the adversary exposes the states of both parties ($S_A \leftarrow$ ExposeA; $S_B \leftarrow$ ExposeB). Consequently none of A's public keys provides any security guarantees for the encapsulation towards the adversary anymore. If the adversary lets B send a ciphertext and thereby propose a new public key to A, A's future session keys are required to be secure again $(C_{B1} \leftarrow_\$ \text{SndB}(\epsilon); \text{RcvA}(C_{B1}, \epsilon))$. Impersonating A towards B and then exposing B to obtain his state has—according to the KIND_R game—no influence on the traceability of A's future session keys $((S'_A, k', C') \leftarrow_\$ \text{snd}_A(S_A, \epsilon); \text{RcvB}(C', \epsilon); S'_B \leftarrow \text{ExposeB})$. However, our construction allows the adversary to impersonate B towards A afterwards: the impersonation of A towards B only *invalidates* the kuKEM secret key in B's state via the key update in B's receive algorithm. The signing key in B's state is still valid for the communication to A since it was not modified at the receipt of the impersonating ciphertext. As such, the adversary may use the signing key and then implant further public keys in A's state by sending these public keys to A $((S''_B, C'') \leftarrow_\$ \text{snd}_B(S'_B, \epsilon); \text{RcvA}(C'', \epsilon))$. These public keys do not provide security with respect to A's session keys since the adversary can freely choose them. As a result, only the public key that B sent in sync before A was impersonated towards B belongs to a secret key that the adversary does not know (public key in C_{B1}). Since A has no indication which public key's secret key is not known by the adversary (note that A and B were exposed at the beginning of the presented scenario and the adversary planted own public keys in A's state at the end of the scenario by sending valid ciphertexts), A needs to encapsulate to all public keys in order to obtain at least one encapsulated key as secret input to the random oracle such that the session key also remains secure $(C_{A1} \leftarrow_\$ \text{SndA}(\epsilon); k_b \leftarrow_\$ \text{Challenge}(A, 1))$.

Observe that the scenario, described above, lacks an argument why also the first public key in A's state needs to be used for the encapsulation if A received further public keys from B afterwards. The reason for also using the first public key, that is always derived from the previous random oracle output, lies within A's sending after becoming out of sync. A became out of sync by receiving C'' (see above). When sending C_{A1}, A derived a new public key for her state. The secret key to this public key was part of the same random oracle output as the session key that is challenged afterwards $(A, 1)$. As argued before, this session key is secure (for all details we refer the reader to the proof in the full version [14]). Consequently the public key in A's state after sending C_{A1} provides security against the adversary regrading encapsulations. However, the adversary can still plant new public keys to A's state $((S'''_B, C''') \leftarrow_\$ \text{snd}_B(S''_B, \epsilon); \text{RcvA}(C''', \epsilon))$. As such, only the first public key in A's state provides security after A became out of sync (and sent once afterwards). All remaining public keys may belong to secret keys chosen by the adversary. Since A will not notice when she became out of sync, she also needs to include the first public key in her state for encapsulating within her send algorithm in order to compute secure session keys $(C_{A2} \leftarrow_\$ \text{SndA}(\epsilon); k_{b2} \leftarrow_\$ \text{Challenge}(A, 2))$.

As a result, A always needs to encapsulate to all public keys in her state such that at least one encapsulated key is a secret input to the random oracle (in case her future session keys were not marked traceable by the KIND_R game).

Optimal Channel Security Against Fine-Grained State Compromise: The Safety of Messaging

Joseph Jaeger[✉] and Igors Stepanovs

Department of Computer Science and Engineering,
University of California San Diego, La Jolla, USA
{jsjaeger,istepano}@eng.ucsd.edu

Abstract. We aim to understand the best possible security of a (bidirectional) cryptographic channel against an adversary that may arbitrarily and repeatedly learn the secret state of either communicating party. We give a formal security definition and a proven-secure construction. This construction provides better security against state compromise than the Signal Double Ratchet Algorithm or any other known channel construction. To facilitate this we define and construct new forms of public-key encryption and digital signatures that update their keys over time.

1 Introduction

End-to-end encrypted communication is becoming a usable reality for the masses in the form of secure messaging apps. However, chat sessions can be extremely long-lived and their secrets are stored on end user devices, so they are particularly vulnerable to having their cryptographic secrets exfiltrated to an attacker by malware or physical access to the device. The Signal protocol [33] by Open Whisper Systems tries to mitigate this threat by continually updating the key used for encryption. Beyond its use in the Signal messaging app, this protocol has been adopted by a number of other secure messaging apps. This includes being used by default in WhatsApp and as part of secure messaging modes of Facebook Messenger, Google Allo, and Skype.

WhatsApp alone has 1 billion daily active users [43]. It is commonly agreed in the cryptography and security community that the Signal protocol is secure. However, the protocol was designed without an explicitly defined security notion. This raises the questions: what security does it achieve and could we do better?

In this work we study the latter question, aiming to understand the best possible security of two-party communication in the face of state exfiltration. We formally define this notion of security and design a scheme that provably achieves it.

Security against compromise. When a party's secret state is exposed we would like both that the security of past messages and (as soon as possible) the security of future messages not be damaged. These notions have been considered in a

© International Association for Cryptologic Research 2018
H. Shacham and A. Boldyreva (Eds.): CRYPTO 2018, LNCS 10991, pp. 33–62, 2018.
https://doi.org/10.1007/978-3-319-96884-1_2

variety of contexts with differing terminology. The systemization of knowledge paper on secure messaging [42] by Unger et al. evaluates and systematizes a number of secure messaging systems. In it they describe a variety of terms for these types of security including "forward secrecy," "backwards secrecy," "self-healing," and "future secrecy" and note that they are "controversial and vague." Cohn-Gordon et al. [15] study the future direction under the term of post-compromise security and similarly discuss the terms "future secrecy," "healing," and "bootstrapping" and note that they are "intuitive" but "not well-defined." Our security notion intuitively captures any of these informal terms, but we avoid using any of them directly by aiming generically for the best possible security against compromise.

Channels. The standard model for studying secure two party communication is that of the (cryptographic) channel. The first attempts to consider the secure channel as a cryptographic object were made by Shoup [39] and Canetti [11]. It was then formalized by Canetti and Krawczyk [13] as a modular way to combine a key exchange protocol with authenticated encryption, which covers both privacy and integrity. Krawczyk [28] and Namprempre [32] study what are the necessary and sufficient security notions to build a secure channel from these primitives.

Modern definitions of channels often draw from the game-based notion of security for stateful authenticated-encryption as defined by Bellare et al. [4]. We follow this convention which assumes initial generation of keys is trusted. In addition to requiring that a channel provides integrity and privacy of the encrypted data, we will require integrity for associated data as introduced by Rogaway [36].

Recently Marson and Poettering [30] closed a gap in the modeling of two-party communication by capturing the bidirectional nature of practical channels in their definitions. We work with their notion of bidirectional channels because it closely models the behavior desired in practice and the bidirectional nature of communication allows us to achieve a fine-grained security against compromise.

Definitional contributions. This paper aims to specify and achieve the best possible security of a bidirectional channel against state compromise. We provide a formal, game-based definition of security and a construction that provably achieves it. We analyze our construction in a concrete security framework [2] and give precise bounds on the advantage of an attacker.

To derive the best possible notion of security against state compromise we first specify a basic input-output interface via a game that describes how the adversary interacts with the channel. This corresponds roughly to combining the integrity and confidentiality games of [30] and adding an oracle that returns the secret state of a specified user to the adversary. Then we specify several attacks that break the security of *any* channel. We define our final security notion by minimally extending the initial interface game to disallow these unavoidable attacks while allowing all other behaviors. Our security definition is consequently the best possible with respect to the specified interface because our attacks rule out the possibility of any stronger notion.

One security notion is an all-in-one notion in the style of [37] that simultaneously requires integrity and privacy of the channel. It asks for the maximal possible security in the face of the exposure of either party's state. A surprising requirement of our definition is that given the state of a user the adversary should not be able to decrypt ciphertexts sent by that user or send forged ciphertexts to that user.

Protocols that update their keys. The OTR (Off-the-Record) messaging protocol [10] is an important predecessor to Signal. It has parties repeatedly exchange Diffie-Hellman elements to derive new keys. The Double Ratchet Algorithm of Signal uses a similar Diffie-Hellman update mechanism and extends it by using a symmetric key-derivation function to update keys when there is no Diffie-Hellman update available. Both methods of updating keys are often referred to as ratcheting (a term introduced by Langley [29]). While the Double Ratchet Algorithm was explicitly designed to achieve strong notions of security against state compromise with respect to privacy, the designers explicitly consider security against a passive eavesdropper [21]; authenticity in the face of compromise is out of scope.

The first academic security analysis of Signal was due to Cohn-Gordan et al. [14]. They only considered the security of the key exchange underlying the Double Ratchet Algorithm and used a security definition explicitly tailored to understanding its security instead of being widely applicable to any scheme.

Work by Bellare et al. [7] sought to formally understand ratcheting as an independent primitive, introducing the notions of (one-directional) ratcheted key exchange and ratcheted encryption. In their model a compromise of the receiving party's secrets permanently and irrevocably disrupts all security, past and future. Further they strictly separate the exchange of key update information from the exchange of messages. Such a model cannot capture a protocol like the Double Ratchet Algorithm for which the two are inextricably combined. On the positive side, they did explicitly model authenticity in the face of compromise.

In [26], Günther and Mazaheri study a key update mechanism introduced in TLS 1.3. Their security definition treats update messages as being out-of-band and thus implicitly authenticated. Their definition is clearly tailored to understand TLS 1.3 specifically.

Instead of analyzing an existing scheme, we strive to understand the best possible security with respect to both privacy and authenticity in the face of state compromise. The techniques we use to achieve this differ from those underlying the schemes discussed above, because all of them rely on exchanging information to create a shared symmetric key that is ultimately used for encryption. Our security notion is not achievable by a scheme of this form and instead requires that asymmetric primitives be used throughout.

Consequently, our scheme is more computationally intensive than those mentioned above. However, as a part of OTR or the Double Ratchet Algorithm, when users are actively sending messages back and forth (the case where efficiency is most relevant), they will be performing asymmetric Diffie-Hellman based key updates prior to most message encryptions. This indicates that the overhead of extra computation with asymmetric techniques is not debilitating in our motivating context of secure messaging. However, the asymmetric techniques we require are likely less efficient than Diffie-Hellman computations so we do not currently know whether our scheme meets realistic efficiency requirements.

Our construction. Our construction of a secure channel is given in Sect. 6.1. It shows how to generically build the channel from a collision-resistant hash function, a public-key encryption scheme, and a digital signature scheme. The latter two require new versions of the primitives that we describe momentarily.

The hash function is used to store transcripts of the communication in the form of hashes of all sent or received ciphertexts. These transcripts are included as part of every ciphertext and a user will not accept a ciphertext with transcripts that do not match those it has stored locally. Every ciphertext sent by a user is signed by their current digital signature signing key and includes the verification key corresponding to their next signing key. Similarly a user will include a new encryption key with every ciphertext they send. The sending user will use the most recent encryption key they have received from the other user and the receiving user will delete all decryption keys that are older than the one most recently used by the sender.

New notions of public-key encryption and digital signatures. Our construction uses new forms of public-key encryption and digital signatures that update their keys over time, which we define in Sect. 3. The former updates its keys with every ciphertext. We refer to it as key-updating public-key encryption. The latter includes extra algorithms that allow the keys to be updated with respect to an arbitrary string. We refer to it as key-updatable digital signature schemes. In our construction a user updates their signing key with their transcript every time they receive a ciphertext.

For public-key encryption we consider encryption with labels and require an IND-CCA style security be maintained even if the adversary is given the decryption key after all challenge ciphertexts have been decrypted or an adversarially generated ciphertext has been decrypted. We show how to construct such scheme from hierarchical identity-based encryption [23].

For digital signatures, security requires that an adversary is unable to forge a signature even given the signing key as long as the sequence of strings used to update it is not a prefix of the sequence of strings used to update the verification key. We additionally require that the scheme has unique signatures (i.e. for any sequence of updates and any message an adversary can only find one signature that will verify). We show how to construct this from a digital signature scheme that is forward secure [5] and has unique signatures.

Related work. Several works [9, 22] extended the definitions of channels to address the stream-based interface provided by channels like TLS, SSH, and QUIC. Our primary motivation is to build a channel for messaging where an atomic interface for messages is more appropriate.

Numerous areas of research within cryptography are motivated by the threat of key compromise. These include key-insulated cryptography [18–20], secret sharing [31,38,41], threshold cryptography [16], proactive cryptography [34], and forward security [17,25]. Forward security, in particular, was introduced in the context of key-exchange [17,25] but has since been considered for a variety of primitives including symmetric [8] and asymmetric encryption [12] and digital signature schemes [5]. Green and Miers [24] propose using puncturable encryption for forward secure asynchronous messaging.

In concurrent and independent work, Poettering and Rösler [35] extend the definitions of ratcheted key exchange from [7] to be bidirectional. Their security definition is conceptually similar to our definition for bidirectional channels because both works aim to achieve strong notions of security against an adversary that can arbitrarily and repeatedly learn the secret state of either communicating party. In constructing a secure ratcheted key exchange scheme they make use of a key-updatable key encapsulation mechanism (KEM), a new primitive they introduce in their work. The key-updatable nature of this is conceptually similar to that of the key-updatable digital signature schemes we introduce in our work. To construct such a KEM they make use of hierarchical identity-based encryption in a manner similar to how we construct key-updating public-key encryption. The goal of their work differs from ours; they only consider security for the exchange of symmetric keys while we do so for the exchange of messages.

2   Preliminaries

Notation and conventions. Let $\mathbb{N} = \{0, 1, 2, \ldots\}$ be the set of non-negative integers. Let ε denote the empty string. If $x \in \{0,1\}^*$ is a string then $|x|$ denotes its length. By $x \parallel y$ we denote the concatenation of strings x and y. If X is a finite set, we let $x \leftarrow_{\$} X$ denote picking an element of X uniformly at random and assigning it to x. By $(X)^n$ we denote the n-ary Cartesian product of X. We let $x_1 \leftarrow x_2 \leftarrow \cdots \leftarrow x_n \leftarrow v$ denote assigning the value v to each variable x_i for $i = 1, \ldots, n$.

If **mem** is a table, we use $\mathbf{mem}[p]$ to denote the element of the table that is indexed by p. By $\mathbf{mem}[0, \ldots, \infty] \leftarrow v$ we denote initializing all elements of **mem** to v. For $a, b \in \mathbb{N}$ we let $v \leftarrow \mathbf{mem}[a, \ldots, b]$ denote setting v equal to the tuple obtained by removing all $\perp$ elements from $(\mathbf{mem}[a], \mathbf{mem}[a+1], \ldots, \mathbf{mem}[b])$. It is the empty vector () if all of these table entries are $\perp$ or if $a > b$. A tuple $\boldsymbol{x} = (x_1, \ldots)$ specifies a uniquely decodable concatenation of strings $x_1, \ldots$. We say $\boldsymbol{x} \sqsubseteq \boldsymbol{y}$ if $\boldsymbol{x}$ is a prefix of $\boldsymbol{y}$. More formally, $(x_1, \ldots, x_n) \sqsubseteq (y_1, \ldots, y_m)$ if $n \leq m$ and $x_i = y_i$ for all $i \in \{1, \ldots, n\}$.

Algorithms may be randomized unless otherwise indicated. Running time is worst case. If A is an algorithm, we let $y \leftarrow A(x_1, \ldots; r)$ denote running A with random coins r on inputs $x_1, \ldots$ and assigning the output to y. Any state maintained by an algorithm will explicitly be shown as input and output of that algorithm. We let $y \leftarrow_{\$} A(x_1, \ldots)$ denote picking r at random and letting $y \leftarrow A(x_1, \ldots; r)$. We omit the semicolon when there are no inputs other than the random coins. We let $[A(x_1, \ldots)]$ denote the set of all possible outputs of A when invoked with inputs $x_1, \ldots$. Adversaries are algorithms. The instruction $\mathbf{abort}(x_1, \ldots)$ is used by an adversary to immediately halt with output $(x_1, \ldots)$.

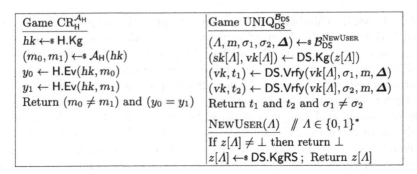

Fig. 1. Games defining collision-resistance of function family H and signature uniqueness of key-updatable digital signature scheme DS.

We use a special symbol $\perp \notin \{0,1\}^*$ to denote an empty table position, and we also return it as an error code indicating an invalid input. An algorithm may not accept $\perp$ as input. If $x_i = \perp$ for some i when executing $(y_1, \dots) \leftarrow A(x_1 \dots)$ we assume that $y_j = \perp$ for all j. We assume that adversaries never pass $\perp$ as input to their oracles.

We use the code based game playing framework of [6]. (See Fig. 1 for an example of a game.) We let $\Pr[G]$ denote the probability that game G returns true. In code, tables are initially empty. We adopt the convention that the running time of an adversary means the worst case execution time of the adversary in the game that executes it, so that time for game setup steps and time to compute answers to oracle queries is included.

Function families. A family of functions H specifies algorithms H.Kg and H.Ev, where H.Ev is deterministic. Key generation algorithm H.Kg returns a key hk. Evaluation algorithm H.Ev takes hk and an input $x \in \{0,1\}^*$ to return an output y, denoted by $y \leftarrow$ H.Ev(hk, x).

Collision-resistant functions. Consider game CR of Fig. 1 associated to a function family H and an adversary $\mathcal{A}_H$. The game samples a random key hk for function family H. In order to win the game, adversary $\mathcal{A}_H$ has to find two distinct messages m_0, m_1 such that H.Ev$(hk, m_0) =$ H.Ev(hk, m_1). The advantage of $\mathcal{A}_H$ in breaking the CR security of H is defined as $\mathsf{Adv}_H^{cr}(\mathcal{A}_H) = \Pr[\mathrm{CR}_H^{\mathcal{A}_H}]$.

Digital signature schemes. A digital signature scheme DS specifies algorithms DS.Kg, DS.Sign and DS.Vrfy, where DS.Vrfy is deterministic. Associated to DS is a key generation randomness space DS.KgRS and signing algorithm's randomness space DS.SignRS. Key generation algorithm DS.Kg takes randomness $z \in$ DS.KgRS to return a signing key sk and a verification key vk, denoted by $(sk, vk) \leftarrow$ DS.Kg(z). Signing algorithm DS.Sign takes sk, a message $m \in \{0,1\}^*$ and randomness $z \in$ DS.SignRS to return a signature σ, denoted by $\sigma \leftarrow$ DS.Sign$(sk, m; z)$. Verification algorithm DS.Vrfy takes vk, σ, and m to return a decision $t \in \{\mathsf{true}, \mathsf{false}\}$ regarding whether σ is a valid signature of m under vk, denoted by $t \leftarrow$ DS.Vrfy(vk, σ, m). The correctness condition for DS requires that $\mathsf{DS.Vrfy}(vk, \sigma, m) = \mathsf{true}$ for all $(sk, vk) \in [\mathsf{DS.Kg}]$, all $m \in \{0,1\}^*$, and all $\sigma \in [\mathsf{DS.Sign}(sk, m)]$.

We define the min-entropy of algorithm $\mathsf{DS.Kg}$ as $\mathrm{H}_\infty(\mathsf{DS.Kg})$, such that

$$2^{-\mathrm{H}_\infty(\mathsf{DS.Kg})} = \max_{vk} \Pr\left[vk^* = vk : (sk^*, vk^*) \leftarrow_\$ \mathsf{DS.Kg}\right].$$

The probability is defined over the random coins used for $\mathsf{DS.Kg}$. Note that the min-entropy is defined with respect to verification keys, regardless of the corresponding values of the secret keys.

3   New Asymmetric Primitives

In this section we define key-updatable digital signatures and key-updating public-key encryption. The former allows its keys to be updated with arbitrary strings. The latter updates its keys with every ciphertext that is sent/received. While in general one would prefer the size of keys, signatures, and ciphertexts to be constant we will be willing to accept schemes for which these grow linearly in the number of updates. As we will discuss later, these are plausibly acceptable inefficiencies for our use cases.

We specify multi-user security definitions for both primitives, because it allows tighter reductions when we construct a channel from these primitives. Single-user variants of these definitions are obtained by only allowing the adversary to interact with one user and can be shown to imply the multi-user versions by a standard hybrid argument. Starting with [1] constructions have been given for a variety of primitives that allow multi-user security to be proven without the factor q security loss introduced by a hybrid argument. If analogous constructions can be found for our primitives then our results will give tight bounds on the security of our channel.

3.1   Key-Updatable Digital Signature Schemes

We start by formally defining the syntax and correctness of a key-updatable digital signature scheme. Then we specify a security definition for it. We will briefly sketch how to construct such a scheme, but leave the details to [27].

Syntax and correctness. A *key-updatable* digital signature scheme is a digital signature scheme with additional algorithms $\mathsf{DS.UpdSk}$ and $\mathsf{DS.UpdVk}$, where $\mathsf{DS.UpdVk}$ is deterministic. Signing-key update algorithm $\mathsf{DS.UpdSk}$ takes a signing key sk and a key update information $\Delta \in \{0,1\}^*$ to return a new signing key sk, denoted by $sk \leftarrow_\$ \mathsf{DS.UpdSk}(sk, \Delta)$. Verification-key update algorithm $\mathsf{DS.UpdVk}$ takes a verification key vk and a key update information $\Delta \in \{0,1\}^*$ to return a new verification key vk, denoted by $vk \leftarrow \mathsf{DS.UpdVk}(vk, \Delta)$.

For compactness, when $\boldsymbol{\Delta} = (\Delta_1, \Delta_2, \ldots)$ we sometimes write $(vk, t) \leftarrow \mathsf{DS.Vrfy}(vk, \sigma, m, \boldsymbol{\Delta})$ to denote updating the key via $vk \leftarrow \mathsf{DS.UpdVk}(vk, \Delta_i)$ for $i = 1, \ldots, n$ and then evaluating $t \leftarrow \mathsf{DS.Vrfy}(vk, \sigma, m)$.

Game $\text{DSCORR}^{\mathcal{C}}_{\text{DS}}$	Game $\text{PKECORR}^{\mathcal{C}}_{\text{PKE}}$
$i \leftarrow 0$; $\nu \leftarrow_{\$} \text{DS.KgRS}$	$\nu \leftarrow_{\$} \text{PKE.KgRS}$
$(sk, vk) \leftarrow \text{DS.Kg}(\nu)$	$(ek, dk) \leftarrow \text{PKE.Kg}(\nu)$
$\mathcal{C}^{\text{UPD,SIGN}}(\nu)$	$\mathcal{C}^{\text{ENC}}(\nu)$
Return **bad**	Return **bad**
$\underline{\text{UPD}(\Delta)}$ // $\Delta \in \{0,1\}^{*}$	$\underline{\text{ENC}(m, \ell)}$ // $m, \ell \in \{0,1\}^{*}$
$i \leftarrow i+1$; $\Delta_i \leftarrow \Delta$	$(ek, c) \leftarrow_{\$} \text{PKE.Enc}(ek, \ell, m)$
$sk \leftarrow_{\$} \text{DS.UpdSk}(sk, \Delta)$	$(dk, m') \leftarrow_{\$} \text{PKE.Dec}(dk, \ell, c)$
Return sk	If $m' \neq m$ then **bad** $\leftarrow$ **true**
$\underline{\text{SIGN}(m)}$ // $m \in \{0,1\}^{*}$	Return (ek, dk)
$\sigma \leftarrow_{\$} \text{DS.Sign}(sk, m)$	
$\boldsymbol{\Delta} \leftarrow (\Delta_1, \ldots, \Delta_i)$	
$(vk^{*}, t) \leftarrow \text{DS.Vrfy}(vk, \sigma, m, \boldsymbol{\Delta})$	
If not t then **bad** $\leftarrow$ **true**	

Fig. 2. Games defining correctness of key-updatable digital signature scheme DS and correctness of key-updating public-key encryption scheme PKE.

The key-update correctness condition requires that signatures must verify correctly as long as the signing and the verification keys are both updated with the same sequence of key update information $\boldsymbol{\Delta} = (\Delta_1, \Delta_2, \ldots)$. To formalize this, consider game DSCORR of Fig. 2, associated to a key-updatable digital signature scheme DS and an adversary $\mathcal{C}$. The advantage of an adversary $\mathcal{C}$ against the correctness of DS is given by $\text{Adv}^{\text{dscorr}}_{\text{DS}}(\mathcal{C}) = \Pr[\text{DSCORR}^{\mathcal{C}}_{\text{DS}}]$. We require that $\text{Adv}^{\text{dscorr}}_{\text{DS}}(\mathcal{C}) = 0$ for all (even unbounded) adversaries $\mathcal{C}$. See Sect. 4 for discussion on game-based definitions of correctness.

Signature uniqueness. We will be interested in schemes for which there is only a single signature that will be accepted for any message m and any sequence of updates $\boldsymbol{\Delta}$. Consider game UNIQ of Fig. 1, associated to a key-updatable digital signature scheme DS and an adversary $\mathcal{B}_{\text{DS}}$. The adversary $\mathcal{B}_{\text{DS}}$ can call the oracle NEWUSER arbitrarily many times with a user identifier Λ and be given the randomness used to generate the keys of Λ. The adversary ultimately outputs a user id Λ, message m, signatures σ_1, σ_2, and key update vector $\boldsymbol{\Delta}$. It wins if the signatures are distinct and both verify for m when the verification key of Λ is updated with $\boldsymbol{\Delta}$. The advantage of $\mathcal{B}_{\text{DS}}$ in breaking the UNIQ security of DS is defined by $\text{Adv}^{\text{uniq}}_{\text{DS}}(\mathcal{B}_{\text{DS}}) = \Pr[\text{UNIQ}^{\mathcal{B}_{\text{DS}}}_{\text{DS}}]$.

Signature unforgeability under exposures. Our main security notion for signatures asks that the adversary not be able to create signatures for any key update vector $\boldsymbol{\Delta}$ unless it was given a signature for that key update vector or given the signing key such that the vector of strings it had been updated with was a prefix of $\boldsymbol{\Delta}$. Consider game UFEXP of Fig. 3, associated to a key-updatable digital signature scheme DS and an adversary $\mathcal{A}_{\text{DS}}$.

Game $\mathrm{UFEXP}_{\mathsf{DS}}^{\mathcal{A}_{\mathsf{DS}}}$	Game $\mathrm{INDEXP}_{\mathsf{PKE}}^{\mathcal{A}_{\mathsf{PKE}}}$
$\mathrm{out} \leftarrow_{\$} \mathcal{A}_{\mathsf{DS}}^{\mathrm{NEWUSER},\mathrm{UPD},\mathrm{SIGN},\mathrm{EXP}}$	$b \leftarrow_{\$} \{0,1\}$
$(\Lambda, \sigma, m, \boldsymbol{\Delta}) \leftarrow \mathrm{out}$	$b' \leftarrow_{\$} \mathcal{A}_{\mathsf{PKE}}^{\mathrm{NEWUSER},\mathrm{ENC},\mathrm{DEC},\mathrm{EXPRAND},\mathrm{EXPDK}}$
$\mathrm{forgery} \leftarrow (\sigma, m, \boldsymbol{\Delta})$	Return $b = b'$
$\mathrm{given} \leftarrow (\sigma^*[\Lambda], m^*[\Lambda], \boldsymbol{\Delta}^*[\Lambda])$	$\underline{\mathrm{NEWUSER}(\Lambda)} \quad /\!\!/ \; \Lambda \in \{0,1\}^*$
$t_1 \leftarrow (\mathrm{forgery} = \mathrm{given})$	If $dk[\Lambda] \neq \bot$ or $\mathsf{nextop} \neq \bot$ then return $\bot$
$t_2 \leftarrow \boldsymbol{\Delta}'[\Lambda] \sqsubseteq \boldsymbol{\Delta}$	$s[\Lambda] \leftarrow 0$; $r[\Lambda] \leftarrow 0$; $\mathsf{exp}[\Lambda] \leftarrow \mathsf{false}$
$\mathsf{cheated} \leftarrow (t_1 \text{ or } t_2)$	$z[\Lambda] \leftarrow_{\$} \mathsf{PKE.EncRS}$; $(ek[\Lambda], dk[\Lambda]) \leftarrow_{\$} \mathsf{PKE.Kg}$
$vk \leftarrow vk[\Lambda]$	Return $ek[\Lambda]$
$(vk, \mathsf{win}) \leftarrow \mathsf{DS.Vrfy}(vk, \sigma, m, \boldsymbol{\Delta})$	$\underline{\mathrm{ENC}(\Lambda, m_0, m_1, \ell)} \quad /\!\!/ \; \Lambda, m_0, m_1, \ell \in \{0,1\}^*$
Return win and not $\mathsf{cheated}$	If $dk[\Lambda] = \bot$ or $|m_0| \neq |m_1|$ then return $\bot$
$\underline{\mathrm{NEWUSER}(\Lambda)} \quad /\!\!/ \; \Lambda \in \{0,1\}^*$	$\mathsf{noch} \leftarrow (\mathbf{ch}[\Lambda][s[\Lambda]+1] = \text{"forbidden"})$
If $vk[\Lambda] \neq \bot$ then return $\bot$	$t \leftarrow (\mathsf{noch} \text{ or } \mathsf{exp}[\Lambda])$
$i[\Lambda] \leftarrow 0$	If $m_0 \neq m_1$ and t then return $\bot$
$\boldsymbol{\Delta}^*[\Lambda] \leftarrow \bot$; $\boldsymbol{\Delta}'[\Lambda] \leftarrow \bot$	$s[\Lambda] \leftarrow s[\Lambda] + 1$
$(sk[\Lambda], vk[\Lambda]) \leftarrow_{\$} \mathsf{DS.Kg}$	$(ek[\Lambda], c) \leftarrow \mathsf{PKE.Enc}(ek[\Lambda], \ell, m_b; z[\Lambda])$
Return $vk[\Lambda]$	$z[\Lambda] \leftarrow_{\$} \mathsf{PKE.EncRS}$; $\mathsf{nextop} \leftarrow \bot$
$\underline{\mathrm{UPD}(\Lambda, \Delta)} \quad /\!\!/ \; \Lambda, \Delta \in \{0,1\}^*$	$\mathbf{ctable}[\Lambda][s[\Lambda]] \leftarrow (c, \ell)$
If $sk[\Lambda] = \bot$ then return $\bot$	If $m_0 \neq m_1$ then $\mathbf{ch}[\Lambda][s[\Lambda]] \leftarrow \text{"done"}$
$i[\Lambda] \leftarrow i[\Lambda] + 1$; $\Delta_{i[\Lambda]}[\Lambda] \leftarrow \Delta$	Return $(ek[\Lambda], c)$
$sk[\Lambda] \leftarrow_{\$} \mathsf{DS.UpdSk}(sk[\Lambda], \Delta)$	$\underline{\mathrm{DEC}(\Lambda, c, \ell)} \quad /\!\!/ \; \Lambda, c, \ell \in \{0,1\}^*$
Return $\bot$	If $dk[\Lambda] = \bot$ or $\mathsf{nextop} \neq \bot$ then return $\bot$
$\underline{\mathrm{SIGN}(\Lambda, m)} \quad /\!\!/ \; \Lambda, m \in \{0,1\}^*$	$r[\Lambda] \leftarrow r[\Lambda] + 1$
If $sk[\Lambda] = \bot$ then return $\bot$	$(dk[\Lambda], m) \leftarrow_{\$} \mathsf{PKE.Dec}(dk[\Lambda], \ell, c)$
$\sigma \leftarrow_{\$} \mathsf{DS.Sign}(sk[\Lambda], m)$	If $(c, \ell) \neq \mathbf{ctable}[\Lambda][r[\Lambda]]$ or $\mathsf{restricted}[\Lambda]$ then
$sk[\Lambda] \leftarrow \bot$	$\quad \mathsf{restricted}[\Lambda] \leftarrow \mathsf{true}$
$(\sigma^*[\Lambda], m^*[\Lambda]) \leftarrow (\sigma, m)$	$\quad$ Return m
$\boldsymbol{\Delta}^*[\Lambda] \leftarrow (\Delta_1[\Lambda], \ldots, \Delta_{i[\Lambda]}[\Lambda])$	Return $\bot$
Return σ	$\underline{\mathrm{EXPRAND}(\Lambda)} \quad /\!\!/ \; \Lambda \in \{0,1\}^*$
$\underline{\mathrm{EXP}(\Lambda)} \quad /\!\!/ \; \Lambda \in \{0,1\}^*$	If $dk[\Lambda] = \bot$ or $\mathsf{nextop} \neq \bot$ then return $\bot$
If $sk[\Lambda] = \bot$ then return $\bot$	$\mathbf{ch}[\Lambda][s[\Lambda]+1] \leftarrow \text{"forbidden"}$
If $\boldsymbol{\Delta}'[\Lambda] = \bot$ then	$\mathsf{nextop} \leftarrow \text{"encrypt"}$; Return $z[\Lambda]$
$\quad \boldsymbol{\Delta}'[\Lambda] \leftarrow (\Delta_1[\Lambda], \ldots, \Delta_{i[\Lambda]}[\Lambda])$	$\underline{\mathrm{EXPDK}(\Lambda)} \quad /\!\!/ \; \Lambda \in \{0,1\}^*$
Return $sk[\Lambda]$	If $dk[\Lambda] = \bot$ or $\mathsf{nextop} \neq \bot$ then return $\bot$
	If $\mathsf{restricted}[\Lambda]$ then return $dk[\Lambda]$
	If $\exists i \in (r[\Lambda], s[\Lambda]]$ s.t. $\mathbf{ch}[\Lambda][i] = \text{"done"}$ then
	$\quad$ Return $\bot$
	$\mathsf{exp}[\Lambda] \leftarrow \mathsf{true}$; Return $dk[\Lambda]$

Fig. 3. Games defining signature unforgeability under exposures of key-updatable digital signature scheme DS, and ciphertext indistinguishability under exposures of key-updating public-key encryption scheme PKE.

The adversary $\mathcal{A}_{DS}$ can call the oracle NEWUSER arbitrarily many times for any user identifier Λ and be given the verification key for that user. Then it can interact with user Λ via three different oracles. Via calls to UPD with a string Δ it requests that the signing key for the specified user be updated with Δ. Via calls to SIGN with message m it asks for a signature of m using the signing key for the specified user. When it does so the signing key is erased so it can no longer interact with that user and $\Delta^*[\Lambda]$ is used to store the vector of strings the key was updated with.[1] Via calls to EXP it can ask to be given the current signing key of the specified user. When it does so $\Delta'[\Lambda]$ is used to store the vector of strings the key was updated with.

At the end of the game the adversary outputs a user id Λ, signature σ, message m, and key update vector Δ. The adversary has cheated if it previously received σ as the result of calling SIGN(Λ, m) and $\Delta = \Delta^*[\Lambda]$, or if it exposed the signing key of Λ and $\Delta'[\Lambda]$ is a prefix of Δ. It wins if it has not cheated and the signature it output verifies for m when the verification key of Λ is updated with Δ. The advantage of $\mathcal{A}_{DS}$ in breaking the UFEXP security of DS is defined by $\mathsf{Adv}_{DS}^{ufexp}(\mathcal{A}_{DS}) = \Pr[\mathrm{UFEXP}_{DS}^{\mathcal{A}_{DS}}]$.

<u>Construction.</u> In [27] we use a forward secure [5] key-evolving signature scheme with unique signatures to construct a signature scheme secure with respect to both of the above definitions. Roughly, a key-evolving signature scheme is like a key-updatable digital signature that can only update with $\Delta = \varepsilon$. In order to enable updates with respect to arbitrary key update information, we sign each update string with the current key prior to evolving the key, and then include these intermediate signatures with our final signature.

3.2   Key-Updating Public-Key Encryption Schemes

We start by formally defining the syntax and correctness of a key-updating public-key encryption. Then we specify a security definition for it. We will briefly sketch how to construct such a scheme, but leave the details to [27]. We consider public-key encryption with labels as introduced by Shoup [40].

<u>Syntax and correctness.</u> A key-updating public-key encryption scheme PKE specifies algorithms PKE.Kg, PKE.Enc, PKE.Dec. Associated to PKE is a key generation randomness space PKE.KgRS and encryption randomness space PKE.EncRS. Key generation algorithm PKE.Kg takes randomness $z \in$ PKE.KgRS to return an encryption key ek and a decryption key dk, denoted by $(ek, dk) \leftarrow$ PKE.Kg(z). Encryption algorithm PKE.Enc takes ek, a label $\ell \in \{0,1\}^*$, a message $m \in \{0,1\}^*$ and randomness $z \in$ PKE.EncRS to return a new encryption key ek and a ciphertext c, denoted by $(ek, c) \leftarrow$ PKE.Enc($ek, \ell, m; z$). Decryption algorithm PKE.Dec takes dk, ℓ, c to return a new decryption key dk and a message $m \in \{0,1\}^*$, denoted by $(dk, m) \leftarrow\!\!{}_\$ $ PKE.Dec(dk, ℓ, c).

---

[1] We are thus defining security for a *one-time* signature scheme, because a particular key will only be used for one signature. This is all we require for our application, but the definition and construction we provide could easily be extended to allow multiple signatures if desired.

The correctness condition requires that ciphertexts decrypt correctly as long as they are received in the same order they were created and with the same labels. To formalize this, consider game PKECORR of Fig. 2, associated to a key-updating public-key encryption scheme PKE and an adversary $\mathcal{C}$. The advantage of an adversary $\mathcal{C}$ against the correctness of PKE is given by $\mathsf{Adv}^{\mathsf{pkecorr}}_{\mathsf{PKE}}(\mathcal{C}) = \Pr[\mathrm{PKECORR}^{\mathcal{C}}_{\mathsf{PKE}}]$. Correctness requires that $\mathsf{Adv}^{\mathsf{pkecorr}}_{\mathsf{PKE}}(\mathcal{C}) = 0$ for all (even computationally unbounded) adversaries $\mathcal{C}$. See Sect. 4 for discussion on game-based definitions of correctness.

Define the min-entropy of algorithms PKE.Kg and PKE.Enc as $\mathrm{H}_{\infty}(\mathsf{PKE.Kg})$ and $\mathrm{H}_{\infty}(\mathsf{PKE.Enc})$, respectively, defined as follows:

$$2^{-\mathrm{H}_{\infty}(\mathsf{PKE.Kg})} = \max_{ek} \Pr\left[ek^* = ek : (ek^*, dk^*) \leftarrow_{\$} \mathsf{PKE.Kg}\right],$$

$$2^{-\mathrm{H}_{\infty}(\mathsf{PKE.Enc})} = \max_{ek,\ell,m,c} \Pr\left[c^* = c : (ek^*, c^*) \leftarrow_{\$} \mathsf{PKE.Enc}(ek, \ell, m)\right].$$

The probability is defined over the random coins used by PKE.Kg and PKE.Enc, respectively. Note that min-entropy does not depend on the output values dk^* (in the former case) and ek^* (in the latter case).

Ciphertext indistinguishability under exposures. Consider game INDEXP of Fig. 3, associated to a key-updating public-key encryption scheme PKE and an adversary $\mathcal{A}_{\mathsf{PKE}}$. Roughly, it requires that PKE maintain CCA security [3] even if $\mathcal{A}_{\mathsf{PKE}}$ is given the decryption key (as long as that decryption key is no longer able to decrypt any challenge ciphertexts).

The adversary $\mathcal{A}_{\mathsf{PKE}}$ can call the oracle NEWUSER arbitrarily many times with a user identifier Λ and be given the encryption key of that user. Then it can interact with user Λ via four oracles. Via calls to ENC with messages m_0, m_1 and label ℓ it requests that one of these messages be encrypted using the specified label (which message is encrypted depends on the secret bit b). It will be given back the new encryption key and the produced ciphertext. If $m_0 \neq m_1$ we remember that a challenge query was done.

Via calls to DEC with ciphertext c and ℓ it requests that the ciphertext be decrypted with the specified label. Adversary $\mathcal{A}_{\mathsf{PKE}}$ will only be given the result of this decryption if the pair (c, ℓ) was not obtained from a call to ENC. Once the adversary queries such pair, the user Λ becomes "restricted" and the oracle will return the true decryption of all future ciphertexts for this user.

Via calls to EXPRAND it asks to be given the next randomness that will be used for encryption. This represents the adversary exposing the randomness while the encryption is taking place so we require that after a call to EXPRAND the adversary immediately makes the corresponding call to ENC. During this call challenges are forbidden so it must choose $m_0 = m_1$.

Via calls to EXPDK it asks to be given the current decryption key of the user. It may not do so if a challenge query was done but the user has not decrypted the corresponding ciphertext yet (unless the user is restricted). Otherwise the decryption key is returned and the user is considered to be exposed. Once a user is exposed challenges are not allowed so for all future calls to ENC the adversary required to choose $m_0 = m_1$.

At the end of the game the adversary outputs a bit b' representing its guess of the secret bit b. The advantage of $\mathcal{A}_{\mathsf{PKE}}$ in breaking the INDEXP security of PKE is defined as $\mathsf{Adv}^{\mathsf{indexp}}_{\mathsf{PKE}}(\mathcal{A}_{\mathsf{PKE}}) = 2\Pr[\mathrm{INDEXP}^{\mathcal{A}_{\mathsf{PKE}}}_{\mathsf{PKE}}] - 1$.

Many of the variables used to track the behavior of the adversary in INDEXP are analogous to variables we use and discuss in detail in Sect. 5 when defining security of a channel. The reader interested in understand the pseudocode of INDEXP in detail is encouraged to read that section first.

Construction. In [27] we use a hierarchical identity-based encryption (HIBE) scheme to construct a secure key-updating encryption scheme. Roughly, a HIBE assigns a decryption key to any identity (vector of strings). A decryption key for an identity $\boldsymbol{I}$ can be used to create decryption keys for an identity of which $\boldsymbol{I}$ is a prefix. Security requires that the adversary be unable to learn about encrypted messages encrypted to an identity $\boldsymbol{I}$ even if given the decryption key for many identities as long as none of them were prefixes of $\boldsymbol{I}$. To create a key-updating encryption scheme we use the vector of ciphertexts and labels a user has received so far as the identity. The security of this scheme then follows from the security of the underlying HIBE in a fairly straightforward manner.

4   Bidirectional Cryptographic Channels

In this section we formally define the syntax and correctness of bidirectional cryptographic channels. Our notion of bidirectional channels will closely match that of Marson and Poettering [30]. Compared to their definition, we allow the receiving algorithm to be randomized and provide an alternative correctness condition. We argue that the new correctness condition is more appropriate for our desired use case of secure messaging. Henceforth, we will omit the adjective "bidirectional" and refer simply to channels.

Syntax of channel. A channel provides a method for two users to exchange messages in an arbitrary order. We will refer to the two users of a channel as the initiator $\mathcal{I}$ and the receiver $\mathcal{R}$. There will be no formal distinction between the two users, but when specifying attacks we follow the convention of having $\mathcal{I}$ send a ciphertext first. We will use u as a variable to represent an arbitrary user and $\bar{u}$ to represent the other user. More formally, when $u \in \{\mathcal{I}, \mathcal{R}\}$ we let $\bar{u}$ denote the sole element of $\{\mathcal{I}, \mathcal{R}\} \setminus \{u\}$.

A channel Ch specifies algorithms Ch.Init, Ch.Send, and Ch.Recv. Initialization algorithm Ch.Init returns initial states $st_{\mathcal{I}} \in \{0,1\}^*$ and $st_{\mathcal{R}} \in \{0,1\}^*$, where $st_{\mathcal{I}}$ is $\mathcal{I}$'s state and $st_{\mathcal{R}}$ is $\mathcal{R}$'s state. We write $(st_{\mathcal{I}}, st_{\mathcal{R}}) \leftarrow_{\$} \mathsf{Ch.Init}$. Sending algorithm Ch.Send takes state $st_u \in \{0,1\}^*$, associated data $ad \in \{0,1\}^*$, and message $m \in \{0,1\}^*$ to return updated state $st_u \in \{0,1\}^*$ and a ciphertext $c \in \{0,1\}^*$. We write $(st_u, c) \leftarrow_{\$} \mathsf{Ch.Send}(st_u, ad, m)$. Receiving algorithm takes state $st_u \in \{0,1\}^*$, associated data $ad \in \{0,1\}^*$, and ciphertext $c \in \{0,1\}^*$ to return updated state $st_u \in \{0,1\}^* \cup \{\bot\}$ and message $m \in \{0,1\}^* \cup \{\bot\}$. We write $(st_u, m) \leftarrow_{\$} \mathsf{Ch.Recv}(st_u, ad, c)$, where $m = \bot$ represents a rejection of ciphertext c and $st_u = \bot$ represents the channel being permanently shut down from the perspective of u (recall our convention regarding $\bot$ as input to an algorithm). One notion of correctness we discuss will require that $st_u = \bot$ whenever $m = \bot$. The other will require that st_u not be changed from its input value when $m = \bot$.

We let Ch.InitRS, Ch.SendRS, and Ch.RecvRS denote the sets of possible random coins for Ch.Init, Ch.Send, and Ch.Recv, respectively. Note that for full generality we allow Ch.Recv to be randomized. Prior work commonly requires this algorithm to be deterministic.

<u>Correctness of channel.</u> In Fig. 4 we provide two games, defining two alternative correctness requirements for a cryptographic channel. Lines labelled with the name of a game are included only in that game. The games differ in whether the adversary is given access to an oracle ROBUST or to an oracle REJECT. Game CORR uses the former, whereas game CORR⊥ uses the latter. The advantage of an adversary $\mathcal{C}$ against the correctness of channel Ch is given by $\mathsf{Adv}^{\mathsf{corr}}_{\mathsf{Ch}}(\mathcal{C}) = \Pr[\mathrm{CORR}^{\mathcal{C}}_{\mathsf{Ch}}]$ in one case, and $\mathsf{Adv}^{\mathsf{corr}\bot}_{\mathsf{Ch}}(\mathcal{C}) = \Pr[\mathrm{CORR}\bot^{\mathcal{C}}_{\mathsf{Ch}}]$ in the other case. Correctness with respect to either notion requires that the advantage is equal 0 for all (even computationally unbounded) adversaries $\mathcal{C}$.

Games $\mathrm{CORR}^{\mathcal{C}}_{\mathsf{Ch}}$, $\mathrm{CORR}\bot^{\mathcal{C}}_{\mathsf{Ch}}$

$s_{\mathcal{I}} \leftarrow r_{\mathcal{I}} \leftarrow s_{\mathcal{R}} \leftarrow r_{\mathcal{R}} \leftarrow 0$; $\nu \leftarrow\!\!\$ \mathsf{Ch.InitRS}$; $(st_{\mathcal{I}}, st_{\mathcal{R}}) \leftarrow \mathsf{Ch.Init}(\nu)$
$\mathcal{C}^{\text{SEND,RECV,ROBUST}}(\nu)$ // $\mathrm{CORR}^{\mathcal{C}}_{\mathsf{Ch}}$
$\mathcal{C}^{\text{SEND,RECV,REJECT}}(\nu)$ // $\mathrm{CORR}\bot^{\mathcal{C}}_{\mathsf{Ch}}$
Return bad

SEND(u, ad, m) // $u \in \{\mathcal{I}, \mathcal{R}\}, (ad, m) \in (\{0,1\}^*)^2$
$s_u \leftarrow s_u + 1$; $z \leftarrow\!\!\$ \mathsf{Ch.SendRS}$; $(st_u, c) \leftarrow \mathsf{Ch.Send}(st_u, ad, m; z)$
$\mathbf{ctable}_{\bar{u}}[s_u] \leftarrow (c, ad)$; $\mathbf{mtable}_{\bar{u}}[s_u] \leftarrow m$; Return z

RECV(u) // $u \in \{\mathcal{I}, \mathcal{R}\}$
If $\mathbf{ctable}_u[r_u + 1] = \bot$ then return $\bot$
$r_u \leftarrow r_u + 1$; $(c, ad) \leftarrow \mathbf{ctable}_u[r_u]$
$\eta \leftarrow\!\!\$ \mathsf{Ch.RecvRS}$; $(st_u, m) \leftarrow \mathsf{Ch.Recv}(st_u, ad, c; \eta)$
If $m \neq \mathbf{mtable}_u[r_u]$ then **bad** $\leftarrow$ **true**
Return η

ROBUST(st, ad, c) // $(st, ad, c) \in (\{0,1\}^*)^3$
$(st', m) \leftarrow\!\!\$ \mathsf{Ch.Recv}(st, ad, c)$
If $m = \bot$ and $st' \neq st$ then **bad** $\leftarrow$ **true**

REJECT(st, ad, c) // $(st, ad, c) \in (\{0,1\}^*)^3$
$(st', m) \leftarrow\!\!\$ \mathsf{Ch.Recv}(st, ad, c)$
If $m = \bot$ and $st' \neq \bot$ then **bad** $\leftarrow$ **true**

Fig. 4. Games defining correctness of channel Ch. Lines labelled with the name of a game are included only in that game. CORR requires that Ch be robust when given an incorrect ciphertext via oracle ROBUST. CORR⊥ requires that Ch permanently returns $\bot$ when given an incorrect ciphertext via oracle REJECT.

Our use of games to define correctness conditions follows the work of Marson and Poettering [30] and Bellare et. al. [7]. By considering unbounded adversaries and requiring an advantage of 0 we capture a typical information-theoretic perfect correctness requirement without having to explicitly quantify over sequences of actions. In this work we require only the perfect correctness because it is achieved by our scheme; however, it would be possible to capture computational correctness by considering a restricted class of adversaries.

Both games require that ciphertexts sent by any user are always decrypted to the correct message by the other user. This is modeled by providing adversary C with access to oracles SEND and RECV. We assume that messages from u to $\bar{u}$ are received in the same order they were sent, and likewise that messages from $\bar{u}$ to u are also received in the correct order (regardless Aof how they are interwoven on both sides, since ciphertexts are being sent in both directions).

The games differ in how the channel is required to behave in the case that a ciphertext is rejected. Game CORR (using oracle ROBUST) requires that the state of the user not be changed so that the channel can continue to be used. Game CORR⊥ (using oracle REJECT) requires that the state of the user is set to ⊥. According to our conventions about the behavior of algorithms given ⊥ as input (see Sect. 2), the channel will then refuse to perform any further actions by setting all subsequent outputs to ⊥. We emphasize that the adversary specifies all inputs to Ch.Recv when making calls to ROBUST and REJECT, so the behavior of those oracles is not related to the behavior of the other two oracles for which the game maintains the state of both users.

Comparison of correctness notions. The correctness required by CORR⊥ is identical to that of Marson and Poettering [30]. The CORR notion of correctness instead uses a form of robustness analogous to that of [7]. In [27] we discuss how these correctness notions have different implications for the *security* of the channel. It is trivial to convert a CORR-correct channel to a CORR⊥-correct channel and vice versa. Thus we will, without loss of generality, only provide a scheme achieving CORR-correctness.

5   Security Notion for Channels

In this section we will define what it means for a channel to be secure in the presence of a strong attacker that can steal the secrets of either party in the communication. Our goal is to give the strongest possible notion of security in this setting, encompassing both the privacy of messages and the integrity of ciphertexts. We take a fine-grained look at what attacks are possible and require that a channel be secure against all attacks that are not syntactically inherent in the definition of a channel.

To introduce our security notion we will first describe a simple interface of how the adversary is allowed to interact with the channel. Then we show attacks that would break the security of *any* channel using this interface. Our final security notion will be created by adding checks to the interface that prevents adversary from performing any sequence of actions that leads to these unpreventable breaches of security. We introduce only the minimal necessary restrictions preventing the attacks, making sure that we allow *all* adversaries that do not trivially break the security as per above.

5.1   Channel Interface Game

Consider game INTER in Fig. 5. It defines the interface between an adversary $\mathcal{D}$ and a channel Ch. A secret bit b is chosen at random and the adversary's goal is to guess this bit given access to a left-or-right sending oracle, real-or-$\perp$ receiving oracle, and an exposure oracle. The sending oracle takes as input a user $u \in \{\mathcal{I}, \mathcal{R}\}$, two messages $m_0, m_1 \in \{0, 1\}^*$, and associated data ad. Then it returns the encryption of m_b with ad by user u. The receiving oracle RECV takes as input a user u, a ciphertext c, and associated data ad. It has user u decrypt this ciphertext using ad, and proceeds as follows. If $b = 0$ holds (along with another condition we discuss momentarily) then it returns the valid decryption of this ciphertext; otherwise it returns $\perp$. The exposure oracle EXP takes as input a user u, and a flag rand. It returns user's state st_u, and it might return random coins that will be used the next time this user runs algorithms Ch.Send or Ch.Recv (depending on the value of rand, which we discuss below). The advantage of adversary $\mathcal{D}$ against channel Ch is defined by $\mathsf{Adv}_{\mathsf{Ch}}^{\mathsf{inter}}(\mathcal{D}) = 2\Pr[\mathrm{INTER}_{\mathsf{Ch}}^{\mathcal{D}}] - 1$.

This interface gives the adversary full control over the communication between the two users of the channel. It may modify, reorder, or block any

Game $\mathrm{INTER}_{\mathsf{Ch}}^{\mathcal{D}}$	$\mathrm{RECV}(u, c, ad)$
$b \leftarrow_\$ \{0, 1\}$	$/\!\!/ \ u \in \{\mathcal{I}, \mathcal{R}\}, (c, ad) \in (\{0,1\}^*)^2$
$s_\mathcal{I} \leftarrow r_\mathcal{I} \leftarrow s_\mathcal{R} \leftarrow r_\mathcal{R} \leftarrow 0$	If nextop $\neq (u, \text{“recv”})$
$(st_\mathcal{I}, st_\mathcal{R}) \leftarrow_\$ \mathsf{Ch.Init}$	and nextop $\neq \perp$ then return $\perp$
$(z_\mathcal{I}, z_\mathcal{R}) \leftarrow_\$ (\mathsf{Ch.SendRS})^2$	$(st_u, m) \leftarrow \mathsf{Ch.Recv}(st_u, ad, c; \eta_u)$
$(\eta_\mathcal{I}, \eta_\mathcal{R}) \leftarrow_\$ (\mathsf{Ch.RecvRS})^2$	nextop $\leftarrow \perp$; $\eta_u \leftarrow_\$ \mathsf{Ch.RecvRS}$
$b' \leftarrow_\$ \mathcal{D}^{\mathrm{SEND}, \mathrm{RECV}, \mathrm{EXP}}$	If $m \neq \perp$ then $r_u \leftarrow r_u + 1$
Return $(b' = b)$	If $b = 0$ and $(c, ad) \neq \mathbf{ctable}_u[r_u]$ then
	Return m
$\mathrm{SEND}(u, m_0, m_1, ad)$	Return $\perp$
$/\!\!/ \ u \in \{\mathcal{I}, \mathcal{R}\}, (m_0, m_1, ad) \in (\{0,1\}^*)^3$	
If nextop $\neq (u, \text{“send”})$	$\mathrm{EXP}(u, \mathsf{rand})$
and nextop $\neq \perp$ then return $\perp$	$/\!\!/ \ u \in \{\mathcal{I}, \mathcal{R}\}, \mathsf{rand} \in \{\varepsilon, \text{“send”}, \text{“recv”}\}$
If $|m_0| \neq |m_1|$ then return $\perp$	If nextop $\neq \perp$ then return $\perp$
$(st_u, c) \leftarrow \mathsf{Ch.Send}(st_u, ad, m_b; z_u)$	$(z, \eta) \leftarrow (\varepsilon, \varepsilon)$
nextop $\leftarrow \perp$	If rand $= \text{“send”}$ then
$s_u \leftarrow s_u + 1$; $z_u \leftarrow_\$ \mathsf{Ch.SendRS}$	nextop $\leftarrow (u, \text{“send”})$; $z \leftarrow z_u$
$\mathbf{ctable}_{\bar{u}}[s_u] \leftarrow (c, ad)$	Else if rand $= \text{“recv”}$ then
Return c	nextop $\leftarrow (u, \text{“recv”})$; $\eta \leftarrow \eta_u$
	Return (st_u, z, η)

Fig. 5. Game defining interface between adversary $\mathcal{D}$ and channel Ch.

communication as it sees fit. The adversary is able to exfiltrate the secret state of either party at any time.

Let us consider the different cases of how a user's secrets might be exposed. They could be exposed while the user is in the middle of performing a Ch.Send operation, in the middle of performing a Ch.Recv operation, or when the user is idle (i.e. not in the middle of performing Ch.Send or Ch.Recv). In the last case we expect the adversary to learn the user's state st_u, but nothing else. If the adversary is exposing the user during an operation, they would potentially learn the state before the operation, any secrets computed during the operation, and the state after the operation. We capture this by leaking the state from before the operation along with the randomness that will be used when the adversary makes its next query to SEND or RECV. This allows the adversary to compute the next state as well. The three possible values of rand are rand = "send" for the first possibility, rand = "recv" for the second possibility, and rand = ε for the third. These exposures represent what the adversary is learning while a particular operation is occurring, so we require (via nextop) that after such an exposure it immediately makes the corresponding oracle query. Without the use of the exposure oracle the game specified by this interface would essentially be equivalent to the combination of the integrity and confidentiality security notions defined by Marson and Poettering [30] in the all-in-one definition style of Rogaway and Shrimpton [37].

The interface game already includes some standard checks. First, we require that on any query (u, m_0, m_1, ad) to SEND the adversary must provide equal length messages. If the adversary does not do so (i.e. $|m_0| \neq |m_1|$) then SEND returns $\perp$ immediately. This prevents the inherent attack where an adversary could distinguish between the two values of b by asking for encryptions of different length messages and checking the length of the output ciphertext. Adversary $\mathcal{D}_1$ in Fig. 6 does just that and would achieve $\mathsf{Adv}_{\mathsf{Ch}}^{\mathsf{inter}}(\mathcal{D}_1) > 1/2$ against any channel Ch if not for that check.

Second, we want to prevent RECV from decrypting ciphertexts that are simply forwarded to it from SEND. So for each user u we keep track of counters s_u and r_u that track how many messages that user has sent and received. Then at the end of a SEND call to u the ciphertext-associated data pair (c, ad) is stored in the table $\mathbf{ctable}_{\overline{u}}$ with index s_u. When RECV is called for user $\overline{u}$ it will compare the pair (c, ad) against $\mathbf{ctable}_{\overline{u}}[r_{\overline{u}}]$ and if the pair matches return $\perp$ regardless of the value of the secret bit. If we did not do this check then for any channel Ch the adversary $\mathcal{D}_2$ shown in Fig. 6 would achieve $\mathsf{Adv}_{\mathsf{Ch}}^{\mathsf{inter}}(\mathcal{D}_2) = 1$.

We now specify several efficient adversaries that will have high advantage for *any* choice of Ch. For concreteness we always have our adversaries immediately start the actions required to perform the attacks, but all of the attacks would still work if the adversary had performed a number of unrelated procedure calls first. Associated data will never be important for our attacks so we will always set it to ε. We will typically set $m_0 = 0$ and $m_1 = 1$. For the following we let Ch be any channel and consider the adversaries shown in Fig. 6.

Adversary $\mathcal{D}_1^{\text{SEND,RECV,EXP}}$	Adversary $\mathcal{D}_{3.1}^{\text{SEND,RECV,EXP}}$	Adversary $\mathcal{D}_4^{\text{SEND,RECV,EXP}}$
$(st, z, \eta) \leftarrow \text{EXP}(\mathcal{I}, \varepsilon)$	$(st, z, \eta) \leftarrow \text{EXP}(\mathcal{I}, \varepsilon)$	$c \leftarrow \text{SEND}(\mathcal{I}, 0, 1, \varepsilon)$
$n \leftarrow \max_{c \in [\text{Ch.Send}(st,\varepsilon,1)]} \lvert c \rvert$	$(st, c) \leftarrow\!\!\$ \text{Ch.Send}(st, \varepsilon, 1)$	$(st, z, \eta) \leftarrow \text{EXP}(\mathcal{R}, \varepsilon)$
$m \leftarrow\!\!\$ \{0,1\}^{n+2}$	$m \leftarrow \text{RECV}(\mathcal{R}, c, \varepsilon)$	$(st, m) \leftarrow\!\!\$ \text{Ch.Recv}(st, \varepsilon, c)$
$c \leftarrow \text{SEND}(\mathcal{I}, m, 1, \varepsilon)$	$(st, c) \leftarrow\!\!\$ \text{Ch.Send}(st, \varepsilon, 1)$	If $m = 1$ then return 1
If $\lvert c \rvert \leq n$ then return 1	$m \leftarrow \text{RECV}(\mathcal{R}, c, \varepsilon)$	Return 0
Return 0	If $m = \bot$ then return 1	

Adversary $\mathcal{D}_2^{\text{SEND,RECV,EXP}}$	Return 0	Adversary $\mathcal{D}_5^{\text{SEND,RECV,EXP}}$
$c \leftarrow \text{SEND}(\mathcal{I}, 1, 1, \varepsilon)$	Adversary $\mathcal{D}_{3.2}^{\text{SEND,RECV,EXP}}$	$(st, z, \eta) \leftarrow \text{EXP}(\mathcal{R}, \varepsilon)$
$m \leftarrow \text{RECV}(\mathcal{R}, c, \varepsilon)$	$(st, z, \eta) \leftarrow \text{EXP}(\mathcal{I}, \varepsilon)$	$c \leftarrow \text{SEND}(\mathcal{I}, 0, 1, \varepsilon)$
If $m = \bot$ then return 1	$(st, c) \leftarrow\!\!\$ \text{Ch.Send}(st, \varepsilon, 1)$	$(st, m) \leftarrow\!\!\$ \text{Ch.Recv}(st, \varepsilon, c)$
Return 0	$m \leftarrow \text{RECV}(\mathcal{R}, c, \varepsilon)$	If $m = 1$ then return 1

Adversary $\mathcal{D}_3^{\text{SEND,RECV,EXP}}$	$c \leftarrow \text{SEND}(\mathcal{R}, 0, 1, \varepsilon)$	Return 0
$(st, z, \eta) \leftarrow \text{EXP}(\mathcal{I}, \varepsilon)$	$(st, m) \leftarrow\!\!\$ \text{Ch.Recv}(st, \varepsilon, c)$	Adversary $\mathcal{D}_6^{\text{SEND,RECV,EXP}}$
$(st, c) \leftarrow\!\!\$ \text{Ch.Send}(st, \varepsilon, 1)$	If $m = 1$ then return 1	$(st, z, \eta) \leftarrow \text{EXP}(\mathcal{I}, \text{"send"})$
$m \leftarrow \text{RECV}(\mathcal{R}, c, \varepsilon)$	Return 0	$(st, c) \leftarrow \text{Ch.Send}(st, \varepsilon, 1; z)$
If $m = \bot$ then return 1		$c' \leftarrow \text{SEND}(\mathcal{I}, 0, 1, \varepsilon)$
Return 0		If $c' = c$ then return 1
		Return 0

Fig. 6. Generic attacks against any channel Ch with interface INTER.

Trivial Forgery. If the adversary exposes the secrets of u it will be able to forge a ciphertext that $\bar{u}$ would accept at least until the future point in time when $\bar{u}$ has received the ciphertext that u creates next. For a simple example of this consider the third adversary, $\mathcal{D}_3$. It exposes the secrets of user $\mathcal{I}$, then uses them to perform its own Ch.Send computation locally, and sends the resulting ciphertext to $\mathcal{R}$. Clearly this ciphertext will always decrypt to a non-$\bot$ value so the adversary can trivially determine the value of b and achieve $\text{Adv}_{\text{Ch}}^{\text{inter}}(\mathcal{D}_3) = 1$.

After an adversary has done the above to trivially send a forgery to $\bar{u}$ it can easily perform further attacks on both the integrity and authenticity of the channel. These are shown by adversaries $\mathcal{D}_{3.1}$ and $\mathcal{D}_{3.2}$. The first displays the fact that the attacker can easily send further forgeries to $\bar{u}$. The second displays the fact that the attacker can now easily decrypt any messages sent by $\bar{u}$. We have $\text{Adv}_{\text{Ch}}^{\text{inter}}(\mathcal{D}_{3.1}) = 1$ and $\text{Adv}_{\text{Ch}}^{\text{inter}}(\mathcal{D}_{3.2}) = 1$.

Trivial Challenges. If the adversary exposes the secrets of u it will necessarily be able to decrypt any ciphertexts already encrypted by $\bar{u}$ that have not already been received by u. Consider the adversary $\mathcal{D}_4$. It determines what message was encrypted by user $\mathcal{I}$ by exposing the state of $\mathcal{R}$, and uses that to run Ch.Recv. We have $\text{Adv}_{\text{Ch}}^{\text{inter}}(\mathcal{D}_4) = 1$.

Similarly, if the adversary exposes the secrets of u it will necessarily be able to decrypt any future ciphertexts encrypted by $\bar{u}$, until $\bar{u}$ receives the ciphertext that u creates next. Consider the adversary $\mathcal{D}_5$. It is essentially the identical to adversary $\mathcal{D}_4$, except it reverses the order of the calls made to SEND and EXP. We have $\mathsf{Adv}_{\mathsf{Ch}}^{\mathsf{inter}}(\mathcal{D}_5) = 1$.

Exposing Randomness. If an adversary exposes user u with rand = "send" then it is able to compute the next state of u by running Ch.Send locally with the same randomness that u will use. So in this case the security game must act as if the adversary exposed both the current and the next state. In particular, the attacks above could only succeed until, first, the exposed user u updated its secrets and, second, user $\bar{u}$ updates its secrets accordingly (which can happen after it receives the next message from u). But if the randomness was exposed, then secrets would need to be updated at least twice until the security is restored.

Exposing user u with rand = "send" additionally allows the attack shown in $\mathcal{D}_6$. The adversary exposes the state and the sending randomness of $\mathcal{I}$, encrypts 1 locally using these exposed values of $\mathcal{I}$, and then calls SEND to get a challenge ciphertext sent by $\mathcal{I}$. The adversary compares whether the two ciphertexts are the same to determine the secret bit. We have $\mathsf{Adv}_{\mathsf{Ch}}^{\mathsf{inter}}(\mathcal{D}_6) = 1$. More broadly, if the adversary exposes the secrets of u with rand = "send" it will always be able to tell what is the next message encrypted by u.

Exposing with rand = "recv" does not generically endow the adversary with the ability to do any additional attacks.

5.2   Optimal Security of a Channel

Our full security game is obtained by adding a minimal amount of code to INTER to disallow the generic attacks just discussed. Consider the game AEAC (authenticated encryption against compromise) shown in Fig. 7. We define the advantage of an adversary $\mathcal{D}$ against channel Ch by $\mathsf{Adv}_{\mathsf{Ch}}^{\mathsf{aeac}}(\mathcal{D}) = 2\Pr[\mathrm{AEAC}_{\mathsf{Ch}}^{\mathcal{D}}] - 1$.

We now have a total of eight variables to control the behavior of the adversary and prevent it from abusing trivial attacks. Some of the variables are summarized in Fig. 8. We have already seen s_u, r_u, nextop, and $\mathbf{ctable}_u$ in INTER. The new variables we have added in AEAC are tables $\mathbf{forge}_u$ and $\mathbf{ch}_u$, number $\mathcal{X}_u \in \mathbb{N}$, and flag $\mathsf{restricted}_u \in \{\mathsf{true}, \mathsf{false}\}$. We now discuss the new variables.

The table $\mathbf{forge}_u$ was added to prevent the type of attack shown in $\mathcal{D}_3$. When the adversary calls EXP on user u we set $\mathbf{forge}_{\bar{u}}$ to "trivial" for the indices of ciphertexts for which this adversary is now necessarily able to create forgeries. If the adversary takes advantage of this to send a ciphertext of its own creation to $\bar{u}$ then the flag $\mathsf{restricted}_{\bar{u}}$ will be set, whose effect we will describe momentarily.

The table $\mathbf{ch}_u$ is used to prevent the types of attacks shown by $\mathcal{D}_4$ and $\mathcal{D}_6$. Whenever the adversary makes a valid challenge query[2] to user u we set $\mathbf{ch}_u[s_u]$ to "done". The game will not allow the adversary to expose $\bar{u}$'s secrets if there are any challenge queries for which u sent a ciphertext that $\bar{u}$ has not received yet. This use of $\mathbf{ch}_u$ prevents an attack like $\mathcal{D}_4$. To prevent an attack like $\mathcal{D}_6$, we set $\mathbf{ch}_u[s_u + 1]$ to "forbidden" whenever the adversary exposes the state and sending randomness of u. This disallows the adversary from doing a challenge

---

[2] We use the term challenge query to refer to a SEND query for which $m_0 \neq m_1$.

Game $\text{AEAC}_{\text{Ch}}^{\mathcal{D}}$

$b \leftarrow_\$ \{0,1\}$; $s_\mathcal{I} \leftarrow 0$; $r_\mathcal{I} \leftarrow 0$; $s_\mathcal{R} \leftarrow 0$; $r_\mathcal{R} \leftarrow 0$

$\text{restricted}_\mathcal{I} \leftarrow \text{false}$; $\text{restricted}_\mathcal{R} \leftarrow \text{false}$

$\mathbf{forge}_\mathcal{I}[0 \ldots \infty] \leftarrow \text{``nontrivial''}$ ; $\mathbf{forge}_\mathcal{R}[0 \ldots \infty] \leftarrow \text{``nontrivial''}$

$\mathcal{X}_\mathcal{I} \leftarrow 0$; $\mathcal{X}_\mathcal{R} \leftarrow 0$; $(st_\mathcal{I}, st_\mathcal{R}) \leftarrow_\$ \text{Ch.Init}$

$(z_\mathcal{I}, z_\mathcal{R}) \leftarrow_\$ (\text{Ch.SendRS})^2$; $(\eta_\mathcal{I}, \eta_\mathcal{R}) \leftarrow_\$ (\text{Ch.RecvRS})^2$

$b' \leftarrow_\$ \mathcal{D}^{\text{SEND,RECV,EXP}}$

Return $(b' = b)$

$\underline{\text{SEND}(u, m_0, m_1, \text{ad})}$ // $u \in \{\mathcal{I}, \mathcal{R}\}, (m_0, m_1, \text{ad}) \in (\{0,1\}^*)^3$

If $\mathbf{nextop} \neq (u, \text{``send''})$ and $\mathbf{nextop} \neq \bot$ then return $\bot$

If $|m_0| \neq |m_1|$ then return $\bot$

If $(r_u < \mathcal{X}_u$ or restricted_u or $\mathbf{ch}_u[s_u + 1] = \text{``forbidden''})$ and $m_0 \neq m_1$ then

    Return $\bot$

$(st_u, c) \leftarrow \text{Ch.Send}(st_u, \text{ad}, m_b; z)$

$\mathbf{nextop} \leftarrow \bot$; $s_u \leftarrow s_u + 1$; $z_u \leftarrow_\$ \text{Ch.SendRS}$

If not restricted_u then $\mathbf{ctable}_{\overline{u}}[s_u] \leftarrow (c, \text{ad})$

If $m_0 \neq m_1$ then $\mathbf{ch}_u[s_u] \leftarrow \text{``done''}$

Return c

$\underline{\text{RECV}(u, c, \text{ad})}$ // $u \in \{\mathcal{I}, \mathcal{R}\}, (c, \text{ad}) \in (\{0,1\}^*)^2$

If $\mathbf{nextop} \neq (u, \text{``recv''})$ and $\mathbf{nextop} \neq \bot$ then return $\bot$

$(st_u, m) \leftarrow \text{Ch.Recv}(st_u, \text{ad}, c; \eta_u)$

$\mathbf{nextop} \leftarrow \bot$; $\eta_u \leftarrow_\$ \text{Ch.RecvRS}$

If $m = \bot$ then return $\bot$

$r_u \leftarrow r_u + 1$

If $\mathbf{forge}_u[r_u] = \text{``trivial''}$ and $(c, \text{ad}) \neq \mathbf{ctable}_u[r_u]$ then

    $\text{restricted}_u \leftarrow \text{true}$

If restricted_u or $(b = 0$ and $(c, \text{ad}) \neq \mathbf{ctable}_u[r_u])$ then

    Return m

Return $\bot$

$\underline{\text{EXP}(u, \text{rand})}$ // $u \in \{\mathcal{I}, \mathcal{R}\}, \text{rand} \in \{\varepsilon, \text{``send''}, \text{``recv''}\}$

If $\mathbf{nextop} \neq \bot$ then return $\bot$

If restricted_u then return (st_u, z_u, η_u)

If $\exists i \in (r_u, s_{\overline{u}}]$ s.t. $\mathbf{ch}_{\overline{u}}[i] = \text{``done''}$ then

    Return $\bot$

$\mathbf{forge}_{\overline{u}}[s_u + 1] \leftarrow \text{``trivial''}$; $(z, \eta) \leftarrow (\varepsilon, \varepsilon)$; $\mathcal{X}_{\overline{u}} \leftarrow s_u + 1$

If $\text{rand} = \text{``send''}$ then

    $\mathbf{nextop} \leftarrow (u, \text{``send''})$; $z \leftarrow z_u$; $\mathcal{X}_{\overline{u}} \leftarrow s_u + 2$

    $\mathbf{forge}_{\overline{u}}[s_u + 2] \leftarrow \text{``trivial''}$ ; $\mathbf{ch}_u[s_u + 1] \leftarrow \text{``forbidden''}$

Else if $\text{rand} = \text{``recv''}$ then

    $\mathbf{nextop} \leftarrow (u, \text{``recv''})$; $\eta \leftarrow \eta_u$

Return (st_u, z, η)

Fig. 7. Game defining AEAC security of channel Ch.

Variable	Set to x when y occurs	Effect
nextop_u	$(u, \text{"send"})$ when z_u is exposed	$-$ u must send next
	$(u, \text{"recv"})$ when η_u is exposed	$-$ u must receive next
forge_u	"trivial" when $\bar{u}$ is exposed	$-$ forgeries to u set restricted_u
ch_u	"done" when challenge from u	$-$ prevents an exposure of $\bar{u}$
	"forbidden" when z_u is exposed	$-$ prevents a challenge from u
$\mathcal{X}_u$	when $\bar{u}$ is exposed	$-$ prevents challenges until $r_u = \mathcal{X}_u$
restricted_u	true when trivial forgery to u	$-$ prevents challenges from u
		$+$ (c, ad) from u not added to $\text{ctable}_{\bar{u}}$
		$\pm$ show decryption of (c, ad) sent to u
		$+$ EXP calls to u always allowed and will not change other variables

Fig. 8. Table summarizing some important variables in game AEAC. A "$-$" indicates a way in which the behavior of the adversary is being restricted. A "$+$" indicates a way in which the behavior of the adversary is being enabled.

query during its next SEND call to u (the call for which the adversary knows the corresponding randomness).

The number $\mathcal{X}_u$ prevents attacks like $\mathcal{D}_5$. When u is exposed $\mathcal{X}_{\bar{u}}$ will be set to a number that is 1 or 2 greater than the current number of ciphertexts u has sent (depending on the value of rand) and challenge queries from $\bar{u}$ will not be allowed until it has received that many ciphertexts. This ensures that the challenge queries from $\bar{u}$ are not issued with respect to exposed keys of u.[3]

Finally the flag restricted_u serves to both allow and disallow some attacks. The flag is initialized to false. It is set to true when the adversary forges a ciphertext to u after exposing $\bar{u}$. Once u has received a different ciphertext than was sent by $\bar{u}$ there is no reason to think that u should be able to decrypt ciphertexts sent by $\bar{u}$ or send its own ciphertexts to $\bar{u}$. As such, if u is restricted (i.e. $\text{restricted}_u = \text{true}$) we will not add its ciphertexts to $\text{ctable}_{\bar{u}}$, we will always show the true output when u attempts to decrypt ciphertexts given to it by the adversary (even if they were sent by $\bar{u}$), and if the adversary asks to expose u we will return all of its secret state without setting any of the other variables that would restrict the actions the adversary is allowed to take.

The above describes how restricted_u allows some attacks. Now we discuss how it prevents attacks like $\mathcal{D}_{3.1}$ and $\mathcal{D}_{3.2}$. Once the adversary has sent its own ciphertext to u we must assume that the adversary will be able to decrypt ciphertexts sent by u and able to send its own ciphertexts to u that will decrypt to non-$\perp$ values. The adversary could simply have "replaced" $\bar{u}$ with itself. To address this we prevent all challenge queries from u, and decryptions performed by u are always given back to the adversary regardless of the secret bit.

Informal description of the security game. In [27] we provide a thorough written description of our security model to facilitate high-level understanding of it. For

---

[3] The symbol chi is meant to evoke the word "challenge" because it stores the next time the adversary may make a challenge query.

intricate security definitions like ours there is often ambiguity or inconsistency in subtle corner cases of the definition when written out fully in text. As such this description should merely be considered an informal aid while the pseudocode of Fig. 7 is the actual definition.

Comparison to recent definitions. The three recents works we studied while deciding how to write our security definition were [7, 14, 26]. Their settings were all distinct, but each presented security models that involve different "stages" of keys. All three works made distinct decisions in how to address challenges in different stages. In [27] we discuss these decisions, noting that they result in qualitatively identical but quantitatively distinct definitions.

6   Construction of a Secure Channel

6.1   Our Construction

We are not aware of any secure channels that would meet (or could easily be modified to meet) our security notion. The "closest" (for some unspecified, informal notion of distance) is probably the Signal Double Ratchet Algorithm. However, it relies on symmetric authenticated encryption for both privacy and integrity so it is inherently incapable of achieving our strong notion of security. Later, we describe an attack against a variant of our proposed construction that uses symmetric primitives to exhibit the sorts of attacks that are unavoidable when using them. A straightforward variant of this attack would also apply against the Double Ratchet Algorithm.

In this section we construct our cryptographic channel and motivate our design decisions by giving attacks against variants of the channel. In Sect. 6.2 we will prove its security by reducing it to that of its underlying components.

The idea of our scheme is as follows. Both parties will keep track of a transcript of the messages they have sent and received, τ_s and τ_r. These will be included as a part of every ciphertext and verified before a ciphertext is accepted. On seeing a new ciphertext the appropriate transcript is updated to be the hash of the ciphertext (note that the old transcript is part of this ciphertext, so the transcript serves as a record of the entire conversation). Sending transcripts (vector of τ_s) are stored until the other party has acknowledged receiving a more recent transcript.

For authenticity, every time a user sends a ciphertext they authenticate it with a digital signature and include in it the verification key for the signing key that they will use to sign the next ciphertext they send. Any time a user receives a ciphertext they will use the new receiving transcript produced to update their current signing key.

For privacy, messages will be encrypted using public-key encryption. With every ciphertext the sender will include the encryption key for a new decryption key they have generated. Decryption keys are stored until the other party has acknowledged receiving a more recent encryption key. The encryption will use as a label all of the extra data that will be included with the ciphertext (i.e. a sending counter, a receiving counter, an associated data string, a new verification key, a new encryption key, a receiving transcript, and a sending transcript). The formal definition of our channel is as follows.

Algorithm SCh.Init

$(sk_\mathcal{I}, vk_\mathcal{R}) \leftarrow_\$ \mathsf{DS.Kg}$; $(ek_\mathcal{I}, \mathbf{dk}_\mathcal{R}[0]) \leftarrow_\$ \mathsf{PKE.Kg}$

$(sk_\mathcal{R}, vk_\mathcal{I}) \leftarrow_\$ \mathsf{DS.Kg}$; $(ek_\mathcal{R}, \mathbf{dk}_\mathcal{I}[0]) \leftarrow_\$ \mathsf{PKE.Kg}$

$hk \leftarrow_\$ \mathsf{H.Kg}$; $\tau_r \leftarrow \varepsilon$; $\boldsymbol{\tau}_s[0] \leftarrow \varepsilon$; $s \leftarrow r \leftarrow r^{ack} \leftarrow 0$

$st_\mathcal{I} \leftarrow (s, r, r^{ack}, sk_\mathcal{I}, vk_\mathcal{I}, ek_\mathcal{I}, \mathbf{dk}_\mathcal{I}, hk, \tau_r, \boldsymbol{\tau}_s)$

$st_\mathcal{R} \leftarrow (s, r, r^{ack}, sk_\mathcal{R}, vk_\mathcal{R}, ek_\mathcal{R}, \mathbf{dk}_\mathcal{R}, hk, \tau_r, \boldsymbol{\tau}_s)$

Return $(st_\mathcal{I}, st_\mathcal{R})$

Algorithm SCh.Send(st, ad, m)

$(s, r, r^{ack}, sk, vk, ek, \mathbf{dk}, hk, \tau_r, \boldsymbol{\tau}_s) \leftarrow st$; $s \leftarrow s + 1$

$(sk', vk') \leftarrow_\$ \mathsf{DS.Kg}$; $(ek', \mathbf{dk}[s]) \leftarrow_\$ \mathsf{PKE.Kg}$

$\ell \leftarrow (s, r, ad, vk', ek', \tau_r, \boldsymbol{\tau}_s[s-1])$

$(ek, c') \leftarrow_\$ \mathsf{PKE.Enc}(ek, \ell, m)$

$v \leftarrow (c', \ell)$; $\sigma \leftarrow_\$ \mathsf{DS.Sign}(sk, v)$

$c \leftarrow (\sigma, v)$; $\boldsymbol{\tau}_s[s] \leftarrow \mathsf{H.Ev}(hk, c)$

$st \leftarrow (s, r, r^{ack}, sk', vk, ek, \mathbf{dk}, hk, \tau_r, \boldsymbol{\tau}_s)$

Return (st, c)

Algorithm SCh.Recv(st, ad, c)

$(s, r, r^{ack}, sk, vk, ek, \mathbf{dk}, hk, \tau_r, \boldsymbol{\tau}_s) \leftarrow st$

$(\sigma, v) \leftarrow c$; $(c', \ell) \leftarrow v$

$(s', r', ad', vk', ek', \tau_r', \tau_s') \leftarrow \ell$

If $s' \neq r+1$ or $\tau_r' \neq \boldsymbol{\tau}_s[r']$ or $\tau_s' \neq \tau_r$ or $ad \neq ad'$ then return $(st, \bot)$

$(vk'', t) \leftarrow \mathsf{DS.Vrfy}(vk, \sigma, v, \boldsymbol{\tau}_s[r^{ack}+1, \ldots, r'])$

If not t then return $(st, \bot)$

$r \leftarrow r+1$; $r^{ack} \leftarrow r'$; $(\mathbf{dk}[r'], m) \leftarrow_\$ \mathsf{PKE.Dec}(\mathbf{dk}[r'], \ell, c')$

$\boldsymbol{\tau}_s[0, \ldots, r'-1] \leftarrow \bot$; $\mathbf{dk}[0, \ldots, r'-1] \leftarrow \bot$

$\tau_r \leftarrow \mathsf{H.Ev}(hk, c)$; $sk \leftarrow_\$ \mathsf{DS.UpdSk}(sk, \tau_r)$

$st \leftarrow (s, r, r^{ack}, sk, vk', ek', \mathbf{dk}, hk, \tau_r, \boldsymbol{\tau}_s)$

Return (st, m)

Fig. 9. Construction of channel SCh = SCH[DS, PKE, H] from function family H, key-updatable digital signature scheme DS, and key-updating public-key encryption scheme PKE.

Cryptographic channel SCH[DS, PKE, H]. Let DS be a key-updatable digital signature scheme, PKE be a key-updating public-key encryption scheme, and H be a family of functions. We build a cryptographic channel SCh = SCH[DS, PKE, H] as defined in Fig. 9.

A user's state st_u, among other values, contains counters s_u, r_u, r_u^{ack}. Here, s_u is the number of messages that u sent to $\bar{u}$, and r_u is the number of messages they received back from $\bar{u}$. The counter r_u^{ack} stores the last value of $r_{\bar{u}}$ in a ciphertext received by u (i.e. the index of the last ciphertext that u believes $\overline{u}$ has received and acknowledged). This counter is used to ensure that prior to running a signature verification algorithm, the verification key vk is updated with respect to the same transcripts as the signing key sk (at the time it was used to produce the signature). Note that algorithm DS.Vrfy returns (vk'', t) where t is the result of verifying that σ is a valid signature for v with respect to verification key vk'' (using the notation convention from Sect. 3).

<u>Inefficiencies of SCh.</u> A few aspects of SCh are less efficient than one would a priori hope. The state maintained by a user u (specifically the tables $\mathbf{dk}_u$ and $\tau_{s,u}$) is not constant in size, but instead grows linearly with the number of ciphertexts that u sent to $\overline{u}$ without receiving a reply back. Additionally, when DS is instantiated with the particular choice of DS that we define in [27] the length of the ciphertext sent by a user u will grow linearly in the number of ciphertexts that u has received since the last time they sent a ciphertext. When PKE is instantiated with the scheme we define in [27] there is an extra state being stored that is linear in the number of ciphertexts that u has sent since it last received a ciphertext. Such inefficiencies would be unacceptable for a protocol like TLS or SSH, but in our motivating context of messaging is it plausible that they are acceptable. Each message is human generated and the state gets "refreshed" regularly if the two users regularly reply to one another. One could additionally consider designing an app to regularly send an empty message whose sole purpose is state refreshing. We leave as interesting future work improving on the efficiency of our construction.

<u>Design decisions.</u> We will now discuss attacks against different variants of SCh. This serves to motivate the decisions made in its design and give intuition for why it achieves the desired security. Several steps in the security proof of this construction can be understood by noting which of these attacks are ruled out in the process.

The attacks are shown in Figs. 10 and 11. The first several attacks serve to demonstrate that Ch.Send must use a sufficient amount of randomness (shown in $\mathcal{D}_a$, $\mathcal{D}_b$, $\mathcal{D}_c$) and that H needs to be collision resistant (shown in $\mathcal{D}_b$, $\mathcal{D}_c$). The next attack shows why our construction would be insecure if we did not use labels with PKE (shown in $\mathcal{D}_d$). Then we provide two attacks showing why the keys of DS and PKE need to be updated (shown in $\mathcal{D}_e$, $\mathcal{D}_f$). Then we show an attack that arises if multiple valid signatures can be found for the same string (shown in $\mathcal{D}_g$). Finally, we conclude with attacks that would apply if we used symmetric instead of asymmetric primitives to build SCh (shown in $\mathcal{D}_h$, $\mathcal{D}_i$).

<u>Scheme with insufficient sending entropy.</u> Any scheme whose sending algorithm has insufficient entropy will necessarily be insecure. For simplicity let SCh_1 be a variant of SCh such that SCh_1.Send is deterministic (the details of how we are making it deterministic do not matter). We can attack both the message privacy and the integrity of such a scheme.

Consider the adversary $\mathcal{D}_a$. It exposes $\mathcal{I}$, encrypts the message 1 locally, and then sends a challenge query to $\mathcal{I}$ asking for the encryption of either 1 or

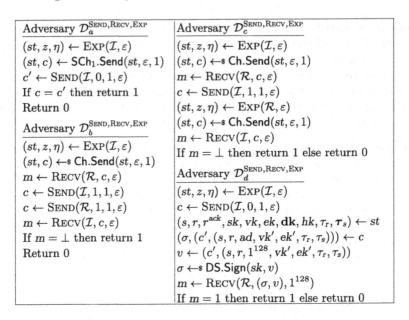

Fig. 10. Attacks against variants of SCh.

0. By comparing the ciphertext it produced to the one returned by SEND it can determine which message was encrypted, learning the secret bit. We have $\mathsf{Adv}^{\mathsf{aeac}}_{\mathsf{SCh}_1}(\mathcal{D}_a) = 1$. This attack is fairly straightforward and will be ruled out by the security of PKE in our proof without having to be addressed directly.

The attacks against integrity are more subtle. They are explicitly addressed in the first game transition of our proof. Let $\mathsf{Ch} = \mathsf{SCh}_1$ and consider adversaries $\mathcal{D}_b$ and $\mathcal{D}_c$. They both start by doing the same sequence of operations: expose $\mathcal{I}$, use its secret state to encrypt and send message 1 to $\mathcal{R}$, then ask $\mathcal{I}$ to produce an encryption of 1 for $\mathcal{R}$ (which will be the same ciphertext as above, because $\mathsf{SCh}_1.\mathsf{Send}$ is deterministic). Now $\mathsf{restricted}_{\mathcal{R}} = \mathsf{true}$ because oracle RECV was called on a trivially fogeable ciphertext that was not produced by oralce SEND. But $\mathcal{R}$ has received the exact same ciphertext that $\mathcal{I}$ sent. Different attacks are possible from this point.

Adversary $\mathcal{D}_b$ just asks $\mathcal{R}$ to send a message and forwards it along to $\mathcal{I}$. Since $\mathcal{R}$ was restricted the ciphertext does not get added to $\mathbf{ctable}_{\mathcal{I}}$ so it can be used to discover the secret bit. We have $\mathsf{Adv}^{\mathsf{aeac}}_{\mathsf{SCh}_1}(\mathcal{D}_b) = 1$. Adversary $\mathcal{D}_c$ exposes $\mathcal{R}$ and uses the state it obtains to create its own forgery to $\mathcal{I}$. It then returns 1 or 0 depending on whether RECV returns the correct decryption or $\perp$. This attack succeeds because exposing $\mathcal{R}$ when it is restricted will not set any of the variables that would typically prevent the adversary from winning by creating a forgery. We have $\mathsf{Adv}^{\mathsf{aeac}}_{\mathsf{SCh}_1}(\mathcal{D}_c) = 1$. We have not shown it, but another message privacy attack at this point (instead of proceeding as $\mathcal{D}_b$ or $\mathcal{D}_c$) could have asked for another challenge query from $\mathcal{I}$, exposed $\mathcal{R}$, and used the exposed state to trivially determine which message was encrypted.

Scheme without collision-resistant hashing. If it is easy to find collisions in H then we can attack the channel by causing both parties to have matching transcripts despite having seen different sequences of ciphertexts. For concreteness let SCh_2 be a variant of our scheme using a hash function that outputs 0^{128} on all inputs. Let $\mathsf{Ch} = \mathsf{SCh}_2$ and again consider adversaries $\mathcal{D}_b$ and $\mathcal{D}_c$. We no longer expect the ciphertexts that they produce locally to match the ciphertexts returned by $\mathcal{I}$. However, they will have the same hash value and thus produce the same transcript $\tau_{r,\mathcal{R}} = 0^{128} = \tau_{s,\mathcal{I}}$. Consequently, $\mathcal{R}$ still updates its signing key in the same way regardless of whether it receives the ciphertext produced by $\mathcal{I}$ or the ciphertext locally generated by adversary. So the messages subsequently sent by $\mathcal{R}$ will still be accepted by $\mathcal{I}$. We have $\mathsf{Adv}^{\mathsf{aeac}}_{\mathsf{SCh}_2}(\mathcal{D}_b) = 1$ and $\mathsf{Adv}^{\mathsf{aeac}}_{\mathsf{SCh}_2}(\mathcal{D}_c) = 1$.

Scheme without PKE labels. Let SCh_3 be a variant of SCh that uses a public-key encryption scheme that does not accept labels and consider adversary $\mathcal{D}_d$. It exposes $\mathcal{I}$ and asks $\mathcal{I}$ for a challenge query. It then uses the state it exposed to trivially modify the ciphertext sent from $\mathcal{I}$ (we chose to have it change ad from ε to 1^{128}) and sends it to $\mathcal{R}$. Since the ciphertext sent to $\mathcal{R}$ has different associated data than the one sent by $\mathcal{I}$ the adversary will be given the decryption of this ciphertext. But without the use of labels this decryption by PKE is independent of the associated data and will thus reveal the true decryption of the challenge ciphertext to $\mathcal{I}$. We have $\mathsf{Adv}^{\mathsf{aeac}}_{\mathsf{SCh}_3}(\mathcal{D}_d) = 1$.

Schemes without key updating. We will now show why it is necessary to define new forms of PKE and DS for our construction.

Let SCh_4 be a variant of SCh that uses a digital signature scheme that does not update its keys. Consider adversary $\mathcal{D}_e$. It exposes $\mathcal{I}$, then queries SEND for $\mathcal{I}$ to send a message to $\mathcal{R}$, but uses the exposed secrets to replace it with a locally produced ciphertext c. It calls RECV for $\mathcal{R}$ with c, which sets $\mathsf{restricted}_{\mathcal{R}} = \mathsf{true}$. Since the signing key is not updated in SCh_4, the adversary now exposes $\mathcal{R}$ to obtain a signing key whose signatures will be accepted by $\mathcal{I}$. It uses this to forge a ciphertext to $\mathcal{I}$ to learn the secret bit. We have $\mathsf{Adv}^{\mathsf{aeac}}_{\mathsf{SCh}_4}(\mathcal{D}_e) = 1$.

Let SCh_5 be a variant of SCh that uses a public-key encryption scheme that does not update its keys. Consider adversary $\mathcal{D}_f$. It exposes $\mathcal{I}$ and uses this to send $\mathcal{R}$ a different ciphertext than is sent by $\mathcal{I}$ (setting $\mathsf{restricted}_{\mathcal{R}} = \mathsf{true}$). Since the decryption key is not updated, the adversary now exposes $\mathcal{R}$ to obtain a decryption key that can be used to decrypt a challenge ciphertext sent by $\mathcal{I}$. We have $\mathsf{Adv}^{\mathsf{aeac}}_{\mathsf{SCh}_5}(\mathcal{D}_f) = 1$.

Scheme with non-unique signatures. Let SCh_6 be a variant of our scheme using a digital signature scheme that does not have unique signatures. For concreteness, assume that $\sigma \,\|\, sk$ is a valid signature whenever σ is. Then consider adversary $\mathcal{D}_g$. It exposes $\mathcal{I}$ and has $\mathcal{I}$ send a challenge ciphertext. Then it modifies the ciphertext by changing the signature and forwards this modified ciphertext on to $\mathcal{R}$. The adversary is given back the true decryption of this ciphertext (because it was changed) which trivially reveals the secret bit of the game (here it is important that the signature is not part of the label used for encryption/decryption). We have $\mathsf{Adv}^{\mathsf{aeac}}_{\mathsf{SCh}_6}(\mathcal{D}_g) = 1$.

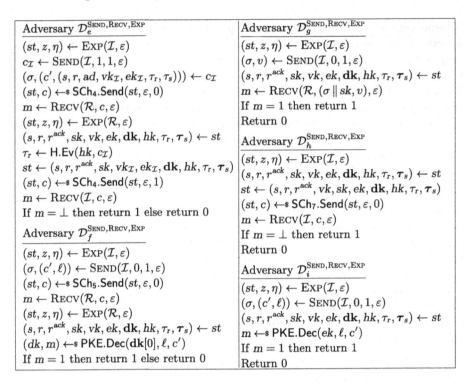

Adversary $\mathcal{D}_e^{\text{SEND},\text{RECV},\text{EXP}}$

$(st, z, \eta) \leftarrow \text{EXP}(\mathcal{I}, \varepsilon)$
$c_{\mathcal{I}} \leftarrow \text{SEND}(\mathcal{I}, 1, 1, \varepsilon)$
$(\sigma, (c', (s, r, ad, vk_{\mathcal{I}}, ek_{\mathcal{I}}, \tau_r, \tau_s)))) \leftarrow c_{\mathcal{I}}$
$(st, c) \leftarrow\!\!\$ \, \text{SCh}_4.\text{Send}(st, \varepsilon, 0)$
$m \leftarrow \text{RECV}(\mathcal{R}, c, \varepsilon)$
$(st, z, \eta) \leftarrow \text{EXP}(\mathcal{R}, \varepsilon)$
$(s, r, r^{ack}, sk, vk, ek, \mathbf{dk}, hk, \tau_r, \tau_s) \leftarrow st$
$\tau_r \leftarrow \text{H.Ev}(hk, c_{\mathcal{I}})$
$st \leftarrow (s, r, r^{ack}, sk, vk_{\mathcal{I}}, ek_{\mathcal{I}}, \mathbf{dk}, hk, \tau_r, \tau_s)$
$(st, c) \leftarrow\!\!\$ \, \text{SCh}_4.\text{Send}(st, \varepsilon, 1)$
$m \leftarrow \text{RECV}(\mathcal{I}, c, \varepsilon)$
If $m = \bot$ then return 1 else return 0

Adversary $\mathcal{D}_f^{\text{SEND},\text{RECV},\text{EXP}}$

$(st, z, \eta) \leftarrow \text{EXP}(\mathcal{I}, \varepsilon)$
$(\sigma, (c', \ell)) \leftarrow \text{SEND}(\mathcal{I}, 0, 1, \varepsilon)$
$(st, c) \leftarrow\!\!\$ \, \text{SCh}_5.\text{Send}(st, \varepsilon, 0)$
$m \leftarrow \text{RECV}(\mathcal{R}, c, \varepsilon)$
$(st, z, \eta) \leftarrow \text{EXP}(\mathcal{R}, \varepsilon)$
$(s, r, r^{ack}, sk, vk, ek, \mathbf{dk}, hk, \tau_r, \tau_s) \leftarrow st$
$(dk, m) \leftarrow\!\!\$ \, \text{PKE.Dec}(\mathbf{dk}[0], \ell, c')$
If $m = 1$ then return 1 else return 0

Adversary $\mathcal{D}_g^{\text{SEND},\text{RECV},\text{EXP}}$

$(st, z, \eta) \leftarrow \text{EXP}(\mathcal{I}, \varepsilon)$
$(\sigma, v) \leftarrow \text{SEND}(\mathcal{I}, 0, 1, \varepsilon)$
$(s, r, r^{ack}, sk, vk, ek, \mathbf{dk}, hk, \tau_r, \tau_s) \leftarrow st$
$m \leftarrow \text{RECV}(\mathcal{R}, (\sigma \,\|\, sk, v), \varepsilon)$
If $m = 1$ then return 1
Return 0

Adversary $\mathcal{D}_h^{\text{SEND},\text{RECV},\text{EXP}}$

$(st, z, \eta) \leftarrow \text{EXP}(\mathcal{I}, \varepsilon)$
$(s, r, r^{ack}, sk, vk, ek, \mathbf{dk}, hk, \tau_r, \tau_s) \leftarrow st$
$st \leftarrow (s, r, r^{ack}, vk, sk, ek, \mathbf{dk}, hk, \tau_r, \tau_s)$
$(st, c) \leftarrow\!\!\$ \, \text{SCh}_7.\text{Send}(st, \varepsilon, 0)$
$m \leftarrow \text{RECV}(\mathcal{I}, c, \varepsilon)$
If $m = \bot$ then return 1
Return 0

Adversary $\mathcal{D}_i^{\text{SEND},\text{RECV},\text{EXP}}$

$(st, z, \eta) \leftarrow \text{EXP}(\mathcal{I}, \varepsilon)$
$(\sigma, (c', \ell)) \leftarrow \text{SEND}(\mathcal{I}, 0, 1, \varepsilon)$
$(s, r, r^{ack}, sk, vk, ek, \mathbf{dk}, hk, \tau_r, \tau_s) \leftarrow st$
$m \leftarrow\!\!\$ \, \text{PKE.Dec}(ek, \ell, c')$
If $m = 1$ then return 1
Return 0

Fig. 11. Attacks against variants of SCh.

Scheme with symmetric primitives. Let SCh_7 be a variant of our scheme that uses a MAC instead of a digital signature scheme (e.g. $vk = sk$ always, and vk is presumably no longer sent in the clear with the ciphertext). Consider adversary $\mathcal{D}_h$. It simply exposes $\mathcal{I}$ and then uses $\mathcal{I}$'s vk to send a message to $\mathcal{I}$. This trivially allows it to determine the secret bit. Here we used that PKE will decrypt any ciphertext to a non-$\bot$ value. We have $\text{Adv}_{\text{SCh}_7}^{\text{aeac}}(\mathcal{D}_h) = 1$.

Similarly let SCh_8 be a variant of our scheme that uses symmetric encryption instead of public-key encryption (e.g. $ek = dk$ always, and ek is presumably no longer sent in the clear with the ciphertext). Adversary $\mathcal{D}_i$ exposes user $\mathcal{I}$ and then uses the corresponding ek to decrypt a challenge message encrypted by $\mathcal{I}$. We have $\text{Adv}_{\text{SCh}_8}^{\text{aeac}}(\mathcal{D}_i) = 1$.

Stated broadly, a scheme that relies on symmetric primitives will not be secure because a user will know sufficient information to send a ciphertext that they would themselves accept or to read a message that they sent to the other user. Our security notion requires that this is not possible.

6.2 Security Theorem

The following theorem bounds the advantage of an adversary breaking the AEAC security of SCh using the advantages of adversaries against the CR security of

H, the UFEXP and UNIQ security of DS, the INDEXP security of PKE, and the min-entropy of DS and PKE.

Theorem 1. *Let* DS *be a key-updatable digital signature scheme,* PKE *be a key-updating public-key encryption scheme, and* H *be a family of functions. Let* SCh = SCH[DS, PKE, H]. *Let* $\mathcal{D}$ *be an adversary making at most* q_{SEND} *queries to its* SEND *oracle,* q_{RECV} *queries to its* RECV *oracle, and* q_{EXP} *queries to its* EXP *oracle. Then we can build adversaries* $\mathcal{A}_{\text{H}}$, $\mathcal{A}_{\text{DS}}$, $\mathcal{B}_{\text{DS}}$, *and* $\mathcal{A}_{\text{PKE}}$ *such that*

$$\mathsf{Adv}^{\text{aeac}}_{\text{SCh}}(\mathcal{D}) \leq 2 \cdot (q_{\text{SEND}} \cdot 2^{-\mu} + \mathsf{Adv}^{\text{cr}}_{\text{H}}(\mathcal{A}_{\text{H}}) + \mathsf{Adv}^{\text{ufexp}}_{\text{DS}}(\mathcal{A}_{\text{DS}}) +$$
$$+ \mathsf{Adv}^{\text{uniq}}_{\text{DS}}(\mathcal{B}_{\text{DS}})) + \mathsf{Adv}^{\text{indexp}}_{\text{PKE}}(\mathcal{A}_{\text{PKE}})$$

where $\mu = \mathrm{H}_\infty(\text{DS.Kg}) + \mathrm{H}_\infty(\text{PKE.Kg}) + \mathrm{H}_\infty(\text{PKE.Enc})$. *Adversary* $\mathcal{A}_{\text{DS}}$ *makes at most* $q_{\text{SEND}} + 2$ *queries to its* NEWUSER *oracle,* q_{SEND} *queries to its* SIGN *oracle, and* q_{EXP} *queries to its* EXP *oracle. Adversary* $\mathcal{B}_{\text{DS}}$ *makes at most* $q_{\text{SEND}} + 2$ *queries to its* NEWUSER *oracle. Adversary* $\mathcal{A}_{\text{PKE}}$ *makes at most* $q_{\text{SEND}} + 2$ *queries to its* NEWUSER *oracle,* q_{SEND} *queries to its* ENC *oracle,* q_{RECV} *queries to its* DEC *oracle,* $q_{\text{SEND}} + 2$ *queries to its* EXPDK *oracle, and* $\min\{q_{\text{EXP}}, q_{\text{SEND}} + 1\}$ *queries to its* EXPRAND *oracle. Adversaries* $\mathcal{A}_{\text{H}}$, $\mathcal{A}_{\text{DS}}$, $\mathcal{B}_{\text{DS}}$, *and* $\mathcal{A}_{\text{PKE}}$ *all have runtime about that of* $\mathcal{D}$.

The proof is in [27]. It broadly consists of two stages. The first stage of the proof (consisting of three game transitions) argues that the adversary will not be able to forge a ciphertext to an unrestricted user except by exposing the other user. This argument is justified by a reduction to an adversary $\mathcal{A}_{\text{DS}}$ against the security of the digital signature scheme. However, care must be taken in this reduction to ensure that $\mathcal{D}$ cannot induce behavior in $\mathcal{A}_{\text{DS}}$ that would result in $\mathcal{A}_{\text{DS}}$ cheating in the digital signature game. Addressing this possibility involves arguing that $\mathcal{D}$ cannot predict any output of SEND (from whence the min-entropy term in the bound arises) and that it cannot find any collisions in the hash function H.

Once this stage is complete the output of RECV no longer depends on the secret bit b, so we move to using the security of PKE to argue that $\mathcal{D}$ cannot use SEND to learn the value of the secret bit. This is the second stage of the proof. But prior to this reduction we have to make one last argument using the security of DS. Specifically we show that, given a ciphertext (σ, v), the adversary will not be able to find a new signature σ' such that (σ', v) will be accepted by the receiver (otherwise since $\sigma \neq \sigma'$, oracle RECV would return the true decryption of this ciphertext which would be the same as the decryption of the original ciphertext and thus allow a trivial attack). Having done this, the reduction to the security of PKE is straightforward.

Acknowledgments. We thank Mihir Bellare for extensive discussion on preliminary versions of this paper. We thank the CRYPTO 2018 reviewers for their comments. Jaeger and Stepanovs were supported in part by NSF grants CNS-1717640 and CNS-1526801.

References

1. Bellare, M., Boldyreva, A., Micali, S.: Public-key encryption in a multi-user setting: security proofs and improvements. In: Preneel, B. (ed.) EUROCRYPT 2000. LNCS, vol. 1807, pp. 259–274. Springer, Heidelberg (2000). https://doi.org/10.1007/3-540-45539-6_18

2. Bellare, M., Desai, A., Jokipii, E., Rogaway, P.: A concrete security treatment of symmetric encryption. In: FOCS 1997 (1997)

3. Bellare, M., Desai, A., Pointcheval, D., Rogaway, P.: Relations among notions of security for public-key encryption schemes. In: Krawczyk, H. (ed.) CRYPTO 1998. LNCS, vol. 1462, pp. 26–45. Springer, Heidelberg (1998). https://doi.org/10.1007/BFb0055718

4. Bellare, M., Kohno, T., Namprempre, C.: Breaking and provably repairing the ssh authenticated encryption scheme: a case study of the encode-then-encrypt-and-mac paradigm. ACM Trans. Inf. Syst. Secur. (TISSEC) **7**(2), 206–241 (2004)

5. Bellare, M., Miner, S.K.: A forward-secure digital signature scheme. In: Wiener, M. (ed.) CRYPTO 1999. LNCS, vol. 1666, pp. 431–448. Springer, Heidelberg (1999). https://doi.org/10.1007/3-540-48405-1_28

6. Bellare, M., Rogaway, P.: The security of triple encryption and a framework for code-based game-playing proofs. In: Vaudenay, S. (ed.) EUROCRYPT 2006. LNCS, vol. 4004, pp. 409–426. Springer, Heidelberg (2006). https://doi.org/10.1007/11761679_25

7. Bellare, M., Singh, A.C., Jaeger, J., Nyayapati, M., Stepanovs, I.: Ratcheted encryption and key exchange: the security of messaging. In: Katz, J., Shacham, H. (eds.) CRYPTO 2017. LNCS, vol. 10403, pp. 619–650. Springer, Cham (2017). https://doi.org/10.1007/978-3-319-63697-9_21

8. Bellare, M., Yee, B.: Forward-security in private-key cryptography. In: Joye, M. (ed.) CT-RSA 2003. LNCS, vol. 2612, pp. 1–18. Springer, Heidelberg (2003). https://doi.org/10.1007/3-540-36563-X_1

9. Boldyreva, A., Degabriele, J.P., Paterson, K.G., Stam, M.: Security of symmetric encryption in the presence of ciphertext fragmentation. In: Pointcheval, D., Johansson, T. (eds.) EUROCRYPT 2012. LNCS, vol. 7237, pp. 682–699. Springer, Heidelberg (2012). https://doi.org/10.1007/978-3-642-29011-4_40

10. Borisov, N., Goldberg, I., Brewer, E.: Off-the-record communication, or, why not to use PGP. In: ACM Workshop on Privacy in the Electronic Society (2004)

11. Canetti, R.: Universally composable security: a new paradigm for cryptographic protocols. In: FOCS 2001 (2001)

12. Canetti, R., Halevi, S., Katz, J.: A forward-secure public-key encryption scheme. J. Cryptol. **20**(3), 265–294 (2007)

13. Canetti, R., Krawczyk, H.: Analysis of key-exchange protocols and their use for building secure channels. In: Pfitzmann, B. (ed.) EUROCRYPT 2001. LNCS, vol. 2045, pp. 453–474. Springer, Heidelberg (2001). https://doi.org/10.1007/3-540-44987-6_28

14. Cohn-Gordon, K., Cremers, C., Dowling, B., Garratt, L., Stebila, D.: A formal security analysis of the Signal messaging protocol. In: Proceedings of IEEE European Symposium on Security and Privacy (EuroS&P) (2017)

15. Cohn-Gordon, K., Cremers, C., Garratt, L.: On post-compromise security. In: IEEE Computer Security Foundations Symposium (CSF) (2016)

16. Desmedt, Y., Frankel, Y.: Threshold cryptosystems. In: Brassard, G. (ed.) CRYPTO 1989. LNCS, vol. 435, pp. 307–315. Springer, New York (1990). https://doi.org/10.1007/0-387-34805-0_28

17. Diffie, W., van Oorschot, P.C., Wiener, M.J.: Authentication and authenticated key exchanges. Des. Codes Crypt. **2**(2), 107–125 (1992)
18. Dodis, Y., Katz, J., Xu, S., Yung, M.: Key-insulated public key cryptosystems. In: Knudsen, L.R. (ed.) EUROCRYPT 2002. LNCS, vol. 2332, pp. 65–82. Springer, Heidelberg (2002). https://doi.org/10.1007/3-540-46035-7_5
19. Dodis, Y., Katz, J., Xu, S., Yung, M.: Strong key-insulated signature schemes. In: Desmedt, Y.G. (ed.) PKC 2003. LNCS, vol. 2567, pp. 130–144. Springer, Heidelberg (2003). https://doi.org/10.1007/3-540-36288-6_10
20. Dodis, Y., Luo, W., Xu, S., Yung, M.: Key-insulated symmetric key cryptography and mitigating attacks against cryptographic cloud software. In: ASIACCS 2012 (2012)
21. Perrin, T. (ed.), Marlinspike, M.: The double ratchet algorithm, 20 November 2016. https://whispersystems.org/docs/specifications/doubleratchet/
22. Fischlin, M., Günther, F., Marson, G.A., Paterson, K.G.: Data is a stream: security of stream-based channels. In: Gennaro, R., Robshaw, M. (eds.) CRYPTO 2015. LNCS, vol. 9216, pp. 545–564. Springer, Heidelberg (2015). https://doi.org/10.1007/978-3-662-48000-7_27
23. Gentry, C., Silverberg, A.: Hierarchical ID-based cryptography. In: Zheng, Y. (ed.) ASIACRYPT 2002. LNCS, vol. 2501, pp. 548–566. Springer, Heidelberg (2002). https://doi.org/10.1007/3-540-36178-2_34
24. Green, M.D., Miers, I.: Forward secure asynchronous messaging from puncturable encryption. In: IEEE Symposium on Security and Privacy (2015)
25. Günther, C.G.: An identity-based key-exchange protocol. In: Quisquater, J.-J., Vandewalle, J. (eds.) EUROCRYPT 1989. LNCS, vol. 434, pp. 29–37. Springer, Heidelberg (1990). https://doi.org/10.1007/3-540-46885-4_5
26. Günther, F., Mazaheri, S.: A formal treatment of multi-key channels. In: Katz, J., Shacham, H. (eds.) CRYPTO 2017. LNCS, vol. 10403, pp. 587–618. Springer, Cham (2017). https://doi.org/10.1007/978-3-319-63697-9_20
27. Jaeger, J., Stepanovs, I.: Optimal Channel Security Against Fine-Grained State Compromise: The Safety of Messaging. Cryptology ePrint Archive, Report 2018/XYZ (2018, To appear)
28. Krawczyk, H.: The order of encryption and authentication for protecting communications (or: how secure is SSL?). In: Kilian, J. (ed.) CRYPTO 2001. LNCS, vol. 2139, pp. 310–331. Springer, Heidelberg (2001). https://doi.org/10.1007/3-540-44647-8_19
29. Langley, A.: Pond. GitHub repository, README.md (2012). https://github.com/agl/pond/commit/7bb06244b9aa121d367a6d556867992d1481f0c8
30. Marson, G.A., Poettering, B.: Security notions for bidirectional channels. IACR Trans. Symm. Cryptol. **2017**(1), 405–426 (2017)
31. Mignotte, M.: How to share a secret? In: Beth, T. (ed.) EUROCRYPT 1982. LNCS, vol. 149, pp. 371–375. Springer, Heidelberg (1983). https://doi.org/10.1007/3-540-39466-4_27
32. Namprempre, C.: Secure channels based on authenticated encryption schemes: a simple characterization. In: Zheng, Y. (ed.) ASIACRYPT 2002. LNCS, vol. 2501, pp. 515–532. Springer, Heidelberg (2002). https://doi.org/10.1007/3-540-36178-2_32
33. Open Whisper Systems. Signal protocol library for Java/Android. GitHub repository (2017). https://github.com/WhisperSystems/libsignal-protocol-java
34. Ostrovsky, R., Yung, M.: How to withstand mobile virus attacks (extended abstract). In: ACM PODC 1991 (1991)

35. Poettering, B., Rösler, P.: Ratcheted key exchange, revisited. Cryptology ePrint Archive, Report 2018/296 (2018). https://eprint.iacr.org/2018/296
36. Rogaway, P.: Authenticated-encryption with associated-data. In: ACM CCS 2002 (2002)
37. Rogaway, P., Shrimpton, T.: A provable-security treatment of the key-wrap problem. In: Vaudenay, S. (ed.) EUROCRYPT 2006. LNCS, vol. 4004, pp. 373–390. Springer, Heidelberg (2006). https://doi.org/10.1007/11761679_23
38. Shamir, A.: How to share a secret. Commun. Assoc. Comput. Mach. **22**(11), 612–613 (1979)
39. Shoup, V.: On formal models for secure key exchange. Cryptology ePrint Archive, Report 1999/012 (1999). http://eprint.iacr.org/1999/012
40. Shoup, V.: A proposal for an ISO standard for public key encryption. Cryptology ePrint Archive, Report 2001/112 (2001). https://eprint.iacr.org/2001/112
41. Tompa, M., Woll, H.: How to share a secret with cheaters. J. Cryptol. **1**(2), 133–138 (1988)
42. Unger, N., Dechand, S., Bonneau, J., Fahl, S., Perl, H., Goldberg, I., Smith, M.: SoK: secure messaging. In: IEEE Symposium on Security and Privacy (2015)
43. WhatsApp Blog. Connecting one billion users every day, 26 July 2017. https://blog.whatsapp.com/10000631/Connecting-One-Billion-Users-Every-Day

Out-of-Band Authentication in Group Messaging: Computational, Statistical, Optimal

Lior Rotem[✉] and Gil Segev

School of Computer Science and Engineering,
Hebrew University of Jerusalem, 91904 Jerusalem, Israel
{lior.rotem,segev}@cs.huji.ac.il

Abstract. Extensive efforts are currently put into securing messaging platforms, where a key challenge is that of protecting against man-in-the-middle attacks when setting up secure end-to-end channels. The vast majority of these efforts, however, have so far focused on securing *user-to-user* messaging, and recent attacks indicate that the security of *group* messaging is still quite fragile.

We initiate the study of out-of-band authentication in the group setting, extending the user-to-user setting where messaging platforms (e.g., Telegram and WhatsApp) protect against man-in-the-middle attacks by assuming that users have access to an external channel for authenticating one short value (e.g., two users who recognize each other's voice can compare a short value). Inspired by the frameworks of Vaudenay (CRYPTO '05) and Naor et al. (CRYPTO '06) in the user-to-user setting, we assume that users communicate over a completely-insecure channel, and that a group administrator can out-of-band authenticate one short message to all users. An adversary may read, remove, or delay this message (for all or for some of the users), but cannot undetectably modify it.

Within our framework we establish tight bounds on the tradeoff between the adversary's success probability and the length of the out-of-band authenticated message (which is a crucial bottleneck given that the out-of-band channel is of low bandwidth). We consider both computationally-secure and statistically-secure protocols, and for each flavor of security we construct an authentication protocol and prove a lower bound showing that our protocol achieves essentially the best possible tradeoff.

In particular, considering groups that consist of an administrator and k additional users, for statistically-secure protocols we show that at least $(k + 1) \cdot (\log(1/\epsilon) - \Theta(1))$ bits must be out-of-band authenticated, whereas for computationally-secure ones $\log(1/\epsilon) + \log k$ bits suffice, where ϵ is the adversary's success probability. Moreover, instantiating our computationally-secure protocol in the random-oracle model

L. Rotem and G. Segev—Supported by the Israel Science Foundation (Grant No. 483/13) and by the Israeli Centers of Research Excellence (I-CORE) Program (Center No. 4/11).

© International Association for Cryptologic Research 2018
H. Shacham and A. Boldyreva (Eds.): CRYPTO 2018, LNCS 10991, pp. 63–89, 2018.
https://doi.org/10.1007/978-3-319-96884-1_3

yields an efficient and practically-relevant protocol (which, alternatively, can also be based on any one-way function in the standard model).

1   Introduction

Instant messaging is gaining extremely-increased popularity as a tool enabling users to communicate with other users either individually or within groups. A variety of available messaging platforms hold an overall user base of more than 1.5 billion active users (e.g., WhatsApp, Signal, Telegram, and many more [Wik]), and recognize user authentication and end-to-end encryption as key ingredients for ensuring secure communication within them.

Extensive efforts are currently put into securing messaging platforms, both commercially (e.g., [PM16, Telb, Wha]) and academically (e.g., [FMB+16, BSJ+17, CCD+17, KBB17]). The vast majority of these efforts, however, have so far focused on securing *user-to-user* messaging, and substantially less attention has been devoted to securing *group* messaging. Unfortunately, it recently turned out that whereas the security of user-to-user messaging is gradually reaching a stable ground, the security of group messaging is still quite fragile [CGCG+17, RMS18, Gre18a, Gre18b].

Out-of-band authentication. A key challenge in securing messaging platforms is that of protecting against man-in-the-middle attacks when setting up secure end-to-end channels. Such attacks are enabled by the inability of users to authenticate their incoming messages given the somewhat ad-hoc nature of messaging platforms.[1] To this end, various messaging platforms enable "out-of-band" authentication, assuming that users have access to an *external* channel for authenticating short values. These values typically correspond to short hash values that are derived, for example, from the public keys of the users, or more generally from the transcript of any key-exchange protocol that the users execute for setting up a secure end-to-end channel.

For example, in the user-to-user setting, some messaging platforms offer users the ability to compare with each other a value that is displayed by their devices (e.g., Telegram [Tela], WhatsApp [Wha] and Viber [Vib]).[2] This may rely on the realistic assumption that by recognizing each other's voice, two users can establish a *low-bandwidth authenticated channel*: A man-on-the-middle adversary

---

[1] Despite the significant threats posed by man-in-the-middle attacks, research on the security of group messaging has so far assumed an initial authenticated setup phase (e.g., [CGCG+17, RMS18]), and did not address this security-critical assumption.

[2] For example, as specified in WhatsApp's security whitepaper [Wha, p. 10]: "WhatsApp users additionally have the option to verify the keys of the other users with whom they are communicating so that they are able to confirm that an unauthorized third party (or WhatsApp) has not initiated a man-in-the-middle attack. This can be done by scanning a QR code, or by comparing a 60-digit number. [...] The 60-digit number is computed by concatenating the two 30-digit numeric fingerprints for each user's Identity Key".

can view, delay or even remove any message sent over this channel, but cannot modify its content in an undetectable manner.

Such an authentication model was initially proposed back in 1984 by Rivest and Shamir [RS84]. They constructed the "Interlock" protocol which enables two users, who recognize each other's voice, to mutually authenticate their public keys in the absence of a trusted infrastructure.[3] More recently, motivated by the task of securely pairing wireless devices (e.g., wireless USB or Bluetooth devices), this model was formalized by Vaudenay [Vau05] in the computational setting and extended by Naor et al. [NSS06, NSS08] to the statistical setting (considering computationally-bounded and computationally-unbounded adversaries, respectively).

Given that the out-of-band channel is of low bandwidth, it is of extreme importance to construct out-of-band authentication protocols with an essentially optimal tradeoff between the length of their out-of-band authenticated value and the adversary's success probability. Vaudenay and Naor et al. provided a complete characterization of this tradeoff, resulting in optimal computationally-secure and statistically-secure protocols.

Out-of-band authentication: The group setting. Motivated by the insufficiently explored security of group messaging, we initiate the study of out-of-band message authentication protocols in the group setting. We extend the user-to-user setting to consider a group of users that consists of a sender (e.g., the group administrator) and multiple receivers (e.g., all other group members): All users communicate over an insecure channel, and we assume that the sender can out-of-band authenticate one short message to all receivers.[4] As in the user-to-user setting, this can be based, for example, on the assumption that each user can identify the administrator's voice, and having the administrator record and broadcast a short voice message. As above, we assume that an adversary may read or remove any message sent over the out-of-band channel for some or all receivers, and may delay it for different periods of time for different receivers, but cannot modify it in an undetectable manner.

Equipped with such an authentication protocol, the users of a group can now authenticate their public keys, or more generally, authenticate the transcript of any group key-exchange protocol of their choice. As in the user-to-user setting, given that the out-of-band channel is of low bandwidth, we aim at identifying the optimal tradeoff between the length of the out-of-band authenticated value and the adversary's success probability, and at constructing protocols that achieve this best-possible tradeoff.

---

[3] Unfortunately, potential attacks on the Interlock protocol were identified later on [BM94, Ell96].

[4] Clearly, one may consider a less-minimal extension where *several* users are allowed to send out-of-band authenticated values (i.e., not only the group administrator that we denote as the sender), but as our results show this is in fact not required.

1.1   Our Contributions

Modeling out-of-band authentication in the group setting. In this work we first put forward a realistic framework and strong notions of security for out-of-band message authentication protocols in the group setting. We consider a group of users that consists of a sender (e.g., the group administrator) and k receivers (e.g., all other group members), where for every $i \in [k]$ the sender would like to authenticate a message m_i to the ith receiver. We assume that all users are connected via an insecure channel (over which a man-in-the-middle adversary has complete control), and via a low-bandwidth "out-of-band" authenticated channel that enables the sender to authenticate one short message to all receivers. Adversaries may read or remove this message for some or all receivers, and may delay it for different periods of time for different receivers, but cannot modify it in an undetectable manner (we refer the reader to Sect. 3 for a formal description of our communication model and notions of security).

Identifying the optimal tradeoff: Protocols and matching lower bounds. Within our framework we then construct out-of-band authentication protocols with an optimal tradeoff between the length of their out-of-band authenticated value and the adversary's success probability. We consider both the computational setting where security is guaranteed against computationally-bounded adversaries, and the statistical setting where security is guaranteed against computationally-unbounded adversaries. In each setting we construct an authentication protocol, and then prove a lower bound showing that our protocol achieves essentially the best possible tradeoff between the length of the out-of-band authenticated value and the adversary's success probability. Our results are briefly summarized in Table 1, and we refer the reader to the following section for a more detailed overview and theorem statements.

Table 1. The length of the out-of-band authenticated value in our protocols and lower bounds. We denote by k the number of receivers (i.e., we consider groups of size $k + 1$), and by ϵ the adversary's forgery probability. Our computationally-secure protocol relies on the existence of any one-way function (see Theorem 1.1), whereas our statistically-secure protocol and our two lower bounds do not rely on any computational assumptions (see Theorems 1.2, 1.3 and 1.4).

	Our Protocols	Our Lower Bounds
Computational Security	$\log(1/\epsilon) + \log k$	$\log(1/\epsilon) + \log k - \Theta(1)$
Statistical Security	$(k + 1) \cdot (\log(1/\epsilon) + \log k + \Theta(1))$	$(k + 1) \cdot \log(1/\epsilon) - k$

Note that our upper bound and lower bound in the computational setting match within an additive constant term, whereas in the statistical setting they match within an additive $(k + 1) \log k + \Theta(k)$ term (however, whenever $\epsilon = o(1/k)$ as one would typically expect when setting a bound on the adversary's forgery probability, this difference becomes a lower-order term).

Computational vs. statistical security. Our tight bounds reveal a significant gap between the possible length of the out-of-band authenticated value in
the computational setting and in the statistical setting: Whereas in the statistical setting we prove a lower bound that depends linearly on the size of the
group, the length of the out-of-band authenticated value in our computationally-
secure protocol depends very weakly on the size of the group. Moreover, when
instantiating its cryptographic building block (a concurrent non-malleable commitment scheme) in the random-oracle model, our approach yields an efficient
and practically-relevant protocol (which, alternatively, can also be based on any
one-way function in the standard model).[5]

1.2   Overview of Our Contributions

A naive approach to constructing an out-of-band authentication protocol in the
group setting is to rely on any such protocol in the user-to-user setting: Given a
sender and k receivers, we can invoke a user-to-user protocol between the sender
and each of the receivers. Thus, if the length of the out-of-band authenticated
value in the underlying user-to-user protocol is $\ell(\epsilon)$ bits (where ϵ is the adversary's forgery probability), then the length of the out-of-band authenticated
value in the resulting group protocol is $k \cdot \ell(\epsilon/k)$ bits.[6] Thus, the naive approach
yields out-of-band authenticated values whose length is linear in the size of the
group, and the key technical challenge underlying our work is understanding
whether or not this is the best possible.

Concretely, the user-to-user protocols of Vaudenay [Vau05] and Naor et al.
[NSS06] have out-of-band authenticated values of lengths $\ell(\epsilon) = \log(1/\epsilon)$ and
$\ell(\epsilon) = 2\log(1/\epsilon) + \Theta(1)$, respectively. Thus, instantiating the naive approach with
their protocols yields computationally-secure and statistically-secure protocols
where the sender out-of-band authenticates $k \cdot (\log(1/\epsilon) + \log k)$ bits and $2k \cdot$
$(\log(1/\epsilon) + \log k + \Theta(1))$ bits, respectively.

Our results show that, unlike in the user-to-user setting, in the group setting
computationally-secure and statistically-secure protocols exhibit completely different behaviors. First, we show that for computationally-secure protocols it is
possible to do dramatically better compared to the naive approach and completely eliminate the linear dependency on the size of the group. We prove the
following two theorems providing an out-of-band authentication protocol and a
matching lower bound:

---

[5] Concretely, when setting the adversary's forgery probability ϵ to 2^{-30} in a group
that consists of $k = 2^{10}$ users, then in any statistically-secure protocol more than
$k \cdot \log(1/\epsilon) = 2^{10} \cdot 30$ bits must be out-of-band authenticated, whereas in our
computationally-secure protocol only $\log(1/\epsilon) + \log k = 40$ bits are out-of-band
authenticated.

[6] Note that if the adversary's forgery probability in the group protocol should be at
most ϵ, then the user-to-user protocol should be parameterized, for example, with ϵ/k
as the adversary's forgery probability (enabling a union bound over the k executions).

Theorem 1.1. *Assuming the existence of any one-way function, for any $k \geq 1$ there exists a computationally-secure constant-round k-receiver out-of-band message authentication protocol in which the sender out-of-band authenticates $\log(1/\epsilon) + \log k$ bits, where ϵ is the adversary's forgery probability.*

Theorem 1.2. *In any computationally-secure k-receiver out-of-band message authentication protocol, the sender must out-of-band authenticate at least $\log(1/\epsilon) + \log k - \Theta(1)$ bits, where ϵ is the adversary's forgery probability.*

Then, we show that for statistically-secure protocols the naive approach is in fact asymptotically optimal, but it can still be substantially improved by a multiplicative constant factor (which is of key importance given that the out-of-band channel is of low bandwidth). We prove the following two theorems, once again providing an out-of-band authentication protocol and a lower bound:

Theorem 1.3. *For any $k \geq 1$ there exists a statistically-secure k-receiver out-of-band message authentication protocol in which the sender out-of-band authenticates $(k + 1) \cdot (\log(1/\epsilon) + \log k + \Theta(1))$ bits, where ϵ is the adversary's forgery probability.*

Theorem 1.4. *In any statistically-secure k-receiver out-of-band message authentication protocol, the sender must out-of-band authenticate at least $(k + 1) \cdot \log(1/\epsilon) - k$ bits, where ϵ is the adversary's forgery probability.*

As discussed above, note that here our upper bound and lower bound differ by an additive $(k + 1) \cdot \log k + \Theta(k)$ term. However, whenever $\epsilon = o(1/k)$ as one would typically expect when setting a bound on the adversary's forgery probability, this difference becomes a lower-order term.

In the remainder of this section we overview the main ideas underlying our protocols and lower bounds, first describing our contributions in the computational setting, and then describing our contributions in the statistical setting.

Computational security: Our protocol. Our computationally-secure protocol is inspired by the user-to-user protocol proposed by Vaudenay [Vau05]. In his protocol the sender S first commits to the value (m, r_S), where m is the message to be authenticated, and r_S is a random ℓ-bit string. The receiver R then replies with a random string r_R, followed by S revealing r_S and out-of-band authenticating $r_S \oplus r_R$. Finally, the receiver R accepts m if and only if the out-of-band authenticated value is consistent with his view of the protocol.

When moving to the group setting, however, a man-in-the-middle adversary has many more possible ways to interleave its interactions with the parties, thus providing security becomes a much more intricate task. For instance, a naive attempt to generalize Vaudenay's protocol to the group setting (while keeping the out-of-band authenticated value short) might naturally rely on the following idea: Have the sender choose a single value r_S and send each receiver a commitment to (m_i, r_S),[7] and then have each receiver R_i reply with a string r_{R_i} to all

---

[7] Of course, a commitment scheme may be interactive, but we use this terminology for ease of presentation in the overview.

other parties.[8] The out-of-band authenticated value is then $r_S \oplus r_{R_1} \oplus \ldots \oplus r_{R_k}$, and each receiver R_i accepts the message m_i if and only if this value is consistent with his view of the protocol. Alas, this protocol is completely insecure – even when considering just one additional receiver. For example, an adversary can send R_1 a commitment to $(\widehat{m_1}, \widehat{r_S})$ for a message $\widehat{m_1} \neq m_1$ and an arbitrary $\widehat{r_S}$. After learning r_S and r_{R_2}, the adversary can simply send R_1 the value $\widehat{r_{R_2}} = r_{R_2} \oplus r_S \oplus \widehat{r_S}$ instead of r_{R_2}. Since $r_S \oplus r_{R_2} = \widehat{r_S} \oplus \widehat{r_{R_2}}$, the attack will go undetected and the receiver R_1 will accept a fraudulent message $\widehat{m_1}$.

To immune our protocol from attacks as the one described above, the receivers in our protocol must avoid sending their random strings in the clear. Rather, they too send *commitments* of these strings at the beginning of the protocol. Informally, our protocol proceeds as follows: (1) Each R_i sends a commitment to a random ℓ-bit string r_{R_i}; (2) S chooses a random string r_S and sends a commitment to (m_i, r_S) to each R_i; (3) The receivers open their commitments; (4) S opens her commitments; (5) S out-of-band authenticates $r_S \oplus r_{R_1} \oplus \ldots \oplus r_{R_k}$. One can verify that the additional commitments indeed prevent the aforementioned attack, but there are clearly many additional attacks to consider given that an adversary has many possible ways to interleave its interactions with the parties.

The multitude of commitments in our protocol, and the many possible synchronizations an adversary may impose on them in the group setting, make proving the security of our protocol a challenging task. Nonetheless, we are able to show that when the commitment scheme being used is a concurrent non-malleable commitment scheme (see Sect. 2 for a formal definition), our protocol is indeed secure: Setting $\ell = \log(1/\epsilon) + \log k$ guarantees that the adversary's forgery probability is at most ϵ.

Technical details omitted, the intuition behind the security of the protocol is the following. An adversary A wishing to cause some R_i to accept a fraudulent message, essentially has to choose between two options. If A delivers all commitments to S and to R_i before R_i reveals r_{R_i}, then R_i accepting a fraudulent message implies breaking the concurrent non-malleability of the commitment scheme: The $2k$ commitments delivered to S and to R_i by the adversary must define values whose exclusive-or is equal to $r_{R_i} \oplus r_S$. These commitments thus satisfy a "non-trivial" relation which violates the concurrent non-malleability of the commitment scheme. On the other hand, if r_{R_i} is revealed before all commitments were delivered to S, then r_S is chosen after all commitments were delivered to S and to R_i. Hence, all other values contributing to the authenticated value sent by S, and to the value R_i is expecting to see as the out-of-band authenticated value, have already been determined, so the exclusive-or of all relevant values guarantees that the probability of the chosen r_S to result in equality is $2^{-\ell}$.

Computational security: Lower bound. Already in the user-to-user setting, at least $\log(1/\epsilon)$ bits must be out-of-band authenticated, where ϵ is the

---

[8] We do not go into details regarding the possible models of insecure communication in this high-level overview, and we refer the reader to Sect. 3 for an in-depth discussion.

adversary's forgery probability. This can be proved, for example, by analyzing the collision entropy of the random variable corresponding to the out-of-band authenticated value (see, for example, [PV06]). We show that such an analysis can be extended to the group setting, resulting in a stronger lower bound which depends on the size of the group (and is in fact optimal given our above-described protocol).

Specifically, we show an efficient attack against any k-receiver protocol that succeeds with probability roughly $k \cdot 2^{-\ell}$, where ℓ is the number of bits the sender authenticates out-of-band. Given such a protocol π involving a sender and k receivers, our attacker runs $k + 1$ independent executions of π, one with each party taking part in the protocol. In each execution, the attacker independently chooses k random messages as the input to the sender (the true sender in the execution with the sender, and the simulated one in the executions with each of the receivers), and honestly simulates the roles of all other parties. Now, if the out-of-band authenticated value in the execution with the sender is equal to the out-of-band authenticated value in one of the k executions with the receivers, then the attacker combines these two executions by forwarding the out-of-band authenticated value that is sent by the true sender for replacing the simulated value in the execution with that receiver.

Observe that the probability of a successful forgery is roughly the probability that the out-of-band authenticated value in the execution with the sender is indeed equal to the out-of-band authenticated value in one of the k executions with the receivers.[9] Hence, in order to analyze the effectiveness of this attack, it is sufficient to bound the probability of this event. We manage to provide a $\Theta(k \cdot 2^{-\ell})$ lower bound on the probability of this event, which yields Theorem 1.2.

Statistical security: Our protocol. The starting point of our statistically-secure protocol is the iterative hashing protocol of Naor et al. [NSS06]. Loosely speaking, in their protocol the parties maintain a joint sequence of values of decreasing length, starting with the input message of the sender and ending up with the out-of-band authenticated value. In each round, the parties apply to the current value a hash function that is *cooperatively chosen* by both parties: Half of the randomness for choosing the function is determined by the sender, and the other half by the receiver.

As noted above, when moving to the group setting, a naive generalization of the Naor et al. protocol in which the sender executes the user-to-user protocol with each receiver independently, will result in a blow-up of factor k in the length of the out-of-band authenticated value. However, we show that it is possible to exploit the specific structure of the Naor et al. protocol, and in particular of the out-of-band authenticated value, in order to cut its length in the group setting roughly by half (compared to the naive generalization). The main observation underlying our approach is that the k executions of the user-to-user

---

[9] A successful forgery also requires that the input message for that particular receiver is different in the two executions, but this has little effect on the probability of forgery when the input messages are not too short.

protocol need not be completely independent. More concretely, we show that if in the last round (before sending the out-of-band authenticated value), the sender contributes *the same randomness* for all k hash functions, then all k executions are "tied together" in a way that permits a significant reduction in the number of bits that are authenticated out-of-band. Security is now of course not trivially guaranteed, as this change introduces heavy dependencies between the executions. We nevertheless manage to prove, carefully adjusting the structure of our protocol, that the resulting protocol provides an essentially optimal tradeoff between the length of the out-of-band authenticated value and its security.

Statistical security: Lower bound. We prove our lower bound in the statistical security setting by providing a lower bound on the Shanon entropy of the random variable corresponding to the out-of-band authenticated value in any out-of-band authentication protocol. Intuitively speaking, at the beginning of any such protocol, the out-of-band authenticated value is completely undetermined, while at the end of the execution it is fully determined. We show that if the forgery probability is to be bounded by ϵ, this decline in entropy must adhere to a specific structure: Each party must decrease the entropy of the out-of-band authenticated value – via the messages it sends during the execution of the protocol – by at least $\log(1/\epsilon) - 1$ bits on average. It follows that $\mathrm{H}(\Sigma) \geq (k+1) \cdot \log(1/\epsilon) - k$, where Σ is the afore-defined random variable and k is the number of receivers.

We formalize and prove this intuition by presenting a collection of $k + 1$ attacks against any k-receiver out-of-band authentication protocol, one per each participating party. Loosely speaking, the attack corresponding to party P (where P may be the sender or any of the receivers) consists of running two executions of the protocol. First, our adversary plays the role of P in an honest execution of the protocol with all other parties, and obtains the out-of-band authenticated value σ to be sent at the end of this execution. Then, the adversary runs an execution of the protocol with P, playing the role of all other parties, while choosing their messages throughout the protocol not only conditioned on their views, but also conditioned on the out-of-band authenticated value being σ. We show in our analysis that if we denote by ϵ_P the success probability of the attack corresponding to party P, then it holds that $\prod_P \epsilon_P \geq 2^{-\mathrm{H}(\Sigma)-k}$. Hence, if the probability of a successful forgery in any attack (and in particular in our $k + 1$ attacks) is at most ϵ, then it holds that

$$2^{-\mathrm{H}(\Sigma)-k} \leq \prod_P \epsilon_P \leq \epsilon^{k+1},$$

and our lower bound follows. Our proof technique is inspired by the lower bound of Naor et al. [NSS06] for statistically-secure user-to-user out-of-band authentication protocols. In the group setting, however, there are many more "independent" attacks to consider, adding to the intricacy of the proof.

1.3 Paper Organization

The remainder of this paper is organized as follows. In Sect. 2 we review the basic notions and tools that are used in this paper. In Sect. 3 we put forward our framework for out-of-band message authentication protocols in the group setting, formally discussing our communication models and notions of security. Then, in Sects. 4 and 5 we present our protocols and prove our corresponding lower bounds in the computational and statistical settings, respectively.

2 Preliminaries

In this section we present the notation and basic definitions that are used in this work. For a distribution X we denote by $x \leftarrow X$ the process of sampling a value x from the distribution X. Similarly, for a set $\mathcal{X}$ we denote by $x \leftarrow \mathcal{X}$ the process of sampling a value x from the uniform distribution over $\mathcal{X}$. For an integer $n \in \mathbb{N}$ we denote by $[n]$ the set $\{1, \ldots, n\}$. A function $\nu : \mathbb{N} \to \mathbb{R}^+$ is *negligible* if for any polynomial $p(\cdot)$ there exists an integer N such that for all $n > N$ it holds that $\nu(n) \leq 1/p(n)$.

Shannon entropy and mutual information. For random variables X, Y and Z we rely the following standard notions:

- The *entropy* of X is defined as $H(X) = -\sum_x \Pr[X = x] \cdot \log \Pr[X = x]$.
- The *conditional entropy* of X given Y is defined as $H(X|Y) = \sum_y \Pr[Y = y] \cdot H(X|Y = y)$.
- The *mutual information* of X and Y is defined as $I(X;Y) = H(X) - H(X|Y)$.
- The *mutual information* of X and Y given Z is defined as $I(X;Y|Z) = H(X|Z) - H(X|Z,Y)$.

Non-malleable commitment schemes. In this paper we rely on the notion of statistically-binding concurrent non-malleable commitments (for basic definitions and background on commitment schemes, we refer the reader to [Gol01]). We follow the indistinguishability-based definition of Lin and Pass [LP11], though we find it convenient to consider non-malleability with respect to content, other than with respect to identities. For simplicity, the definition below only addresses the one-many setting (which is equivalent to the general many-many setting [PR05]), as this is enough for our needs. Lin and Pass [LP11] and Goyal [Goy11] have shown that constant-round concurrent non-malleable commitment schemes can be constructed from any one-way function (the round complexity was further improved by Ciampi et al. [COS+17] to just 4 rounds). From a more practical perspective, such schemes can be constructed efficiently in the random-oracle model [BR93]. For further information regarding non-malleable and concurrent non-malleable commitment schemes see, for example, [DDN00, CIO98, FF00, CF01, PR05, PR08, LPV08] and the references therein.

Intuitively speaking, a (one-many) concurrent non-malleable commitment scheme has the following guarantee: Any efficient adversary cannot use a commitment to some value v in order to produce commitments to values $\widehat{v_1}, \ldots, \widehat{v_k}$

that are "non-trivially" related to v. More formally, Let $\mathsf{Com} = (C, R)$ be a statistically-binding commitment scheme, and let $k = k(\cdot)$ be a function of the security parameter $\lambda \in \mathbb{N}$, bounded by some polynomial. Consider an efficient adversary A that gets an auxiliary input $z \in \{0, 1\}^*$ (in addition to the security parameter) and participates in the following "man-in-the-middle" experiment. A takes part in a single "left" interaction and in k "right" interactions: In the left interaction, A interacts with the committer C, and receives a commitment to a value v. Denote the resulting commitment (transcript of the interaction) by c. In the right interactions, A interacts with the receiver R, resulting in k commitments $\widehat{c}_1, \ldots, \widehat{c}_k$. We define k related values $\widehat{v}_1, \ldots, \widehat{v}_k$ in the following manner. For every $i \in [k]$, if $\widehat{c}_i = c$, if $\widehat{c}_i$ is not a valid commitment, or if $\widehat{c}_i$ can be opened to more than one value, we let $\widehat{v}_i = \perp$ (note that by the statistical binding property of Com, the latter case only happens with negligible probability). Otherwise, $\widehat{v}_i$ is the unique value to which $\widehat{c}_i$ may be opened. Let $\mathsf{mim}^A_{\mathsf{Com}}(v, z)$ denote the random variable that includes the values $\widehat{v}_1, \ldots, \widehat{v}_k$ and A's view at the end of the afore-described experiment.

Definition 2.1. *Let A and D be a pair of algorithms. We define the advantage of (A, D) with respect to security parameter $\lambda \in \mathbb{N}$ as*

$$\mathsf{Adv}^{A,D}_{\mathsf{Com}}(\lambda) \overset{\text{def}}{=} \max_{v, v' \in \{0,1\}^\lambda} \left\{ \Pr\left[D(1^\lambda, \mathsf{mim}^A_{\mathsf{Com}}(v, z)) = 1 \right] \right.$$
$$\left. - \Pr\left[D(1^\lambda, \mathsf{mim}^A_{\mathsf{Com}}(v', z)) = 1 \right] \right\}.$$

We say that a statistically-binding commitment scheme is concurrent non-malleable if for any pair of probabilistic polynomial-time algorithms (A, D) there exists a negligible function $\nu = \nu(\cdot)$ such that $\mathsf{Adv}^{A,D}_{\mathsf{Com}}(\lambda) \leq \nu(\lambda)$ for all sufficiently large $\lambda \in \mathbb{N}$.

3   The Communication Model and Notions of Security

We consider the message authentication problem in a setting involving a group of $k + 1$ users: A sender S and k receivers $R_1, \ldots, R_k$. For each $i \in [k]$ the sender would like to authenticate a message m_i to the ith receiver R_i. We assume that the users communicate over two channels: An insecure channel over which a man-in-the-middle adversary has complete control, and a low-bandwidth "out-of-band" authenticated channel, enabling the sender to authenticate one short message to all receivers. In what follows we formally specify the underlying communication model as well as the notions of security that we consider in this work (generalizing those of Vaudenay [Vau05] and Naor et al. [NSS06] to the group setting).

3.1   Communication Model

Our starting point is the framework of Vaudenay [Vau05] and Naor et al. [NSS06] which considers a sender who wishes to authenticate a single message to a single receiver using out-of-band authentication. They modeled this interaction by providing the sender and the receiver with two types of channels: A bidirectional insecure channel that is completely vulnerable to man-in-the middle attacks, and an authenticated unidirectional low-bandwidth channel from the sender to the receiver (an "out-of-band" channel).

We extend this model to the group setting in the following manner. Similarly to the framework of Vaudenay and Naor et al. we assume that the parties are connected via two types of communication channels: An insecure channel and an authenticated low-bandwidth channel. As for the authenticated channel, we assume that the sender S is equipped with an out-of-band channel, through which S may send a short message visible to all receivers in an authenticated manner (e.g., a voice message in group messaging). The adversary may read or remove this message for some or all receivers, and may delay it for different periods of time for different receivers, but cannot modify it in an undetectable manner. One may also consider a scenario where S, as well as the receivers, may send multiple messages over the out-of-band authenticated channel throughout the protocol. However, this is less desirable from a practical standpoint, and in any case, will not be necessary in our protocols. Furthermore, our lower bounds readily capture this more general case as well, providing a lower bound on the total number of bits sent over the authenticated channel throughout the protocol.

As mentioned above, we also assume that the parties are connected among themselves in a network of insecure channels. These channels are vulnerable to man-in-the-middle attacks, and the adversary is assumed to have complete control over them: The adversary can read, delay and stop messages sent by the parties, as well as insert new messages at any point in time. In particular, this provides the adversary with considerable control over the synchronization of the protocol's execution. Nonetheless, the execution is still guaranteed to be "marginally synchronized": Each party sends her messages in the ith round of the protocol only upon receiving all due messages of round $i - 1$.

One may consider various possible networks to define the topology of the insecure channels. Two extremes of that spectrum are the following:

- **The star network model:** In this model each receiver R_i is connected to the sender S via a bidirectional insecure channel. In particular, the receivers cannot send messages directly to each other, and any communication among them must pass through the sender S.
- **The complete network model:** In this model every pair of parties (sender and receiver as well as two receivers) is connected through an insecure channel.

In that respect, our results – both in the computational setting and in the statistical setting – will be of the strongest form possible. Our protocols will be articulated, and their correctness and security proven, in the restrictive "star" network model, which in particular means that they can be implemented in models richer in channels, and namely in the complete network model (in that case, some communication efficiency optimizations are possible). Our lower bounds on the other hand, will assume complete communication networks, and will hence apply to weaker network models as well.

3.2   Notions of Security

In what follows we define the security and correctness requirements of out-of-band authentication protocols, essentially extending those of Vaudenay [Vau05] and Naor et al. [NSS06] to the group setting in an intuitive manner. In such protocols, the input to the sender S is a vector of message $m_1, \ldots, m_k$ which may be chosen by the adversary. At the end of the execution, each receiver R_i outputs either a message $\widehat{m_i}$ or the unique symbol $\bot$, implying rejection. Informally, correctness states that in an honest execution, with high probability all receivers output the correct message; i.e., $\widehat{m_i} = m_i$ for every $i \in [k]$. As for security, we demand that an adversary (which is efficient in the computational setting and unbounded in the statistical setting) cannot convince a receiver to output an incorrect message; i.e., the probability that $\widehat{m_i} \notin \{m_i, \bot\}$ is bounded by a pre-specified parameter.

For the sake of generality, Definitions 3.1 and 3.2 below are articulated without specific reference to an underlying communication model, and may be applied to any of the group communication models discussed above. We begin with a formal definition of out-of-band authentication in the statistical setting.

Definition 3.1. *A statistically-secure out-of-band $(n, \ell, k, r, \epsilon)$-authentication protocol is a $(k + 1)$-party r-round protocol in which the sender S is invoked on a k-tuple of n-bit messages, and sends at most ℓ bits over the authenticated out-of-band channel. The following requirements must hold:*

- **Correctness:** *In an honest execution of the protocol, for all input messages $m_1, \ldots, m_k \in \{0,1\}^n$ to S and for every $i \in [k]$, receiver R_i outputs m_i with probability 1.*
- **Unforgeability:** *For any adversary and for every adversarially-chosen input messages $m_1, \ldots, m_k$ on which S is invoked, the probability that there exists some $i \in [k]$ for which receiver R_i outputs some message $\widehat{m_i} \notin \{m_i, \bot\}$ is at most ϵ.*

A computationally-secure out-of-band authentication protocol is defined similarly, except that security need only hold against efficient adversaries, and the probability of forgery is also allowed to additively grow (with respect to the statistical setting) by a negligible function of the security parameter $\lambda \in \mathbb{N}$.

Definition 3.2. *Let $n = n(\lambda), \ell = \ell(\lambda), k = k(\lambda), r = r(\lambda)$ and $\epsilon = \epsilon(\lambda)$ be functions of the security parameter $\lambda \in \mathbb{N}$. A computationally-secure out-of-band $(n, \ell, k, r, \epsilon)$-authentication protocol is a $(k + 1)$-party r-round protocol in which the sender S is invoked on a k-tuple of n-bit messages, and sends at most ℓ bits over the authenticated out-of-band channel. The following requirements must hold:*

- **Correctness:** *In an honest execution of the protocol, for all input messages $m_1, \ldots, m_k \in \{0,1\}^n$ to S and for every $i \in [k]$, receiver R_i outputs m_i with probability 1.*

- **Unforgeability:** *For any probabilistic polynomial-time adversary there exists a negligible function $\nu = \nu(\cdot)$ such that the following holds: For every input messages $m_1, \ldots, m_k$ chosen by the adversary and on which S is invoked, the probability that there exists some $i \in [k]$ for which receiver R_i outputs some message $\widehat{m_i} \notin \{m_i, \bot\}$ is at most $\epsilon + \nu(\lambda)$ for all sufficiently large $\lambda \in \mathbb{N}$.*

4   The Computational Setting

In this section we prove tight bounds for computationally-secure out-of-band authentication in the group setting. In Sect. 4.1 we present our computationally-secure protocol and discuss its possible instantiations (both in the standard model and in the random-oracle model). The security proof of our protocol is provided in the full version of the paper [RS18]. In Sect. 4.2 we prove a matching lower bound on the length of the out-of-band authenticated value in any computationally-secure protocol.

4.1   Our Protocol and Its Instantiations

Let $\mathsf{Com} = (C_{\mathsf{Com}}, R_{\mathsf{Com}})$ be a concurrent non-malleable commitment scheme that is statistically binding (see Sect. 2 and Definition 2.1). Our protocol, denoted π_{Comp}, is parameterized by the security parameter $\lambda \in \mathbb{N}$, by the number $k = k(\lambda)$ of receivers, by the length $\ell = \ell(\lambda)$ of the out-of-band authenticated value, and by the length $n = n(\lambda)$ of the messages that the user would like to authenticate. The protocol is defined as follows: ·

1. For every $i \in [k]$ the receiver R_i chooses a random ℓ-bit string $r_i \leftarrow \{0, 1\}^\ell$, and commits to it to the sender S using Com. For every $i \in [k]$ denote the resulting commitment according to the view of R_i by c_i, and denote the commitments received by S by $\widehat{c_i}$.[10]
2. The sender S chooses a random string $r_s \leftarrow \{0, 1\}^\ell$, and executes k (possibly parallel) executions of Com to commit to the message (m_i, r_s) to the receiver R_i for every $i \in [k]$. Denote the resulting commitments, as seen by the sender S by c_s^i, and denote the commitment received by R_i by $\widehat{c_s^i}$.

   For every $i \in [k]$ the sender S also explicitly appends the following information to the first message it sends R_i as part of the commitment: (1) The message m_i, and (2) the (possibly tampered with) commitments $(\widehat{c_j})_{j \in [k] \setminus \{i\}}$ received from the other receivers in Step 1 of the protocol. We let $\widehat{m_i}$ and $(\widehat{c_{j \to i}})_{j \in [k] \setminus \{i\}}$ denote the message and the forwarded commitments as received by R_i.

---

[10] As a commitment scheme may be interactive, when referring to a commitment, we mean the transcript of the interaction between the committer and the receiver during an execution of the commit phase of the commitment scheme. When the scheme is non-interactive, a commitment is simply a single string sent from the committer to the receiver.

3. For every $i \in [k]$ the receiver R_i sends a decommitment d_i of her commitment from Step 1 of the protocol to reveal r_i to the sender S. Let $\widehat{d_i}$ denote the decommitment received by S from R_i. For every $i \in [k]$ the sender S then checks whether $\widehat{d_i}$ is a valid decommitment to $\widehat{c_i}$. If so, let $\widehat{r_i}$ denote the committed value. Otherwise, S sends $\bot$ over the authenticated channel, in which case all receivers output $\bot$.

4. For every $i \in [k]$, the sender S sends receiver R_i a decommitment d_s^i to the corresponding commitment from Step 2 of the protocol, and reveals r_s to R_i. Denote by $\widehat{d_s^i}$ the decommitment received by R_i. For every $i \in [k]$ the receiver R_i checks if $\widehat{d_s^i}$ is a valid decommitment to $\widehat{c_s^i}$. If it is, denote the committed value by $(\widehat{m_i'}, \widehat{r_s^i})$. If it is not a valid decommitment or if $\widehat{m_i'} \neq \widehat{m_i}$ (where $\widehat{m_i}$ was received in Step 2), then R_i outputs $\bot$ and terminates.

   For every $i \in [k]$ the sender S also sends R_i the (possibly tampered with) decommitments $(\widehat{d_j})_{j \in [k] \setminus \{i\}}$ she received in Step 3. We let $(\widehat{d_{j \to i}})_{j \in [k] \setminus \{i\}}$ denote the decommitments received by R_i. If for some $j \in [k] \setminus \{i\}$ it holds that $\widehat{d_{j \to i}}$ is not a valid decommitment to $\widehat{c_{j \to i}}$ received by R_i is Step 2, then R_i outputs $\bot$ and terminates. Otherwise, denote by $(\widehat{r_{j \to i}})_{j \in [k] \setminus \{i\}}$ the values obtained by opening the commitments.

5. S computes $\sigma = r_s \oplus \widehat{r_1} \oplus \ldots \oplus \widehat{r_k}$ and sends σ over the authenticated out-of-band channel. Every receiver R_i computes $\widehat{\sigma_i} = r_s^i \oplus \widehat{r_{1 \to i}} \oplus \ldots \oplus \widehat{r_{i-1 \to i}} \oplus r_i \oplus \widehat{r_{i+1 \to i}} \oplus \ldots \widehat{r_{k \to i}}$, and then outputs $\widehat{m_i}$ (received in Step 2) if $\widehat{\sigma_i} = \sigma$ and outputs $\bot$ otherwise.

Theorem 4.1 (when combined with the existence of a constant-round concurrent non-malleable statistically-binding commitment scheme based on any one-way function – see Sect. 2) implies Theorem 1.1 as an immediate corollary:

Theorem 4.1. *Let $k = k(\cdot), \ell = \ell(\cdot), r = r(\cdot)$ and $n = n(\cdot)$ be functions of the security parameter $\lambda \in \mathbb{N}$ and let* Com *be an r-round concurrent non-malleable commitment scheme. Then, protocol π_{Comp} is a computationally-secure out-of-band $(n, \ell, k, O(r), k \cdot 2^{-\ell})$-authentication protocol.*

The correctness and round complexity of π_{Comp} are straightforward. The unforgeability of the protocol (according to the parameters of Theorem 4.1) is proven in the full version of this paper [RS18].

Possible instantiations. Our protocol π_{Comp} can be instantiated with Com being any concurrent non-malleable statistically-biding commitment scheme. From a theoretical point of view, Lin and Pass [LP11] and Goyal [Goy11] gave constant-round constructions of such schemes from any one-way function (and the round complexity was further improved by [COS+17]). Hence, our protocol can also be instantiated as a constant-round protocol, assuming only the existence of one-way functions. This assumption is minimal and necessary, since Naor et al. [NSS06] showed that even in the user-to-user setting, any computationally-secure out-of-band authentication protocol for which $\ell < 2 \log 1/\epsilon - \Theta(1)$ implies the existence of one-way functions.

From a more practical standpoint, a *non-interactive* concurrent non-malleable statistically-biding commitment scheme can be very efficiently constructed in the random oracle model [BR93]. Thus, instantiating π_{Comp} with a cryptographic hash function (e.g., SHA-2) as the random oracle yields a highly efficient protocol. Given a random oracle H, in order to commit to a value v, one simply has to send $c = H(v, r)$ for a sufficiently long random string r. Decommitment is done by revealing v and r, and the receiver asserts that $c = H(v, r)$. Consider a pair of poly-query algorithms (A, D), where A is the man-in-the-middle adversary and D is the distinguisher (see Definition 2.1). Informally speaking, assume H is sufficiently length-increasing (say, length-doubling) so that it is difficult to find an element y in its image without querying H on a pre-image of y. So the algorithm A, that receives $c = H(v, r)$ and produces $c_1 = H(v_1, r_1), \ldots, c_k = H(v_k, r_k)$, knows $v_1, \ldots, v_k$ with overwhelming probability. Hence, it can distinguish between the case that $c = H(v, r)$, and the case that $c = H(v', r')$ where the value v' – when taken together with $v_1, \ldots, v_k$ and the view of A – does not satisfy the polynomial time relation defined by the distinguisher D. By a standard argument, this is hard for any adversary making a polynomial number of queries to the random oracle.

Non-malleable commitment schemes also exist in the common reference string (CRS) model (see, for example, [CIO98, CKO+01, FF00, CF01, DG03]). However, assuming a trusted CRS may be somewhat incompatible with the ad-hoc nature of instant messaging platforms and applications.

4.2   Lower Bound

In this section, we prove a lower bound on the length of out-of-band authenticated value in any out-of-band authentication protocol, as a function of the desired security level ϵ and of the number of receivers k. Our bound shows that the length of the out-of-band authenticated value in our protocol π_{Comp} of Sect. 4.1 is optimal (up to an additive constant). The lower bound is stated by the following Theorem, which yields Theorem 1.2.

Theorem 4.2. *For any computationally-secure $(n, \ell, k, r, \epsilon)$-authentication protocol where $n \geq \log(1/\epsilon) + \log k + 3$ and $\epsilon < 1/6$, it holds that $\ell \geq \log 1/\epsilon + \log k - 3$.*

Proof. Let $\pi = (S, R_1, \ldots, R_k)$ be a k-receiver out-of-band authentication protocol for messages of length n in the complete network communication model. We present an efficient adversary A that succeeds in fooling at least one of the reveivers with probability at least $k \cdot 2^{-\ell-3}$, and the theorem follows (for an intuitive overview of the attack and analysis, see Sect. 1.2).

On input 1^λ, A runs the following steps:

1. A samples k input messages $(m_1, \ldots, m_k) \leftarrow \{0, 1\}^{m \times k}$ as the input to the sender S, and runs an execution with S in which A plays the role of all receivers. Denote by $\sigma \in \{0, 1\}^\ell$ the value that S sends over the authenticated channel at the end of this execution.

2. For every $i \in [k]$, A samples k input messages $(\widehat{m_1^i}, \ldots, \widehat{m_k^i}) \leftarrow \{0, 1\}^{m \times k}$ uniformly at random (independently from the messages sampled in the other executions), and runs an execution of π with R_i in which A plays the role of the sender (with input $(\widehat{m_1^i}, \ldots, \widehat{m_k^i})$) and all other receivers. For every $i \in [k]$ denote the out-of-band authenticated value the (simulated) sender sends in the end of the execution with the true receiver R_i by $\widehat{\sigma}_i$.

We first wish to lower bound the probability that there exists some receiver R_i that outputs $\widehat{m_i^i}$. By the correctness of π, this is at least the probability that $\widehat{\sigma}_i = \sigma$. Thus, for every $i \in [k]$, it holds that

$$\Pr\left[R_i \text{ outputs } \widehat{m_i^i}\right] \geq \Pr[\widehat{\sigma}_i = \sigma] = \sum_{v \in \{0,1\}^\ell} \Pr[\sigma = v] \cdot \Pr[\widehat{\sigma}_i = v].$$

More generally, for any subset $\mathcal{I} \subseteq [k]$ of the receivers, it holds that

$$\Pr\left[\forall i \in \mathcal{I} : R_i \text{ outputs } \widehat{m_i^i}\right] \geq \sum_{v \in \{0,1\}^\ell} \Pr[\sigma = v] \cdot \prod_{i \in \mathcal{I}} \Pr[\widehat{\sigma}_i = v]$$

$$= \sum_{v \in \{0,1\}^\ell} (\Pr[\sigma = v])^{|\mathcal{I}|+1}.$$

The inequality follows by the fact that the executions A conducts with the receivers are independent from each other, and the equality holds since σ and $\widehat{\sigma}_i$ are identically distributed for every $i \in [k]$. The inclusion-exclusion principle now yields that the probability that for at least one receiver it holds that $\widehat{\sigma}_i = \sigma$ is

$$\Pr\left[\exists i \in [k] \text{ s.t. } R_i \text{ outputs } \widehat{m_i^i}\right] \geq \sum_{i=1}^{k} (-1)^{i+1} \cdot \binom{k}{i} \cdot \left(\sum_{v \in \{0,1\}^\ell} (\Pr[\sigma = v])^{i+1}\right).$$

The above probability is minimized when the distribution of σ over a random execution of the protocol as described above is uniform; i.e., when $\Pr[\sigma = v] = 2^{-\ell}$ for all $v \in \{0, 1\}^\ell$. Hence, it holds that

$$\Pr\left[\exists i \in [k] \text{ s.t. } R_i \text{ outputs } \widehat{m_i^i}\right] \geq \sum_{i=1}^{k} (-1)^{i+1} \cdot \binom{k}{i} \cdot 2^{-i \cdot \ell}.$$

In what follows, we make use of the following claim, which bounds the above expression. For the proof of Claim 4.3, see the full version of this paper [RS18].

Claim 4.3. $\sum_{i=1}^{k} (-1)^{i+1} \cdot \binom{k}{i} \cdot 2^{-i \cdot \ell} \geq \min\{1/3, k \cdot 2^{-\ell}/4\}$.

Let Forge_A denote the event in which for some $i \in [k]$, R_i outputs $\widehat{m_i^i} \neq m_i$. By Claim 4.3,

$$
\begin{aligned}
\Pr\left[\mathsf{Forge}_A\right] &= \Pr\left[\exists i \in [k] \text{ s.t. } m_i \neq \widehat{m_i^i} \wedge R_i \text{ outputs } \widehat{m_i^i}\right] \\
&\geq \Pr\left[\forall j \in [k], m_j \neq \widehat{m_j^j} \wedge \exists i \in [k] \text{ s.t. } R_i \text{ outputs } \widehat{m_i^i}\right] \\
&\geq \Pr\left[\exists i \in [k] \text{ s.t. } R_i \text{ outputs } \widehat{m_i^i}\right] - \Pr\left[\exists j \in [k] \text{ s.t. } m_j = \widehat{m_j^j}\right] \\
&\geq \min\left\{\frac{1}{3}, \frac{k}{4} \cdot 2^{-\ell}\right\} - k \cdot 2^{-n} \\
&\geq \min\left\{\frac{1}{6}, k \cdot 2^{-\ell-2} - k \cdot 2^{-n}\right\}.
\end{aligned}
$$

The last inequality holds since $n \geq \log k + \log 1/\epsilon + 3 > \log k + 3$ and thus $k \cdot 2^{-n} < 1/6$. Finally, since $\epsilon < 1/6$ and $n \geq \log k + \log 1/\epsilon$, it holds that

$$
\epsilon \geq k \cdot 2^{-\ell-2} - k \cdot 2^{-n} \geq k \cdot 2^{-\ell-2} - \epsilon.
$$

Equivalently, $\epsilon \geq k \cdot 2^{-\ell-3}$, which implies $\ell \geq \log 1/\epsilon + \log k - 3$. ∎

5   The Statistical Setting

In this section we prove tight bounds for statistically-secure out-of-band authentication protocols in the group setting. First, in Sect. 5.1 we present our statistically-secure protocol. Then, in Sect. 5.2 we prove the security of our protocol, and in Sect. 5.3 we prove a matching lower bound on the length of the out-of-band authenticated value in any statistically-secure protocol.

5.1   Our Protocol

Our protocol, denoted π_{Stat}, is parametrized by the maximal forgery probability $\epsilon \in (0, 1)$, integers $n, k \in \mathbb{N}$ denoting the length of each message and the number of receivers, respectively, and an odd integer $r \in \mathbb{N}$ denoting the number of rounds (we refer the reader to Sect. 1.2 for an intuitive overview of the protocol).

Notation. Denote the Galois field with q elements by $GF(q)$. Then, a message m of length n can be parsed as a polynomial of degree at most $\lceil n/\log q \rceil$ over $GF(q)$. Namely, a message $m = m_1, \ldots, m_t \in GF(q)^t$ defines a polynomial in the following manner: For every $x \in GF(q)$, we let $m(x) = \sum_{i=1}^{t} m_i \cdot x^i$. Then, for two distinct messages $m, \widehat{m} \in GF(q)^t$ and any two field elements $y, \widehat{y} \in GF(q)$, it holds that the polynomials $m(\cdot) + y$ and $\widehat{m}(\cdot) + \widehat{y}$ are distinct and thus $\Pr_{x \leftarrow GF(q)}[m(x) + y = \widehat{m}(x) + \widehat{y}] \leq t/q$. Let $\epsilon' = \epsilon/k$, and let $n_1 = n$. For every $j \in [r-1]$ let q_j be a prime number chosen in a deterministically and agreed upon manner in the interval $\left[\frac{2^{r-j} \cdot n_j}{\epsilon'}, \frac{2^{r-j+1} \cdot n_j}{\epsilon'}\right)$, and let $n_{j+1} = \lceil 2\log q_j \rceil$.

Our protocol π_{Stat} is then defined by the following steps:

1. For every $i \in [k]$, S sends $m^1_{S,i} = m_i$ to R_i. Denote by $m^1_{R_i}$ the string received by R_i.
2. For $j = 1$ to $r - 2$:
   (a) If j is odd, then for every $i \in [k]$:
       i. S chooses $y^j_i \leftarrow GF(q_j)$ and sends it to R_i.
       ii. R_i receives $\widehat{y^j_i}$, chooses $x^j_i \leftarrow GF(q_j)$ and sends it to S.
       iii. S receives $\widehat{x^j_i}$ and computes $m^{j+1}_{S,i} = \widehat{x^j_i} \| m^j_{S,i}(\widehat{x^j_i}) + y^j_i$.
       iv. R_i computes $m^{j+1}_{R_i} = x^j_i \| m^j_{R_i}(x^j_i) + \widehat{y^j_i}$.
   (b) if j is even, then for every $i \in [k]$:
       i. R_i chooses $y^j_i \leftarrow GF(q_j)$ and sends it to S.
       ii. S receives $\widehat{y^j_i}$, chooses $x^j_i \leftarrow GF(q_j)$ and sends it to R_i.
       iii. R_i receives $\widehat{x^j_i}$ and computes $m^{j+1}_{R_i} = \widehat{x^j_i} \| m^j_{R_i}(\widehat{x^j_i}) + y^j_i$.
       iv. S computes $m^{j+1}_{S,i} = x^j_i \| m^j_{S,i}(x^j_i) + \widehat{y^j_i}$.
3. For every $i \in [k]$, R_i chooses $y^{r-1}_i \leftarrow GF(q_j)$ and sends it to S.
4. S receives $\widehat{y^{r-1}_1}, \ldots, \widehat{y^{r-1}_k}$, chooses $x^{r-1} \leftarrow GF(q_{r-1})$, and for every $i \in [k]$ sends $x^{r-1}_i = x^{r-1}$ to R_i.
5. For every $i \in [k]$, R_i receives $\widehat{x^{r-1}_i}$ and computes $\widehat{\sigma}_i = m^{r-1}_{R_i}(\widehat{x^{r-1}_i}) + y^{r-1}_i$. Denote $m^r_{R_i} = \widehat{x^{r-1}_i} \| \widehat{\sigma}_i$.
6. For every $i \in [k]$, S computes $\sigma_i = m^{r-1}_{S,i}(x^{r-1}) + \widehat{y^{r-1}_i}$. Denote $m^r_{S,i} = x^{r-1} \| \sigma_i$. S sends $x^{r-1} \| \sigma_1 \| \ldots \| \sigma_k$ over the authenticated channel.
7. For every $i \in [k]$, if $m^r_{S,i} = m^r_{R_i}$ (i.e., if $x^{r-1} = \widehat{x^{r-1}_i}$ and $\sigma_i = \widehat{\sigma}_i$), R_i outputs $m^1_{R_i}$. Otherwise, R_i outputs $\perp$.

The following theorem (when the protocol is invoked with at least $\log^* n$ rounds) implies Theorem 1.3 as an immediate corollary:

Theorem 5.1. *Let $n, k \in \mathbb{N}$, let $r \geq 3$, and let $\epsilon \in (0, 1)$. Then, protocol π_{Stat} is a statistically-secure out-of-band $(n, \ell, k, r, \epsilon)$-authentication protocol, where $\ell = (k + 1) \cdot \left(\log \frac{1}{\epsilon} + \log k + \log^{(r-1)} n + O(1) \right)$.*

The correctness of our protocol is straightforward. In Lemma 5.2 we bound the length ℓ of the out-of-band authenticated value as stated in Theorem 5.1, and the proof of unforgeability is given in Sect. 5.2, yielding Theorem 5.1. A corollary of Lemma 5.2 is that when invoked with $r = \Omega(\log^* n)$, the sender in protocol π_{Stat} has to authenticate at most $(k + 1) \cdot (\log(1/\epsilon) + \log k + O(1))$ bits.

Lemma 5.2. *Let $n, k \in \mathbb{N}$, let $r \geq 3$, and let $\epsilon \in (0, 1)$. Then, in protocol π_{Stat} it holds that $\ell \leq (k + 1) \cdot \left(\log \frac{1}{\epsilon} + \log k + \log^{(r-1)} n + O(1) \right)$.*

The proof of Lemma 5.2 will make use of the following two claims.

Claim 5.3. *If $n_j > 2^{r-j}/\epsilon'$ for every $j \in [r-2]$, then $n_{j+1} \leq \max\{4\log^{(j)} n$*
$+4\log 5 + 3, 27\}$ for every $j \in [k-2]$.

Proof. The proof is by induction on j. Since $n_j > 2^{r-j}/\epsilon'$ for every $j \in [r-2]$, it holds that for every $j \in [r-2]$,

$$q_j < \frac{2^{r-j+1}}{\epsilon'} \cdot n_j \leq 2n_j^2.$$

This implies that for every $j \in [r-2]$, it holds that

$$n_{j+1} = \lceil 2\log q_j \rceil < \lceil 2\log\left(2n_j^2\right) \rceil \leq 4\log n_j + 3.$$

For $j = 1$, the claim indeed yields: $n_2 < 4\log n + 3$. For $2 \leq j \leq r-2$, if $n_j \leq 27$, then $n_{j+1} < 4\log 27 + 3 < 23$. Otherwise, by the induction hypothesis, it holds that

$$n_{j+1} \leq 4\log n_j + 3 \leq 4\log\left(4\log^{(j-1)} n + 4\log 5 + 3\right) + 3.$$

Consider the following two cases:

1. If $\log^{(j-1)} n \leq 4\log 5 + 3$, then $n_{j+1} \leq 4\log(20\log 5 + 15) + 3 < 27$.
2. If $\log^{(j-1)} n > 4\log 5 + 3$, then $n_{j+1} \leq 4\log\left(5\log^{(j-1)} n\right) + 3 = 4\log^{(j)} n + 4\log 5 + 3$.

∎

Claim 5.4. *If $n_j \leq 2^{r-j}/\epsilon'$ for some $j \in [r-2]$, then for every $j' \in \{j, \ldots, r-2\}$, it holds that $n_{j'} \leq 2^{r-j'}/\epsilon'$.*

Proof. Assume $n_j \leq 2^{r-j}/\epsilon'$ for some $j \in [r-3]$. We prove $n_{j+1} \leq 2^{r-j-1}/\epsilon'$ and the claim follows. By the assumption on n_j, it holds that

$$n_{j+1} = \lceil 2\log q_j \rceil$$

$$\leq \left\lceil 2\log\left(\frac{2^{r-j}}{\epsilon'} \cdot n_j\right) \right\rceil$$

$$\leq \left\lceil 4\log\left(\frac{2^{r-j}}{\epsilon'}\right) \right\rceil$$

$$\leq 4 \cdot \left(r - j + \log\frac{1}{\epsilon'}\right) + 1$$

$$\leq 2^{r-j+\log\frac{1}{\epsilon'}-1}$$

$$= \frac{2^{r-j-1}}{\epsilon'}.$$

The last inequality follows by the fact that $4x + 1 \leq 2^{x-1}$ for any $x \geq 6$ (if $r - j + \log(1/\epsilon') < 6$ then the parties can jump to Step 3 of the protocol and complete it, while S only has to send $(k+1) \cdot O(1)$ bits over the out-of-band channel, which implies Lemma 5.2). ∎

We are now ready to prove Lemma 5.2.

Proof of Lemma 5.2. Informally speaking, we prove that q_{r-1} is at most roughly $1/\epsilon'$, and then the lemma follows, since S authenticates to $k+1$ elements in $GF(q_{r-1})$, which can be encoded using $\lceil (k+1) \cdot \log q_{r-1} \rceil$ bits.

More formally, we consider two separate cases. First we consider the case where $n_j > 2^{r-j}/\epsilon'$ for every $j \in [r-2]$. By Claim 5.3, it holds that $n_{r-1} \leq \max\left\{4\log^{(r-2)} n + 4\log 5 + 3, 27\right\}$. If $n_{r-1} \leq 27$, then $q_{r-1} < 4 \cdot 27/\epsilon'$, and then

$$
\begin{aligned}
\ell &= \lceil (k+1) \cdot \log q_{r-1} \rceil \\
&\leq (k+1) \cdot \left(\log \frac{1}{\epsilon'} + O(1) \right) \\
&= (k+1) \cdot \left(\log \frac{1}{\epsilon} + \log k + O(1) \right).
\end{aligned}
$$

Otherwise, it holds that $n_{r-1} \leq 4\log^{(r-2)} n + 4\log 5 + 3$. Hence,

$$
\begin{aligned}
\ell &= \lceil (k+1) \cdot \log q_{r-1} \rceil \\
&= \left\lceil (k+1) \cdot \log \left(\frac{4}{\epsilon'} \cdot n_{r-1} \right) \right\rceil \\
&\leq (k+1) \cdot \left(\log \frac{1}{\epsilon} + \log k + \log^{(r-1)} n + O(1) \right).
\end{aligned}
$$

We now turn to consider the case where there exists some $j \in [r-2]$ such that $n_j \leq 2^{r-j}/\epsilon'$. By Claim 5.4, this means that $n_{r-2} \leq 4/\epsilon'$. Therefore,

$$
n_{r-1} = \lceil 2\log q_{r-2} \rceil \leq \left\lceil 2\log \frac{2^3}{\epsilon'} \cdot n_{r-2} \right\rceil \leq 4\log \frac{1}{\epsilon'} + 11.
$$

Where this is the case, the parties can set $q_{r-1} = \Theta(1/\epsilon')$, and the security of the protocol is preserved. This is due to the fact that our proof of security (see Sect. 5.2) only relies on the fact that two distinct polynomials over $GF(q_{r-1})$ defined by n_{r-1}-bit strings evaluate to the same value on at most $\epsilon'/2$ field elements; i.e., $q_{r-1}^{-1} \cdot \lceil n_{r-1}/\log(1/\epsilon') \rceil \leq \epsilon'/2$. If $q_{r-1} = \Theta(1/\epsilon')$, then indeed

$$
\ell \leq (k+1) \cdot \left(\log \frac{1}{\epsilon} + \log k + \log^{(r-1)} n + O(1) \right),
$$

concluding the proof. ∎

5.2 Proof of Security

In this section, we prove the unforgeability of our protocol π_{Stat}, proving Theorem 5.1. For an adversary A, let $\mathsf{Forge}_{A,i}$ denote the event in which R_i outputs $\widehat{m_i} \notin \{m_i, \perp\}$ in an execution of π_{Stat} with A, and let $\mathsf{Forge}_A = \bigcup_{i \in [k]} \mathsf{Forge}_{A,i}$. The following Lemma captures the unforgeability of π_{Stat}.

Lemma 5.5. *For any computationally unbounded adversary A, it holds that* $\Pr\left[\mathsf{Forge}_A\right] \leq \epsilon$.

Proof. We prove that for every $i \in [k]$, any computationally unbounded adversary A succeeds in making R_i output a fraudulent message with probability at most $\epsilon' = \epsilon/k$ and the theorem thus follows by union bound. Note that if A fools R_i this in particular means that $m_{S,i}^1 \neq m_{R_i}^1$ but $m_{S,i}^r = m_{R_i}^r$. Hence, there exists a round $j \in [r-1]$ such that $m_{S,i}^j \neq m_{R_i}^j$ but $m_{S,i}^{j+1} = m_{R_i}^{j+1}$; denote this event by Coll_i^j. We will prove that for every j, $\Pr\left[\mathsf{Coll}_i^j\right] \leq \epsilon'/2^{r-j}$, and then by taking a union bound over all rounds, the probability of $\mathsf{Forge}_{A,i}$ is at most $\sum_{j=1}^{r-1} \Pr\left[\mathsf{Coll}_i^j\right] \leq \sum_{j=1}^{r-1} \epsilon'/2^{r-j} < \epsilon'$.

We denote by $T(v)$ the time in which a message v in the protocol is sent and fixed. We analyze separately the case where the round index j is odd, and the case that it is even. We start by bounding $\Pr\left[\mathsf{Coll}_i^j\right]$ in case j is odd (R_i picks the evaluation point of the polynomial and S chooses the shift), and consider three possible attack timings:

1. $T(\widehat{x_i^j}) < T(x_i^j)$: In this case, R_i chooses x_i^j at random from the field only after $\widehat{x_i^j}$ was fixed and sent to S. Recall that $\widehat{x_i^j}$ is the first part of $m_{S,i}^{j+1}$ and x_i^j is the first part of $m_{R_i}^{j+1}$. Hence,

$$\Pr\left[\mathsf{Coll}_i^j\right] \leq \Pr_{x_i^j \leftarrow GF(q_j)}\left[x_i^j = \widehat{x_i^j}\right] = \frac{1}{q_j} \leq \frac{\epsilon'}{2^{r-j}}.$$

2. $T(\widehat{x_i^j}) \geq T(x_i^j)$ and $T(\widehat{y_i^j}) \geq T(y_i^j)$: In this case, if the adversary chooses $\widehat{x_i^j} \neq x_i^j$, then $\Pr\left[\mathsf{Coll}_i^j\right] = \Pr\left[m_{S,i}^{j+1} = m_{R_i}^{j+1}\right] = 0$. So for the remainder of the analysis of this case, we assume $\widehat{x_i^j} = x_i^j$. Since j is odd, it is always the case that $T(x_i^j) > T(\widehat{y_i^j})$; i.e., R_i chooses x_i^j after receiving $\widehat{y_i^j}$. Since we are also in the case where $T(\widehat{y_i^j}) \geq T(y_i^j)$, this means that R_i chooses x_i^j when $m_{S,i}^j, m_{R_i}^j, y_i^j$ and $\widehat{y_i^j}$ are all fixed. In particular, if $m_{S,i}^j \neq m_{R_i}^j$, then the polynomials $m_{S,i}^j(\cdot) + y_i^j$ and $m_{R_i}^j(\cdot) + \widehat{y_i^j}$ are two distinct polynomials of degree at most $\lceil n_j/\log q_j \rceil$. Hence,

$$\Pr\left[\mathsf{Coll}_i^j\right] = \Pr_{x_i^j \leftarrow GF(q_j)}\left[m_{S,i}^j \neq m_{R_i}^j \wedge m_{S,i}^j(x_i^j) + y_i^j = m_{R_i}^j(x_i^j) + \widehat{y_i^j}\right]$$
$$\leq \frac{1}{q_j} \cdot \left\lceil \frac{n_j}{\log q_j} \right\rceil \leq \frac{\epsilon'}{2^{r-j}}.$$

3. $T(\widehat{x_i^j}) \geq T(x_i^j)$ and $T(\widehat{y_i^j}) < T(y_i^j)$: As before, if $\widehat{x_i^j} \neq x_i^j$, then $\Pr\left[\mathsf{Coll}_i^j\right] = 0$, so we assume $\widehat{x_i^j} = x_i^j$. In this case, S chooses y_i^j and R_i chooses x_i^j when the

adversary has already chosen $\widehat{y_i^j}$. Since y_i^j and x_i^j are chosen independently, we may assume without loss of generality that $T(y_i^j) > T(x_i^j)$, meaning y_i^j is chosen when $m_{S,i}^j, m_{R_i}^j, \widehat{y_i^j}$ and x_i^j are already fixed (and thus also $\widehat{x_i^j}$, since we assume $\widehat{x_i^j} = x_i^j$). It follows that

$$\Pr\left[\mathsf{Coll}_i^j\right] = \Pr_{y_i^j \leftarrow GF(q_j)}\left[y_i^j = m_{R_i}^j(x_i^j) + \widehat{y_i^j} - m_{S,i}^j(x_i^j)\right] \leq \frac{1}{q_j} \leq \frac{\epsilon'}{2^{r-j}}.$$

We now turn to bound $\Pr\left[\mathsf{Coll}_i^j\right]$ in case j is even (S picks the evaluation point of the polynomial and R_i chooses the shift). The proof is very similar to the case where j is odd, and considers the same three cases:

1. $T(\widehat{x_i^j}) < T(x_i^j)$: In this case, S chooses x_i^j at random when $\widehat{x_i^j}$ is fixed. Therefore,

$$\Pr\left[\mathsf{Coll}_i^j\right] \leq \Pr_{x_i^j \leftarrow GF(q_j)}\left[x_i^j = \widehat{x_i^j}\right] = \frac{1}{q_j} \leq \frac{\epsilon'}{2^{r-j}}.$$

2. $T(\widehat{x_i^j}) \geq T(x_i^j)$ and $T(\widehat{y_i^j}) \geq T(y_i^j)$: As in the analysis for odd values of j, we can assume $\widehat{x_i^j} = x_i^j$, and we know that S chooses x_i^j when $m_{S,i}^j, m_{R_i}^j, y_i^j$ and $\widehat{y_i^j}$ are all fixed (in the last round, this follows also by the fact that S chooses x^{r-1} after receiving all $\widehat{y_i^{r-1}}$'s). In particular, if $m_{S,i}^j \neq m_{R_i}^j$, then the polynomials $m_{S,i}^j(\cdot) + y_i^j$ and $m_{R_i}^j(\cdot) + y_i^j$ are two distinct polynomials of degree at most $\lceil n_j / \log q_j \rceil$. Hence,

$$\Pr\left[\mathsf{Coll}_i^j\right] = \Pr_{x_i^j}\left[m_{S,i}^j \neq m_{R_i}^j \wedge m_{S,i}^j(x_i^j) + \widehat{y_i^j} = m_{R_i}^j(x_i^j) + y_i^j\right] \leq \frac{\epsilon'}{2^{r-j}}.$$

3. $T(\widehat{x_i^j}) \geq T(x_i^j)$ and $T(\widehat{y_i^j}) < T(y_i^j)$: As before, we assume $\widehat{x_i^j} = x_i^j$, and we know that R_i chooses y_i^j and S chooses x_i^j when the adversary has already chosen $\widehat{y_i^j}$. Since y_i^j and x_i^j are chosen independently, we may assume without loss of generality that $T(y_i^j) > T(x_i^j)$, meaning y_i^j is chosen when $m_{S,i}^j, m_{R_i}^j, \widehat{y_i^j}$ and x_i^j are already fixed. Hence,

$$\Pr\left[\mathsf{Coll}_i^j\right] = \Pr_{y_i^j \leftarrow GF(q_j)}\left[y_i^j = m_{S,i}^j(x_i^j) + \widehat{y_i^j} - m_{R_i}^j(x_i^j)\right] \leq \frac{\epsilon'}{2^{r-j}}.$$

Let $\mathsf{Coll}_i = \bigcup_{j \in [r-1]} \mathsf{Coll}_i^j$. By taking a union bound over all rounds, it follows that for every $i \in [k]$,

$$\Pr[\mathsf{Coll}_i] \leq \sum_{j=1}^{r-1} \Pr\left[\mathsf{Coll}_i^j\right] \leq \sum_{j=1}^{r-1} \frac{\epsilon'}{2^{r-j}} < \epsilon'.$$

Since for every $i \in [k]$, it is the case that $\mathsf{Forge}_{A,i}$ implies Coll_i, it holds that for every $i \in [k]$, $\Pr\left[\mathsf{Forge}_{A,i}\right] \leq \Pr\left[\mathsf{Coll}_i\right] \leq \epsilon'$. By taking a union bound over all receivers it holds that $\Pr\left[\mathsf{Forge}_A\right] \leq k \cdot \epsilon' = \epsilon$. ∎

5.3   Lower Bound

In this section we present a lower bound on the number of bits the sender has to out-of-band authenticate in the group setting. We prove the following theorem:

Theorem 5.6. *For   any   statistically-secure   out-of-band $(n, \ell, k, r, \epsilon)-$ authentication protocol, if $n \geq (k + 2) \cdot \log(1/\epsilon)$ then $\ell \geq (k + 1) \cdot \log(1/\epsilon) - k$.*

Proof. Let $\pi = (S, R_1, \ldots, R_k)$ be a statistically-secure out-of-band $(n, \ell, k, r, \epsilon)$-authentication protocol. We assume without loss of generality that $r \equiv 1 \mod (k + 1)$ and that π has the following structure. For every $j \in [r - 1]$, in round j there exists a single "active" party that sends a message (over the insecure channels) to each of the other parties, and all other parties do not send any messages in that round. If $j \equiv 1 \mod (k + 1)$, then the sender S is the active party in round j. Otherwise, if $j \equiv i + 1 \mod (k + 1)$ for some $i \in [k]$, then receiver R_i is the active user in round j. Denote the vector of messages sent in round j by x_{j-1} and the random variable describing that vector by X_{j-1} (so the vectors of messages sent over the insecure channels are $x_0, \ldots, x_{r-2}$). Finally, in round r, the sender S sends the short out-of-band authenticated value σ, and we denote the random variable describing it by Σ. We also denote the random variable describing the vector of input messages to S by M.

Observe, that we can write the Shannon entropy of Σ as

$$
\begin{aligned}
\mathrm{H}(\Sigma) &= \mathrm{H}(\Sigma) - \mathrm{H}(\Sigma|M, X_0) + \sum_{j \in [r-2]} (\mathrm{H}(\Sigma|M, X_0, \ldots, X_{j-1}) \\
&\quad - \mathrm{H}(\Sigma|M, X_0, \ldots, X_j)) + \mathrm{H}(\Sigma|M, X_0, \ldots, X_{r-2}) \\
&= \mathrm{I}(\Sigma; M, X_0) + \sum_{j \in [r-2]} \mathrm{I}(\Sigma; X_j|M, X_0, \ldots, X_{j-1}) \\
&\quad + \mathrm{H}(\Sigma|M, X_0, \ldots, X_{r-2}) \\
&= \mathrm{I}(\Sigma; M, X_0) + \sum_{\substack{i \in \{0,\ldots,k\}}} \sum_{\substack{j \in [r-2]: \\ j \equiv i \mod (k+1)}} \mathrm{I}(\Sigma; X_j|M, X_0, \ldots, X_{j-1}) \\
&\quad + \mathrm{H}(\Sigma|M, X_0, \ldots, X_{r-2}).
\end{aligned}
$$

To bound the above expression, we make use of the following two lemmata, proofs for which are provided in the full version of the paper [RS18]. Intuitively speaking, Lemma 5.7 shows that the messages of the sender S during the execution of π need to reduce, on average, roughly $\log(1/\epsilon)$ bits of entropy from the out-of-band authenticated value.

Lemma 5.7. *If $n \geq 1/k \cdot \log(1/\epsilon)$, then*

$$I(\Sigma; M, X_0) + \sum_{\substack{j \in [r-2]: \\ j \equiv 0 \mod (k+1)}} I(\Sigma; X_j | M, X_0, \ldots, X_{j-1})$$

$$+ H(\Sigma | M, X_0, \ldots, X_{r-2}) \geq \log(1/\epsilon) - 1.$$

In a similar fashion, Lemma 5.8 shows that for any $i \in [k]$, the messages of receiver R_i during the execution of π need to reduce, on average, roughly $\log(1/\epsilon)$ bits of entropy from the out-of-band authenticated value.

Lemma 5.8. *If $n \geq (k+2) \cdot \log(1/\epsilon)$ and $\ell \leq (k+1) \cdot \log(1/\epsilon)$, then for every $i \in [k]$,*

$$\sum_{\substack{j \in [r-2]: \\ j \equiv i \mod (k+1)}} I(\Sigma, X_j | M, X_0, \ldots, X_{j-1}) \geq \log(1/\epsilon) - 1.$$

Now, if $\ell > (k+1) \cdot \log(1/\epsilon)$, then the theorem follows. Otherwise, by Lemmata 5.7 and 5.8 it holds that, $\ell \geq H(\Sigma) \geq (k+1) \cdot \log(1/\epsilon) - k$, concluding the proof of Theorem 5.6.

References

[BM94] Bellovin, S.M., Merritt, M.: An attack on the Interlock protocol when used for authentication. IEEE Trans. Inf. Theory **40**(1), 273–275 (1994)

[BR93] Bellare, M., Rogaway, P.: Random oracles are practical: a paradigm for designing efficient protocols. In: Proceedings of the 1st ACM Conference on Computer and Communications Security, pp. 62–73 (1993)

[BSJ+17] Bellare, M., Singh, A.C., Jaeger, J., Nyayapati, M., Stepanovs, I.: Ratcheted encryption and key exchange: the security of messaging. In: Katz, J., Shacham, H. (eds.) CRYPTO 2017. LNCS, vol. 10403, pp. 619–650. Springer, Cham (2017). https://doi.org/10.1007/978-3-319-63697-9_21

[CCD+17] Cohn-Gordon, K., Cremers, C.J.F., Dowling, B., Garratt, L., Stebila, D.: A formal security analysis of the Signal messaging protocol. In: Proceedings of the 2nd IEEE European Symposium on Security and Privacy (EuroS&P), pp. 451–466 (2017)

[CF01] Canetti, R., Fischlin, M.: Universally composable commitments. In: Kilian, J. (ed.) CRYPTO 2001. LNCS, vol. 2139, pp. 19–40. Springer, Heidelberg (2001). https://doi.org/10.1007/3-540-44647-8_2

[CGCG+17] Cohn-Gordon, K., Cremers, C., Garratt, L., Millican, J., Milner, K.: On ends-to-ends encryption: Asynchronous group messaging with strong security guarantees. Cryptology ePrint Archive, Report 2017/666 (2017)

[CIO98] Crescenzo, G.D., Ishai, Y., Ostrovsky, R.: Non-interactive and non-malleable commitment. In: Proceedings of the 30th Annual ACM Symposium on Theory of Computing, pp. 141–150 (1998)

[CKO+01] Di Crescenzo, G., Katz, J., Ostrovsky, R., Smith, A.: Efficient and non-interactive non-malleable commitment. In: Pfitzmann, B. (ed.) EUROCRYPT 2001. LNCS, vol. 2045, pp. 40–59. Springer, Heidelberg (2001). https://doi.org/10.1007/3-540-44987-6_4

[COS+17] Ciampi, M., Ostrovsky, R., Siniscalchi, L., Visconti, I.: Four-round concurrent non-malleable commitments from one-way functions. In: Katz, J., Shacham, H. (eds.) CRYPTO 2017. LNCS, vol. 10402, pp. 127–157. Springer, Cham (2017). https://doi.org/10.1007/978-3-319-63715-0_5

[DDN00] Dolev, D., Dwork, C., Naor, M.: Non-malleable cryptography. SIAM J. Comput. **30**(2), 391–437 (2000)

[DG03] Damgard, I., Groth, J.: Non-interactive and reusable non-malleable commitment schemes. In: Proceedings of the 35th Annual ACM Symposium on Theory of Computing, pp. 426–437 (2003)

[Ell96] Ellison, C.M.: Establishing identity without certification authorities. In: Proceedings of the 6th USENIX Security Symposium, p. 7 (1996)

[FF00] Fischlin, M., Fischlin, R.: Efficient non-malleable commitment schemes. In: Bellare, M. (ed.) CRYPTO 2000. LNCS, vol. 1880, pp. 413–431. Springer, Heidelberg (2000). https://doi.org/10.1007/3-540-44598-6_26

[FMB+16] Frosch, T., Mainka, C., Bader, C., Bergsma, F., Schwenk, J., Holz, T.: How secure is TextSecure? In: Proceedings of the 1st IEEE European Symposium on Security and Privacy (EuroS&P), pp. 457–472 (2016)

[Gol01] Goldreich, O.: Foundations of Cryptography – Volume 1: Basic Techniques. Cambridge University Press, Cambridge (2001)

[Goy11] Goyal, V.: Constant round non-malleable protocols using one way functions. In: Proceedings of the 43rd Annual ACM Symposium on Theory of Computing, pp. 695–704 (2011)

[Gre18a] Green, M.: Attack of the week: Group messaging in WhatsApp and Signal. A Few Thoughts on Cryptographic Engineering (2018). https://blog. cryptographyengineering.com/2018/01/10/attack-of-the-week-group-messaging

[Gre18b] Greenberg, A.: WhatsApp security flaws could allow snoops to slide into group chats. Wired Mag. (2018). https://www.wired.com/story/whatsapp-security-flaws-encryption-group-chats

[KBB17] Kobeissi, N., Bhargavan, K., Blanchet, B.: Automated verification for secure messaging protocols and their implementations: a symbolic and computational approach. In: Proceedings of the 2nd IEEE European Symposium on Security and Privacy (EuroS&P), pp. 435–450 (2017)

[LP11] Lin, H., Pass, R.: Constant-round non-malleable commitments from any one-way function. In: Proceedings of the 43rd Annual ACM Symposium on Theory of Computing, pp. 705–714 (2011)

[LPV08] Lin, H., Pass, R., Venkitasubramaniam, M.: Concurrent non-malleable commitments from any one-way function. In: Canetti, R. (ed.) TCC 2008. LNCS, vol. 4948, pp. 571–588. Springer, Heidelberg (2008). https://doi. org/10.1007/978-3-540-78524-8_31

[NSS06] Naor, M., Segev, G., Smith, A.: Tight bounds for unconditional authentication protocols in the manual channel and shared key models. In: Dwork, C. (ed.) CRYPTO 2006. LNCS, vol. 4117, pp. 214–231. Springer, Heidelberg (2006). https://doi.org/10.1007/11818175_13

[NSS08] Naor, M., Segev, G., Smith, A.D.: Tight bounds for unconditional authentication protocols in the manual channel and shared key models. IEEE Trans. Inf. Theory **54**(6), 2408–2425 (2008)

[PM16] Perrin, T., Marlinspike, M.: The double ratchet algorithm (2016). https://signal.org/docs/specifications/doubleratchet/doubleratchet.pdf. Accessed 16 May 2018

[PR05] Pass, R., Rosen, A.: Concurrent non-malleable commitments. In: Proceedings of the 46th Annual IEEE Symposium on Foundations of Computer Science, pp. 563–572 (2005)

[PR08] Pass, R., Rosen, A.: New and improved constructions of nonmalleable cryptographic protocols. SIAM J. Comput. **38**(2), 702–752 (2008)

[PV06] Pasini, S., Vaudenay, S.: An optimal non-interactive message authentication protocol. In: Pointcheval, D. (ed.) CT-RSA 2006. LNCS, vol. 3860, pp. 280–294. Springer, Heidelberg (2006). https://doi.org/10.1007/11605805_18

[RMS18] Rösler, P., Mainka, C., Schwenk, J.: More is less: on the end-to-end security of group chats in Signal, WhatsApp, and Threema. In: Proceedings of the 3rd IEEE European Symposium on Security and Privacy (EuroS&P) (2018)

[RS84] Rivest, R.L., Shamir, A.: How to expose an eavesdropper. Commun. ACM **27**(4), 393–395 (1984)

[RS18] Rotem, L., Segev, G.: Out-of-band authentication in group messaging: computational, statistical, optimal. Cryptology ePrint Archive, Report 2018/493 (2018)

[Tela] Telegram. End-to-end encrypted voice calls - key verification. https://core.telegram.org/api/end-to-end/voice-calls#key-verification. Accessed 16 May 2018

[Telb] Telegram. End-to-end encryption. https://core.telegram.org/api/end-to-end. Accessed 16 May 2018

[Vau05] Vaudenay, S.: Secure communications over insecure channels based on short authenticated strings. In: Shoup, V. (ed.) CRYPTO 2005. LNCS, vol. 3621, pp. 309–326. Springer, Heidelberg (2005). https://doi.org/10.1007/11535218_19

[Vib] Viber encryption overview. https://www.viber.com/app/uploads/Viber-Encryption-Overview.pdf. Accessed 16 May 2018

[Wha] WhatsApp encryption overview. https://www.whatsapp.com/security/WhatsApp-Security-Whitepaper.pdf. Accessed 16 May 2018

[Wik] Wikipedia. Instant messaging. https://en.wikipedia.org/wiki/Instant_messaging. Accessed 16 May 2018

Implementations and Physical Attacks Prevention

Faster Homomorphic Linear Transformations in **HElib**

Shai Halevi[1]([⊠]) and Victor Shoup[1,2]

[1] IBM Research, Yorktown Heights, NY, USA
shaih@alum.mit.edu
[2] New York University, New York, USA
shoup@cs.nyu.edu

Abstract. **HElib** is a software library that implements homomorphic encryption (HE), with a focus on effective use of "packed" ciphertexts. An important operation is applying a known linear map to a vector of encrypted data. In this paper, we describe several algorithmic improvements that significantly speed up this operation: in our experiments, our new algorithms are 30–75 times faster than those previously implemented in **HElib** for typical parameters.

One application that can benefit from faster linear transformations is bootstrapping (in particular, "thin bootstrapping" as described in [Chen and Han, Eurocrypt 2018]). In some settings, our new algorithms for linear transformations result in a 6× speedup for the entire thin bootstrapping operation.

Our techniques also reduce the size of the large public evaluation key, often using 33%–50% less space than the previous **HElib** implementation. We also implemented a new tradeoff that enables a drastic reduction in size, resulting in a 25× factor or more for some parameters, paying only a penalty of a 2–4× times slowdown in running time (and giving up some parallelization opportunities).

Keywords: Homomorphic encryption · Implementation
Linear transformations

1 Introduction

Homomorphic encryption (HE) [5,13] enables performing arithmetic operations on encrypted data even without knowing the secret key. All contemporary HE schemes roughly follow the outline of Gentry's first candidate, where fresh ciphertexts are "noisy" to ensure security. This noise grows with every operation, until it becomes so large so as to cause decryption errors. This results in a "somewhat homomorphic" encryption scheme (SWHE) that can only evaluate low-depth circuits, such a scheme can be converted to a "fully homomorphic"

Supported by the Defense Advanced Research Projects Agency (DARPA) and Army Research Office (ARO) under Contract No. W911NF-15-C-0236.

H. Shacham and A. Boldyreva (Eds.): CRYPTO 2018, LNCS 10991, pp. 93–120, 2018.
https://doi.org/10.1007/978-3-319-96884-1_4

encryption scheme (FHE) using bootstrapping. The most asymptotically efficient SWHE schemes are based on the hardness of ring-LWE. Most of these scheme use $R_p = \mathbb{Z}[X]/(F(X), p)$ as their native plaintext space, with F a cyclotomic polynomial and p an integer (usually a prime or prime power).

Smart and Vercauteren observed [15] that (for a prime p) an element in this native plaintext space can be used to encode (via Chinese Remaindering) a vector of values from a finite field $\mathbb{F}_{p^d}$, for some integer d that depends on F and p, and that operations on elements in R_p induce the corresponding entry-wise operation on the encoded vectors. This technique of encoding many plaintext elements from $\mathbb{F}_{p^d}$ in a single R_p element, which is then encrypted and manipulated homomorphically, is called "ciphertext packing", and the entries in the vector are called "plaintext slots." Gentry, Halevi, and Smart showed in [6] how to use special automorphisms on R_p (which were used for different purposes in [2, 10]) to enable data movement between the slots.

HElib [7–9] is an open-source C++ library that implements the ring variant of the scheme due to Brakerski-Gentry-Vaikuntanathan [2], focusing on effective use of ciphertext packing. It includes an implementation of the BGV scheme itself with all its basic homomorphic operations, as well as higher-level procedures for data-movement, simple linear algebra, bootstrapping, etc. One can think of the lower levels of HElib as providing a "hardware platform", defining a set of operations that can be applied homomorphically. These operations include entry-wise addition and multiplication operations on the vector of plaintext values, as well as data movement, making this "platform" a SIMD environment.

Our Results. In this work, we improve performance of core linear algebra algorithms in HElib that apply publicly known linear transformations to encrypted vectors. These improvements are now integrated into HElib. For typical, realistic parameter settings, our new algorithms can run 30–75 times faster than those in the previous implementation of HElib, where the exact speedup depends on myriad details.[1] Our implementation also exploits multiple cores, when available, to get even further speedups.

Our techniques also reduce the size of the large public evaluation key. In the old HElib implementation, the evaluation key typically consists of a large number of large "key switching matrices": Each of these "matrices" can take 1–4 MB of space, and the implementation uses close to a hundred of them. Our new implementation reduces the number of key-switching matrices by 33–50% in some parameter settings (that arise fairly often in practice), while at the same time improves the running time. Moreover, a new tradeoff that we implemented enables a drastic reduction in the number of matrices (sometimes as few as four or six matrices overall), for a small price of only 2–4× in performance. This space efficient variation, however, is inherently sequential, as opposed to our other procedure than can be easily parallelized.

---

[1] One could also consider algorithms that apply encrypted linear transformations to encrypted vectors; some of our new algorithmic techniques may apply to that problem as well; however, we have not yet implemented this in HElib.

Applications. Linear transformations of encrypted vectors is a manifestly fundamental operation with many applications. For one example, HElib itself makes critical use of such transformations in its bootstrapping logic. As reported in [8], the bootstrapping routine can typically spend 25–40% of its time performing such transformations. In addition, a new "thin bootstrapping" technique, due to Chen and Han [4], is useful to bootstrap encrypted vectors whose entries are in the base field, rather than an extension field. In practice, this is an important special case of bootstrapping, and our faster algorithms for linear transformations play an even more significant role here. Our timing results in Sect. 9 show that for large vectors, these faster algorithms are essential to make "thin bootstrapping" practical.

As another example, consider a *private information retrieval* protocol in which a client selects one value from a database of values held by a server, while hiding from the server which value was accessed. Using HE, one way to do this is for the server to encode each value as a column vector. The collection of all such values held by the server is thus encoded as a matrix M, where each column in M corresponds to one value. To access the ith value, the client can send to the server an encrypted unit vector v with 1 in the ith entry (or some other encrypted information from which the server can homomorphically compute such an encrypted unit vector). The server then homomorphically computes $M \times v$, which is an encryption of the selected column of M. The server sends the result to the client, who can decrypt it and recover the selected value.

Techniques. In the linear transformation algorithms previously implemented in HElib, the bulk of the time is spent moving data among the slots in the encrypted vector. As mentioned above, this is accomplished by using special automorphisms. The main cost of applying such an automorphism to a ciphertext is actually that of "key switching": after applying the automorphism to each ring element in the ciphertext (which is actually a very cheap operation), we end up with an encryption relative to the "wrong" secret key; we can recover a ciphertext relative to the "right" secret key by using data in the public key specific to this particular automorphism — a so-called "key switching matrix."

The main goals in improving performance are therefore to reduce the number of automorphisms, and to reduce the cost of each automorphism.

– To reduce the number of automorphisms, we introduce a "baby-step/giant-step" strategy for computing all of the required automorphisms. This strategy generalizes a similar idea that was used in [8] in the context of bootstrapping. This strategy by itself speeds up the computation by a factor of 15–20 in typical settings. See Sect. 4.1.
– We further reduce the number of automorphisms by refactoring a number of computations, more aggressively exploiting the algebraic properties of the automorphisms that we use. See Sect. 4.4.

- To reduce the cost of each automorphism, we introduce a new technique for "hoisting" the expensive parts of these operations out of the main loop.[2] Our main observation is that applying many automorphisms to the same ciphertext v can be done faster than applying each one separately. Instead, we can perform an expensive pre-computation that depends only on v (but not the automorphisms themselves), and this pre-computation makes each automorphism much cheaper (typically, 6–8 times faster). See Sects. 4.2 and 5.
- Recall that key switching matrices are a part of the public key and consume quite a lot of space (typically several megabytes per matrix), so keeping their numbers down is desirable. In the previous implementation of HElib, there can easily be several hundred such matrices in the public key. We introduce a new technique that reduces the number of key-switching matrices by 33–50% in some parameter settings (that arise fairly often in practice), while at the same time improves the running time of our algorithms. See Sect. 4.3.
- We introduce yet another technique that drastically reduces the number of key-switching matrices to a very small number (less than 10), but comes at a cost in running time (typically 2–4 times more slowly as our fastest algorithms), and cannot be parallelized.[3] Achieving this reduction in key-switching storage without too much degradation in running time requires some new algorithmic ideas. See Sect. 4.5.

Outline. The rest of the paper is organized as follows.

- In Sect. 2, we introduce notation and terminology, and review the basics of the BGV cryptosystem, including ciphertext packing and automorphisms.
- In Sect. 3, we review the basic ideas underlying the previous algorithms in HElib for applying linear transformations homomorphically. We focus on restricted linear transformations, the "one-dimensional" transformations Mat-Mul1D and BlockMatMul1D. It turns out that considering these restricted transformations is sufficient: they can be used directly in applications such as bootstrapping, and can be easily be used to implement more general linear transformations.
- In Sect. 4, we give a more detailed overview of our new techniques.
- In Sect. 5, we give more of the details of our new hoisting technique.
- In Sect. 6, we present all of our new algorithms for MatMul1D and BlockMat-Mul1D in detail.
- In Sect. 7, we describe how to use algorithms for MatMul1D and BlockMat-Mul1D for more general linear transformations.

---

[2] "Hoisting" is a term used in compiler optimization to describe the action of "hoisting" a computation out of a loop, so that it is only performed once, instead of in every loop iteration.

[3] While the "top level" operations in our linear transformations are inherently sequential when using this technique, lower-level routines in HElib will still exploit multiple cores, if available. Such low-level parallelism are usually less effective, however.

- In Sect. 8, we review the bootstrapping procedure from [8], and discuss how those techniques can be adapted to the "thin bootstrapping" technique of Chen and Han [4].
- In Sect. 9, we report on the performance of the implementation of our new algorithms (and their application to bootstrapping).

2   Notations and Background

For a positive modulus $q \in \mathbb{Z}_{>0}$, we identify the ring $\mathbb{Z}_q$ with its representation as integers in $[-q/2, q/2)$ (except for $q = 2$ where we use $\{0, 1\}$). For integer z, we denote by $[z]_q$ the reduction of z modulo q into the same interval. This notation extends to vectors and matrices coordinate-wise, and to elements of other algebraic groups/rings/fields by considering their coefficients in some convenient basis (e.g., the coefficient of polynomials in the power basis when talking about $\mathbb{Z}[X]$). The norm of a ring element $\|a\|$ is defined as the norm of its coefficient vector in that basis.[4]

2.1   The BGV Cryptosystem

The BGV ring-LWE-based scheme [3] is defined over a ring $R \stackrel{\text{def}}{=} \mathbb{Z}[X]/(\Phi_m(X))$, where $\Phi_m(X)$ is the mth cyclotomic polynomial. For an arbitrary integer modulus N (not necessarily prime) we denote the ring $R_N \stackrel{\text{def}}{=} R/NR$.

As implemented in HElib, the native plaintext space of the BGV cryptosystem is R_{p^r} for a prime power p^r. The scheme is parametrized by a sequence of decreasing moduli $q_L \gg q_{L-1} \gg \cdots \gg q_0$, and an "$i$th level ciphertext" in the scheme is a vector $v \in R_{q_i}^2$. Secret keys are elements $s \in R$ with "small" coefficients (chosen in $\{0, \pm 1\}$ in HElib), and we view s as the second element of the 2-vector $\mathsf{sk} = (1, s) \in R^2$. A level-$i$ ciphertext $v = (p_0, p_1)$ encrypts a plaintext element $\alpha \in R_{p^r}$ with respect to $\mathsf{sk} = (1, s)$ if $[\langle \mathsf{sk}, v \rangle]_{q_i} = [p_0 + s \cdot p_1]_{q_i} = \alpha + p^r \cdot \epsilon$ (in R) for some "small" error term, $\|\epsilon\| \ll q_i/p^r$.

The error term grows with homomorphic operations of the cryptosystem, and switching from q_{i+1} to q_i is used to decrease the error term roughly by the ratio q_{i+1}/q_i. Once we have a level-0 ciphertext v, we can no longer use that technique to reduce the noise. To enable further computation, we need to use Gentry's bootstrapping technique [5]. In HElib, each q_i is a product of small (machine-word sized) primes.

2.2   Encoding Vectors in Plaintext Slots

As observed by Smart and Vercauteren [15], an element of the native plaintext space $\alpha \in R_{p^r}$ can be viewed as encoding a vector of "plaintext slots" containing elements from some smaller ring extension of $\mathbb{Z}_{p^r}$ via Chinese remaindering.

---

[4] The difference between the norm in the different bases is not very important for the current work.

In this way, a single arithmetic operation on α corresponds to the same operation applied component-wise to all the slots.

Specifically, suppose the factorization of $\Phi_m(X)$ modulo p^r is $\Phi_m(X) \equiv F_1(X) \cdots F_\ell(X) \pmod{p^r}$, where each F_i has the same degree d, which is equal to the order of p modulo m, so that $\ell = \phi(m)/d$. (This factorization can be obtained by factoring $\Phi_m(X)$ modulo p, followed by Hensel lifting.) Then we have the isomorphism $R_{p^r} \cong \bigoplus_{i=1}^\ell (\mathbb{Z}[X]/(p^r, F_i(X)))$.

Let us now denote $E \overset{\text{def}}{=} \mathbb{Z}[X]/(p^r, F_1(X))$, and let ζ be the residue class of X in E, which is a principal mth root of unity, so that $E = \mathbb{Z}/(p^r)[\zeta]$. The rings $\mathbb{Z}[X]/(p^r, F_i(X))$ for $i = 1, \ldots, \ell$ are all isomorphic to E, and their direct product is isomorphic to R_{p^r}, so we get an isomorphism between R_{p^r} and E^ℓ. HElib makes extensive use of this isomorphism, using it to encode an ℓ-vector of elements in E as an element of the native plaintext space R_{p^r}. Addition and multiplication of ciphertexts act on all ℓ slots of the corresponding plaintext in parallel.

2.3   Hypercube Structure and One-Dimensional Rotations

Beyond addition and multiplications, we can also manipulate elements in R_{p^r} using a set of automorphisms on R_{p^r} of the form

$$\theta_t : R_{p^r} \longrightarrow R_{p^r}, \quad a(X) \longmapsto a(X^t) \pmod{(p^r, \Phi_m(X))}.$$

for $t \in \mathbb{Z}_m^*$. Since each θ_t is an automorphism, it distributes over addition and multiplication, i.e., $\theta_t(\alpha+\beta) = \theta_t(\alpha)+\theta_t(\beta)$ and $\theta_t(\alpha\beta) = \theta_t(\alpha)\theta_t(\beta)$. Also, these automorphisms commute with one another, i.e., $\theta_t\theta_{t'} = \theta_{tt'} = \theta_{t'}\theta_t$. Moreover, for any integer i, we have $\theta_t^i = \theta_{t^i}$.

We can homomorphically apply such an automorphism by applying it to the individual ciphertext components and then performing "key switching" (see [3,6]). In somewhat more detail, a ciphertext in HElib consists of two "parts," each an element of R_q for some q. Applying the same automorphism (defined in R_q) to the two parts, we get a ciphertext with respect to a different secret key. In order to do anything more with this ciphertext, we usually have to convert it back to a ciphertext with respect to the original secret key. In order to do this, the public-key must contain data specific to the automorphism θ_t, called a "key switching matrix".[5] We will discuss this key-switching operation in more detail below in Sect. 5.

As discussed in [6], these automorphisms induce a hypercube structure on the plaintext slots, that depends on the structure of the group $\mathbb{Z}_m^*/\langle p \rangle$. Specifically, HElib keeps a hypercube basis $g_1, \ldots, g_n \in \mathbb{Z}_m^*$ with orders $D_1, \ldots, D_n \in \mathbb{Z}_{>0}$, and then defines the set of representatives for $\mathbb{Z}_m^*/\langle p \rangle$ as

$$\{g_1^{e_1} \cdots g_n^{e_n} : 0 \le e_s < D_s, \ s = 1, \ldots, n\}.$$

---

[5] Note that this "key switching" technique is a generalization of that used to allow multiplication of ciphertexts.

More precisely, D_s is the order of g_s in $\mathbb{Z}_m^*/\langle p, g_1, \ldots, g_{s-1}\rangle$. Thus, the slots are in one-to-one correspondence with tuples $(e_1, \ldots, e_n)$ with $0 \leq e_s < D_s$. This induces an n-dimensional hypercube structure on the plaintext space. If we fix $e_1, \ldots, e_{s-1}, e_{s+1}, \ldots, e_n$, and let e_s range over $0, \ldots, D_s - 1$, we get a set of D_s slots, which we refer to as a *hypercolumn in dimension s* (and there are ℓ/D_s such hypercolumns).

Using automorphisms, we can efficiently perform rotations in any dimension; a *rotation by i in dimension s* maps a slot corresponding to $(e_1, \ldots, e_s, \ldots, e_n)$ to the slot corresponding to $(e_1, \ldots, e_s + i \bmod D_s, \ldots, e_n)$. In other words, it rotates each hypercolumn in dimension s by i. We denote by ρ_s the rotation-by-1 operation in dimension s. Observe that ρ_s^i is the rotation-by-i operation in dimension s.

We can implement ρ_s^i by applying either one or two of the automorphisms $\{\theta_t\}_{t \in \mathbb{Z}_m^*}$ defined above. If the order of g_s in $\mathbb{Z}_m^*$ is D_s, then we get by with just a single automorphism, since

$$\rho_s^i(\alpha) = \theta_{g_s^i}(\alpha). \tag{1}$$

In this case, we call s a "good dimension".

If the order of g_s in $\mathbb{Z}_m^*$ is different from D_s, then we call s a "bad dimension", and we need to implement this rotation using two automorphisms. Specifically, we use a constant "0–1 mask value" μ that selects some slots and zeros-out the others, and use the two automorphisms $\psi \overset{\text{def}}{=} \theta_{g_s^i}$ and $\psi^* \overset{\text{def}}{=} \theta_{g_s^{i-D}}$. Then we have

$$\rho_s^i(\alpha) = \psi(\mu \cdot \alpha) + \psi^*((1 - \mu) \cdot \alpha). \tag{2}$$

The idea is roughly as follows. Even though ψ does not act as a rotation by i in dimension s, it does act as the desired rotation if we restrict it to inputs with zeros in each slot whose coordinate in dimension s is at least $D - i$. Similarly, ψ^* acts as the desired rotation if we restrict it to inputs with zeros in each slot whose coordinate in dimension s is less than $D - i$. This tells us that μ should have a 1 in all slots whose coordinate in dimension s is less than $D - i$, and a 0 in all other slots. Note also that

$$\rho_s^i(\alpha) = \mu' \cdot \psi(\alpha) + (1 - \mu') \cdot \psi^*(\alpha), \tag{3}$$

where $\mu' = \psi(\mu)$ is a mask with a 1 is all slots whose coordinate in dimension s is at least i, and a 0 in all other slots. This formulation will be convenient in some of the algorithms we present.

2.4 Frobenius and Linearized Polynomials

We define the automorphism $\sigma \overset{\text{def}}{=} \theta_p$, which is the Frobenius map on R_{p^r} (where θ_p is one of the automorphisms defined in Sect. 2.3). It acts on each slot independently as the Frobenius map σ_E on E, which sends ζ to ζ^p and leaves elements of $\mathbb{Z}_{p^r}$ fixed. (When $r = 1$, σ is the same as the pth power map on E.)

For any $\mathbb{Z}_{p^r}$-linear transformation on E, denoted M, there exist unique constants $\lambda_0, \ldots, \lambda_{d-1} \in E$ such that $M(\eta) = \sum_{j=0}^{d-1} \lambda_j \sigma_E^j(\eta)$ for all $\eta \in E$. When $r = 1$, this follows from the general theory of linearized polynomials (see, e.g., Theorem 10.4.4 on p. 237 of [14]), but the same results are easily seen to hold for $r > 1$ as well. These constants are readily computable by solving a system of equations mod p^r.

Using linearized polynomials, we may effectively apply a fixed linear map to each slot of a plaintext element $\alpha \in R_{p^r}$ (either the same or different maps in each slot) by computing $\sum_{j=0}^{d-1} \kappa_j \sigma^j(\alpha)$, where the κ_j's are R_{p^r}-constants obtained by embedding appropriate E-constants in the slots.

2.5 Key Switching Strategies

The total number of automorphisms is $\phi(m)$, which is typically many thousands, so it is not very practical to store all possible key switching matrices in the public key: each such matrix typically occupies a few megabytes of storage, and storing all of them will consume hundreds of gigabytes. Therefore, we consider strategies that trade off space for time with respect to key switching matrices.

For almost all applications, we only need the key switching matrices for one-dimensional rotations in each dimension, as well as for the Frobenius map (and its powers). For a fixed dimension $s = 1, \ldots, n$ of size $D \stackrel{\text{def}}{=} D_s$ with generator $g \stackrel{\text{def}}{=} g_s$, consider the automorphism $\theta \stackrel{\text{def}}{=} \theta_{g_s}$. In the original implementation of HElib, one of two key switching strategies for dimension s are used.

Full: We store key switching matrices for θ^i for $i = 0, \ldots, D - 1$. If s is a "bad dimension", we additionally store key switching matrices for θ^{-i} for $i = 1, \ldots, D - 1$.

Baby-step/giant-step: We store key switching matrices for θ^j with $j = 1, \ldots, g - 1$, where $g \stackrel{\text{def}}{=} \lceil \sqrt{D} \rceil$ (the "baby steps"), as well as for θ^{gk} with $k = 1, \ldots, h - 1$, where $h \stackrel{\text{def}}{=} \lceil D/g \rceil$ (the "giant steps"). If s is a "bad dimension", we additionally store key switching matrices for θ^{-gk} with $k = 1, \ldots, h$ (negative "giant steps").

Using the full strategy, any rotation in dimension s can be implemented using a single automorphism and key switching if s is a good dimension, and using two automorphisms and key switchings if s is a bad dimension.

Using the baby-step/giant-step strategy, any rotation in dimension s can be implemented using at most two automorphisms and key switchings if s is a good dimension, and using at most four automorphisms and key switchings if s is a bad dimension. The idea is that to compute $\theta^i(v)$, for a given $i = 0, \ldots, D - 1$, we can write $i = j + gk$, so that to compute $\theta^i(v)$, we first compute $w = \theta^{gk}(v)$, which takes one automorphism and a key switching, and then compute $\theta^j(w)$, which takes another automorphism and key switching.

These two strategies give us a time/space trade-off: although it slows down the computation time by a factor of two, the baby-step/giant-step strategy requires space for just $O(\sqrt{D})$ key switching matrices, rather than the $O(D)$ key switching matrices required by the full strategy.

The same two strategies can be used to store key switching matrices for powers of the Frobenius map, so that any power of the Frobenius map can be computed using either one or two automorphisms. Indeed, it is convenient to think of the powers of the Frobenius map as defining an additional (effectively "good") dimension.

The default behavior of HElib is to use the full key-switching strategy for "small" dimensions (of size at most 50), and the baby-step/giant-step strategy for larger dimensions.

3 Matrix Multiplication — Basic Ideas

In [7], it is observed that we can multiply a matrix $M \in E^{\ell \times \ell}$ by a column vector $v \in E^{\ell \times 1}$ by computing

$$Mv = M_0 v_0 + \cdots + M_{\ell-1} v_{\ell-1}, \tag{4}$$

where each v_i is the vector obtained by rotating the entries of v by i positions, and each M_i is a diagonal matrix containing one diagonal of M.

3.1 MatMul1D: One-Dimensional E-Linear Transformations

In many applications, such as the recryption procedure in [8], instead of a general E-linear transformation on R_{p^r}, we only need to work with a *one-dimensional* E-linear transformation that acts independently on the individual hypercolumns of a single dimension $s = 1, \ldots, n$. We can adapt the diagonal decomposition of Eq. (4) to this setting using appropriate rotation maps on the slots of R_{p^r}. Let $\rho \overset{\text{def}}{=} \rho_s$ be the rotation-by-1 map in dimension s, and let $D \overset{\text{def}}{=} D_s$ be the size of dimension s. If T is a one-dimensional E-linear transformation on R_{p^r}, then for every $v \in R_{p^r}$, we have

$$T(v) = \sum_{i=0}^{D-1} \kappa_i \cdot \rho^i(v), \tag{5}$$

where the κ_i's are constants in R_{p^r} determined by T, obtained by embedding appropriate constants in E in each slot. Equation (5) translates directly into a simple homomorphic evaluation algorithm, just by applying the same operations to a ciphertext encrypting v. In a straightforward implementation, in a good dimension, the computational cost is about D automorphisms and D constant-ciphertext multiplications, and the noise cost is a single constant-ciphertext multiplication. In bad dimensions, all of these costs would essentially double. In practice, if the constants have been pre-computed, the computation cost of the constant-ciphertext multiplications is negligible compared to that of the automorphisms.

One of our main goals in this paper is to dramatically improve upon the computational cost for performing such a MatMul1D operation.

3.2 BlockMatMul1D: One-Dimensional $\mathbb{Z}_{p^r}$-Linear Transformations

In some applications (again, including the recryption procedure in [8]), instead of applying an E-linear transformation, we need to apply a $\mathbb{Z}_{p^r}$-linear map. Again, we focus on *one-dimensional* $\mathbb{Z}_{p^r}$-linear maps that act independently on the hypercolumns of a single dimension.

We can still use the same diagonal decomposition as in Eq. (4), except that the entries in the diagonal matrices are no longer elements of E, but rather, $\mathbb{Z}_{p^r}$-linear maps on E. These maps may be encoded using linearized polynomials, as in Sect. 2.4. Therefore, if T is a one-dimensional $\mathbb{Z}_{p^r}$-linear transformation on R_{p^r}, then for every $v \in R_{p^r}$, we have

$$T(v) = \sum_{i=0}^{D-1} \sum_{j=0}^{d-1} \kappa_{i,j} \cdot \sigma^j \big(\rho^i(v) \big), \tag{6}$$

where the $\kappa_{i,j}$'s are constants in R_{p^r} determined by T.

A naive homomorphic implementation of the formula from Eq. (6) takes $O(dD)$ automorphisms, but as shown in [8], this can be reduced to $O(d + D)$ automorphisms. In this paper, we will also present significant improvements to the BlockMatMul1D algorithm in [8], although they are not as dramatic as our improvements to the MatMul1D algorithm.

4 Overview of Algorithmic Improvements

4.1 Baby-Step/Giant-Step Multiplication

As already mentioned, [8] introduces a technique that reduces the number of automorphisms needed to implement BlockMatMul1D in dimension s from $O(dD)$ to $O(d + D)$, where $D \overset{\text{def}}{=} D_s$ is the size of the dimension, and d is the order of $p \bmod m$. A very similar idea, essentially a baby-step/giant-step technique, can be used to reduce the number of automorphisms needed to implement MatMul1D in dimension s from $O(D)$ to $O(\sqrt{D})$. See Sect. 6 for details.

This technique is distinct from the baby-step/giant-step key switching strategy discussed above in Sect. 2.5. However, for best results, the two techniques should be combined in a way that harmonizes the baby-step/giant-step thresholds.

4.2 Hoisting

As we have seen, in many situations, we want to compute $\psi(v)$ for a fixed ciphertext v and many automorphisms ψ. Assuming we have key switching matrices for each automorphism ψ, the dominant cost of computing all of these values is that of performing one key-switching operation for each ψ. Our "hoisting" technique is a method that refactors the computation, performing a pre-computation that only depends on v, and whose computational cost is roughly equivalent to a single key-switching operation. After performing this pre-computation, computing

$\psi(v)$ for any individual ψ is much faster than a single key-switching operation (typically, around 6–8 times faster). We describe this idea in more detail below in Sect. 5.

4.3   Better Key Switching Strategies in Bad Dimensions

Recall from Sect. 2.5 that with the "full" key-switching strategy, in a bad dimension, we stored key-switching matrices for the automorphisms θ^i, with $i = -(D-1), \ldots, -1, 1, \ldots, D-1$. To perform a rotation by i on v in the given dimension, we need to compute $\theta^i(v)$ and $\theta^{i-D}(v)$, and so with these key-switching matrices available, we need to perform two automorphisms and key switchings. However, we do not really need all of these negative-power key switching matrices. In fact, we can get by with key-switching matrices just for θ^i, with $i = 1, \ldots, D-1$, and for θ^{-D}. To perform a rotation by i on v in the given dimension, we can compute $w = \theta^i(v)$ and $\theta^{-D}(w) = \theta^{i-D}(v)$. So again, we need to perform two automorphisms and key switchings. This cuts the number of key-switching matrices in half without a significant increase in running time. Moreover, this key-switching strategy aligns well with the strategy discussed below for decoupling rotations and automorphisms in bad dimensions.

Similarly, for the baby-step/giant-step key-switching strategy in a bad dimension, we just store a key-switching matrix for θ^{-D}, rather than for all the negative "giant steps". This cuts down the number of key-switching matrices by a third. Moreover, the number of key switchings we need to perform per rotation is only 3 (instead of 4).

4.4   Decoupling Rotations and Automorphisms in Bad Dimensions

Recall that by Eq. (3), a rotation by i on a ciphertext v in a given bad dimension can be implemented as $\mu\theta^i(v) + (1-\mu)\theta^{i-D}(v)$, where μ is a "mask" (a constant with a 0 or 1 encoded in each slot). It turns out that in our matrix-vector computations, it is best to work directly with this implementation, and algebraically refactor the computation to improve both running time and noise. This refactoring exploits the fact that θ is an automorphism. See Sect. 6 for details.

4.5   A Horner-Like Rule with Application to a Minimal Key-Switching Strategy

We introduce a new key-switching strategy that reduces the storage requirements even further, to just 1, 2, or 3 key-switching matrices per dimension. This, combined with a simple algorithmic idea, allow us to implement a variant of the baby-step/giant-step multiplication strategy that does not run too much more slowly than when using the full or baby-step/giant-step key-switching strategy. To do this, we observe that if we need to compute $\sum_{i=0}^{h-1} \psi^i(v_i)$, where ψ is some automorphism and the v_i's are ciphertexts, we can do this using Horner's rule, provided we have a key-switching matrix just for ψ. Specifically, we can compute

$$\sum_{i=0}^{h-1} \psi^i(v_i) = \psi\big(\cdots\psi\big(\psi(v_{h-1}) + v_{h-2}\big) + \cdots\big) + v_0.$$

That is, we set $w_{h-1} \leftarrow v_{h-1}$, then $w_{i-1} \leftarrow \psi(w_i) + v_{h-1}$ for $i = h - 1, \ldots, 1$, and finally we output w_0.

4.6   Exploiting Multi-core Platforms

With the exception of the minimal key-switching strategy discussed above, all our other algorithms are very amenable to parallelization. We thus implemented them so as to exploit multiple cores, when available.

5   Hoisting

A ciphertext in HElib is a vector $v = (p_0, p_1) \in R_q^2$, with each "part" p_0, p_1 represented in a DoubleCRT format (i.e., both integer and polynomial CRT) [9]. We recall the steps in the computation of each $\psi(v)$, as implemented in HElib.

1. **Automorphism:** We first apply the automorphism to each part of v, computing $p_j' \leftarrow \psi(p_j)$ for $j = 0, 1$.

   Applying an automorphism to a DoubleCRT object is a fast, linear time operation, so this step is cheap. If $v = (p_0, p_1)$ decrypts to α under the secret key $\mathsf{sk} = (1, s)$, then $v' = (p_0', p_1')$ decrypts to $\psi(\alpha)$ under the secret key $\mathsf{sk}' = (1, \psi(s))$. We next have to perform a "relinearization" operation which converts v' back to a ciphertext that decrypts to $\psi(\alpha)$ under the original secret key sk. This operation can itself be broken down into two steps:

2. **Break into digits:** decompose p_1' into "small" pieces: $p_1' = \sum_k q_k' \Delta_k$. Here, the Δ_k's are integer constants, and the pieces q_k' are elements of R of small norm. This operation is rather expensive, as it requires conversions between DoubleCRT and coefficient representations of elements in R_q.

3. **Key switching:** compute the ciphertext $(p_0' + p_0'', p_1'')$, where

$$p_j'' = \sum_k q_k' A_{jk}, \quad (j = 0, 1).$$

Here, the A_{jk}'s are the "key switching matrices", namely, pre-computed elements in R_Q (for some larger Q) which are stored in the public key. The A_{jk}'s are stored in DoubleCRT format, so if we have the q_k' in the same DoubleCRT format then this operation is also a fast, linear time operation.

The key observation to our new technique is that we can reverse the order of the first two steps above, without affecting the correctness of the procedure. Namely our new procedure is as follows:

1. **Break into digits:** decompose the original p_1 *before applying the automorphism* into "small" pieces: $p_1 = \sum_k q_k \Delta_k$.
2. **Automorphism:** compute $p'_0 \leftarrow \psi(p_0)$, and $q'_k \leftarrow \psi(q_k)$ for each q_k. Namely, p'_0 is computed just as before, but we apply the automorphism to the pieces q_k from above rather than to p_1 itself.
3. **Key switching:** compute the ciphertext $(p'_0 + p''_0, p''_1)$, where

$$p''_j = \sum_k q'_k A_{jk}, \quad (j = 0, 1).$$

This is exactly the same computation as before.

The reasons that this works, is that (i) ψ is an automorphism (so it distributes over addition and multiplication), and (ii) applying ψ does not significantly change the norm of an element (cf. [10]). In a little more detail, correctness of the key-switching step depends only on the following two conditions on the q'_k's:

(a) $\sum_k q'_q \Delta_k = \psi(p_1)$, and
(b) the q'_k's have low norm.

Condition (a) is satisfied in our new procedure since ψ is an automorphism (which acts as the identity on integers), and so

$$\psi(p_1) = \psi\left(\sum_k q_k \Delta_k\right) = \sum_k \psi(q_k) \Delta_k = \sum_k q'_k \Delta_k.$$

Condition (b) is satisfied since the pieces q_k have small norm, and applying ψ to a ring element does not increase its norm significantly.

The new procedure is therefore just as effective as the old one, but now the expensive break-into-digits step can be preformed only once, as a pre-computation that depends only on v, rather than having to perform it for every automorphism ψ. The flip side is that we need to apply ψ to each one of the parts q_k instead of only once to p_1. But as we mentioned, this is a cheap operation.

5.1   Interaction with Key-Switching Strategy

If we want to compute $\psi(v)$ for various automorphisms ψ, and we have key-switching matrices for all of the ψ's. Then we can apply the above hoisting strategy directly. In some situations, what we want to do is compute $\theta^i(v)$ for $i = 0, \ldots, D - 1$, where $\theta = \theta_{g_s}$ for some dimension s with generator $g_s \in \mathbb{Z}_m^*$, and where $D = D_s$ is the size of the dimension. If we are employing the baby-step/giant-step strategy for storing key-switching matrices, then we do not have all of the requisite key-switching matrices, so we cannot use the hoisting strategy directly. Instead, what we can do is the following. Since we have key-switching matrices for all of the giant steps θ^{gj}, for $j = 1, \ldots, h - 1$, we can use hoisting to compute $\theta^{gj}(v)$ for all of the giant steps, and for each of these values, we perform the pre-computation (i.e., the break-into-digits step). Then, since we have key-switching matrices for all of the baby steps θ^k, for $k = 1, \ldots, g - 1$, we can compute any value $\theta^{gj+k}(v)$ as $\theta^k(\theta^{gj}(v))$, using the precomputed data for $\theta^{gj}(v)$ and the key-switching matrix for θ^k.

6   Algorithms for One-Dimensional Linear Transformations

In this section, we describe in detail our algorithms for applying one-dimensional linear transformations to a ciphertext v. We fix a dimension $s = 1, \ldots, n$. Recall from Sect. 2.3 that $\rho \overset{\text{def}}{=} \rho_s$ is the rotation-by-1 map in dimension s, and that $D \overset{\text{def}}{=} D_s$ is the size of dimension s.

6.1   Logic for Basic MatMul1D

Recall from Sect. 3.1 that for the MatMul1D calculation, we need to compute

$$w = \sum_{i \in [D]} \kappa(i) \rho^i(v),$$

where the $\kappa(i)$'s are constants in R_{p^r} that depend on the matrix.

If s is a good dimension, then ρ is realized with a single automorphism, $\rho = \theta \overset{\text{def}}{=} \theta_{g_s}$ where $g_s \in \mathbb{Z}_m^*$ is the generator for dimension s. We can easily implement this in a number of ways. For example, we can use the hoisting technique from Sect. 5 to compute all of the values $\theta^i(v)$ for $i \in [D]$. Alternatively, if we are using a minimal key-switching strategy (see Sect. 4.5), then with just a key-switching matrix for θ, we can compute the values $\theta^i(v)$ iteratively, computing $\theta^{i+1}(v)$ from $\theta^i(v)$ as $\theta(\theta^i(v))$.

6.2   Revised Logic for Bad Dimensions

From Eq. (3), if s is a bad dimension, then we have

$$\rho^i(v) = \mu(i)\theta^i(v) + \mu'(i)\theta^{i-D}(v), \tag{7}$$

where $\mu(i)$ is a "0–1 mask" and $\mu'(i) = 1 - \mu(i)$. As discussed in Sect. 4.4, it is useful to algebraically decouple the rotations and automorphisms in a bad dimension, which we can do as follows:

$$
\begin{aligned}
w &= \sum_{i \in [D]} \kappa(i)\rho^i(v) \\
&= \sum_{i \in [D]} \kappa(i)\{\mu(i)\theta^i(v) + \mu'(i)\theta^{i-D}(v)\} \\
&= \sum_{i \in [D]} \kappa'(i)\theta^i(v) \;+\; \theta^{-D}\left[\sum_{i \in [D]} \kappa''(i)\theta^i(v)\right],
\end{aligned}
$$

where

$$\kappa'(i) = \mu(i)\kappa(i) \quad \text{and}$$
$$\kappa''(i) = \theta^D\{\mu'(i)\kappa(i)\}.$$

To implement this, we have to compute $\theta^i(v)$ for all $i \in [D]$. This can be done using the same strategies as were discussed above in a good dimension, using either hoisting or iteration. The only other automorphism we need to compute is one evaluation of θ^{-D}. Note that with our new key-switching strategy (see Sect. 4.3), we always have available a key-switching matrix for θ^{-D}.

If we ignore the cost of pre-computing all the constants in DoubleCRT format, we see that the computational cost is roughly the same in both good and bad dimensions. This is because the time needed to perform all the constant-ciphertext multiplications is very small in comparison to the time needed to perform all the automorphisms. The cost in noise is also about the same, essentially, one constant-ciphertext multiplication.

6.3   Baby-Step/Giant-Step Logic

We now present the logic for a new baby-step/giant-step multiplication algorithm. As discussed above in Sect. 4.1, this idea is very similar to the Block-MatMul1D implementation described in [8]. Set $g = \lceil\sqrt{D}\rceil$ and $h = \lceil D/g\rceil$. We have:

$$w = \sum_{i\in[D]} \kappa(i)\rho^i(v)$$
$$= \sum_{j\in[g]}\sum_{k\in[h]} \kappa(j + gk)\rho^{j+gk}(v)$$
$$= \sum_{k\in[h]} \rho^{gk}\left[\sum_{j\in[g]} \kappa'(j + gk)\rho^j(v)\right],$$

where $\kappa'(j + gk) = \rho^{-gk}(\kappa(j + gk))$.

Algorithm 1. In a good dimension, where $\rho = \theta$, we can implement the above logic using the following algorithm.

1. For each $j \in [g]$, compute $v_j = \theta^j(v)$.
2. For each $k \in [h]$, compute

$$w_k = \sum_{j\in[g]} \kappa'(j + gk)v_j.$$

3. Compute

$$w = \sum_{k\in[h]} \theta^{gk}(w_k).$$

Step 1 of the algorithm can be implemented by hoisting, or if we are using a minimal key-switching strategy, by iteration. Also, if we employ the minimal key-switching strategy, then Step 3 can be implemented using the Horner-rule idea discussed in Sect. 4.5 — for this, we just need a key-switching matrix for θ^g. Otherwise, if we have key switching matrices for all of the ρ^{gk}'s, it is somewhat faster to apply all of these automorphisms independently, which is also amenable to parallelization.

6.4   Revised Baby-Step/Giant-Step Logic for Bad Dimensions

Set $g = \lceil \sqrt{D} \rceil$ and $h = \lceil D/g \rceil$. Again, using Eq. (7), and the idea of algebraically decoupling the rotations and automorphisms in a bad dimension, we have:

$$
\begin{aligned}
w &= \sum_{i\in[D]} \kappa(i)\rho^i(v) \\
&= \sum_{i\in[D]} \kappa(i)\{\mu(i)\theta^i(v) + \mu'(i)\theta^{i-D}(v)\} \\
&= \sum_{j\in[g]}\sum_{k\in[h]} \kappa(j+gk)\{\mu(j+gk)\theta^{j+gk}(v) + \mu'(j+gk)\theta^{j+gk-D}(v)\} \\
&= \sum_{k\in[h]} \theta^{gk}\left[\sum_{j\in[g]}\{\kappa'(j+gk)\theta^j(v) + \kappa''(j+gk)\theta^{j-D}(v)\}\right],
\end{aligned}
$$

where

$$
\kappa'(j+gk) = \theta^{-gk}\{\mu(j+gk)\kappa(j+gk)\} \quad \text{and}
$$
$$
\kappa''(j+gk) = \theta^{-gk}\{\mu'(j+gk)\kappa(j+gk)\}.
$$

Based on this, we derive the following:

Algorithm 2.

1. Compute $v' = \theta^{-D}(v)$.
2. For each $j \in [g]$, compute $v_j = \theta^j(v)$ and $v'_j = \theta^j(v')$
3. For each $k \in [h]$, compute

$$
w_k = \sum_{j\in[g]} \{\kappa'(j+gk)v_j + \kappa''(j+gk)v'_j\}.
$$

4. Compute

$$
w = \sum_{k\in[h]} \theta^{gk}(w_k).
$$

Step 2 of the algorithm can be implemented by hoisting, or if we are using a minimal key-switching strategy, by iteration. Also, if we employ the minimal key-switching strategy, then Step 4 can be implemented using Horner's rule.

As before, if we have key switching matrices for all of the ρ^{gk}'s, it is somewhat faster to apply all of these automorphisms independently, which is also amenable to parallelization.

Based on experimental data, we find that using the baby-step/giant-step multiplication algorithms are faster in dimensions for which we are using a baby-step/giant-step key-switching strategy. Moreover, even if we are using the full key-switching strategy, and we have all key-switching matrices for that dimensions available, the baby-step/giant-step multiplication algorithms are still faster in very large dimensions (say, on the order of several hundred).

6.5 Alternative Revised Baby-Step/Giant-Step Logic for Bad dimensions

We considered, implemented, and tested an alternative algorithm, which was found to be slightly slower and was hence disabled. It proceeds as follows: Set $g = \lceil\sqrt{D}\rceil$ and $h = \lceil D/g\rceil$.

$$w = \sum_{i\in[D]} \kappa(i)\rho^i(v)$$

$$= \sum_{i\in[D]} \kappa(i)\{\mu(i)\theta^i(v) + \mu'(i)\theta^{i-D}(v)\}$$

$$= \sum_{j\in[g]}\sum_{k\in[h]} \kappa(i)\{\mu(j+gk)\theta^{j+gk}(v) + \mu'(j+gk)\theta^{j+gk-D}(v)\}$$

$$= \sum_{k\in[h]} \theta^{gk}\left[\sum_{j\in[g]} \kappa'(j+gk)\theta^j(v)\right] + \theta^{-D}\left(\sum_{k\in[h]} \theta^{gk}\left[\kappa''(j+gk)\theta^j(v)\right]\right),$$

where

$$\kappa'(j+gk) = \theta^{-gk}\{\mu(j+gk)\kappa(j+gk)\} \text{ and}$$
$$\kappa''(j+gk) = \theta^{D-gk}\{\mu'(j+gk)\kappa(j+gk)\}.$$

Based on this, we derive the following:

Algorithm 3.

1. For each $j \in [g]$, compute $v_j = \theta^j(v)$
2. For each $k \in [h]$, compute

$$u_k = \sum_{j\in[g]} \kappa'(j+gk)v_j \text{ and} u'_k = \sum_{j\in[g]} \kappa''(j+gk)v'_j.$$

3. Compute

$$u = \sum_{k\in[h]} \theta^{gk}(u_k) \text{ and} u' = \sum_{k\in[h]} \theta^{gk}(u'_k).$$

4. Compute

$$w = u + \theta^{-D}(u').$$

6.6   BlockMatMul1D Logic

Recall from Sect. 3.2 that for the BlockMatMul1D calculation, we need to compute

$$w = \sum_{j \in [d]} \sum_{i \in [D]} \kappa(i,j) \sigma^j (\rho^i(v))$$

$$= \sum_{j \in [d]} \sigma^j \left[\sum_{i \in [D]} \kappa'(i,j) \rho^i(v) \right],$$

where $\kappa'(i,j) = \sigma^{-j}(\kappa(i,j))$. Here, σ is the Frobenius automorphism. This strategy is very similar to the baby-step/giant-step strategy used for the MatMul1D computation.

Algorithm 4. In a good dimension, where $\rho = \theta$, we can implement the above logic using the following algorithm.

1. Initialize an accumulator $w_j = 0$ for each $j \in [d]$.
2. For each $i \in [D]$:
   (a) compute $v_i = \theta^i(v)$;
   (b) for each $j \in [d]$, add $\kappa'(i,j)v_i$ to w_j.
3. Compute

$$w = \sum_{j \in [d]} \sigma^j(w_j).$$

Step 2(a) of the algorithm can be implemented by hoisting, or if we are using a minimal key-switching strategy, by iteration. Also, if we employ the minimal key-switching strategy, then Step 3 can be implemented using Horner's rule, using just a key-switching matrix for σ. If we have key switching matrices for all of the σ^j's, it is somewhat faster to apply all of these automorphisms independently, which is also amenable to parallelization.

Often, D is much larger than d. Assuming we are using the hoisting technique in Step 2(a), it is much faster to perform Step 2(a) on the dimension of larger size D, and to perform Step 3 on the dimension of smaller size d. Indeed, the amortized cost of computing each of the d automorphisms in Step 3 is much greater than the amortized cost of computing each of the D automorphisms (via hoisting) in Step 2(a). Note that in our actual implementation, if it turns out that D is in fact smaller than d, then we switch the roles of θ and σ.

Observe that we store d accumulators $w_0, \ldots, w_{d-1}$, rather than store the intermediate values $v_0, \ldots, v_{D-1}$. Either strategy would work, but assuming D is much larger than d, we save space with this strategy (even though it is slightly more challenging to parallelize).

6.7   Revised BlockMatMul1D Logic for Bad Dimensions

Again, using Eq. (7) and the idea of algebraically decoupling rotations and automorphism, we have:

$$w = \sum_{j \in [d]} \sum_{i \in [D]} \kappa(i,j)\sigma^j(\rho^i(v))$$

$$w = \sum_{j \in [d]} \sum_{i \in [D]} \kappa(i,j)\sigma^j \left\{ \mu(i)\theta^i(v) + \mu'(i)\theta^{i-D}(v) \right\}$$

$$= \sum_{j \in [d]} \sigma^j \left[\sum_{i \in [D]} \kappa'(i,j)\theta^i(v) \right] + \theta^{-D} \left(\sum_{j \in [d]} \sigma^j \left[\sum_{i \in [D]} \kappa''(i,j)\theta^i(v) \right] \right),$$

where

$$\kappa'(i,j) = \sigma^{-j}(\kappa(i,j))\mu(i) \quad \text{and}$$
$$\kappa''(i,j) = \theta^D \left\{ \sigma^{-j}(\kappa(i,j))\mu'(i) \right\}.$$

Based on this, we derive the following:

Algorithm 5.

1. Initialize accumulators $u_j = 0$ and $u'_j = 0$ for each $j \in [d]$.
2. For each $i \in [D]$:
   (a) compute $v_i = \rho^i(v)$;
   (b) for each $j \in [d]$, add $\kappa'(i,j)v_i$ to u_j and add $\kappa''(i,j)v_i$ to u'_j
3. Compute

$$u = \sum_{j \in [d]} \sigma^j(u_j) \quad \text{and} \quad u' = \sum_{j \in [d]} \sigma^j(u'_j).$$

4. Compute

$$w = u + \theta^{-D}(u').$$

As above, Step 2(a) of the algorithm can be implemented by hoisting, or if we are using a minimal key-switching strategy, by iteration. Also, if we employ the minimal key-switching strategy, then Step 3 can be implemented using Horner's rule, using just a key-switching matrix for σ. Again, if it turns out that D is in fact smaller than d, then we switch the roles of θ and σ.

7   Algorithms for Arbitrary Linear Transformations

So far, we have described algorithms for applying one-dimensional linear transformations to an encrypted vector, that is, E- or $\mathbb{Z}_{p^r}$-linear transformations that act independently on the hypercolumns in a single dimension (i.e., the MatMul1D and BlockMatMul1D operations introduced in Sect. 3). Many of the techniques we have introduced can be adapted to arbitrary linear transformations. However, from a software design point of view, we adopted a strategy of designing a simple reduction from the general case to the one-dimensional case. For some parameter settings, this approach may not be optimal, but it is almost always much faster than the previous implementations of these operations in HElib.

We first consider the MatMulFull operation, which applies a general E-linear transformation to an encrypted vector. Here, an encrypted vector is a ciphertext whose corresponding plaintext is a vector with $\ell = \phi(m)/d$ slots. One can easily extend the MatMulFull operation to E-linear transformations on larger encrypted vectors that comprise several ciphertexts, although we have not yet implemented such an extension.

Recall from Sect. 2.3 that $\ell = D_1 \cdots D_n$, where for $s = 1, \ldots, n$, the size of dimension s is D_s, and ρ_s is the rotation-by-1 map on dimension s. In [7], it was observed that we can apply the MatMulFull operation to a ciphertext v by using a generalization of the simple rotation strategy we presented above in Eq. (4). More specifically, if T is an E-linear transformation on R_{p^r}, then for every $v \in R_{p^r}$, we have

$$T(v) = \sum_{i_1 \in [D_1]} \cdots \sum_{i_n \in [D_n]} \kappa_{i_1, \ldots, i_n} \cdot (\rho_n^{i_n} \cdots \rho_1^{i_1})(v), \tag{8}$$

where the $\kappa_{i_1, \ldots, i_n}$'s are constants in R_{p^r} determined by the linear transformation. For each $(i_1, \ldots, i_{n-1})$, there is a one-dimensional E-linear transformation $T'_{i_1, \ldots, i_{n-1}}$ that acts on dimension n, such that for every $w \in R_{p^r}$, we have

$$T'_{i_1, \ldots, i_{n-1}}(w) = \sum_{i_n \in [D_n]} \kappa_{i_1, \ldots, i_n} \cdot (\rho_n^{i_n} \cdots \rho_1^{i_1})(w).$$

Therefore, we can refactor Eq. (8) as follows:

$$T(v) = \sum_{i_1 \in [D_n]} \cdots \sum_{i_{n-1} \in [D_{n-1}]} T'_{i_1, \ldots, i_{n-1}} \{ (\rho_{n-1}^{i_{n-1}} \cdots \rho_1^{i_1})(v) \}. \tag{9}$$

To implement Eq. (9), we compute all of the rotations $(\rho^{i_{n-1}} \cdots \rho^{i_1})(v)$ using a simple recursive algorithm. The main type of operation performed here is to compute all of the rotations $\rho_s^{i_s}(w)$ for a given w, a given dimension, and for all $i_s \in [D_s]$. In a good dimension, where $\rho_s = \theta_{g_s}$, we can use hoisting (see Sect. 5) to speed things up, provided the required key-switching matrices are available, or sequentially if not. For bad dimensions, we can use the decoupling idea discussed in Sect. 4.4. Specifically, using Eq. (7), if $\theta \stackrel{\text{def}}{=} \theta_{g_s}$, then

$$\rho_s^{i_s}(w) = \mu_{i_s} \theta^{i_s}(w) + (1 - \mu_{i_s}) \theta^{i_s - D_s}$$

for an appropriate mask μ_{i_s}. Then we can compute $w' = \theta^{-D_s}$, which requires a single key-switching using our new key-switching strategy (see Sect. 4.3). After this, we need to compute $\theta^{i_s}(w)$ and $\theta^{i_s}(w')$ for all $i_s \in [D_s]$, which again, can be done by hoisting or iteration, as appropriate.

The other main type of operation needed to implement Eq. (9) is the application of all of the one-dimensional transformations $T'_{i_1,\ldots,i_{n-1}}$ in dimension n, for which we can use our improved implementation of MatMul1D.

The speedup over the previous implementation in HElib will be roughly equal to the speedup of our new implementation of MatMul1D in dimension n. So to get the best performance, our implementation orders the dimensions so that D_n is the largest dimension size. If dimension n is a bad dimension, we also save on noise as well (we save noise equal to that of one constant-ciphertext multiplication). In many applications, it is desirable to choose parameters so that there is one very large dimension, and zero, one, or two very small dimensions — indeed, by default, HElib will choose parameters in this way. In this typical setting, the speedup for MatMulFull will be very significant.

Finally, we mention that the above techniques carry over in an obvious way to general $\mathbb{Z}_{p^r}$-linear transformations on R_{p^r}. As above, there is a simple reduction from the general BlockMatMulFull operation to the one-dimensional Block-MatMul1D operation. The previous implementation of BlockMatMulFull was not particularly well optimized, and because of this, the speedup we get is roughly equal to n times the speedup of our implementation of BlockMatMul1D, where, again, n is the number of dimensions in the underlying hypercube.

8 Application to "thin" Bootstrapping

HElib implements a general bootstrapping algorithm, which will convert an arbitrary noisy ciphertext into an equivalent ciphertext with less noise. However, in some applications, ciphertexts are not completely arbitrary. Recall that plaintexts can be viewed as vectors of slots, where each slot contains an element of $E = \mathbb{Z}_{p^r}[\zeta]$, where ζ is a root of a polynomial over $\mathbb{Z}_{p^r}$ of degree d. In some applications, one sometimes works with "thin" plaintexts, where the slots contain "constants", i.e., elements of the subring $\mathbb{Z}_{p^r}$ of E.

One could of course apply the HElib bootstrapping algorithm directly to such "thin" ciphertexts, but that would be quite wasteful. We can get more efficient implementation (in an amortized sense) by bootstrapping "batches" of d ciphertexts at a time: We can take d thin ciphertexts, pack them together to form a single ciphertext where each slot is fully packed, bootstrap this fully packed ciphertext, and then unpack it back to d thin ciphertexts. This approach, however, is only applicable when we have many ciphertexts to bootstrap, and it is not very convenient from a software engineering perspective. Moreover it also introduces some additional noise in the packing/unpacking steps.

Recently, Chen and Han devised an approach for more efficient and direct bootstrapping of thin ciphertexts [4], and we adapted their approach to HElib. We combined Chen and Han's ideas with numerous optimizations for the linear algebra part of the bootstrapping from [8], reducing the bulk computation to a sequence of MatMul1D operations, where our improved algorithms for these operations yield great performance dividends. We implemented this new thin bootstrapping, and report on its performance below in Sect. 9.

Let us review the bootstrapping procedure of [8], which has been implemented in HElib, and then outline how to adapt it to incorporate Chen and Han's technique.

A plaintext element $\alpha \in R_{p^r}$ can be viewed in a couple of different ways. It can be viewed as a vector of plaintext slots:

$$\alpha = \left(\sum_j a_{1j}\zeta^j, \ldots, \sum_j a_{\ell j}\zeta^j \right),$$

where the a_{ij}'s are scalars in $\mathbb{Z}_{p^r}$. Here, $\sum_j a_{ij}\zeta^j \in E$ is the content of the ith slot of α. For a thin plaintext, only the a_{i0}'s are non-zero elements in E.

The above representation corresponds to some $\mathbb{Z}_{p^r}$-basis of R_{p^r}, namely $\alpha = \sum_{ij} a_{ij}\lambda_{ij}$ (with $\lambda_{ij} \in R_{p^r}$ being the element with ζ^j in the ith slot and zero elsewhere). But we can express the same α on an arbitrary $\mathbb{Z}_{p^r}$-basis $\{\beta_{ij}\}$ of R_{p^r},

$$\alpha = \sum_{ij} b_{ij}\beta_{ij} \quad (\text{where } b_{ij} \in \mathbb{Z}_{p^r}).$$

For example, for the *power basis*, the β_{ij}'s are powers of X modulo $(p^r, \phi_m(X))$. As it turns out, for bootstrapping it is more convenient to use the *powerful basis*, introduced by Lyubashevsky et al. [11,12] and developed further by Alperin-Sheriff and Peikert [1]. The bootstrapping algorithms in HElib make use of the powerful basis, as it allows us to decompose the required linear transformations into a sequence of one-dimensional linear transformations.

Here is a rough outline of HElib's bootstrapping procedure for fully packed ciphertexts. We start out with a ciphertext encrypting a plaintext $\beta = \sum_{ij} b_{ij}\beta_{ij}$.

1. Perform a modulus switching and homomorphic inner product, obtaining a ciphertext with very little noise that encrypts some $\beta^* = \sum_{ij} b_{ij}^*\beta_{ij}$. The b_{ij}^* coefficients are actually in $\mathbb{Z}_{p^s}$ for some $s > r$, and have the property that there is a (non-linear) "digit extraction" procedure that computes b_{ij} from b_{ij}^*. (In more detail, it computes $b_{ij} = \lfloor b_{ij}^*/p^{s-r} \rfloor$.)
2. Perform a linear "coefficient to slot" operation that transforms the ciphertext encrypting β^* to one encrypting $\alpha^* = \left(\sum_j b_{1j}^*\zeta^j, \ldots, \sum_j b_{\ell j}^*\zeta^j \right)$.
3. Unpack the ciphertext encrypting α^* into d thin ciphertexts, where for $j = 0, \ldots, d-1$, the jth unpacked ciphertext encrypts $(b_{1j}^*, \ldots, b_{\ell j}^*)$.
4. Apply the above-mentioned "digit extraction" procedure to each unpacked thin ciphertext, obtaining d thin ciphertexts, where the jth ciphertext is an encryption of $(b_{1j}, \ldots, b_{\ell j})$.
5. Repack the thin ciphertexts from the previous step, obtaining an encryption of $\alpha = \left(\sum_j b_{1j}\zeta^j, \ldots, \sum_j b_{\ell j}\zeta_j \right)$.
6. Perform a linear "slot to coefficient" operation, which is the inverse of the "coefficient to slot" operation in Step 2, to transform the encryption of α in the previous step to an encryption of $\beta = \sum_{ij} b_{ij}\beta_{ij}$.

By careful usage of the powerful basis for $\{\beta_{ij}\}_{ij}$, each of the linear operations, "coefficient to slot" and "slot to coefficient", can be implemented using one BlockMatMul1D operation and a small number (typically one or two) Mat-Mul1D operations. More specifically, the "slot to coefficient" transformation L can be decomposed as $L = L_t \cdots L_2 L_1$, where L_1 is a one-dimensional $\mathbb{Z}_{p^r}$-linear transformation (i.e., a BlockMatMul1D operation), and $L_2, \ldots, L_n$ are one-dimensional E-linear transformations (i.e., MatMul1D operations). The inverse "coefficient to slot" transformation can therefore also be decomposed as $L^{-1} = L_1^{-1} L_2^{-1} \cdots L_n^{-1}$. See [8] for details of the definitions of the maps $L_1, \ldots, L_n$.

We now review Chen and Han's technique from [4], adapted to HElib's strategy to dealing with linear transformations. We start with a ciphertext encrypting a thin plaintext $\alpha = (a_{10}, \ldots, a_{\ell 0})$.

1. First apply the "slot to coefficient" transformation, obtaining an encryption of $\beta = \sum_i a_{i0} \beta_{i0}$.
2. Perform the modulus switching and homomorphic inner product, obtaining a ciphertext with less noise that encrypts $\beta^* = \sum_{ij} a_{ij}^* \beta_{ij}$.
3. Apply the "coefficient to slot" transformation, which places $\sum_{ij} a_{ij}^* \zeta^j$ in the ith slot, followed by a slot-wise projection function π that maps each $\sum_{ij} a_{ij}^* \zeta^j$ to a_{i0}^*, obtaining a ciphertext that encrypts $\alpha^* = (a_{10}^*, \ldots, a_{\ell 0}^*)$.
4. Apply the "digit extraction" procedure, obtaining a ciphertext encrypting $\alpha = (a_{10}, \ldots, a_{\ell 0})$.

Clearly, this procedure only performs a single digit extraction operation, versus the d digit extraction operations that are required for fully packed bootstrapping.

As another benefit, observe that in Step 1 we are applying the linear transformation $L = L_t \cdots L_2 L_1$ to a thin plaintext. It turns out, that the restriction of L_1 to the subspace of thin plaintexts is in fact an E-linear transformation (this is easily seen from the definition of L_1 in [8]). Therefore, we can implement L_1 as a MatMul1D operation, rather than as a more expensive BlockMatMul1D operation. (The other transformations $L_2, \ldots, L_n$ are already implemented as MatMul1D operations.)

Moreover, in Step 3, we are computing

$$\pi L^{-1} = (\pi L_1^{-1}) L_2^{-1} \cdots L_n^{-1}.$$

We can rewrite πL_1^{-1} as τK, where τ is the slot-wise trace map and K is a certain E-linear transformation derived from L_1^{-1}. The trace map τ_E on E sends $\eta \in E$ to $\sum_{j=0}^{d-1} \sigma_E^j(\eta)$, where σ_E is the Frobenius map on E. The decomposition of πL_1^{-1} as τK follows from the general fact that for every $\mathbb{Z}_{p^r}$-linear map M from E to $\mathbb{Z}_{p^r}$, there exists $\lambda_M \in E$ such that $M(\eta) = \tau_E(\lambda_M \eta)$ for all $\eta \in E$. Indeed, L_1^{-1} can be represented by a matrix whose entries are themselves $\mathbb{Z}_{p^r}$-linear maps on E, and so πL_1^{-1} can be represented by a matrix whose entries are $\mathbb{Z}_{p^r}$-linear maps from E to $\mathbb{Z}_{p^r}$. If we replace each such map M with the multiplication-by-λ_M map, we obtain the matrix for the E-linear map K, and we have $\pi L_1^{-1} = \tau K$.

Thus, we can implement πL_1^{-1} using one MatMul1D operation and one application of the slot-wise trace map τ. We can quickly compute the slot-wise trace using one of several strategies. If we have key switching matrices for σ^j, for all $j = 1, \ldots, d-1$, where $\sigma \overset{\text{def}}{=} \theta_p$, we can compute the trace of a ciphertext v via hosting by first computing $\sigma^j(v)$ for $j = 0, \ldots, d-1$, and then adding these up. Alternatively, if $v^{(s)} \overset{\text{def}}{=} \sum_{j=0}^{s-1} \sigma^j(v)$, we can the relation $v^{(s+t)} = \sigma^t(v^{(s)}) + v^{(t)}$. If we are using the baby-step/giant-step key switching strategy, then we can compute the trace of v using $O(\log d)$ key-switching operations via a "repeated doubling" computation strategy. If we are using the minimal key-switching strategy, we can use this same relation to compute the trace of v using $O(\sqrt{d})$ key-switching operations via a baby-step/giant-step computation strategy; for this to work, we just need key-switching matrices for σ and σ^g, where $g \approx \sqrt{d}$.

9 Timings

We now present some timing data that demonstrates the effectiveness of our new techniques. All of our testing was done on a machine with an Intel Xeon CPU, E5-2698 v3 @2.30 GHz (which is a Haswell processor), featuring 32 cores and 250 GB of main memory. The compiler was GCC version 4.8.5, and we used NTL version 10.5.0 and GMP version 6.0.

Table 1 shows the running time (in seconds) for the old default behavior ("old def") and the new default behavior ("new def") for MatMul1D computations (see Sect. 3.1). We do this for various values of m defining a cyclotomic polynomial of degree $\phi(m)$. The quantity d is the order of p mod m (which represents the "size" of each slot), while the quantity D is the size of the dimension. We worked with plaintext spaces modulo $p^r = 2$ in all of these examples. A value of D marked with "*" denotes a "bad" dimension. Table 1 does not show the time taken to build the constants associated with a matrix or to convert them to DoubleCRT representation. One sees that for the large dimension of size 682 (which is a typical size for many applications), we get a speedup of 30 if it is a good dimension, and a speedup of 75 if it is bad. Speedups for smaller dimensions are less dramatic, but still quite significant.

Table 2 shows more detailed information on various implementation strategies, as well as the cost of precomputing matrix constants. The "build" column shows the time to build the constants associated with the matrix in a polynomial representation. The "conv" column shows that time required to convert these constants to DoubleCRT representation. The following columns show that time required to perform the matrix-vector multiplication, based on a variety of key switching and algorithmic strategies. The columns are labeled as "[MBF]/[BF][HN]", where

MBF: **M** is for Min KS strategy, **B** is for Baby-step/giant-step key-switching strategy, **F** is for Full key-switching strategy,

BF: **B** is for Baby-step/giant-step multiplication strategy, **F** is for Full multiplication strategy,

HN: **H** is for Hoisting, **N** is for No hoisting.

As one can see from the data, the cost of converting constant to DoubleCRT representation can easily exceed the cost of the remaining operations, so it is essential that these conversions are done as precomputations, if at all possible.

Consider the first line in Table 2. Column B/BH represents the default behavior: baby-step/giant-step key switching (since it is a large dimension of size 682), baby-step/step-step multiplication, and hoisting (only the baby steps are subject to hoisting). The next column (B/BN) is the same, except the baby steps are not hoisted, which is why it is slower. Column B/FH shows what happens if we do not use baby-step/giant-step multiplication, and rely exclusively on hoisting (as in Sect. 5.1). One can see that for such a large dimension, this is not an optimal strategy. Column M/B shows what happens when we use the minimal key switching strategy (with baby-step/step-step multiplication). Even though it needs only two key switching matrices (rather than about 50), it is less than twice as slow as the best strategy (although it does not parallelize very well). The algorithm represented by column B/FN corresponds directly to the algorithm originally implemented in HElib. The next line in the table represents a bad dimension. We note that for bad dimensions, the algorithm originally implemented in HElib is about twice as slow as the one represented by column B/FN (this is why the timing data in Table 1 for bad dimensions is not equal to the numbers in column B/FN of Table 2).

Table 3 shows corresponding timing data for BlockMatMul1D computations (see Sect. 3.2). For good dimensions, the previous implementation in HElib roughly corresponds to the non-hoisting strategy in our new implementation. So one can see that with hoisting we get a speedup of up to 4 times over the previous implementation for large dimensions (but only about 1.5 for small dimensions). For large, bad dimensions, in the previous implementation in HElib, the running time will be close to twice that of the non-hoisting strategy in our new implementation; therefore, the speedup in such dimensions is close to a factor of about 5.

Table 4 shows the effectiveness of parallelization using multiple cores. We show times for both MatMul1D and BlockMatMul1D, using 1, 4, and 16 threads. These times are for the default strategies, and do not show the time required to build the matrix constants or convert them to DoubleCRT representation. While the speedups do not quite scale linearly with the number of cores, they are clearly significant, with 16 cores yielding roughly an 8× speedup in large dimensions and 4× speedup in small ones.

We do not present detailed results for the running times of our new implementation of MatMulFull and BlockMatMulFull, discussed in Sect. 7. However, our experiments indicate that the speedups predicted in Sect. 7 closely align with practice: the speedup for MatMulFull is about the same as our speedup for MatMul1D in the largest dimension; the speedup for BlockMatMulFull is roughly our speedup for BlockMatMul1D in the largest dimension, times the number of dimensions in the hypercube.

Finally, we present some timing results to demonstrate the efficacy of our new algorithms in the context of bootstrapping, as discussed in Sect. 8. We chose large parameters that demonstrate well the potential saving with our new implementation. Specifically, we used $m = 49981$ and $p^r = 2$, for which we have $\phi(m) = 49500$ and $d = 30$. The hypercube structure for $\mathbb{Z}_m^*/(p^r)$ has two dimensions, one of size 150 and one of size 11, for a total of 1650 slots. We note that most parameter choices in [8] attempted to balance the size of the different dimensions, specifically because the linear transformations would take too long otherwise. One of the benefits of our faster algorithms is thus to free us from having to consider that aspect; indeed, our timing shows that the linear transformations are now quite fast even for this "unbalanced" setting.

We ran our tests with ciphertexts with 55 "levels" (for an estimated security parameter of about 80). For these parameters, the bootstrapping procedure consumes about 10 levels, leaving about 45 levels for other computations. Table 5 shows the running time (in seconds) for both the thin bootstrapping and packed bootstrapping routines with both the old and new matrix multiplication algorithms. These results make it clear that for such large hypercubes, thin bootstrapping must be done using our new, faster matrix multiplication to be truly practical.

Table 1. MatMul1D: summary of old vs new, time in seconds

m	$\phi(m)$	d	D	old def	new def	speedup
15709	15004	22	682	69.28	2.22	31.20
15709	15004	22	682*	138.20	3.14	75.86
18631	18000	25	120	20.27	1.38	14.69
18631	18000	25	120*	39.97	1.69	23.65
24295	18816	28	42	3.18	0.51	6.24
24295	18816	28	42*	6.20	0.55	11.27

Table 2. Different strategies for MatMul1D, time in seconds

m	$\phi(m)$	d	D	build	conv	M/B	M/F	B/BH	B/BN	B/FH	B/FN	F/FH	F/FN
15709	15004	22	682	0.47	5.54	3.80	44.81	2.22	3.19	6.46	69.28	5.30	28.30
15709	15004	22	682*	0.56	11.07	5.93	44.86	3.14	5.03	7.33	69.70	5.94	29.16
18631	18000	25	120	0.08	1.96	2.43	13.81	1.38	2.04	2.36	20.27	1.29	8.70
18631	18000	25	120*	0.10	3.91	3.68	13.95	1.69	2.89	2.45	20.27	1.29	8.78
24295	18816	28	42	0.03	0.70	1.39	5.09	0.82	1.17	1.11	6.87	0.51	3.18
24295	18816	28	42*	0.04	1.39	2.17	5.09	0.95	1.64	1.20	6.94	0.55	3.20

Table 3. Different strategies for BlockMatMul1D, time in seconds

m	$\phi(m)$	d	D	build	conv	M/	B/H	B/N	F/H	F/N
15709	15004	22	682	15.47	122.62	54.73	21.03	84.42	18.15	42.67
15709	15004	22	682*	17.31	246.89	64.98	36.81	99.84	32.41	57.07
18631	18000	25	120	2.44	49.59	18.83	9.84	27.90	6.88	14.66
18631	18000	25	120*	2.96	98.79	23.83	17.62	35.80	12.73	20.58
24295	18816	28	42	0.95	19.73	9.25	7.84	13.64	5.01	7.70
24295	18816	28	42*	1.15	39.72	13.49	14.73	20.45	9.65	12.47

Table 4. Multithreading for MatMul1D/BlockMatMul1D, time in seconds

				MatMul1D			BlockMatMul1D		
m	$\phi(m)$	d	D	$nt = 1$	$nt = 4$	$nt = 16$	$nt = 1$	$nt = 4$	$nt = 16$
15709	15004	22	682	2.18	0.67	0.29	20.21	7.60	2.47
15709	15004	22	682*	3.14	0.97	0.42	35.50	12.17	4.70
18631	18000	25	120	1.35	0.49	0.20	7.97	2.49	1.03
18631	18000	25	120*	1.65	0.58	0.29	13.89	4.30	1.67
24295	18816	28	42	0.47	0.23	0.15	4.98	1.37	0.61
24295	18816	28	42*	0.51	0.22	0.14	9.51	2.67	1.08

Table 5. Bootstrapping, time in seconds

	old		new	
	total	linear	total	linear
thin bootstrap	474.18	428.76	80.31	36.17
packed bootstrap	2120.05	804.30	1413.02	102.65

References

1. Alperin-Sheriff, J., Peikert, C.: Practical bootstrapping in quasilinear time. In: Canetti, R., Garay, J.A. (eds.) CRYPTO 2013. LNCS, vol. 8042, pp. 1–20. Springer, Heidelberg (2013). https://doi.org/10.1007/978-3-642-40041-4_1
2. Brakerski, Z., Gentry, C., Vaikuntanathan, V.: Fully homomorphic encryption without bootstrapping. In: Innovations in Theoretical Computer Science (ITCS 2012) (2012). http://eprint.iacr.org/2011/277
3. Brakerski, Z., Gentry, C., Vaikuntanathan, V.: (Leveled) fully homomorphic encryption without bootstrapping. ACM Trans. Comput. Theory 6(3), 13 (2014)
4. Chen, H., Han, K.: Homomorphic lower digits removal and improved FHE bootstrapping. In: Nielsen, J.B., Rijmen, V. (eds.) EUROCRYPT 2018. LNCS, vol. 10820, pp. 315–337. Springer, Cham (2018). https://doi.org/10.1007/978-3-319-78381-9_12

5. Gentry, C.: Fully homomorphic encryption using ideal lattices. In: Proceedings of the 41st ACM Symposium on Theory of Computing - STOC 2009, pp. 169–178. ACM (2009)
6. Gentry, C., Halevi, S., Smart, N.P.: Fully homomorphic encryption with polylog overhead. In: Pointcheval, D., Johansson, T. (eds.) EUROCRYPT 2012. LNCS, vol. 7237, pp. 465–482. Springer, Heidelberg (2012). https://doi.org/10.1007/978-3-642-29011-4_28
7. Halevi, S., Shoup, V.: Algorithms in HElib. In: Garay, J.A., Gennaro, R. (eds.) CRYPTO 2014, Part I. LNCS, vol. 8616, pp. 554–571. Springer, Heidelberg (2014). https://doi.org/10.1007/978-3-662-44371-2_31
8. Halevi, S., Shoup, V.: Bootstrapping for HElib. In: Oswald, E., Fischlin, M. (eds.) EUROCRYPT 2015, Part I. LNCS, vol. 9056, pp. 641–670. Springer, Heidelberg (2015). https://doi.org/10.1007/978-3-662-46800-5_25
9. Halevi, S., Shoup, V.: HElib - an implementation of homomorphic encryption, September 2014. https://github.com/shaih/HElib/
10. Lyubashevsky, V., Peikert, C., Regev, O.: On ideal lattices and learning with errors over rings. In: Gilbert, H. (ed.) EUROCRYPT 2010. LNCS, vol. 6110, pp. 1–23. Springer, Heidelberg (2010). https://doi.org/10.1007/978-3-642-13190-5_1
11. Lyubashevsky, V., Peikert, C., Regev, O.: A toolkit for ring-LWE cryptography. In: Johansson, T., Nguyen, P.Q. (eds.) EUROCRYPT 2013. LNCS, vol. 7881, pp. 35–54. Springer, Heidelberg (2013). https://doi.org/10.1007/978-3-642-38348-9_3
12. Lyubashevsky, V., Peikert, C., Regev, O.: On ideal lattices and learning with errors over rings. J. ACM 60(6), 43 (2013). Early version in EUROCRYPT 2010
13. Rivest, R., Adleman, L., Dertouzos, M.: On data banks and privacy homomorphisms. In: Foundations of Secure Computation, pp. 169–177. Academic Press (1978)
14. Roman, S.: Field Theory, 2nd edn. Springer, New York (2006). https://doi.org/10.1007/0-387-27678-5
15. Smart, N.P., Vercauteren, F.: Fully homomorphic SIMD operations. Des. Codes Cryptogr. 71(1), 57–81 (2014). Early verion at http://eprint.iacr.org/2011/133

CAPA: The Spirit of Beaver Against Physical Attacks

Oscar Reparaz[1,2(✉)], Lauren De Meyer[1], Begül Bilgin[1], Victor Arribas[1], Svetla Nikova[1], Ventzislav Nikov[3], and Nigel Smart[1,4]

[1] KU Leuven, imec - COSIC, Leuven, Belgium
{oscar.reparaz,lauren.demeyer,begul.bilgin,victor.arribas,
svetla.nikova,nigel.smart}@esat.kuleuven.be
[2] Square Inc., San Francisco, USA
[3] NXP Semiconductors, Leuven, Belgium
venci.nikov@gmail.com
[4] University of Bristol, Bristol, UK

Abstract. In this paper we introduce two things: On one hand we introduce the Tile-Probe-and-Fault model, a model generalising the wire-probe model of Ishai et al. extending it to cover both more realistic side-channel leakage scenarios on a chip and also to cover fault and combined attacks. Secondly we introduce CAPA: a combined Countermeasure Against Physical Attacks. Our countermeasure is motivated by our model, and aims to provide security against higher-order SCA, multiple-shot FA and combined attacks. The tile-probe-and-fault model leads one to naturally look (by analogy) at actively secure multi-party computation protocols. Indeed, CAPA draws much inspiration from the MPC protocol SPDZ. So as to demonstrate that the model, and the CAPA countermeasure, are not just theoretical constructions, but could also serve to build practical countermeasures, we present initial experiments of proof-of-concept designs using the CAPA methodology. Namely, a hardware implementation of the KATAN and AES block ciphers, as well as a software bitsliced AES S-box implementation. We demonstrate experimentally that the design can resist second-order DPA attacks, even when the attacker is presented with many hundreds of thousands of traces. In addition our proof-of-concept can also detect faults within our model with high probability in accordance to the methodology.

1 Introduction

Side-channel analysis attacks (SCA) [41] are cheap and scalable methods to extract secrets, such as cryptographic keys or passwords, from embedded electronic devices. They exploit unintended signals (such as the instantaneous power consumption [42] or the electromagnetic radiation [24]) stemming from a cryptographic implementation. In the last twenty years, plenty of countermeasures to mitigate the impact of side-channel information have been developed. Masking [15,26] is an established solution that stands out as a provably secure yet practically useful countermeasure.

© International Association for Cryptologic Research 2018
H. Shacham and A. Boldyreva (Eds.): CRYPTO 2018, LNCS 10991, pp. 121–151, 2018.
https://doi.org/10.1007/978-3-319-96884-1_5

Fault analysis (FA) is another relevant attack vector for embedded cryptography. The basic principle is to disturb the cryptographic computation somehow (for example, by under-powering the cryptographic device, or by careful illumination of certain areas in the silicon die). The result of a faulty computation can reveal a wealth of secret information: in the case of RSA or AES, a single faulty ciphertext pair makes key recovery possible [10,48]. Countermeasures are essentially based on adding some redundancy to the computation (in space or time). In contrast to masking, the countermeasures for fault analysis are mostly heuristic and lack a formal background.

However, there is a tension between side-channel countermeasures and fault analysis countermeasures. On the one hand, fault analysis countermeasures require redundancy, which can give out more leakage information to an adversary. On the other hand, a device that implements first-order masking offers an adversary double the attack surface to insert a fault in the computation. A duality relation between SCA and FA was pointed out in [23]. There is clearly a need for a combined countermeasure that tackles both problems simultaneously.

In this work we introduce a new attack model to capture this combined attack surface which we call the *tile-probe-and-fault* model. This model naturally extends the wire-probe model of [34]. In the wire-probe model individual wires of a circuit may be targetted for probing. The goal is then to protect against a certain fixed set of wire-probes. In our model, inspired by modern processor designs, we allow whole areas (or tiles) to be probed, and in addition we add the possibility of the attacker inducing faults on such tiles.

Protection against attacks in the wire-probe model is usually done via masking; which is in many cases the extension of ideas from *passively secure* secret sharing based Multi-Party Computation (MPC) to the side-channel domain. It is then natural to look at *actively secure* MPC protocols for the extension to fault attacks. The most successful modern actively secure MPC protocols are in the SPDZ family [20]. These use a pre-processing or *preparation* phase to produce so called *Beaver triples*, named after Beaver [6]. These auxiliary data values, which will be explained later, are prepared either before a computation, or in a just-in-time manner, so as to enable an efficient protocol to be executed. This use of *prepared* Beaver triples also explains, partially, the naming of our system, CAPA (a Combined countermeasure Against Physical Attacks), since Capa is also the beaver spirit in Lakota mythology. In this mythology, Capa is the lord of domesticity, labour and *preparation*.

1.1 Previous Work

Fault Attack Models and Countermeasures: Faults models typically describe the characterization of an attacker's ability. That is, the fault model is constructed as a combination of the following: the precision of the fault location and time, the number of affected bits which highly depends on the architecture, the effect of the fault (flip/set/reset/random) and its duration (transient/permanent). Moreover, the fault can target the clock or power line, storage units, combinational or control logic.

When it comes to countermeasures, one distinguishes between protection of the algorithm on the one hand and protection of the device itself by using, for example, active or passive shields on the other. No countermeasure provides perfect security at a finite cost; it is the designer's responsibility to strive for a balance between high-level (algorithmic) countermeasures and low-level ones that work at the circuit level and complement each other. In this paper, we discuss the former.

One algorithmic technique is to replicate the calculation m times in either time or space and only complete if all executions return the same result [54]. This countermeasure has the important caveat that there are conceptually simple attacks, such as m identical fault injections in each execution, that break the implementation with probability one. However, it should be stated that these attacks are not trivial to mount in practice when the redundancy is in space.

A second method is to use an error correcting or detecting code [8,12,13,32, 35–39,46]. This means one performs all calculations on both data and checksum. A drawback is that error correcting/detecting codes only work in environments in which errors are randomly generated, as opposed to maliciously generated. Thus, a skilled attacker may be able to carefully craft a fault that results in a valid codeword and is thus not detected. A detailed cost comparison between error detection codes and doubling is given in [44].

Another approach is that of infective computation [25,43], where any fault injected will affect the ciphertext in a way that no secret information can be extracted from it. This method ensures the ciphertext can always be returned without the need for integrity checks. While infective methods are very efficient, the schemes proposed so far have all been broken [5].

Side-Channel Attack Models and Countermeasures: A side-channel adversary typically uses the noisy leakage model [55], where side-channel analysis (SCA) attacks are bounded by the statistical moment of the attack due to a limited number of traces and noisy leakages. Given enough noise and an independent leakage assumption of each wire, this model, when limited to the tth-order statistical moment, is shown to be comparable to the t-probing model introduced in [34], where an attacker is allowed to probe, receive and combine the noiseless information about t wires within a time period [21]. Finally, it has been shown in [4] that a (semi-)parallel implementation is secure in the tth-order bounded moment model if its complete serialization is secure at the t-probing model.

While the countermeasures against fault attacks are limited to resist only a small subset of the real-world adversaries and attack models, protection against side-channel attacks stands on much more rigorous grounds and generally scales well with the attacker's powers. A traditional solution is to use masking schemes [9,29,34,51,56,58,59] to implement a function in a manner in which higher-order SCA is needed to extract any secret information, *i.e.* the attacker must exploit the joint leakage of several intermediate values. Masking schemes are analogues of the passively secure threshold MPC protocols based on secret sharing. One can thus justify their defence by appealing to the standard MPC literature. In MPC, a number of parties can evaluate a function on shared data, even in the presence of adversaries amongst the computing parties. The maximum number of dishonest parties which can be tolerated is called the threshold. In an embedded scenario, the basic idea is that different parts of a chip simulate the parties in an MPC protocol.

Combining Faults and Side-Channels Models and Countermeasures. The importance of combined countermeasures becomes more apparent as attacks such as [2] show the feasibility of combined attacks. Being a relatively new threat, combined adversarial models lack a joint description and are typically limited to the combination of a certain side-channel model and a fault model independently.

One possible countermeasure against combined attacks is found in leakage resilient schemes [45], although none of these constructions provide *provable* security against FA. Typical leakage resilient schemes rely on a relatively simple and easy to protect key derivation function in order to update the key that is used by the cryptographic algorithm within short periods. That is, a leakage resilient scheme acts as a specific "mode of operation". Thus, it cannot be a drop-in replacement for a standard primitive such as the AES block cipher. The aforementioned period can be as short as one encryption per key in order to eliminate fault attacks completely. However, the synchronization burden this countermeasure brings, makes it difficult to integrate with deployed protocols.

There are a couple of alternative countermeasures proposed for embedded systems in recent years. In private circuits II [16,33], the authors use redundancy on top of a circuit that already resists SCA (private circuits I [34]) to add protection against FA. In ParTI [62], threshold implementations (TI) are combined with concurrent error detection techniques. ParTI naturally inherits the drawbacks of using an error correction/detection code. Moreover, the detectable faults are limited in hamming weight due to the choice of the code. Finally, in [63], infective computation is combined with error-preserving computation to obtain a side-channel and fault resistant scheme. However, combined attacks are not taken into account.

Given the above introduction, it is clear that both combined attack models and countermeasures are not mature enough to cover a significant part of the attack surface.

Actively Secure MPC. Modern MPC protocols obtain active security, *i.e.* security against malicious parties which can actively deviate from the protocol. By mapping such protocols to the on-chip side-channel countermeasures, we would be able to protect against an eavesdropping adversary that inserts faults into a subset of the simulated parties. An example of a practical attack that fits this model is the combined attack of Amiel et al. [2]. We place defences against faults on the same theoretical basis as defences against side-channels.

To obtain maliciously secure MPC protocols in the secret-sharing model, there are a number of approaches. The traditional approach is to use Verifiable Secret Sharing (VSS), which works in the information theoretic model and requires that strictly less than $n/3$ parties can be corrupt. The modern approach, adopted by protocols such as BODZ, SPDZ, Tiny-OT, MASCOT etc. [7,20,40,50], is to work in a full threshold setting (*i.e.* all but one party can be corrupted) and attach information theoretic MACs to each data item. This approach turns out to be very efficient in the MPC setting, apart from its usage of public-key primitives. The computational efficiency of the use of information theoretic MACs and the active adversarial model of SPDZ lead us to adopt this philosophy.

1.2   Our Contributions

Our contributions are threefold. We first introduce the *tile-probe-and-fault* model, a new adversary model for physical attacks on embedded systems. We then use the analogy between masking and MPC to provide a methodology, which we call CAPA, to protect against such a tile-probe-and-fault attacker. Finally, we illustrate that the CAPA methodology can be prototyped by describing specific instantiations of the CAPA methodology, and our experimental results.

Tile-probe-and-fault model. We introduce a new adversary model that expands on the wire-probe model and brings it closer to real-world processor designs. Our model is set in an architecture that mimics the actively secure MPC setting that inspires our countermeasures (see Fig. 1). Instead of individual wires at the foundation of the model, we visualize a separation of the chip (integrated circuit) into areas or tiles, consisting of not only many wires, but also complete blocks of combinational and sequential logic. Such tiled designs are inherent in many modern processor architectures, where the tiles correspond to "cores" and the wires correspond to the on-chip interconnect. This can easily be related to a standard MPC architecture where each tile behaves like a separate party. The main difference between our architecture and the MPC setting is that in the latter, parties are assumed to be connected by a complete network of authenticated channels. In our architecture, we know exactly how the wires are connected in the circuit.

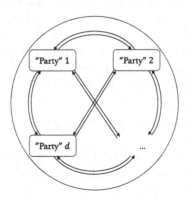

Fig. 1. Partition of the integrated circuit area into tiles, implementing MPC "parties"

The tile architecture satisfies the independent leakage assumption [21] amongst tiles. That is, leakage is local and thus observing the behaviour of a tile by means of probing, faulting or observing its side-channel leakage, does not give unintended information about another tile through, for example, coupling.

As the name implies, the adversary in our model exploits side-channels and introduces faults. We stress that our goal is to *detect* faults as opposed to *tolerate* or *correct* them. That is, if an adversary interjects a fault, we want our system to abort without revealing any of the underlying secrets.

CAPA Methodology. We introduce CAPA, a countermeasure against the tile-probe-and-fault-attacker, which is suitable for implementation in both hardware and software. CAPA inherits theoretical aspects of the MPC protocol SPDZ [20] by similarly computing on shared values, along with corresponding shared MAC tags. The former prevents the adversary from learning sensitive values, while the latter allows for detection of any faults introduced. Moreover, having originated from the MPC protocol SPDZ, CAPA is the first countermeasure with *provable* security against *combined* attacks. The methodology can be scaled to achieve an arbitrary fault detection probability and is suitable for implementation in both hardware and software.

Experimental Results. We provide examples of CAPA designs in hardware of the KATAN and AES block ciphers as well as a software bitsliced implementation of the AES S-box. Our designs show that our methodology is feasible to implement, and in addition our attack experiments confirm our theoretical claims. For example, we implemented a second-order secure hardware implementation of KATAN onto a Spartan-6 FPGA and perform a non-specific leakage detection test, which does not show evidence of first- or second-order leakage with up to 100 million traces. Furthermore, we deploy a second-order secure software based CAPA implementation of the AES S-box on an ARM Cortex-M4 and take electromagnetic measurements; for this implementation neither first-nor second-order leakage is detected with up to 200 000 traces. Using toy parameters, we verify our claimed fault detection probability for the AES S-box software implementation. It should be noted that our experimental implementations are to be considered only proof-of-concept; they are currently too expensive to be used in practice. But the designs demonstrate that the overall methodology can provide significant side-channel and fault protection, and they provide a benchmark against which future improvements can be measured.

2   The Tile-Probe-and-Fault Model

The purpose of this section is to introduce a new adversarial model in which our security guarantees are based. This model is strictly more powerful than the traditional DPA or DFA models.

Tile Architecture. Consider a partition of the chip in a number of tiles T_i, with wires running between each pair of tiles as shown in Fig. 1. We call the set of all tiles T. Each tile $T_i \in T$ possesses its own combinational logic, control logic (or program code) and pseudo-random number generator needed for the calculations of one share. In the abstract setting, we consider each tile as the set composed of all input and intermediate values on the wires and memory elements of those blocks. A probe-and-fault attacker may obtain, for a given subset of tiles, *all* the internal information at given time intervals on this set of tiles. He may also inject faults (known or random) into each tile in this set.

In our model, each sensitive variable is split into d shares through secret sharing. Without loss of generality, we use Boolean sharing in this paper.

We define each tile such that it stores and manipulates at most one share of each intermediate variable. Any wire running from one tile to another carries only blinded versions of a sensitive variables' share used by T_i. We make minimal assumptions on the security of these wires. Instead, we include all the information on the unidirectional wires in Fig. 1 in the tile on the *receiving* and not the *sending* end. We thus assume only one tile is affected by an integrity failure of a wire. We assume that shared calculations are performed in parallel without loss of generality. The redundancy of intermediate variables and logic makes the tiles completely independent apart from the communication through wires.

Probes. Throughout this work, we assume a powerful d_p-probing adversary where we give an attacker information about all intermediate values possessed by d_p specified tiles, *i.e.* $\cup_{i \in i_1,...,i_{d_p}} T_i$. The attacker obtains *all* the intermediate values on the tile (such as internal wire and register values) with probability one and obtains these values from the start of the computation until the end. Note that this is stronger than both the standard t-probing adversary which gives access to only t intermediate *values* within a certain amount of time [34] and ϵ-probing adversary where the information about t intermediate *values* is gained with certain probability. In our d_p-probing model, the adversary gets information from n intermediate values from d_p tiles where $n \gg d_p$. Therefore, our d_p-probing model is more generic and covers realistic scenarios including an attacker with a limited number of EM probes which enable observation of multiple intermediate values simultaneously within arbitrarily close proximity on the chip.

Faults. We also consider two types of fault models. Firstly, a d_f-faulting adversary which can induce chosen-value faults in any number of intermediate bits/values within d_f tiles, *i.e.* from the set $\cup_{i \in i_1,...,i_{d_f}} T_i$. These faults can have the nature of *either* flipping the intermediate values with a pre-calculated (adversarially chosen) offset *or* setting the intermediate values to a chosen fixed value. In particular, the faults are not limited in hamming weight. One can relate this type of faults with, for example, very accurate laser injections.

Secondly, we consider an ϵ-faulting adversary which is able to insert a random-value fault in *any* variable belonging to *any* party. This is a somehow new MPC model, and essentially means that all parties are randomly corrupted.

The ϵ-adversary may inject the random-value fault according to some distribution (for example, flip each bit with certain probability), but he cannot set all intermediates to a chosen fixed value. This adversary is different from the d_f-faulting adversary. One can relate the ϵ-faulting adversary to a certain class of non-localised EM attacks.

Time Periods. We assume a notion of time periods; where the period length is at least one clock cycle. We require that a d_f-fault to an adversarially chosen value cannot be preceded by a probe within the same time period. Thus adversarial faults can only depend on values from previous time periods. This time restriction is justified by practical experimental constraints; where the time period is naturally upper bounded by the time it takes to set up such a specific laser injection.

Adversarial Models. Given the aforementioned definitions, we consider on the one hand an active adversary $\mathscr{A}_1$ with both d_p-probing and d_f-faulting capabilities simultaneously. We define $\mathcal{P}_1$ the set of up to d_p tiles that can be probed and $\mathcal{F}_1$ the set of up to d_f tiles that can be faulted by $\mathscr{A}_1$. Since each tile potentially sees a different share of a variable and we use a d-sharing for each variable, we constrain the attack surface (the sets of adversarially probed and potentially modified tiles) as follows:

$$(\mathcal{F}_1 \cup \mathcal{P}_1) \subseteq \cup_{j=1}^{d-1} \mathcal{T}_{i_j}$$

The constraint implies that at least one share remains unaccessed/honest and thus $|\mathcal{F}_1 \cup \mathcal{P}_1| \leq d-1$. Within those $d-1$ tiles, the adversary can probe and fault arbitrarily many wires, including the wires *arriving* at each tile. The adversary's d_f-faulting capabilities are limited in time by our definition of time periods, which implies that any d_f-fault cannot be preceded by another probe within the same time period.

We also consider an active adversary $\mathscr{A}_2$ that has d_p-probing and ϵ-faulting capabilities simultaneously. In this case, the constraint on the set of probed tiles $\mathcal{P}_2$ remains the same:

$$\mathcal{P}_2 \subseteq \cup_{j=1}^{d-1} \mathcal{T}_{i_j}$$

but the set of faulted tiles is no longer constrained:

$$\mathcal{F}_2 \subseteq \mathcal{T}$$

Moreover, as ϵ-faults do not require the same set-up time as d_f-faults, they are not limited in time. Note that, ϵ-faults do not correspond to a standard adversary model in the MPC literature; thus this part of our model is very much an aspect of our side-channel and fault analysis focus. A rough equivalent model in the MPC literature would be for an honest-but-curious adversary who is able to replace the share or MAC values of honest players with values selected from a given random distribution. Whilst such an attack makes sense in the hardware model we consider, in the traditional MPC literature this model is of no interest due to the supposed isolated nature of computing parties.

As our constructions are based on MPC protocols which are *statically secure* we make the same assumptions in our tile-probe-and-fault model, *i.e.* the selection of tiles attacked must be fixed beforehand and cannot depend on information gathered during computation. This model reflects realistic attackers since it is infeasible to move a probe or a laser during a computation with today's resources. We thus assume that both adversaries $\mathscr{A}_1$ and $\mathscr{A}_2$ are static.

3   The CAPA Design

The CAPA methodology consists of two stages. A preprocessing step generates auxiliary data, which is used to perform the actual cryptographic operation in the evaluation step. We first present some notation, then the building blocks for the main evaluation, and finally the preprocessing components.

Notation. Although generalization to any finite field holds, in this paper we work over a field $\mathbb{F}_q$ with characteristic 2, for example $GF(2^k)$ for a given k, as this is sufficient for application to most symmetric ciphers. We use $\cdot$ and $+$ to describe multiplication and addition in $\mathbb{F}_q$ respectively. We use upper case letters for constants. The lower case letters x, y, z are reserved for the variables used only in the evaluation stage (e.g. sensitive variables) whereas $a, b, c, \ldots$ represent auxiliary variables generated from randomness in the preprocessing stage. The kronecker delta function is denoted by $\delta_{i,j}$. We use $L(.)$ to denote an additively homomorphic function and $A(.) = C + L(.)$ with C some constant.

Information Theoretic MAC Tags and the MAC Key α. We represent a value $a \in \mathbb{F}_q$ (similarly $x \in \mathbb{F}_q$) as a pair $\langle a \rangle = (\boldsymbol{a}, \boldsymbol{\tau}^a)$ of data and tag shares in the masked domain. The data shares $\boldsymbol{a} = (a_1, \ldots, a_d)$ satisfy $\sum a_i = a$. For each $a \in \mathbb{F}_q$, there exists a corresponding MAC tag τ^a computed as $\tau^a = \alpha \cdot a$, where α is a MAC key, which is secret-shared amongst the tiles as $\alpha = \sum \alpha_i$.

Analogously to the data, the MAC tag is shared $\boldsymbol{\tau}^a = (\tau_1^a, \ldots, \tau_d^a)$, such that it satisfies $\sum \tau_i^a = \tau^a$, but the MAC key itself does not carry a tag. Depending on a security parameter m, there can be m independent MAC keys $\alpha[j] \in \mathbb{F}_q$ for $j \in \{1, \ldots, m\}$. In that case, α as well as τ^a are in $\mathbb{F}_q^m$ and the tag shares satisfy $\sum \tau_i^a[j] = \tau^a[j] = \alpha[j] \cdot a, \forall j \in \{1, \ldots, m\}$. Further we assume $m = 1$ unless otherwise mentioned.

3.1   Evaluation Stage

We let each tile $\mathcal{T}_i$ hold the ith share of each sensitive and auxiliary variable $(x_i, \ldots, a_i, \ldots)$ and the MAC key share α_i. We first describe operations that do not require communication between tiles.

Addition. To compute the addition (z, τ^z) of (x, τ^x) and (y, τ^y), each tile performs local addition of their data shares $z_i = x_i + y_i$ and their tag shares $\tau_i^z = \tau_i^x + \tau_i^y$. When one operand is public (for example, a cipher constant $C \in \mathbb{F}_q$), the sum can be computed locally as $z_i = x_i + C \cdot \delta_{i,1}$ for value shares and $\tau_i^z = \tau_i^x + C \cdot \alpha_i$ for tag shares.

Multiplication by a Public Constant. Given a public constant $C \in \mathbb{F}_q$, the multiplication (z, τ^z) of (x, τ^x) and C is obtained locally by setting $z_i = C \cdot x_i$ and $\tau_i^z = C \cdot \tau_i^x$.

The following operations, on the other hand, require auxiliary data generated in a preprocessing stage and also communication between the tiles.

Multiplication. Multiplication of (x, τ^x) and (y, τ^y) requires as auxiliary data a Beaver *triple* $(\langle a \rangle, \langle b \rangle, \langle c \rangle)$, which satisfies $c = a \cdot b$, for random a and b. The multiplication itself is performed in four steps.

- Step A. In the *blinding* step, each tile $\mathcal{T}_i$ computes locally a randomized version of its share of the secret: $\varepsilon_i = x_i + a_i$ and $\eta_i = y_i + b_i$.
- Step B. In the *partial unmasking* step, each tile $\mathcal{T}_i$ broadcasts its own shares ε_i and η_i to other tiles, such that each tile can construct and store locally the values $\varepsilon = \sum \varepsilon_i$ and $\eta = \sum \eta_i$. The value ε (resp. η) is the *partial unmasking* of $(\varepsilon, \tau^\varepsilon)$ (resp. (η, τ^η)), *i.e.* the value ε (resp. η) is unmasked but its tag τ^ε (τ^η) remains shared. These values are blinded versions of the secrets x and y and can therefore be made public.
- Step C. In the *MAC-tag checking* step, the tiles check whether the tags τ^ε (τ^η) are consistent with the public values ε and η, using a method which we will explain later in this section.
- Step D. In the *Beaver computation* step, each tile locally computes

$$z_i = c_i + \varepsilon \cdot b_i + \eta \cdot a_i + \varepsilon \cdot \eta \cdot \delta_{i,1}$$
$$\tau_i^z = \tau_i^c + \varepsilon \cdot \tau_i^b + \eta \cdot \tau_i^a + \varepsilon \cdot \eta \cdot \alpha_i.$$

It can be seen easily that the sharing (z, τ^z) corresponds to $z = x \cdot y$ unless faults occurred. Step B and C are the only steps that require communication among tiles. Step A and D are completely local. Note that to avoid leaking information on the sensitive data x and y, the shares ε_i and η_i must be synchronized using memory elements after step A, before being released to other tiles in step B. Moreover, we remark that step C does not require the result of step B and can thus be performed in parallel.

Squaring. Squaring is a linear operation in characteristic 2 fields. Hence, the output shares of a squaring operation can be computed locally using the input shares. However, obtaining the corresponding tag shares is non-trivial. To square (x, τ^x) into (z, τ^z), we therefore require an auxiliary *tuple* $(\langle a \rangle, \langle b \rangle)$ such that $b = a^2$. The procedure to obtain (z, τ^z) mimics that of multiplication with some modifications: there is only one partially unmasked value $\varepsilon = x + a$, whose tag needs to be checked, and each tile calculates $z_i = b_i + \varepsilon^2 \cdot \delta_{i,1}$ and $\tau_i^z = \tau_i^b + \varepsilon^2 \cdot \alpha_i$.

Following the same spirit, we can also perform the following operations.

Affine Transformation. Provided that we have access to a tuple $(\langle a \rangle, \langle b \rangle)$ such that $b = A(a)$, we can compute (z, τ^z) satisfying $z = A(x) = C + L(x)$, where $L(x)$ is an additively homomorphic function over the finite field, by computing the output sharing as $z_i = b_i + L(\varepsilon) \cdot \delta_{i,1}$ and $\tau_i^z = \tau_i^b + L(\varepsilon) \cdot \alpha_i$.

Multiplication following Linear Transformations. The technique used for the above additively homomorphic operations can be generalized even further to compute $z = L_1(x) \cdot L_2(y)$ in shared form, where L_1 and L_2 are additively homomorphic functions. A trivial methodology would require two tuples $(\langle a_i \rangle, \langle b_i \rangle)$ with $b_i = L_i(a_i)$ for $i \in \{1, 2\}$, plus a standard Beaver triple (*i.e.* requiring seven pre-processed data items). We see that we can do the same operation with five pre-processed items $(\langle a \rangle, \langle b \rangle, \langle c \rangle, \langle d \rangle, \langle e \rangle)$, such that $c = L_1(a)$, $d = L_2(b)$ and $e = L_1(a) \cdot L_2(b)$. The tiles partially unmask $x + a$ (resp. $y + b$) to obtain ε (resp. η) and verify them. Each tile computes its value share and tag share of z as $z_i = e_i + L_1(\varepsilon) \cdot d_i + L_2(\eta) \cdot c_i + L_1(\varepsilon) \cdot L_2(\eta) \cdot \delta_{i,1}$ and $\tau_i^z = \tau_i^e + L_1(\varepsilon) \cdot \tau_i^d + L_2(\eta) \cdot \tau_i^c + L_1(\varepsilon) \cdot L_2(\eta) \cdot \alpha_i$, respectively. We refer to $(\langle a \rangle, \langle b \rangle, \langle c \rangle, \langle d \rangle, \langle e \rangle)$ as a *quintuple*.

Proof

$$\sum_{i=1}^{d} z_i = \sum_{i=1}^{d} (e_i + L_1(\varepsilon) \cdot d_i + L_2(\eta) \cdot c_i) + L_1(\varepsilon) \cdot L_2(\eta)$$

$$= \sum_{i=1}^{d} e_i + L_1(\varepsilon) \cdot \sum_{i=1}^{d} d_i + L_2(\eta) \cdot \sum_{i=1}^{d} c_i + L_1(\varepsilon) \cdot L_2(\eta)$$

$$= L_1(a) \cdot L_2(b) + L_1(x + a) \cdot L_2(b) + L_2(y + b) \cdot L_1(a) + L_1(x + a) \cdot L_2(y + b)$$

$$= L_1(a) \cdot L_2(b) + L_1(x) \cdot L_2(b) + L_1(a) \cdot L_2(b) + L_1(a) \cdot L_2(y) + L_1(a) \cdot L_2(b)$$

$$+ L_1(x) \cdot L_2(y) + L_1(x) \cdot L_2(b) + L_1(a) \cdot L_2(y) + L_1(a) \cdot L_2(b)$$

$$= L_1(x) \cdot L_2(y)$$

$$\sum_{i=1}^{d} \tau_i^z = \sum_{i=1}^{d} (\tau_i^e + L_1(\varepsilon) \cdot \tau_i^d + L_2(\eta) \cdot \tau_i^c + L_1(\varepsilon) \cdot L_2(\eta) \cdot \alpha_i)$$

$$= \sum_{i=1}^{d} \tau_i^e + L_1(\varepsilon) \cdot \sum_{i=1}^{d} \tau_i^d + L_2(\eta) \cdot \sum_{i=1}^{d} \tau_i^c + L_1(\varepsilon) \cdot L_2(\eta) \cdot \sum_{i=1}^{d} \alpha_i$$

$$= \alpha \cdot e + L_1(\varepsilon) \cdot \alpha \cdot d + L_2(\eta) \cdot \alpha \cdot c + L_1(\varepsilon) \cdot L_2(\eta) \cdot \alpha$$

$$= \alpha \cdot (e + L_1(\varepsilon) \cdot d + L_2(\eta \cdot c + L_1(\varepsilon) \cdot L_2(\eta)))$$

$$= \alpha \cdot L_1(x) \cdot L_2(y)$$

Checking the MAC Tag of Partially Unmasked Values. Consider a public value $\varepsilon = x + a$, calculated in the partial unmasking step of the Beaver multiplication operation. Recall that we obtain its MAC-tag shares as follows: $\tau_i^\varepsilon = \tau_i^a + \tau_i^x$. During the MAC-tag checking step of the Beaver operation, the authenticity of τ^ε corresponding to ε is tested. As ε is public, each tile can calculate and broadcast the value $\varepsilon \cdot \alpha_i + \tau_i^\varepsilon$. For a correct tag, we expect $\sum \tau_i^\varepsilon = \alpha \cdot \varepsilon$, thus each tile computes $\sum (\varepsilon \cdot \alpha_i + \tau_i^\varepsilon)$ and proceeds if the result is zero. Recall that the broadcasting must be preceded by a synchronization of the shares.

Note on Unmasked Values/Calculations. There are several components in a cipher which do not need to be protected against SCA (*i.e.* masked), because their specific values are not sensitive. One prominent example is the control unit which decides what operations should be performed (*e.g.* the round counter). Other examples are constants such as the AES affine constant 0x63 or public values such as ε in a Beaver calculation and the difference $\varepsilon \cdot \alpha + \tau^\varepsilon$ during the MAC-tag checking phase.

While these *public* components are not sensitive in a SCA context, they *can* be targeted in a fault attack. It is therefore important to introduce some redundancy. Each tile should have its own control logic and keep a local copy of all public values to avoid single points of attack. The shares ε_i are distributed to *all* tiles so that ε can be unmasked by each tile separately and any subsequent computation performed on these public values is repeated by each tile. Finally, each tile also keeps its own copy of the abort status. This is in fact completely analogous to the MPC scenario.

3.2   Preprocessing Stage

The auxiliary data $(\langle a \rangle, \langle b \rangle, \ldots)$ required in the Beaver evaluations, is generated in a preprocessing stage. This preparation corresponds to the *offline* phase in SPDZ. However, CAPA's preprocessing stage is lighter and does not require a public key calculation due to the differences in adversary model. As in SPDZ, this stage is completely independent from the sensitive data of the main evaluation. Below, we describe the generation of a Beaver triple used in multiplication. This can trivially be generalized to tuples and quintuples.

Auxiliary Data Generation. To generate a triple $(\langle a \rangle, \langle b \rangle, \langle c \rangle)$ satisfying $c = a \cdot b$, we draw random shares $a = (a_1, \ldots, a_d)$ and $b = (b_1, \ldots, b_d)$ and use a passively secure shared multiplier to compute c s.t. $c = a \cdot b$. We then use another such multiplication with the shared MAC key α to generate tag shares τ^a, τ^b, τ^c. We note that the shares a_i, b_i are randomly generated by tile $\mathcal{T}_i$. There are thus d separate PRNG's on d distinct tiles.

Passively Secure Shared Multiplier. For a secure implementation of a shared multiplication, no subset of $d - 1$ tiles should have access to all shares of any variable. This concept, which is used in the context of secure implementations against SCA on hardware, is precisely called $d - 1^{\text{th}}$-order non-completeness in [9,52]. In the last decade, there has been significant improvement on passively secure shared multipliers that can be used in both hardware and software [9, 27,29,51,56]. In principle, CAPA can use any such multiplier as long as the tile structure still holds.

A close inspection of existing multipliers show that they require the calculation of the cross products $a_i b_j$. In order to make these multipliers compatible with the CAPA tile architecture, we define tiles $\mathcal{T}_{i,j}$ which receive a_i from $\mathcal{T}_i$ and b_j from $\mathcal{T}_j$ where $i \neq j$ in order to handle the pair (a_i, b_j) to be used during tuple, triple and quintuple generation. This implies $d(d-1)$ smaller tiles used only during auxiliary data generation in addition to d tiles used for both auxiliary data generation and evaluation. The output wires from $\mathcal{T}_{i,j}$ are only connected to $\mathcal{T}_i$ and carry randomized information.

The multipliers used in the preprocessing phase are only *passively* secure. We also ensure resistance against *active* adversaries because on the one hand, deterministic faults are limited to $d-1$ tiles and on the other, because of a relation verification step, which is explained in the next section.

3.3 Relation Verification of Auxiliary Data

The information theoretic MAC tags provide security against faults induced in the evaluation stage. To detect faults in the preprocessing stage, we perform a relation verification of the auxiliary data. This relation verification step is done for each generated triple that is passed from the preprocessing to the evaluation stage and ensures that the triple is functionally correct (*i.e.* $c = a \cdot b$) by sacrificing another triple. That is, we take as input two triples $(\langle a \rangle, \langle b \rangle, \langle c \rangle)$ and $(\langle d \rangle, \langle e \rangle, \langle f \rangle)$, that should satisfy the same relation, in this example $c = a \cdot b$ and $f = d \cdot e$. The following Beaver computation holds if and only if both relations are satisfied:

- Draw a random $r_1 \in \mathbb{F}_q$
- Use triple $(\langle d \rangle, \langle e \rangle, \langle f \rangle)$ to calculate the multiplication of $r_1 \cdot \langle a \rangle$ and $\langle b \rangle$ using a constant multiplication with r_1, followed by the Beaver equation for multiplication described above. The result $\langle \tilde{c} \rangle$ is a shared representation of $\tilde{c} = r_1 \cdot a \cdot b$.
- For each share i, calculate the difference with the shares and tags of $r_1 \cdot c$: $\Delta_i = r_1 \cdot c_i + \tilde{c}_i$ and $\tau_i^\Delta = r_1 \cdot \tau_i^c + \tau_i^{\tilde{c}}$.
- Unmask the resulting differences Δ and τ^Δ.
- If a difference is nonzero, reject $(\langle a \rangle, \langle b \rangle, \langle c \rangle)$ as a valid triple.
- Pick another $r_2 \in \mathbb{F}_q$ such that $r_2 \neq r_1$ and repeat a second time.

Note that this relation verification ensures that the second triple is functionally correct too. However, it is burnt (or "sacrificed") in this process in order to ensure that the first triple can be used securely further on. Note that this relation verification or "sacrificing" step is mandatory in each Beaver-like operation.

Why We Need Randomization. This sacrificing step involves two values r_1 and r_2. We present the following attack to illustrate why this randomization is needed. Again, we elaborate on triples, but the same can be said for tuples and quintuples. As the security does not rely on the secrecy of r_1 and r_2, we assume for simplicity that they are known to the attacker. We only stress that they are different: $r_1 \neq r_2$.

Consider two triples $(\langle a \rangle, \langle b \rangle, \langle c' \rangle)$ and $(\langle d \rangle, \langle e \rangle, \langle f' \rangle)$ at the input of the sacrificing stage. We assume that the adversary has introduced an additive difference into one share of c' and f' such that $c' = a \cdot b + \Delta_c$ and $f' = d \cdot e + \Delta_f$. This fault is injected before the MAC tag calculation, so that $\tau^{c'}$ and $\tau^{f'}$ are valid tags for the faulted values c' and f' respectively. In particular, this means we have $\tau^{c'} = \tau^c + \alpha \cdot \Delta_c$ and $\tau^{f'} = \tau^f + \alpha \cdot \Delta_f$.

The sacrificing step calculates the following four differences (for $r_j = r_1$ and r_2) and only succeeds if all are zero.

$$\Delta_j = \sum_{i=1}^{d} \left(r_j \cdot c_i' + f_i' + \varepsilon \cdot e_i + \eta \cdot d_i \right) + \varepsilon \cdot \eta$$

$$= r_j \cdot \Delta_c + \Delta_f \overset{?}{=} 0$$

$$\tau^{\Delta_j} = \sum_{i=1}^{d} \left(r_j \cdot \tau_i^{c'} + \tau_i^{f'} + \varepsilon \cdot \tau_i^e + \eta \cdot \tau_i^d + \varepsilon \cdot \eta \cdot \alpha_i \right)$$

$$= r_j \cdot \alpha \cdot \Delta_c + \alpha \cdot \Delta_f \overset{?}{=} 0$$

Without randomization (*i.e.* $r_1 = r_2 = 1$), the attacker only has to match the differences $\Delta_f = \Delta_c$ to pass verification. With a random r_1, the attacker can fix $\Delta_f = r_1 \cdot \Delta_c$ to automatically force Δ_1 and τ^{Δ_1} to zero. Even if he does not know r_1, he has probability as high as 2^{-k} to guess it correctly.

Only thanks to the repetition of the relation verification with r_2, the adversary is detected with a probability $1 - 2^{-km}$. Assuming he fixed $\Delta_f = r_1 \cdot \Delta_c$, it is impossible to also achieve $\Delta_f = r_2 \cdot \Delta_c$. Even if the attacker manages to force Δ_2 to zero with an additive injection (since he knows all components r_2, Δ_c and Δ_f), he cannot get rid of the difference $\tau^{\Delta_2} = r_2 \cdot \alpha \cdot \Delta_c + \alpha \cdot \Delta_f$ without knowing the MAC key. Since α remains secret, the attacker only has a success probability of 2^{-km} to succeed.

4   Discussion

4.1   Security Claims

With both described adversaries $\mathscr{A}_1$ and $\mathscr{A}_2$, our design CAPA claims provable security against the following types of attacks as well as a combined attack of the two

1. Side-Channel Analysis (*i.e.* against $d - 1$ tile probing adversary).
2. Fault Attacks (*i.e.* an adversary introducing either known faults into $d - 1$ tiles or random faults everywhere).

Side-channel Analysis. One can check that no union of $d - 1$ tiles $\cup_{j \in j_1, \ldots, j_{d-1}} \mathcal{T}_j$ has all the shares of a sensitive value. Very briefly, we can reason to this $d - 1^{\text{th}}$-order non-completeness as follows. All computations are local with the exception of the unmasking of public values such as ε. However, the broadcasting of all shares of ε does not break non-completeness since $\varepsilon = x + a$ is not sensitive itself but rather a blinded version of a sensitive value x, using a random a that is shared across all tiles. Unmasking the public value ε therefore gives each tile T_i only one share $\varepsilon + a_i$ of a new sharing of the secret x:

$$x = (a_1, \ldots, a_{i-1}, \varepsilon + a_i, a_{i+1}, \ldots, a_d)$$

In this sharing, no union of $d - 1$ shares suffices to recover the secret. Our architecture thus provides non-completeness for all sensitive values. As a result, our d-share implementation is secure against $d - 1$-probing attacks. Any number of probes following the adversaries' restrictions leak no sensitive data. Our model is related to the wire-probe model, but with wires replaced by entire tiles. We can thus at least claim security against $d - 1^{\text{th}}$ order SCA.

Fault Attacks. A fault is only undetected if both value and MAC tag shares are modified such that they are consistent. Adversary $\mathcal{A}_1$ can fault at most $d_f < d$ tiles, which means he requires knowledge of the MAC key $\alpha \in GF(2^{km})$ to forge a valid tag for a faulty value. Since α is secret, his best option is to guess the MAC key. This guess is correct with probability 2^{-km}. Adversary $\mathcal{A}_2$ has ϵ-faulting abilities only and will therefore only avoid detection if the induced faults in value and tag shares *happen* to be consistent. This is the case with probability 2^{-km}. We can therefore claim an error detection probability (EDP) of $1 - \frac{1}{2^{km}}$. The EDP does not depend on the number of faulty bits (or the hamming weight of the injected fault).

Combined Attacks. In a combined attack, an adversary with d_f-faulting capabilities can mount an attack where he uses the knowledge obtained from probing some tiles $\in \mathcal{P}_1$ to carefully forge the faults. In SPDZ, commitments are used to avoid the so called "rushing adversary". CAPA does not need commitments as the timing limitation on $\mathscr{A}_1$ adversary ensures a d_f-fault cannot be preceded by a probe in the same clock cycle. As a result, we inherit the security claims of SPDZ and the claimed EDP is not affected by probing or SCA. Also, the injection of a fault in CAPA does not change the side-channel security. Performing a side-channel attack on a perturbed execution does not reveal any additional information because the Beaver operations do not allow injected faults to propagate through a calculation into a difference that depends on sensitive information. We can claim this security, because of the aspects inherited from MPC. CAPA is essentially secure against a very powerful adversary that has complete control (hence *combined* attacks) over all but one of the tiles.

What Does Our MAC Security Mean? We stress that CAPA provides significantly higher security than existing approaches against faults. An adversary that injects errors in up to d_f tiles cannot succeed with more than the claimed detection probability. This means that our design can stand $d'_f \gg d_f$ shots if they affect at most d_f tiles. This is the case even if those d_f tiles leak their entire state; hence our resistance against combined attacks. The underlying reason for this is that to forge values, an attacker needs to know the MAC key, but since this is also shared, the attacker does not gain any information on the MAC key and their best strategy is to insert a random fault, which is detected with probability $1 - 2^{-km}$. Moreover, our solution is incredibly scalable compared to for example error detection code solutions.

How Much Do Tags Leak? The tag shares τ_i^a form a Boolean masking of a variable τ^a. This variable τ^a itself is an information theoretic MAC tag of the underlying value a and can be seen as a multiplicative share of a. We therefore require the MAC key to change for each execution. Hence MAC tag shares are a Boolean masking of a multiplicative share and are expected to leak very little information in comparison with the value shares themselves.

Forbidding the All-0s MAC Key. If the MAC key size mk is small, we should forbid the all-0 MAC key. This ensures that tags are injective: if an attacker changes a value share, he must change the tag share. We only pay with a slight decrease in the claimed detection probability. By excluding 1 of the 2^{km} MAC key possibilities, we reduce the fault detection probability to $1 - 2^{-\kappa}$, where $\kappa = \log_2(2^{km} - 1)$.

4.2 Attacks

The Glitch Power Supply or Clock Attack. The solution presented in this paper critically depends on the fact that there is no single point where an attacker can insert a fault that affects all d tiles deterministically. An attacker may try to glitch the chip clock line that is shared among all tiles. In this case, the attacker could try to carefully insert a glitch so that writing to the abort register is skipped or a test instruction is skipped. Since all tiles share the same clock, the attacker can bypass in this way the tag verification step. Similar comments apply, for example, to glitches in the power line. The bottom line is that one should design the hardware architecture accordingly, that is, deploy low-level circuit countermeasures that detect or avoid this attack vector.

Skipping Instructions. In software, when each tile is a separate processor (with its own program counter, program memory and RAM memory), skipping one instruction in up to $d - 1$ shares would be detected. The unaffected tiles will detect this misbehavior when checking partially unmasked values.

Safe Error Attack. We point out a specific attack that targets any countermeasure against a probing and faulting adversary. In a safe error attack [65], the attacker perturbs the implementation in a way that the output is only affected if a sensitive variable has a certain value. The attacker learns partial secret information by merely observing whether or not the computation succeeds (*i.e.* does not abort). Consider for example a shared multiplication of a variable x and a secret y and call the resulting product $z = xy$. The adversary faults one of the inputs with an additive nonzero difference such that the multiplication is actually performed on $x' = x + \Delta$ instead of x. Such an additive fault can be achieved by affecting only one share/tile. The multiplication results in the faulty product $z' = z + \Delta \cdot y$. The injected fault has propagated into a difference that depends on sensitive data (y). As a result, the success or failure of any integrity check following this multiplication depends on y. In particular, if nothing happens (all checks pass), the attacker learns that y must be 0.

Among existing countermeasures against combined attacks, none provide protection against this kind of selective failure attack as they cannot detect the initial fault Δ. The attacker can always target the wire running from the last integrity check on x to the multiplication with y. We believe CAPA is currently unique in preventing this type of attack. One can verify that the MAC-tag checking step in a Beaver operation successfully prevents Δ from propagating to the output. This integrity check only passes if all tiles have a correct copy of the public value ε. Any faults injected after this check have a limited impact as the calculation finishes locally. That is, once the correct public values are established between the tiles, the shares of the multiplication output z are calculated without further communication among tiles. The adversary is thus unable to elicit a fault that depends on sensitive data.

PACA. We claim security against the passive and active combined attack (PACA) on masked AES described in [2] because CAPA does not output faulty ciphertexts. A second attack in this work uses another type of safe errors (or ineffective faults as they are called in this work) which are impossible to detect. The attacker fixes a specific wire to the value zero (this requires the d_f-faulting capability) and collects power traces of the executions that succeed. This means the attacker only collects traces of encryptions in which that specific wire/share was already zero. The key is then extracted using $d - 1^{\text{th}}$-order SCA on the remaining $d - 1$ shares. This safe error attack however falls outside our model since the adversary gets access (either by fault or SCA) to all d shares and thus $(\mathcal{F}_1 \cup \mathcal{P}_1) = \mathcal{T}$.

Advanced Physical Attacks. In our description we are assuming that during the broadcast phase there are no "races" between tiles: by design, each tile sets its share to be broadcasted at clock cycle t and captures other tiles' share in the same clock cycle t. We are implicitly assuming that tiles cannot do much work between these two events. If this assumption is violated (for example, using advanced circuit editing tools), a powerful adversary could bypass any verification. This is why in the original SPDZ protocol there are commitments prior to broadcasting operations; if this kind of attack is a concern one could adapt the same principles of commitments to CAPA. This is a very strong adversarial model that we consider out of scope for this paper.

4.3   Differences with SPDZ

Offline Phase. In SPDZ, the auxiliary data is generated using a somewhat homomorphic encryption scheme. The mapping onto a chip environment thus seems prohibitive due to the need for this expensive public-key machinery to obtain full threshold and the large storage required. We avoid this by generating the Beaver triples using passively secure shared multipliers. Furthermore, to avoid the large storage requirement, we produce the auxiliary data on the fly whenever required.

MAC Tag Checking. SPDZ delays the tag checking of public values until the very end of the encryption by using commitments. For this, each party keeps track of publicly opened values. This is to avoid a slowdown of the computation and because in the MPC setting, local memory is cheaper than communication costs. In an embedded scenario the situation is opposite so we check the opened values on the fly at the cost of additional dedicated circuit. In hardware, we "simulate" the broadcast channel by wiring between all tiles. Each tile keeps a local copy of those broadcasted values.

Adversary. Although MPC considers mainly the "synchronous" communication model, the SPDZ adversary model also includes the so-called "rushing" adversary, which first collects all inputs from the other parties and only then decides what to send in reply. In our embedded setting, as already pointed out, the "rushing" adversary is impossible. Due to the nature of the implementation, the computational environment and storage is very much restricted. On the other hand, communication channels are very efficient and can be assumed to be automatically synchronous with all tiles progressing in-step in the computation.

4.4   Cost Analysis and Scalability

The computation as described in Sect. 3.1 scales nicely with the masking order d and the security parameter m. For any fixed number of shares d, the circuit area scales linearly in m (see for example Table 2). Storage increases with a factor $(m+1)d$ compared to a plain implementation. We note that our implementations run in almost the same amount of cycles as that of a plain implementation. There is almost no loss in throughput and only negligible in latency. In software as well, the timing scales linearly if tiles run in parallel (Table 1).

Table 1. Overview of the number of $\mathbb{F}_q$ multiplications (.), $\mathbb{F}_q$ additions (+) and linear operations in $GF(2)$ $(L(.))$ required to calculate all building blocks with d shares and m tags

| | Public Values | | | Output calculation | | | | MAC check | |
| | | | | Value | | Tags | | | |
	.	+	$L(.)$	.	+	.	+	.	+
Add.					d		dm		
Add. with C					1	dm	dm		
Multip. with C				d		dm			
Multip.	d	$2d + 2(d-1)d$		$2d$	$2d+1$	$3dm$	$3dm$	$2dm$	$4dm + 2(d-1)dm$
Square/Affine	$d + (d-1)d$		d		1	dm	dm	dm	$2dm + (d-1)dm$
$L_1(x) \cdot L_2(y)$	d	$2d + 2(d-1)d$	$d+d$	$2d$	$2d+1$	$3dm$	$3dm$	$2dm$	$4dm + 2(d-1)dm$

This efficiency does not come for free. The complexity is shifted to the pre-processing stage; indeed the generation of auxiliary triples is the most expensive part of the implementation. There is a trade-off to be made here between the online and offline complexity. The more auxiliary data we prepare "offline", the more efficient the online computation.

Complexity for Passive Attacker Scenario. It is remarkable that if active attackers are ruled out, and only SCA is a concern, then the complexity of the principal computation is linear in d. This may seem like a significant improvement over previous masking schemes which have quadratic complexity on the security order [18,34,58]. However, this complexity is again pushed into the preprocessing stage. Nevertheless, this can be interesting especially for software implementations in platforms where a large amount of RAM is available to store the auxiliary data generated in Sect. 3.2. The same comments apply to FPGAs with plenty of BlockRAM.

Optimization of Preprocessing. It may be beneficial to store the output of the preprocessing stage Sect. 3.2 in a table for later usage. One could optimize this process by recycling auxiliary data (sample elements with replacement from the table). Of course, this would void the provable security claims; but if performed with care (with appropriate table shuffling and table elements refresh), this can give rise to an implementation that is secure in practice.

5   Proof-of-Concept

In this section we detail a proof-of-concept implementation of the CAPA methodology in both a hardware and a software environment. We emphasize specific concepts for hardware and software implementations and provide case studies of KATAN-32 [14] and AES [1], which cover operations in different fields, possibility of bitsliced implementations, specific timing and memory optimizations, and performance results.

5.1   Hardware Implementations

We now describe two case studies for applying CAPA in hardware. Our implementations are somewhat optimized for latency rather than area with d tiles spatially separated and operating in parallel, each with its own combinational and control logic and auxiliary data preparation module. These preparation modules are equipped with a passively secure shared multiplication with higher-order non-completeness. Literature provides us with a broad spectrum of multipliers to choose from [9,27,29,51,56]. In order to minimize the randomness requirement, our implementation uses the one from [29], hereafter referred to as DOM.

Library. For synthesis, we use Synopsis Design Compiler Version I-2013.12 using the NanGate 45 nm Open Cell library [49] for ease of future comparisons. We choose the compile option – exact_map to prevent optimization across tiles. The area results are provided in 2-input NAND-gate equivalents (GE).

Table 2. Area (GE) of 2-share KATAN-32 implementations with m MAC keys $\alpha[j] \in \mathbb{F}_q$

	No tags	$m = 1$	$m = 8$	Any m
- Evaluation	2 315	4 708	21 404	$\approx 2\,315 + 2\,390\,m$
* Shift Register	888	1 823	8 419	$\approx 888 + 935\,m$
* Key Schedule	1 427	2 885	12 985	$\approx 1\,427 + 1\,455\,m$
- Preprocessing (x3)	363	679	2 727	$\approx 363 + 315\,m$
* Two triple generation	237	431	1 786	$\approx 237 + 195\,m$
* Relation verification	126	248	941	$\approx 126 + 120\,m$
Total	3 672	7 103	30 596	$\approx 3\,672 + 3\,430\,m$

Case Study: KATAN-32. KATAN-32 is a shift register based block cipher, which has a 80-bits key and processes 32-bit plaintext input. It is designed specifically for efficient hardware implementations and performs 254 cycles of four AND-XOR operations. Hence, its natural shared data representation is in the field $\mathbb{F}_q = GF(2)$, which makes the mapping into CAPA operations relatively straightforward. However, the small finite field means that we need to utilize a vectorized MAC-tag operation ($m > 1$) to ensure a good probability of detecting errors. Our implementation is round based, as in [14] with three AND-XOR Beaver operations and one constant AND-XOR calculated in parallel. Each Beaver AND-XOR operation requires two cycles, and is implemented in a pipelined fashion such that the latency of the whole computation increases only by one clock cycle.

Implementation Cost. Tables 2 and 3 summarize the area of our KATAN implementations. Naturally, compared to a shared implementation without MAC tags, the state registers grow with a factor $m + 1$ as the MAC-key size increases. In the last columns, we extrapolate the area results for any m.

Each Beaver multiplication in GF(2) requires one triple, and each triple needs $2d$ random bits for generating $\boldsymbol{a}$ and $\boldsymbol{b}$. A d-share DOM multiplication requires $\binom{d}{2}$ units of randomness. The construction of one triple requires $1 + 3m$ masked

Table 3. Area (GE) of 3-share KATAN-32 implementations with m MAC keys $\alpha[j] \in \mathbb{F}_q$

	No tags	$m = 1$	$m = 8$	Any m
- Evaluation	3 560	7 139	32 368	$\approx 3\,560 + 3\,580\,m$
* Shift Register	1 363	2 812	12 890	$\approx 1\,363 + 1\,450\,m$
* Key Schedule	2 197	4 327	19 478	$\approx 2\,197 + 2\,130\,m$
- Preprocessing (x3)	638	1 468	7 124	$\approx 638 + 830\,m$
* Two triple generation	428	952	4 694	$\approx 428 + 524\,m$
* Relation verification	210	516	2 430	$\approx 210 + 306\,m$
Total	5 971	12 083	55 254	$\approx 5\,971 + 6\,112\,m$

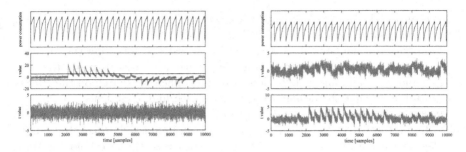

Fig. 2. Non-specific leakage detection on the first 31 rounds of first-order KATAN. Left column: PRNG off (24K traces). Right column: PRNG on (100M traces). Rows (top to bottom): exemplary power trace; first-order t-test; second-order t-test

multiplications: one to obtain the multiplication c of a and b; and $3m$ to obtain the m tags τ^a, τ^b and τ^c. Due to the relation verification through the sacrificing of another triple, the randomness must be doubled. Hence, the total required number of random bits per round of KATAN is $3 \cdot 2 \cdot (2d + (1 + 3m)\frac{d(d-1)}{2}))$.

Experimental Validation. The goal of the prior proof-of-concept implementation is to experimentally validate the protection against side-channel attacks offered by the CAPA methodology. We deploy a first- and second-order secure KATAN instance onto a Xilinx Spartan-6 FPGA. Our platform is a Sakura-G board specifically designed for side-channel evaluation with two FPGA's to minimize platform noise: a control FPGA handles I/O with the host computer and supplies masked data to the crypto FPGA, which implements both the preprocessing and evaluation. The KATAN implementations use $d = 2$ (resp. $d = 3$) shares and $m = 2$ MAC keys. The parameter $m = 2$ is insufficient in practice, but serves for this experiment since m has no influence on SCA security. The designs are

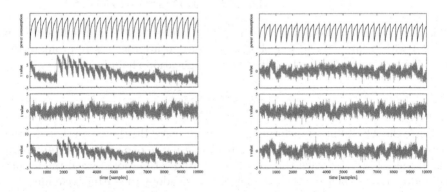

Fig. 3. Non-specific leakage detection on the first 31 rounds of second-order KATAN. Left column: PRNG off (24K traces). Right column: PRNG on (100M traces). Rows (top to bottom): exemplary power trace; first-order t-test; second-order t-test; third-order t-test

clocked at 3 MHz and we sample power traces of 10 000 time samples each at 1 GS/s. Exemplary traces are shown in Fig. 2, top.

We perform a non-specific leakage detection test [17] following the methodology from [57,61]. First, we test the designs without masks to verify that our setup is indeed sound and able to detect leakage. Then we switch on the PRNG and corroborate that the design does not leak with high confidence.

In Fig. 2, we show the results for the first-order secure design ($d = 2$). In the left column, the PRNG is turned off, emulating an unmasked design. Indeed, we see clear leakage at first order, since the t-statistics cross the threshold 4.5. With the PRNG on (right column), no first-order leakage is detected with up to 100 million traces. As expected, we do see second-order leakage. Figure 3 exhibits the results for the second-order secure design ($d = 3$). The left column shows clear leakage at first, second and third order when the PRNG is turned off. In the right column, we repeat the procedure with PRNG on and no univariate leakage is detected with up to 100 million traces.[1]

Case Study: AES. There has been a great deal of work on MPC and masked implementations of the basic AES operations. We take what has now become the traditional approach and work in the field $GF(2^8)$ with $m = 1$ for AES, *i.e.* the MAC key, data and tag shares α_i, a_i and τ_i^a are $\in GF(2^8)$. The ShiftRows and MixColumns operations are linear in $GF(2^8)$, hence are straightforward. Here, we only describe the S-box calculation.

Design Choices. The AES S-box consists of an inversion in $GF(2^8)$, followed by an affine transformation over bits. We distinguish two methodologies for the S-box implementation: It is well known that the combination of the two operations can be expressed by the following polynomial in $GF(2^8)$ [19]:

$$\text{S-box}(x) = \texttt{0x63} + \texttt{0x8F} \cdot x^{127} + \texttt{0xB5} \cdot x^{191} + \texttt{0x01} \cdot x^{223} + \texttt{0xF4} \cdot x^{239}$$
$$+ \texttt{0x25} \cdot x^{247} + \texttt{0xF9} \cdot x^{251} + \texttt{0x09} \cdot x^{253} + \texttt{0x05} \cdot x^{254} \quad (1)$$

This polynomial can be implemented using 6 squares and 7 multiplications in $GF(2^8)$ with a latency of 13 clock cyles. A second approach is to evaluate the inversion $x \longrightarrow x^{254}$ using the following multiplication chain from [30]:

$$x^{254} = x^4 \cdot \left(\left((x^5)^5 \right)^5 \right)^2$$

Since the AES affine transform $A(x)$ is linear over $GF(2)$, we can then use the Beaver operation described in Sect. 3.1 to evaluate it in one cycle, using auxiliary affine tuples $(\langle a \rangle, \langle b \rangle)$ such that $b = A(a)$. Initial estimations reveal the former method is more expensive than the latter, so we adopt the latter technique.

---

[1] Since our implementation handles 3 shares, we expect to detect leakage in the third order. Due to platform noise, this is not visible.

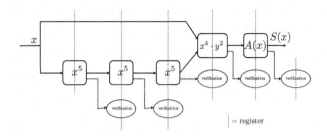

Fig. 4. AES S-box pipeline

Multiplication Chain. Our implementation of the proposed multiplication chain uses two types of operations: x^5 and $x^4 \cdot y^2$, which can both be computed as described in Sect. 3.1 (Multiplication following Linear Transformations). Given an input $\langle x \rangle$ and a triple $(\langle a \rangle, \langle b \rangle, \langle c \rangle)$ such that $b = a^4$ and $c = a^5$, we calculate the CAPA exponentiation to the power five. Likewise, we perform the map $x^4 \cdot y^2$ (with $y = x^{125}$) in one cycle, using quintuples $(\langle a \rangle, \langle b \rangle, \langle c \rangle, \langle d \rangle, \langle e \rangle)$ such that $c = a^4$, $d = b^2$ and $e = c \cdot d = a^4 \cdot b^2$. As a result, an inversion in $GF(2^8)$ costs only 4 cycles, using 3 exponentiation triples and 1 quintuple. Combined with the affine stage, we obtain the S-box output in 5 cycles (see Fig. 4). This approach does not only optimize the number of cycles but also the amount of required randomness. The S-box is implemented as a five stage pipeline.

Implementation Cost. We use a serialized AES architecture, based on that in [28]. One round of the cipher requires 21 clock cycles, making the latency of one complete encryption 226 clock cycles. Since the unprotected serialised implementation of [47] also requires 226 cycles, the timing performance is very good.

Table 4 presents the area for the different blocks that make up our AES implementation. We can see a significant difference between the preprocessing and evaluation stages, *i.e.* the efficient calculation phase comes at the cost of expensive resource generation machinery.

Table 5 summarizes the required number of random bytes for the generation of the triples/tuples for the AES S-box as a function of the number of MAC keys m and the number of shares d. Recall that the S-box needs three exponentiation triples, one quintuple and one affine tuple per cycle (doubled for the sacrificing). Each of these uses d initial bytes of randomness per input for the shares of a (and b). Furthermore, recall that each masked multiplication requires $\binom{d}{2}$ bytes or randomness. That is, for $d = 3$ and $m = 1$, we need 156 bytes of randomness per S-box evaluation.

Table 4. Areas for first- and second-order AES implementations with $m = 1$ in 2-NAND Gate Equivalents (GE)

Evaluation	$d = 2$	$d = 3$	Preprocessing	$d = 2$	$d = 3$
S-box	18 810	28 234	Quintuples	29 147	53 212
* Beaver x^5 (x3)	3 914	5 875	* Generation	15 092	32 241
* Beaver $x^4 y^2$	4 944	7 427	* Sacrificing	14 055	20 971
* Beaver Affine	1 563	2 344	Triples (x3)	19 106	34 954
State array	4 962	7 466	* Generation	9 804	21 112
* MixColumns	1 056	1 584	* Sacrificing	9 302	13 842
Key array	3 225	4 835	Affine tuples	7 603	14 657
Others	1 296	1 839	* Generation	4 821	10 444
			* Sacrificing	2 782	4 213
Total	28 293	42 374	Total	94 068	172 731
TOTAL				122 361	215 105

Table 5. The number of randomness in bytes for the initial sharing, shared multiplication and the sacrifice required for AES S-box

	Initial sharing	Shared mult.	Total
Exp. triple	d	$1 + 3m$	$2(d + (1 + 3m)\frac{d(d-1)}{2})$
Quintuple	$2d$	$1 + 5m$	$2(2d + (1 + 5m)\frac{d(d-1)}{2})$
Affine tuple	d	$2m$	$2(d + 2m\frac{d(d-1)}{2})$
Total			$12d + 2(4 + 16m)\frac{d(d-1)}{2}$

5.2 Software Implementation

CAPA is a suitable technique for software implementations if we map different tiles to different processors/cores. We do, however, need to place some constraints on the underlying hardware architecture; namely each processor should have an independent memory bank. Otherwise, a single affected tile (processor) could compromise the security of the whole system by for example dumping the entire memory contents (including all shares for sensitive variables).

This model therefore does not perfectly fit commercial off-the-shelf multi-core architectures, but we think isolated memory regions is a reasonable assumption for future micro-processors. While we do not have access to such architecture, as a proof of concept we emulate the proposed multi-processor architecture by time-sharing a 32-bit single-core ARM Cortex-M4 processor. This proof-of-concept does not provide resistance against attacks such as the memory dump example above.

Case Study: AES S-box. Even though it is possible to implement the AES S-box using $GF(2^8)$ operations in SW also, we base our bitsliced software implementation on the principles of gate-level masking and we use the depth-16 AES S-box circuit by Boyar et al. [11] in order to provide competitive through-put. Our high-level implementation processes 32 blocks simultaneously which is compatible with the word size of our processor and can naturally be reduced. As the circuit boils down to a series of XOR and AND operations over pairs of value and tag shares, we redefine these elementary operations in the same way as previous works [3, Sect. 4]. We note that this technique is independent from the concrete design, and one could apply the same principles to different ciphers.

We create a prototype implementation in C99. This is an unoptimized implementation meant for functionality and security testing. We compile with `gcc-arm` `4.8.4`. The 32 parallel SubBytes operations are performed in 2.52 million cycles (15ms) at 168MHz with $m = 8$ MAC tags and $d = 3$ shares. The implementation holds 41 intermediate variables in the stack (but this can be optimized); each takes $d \cdot w$ bytes for value shares and $m \cdot d \cdot w$ bytes for tag shares ($w = 4$ is number of bytes per word).

Experimental Validation of DPA Security. We use an STM32F407 32-bit ARM Cortex-M4 processor running the C99 implementation. We take EM measurements with an electromagnetic probe on top of a decoupling capacitor. This platform is very low noise: a DPA attack on the unprotected byte-oriented AES implementation succeeds with only 15 traces. Each trace is slightly above 500 000 time samples long and covers the entire execution of SubBytes. An exemplary trace is depicted at the top of Fig. 5.

Following the same procedure as in Sect. 5.1, we first perform a non-specific leakage detection test with the masking PRNG turned off. The results of the first-, second- and third-order leakage tests are shown on the left side of Fig. 5. Severe leakage is detected, which confirms that the setup is sound. When we plug in the PRNG, no leakage is detected with up to 200 000 traces (the statistic does not surpass the threshold $C = \pm 4.5$). This serves to confirm that the implementation effectively masks all intermediates, and that first- nor second-order DPA is not possible on this implementation. SPA features within an electromagnetic trace are better visible in the cross-correlation matrix shown in Fig. 6.

Experimental Validation of DFA Security. For the purposes of validating our theoretical security claims on CAPA's protection against fault attacks, we scale down our software AES SubBytes implementation, reducing the MAC key size to $m = 2$ and scaling down words to bits ($k = 1$). Note that this parameter choice lowers the detection probability; the point of using these toy parameters is only to verify more comfortably that the detection probability works as expected. It is easier to verify that the detection probability is $1 - 2^{-2}$ rather than $1 - 2^{-40}$. This concrete parameter choice is naturally not to be used in a practical deployment.

When barring the all zeroes key, we expect the attacker to succeed with probability at most $\frac{1}{2^{mk}-1} = \frac{1}{2^2-1} = 33\%$. The instrumented implementation conditionally inserts faults in value and/or tag shares. We repeat the SubBytes execution 1000 times, each iteration with a fresh MAC key. Faults are inserted in a random location during the execution of the S-box.

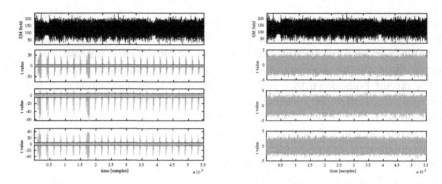

Fig. 5. Non-specific leakage detection on second-order SubBytes. Left column: masks off. Right column: masks on (200K traces). Rows (top to bottom): one exemplary EM trace, first-order t-test; second-order t-test; third-order t-test

Fig. 6. Cross-correlation for second-order SubBytes. One can identify the 34 AND gates in the SubBytes circuit of Boyar et al. [11].

We verify that single faults on only values or only tags are detected unconditionally when we bar the all-0s key. When a single-bit offset (fault) is inserted in a single tile in both the value and tag share, it is indeed detected in approximately 66% of the iterations. Inserting a single-bit offset in value share and a random-bit offset in tag share is a worse attack strategy and is detected in around 83% of the experiments. The same results hold when faults are inserted in up to $d - 1$ tiles. When the value and tag shares in all d tiles are modified and fixed to a known value, the fault escapes detection with probability one, as expected.

6   Conclusion

In this paper, we introduced the first adversary model that jointly considers side-channels and faults in a unified and formal way. The *tile-probe-and-fault* security model extends the more traditional wire-probe model and accounts for a more realistic and comprehensive adversarial behavior. Within this model, we developed the methodology CAPA: a new combined countermeasure against physical attacks. CAPA provides security against higher-order DPA, multiple-shot DFA and combined attacks. CAPA scales to arbitrary security orders and borrows concepts from SPDZ, an MPC protocol. We showed the feasibility of implementing CAPA in embedded hardware and software by providing prototype implementations of established block ciphers. We hope CAPA provides an interesting addition to the embedded designer's toolbox, and stimulates further research on combined countermeasures grounded on more formal principles.

Acknowledgements. This work was supported in part by the Research Council KU Leuven: C16/15/058 and OT/13/071, by the NIST Research Grant 60NANB15D346 and the EU H2020 project FENTEC. Oscar Reparaz and Begül Bilgin are postdoctoral fellows of the Fund for Scientific Research - Flanders (FWO) and Lauren De Meyer is funded by a PhD fellowship of the FWO. The work of Nigel Smart has been supported in part by ERC Advanced Grant ERC-2015-AdG-IMPaCT, by the Defense Advanced Research Projects Agency (DARPA) and Space and Naval Warfare Systems Center, Pacific (SSC Pacific) under contract No. N66001-15-C-4070, and by EPSRC via grants EP/M012824 and EP/N021940/1.

References

1. Advanced Encryption Standard (AES): National Institute of Standards and Technology (NIST), FIPS PUB 197, U.S. Department of Commerce, November 2001
2. Amiel, F., Villegas, K., Feix, B., Marcel, L.: Passive and active combined attacks: combining fault attacks and side channel analysis. In: Breveglieri, L., Gueron, S., Koren, I., Naccache, D., Seifert, J. (eds.) FDTC 2007, pp. 92–102. IEEE Computer Society (2007)
3. Balasch, J., Gierlichs, B., Reparaz, O., Verbauwhede, I.: DPA, bitslicing and masking at 1 GHz. In: Güneysu and Handschuh [31], pp. 599–619
4. Barthe, G., Dupressoir, F., Faust, S., Grégoire, B., Standaert, F.-X., Strub, P.-Y.: Parallel implementations of masking schemes and the bounded moment leakage model. In: Coron, J.-S., Nielsen, J.B. (eds.) EUROCRYPT 2017, Part I. LNCS, vol. 10210, pp. 535–566. Springer, Cham (2017). https://doi.org/10.1007/978-3-319-56620-7_19
5. Battistello, A., Giraud, C.: Fault analysis of infective AES computations. In: Fischer, W., Schmidt, J., (eds.) FDTC 2013, pp. 101–107. IEEE Computer Society (2013)
6. Beaver, D.: Precomputing oblivious transfer. In: Coppersmith, D. (ed.) CRYPTO 1995. LNCS, vol. 963, pp. 97–109. Springer, Heidelberg (1995). https://doi.org/10.1007/3-540-44750-4_8
7. Bendlin, R., Damgård, I., Orlandi, C., Zakarias, S.: Semi-homomorphic encryption and multiparty computation. In: Paterson [53], pp. 169–188
8. Bertoni, G., Breveglieri, L., Koren, I., Maistri, P., Piuri, V.: Error analysis and detection procedures for a hardware implementation of the advanced encryption standard. IEEE Trans. Comput. **52**(4), 492–505 (2003)

9. Bilgin, B., Gierlichs, B., Nikova, S., Nikov, V., Rijmen, V.: Higher-order threshold implementations. In: Sarkar, P., Iwata, T. (eds.) ASIACRYPT 2014, Part II. LNCS, vol. 8874, pp. 326–343. Springer, Heidelberg (2014). https://doi.org/10.1007/978-3-662-45608-8_18

10. Boneh, D., DeMillo, R.A., Lipton, R.J.: On the importance of eliminating errors in cryptographic computations. J. Cryptol. 14(2), 101–119 (2001)

11. Boyar, J., Matthews, P., Peralta, R.: Logic minimization techniques with applications to cryptology. J. Cryptol. 26(2), 280–312 (2013)

12. Bringer, J., Carlet, C., Chabanne, H., Guilley, S., Maghrebi, H.: Orthogonal direct sum masking- a smartcard friendly computation paradigm in a code, with builtin protection against side-channel and fault attacks. In: Naccache, D., Sauveron, D. (eds.) WISTP 2014. LNCS, vol. 8501, pp. 40–56. Springer, Heidelberg (2014). https://doi.org/10.1007/978-3-662-43826-8_4

13. Bringer, J., Chabanne, H., Le, T.: Protecting AES against side-channel analysis using wire-tap codes. J. Cryptogr. Eng. 2(2), 129–141 (2012)

14. De Cannière, C., Dunkelman, O., Knežević, M.: KATAN and KTANTAN — a family of small and efficient hardware-oriented block ciphers. In: Clavier, C., Gaj, K. (eds.) CHES 2009. LNCS, vol. 5747, pp. 272–288. Springer, Heidelberg (2009). https://doi.org/10.1007/978-3-642-04138-9_20

15. Chari, S., Jutla, C.S., Rao, J.R., Rohatgi, P.: Towards sound approaches to counteract power-analysis attacks. In: Wiener [64], pp. 398–412

16. Cnudde, T.D., Nikova, S.: More efficient private circuits II through threshold implementations. In: FDTC 2016, pp. 114–124. IEEE Computer Society (2016)

17. Cooper, J., DeMulder, E., Goodwill, G., Jaffe, J., Kenworthy, G., Rohatgi, P.: Test Vector Leakage Assessment (TVLA) methodology in practice. In: International Cryptographic Module Conference (2013)

18. Coron, J.-S.: Higher order masking of look-up tables. In: Nguyen, P.Q., Oswald, E. (eds.) EUROCRYPT 2014. LNCS, vol. 8441, pp. 441–458. Springer, Heidelberg (2014). https://doi.org/10.1007/978-3-642-55220-5_25

19. Daemen, J., Rijmen, V.: The Design of Rijndael: AES - The Advanced Encryption Standard. Information Security and Cryptography. Springer, Heidelberg (2002). https://doi.org/10.1007/978-3-662-04722-4

20. Damgård, I., Pastro, V., Smart, N.P., Zakarias, S.: Multiparty computation from somewhat homomorphic encryption. In: Safavi-Naini and Canetti [60], pp. 643–662

21. Duc, A., Faust, S., Standaert, F.-X.: Making masking security proofs concrete. In: Oswald, E., Fischlin, M. (eds.) EUROCRYPT 2015, Part I. LNCS, vol. 9056, pp. 401–429. Springer, Heidelberg (2015). https://doi.org/10.1007/978-3-662-46800-5_16

22. Fischer, W., Homma, N. (eds.): CHES 2017. LNCS, vol. 10529. Springer, Cham (2017). https://doi.org/10.1007/978-3-319-66787-4

23. Gammel, B.M., Mangard, S.: On the duality of probing and fault attacks. J. Electron. Test. 26(4), 483–493 (2010)

24. Gandolfi, K., Mourtel, C., Olivier, F.: Electromagnetic analysis: concrete results. In: Koç, Ç.K., Naccache, D., Paar, C. (eds.) CHES 2001. LNCS, vol. 2162, pp. 251–261. Springer, Heidelberg (2001). https://doi.org/10.1007/3-540-44709-1_21

25. Gierlichs, B., Schmidt, J.-M., Tunstall, M.: Infective computation and dummy rounds: fault protection for block ciphers without check-before-output. In: Hevia, A., Neven, G. (eds.) LATINCRYPT 2012. LNCS, vol. 7533, pp. 305–321. Springer, Heidelberg (2012). https://doi.org/10.1007/978-3-642-33481-8_17

26. Goubin, L., Patarin, J.: DES and differential power analysis the "Duplication" method. In: Koç, Ç.K., Paar, C. (eds.) CHES 1999. LNCS, vol. 1717, pp. 158–172. Springer, Heidelberg (1999). https://doi.org/10.1007/3-540-48059-5_15

27. Groß, H., Mangard, S.: Reconciling d+1 masking in hardware and software. In: Fischer and Homma [22], pp. 115–136

28. Groß, H., Mangard, S., Korak, T.: Domain-oriented masking: compact masked hardware implementations with arbitrary protection order. IACR Cryptology ePrint Archive, 2016:486 (2016)

29. Gross, H., Mangard, S., Korak, T.: An efficient side-channel protected AES implementation with arbitrary protection order. In: Handschuh, H. (ed.) CT-RSA 2017. LNCS, vol. 10159, pp. 95–112. Springer, Cham (2017). https://doi.org/10.1007/978-3-319-52153-4_6

30. Grosso, V., Prouff, E., Standaert, F.-X.: Efficient masked S-boxes processing – a step forward –. In: Pointcheval, D., Vergnaud, D. (eds.) AFRICACRYPT 2014. LNCS, vol. 8469, pp. 251–266. Springer, Cham (2014). https://doi.org/10.1007/978-3-319-06734-6_16

31. Güneysu, T., Handschuh, H. (eds.): CHES 2015. LNCS, vol. 9293. Springer, Heidelberg (2015). https://doi.org/10.1007/978-3-662-48324-4

32. Guo, X., Mukhopadhyay, D., Jin, C., Karri, R.: Security analysis of concurrent error detection against differential fault analysis. J. Cryptogr. Eng. 5(3), 153–169 (2015)

33. Ishai, Y., Prabhakaran, M., Sahai, A., Wagner, D.: Private circuits II: keeping secrets in tamperable circuits. In: Vaudenay, S. (ed.) EUROCRYPT 2006. LNCS, vol. 4004, pp. 308–327. Springer, Heidelberg (2006). https://doi.org/10.1007/11761679_19

34. Ishai, Y., Sahai, A., Wagner, D.: Private circuits: securing hardware against probing attacks. In: Boneh, D. (ed.) CRYPTO 2003. LNCS, vol. 2729, pp. 463–481. Springer, Heidelberg (2003). https://doi.org/10.1007/978-3-540-45146-4_27

35. Joshi, N., Wu, K., Karri, R.: Concurrent error detection schemes for involution ciphers. In: Joye, M., Quisquater, J.-J. (eds.) CHES 2004. LNCS, vol. 3156, pp. 400–412. Springer, Heidelberg (2004). https://doi.org/10.1007/978-3-540-28632-5_29

36. Joye, M., Manet, P., Rigaud, J.: Strengthening hardware AES implementations against fault attacks. IET Inf. Secur. 1(3), 106–110 (2007)

37. Karpovsky, M., Kulikowski, K.J., Taubin, A.: Differential fault analysis attack resistant architectures for the advanced encryption standard. In: Quisquater, J.J., Paradinas, P., Deswarte, Y., El Kalam, A.A. (eds.) Smart Card Research and Advanced Applications VI. IFIP, vol. 153, pp. 177–192. Springer, Boston (2004). https://doi.org/10.1007/1-4020-8147-2_12

38. Karri, R., Kuznetsov, G., Goessel, M.: Parity-based concurrent error detection of substitution-permutation network block ciphers. In: Walter, C.D., Koç, Ç.K., Paar, C. (eds.) CHES 2003. LNCS, vol. 2779, pp. 113–124. Springer, Heidelberg (2003). https://doi.org/10.1007/978-3-540-45238-6_10

39. Karri, R., Wu, K., Mishra, P., Kim, Y.: Concurrent error detection schemes for fault-based side-channel cryptanalysis of symmetric block ciphers. IEEE Trans. CAD Integr. Circ. Syst. 21(12), 1509–1517 (2002)

40. Keller, M., Orsini, E., Scholl, P.: MASCOT: faster malicious arithmetic secure computation with oblivious transfer. In: Weippl, E.R., Katzenbeisser, S., Kruegel, C., Myers, A.C., Halevi, S. (eds.) ACM CCS 2016, pp. 830–842. ACM Press, October 2016

41. Kocher, P.C.: Timing attacks on implementations of Diffie-Hellman, RSA, DSS, and other systems. In: Koblitz, N. (ed.) CRYPTO 1996. LNCS, vol. 1109, pp. 104–113. Springer, Heidelberg (1996). https://doi.org/10.1007/3-540-68697-5_9

42. Kocher, P.C., Jaffe, J., Jun, B.: Differential power analysis. In: Wiener [64], pp. 388–397

43. Lomné, V., Roche, T., Thillard, A.: On the need of randomness in fault attack countermeasures - application to AES. In: Bertoni, G., Gierlichs, B. (eds.) FDTC 2012, pp. 85–94. IEEE Computer Society (2012)

44. Malkin, T.G., Standaert, F.-X., Yung, M.: A comparative cost/security analysis of fault attack countermeasures. In: Breveglieri, L., Koren, I., Naccache, D., Seifert, J.-P. (eds.) FDTC 2006. LNCS, vol. 4236, pp. 159–172. Springer, Heidelberg (2006). https://doi.org/10.1007/11889700_15

45. Medwed, M., Standaert, F.-X., Großschädl, J., Regazzoni, F.: Fresh re-keying: security against side-channel and fault attacks for low-cost devices. In: Bernstein, D.J., Lange, T. (eds.) AFRICACRYPT 2010. LNCS, vol. 6055, pp. 279–296. Springer, Heidelberg (2010). https://doi.org/10.1007/978-3-642-12678-9_17

46. Mitra, S., McCluskey, E.J.: Which concurrent error detection scheme to choose? In: Proceedings IEEE International Test Conference 2000, Atlantic City, NJ, USA, October 2000, pp. 985–994. IEEE Computer Society (2000)

47. Moradi, A., Poschmann, A., Ling, S., Paar, C., Wang, H.: Pushing the limits: a very compact and a threshold implementation of AES. In: Paterson [53], pp. 69–88

48. Mukhopadhyay, D.: An improved fault based attack of the advanced encryption standard. In: Preneel, B. (ed.) AFRICACRYPT 2009. LNCS, vol. 5580, pp. 421–434. Springer, Heidelberg (2009). https://doi.org/10.1007/978-3-642-02384-2_26

49. NANGATE: The NanGate 45nm Open Cell Library. http://www.nangate.com

50. Nielsen, J.B., Nordholt, P.S., Orlandi, C., Burra, S.S.: A new approach to practical active-secure two-party computation. In: Safavi-Naini and Canetti [60], pp. 681–700

51. Nikova, S., Rechberger, C., Rijmen, V.: Threshold implementations against side-channel attacks and glitches. In: Ning, P., Qing, S., Li, N. (eds.) ICICS 2006. LNCS, vol. 4307, pp. 529–545. Springer, Heidelberg (2006). https://doi.org/10.1007/11935308_38

52. Nikova, S., Rijmen, V., Schläffer, M.: Secure hardware implementation of non-linear functions in the presence of glitches. In: Lee, P.J., Cheon, J.H. (eds.) ICISC 2008. LNCS, vol. 5461, pp. 218–234. Springer, Heidelberg (2009). https://doi.org/10.1007/978-3-642-00730-9_14

53. Paterson, K.G. (ed.): EUROCRYPT 2011. LNCS, vol. 6632. Springer, Heidelberg (2011). https://doi.org/10.1007/978-3-642-20465-4

54. Patranabis, S., Chakraborty, A., Nguyen, P.H., Mukhopadhyay, D.: A biased fault attack on the time redundancy countermeasure for AES. In: Mangard, S., Poschmann, A.Y. (eds.) COSADE 2014. LNCS, vol. 9064, pp. 189–203. Springer, Cham (2015). https://doi.org/10.1007/978-3-319-21476-4_13

55. Prouff, E., Rivain, M.: Masking against side-channel attacks: a formal security proof. In: Johansson, T., Nguyen, P.Q. (eds.) EUROCRYPT 2013. LNCS, vol. 7881, pp. 142–159. Springer, Heidelberg (2013). https://doi.org/10.1007/978-3-642-38348-9_9

56. Reparaz, O., Bilgin, B., Nikova, S., Gierlichs, B., Verbauwhede, I.: Consolidating masking schemes. In: Gennaro, R., Robshaw, M. (eds.) CRYPTO 2015, Part I. LNCS, vol. 9215, pp. 764–783. Springer, Heidelberg (2015). https://doi.org/10.1007/978-3-662-47989-6_37

57. Reparaz, O., Gierlichs, B., Verbauwhede, I.: Fast leakage assessment. In: Fischer and Homma [22], pp. 387–399
58. Rivain, M., Prouff, E.: Provably secure higher-order masking of AES. In: Mangard, S., Standaert, F.-X. (eds.) CHES 2010. LNCS, vol. 6225, pp. 413–427. Springer, Heidelberg (2010). https://doi.org/10.1007/978-3-642-15031-9_28
59. Roche, T., Prouff, E.: Higher-order glitch free implementation of the AES using secure multi-party computation protocols - extended version. J. Cryptogr. Eng. **2**(2), 111–127 (2012)
60. Safavi-Naini, R., Canetti, R. (eds.): CRYPTO 2012. LNCS, vol. 7417. Springer, Heidelberg (2012). https://doi.org/10.1007/978-3-642-32009-5
61. Schneider, T., Moradi, A.: Leakage assessment methodology - a clear roadmap for side-channel evaluations. In: Güneysu and Handschuh [31], pp. 495–513
62. Schneider, T., Moradi, A., Güneysu, T.: ParTI – towards combined hardware countermeasures against side-channel and fault-injection attacks. In: Robshaw, M., Katz, J. (eds.) CRYPTO 2016, Part II. LNCS, vol. 9815, pp. 302–332. Springer, Heidelberg (2016). https://doi.org/10.1007/978-3-662-53008-5_11
63. Seker, O., Eisenbarth, T., Steinwandt, R.: Extending glitch-free multiparty protocols to resist fault injection attacks. IACR Cryptology ePrint Archive, 2017:269 (2017)
64. Wiener, M. (ed.): CRYPTO 1999. LNCS, vol. 1666. Springer, Heidelberg (1999). https://doi.org/10.1007/3-540-48405-1
65. Yen, S., Joye, M.: Checking before output may not be enough against fault-based cryptanalysis. IEEE Trans. Comput. **49**(9), 967–970 (2000)

Authenticated and Format-Preserving Encryption

Fast Message Franking: From Invisible Salamanders to Encryptment

Yevgeniy Dodis[1], Paul Grubbs[2]([envelope]), Thomas Ristenpart[2], and Joanne Woodage[3]

[1] New York University, New York, USA
dodis@cs.nyu.edu
[2] Cornell Tech, New York, USA
pag225@cornell.edu
[3] Royal Holloway, University of London, Egham, UK

Abstract. Message franking enables cryptographically verifiable reporting of abusive messages in end-to-end encrypted messaging. Grubbs, Lu, and Ristenpart recently formalized the needed underlying primitive, what they call compactly committing authenticated encryption (AE), and analyze security of a number of approaches. But all known secure schemes are still slow compared to the fastest standard AE schemes. For this reason Facebook Messenger uses AES-GCM for franking of attachments such as images or videos.

We show how to break Facebook's attachment franking scheme: a malicious user can send an objectionable image to a recipient but that recipient cannot report it as abuse. The core problem stems from use of fast but non-committing AE, and so we build the fastest compactly committing AE schemes to date. To do so we introduce a new primitive, called encryptment, which captures the essential properties needed. We prove that, unfortunately, schemes with performance profile similar to AES-GCM won't work. Instead, we show how to efficiently transform Merkle-Damgård-style hash functions into secure encryptments, and how to efficiently build compactly committing AE from encryptment. Ultimately our main construction allows franking using just a single computation of SHA-256 or SHA-3. Encryptment proves useful for a variety of other applications, such as remotely keyed AE and concealments, and our results imply the first single-pass schemes in these settings as well.

1 Introduction

End-to-end encrypted messaging systems including WhatsApp [40], Signal [38], and Facebook Messenger [13] have increased in popularity — billions of people now rely on them for security. In these systems, intermediaries including the messaging service provider should not be able to read or modify messages. Providers simultaneously want to provide abuse reporting: should one user send another a harmful message, image, or video, the recipient should be able to report the content to the service provider. End-to-end encryption would seem to prevent the provider from verifying that the reported message was the one sent.

© International Association for Cryptologic Research 2018
H. Shacham and A. Boldyreva (Eds.): CRYPTO 2018, LNCS 10991, pp. 155–186, 2018.
https://doi.org/10.1007/978-3-319-96884-1_6

Facebook suggested a way to navigate this tension in the form of message franking [14,30]. The idea is to enable the recipient to cryptographically prove to the service provider that the reported message was the one sent. Grubbs, Lu, and Ristenpart (GLR) [17] provided the first formal treatment of the problem, and introduced compactly committing authenticated encryption with associated data (ccAEAD) as the key primitive. A secure ccAEAD scheme is symmetric encryption for which a short portion of the ciphertext serves as a cryptographic commitment to the underlying message (and associated data). They detailed appropriate security notions and security proofs that provide validation of the main Facebook message franking approach and a faster custom ccAEAD scheme called Committing Encrypt-and-PRF (CEP).

The Facebook scheme composes HMAC (serving the role of a commitment) with a standard encrypt-then-MAC AEAD scheme. Their scheme therefore requires a full three cryptographic passes over messages. The CEP construction gets this down to two. But even that does not match the fastest standard AE schemes such as AES-GCM [28] and OCB [32]. These require at most one blockcipher call (on the same key) per block of message and some arithmetic operations in $GF(2^n)$, which are faster than a blockcipher invocation. As observed by GLR, however, these schemes are not compactly committing: one can find two distinct messages and two encryption keys that lead to the same tag. This violates what they call receiver binding, and could in theory allow a malicious recipient to report a message that was never sent.

Existing ccAEAD schemes are not considered fast enough for all applications of message franking by practitioners [30]. Facebook Messenger does not use the ccAEAD scheme mentioned above to directly encrypt attachments, rather using a kind of hybrid encryption combining ccAEAD of a symmetric key that is in turn used with AES-GCM to encrypt the attachment. Use of AES-GCM does not necessarily seem problematic despite the GLR results; the latter do not imply any concrete attack on Facebook's system.

Breaking Facebook's attachment franking. Our first contribution is to show an attack against Facebook's attachment franking scheme. The attack enables a malicious sender to transmit an abusive attachment (e.g., an objectionable image or video) to a receiver so that: (1) the recipient receives the attachment (it decrypts correctly), yet (2) reporting the abusive message fails — Facebook's systems essentially "lose" the abusive image, rendering them invisible from the abuse handling team. Instead what gets reported to Facebook is a different, innocuous image. See Fig. 3.

Perhaps confusingly, our attack does not violate the primary reason for requiring receiver binding in committing AE (preventing a malicious recipient from framing a user as having sent a message they didn't send). Instead it violates what GLR call sender binding security: a malicious sender should not be able to force an abusive message to be received by the recipient, yet that recipient can't report it properly. Nevertheless, the root cause of this vulnerability in Facebook's case is the use of an AE scheme that is not a binding commitment to its message or, equivalently in this context, that is not a robust encryption scheme [1,15,16].

Briefly, Facebook uses a cryptographic hash of the AES-GCM ciphertext, along with a randomly-generated value, as an identifier for the attachment. For a given abusive message, our attack efficiently finds two keys and a ciphertext, such that the first key decrypts the ciphertext to the abusive attachment while the other key successfully decrypts the same ciphertext, but to another innocuous attachment. The malicious sender transmits two messages with the different keys but the same attachment ciphertext. Facebook's systems deduplicate the two attachments, and the report will only include the non-abusive image.

We responsibly disclosed this vulnerability to Facebook, and in fact they helped us understand how our attack works against their systems (much of the abuse handling code is server-side and closed source). The severity of the issue led them to patch their (server-side) systems and to award us a bug bounty. Their fix is ad hoc and involves deduplicating more carefully. But the vulnerability would have been avoided in the first place by using a fast ccAEAD scheme that provided the binding security properties implicitly assumed of, but not actually provided by, AES-GCM.

Towards faster ccAEAD schemes: encryptment.   This message franking failure motivates the need for faster schemes. As mentioned, the best known secure ccAEAD scheme from GLR is two pass, requiring computing both HMAC and AES-CTR mode (or similar) over the message. The fastest standard AE schemes [22,28,32], however, require just a single pass using a blockcipher with a single key. Can we build ccAEAD schemes that match this performance?

To tackle this question we first abstract out the core technical challenge underlying ccAEAD: building a one-time encryption mechanism that simultaneously encrypts and compactly commits to the message. We formalize this in a new primitive that we call *encryptment*. An encryptment of a message using a key K_{EC} is a pair (C_{EC}, B_{EC}) where C_{EC} is a ciphertext and B_{EC} is a binding tag. By compactness we require that $|B_{EC}|$ is independent of the length of the message. Decryption takes as input K_{EC}, C_{EC}, B_{EC} and returns a message (or $\perp$). Finally, there is a verification algorithm that takes a key, a message, and a binding tag, and determines whether the tag is a commitment to the message. Encryptment supports associated data also, but we defer the details to the body.

We introduce security notions for encryptment. These include a real-or-random style confidentiality goal in which the adversary must distinguish between a single encryptment and an appropriate-length sequence of random bits. Additionally we require sender binding and receiver binding notions like those from GLR (but adapted to the encryptment syntax), and finally a strong correctness property that is easy to meet. Comparatively, GLR require many-time confidentiality and integrity notions in addition to various binding notions.

Therefore encryptment is substantially simpler than ccAEAD, making analyses easier and, we think, design of constructions more intuitive. At the same time, we will be able to build ccAEAD from encryptment using simple, efficient transforms. In the other direction, we show that one can also build encryptment from ccAEAD, making the two primitives equivalent from a theoretical perspective. Encryptment also turns out to be the "right" primitive for a number of other applications: robust authenticated-encryption [1,15,16], concealments [12], remotely keyed authenticated encryption [12], and perhaps even more.

Fast encryptment from fixed-key blockciphers?   Given a simpler formulation in hand, we turn to building fast schemes. First, we show a negative result: encryptment schemes cannot match the efficiency profile of OCB or AES-GCM. In fact we rule out any scheme that uses just a single blockcipher invocation for each block of message, with some fixed small set of keys.

The negative result makes use of a connection between encryption and collision-resistant (CR) hashing. Because encryptment schemes are deterministic, we can think of the computation of a binding tag B_{EC} as a deterministic function $F(K_{EC}, M)$ applied to the key and message; verification simply checks that $F(K_{EC}, M) = B_{EC}$. Then, receiver binding is achieved if and only if F is CR: the adversary shouldn't be able to find $(K_{EC}, M) \neq (K'_{EC}, M')$ such that $F(K_{EC}, M) = F(K'_{EC}, M')$.

Given this connection, we can exploit previous work on ruling out fixed-key blockcipher-based CR hashing [34,35,37]. A simple corollary of [35, Theorem 1] is that one cannot prove receiver binding security for any rate-1 fixed-key blockcipher-based encryptment. (Rate-1 meaning one blockcipher call per block of message.) Since OCB and AES-GCM fall into this category of rate-1, they don't work, but neither do other similar blockcipher-based schemes. Our negative result also rules out rate-1 ccAEAD, due to our aforementioned result that (fast) ccAEAD implies (fast) encryptment.

One-pass encryptment from hashing.   Given the connection just mentioned, it is natural to turn to CR hashing as a starting point for building as-fast-as-possible encryptment. We do so and show how to achieve secure encryptment using just a single pass of a secure cryptographic hash function. The encryptment can be viewed as a mode of operation of a fixed-input-length compression function, such as the one underlying SHA-256 or other Merkle-Damgård style constructions.

Let $f(x, y)$ be a compression function on two n-bit inputs and with output an n-bit string. Then our HFC (hash function chaining) encryptment works as shown in Fig. 8. Basically one hashes $K_{EC} \| (M_1 \oplus K_{EC}) \| \cdots \| (M_2 \oplus K_{EC})$ using a standard iteration of f. But, additionally, one uses the intermediate chaining values as pads to encrypt the message blocks. Decryption simply computes the hash, recovering message blocks as it goes.

We prove that our HFC scheme is a secure encryptment. Binding is inherited from the CR of the underlying hash function. We show confidentiality assuming $f(x, y \oplus K_{EC})$ is a related-key-attack-secure pseudorandom function (RKA-PRF) [3] when keyed by K_{EC}. For standard designs, such as the Davies-Meyer construction $f(x, y \oplus K_{EC}) = E(y \oplus K_{EC}, x) \oplus x$, we can reduce RKA-PRF security to RKA-PRP security of the underlying blockcipher E. This property is already an active target of cryptographic analysis for standard E (such as AES), giving us confidence in the assumption. Because SHA-256 uses a DM-style compression function, this also gives confidence for using SHA-256 (or SHA-384, SHA-512).

From a theoretical perspective, one might want to avoid relying on RKA security (compared to standard PRF security). We discuss approaches for doing so in the body, but the resulting constructions are not as fast or elegant as HFC.

HFC has some features in common with the Duplex authenticated-encryption mode [6] using Keccak (SHA-3) [5]. In fact the Duplex mode gives rise to a secure encryption scheme as well. See the full version for a discussion. The way we key in HFC is also similar to the Halevi-Krawczyk construction for reducing the assumptions needed on hash functions in digital signature settings [20], but the keying serves a different role here and their analysis techniques are not applicable.

From encryption to ccAEAD. We show several efficient transforms for building a ccAEAD scheme given a secure encryption. First consider doing so given also a secure (standard) AE scheme. To encrypt a message M, first generate a random key K_{EC} and then compute an encryption (C_{EC}, B_{EC}) for K_{EC}, M. Encrypt K_{EC} under the long-lived AE key K using as associated data the binding tag B_{EC}. The resulting ciphertext is the AE ciphertext (including its authentication tag) along with C_{EC}, B_{EC}. We prove that this transformation provides the multi-opening confidentiality and integrity goals for ccAEAD of GLR, assuming the standard security of the AE scheme and the aforementioned security goals are met for the encryption scheme.

One can instead use just two additional PRF calls to securely convert an encryption scheme to a ccAEAD scheme. One can, for example, instantiate the PRF with the SHA-256 compression function, to have a total cost of at most $m + 4$ SHA-256 compression function calls for a message that can be parsed into m blocks of 256 bits. Another transform uses a single tweakable blockcipher call in addition to the encryption. See the full version for details.

Our approach of hashing-based ccAEAD has a number of attractive features. HFC works with any hash function that iterates a secure compression function, giving us a wide variety of options for instantiation. Because of our simplified formalization via encryption, the security proofs are modular and conceptually straightforward. As already mentioned it is fast in terms of the number of underlying primitive calls. If instantiated using SHA-256, one can use the SHA hardware instructions [18] now supported on some AMD and ARM processors, and that are likely to be incorporated in future Intel processors. Finally, HFC-based ccAEAD is simple to implement.

Other applications. Encryption proves a useful abstraction for other applications as well. In the full version of this work, we show how it suffices for building concealments [12] (a conceptually similar, but distinct, primitive) which, in turn, can be used to build remotely keyed AE [12]. Previous constructions of these required two passes over the message. Our new encryption-based approach gives the first single-pass concealments and remotely keyed AE. Finally, encryption schemes give rise to robust AE [15] via some of our transforms mentioned above. We expect that encryption will find further applications in the future.

2 Definitions and Preliminaries

Preliminaries. For an alphabet Σ, we let Σ^* denote the set of all strings of symbols from that alphabet, and let Σ^n denote the set of all such strings of length n. For a string $x \in \Sigma^*$, we write $|x|$ to denote the length of the string x. We let ε denote the empty string, and $\perp$ denote the distinguished error symbol. We write $x \leftarrow_\$ \mathcal{X}$ to denote choosing an element at random from the set $\mathcal{X}$.

We define the XOR of two strings of different lengths to return the XOR of the shorter string and the truncation of the longer string to the length of the shorter string. Our proofs assume a RAM model of computation where most operations are unit cost. We use big-O notation $\mathcal{O}(\cdot)$ to hide small constants related to the internal data structures (e.g., tables of queries) used by reductions.

For a deterministic algorithm A, we write $y \leftarrow A(x_1, \dots)$ to denote running A on inputs $x_1, \dots$ to produce output y. For a probabilistic algorithm A with associated coin space $\mathcal{C}$, we write $y \leftarrow_\$ A(x_1, \dots)$ to denote choosing coins $c \leftarrow_\$ \mathcal{C}$ and returning $y \leftarrow A(x_1, \dots; c)$, where $y \leftarrow A(x_1, \dots; c)$ denotes running A on the given inputs with coins c fixed, to deterministically produce output y.

Collision-resistant functions. Let $\mathcal{H} : Dom \to \{0, 1\}^n$ be a function on some domain $Dom \subset \{0, 1\}^*$. The collision resistance game CR has $\mathcal{A}$ run and output a pair of messages X, X'. If analysis is with respect to an ideal primitive such as an ideal cipher, then $\mathcal{A}$ is given oracle access to this primitive also. The game outputs true if $\mathcal{H}(X) = \mathcal{H}(X')$ and $X \neq X'$. The CR advantage of an adversary $\mathcal{A}$ against $\mathcal{H}$ is defined $\mathbf{Adv}_{\mathcal{H}}^{\mathrm{cr}}(\mathcal{A}) = \Pr\left[\, \mathrm{CR}_{\mathcal{H}}^{\mathcal{A}} \Rightarrow \mathsf{true} \,\right]$, where the probability is over the coins of $\mathcal{A}$ and those of any ideal primitive. We measure the efficiency of the attacker in terms of their resources, e.g. run time or number of queries made to some underlying primitive.

For space reasons, we direct the reader to [33] for syntax and correctness notions for AEAD. We require that AEAD schemes offer both real-or-random confidentiality and ciphertext integrity. These will be formalized in Sect. 7.

3 Invisible Salamanders: Breaking Facebook's Franking

In this section we demonstrate an attack against Facebook's message franking. Facebook uses AES-GCM to encrypt attachments sent via Secret Conversations. The attack creates a "colliding" GCM ciphertext which decrypts to an abusive attachment via one key and an innocuous attachment via the other. This combined with the behavior of Facebook's server-side abuse report generation code prevents abusive messages from being reported to Facebook. Since messages in Secret Conversations are called "salamanders" by Facebook (perhaps inspired by the Axolotl ratchet used in Signal, named for an endangered salamander), ensuring Facebook does not see a message essentially makes it an *invisible salamander*. We responsibly disclosed the vulnerability to Facebook. They have remediated it and have given us a bug bounty for reporting the issue.

Facebook's attachment franking. A diagram of Facebook's franking protocol for attachments (e.g., images and videos) is in Fig. 1. The protocol uses CtE2, Facebook's ccAEAD scheme for chat messages described in [14,30] and analyzed in [17], as a subroutine. Some encryption and HMAC keys, as well as some other details like headers and associated data not important to the presentation of the protocol, have been removed for simplicity in the diagram and prose below. Consult [14,17] for additional details. For ease of exposition we divide the protocol into three phases: the *sending phase* involving the sender Alice and Facebook, the *receiving phase* involving the receiver Bob and Facebook, and the *reporting phase* between Bob and Facebook.

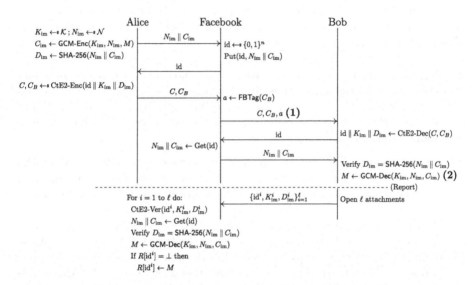

Fig. 1. Facebook's attachment franking protocol [29,30]. The sending phase consists of everything from the upper-left corner to the message marked (1). The receiving phase consists of everything strictly after (1) and before (2). The reporting phase is below the dashed line. The descriptions of Facebook's behavior during the reporting phase were paraphrased (with permission) from conversations with Jon Millican, whom the authors thank profusely.

Sending phase: In the first part of the sending phase, Alice generates a key K_{im} and nonce N_{im} and encrypts M_a using AES-GCM (described in pseudocode in Fig. 2) to obtain a ciphertext C_{im}. The sender computes the SHA-256 digest D_{im} of $N_{\mathrm{im}} \| C_{\mathrm{im}}$ and sends Facebook $N_{\mathrm{im}} \| C_{\mathrm{im}}$ for storage. Facebook generates a random identifier id and puts $N_{\mathrm{im}} \| C_{\mathrm{im}}$ in a key-value data structure with key id. Facebook then sends id to Alice. In the second part of the sending phase, Alice encrypts the message $\mathrm{id} \| K_{\mathrm{im}} \| D_{\mathrm{im}}$ using CtE2 to obtain the ccAEAD ciphertext C, C_B. Below, we will call a message containing an identifier, key and digest an "attachment metadata" message. Alice sends C, C_B to Facebook, which runs FBTag on C_B (this amounts to HMAC-SHA256 with an internal

Facebook key and some metadata) as in the standard message franking protocol to obtain a. Facebook sends C, C_B, a to the receiver.

Receiving phase: Upon receiving a message C, C_B, a from Alice (via Facebook), Bob runs CtE2-Dec on C, C_B to obtain $\text{id} \,\|\, K_{\text{im}} \,\|\, D_{\text{im}}$. Bob then sends id to Facebook, which gets the value $N_{\text{im}} \,\|\, C_{\text{im}}$ associated with id in its key-value store and sends it to Bob. Bob verifies that $D_{\text{im}} = \text{SHA-256}(N_{\text{im}} \,\|\, C_{\text{im}})$ and decrypts C_{im} to obtain the attachment content M_a.

Reporting phase: Bob sends all recent messages to Facebook along with their commitment openings and a values (not pictured in the diagram). For each message, Facebook verifies the commitment using CtE2-Ver and the authentication tag a using its internal HMAC key. Then, if the commitment verifies correctly and the message contains attachment metadata, Facebook gets the attachment ciphertext and nonce $N_{\text{im}} \,\|\, C_{\text{im}}$ from its key-value store using its identifier id. Facebook verifies that $D_{\text{im}} = \text{SHA-256}(N_{\text{im}} \,\|\, C_{\text{im}})$ and decrypts C_{im} with K_{im} and N_{im} to obtain the attachment content M_a. If no other attachment metadata message containing identifier id has already been seen, the plaintext M_a is added to the abuse report R. (Looking ahead, this is the application-level behavior that enables the attack, which will violate the one-to-one correspondence between id and plaintext that is assumed here.)

Attack intuition. The threat model of this attack is a malicious Alice who wants to send an abusive attachment to Bob, but prevent Bob from reporting it to Facebook. The attachment can be an offensive image (e.g., a picture of abusive text or of a gun) or video. We focus our discussion below on images.

The attack has two main steps: (1) generating the colliding ciphertext and (2) sending it twice to Bob. In step (1), Alice creates two GCM keys and a single GCM ciphertext which decrypts (correctly) to the abusive attachment under one key and to a different attachment under the other key. In step (2), Alice sends the ciphertext to Facebook and gets an identifier back. Alice then sends the identifier to Bob twice, once with each key.

On receiving the two messages, Bob decrypts the image twice and sees both the abusive attachment and the other one. When Bob reports the conversation to Facebook, its server-side code verifies both decryptions of the image ciphertext but only inserts the other decryption into the abuse report—the human making the abusive-or-not judgment will have no idea Bob saw the abusive attachment.

We will describe two variants of the attack. We will begin with the case where the second decryption of the colliding ciphertext is junk bytes with no particular structure. This variant is simple but easily detectable, since the junk bytes will not display correctly. Then we give a more advanced variant where the second decryption correctly displays an innocuous attachment, like a picture of a kitten.

Generating the colliding ciphertext—simple variant. Alice begins the attack with an abusive attachment M_a^{ab}. Alice chooses two distinct 128-bit GCM keys K_1 and K_2 and a nonce N_{im}, then computes a ciphertext C_a via $\text{CTR-Enc}(K_1, N_{\text{im}} + 2, M_a^{\text{ab}})$, where CTR-Enc denotes CTR-mode encryption with the given key and nonce. The nonce is $N_{\text{im}} + 2$ to match GCM, see Fig. 2.

In Facebook's scheme Alice can choose the keys and the nonce, but this is not necessary—any combination of two keys and a nonce will work.

The ciphertext C_a is almost, but not quite, the ciphertext Alice will use in the attack. To ensure GCM decryption is correct for both keys, Alice generates the colliding GCM tag and final ciphertext block using Collide-GCM(K_1, K_2, N_{im}, C_a) (described in Fig. 2). The function Collide-GCM works by computing the tags for the two keys then solving a linear equation to find the value of the last ciphertext block. We use the final ciphertext block as the variable, but a different ciphertext block or a block of associated data could be used instead. The output $N_{im} \| C_{im} \| T$ correctly decrypts to M_a^{ab} under K_1 and to another plaintext M_j under K_2. However, the plaintext M_j will be random bytes with no structure.

GCM(K, N, AD, M):	Collide-GCM(K_1, K_2, N_{im}, C_a):
$H \leftarrow E_K(0^{128})$	$H_1 \leftarrow E_{K_1}(0^{128})\,;\,H_2 \leftarrow E_{K_2}(0^{128})$
$P \leftarrow E_K(N+1)$	$P_1 \leftarrow E_{K_1}(N_{im}+1)\,;\,P_2 \leftarrow E_{K_2}(N_{im}+1)$
lens $\leftarrow$ encode$_{64}$($\|AD\|$) $\|$ encode$_{64}$($\|M\|$)	mlen $\leftarrow |M_a|/128$
$T \leftarrow$ lens $\cdot H + P$	lens $\leftarrow$ encode$_{64}$(0) $\|$ encode$_{64}$($|M_a|$)
mlen $\leftarrow |M|/128$	acc $\leftarrow$ lens $\cdot (H_1 + H_2) + P_1 + P_2$
adlen $\leftarrow |AD|/128$	For $i = 1$ to mlen $- 1$:
blen $\leftarrow$ mlen $+$ adlen	$\quad$ acc $\leftarrow$ acc $+ C_i \cdot (H_1^{mlen+2-i} + H_2^{mlen+2-i})$
For $i = 1$ to adlen:	$\quad C \leftarrow C \| C_i$
$\quad T \leftarrow T + AD[i] \cdot H^{blen+2-i}$	inv $\leftarrow (H_1^2 + H_2^2)^{-1}$
For $i = 1$ to mlen:	$C_{mlen} \leftarrow$ acc $\cdot$ inv
$\quad C_i \leftarrow E_K(N_{im} + 1 + i) + M[i]$	$C_{im} \leftarrow C \| C_{mlen}$
$\quad T \leftarrow T + C_i \cdot H^{blen+2-i-adlen}$	$T \leftarrow$ GHASH(H_1, C_{im}) $+ P_1$
$\quad C \leftarrow C \| C_i$	Return $N_{im} \| C_{im} \| T$
Return $AD \| N_{im} \| C_{im} \| T$	

Fig. 2. (Left) The Galois/Counter block cipher mode. **(Right)** The Collide-GCM algorithm. Array indexing is done in terms of 128-bit blocks. We assume all input bit lengths are multiples of 128 for simplicity, and that the input M_a to Collide-GCM is at least two blocks in length. The function GHASH is the standard GCM polynomial hash (the lines which assign to T on the left). The function encode$_{64}(\cdot)$ returns a 64-bit representation of its input. Arithmetic is in GF(2^{128}). The function Collide-GCM can take arbitrary headers, but we elide them for simplicity.

Sending the colliding ciphertext. Alice continues the sending phase with Facebook, obtaining an identifier id for the ciphertext $N_{im} \| C_{im}$. Alice then creates two attachment metadata messages: $MD_1 = $ id $\| K_2 \| D_{im}$ and $MD_2 = $ id $\| K_1 \| D_{im}$. Alice completes the remainder of the sending phase twice, first with MD_1 and then with MD_2. (The first message sent is associated to the junk message.) After finishing the receiving phase for MD_1, Bob will decrypt C_{im} with K_2, giving M_j. After finishing the receiving phase with MD_2, Bob will decrypt C_{im} with K_1 and see M_a^{ab}. We emphasize that both attachment metadata messages are valid, and no security properties of CtE2 are violated.

When Bob reports the recent messages, Facebook will verify both MD_1 and MD_2 and check the digest D_{im} matches the value $N_{im} \| C_{im}$ stored with identifier id. *However, it will only insert the first decryption, the plaintext* M_j, *into the abuse report.* The system sees the second ciphertext has the same SHA-256 hash and identifier, and assumes it's a duplicate: the human viewing the report will have no idea Bob ever saw the message M_a^{ab}.

3.1 Advanced Variant and Proof of Concept

Next we will describe the advanced variant of the attack (in which both decryptions correctly display as attachments) and our proof-of-concept implementation. Ensuring both decryptions are valid attachments is important because the simple variant (where one decryption is random bytes) may not have sufficed for a practical exploit if Facebook only inserted valid images into their abuse reports. We implemented the advanced variant and crafted a colliding ciphertext for which the "abusive" decryption M_a^{ab} is the image of an Axolotl salamander in Fig. 3. The innocuous decryption M_j is the image of a kitten in that figure. We verified both display correctly in Facebook Messenger's browser client.

Fig. 3. Two images with the same GCM ciphertext $C_{im} \parallel T$ when encrypted using 16-byte key $K_1 = (03)^{16}$ or $K_2 = (02)^{16}$, nonce $N_{im} = 10606665379$, and associated data $H = (ad)^{32}$ (all given in hex where exponentiation indicates repetition). **(Left)** The titular invisible salamander, which is the image delivered to the recipient. **(Right)** An image of a kitten that is put in the recipient's abuse report instead of the salamander.

The only difference between the advanced variant and the one described above is the way Alice generates the ciphertext C_a which is input to Collide-GCM. Instead of simply encrypting the abusive attachment M_a^{ab}, Alice first merges M_a^{ab} and another innocuous attachment M_j using a function Att-Merge(K_1, K_2, M_a^{ab}, M_j) which takes the two keys and attachments and outputs a nonce N_{im} and C_a so that CTR-Dec($K_1, N_{im} + 2, C_a$) displays M_a^{ab} and CTR-Dec($K_2, N_{im} + 2, C_a$) displays M_j. The exact implementation of Att-Merge is file-format-specific, but for most formats Att-Merge has two main steps: (1) a *nonce search* yielding a nonce which gives a collision on some region of the ciphertext, and (2) a *plaintext restructuring* that expands the plaintexts with random bytes in locations that are ignored by parsers for their respective file formats. We implemented Att-Merge for JPEG and BMP images (the salamander image and the kitten image, respectively), so our discussion will focus on these formats.

Before discussing our implementation of Att-Merge we will briefly describe the JPEG and BMP file formats. JPEG files *must* begin with the two-byte sequence ffd8 and end with ffd9. JPEGs can have comments. They are indicated with the two-byte sequence fffe followed by a big-endian two-byte encoding of the comment length. BMP files *must* begin with 424d, and the next four bytes *must* be the length block. The length block in a BMP file is a four-byte (little-endian) encoding of the file length. All the BMP parsers we used only read the number of bytes indicated in the header and ignore trailing bytes.

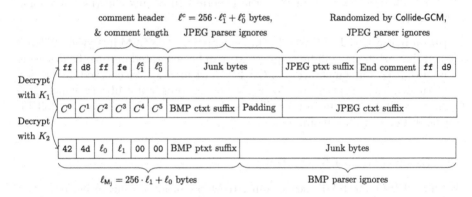

Fig. 4. Diagram of the JPEG M_a^{ab} **(top)** and BMP M_j **(bottom)** plaintexts output by the plaintext restructuring step, and their ciphertext **(middle)**. The leftmost block of each file is the first byte. The "BMP ptxt suffix" is the suffix of the original BMP starting at byte 6. The "JPEG ptxt suffix" is the bytes of the original JPEG starting at byte 2 and ending before the final two bytes. The region marked "End comment" begins with the comment header and comment length bytes (which are *not* randomized by Collide-GCM), but we do not depict them for simplicity.

Nonce search. Since file formats generally have some internal structure (like having a fixed byte sequence at the beginning or end) Att-Merge must choose a nonce so that the keystreams for the two keys respect this structure. JPEG and BMP files must begin with different fixed two-byte sequences, so the keystreams XORed with those sequences must result in a collision for the first two bytes. The plaintext restructuring step will need the JPEG to have a comment header in the next two bytes, which in the BMP plaintext contain the file length. Thus, the nonce output by Att-Merge must produce a collision in the first four bytes of the ciphertext (marked C^0 through C^4 in Fig. 4), which happens for about one in 2^{32} nonces. We wrote a simple Python script to search through nonces until we found 10606665379, which produces the required collision. Finding that nonce took roughly three hours on a 3.4 GHz quad-core Intel i7.

Plaintext restructuring. After the nonce search, the two plaintexts can be restructured. For JPEG and BMP images Att-Merge performs the following steps: (1) inserting the decryption (under K_1) of the BMP ciphertext into a comment region at the beginning of the JPEG, (2) inserting an additional comment at the end of the JPEG so the bytes randomized by Collide-GCM are ignored by the JPEG parser, and (3) appending the decryption (under K_2) of the JPEG ciphertext to the end of the BMP plaintext. See Fig. 4 for a diagram of the JPEG and BMP plaintexts after restructuring.

One important subtlety is that JPEG comments are at most 2^{16} bytes in length, so the BMP image must be smaller than 2^{16} bytes. In fact, it is advantageous for the BMP to be as small as possible because the comment length bytes in the JPEG are not fixed by the nonce search. A more detailed explanation of this issue and plaintext restructuring in general will be given in the full version of this work.

Implementing Collide-GCM. We implemented Collide-GCM in Python 2.7 and verified that arbitrary colliding ciphertexts can be generated in roughly 45 s using an unoptimized implementation of $GF(2^{128})$ arithmetic. We checked decryption correctness using cryptography.io, a Python cryptography library which uses OpenSSL's GCM implementation. This sufficed as a proof-of-concept exploit for Facebook's engineering team.

3.2 Discussion and Mitigation

We chose JPEG and BMP files for our Att-Merge proof of concept because their formats can tolerate random bytes in different regions of the file (the beginning and the end, respectively). We did not try to extend the Att-Merge to other common image formats but it is possible. We did not try to implement Att-Merge for video file formats. Such formats are substantially more complex than image formats, but we conjecture it is possible to extend the attack to video files.

Relation to GLR. In [17] GLR proved CtE2 is a ccAEAD scheme, and one may wonder whether this attack shows their proof is incorrect. Their proof only applies to CtE2 itself, not to the composition of CtE2 and GCM. Concretely, GLR analyzed CtE2 as it is used for text chat messages in Messenger, but did not analyze how it is used for attachments. This attack points to a gap between GLR's analysis and what Facebook actually uses, but it does not mean GLR's proof is incorrect. Indeed, the fact that the attack works without breaking CtE2's binding highlights the surprising subtlety of security notions for this setting.

The Collide-GCM algorithm in Fig. 2 is related to the r-BIND attack against GCM given by GLR [17]. However, their attack is insufficient to exploit Facebook's attachment franking—it only creates ciphertexts with colliding tags, but not the same ciphertext. Thus using it against Facebook wouldn't work, because the SHA-256 hashes of the two images would not collide. The Collide-GCM algorithm works even if the entire ciphertext, including any headers and the nonce, act as the commitment and the only opening is the encryption key.

Mitigating the attack. There are two main ways this attack can be mitigated. The first is a "server-software-only" patch that ensures abuse reports containing attachments are not deduplicated by attachment identifier. The second is changing the Messenger clients to use a ccAEAD scheme instead of GCM to encrypt attachments. In response to our bug report, Facebook deployed the first mitigation, primarily because it did not require patching the Messenger clients (an expensive and time-consuming process). Despite requiring less engineering effort, we believe this mitigation has some important drawbacks. Most notably, it leaves the underlying cryptographic issue intact: attachments are still encrypted using GCM. This means future changes to either the Messenger client or Facebook's server-side code could re-expose the vulnerability. Using a ccAEAD in place of GCM for attachment encryption would immediately prevent any deduplication behavior from being exploited, since the binding security of ccAEAD implies attachment identifiers uniquely identify the attachment *plaintexts*.

4   A New Primitive: Encryptment

In this section, we introduce a new primitive called an *encryptment scheme*. Encryptment schemes allow both encryption of, and commitment to[1], a message. Moreover, the schemes which we target and ultimately build achieve both security goals with only a *single* pass over the underlying data.

While the syntax of encryptment schemes is similar to that of the ccAEAD schemes we ultimately look to build, the key difference is that we expect far more minimal security notions from encryptment schemes (see Sect. 7 for a more detailed discussion). Looking ahead, we shall see that a secure encryptment scheme is the key building block for more complex primitives such as ccAEAD schemes, robust encryption [1,15,16], cryptographic concealments [12], and domain extension for authenticated encryption and remotely keyed AE [12], facilitating the construction of very efficient instantiations of these primitives. In Sect. 7.3 we show how to build ccAEAD from encryptment. The other primitives are deferred to the full version of this work.

Encryptment schemes. Applying the encryptment algorithm to a given key, header and message tuple (K_{EC}, H, M) returns a pair (C_{EC}, B_{EC}) which we call an *encryptment*. We refer to encryptment component C_{EC} as the *ciphertext*, and to B_{EC} as the *binding tag*. Together the ciphertext/binding tag pair (C_{EC}, B_{EC}) function as an encryption of M under key K_{EC}, so that given $(K_{EC}, H, C_{EC}, B_{EC})$, the opening algorithm DO can recover the underlying message M. The binding tag B_{EC} simultaneously acts as a commitment to the underlying header and message, with opening K_{EC}; the validity of this commitment to a given pair (H, M) is checked by the verification algorithm EVer. Looking ahead, we will

---

[1] A secure *commitment* allows a user to commit to a message without revealing its content; see [10] for further discussion.

actually require that B_{EC} acts as a commitment to the opening K_{EC} also, in that it should be infeasible to find $K_{EC} \neq K'_{EC}$ which verify the same B_{EC}.

Formally an encryptment scheme is a tuple $EC = (EKg, EC, DO, EVer)$ defined as follows. Associated to the scheme is a key space $\mathcal{K}_{EC} \subseteq \Sigma^*$, header space $\mathcal{H}_{EC} \subseteq \Sigma^*$, message space $\mathcal{M}_{EC} \subseteq \Sigma^*$, ciphertext space $\mathcal{C}_{EC} \subseteq \Sigma^*$, and binding tag space $\mathcal{T}_{EC} \subseteq \Sigma^*$.

- The randomized key generation EKg algorithm takes no input, and outputs a key $K_{EC} \in \mathcal{K}_{EC}$.
- The encryptment algorithm EC is a deterministic algorithm which takes as input a key $K_{EC} \in \mathcal{K}_{EC}$, a header $H \in \mathcal{H}_{EC}$, and a message $M \in \mathcal{M}_{EC}$, and outputs an encryptment $(C_{EC}, B_{EC}) \in \mathcal{C}_{EC} \times \mathcal{T}_{EC}$.
- The decryptment algorithm DO is a deterministic algorithm which takes as input a key $K_{EC} \in \mathcal{K}_{EC}$, a header $H \in \mathcal{H}_{EC}$, and an encryptment $(C_{EC}, B_{EC}) \in \mathcal{C}_{EC} \times \mathcal{T}_{EC}$, and outputs a message $M \in \mathcal{M}_{EC}$ or the error symbol $\perp$. We assume that if $(K_{EC}, H, C_{EC}, B_{EC}) \notin \mathcal{K}_{EC} \times \mathcal{H}_{EC} \times \mathcal{C}_{EC} \times \mathcal{T}_{EC}$, then $\perp \leftarrow DO(K_{EC}, H, C_{EC}, B_{EC})$.
- The verification algorithm $EVer$ is a deterministic algorithm which takes as input a header $H \in \mathcal{H}_{EC}$, a message $M \in \mathcal{M}_{EC}$, a key $K_{EC} \in \mathcal{K}_{EC}$, and a binding tag $B_{EC} \in \mathcal{T}_{EC}$, and returns a bit b. We assume that if $(H, M, K_{EC}, B_{EC}) \notin \mathcal{H}_{EC} \times \mathcal{M}_{EC} \times \mathcal{K}_{EC} \times \mathcal{T}_{EC}$ then $0 \leftarrow EVer(H, M, K_{EC}, B_{EC})$.

Length regularity and compactness. We impose two requirements on the lengths of the encryptments output by encryptment schemes. First, we require *compactness*: that the binding tags B_{EC} output by an encryptment scheme are of constant length btlen *regardless* of the length of the underlying message, and that btlen is linear in the key size. Second, we require *length regularity*: that the length of ciphertexts C_{EC} depend only on the length of the underlying message. Formally, we require there exists a function $\mathsf{clen} \colon \mathbb{N} \to \mathbb{N}$ such that for all $(H, M) \in \mathcal{H}_{EC} \times \mathcal{M}_{EC}$ it holds that $|C_{EC}| = \mathsf{clen}(|M|)$ with probability one for the sequence of algorithm executions: $K_{EC} \leftarrow_\$ EKg$; $(C_{EC}, B_{EC}) \leftarrow EC(K_{EC}, H, M)$.

Correctness. We define two correctness notions for encryptment schemes, which we formalize via the games COR and S-COR shown in Fig. 5. We require that *all* encryptment schemes satisfy our all-in-one *correctness* notion, which requires that honestly generated encryptments both decrypt to the correct underlying message, and successfully verify, with probability one. Formally, we say that an encryptment scheme $EC = (EKg, EC, DO, EVer)$ is correct if for all header/message pairs $(H, M) \in \mathcal{H}_{EC} \times \mathcal{M}_{EC}$, it holds that $\Pr[\,COR_{EC}(H, M) \Rightarrow 1\,] = 1$, where the probability is over the coins of EKg.

$\underline{COR_{EC}(H, M)}$
$K_{EC} \leftarrow_\$ EKg$
$(C_{EC}, B_{EC}) \leftarrow EC(K_{EC}, H, M)$
$M' \leftarrow DO(K_{EC}, H, C_{EC}, B_{EC})$
$b \leftarrow EVer(H, M', K_{EC}, B_{EC})$
Return $(M = M' \wedge b = 1)$

$\underline{S\text{-}COR_{EC}(K_{EC}, H, C_{EC}, B_{EC})}$
$M \leftarrow DO(K_{EC}, H, C_{EC}, B_{EC})$
$(C'_{EC}, B'_{EC}) \leftarrow EC(K_{EC}, H, M)$
Return $((C_{EC}, B_{EC}) = (C'_{EC}, B'_{EC}))$

Fig. 5. Correctness games for an encryptment scheme $EC = (EKg, EC, DO, EVer)$.

We additionally define *strong correctness*, which requires that for each tuple $(K_{EC}, H, M) \in \mathcal{K}_{EC} \times \mathcal{H}_{EC} \times \mathcal{M}_{EC}$ there is a unique encryption (C_{EC}, B_{EC}) such that $M \leftarrow DO(K_{EC}, H, C_{EC}, B_{EC})$. We formalize this in game S-COR, and say that an encryption scheme $EC = (EKg, EC, DO, EVer)$ is strongly correct if for all tuples $(K_{EC}, H, C_{EC}, B_{EC}) \in \mathcal{K}_{EC} \times \mathcal{H}_{EC} \times \mathcal{M}_{EC} \times \mathcal{C}_{EC} \times \mathcal{T}_{EC}$, it holds that $\Pr[\text{S-COR}_{EC}(K_{EC}, H, C_{EC}, B_{EC}) \Rightarrow 1] = 1$. While we only require that encryption schemes satisfy correctness, the schemes we build will also possess the stronger property (which simplifies their security proofs). We note that strong correctness can be added to any encryption scheme by making DO recompute a ciphertext after decrypting, and returning $\perp$ if the two do not match; however for efficiency we target schemes which achieve strong correctness without this.

4.1 Security Goals for Encryption

We require encryption schemes to satisfy both one-time real-or-random (otROR) security, and a variant of one-time ciphertext integrity (SCU) which requires forging a ciphertext for a given binding tag with a known key; we motivate this variant below. The security games for both notions are shown in Fig. 6.

otROR0$_{EC}^{\mathcal{A}}$:	otROR1$_{EC}^{\mathcal{A}}$:	SCU$_{EC}^{\mathcal{A}}$:
$K_{EC} \leftarrow^\$ EKg$	query-made $\leftarrow$ false	$K_{EC} \leftarrow^\$ EKg$
query-made $\leftarrow$ false	$b \leftarrow^\$ \mathcal{A}^{\$(\cdot,\cdot)}$	win $\leftarrow$ false
$b \leftarrow^\$ \mathcal{A}^{enc(\cdot,\cdot)}$	Return b	query-made $\leftarrow$ false
Return b		$\varepsilon \leftarrow^\$ \mathcal{A}^{enc(\cdot,\cdot),dec(\cdot,\cdot)}$
	$\$(H, M)$:	Return win
enc(K_{EC}, H, M):	If query-made = true then	
If query-made = true then	Return $\perp$	enc(H, M):
Return $\perp$	query-made $\leftarrow$ true	If query-made = true then
query-made $\leftarrow$ true	$C_{EC} \leftarrow^\$ \{0,1\}^{clen(|M|)}$	Return $\perp$
$(C_{EC}, B_{EC}) \leftarrow EC(K_{EC}, H, M)$	$B_{EC} \leftarrow^\$ \{0,1\}^{btlen}$	query-made $\leftarrow$ true
Return (C_{EC}, B_{EC})	Return (C_{EC}, B_{EC})	$(C_{EC}, B_{EC}) \leftarrow EC(K_{EC}, H, M)$
		Return $((C_{EC}, B_{EC}), K_{EC})$
sr-BIND$_{CE}^{\mathcal{A}}$:	s-BIND$_{CE}^{\mathcal{A}}$:	
$(V_1, V_2, B_{EC}) \leftarrow^\$ \mathcal{A}$	$(K_{EC}, H, C_{EC}, B_{EC}) \leftarrow^\$ \mathcal{A}$	dec(H', C_{EC}'):
$(H, M, K_{EC}) \leftarrow V_1$	$M' \leftarrow DO(K_{EC}, H, C_{EC}, B_{EC})$	If query-made = false then
$(H', M', K_{EC}') \leftarrow V_2$	If $M' = \perp$ then Return false	Return $\perp$
$b \leftarrow EVer(H, M, K_{EC}, B_{EC})$	$b \leftarrow EVer(H, M', K_{EC}, B_{EC})$	If $(H', C_{EC}') = (H, C_{EC})$ then
$b' \leftarrow EVer(H', M', K_{EC}', B_{EC})$	If $b = 1$ then	Return $\perp$
If $V_1 = V_2$ then	Return true	$M' \leftarrow DO(K_{EC}, H', C_{EC}', B_{EC})$
Return false	Return false	If $M' \neq \perp$ then win $\leftarrow$ true
Return $(b = b' = 1)$		Return M'

Fig. 6. One-time real-or-random (otROR), second-ciphertext unforgeability (SCU), and binding notions for an encryption scheme $EC = (EKg, EC, DO, EVer)$.

Confidentiality. We define otROR security for an encryption scheme $EC = (EKg, EC, DO, EVer)$ in terms of games otROR0 and otROR1. Each game allows an attacker $\mathcal{A}$ to make one query of the form (H, M) to his real-or-random encryption oracle; in game otROR0 he receives back the real encryption $(C_{\mathsf{EC}}, B_{\mathsf{EC}})$ encrypting the input under a secret key, and in game otROR1 he receives back random bit strings. For an encryption scheme EC and adversary $\mathcal{A}$, we define the otROR advantage of $\mathcal{A}$ against EC as

$$\mathbf{Adv}_{\mathsf{EC}}^{\text{ot-ror}}(\mathcal{A}) = \left| \Pr\left[\, \text{otROR0}_{\mathsf{EC}}^{\mathcal{A}} \Rightarrow 1 \,\right] - \Pr\left[\, \text{otROR1}_{\mathsf{EC}}^{\mathcal{A}} \Rightarrow 1 \,\right] \right|,$$

where the probability is over the coins of EKg and $\mathcal{A}$.

Second-ciphertext unforgeability. We also ask that encryption schemes meet an unforgeability goal that we call second-ciphertext unforgeability (SCU). In this game, the attacker first learns an encryption $(C_{\mathsf{EC}}, B_{\mathsf{EC}})$ corresponding to a chosen header/message pair (H, M) under key K_{EC}. We then require that the attacker shouldn't be able to find a *distinct* header and ciphertext pair $(H', C'_{\mathsf{EC}}) \neq (H, C_{\mathsf{EC}})$ such that $\mathsf{DO}(K_{\mathsf{EC}}, H', C'_{\mathsf{EC}}, B_{\mathsf{EC}})$ does not return an error. This should hold even if the attacker knows K_{EC}. Looking ahead, this is a necessary and sufficient condition needed from encryption when using it to build ccAEAD schemes from fixed domain authenticated encryption.

Formally, the game SCU is shown in Fig. 6. To an encryption scheme EC and adversary $\mathcal{A}$, we define the second-ciphertext unforgeability (SCU) advantage to be $\mathbf{Adv}_{\mathsf{EC}}^{\text{scu}}(\mathcal{A}) = \Pr\left[\, \text{SCU}_{\mathsf{EC}}^{\mathcal{A}} \Rightarrow \mathsf{true} \,\right]$, where the probability is again over the coins of EKg and $\mathcal{A}$.

Binding security. We finally require that encryption schemes satisfy certain binding notions. We start by generalizing the receiver binding notion r-BIND for ccAEAD schemes from [17], and adapting the syntax to the encryption setting. r-BIND security requires that no computationally efficient adversary can find two keys, message, header triples $(K_{\mathsf{EC}}, H, M), (K'_{\mathsf{EC}}, H', M')$ and a binding tag B_{EC} such that $(H, M) \neq (H', M')$ and $\mathsf{EVer}(H, M, K_{\mathsf{EC}}, B_{\mathsf{EC}}) = \mathsf{EVer}(H', M', K'_{\mathsf{EC}}, B_{\mathsf{EC}}) = 1$. A simple strengthening of this notion — which we denote sr-BIND (for *strong* receiver binding) — allows the adversary to instead win if $(H, M, K_{\mathsf{EC}}) \neq (H', M', K'_{\mathsf{EC}})$. The pseudocode game sr-BIND is shown in Fig. 6, where we define the sr-BIND advantage of an adversary $\mathcal{A}$ against EC as $\mathbf{Adv}_{\mathsf{EC}}^{\text{sr-bind}}(\mathcal{A}) = \Pr\left[\, \text{sr-BIND}_{\mathsf{EC}}^{\mathcal{A}} \Rightarrow \mathsf{true} \,\right]$. The corresponding game and advantage term for r-BIND security are defined analogously. The stronger receiver binding notion implies the prior notion, and indeed is strictly stronger. We defer the details to the full version. For our purposes, it will simplify our negative results about rate-1 blockcipher-based encryption.

We additionally define the notion of sender binding. It ensures that a sender must itself commit to the message underlying an encryption, by requiring that it is infeasible to find an encryption which decrypts correctly but for which verification fails. Without this requirement, a malicious sender may be able to send an abusive message to a receiver with a faulty commitment such that a receiver is unable to report it. We define sender binding security formally via the game s-BIND in Fig. 6. We define the s-BIND advantage of an adversary $\mathcal{A}$ against an encryption scheme EC as $\mathbf{Adv}_{\mathsf{EC}}^{\text{s-bind}}(\mathcal{A}) = \Pr\left[\, \text{s-BIND}_{\mathsf{EC}}^{\mathcal{A}} \Rightarrow \mathsf{true} \,\right]$.

Binding notions and the Facebook attack.   Looking ahead, the analogous strong receiver binding notion for ccAEAD schemes is the property that would have prevented the Facebook attack, had they used a scheme that enjoyed it. This is because receiver binding implies that it is computationally intractable for an attacker to find two distinct keys that verify the same binding tag. In the Facebook attack, the sender was able to exploit this weakness to violate a security property similar to GLR's sender binding notion [17], which ensures decryption can only succeed if the binding tag commits to the underlying plaintext. Canonically, however, receiver binding is modeling the ability of a malicious receiver to frame the sender as having sent a message they did not, in fact, send. Such an attack doesn't work against Facebook's attachment franking scheme because the encryption of the AES-GCM key enjoys receiver binding, and prevents the recipient from forging an abuse report for an image that wasn't sent.

Relation to ccAEAD.   Given the simpler security properties expected of them, building highly efficient secure encryption schemes is a more straightforward task than constructing a ccAEAD scheme directly. However, as we shall see, encryption isolates the core complexity of building ccAEAD schemes with multi-opening security. In particular, in Sect. 7.3 we give a generic transform which allows one to build a multi-opening secure ccAEAD schemes from a secure encryption scheme and secure AEAD scheme. Armed with this transform, in Sect. 6 we show how to construct a secure encryption scheme from cryptographic hash functions. Together, our results will yield the first single-pass, single-primitive constructions of ccAEAD.

Binding and correctness imply ciphertext integrity.   One reason we have introduced encryption as a standalone primitive (instead of directly working with the ccAEAD formulation from GLR) is that it simplifies security analyses. One useful tool towards this is that we can show the following lemma, which states that for any encryption scheme EC that enjoys strong correctness, the combination of r-BIND and s-BIND security suffice to prove the SCU security.

Lemma 1. *Let* $\mathsf{EC} = (\mathsf{EKg}, \mathsf{EC}, \mathsf{DO}, \mathsf{EVer})$ *be a strongly correct encryption scheme, and consider an attacker $\mathcal{A}$ in the* SCU *game against* EC. *Then there exist attackers $\mathcal{B}$ and $\mathcal{C}$ such that* $\mathbf{Adv}_{\mathsf{EC}}^{\mathrm{scu}}(\mathcal{A}) \leq \mathbf{Adv}_{\mathsf{EC}}^{\mathrm{s\text{-}bind}}(\mathcal{B}) + \mathbf{Adv}_{\mathsf{EC}}^{\mathrm{r\text{-}bind}}(\mathcal{C})$, *and moreover $\mathcal{B}$ and $\mathcal{C}$ both run in the same time as $\mathcal{A}$.*

We give a proof sketch and defer details to the full version. Let $((C_{\mathsf{EC}}, B_{\mathsf{EC}}), K_{\mathsf{EC}})$ be the tuple corresponding to $\mathcal{A}$'s single encryption query (H, M) in the SCU game, and suppose that $\mathcal{A}$ subsequently wins the game with decryption oracle query (H', C'_{EC}), meaning that $\mathsf{DO}(K_{\mathsf{EC}}, H', C'_{\mathsf{EC}}, B_{\mathsf{EC}}) = M' \neq \perp$ and $(H', C'_{\mathsf{EC}}) \neq (H, C_{\mathsf{EC}})$. The proof first argues that if the scheme is s-BIND-secure, then any ciphertext which decrypts correctly must also verify correctly. As such, it follows that if $(H, M) \neq (H', M')$ for the winning query, then this can be used to construct a winning tuple for an attacker in the r-BIND game against EC; we bound the probability that this occurs with a reduction to r-BIND security. On the other hand, if $(H, M) = (H', M')$, then it must be the case that

$C_{\mathsf{EC}} \neq C'_{\mathsf{EC}}$ — but this in turn implies that we have found two distinct encrypt-ments which decrypt to the same header and message under K_{EC}, violating strong correctness.

A simple encryption construction. It is straightforward to construct an encryption scheme by composing a secure encryption scheme and a commit-ment scheme. One can just use a simple adaptation of the CtE2 ccAEAD scheme from [17]. We defer the details to the full version. But such generic compositions are inherently two pass and we seek faster schemes.

5   On Efficient Fixed-Key Blockcipher-Based Encryption

We are interested in building encryption schemes — and ultimately, more complex primitives such as ccAEAD schemes — from just a blockcipher used on a small number of keys and other primitive arithmetic operations (XOR, finite field arithmetic, etc.). Beyond being an interesting theoretical question, there is the practical motivation that the current fastest AEAD schemes, such as OCB [32], fall into this category.

As a simple motivating example illustrating the challenging nature of this task, we note that OCB does *not* satisfy r-BIND security (see Sect. 4) when reframed as an encryption scheme in the natural way. The high level reason for this (modulo a number of details), is that in OCB the binding tag is computed as a function over the XOR of the message blocks. As such, it is straightforward to construct two distinct messages such that the blocks XOR to the same value (and thus produce the same binding tag), thereby violating r-BIND security. Full details of the scheme and attack are given in the full version.

For the remainder of this section, we formally define high-rate encryption schemes, and show how prior results on the impossibility of high-rate CR func-tions can be used to rule out high-rate encryption schemes as well.

A connection between hashing and encryption. Towards showing neg-ative results, we must first define more carefully what we mean by the rate of encryption schemes. We are inspired by (and will later exploit connec-tions to) the definitions of rate from the blockcipher-based hash function lit-erature [9,34,35]. Consider a compression function $\mathcal{H}\colon \{0,1\}^{mn} \to \{0,1\}^{rn}$ for $m > r \geq 1$ and $n \geq 1$, which uses $k \geq 1$ calls of a blockcipher $E\colon \{0,1\}^{\kappa} \times \{0,1\}^n \to \{0,1\}^n$ $(m,r,n,k,\kappa \in \mathbb{N})$. Then following [35], we may write $\mathcal{H}$ as shown in Fig. 7, where we let $K_1, \ldots, K_k$ be any *fixed* strings[2], and $f_i\colon \{0,1\}^{(m+(i-1))n} \to \{0,1\}^n$ $(i = 1, \ldots, k)$, $g\colon \{0,1\}^{(m+k)n} \to \{0,1\}^{rn}$ are functions.

---

[2] One can modify our definitions so keys can be picked from a set as a function of the current round and messages, what Rogaway and Steinberger refer to as the no-fixed order model, and as first done in [9]. A negative result based on [9, Theorem 5] would rule out encryption using any rate-1 no-fixed order verification algorithm.

The *rate* of $\mathcal{H}$ is defined to be m/k; so a rate-$\frac{1}{\beta}$ function $\mathcal{H}$ makes β blockcipher calls per n-bits of input. For example, a rate-1 $\mathcal{H}$ would achieve a single blockcipher call per n-bit block of input. A consequence of the more general results of [35] (see below) is that they rule out rate-1 functions achieving security past $2^{n/4}$ queries to E by an adversary, when modeling E as an ideal cipher. We would like to exploit their negative results to similarly rule out rate-1 encryption schemes.

$$\boxed{\begin{array}{l} \mathcal{H}(V) \\ \text{For } i = 1 \text{ to } k \text{ do} \\ \quad X_i \leftarrow f_i(V, Y_1, \dots, Y_{i-1}) \\ \quad Y_i \leftarrow E_{K_i}(X_i) \\ W \leftarrow g(V, Y_1, \dots, Y_k) \\ \text{Return } W \end{array}}$$

Fig. 7. A blockcipher-based compression function.

We now focus attention on encryption schemes that fall into a certain form. Consider an encryption scheme $\mathsf{EC} = (\mathsf{EKg}, \mathsf{EC}, \mathsf{DO}, \mathsf{EVer})$. Because EC is deterministic, we can view computing the binding tag as a function $F(K_{\mathsf{EC}}, H, M)$ defined by computing $(C_{\mathsf{EC}}, B_{\mathsf{EC}}) = \mathsf{EC}(K_{\mathsf{EC}}, H, M)$ and outputting B_{EC}. The verification algorithm $\mathsf{EVer}(H, M, K_{\mathsf{EC}}, B_{\mathsf{EC}})$ checks that $F(K_{\mathsf{EC}}, H, M) = B_{\mathsf{EC}}$. (One can generalize this definition by allowing EC and EVer to use different functions F, F' to compute the binding tag; the lower bounds given in this section on the rate of such functions readily extend to this case also.)

With this in place, we can define the rate of verification for encryption analogously to defining the rate of a hash function $\mathcal{H}$, by saying that an encryption scheme has rate-$\frac{1}{\beta}$ if the associated function F makes β blockcipher calls per n-bits of header and message data (or equivalently, can process (H, M) of combined length mn-bits using βm blockcipher calls).

Now we can give a generic, essentially syntactic, transform from an encryption scheme to a hash function. For an encryption scheme EC, let F be the associated binding tag computation function as per above. Let $\mathcal{H}: \{0,1\}^* \to \{0,1\}^n$ be the function defined as $\mathcal{H}(X) = F(K_{\mathsf{EC}}, \varepsilon, X)$ for K_{EC} an arbitrary, fixed bit string. (Here we take $H = \varepsilon$, so that the number of block cipher calls required to compute F is solely determined by the length of the input X). The following is simple to prove.

Theorem 1. *Let* EC *be a encryption scheme with binding codes, and let* $\mathcal{H}$ *be defined as in the previous paragraph. For any collision-resistance adversary* $\mathcal{A}$, *we give an* r-BIND *adversary* $\mathcal{B}$ *so that* $\mathbf{Adv}_{\mathcal{H}}^{\mathrm{cr}}(\mathcal{A}) \leq \mathbf{Adv}_{\mathsf{EC}}^{\mathrm{r\text{-}bind}}(\mathcal{B})$. *The adversary* $\mathcal{B}$ *runs in the same amount of time as* $\mathcal{A}$.

Theorem 1 allows us to apply known negative results about efficient CR-hashing. For example, we have the following corollary of Theorem 1 and [35, Theorem 1]:

Corollary 1. *Fix* $m > r \geq 1$ *and* $n > 0$ $(m, r, n \in \mathbb{N})$. *Let* $N = 2^n$. *Let* EC *be an encryption scheme with ideal-cipher-based binding codes of length* rn *and that has message space including strings of length* mn. *Then there is a runnable adversary* $\mathcal{A}$ *making* $q = k(N^{1-(m-r)/k} + 1)$ *ideal cipher queries and achieving* $\mathbf{Adv}_{\mathsf{EC}}^{\mathrm{r\text{-}bind}}(\mathcal{A}) = 1$, *where* $k \in \mathbb{N}$ *denotes the number of permutation calls required to compute the binding code for an* mn-bit input.

This immediately rules out security of rate-1 schemes that achieve the efficiency of OCB, i.e., having $k = m$, m arbitrarily large, and $r = 1$. Consider the minimal case that $m = 2$ (two block messages), then $\mathcal{A}$ only requires $q = 2$ queries to succeed. Stronger results ruling out rate-$\frac{1}{2}$ verification can be similarly lifted from [35, Theorem 2] under some technical conditions about the verification function and the adversary. The results above were cast in terms of r-BIND security, but extend to sr-BIND security because the latter implies the former.

Ultimately these negative results indicate that for an r-BIND-secure encryptment scheme, the best we can hope for is either a rate-$\frac{1}{3}$ construction with a small set of keys, or to allow rekeying with each block of message. We therefore turn to building as efficient-as-possible constructions.

In Sect. 7, we will describe how the existence of an r-BIND-secure ccAEAD scheme of a given rate implies the existence of a given r-BIND-secure encryptment scheme of the same rate, and so the results of this section exclude the existence of rate-1 or rate-$\frac{1}{2}$ ccAEAD schemes also.

6 Encryptment from Hashing

In this section, we turn our attention to building secure and efficient encryptment schemes. As we shall see in Sect. 7, these can be lifted to multi-opening, many-time secure ccAEAD via simple and efficient transforms.

As one might expect given the close relationship between binding and CR hashing discussed previously in Sect. 5, our starting point will be cryptographic hashing. A slightly simplified version of the construction is shown in Fig. 8 (padding details are omitted), where f is a compression function. In summary, the scheme hashes the key, associated data and message data (the latter two of which are repeatedly XOR'd with the key). Intermediate chaining variables from the hash computation are used as pads to encrypt the message data, while the final chaining variable constitutes the binding tag.

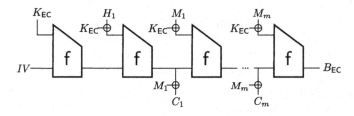

Fig. 8. Encryptment in the HFC scheme for a 1-block header and m-block message. For simplicity the diagram does not show the details of padding.

Intuitively, (strong) receiver binding derives from the collision resistance of the underlying hash function. We XOR the key into all the associated data and message blocks to ensure that every application of the compression function is keyed. This is critical; just prepending (or both prepending and appending) the key to the data leads to a scheme whose confidentiality is easily broken.

Likewise one cannot dispense with the additional initial block that simply processes the key, otherwise the encoding of the key, associated data, and message would not be injective and binding attacks result.

Some notation. Before defining the full scheme, we first give some additional notation which will simplify the presentation. The algorithm Parse_d is used to partition a string into d-bit blocks. Formally, we define Parse_d to be the algorithm which on input X outputs $(X_1, \ldots, X_\ell)$ such that $|X_i| = d$ for $1 \le i \le \ell - 1$ and $|X_\ell| = |X| \mod d$. For correctness, we require that $X = X_1 \| \ldots \| X_\ell$. Similarly, we define Trunc_r to be the algorithm which on input X outputs the r leftmost bits of X. We write $\langle y \rangle_{64}$ to be the encoding of y as a 64-bit string.

Our scheme utilizes a padding scheme $\mathsf{PadS} = (\mathsf{PadH}, \mathsf{PadM}, \mathsf{PadSuf}, \mathsf{Pad})$. The padding scheme is parameterized by a pair of numbers d, n, but we omit these in the notation for simplicity. We assume $d \ge n \ge 128$. The algorithms $\mathsf{PadH}, \mathsf{PadM}$, and PadSuf are shown in Fig. 9. Notice that for all header and message pairs (H, M), it holds that if $|M| \mod n = r$, then $r + |\mathsf{PadSuf}(|H|, |M|)|$ will be equal to either d or $2d$. The full padding function is then defined to be $\mathsf{Pad}(H, M) = \mathsf{PadH}(H)\|\mathsf{PadM}(M)\|\mathsf{PadSuf}(|H|, |M|)$. Note that $|\mathsf{Pad}(H, M)|$ is a multiple of d

$\mathsf{PadH}(H)$
$(H_1, \ldots, H_h) \leftarrow \mathsf{Parse}_d(H \| 0^{d - |H| \mod d})$
Return $(H_1, \ldots, H_h)$
$\mathsf{PadM}(M)$
$(M'_1, \ldots, M'_m) \leftarrow \mathsf{Parse}_n(M)$
For $i = 1, \ldots, m - 1$
$\quad M_i \leftarrow M'_i \| 0^{d-n}$
Return $(M_1, \ldots, M_{m-1}, M'_m)$
$\mathsf{PadSuf}(\ell_H, \ell_M)$
$p \leftarrow \min\{i \in \mathbb{N} : d \,|\, ((\ell_M \mod n) + i + 128)\}$
Return $0^p \| \langle \ell_H \rangle_{64} \| \langle \ell_M \rangle_{64}$

Fig. 9. Padding scheme $\mathsf{PadS} = (\mathsf{PadH}, \mathsf{PadM}, \mathsf{PadSuf}, \mathsf{Pad})$. We require that $\ell_H, \ell_M \in \mathbb{N}^2$.

and that the function $\mathsf{Pad}(H, M)$ is injective, i.e., for all pairs $(H, M), (H', M')$, $\mathsf{Pad}(H, M) = \mathsf{Pad}(H', M')$ only if $(H, M) = (H', M')$.

Next we define iterated functions. Let $\mathsf{f} \colon \{0,1\}^n \times \{0,1\}^d \to \{0,1\}^n$ be a function for some $d \ge n \ge 128$, let $D^+ = \cup_{i \ge 1}\{0,1\}^{id}$ and let $V_0 \in \{0,1\}^n$. Then $\mathsf{f}^+ \colon \{0,1\}^n \times D^+ \to \{0,1\}^n$ denotes the *iteration* of f, where $\mathsf{f}^+(V_0, X_1 \| \cdots \| X_m) = V_m$ is computed via $V_i = \mathsf{f}(V_{i-1}, X_i)$ for $1 \le i \le m$.

The HFC encryption scheme. The hash-function-chaining encryption scheme $\mathsf{HFC} = (\mathsf{HFCKg}, \mathsf{HFCEnc}, \mathsf{HFCDec}, \mathsf{HFCVer})$ is based on a compression function $\mathsf{f} \colon \{0,1\}^n \times \{0,1\}^d \to \{0,1\}^n$. The pseudocode for the encryption and decryption algorithms is presented in Fig. 10.

Key generation HFCKg simply chooses $K_{\mathsf{EC}} \leftarrow_\$ \{0,1\}^d$. Encryption first pads the header and message using the padding functions PadH and PadM respectively. We let $IV \in \{0,1\}^n$ be a fixed constant value (also called an initialization vector). The scheme computes an initial chaining variable as $V_0 = \mathsf{f}(IV, K_{\mathsf{EC}})$. It then hashes $\mathsf{PadH}(H) \| \mathsf{PadM}(M) \| \mathsf{PadSuf}(|H|, |M|)$ with f^+, the iteration of the compression function f, where the secret encryption key K_{EC} is XORed into each d-bit block prior to hashing. The final chaining variable produced by this process forms the binding tag B_{EC}. Notice that while the compression function takes d-bit inputs, the way in which the message

HFCEnc(K_{EC}, H, M):	HFCDec($K_{EC}, H, C_{EC}, B_{EC}$):
$H_1, \ldots, H_h \leftarrow \text{PadH}(H)$	$H_1, \ldots, H_h \leftarrow \text{PadH}(H)$
$M_1, \ldots, M_m \leftarrow \text{PadM}(M)$	$C_1, \ldots, C_m \leftarrow \text{Parse}_n(C_{EC})$
$V_0 \leftarrow f(IV, K_{EC})$	$V_0 \leftarrow f(IV, K_{EC})$
$V_h \leftarrow f^+(V_0, (K_{EC} \oplus H_1) \| \cdots \| (K_{EC} \oplus H_h))$	$V_h \leftarrow f^+(V_0, (K_{EC} \oplus H_1) \| \cdots \| (K_{EC} \oplus H_h))$
$C_{EC} \leftarrow \varepsilon$	For $i = 1, \ldots, m - 1$ do
For $i = 1, \ldots, m - 1$ do	$\quad M_i \leftarrow V_{h+i-1} \oplus C_i$
$\quad C_{EC} \leftarrow C_{EC} \| (V_{h+i-1} \oplus M_i)$	$\quad V_{h+i} \leftarrow f(V_{h+i-1}, (K_{EC} \oplus M_i \| 0^{d-n}))$
$\quad V_{h+i} \leftarrow f(V_{h+i-1}, (K_{EC} \oplus M_i))$	$M_m \leftarrow V_{h+m-1} \oplus C_m$
$C_{EC} \leftarrow C_{EC} \| (V_{h+m-1} \oplus M_m)$	$M'_m, M'_{m+1} \leftarrow \text{Parse}_d(M_m \| \text{PadSuf}(|H|, |C_{EC}|))$
$M'_m, M'_{m+1} \leftarrow \text{Parse}_d(M_m \| \text{PadSuf}(|H|, |M|))$	$B'_{EC} \leftarrow f^+(V_{h+m-1}, (K_{EC} \oplus M'_m) \| (K_{EC} \oplus M'_{m+1}))$
$B_{EC} \leftarrow f^+(V_{h+m-1}, (K_{EC} \oplus M'_m) \| (K_{EC} \oplus M'_{m+1}))$	If $B'_{EC} \neq B_{EC}$ then
Return (C_{EC}, B_{EC})	$\quad$ Return $\perp$
	Return $M_1 \| \cdots \| M_m$

Fig. 10. The HFC encryptment scheme HFC built from a compression function $f \colon \{0,1\}^n \times \{0,1\}^d \to \{0,1\}^n$ and padding scheme $\text{PadS} = (\text{PadH}, \text{PadM}, \text{PadSuf}, \text{Pad})$. Here $K_{EC} \in \{0,1\}^d$, and $IV \in \{0,1\}^n$ is a fixed public constant.

data is padded means we only process n-bits of message in each compression function call. We will see that the collision resistance of the iterated hash function when instantiated with an appropriate compression function implies the sr-BIND security of the construction.

Rather than running a separate encryption algorithm alongside this process to encrypt the message, we instead generate ciphertext blocks by XORing the message blocks M_i with intermediate chaining variables, yielding $C_i = V_{h+i-1} \oplus M_i$ for $1 \leq i \leq m$ where h denotes the number of header blocks. Recall that in our notation $X \oplus Y$ silently truncates the longer string to the length of the shorter string, and so only the n-bits of message data in each d-bit padded message block is XORed with the n-bit chaining variable; similarly, if message M is such that $|M| \bmod n = r$, then the final ciphertext block produced by this process is truncated to the leftmost r-bits. The properties of the compression function ensure that the chaining variables are pseudorandom, thus yielding the required otROR security. By 'reusing' chaining variables as random pads we can achieve encryption with no additional overhead over just computing the binding tag, incurring a significant efficiency saving (see further discussion below).

Decryption $\text{DO}(K_{EC}, H, C_{EC})$ begins by padding H into d-bit blocks via $\text{PadH}(H)$ and parsing C_{EC} into n-bit blocks. The algorithm computes the initial chaining variable as $V_0 = f(IV, K_{EC})$, then hashes the padded header as in encryption. The scheme then recovers the first message block M_1 by XORing the chaining variable into the first ciphertext block C_1. This is then used to compute the next chaining variable via application of f, and so on. Notice how at most n-bits of message data is recovered in each such step; this is why we must process only n-bits of message data in each compression function call, else the decryptor would be unable to compute the next chaining variable. Finally, DO recomputes and verifies the binding tag, returning the message only if verification succeeds.

The verification algorithm (not shown), on input (K_{EC}, H, M, B_{EC}), pads the message to $\text{PadH}(H) \| \text{PadM}(M) \| \text{PadSuf}(|H|, |M|)$, XORs K_{EC} into every block, and hashes the resulting string with f^+ with initial chaining variable $V_0 = f(IV, K_{EC})$, checking that the output matches the binding tag B_{EC}.

Our padding scheme is a variant of MD strengthening. We will not rely on the strengthening for its traditional purpose of forming a suffix-free padding scheme; we use strengthening only for injectivity and will assume more of f.

Efficiency. The efficiency of the scheme (in terms of throughput) depends on the parameters d, n, where recall that $f: \{0,1\}^n \times \{0,1\}^d \to \{0,1\}^n$. As discussed previously, at most n-bits of message data can be processed in each compression function call. As such, the HFC encryption scheme achieves optimal throughput when $d = n$. In this case no padding is applied to the message blocks, and so computing the full encryption incurs *no overhead* over simply computing the binding tag. If $d > n$, then some throughput is lost due to the padding. In the full version we present an alternative padding scheme for this case, which recovers some throughput by padding message blocks with header data.

6.1   Analyzing the HFC Encryption Scheme

In this section, we analyze the security of the HFC encryption scheme, relative to the security goals detailed in Sect. 4. We also discuss some of the options for instantiating the compression function f.

Strong receiver binding. We begin by proving that the HFC encryption scheme satisfies strong receiver binding. Observe that the binding tag computation performed by HFCEnc on input tuple (K_{EC}, H, M) is equivalent to XORing K_{EC} into each d-bit block of $0^d \parallel \text{Pad}(H, M)$ (we refer to this as 'encoding' the tuple), and hashing the resulting string with f^+. Moreover, it is straightforward to verify that the injectivity of Pad implies that the encoding map is injective also. So any tuple breaking the sr-BIND security of HFC is a collision against f^+.

A well-known folklore result (see [2]) gives that f^+ is collision-resistant provided the underlying compression function is collision-resistant, and that it is hard to find an input which hashes to the IV. Standard compression functions satisfy both properties. The full proof of the following is given in the full version. The conditions on d, n below are due to the padding scheme and can be relaxed.

Theorem 2. *Let* HFC *be as shown in Fig. 10, using compression function* $f: \{0,1\}^n \times \{0,1\}^d \to \{0,1\}^n$ *where* $d \geq n \geq 128$. *Then for any adversary* $\mathcal{A}$ *in the* sr-BIND *game against* HFC, *there exists an adversary* $\mathcal{B}$ *such that* $\mathbf{Adv}_{\mathsf{HFC}}^{\text{sr-bind}}(\mathcal{A}) \leq \mathbf{Adv}_{f^+}^{\text{cr}}(\mathcal{B})$, *where adversary* $\mathcal{B}$ *runs in the same time as* $\mathcal{A}$.

Sender binding and correctness. The s-BIND security of HFC is immediate because decryption verifies the binding tag. Similarly, it is straightforward to verify that the scheme is strongly correct. Therefore Lemma 1 allows us to bound the SCU security of HFC as an immediate consequence of these observations coupled with Theorem 2.

One-time confidentiality. All that remains is to bound the otROR security of HFC. We do this in the next theorem, by reducing otROR security of HFC to the related-key attack (RKA) PRF security [3] of f for a specific class of related-key deriving functions.

Let $F \colon \{0,1\}^n \times \{0,1\}^d \to \{0,1\}^n$ be a function, and consider the games RKA-PRF0 and RKA-PRF1. In both games a key $K_{prf} \leftarrow^{\$} \{0,1\}^d$ is chosen. The attacker is given access to an oracle to which he may submit queries of the form $(X, Y) \in \{0,1\}^n \times \{0,1\}^d$. In game RKA-PRF0, the oracle returns $F(X, Y \oplus K_{prf})$. In game RKA-PRF1, the oracle returns a random bit string for each query, answering consistently if $(X, Y \oplus K_{prf})$ collides with a previous query. The *linear-only* RKA-PRF advantage of an adversary $\mathcal{A}$ is defined as

$$\mathbf{Adv}_F^{\oplus\text{-prf}}(\mathcal{A}) = \left| \Pr\left[\text{RKA-PRF0}_F^{\mathcal{A}} \Rightarrow 1 \right] - \Pr\left[\text{RKA-PRF1}^{\mathcal{A}} \Rightarrow 1 \right] \right| ,$$

where the probabilities are over the coins used in the games.

The proof of the following theorem then follows from a reduction to the RKA-PRF security of f, coupled with a birthday bound to account for collisions during the challenge ciphertext computation. The proof is given in the full version.

Theorem 3. *Let* HFC *be as shown in Fig. 10, using compression function* $f \colon \{0,1\}^n \times \{0,1\}^d \to \{0,1\}^n$ *where* $d \geq n \geq 128$. *Then for any adversary* $\mathcal{A}$ *in the otROR game against* HFC, *there exists an adversary* $\mathcal{B}$ *such that* $\mathbf{Adv}_{\mathsf{HFC}}^{\text{ot-ror}}(\mathcal{A}) \leq \mathbf{Adv}_{\mathsf{f}}^{\oplus\text{-prf}}(\mathcal{B}) + \frac{\ell^2}{2^n}$, *where* $\ell \cdot d$ *denotes the length of* $\mathcal{A}$*'s encryption query after padding. The adversary* $\mathcal{B}$ *runs in time that of* $\mathcal{A}$ *plus an* $\mathcal{O}(\ell)$ *overhead and makes at most* ℓ *queries.*

Instantiations. The obvious (and probably best) choice to instantiate f is the SHA-256 or SHA-512 compression function. These provide good software performance, and there is a shift towards widespread hardware support in the form of the Intel SHA instructions [11,18,39]. Extensive cryptanalysis for the CR (e.g., [23,26,36]), preimage resistance (e.g., [19,23]), and RKA-PRP of the associated SHACAL-2 blockcipher (e.g., [21,24,25,27]) gives confidence in its security. Another approach would be to use AES via a PGV compression function [31] like Davies-Meyer (DM). Security of AES has been studied extensively, and known attacks do not falsify the assumptions we need [7,8]. On systems with AES-NI, HFC instantiated with DM-AES will have very good performance. More problematic is that binding can only hold up 2^{64}, which is in general insufficient in practice. Other options, although in some cases less well-studied cryptanalytically, include SHA-3 finalists. In particular, a variant of the HFC construction using a sponge-based mode such as Keccak, in which the key is fed to the sponge prior to hashing the message blocks, would allow us to avoid the RKA assumption. We could also remove the assumption by using a compression function with a dedicated key input such as LP231 [34]. We discuss both cases, and include a more thorough discussion of instantiations, in the full version.

7   Compactly Committing AEAD from Encryption

In this section we recall the formal notions for compactly committing AEAD schemes (ccAEAD schemes), following the treatment given by GLR [17], and compare these to encryption. With this in place, we show in Sect. 7.3 how to build ccAEAD from encryption with very efficient transforms. In the full version, we will show how to construct a secure encryption scheme from a ccAEAD scheme in a way that transfers our negative results from Sect. 5 to ccAEAD; this result does not appear here for space reasons.

7.1   ccAEAD Syntax and Correctness

Encryptment can be viewed as a one-time secure, deterministic variant of ccAEAD. We discuss further the differences between the two primitives later in the section.

ccAEAD schemes.   Formally, a ccAEAD scheme is a tuple of algorithms $\mathsf{CE} = (\mathsf{Kg}, \mathsf{Enc}, \mathsf{Dec}, \mathsf{Ver})$ with associated key space $\mathcal{K} \subseteq \Sigma^*$, header space $\mathcal{H} \subseteq \Sigma^*$, message space $\mathcal{M} \subseteq \Sigma^*$, ciphertext space $\mathcal{C} \subseteq \Sigma^*$, opening space $\mathcal{K}_f \subseteq \Sigma^*$, and binding tag space $\mathcal{T} \subseteq \Sigma^*$, defined as follows. The randomized key generation algorithm Kg takes no input, and outputs a secret key $K \in \mathcal{K}$. The randomized encryption algorithm Enc takes as input a tuple $(K, H, M) \in \mathcal{K} \times \mathcal{H} \times \mathcal{M}$ and outputs a ciphertext/binding tag pair $(C, C_B) \in \mathcal{C} \times \mathcal{T}$. The deterministic decryption algorithm Dec takes as input a tuple $(K, H, C, C_B) \in \mathcal{H} \times \mathcal{M} \times \mathcal{C} \times \mathcal{T}$, and outputs a message/opening pair $(M, K_f) \in \mathcal{M} \times \mathcal{K}_f$ or the error symbol $\bot$. The deterministic verification algorithm Ver takes as input a tuple $(H, M, K_f, C_B) \in \mathcal{H} \times \mathcal{M} \times \mathcal{K}_f \times \mathcal{T}$, and outputs a bit b. We assume that if Dec and Ver are queried on inputs which do not lie in their defined input spaces, then they return $\bot$ and 0 respectively.

Correctness and compactness.   Correctness for ccAEAD schemes is defined identically to the COR correctness notion for encryptment schemes (Fig. 5), except in the ccAEAD case the probability is now over the coins of Enc also. We require that the structure of ciphertexts C depend only on the length of the underlying message. Formally, let $M^* = \{i \mid \exists m \in \mathcal{M} : |m| = i\}$. Then we require that the ciphertext space $\mathcal{C}$ can be partitioned into disjoint sets $\mathcal{C}(i) \subseteq \mathcal{C}, i \in M^*$, such that for all $(H, M) \in \mathcal{H} \times \mathcal{M}$ it holds that $C \in \mathcal{C}(|M|)$ with probability one for the sequence of algorithm executions: $K \leftarrow\!\!\text{\tiny \$}\ \mathsf{Kg}$; $(C, C_B) \leftarrow\!\!\text{\tiny \$}\ \mathsf{Enc}(K, H, M)$. Finally, we require that the binding tags C_B are *compact*, by which we mean that all C_B returned by a ccAEAD scheme are of constant length blen which is linear in the key size.

Comparison with encryptment.   With this in place, we highlight the key differences between encryptment and ccAEAD schemes. The overarching difference is that encryptment schemes are single-use (a key is only ever used to encrypt a single message), whereas ccAEAD schemes are multi-use. To support this, the encryption algorithm for ccAEAD schemes is randomized, whereas for encryptment this algorithm is deterministic. This is necessary for achieving schemes that enjoy security in the face of attackers that can obtain multiple encryptions. Moreover, while encryptment schemes are restricted to use the same key for verification as they use for encryption, ccAEAD schemes output an explicit opening key K_f during decryption. There is no requirement that this equal the secret key used for encryption. Again, outputting an opening key distinct from the encryption key allows for ccAEAD schemes that maintain confidentiality and integrity even after some ciphertexts produced under a given encryption key have been opened.

AEAD schemes. The usual definition of AEAD schemes (see Sect. 2) can be recovered from the above definition of ccAEAD schemes by noticing that the tuple of AEAD algorithms AEAD = (AEAD.kg, AEAD.enc, AEAD.dec) can be defined identically to their ccAEAD variants, except we view the ciphertext/binding tag pair as a single ciphertext, and modify decryption to no longer output the opening, in the AEAD case. This framing allows us to define security notions for AEAD schemes as a special case of those notions for ccAEAD schemes for conciseness and ease of comparison. Similarly regular AE schemes are defined to be the same as AEAD schemes but with all references to the header removed.

7.2 Security Notions for Compactly Committing AEAD

We now define the security notions for ccAEAD schemes, following GLR. They adapt the familiar security notions of real-or-random (ROR) ciphertext indistinguishability [33], and ciphertext integrity (CTXT) [4] for AE schemes to the ccAEAD setting. We focus on GLR's *multi-opening* (MO) security notions. MO-ROR (resp. MO-CTXT) requires that if multiple messages are encrypted under the same key, then learning the message/opening pair (M, K_f) for some of the resulting ciphertexts does not compromise the ROR (resp. CTXT) security of the remaining unopened ciphertexts. This precludes schemes which for example have the opening key K_f equal to the secret encryption key K.

Confidentiality. Games MO-REAL and MO-RAND are shown in Fig. 11. In both variants, the attacker is given access to an oracle **ChalEnc** to which he may submit message/header pairs. This oracle returns real (resp. random) ciphertext/binding tag pairs in game MO-REAL (resp. MO-RAND). The attacker is then challenged to distinguish between the two games. To model multi-opening security, the attacker is also given a pair of encryption/decryption oracles, **Enc** and **Dec**, and may submit the (real) ciphertexts generated via a query to the former to the latter, learning the openings of these ciphertexts in the process. The challenge decryption oracle will return $\perp$ for any ciphertext not generated via the encryption oracle, to prevent the attacker trivially winning by decrypting a ciphertext returned by **ChalEnc**. We define the advantage of an attacker $\mathcal{A}$ in game MO-ROR against a ccAEAD scheme CE as

$$\mathbf{Adv}_{\mathsf{CE}}^{\mathrm{mo\text{-}ror}}(\mathcal{A}) = \left| \Pr\left[\,\mathrm{MO\text{-}REAL}_{\mathsf{CE}}^{\mathcal{A}} \Rightarrow 1\,\right] - \Pr\left[\,\mathrm{MO\text{-}RAND}_{\mathsf{CE}}^{\mathcal{A}} \Rightarrow 1\,\right] \right| .$$

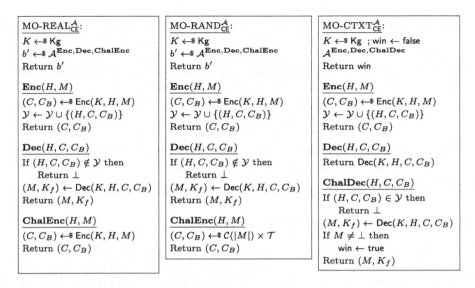

Fig. 11. Confidentiality (left two games) and ciphertext integrity (rightmost) games for ccAEAD.

Ciphertext integrity. Ciphertext integrity guarantees that an attacker cannot produce a fresh ciphertext which will decrypt correctly. The multi-opening adaptation to the ccAEAD setting MO-CTXT is shown in Fig. 11. The attacker $\mathcal{A}$ is given access to encryption oracle **Enc** and a challenge decryption oracle **ChalDec**. The attacker wins if he submits a ciphertext to **ChalDec** which decrypts correctly and which wasn't the result of a previous query to the encryption oracle. To model multi-opening security, the attacker is given access to a further oracle **Dec** via which he may decrypt ciphertexts and learn the corresponding openings. The advantage of an attacker $\mathcal{A}$ in game MO-CTXT against a ccAEAD scheme CE is then defined

$$\mathbf{Adv}_{\mathsf{CE}}^{\mathrm{mo\text{-}ctxt}}(\mathcal{A}) = \Pr\left[\,\mathrm{MO\text{-}CTXT}_{\mathsf{CE}}^{\mathcal{A}} \Rightarrow \mathsf{true}\,\right]\,.$$

Security for standard AEAD. We note that the familiar ROR and CTXT notions for AEAD schemes can be recovered from the corresponding ccAEAD games in Fig. 11 by reframing the ccAEAD scheme as an AEAD scheme as described previously, removing access to oracle **Dec** in all games, and removing **Enc** in MO-REAL and MO-RAND. Advantage functions are defined analogously. Since here we are removing attacker capabilities, it follows that security for a ccAEAD scheme with respect to these notions implies security for the derived AEAD scheme also.

Receiver and sender binding. Strong receiver binding for ccAEAD schemes is the same as for encryption (Fig. 6), except the attacker outputs openings K_f, K_f' rather than secret keys K, K' as part of his guess. The sender binding game for a ccAEAD scheme challenges an attacker $\mathcal{A}$ to output a tuple

(K, H, C, C_B) such that $(K_f, M) \leftarrow \mathsf{Dec}(K, H, C, C_B)$ does not equal $\perp$ but $\mathsf{Ver}(H, M, K_f, C_B) = 0$. This is the same as the associated game for encryption, except that the opening K_f recovered during decryption is used for verification rather than the key output by $\mathcal{A}$. Given the similarities, we abuse notation by using the same names for ccAEAD binding notion games and advantage terms as in the encryption case; which version will be clear from the context.

Given that both target certain binding notions, a natural question is whether an sr-BIND secure ccAEAD scheme is also robust [16], and vice versa. In the full version, we show that neither notion implies the other in generality. We also discuss the conditions under which the ccAEAD schemes we build from secure encryption are robust.

7.3   Encryption to ccAEAD Transforms

We now turn to building ccAEAD from encryption. Fix an encryption scheme $\mathsf{EC} = (\mathsf{EKg}, \mathsf{EC}, \mathsf{DO}, \mathsf{EVer})$ and a standard AEAD scheme $\mathsf{AEAD} = (\mathsf{AEAD.Kg}, \mathsf{AEAD.enc}, \mathsf{AEAD.dec})$. Let $\mathsf{CE}[\mathsf{EC}, \mathsf{AEAD}] = (\mathsf{Kg}, \mathsf{Enc}, \mathsf{Dec}, \mathsf{Ver})$ be the ccAEAD scheme whose encryption, decryption, and verification algorithms are shown in Fig. 12. Key generation Kg runs $K \leftarrow_\$ \mathsf{AEAD.Kg}$ and outputs K.

To encrypt a header/message (H, M), Enc uses the key generation algorithm of the encryption scheme to generate a one-time encryption key $K_{\mathsf{EC}} \leftarrow_\$ \mathsf{EKg}$, and computes the encryption of the header and message via $(C_{\mathsf{EC}}, B_{\mathsf{EC}}) \leftarrow \mathsf{EC}(K_{\mathsf{EC}}, H, M)$. The scheme then uses the encryption algorithm of the AEAD scheme to encrypt the one-time key K_{EC} with header B_{EC}, producing $C_{\mathsf{AE}} \leftarrow_\$ \mathsf{AEAD.enc}(K, B_{\mathsf{EC}}, K_{\mathsf{EC}})$, and outputs $((C_{\mathsf{EC}}, C_{\mathsf{AE}}), B_{\mathsf{EC}})$. On input $(K, (C_{\mathsf{EC}}, C_{\mathsf{AE}}), B_{\mathsf{EC}})$, Dec computes $K_{\mathsf{EC}} \leftarrow \mathsf{AEAD.dec}(K, B_{\mathsf{EC}}, C_{\mathsf{AE}})$ and if $K_{\mathsf{EC}} = \perp$ returns $\perp$ since this clearly indicates that C_{AE} is invalid. The recovered key K_{EC} is in turn used to recover the message via $M \leftarrow \mathsf{DO}(K_{\mathsf{EC}}, H, C_{\mathsf{EC}}, B_{\mathsf{EC}})$. If $M = \perp$, the scheme returns $\perp$; otherwise, EC returns (M, K_{EC}) as the message/opening pair. Ver simply applies the verification algorithm EVer of the underlying encryption scheme to the input tuple and returns the result.

Notice that by including the binding tag B_{EC} as the header in the authenticated encryption, this ensures the integrity of B_{EC}. If we did not authenticate B_{EC} then an attacker could trivially break the MO-CTXT-security of the scheme by using an **Enc** query to obtain ciphertext $((C_{\mathsf{EC}}, C_{\mathsf{AE}}), B_{\mathsf{EC}})$ for a pair (H, M), submitting that ciphertext to **Dec** to recover the opening/key K_{EC}, with which he can easily create a valid forgery by computing $(C'_{\mathsf{EC}}, B'_{\mathsf{EC}}) \leftarrow \mathsf{EC}(K_{\mathsf{EC}}, H', M')$ for some distinct header/message pair and outputting $((C'_{\mathsf{EC}}, C_{\mathsf{AE}}), B'_{\mathsf{EC}})$. Including the binding tag as the header in the AEAD ciphertext means that an attacker trying to replicate the above mix-and-match attack must create a forgery for an encryption binding tag and key already returned as the result of an **Enc** query, thus violating the SCU security of the underlying encryption scheme.

Security of the transform. Next, we analyze the security of the ccAEAD scheme $\mathsf{CE}[\mathsf{EC}, \mathsf{AEAD}]$ shown in Fig. 12. We begin with confidentiality. The proof of the following theorem follows from reductions to the ROR security of the underlying encryption and AEAD schemes, and is given in the full version.

Theorem 4. *Let* EC *be an encryption scheme,* AEAD *be an authenticated encryption scheme, and let* $\mathsf{CE}[\mathsf{EC}, \mathsf{AEAD}]$ *be the ccAEAD scheme built from* EC *according to Fig. 12. Then for any adversary* $\mathcal{A}$ *in the MO-ROR game against* CE *making a total of* q *queries, of which* q_c *are to* **ChalEnc** *and* q_e *are to* **Enc***, there exists adversaries* $\mathcal{B}$ *and* $\mathcal{C}$ *such that*

$$\mathbf{Adv}_{\mathsf{CE}}^{\text{mo-ror}}(\mathcal{A}) \leq 2 \cdot \mathbf{Adv}_{\mathsf{AEAD}}^{\text{ror}}(\mathcal{B}) + q_c \cdot \mathbf{Adv}_{\mathsf{EC}}^{\text{ot-ror}}(\mathcal{C}) \ .$$

Adversaries $\mathcal{B}$ *and* $\mathcal{C}$ *run in the same time as* $\mathcal{A}$ *with an* $\mathcal{O}(q)$ *overhead, and adversary* $\mathcal{B}$ *makes at most* $q_c + q_e$ *encryption oracle queries.*

$\mathsf{Enc}(K, H, M)$:
  $K_{\mathsf{EC}} \leftarrow^\$ \mathsf{EKg}$
  $(C_{\mathsf{EC}}, B_{\mathsf{EC}}) \leftarrow \mathsf{EC}(K_{\mathsf{EC}}, H, M)$
  $C_{\mathsf{AE}} \leftarrow^\$ \mathsf{AEAD.enc}(K, B_{\mathsf{EC}}, K_{\mathsf{EC}})$
  Return $((C_{\mathsf{EC}}, C_{\mathsf{AE}}), B_{\mathsf{EC}})$

$\mathsf{Dec}(K, H, (C, C_B))$:
  $(C_{\mathsf{EC}}, C_{\mathsf{AE}}) \leftarrow C \ ; B_{\mathsf{EC}} \leftarrow C_B$
  $K_{\mathsf{EC}} \leftarrow \mathsf{AEAD.dec}(K, B_{\mathsf{EC}}, C_{\mathsf{AE}})$
  If $K_{\mathsf{EC}} = \bot$ then Return $\bot$
  $M \leftarrow \mathsf{DO}(K_{\mathsf{EC}}, H, C_{\mathsf{EC}}, B_{\mathsf{EC}})$
  If $M = \bot$ then Return $\bot$
  Return (M, K_{EC})

Fig. 12. A generic transform from an encryption scheme EC and a standard authenticated encryption scheme AEAD to a multi-opening ccAEAD scheme $\mathsf{CE}[\mathsf{EC}, \mathsf{AEAD}]$. Verification simply runs EVer.

Next we bound the MO-CTXT advantage of any adversary against $\mathsf{CE}[\mathsf{EC}, \mathsf{AEAD}]$, via a reduction to the CTXT security of the underlying AEAD scheme, and the SCU security of the encryption scheme; we defer the proof to the full version.

Theorem 5. *Let* EC *be an encryption scheme,* AEAD *be an authenticated encryption scheme, and let* $\mathsf{CE}[\mathsf{EC}, \mathsf{AEAD}]$ *be the ccAEAD scheme built from* EC *according to Fig. 12. Then for any adversary* $\mathcal{A}$ *in the MO-CTXT game against* CE *making a total of* q *queries, of which* q_e *are to* **Enc***, there exists adversaries* $\mathcal{B}$ *and* $\mathcal{C}$ *such that*

$$\mathbf{Adv}_{\mathsf{CE}}^{\text{mo-ctxt}}(\mathcal{A}) \leq \mathbf{Adv}_{\mathsf{AEAD}}^{\text{ctxt}}(\mathcal{B}) + q_e \cdot \mathbf{Adv}_{\mathsf{EC}}^{\text{scu}}(\mathcal{C}) \ .$$

Adversaries $\mathcal{B}$ *and* $\mathcal{C}$ *run in the same time as* $\mathcal{A}$ *with an* $\mathcal{O}(q)$ *overhead, and adversary* $\mathcal{B}$ *makes at most as many queries as* $\mathcal{A}$*.*

We omit bounding the s-BIND and sr-BIND security of $\mathsf{CE}[\mathsf{EC}, \mathsf{AEAD}]$, since CE inherits these properties directly from EC. By reframing CE as a regular AEAD scheme, our transform yields a ROR and CTXT secure single-pass AEAD scheme. To implement the transform, the fixed-input-length AE scheme must be instantiated. One can use, for example, AES-GCM or OCB. In the full version of the paper, we provide two other approaches for building ccAEAD from encryption, which use a PRF and a tweakable block cipher respectively.

Acknowledgments. The authors thank Jon Millican for his help on understanding Facebook's message franking systems. Dodis is partially supported by gifts from

VMware Labs and Google, and NSF grants 1619158, 1319051, 1314568. Grubbs is supported by an NSF Graduate Research Fellowship. A portion of this work was completed while Grubbs visited Royal Holloway University, and he thanks Kenny Patterson for generously hosting him. Ristenpart is supported in part by NSF grants 1704527 and 1514163, as well as a gift from Microsoft. Woodage is supported by the EPSRC and the UK government as part of the Centre for Doctoral Training in Cyber Security at Royal Holloway, University of London (EP/K035584/1).

References

1. Abdalla, M., Bellare, M., Neven, G.: Robust encryption. In: Micciancio, D. (ed.) TCC 2010. LNCS, vol. 5978, pp. 480–497. Springer, Heidelberg (2010). https://doi.org/10.1007/978-3-642-11799-2_28
2. Bellare, M., Jaeger, J., Len, J.: Better than advertised: improved collision-resistance guarantees for MD-based hash functions. In: ACM CCS (2017)
3. Bellare, M., Kohno, T.: A theoretical treatment of related-key attacks: RKA-PRPs, RKA-PRFs, and applications. In: Biham, E. (ed.) EUROCRYPT 2003. LNCS, vol. 2656, pp. 491–506. Springer, Heidelberg (2003). https://doi.org/10.1007/3-540-39200-9_31
4. Bellare, M., Namprempre, C.: Authenticated encryption: relations among notions and analysis of the generic composition paradigm. In: Okamoto, T. (ed.) ASIACRYPT 2000. LNCS, vol. 1976, pp. 531–545. Springer, Heidelberg (2000). https://doi.org/10.1007/3-540-44448-3_41
5. Bertoni, G., Daemen, J., Peeters, M., Van Assche, G.: Keccak sponge function family main document. Submission to NIST SHA3 (2009)
6. Bertoni, G., Daemen, J., Peeters, M., Van Assche, G.: Duplexing the sponge: single-pass authenticated encryption and other applications. In: Miri, A., Vaudenay, S. (eds.) SAC 2011. LNCS, vol. 7118, pp. 320–337. Springer, Heidelberg (2012). https://doi.org/10.1007/978-3-642-28496-0_19
7. Biryukov, A., Khovratovich, D.: Related-key cryptanalysis of the full AES-192 and AES-256. In: Matsui, M. (ed.) ASIACRYPT 2009. LNCS, vol. 5912, pp. 1–18. Springer, Heidelberg (2009). https://doi.org/10.1007/978-3-642-10366-7_1
8. Biryukov, A., Khovratovich, D., Nikolić, I.: Distinguisher and related-key attack on the full AES-256. In: Halevi, S. (ed.) CRYPTO 2009. LNCS, vol. 5677, pp. 231–249. Springer, Heidelberg (2009). https://doi.org/10.1007/978-3-642-03356-8_14
9. Black, J., Cochran, M., Shrimpton, T.: On the impossibility of highly-efficient blockcipher-based hash functions. In: Cramer, R. (ed.) EUROCRYPT 2005. LNCS, vol. 3494, pp. 526–541. Springer, Heidelberg (2005). https://doi.org/10.1007/11426639_31
10. Brassard, G., Chaum, D., Crépeau, C.: Minimum disclosure proofs of knowledge. JCSS **37**, 156–189 (1988)
11. Advanced Micro Devices: The ZEN microarchitecture (2016). https://www.amd.com/en/technologies/zen-core
12. Dodis, Y., An, J.H.: Concealment and its applications to authenticated encryption. In: Biham, E. (ed.) EUROCRYPT 2003. LNCS, vol. 2656, pp. 312–329. Springer, Heidelberg (2003). https://doi.org/10.1007/3-540-39200-9_19
13. Facebook: Facebook Messenger app (2016). https://www.messenger.com/
14. Facebook: Messenger Secret Conversations Technical Whitepaper (2016)

15. Farshim, P., Libert, B., Paterson, K.G., Quaglia, E.A.: Robust encryption, revisited. In: Kurosawa, K., Hanaoka, G. (eds.) PKC 2013. LNCS, vol. 7778, pp. 352–368. Springer, Heidelberg (2013). https://doi.org/10.1007/978-3-642-36362-7_22

16. Farshim, P., Orlandi, C., Rosie, R.: Security of symmetric primitives under incorrect usage of keys. In: FSE (2017)

17. Grubbs, P., Lu, J., Ristenpart, T.: Message franking via committing authenticated encryption. In: Katz, J., Shacham, H. (eds.) CRYPTO 2017. LNCS, vol. 10403, pp. 66–97. Springer, Cham (2017). https://doi.org/10.1007/978-3-319-63697-9_3

18. Gulley, S., Gopal, V., Yap, K., Feghali, W., Guilford, J.: Intel SHA extensions (2013). https://software.intel.com/en-us/articles/intel-sha-extensions

19. Guo, J., Ling, S., Rechberger, C., Wang, H.: Advanced meet-in-the-middle preimage attacks: first results on full tiger, and improved results on MD4 and SHA-2. In: Abe, M. (ed.) ASIACRYPT 2010. LNCS, vol. 6477, pp. 56–75. Springer, Heidelberg (2010). https://doi.org/10.1007/978-3-642-17373-8_4

20. Halevi, S., Krawczyk, H.: Strengthening digital signatures via randomized hashing. In: Dwork, C. (ed.) CRYPTO 2006. LNCS, vol. 4117, pp. 41–59. Springer, Heidelberg (2006). https://doi.org/10.1007/11818175_3

21. Hong, S., Kim, J., Lee, S., Preneel, B.: Related-key rectangle attacks on reduced versions of SHACAL-1 and AES-192. In: Gilbert, H., Handschuh, H. (eds.) FSE 2005. LNCS, vol. 3557, pp. 368–383. Springer, Heidelberg (2005). https://doi.org/10.1007/11502760_25

22. Jutla, C.S.: Encryption modes with almost free message integrity. In: Pfitzmann, B. (ed.) EUROCRYPT 2001. LNCS, vol. 2045, pp. 529–544. Springer, Heidelberg (2001). https://doi.org/10.1007/3-540-44987-6_32

23. Khovratovich, D., Rechberger, C., Savelieva, A.: Bicliques for preimages: attacks on Skein-512 and the SHA-2 family. In: Canteaut, A. (ed.) FSE 2012. LNCS, vol. 7549, pp. 244–263. Springer, Heidelberg (2012). https://doi.org/10.1007/978-3-642-34047-5_15

24. Kim, J., Kim, G., Hong, S., Lee, S., Hong, D.: The related-key rectangle attack – application to SHACAL-1. In: Wang, H., Pieprzyk, J., Varadharajan, V. (eds.) ACISP 2004. LNCS, vol. 3108, pp. 123–136. Springer, Heidelberg (2004). https://doi.org/10.1007/978-3-540-27800-9_11

25. Kim, J., Kim, G., Lee, S., Lim, J., Song, J.: Related-key attacks on reduced rounds of SHACAL-2. In: Canteaut, A., Viswanathan, K. (eds.) INDOCRYPT 2004. LNCS, vol. 3348, pp. 175–190. Springer, Heidelberg (2004). https://doi.org/10.1007/978-3-540-30556-9_15

26. Lamberger, M., Mendel, F.: Higher-order differential attack on reduced SHA-256. IACR ePrint, Report 2011/037 (2011)

27. Lu, J., Kim, J., Keller, N., Dunkelman, O.: Related-key rectangle attack on 42-round SHACAL-2. In: Katsikas, S.K., López, J., Backes, M., Gritzalis, S., Preneel, B. (eds.) ISC 2006. LNCS, vol. 4176, pp. 85–100. Springer, Heidelberg (2006). https://doi.org/10.1007/11836810_7

28. McGrew, D., Viega, J.: The Galois/counter mode of operation (GCM). In: NIST Modes of Operation (2004)

29. Millican, J.: Personal communication, Feb 2018

30. Millican, J.: Challenges of E2E Encryption in Facebook Messenger. RWC (2017)

31. Preneel, B., Govaerts, R., Vandewalle, J.: Hash functions based on block ciphers: a synthetic approach. In: Stinson, D.R. (ed.) CRYPTO 1993. LNCS, vol. 773, pp. 368–378. Springer, Heidelberg (1994). https://doi.org/10.1007/3-540-48329-2_31

32. Rogaway, P., Bellare, M., Black, J.: OCB: a block-cipher mode of operation for efficient authenticated encryption. ACM TISSEC **6**, 365–403 (2003)

33. Rogaway, P., Shrimpton, T.: A provable-security treatment of the key-wrap problem. In: Vaudenay, S. (ed.) EUROCRYPT 2006. LNCS, vol. 4004, pp. 373–390. Springer, Heidelberg (2006). https://doi.org/10.1007/11761679_23

34. Rogaway, P., Steinberger, J.: Constructing cryptographic hash functions from fixed-key blockciphers. In: Wagner, D. (ed.) CRYPTO 2008. LNCS, vol. 5157, pp. 433–450. Springer, Heidelberg (2008). https://doi.org/10.1007/978-3-540-85174-5_24

35. Rogaway, P., Steinberger, J.: Security/efficiency tradeoffs for permutation-based hashing. In: Smart, N. (ed.) EUROCRYPT 2008. LNCS, vol. 4965, pp. 220–236. Springer, Heidelberg (2008). https://doi.org/10.1007/978-3-540-78967-3_13

36. Sanadhya, S.K., Sarkar, P.: New collision attacks against up to 24-step SHA-2. In: Chowdhury, D.R., Rijmen, V., Das, A. (eds.) INDOCRYPT 2008. LNCS, vol. 5365, pp. 91–103. Springer, Heidelberg (2008). https://doi.org/10.1007/978-3-540-89754-5_8

37. Shrimpton, T., Stam, M.: Building a collision-resistant compression function from non-compressing primitives. In: Aceto, L., Damgård, I., Goldberg, L.A., Halldórsson, M.M., Ingólfsdóttir, A., Walukiewicz, I. (eds.) ICALP 2008. LNCS, vol. 5126, pp. 643–654. Springer, Heidelberg (2008). https://doi.org/10.1007/978-3-540-70583-3_52

38. Open Whisper Systems: Signal (2016). https://signal.org/

39. van der Linde, W.: Parallel SHA-256 in NEON for use in hash-based signatures. BSc thesis, Radboud University (2016)

40. Whatsapp: Whatsapp (2016). https://www.whatsapp.com/

Indifferentiable Authenticated Encryption

Manuel Barbosa[1] and Pooya Farshim[2,3]([⊠])

[1] INESC TEC and FC University of Porto, Porto, Portugal
mbb@dcc.fc.up.pt
[2] DI/ENS, CNRS, PSL University, Paris, France
pooya.farshim@gmail.com
[3] Inria, Paris, France

Abstract. We study Authenticated Encryption with Associated Data (AEAD) from the viewpoint of composition in arbitrary (single-stage) environments. We use the indifferentiability framework to formalize the intuition that a "good" AEAD scheme should have random ciphertexts subject to decryptability. Within this framework, we can then apply the indifferentiability composition theorem to show that such schemes offer extra safeguards wherever the relevant security properties are not known, or cannot be predicted in advance, as in general-purpose crypto libraries and standards.

We show, on the negative side, that generic composition (in many of its configurations) and well-known classical and recent schemes fail to achieve indifferentiability. On the positive side, we give a provably indifferentiable Feistel-based construction, which reduces the round complexity from at least 6, needed for blockciphers, to only 3 for encryption. This result is not too far off the theoretical optimum as we give a lower bound that rules out the indifferentiability of *any* construction with less than 2 rounds.

Keywords: Authenticated encryption · Indifferentiability
Composition · Feistel · Lower bound · CAESAR

1 Introduction

Authenticated Encryption with Associated Data (AEAD) [10,54] is a fundamental building block in cryptographic protocols, notably those enabling secure communication over untrusted networks. The syntax, security, and constructions of AEAD have been studied in numerous works. Recent, ongoing standardization processes, such as the CAESAR competition [14] and TLS 1.3, have revived interest in this direction. Security notions such as misuse-resilience [38,43,52,56], robustness [2,6,41], multi-user security [19], reforgeability [36], and unverified plaintext release [5], as well as syntactic variants such as online operation [43] and variable stretch [41,57] have been studied in recent works.

Building on these developments, and using the indifferentiability framework of Maurer, Renner, and Holenstein [48], we propose new definitions that bring a

© International Association for Cryptologic Research 2018
H. Shacham and A. Boldyreva (Eds.): CRYPTO 2018, LNCS 10991, pp. 187–220, 2018.
https://doi.org/10.1007/978-3-319-96884-1_7

new perspective to the design of AEAD schemes. In place of focusing on specific property-based definitions, we formalize when an AEAD behaves like a *random* one. A central property of indifferentiable schemes is that they offer security with respect to a wide class of games. This class includes all the games above plus many others, including new unforeseen ones. Indifferentiability has been used to study the security of hash functions [15, 21] and blockciphers [4, 24, 33, 44], where constructions have been shown to behave like random oracles or ideal ciphers respectively. We investigate this question for authenticated encryption and ask if, and how efficiently, can indifferentiable AEAD schemes be built.

Our main contributions are as follows.

Definitions: We define ideal authenticated-encryption as one that is indifferentiable from a *random keyed injection*. This definition gives rise to a new model that is intermediate between the random-oracle and the ideal-cipher models. Accordingly, the random-injection model offers new efficiency and security trade-offs when compared to the ideal-cipher model.

Constructions: We obtain both positive and negative results for indifferentiable AEAD schemes. For most well-known constructions our results are negative. However, our main positive result is a Feistel construction that reduces the number of rounds from eight for ideal ciphers to only *three* for ideal keyed injections. This result improves the concrete parameters involved as well. We also give a transformation from offline to online ideal AEADs.

Lower bounds: Three rounds of Feistel are necessary to build injections. However, we prove a stronger result that lower bounds the number of primitive queries as a function of message blocks in *any* construction. This, in turn, shows that the *rate* of our construction is not too far off the optimal solution. For this we combine two lower bound techniques, one for collision resistance and the other for pseudorandomness, which may be of independent interest.

1.1   Background on Indifferentiability

A common paradigm in the design of symmetric schemes is to start from some simple primitive, such as a public permutation or a compression function, and through some "mode of operation" build a more complex scheme, such as a blockcipher or a variable-length hash function. The provable-security of such constructions has been analyzed mainly through two approaches. One is to formulate specific game-based properties, and then show that the construction satisfies them if its underlying primitives are secure. This methodology has been successfully applied to AEAD schemes. (See works cited in the opening paragraph of the paper.) Following this approach, higher-level protocols need to choose from a catalog of explicit properties offered by various AEAD schemes. For example, one would use an MRAE scheme whenever nonce-reuse cannot be excluded [38, 43, 52, 56] or a key-dependent message (KDM) secure one when the scheme is required to securely encrypt its own keys [7, 18].

The seminal work of Maurer, Renner, and Holenstein (MRH) on the indifferentiability of random systems [48] provides an alternative path to study the security of symmetric schemes. In this framework, a public primitive f is available.

The goal is to build another primitive F from f via a construction C^f. Indifferentiability formalizes a set of necessary and sufficient conditions for the construction C^f to securely replace its ideal counterpart F in a wide range of environments: there exists a *simulator* S such that the systems (C^f, f) and (F, S^F) are indistinguishable, even when the distinguisher has access to f. Indeed, the composition theorem proved by MRH states that, if C^f is indifferentiable from F, then C^f can securely replace F in *arbitrary* (single-stage) contexts. Thus, proving that a construction C is indifferentiable from an ideal object F amounts to proving that C^f *retains essentially all security properties implicit in F*. This approach has been successfully applied to the analysis of many symmetric cryptographic constructions in various ideal-primitive models; see, e.g., [21,26,33,44]. Our work is motivated by this composition property.

1.2   Motivation

Maurer, Renner, and Holenstein proposed indifferentiability as an alternative to the Universal Composability (UC) framework [20] for compositional reasoning in idealized models of computation such as the random-oracle (RO) and the ideal-cipher (IC) models. Indifferentiability permits finding constructions that can safely replace ideal primitives (e.g., the random oracle) in various schemes.

The UC framework provides another general composition theorem, which has motivated the study of many UC-secure cryptographic protocols. Küsters and Tuengerthal [47] considered UC-secure symmetric encryption and defined an ideal functionality on par with standard notions of symmetric encryption security. This, however, resulted in an intricate functionality definition that adds complexity to the analysis of higher-level protocols. By adopting indifferentiability for the study of AEADs, we follow an approach that has been successfully applied to the study of other *symmetric* primitives. As random oracles formalize the intuition that well-designed hash functions have random-looking outputs, ideal encryption formalizes random-looking ciphertexts subject to decryptability. This results in a simple and easy-to-use definition. We discuss the benefits of this approach next and give limitations and open problems at the end of the section.

Once a primitive is standardized for general use, it is hard to predict in which environments it will be deployed, and which security properties may be intuitively expected from it. For example, consider a setting where a protocol designer follows the intuition that an AEAD scheme "essentially behaves randomly" and, while not knowing that AE security does not cover key-dependent message attacks [7,18,40] (KDM), uses a standardized general-purpose scheme for disk encryption. In other settings, a designer might create correlations among keys (as in 3GPP) expecting the underlying scheme to offer security against related-key attacks [8] (RKAs). Certain protocols rely on AE schemes that need to be committing against malicious adversaries, which can choose all inputs and thus also the keys. This has lead to the formalizations of committing [39] and key-robust [34] authenticated encryption. When there is leakage, parts of the key and/or randomness might be revealed [9]. All of these lie beyond standard notions of AE security, so the question is how should one deal with such a multitude of security properties.

One approach would be to formulate a new "super" notion that encompasses all features of the models above. This is clearly not practical. The model (and analyses using it) will be error-prone and, moreover, properties that have not yet been formalized will not be accounted for. Instead, and as mentioned above, we consider the following approach: a good AEAD scheme should behave like a random oracle, except that its ciphertexts are invertible. We formulate this in the language of indifferentiability, which results in a simple, unified, and easy to use definition. In indifferentiability, all inputs are under the control of the adversary. This means that the security guarantees offered extend to notions that allow for tampering with keys or creation of dependencies among the inputs. Once indifferentiability is proved, security with respect to *all* these games, combinations thereof and new unforeseen ones, jointly follows from the composition theorem.

Therefore one use-case for indifferentiable schemes would be to provision additional safeguards against *primitive misuse* in various deployment scenarios, such as general-purpose crypto libraries or standards, where the relevant security properties for target applications are complex or not known. We discussed some of these in the paragraph above. Protocol designers can rely on the intuition given by an ideal view of AEADs when integrating schemes into higher-level protocols, keeping game-based formulations implicit. Other applications include symbolic protocol analysis, where such idealizations are intrinsic [49] and security models where proof techniques such as programmability may be required [59].

A CONCRETE EXAMPLE. In Facebook's message-franking protocol, an adversary attempts to compute a ciphertext that it can later open in two ways by revealing different keys, messages and header information. (Facebook sees one (harmless) message, whereas the receiver gets another (possibly abusive) message.) Grubbs, Lu, and Ristenpart [39] formalize the security of such protocols and show that a standard AEAD can be used here, provided that it satisfies an additional security property called r-BIND [39, Fig. 17 (left)].

One important feature of this definition is that it relies on a single-stage game in the sense of [53]. The single-stage property immediately implies that any indifferentiable scheme is r-BIND if the ideal encryption scheme itself satisfies the r-BIND property. In contrast, not every AE-secure scheme is r-BIND secure [39]. Interestingly, it is easy to see that the ideal encryption scheme (a keyed random injection) indeed satisfies r-BIND and this is what, intuitively, the protocol designers seem to have assumed: that ciphertexts *look random* and thus collisions are hard to find, even if keys are adversarially chosen.

Indifferentiable AEADs therefore allow designers to rely on the above (arguably pragmatic) random-behavior intuition much in the same way as they do when using hash functions as random oracles. As the practicality of random oracles stems from their random output behavior (beyond PRF security or collision resistance) indifferentiable AEAD offers similar benefits: instead of focusing on a specific game-based property, it considers a fairly wide *class* of games for which the random behavior provably holds.

Thus an indifferentiable AE can be used as a safety net to ensure any existing or future single-stage assumptions one may later need is satisfied (with the caveat of possibly weaker bounds). However, we note that for RO indifferentiability there is the additional motivation that a fair number of security proofs involving hash functions rely on modeling the hash as RO. Our work also unlocks the possibility to use the *full* power of random injections in a similar way (see [46] and Footnote 2).

To summarize, in the context of Facebook's protocol, if an indifferentiable scheme was used from the start, it would have automatically met the required binding property. The same holds for RKA security (in 3GPP), KDM security (in disk encryption), and other single-stage AEAD applications.

1.3   Overview of Technical Contributions

DEFINITIONS. The MRH framework has been formulated with respect to a general class of random systems. We make this definition explicit for AEAD schemes by formulating an adequate *ideal reference object*. This object has been gradually emerging through the notion of a pseudorandom injection (PRI) in a number of works [41,43,56], and has been used to study the security of offline and online AEADs [41,43]. We lift these notions to the indifferentiability setting by introducing offline and online *random injections*, which may be also keyed or tweaked. As a result, we obtain a new idealized model of computation: the ideal-encryption (or ideal-injection) model, which is intermediate between the RO and IC models. Along the way, we give an extension of the composition theorem to include game-based properties with multiple adversaries.

ANALYSIS OF KNOWN SCHEMES. We examine generic and specific constructions of AEADs that appear in the literature. Since indifferentiability implies security in the presence of nonce-misuse (MRAE) as well as its recent strengthening to variable ciphertext stretch, RAE security,[1] we rule out the indifferentiability of a number of (classical) schemes that do not achieve these levels of security. This includes OCB [55], CCM, GCM, and EAX [13], and all but two of the third-round CAESAR candidates [14]. The remaining two candidates, AEZ [42] and DEOXYS-II [45], are also ruled out, but only using specific indifferentiability attacks. We discuss our conclusions for CAESAR submissions in [1, Sect. 4.2].

We then turn our attention to generic composition [10,51]. We study the well-known Encrypt-then-MAC and MAC-then-Encrypt constructions via the composition patterns of Namprempre, Rogaway and Shrimpton [51]. These include Synthetic Initialization Vector (SIV) [56] and EAX [13]. To simplify and generalize the analysis, we start by presenting a template for generic composition, consisting of a preprocessing and a post-processing phase, that encompasses a

---

[1] The notion of RAE security that we use deviates from the original notion proposed in [41] by not considering benevolous leakage of information during decryption. This is because all indifferentiable constructions must guarantee that, like the ideal object, decryption gives the stronger guarantee that $\perp$ is returned for all invalid ciphertexts.

number of schemes that we have found in the literature. We show that if there is an insufficient flow of information in a scheme—a notion that we formalize—differentiating attacks exist. Our attacks render all of these constructions except A8 and *key reusing* variants of A2 and A6 as indifferentiability candidates.

In short, contrarily to our expectations based on known results for hash functions and permutations, we could not find a well-known AEAD construction that meets the stronger notion of indifferentiability. We stress that these findings do *not* contradict existing security claims. However, an indifferentiability attack can point to environments in which the scheme will not offer the expected levels of security. For example, some of our differentiators stem from the fact that ciphertexts do not depend on all keying material, giving way to related-key attacks. In others, the attacks target intermediate computation values and are reminiscent of padding oracles. For these reasons, and even though our results do not single out any of the CAESAR candidates as being better or worse than the others, we pose that our results are aligned with the fundamental goal of CAESAR and prior competitions such as AES and SHA-3, to "boost to the cryptographic research community's understanding" of the primitive [14].

BUILDING INJECTIONS. We revisit the classical Encode-then-Encipher (EtE) transform [11]. Given expansion τ, which indicates the required level of authenticity, EtE pads the input message with 0^τ and enciphers it with a variable-input-length (VIL) blockcipher. Decryption checks the consistency of the padding after recovering the message. We show that EtE is indifferentiable from a random injection in the VIL ideal-cipher model for any (possibly small) value of τ.

The ideal cipher underlying EtE can be instantiated via the Feistel construction [23] in the random-oracle model or via the confusion-diffusion construction [33] in the random-permutation model. In a series of works, the number of rounds needed for indifferentiability of Feistel has been gradually reduced from 14 [23,44] to 10 [25,27] and recently to 8 [28]. Due to the existence of differentiators [23,24], the number of rounds must be at least 6. For confusion-diffusion, 7 rounds are needed for good security bounds [33]. This renders the above approach to construct random injections somewhat suboptimal in terms of queries per message block to their underlying ideal primitives (i.e., their *rate*).

Our main positive result is the indifferentiability of *three*-round Feistel for large (but variable) expansion values τ. Three rounds are also necessary, as we give a differentiator against the 2-round Feistel network for any τ. In light of the above results, and state-of-the-art 2.5-round constructions such as AEZ, this is a surprisingly small price to pay to achieve indifferentiability. Our results, therefore, support inclusion of redundancy for achieving authenticity (as opposed to generic composition). Furthermore, when using a blockcipher for encryption with redundancy, a significantly reduced number of rounds may suffice.

THE SIMULATOR. Our main construction is an unbalanced 3-round Feistel network Φ_3 with independent round functions where an input X_1 is encoded with redundancy as $(0^\tau, X_1)$ (see Fig. 1). The main task of our indifferentiability simulator is to consistently respond to round-function oracle queries that correspond to those that the construction makes for some (possibly unknown) input X_1. We show that with overwhelming probability the simulator can detect when consistency with the construction must be enforced; the remaining isolated queries can be simulated using random and independent values.

Take, for example, a differentiator that computes $(X_3', X_4') := \Phi_3(X_1)$ for some random X_1, then computes the corresponding round-function outputs $X_2 := F_1(X_1)$, $Y_2 := F_2(X_2)$, $Y_3 := F_3(X_1 \oplus Y_2)$, and finally checks if $(X_3', X_4') = (X_1 \oplus Y_2, X_2 \oplus Y_3)$. Note that these queries need *not* arrive in this particular order. Indeed, querying $F_1(X_1)$ first gives the simulator an advantage as it can preemptively complete this chain of queries and use its ideal injection to give consistent responses. A better (and essentially the only) alternative for the differentiator would be to check the consistency of outputs by going through the construction in the backward direction. We show, however, that whatever query strategy is adopted by the differentiator, the simulator can take output values fixed by the ideal injection and work out answers for the round function oracles that are consistent with the construction in the real world.

A crucial part of this analysis hinges on the fact that the output of the first round function is directly fed as input to the second round function as a consequence of fixing parts of the input to 0^τ.[2] As corollaries of our results we obtain efficient and (simultaneously) RKA and KDM-secure offline (and, as we shall see, online) AEAD schemes in the random-permutation model under natural, yet practically relevant restrictions

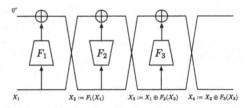

Fig. 1. Injection from 3-round Feistel.

on these security models. For example, if the ideal AEAD $\mathcal{AE}$ is secure under encryptions of $\phi^{\mathcal{AE}}(K)$ for some oracle machine $\phi^{\mathcal{AE}}$, then so is an indifferentiable construction C^π in presence of encryptions of $\phi^{C^\pi}(K)$, the restriction being that ϕ does not directly access π.

BOUNDS. Security bounds, including simulator query complexity, are important considerations for practice. Our bound for the Encode-then-Encipher construction is essentially tight. Our simulator for the 3-round Feistel construction has a quadratic query complexity and overall bounds are birthday-type. Improving these bounds, or proving lower bounds for them [32], remain open for subsequent work.

Our construction of an ideal encryption scheme from a non-keyed ideal injection introduces an additional multiplicative factor related to the number of

---

[2] Padding with 0^τ has also been used by Kiltz, Pietrzak, and Szegedy [46] who study the *public* indifferentiability of injections while building digital signature schemes with message recovery. The motivation there is to design schemes with optimal overhead that also come with tight security reductions. However, this level of indifferentiability is not sufficient in the AEAD setting as it does not even imply CPA security.

different ideal injection keys queried by the differentiator, resulting from a hybrid argument over keys. Furthermore, the number of ideal injection keys used in the construction is bound to the number of encryption and decryption operations that are carried out. This means that the overall bound for our authenticated encryption construction includes a multiplicative factor of q^3 (see Sect. 5.3).

We note that the concrete constructions that we analyze may satisfy (R)AE, RKA or KDM security with improved bounds (via game-specific security analyses), while remaining compatible with the single proof and bound that we present for all single-stage games.

ONLINE AEADS. We give simple solutions to the problem of constructing an indifferentiable *segment-oriented* online AEAD scheme from an offline AEAD. Following [43], we define ideal online AEAD scheme via initialization, next-segment encryption/decryption, and last-segment encryption/decryption procedures. The difference between next-segment and last-segment operations is that the former propagates *state values*, whereas the latter does not. Since a differentiator typically has access to *all* interfaces of a system, the state values become under its control/view. For this we restrict the state size of the ideal object to be finite and hence definitionally deviate from [43] in this aspect. Therefore our constructions have the extra security property that the state value hides all information about past segments.

The most natural way to construct an ideal online AEAD would be to *chain* encryptions of the segments by *tweaking* the underlying encryption primitive with the input history so far, as in the **CHAIN** transform of HRRV [43, Fig. 8]. We show, however, that standard XOR-based tweaking techniques are not sound in the indifferentiability setting and, in particular, we present a differentiating attack on **CHAIN**. However, by decomposing the ideal object for online AEAD into simpler ones [29, 48], we recover an indifferentiable variant of the construction called **HashCHAIN**, where a random oracle is used to prepare the state for the next segment. Via optimizations specific to 3-round Feistel, we reduce overheads to a *constant* number of hashes per segment.

LOWER BOUNDS. The indifferentiability of Sponge [15] allows us to instantiate the round functions in 3-round Feistel with this construction and derive a random injection in the random-permutation model.[3] This construction requires roughly $3w$ calls to its underlying (one-block) permutation, where w is the total number of input blocks. This is slightly higher than $2.5w$ for AEZ (which shares some of its design principles with us, but does not offer indifferentiability). This leads us to ask whether or not an indifferentiable construction with rate *less than* 3 is achievable. Our second main result is a lower bound showing the impossibility of *any* such construction with rate (strictly) less than 2. To prove this lower bound, we combine negative results for constructions of collision-resistant hash functions [17, 58] and pseudorandom number generators by Gennaro and Trevisan [37], and put critical use to the existence of an indifferentiability simulator. To the best of our knowledge, this is the first impossibility result

---

[3] The intermediate (expanding) round function can alternatively be fully parallelized.

that exploits indifferentiability, so the proof technique may be of independent interest.

LIMITATIONS AND FUTURE WORK. As clarified by Ristenpart, Shacham, and Shrimpton [53], the indifferentiability composition theorem may not apply to multi-stage games where multiple adversaries cannot be collapsed into a single central adversary. Indifferentiable AEAD schemes come with similar limitations.

Indifferentiability typically operates in an ideal model of computation. This leaves open the question of standard-model security. However, it does not exclude a "best of the two worlds" construction, which is both indifferentiable and is RAE secure in the standard-model. For example, chop-Merkle–Damgård [21] can be proven both indifferentiable from a random oracle *and* collision resistant in the standard model. We leave exploring this for future work.

2   Basic Definitions

We let $\{0,1\}^*$ denote the set of all finite-length bit strings, including the empty string ε. For bit strings X and Y, $X|Y$ denotes concatenation and (X,Y) denotes a decodable encoding of X and Y. The length of a string X is denoted by $|X|$.

GAMES. An n-adversary game $\mathbf{G}$ is a Turing machine $\mathbf{G}^{\Sigma,\mathscr{A}_1,\dots,\mathscr{A}_n}$ where Σ is a system (or functionality) and $\mathscr{A}_i$ are adversarial procedures that can keep full local state but may only communicate with each other through $\mathbf{G}$. We say an n-adversary game $\mathbf{G}_n$ is reducible to an m-adversary game if there is a $\mathbf{G}_m$ such that for any $(\mathscr{A}_1,\dots,\mathscr{A}_n)$ there are $(\mathscr{A}'_1,\dots,\mathscr{A}'_m)$ such that for all Σ we have that $\mathbf{G}_n^{\Sigma,\mathscr{A}_1,\dots,\mathscr{A}_n} = \mathbf{G}_m^{\Sigma,\mathscr{A}'_1,\dots,\mathscr{A}'_m}$. Two games are equivalent if they are reducible in both directions. An n-adversary game is called n-stage [53] if it is not equivalent to any m-adversary game with $m < n$. Any single-stage game $\mathbf{G}^{\Sigma,\mathscr{A}}$ can be also written as $\mathscr{A}^{\overline{\mathbf{G}}^\Sigma}$ for some oracle machine $\overline{\mathbf{G}}$ and a class of adversarial procedures $\mathscr{A}$ compatible with a modified syntax in which the game is called as an oracle.

REFERENCE OBJECTS. Underlying the security definition for a cryptographic primitive there often lies an *ideal* primitive that is used as a *reference object* to formalize security. For instance, the security of PRFs is defined with respect to a random oracle, PRPs with respect to an ideal cipher, and as mentioned above, AEADs with respect to a random *injection*. Given the syntax and the correctness condition of a cryptographic primitive, we will define its ideal counterpart as the uniform distribution over the set of all functions that meet these syntactic and correctness requirements (but without any efficiency requirements). We start by formalizing a general class of ideal functions—that may be keyed, admit auxiliary data (such as nonces or authenticated data), or allow for variable-length outputs—and derive distributions of interest to us by imposing structural restrictions over the class of considered functions. This approach has also been used in [16].

IDEAL FUNCTIONS. A variable-output-length (VOL) function $\mathcal{F}$ with auxiliary input has signature $\mathcal{F} : \mathcal{A} \times \mathcal{M} \times \mathcal{X} \longrightarrow \mathcal{R}$, where $\mathcal{A}$ is the auxiliary-input space, $\mathcal{M}$ is the message space, $\mathcal{X} \subseteq \mathbb{N}$ is the expansion space, and $\mathcal{R}$ is the range. We let $\mathrm{Fun}[\mathcal{A} \times \mathcal{M} \times \mathcal{X} \longrightarrow \mathcal{R}]$ be the set of all such functions satisfying $\forall(A, M, \tau) \in \mathcal{A} \times \mathcal{M} \times \mathcal{X} : |\mathcal{F}(A, M, \tau)| = \tau$, We endow the above set with the uniform distribution and denote the action of sampling a uniform function $\mathcal{F}$ via $\mathcal{F} \twoheadleftarrow \mathrm{Fun}[\mathcal{A} \times \mathcal{M} \times \mathcal{X} \longrightarrow \mathcal{R}]$ (and analogously for expanding functions). To ease notation, given a function $\mathcal{F}$, we define $\mathrm{Fun}[\mathcal{F}]$ to be the set of all functions with signature identical to that of $\mathcal{F}$. Granting oracle access to $\mathcal{F}$ to all parties (honest or otherwise) results in an ideal model of computation.

INJECTIONS. We define $\mathrm{Inj}[\mathcal{A} \times \mathcal{M} \times \mathcal{X} \longrightarrow \mathcal{R}]$ to be the set of all expanding functions that are injective on $\mathcal{M}$: $\forall(A, M, \tau), (A, M', \tau) \in \mathcal{A} \times \mathcal{M} \times \mathcal{X}$: $M \neq M' \implies \mathcal{F}(A, M, \tau) \neq \mathcal{F}(A, M', \tau)$, and satisfy the length restriction $\forall(A, M, \tau) \in \mathcal{A} \times \mathcal{M} \times \mathcal{X} : |\mathcal{F}(A, M, \tau)| = |M| + \tau$. Each injective function defines a unique inverse function $\mathcal{F}^-$ that maps (A, C, τ) to either a unique M if and only if C is within the range of $\mathcal{F}(A, \cdot, \tau)$, or to $\bot$ otherwise. (Such functions are therefore *tidy* in the sense of [51].) This gives rise to a *strong* induced model for injections where oracle access is extended to include $\mathcal{F}^-$, which we always assume to be the case when working with injections.

When $k = 0$ the key space contains the single ε key and we recover unkeyed functions. We use the following abbreviations: $\mathrm{Fun}[n, m]$ is the set of functions mapping n bits to m bits and $\mathrm{Perm}[n]$ is the set of permutations over n bits.

LAZY SAMPLERS. Various ideal objects (such as random oracles) often appear as algorithmic procedures that lazily sample function values at each point. These procedures can be extended to admit auxiliary data and respect either of our length-expansion requirements above. Furthermore, given a list L of input-output pairs, these samplers can be modified to sample a function that is also consistent with the points defined in L (i.e., the conditional distribution given L is also samplable). We denote the lazy sampler for random oracles with $(Y; L) \twoheadleftarrow \mathrm{LazyRO}(A, X, \tau; L)$ and that for ideal ciphers with $(Y; L) \twoheadleftarrow \mathrm{LazyIC}^{\pm}(A, X; L)$. The case of random injection is less well known, but such a procedure appears in [56, Fig. 6]. We denote this sampler with $(Y; L) \twoheadleftarrow \mathrm{LazyRI}^{\pm}(A, X, \tau; L)$.

2.1 Authenticated-Encryption with Associated-Data

We follow [43] in formalizing the syntax of (offline) AEAD schemes.[4] We allow for arbitrary plaintexts and associated data, and also include an explicit expansion parameter τ specifying the level of authenticity. Associated data may contain information that may be needed in the clear by a higher-level protocol that nevertheless should be authentic. We also only allow for public nonces as the benefits of the AE5 syntax with a private nonce are unclear [50].

SYNTAX AND CORRECTNESS. An AEAD scheme is a triple of algorithms $\Pi :=$ $(\mathcal{K}, \mathcal{AE}, \mathcal{AD})$ where: (1) $\mathcal{K}$ is the randomized key-generation algorithm which

---

[4] When referring to an AEAD without specifying its type, we mean an *offline* AEAD.

GAME **RAE-Real$_\Pi^{\mathscr{A}}$**	GAME **RAE-Ideal$_\Pi^{\mathscr{A}}$**
$K \twoheadleftarrow \mathcal{K}$	$(\mathcal{AE}', \mathcal{AD}') \twoheadleftarrow \mathrm{AE}[\Pi]$
$b \twoheadleftarrow \mathscr{A}^{\mathcal{AE}(K,\cdot,\cdot,\cdot,\cdot),\mathcal{AD}(K,\cdot,\cdot,\cdot,\cdot)}$	$K \twoheadleftarrow \mathcal{K}$
return b	$b \twoheadleftarrow \mathscr{A}^{\mathcal{AE}'(K,\cdot,\cdot,\cdot,\cdot),\mathcal{AD}'(K,\cdot,\cdot,\cdot,\cdot)}$
	return b

Fig. 2. Games defining RAE security. The adversary queries its oracles on inputs that belong to appropriate spaces.

returns a key K. This algorithm defines a non-empty set, the support of $\mathcal{K}$, and an associated distribution on it. Slightly abusing notation, we denote all these by $\mathcal{K}$. (2) $\mathcal{AE}$ is the deterministic encryption algorithm with signature $\mathcal{AE} : \mathcal{K} \times \mathcal{N} \times \mathcal{H} \times \{0,1\}^* \times \mathcal{X} \longrightarrow \{0,1\}^*$. Here $\mathcal{N} \subseteq \{0,1\}^*$ is the nonce space, $\mathcal{H} \subseteq \{0,1\}^*$ is the associated data space, and $\mathcal{X} \subseteq \mathbb{N}$ is the set of allowed expansion values. We typically have that $\mathcal{K} = \{0,1\}^k$, $\mathcal{N} = \{0,1\}^n$ for $k, n \in \mathbb{N}$, $\mathcal{H} = \{0,1\}^*$, and the expansion space contains a single value. (3) $\mathcal{AD}$ is the deterministic decryption algorithm with signature $\mathcal{AD} : \mathcal{K} \times \mathcal{N} \times \mathcal{H} \times \{0,1\}^* \times \mathcal{X} \longrightarrow \{0,1\}^* \cup \{\bot\}$. As usual we demand that $\mathcal{AD}(K, N, A, \mathcal{AE}(K, N, A, M, \tau), \tau) = M$ for all inputs from the appropriate spaces. We also impose the ciphertext expansion restriction that for all inputs from the appropriate spaces $|\mathcal{AE}(K, N, A, M, \tau)| - |M| = \tau$.

IDEAL AEAD. An ideal AEAD is an injection with signature $(\mathcal{K} \times \mathcal{N} \times \mathcal{H}) \times \mathcal{M} \times \mathcal{X} \longrightarrow \mathcal{C}$ and satisfying the ciphertext-expansion restriction. Therefore an ideal AEAD is a random injection in $\mathrm{Inj}[(\mathcal{K} \times \mathcal{N} \times \mathcal{H}) \times \mathcal{M} \times \mathcal{X} \longrightarrow \mathcal{C}]$. Given a concrete AEAD scheme Π with signature $\mathcal{K} \times \mathcal{N} \times \mathcal{H} \times \mathcal{M} \times \mathcal{X} \longrightarrow \mathcal{C}$ we associate the space $\mathrm{AE}[\Pi] := \mathrm{Inj}[(\mathcal{K} \times \mathcal{N} \times \mathcal{H}) \times \mathcal{M} \times \mathcal{X} \longrightarrow \mathcal{C}]$ to it.

NAMING CONVENTIONS. When referring to AEAD schemes we use $(\mathcal{AE}, \mathcal{AD})$ instead of $(\mathcal{F}, \mathcal{F}^-)$. When the associated-data space is empty, we use $(\mathcal{E}, \mathcal{D})$ for (encryption without associated data), when the nonce space is also empty we use $(\mathcal{F}, \mathcal{F}^-)$ (for keyed injection), when $\tau = 0$ as well we use $(\mathsf{E}, \mathsf{E}^-)$ (for block-cipher), and if these are also unkeyed we use (ρ, ρ^-) and (π, π^-) respectively. For a random function (without inverse) we use $\mathcal{H}$.

RAE SECURITY. Robust AE (RAE) security [41,43] requires that an AEAD scheme behaves indistinguishably from an ideal AEAD under a random key. Formally, for scheme $\Pi = (\mathcal{K}, \mathcal{AE}, \mathcal{AD})$ and adversary $\mathscr{A}$ we define

$$\mathbf{Adv}_\Pi^{\mathrm{rae}}(\mathscr{A}) := \Pr\left[\mathbf{RAE\text{-}Real}_\Pi^{\mathscr{A}}\right] - \Pr\left[\mathbf{RAE\text{-}Ideal}_\Pi^{\mathscr{A}}\right],$$

where games **RAE-Real$_\Pi^{\mathscr{A}}$** and **RAE-Ideal$_\Pi^{\mathscr{A}}$** are defined in Fig. 2. Informally, we say Π is RAE secure if $\mathbf{Adv}_\Pi^{\mathrm{rae}}(\mathscr{A})$ is "small" for any "reasonable" $\mathscr{A}$. Misuse-resilient AE (MRAE) security [56] weakens RAE security by constraining the adversary to a fixed and sufficiently large value of expansion τ. AE security [54] weakens MRAE security and requires that the adversary does not repeat nonces in its queries to either oracle. These definitions lift to idealized models of computation where, for example, access to an ideal injection in both the forward and backward directions is provided.

The proposition below formalizes the intuition that the ideal AEAD, i.e., the trivial AEAD scheme in the ideal AEAD model, is RAE secure. This fact will be used when studying the relation between indifferentiability and RAE security. The proof follows from the fact that unless the attacker can discover the secret key, the construction oracle behaves independently from the ideal AEAD oracle.

Proposition 1 (Ideal AEAD is RAE secure). *For any q-query adversary $\mathscr{A}$ attacking the trivial ideal AEAD Π in the ideal AEAD model we have that* $\mathbf{Adv}_{\Pi}^{\mathrm{rae}}(\mathscr{A}) \leq q/2^k$.

3   AEAD Indifferentiability

The indifferentiability framework of Maurer, Renner, and Holenstein (MRH) [48] formalizes a set of necessary and sufficient conditions for one system to securely replace another in a wide class of environments. This framework has been successfully used to justify the structural soundness of a number of cryptographic constructions, including hash functions [21,31], blockciphers [4,23,33], and domain extenders for them [22]. The indifferentiability framework is formulated with respect to general systems. When the ideal AEAD object defined in Sect. 2.1 is used, a notion of indifferentiability for AEAD schemes emerges. In this section, we recall indifferentiability of systems and make it explicit for AEAD schemes. We will then discuss some of its implications that motivate our work.

3.1   Definition

A random system or functionality $\Sigma := (\Sigma.\mathrm{hon}, \Sigma.\mathrm{adv})$ is accessible via two interfaces $\Sigma.\mathrm{hon}$ and $\Sigma.\mathrm{adv}$. Here, $\Sigma.\mathrm{hon}$ provides a public interface through which the system can be accessed. $\Sigma.\mathrm{adv}$ corresponds to a (possibly extended) interface that models adversarial access to the inner workings of the system, which may be exploited during an attack on constructions. A system typically implements some ideal object $\mathcal{F}$, or it is itself a construction $\mathrm{C}^{\mathcal{F}'}$ relying on some underlying (lower-level) ideal object $\mathcal{F}'$.

INDIFFERENTIABILITY [48]. Let Σ_1 and Σ_2 be two systems and $\mathcal{S}$ be an algorithm called the simulator. The (strong) indifferentiability advantage of a (possibly unbounded) differentiator $\mathscr{D}$ against (Σ_1, Σ_2) with respect to $\mathcal{S}$ is

$$\mathbf{Adv}_{\Sigma_1,\Sigma_2,\mathcal{S}}^{\mathrm{indiff}}(\mathscr{D}) := \Pr\left[\mathbf{Diff\text{-}Real}_{\Sigma_1}^{\mathscr{D}}\right] - \Pr\left[\mathbf{Diff\text{-}Ideal}_{\Sigma_2,\mathcal{S}}^{\mathscr{D}}\right] ,$$

where games $\mathbf{Diff\text{-}Real}_{\Sigma_1}^{\mathscr{D}}$ and $\mathbf{Diff\text{-}Ideal}_{\Sigma_2,\mathcal{S}}^{\mathscr{D}}$ are defined in Fig. 3. Informally, we call Σ_1 indifferentiable from Σ_2 if, for an "efficient" $\mathcal{S}$, the advantage above is "small" for all "reasonable" $\mathscr{D}$.

GAME **Diff-Real**$_{\Sigma_1}^{\mathscr{D}}$		GAME **Diff-Ideal**$_{\Sigma_2,\mathcal{S}}^{\mathscr{D}}$	
$b \twoheadleftarrow \mathscr{D}^{\text{CONST,PRIM}}$; return b		$b \twoheadleftarrow \mathscr{D}^{\text{CONST,PRIM}}$; return b	
PROC. CONST(X)	PROC. PRIM(X)	PROC. CONST(X)	PROC. PRIM(X)
return $\Sigma_1.\text{hon}(X)$	return $\Sigma_1.\text{adv}(X)$	return $\Sigma_2.\text{hon}(X)$	return $\mathcal{S}^{\Sigma_2.\text{adv}}(X)$

Fig. 3. Games defining the indifferentiability of two systems.

In the rest of the paper we consider a specific application of this definition to two systems with interfaces $(\Sigma_1.\text{hon}(X), \Sigma_1.\text{adv}(x)) := (C^{\mathcal{F}_1}(X), \mathcal{F}_1(x))$ and $(\Sigma_2.\text{hon}(X), \Sigma_2.\text{adv}(x)) := (\mathcal{F}_2(X), \mathcal{F}_2(x))$, where $\mathcal{F}_1$ and $\mathcal{F}_2$ are two ideal cryptographic objects sampled from their associated distributions and $C^{\mathcal{F}_1}$ is a construction of $\mathcal{F}_2$ from $\mathcal{F}_1$. To ease notation, we denote the advantage function by $\mathbf{Adv}_{C,\mathcal{S}}^{\text{indiff}}(\mathscr{D})$ when $\mathcal{F}_1$ and $\mathcal{F}_2$ are clear from context. Typically $\mathcal{F}_2$ will be an ideal AEAD and $\mathcal{F}_1$ a random oracle or an ideal cipher.

3.2 Consequences

MRH [48] prove the following composition theorem for indifferentiable systems. Here we state a game-based formulation from [53].

Theorem 1 (Indifferentiability composition). *Let $\Sigma_1 := (C^{\mathcal{F}_1}, \mathcal{F}_1)$ and $\Sigma_2 := (\mathcal{F}_2, \mathcal{F}_2)$ be two indifferentiable systems with simulator $\mathcal{S}$. Let $\mathbf{G}$ be a single-stage game. Then for any adversary $\mathscr{A}$ there exist an adversary $\mathscr{B}$ and a differentiator $\mathscr{D}$ such that*

$$\Pr\left[\mathbf{G}^{C^{\mathcal{F}_1}, \mathscr{A}^{\mathcal{F}_1}}\right] \leq \Pr\left[\mathbf{G}^{\mathcal{F}_2, \mathscr{B}^{\mathcal{F}_2}}\right] + \mathbf{Adv}_{C,\mathcal{S}}^{\text{indiff}}(\mathscr{D}) .$$

As discussed in [53], the above composition does not necessarily extend to multi-stage games since the simulator often needs to keep *local* state in order to guarantee consistency. However, some (seemingly) multi-stage games can be written as equivalent single-stage games (see Sect. 2 for a definition of game equivalence). Indeed, any n-adversary game where only one adversary can call the primitive directly and the rest call it indirectly *via the construction* can be written as a single-stage game as the game itself has access to the construction. We summarize this observation in the following theorem, which generalizes a result for related-key security in [35].

Theorem 2. *Let $\Sigma_1 := (C^{\mathcal{F}_1}, \mathcal{F}_1)$ and $\Sigma_2 := (\mathcal{F}_2, \mathcal{F}_2)$ be two indifferentiable systems with simulator $\mathcal{S}$. Let $\mathbf{G}$ be an n-adversary game and $\mathscr{A} := (\mathscr{A}_1, \ldots, \mathscr{A}_n)$ be an n-tuple of adversaries where $\mathscr{A}_1$ can access $\mathcal{F}_1$ but $\mathscr{A}_i$ for $i > 1$ can only access $C^{\mathcal{F}_1}$. Then there is an n-adversary $\mathscr{B}$ and a differentiator $\mathscr{D}$ such that*

$$\Pr\left[\mathbf{G}^{C^{\mathcal{F}_1}, \mathscr{A}_1^{\mathcal{F}_1}, \mathscr{A}_2^{C^{\mathcal{F}_1}}, \ldots, \mathscr{A}_n^{C^{\mathcal{F}_1}}}\right] \leq \Pr\left[\mathbf{G}^{\mathcal{F}_2, \mathscr{B}_1^{\mathcal{F}_2}, \mathscr{B}_2^{\mathcal{F}_2}, \ldots, \mathscr{B}_n^{\mathcal{F}_2}}\right] + \mathbf{Adv}_{C,\mathcal{S}}^{\text{indiff}}(\mathscr{D}) .$$

REMARK 1. There is a strong practical motivation for the restriction imposed on the class of games above. Consider, for example, security against related-key attacks (RKAs) where the related-key deriving (RKD) function $\phi^{\mathcal{F}_1}$ may depend on the ideal primitive [3]. The RKA game is *not* known to be equivalent to a single-stage game. The authors in [35] consider a restricted form of this game where dependence of ϕ on the ideal primitive $\mathcal{F}_1$ is constrained to be through the construction $\mathsf{C}^{\mathcal{F}_1}$ only. In other words, an RKD function takes the form $\phi^{\mathsf{C}^{\mathcal{F}_1}}$ rather than $\phi^{\mathcal{F}_1}$. When comparing the RKA security of a construction $\mathsf{C}^{\mathcal{F}_1}$ to the RKA security of its ideal counterpart, one would expect the set of RKD functions from which ϕ is drawn in two games to be syntactically fixed and hence comparable. Since no underlying ideal primitive for $\mathcal{F}_2$ exists, RKD functions take the form $\phi^{\mathcal{F}_2}$ and hence it is natural to consider RKD functions of the form $\phi^{\mathsf{C}^{\mathcal{F}_1}}$ with respect to $\mathsf{C}^{\mathcal{F}_1}$. The same line of reasoning shows that an indifferentiable construction would resist key-dependent message (KDM) attacks for key-dependent deriving functions that depend on the underlying ideal primitive via the construction only. Other (multi-stage) security notions that have a practically relevant single-stage formulation include security against bad-randomness attacks, where malicious random coins are computed using the construction, and leakage-resilient encryption where leakage functions may rely on the construction. Therefore from a practical point of view, composition extends well beyond 1-adversary games.

REMARK 2. Theorem 1 reduces the security of one system to that of another. For instance, one can deduce the RKA (resp., KDM or leakage-resilient) security of an indifferentiable construction $\mathsf{C}^{\mathcal{F}_1}$ of $\mathcal{F}_2$ *if* $\mathcal{F}_2$ itself can be proven to be RKA (resp., KDM or leakage-resilient) secure. We have seen an example of the latter in Proposition 1, where the ideal AEAD scheme is shown to be RAE secure. Hence Theorem 1 and Proposition 1 immediately allow us to deduce that an indifferentiable AEAD construction $\mathsf{C}^{\mathcal{F}_1}$ will be RAE secure in the idealized model of computation induced by its underlying ideal primitive $\mathcal{F}_1$. Analogous propositions for RKA, KDM, leakage resilience of the ideal AEAD scheme (for quantified classes of related-key deriving functions, key-dependent deriving, and leakage functions) can be formulated. This in turn implies that an indifferentiable AEAD scheme will resist strong forms of related-key, KDM, and leakage attacks.

4   Differentiators

Having defined AEAD indifferentiability, we ask whether or not (plausibly) indifferentiable constructions of AEAD schemes in the literature exist. In this section we present a number of generic and specific attacks that essentially rule out the indifferentiability of many constructions that we have found in the literature. We emphasize that existing schemes were not designed with the goal of meeting indifferentiability, and our attacks do not contradict any security claims made under the standard RAE, MRAE, or AE models. Indeed, many AEAD schemes are designed with the goal of maximizing efficiency, forsaking stronger security goals such as misuse resilience or robustness.

4.1   Generic Composition

Any construction that is not (M)RAE secure (in the sense of [41,56]) can be immediately excluded as one that is indifferentiable: the ideal AEAD is RAE secure (Proposition 1), furthermore RAE is a single-stage game and hence implied by indifferentiability (Theorem 1). This simple observation rules out the indifferentiability of a number notable AEAD schemes such as OCB [55], CCM, GCM, EAX [13], and many others. The MRAE insecurity of these schemes are discussed in the respective works.

RAE insecurity can be used to also rule out the indifferentiability of some generic AEAD constructions. In this section, we present a more general result by giving differentiators against a wide class of generically composed schemes, some of which have been proven to achieve RAE security. This class consists of schemes built from a hash function $\mathcal{H}$, which we treat as a random oracle, and an encryption scheme $(\mathcal{E}, \mathcal{D})$, which we consider to be an ideal AEAD *without* associated data. We assume that the encryption algorithm of the composed scheme operates as follows. An initialization procedure $\mathcal{I}_e$ is used to prepare the inputs to a preprocessing algorithm $\mathcal{E}_0^{\mathcal{H}}$ and a post-processing algorithm $\mathcal{E}_1^{\mathcal{H}}$. The preprocessing algorithm prepares the inputs to the underlying $\mathcal{E}$ algorithm. The post-processing algorithm gets the output ciphertext and completes encryption (e.g., by appending a tag value). The decryption algorithm operates analogously by reversing this process via an initialization procedure $\mathcal{I}_d$, a preprocessing algorithm $\mathcal{D}_0^{\mathcal{H}}$ and a post-processing algorithm $\mathcal{D}_1^{\mathcal{H}}$. See Fig. 4 for the details.

ALGO. $\mathcal{AE}(K, N, A, M, \tau)$	ALGO. $\mathscr{D}_1^{\mathrm{CONST}^+, \mathrm{PRIM2}}(\tau)$
$(est_0, est_1) \leftarrow \mathcal{I}_e(K, N, A, M, \tau)$	$(K, N, A, M) \twoheadleftarrow \{0,1\}^{4n}$
$(K', N', M', \tau') \leftarrow \mathcal{E}_0^{\mathcal{H}}(est_0)$	$C \leftarrow \mathrm{CONST}^+(K, N, A, M, \tau)$
$C' \leftarrow \mathcal{E}(K', N', \varepsilon, M', \tau')$	$(est_0, est_1) \leftarrow \mathcal{I}_e(K, N, A, M, \tau)$
$C \leftarrow \mathcal{E}_1^{\mathcal{H}}(C', est_1)$	$C' \leftarrow \mathcal{R}_1(C)$
return C	$\tilde{C} \leftarrow \mathcal{E}_1^{\mathrm{PRIM2}}(C', est_1)$
	return $(\tilde{C} = C)$

Fig. 4. Template for generically composed AEAD $(\mathcal{AE}, \mathcal{AD})$ (left) and a differentiator for type-I schemes (right).

The next theorem shows that this class of schemes are differentiable if certain conditions on information passed between the above sub-procedures are met.

Theorem 3 (Differentiability of generic composition). *Let Π be a generically composed AEAD scheme from an encryption scheme (without associated data) $(\mathcal{E}, \mathcal{D})$ and a hash function $\mathcal{H}$ following the structure shown in Fig. 4 for some algorithms $(\mathcal{I}_e, \mathcal{E}_0, \mathcal{E}_1, \mathcal{I}_d, \mathcal{D}_0, \mathcal{D}_1)$. Let $\Delta_C := |C| - |C'|$ denote the ciphertext overhead. Suppose that the following condition holds.*

Type-I: *Let est_1 be the state passed to $\mathcal{E}_1$. We require that for all inputs (K, N, A, M) and for a sufficiently large Δ_1 we have that $|(K, N, A, M)| - |est_1| \geq \Delta_1$.[5] Furthermore, there is a recovery algorithm $\mathcal{R}_1$ (with no oracle access) that on input C recovers C', the internal ciphertext output by $\mathcal{E}$.*

Then Π is differentiable. More precisely, for any type-I scheme Π there exists a differentiator $\mathscr{D}_1$ such that for any simulator S making at most q queries in total to its ideal AEAD oracles

$$\mathbf{Adv}_{\Pi,S}^{\mathrm{indiff}}(\mathscr{D}_1) \geq 1 - q/2^{\Delta_1} - (q+1)/2^{\Delta_C} \; .$$

The complete version of this theorem in [1, Sect. 4.1] covers also type-II schemes, where decryption omits Δ_2 bits of information about (K, N, A, C) from the partial information used to recover plaintexts.

Proof. We give the proof for type-I schemes. The differentiator computes a ciphertext for a random set of inputs using the construction in the forward direction and then checks if the result matches that computed via the generic composition using the provided primitive oracles. To rule out the existence of successful simulators the differentiator must ensure that it does not reveal information that allows the simulator to use its ideal construction oracles to compute a correct ciphertext. The restriction on the size of est_1 (and the ability to recompute the internal ciphertext C' via $\mathcal{R}_1$) will be used to show this. The pseudocode for the differentiator, which we call $\mathscr{D}_1$, is shown in Fig. 4 (left). The attack works for any given value of τ and to simplify the presentation, we have assumed all spaces consist of bit strings of length n.

ANALYSIS OF $\mathscr{D}_1$. It is easy to see that when $\mathscr{D}_1$ is run in the real world its output will be always 1. This follows from the fact that $\mathcal{R}_1(C)$ will correctly recover the internal ciphertext C' and hence $\mathcal{E}_1^{\mathrm{PRIM2}}(C', est_1)$, being run with respect to correct inputs and hash oracle, will also output C.

We now consider the ideal world. We first modify the ideal game so that the ideal object presented to the simulator is independent of that used to answer construction queries placed by the differentiator. This game is identical to the ideal world unless S queries the forward construction oracle on (K, N, A, M, τ) (call this event E_1) or the backward construction oracle on (K, N, A, C, τ) (call this event E_2). We will bound the probability of each of these events momentarily. In the modified game, we claim that no algorithm S can compute C from (C', est_1). This is the only information about C that is revealed to a simulator and this claim in particular means that running $\mathcal{E}_1^{\mathrm{PRIM2}}(C', est_1)$ within $\mathscr{D}_1$ won't output the correct C either. The answers to oracle queries placed by S can be computed independently of the ideal construction oracles. Furthermore, (C', est_1) misses at least Δ_C bits of information about C as est_1 is computed independently of C. The simulator therefore has at most a probability of $1/2^{\Delta_C}$ of outputting C in this game. The bound in the theorem statement follows from a simple analysis of the probabilities of events E_1 and E_2 in the modified game.

---

[5] We do not count the length of τ as our attack also works for fixed values of τ.

The proof for type-II schemes follows along the same lines and yields similar bounds. The full details for schemes of both types are given in [1, Sect. 4.1].   □

CONSEQUENCES FOR GENERIC COMPOSITION. Namprempre, Rogaway, and Shrimpton [51] explore various methods to generically compose an AEAD scheme from a nonce-based AE scheme (without associated data) and a MAC. In their analysis the authors single out eight favored schemes A1–A8. Roughly speaking, schemes A1, A2, and A3 correspond to Encrypt-and-MAC where, respectively, N, (N, A), and (N, A, M) are used in the preparation of the input IV to the base AE scheme. Scheme A4 is the Synthetic Initialization Vector (SIV) mode of operation [56, Fig. 5], which is misuse resilient. Schemes A5 and A6 correspond to Encrypt-then-MAC, where IV is computed using N and (N, A), respectively. Schemes A7 and A8 correspond to MAC-then-Encrypt, where IV is computed using N and (N, A) respectively. The MAC component in all these schemes is computed over (N, A, M). Key L is used for IV and MAC generation, and an independent key K is used in encryption. We refer the reader to the original paper [51, Fig. 2] for further details. For convenience, we have also included the diagrams for the A (as well as B and N) schemes in [1, Appendix A] with the authors' permission.

In [1, Sect. 4.1] we give an analysis of how each of these schemes, as well as all the others discussed in [56], are affected by the generic attacks given in Theorem 3. We find that all A schemes except A8 (which generalizes the structure of the constructions we give in the next section) are differentiable. When looking at the same schemes but assuming that the encryption and authentication keys are identical (i.e., under *key reuse*), schemes A2, A6, and A8 no longer fall prey to our generic attacks. We leave analyzing their indifferentiability as an open problem. Finally, all B-schemes and N-schemes are found to be differentiable as well. In the literature, we also found a recent scheme called Robust Initialization Vector (RIV) [2] that is MRAE secure and bears similarities to our constructions. We show in [1, Appendix C] that RIV is type-I and hence differentiable.

5   Ideal Offline AEAD

We now give two constructions of ideal AEAD from simpler ideal primitives. The first is based on a VIL blockcipher, it enjoys a simpler analysis and supports any expansion τ. The second is based on the unbalanced 3-round Feistel network, where round functions are alternatively compressing and expanding random oracles. It achieves higher efficiency, but here τ must be sufficiently large.

We present our proofs in a modular way. We first build ideal AEADs that achieve indifferentiability in a restricted setting where all parameters except the input message are fixed. More precisely, we first show that there is a simulator S that for any *arbitrary but fixed* value of $K' := (K, N, A, \tau)$ is successful against all differentiators that are K'-bound in the sense that they only query the construction and primitive oracles on values specified by K'. To this end, we also begin with the simplifying assumption that the underlying ideal objects can be

Algo. $\mathcal{AE}(K,N,A,M,\tau)$	Algo. $\mathcal{AD}(K,N,A,C,\tau)$
$K' \leftarrow (K,N,A,\tau)$	$K' \leftarrow (K,N,A,\tau)$
$C \leftarrow \mathsf{E}(K',0^\tau\|M)$	$T\|M \leftarrow \mathsf{E}^-(K',C)$, where $|T| = \tau$
return C	if $T \neq 0^\tau$ return $\perp$ else return M

Fig. 5. The (un-hashed) Encode-then-Encipher construction. In the full scheme we set $K' \leftarrow \mathcal{H}(K,N,A,\tau)$ for a random oracle $\mathcal{H}$.

keyed with keys of arbitrary length. We then show how these restrictions and simplifying assumptions can be removed to obtain fully indifferentiable AEADs.

5.1 Indifferentiability of Encode-then-Encipher

Our first construction transforms a VIL ideal cipher with arbitrary key space into an ideal AEAD. It follows the Encode-then-Encipher (EtE) transform of Bellare and Rogaway [11]. In its most simple form, EtE fixes τ bits of the input to 0^τ and checks the correctness of the included redundancy upon inversion (see Fig. 5).[6] The domain of the underlying blockcipher should therefore be at least τ bits longer than that needed for the injection. This, in particular, is the case when both objects have variable input lengths. The results of this section (in contrast to the attacks against other generic schemes) support the soundness of EtE-based schemes from an indifferentiability perspective.

Theorem 4 (EtE is indifferentiable). *The EtE construction in Fig. 5 is indifferentiable from an ideal AEAD for any fixed $K' := (K,N,A,\tau)$ when instantiated with a VIL ideal cipher $(\mathsf{E},\mathsf{E}^-)$. More precisely, there is an expected $4q$-query simulator $\mathcal{S}(\cdot\,;K')$ that presents a perfect simulation of the underlying permutation for any K'-bound $q/2$-query differentiator $\mathcal{D}$ for $q/2 \leq 2^{n+\tau}/8$.*

Proof (Sketch). Since the key values are fixed, we denote $(\mathsf{E},\mathsf{E}^-)$ with (ρ,ρ^-), an unkeyed VIL random injection. The simulator will simulate the permutation on inputs of the form $0^\tau\|M$ via the ideal AEAD oracle ρ and will use a lazily sampled injection *disjoint* from ρ (i.e., one whose domain and range are disjoint from those of ρ) for inputs of the form $T\|M$ with $T \neq 0^\tau$. The simulator can always detect when a query must be consistent with the ideal AEAD oracle: such queries will always correspond to inputs of the form $0^\tau\|M$ in forward queries and outputs that are invertible under ρ^- in backward queries. All other queries are answered by lazily sampling the disjoint injection. However, in order to offer a perfect simulation, the simulator must condition this lazy sampling by rejecting

---

[6] In both the EtE construction and the Feistel construction in the next section, the 0^τ constant can be replaced by any fixed constant Δ of the same length. For EtE the indifferentiability proof is the same. For the Feistel construction the proof can be easily adapted. To see this, note that any round function $F_1(X)$ can be replaced with an indifferentiable one $F_1'(X) = \Delta \oplus F_1(X)$. The resulting construction becomes identical to the one using 0^τ by cancellation.

any sampled inverses of the form $0^\tau | M$ and sampled outputs that are invertible under ρ^-. This rejection sampling yields a simulator that runs in expected polynomial time as stated in the theorem. This simulator can be converted into one that runs in strict polynomial time in the standard way by capping the number of samples to t tries. With $q \leq 2^{n+\tau}/4$, this simulator fails with probability at most $(2/3)^t$ for each differentiator query, and hence introduces a statistical distance of $q(2/3)^t$. The full proof and the simulator are given in [1, Sect. 5.1]. $\square$

5.2 Indifferentiability of 3-round Feistel

A variable-input-length (VIL) permutation can be constructed via the Feistel construction [23] from a VIL/VOL random oracle, or via the confusion-diffusion construction [33] from a fixed-input-length (FIL) random permutation.[7] The number of rounds needed for indifferentiability of Feistel from an ideal cipher has been gradually reduced to 8 [28]; whereas for confusion-diffusion 7 rounds are needed for good security bounds [33]. This state of affairs

$$
\begin{array}{l}
\text{ALGO. } \mathscr{D}^{\text{Const}^+,\text{Prim}_2} \\
\hline
X_1 \twoheadleftarrow \{0,1\}^n \\
(X_2, X_3) \leftarrow \text{Const}^+(X_1) \\
Y_2 \leftarrow \text{Prim}_2(X_2) \\
\text{Return } (X_1 \oplus Y_2 = X_3)
\end{array}
$$

Fig. 6. The 2-round Feistel differentiator.

leaves the above approach to the design of random injections somewhat suboptimal in terms of the number of queries per message block to a random permutation.

We ask whether this rate can be improved for random *injections*. We start from the observation that indifferentiability attacks against 5-round Feistel do not necessarily translate to those that fix parts of the input to 0^τ. Despite this, we show that differentiating attacks against 2-round Feistel still exist.

Proposition 2 (Differentiability of 2-round Feistel). *The 2-round unbalanced Feistel construction Φ_2 (cf. Fig. 1) with the left part of the input fixed to 0^τ is differentiable from an ideal injection.*

Proof (Sketch). Consider the differentiator $\mathscr{D}$ in Fig. 6 that checks the consistency of simulated output against the construction on a random input X. In the real world, $\mathscr{D}$ will output 1 with probability 1. In the ideal world the simulator has to guess value Y_2, which it won't be able to do except with probability negligible in n as the query placed by $\mathscr{D}$ is hidden from its view. $\square$

The simplicity of the above attack and the necessity for large number of rounds in building indifferentiable permutations raise the undesirable possibility

---

[7] Using a hybrid argument the indifferentiability of the Feistel and confusion-diffusion constructions carry over to variable input lengths. The VIL/VOL hash function in Feistel can itself be instantiated with the Sponge construction [15] in the random-permutation model. Note that, when dealing with domain and range extension for Sponge one needs to take care of encoding the lengths of inputs and outputs as part of the inputs fed to the random oracle [29].

that many rounds would also be needed for building random injections. We show, perhaps surprisingly, that this is not the case and adding only one extra round results in indifferentiability as long as τ and the input size are sufficiently large. This means, somewhat counter-intuitively, that the efficiency of constructions of ideal injections can be increased when a *higher* level of security is required. The 3-round Feistel construction and variable names are shown in Fig. 1.

We present the more intricate part of the proof of the following theorem in the code-based game-playing framework [12] to help its readability and verifiability.

Theorem 5 (Indifferentiability of 3-round Feistel). *Take the 3-round Feistel construction* Φ_3 *shown in Fig. 1 when it is instantiated with three independent keyed random oracles (the round functions are all keyed with the same key). This construction is indifferentiable from an ideal AEAD scheme for any fixed key of the form* $K' := (K, N, A, \tau)$. *More precisely, there is a simulator* S *such that for all* $(q_e, q_d, q_1, q_2, q_3)$-*query* K'-*bound differentiators* $\mathcal{D}$ *with* $q_e + q_d + 2q_1 + q_2 + q_3 \leq q$ *we have*

$$\mathbf{Adv}^{\mathrm{indiff}}_{\Phi_3, S}(\mathcal{D}) \leq 9q^2/2^\tau \ ,$$

as long as $q_2(q_1 + q_2 + q_3) \leq 2^{n+\tau}/2$ *and* $q_e + q_1 \leq 2^n/2$. *The simulator places at most* q^2 *queries to its oracles.*

Proof. To make the notation lighter we omit the key input to the various ideal objects (as we are dealing with K'-bound differentiators) and indicate forward/backward queries to the construction or ideal AEAD by C/C^-, and queries to the real or simulated round functions by F_1, F_2, and F_3. To simplify the analysis, we consider a *restricted* class of differentiators that (1) query $C(X_1)$ before any query $F_1(X_1)$, and (2) never query C^-. We also call a simulator C-respecting if it calls C only when simulating $F_1(X_1)$, in which case it places a single query $C(X_1)$. The following lemma deals with this simplification.

Lemma 1 (Restricting $\mathcal{D}$). *For any* $(q_e, q_d, q_1, q_2, q_3)$-*query differentiator* $\mathcal{D}$ *there is a restricted* $(q_e + q_1, 0, q_1, q_2, q_3)$-*query differentiator* $\mathcal{D}'$ *such that for any C-respecting simulator* S, *and as long as* $q_e + q_1 \leq 2^n/2$, *we have*

$$|\mathbf{Adv}^{\mathrm{indiff}}_{\Phi_3, S}(\mathcal{D}') - \mathbf{Adv}^{\mathrm{indiff}}_{\Phi_3, S}(\mathcal{D})| \leq 3q_d/2^\tau \ .$$

We give the proof of this auxilliary lemma in [1, Sect. 5.2]. Intuitively, we can convert any distinguisher $\mathcal{D}$ into a restricted $\mathcal{D}'$ that always calls the construction before it answers a query to F_1 and intercepts all queries to the inverse construction oracle and returns $\perp$ if the queried value was never computed by the construction in the forward direction. The lemma follows from bounding the probability that $\mathcal{D}'$ provides a wrong answer in either world. The C-respecting restriction is used to upper-bound the total number of forward construction queries in the ideal world (including simulator calls).

We prove indifferentiability with respect to restricted differentiators via a sequence of games as follows. We start with the real game, which includes oracles for the construction and the round functions, and gradually modify the implementations of these oracles until: (1) the construction no longer places any queries to the round functions and is implemented as an ideal injection; and (2) the round functions use this (ideal) construction oracle. We now describe these games. We give the pseudocode in Figs. 7 and 8.

$\mathbf{G}_0$: This game is identical to the (restricted) real game. Here the construction oracle C calls F_1, F_2 and F_3 and adds entries to lists L_1, L_2, and L_3.

$\mathbf{G}_1$: This game introduces flag_1. The game sets flag_1 if F_1 chooses an output value that was already queried to F_2. As we will see, we can easily bound the probability of this flag getting set via the birthday bound.[8]

$\mathbf{G}_2$: This game explicitly samples fresh values that are added to L_1 and L_2 as a result of a non-repeat query X_1 to C within the code of C rather than under the corresponding round functions. This is a conceptual modification and the game is identical to $\mathbf{G}_1$. Indeed, the sampled L_1 entry is always guaranteed to be fresh assuming a non-repeat value X_1, and the L_2 entry will be also non-repeat or flag_1 is set. List L_C is used to deal with repeat queries and avoid spurious samplings.

$\mathbf{G}_3$: This game introduces a (conceptual) *change of random variables*. Instead of choosing Y_1 and Y_2 (i.e., the outputs of F_1 and F_2) randomly and computing the outputs (X_3, X_4) of the construction, it first chooses (X_3, X_4) and sets Y_1 and Y_2 based on these, the input, and Y_3. This is done via a linear change of variables that will not affect the distributions of Y_1 and Y_2, as we show below. This game constitutes our first step in constructing the simulator by defining the outputs of F_1 and F_2 in terms of those for C. The proof, however, is not yet complete: although C is implemented independently of the round functions, F_2 and F_3 need access to the list of queries made to C.

$\mathbf{G}_4$: This game removes flag_1 (which allowed the previous transitions to be carried out in a conservative way) as we wish to gradually construct the code of the simulator, and this code is not needed in the final simulation.[9]

$\mathbf{G}_5$: This game shifts most of the code from the C oracle to the F_1 oracle. In particular, the manipulations of L_1 and L_2 are now done within F_1. The outputs of C are still sampled within the construction procedure and C makes a call to F_1. Procedure F_1 retrieves the necessary (X_3, X_4) values by calling back the construction (note these are now added to L_C prior to calling F_1). This modification is conceptual since (1) restricted differentiators *always* call the construction oracle before calling F_1 and hence the entry for X_1 will already be in the list L_C, and (2) although some queries to F_2 and F_3 may no longer be done, these oracles behave as random oracles and hence performing such queries earlier or later does not affect the view of the adversary.

$\mathbf{G}_6$: This game removes the query to F_1 from C and adds a bad event based on flag_2 to F_2 that guarantees that this game is identical to $\mathbf{G}_5$ until flag_2. Removing the call to F_1 from C has implications for F_2, since the operation

---

[8] As usual, once a flag is set, nothing matters. E.g., we can assume the game returns 0.

[9] We need not introduce additional terms here. Suppose games $\mathbf{G}$ and $\mathbf{G}''$ never set flag, but game $\mathbf{G}'$ does. If these games are identical until flag is set, then the distance between $\mathbf{G}$ and $\mathbf{G}''$ is bounded by the probability of flag being set in any game.

of this oracle depends on entries that were added to L_2 whenever a call to C (and therefore a call to F_1) occurred. For each F_2 query, we therefore need to ensure that processing left undone in this modified construction oracle (which may influence the view of the adversary) is carried out as before. To this end, we go through the entries in L_C and check if an entry $(X_1, (X_3, X_4))$ occurred that might have set the value of Y_2. If more than one such entry exists, then this is detected as a collision at the output of F_1 and flag$_2$ is set. If only one candidate is found, this corresponds exactly to the query that would have been made by the removed F_1 call. If no candidate is found, then the oracle simply samples a fresh value as before. The games are therefore identical until flag$_2$ is set, the probability of which we bound below.

G_7: This game introduces a conceptual change in the way the loops in F_2 are executed. First, all X_3 values corresponding to entries in L_C are queried to F_3 if they were not previously done so. This means that the subsequent search for a good Y_3 can be equivalently made by going through those entries in L_C whose X_3 value is *already* present in L_3. This change sets the ground for the next game where we drop the first loop completely.

G_8: We now remove the code that corresponds to the first loop in F_2 completely and argue that there is a rare event that allows us to prove the games identical until bad and bound the statistical distance between the two. This rare event is explicitly shown, for convenience, as a dummy flag$_3$: it is activated whenever the first loop was adding to list L_3 a freshly sampled entry (X_3, Y_3), which is used by the second loop. Again we can bound the probability of this event easily, as F_3 implements a random oracle.

G_9: This game rewrites the loops in F_2 and only looks in L_C for values that will be used by F_2, i.e., only those entries with $X_4 = X_2 \oplus Y_3$ will be searched over. This is a conceptual change.

G_{10}: This game introduces flag$_C$, which is set if collisions in the outputs of C are found. This prepares us to modify the implementation of C from a random function to a random injection. We bound this via a standard RF/RI switching lemma. This game also introduces a (partial and so far unused) inverse C^- to C that returns the preimage to (X_3, X_4) if this value was queried to C. This will allow us to remove the dependency on the L_C next. (Recall that the differentiator is restricted and it cannot call C^- at all.)

G_{11}: In this game F_2 no longer uses L_C; instead it uses C^- to check if a value was queried to C. Since this partial inverse oracle always returns $\bot$ for inputs that are not on L_C, this game is identical to the previous game. (Note also that we may also omit the re-computation of (X_3, X_4).)

G_{12}: This game modifies C to the forward direction of a random injection oracle and C^- to its backward direction (which could return a non-$\bot$ value even if an inverse is not found in L_C). This modification can be bounded by looking at the probability that the simulator places an inverse query that was not previously obtained from the forward construction oracle.

Now observe that G_{12} is the ideal game where procedures F_1, F_2 and F_3 make use of random injection oracles (C, C^-) but *not* its internal list L_C. By viewing

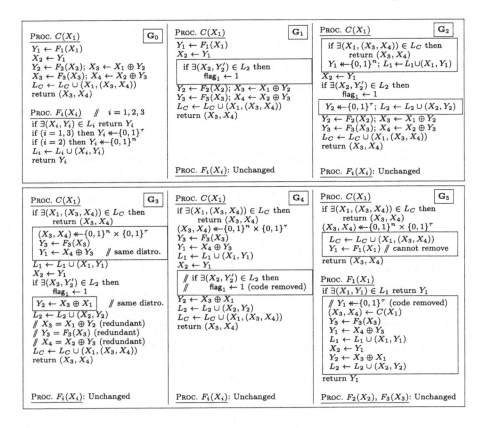

Fig. 7. Games G_0 to G_5.

the implementations of these procedures as three (sub-)simulators S_1, S_2 and S_3 we arrive at our simulator. We note that S_2 can omit flag_2 in F_2 with no loss in advantage (cf. footnote in the conservative jump to G_4 above). We also note that this simulator is C-respecting as needed in Lemma 1 above, and that it places at most q^2 oracle queries (it is quadratic due to the loop in $S_2^{C^-}$). The remainder of the proof consists of bounding the probabilities of setting the four flags in the game sequence above. The details of this analysis and the extracted code for the simulator can be found in [1, Sect. 5.2].   □

5.3   Removing Restrictions and Simplifications

Our AEAD schemes were analyzed with respect to differentiators that were bound to a fixed (K, N, A, τ). We deal with arbitrary (K, N, A, τ) by applying a hybrid argument. For this argument to hold, it is important to ensure that the simulators do not "interfere" with each other: not only should they be run on independent coins, but also their ideal AEAD oracles should be independent. We formalize this argument in a more general form.

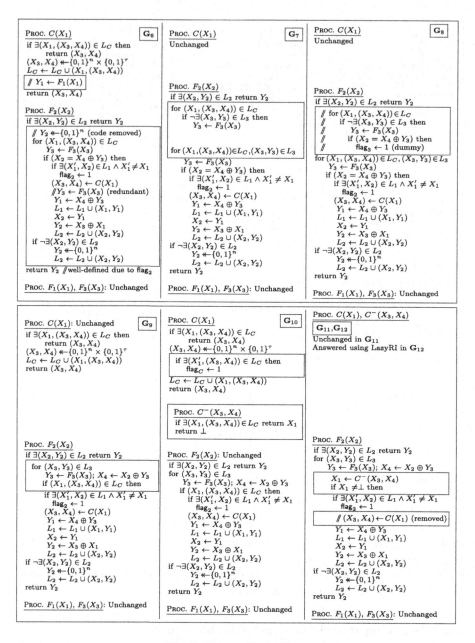

Fig. 8. Games $\mathbf{G}_6$ to $\mathbf{G}_{12}$.

FROM KEY-WISE TO FULL INDIFFERENTIABILITY. We call a keyed ideal object $\mathcal{F}$ *uniformly keyed* if $\mathcal{F}(K, X)$ and $\mathcal{F}(K', X)$ are identically and independently distributed for any X and distinct keys K and K'. Let $\mathsf{C}^{\mathcal{F}_1}$ be a construction of a uniformly keyed object $\mathcal{F}_2$ from a uniformly keyed object $\mathcal{F}_1$. We call the construction key-respecting if for all inputs (K, X) it queries $\mathcal{F}_1$ on K only. We call a simulator (for $\mathcal{F}_1$) key-respecting if for all inputs (K, X) it queries $\mathcal{F}_2$ on K only. We call a differentiator key-respecting if it always queries both the construction and the primitive oracles on K only. We call the construction key-wise indifferentiable if it is indifferentiable with a key-respecting simulator against all key-respecting differentiators. The following lemma follows from a standard hybrid argument (see [1, Appendix D]).

Lemma 2 (Hybrid over keys). *Let $\mathcal{F}_1$ and $\mathcal{F}_2$ be two uniformly keyed objects and $C^{\mathcal{F}_1}$ be a key-respecting construction of $\mathcal{F}_2$ from $\mathcal{F}_1$. Then if $C^{\mathcal{F}_1}$ is key-wise indifferentiable, it is also (fully) indifferentiable. More precisely, for any key-respecting simulator $\mathcal{S}$ and any q-query (unrestricted) differentiator $\mathcal{D}$ there is a key-respecting differentiator $\mathcal{D}'$ such that*

$$\mathbf{Adv}_{C,\mathcal{S}}^{\mathrm{indiff}}(\mathcal{D}) \leq q \cdot \mathbf{Adv}_{C,\mathcal{S}}^{\mathrm{indiff}}(\mathcal{D}') .$$

In order to apply this result to the EtE and 3-round Feistel it suffices to syntactically express all underlying ideal objects as a *single* keyed primitive and then show that they are key respecting. We note that the key-respecting restriction forces the use of the *same* key on all underlying ideal objects, which agrees with our observations on the benefits of key reuse in Sect. 4.1.

DEALING WITH KEYS OF ARBITRARY SIZE. Objects with an arbitrarily large key space can be indifferentiably built from those with a smaller key space in the standard way by hashing the key using a random oracle. This means we can remove the assumption of variable key lengths on the VIL ideal cipher in our construction. We prove the following result in [1, Sect. 5.3].

Proposition 3 (Key extension via hashing). *Let $\mathcal{F}_1$ and $\mathcal{F}_2$ be two uniformly keyed ideal objects with key spaces $\mathcal{K}_1$ and $\mathcal{K}_2$ respectively. Let $\mathcal{H} : \mathcal{K}_2 \longrightarrow \mathcal{K}_1$ be a random oracle. Suppose further that for some (and hence any) $K_1 \in \mathcal{K}_1$ and $K_2 \in \mathcal{K}_2$ we have that $\mathcal{F}_1(K_1, X)$ is identically distributed to $\mathcal{F}_2(K_2, X)$. Then $C^{\mathcal{F}_1, \mathcal{H}}(K, X) := \mathcal{F}_1(\mathcal{H}(K), X)$ is indifferentiable from $\mathcal{F}_2$. More precisely, there is a simulator $\mathcal{S}$ such that for any $q/3$-query differentiator $\mathcal{D}$,*

$$\mathbf{Adv}_{C,\mathcal{F}_2}^{\mathrm{indiff}}(\mathcal{D}) \leq 2q^2/|\mathcal{K}_1| .$$

THE FULL CONSTRUCTION. Our final AEAD construction can be written as $\mathcal{AE}(K, N, A, M, \tau) = \Phi_3(K', M)$, where $K' = \mathcal{H}(K, N, A, \tau)$ and Φ_3 is the ideal injection instantiated with 3-round Feistel. The latter uses independent keyed random oracles $\mathcal{F}_i$ all with key space $\mathcal{K}$ matching the co-domain of $\mathcal{H}$. Combining Theorem 5 with Lemmas 2 and 3 we obtain an overall bound $9q^3/2^\tau + 2q^2/|\mathcal{K}|$, where q is an upper bound on the number of oracles queries.

5.4   Ideal Online AEAD

Offline AEAD schemes can fall short of providing adequate levels of functionality or efficiency in settings where data arrives one segment at a time and should be processed immediately without the knowledge of future segments. In an *online* AEAD scheme, the encryption and decryption algorithms are replaced by *stateful* segment-oriented ones that process the inputs one segment at a time. We formalize ideal online AEAD next and briefly present our results in indifferentiably constructing online AEAD schemes.

ONLINE FUNCTIONS AND IDEAL ONLINE AEAD. An online function(ality) is a triple of functions with signatures

$$\mathcal{F}_0 : \mathcal{A}_0 \longrightarrow \mathcal{S}, \quad \mathcal{F}_1 : \mathcal{S} \times \mathcal{A} \times \mathcal{M} \times \mathcal{X} \longrightarrow \mathcal{R}_1 \times \mathcal{S}, \quad \mathcal{F}_2 : \mathcal{S} \times \mathcal{A} \times \mathcal{M} \times \mathcal{X} \longrightarrow \mathcal{R}_2 .$$

We define $\mathrm{Onj}^+[\mathcal{A}_0, \mathcal{A}, \mathcal{M}, \mathcal{X}, \mathcal{S}, \mathcal{R}_1, \mathcal{R}_2]$ as the set of online functions for which $\mathcal{F}_1$ and $\mathcal{F}_2$ are injective over $\mathcal{M}$ and respect the length-expansion requirement. An ideal online AEAD is a uniform function in $\mathrm{Onj}^+[\mathcal{A}_0, \mathcal{A}, \mathcal{M}, \mathcal{X}, \mathcal{S}, \mathcal{R}_1, \mathcal{R}_2]$ where $\mathcal{A}_0 := \mathcal{K} \times \mathcal{N}$, $\mathcal{A} := \mathcal{H}$, and $\mathcal{R}_1 := \mathcal{R}_2 := \mathcal{C}$.

INDIFFERENTIABLE ONLINE AEAD. The **CHAIN** construction of [43] is trivially differentiable from an ideal online AEAD as its initialization procedure $\mathcal{AE}$.init and state-update procedures are not random. Indeed, we need to modify this and other aspects of its design (cf. [1, Sect. 7.2]) to achieve indifferentiability. Intuitively, the computation of a ciphertext/state pair must be done in a way that forces the differentiator to reveal all necessary information that is needed to recompute them via the ideal objects accessible to the simulator. Following this, we propose a new construction in Fig. 9, which we call **HashCHAIN**. Here, $\mathcal{E}$ is an offline ideal AEAD with key length k, and $\mathcal{H}_i$ are VIL/VOL keyed random oracles with key size k that admit outputs of lengths k and $2k$. These are implemented from a single random oracle via domain separation. The nonce and associated-data spaces of the online scheme are arbitrary. Its message, expansion and ciphertext spaces match those of the offline scheme. The state space is $\mathcal{S} := \mathcal{K}$. A formal statement and proof of the following theorem are given in [1, Sect. 7.2]. In the proof we apply parallel composition of indifferentiability, which permits modifying the ideal AEAD reference object until we arrive at **HashCHAIN**.

Theorem 6 (HashCHAIN is indifferentiable). *The* **HashCHAIN** *construction in Fig. 9 is indifferentiable from an ideal online AEAD.*

6   Efficiency Lower Bounds

Suppose we instantiate the random oracles underlying our Feistel-based construction with the Sponge construction. Suppose also that the underlying Sponges absorb inputs and expand outputs in blocks of n bits (i.e., the Sponge has bit-rate n). Finally, assume that our input message is w blocks long. This means that in both of our constructions roughly w primitive calls are used in each round of Feistel. This adds up to $3w$ overall primitive calls for the second construction and $8w$ calls for the first one. Our second construction is therefore

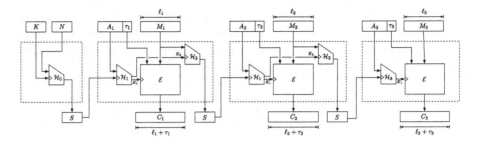

Fig. 9. The **HashCHAIN** transform.

almost 3 times faster than the first. We next show that our more efficient construction is not too far from the theoretically optimal solution by proving that at least $2w$ calls are necessary for *any* indifferentiable construction. We do this by first giving a lower bound for indifferentiable constructions of random oracles (which is tight as it is essentially matched by Sponge) and then show how to derive the lower bound for random injections from it.

Theorem 7 (Efficiency lower bound). *Any indifferentiable construction of a random function* $C^\pi : \{0,1\}^{wn} \longrightarrow \{0,1\}^{wn}$ *from a random permutation* $\pi : \{0,1\}^n \longrightarrow \{0,1\}^n$ *must place at least* $q \geq 2w-2$ *queries to* π. *More precisely, for any such q-query construction* C^π *and any* q_S-*query indifferentiability simulator* S *there is a w-query differentiator* $\mathscr{D}$ *such that*

$$2 \cdot \mathbf{Adv}^{\text{indiff}}_{C,S}(\mathscr{D}) \geq 1 - 1/2^{(q-(2w-2))n} - (q^2 + q_S)/2^n .$$

Proof. We prove this result by constructing a differentiator against any construction C^π that places $q < 2w - 2$ queries to π. Any such C^π can be written using (π-independent) functions $f_1, \ldots f_{q+1}$ where

$$f_i : \{0,1\}^{(w+i-1)n} \longrightarrow \{0,1\}^{(w+i)n} \quad \text{for} \quad 1 \leq i \leq q ,$$
$$f_{q+1} : \{0,1\}^{(w+q)n} \longrightarrow \{0,1\}^{wn} .$$

This reflects the fact that each f_i can recompute everything that depends only on the initial inputs, but also needs to take as additional inputs the values returned by π at each of the previous calls. See [1, Sect. 6] for a schematic diagram.

Consider the first $w - 1$ calls to π. There are $2^{(w-1)n}$ possible tuples $P = (P_1, \ldots, P_{w-1})$ that can define the inputs to such queries. Since in total there are 2^{wn} possible inputs, by a counting argument, a subset $D[C, \pi]$ of the input values of size at least $2^n = 2^{wn}/2^{(w-1)n}$ will be mapped by a construction C to the same $P[C, \pi]$, for any given π. Set $D[C, \pi]$ and points $P[C, \pi]$ can be found by a (possibly unbounded) attacker $\mathscr{D}$ using only $w - 2$ queries to π. Algorithm $\mathscr{D}$ proceeds in rounds as follows. There is at least one point $P_1 \in \{0,1\}^n$ such that f_1 always chooses P_1 for at least $2^{wn}/2^n = 2^{(w-1)n}$ of its inputs. No queries to π are needed to find P_1 and we set $D[C, \pi]$ to a corresponding set of colliding inputs. We then get $Z_1 := \pi(P_1)$ and we use it to analyze the operation of f_2.

Given Z_1 and $D[C, \pi]$, at least $2^{(w-1)n}/2^n$ of the inputs in $D[C, \pi]$ are such that f_2 always chooses the same query point P_2 to π. We update $D[C, \pi]$ to this subset. Continuing in this manner, we obtain a set $D[C, \pi]$ of at least 2^n points such that f_{w-1} chooses a point P_{w-1} for all inputs in $D[C, \pi]$.

Put together, the restriction of C^π to inputs in $D[C, \pi]$ guarantees that the construction always queries π at P_i for queries $i = 1, \ldots, w - 1$ and then places an arbitrary sequence of $q - (w - 1)$ queries to π. Furthermore, from the previous discussion we can assume that differentiator $\mathscr{D}$ knows the description of set $D[C, \pi]$ and values $Z[C, \pi] := (Z_1, \ldots, Z_{w-1}) = (\pi(P_1), \ldots, \pi(P_{w-1}))$.

Now consider a pseudorandom generator $PRG : D[C, \pi] \times \{0, 1\}^{(q-(w-1))n} \longrightarrow \{0, 1\}^{wn}$ that has $Z[C, \pi]$ hardwired in and operates as

$$PRG[Z[C, \pi]](X, Z_w, \ldots, Z_q) := C^{Z_1, \ldots, Z_q}(X) \, ,$$

where $C^{Z_1, \ldots, Z_q}(X)$ denotes running $C^\pi(X)$, answering the i-th query with Z_i. It is at this step that we follow the techniques of Gennaro and Trevisan [37]. If $\mathscr{D}$ can distinguish the output of PRG from a random string, this will allow differentiating C^π from a random function. We now show that such an attack is guaranteed to exist if C does not make a sufficient number of queries to π.

Our first claim is that if C^π is indifferentiable then $PRG[Z[C, \pi]]$ is a secure pseudorandom generator over a random choice of π. More precisely, our goal is to show that under the indifferentiability of C^π, the distribution $\{ (Y, Z[C, \pi]) : \pi \twoheadleftarrow Perm[n]; Y \twoheadleftarrow \{0, 1\}^{wn} \}$ is statistically close to

$$(PRG[Z[C, \pi]](X, Z_w, \ldots, Z_q), Z[C, \pi]) :$$
$$\pi \twoheadleftarrow Perm[n]; \ X \twoheadleftarrow D[C, \pi]; \ Z_w, \ldots, Z_q \twoheadleftarrow \{0, 1\}^{(q-(w-1))n}$$

The points in $Z[C, \pi]$ are computed using oracle access to π at the onset and, being part of the description of the PRG, are in the view of a PRG distinguisher. Take distribution $\{ C^\pi(X) : \pi \twoheadleftarrow Perm[n]; X \twoheadleftarrow D[C, \pi] \}$. We first argue this is statistically close to

$$PRG[Z[C, \pi]](X, Z_w, \ldots, Z_q) : \pi \twoheadleftarrow Perm[n];$$
$$X \twoheadleftarrow D[C, \pi]; \ Z_w, \ldots, Z_q \twoheadleftarrow \{0, 1\}^{(q-(w-1))n}$$

To see this, note that the simulation of π using Z_i is fully consistent for queries $i = 1, \ldots, w - 1$. This is also the case for $i \geq w$ unless $Z_1, \ldots, Z_q$ are not all distinct, which by the birthday bound occurs with probability at most $q^2/2^n$. We are left with proving the following distributions statistically close.

$$(C^\pi(X), Z[C, \pi]) : \pi \twoheadleftarrow Perm[n]; \ X \twoheadleftarrow D[C, \pi]$$
$$(Y, Z[C, \pi]) : \pi \twoheadleftarrow Perm[n]; \ Y \twoheadleftarrow \{0, 1\}^{wn}$$

Here we cannot directly apply indistinguishability of $C^\pi(X)$ from a truly random wn-bit function $\mathcal{H}(X)$ (which follows from indifferentiability) as the hardwired values $Z[C, \pi]$ are in the distinguisher's view. Instead we proceed via a sequence of games as follows. First, we use the indifferentiability simulator $\mathcal{S}$ to deduce that the following distributions are statistically close.

$$(\mathrm{C}^{\pi}(X), \mathrm{Z}[\mathrm{C}, \pi]) : \pi \twoheadleftarrow \mathrm{Perm}[n]; \; X \twoheadleftarrow \mathrm{D}[\mathrm{C}, \pi]$$

$$(\mathcal{H}(X), \mathrm{Z}[\mathrm{C}, \mathcal{S}^{\mathcal{H}}]) : \mathcal{H} \twoheadleftarrow \mathrm{Fun}[wn, wn]; \; X \twoheadleftarrow \mathrm{D}[\mathrm{C}, \mathcal{S}^{\mathcal{H}}]$$

This follows directly from the definition of indifferentiability. Consider a differentiator that constructs $\mathrm{Z}[\mathrm{C}, \mathrm{PRIM}]$ and $\mathrm{D}[\mathrm{C}, \mathrm{PRIM}]$ using the real or simulated π-oracle PRIM, then queries its real or ideal construction oracle on $X \twoheadleftarrow \mathrm{D}[\mathrm{C}, \mathrm{PRIM}]$ to obtain the first component above. Any successful distinguisher for the above distributions could be used by this differentiator to contradict the indifferentiability assumption with the same advantage. This differentiator places exactly w queries ($w - 1$ queries to the real or simulated π-oracle PRIM to construct $\mathrm{Z}[\mathrm{C}, \mathrm{PRIM}]$ and one extra query to the real or ideal construction oracle). Note that this argument also shows that $\mathrm{D}[\mathrm{C}, \mathcal{S}^{\mathcal{H}}]$ must also have at least 2^n points.

The next step is to show that we can replace $\mathcal{H}(X)$ with Y for an *independently sampled* random string Y that is *not* computed via the random oracle. More precisely, we argue that the following distributions are statistically close.

$$(\mathcal{H}(X), \mathrm{Z}[\mathrm{C}, \mathcal{S}^{\mathcal{H}}]) : \mathcal{H} \twoheadleftarrow \mathrm{Fun}[wn, wn]; \; X \twoheadleftarrow \mathrm{D}[\mathrm{C}, \mathcal{S}^{\mathcal{H}}]$$

$$(Y, \mathrm{Z}[\mathrm{C}, \mathcal{S}^{\mathcal{H}}]) : \mathcal{H} \twoheadleftarrow \mathrm{Fun}[wn, wn]; \; Y \twoheadleftarrow \{0,1\}^{wn}$$

Suppose $\mathcal{S}$ places at most q_S queries to $\mathcal{H}$. The set $\mathrm{D}[\mathrm{C}, \pi]$ has size at least 2^n and hence so does the set $\mathrm{D}[\mathrm{C}, \mathcal{S}^{\mathcal{H}}]$. Now since X is chosen uniformly at random from $\mathrm{D}[\mathrm{C}, \mathcal{S}^{\mathcal{H}}]$, the simulator $\mathcal{S}$ will query $\mathcal{H}$ on X with probability at most $q_S/2^n$. Hence $\mathcal{H}(X)$ is independent of the simulators view and we may replace it with independent random value Y.

Finally, we use indifferentiability once more to show that we can replace $\mathrm{Z}[\mathrm{C}, \mathcal{S}^{\mathcal{H}}]$ back by $\mathrm{Z}[\mathrm{C}, \pi]$ in the presence of the independently sampled random string Y. The differentiator we construct uses the real or simulated π-oracle PRIM to construct set $\mathrm{Z}[\mathrm{C}, \pi]$ or $\mathrm{Z}[\mathrm{C}, \mathcal{S}^{\mathcal{H}}]$, respectively, and then samples value Y. Again, any successful distinguisher for the above distributions will be translated into a differentiating attack with the same advantage, resulting in a successful differentiator that places exactly $w - 1$ queries.

This concludes the proof of our claim that PRG is secure over seed space $\mathrm{D}'[\mathrm{C}, \pi] := \mathrm{D}[\mathrm{C}, \pi] \times \{0,1\}^{(q-(w-1))n}$ (of overall size at least $2^{(q-w+2)n}$) and range $\mathrm{R} := \{0,1\}^{wn}$ with advantage at most $(q^2 + q_S)/2^n + 2\delta$, where δ is the maximum advantage $\mathbf{Adv}_{\mathrm{C}, \mathcal{S}}^{\mathrm{indiff}}(\mathscr{D})$ over all $\mathscr{D}$ placing at most w queries.

We now show that, unless C^{π} makes a large number of queries to π the above PRG cannot be secure. The queries of C^{π} translate to the size of the seed space of PRG as this does not make any queries to π beyond the initial $w - 1$ queries used to hardwire the fixed $\mathrm{Z}[\mathrm{C}, \pi]$ values. However, the outputs of any PRG with domain $\mathrm{D}'[\mathrm{C}, \pi]$ and range R can be information-theoretically distinguished from random with advantage $1 - |\mathrm{D}'[\mathrm{C}, \pi]|/|\mathrm{R}|$. We therefore must have that

$$1 - |\mathrm{D}[\mathrm{C}, \pi] \times \{0,1\}^{(q-(w-1))n}|/|\{0,1\}^{wn}| \leq (q^2 + q_S)/2^n + 2\delta \;.$$

If C^{π} is indifferentiable, we get $q \geq 2w - 2$, when $q^2 + q_S \leq 2^n/2$ and $\delta = 1$. $\quad\square$

The above lower bound is essentially tight for random functions as the Sponge construction meets it up to constant terms. The proof, however, does not directly apply to random injections ρ, as the inverse oracle ρ^- would allow an adversary to invert the outputs of the PRG. The next proposition shows that by chopping sufficiently many bits of the outputs of ρ, a random function can be *indifferentiably* obtained from a random injection in a *single* query. Together with the above result this extends the lower bound to random injections as well.

Proposition 4. *Let $\rho : \{0,1\}^{wn} \longrightarrow \{0,1\}^{wn+n}$ be a random injection with inverse ρ^-. Let $C^\rho(X) := \rho(X)[1..wn]$ be the construction that chops n bits of $\rho(X)$. Then C^ρ is indifferentiable from a length-preserving random function.*

The proof is given in [1, Sect. 6] where we construct a simulator that uses the random oracle output and samples the extension bits independently, keeping a list for consistency. Our construction of random injections via the 3-round Feistel construction places $3w + O(1)$ queries to π. This is somewhat higher than the $2w - 2$ required by the lower bound. We leave bridging this gap for random injections (and indeed also permutations) as the main open problem in this area.

Acknowledgments. The authors would like to thank Phillip Rogaway, Martijn Stam, and Stefano Tessaro for their comments. Barbosa was supported in part by Project NORTE-01-0145-FEDER-000020, financed by the North Portugal Regional Operational Programme (NORTE 2020) under the PORTUGAL 2020 Partnership Agreement, and through the European Regional Development Fund (ERDF). Farshim was supported in part by the European Research Council under the European Community's Seventh Framework Programme (FP7/2007-2013 Grant Agreement no. 339563 - CryptoCloud). This work was initiated during a short-term scientific mission sponsored by the COST CryptoAction (IC1306).

References

1. Barbosa, M., Farshim, P.: Indifferentiable Authenticated Encryption. Cryptology ePrint Archive (2018)
2. Abed, F., Forler, C., List, E., Lucks, S., Wenzel, J.: RIV for robust authenticated encryption. In: Peyrin, T. (ed.) FSE 2016. LNCS, vol. 9783, pp. 23–42. Springer, Heidelberg (2016). https://doi.org/10.1007/978-3-662-52993-5_2
3. Albrecht, M.R., Farshim, P., Paterson, K.G., Watson, G.J.: On cipher-dependent related-key attacks in the ideal-cipher model. In: Joux, A. (ed.) FSE 2011. LNCS, vol. 6733, pp. 128–145. Springer, Heidelberg (2011). https://doi.org/10.1007/978-3-642-21702-9_8
4. Andreeva, E., Bogdanov, A., Dodis, Y., Mennink, B., Steinberger, J.P.: On the indifferentiability of key-alternating ciphers. In: Canetti, R., Garay, J.A. (eds.) CRYPTO 2013. LNCS, vol. 8042, pp. 531–550. Springer, Heidelberg (2013). https://doi.org/10.1007/978-3-642-40041-4_29
5. Andreeva, E., Bogdanov, A., Luykx, A., Mennink, B., Mouha, N., Yasuda, K.: How to securely release unverified plaintext in authenticated encryption. In: Sarkar, P., Iwata, T. (eds.) ASIACRYPT 2014. LNCS, vol. 8873, pp. 105–125. Springer, Heidelberg (2014). https://doi.org/10.1007/978-3-662-45611-8_6

6. Ashur, T., Dunkelman, O., Luykx, A.: Boosting authenticated encryption robustness with minimal modifications. In: Katz, J., Shacham, H. (eds.) CRYPTO 2017. LNCS, vol. 10403, pp. 3–33. Springer, Cham (2017). https://doi.org/10.1007/978-3-319-63697-9_1

7. Bellare, M., Keelveedhi, S.: Authenticated and misuse-resistant encryption of key-dependent data. In: Rogaway, P. (ed.) CRYPTO 2011. LNCS, vol. 6841, pp. 610–629. Springer, Heidelberg (2011). https://doi.org/10.1007/978-3-642-22792-9_35

8. Bellare, M., Kohno, T.: A theoretical treatment of related-key attacks: RKA-PRPs, RKA-PRFs, and applications. In: Biham, E. (ed.) EUROCRYPT 2003. LNCS, vol. 2656, pp. 491–506. Springer, Heidelberg (2003). https://doi.org/10.1007/3-540-39200-9_31

9. Barwell, G., Martin, D.P., Oswald, E., Stam, M.: Authenticated encryption in the face of protocol and side channel leakage. In: Takagi, T., Peyrin, T. (eds.) ASIACRYPT 2017. LNCS, vol. 10624, pp. 693–723. Springer, Cham (2017). https://doi.org/10.1007/978-3-319-70694-8_24

10. Bellare, M., Namprempre, C.: Authenticated encryption: relations among notions and analysis of the generic composition paradigm. In: Okamoto, T. (ed.) ASIACRYPT 2000. LNCS, vol. 1976, pp. 531–545. Springer, Heidelberg (2000). https://doi.org/10.1007/3-540-44448-3_41

11. Bellare, M., Rogaway, P.: Encode-then-encipher encryption: how to exploit nonces or redundancy in plaintexts for efficient cryptography. In: Okamoto, T. (ed.) ASIACRYPT 2000. LNCS, vol. 1976, pp. 317–330. Springer, Heidelberg (2000). https://doi.org/10.1007/3-540-44448-3_24

12. Bellare, M., Rogaway, P.: The security of triple encryption and a framework for code-based game-playing proofs. In: Vaudenay, S. (ed.) EUROCRYPT 2006. LNCS, vol. 4004, pp. 409–426. Springer, Heidelberg (2006). https://doi.org/10.1007/11761679_25

13. Bellare, M., Rogaway, P., Wagner, D.: The EAX mode of operation. In: Roy, B., Meier, W. (eds.) FSE 2004. LNCS, vol. 3017, pp. 389–407. Springer, Heidelberg (2004). https://doi.org/10.1007/978-3-540-25937-4_25

14. Bernstein, D.J.: Cryptographic competitions (2014). https://competitions.cr.yp.to/index.html

15. Bertoni, G., Daemen, J., Peeters, M., Van Assche, G.: On the indifferentiability of the sponge construction. In: Smart, N. (ed.) EUROCRYPT 2008. LNCS, vol. 4965, pp. 181–197. Springer, Heidelberg (2008). https://doi.org/10.1007/978-3-540-78967-3_11

16. Bellare, M., Bernstein, D.J., Tessaro, S.: Hash-function based PRFs: AMAC and its multi-user security. In: Fischlin, M., Coron, J.-S. (eds.) EUROCRYPT 2016. LNCS, vol. 9665, pp. 566–595. Springer, Heidelberg (2016). https://doi.org/10.1007/978-3-662-49890-3_22

17. Black, J., Cochran, M., Shrimpton, T.: On the impossibility of highly-efficient blockcipher-based hash functions. In: Cramer, R. (ed.) EUROCRYPT 2005. LNCS, vol. 3494, pp. 526–541. Springer, Heidelberg (2005). https://doi.org/10.1007/11426639_31

18. Black, J., Rogaway, P., Shrimpton, T.: Encryption-scheme security in the presence of key-dependent messages. In: Nyberg, K., Heys, H. (eds.) SAC 2002. LNCS, vol. 2595, pp. 62–75. Springer, Heidelberg (2003). https://doi.org/10.1007/3-540-36492-7_6

19. Bellare, M., Tackmann, B.: The multi-user security of authenticated encryption: AES-GCM in TLS 1.3. In: Robshaw, M., Katz, J. (eds.) CRYPTO 2016. LNCS, vol. 9814, pp. 247–276. Springer, Heidelberg (2016). https://doi.org/10.1007/978-3-662-53018-4_10

20. Canetti, R.: Universally composable security: a new paradigm for cryptographic protocols. In: FOCS 2001. IEEE Computer Society Press (2001)

21. Coron, J.-S., Dodis, Y., Malinaud, C., Puniya, P.: Merkle-Damgård revisited: how to construct a hash function. In: Shoup, V. (ed.) CRYPTO 2005. LNCS, vol. 3621, pp. 430–448. Springer, Heidelberg (2005). https://doi.org/10.1007/11535218_26

22. Coron, J.-S., Dodis, Y., Mandal, A., Seurin, Y.: A domain extender for the ideal cipher. In: Micciancio, D. (ed.) TCC 2010. LNCS, vol. 5978, pp. 273–289. Springer, Heidelberg (2010). https://doi.org/10.1007/978-3-642-11799-2_17

23. Coron, J.-S., Holenstein, T., Künzler, R., Patarin, J., Seurin, Y., Tessaro, S.: How to build an ideal cipher: the indifferentiability of the Feistel construction. J. Cryptol. **29**(1), 61–114 (2016)

24. Coron, J.-S., Patarin, J., Seurin, Y.: The random Oracle model and the ideal cipher model are equivalent. In: Wagner, D. (ed.) CRYPTO 2008. LNCS, vol. 5157, pp. 1–20. Springer, Heidelberg (2008). https://doi.org/10.1007/978-3-540-85174-5_1

25. Dachman-Soled, D., Katz, J., Thiruvengadam, A.: 10-round Feistel is indifferentiable from an ideal cipher. In: Fischlin, M., Coron, J.-S. (eds.) EUROCRYPT 2016. LNCS, vol. 9666, pp. 649–678. Springer, Heidelberg (2016). https://doi.org/10.1007/978-3-662-49896-5_23

26. Dai, Y., Seurin, Y., Steinberger, J., Thiruvengadam, A.: Indifferentiability of iterated Even-Mansour ciphers with non-idealized key-schedules: five rounds are necessary and sufficient. In: Katz, J., Shacham, H. (eds.) CRYPTO 2017. LNCS, vol. 10403, pp. 524–555. Springer, Cham (2017). https://doi.org/10.1007/978-3-319-63697-9_18

27. Dai, Y., Steinberger, J.: Indifferentiability of 10-round Feistel networks. Cryptology ePrint Archive, Report 2015/874

28. Dai, Y., Steinberger, J.: Indifferentiability of 8-round Feistel networks. In: Robshaw, M., Katz, J. (eds.) CRYPTO 2016. LNCS, vol. 9814, pp. 95–120. Springer, Heidelberg (2016). https://doi.org/10.1007/978-3-662-53018-4_4

29. Demay, G., Gaži, P., Hirt, M., Maurer, U.: Resource-restricted indifferentiability. In: Johansson, T., Nguyen, P.Q. (eds.) EUROCRYPT 2013. LNCS, vol. 7881, pp. 664–683. Springer, Heidelberg (2013). https://doi.org/10.1007/978-3-642-38348-9_39

30. Dodis, Y., Reyzin, L., Rivest, R.L., Shen, E.: Indifferentiability of permutation-based compression functions and tree-based modes of operation, with applications to MD6. In: Dunkelman, O. (ed.) FSE 2009. LNCS, vol. 5665, pp. 104–121. Springer, Heidelberg (2009). https://doi.org/10.1007/978-3-642-03317-9_7

31. Dodis, Y., Ristenpart, T., Shrimpton, T.: Salvaging Merkle-Damgård for practical applications. In: Joux, A. (ed.) EUROCRYPT 2009. LNCS, vol. 5479, pp. 371–388. Springer, Heidelberg (2009). https://doi.org/10.1007/978-3-642-01001-9_22

32. Dodis, Y., Ristenpart, T., Steinberger, J., Tessaro, S.: To hash or not to hash again? (In)differentiability results for H^2 and HMAC. In: Safavi-Naini, R., Canetti, R. (eds.) CRYPTO 2012. LNCS, vol. 7417, pp. 348–366. Springer, Heidelberg (2012). https://doi.org/10.1007/978-3-642-32009-5_21

33. Dodis, Y., Stam, M., Steinberger, J., Liu, T.: Indifferentiability of confusion-diffusion networks. In: Fischlin, M., Coron, J.-S. (eds.) EUROCRYPT 2016. LNCS, vol. 9666, pp. 679–704. Springer, Heidelberg (2016). https://doi.org/10.1007/978-3-662-49896-5_24

34. Farshim, P., Orlandi, C., Roşie, R.: Security of symmetric primitives under incorrect usage of keys. IACR Trans. Symm. Cryptol. **2017**(1), 449–473 (2017)

35. Farshim, P., Procter, G.: The related-key security of iterated Even–Mansour ciphers. In: Leander, G. (ed.) FSE 2015. LNCS, vol. 9054, pp. 342–363. Springer, Heidelberg (2015). https://doi.org/10.1007/978-3-662-48116-5_17

36. Forler, C., List, E., Lucks, S., Wenzel, J.: Reforgeability of authenticated encryption schemes. In: Pieprzyk, J., Suriadi, S. (eds.) ACISP 2017. LNCS, vol. 10343, pp. 19–37. Springer, Cham (2017). https://doi.org/10.1007/978-3-319-59870-3_2

37. Gennaro, R., Trevisan, L.: Lower bounds on the efficiency of generic cryptographic constructions. In: 41st FOCS. IEEE (2000)

38. Gueron, S., Lindell, Y.: GCM-SIV: full nonce misuse-resistant authenticated encryption at under one cycle per byte. In: ACM CCS 2015. ACM (2015)

39. Grubbs, P., Lu, J., Ristenpart, T.: Message franking via committing authenticated encryption. In: Katz, J., Shacham, H. (eds.) CRYPTO 2017. LNCS, vol. 10403, pp. 66–97. Springer, Cham (2017). https://doi.org/10.1007/978-3-319-63697-9_3

40. Halevi, S., Krawczyk, H.: Security under key-dependent inputs. In: ACM CCS 2007. ACM Press (2007)

41. Hoang, V.T., Krovetz, T., Rogaway, P.: Robust authenticated-encryption AEZ and the problem that it solves. In: Oswald, E., Fischlin, M. (eds.) EUROCRYPT 2015. LNCS, vol. 9056, pp. 15–44. Springer, Heidelberg (2015). https://doi.org/10.1007/978-3-662-46800-5_2

42. Hoang, V.T., Krovetz, T., Rogaway, P.: AEZ v5: authenticated encryption by enciphering (2017). https://competitions.cr.yp.to/round3/aezv5.pdf

43. Hoang, V.T., Reyhanitabar, R., Rogaway, P., Vizár, D.: Online authenticated-encryption and its nonce-reuse misuse-resistance. In: Gennaro, R., Robshaw, M. (eds.) CRYPTO 2015. LNCS, vol. 9215, pp. 493–517. Springer, Heidelberg (2015). https://doi.org/10.1007/978-3-662-47989-6_24

44. Holenstein, T., Künzler, R., Tessaro, S.: The equivalence of the random oracle model and the ideal cipher model, revisited. In: 43rd ACM STOC. ACM (2011)

45. Jean, J., Nikolić, I., Peyrin, T., Seurin, Y.: Deoxys v1.41 (2016). https://competitions.cr.yp.to/round3/deoxysv141.pdf

46. Kiltz, E., Pietrzak, K., Szegedy, M.: Digital signatures with minimal overhead from indifferentiable random invertible functions. In: Canetti, R., Garay, J.A. (eds.) CRYPTO 2013. LNCS, vol. 8042, pp. 571–588. Springer, Heidelberg (2013). https://doi.org/10.1007/978-3-642-40041-4_31

47. Küsters, R., Tuengerthal, M.: Universally composable symmetric encryption. In: CSF 2009. IEEE Computer Society (2009)

48. Maurer, U., Renner, R., Holenstein, C.: Indifferentiability, impossibility results on reductions, and applications to the random oracle methodology. In: Naor, M. (ed.) TCC 2004. LNCS, vol. 2951, pp. 21–39. Springer, Heidelberg (2004). https://doi.org/10.1007/978-3-540-24638-1_2

49. Micciancio, D., Warinschi, B.: Soundness of formal encryption in the presence of active adversaries. In: Naor, M. (ed.) TCC 2004. LNCS, vol. 2951, pp. 133–151. Springer, Heidelberg (2004). https://doi.org/10.1007/978-3-540-24638-1_8

50. Namprempre, C., Rogaway, P., Shrimpton, T.: AE5 security notions: definitions implicit in the CAESAR call. Cryptology ePrint Archive, Report 2013/242

51. Namprempre, C., Rogaway, P., Shrimpton, T.: Reconsidering generic composition. In: Nguyen, P.Q., Oswald, E. (eds.) EUROCRYPT 2014. LNCS, vol. 8441, pp. 257–274. Springer, Heidelberg (2014). https://doi.org/10.1007/978-3-642-55220-5_15

52. Peyrin, T., Seurin, Y.: Counter-in-tweak: authenticated encryption modes for tweakable block ciphers. In: Robshaw, M., Katz, J. (eds.) CRYPTO 2016. LNCS, vol. 9814, pp. 33–63. Springer, Heidelberg (2016). https://doi.org/10.1007/978-3-662-53018-4_2

53. Ristenpart, T., Shacham, H., Shrimpton, T.: Careful with composition: limitations of the indifferentiability framework. In: Paterson, K.G. (ed.) EUROCRYPT 2011. LNCS, vol. 6632, pp. 487–506. Springer, Heidelberg (2011). https://doi.org/10.1007/978-3-642-20465-4_27

54. Rogaway, P.: Authenticated-encryption with associated-data. In: ACM CCS 2002. ACM (2002)

55. Rogaway, P., Bellare, M., Black, J., Krovetz, T.: OCB: a block-cipher mode of operation for efficient authenticated encryption. In: ACM CCS 2001. ACM (2001)

56. Rogaway, P., Shrimpton, T.: A provable-security treatment of the key-wrap problem. In: Vaudenay, S. (ed.) EUROCRYPT 2006. LNCS, vol. 4004, pp. 373–390. Springer, Heidelberg (2006). https://doi.org/10.1007/11761679_23

57. Reyhanitabar, R., Vaudenay, S., Vizár, D.: Authenticated encryption with variable stretch. In: Cheon, J.H., Takagi, T. (eds.) ASIACRYPT 2016. LNCS, vol. 10031, pp. 396–425. Springer, Heidelberg (2016). https://doi.org/10.1007/978-3-662-53887-6_15

58. Stam, M.: Beyond uniformity: better security/efficiency tradeoffs for compression functions. In: Wagner, D. (ed.) CRYPTO 2008. LNCS, vol. 5157, pp. 397–412. Springer, Heidelberg (2008). https://doi.org/10.1007/978-3-540-85174-5_22

59. Unruh, D.: Programmable encryption and key-dependent messages. Cryptology ePrint Archive, Report 2012/423

The Curse of Small Domains: New Attacks on Format-Preserving Encryption

Viet Tung Hoang[1]([✉]), Stefano Tessaro[2], and Ni Trieu[3]

[1] Department of Computer Science, Florida State University, Tallahassee, USA
tvhoang@cs.fsu.edu
[2] Department of Computer Science, University of California Santa Barbara, Santa Barbara, USA
[3] Department of Computer Science, Oregon State University, Corvallis, USA

Abstract. Format-preserving encryption (FPE) produces ciphertexts which have the same format as the plaintexts. Building secure FPE is very challenging, and recent attacks (Bellare, Hoang, Tessaro, CCS '16; Durak and Vaudenay, CRYPTO '17) have highlighted security deficiencies in the recent NIST SP800-38G standard. This has left the question open of whether practical schemes with high security exist.

In this paper, we continue the investigation of attacks against FPE schemes. Our first contribution are new known-plaintext message recovery attacks against Feistel-based FPEs (such as FF1/FF3 from the NIST SP800-38G standard) which improve upon previous work in terms of amortized complexity in multi-target scenarios, where multiple ciphertexts are to be decrypted. Our attacks are also qualitatively better in that they make no assumptions on the correlation between the targets to be decrypted and the known plaintexts. We also surface a new vulnerability specific to FF3 and how it handles odd length domains, which leads to a substantial speedup in our attacks.

We also show the first attacks against non-Feistel based FPEs. Specifically, we show a strong message-recovery attack for FNR, a construction proposed by Cisco which replaces two rounds in the Feistel construction with a pairwise-independent permutation, following the paradigm by Naor and Reingold (JoC, '99). We also provide a strong ciphertext-only attack against a variant of the DTP construction by Brightwell and Smith, which is deployed by Protegrity within commercial applications. All of our attacks show that existing constructions fall short of achieving desirable security levels. For Feistel and the FNR schemes, our attacks become feasible on small domains, e.g., 8 bits, for suggested round numbers. Our attack against the DTP construction is practical even for large domains. We provide proof-of-concept implementations of our attacks that verify our theoretical findings.

Keywords: Format-preserving encryption · Attacks

© International Association for Cryptologic Research 2018
H. Shacham and A. Boldyreva (Eds.): CRYPTO 2018, LNCS 10991, pp. 221–251, 2018.
https://doi.org/10.1007/978-3-319-96884-1_8

1   Introduction

A *format-preserving encryption* (FPE) scheme is a deterministic symmetric encryption mechanism which preserves the format of the data, i.e., the ciphertext has the same format as the plaintext. For instance, a valid SSN is encrypted into a valid SSN, a valid credit-card number is encrypted into a valid credit-card number, etc. The first known constructions date back to Brightwell and Smith [6] and Black and Rogaway [4], and a formal treatment was later given by Bellare, Ristenpart, Rogaway, and Stegers [2]. The widespread interest in FPE from industry stems for its usage in the financial sector to encrypt credit-card numbers, as well as its ability to add encryption to legacy databases and applications without violating existing format constraints. FPE has been used and deployed by several companies, e.g., Voltage, Veriphone, Ingenico, Protegrity, Cisco, as well as by major credit-card payment organizations. While precise numbers are not known, it is safe to assume that vast amounts of data are currently encrypted with FPE in industrial settings.

However, building secure FPE is a challenging question, largely because (1) the domain is usually non-binary, and standard cryptographic primitives, e.g., AES, operate on fixed-length binary domains, and (2) the domain can be *small*, and it is hard to devise schemes where the domain size is not a security parameter. For example, the ANSI ASC X9.124 standard adopted by the financial industry envisions applications with domains as small as two decimal digits. While provably-secure schemes *do* exist [11,13,15], they consistently fail to meet practical efficiency demands. Consequently, practical designs have been validated via cryptanalysis only, and NIST has recently standardized [9] two constructions, FF1 [3] and FF3 [5], both based on Feistel networks. Recent works have however cast some doubt on the security of these constructions, which appear to be far from the initial desiderata set by NIST's selection process, which required 128 bits of security. (Indeed, one construction, FF2 [16], was dropped for far less severe attacks [10] than those by now known to exist against all Feistel-based constructions.) This state of affairs is particularly alarming, given the large-scale usage of FPE.

In a nutshell, this paper will take FPE cryptanalysis even further, providing more evidence that practical FPE constructions with high security are still beyond reach. This is particularly important as existing standards (NIST SP 800-38G, ANSI ASC X9.124) are being revised in view of recent attacks. We will strengthen prior attacks, and also present new attacks against practical constructions (employed in industry) which do not follow the Feistel paradigm.

EXISTING CRYPTANALYSIS. Let us first review recent cryptanalytic attacks against FPE. Formally, an FPE scheme F is a pair of deterministic algorithms (F.E, F.D), where F.E : F.Keys × F.Twk × F.Dom → F.Dom is the encryption algorithm, F.D : F.Keys × F.Twk × F.Dom → F.Dom the decryption algorithm, F.Keys the key space, F.Twk the tweak space, and F.Dom the domain. For every key $K \in$ F.Keys and tweak $T \in$ F.Twk, the map F.E$(K, T, \cdot)$ is a permutation over F.Dom, and F.D$(K, T, \cdot)$ reverses F.E$(K, T, \cdot)$.

Table 1. Attack parameters and effectiveness. This is for balanced-Feistel FPE with domain $\{0,1\}^{2n}$ ($n \geq 3$) and r rounds, with $N = 2^n$. Our attack LD does not limit the number of targets p, and thus p can be $O(N^2)$. In contrast, BHT's attack can only handle a single target. Both attacks achieve high advantage, as shown in the second row. The third and fourth rows respectively show the running time and the number of ciphertexts for the attacks, with a generic number p of targets for LD, and a single target for BHT's attack. The fifth and sixth row shows the amortized time and the number of ciphertexts per target, if $p = \Omega(N^2)$. The seventh row shows the maximum number of ciphertexts per tweak that each attack requires, and the last row shows the needed correlation between known messages and the target messages for each attack.

	Our attack LD	BHT's attack [1]
Advantage	$1 - 1/N$	$1 - 2/N$
Running time	$O(n^{1.5}N^{r-2} + N^{r-2}np)$	$O(n \cdot N^{r-2})$
Total ciphertexts	$O(n^{1.5}N^{r-2} + N^{r-3}np)$	$O(n \cdot N^{r-2})$
Time per target	$O(n \cdot N^{r-2})$	$O(n \cdot N^{r-2})$
Ciphertexts per target	$O(n \cdot N^{r-3})$	$O(n \cdot N^{r-2})$
Ciphertexts per tweak	$O(\sqrt{n} \cdot N)$	3
Known msg vs target	No correlation	Same right half

Bellare, Hoang, and Tessaro (BHT) [1] recently introduced a framework for known-plaintext message-recovery attacks on FPE. More concretely, they introduce the notion of a *message sampler*, an algorithm XS that returns a tuple $((T_1, X_1), \ldots, (T_Q, X_Q), Z^*, a)$ that consists of Q *distinct* tweak-message pairs (T_i, X_i), a *target message* Z^*, and (possibly) some *auxiliary information* $a \in \{0,1\}^*$. Then, an attacker against XS attempts to recover Z^* given

$$(T_1, \mathsf{F.E}(K, T_1, X_1)), \ldots (T_Q, \mathsf{F.E}(K, T_Q, X_Q)), a$$

for a secret key K. The attacker's advantage is obtained by subtracting from its success probability that of the best possible trivial attacker that only gets $T_1, \ldots, T_Q$ and a. Therefore, *any* message sampler with a corresponding attacker achieving substantial advantage within feasible computational constraints is effectively a break, since the scheme fails to satisfy some ideal property to be expected.

For example, for the *balanced* r-round Feistel construction with domain $\mathbb{Z}_N \times \mathbb{Z}_N$ (meaning the domain size is N^2), where $N = 2^n$, BHT provide a sampler and an attack which succeed with $O(n \cdot N^{r-2})$ ciphertexts, where in particular these ciphertexts consist of the encryption of three messages (one of which is the target one) under $O(n \cdot N^{r-2})$ distinct tweaks.[1] (The attack is summarized in Table 1.) While the attack is generic, when applied to the setting of NIST's standardized

---

[1] BHT actually give three attacks with different complexity, but only one of them can fully recover the target message; the other two can only recover a half of the target. Since our attack can recover all target messages in their entirety, here we only compare our attack with the Full-Message Recovery attack of BHT.

constructions FF1/FF3, which use $r = 10$ and $r = 8$, respectively, the attack becomes particularly threatening for small domains. The fact that the number of ciphertexts is larger than the domain size N is no contradiction – the point is that the number of ciphertexts *per tweak* is small, and this makes a generic message recovery without the ciphertexts only possible with small probability.

We also point out the work by Durak and Vaudenay (DV) [8]. They give a message-recovery attack against FF3 which uses only *two* tweaks, yet their attack is due to a flaw in the tweaking mechanism used in FF3, rather than being a generic issue of Feistel. In contrast, BHT's attacks succeed even if the flaw behind DV's attack is fixed.

NIST has temporarily discouraged the use of FF3 as the result of DV's attack[2], whereas a draft update of the ANSI ASC X9.124 standard additionally suggests double encryption on small domains as a result of BHT's attacks.

OUR CONTRIBUTIONS. The BHT attacks can be mitigated by increasing the number of rounds of the constructions. However, this raises the question of whether the attacks are the best possible, and whether new, stronger attacks, are possible. Similarly, plain Feistel is not the only approach used in practice for FPE. For example, Cisco presented a variant of Feistel, called FNR [7], which appears to bypass the BHT attacks. Protegrity is another very active company in the FPE domain and uses a different construction [12], called DTP (from "Data-type preserving" encryption), based on Brightwell and Smith's [6] construction. It is well possible that these constructions are not affected by attacks, and may end up being superior to NIST-standardized constructions.

Our first contribution will be new attacks against Feistel-based FPE that improve upon BHT in settings where multiple messages can be recovered, as well as only requiring weaker correlations in the known messages for which the FPE construction is evaluated. We will then provide an attack against FNR, thus showing it too fails to provide sufficient security. Finally, we provide a strong ciphertext-only attack against DTP. In particular, while our attacks against Feistel and FNR relies on weaknesses for small domains, our attack against DTP works even on large domains.

We complement our attacks with proof-of-concept implementations that validate experimentally our theoretical findings.

NEW ATTACKS AGAINST FEISTEL-BASED FPE. We strengthen the attacks from BHT by considering the setting where the attacker is given *multiple* target messages $Z_1^*, \ldots, Z_p^*$ it is trying to recover. This captures for example an attempt by the attacker to compromise a large fraction of an FPE-encrypted database, as opposed to an individual record in it. Clearly, this task should be harder than recovering a single target, and a good FPE scheme should guarantee that the cost of recovering p messages is roughly p times that of recovering one message. Indeed, this is true when mounting BHT's attacks, as the only option is to apply the attack to each target.

---

[2] https://csrc.nist.gov/News/2017/Recent-Cryptanalysis-of-FF3.

We will show however that for the r-round Feistel construction with domain $\mathbb{Z}_M \times \mathbb{Z}_N$, multiple targets can be recovered much faster, in fact with a number of ciphertexts comparable to what is needed for a single target. As summarized in Table 1, for the special case $M = N = 2^n$, the amortized number of ciphertexts *per target* is only $O(n \cdot N^{r-3})$, as opposed to $O(n \cdot N^{r-2})$ when using BHT repeatedly. A further advantage of our attack is that the known plaintexts revealed to the attacker are not correlated with the target messages – whereas BHT assumed a fairly artificial setting where (partially) known plaintexts exhibit strong correlations with the target message.

More concretely, the attacker is supplied τ *known* distinct messages $X_1, \ldots, X_\tau$, and we have p targets $Z_1, \ldots, Z_p$. Then, the attacker gets encryptions of these $\tau + p$ messages (assumed to be distinct) under q known tweaks $T_1, \ldots, T_q$ (thus, the attacker sees $q \times (\tau + p)$ ciphertexts). The goal is to recover all of $Z_1, \ldots, Z_p$. The only assumptions here are that (1) The right halves of $X_1, \ldots, X_\tau$ cover all of $\mathbb{Z}_N$, and (2) $Z_1, \ldots, Z_p$ have (as a tuple) sufficient min-entropy conditioned on $X_1, \ldots, X_\tau, T_1, \ldots, T_q$, say at least θ. Because of this, the probability that an ideal adversary that does not learn the ciphertexts recovers all of $Z_1, \ldots, Z_p$ here is at most $2^{-\theta}$. In contrast, we give an attack which recovers them with high probability whenever q is large enough. See Table 1 for the exact complexities when $M = N = 2^n$.

We stress that unlike the BHT attacks, the attacker is not aware of any correlation between the known plaintexts $X_1, \ldots, X_\tau$ and the target plaintexts $Z_1, \ldots, Z_p$. Of course, every right half of $Z_1, \ldots, Z_p$ will appear among $X_1, \ldots, X_\tau$, but the attacker does not know which of the inputs have matching right halves. Also, we point out that the restriction of all right halves appearing in $X_1, \ldots, X_\tau$ is not as artificial as it may at first appear. If these inputs are drawn uniformly at random (under the constraint of being distinct), and $\tau = \Theta(N\sqrt{n})$, then we can show that all right halves are going to appear with high probability by a variant of the so-called "coupon collector" argument. Even more importantly, if they do not cover all of $\mathbb{Z}_N$, our attacks recovers all of the $Z_1, \ldots, Z_p$ whose right halves overlap with those of $X_1, \ldots, X_\tau$.

THE DANGER OF ASYMMETRY. We note that the complexity of our attack is not symmetric in M and N. In particular, the attack's performance improves with a smaller N and a larger M. This is particularly problematic for FF3, which in the case of odd-length domains (e.g., $\{0, \ldots, 9\}^3$) would exactly create such a convenient asymmetry, setting $M = 100$ and $N = 10$. This feature was already present in the left-half attack of BHT, but went unobserved.

THE FNR CONSTRUCTION. Cisco proposed the FNR construction [7] as an approach to encrypt IP addresses. While we are not aware whether FNR was indeed used, it adopts a potentially interesting idea which seemingly prevents our and BHT's attacks against Feistel. Essentially, it uses Naor and Reingold's [14] idea of replacing the two outer rounds of the Feistel construction with a pairwise independent permutation while retaining security.

Initially, it is not clear how existing attacks against Feistel can be used when a pairwise-independent permutation is used. We show however that this approach too fails, and in fact, in terms of our attacks, FNR with r-rounds appear to be as secure as plain Feistel with $r + 2$ rounds, somehow matching (though in a different and unexplored context) the initial intuition by Naor and Reingold.

THE DTP SCHEME AND ITS INSECURITY. Another solution is the DTP scheme put forward by Protegrity [12], which is a variation of the scheme by Smith and Brightwell [6] and which has been argued to be potentially superior to FPE.[3] In particular, reframing it in our language, DTP requires a distinct tweak per encryption, thus potentially achieving higher security by preventing detection of equal plaintexts being encrypted. However, we give an attack that only requires multiple encryptions of the same target message with different tweaks (and is thus compatible with the envisioned usage scenario). The attack differs from those against Feistel-based FPE, but again is in the same spirit of using encryptions under multiple tweaks to amplify subtle statistical deviations. We have confirmed that a variant of this scheme, called DTP-2, is still deployed by Protegrity, even though it is being phased out to be replaced with FF1.[4]

Abstractly, the main issue of DTP is that it encrypts individual digits of the plaintext $x_1 x_2 \ldots x_n$ (where $x_i \in \mathbb{Z}_d$) as $c_i \leftarrow x_i + z_i \pmod{d}$, where the z_i's are pseudorandom elements of $\mathbb{Z}_D$. For example, one could use $d = 10$ (to encrypt decimal numbers) and $D = 256$ (e.g., the z_i's are individual bytes from an AES output). Then, it is not hard to see that the c_i values are not pseudorandom anymore, and there is in fact a noticeable statistical deviation. This is because $z_i \in \{0, 1, \ldots, 5\}$ is more likely to occur than $z_i \in \{6, \ldots, 9\}$. Our recent interactions with Protegrity indicate that $d = 62$ is more commonly used (to accommodate for the alphabet $\{a, \ldots, z, A, \ldots, Z, 0, \ldots, 9\}$), and this introduces even more important biases. As we show below in Table 4, there is a factor 10 improvement in the number of ciphertexts required by our attack when switching from $d = 10$ to $d = 62$.

Our attack is stronger than those against Feistel and FNR as it also works on large input spaces – the problem being exploited here is the mapping between binary outputs (corresponding to the choice of D) to elements in another alphabet (by reducing mod d). The observation that encryptions are biased is not novel (cf. e.g. https://en.wikipedia.org/wiki/Format-preserving_encryption), but our attacks highlights how such biases can be exploited for full-message recovery in a multi-tweak scenario.

We note that the spec (as well as the original description in [6]) allow for some key-dependent pre-processing of the plaintext which Protegrity makes *explicitly optional* if tweaks are chosen uniformly at random. The version without pre-processing is the version we attack here. With pre-processing, our attack does not apply, but note that [6] acknowledges the pre-processing itself only suffices to deter "casual attacks" and this is unlikely a strong countermeasure.

---

[3] http://www.protegrity.com/role-of-standards-nist-data-security/.

[4] The findings of this paper have been in particular shared with Protegrity.

2   Preliminaries

NOTATION. We let ε denote the empty string. If y is a string then $|y|$ denotes its length and $y[i]$ denotes its i-th bit for $1 \leq i \leq |y|$. If X is a finite set, we let $x \leftarrow_\$ X$ denote picking an element of X uniformly at random and assigning it to x. Algorithms may be randomized unless otherwise indicated. Running time is worst case. If A is an algorithm, we let $y \leftarrow A(x_1,\ldots;r)$ denote running A with random coins r on inputs $x_1,\ldots$ and assigning the output to y. We let $y \leftarrow_\$ A(x_1,\ldots)$ be the result of picking r at random and letting $y \leftarrow A(x_1,\ldots;r)$. We let $[A(x_1,\ldots)]$ denote the set of all possible outputs of A when invoked with inputs $x_1,\ldots$. By $\Pr[G]$ we denote the probability of the event that the execution of game G results in the game returning true. If D is a set then $\mathrm{Perm}(D)$ denotes the set of all permutations on D. Let $\exp(x)$ denote e^x, where e is the base of the natural logarithm.

FPE. An FPE scheme F specifies a pair of deterministic algorithms $(\mathsf{F.E}, \mathsf{F.D})$, where $\mathsf{F.E} : \mathsf{F.Keys} \times \mathsf{F.Twk} \times \mathsf{F.Dom} \rightarrow \mathsf{F.Dom}$ is the encryption algorithm, $\mathsf{F.D} : \mathsf{F.Keys} \times \mathsf{F.Twk} \times \mathsf{F.Dom} \rightarrow \mathsf{F.Dom}$ the decryption algorithm, $\mathsf{F.Keys}$ the key space, $\mathsf{F.Twk}$ the tweak space, and $\mathsf{F.Dom}$ the domain. For every key $K \in \mathsf{F.Keys}$ and tweak $T \in \mathsf{T}$, the map $\mathsf{F.E}(K,T,\cdot)$ is a permutation over $\mathsf{F.Dom}$, and $\mathsf{F.D}(K,T,\cdot)$ reverses $\mathsf{F.E}(K,T,\cdot)$.

CHERNOFF BOUND. Our results heavily rely on the well-known Chernoff bounds. We recall the details of Chernoff bounds below.

Lemma 1 (Chernoff bounds). *Let* $Y_1,\ldots,Y_\ell$ *be independent Bernoulli random variables with* $\Pr[Y_1 = 1] = \cdots = \Pr[Y_\ell = 1] = \mu$. *Then,*

$$\Pr\left[Y_1 + \cdots + Y_\ell \geq (1+\epsilon)\ell\mu\right] \leq \exp\left(\frac{-\epsilon^2 \ell \mu}{2+\epsilon}\right) \text{ for any } \epsilon > 0, \text{ and}$$

$$\Pr\left[Y_1 + \cdots + Y_\ell \leq (1-\epsilon)\ell\mu\right] \leq \exp\left(\frac{-\epsilon^2 \ell \mu}{2}\right) \text{ for any } 0 < \epsilon < 1.$$

3   Message Recovery Framework

Here we give a new formalization of message-recovery attacks, generalizing the definition of Bellare, Hoang, and Tessaro (BHT) [1] for attacking multiple target messages.

A HIGH-LEVEL INTUITION. Under our framework, there are τ known messages and p target messages. An adversary $\mathcal{A}$ will receive the ciphertexts of those, each under multiple tweaks, and has to recover at least $d \leq p$ targets to win the game, where d is a parameter of the message-recovery game. For example $d = 1$ means that as long as the adversary recovers a single target message, it wins the game, and $d = p$ means that the adversary has to recover all targets to win.

Following BHT, we aim for a generalized framework that can capture BHT's attack, where known messages are correlated with the targets. Thus in our notion, the known messages and the target messages, and also the tweaks, are generated via a message sampler XS. The adversary $\mathcal{A}$ receives the tweaks and the ciphertexts, and some auxiliary information that contains information about the known messages, and possibly some partial information about the targets. We stress that only the sampler knows the target messages, and the adversary $\mathcal{A}$ just knows some partial information of the target messages that the auxiliary information reveals.

The framework above allows samplers that output target messages that are trivial to guess. Thus for any FPE scheme, there is an adversary that with high probability can recover target messages produced by those degenerate samplers by merely guessing, but of course this does not imply a vulnerability of the FPE scheme. Following BHT, we define the d-target advantage $\mathbf{Adv}^{mr}_{F,XS,d}(\mathcal{A})$ of adversary $\mathcal{A}$ against FPE scheme F and sampler XS as the difference between (i) the chance that $\mathcal{A}$ can recover at least d targets, and (ii) the probability of the best strategy of guessing that many targets given just the auxiliary information (but not the ciphertexts). Hence for an FPE scheme F, if one can construct an efficient adversary $\mathcal{A}$ and an efficient sampler XS such that $\mathbf{Adv}^{mr}_{F,XS,d}(\mathcal{A})$ is large, it means that this particular FPE scheme F is indeed vulnerable.

Our notion only models non-adaptive attacks and requires adversaries to recover at least d targets. However, recall that here we are giving an attack notion, and thus these restrictions only make our attacks better. On the other hand, if an FPE scheme meets our notion, it does not necessarily mean that the scheme is secure for real-world usage. Below, we will formalize our framework.

SAMPLERS AND GUESSING PROBABILITY. A *message sampler* is an algorithm XS that returns $((T_1, X_1), \ldots, (T_Q, X_Q), Z_1, \ldots, Z_p, a)$ that consists of Q tweak-message pairs (T_i, X_i), p *target messages* Z_j, and some *auxiliary information* $a \in \{0, 1\}^*$. Note that encryption schemes of FPEs are deterministic, and thus it is trivial to detect repetition among the pairs $(T_1, X_1), \ldots, (T_Q, X_Q)$ given their ciphertexts. Therefore, following BHT, we require the *distinctness* condition that the Q pairs $(T_1, X_1), \ldots, (T_Q, X_Q)$ be distinct. Define the d-target message-guessing (mg) advantage against a sampler XS as

$$\mathbf{Adv}^{mg}_{XS,d} = \max_{\mathcal{S}} \Pr[\mathbf{G}^{mg}_{XS,d}(\mathcal{S})],$$

where game $\mathbf{G}^{mg}_{XS}(\mathcal{S})$ is defined in the top panel of Fig. 1. This is the probability of the best possible way at guessing at least d target messages given the tweaks and auxiliary information. For the special case $d = p$, meaning that one has to guess all target messages, we write $\mathbf{Adv}^{mg}_{XS}$ instead of $\mathbf{Adv}^{mg}_{XS,p}$. To account for the efficiency of attacks, besides the number of ciphertexts Q, we also consider the *number of ciphertexts per recovered target* $q_t = Q/d$. This is the amortized data complexity.

MESSAGE-RECOVERY NOTION. Let F be an FPE scheme. Let XS be a message sampler such that $T_1, \ldots, T_Q \in$ F.Twk and $X_1, \ldots, X_Q, Z_1, \ldots, Z_p \in$ F.Dom for any $((T_1, X_1), \ldots, (T_Q, X_Q), Z_1, \ldots, Z_p, a)$ in [XS]. Define the d-target message-recovery (mr) advantage of $\mathcal{A}$ against F, XS as

$$\mathbf{Adv}^{\mathsf{mr}}_{\mathsf{F},\mathsf{XS},d}(\mathcal{A}) = \Pr[\mathbf{G}^{\mathsf{mr}}_{\mathsf{F},\mathsf{XS},d}(\mathcal{A})] - \mathbf{Adv}^{\mathsf{mg}}_{\mathsf{XS},d} \; .$$

The mr game $\Pr[\mathbf{G}^{\mathsf{mr}}_{\mathsf{F},\mathsf{XS},d}(\mathcal{A})]$ is defined in the bottom panel of Fig. 1, measuring $\mathcal{A}$'s advantage at recovering at least d target messages given the tweaks, ciphertexts, and auxiliary information. For $d = p$, meaning that the adversary has to recover all targets, we write $\mathbf{Adv}^{\mathsf{mr}}_{\mathsf{F},\mathsf{XS}}(\mathcal{A})$ instead of $\mathbf{Adv}^{\mathsf{mr}}_{\mathsf{F},\mathsf{XS},p}(\mathcal{A})$.

Game $\mathbf{G}^{\mathsf{mg}}_{\mathsf{XS},d}(\mathcal{S})$

$((T_1, X_1), \ldots, (T_Q, X_Q)), Z_1, \ldots, Z_p, a) \leftarrow\!\!\text{\$}\; \mathsf{XS}$

$(Z_1^*, \ldots, Z_p^*) \leftarrow\!\!\text{\$}\; \mathcal{S}(T_1, \ldots, T_Q, a); \;\; t \leftarrow \min\{d, p\}$

Return $(\exists\, i_1 < \cdots < i_t$ such that $(Z_{i_1} = Z_{i_1}^*) \wedge \cdots \wedge (Z_{i_t} = Z_{i_t}^*))$

Game $\mathbf{G}^{\mathsf{mr}}_{\mathsf{F},\mathsf{XS},d}(\mathcal{A})$

$K \leftarrow\!\!\text{\$}\; \mathsf{F}.\mathsf{Keys}; \;\; ((T_1, X_1), \ldots, (T_Q, X_Q)), Z_1, \ldots, Z_p, a) \leftarrow\!\!\text{\$}\; \mathsf{XS}$

For $i = 1, \ldots, Q$ do $Y_i \leftarrow \mathsf{F}.\mathsf{E}(K, T_i, X_i)$

$(Z_1^*, \ldots, Z_p^*) \leftarrow\!\!\text{\$}\; \mathcal{A}((T_1, Y_1), \ldots, (T_Q, Y_Q), a); \;\; t \leftarrow \min\{d, p\}$

Return $(\exists\, i_1 < \cdots < i_t$ such that $(Z_{i_1} = Z_{i_1}^*) \wedge \cdots \wedge (Z_{i_t} = Z_{i_t}^*))$

Fig. 1. Games defining message-recovery notion of an FPE scheme F, parameterized by a message sampler XS.

RELATION TO BHT'S NOTION. BHT's notion is the special case of the definition above where the number of target message p is 1. However, in practice, it is not economical to collect a lot of known message-ciphertext pairs to recover just a single target message. If we can instead spend the same amount of resource but recover multiple messages, the cost will be amortized by the number of recovered targets, cheapening the attack. Thus compared to BHT's definition, ours gives a more realistic attack model.

REMARKS. Most existing notions in the cryptanalytic literature only define codebook-recovery attacks, but our attacks or BHT's attack do not fit into this category. Bellare, Ristenpart, Rogaway, and Stegers (BRRS) [2] define a message-recovery notion for FPEs, but again (i) this notion considers just a single target message, and (ii) more importantly, the number of ciphertexts under this notion cannot exceed the domain size. Thus BRRS's notion also fails to capture our attack or BHT's attack.

4 Attacking Feistel-Based FPE

In this section, we first recall the Feistel-based FPE constructions, as in NIST standards FF1 or FF3, and then give a message-recovery attack on a generic

FPE scheme. Compared to BHT's attacks [1], our attack can deal with a general number of target messages and recover all of them, and thus have better amortized cost. Moreover, we do not require any correlation between the known messages and the targets.

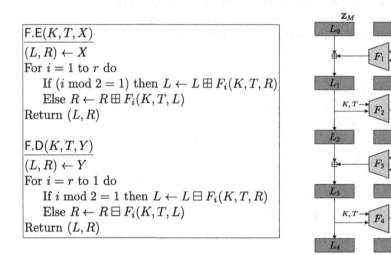

$\underline{\mathsf{F.E}(K, T, X)}$

$(L, R) \leftarrow X$
For $i = 1$ to r do
    If $(i \bmod 2 = 1)$ then $L \leftarrow L \boxplus F_i(K, T, R)$
    Else $R \leftarrow R \boxplus F_i(K, T, L)$
Return (L, R)

$\underline{\mathsf{F.D}(K, T, Y)}$

$(L, R) \leftarrow Y$
For $i = r$ to 1 do
    If $i \bmod 2 = 1$ then $L \leftarrow L \boxminus F_i(K, T, R)$
    Else $R \leftarrow R \boxminus F_i(K, T, L)$
Return (L, R)

Fig. 2. Left: The code for the encryption and decryption algorithms of $\mathsf{F} =$ **Feistel**$[r, M, N, \boxplus, \mathrm{PL}]$, where $\mathrm{PL} = (\mathcal{T}, \mathcal{K}, F_1, \dots, F_r)$. **Right:** An illustration of encryption with $r = 4$ rounds.

FEISTEL-BASED CONSTRUCTIONS. Most existing FPE schemes, including the FF1 and FF3 standards [9], are based on Feistel networks. Following BHT, we specify Feistel-based FPE in a general, parameterized way. This allows us to refer to both schemes of ideal round functions for the analysis, and schemes of some concrete round functions for realizing the standards.

We associate to parameters $r, M, N, \boxplus, \mathrm{PL}$ an FPE scheme $\mathsf{F} = $ **Feistel**$[r, M, N, \boxplus, \mathrm{PL}]$. Here $r \geq 2$ is an integer, the number of rounds, and $\boxplus$ is an operation for which $(\mathbb{Z}_M, \boxplus)$ and $(\mathbb{Z}_N, \boxplus)$ are Abelian groups. We let $\boxminus$ denote the inverse operator of $\boxplus$, meaning that $(X \boxplus Y) \boxminus Y = X$ for every X and Y. Integers $M, N \geq 1$ define the domain of F as $\mathsf{F.Dom} = \mathbb{Z}_M \times \mathbb{Z}_N$. The parameter $\mathrm{PL} = (\mathcal{T}, \mathcal{K}, F_1, \dots, F_r)$ specifies the set $\mathcal{T}$ of tweaks and a set $\mathcal{K}$ of keys, meaning $\mathsf{F.Twk} = \mathcal{T}$ and $\mathsf{F.Keys} = \mathcal{K}$, and the round functions $F_1, \dots, F_r$ such that $F_i : \mathcal{K} \times \mathcal{T} \times \mathbb{Z}_N \to \mathbb{Z}_M$ if i is odd, and $F_i : \mathcal{K} \times \mathcal{T} \times \mathbb{Z}_M \to \mathbb{Z}_N$ if i is even. The code of $\mathsf{F.E}$ and $\mathsf{F.D}$ is shown in Fig. 2.

Classical Feistel schemes correspond to the boolean case, where $M = 2^m$ and $N = 2^n$ are powers of two, and $\boxplus$ is the bitwise xor operator $\oplus$. The scheme is balanced if $M = N$ and unbalanced otherwise. For $X = (L, R) \in \mathbb{Z}_M \times \mathbb{Z}_N$, we call L and R the *left segment* and *right segment* of X, respectively. We write $\mathbf{Left}(X)$ and $\mathbf{Right}(X)$ to refer to the left and right segments of X respectively.

For simplicity, we assume that 0 is the zero element of the groups $(\mathbb{Z}_M, \boxplus)$ and $(\mathbb{Z}_N, \boxplus)$.

For analysis, the round functions are modeled as truly random. Formally, let $\mathcal{T} = \{0, 1\}^*$, and let $\mathcal{K}$ be the set $\mathbf{RF}(\mathcal{T}, r, M, N)$ of all tuples of functions $(G_1, \ldots, G_r)$ such that $G_i : \mathcal{T} \times \mathbb{Z}_N \rightarrow \mathbb{Z}_M$ if i is odd, and $G_i : \mathcal{T} \times \mathbb{Z}_M \rightarrow \mathbb{Z}_N$ if i is even. Then for $1 \le i \le r$ define $F_i(K, \cdot, \cdot) = G_i(\cdot, \cdot)$, where $(G_1, \ldots, G_r) \leftarrow K$. We write $\mathbf{Feistel}[r, M, N, \boxplus]$ to denote $\mathbf{Feistel}[r, M, N, \boxplus, \mathrm{PL}]$, for the particular choice $\mathrm{PL} = (\mathcal{T}, \mathcal{K}, F_1, \ldots, F_r)$ above.

Schemes in the standards [9] specify the round functions using AES. Using the standard assumption that AES is a PRF, one can focus on attacking Feistel-based schemes of ideal round functions, with small differences in the advantage.

SETUP. We give a message-recovery attack on a generic Feistel-based FPE $\mathsf{F} = \mathbf{Feistel}[r, M, N, \boxplus, \mathrm{PL}]$. Like the prior work of BHT [1], we only consider the case that r is even, as NIST standards only use $r = 8$ (for FF3) or $r = 10$ (for FF1). Under our attack, there are τ known messages $X_1, \ldots, X_\tau$ and p targets $Z_1, \ldots, Z_p$. The adversary is given the encryption of those $\tau + p$ distinct messages under q tweaks $T_1, \ldots, T_q$, for an appropriately large q. Due to the distinctness requirement, $X_1, \ldots, X_\tau, Z_1, \ldots, Z_p$ must be distinct. The auxiliary information is $(X_1, \ldots, X_\tau, p, q)$. The only requirement in our attack is that with high probability, the right halves of the known messages $X_1, \ldots, X_\tau$ cover at least d of the right halves of the targets. We have no restriction on the number p of targets or the parameter d, (except the unavoidable constraint that $d \le p$) so potentially p can be as large as $MN - \tau$. Our attack will recover d targets out of $Z_1, \ldots, Z_p$.

A special important case in our attack is that the right halves of $X_1, \ldots, X_\tau$ cover everything in $\mathbb{Z}_N$; in this case we can recover all targets. At the first glance, this requirement seems contrived, and thus it is unclear how the adversary can mount such an attack. However, we will show that for $\tau = \lceil \min\{2\sqrt{MN \ln(N)}, 2N \ln(N)\} \rceil$, if the known messages are sampled uniformly without replacement from $\mathbb{Z}_M \times \mathbb{Z}_N$ then they will meet the requirement above. Concretely, if we want to recover PINs, meaning $M = N = 100$, we need to obtain $\lceil 2N\sqrt{\ln(N)} \rceil = 430$ random known messages. In contrast, BHT's attack needs to obtain two known messages, but one of those must have the same right half as the target.

To explain the bound $\lceil \min\{2\sqrt{MN \ln(N)}, 2N \ln(N)\} \rceil$ above, note that this is the well-known coupon collector's problem: there are N types of coupons and a collector wishes to collect all of them. In the classical setting, in each draw, the collector is given a uniformly random type of coupon, and it will take $\Theta(N \ln(N))$ draws, with very high probability, for the collector to get all N types. In our setting, the coupons are the values of the right halves of the known messages, but in each draw, the type of the given coupon is *not* exactly uniformly random. In fact, since known messages must be distinct, each draw is slightly biased towards new types of coupons. Thus in our setting, to get all types of coupons with high probability, the number of draws is smaller than the classical result, about $O(N\sqrt{\ln(N)})$ in the balanced case $M = N$. This intuition is formalized in Lemma 2 below; the proof is in the full version.

Lemma 2 (Biased coupon collector's problem). *Let $M \geq 2$ and $N \geq 2$ be integers and let $\tau = \lceil \min\{2\sqrt{MN \ln(N)}, 2N \ln(N)\} \rceil$. Let $X_1, \ldots, X_\tau$ be sampled uniformly without replacement from the set $\mathbb{Z}_M \times \mathbb{Z}_N$. Then we have $\{\mathbf{Right}(X_1), \ldots, \mathbf{Right}(X_\tau)\} = \mathbb{Z}_N$ with probability at least $1 - 1/N$.*

From Lemma 2 above, the requirement of our attack is quite mild, yet it is powerful, recovering as many targets as possible. In contrast, in BHT's attack, there is only a single target (meaning $p = 1$), and the first known message must have the same right half as the target message. Of course in our attack, for each target Z_i, there is some known message X_j of the same right half as Z_i, but the adversary does not know what is j.

THE ATTACK. We formalize the attack via the message-recovery framework, by specifying a class $\mathsf{SC1}_{p,q,\delta,\theta}$ of samplers, and then giving a lower bound on the mr-advantage of the attack for any sampler in this class. First, let $\mathsf{DC1}_{p,q,d,\delta,\theta}$ be the class of all algorithms D that outputs q distinct tweaks $T_1, \ldots, T_q \in \{0,1\}^*$, and distinct $X_1, \ldots, X_\tau, Z_1, \ldots, Z_p \in \mathbb{Z}_M \times \mathbb{Z}_N$ such that (1) with probability at least $1 - \delta$, there are d or more indices k such that $Z_k \in \{\mathbf{Right}(X_1), \ldots, \mathbf{Right}(X_\tau)\}$ and (2) given $X_1, \ldots, X_\tau, T_1, \ldots, T_q$, for any subset $\{r_1, \ldots, r_d\} \subseteq \{1, \ldots, \tau\}$, for any $Z_1^*, \ldots, Z_d^* \in \mathbb{Z}_M \times \mathbb{Z}_N \setminus \{X_1, \ldots, X_\tau\}$, the conditional probability that $Z_{r_1} = Z_1^*, \ldots, Z_{r_d} = Z_d^*$ is at most $2^{-\theta}$.[5] To any such D, we associate the sampler

> Sampler $\mathsf{XS}[\mathsf{D}]$
> ───────────────────────
> $(T_1, \ldots, T_q, X_1, \ldots, X_\tau, Z_1, \ldots, Z_p) \twoheadleftarrow\!\!\$\ \mathsf{D}$
> $a \leftarrow (X_1, \ldots, X_\tau, p, q)$
> Return $\big(\{(T_i, X_j), (T_i, Z_k) \mid i \leq q, j \leq \tau, k \leq p\}, Z_1, \ldots, Z_p, a\big)$

The sampler above returns the pairs (T_i, X_j) and (T_i, Z_k) for every $i \leq q$ and every $j \leq \tau$, and $k \leq p$, where the targets are $Z_1, \ldots, Z_p$. The number of ciphertexts Q is $(\tau + p)q$, and the number of ciphertexts per recovered target q_t is $(\tau + p)q/d$. Let $\mathsf{SC1}_{p,q,d,\delta,\theta} = \{\mathsf{XS}[\mathsf{D}] \mid \mathsf{D} \in \mathsf{DC1}_{p,q,d,\delta,\theta}\}$. We would expect that adversaries will have low mr-advantage, even if q is big. However, the Left-half Differential (LD) attack, given in Fig. 3, can recover d targets out of $Z_1, \ldots, Z_p$ in $O(pqN)$ time. Theorem 3 below gives a lower bound on the mr-advantage of LD.

The bound in Theorem 3, for the special case $d = p$, is illustrated in Fig. 4. For example, for FF1, the attack is only reasonably feasible in very few domains, say one-byte strings ($M = N = 16$) or two-decimal strings ($M = N = 10$), but recall that FF1 and FF3 are supposed to provide 128-bit security whenever the domain size MN is at least 100. For FF3, since there are fewer rounds, the attack is faster, and thus becomes feasible in more domains.

───────────────────────

[5] For the special case where $Z_1, \ldots, Z_p$ are sampled uniformly without replacement from $(\mathbb{Z}_M \times \mathbb{Z}_N) \setminus \{X_1, \ldots, X_\tau\}$, then $\theta = \Theta(d \cdot \log(MN))$.

Adversary $\mathsf{LD}(\{(T_i, C_{i,j}), (T_i, C'_{i,k})\}_{i,j,k}, a)$

$/\!/\ 1 \le i \le q, 1 \le j \le \tau, 1 \le k \le p$

$(X_1, \ldots, X_\tau, p, q) \leftarrow a; \quad S, \mathrm{Dom} \leftarrow \emptyset$

For $j = 1, \ldots, \tau$ do

    If $\mathbf{Right}(X_j) \notin \mathrm{Dom}$ then $S \leftarrow S \cup \{j\}; \quad \mathrm{Dom} \leftarrow \mathrm{Dom} \cup \{\mathbf{Right}(X_j)\}$

For $k \leftarrow 1$ to p do $/\!/$ Recover target Z_k

    For $j \in S, s \in \mathbb{Z}_M$ do $V_{j,s} \leftarrow 0$

    For $i \leftarrow 1$ to $q, j \in S$ do

        $s \leftarrow \mathbf{Left}(C'_{i,k}) \boxminus \mathbf{Left}(C_{i,j}) \boxplus \mathbf{Left}(X_j); \quad V_{j,s} \leftarrow V_{j,s} + 1$

    Let $V_{j^*,s^*} = \max\{V_{j,s} \mid j \in S, s \in \mathbb{Z}_M\}; \quad Z_k \leftarrow (s^*, \mathbf{Right}(X_{j^*}))$

Return $(Z_1, \ldots, Z_p)$

Fig. 3. The Left-half Differential attack.

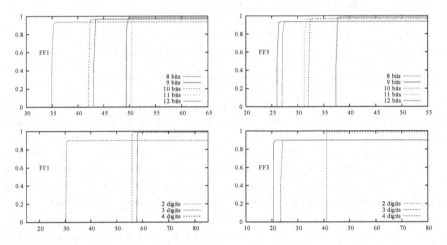

Fig. 4. The mr advantage of the Left-half Differential attack for binary strings of 8–12 bits (top) and decimal strings of 2–4 digits (bottom). The x-axis shows the log, base 2, of the number q of tweaks (which is also roughly q_t, the number of ciphertexts per recovered target), and the y-axis shows $\mathbf{Adv}^{\mathrm{mr}}_{\mathbf{Feistel}[r,M,N,\boxplus],\mathsf{XS}}(\mathsf{LD})$, for XS that outputs $\tau = \lceil \min\{2\sqrt{MN \ln(N)}, 2N \ln(N)\} \rceil$ known messages $X_1, \ldots, X_\tau$ and $p = MN - \tau$ targets; those MN messages are sampled uniformly without replacement from $\mathbb{Z}_M \times \mathbb{Z}_N$. Here we aim to recover all targets, namely $d = p$. On the left, we use the parameters of the FF1 standard. On the right, we use parameters of FF3.

Theorem 3. *Let $M, N \ge 4$ and let $p, q \ge 1$ be integer. Let $r \ge 4$ be an even integer such that $N^{(r-2)/2} \ge 2M$, and let d be an integer such that $1 \le d \le p$. Let $\mathsf{F} = \mathbf{Feistel}[r, M, N, \boxplus]$, and let $\lambda = \left(1 - \frac{1}{M-1}\right)^2 \left(1 - \frac{1}{MN}\right)$. Then for any*

$0 \leq \delta \leq 1$ *and any* $\theta \geq 0$, *and for any sampler* XS *in the class* $SC1_{p,q,d,\delta,\theta}$,

$$\mathbf{Adv}^{mr}_{F,XS,d}(LD) \geq 1 - \delta - d \cdot \exp\Big(\frac{-\lambda Mq}{12 \cdot N^{r-2}}\Big) - MNd \cdot \exp\Big(\frac{-\lambda Mq}{9 \cdot N^{r-2}}\Big) - 2^{-\theta} \ .$$

IDEAS OF THE ATTACK. Our attack is based on an observation by BHT that for any two messages X and X' of the same right half, if we encrypt them under the same tweak to obtain ciphertexts C and C' respectively, then $\mathbf{Left}(C) \boxminus \mathbf{Left}(C')$ is most likely to be $\mathbf{Left}(X) \boxminus \mathbf{Left}(X')$. This observation is formally stated in Lemma 4 below.

Lemma 4 ([1]). *Let* $F = \mathbf{Feistel}[r, M, N, \boxplus]$. *Fix distinct* $X, X' \in \mathbb{Z}_M \times \mathbb{Z}_N$ *of the same right segment, a tweak* $T \in F.\mathsf{Twk}$, *and an even integer* $t \in \{2, 4, \ldots, r\}$. *Pick* $K \leftarrow_\$ F.\mathsf{Keys}$. *Let* L_t *and* L'_t *be the the left segment of the round-t output of* X *and* X' *under* $F(K, T, \cdot)$, *respectively. Then*

(a) $\Pr[L_t \boxminus L'_t = L_0 \boxminus L'_0] \geq \frac{N}{MN-1} + \frac{1-1/(M-1)}{N^{(t-2)/2}}$.
(b) $\Pr[L_t \boxminus L'_t = Z] \leq \frac{N}{MN-1}$, *for any* $Z \in \mathbb{Z}_M \backslash \{L_0 \boxminus L'_0\}$.

The probabilities above are taken over a sampling $K \leftarrow_\$ F.\mathsf{Keys}$.

Consider a target Z_k such that $\mathbf{Right}(Z_k) \in \{\mathbf{Right}(X_1), \ldots, \mathbf{Right}(X_\tau)\}$.[6] Among the known messages $X_1, \ldots, X_\tau$, there will be some X_{j^*} of the same right segment as Z_k. Suppose that somehow we know j^*. Then obviously we can recover the right segment of Z_k. To recover the left segment of Z_k, we will use the above observation of BHT. For all ciphertexts C and C' of X_{j^*} and Z_k under the same tweak respectively, one can guess $\mathbf{Left}(Z_k)$ as $\mathbf{Left}(C') \boxminus \mathbf{Left}(C) \boxplus \mathbf{Left}(X_{j^*})$. However, compared to a random guessing, this is only slightly better; the improvement in the advantage is about $\frac{1-1/(M-1)}{N^{(r-2)/2}}$. To amplify the advantage, we consider ciphertexts C_i and C'_i of X_{j^*} and Z_k under many tweaks T_i, and output the majority value of those $\mathbf{Left}(C'_i) \boxminus \mathbf{Left}(C_i) \boxplus \mathbf{Left}(X_{j^*})$.

Since the algorithm above *assumes* that we are given the index j^*, we are left with the task of finding j^*. We first narrow down our search by considering a smallest possible subset S of $\{1, \ldots, \tau\}$ such that $\{\mathbf{Right}(X_j) \mid j \in S\} = \{\mathbf{Right}(X_1), \ldots, \mathbf{Right}(X_\tau)\}$. Such a set S will contain j^*, but we still do not know which is the right one, among $|S|$ possible values. Next, we try the strategy above for *every* $j \in S$ to see which gives us the best majority value. Specifically, for every $j \in S$, we consider ciphertexts $C_{i,j}$ and $C'_{i,k}$ of X_j and Z_k under tweaks T_i respectively. For every $i \in \{1, \ldots, q\}$, let $U_{i,j} \leftarrow \mathbf{Left}(C'_i) \boxminus \mathbf{Left}(C_i) \boxplus \mathbf{Left}(X_j)$. We then find the majority value of $U_{1,j}, \ldots, U_{q,j}$ together with the number V_j of its occurrences among those q values. Finally, in the election for j^*, each candidate j has V_j votes. The winner is the candidate of the most votes.

---

[6] We stress that the adversary does not need to know that $\mathbf{Right}(Z_k) \in \{\mathbf{Right}(X_1), \ldots, \mathbf{Right}(X_\tau)\}$; it will blindly use the same algorithm for all targets, but will happen to recover Z_k correctly.

The code in Fig. 3 implements the algorithm above as follows. For each $s \in \mathbb{Z}_N$ and each $j \in S$, we count the number $V_{j,s}$ of the occurrences of s in $U_{1,j}, \ldots, U_{q,j}$. We then find (j^*, s^*) such that $V_{j^*,s^*} = \max\{V_{j,s} \mid j \in S, s \in \mathbb{Z}_N\}$. The value s^* is the left segment of Z_k, and the right segment of X_{j^*} is also the right segment of Z_k.

To justify the way we pick j^* above, we need to understand the distribution of $V_{j,s}$, for every $j \in \mathbb{Z}_N \backslash \{j^*\}$ and $s \in \mathbb{Z}_N$. Each such message X_j will have a different right segment from Z_k. The following Lemma 5 tells us that if we encrypt X_j and Z_k under the same tweak to get ciphertexts C and C' respectively, then $\mathbf{Left}(C') \boxminus \mathbf{Left}(C)$ is uniformly distributed over $\mathbb{Z}_M$. The proof is given in the full version.

Lemma 5. *Let* $\mathsf{F} = \mathbf{Feistel}[r, M, N, \boxplus]$. *Fix distinct* $X, X' \in \mathbb{Z}_M \times \mathbb{Z}_N$ *of different right segments, a tweak* $T \in \mathsf{F.Twk}$, *and an even integer* $t \in \{2, 4, \ldots, r\}$. *Pick* $K \leftarrow_\$ \mathsf{F.Keys}$. *Let* L_t *and* L'_t *be the the left segment of the round-t output of* X *and* X' *under* $\mathsf{F}(K, T, \cdot)$, *respectively. Then for any* $Z \in \mathbb{Z}_M$, *we have* $\Pr[L_t \boxminus L'_t = Z] = \frac{1}{M}$, *where the probability is taken over a random sampling* $K \leftarrow_\$ \mathsf{F.Keys}$.

On the one hand, from Lemma 4, the expected value of V_{j^*,s^*} is at least $q(\mu + \Delta)$, where $\mu = \frac{N}{MN-1}$ and $\Delta = \frac{1-1/(M-1)}{N^{(t-2)/2}}$. On the other hand, by using Lemma 5, the expected value of each other $V_{j,s}$ is at most $q\mu$. We will show that it is unlikely for V_{j^*,s^*} to get below the threshold $q(\mu + \Delta/2)$, and any other $V_{j,s}$ is unlikely to get beyond that threshold.

Table 2. Comparison of our Left-half Differential attack, and BHT's attack on Feistel$[r, M, N, \boxplus]$ on parameters of FF1 and FF3. The first column shows the domain $\mathbb{Z}_M \times \mathbb{Z}_N$. The second and third columns show estimated values of q_t— the number of ciphertexts per recovered target—needed for our attack, for FF1 and FF3, respectively, to achieve advantage 0.9. (For our attack, q_t is also approximately q, the number of tweaks.) We use $\tau = \lceil \min\{2\sqrt{MN \ln(N)}, 2N \ln(N)\} \rceil$ known messages $X_1, \ldots, X_\tau$ and $p = MN - \tau$ targets; those MN messages are sampled uniformly without replacement from $\mathbb{Z}_M \times \mathbb{Z}_N$. Our attack aims to recover all targets, namely $d = p$. The fourth and fifth columns show estimated values of q_t needed for BHT's attack, for FF1 and FF3, respectively, to achieve advantage 0.9.

Domain	Our cost q_t (for FF1)	Our cost q_t (for FF3)	BHT's cost q_t (for FF1)	BHT's cost q_t (for FF3)
$\{0,1\}^8$	2^{35}	2^{27}	2^{38}	2^{30}
$\{0,1\}^9$	2^{44}	2^{26}	2^{44}	2^{38}
$\{0,\ldots,9\}^2$	2^{30}	2^{24}	2^{34}	2^{27}
$\{0,\ldots,9\}^3$	2^{56}	2^{21}	2^{56}	2^{49}

DISCUSSION. A concrete comparison of our attack and BHT's attack is shown in Table 2. When the domain length is odd, FF1 and FF3 have different ways to interpret what are M and N. For example, for domain $\{0, \ldots, 9\}^3$ (namely 3-digit numbers), FF1 uses $M = 10$ and $N = 100$, whereas FF3 uses $M = 100$ and $N = 10$. An interesting observation is that in those odd domains, our attack does not improve BHT's attack for FF1, but significantly improves BHT's attack for FF3. For example, for domain $\{0, \ldots, 9\}^3$ above, both attacks use $q_t = 2^{56}$ for FF1, but for FF3, our attack only needs $q_t = 2^{21}$, whereas BHT's attack requires $q_t = 2^{49}$. Thus our attack (i) shows that FF3's way of partitioning odd domains is inferior to that of FF1, and (ii) underscores that for tiny domains, the round counts that FF1 and FF3 use are not enough, as BHT's attack already pointed out. In other words, our attack surfaces weaknesses which might have eliminated these algorithms from consideration during standardization,[7] and they significantly reduce confidence in these algorithms, which are widely deployed.

The recent FF3 attack by Durak and Vaudenay (DV) [8] can recover the entire codebook for quite bigger domains, such as PINs ($M = N = 100$). However, this attack is adaptive, meaning that the adversary must choose the next known message based on prior ciphertexts, which is very hard to mount in practice. Moreover, DV's attack can be easily fixed without performance penalty by restricting the tweak space. In contrast, to thwart our attack or BHT's attack, for tiny domains one has to add a few more rounds, which is widely perceived as a drawback for performance-hungry applications.

EXPERIMENTS. As a proof of concept, we implement our Left-half Differential attack, and evaluate its message-recovery rate against FF3. Each experiment was run using 64 threads in a server of Intel(R) Xeon(R) CPU E5-2699 v3 2.30 GHz CPU and 256 GB RAM. Our implementation, written in Go, uses FF3 source code from Capital One.[8] We evaluate our attack on three domains: $\{0, 1\}^7$ (namely $M = 16$ and $N = 8$), $\{0, \ldots, 9\}^2$ (namely $M = N = 10$), and $\{0, \ldots, 9\}^3$ (namely $M = 100$ and $N = 10$); each on several values of q, the number of tweaks. For each domain $\mathbb{Z}_M \times \mathbb{Z}_N$ and each choice of q, we fix $\tau = \lceil \min\{2\sqrt{MN \ln(N)}, 2N \ln(N)\} \rceil$ known messages whose right segments cover $\mathbb{Z}_N$, and run the attack for 100 trials. In particular, we use $\tau = 33$ for $\{0, 1\}^7$, $\tau = 31$ for $\{0, \ldots, 9\}^2$, and $\tau = 96$ for $\{0, \ldots, 9\}^3$. While the known messages are fixed for all 100 trials, we use $p = MN - \tau$ target messages, and randomly shuffle the targets for each trial. Here we aim to recover all targets, namely $d = p$.

The results of our experiments, given in Table 3, are consistent with (and even slightly better than) Theorem 3. For example, for domain $\{0, \ldots, 9\}^2$, theoretically, one would need to use about $q = 2^{24}$ tweaks to recover all targets with probability nearly 1, and our experiments confirm that using $q = 2^{24}$ indeed gives 100% recovery rate. However, even for $q = 2^{23}$, in every trial we can recover all targets, and the average running time to recover target messages for each trial is about 5.92 min. If one instead uses $q = 2^{22}$, then the recovery rate drops to 86%, meaning that in 86 out of 100 trials, we can recover all targets.

---

[7] Recall that FF2 was eliminated due to a theoretical attack using 2^{64} ciphertexts.

[8] Capital One's code is available at https://github.com/capitalone/fpe.

Table 3. Empirical results of our Left-half Differential attack against FF3.
For each domain (shown in the first column), we run experiments with two values of q
(the number of tweaks) as indicated in the second and fifth columns. The recovery
rates corresponding to these two values of q are given in the third and sixth columns,
respectively. Finally, the average running time (in minutes) of each experiment is given
in the fourth and seventh columns.

Domain	Number of tweaks, q	Recovery rate	Time (min)	Number of tweaks, q	Recovery rate	Time (min)
$\{0,1\}^7$	2^{20}	100%	0.9	2^{19}	66%	0.46
$\{0,\ldots,9\}^2$	2^{23}	100%	5.92	2^{22}	86%	3.06
$\{0,\ldots,9\}^3$	2^{20}	100%	8.72	2^{19}	66%	5.3

Our experiments above empirically confirm the correctness of our attack for
tiny domains. Below, we will give a formal proof to rigorously justify our attack
for all domains.

PROOF OF THEOREM 3. First we show that $\mathbf{Adv}^{\mathrm{mg}}_{\mathsf{XS}} \leq 2^{-\theta}$. Consider an arbitrary simulator $\mathcal{S}$. To win the game, $\mathcal{S}$ must find the first target Z_1. The simulator
is only given the tweaks and the auxiliary information $(X_1, \ldots, X_\tau, p, q)$, and has
to guess correctly at least d components of $(Z_1, \ldots, Z_p)$. From the definition of θ,
the chance that the simulator's guess is correct is at most $2^{-\theta}$. Next, we show
that

$$\Pr[\mathbf{G}^{\mathrm{mr}}_{\mathsf{F,XS}}(\mathsf{LD})] \geq 1 - \delta - d \cdot \exp\left(\frac{-\lambda M q}{12 \cdot N^{r-2}}\right) - MNd \cdot \exp\left(\frac{-\lambda M q}{9 \cdot N^{r-2}}\right) .$$

Let $S \subseteq \{1, \ldots, \tau\}$ be a set such that $\{\mathbf{Right}(X_j) \mid j \in S\} = \{\mathbf{Right}(X_1), \ldots, \mathbf{Right}(X_\tau)\}$. With probability at least $1 - \delta$, at least d targets will have their right halves in $\{\mathbf{Right}(X_j) \mid j \in S\}$. Fix a target Z_k such
that $\mathbf{Right}(Z_k) \in \{\mathbf{Right}(X_j) \mid j \in S\}$. By union bound, it suffices to show
that the chance the adversary fails to recover Z_k is at most

$$\exp\left(\frac{-\lambda M q}{12 \cdot N^{r-2}}\right) + MN \cdot \exp\left(\frac{-\lambda M q}{9 \cdot N^{r-2}}\right) .$$

Recall that for every $j \in S$ and every $s \in \mathbb{Z}_N$, we keep track of the number $V_{j,s}$ of
the occurrences of s among the values $U_{1,j}, \ldots, U_{q,j}$, where $U_{i,j} \leftarrow \mathbf{Left}(C'_{i,k}) \boxminus$
$\mathbf{Left}(C_{i,j}) \boxplus \mathbf{Left}(X_j)$. Let j^* be the element of S such that $\mathbf{Right}(X_{j^*}) = \mathbf{Right}(Z_k)$, and let $s^* \leftarrow \mathbf{Left}(Z_k)$. The adversary can recover Z_k if V_{j^*,s^*} is the
maximum of $\{V_{j,s} \mid j \in S, s \in \mathbb{Z}_N\}$. Let $\mu \leftarrow \frac{N}{MN-1}$ and $\Delta \leftarrow \frac{1-1/(M-1)}{N^{(r-2)/2}}$. We
will give (i) an upper bound for the probability that $V_{j,s}$, with $(j, s) \neq (j^*, s^*)$, is
bigger than the threshold $q(\mu + \Delta/2)$, and (ii) an upper bound for the probability
that V_{j^*,s^*} is smaller than that threshold. Both (i) and (ii) are handled using
Chernoff bounds.

Proceeding into details, fix $(j, s) \neq (j^*, s^*)$. For each $i \leq q$, let Y_i be the Bernoulli
random variable such that $Y_i = 1$ if and only if $U_{i,j} = s$. The random variables

$Y_1, \ldots, Y_q$ are independent and identically distributed (as they are produced from a Feistel network of ideal round functions, under distinct tweaks), and $V_{j,s} = Y_1 + \cdots + Y_q$. Let $\nu = \Pr[Y_1 = 1] \le \mu$ and $\epsilon = \frac{\Delta}{2\nu} \ge \frac{\Delta}{2\mu}$. Note that $\Delta/\mu \le M/N^{(r-2)/2} \le 1/2$, and $\Delta^2/\mu = \lambda M/N^{r-2}$. Then

$$\frac{\epsilon^2 \nu}{2 + \epsilon} = \frac{\Delta}{4/\epsilon + 2} \ge \frac{\Delta}{8\mu/\Delta + 2} = \frac{\Delta^2/\mu}{8 + 2\Delta/\mu} \ge \frac{\lambda M}{9 \cdot N^{r-2}} .$$

Since $(1 + \epsilon)\nu = \nu + \Delta/2 \le \mu + \Delta/2$, by Chernoff bound,

$$\Pr[V_{j,s} \ge q(\mu + \Delta/2)] \le \Pr[Y_1 + \cdots + Y_q \ge q(1 + \epsilon)\nu]$$
$$\le \exp\left(\frac{-\epsilon^2 \nu q}{2 + \epsilon}\right) \le \exp\left(\frac{-\lambda M q}{9 \cdot N^{r-2}}\right) . \tag{1}$$

Next, for each $i \le q$, let Y_i^* be the Bernoulli random variable such that $Y_i^* = 1$ if and only if $U_{i,j^*} = s^*$. Again, the random variables $Y_1^*, \ldots, Y_q^*$ are independent and identically distributed, and $V_{j^*,s^*} = Y_1^* + \cdots + Y_q^*$. Let $\nu^* = \Pr[Y_1^* = 1] \ge \Delta + \mu$ and let $\epsilon^* = \frac{\Delta}{2(\mu + \Delta)}$. Then $0 < \epsilon^* < 1$. Moreover,

$$(\epsilon^*)^2 \nu^* \ge \frac{\Delta^2 q}{4(\mu + \Delta)} = \frac{\Delta^2/\mu}{4(1 + \Delta/\mu)} \ge \frac{\Delta^2/\mu}{6} = \frac{\lambda M}{6 \cdot N^{r-2}} .$$

Since $(1 - \epsilon^*)\nu^* \ge \left(1 - \frac{\Delta}{2(\mu+\Delta)}\right)(\Delta + \mu) = \mu + \Delta/2$, by Chernoff bound,

$$\Pr[V_{j^*,s^*} \le q(\mu + \Delta/2)] \le \Pr[Y_1^* + \cdots + Y_q^* \le q(1 - \epsilon^*)\nu^*]$$
$$\le \exp\left(\frac{-(\epsilon^*)^2 \nu^* q}{2}\right) \le \exp\left(\frac{-\lambda M q}{12 \cdot N^{r-2}}\right) . \tag{2}$$

From Eqs. (1) and (2), the chance that the adversary LD fails to recover Z_k is at most

$$\Pr[V_{j^*,s^*} \le q(\mu + \Delta/2)] + \sum_{(j,s)\ne(j^*,s^*)} \Pr[V_{j,s} \ge q(\mu + \Delta/2)]$$
$$\le \exp\left(\frac{-\lambda M q}{12 \cdot N^{r-2}}\right) + MN \cdot \exp\left(\frac{-\lambda M q}{9 \cdot N^{r-2}}\right) .$$

5   Attacking FNR

In this section, we attack the Flexible Naor-Reingold (FNR) scheme proposed by Cisco [7], which is defined only for the boolean case.[9] It is based on Naor-Reingold generalization of Feistel networks [14], using a pairwise independent permutation and a boolean Feistel-based FPE scheme.

---

[9] While the FNR paper [7] mentions that the scheme can be used to encrypt credit-card numbers, it is unclear how this is possible, as the specific instantiation there only works for binary data.

FNR CONSTRUCTION. Recall that a family $\mathcal{P}$ of permutations on $\{0,1\}^\ell$ is *pairwise independent* if for any $X, X', Y, Y' \in \{0,1\}^\ell$ such that $X \neq X'$ and $Y \neq Y'$,

$$\Pr_{\pi \twoheadleftarrow \mathcal{P}}[(\pi(X) = Y) \wedge (\pi(X') = Y')] = \frac{1}{2^\ell(2^\ell - 1)} .$$

In FNR, the family $\mathcal{P}$ is instantiated as $\mathcal{B}_\ell$, the set of all pairs (B_0, B_1) such that B_0 is an invertible binary matrix of size $\ell \times \ell$, and B_1 is a binary vector of length ℓ. For each $\pi \in \mathcal{P}$, $\pi(X) = (B_0 \cdot X) \oplus B_1$, where the input X is viewed as a binary vector of length ℓ, (B_0, B_1) is the matrix representation of π, and the multiplication $B_0 \cdot X$ is in $\mathsf{GF}(2)$.

In an FNR scheme $\mathsf{F} = \mathbf{FNR}[r, m, n, \mathrm{PL}]$, the domain is $\{0,1\}^m \times \{0,1\}^n$. The parameter $\mathrm{PL} = (\mathcal{T}, \mathcal{K}, F_1, \ldots, F_r)$ specifies the tweak space $\mathcal{T}$ and a Feistel-based FPE scheme $\mathsf{F} = \mathbf{Feistel}[r, 2^m, 2^n, \oplus, \mathrm{PL}]$ as defined in Sect. 4. The key space is $\mathcal{B}_{m+n} \times \mathcal{K}$. On key $K = (B_0, B_1, \widetilde{K})$ and tweak T, to encrypt a message X, one first interprets (B_0, B_1) as a permutation $\pi : \{0,1\}^{m+n} \to \{0,1\}^{m+n}$, computes $U \leftarrow \pi(X)$ and $V \leftarrow \widetilde{\mathsf{F}}.\mathsf{E}(\widetilde{K}, T, U)$, and returns $\pi^{-1}(V)$. Decryption is defined likewise. The code of the encryption and decryption schemes of $\mathbf{FNR}[r, m, n, \mathrm{PL}]$ is given in Fig. 5. If the underlying Feistel-based FPE scheme is $\mathbf{Feistel}[r, 2^m, 2^n, \oplus]$ (meaning ideal round functions), then we write $\mathbf{FNR}[r, m, n]$ for the corresponding FNR scheme. For input length ℓ, the FNR specification only uses the $m = \lceil \ell/2 \rceil$ and $n = \ell - m$, meaning that the Feistel network is a (near)-balanced one. The suggested instantiation in [7] uses $r = 7$.

The FNR spec [7] specifies the round functions using AES. Again, using the standard assumption that AES is a good PRF, one can focus on attacking FNR schemes of ideal round functions, with small differences in the advantage.

THE ATTACK. We now attack the scheme $\mathbf{FNR}[r, m, n]$ scheme for an odd integer $r \geq 7$, with $|m - n| \leq 1$. This is exactly the setting specified by the FNR spec. While FNR also uses a Feistel network, at the first glance, it is unclear how to use the ideas in Sect. 4, because the pairwise independent permutation in FNR will hide the pairwise bias described in Lemma 4. However, we will exploit the fact that the FNR scheme uses *the same* pairwise independent permutation across different tweaks.

Under our attack, there are $\tau = \left\lceil \min\{2 \cdot 2^{(m+n)/2}\sqrt{\ln(2)n}, 2^{n+1}\ln(2)n\} \right\rceil$ known messages $X_1, \ldots, X_\tau$ sampled uniformly without replacement from $\{0,1\}^{m+n}$, and there are p targets $Z_1, \ldots, Z_p$. The adversary is given the encryption of those $\tau + p$ messages under q tweaks $T_1, \ldots, T_q$, for an appropriately large q, and the auxiliary information is $(X_1, \ldots, X_\tau, p, q)$. From the distinctness requirement, these $\tau + p$ messages must be distinct. We have no other restriction on the number p of targets, so potentially p can be as large as $2^{m+n} - \tau$. Our attack will recover all of $Z_1, \ldots, Z_p$, meaning $d = p$. The number of examples Q is $(\tau + p)q$, and the number of examples per target q_t is $(\tau/p + 1)q$.

We formalize the attack via the message-recovery framework, by specifying a class $\mathsf{SC2}_{p,q,\theta}$ of samplers, and then giving a lower bound on the mr-advantage of the attack for any sampler in this class. First, let $\mathsf{DC2}_{p,q,\theta}$ be the

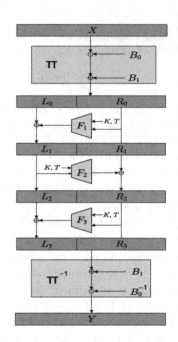

```
F.E(K, T, X)

(B_0, B_1, \widetilde{K}) \leftarrow K
(L, R) \leftarrow U \leftarrow (B_0 \cdot X) \oplus B_1
For i = 1 to r do
    If (i mod 2 = 1) then
        L \leftarrow L \oplus F_i(\widetilde{K}, T, R)
    Else R \leftarrow R \oplus F_i(\widetilde{K}, T, L)
V \leftarrow (L, R); Y \leftarrow B_0^{-1} \cdot (V \oplus B_1)
Return Y
```

```
F.D(K, T, Y)

(B_0, B_1, \widetilde{K}) \leftarrow K
(L, R) \leftarrow V \leftarrow (B_0 \cdot Y) \oplus B_1
For i = r to 1 do
    If (i mod 2 = 1) then
        L \leftarrow L \oplus F_i(\widetilde{K}, T, R)
    Else R \leftarrow R \oplus F_i(\widetilde{K}, T, L)
U \leftarrow (L, R); X \leftarrow B_0^{-1} \cdot (U \oplus B_1)
Return X
```

Fig. 5. Left: The code for the encryption and decryption algorithms of $F = $ **FNR**$[r, m, n, \mathrm{PL}]$, where $\mathrm{PL} = (\mathcal{T}, \mathcal{K}, F_1, \ldots, F_r)$. In implementation, for $(L, R) \leftarrow U$, typically L is the leftmost m-bit substring of U, and R is the rightmost n-bit substring of U. However, in Cisco implementation, L and R are the strings obtained via the odd and even bits of U, respectively. **Right:** An illustration of encryption with $r = 3$ rounds, where $\odot$ denotes the matrix multiplication.

class of all algorithms D that outputs q distinct tweaks $T_1, \ldots, T_q \in \{0, 1\}^*$, and distinct $X_1, \ldots, X_\tau, Z_1, \ldots, Z_p \in \{0, 1\}^{m+n}$ such that (1) $X_1, \ldots, X_\tau$ are sampled uniformly without replacement from $\{0, 1\}^{m+n}$, and (2) given $X_1, \ldots, X_\tau, T_1, \ldots, T_q$, for any fixed $Z_1^*, \ldots, Z_p^*$, the conditional probability that $Z_1 = Z_1^*, \ldots, Z_p = Z_p^*$ is at most $2^{-\theta}$. To any such D, we associate the sampler

> Sampler $\mathsf{XS}[\mathsf{D}]$
> _____
> $(T_1, \ldots, T_q, X_1, \ldots, X_\tau, Z_1, \ldots, Z_p) \leftarrow_\$ \mathsf{D}$
> $a \leftarrow (X_1, \ldots, X_\tau, p, q)$
> Return $\big(\{(T_i, X_j), (T_i, Z_k) \mid i \leq q, j \leq \tau, k \leq p\}, Z_1, \ldots, Z_p, a\big)$

The sampler above return the pairs (T_i, X_j) and (T_i, Z_k) for every $i \leq q, j \leq \tau$, and $k \leq p$, where the targets are $Z_1, \ldots, Z_p$. Let $\mathsf{SC2}_{p,q,\theta} = \{\mathsf{XS}[\mathsf{D}] \mid \mathsf{D} \in \mathsf{DC2}_{p,q,\theta}\}$. The Full-message Differential (FD) attack, given in Fig. 6, can recover all targets $Z_1, \ldots, Z_p$ in $O(pq\tau)$ time. Theorem 6 below gives a lower bound on the mr-advantage of LD; the proof is postponed further below. The bound is illustrated in Fig. 7.

Adversary $\mathsf{FD}(\{(T_i, C_{i,j}), (T_i, C'_{i,k}) \mid i \leq q, j \leq \tau, k \leq p\}, a)$

$(X_1, \ldots, X_\tau, p, q) \leftarrow a; \quad \mu \leftarrow 1/2^{m+n}; \quad \Delta \leftarrow \frac{1}{2 \cdot 2^{2(m+n)}}$

For $k \leftarrow 1$ to p do $/\!/$ Recover target Z_k

    For $j \in \{1, \ldots, \tau\}, s \in \{0,1\}^{m+n}$ do $V_{j,s} \leftarrow 0$

    For $i \leftarrow 1$ to q, $j \leftarrow 1$ to τ do $s \leftarrow C_{i,j} \oplus C'_{i,k}; \quad V_{j,s} \leftarrow V_{j,s} + 1$

    Find smallest j^* s.t. there is only one $s \in \{0,1\}^{m+n}$ with $V_{j^*,s} \leq q(\mu + \Delta/2)$.

    Let $V_{j^*,s^*} = \max\{V_{j^*,s} \mid s \in \{0,1\}^{m+n}\}; \quad Z_k \leftarrow s^* \oplus X_{j^*}$

Return $(Z_1, \ldots, Z_p)$

Fig. 6. The Full-message Differential attack.

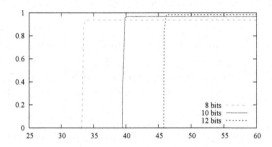

Fig. 7. The mr advantage of the Full-message Differential attack on FNR$[r, n, n]$ for $r = 7$ and $n = 4, 5, 6$. This is the balanced setting $m = n$. The x-axis shows the log, base 2, of the number q of tweaks (which is also roughly q_t, the number of ciphertexts per recovered target), and the y-axis shows $\mathbf{Adv}^{\mathrm{mr}}_{\mathbf{FNR}[r,n,n],\mathsf{XS}}(\mathsf{FD})$, for XS that outputs $\tau = \left\lceil 2^{n+1}\sqrt{\ln(2)n} \right\rceil$ known messages and $p = 2^{2n} - \tau$ targets; those 2^{2n} messages are sampled uniformly without replacement from $\{0,1\}^{2n}$.

Theorem 6. *Let $m, n \geq 3$ and $q \geq 1$ be integers such that $|m - n| \leq 1$, and let $r \geq 7$ be an odd integer. Let $\mathsf{F} = \mathbf{FNR}[r, m, n]$. Let $\lambda = \left(1 - \frac{1}{2^n - 1}\right)^2 \left(1 - \frac{1}{2^{m+n}}\right)$. Then for any $\theta \geq 0$ and for any sampler XS in the class $\mathsf{SC2}_{p,q,\theta}$,*

$$\mathbf{Adv}^{\mathrm{mr}}_{\mathsf{F},\mathsf{XS}}(\mathsf{FD}) \geq 1 - \frac{1}{2^n} - 2^{m+n}p \cdot \exp\left(\frac{-q}{32 \cdot 2^{3(m+n)}}\right) - 2^{m+n}p \cdot \exp\left(\frac{-q}{48 \cdot 2^{3(m+n)}}\right)$$
$$- 2^{m+n}p \cdot \exp\left(\frac{-\lambda q}{9 \cdot 2^{n+(r-2)m}}\right) - p \cdot \exp\left(\frac{-\lambda q}{12 \cdot 2^{n+(r-2)m}}\right) - 2^{-\theta}.$$

IDEAS OF THE ATTACK. For a random variable $W \in \{0,1\}^{m+n}$, we say that it has a *singular* distribution if there is exactly one string $Z \in \{0,1\}^{m+n}$ such that $\Pr[W = Z] \leq 1/2^{m+n}$; otherwise the distribution is *non-singular*. Let $\pi = (B_0, B_1)$ be the pairwise independent permutation in the key of the FNR scheme. Suppose that one encrypts distinct messages X and X' on a tweak T. Then the strings $Y \leftarrow \pi(X)$ and $Y' \leftarrow \pi(X')$ become inputs to a near-balanced, boolean Feistel network, and let U and U' be the corresponding outputs of the Feistel network. Our attack is based on the following observation that is formalized in Lemma 7 below; see the full version for the proof. Specifically, if Y and Y' have the different right segments then the distribution of $U \oplus U'$ is non-singular; in fact, there are 2^m values $Z \in \{0,1\}^{m+n}$ such that $\Pr[U \oplus U' = Z] \le 1/2^{m+n}$. Let C and C' be the ciphertexts of Y and Y' under the FNR scheme, respectively. Then $C \leftarrow \pi^{-1}(U)$ and $C' \leftarrow \pi^{-1}(U')$, and $C \oplus C' = B_0^{-1} \cdot (U \oplus U')$. Thus the distribution of $C \oplus C'$ is also non-singular.

In contrast, suppose that Y and Y' have the same right segments. Then $\Pr[U \oplus U' = Z]$ is significantly larger than $1/2^{m+n}$ for every $Z \in \{0,1\}^{m+n} \setminus \{0^{m+n}\}$, and thus the distribution of $U \oplus U'$, and also that of $C \oplus C'$, are singular in this case. Moreover, the distribution of $U \oplus U'$ peaks at $Y \oplus Y' = B_0 \cdot (X \oplus X')$, and consequently, the distribution of $C \oplus C'$ peaks at $B_0^{-1} \cdot B_0 \cdot (X \oplus X') = X \oplus X'$.

Lemma 7. *Let $r \ge 7$ be an odd integer and let $m, n \ge 2$ be integers such that $|m - n| \le 1$. Let $\mathsf{F} = \mathbf{Feistel}[r, 2^m, 2^n, \oplus]$. Fix distinct $X, X' \in \{0,1\}^{m+n}$, a tweak $T \in \mathsf{F.Twk}$. Pick $K \leftarrow_\$ \mathsf{F.Keys}$. For each integer t, let X_t and X_t' be the the round-t output of X and X' under $\mathsf{F}(K, T, \cdot)$, respectively. Then for any odd integer $t \ge 7$,*

(a) *If X and X' have different right segments then for any non-zero $Z \in \{0,1\}^{m+n}$,*

$$\Pr[X_t \oplus X_t' = Z] = \frac{1}{2^{m+n}} \text{ if } \mathbf{Right}(Z) = 0^n \;,$$

$$\Pr[X_t \oplus X_t' = Z] \ge \frac{1}{2^{m+n}} + \frac{1}{2 \cdot 2^{2(m+n)}} \text{ otherwise} \;.$$

(b) *If X and X' have the same right segments then for any non-zero $Z \in \{0,1\}^{m+n}$,*

$$\Pr[X_t \oplus X_t' = Z] \ge \frac{1}{2^{m+n}} + \frac{1}{2 \cdot 2^{2(m+n)}} \;.$$

Moreover,

$$\Pr[X_t \oplus X_t' = Z] \le \frac{1}{2^{m+n} - 1} + \frac{1}{(2^m - 1)2^{(t-1)(m+n)/2}} \text{ if } Z \ne X \oplus X',$$

$$\Pr[X_t \oplus X_t' = Z] \ge \frac{1}{2^{m+n} - 1} + \frac{1 - 1/(2^m - 1)}{2^n \cdot 2^{(t-1)m/2}} \text{ otherwise} \;.$$

The probabilities above are taken over a sampling $K \leftarrow_\$ \mathsf{F.Keys}$.

Based on the observation above, we can attack the FNR scheme as follows. The adversary receives the encryptions of known messages $X_1, \dots, X_\tau$ and targets $Z_1, \dots, Z_p$, under tweaks $T_1, \dots, T_q$. Fix $k \le p$; we now explain how to recover Z_k. Let $C_{i,j}$ and $C_{i,k}'$ be the ciphertexts of X_j and Z_k under tweak T_i, respectively. To recover a target Z_k, for each $j \le \tau$, we plot the frequency histogram for the values $C_{i,j} \oplus C_{i,k}'$, for every $i = 1, \dots, q$, and call it the histogram of X_j. From the observation above, if $\pi(X_j)$ and $\pi(Z_k)$ have different right segments and q is big enough then the histogram for X_j is *non-singular*, meaning that it has multiple short columns, relative to the height $q/2^{m+n}$. In contrast, if $\pi(X_j)$ and $\pi(Z_k)$ have the same right segments then the histogram for X_j is singular, containing exactly one short column (of height 0). Moreover, in this case, the tallest column corresponds to the value $X_j \oplus Z_k$.

Since $X_1, \ldots, X_\tau$ are sampled uniformly without replacement from $\{0,1\}^{m+n}$ and π is a permutation on $\{0,1\}^{m+n}$, the strings $Y_1 \leftarrow \pi(X_1), \ldots, Y_\tau \leftarrow \pi(X_\tau)$ are also sampled uniformly without replacement from $\{0,1\}^{m+n}$. From the Biased Coupon Collector's problem (Lemma 2), $\{\mathbf{Right}(Y_1), \ldots, \mathbf{Right}(Y_\tau)\} = \{0,1\}^n$ with probability at least $1 - 1/2^n$. Hence there must be some j^* such that Y_{j^*} and $\pi(Z_k)$ have the same right segment. We can find such a j^* by checking if its histogram is singular. Let s^* be the value for the tallest column in the histogram of X_{j^*}. We then can recover Z_k by way of $Z_k \leftarrow s^* \oplus X_{j^*}$.

PROOF OF THEOREM 6. First we show that $\mathbf{Adv}^{\mathsf{mg}}_{\mathsf{XS}} \leq 2^{-\theta}$. Consider an arbitrary simulator $\mathcal{S}$. To win the game, $\mathcal{S}$ must guess all targets, given the tweaks and the auxiliary information. From the definition of θ, the chance that the simulator's guess is correct is at most $2^{-\theta}$. Next, we show that

$$\Pr[\mathbf{G}^{\mathsf{mr}}_{\mathsf{F,XS}}(\mathsf{FD})]$$

$$\geq 1 - 1/2^n - 2^{m+n}p \cdot \exp\left(\frac{-q}{32 \cdot 2^{3(m+n)}}\right) - 2^{m+n}p \cdot \exp\left(\frac{-q}{48 \cdot 2^{3(m+n)}}\right)$$

$$-2^{m+n}p \cdot \exp\left(\frac{-\lambda q}{9 \cdot 2^{n+(r-2)m}}\right) - p \cdot \exp\left(\frac{-\lambda q}{12 \cdot 2^{n+(r-2)m}}\right) .$$

Let $Y \leftarrow \pi(X_1), \ldots, Y_\tau \leftarrow \pi(X_\tau)$. Since $X_1, \ldots, X_\tau$ are sampled uniformly without replacement from $\{0,1\}^{m+n}$ and π is a permutation on $\{0,1\}^{m+n}$, the strings $Y_1, \ldots, Y_\tau$ are also sampled uniformly without replacement from $\{0,1\}^{m+n}$. From the Biased Coupon Collector's problem, $\{\mathbf{Right}(Y_1), \ldots, \mathbf{Right}(Y_\tau)\} = \{0,1\}^n$, with probability at least $1 - 1/2^n$. By union bound, it suffices to prove that for any $k \leq p$, the FD attack fails to recover the target Z_k with probability at most

$$2^{m+n} \cdot \exp\left(\frac{-q}{32 \cdot 2^{3(m+n)}}\right) + 2^{m+n} \cdot \exp\left(\frac{-q}{48 \cdot 2^{3(m+n)}}\right)$$

$$+2^{m+n} \cdot \exp\left(\frac{-\lambda q}{9 \cdot 2^{n+(r-2)m}}\right) + \exp\left(\frac{-\lambda q}{12 \cdot 2^{n+(r-2)m}}\right) .$$

Let $C_{i,j}$ and $C'_{i,k}$ be the ciphertexts for known messages X_i and target Z_k under tweak T_i, respectively. Let $B_{j,i,s}$ be the Bernoulli random variable such that $B_{i,j,s} = 1$ if and only if $C_{i,j} \oplus C'_{i,k} = s$. Now in the histogram for X_j, the height of the column for each value s is $V_{j,s} = B_{1,j,s} + \cdots + B_{q,j,s}$. Note that for each fixed (j, s), the random variables $B_{1,j,s}, \ldots, B_{q,j,s}$ are independent and identically distributed. Let $\mu \leftarrow 1/2^{m+n}$ and $\Delta \leftarrow \frac{1}{2 \cdot 2^{2(m+n)}}$. From Chernoff bound,

(i) For every (j, s), if $\Pr[B_{1,j,s} = 1] \leq \mu$ then $V_{j,s} \geq q(\mu + \Delta/2)$ with probability at most $\exp\left(\frac{-q}{32 \cdot 2^{3(m+n)}}\right)$. That is, a supposedly short column is likely to remain short.

(ii) For every (j, s), if $\Pr[B_{1,j,s} = 1] \geq \mu + \Delta$, we have $V_{j,s} \leq q(\mu + \Delta/2)$ with probability at most $\exp\left(\frac{-q}{48 \cdot 2^{3(m+n)}}\right)$. That is, a supposedly tall column will be likely to remain tall.

Now, consider j such that $\pi(X_j)$ and $\pi(Z_k)$ have different right segments. Since $X_j \neq Z_k$ and FNR is a permutation, the histogram for X_j will surely have one column of height 0, namely the column corresponding to $\pi(0^{m+n})$. To correctly identify the histogram as non-singular, we need one more supposedly short column of this histogram to remain short. From the claim (i) above and from Lemma 7, this happens for every such j with probability at least

$$1 - \tau \cdot \exp\left(\frac{-q}{32 \cdot 2^{3(m+n)}}\right) \geq 1 - 2^{m+n} \cdot \exp\left(\frac{-q}{32 \cdot 2^{3(m+n)}}\right) .$$

Next, consider the smallest j^* such that $\pi(X_{j^*})$ and $\pi(Z_k)$ have the same right segment. Since $X_{j^*} \neq Z_k$ and FNR is a permutation, the histogram for X_{j^*} will surely have one column of height 0, namely the column corresponding to $\pi(0^{m+n})$. To correctly identify the histogram as singular, we need every supposedly tall column of this histogram to remain tall. From the claim (ii) above and from Lemma 7, this happens with probability at least

$$1 - 2^{m+n} \cdot \exp\left(\frac{-q}{48 \cdot 2^{3(m+n)}}\right) .$$

By a union bound, we can realize j^* via checking the singularity of histograms with probability at least

$$1 - 2^{m+n} \cdot \exp\left(\frac{-q}{32 \cdot 2^{3(m+n)}}\right) - 2^{m+n} \cdot \exp\left(\frac{-q}{48 \cdot 2^{3(m+n)}}\right) . \tag{3}$$

Now, once we find j^*, we need to ensure that the peak column indeed corresponds to the value $X_{j^*} \oplus Z_k$. Let $\mu^* = \frac{1}{2^{m+n}-1} + \frac{1/(2^m-1)}{2^{(r-1)(m+n)/2}}$ and $\Delta^* = \frac{1-1/(2^m-2)}{2^n \cdot 2^{(r-1)m/2}}$. From Chernoff bound and Lemma 7,

(iii) For every $s \neq Z_k \oplus X_{j^*}$, $\Pr[B_{1,j^*,s} = 1] \leq \mu^*$, and thus the probability that $V_{j^*,s} \geq q(\mu^* + \Delta^*/2)$ is at most $\exp\left(\frac{-\lambda q}{9 \cdot 2^{n+(r-2)m}}\right)$. That is, it is unlikely that the column corresponding to s is the peak, as it remains lower than $q(\mu^* + \Delta^*/2)$.

(iv) For $s^* = Z_k \oplus X_{j^*}$, $\Pr[B_{1,j^*,s^*} = 1] \geq \mu^* + \Delta^*$, and thus $V_{j^*,s^*} \leq q(\mu^* + \Delta^*/2)$ with probability at most $\exp\left(\frac{-\lambda q}{12 \cdot 2^{n+(r-2)m}}\right)$. That is, the column corresponding to $Z_k \oplus X_{j^*}$ is likely to be the peak, as it remains higher than $q(\mu^* + \Delta^*/2)$.

From (iii) and (iv), the chance that in the histogram of X_{j^*}, the peak column indeed corresponds to $X_{j^*} \oplus Z_k$ is at least

$$1 - 2^{m+n} \cdot \exp\left(\frac{-\lambda q}{9 \cdot 2^{n+(r-2)m}}\right) - \exp\left(\frac{-\lambda q}{12 \cdot 2^{n+(r-2)m}}\right) . \tag{4}$$

From Eqs. (3) and (4), the chance that the attack can recover the target Z_k is at least

$$1 - 2^{m+n} \cdot \exp\left(\frac{-q}{32 \cdot 2^{3(m+n)}}\right) - 2^{m+n} \cdot \exp\left(\frac{-q}{48 \cdot 2^{3(m+n)}}\right)$$
$$-2^{m+n} \cdot \exp\left(\frac{-\lambda q}{9 \cdot 2^{n+(r-2)m}}\right) - \exp\left(\frac{-\lambda q}{12 \cdot 2^{n+(r-2)m}}\right) \ .$$

This completes the proof.

6   Attacking DTP

In this section, we will attack the DTP scheme, by Protegrity Corp. [12], which resembles the seminal FPE construction by Brightwell and Smith [6].

DTP CONSTRUCTION. The DTP scheme has several variants, but here we only consider the simplest and also the most efficient one. Under this version, it requires that each time we encrypt a message, we need to pick a fresh random tweak. Thus in this setting, tweaks serve the same role as initialization vectors in traditional modes of encryption like CBC.

The scheme $\mathsf{F} = \mathbf{DTP}[r, d, D, m, n, \mathrm{PL}]$ has message space $\mathbb{Z}_d^m$ and tweak space $\mathbb{Z}_D^n$, with $d \leq D$ and $n \geq r$. The parameter $\mathrm{PL} = (\mathcal{K}, F)$ specifies the key space $\mathcal{K}$ and the round function $F : \mathcal{K} \times \mathbb{Z}_D^n \rightarrow \mathbb{Z}_D^n$. For example, if we want to encrypt credit-card numbers (CCNs) then $m = 16$, and there are two possible values for d:

 (i) Conventionally, one views CCNs as a sequence of decimal digits, and thus $d = 10$.
(ii) Protegrity prefers to interpret CCNs as a sequence of (case-sensitive) alphanumeric characters for seemingly better security, and thus $d = 62$.

Under the specification in [12], one then instantiates the round function F from AES, interpreting $\{0, 1\}^{128}$ as $\mathbb{Z}_{256}^{16}$ (meaning $n = 16$ and $D = 256$). The code for the encryption and decryption of F is given in Fig. 8. The DTP specification always uses $D = 256$ if $d \leq 256$, and $D = 2^{16}$ if d is bigger. The parameter r specifies how many input characters that one encrypts per one call to the round function F. Initially, Protegrity used $r = 1$; this version is known internally as DTP-1. Eventually, they moved to $r = 3$ for faster speed, and also claimed better security; this is the current version, known as DTP-2 (Fig. 9).

If we consider an ideal round function then $\mathcal{K}$ is the set of all functions $G : \mathbb{Z}_D^n \rightarrow \mathbb{Z}_D^n$, and $F_K(\cdot)$ is defined as the function $G(\cdot)$ that the key K encodes. We write $\mathbf{DTP}[r, d, D, m, n]$ to denote the DTP construction of this particular choice of $\mathrm{PL} = (\mathcal{K}, F)$. As mentioned above, since the DTP spec instantiates the round function via AES, using the standard assumption that AES is a good PRF, one can focus on attacking DTP schemes of ideal round functions, with small differences in the advantage.

F.E(K, T, X)	F.D(K, T, Y)
$x_1 \cdots x_m \leftarrow X$; $T_1 \leftarrow T$; $t \leftarrow \lfloor m/r \rfloor$	$y_1 \cdots y_m \leftarrow Y$; $T_1 \leftarrow T$; $t \leftarrow \lfloor m/r \rfloor$
For $i = 1$ to t do	For $i = 1$ to t do
$\quad z_1 \cdots z_n \leftarrow F_K(T_i)$; $k \leftarrow (i-1)r$	$\quad z_1 \cdots z_n \leftarrow F_K(T_i)$; $k \leftarrow (i-1)r$
$\quad$ For $j = 1$ to r do	$\quad$ For $j = 1$ to r do
$\quad\quad y_{k+j} \leftarrow (x_{k+j} + z_j) \bmod d$	$\quad\quad x_{k+j} \leftarrow (y_{k+j} - z_j) \bmod d$
$\quad T_{i+1} \leftarrow z_{r+1} \cdots z_n x_{k+1} \cdots x_{k+r}$	$\quad T_{i+1} \leftarrow z_{r+1} \cdots z_n x_{k+1} \cdots x_{k+r}$
// Encrypt the trailing digits	// Decrypt the trailing digits
$z_1 \cdots z_n \leftarrow F_K(T_{t+1})$	$z_1 \cdots z_n \leftarrow F_K(T_{t+1})$
For $j = 1$ to $(m \bmod r)$ do	For $j = 1$ to $(m \bmod r)$ do
$\quad y_{tr+j} \leftarrow (x_{tr+j} + z_j) \bmod d$	$\quad x_{tr+j} \leftarrow (y_{tr+j} - z_j) \bmod d$
Return $y_1 \cdots y_m$	Return $x_1 \cdots x_m$

Fig. 8. Code for the encryption and decryption algorithms of F $=$ **DTP**$[r, d, D, m, n, \mathrm{PL}]$, where $\mathrm{PL} = (\mathcal{K}, F)$.

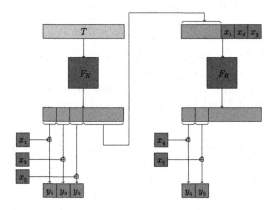

Fig. 9. Illustration of the encryption scheme of F $=$ **DTP**$[r, d, D, m, n, \mathrm{PL}]$, where $\mathrm{PL} = (\mathcal{K}, F)$, for $r = 3$ and $m = 5$, and $\boxplus$ means the addition in mod d.

THE ATTACK. We now give an attack on a general **DTP**$[r, d, D, m, n]$ scheme in which d is *not* a divisor of D. Many applications of DTP use $d = 10$ or $d = 62$ (for examples, encrypting credit-card numbers, social-security numbers, or PINs), and in that case, $D = 256$, falling into our setting. In this attack, we consider only a single target Z. There is no known message, and the auxiliary information is null. The adversary is given the encryption of Z under tweaks $T_1, \ldots, T_q$, for an appropriately large q. The number Q of ciphertexts is q, and so is the number of ciphertexts per recovered target. We assume that Z is uniformly random, independent of the tweaks, so that the message-guessing advantage is low.

Formally, let $\mathsf{DC3}_q$ be the class of all algorithms D that outputs distinct tweaks $T_1, \ldots, T_q \in (\mathbb{Z}_D)^n$. To any such D, we associate the following sampler $\mathsf{XS[D]}$

Adversary $\mathsf{DD}((T_1, C_1), \ldots, (T_q, C_q), a)$

For $i \leftarrow 1$ to m do

    For $k \in \mathbb{Z}_d$ do $V_k \leftarrow 0$

    For $j \leftarrow 1$ to q do $c_1 \cdots c_m \leftarrow C_j$; $V_{c_i} \leftarrow V_{c_i} + 1$

    $r \leftarrow D \bmod d$

    Find the r largest numbers $V_{s_1}, \ldots, V_{s_r}$ in $\{V_k \mid k \in \mathbb{Z}_d\}$

    Find $z_i \in \mathbb{Z}_d$ such that $\{s_j \mid 1 \leq j \leq r\} = \{(z_i + j) \bmod d \mid 1 \leq j \leq r\}$

$Z \leftarrow z_1 \cdots z_m$; Return Z

Fig. 10. The Digit-wise Differential attack.

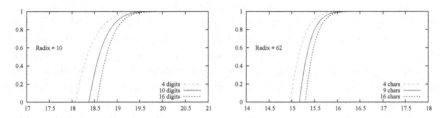

Fig. 11. The mr advantage of the Digit-wise Differential attack on DTP$[3, 10, 256, m, 16]$ (left) and DTP$[3, 62, 256, m, 16]$ (right) for $m = 4, 9, 16$. These are parameter choices for PINs, social security numbers, and credit-card numbers. The x-axis shows the log, base 2, of the number q of ciphertexts, and the y-axis shows $\mathbf{Adv}^{\mathrm{mr}}_{\mathrm{DTP}[3,d,256,m,16],\mathsf{XS}}(\mathsf{DD})$ for $\mathsf{XS} \in \mathsf{SC3}_q$.

Sampler $\mathsf{XS}[\mathsf{D}]$

$(T_1, \ldots, T_q) \leftarrow_\$ \mathsf{D}$; $a \leftarrow \bot$; $Z \leftarrow_\$ (\mathbb{Z}_d)^m$

Return $((T_1, Z), \ldots, (T_q, Z), Z, a)$

The sampler above runs D to generate the tweaks, and then samples a uniformly random target. Define $\mathsf{SC3}_q = \{\mathsf{XS}[\mathsf{D}] \mid \mathsf{D} \in \mathsf{DC3}_q\}$. Since the target is uniformly random and the auxiliary information is null, one would expect that the adversary has low mr-advantage, even if q is big. However, our Digit-wise Differential (DD) attack, given in Fig. 10, will recover the target message for any sampler in $\mathsf{SC3}_q$ within $O(md \log(d) + qm)$ time. Theorem 8 below gives a lower bound on the mr-advantage of DD; the proof is in the full version. The bound is illustrated in Fig. 11.

Theorem 8. *Let $D > d > 1$ be integers such that d is not a divisor of D. Let $m, n, r \geq 1$ be integers such that $n \geq r$, and let $\mathsf{F} = \mathbf{DTP}[r, d, D, m, n]$. Let $s = D \bmod d$. Then for any sampler XS in $\mathsf{SC3}_q$,*

$$\mathbf{Adv}^{\mathrm{mr}}_{\mathsf{F},\mathsf{XS}}(\mathsf{DD}) \geq 1 - \frac{(q \cdot \lceil m/r \rceil)^2}{2 \cdot D^{n-r}} - ms \cdot \exp\left(\frac{-q(d-s)^2}{2Dd(D+d-s)}\right)$$

$$- m(d-s) \cdot \exp\left(\frac{-qs^2}{3Dd(D-s)}\right) - \frac{1}{d^m} .$$

IDEAS OF THE ATTACK. For simplicity, let us start with the special important case $d = 10$ and $D = 256$. Let $Z = z_1 \cdots z_m$, where each z_i is a number in $\{0, \ldots, 9\}$. For simplicity, assume that the $q \cdot \lceil m/r \rceil$ inputs to F are distinct, so that the outputs of F are independent, which holds with high probability. We now explain how the attack can recover, say the first digit z_1 of the target Z, but the same idea works for any digit z_i of Z. The way the encryption works is to pick a random number $B \leftarrow_\$ \{0, \ldots, 255\}$, and then outputs $c_1 \leftarrow z_1 + (B \bmod 10)$ as the first digit of the ciphertext. The problem here is that $B \bmod 10$ is not uniformly distributed in $\{0, 1, \ldots, 9\}$. In fact, for $a \in \{0, 1, \ldots, 9\}$, the probability that $B = a$ is exactly $\frac{\lceil 256/10 \rceil}{256} = \frac{26}{256}$ if $a < 6$, and this probability however is only $\frac{\lfloor 256/10 \rfloor}{256} = \frac{25}{256}$ otherwise. Hence for any fixed number $z_1 \in \{0, 1, \ldots, 9\}$ and any number $a \in \{0, 1, \ldots, 9\}$, the probability that $c_1 \leftarrow z_1 + (B \bmod 10)$ is a is exactly $\frac{26}{256}$ if $a \in \{z_1 \bmod 10, z_1 + 1 \bmod 10, \ldots, z_1 + 5 \bmod 10\}$, and is $\frac{25}{256}$ otherwise. Thus if we encrypt the target Z with a large enough number of times and plot the frequency histogram of the first digit of the ciphertexts, then what we obtain is a 10-column histogram, with 6 tall columns and 4 short ones. These 6 tall columns will be consecutive (possibly with a wrap-around), and the first one corresponds to the value z_1.

Now suppose that we want to deal with generic D and d, but d is not a divisor of D. Let $Z = z_1 \cdots z_m$, where each z_i is a number in $\mathbb{Z}_d$. Consider, say the first digit z_1 of Z. The encryption works by picking a random number $B \leftarrow_\$ \mathbb{Z}_D$ and then outputs $c_1 \leftarrow z_1 + (B \bmod d)$ as the first digit of the ciphertext. Again because d is not a divisor of D, the random variable $B \bmod d$ is not uniformly distributed in $\mathbb{Z}_d$. In fact, for $a \in \mathbb{Z}_d$, the probability that $B = a$ is exactly $\frac{\lceil D/d \rceil}{D}$ if $a < D \bmod d$, and this probability however is only $\frac{\lfloor D/d \rfloor}{D}$ otherwise. By the same argument as the special case above, if we encrypt the target Z with a large enough number of times and plot the frequency histogram of the first digit of the ciphertexts, then what we obtain is a histogram, with $D \bmod d$ tall columns. These tall columns will be consecutive (possibly with a wrap-around), and the first one corresponds to the value z_1.

DISCUSSION. As Theorem 8 suggests, the security of DTP-2 (namely $r = 3$) is not better than that of DTP-1 (namely $r = 1$). Moreover, Protegrity's decision to prefer $d = 62$ over $d = 10$ actually makes security *worse*. As shown in Table 4, if one interprets a CCN as a sequence of 16 decimal digits, then one would need to obtain roughly $575,000$ ciphertexts to recover a CCN with advantage at least 0.9. In contrast, if one interprets a CCN as a sequence of 16 alphanumeric characters, then one would only need about $53,000$ ciphertexts to recover a CCN with advantage at least 0.9.

EXPERIMENTS. We implement our Digit-wise Differential attack in C++ and evaluate its message-recovery rate against both DTP-1 and DTP-2, for domains $\mathbb{Z}_d^m$, with $m \in \{4, 9, 16\}$ and $d \in \{10, 62\}$. (For DTP-1, we only use $d = 10$.) Each experiment for domain $\mathbb{Z}_d^m$ was run using m threads in a server of Intel(R) Xeon(R) CPU E5-2699 v3 2.30 GHz CPU and 256 GB RAM. For each setting,

Table 4. Comparison of security of DTP-2 over the choice of the radix d, on PINs, social security numbers, and credit-card numbers. The first column shows the value of d. The other columns show the estimated number of ciphertexts needed for our attack to achieve advantage 0.9 as suggested by Theorem 8.

Radix d	PINs ($m = 4$)	SSNs ($m = 9$)	CCNs ($m = 16$)
10	460,000	525,000	575,000
62	46,000	51,000	53,000

Table 5. Empirical results of the Digit-wise Differential attack on DTP-1. For each domain (shown in the first column), we run experiments with two values of q (the number of tweaks) as indicated in the second and fifth columns. The recovery rates corresponding to these two values of q are given in the third and sixth columns, respectively. Finally, the average running time (in milliseconds) of each experiment is given in the fourth and seventh columns.

Domain	Number of tweaks, q	Recovery rate	Time (ms)	Number of tweaks, q	Recovery rate	Time (ms)
$\mathbb{Z}_{10}^4$	2^{18}	100%	2.9	2^{17}	98%	1
$\mathbb{Z}_{10}^9$		100%	3		91%	1.49
$\mathbb{Z}_{10}^{16}$		100%	3.5		83%	1.87

Table 6. Empirical results of the Digit-wise Differential attack on DTP-2.

Domain	Number of tweaks, q	Recovery rate	Time (ms)	Number of tweaks, q	Recovery rate	Time (ms)
$\mathbb{Z}_{10}^4$	2^{18}	100%	3	2^{17}	95%	1
$\mathbb{Z}_{10}^9$		100%	3.08		90%	1.53
$\mathbb{Z}_{10}^{16}$		100%	3.58		83%	1.97
$\mathbb{Z}_{62}^4$	2^{16}	100%	0.01	2^{15}	91%	0.01
$\mathbb{Z}_{62}^9$		100%	1.03		78%	0.02
$\mathbb{Z}_{62}^{16}$		100%	1.17		68%	1

we run our attack for several choices of q (the number of tweaks), each for 100 trials, and report the average running time and the empirical recovery rate.

Our experimental results for DTP-1, given in Table 5, are quite consistent with Theorem 8. For example, for domain $\mathbb{Z}_{10}^{16}$ (namely CCNs), theoretically one would need $q = 2^{19}$ tweaks to recover the target with probability nearly 1, and our experiments confirm that using $q = 2^{19}$ indeed gives 100% recovery rate. However, empirically, we find that $q = 2^{18}$ is enough to achieve 100% recovery rate, and each trial takes just 3.5 ms on average. If one instead uses $q = 2^{17}$, the recovery rate drops to 83%.

The experimental results for DTP-2 are given in Table 6, confirming the theoretical observations in Table 4: (1) DTP-2 is just as insecure as DTP-1, and (2) Using radix $d = 62$ instead of $d = 10$ exacerbates the insecurity: for example, for $\mathbb{Z}_{62}^{16}$ (namely CCNs), using $q = 2^{15}$ is already enough to achieve 68% recovery rate, and using $q = 2^{16}$ results in 100% recovery rate.

Acknowledgments. We thank Mihir Bellare and the anonymous CCS and CRYPTO reviewers for insightful feedback. We also thank Michael Maloney and Clyde Williamson of Protegrity Corp. for providing the information of the DTP scheme.

Viet Tung Hoang was supported by NSF grants CICI-1738912 and CRII-1755539. Stefano Tessaro was supported by NSF grants CNS-1553758 (CAREER), CNS-1423566, CNS-1719146, CNS-1528178, and IIS-1528041, and by a Sloan Research Fellowship. Ni Trieu was supported by NSF award #1617197.

References

1. Bellare, M., Hoang, V.T., Tessaro, S.: Message-recovery attacks on Feistel-based format preserving encryption. In: ACM CCS 2016, pp. 444–455. ACM Press (2016)
2. Bellare, M., Ristenpart, T., Rogaway, P., Stegers, T.: Format-preserving encryption. In: Jacobson, M.J., Rijmen, V., Safavi-Naini, R. (eds.) SAC 2009. LNCS, vol. 5867, pp. 295–312. Springer, Heidelberg (2009). https://doi.org/10.1007/978-3-642-05445-7_19
3. Bellare, M., Rogaway, P., Spies, T.: The FFX mode of operation for format-preserving encryption. Submission to NIST, February 2010. http://csrc.nist.gov/groups/ST/toolkit/BCM/documents/proposedmodes/ffx/ffx-spec.pdf
4. Black, J., Rogaway, P.: Ciphers with arbitrary finite domains. In: Preneel, B. (ed.) CT-RSA 2002. LNCS, vol. 2271, pp. 114–130. Springer, Heidelberg (2002). https://doi.org/10.1007/3-540-45760-7_9
5. Brier, E., Peyrin, T., Stern, J.: BPS: a format-preserving encryption proposal. Submission to NIST (2010)
6. Brightwell, M., Smith, H.: Using datatype-preserving encryption to enhance data warehouse security. In: 20th National Information Systems Security Conference Proceedings (NISSC), pp. 141–149 (1997)
7. Dara, S., Fluhrer, S.: FNR: arbitrary length small domain block cipher proposal. In: Chakraborty, R.S., Matyas, V., Schaumont, P. (eds.) SPACE 2014. LNCS, vol. 8804, pp. 146–154. Springer, Cham (2014). https://doi.org/10.1007/978-3-319-12060-7_10
8. Durak, F.B., Vaudenay, S.: Breaking the FF3 format-preserving encryption standard over small domains. In: Katz, J., Shacham, H. (eds.) CRYPTO 2017. LNCS, vol. 10402, pp. 679–707. Springer, Cham (2017). https://doi.org/10.1007/978-3-319-63715-0_23
9. Dworkin, M.: Recommendation for Block Cipher Modes of Operation: Methods for Format-Preserving Encryption. NIST Special Publication 800-38G, March 2016. https://doi.org/10.6028/NIST.SP.800-38G
10. Dworkin, M., Perlner, R.: Analysis of VAES3 (FF2). Cryptology ePrint Archive, Report 2015/306 (2015). http://eprint.iacr.org/2015/306
11. Hoang, V.T., Morris, B., Rogaway, P.: An enciphering scheme based on a card shuffle. In: Safavi-Naini, R., Canetti, R. (eds.) CRYPTO 2012. LNCS, vol. 7417, pp. 1–13. Springer, Heidelberg (2012). https://doi.org/10.1007/978-3-642-32009-5_1

12. Mattsson, U.: Format controlling encryption using datatype preserving encryption. Cryptology ePrint Archive, Report 2009/257 (2009). http://eprint.iacr.org/2009/257
13. Morris, B., Rogaway, P.: Sometimes-Recurse shuffle: almost-random permutations in logarithmic expected time. In: Nguyen, P.Q., Oswald, E. (eds.) EUROCRYPT 2014. LNCS, vol. 8441, pp. 311–326. Springer, Heidelberg (2014). https://doi.org/10.1007/978-3-642-55220-5_18
14. Naor, M., Reingold, O.: On the construction of pseudorandom permutations: Luby-Rackoff revisited. J. Cryptol. **12**(1), 29–66 (1999)
15. Ristenpart, T., Yilek, S.: The Mix-and-Cut shuffle: small-domain encryption secure against N queries. In: Canetti, R., Garay, J.A. (eds.) CRYPTO 2013, Part I. LNCS, vol. 8042, pp. 392–409. Springer, Heidelberg (2013). https://doi.org/10.1007/978-3-642-40041-4_22
16. Vance, J.: VAES3 scheme for FFX: An addendum to The FFX mode of operation for Format Preserving Encryption. Submission to NIST, May 2011

Cryptoanalysis

Cryptanalysis

Cryptanalysis via Algebraic Spans

Adi Ben-Zvi, Arkadius Kalka, and Boaz Tsaban[(✉)]

Department of Mathematics, Bar-Ilan University, Ramat Gan, Israel
adi2lugassy@gmail.com, tschussle@gmail.com, tsaban@math.biu.ac.il

Abstract. We introduce a method for obtaining provable polynomial time solutions of problems in nonabelian algebraic cryptography. This method is widely applicable, easier to apply, and more efficient than earlier methods. After demonstrating its applicability to the major classic nonabelian protocols, we use this method to cryptanalyze the Triple Decomposition key exchange protocol, the only classic group theory based key exchange protocol that could not be cryptanalyzed by earlier methods.

1 Introduction

Since Diffie and Hellman's 1976 key exchange protocol, few alternative protocols withstood cryptanalysis, all based on abelian algebraic structures. In the years 1999 and 2000 [2,12], two general key exchange protocols based on *nonabelian* algebraic structures were introduced. The security of these protocols is based on variations of the conjugacy problem in nonabelian groups. The *Triple Decomposition* key exchange protocol was introduced in 2006 [14,15], and subsequently included in textbooks on nonabelian cryptography [10,16,17]. Its security is based on a problem that is very different from those of the earlier nonabelian key exchange protocols, and it stood out as the only nonabelian group theoretic protocol resisting known cryptanalyses [28].

All mentioned protocols were implemented over Artin's braid group $\mathbf{B}_N$. For the main part of this paper, it suffices to know that this group has an efficient, faithful representation as a group of matrices, and the computational problems on which the security of the above-mentioned protocols are based reduce to the same problems in groups of matrices over finite fields. The details of the reduction are available in Sect. 5.

Our main contribution is the introduction of *algebraic span cryptanalysis*, a general approach for provable polynomial time solutions of computational problems in groups of matrices, and thus in all groups with efficient matrix representations. Algebraic span cryptanalysis improves upon earlier ones (such as the Cheon–Jun method and the linear centralizer method [8,28]), in its wider applicability, simplicity, and efficiency. It solves all problems that were solved by earlier provable methods.

A true challenge for the novelty of a new method is to cryptanalyze a protocol that could not be cryptanalyzed by earlier methods, and the Triple Decomposition key exchange protocol is such. With a novel view at the public information

© International Association for Cryptologic Research 2018
H. Shacham and A. Boldyreva (Eds.): CRYPTO 2018, LNCS 10991, pp. 255–274, 2018.
https://doi.org/10.1007/978-3-319-96884-1_9

provided by this protocol, algebraic span cryptanalysis provides the first crypt-analysis of this protocol.

Previously, provable cryptanalyses of this type were considered theoretical only. Using some algorithmic speed-ups, we also provide the first implementation of a cryptanalysis of this type. All of our experiments, with a wide range of parameters, succeeded in extracting the shared key out of the public information of the protocol.

Related work. The Commutator and the Braid Diffie–Hellman protocols were cryptanalyzed in a number of heuristic ways, but these attacks were foiled by changing the distributions on the group [9,27, and references therein]. In two breakthrough papers [8,28], provable polynomial time algorithms were found for the precise computational problems on which these, and some related proto-cols, are based. In a series of works [18–22, and references therein], Roman'kov and others developed a provable polynomial time method that applies to key exchange protocols with certain commuting substructures. All protocols treated in these papers can be cryptanalyzed using the method presented here. On the other hand, only our method applies to the Triple Decomposition protocol. Alge-braic span cryptanalysis was also applied in a cryptanalysis of the Algebraic Eraser [3].

This paper is a composition of an earlier, hitherto unpublished work by the third named author (Sects. 1–3), and a recent joint work of all three authors (Sects. 4–5).

Paper outline. Section 2 introduces the new method, in general terms. The exact application of this method depends on the specific protocol we wish to cryptanalyze. Section 3 demonstrates the applicability of this method to the classic nonabelian key exchange protocols. In addition to demonstrating the flexibility of this method, this section aims to make the reader comfortable with this method, and thus make it easier to proceed to the next section. Section 4 addresses a hitherto resistant challenge, where the application of our method is more involved. Section 5 provides the details of the Triple Decomposition key exchange protocol and its reduction to a matrix group over a finite field, together with detailed complexity estimates and experimental results.

2 Algebraic Span Cryptanalysis in a Nutshell

Let $\mathbb{F}$ be a finite field, and $M_n(\mathbb{F})$ be the set of $n \times n$ matrices with entries in $\mathbb{F}$. An *algebra* of matrices is a family of matrices $A \subseteq M_n(\mathbb{F})$ that is closed under linear combinations and multiplications. For a set $S \subseteq M_n(\mathbb{F})$, let $\mathrm{Alg}(S)$ be the algebra generated by S, that is, the smallest Algebra $A \subseteq M_n(\mathbb{F})$ that contains S as a subset. Every subalgebra of $M_n(\mathbb{F})$ is also a vector space over the field $\mathbb{F}$. Let $\mathrm{GL}_n(\mathbb{F})$ be the group of invertible matrices in $M_n(\mathbb{F})$. For a subgroup $G \leq \mathrm{GL}_n(\mathbb{F})$, we have $\mathrm{Alg}(G) = \mathrm{span}(G)$, the vector space spanned by G.

For simplicity we assume, throughout, that the dimension of the vector space $\mathrm{Alg}(G)$ is at least a positive constant times n. Notice that even for cyclic groups G, this is typically the case.

Throughout, let ω be the linear algebra constant, the minimal real number such that the complexity of $n \times n$ matrix multiplication is $O(n^\omega)$ field operations.

Proposition 1. *Let $G = \langle g_1, \ldots, g_k \rangle \leq \mathrm{GL}_n(\mathbb{F})$ be a group, and $d \leq n^2$ be the dimension of the vector space $\mathrm{Alg}(G)$. A basis for the vector space $\mathrm{Alg}(G)$ can be computed using $O(kd^2n^2)$ field operations.*

Proof. Initialize a sequence $s = (I)$, the identity matrix, and $i := 1$. Repeat the following as long as there is an element in position i of the sequence s:

1. For $j = 1, \ldots, k$, if $s_i g_j \notin \mathrm{span}\, S$, append $s_i g_j$ at the end of the sequence s.
2. $i := i + 1$.

The resulting sequence s is a basis for $\mathrm{span}\, G$. For each i and each j, the complexity of computing the products $s_i g_j$ is n^ω field operations. Assume that the matrices are stored in s in a vector form, and the matrix s is kept in Echelon normal form throughout the process. Since there are at most d vectors in s, each of length n^2, the complexity of checking whether a vector is in $\mathrm{span}\, s$ is at most $O(dn^2)$. Thus, the overall complexity is $O(kd(n^\omega + dn^2))$ field operations. Since we assume that d is at least a constant multiple of n, the second term dominates the first one. □

Proposition 1 holds, more generally, for semigroups of matrices; but this will not be used here. There are advanced methods, via representation theory, to slightly reduce the complexity of this computation [11] but, for our purposes, Proposition 1 suffices as it is.

2.1 The Method

Algebraic span cryptanalysis is applied as follows. Let $G_1, \ldots, G_k$ be given, publicly known subgroups of $\mathrm{GL}_n(\mathbb{F})$. Assume that a secret $f(g_1, \ldots, g_k)$ is computed from unknown matrices $g_1 \in G_1, \ldots, g_k \in G_k$, by means of a prescribed, public function f. Assume that we can extract, from a protocol transaction, a system of linear equations (or constraints) on the entries of the unknown matrices $g_1, \ldots, g_k$, and we wish to find the secret $f(g_1, \ldots, g_k)$.

Instead of solving the given linear equations subject to the restrictions $g_1 \in G_1, \ldots, g_k \in G_k$ (which may be computationally infeasible), we solve these linear equations subject to the *linear* constraints $g_1 \in \mathrm{Alg}(G_1), \ldots, g_k \in \mathrm{Alg}(G_k)$. We then try to prove (or at least verify by experiments) that, for each solution $\tilde{g}_1, \ldots, \tilde{g}_k$, we have $f(\tilde{g}_1, \ldots, \tilde{g}_k) = f(g_1, \ldots, g_k)$.

Strikingly, this simple method applies, provably, in all cases of nonabelian algebraic cryptography where polynomial-time algorithms are known [12,18–20,28], and in a case that was not cryptanalyzed thus far. We provide these details in the following sections.

The equations do not have to be given as linear. For example, an equation $g_1 a g_2 = b$ with a and b known can be transformed to the equation $a g_2 = g_1^{-1} b$, which is linear in the entries of the matrices g_1^{-1} and g_2.

2.2 Invertibility

Often, as in the latter example, we need some elements in our solution to be invertible. Since there *is* an invertible solution, namely, $(g_1, \ldots, g_k)$, the following lemma (Invertibility Lemma [28, Lemma 9]) guarantees that random solutions are invertible with probability bounded away from zero, provided that the field is not too small. This will be the situation in all of our applications. Thus, we may pick random solutions until they are invertible.

Lemma 1. *For a finite field* $\mathbb{F}$, *let* $a_1, \ldots, a_m \in \mathrm{M}_n(\mathbb{F})$, *such that some linear combination of these matrices is invertible. If* $\alpha_1, \ldots, \alpha_m$ *are chosen uniformly and independently from* $\mathbb{F}$, *then the probability that the linear combination* $\alpha_1 a_1 + \cdots + \alpha_m a_m$ *is invertible is at least* $1 - \frac{n}{|\mathbb{F}|}$.

2.3 Complexity

The next section provides concrete applications of this approach to several problems in the field of nonabelian algebraic cryptography. Enough examples are provided so that the reader can apply this method to additional problems in the field, including essentially all known key exchange protocols based on groups with efficient representations as matrix groups.

In these examples, the proposed platform group is Artin's braid group $\mathbf{B}_N$. However, these problems are known to transform into a matrix group $G \leq \mathrm{GL}_n(\mathbb{F})$ [12, 28]. The reduction uses the the Lawrence–Krammer representation, and thus the matrices are of rank $n = \binom{N}{2}$. Roughly, in this reduction, the cardinality of the field $\mathbb{F}$ is p^d, with $p \approx 2^{N^2 M}$ and $d \approx M$ for some length parameter M (Sect. 5.3). We may assume that $M \approx N$. Then the cost of field multiplication is $d^2 \log p \approx N^5$, ignoring a logarithmic factor. Tighter scrutiny of this reduction is likely to lead to substantially smaller field sizes; the extra factor of N^5 should not be considered definite. Additional details are provided in Sect. 5.

3 Sample Applications

In this section, we apply the algebraic span method to the major classic nonabelian key exchange protocols. The application in these cases is not difficult. This demonstrates the applicability of the method, and also serves as a good preparation for the next section, where a more involved application is made. The protocols are presented for general groups, but they were all proposed to use groups that have efficient representations as matrix groups. We thus assume here that the groups are matrix groups. The exact parameters originally suggested for these protocols are not important: The cryptanalyses we provide are provable, and their complexity is unaffected by the exact settings or distributions used by the protocol. For the main application, we will provide details in Sect. 5.

The protocols to which we apply our method are described succinctly by diagrams. In these diagrams, green letters indicate publicly known elements, and red ones indicate secret elements, known only to their holders. Results of computations involving elements of both colors may be either publicly known, or secret, depending on the context.

Most of these protocols, and their analyzes, use the following notation. For a nonabelian group G and group elements $g, x \in G$, we define $g^x := x^{-1}gx$. Useful identities involving this notation include $g^{xy} = (g^x)^y$, and $g^c = g$ for every element $c \in G$ that commutes with g, that is, when $cg = gc$.

3.1   The Commutator Key Exchange Protocol

A free group word in the variables $x_1, \ldots, x_k$ is a product of the form

$$x_{i_1}^{\epsilon_1} x_{i_2}^{\epsilon_2} \cdots x_{i_m}^{\epsilon_m},$$

with $i_1, \ldots, i_m \in \{1, \ldots, k\}$ and $\epsilon_1, \ldots, \epsilon_m \in \{1, -1\}$, and with no subproduct of the form $x_i x_i^{-1}$ or $x_i^{-1} x_i$. The *Commutator key exchange protocol* [2] is described in Fig. 1 below. In some detail:

1. A nonabelian group G and elements $a_1, \ldots, a_k, b_1, \ldots, b_k \in G$ are publicly given.
2. Alice and Bob choose free group words in the variables $x_1, \ldots, x_k$, $v(x_1, \ldots, x_k)$ and $w(x_1, \ldots, x_k)$, respectively.
3. Alice substitutes $a_1, \ldots, a_k$ for $x_1, \ldots, x_k$, to obtain a secret element $a = v(a_1, \ldots, a_k) \in G$. Similarly, Bob computes a secret element $b = w(b_1, \ldots, b_k)$ in G.
4. Alice sends the conjugated elements $b_1^a, \ldots, b_k^a$ to Bob, and Bob sends the conjugated elements $a_1^b, \ldots, a_k^b$ to Alice.
5. The shared key is the *commutator* $a^{-1}b^{-1}ab$.

As conjugation is a group isomorphism, we have

$$v(a_1^b, \ldots, a_k^b) = v(a_1, \ldots, a_k)^b = a^b = b^{-1}ab.$$

Thus, Alice can compute the shared key $a^{-1}b^{-1}ab$ as $a^{-1}v(a_1^b, \ldots, a_k^b)$, using her secret $a, v(x_1, \ldots, x_k)$ and the public elements $a_1^b, \ldots, a_k^b$. Similarly, Bob computes $a^{-1}b^{-1}ab$ as $w(b_1^a, \ldots, b_k^a)^{-1}b$.

The security of the Commutator key exchange protocol is determined by the difficulty of the following problem. As usual, for a group G and elements $g_1, \ldots, g_k \in G$, the subgroup of G generated by the elements $g_1, \ldots, g_k$ is denoted $\langle g_1, \ldots, g_k \rangle$.

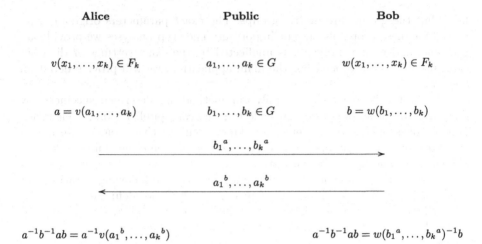

Fig. 1. The Commutator key exchange protocol

Problem 2. *Let G be a group, $a_1, \ldots, a_k, b_1, \ldots, b_k \in G$, $a \in \langle a_1, \ldots, a_k \rangle$, and $b \in \langle b_1, \ldots, b_k \rangle$.*
Given the group elements $a_1, \ldots, a_k, b_1, \ldots, b_k, a_1^b, \ldots, a_k^b, b_1^a, \ldots, b_k^a$, compute the commutator $a^{-1}b^{-1}ab$.

The Commutator key exchange protocol uses Artin's braid group as its platform group, but it is known that the problem reduces, polynomially, to the same problem in matrix groups over finite fields [28]. Thus, we need to solve Problem 2 in matrix groups.

Lemma 3. *Let $x, \tilde{x} \in \mathrm{GL}_n(\mathbb{F})$ and $G = \langle g_1, \ldots, g_k \rangle \le \mathrm{GL}_n(\mathbb{F})$. If $g_i^x = g_i^{\tilde{x}}$ for all $i = 1, \ldots, k$, then $g^x = g^{\tilde{x}}$ for all $g \in \mathrm{Alg}(G)$.*

Proof. Conjugation is an automorphism of the matrix algebra. □

We apply the algebraic span method to Problem 2, as follows:

1. Compute bases for the vector spaces $\mathrm{Alg}(A)$ and $\mathrm{Alg}(B)$. Let d be the maximum of the sizes of these bases.
2. Solve the following homogeneous system of linear equations in the unknown matrix $x \in \mathrm{Alg}(A)$:

$$b_1 \cdot x = x \cdot b_1^a$$

$$\vdots$$

$$b_k \cdot x = x \cdot b_k^a,$$

a system of linear equations on the d coefficients determining the matrix x, as a linear combination of the basis of the space $\mathrm{Alg}(A)$.

3. Fix a basis for the solution space, and pick random solutions until the picked solution $\tilde{a}$ is invertible.
4. Solve the following homogeneous system of linear equations in the unknown matrix $y \in \text{Alg}(B)$:

$$a_1 \cdot y = y \cdot a_1{}^b$$

$$\vdots$$

$$a_k \cdot y = y \cdot a_k{}^b,$$

a system of linear equations on the d coefficients determining y.
5. Fix a basis for the solution space, and pick random solutions until the picked solution $\tilde{b}$ is invertible.
6. *Output:* $\tilde{a}^{-1}\tilde{b}^{-1}\tilde{a}\tilde{b}$.

That step (3) terminates quickly follows from the Invertibility Lemma [28]. We prove that the output is correct. As $\tilde{b} \in \text{Alg}(B)$, we have by Lemma 3 that $\tilde{b}^{\tilde{a}} = \tilde{b}^a$, and therefore

$$(\tilde{b}^{-1})^{\tilde{a}} = (\tilde{b}^{\tilde{a}})^{-1} = (\tilde{b}^a)^{-1} = (\tilde{b}^{-1})^a.$$

It follows that

$$\tilde{a}^{-1}\tilde{b}^{-1}\tilde{a}\tilde{b} = (\tilde{b}^{-1})^{\tilde{a}}\tilde{b} = (\tilde{b}^{-1})^a\tilde{b} = a^{-1}\tilde{b}^{-1}a\tilde{b} = a^{-1}a^{\tilde{b}}.$$

As $a \in \text{Alg}(A)$, we have by Lemma 3 that $a^{\tilde{b}} = a^b$, and thus

$$\tilde{a}^{-1}\tilde{b}^{-1}\tilde{a}\tilde{b} = a^{-1}a^b = a^{-1}b^{-1}ab.$$

Complexity. The step with linear equations computes the nullspace of a $kn^2 \times d$ matrix. Thus, its complexity is $O(\frac{kn^2}{d}d^\omega) = O(kn^2d^{\omega-1})$, which is dominated by the complexity $O(kd^2n^2)$ of computing the algebraic spans. In the actual proposal [2], the dimension d is $O(n^2)$, and the complexity becomes $O(kn^6)$. The parameter k is typically $\sqrt{n}$ (the number of Artin generators in the braid group $\mathbf{B}_N$).

Strikingly, the new cryptanalysis is not only simpler and more general than the previous cryptanalysis (by the linear centralizer method); it is also more efficient. The complexity of the previous cryptanalysis is much larger: $O(n^8 + kn^6)$, that is typically $O(n^8)$ [28].

3.2   The Centralizer Key Exchange Protocol

For a group G and an element $g \in G$, the *centralizer of g in G* is the set

$$C_G(g) := \{h \in G \,:\, gh = hg\}.$$

The *Centralizer key exchange protocol*, introduced by Shpilrain and Ushakov in 2006 [25], is described in Fig. 2. In this protocol, a_1 commutes with b_1 and

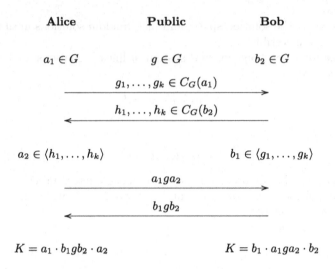

Fig. 2. The Centralizer key exchange protocol

a_2 commutes with b_2. Consequently, the keys computed by Alice and Bob are identical, and they are equal to the group element $a_1 b_1 g a_2 b_2$.

The security of the Centralizer key exchange protocol is determined by the difficulty of the following problem.

Problem 4. Let $G \leq \mathrm{GL}_n(\mathbb{F})$, $g, a_1, b_2 \in G$, $g_1, \ldots, g_k \in C_G(a_1)$, $h_1, \ldots, h_k \in C_G(b_2)$, $a_2 \in \langle h_1, \ldots, h_k \rangle$, and $b_1 \in \langle g_1, \ldots, g_k \rangle$.
Given the group elements $g, g_1, \ldots, g_k, h_1, \ldots, h_k, a_1 g a_2, b_1 g b_2$, compute the product $a_1 b_1 g a_2 b_2$.

The algebraic span method applies, provably, to this problem: We note that $a_1^{-1}(a_1 g a_2) = g a_2$. Find a solution to the system

$$x(a_1 g a_2) = gy$$
$$x g_1 = g_1 x$$
$$\vdots$$
$$x g_k = g_k x$$

with x invertible and $y \in \mathrm{Alg}(\{h_1, \ldots, h_k\})$. In practice, we may start with y which has d variables, and this determines x and then we solve for x.

Let $(\tilde{a}_1, \tilde{a}_2) = (x^{-1}, y)$. Then $\tilde{a}_1 g \tilde{a}_2 = x^{-1} g y = a_1 g a_2$. As the element $\tilde{a}_1 = x^{-1}$ commutes with all elements $g_1, \ldots, g_k$, it also commutes with b_1. As b_2 commutes with $h_1, \ldots, h_k$ and $\tilde{a}_2 \in \mathrm{Alg}(\{h_1, \ldots, h_k\})$, we have $b_2 \tilde{a}_2 = \tilde{a}_2 b_2$. Thus,

$$\tilde{a}_1 b_1 g b_2 \tilde{a}_2 = b_1 \tilde{a}_1 g \tilde{a}_2 b_2 = b_1 a_1 g a_2 b_2.$$

Here, too, the complexity is $O(kd^2 n^2)$.

3.3   The Braid Diffie–Hellman Key Exchange Protocol and the Double Coset Key Exchange Protocol

The *Braid Diffie–Hellman key exchange protocol*, introduced by Ko et al. [12], is illustrated in Fig. 3. For subsets A, B of a group G, that notation $[A, B] = 1$ means that the sets A and B commute elementwise, that is, a and b commute ($ab = ba$) for all elements $a \in A$ and $b \in B$. Since, in the Braid Diffie–Hellman key exchange protocol, the subgroups A and B of G commute element-wise, the keys computed by Alice and Bob are identical.

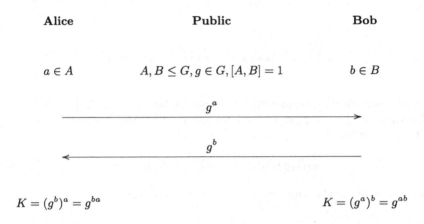

Fig. 3. The Braid Diffie–Hellman key exchange protocol

The security of the Braid Diffie–Hellman key exchange protocol for a platform group G (Fig. 3) is captured by the following problem.

Problem 5. *Let A and B be subgroups of $\mathrm{GL}_n(\mathbb{F})$ with $[A, B] = 1$, and an element $g \in \mathrm{GL}_n(\mathbb{F})$ be given.*
Given a pair (g^a, g^b) where $a \in A$ and $b \in B$, find g^{ab}.

As with all problems in this paper, the original problem is stated for Artin's Braid group, and it is known that it reduces to the same problem in matrix groups over finite fields [8]. We solve it for matrix groups.

To apply the algebraic span method to this problem, solve the equation $g^x = g^a$ subject to the linear constraint $x \in \mathrm{Alg}(A)$, and pick an invertible solution $\tilde{a}$. Then

$$(g^b)^{\tilde{a}} = g^{b\tilde{a}} = g^{\tilde{a}b} = (g^{\tilde{a}})^b = (g^a)^b = g^{ab}.$$

Again, the complexity of the solution is dominated by the computation of $\mathrm{Alg}(A)$.

A generalization of the Braid Diffie–Hellman key exchange protocol was proposed by Cha et al. [7]. A variation of this protocol was proposed in 2005, by Shpilrain and Ushakov [24]. These protocols are both special cases of the *Double Coset key exchange protocol*, illustrated in Fig. 4.

264   A. Ben-Zvi et al.

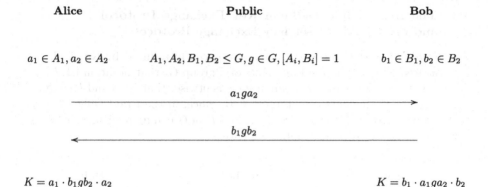

Alice	Public	Bob
$a_1 \in A_1, a_2 \in A_2$	$A_1, A_2, B_1, B_2 \leq G, g \in G, [A_i, B_i] = 1$	$b_1 \in B_1, b_2 \in B_2$

$$a_1 g a_2 \longrightarrow$$

$$\longleftarrow b_1 g b_2$$

| $K = a_1 \cdot b_1 g b_2 \cdot a_2$ | | $K = b_1 \cdot a_1 g a_2 \cdot b_2$ |

Fig. 4. The Double Coset key exchange protocol

One may state the underlying problem as before. Here is how to solve it:
Solve the equation $x_1(a_1 g a_2) = g x_2$ subject to $x_1 \in \mathrm{Alg}(A_1)$ and $x_2 \in \mathrm{Alg}(A_2)$,
with x_1 invertible. Let $(\tilde{a}_1, a_2) = (x_1^{-1}, x_2)$. Then

$$\tilde{a}_1(b_1 g b_2)\tilde{a}_2 = b_1 \tilde{a}_1 g \tilde{a}_2 b_2 = b_1 a_1 g a_2 b_2.$$

The complexity is the same as in the previous solutions.

3.4 Stickel's Key Exchange Protocol

We conclude with an example where the complexity of the cryptanalysis is surprisingly small. The key exchange protocol described in Fig. 5 was introduced by Stickel in 2005 [26].

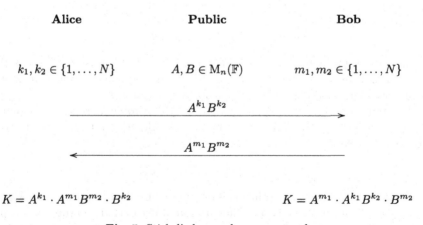

Alice	Public	Bob
$k_1, k_2 \in \{1, \ldots, N\}$	$A, B \in \mathrm{M}_n(\mathbb{F})$	$m_1, m_2 \in \{1, \ldots, N\}$

$$A^{k_1} B^{k_2} \longrightarrow$$

$$\longleftarrow A^{m_1} B^{m_2}$$

| $K = A^{k_1} \cdot A^{m_1} B^{m_2} \cdot B^{k_2}$ | | $K = A^{m_1} \cdot A^{k_1} B^{k_2} \cdot B^{m_2}$ |

Fig. 5. Stickel's key exchange protocol

A successful heuristic cryptanalysis of complexity roughly $n^{2\omega}$ was presented by Shpilrain [23]. Shpilrain's cryptanalysis turned out provable [28]. The algebraic span method provides a simple alternative, of smaller complexity.

The dimension of the *algebras* spanned by the matrices A and B is, by the Cayley–Hamilton Theorem, at most n. Find a matrix $\tilde{A} \in \mathrm{Alg}(\{A\})$ and an invertible matrix $D \in \mathrm{Alg}(\{B\})$ satisfying the linear equation $\tilde{A} = A^{k_1}B^{k_2}D$. Since the dimension is $O(n)$, the complexity is $O(n^4)$. Let $\tilde{B} = D^{-1}$. A cyclic algebra is abelian. Moreover, the matrix $\tilde{B}$ is a finite power of D, and is thus in $\mathrm{Alg}(\{B\})$. Thus,

$$\tilde{A} \cdot A^{m_1}B^{m_2} \cdot \tilde{B} = A^{m_1}\tilde{A}\tilde{B}B^{m_2} = A^{m_1}A^{k_1}B^{k_2}B^{m_2} = K.$$

The overall complexity is just $O(n^4)$.

Variations of this key exchange protocol are proposed every now and then (for example, [1,23]), and are all subject to the cryptanalysis presented here (cf. [5]).

4   Cryptanalysis of the Triple Decomposition Key Exchange Protocol

Kurt's *Triple Decomposition* key exchange protocol [15,17] is described in Fig. 6. In this figure, uppercase letters denote subgroups. An edge between two subgroups means that these subgroups commute elementwise. This ensures that the keys computed by Alice and Bob are both equal to $ab_1a_1b_2a_2b$.

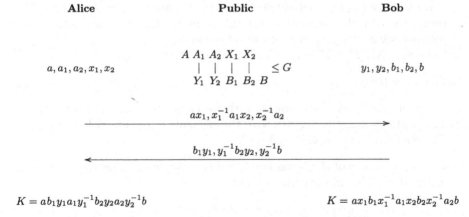

Fig. 6. The Triple Decomposition key exchange protocol

Let $c := x_1^{-1}a_1x_2$. By moving the matrix x_1 or x_2 to the other side of the equation, the public information $x_1^{-1}a_1x_2$ provides a quadratic equation, and similarly for the public information $y_1^{-1}b_2y_2$. Solving quadratic equations may be very difficult. This prevented the application of earlier methods to this key exchange protocol. The natural approach would be to ignore this part of the pubic information, and solve the linear equations provided by the other public items. This works for generic matrix groups, but fails, according to our experiments, for the actual groups proposed in Kurt's paper [15]. We provide here a way that takes the triple products into account, in a linear way, which still provably obtains the correct key. In the framework of algebraic spans, this solution is natural.

The following sets can be computed from the public information:

$$\mathrm{Alg}(B_1)y_1 = \mathrm{Alg}(B_1) \cdot b_1 y_1$$
$$\mathrm{Alg}(B_2 \cup Y_2)y_1 = \mathrm{Alg}(B_2 \cup Y_2) \cdot y_2^{-1} b_2^{-1} y_1 = \mathrm{Alg}(B_2 \cup Y_2) \cdot (y_1^{-1} b_2 y_2)^{-1}$$
$$\mathrm{Alg}(A_2)x_2 = \mathrm{Alg}(A_2) \cdot a_2^{-1} x_2$$
$$\mathrm{Alg}(A_1 \cup X_1)x_2 = \mathrm{Alg}(A_1 \cup X_1) \cdot x_1^{-1} a_1 x_2$$

The invertible matrices y_1 and x_2 are, respectively, in the following intersections of subspaces of $\mathrm{M}_n(\mathbb{F})$:

$$\mathrm{Alg}(Y_1) \cap \mathrm{Alg}(B_1)y_1 \cap \mathrm{Alg}(B_2 \cup Y_2)y_1;$$
$$\mathrm{Alg}(X_2) \cap \mathrm{Alg}(A_2)x_2 \cap \mathrm{Alg}(A_1 \cup X_1)x_2.$$

By the Invertibility Lemma [28, Lemma 9], we can pick invertible elements $\tilde{y}_1$ and $\tilde{x}_2$ in these intersections, respectively. Then:

1. Since the elements y_1 and $\tilde{y}_1$ are in $\mathrm{Alg}(Y_1)$, they commute with the elements of the subgroup A_1.
2. Since $\tilde{y}_1 \in \mathrm{Alg}(B_1)y_1$, we have $\tilde{y}_1 y_1^{-1} \in \mathrm{Alg}(B_1)$, and thus the quotient $\tilde{y}_1 y_1^{-1}$ commutes with the elements of the subgroup X_1. By (1), this quotient also commutes with the elements of the subgroup A_1.
3. Since $\tilde{y}_1 \in \mathrm{Alg}(B_2 \cup Y_2)y_1$, we have $\tilde{y}_1 y_1^{-1} \in \mathrm{Alg}(B_2 \cup Y_2)$.

Similarly, we have:

1. The elements x_2 and $\tilde{x}_2$ commute with the elements of the subgroup B_2.
2. The quotient $\tilde{x}_2 x_2^{-1}$ commutes with the elements of the union $Y_2 \cup B_2$.
3. The quotient $\tilde{x}_2 x_2^{-1}$ is in $\mathrm{Alg}(A_1 \cup X_1)$.

It suffices to use one of the items numbered (3). We will use here the former.
Using the public information, compute

$$\tilde{K} := a x_1 \cdot b_1 y_1 \cdot \tilde{y}_1^{-1} \cdot x_1^{-1} a_1 x_2 \cdot \tilde{x}_2^{-1} \cdot \tilde{y}_1 \cdot y_1^{-1} b_2 y_2 \cdot \tilde{x}_2 \cdot x_2^{-1} a_2 \cdot y_2^{-1} b.$$

We claim that $\tilde{K} = K = a b_1 a_1 b_2 a_2 b$, the key that Alice and Bob established.

Since the subgroups X_1 and B_1 commute elementwise, and $\tilde{y}_1 y_1^{-1} \in \mathrm{Alg}(B_1)$, we have

$$x_1 \cdot b_1 \cdot y_1 \tilde{y}_1^{-1} \cdot x_1^{-1} = b_1 y_1 \tilde{y}_1^{-1}.$$

Since the quotient $\tilde{x}_2 x_2^{-1}$ commutes with the elements of the union $Y_2 \cup B_2$ and $\tilde{y}_1 y_1^{-1} \in \mathrm{Alg}(B_2 \cup Y_2)$, we have

$$x_2 \tilde{x}_2^{-1} \cdot \tilde{y}_1 y_1^{-1} \cdot b_2 \cdot y_2 \cdot \tilde{x}_2 x_2^{-1} = \tilde{y}_1 y_1^{-1} b_2 y_2.$$

Thus,

$$\tilde{K} = ab_1 y_1 \tilde{y}_1^{-1} a_1 \tilde{y}_1 y_1^{-1} b_2 y_2 a_2 y_2^{-1} b.$$

Since the subgroups Y_2 and A_2 commute elementwise, we have

$$y_2 a_2 y_2^{-1} = a_2.$$

Since the quotient $\tilde{y}_1 y_1^{-1}$ commutes with the elements of the subgroup A_1, we have

$$y_1 \tilde{y}_1^{-1} \cdot a_1 \cdot \tilde{y}_1 y_1^{-1} = a_1.$$

It follows that

$$\tilde{K} = ab_1 a_1 b_2 a_2 b,$$

as required.

As in all of our previous examples, the complexity of this cryptanalysis is dominated by the calculation of the algebraic spans, which is $O(kd^2 n^2)$, where k the maximum number of generators of the given subgroups, and d is the maximum dimension of the Algebra generated by them. In particular, it is not greater than $O(kn^6)$.

5 Specifications and Implementation

5.1 Artin's Braid Group $\mathbf{B}_N$

All key exchange protocols addressed in this paper use Artin's *braid group* $\mathbf{B}_N$ as the underlying group. This group is parameterized by a natural number N. Elements of $\mathbf{B}_N$ can be identified with braids on N strands. Braid group multiplication is motivated geometrically, but the details play no role in the present paper. We provide here the necessary details, following the earlier paper [28].

Let S_N be the symmetric group of permutations on N symbols. For our purposes, the braid group $\mathbf{B}_N$ is a group of elements of the form $(i, \mathbf{p})$, where i is an integer, and $\mathbf{p}$ is a finite (possibly, empty) sequence of elements of S_N. In other words, $\mathbf{p} = (p_1, \dots, p_\ell)$ for some $\ell \geq 0$ and $p_1, \dots, p_\ell \in S_N$. The sequence $\mathbf{p} = (p_1, \dots, p_\ell)$ is requested to be *left weighted* (a property whose definition will not be used here), and p_1 must not be the involution $p(k) = N - k + 1$.[1]

For "generic" braids $(i, (p_1, \dots, p_\ell)) \in \mathbf{B}_N$, i is negative and $|i|$ is $O(\ell)$, but this is not always the case. Note that the bit-length of an element $(i, (p_1, \dots, p_\ell)) \in \mathbf{B}_N$ is $O(\log |i| + \ell N \log N)$.

Multiplication is defined on $\mathbf{B}_N$ by an algorithm of complexity $O(\ell^2 N \log N + \log |i|)$. Inversion is of linear complexity. Explicit implementations are provided, for example, in [7].

---

[1] For readers familiar with the braid group, we point out that the sequence $(i, (p_1, \dots, p_\ell))$ encodes the left normal form $\Delta^i p_1 \cdots p_\ell$ of the braid, in Artin's presentation, with Δ being the fundamental, half twist braid on N strands.

5.2   Infimum Reduction

The *infimum* of a braid $b = (i, \mathbf{p})$ is the integer $\inf(b) := i$. As the bit-length of b is $O(\log|i| + \ell N \log N)$, an algorithm polynomial in $|i|$ would be at least *exponential* in the bit-length. This obstacle is eliminated by reducing the infimum [28]. We demonstrate this for the Triple Decomposition key exchange protocol (Sect. 4).

In cases where $\mathbf{p}$ is the empty sequence, we write (i) instead of $(i, \mathbf{p})$. The properties of the braid group $\mathbf{B}_N$ include, among others, the following ones.

(a) $(i) \cdot (j, \mathbf{p}) = (i + j, \mathbf{p})$ for all integers i and all $(j, \mathbf{p}) \in \mathbf{B}_N$.
   In particular, $(i) = (1)^i$ for all i.
(b) $(2) \cdot (i, \mathbf{p}) = (i, \mathbf{p}) \cdot (2)$ for all for all $(i, \mathbf{p}) \in \mathbf{B}_N$.

Thus, $(2j)$ is a central element of $\mathbf{B}_N$ for each integer j. If follows that, for each $(i, \mathbf{p}) \in \mathbf{B}_N$,
$$(i, \mathbf{p}) = (i - (i \bmod 2)) \cdot (i \bmod 2, \mathbf{p}).$$

This way, every braid $x \in \mathbf{B}_N$ decomposes to a unique product $c_x \underline{x}$, where c_b is of the form $(2j)$ (and thus *central*), and $\inf(\underline{b}) \in \{0, 1\}$.

Consider the information in Fig. 6. Since
$$K = ab_1 y_1 a_1 y_1^{-1} b_2 y_2 a_2 y_2^{-1} b = a x_1 b_1 x_1^{-1} a_1 x_2 b_2 x_2^{-1} a_2 b,$$

The central elements $c_{y_1}, c_{y_2}, c_{x_1}, c_{x_2}$ get canceled and do not affect the shared key. Thus, we may assume that the infimum of the braids y_1, y_2, x_1, x_2 is 0 or 1. Decompose the central parts out of the public information:

$$
\begin{aligned}
ax_1 &= c_1 \underline{ax_1}, & y_2^{-1} b &= d_1 \underline{y_2^{-1} b}, \\
x_1^{-1} a_1 x_2 &= c_2 \underline{x_1^{-1} a_1 x_2}, & y_1^{-1} b_2 y_2 &= d_2 \underline{y_1^{-1} b_2 y_2}, \\
x_2^{-1} a_2 &= c_3 \underline{x_2^{-1} a_2}, & b_1 y_1 &= d_3 \underline{b_1 y_1}.
\end{aligned}
$$

The central elements are known, given the public information. Then

$$
\begin{aligned}
K &= ax_1 \cdot b_1 y_1 \cdot y_1^{-1} \cdot x_1^{-1} a_1 x_2 \cdot x_2^{-1} y_1 \cdot y_1^{-1} b_2 y_2 \cdot x_2 \cdot x_2^{-1} a_2 \cdot y_2^{-1} b \\
&= c_1 \underline{ax_1} \ d_3 \underline{b_1 y_1} \ y_1^{-1} \ c_2 x_1^{-1} a_1 x_2 \ x_2^{-1} \ y_1 \ d_2 y_1^{-1} b_2 y_2 \ x_2 \ c_3 x_2^{-1} a_2 \ d_1 y_2^{-1} b \\
&= c_1 c_2 c_3 d_1 d_2 d_3 \ \underline{ax_1} \ \underline{b_1 y_1} \ y_1^{-1} \ \underline{x_1^{-1} a_1 x_2} \ x_2^{-1} \ y_1 \ y_1^{-1} b_2 y_2 \ x_2 \ \underline{x_2^{-1} a_2} \ \underline{y_2^{-1} b} \\
&=: c_1 c_2 c_3 d_1 d_2 d_3 K'.
\end{aligned}
$$

Assume that we have an algorithm for computing the shared key out of the public information, that succeeds when the public braids have infimum 0 or 1. Applying this algorithm to the reduced public braids, we obtain the braid K'. Multiplying by the known central braid $c_1 c_2 c_3 d_1 d_2 d_3$, we obtain the original key K. Thus, we may assume that all public braids, as well as the secret braids y_1, y_2, x_1, x_2 have infimum 0 or 1. Assume that, henceforth.

For a braid $x = (i, \mathbf{p})$, let $\ell(x)$ be the number of permutations in the sequence $\mathbf{p}$. For integers i, s, let

$$[i, s] = \{x \in \mathbf{B}_N : i \le \inf(x) \le \inf(x) + \ell(x) \le s\}.$$

We use the following basic facts about $\mathbf{B}_N$:

1. If $x_1 \in [i_1, s_1]$ and $x_2 \in [i_2, s_2]$, then $x_1 x_2 \in [i_1 + i_2, s_1 + s_2]$.
2. If $x \in [i, s]$, then $x^{-1} \in [-s, -i]$.

By our assumption, the key K is a product of 10 braids with infimum 0 or 1, and thus
$$0 \leq \inf(K) \leq 10.$$

Let ℓ be the maximum of the lengths of the private braids in Fig. 6. In the above reduction, we had $K = c_1 c_2 c_3 d_1 d_2 d_3 K'$, and thus
$$\ell(K') = \ell(K) = \ell(ab_1 a_1 b_2 a_2 b) \leq 6\ell.$$

5.3   Reducing to a Matrix Group over a Finite Field

Let n be a natural number. As usual, we denote the algebra of all $n \times n$ matrices over a field $\mathbb{F}$ by $\mathrm{M}_n(\mathbb{F})$, and the group of invertible elements of this algebra by $\mathrm{GL}_n(\mathbb{F})$. A *matrix group* is a subgroup of $\mathrm{GL}_n(\mathbb{F})$. A *faithful representation* of a group G in $\mathrm{GL}_n(\mathbb{F})$ is a group isomorphism from G onto a matrix group $H \leq \mathrm{GL}_n(\mathbb{F})$. A group is *linear* if it has a faithful representation.

Bigelow and Krammer, established in their breakthrough papers [4,13] that the braid group $\mathbf{B}_N$ is linear, by proving that the so-called *Lawrence–Krammer representation*
$$\mathrm{LK} \colon \mathbf{B}_N \longrightarrow \mathrm{GL}_{\binom{N}{2}}\left(\mathbb{Z}[t^{\pm 1}, \tfrac{1}{2}]\right),$$
whose dimension is
$$n := \binom{N}{2},$$
is injective. The Lawrence–Krammer representation of a braid can be computed in polynomial time. This representation is also invertible in (similar) polynomial time [8,13].

Theorem 6 (Cheon–Jun [8]). *Let $x \in [i, s]$ in $\mathbf{B}_N$. Let $M \geq \max(|i|, |s|)$. Then:*

1. *The degrees of t in $\mathrm{LK}(x) \in \mathrm{GL}_n(\mathbb{Z}[t^{\pm 1}, \tfrac{1}{2}])$ are in $\{-M, -M+1, \ldots, M\}$.*
2. *The rational coefficients $\frac{c}{2^d}$ in $\mathrm{LK}(x)$ (c integer, d nonnegative integer) satisfy: $|c| \leq 2^{N^2 M}, |d| \leq 2NM$.*

In the notation of Theorem 6, Theorem 2 in Cheon–Jun [8] implies that inversion of $\mathrm{LK}(x)$ is of order $N^6 \log M$ multiplications of entries. Ignoring logarithmic factors and thus assuming that each entry multiplication costs $NM \cdot N^2 M = N^3 M^2$, this accumulates to $N^8 M^2$. We also invert the function LK as part of our cryptanalysis. However, the complexity of the computation of the algebraic spans dominates the complexity of these transformations, that are applied only at the beginning and at the end of the cryptanalysis.

Let us return to the Triple Decomposition key exchange protocol. After infimum reduction (who's complexity is negligible), we have

$$K \in [0, 10 + 6\ell].$$

Let $M := 10 + 6\ell$. By the Cheon–Jun Theorem, we have

$$(2^{2NM}t^M) \cdot \mathrm{LK}(K) \in \mathrm{GL}_n(\mathbb{Z}[t]),$$

the absolute values of the coefficients in this matrix are bounded by $2^{N^2(M+1)}$, and the maximal degree of t in this matrix is bounded by $2M$.

Let p be a prime slightly greater than 2^{N^2M+2NM}, and $f(t)$ be an irreducible polynomial over $\mathbb{Z}_p$, of degree d slightly larger than $2M$. Then

$$(2^{2NM}t^M) \cdot \mathrm{LK}(K) = (2^{2NM}t^M) \cdot \mathrm{LK}(K) \bmod (p, f(t)) \in \mathrm{GL}_n(\mathbb{Z}[t]/\langle p, f(t)\rangle),$$

under the natural identification of $\{-(p-1)/2, \ldots, (p-1)/2\}$ with $\{0, \ldots, p-1\}$.

Let $\mathbb{F} = \mathbb{Z}[t]/\langle p, f(t)\rangle = \mathbb{Z}[t^{\pm 1}, \frac{1}{2}]/\langle p, f(t)\rangle$. $\mathbb{F}$ is a finite field of cardinality p^d, where d is the degree of $f(t)$. It follows that the complexity of field operations in $\mathbb{F}$ is, up to logarithmic factors, of order

$$d^2 \log p = O(M^3 N^2).$$

Thus, the key K can be recovered as follows:

1. Apply the composed function $\mathrm{LK}(x) \bmod (p, f(t))$ to the input braids, to obtain a version of this problem in $\mathrm{GL}_n(\mathbb{F})$.
2. Solve the problem there, to obtain $\mathrm{LK}(K) \bmod (p, f(t))$.
3. Compute $(2^{2NM}t^M) \cdot \mathrm{LK}(K) \bmod (p, f(t)) = (2^{2NM}t^M) \cdot \mathrm{LK}(K)$.[2]
4. Divide by $(2^{2NM}t^M)$ to obtain $\mathrm{LK}(K)$.
5. Compute K using the Cheon–Jun inversion algorithm.

The complexity of this *preliminary* cryptanalysis is $O(kn^6)$ field operations, where k is the maximum number of generators in the given subgroups, and n is of order N^2. Roughly, this is $kN^{12} \cdot M^3N^2 = kN^{14}(10 + 6\ell)^3$.

5.4   Reducing the Complexity

To make this cryptanalysis feasible, at least for mildly large parameters, we can improve upon the field multiplication complexity. We do this by applying the Chinese Remainder Theorem (CRT) on both the integer part and the polynomial part. Let

$$p_1, p_2, \cdots = 2, 3, 5, \ldots,$$

the sequence of prime numbers. Also, consider the relatively prime polynomials, or degree 1,

$$x, x \pm 1, x \pm 2, \ldots$$

---

[2] The equality here is over the integers.

We take just enough primes so that their products exceeds $2^{N^2(M+1)}$, and we take the first $2M$ polynomials in our list.

For each pair $(p, f(t))$ of a prime and a polynomial in our lists, we reduce modulo the prime and the polynomial, and apply the cryptanalysis, over the resulting p-element field. In the end, we combine all results using the CRT. Since CRT is done only once, its complexity is dominated by the complexity of the linear span calculations.

Since we ignore logarithmic factors, it suffices to estimate the complexity of the same algorithm, but using just a single prime $p \approx 2^{N^2 M}$. For each linear polynomial, the obtained field size is p, and thus field multiplication is, up to logarithmic factors, of complexity $N^2 M$. We need to repeat this M times. The complexity of each of the M steps is dominated by the computation of the algebraic span, which is of complexity $O(kn^6) = O(kN^{12})$. Thus, the overall complexity is roughly of order

$$M \cdot kN^{12} \cdot N^2 M = kN^{14}M^2 \approx kN^{14}\ell^2.$$

5.5 Specifications of the Triple Decomposition Key Exchange Protocol

We now describe the groups proposed for the actual specification of the Triple Decomposition key exchange protocol. Let $m \geq 2$ be a natural number, and $N := 3m+1$. The braid group $\mathbf{B}_N$ is generated by $N-1$ generators, $\sigma_1, \ldots, \sigma_{N-1}$. One of their defining relations is that σ_i and σ_j commute whenever $|i - j| > 1$. The groups in Fig. 6 are chosen as follows: Fix "generic" braids $g_1, g_2, h_1, h_2 \in \mathbf{B}_N$. Then:

$$A = B = \mathbf{B}_N$$
$$A_1 = \langle \sigma_1^{g_1}, \ldots, \sigma_{m-1}^{g_1} \rangle; \quad Y_1 = \langle \sigma_{m+1}^{g_1}, \ldots, \sigma_{mk}^{g_1} \rangle$$
$$A_2 = \langle \sigma_1^{g_2}, \ldots, \sigma_{m-1}^{g_2} \rangle; \quad Y_2 = \langle \sigma_{m+1}^{g_2}, \ldots, \sigma_{3m}^{g_2} \rangle$$
$$X_1 = \langle \sigma_1^{h_1}, \ldots, \sigma_{2m-1}^{h_1} \rangle; \quad B_1 = \langle \sigma_{2m+1}^{h_1}, \ldots, \sigma_{3m}^{h_1} \rangle$$
$$X_2 = \langle \sigma_1^{h_2}, \ldots, \sigma_{2m-1}^{h_2} \rangle; \quad B_2 = \langle \sigma_{2m+1}^{h_2}, \ldots, \sigma_{3m}^{h_2} \rangle$$

The conjugations prevent an otherwise trivial cryptanalysis [15]. It follows that the parameter k in the complexity estimation is $2m$. This is the same order as $N = 3m$. Thus, the complexity of the cryptanalysis is, roughly, of order $N^{15}\ell^2$.

The value ℓ depends on the way the secret braids are generated. This was never specified exactly. Comparing to more detailed proposals, it is fair to estimate that ℓ is of order much smaller than N.

5.6 Implementation

This type of provable cryptanalyses is generally considered of theoretical interest only [8, 28]. The algorithmic shortcuts described above made it possible, for the first time, to launch our attack on concrete instances, including ones where brute force or naive attacks are infeasible. Being provable, the attacks must find the shared key in all tests, and this provides a "sanity check" for our mathematical reasoning. The attacks were implemented on the computational algebra software MAGMA [6], with no optimizations beyond those specified above. Infimum reduction was not implemented, since for generic braids it has little effect.

For a length parameter l, we chose the braids g_1, g_2, h_1, h_2 as products of l random elements from the set $\{\sigma_1^{\pm 1}, \ldots, \sigma_{N-1}^{\pm 1}\}$. The braids in the subgroup were generated as products of l random generators of that subgroup. Since the running time was long, and we already know that the attacks provable succeed, we conducted only one attack for each set of parameters. Each attack was launched on a single core of a standard desktop CPU. The results are summarized in Tables 1 and 2.

Table 1. Experimental results with MAGMA, single CPU core, $N = 10 = 3 \cdot 3 + 1$

Length	Time	Memory (MB)	Key recovered?
2	114 s	30	Yes
4	10 min	33	Yes
8	38 min	38	Yes
16	3.5 h	108	Yes
32	49 h	249	Yes
64	629 h	640	Yes

Table 2. Experimental results with MAGMA, single CPU core, $N = 13 = 3 \cdot 4 + 1$

Length	Time	Memory (MB)	Key recovered?
2	9 min	195	Yes
4	12 min	201	Yes
8	5 h	215	Yes
16	19 h	548	Yes
32	298 h	1289	Yes

The attacks are highly parallelizable. Larger parameters would necessitate parallel implementations over large grids.

6   Conclusions

We have introduced algebraic span cryptanalysis, a provable method for cryptanalysing nonabelian cryptographic protocols and, more generally, solving computational problems in groups. This method applies to all groups with efficient, faithful representations as matrix groups. The examples provided demonstrate the power, generality, and simplicity of this method.

The novelty of this method is demonstrated by showing that it applies to a protocol that was not approachable by earlier methods. The new method cleared out much of the difficulty of the computational problem behind the Triple Decomposition key exchange protocol, and made it possible for us to find the extra idea to make it work.

Initially considered of theoretical interest only, provable cryptanalysis is now a feasible threat to nonabelian cryptographic protocols. It seems very challenging to devise a nonabelian key exchange protocol that cannot be cryptanalyzed by the algebraic span method.

Acknowledgments. We thank Avraham (Rami) Eizenbud and Craig Gentry for intriguing discussions. A part of this work was carried out while the third named author was on Sabbatical at the Weizmann Institute of Science. This author thanks his hosts for their kind hospitality. The research of the first and third named authors was partially supported by the European Research Council under the ERC starting grant n. 757731 (LightCrypt), and by the BIU Center for Research in Applied Cryptography and Cyber Security, in conjunction with the Israel National Cyber Bureau in the Prime Minister's Office.

References

1. Andrecut, M.: A matrix public key cryptosystem, arXiv eprint 1506.00277 (2015)
2. Anshel, I., Anshel, M., Goldfeld, D.: An algebraic method for public-key cryptography. Math. Res. Lett. **6**, 287–291 (1999)
3. Ben-Zvi, A., Blackburn, S.R., Tsaban, B.: A practical cryptanalysis of the algebraic eraser. In: Robshaw, M., Katz, J. (eds.) CRYPTO 2016. LNCS, vol. 9814, pp. 179–189. Springer, Heidelberg (2016). https://doi.org/10.1007/978-3-662-53018-4_7
4. Bigelow, S.: Braid groups are linear. J. Am. Math. Soc. **14**, 471–486 (2001)
5. Mullan, C.: Cryptanalysing variants of Stickel's key agreement scheme. J. Math. Cryptol. **4**, 365–373 (2011)
6. Bosma, W., Cannon, J., Playoust, C.: The Magma algebra system. I. The user language. J. Symb. Comput. **24**, 235–265 (1997)
7. Cha, J.C., Ko, K.H., Lee, S.J., Han, J.W., Cheon, J.H.: An efficient implementation of braid groups. In: Boyd, C. (ed.) ASIACRYPT 2001. LNCS, vol. 2248, pp. 144–156. Springer, Heidelberg (2001). https://doi.org/10.1007/3-540-45682-1_9
8. Cheon, J.H., Jun, B.: A polynomial time algorithm for the braid Diffie-Hellman conjugacy problem. In: Boneh, D. (ed.) CRYPTO 2003. LNCS, vol. 2729, pp. 212–225. Springer, Heidelberg (2003). https://doi.org/10.1007/978-3-540-45146-4_13
9. Gilman, R., Myasnikov, A., Myasnikov, A., Ushakov, A.: New developments in commutator key exchange. In: Proceedings of the First International Conference on Symbolic Computation and Cryptography, Beijing, pp. 146–150 (2008). http://www-calfor.lip6.fr/~jcf/Papers/scc08.pdf
10. González-Vasco, M., Steinwandt, R.: Group Theoretic Cryptography. Cryptography and Network Security Series. Chapman and Hall/CRC Press, Boca Raton (2015)
11. Holt, D.: Answer to MathOverflow question. http://mathoverflow.net/questions/154761

12. Ko, K.H., Lee, S.J., Cheon, J.H., Han, J.W., Kang, J., Park, C.: New public-key cryptosystem using braid groups. In: Bellare, M. (ed.) CRYPTO 2000. LNCS, vol. 1880, pp. 166–183. Springer, Heidelberg (2000). https://doi.org/10.1007/3-540-44598-6_10

13. Krammer, D.: Braid groups are linear. Ann. Math. **155**, 131–156 (2002)

14. Kurt, Y.: A new key exchange primitive based on the triple decomposition problem, IACR eprint 2006/378

15. Peker, Y.K.: A new key agreement scheme based on the triple decomposition problem. Int. J. Netw. Secur. **16**, 340–350 (2014)

16. Myasnikov, A., Shpilrain, V., Ushakov, A.: Group-Based Cryptography. Birkhäuser, Basel (2008). https://doi.org/10.1007/978-3-7643-8827-0

17. Myasnikov, A., Shpilrain, V., Ushakov, A.: Non-commutative Cryptography and Complexity of Group-Theoretic Problems, vol. 177. American Mathematical Society Surveys and Monographs, Providence (2011)

18. Myasnikov, A., Roman'kov, V.: A linear decomposition attack. Groups Complex. Cryptol. **7**, 81–94 (2015)

19. Roman'kov, V.: Algebraic Cryptography. Omsk State Dostoevsky University, Omsk (2013). (In Russian)

20. Roman'kov, V.: Cryptanalysis of some schemes applying automorphisms. Prikladnaya Discretnaya Matematika **3**, 35–51 (2013). (In Russian)

21. Roman'kov, V.: A nonlinear decomposition attack. Groups Complex. Cryptol. **8**, 197–207 (2016)

22. Roman'kov, V., Obzor, A.: A general encryption scheme using multiplications with cryptanalysis. Prikladnaya Discretnaya Matematika **37**, 52–61 (2017). (In Russian)

23. Shpilrain, V.: Cryptanalysis of Stickel's key exchange scheme. In: Hirsch, E.A., Razborov, A.A., Semenov, A., Slissenko, A. (eds.) CSR 2008. LNCS, vol. 5010, pp. 283–288. Springer, Heidelberg (2008). https://doi.org/10.1007/978-3-540-79709-8_29

24. Shpilrain, V., Ushakov, A.: Thompson's group and public key cryptography. In: Ioannidis, J., Keromytis, A., Yung, M. (eds.) ACNS 2005. LNCS, vol. 3531, pp. 151–163. Springer, Heidelberg (2005). https://doi.org/10.1007/11496137_11

25. Shpilrain, V., Ushakov, A.: A new key exchange protocol based on the decomposition problem. In: Gerritzen, L., Goldfeld, D., Kreuzer, M., Rosenberger, G., Shpilrain, V. (eds.) Algebraic Methods in Cryptography. Contemporary Mathematics, vol. 418, pp. 161–167 (2006)

26. Stickel, E.: A new method for exchanging secret keys. In: Proceedings of the Third International Conference on Information Technology and Applications (ICITA 2005), pp. 426–430 (2005)

27. Tsaban, B.: The Conjugacy Problem: cryptoanalytic approaches to a problem of Dehn. minicourse, Düsseldorf University, Germany, July–August 2012. http://reh.math.uni-duesseldorf.de/~gcgta/slides/Tsaban_minicourses.pdf

28. Tsaban, B.: Polynomial-time solutions of computational problems in noncommutative-algebraic cryptography. J. Cryptol. **28**, 601–622 (2015)

Improved Division Property Based Cube Attacks Exploiting Algebraic Properties of Superpoly

Qingju Wang[1,2,3], Yonglin Hao[4(✉)], Yosuke Todo[5(✉)], Chaoyun Li[6(✉)], Takanori Isobe[7], and Willi Meier[8]

[1] Shanghai Jiao Tong University, Shanghai, China
[2] Technical University of Denmark, Kongens Lyngby, Denmark
[3] SnT, University of Luxembourg, Esch-sur-Alzette, Luxembourg
qingju.wang@uni.lu
[4] State Key Laboratory of Cryptology, Beijing, China
haoyonglin@yeah.net
[5] NTT Secure Platform Laboratories, Tokyo, Japan
todo.yosuke@lab.ntt.co.jp
[6] imec-COSIC, Department of Electrical Engineering (ESAT),
KU Leuven, Leuven, Belgium
chaoyun.li@esat.kuleuven.be
[7] University of Hyogo, Kobe, Japan
takanori.isobe@ai.u-hyogo.ac.jp
[8] FHNW, Windisch, Switzerland
willi.meier@fhnw.ch

Abstract. The cube attack is an important technique for the cryptanalysis of symmetric key primitives, especially for stream ciphers. Aiming at recovering some secret key bits, the adversary reconstructs a superpoly with the secret key bits involved, by summing over a set of the plaintexts/IV which is called a cube. Traditional cube attack only exploits linear/quadratic superpolies. Moreover, for a long time after its proposal, the size of the cubes has been largely confined to an experimental range, e.g., typically 40. These limits were first overcome by the division property based cube attacks proposed by Todo et al. at CRYPTO 2017. Based on MILP modelled division property, for a cube (index set) I, they identify the small (index) subset J of the secret key bits involved in the resultant superpoly. During the precomputation phase which dominates the complexity of the cube attacks, $2^{|I|+|J|}$ encryptions are required to recover the superpoly. Therefore, their attacks can only be available when the restriction $|I| + |J| < n$ is met.

In this paper, we introduced several techniques to improve the division property based cube attacks by exploiting various algebraic properties of the superpoly.

1. We propose the "flag" technique to enhance the preciseness of MILP models so that the proper non-cube IV assignments can be identified to obtain a non-constant superpoly.

© International Association for Cryptologic Research 2018
H. Shacham and A. Boldyreva (Eds.): CRYPTO 2018, LNCS 10991, pp. 275–305, 2018.
https://doi.org/10.1007/978-3-319-96884-1_10

2. A degree evaluation algorithm is presented to upper bound the degree of the superpoly. With the knowledge of its degree, the superpoly can be recovered without constructing its whole truth table. This enables us to explore larger cubes I's even if $|I| + |J| \geq n$.
3. We provide a term enumeration algorithm for finding the monomials of the superpoly, so that the complexity of many attacks can be further reduced.

As an illustration, we apply our techniques to attack the initialization of several ciphers. To be specific, our key recovery attacks have mounted to 839-round TRIVIUM, 891-round Kreyvium, 184-round Grain-128a and 750-round ACORN respectively.

Keywords: Cube attack · Division property · MILP · TRIVIUM Kreyvium · Grain-128a · ACORN · Clique

1   Introduction

Cube attack, proposed by Dinur and Shamir [1] in 2009, is one of the general cryptanalytic techniques of analyzing symmetric-key cryptosystems. After its proposal, cube attack has been successfully applied to various ciphers, including stream ciphers [2–6], hash functions [7–9], and authenticated encryptions [10,11]. For a cipher with n secret variables $\boldsymbol{x} = (x_1, x_2, \ldots, x_n)$ and m public variables $\boldsymbol{v} = (v_1, v_2, \ldots, v_m)$, we can regard the algebraic normal form (ANF) of output bits as a polynomial of $\boldsymbol{x}$ and $\boldsymbol{v}$, denoted as $f(\boldsymbol{x}, \boldsymbol{v})$. For a randomly chosen set $I = \{i_1, i_2, \ldots, i_{|I|}\} \subset \{1, \ldots, m\}$, $f(\boldsymbol{x}, \boldsymbol{v})$ can be represented uniquely as

$$f(\boldsymbol{x}, \boldsymbol{v}) = t_I \cdot p(\boldsymbol{x}, \boldsymbol{v}) + q(\boldsymbol{x}, \boldsymbol{v}),$$

where $t_I = v_{i_1} \cdots v_{i_{|I|}}$, $p(\boldsymbol{x}, \boldsymbol{v})$ only relates to v_s's ($s \notin I$) and the secret key bits $\boldsymbol{x}$, and $q(\boldsymbol{x}, \boldsymbol{v})$ misses at least one variable in t_I. When v_s's ($s \notin I$) and $\boldsymbol{x}$ are assigned statically, the value of $p(\boldsymbol{x}, \boldsymbol{v})$ can be computed by summing the output bit $f(\boldsymbol{x}, \boldsymbol{v})$ over a structure called *cube*, denoted as C_I, consisting of $2^{|I|}$ different $\boldsymbol{v}$ vectors with $v_i, i \in I$ being active (traversing all 0-1 combinations) and non-cube indices $v_s, s \notin I$ being static constants. Traditional cube attacks are mainly concerned about linear or quadratic superpolies. By collecting linear or quadratic equations from the superpoly, the attacker can recover some secret key bits information during the online phase. Aiming to mount distinguishing attack by property testing, cube testers are obtained by evaluating superpolies of carefully selected cubes. In [2], probabilistic tests are applied to detect some algebraic properties such as constantness, low degree and sparse monomial distribution. Moreover, cube attacks and cube testers are acquired experimentally by summing over randomly chosen cubes. So the sizes of the cubes are largely confined. Breakthroughs have been made by Todo et al. in [12] where they introduce the bit-based division property, a tool for conducting integral attacks[1],

---

[1] Integral attacks also require to traverse some active plaintext bits and check whether the summation of the corresponding ciphertext bits have zero-sum property, which equals to check whether the superpoly has $p(\boldsymbol{x}, \boldsymbol{v}) \equiv 0$.

to the realm of cube attack. With the help of mixed integer linear programming (MILP) aided division property, they can identify the variables excluded from the superpoly and explore cubes with larger size, e.g., 72 for 832-round TRIVIUM. This enables them to improve the traditional cube attack.

Division property, as a generalization of the integral property, was first proposed at EUROCRYPT 2015 [13]. With division property, the propagation of the integral characteristics can be deduced in a more accurate manner, and one prominent application is the first theoretic key recovery attack on full MISTY1 [14].

The original division property can only be applied to word-oriented primitives. At FSE 2016, bit-based division property [15] was proposed to investigate integral characteristics for bit-based block ciphers. With the help of division property, the propagation of the integral characteristics can be represented by the operations on a set of 0-1 vectors identifying the bit positions with the zero-sum property. Therefore, for the first time, integral characteristics for bit-based block ciphers SIMON32 and Simeck32 have been proved. However, the sizes of the 0-1 vector sets are exponential to the block size of the ciphers. Therefore, as has been pointed out by the authors themselves, the deduction of bit-based division property under their framework requires high memory for block ciphers with larger block sizes, which largely limits its applications. Such a problem has been solved by Xiang et al. [16] at ASIACRYPT 2016 by utilizing the MILP model. The operations on 0-1 vector sets are transformed to imposing division property values (0 or 1) to MILP variables, and the corresponding integral characteristics are acquired by solving the models with MILP solvers like Gurobi [17]. With this method, they are able to give integral characteristics for block ciphers with block sizes much larger than 32 bits. Xiang et al.'s method has now been applied to many other ciphers for improved integral attacks [18–21].

In [12], Todo et al. adapt Xiang et al.'s method by taking key bits into the MILP model. With this technique, a set of key indices $J = \{j_1, j_2, \ldots, j_{|J|}\} \subset \{1, \ldots, n\}$ is deduced for the cube C_I s.t. $p(\boldsymbol{x}, \boldsymbol{v})$ can only be related to the key bits x_j's ($j \in J$). With the knowledge of I and J, Todo et al. can recover 1-bit of secret-key-related information by executing two phases. In the *offline phase*, a proper assignment to the non-cube IVs, denoted by $\boldsymbol{IV} \in \mathbb{F}_2^m$, is determined ensuring $p(\boldsymbol{x}, \boldsymbol{IV})$ non-constant. Also in this phase, the whole truth table of $p(\boldsymbol{x}, \boldsymbol{IV})$ is constructed through cube summations. In the *online phase*, the exact value of $p(\boldsymbol{x}, \boldsymbol{IV})$ is acquired through a cube summation and the candidate values of x_j's ($j \in J$) are identified by checking the precomputed truth table. A proportion of wrong keys are filtered as long as $p(\boldsymbol{x}, \boldsymbol{IV})$ is non-constant.

Due to division property and the power of MILP solver, cubes of larger dimension can now be used for key recoveries. By using a 72-dimensional cube, Todo et al. propose a theoretic cube attack on 832-round TRIVIUM. They also largely improve the previous best attacks on other primitives namely ACORN, Grain-128a and Kreyvium [12,22]. It is not until recently that the result on TRIVIUM has been improved by Liu et al. [6] mounting to 835 rounds with a new method called the *correlation* cube attack. The correlation attack is based on the *numeric mapping* technique first appeared in [23] originally used for constructing zero-sum distinguishers.

1.1 Motivations

Due to [12,22], the power of cube attacks has been enhanced significantly, however, there are still problems remaining unhandled that we will reveal explicitly.

Finding Proper IV's May Require Multiple Trials. As is mentioned above, the superpoly can filter wrong keys only if a *proper IV assignment* $IV \in \mathbb{F}_2^m$ in the constant part of IVs is found such that the corresponding superpoly $p(x, IV)$ is non-constant. The MILP model in [12,22] only proves the existence of the proper IV's but finding them may not be easy. According to practical experiments, there are quite some IV's making $p(x, IV) \equiv 0$. Therefore, $t \geq 1$ different IV's might be trailed in the precomputation phase before finding a proper one. Since each IV requires to construct a truth table with complexity $2^{|I|+|J|}$, the overall complexity of the offline phase can be $t \times 2^{|I|+|J|}$. When large cubes are used ($|I|$ is big) or many key bits are involved ($|J|$ is large), such a complexity might be at the risk of exceeding the brute-force bound 2^n. Therefore, two assumptions are made to validate their cube attacks as follows.

Assumption 1 (Strong Assumption). *For a cube C_I, there are many values in the constant part of IV whose corresponding superpoly is balanced.*

Assumption 2 (Weak Assumption). *For a cube C_I, there are many values in the constant part of IV whose corresponding superpoly is not a constant function.*

These assumptions are proposed to guarantee the validity of the attacks as long as $|I| + |J| < n$, but the rationality of such assumptions is hard to be proved, especially when $|I| + |J|$ are so close to n in many cases. The best solution is to evaluate different IVs in the MILP model so that the proper IV of the constant part of IVs and the set J are determined simultaneously before implementing the attack.

Restriction of $|I| + |J| < n$. The superpoly recovery has always been dominating the complexity of the cube attack, especially in [12], the attacker knows no more information except which secret key bits are involved in the superpoly. Then she/he has to first construct the whole truth table for the superpoly in the offline phase. In general, the truth-table construction requires repeating the cube summation $2^{|J|}$ times, and makes the complexity of the offline phase about $2^{|I|+|J|}$. Apparently, such an attack can only be meaningful if $|I|+|J| < n$, where n is the number of secret variables. The restriction of $|I|+|J| < n$ barricades the adversary from exploiting cubes of larger dimension or mounting more rounds (where $|J|$ may expand). This restriction can be removed if we can avoid the truth table construction in the offline phase.

1.2 Our Contributions

This paper improves the existing cube attacks by exploiting the algebraic properties of the superpoly, which include the (non-)constantness, low degree and sparse monomial distribution properties. Inspired by the division property based cube attack work of Todo et al. in [12], we formulate all these properties in one framework by developing more precise MILP models, thus we can reduce the complexity of superpoly recovery.

This also enables us to attack more rounds, or employ even larger cubes. Similar to [12], our methods regard the cryptosystem as a non-blackbox polynomial and can be used to evaluate cubes with large dimension compared with traditional cube attack and cube tester. In the following, our contributions are summarized into five aspects.

Flag Technique for Finding Proper IV Assignments. The previous MILP model in [12] has not taken the effect of constant 0/1 bits of the constant part of IVs into account. In their model, the active bits are initialized with division property value 1 and other non-active bits are all initialized to division property value 0. The non-active bits include constant part of IVs, together with some secret key bits and state bits that are assigned statically to 0/1 according to the specification of ciphers. It has been noticed in [22] that constant 0 bits can affect the propagation of division property. But we should pay more attention to constant 1 bits since constant 0 bits can be generated in the updating functions due to the XOR of even number of constant 1's. Therefore, we propose a formal technique which we refer as the "flag" technique where the constant 0 and constant 1 as well as other non-constant MILP variables are treated properly. With this technique, we are able to find proper assignments to constant IVs $(\boldsymbol{IV})$ that makes the corresponding superpoly $(p(\boldsymbol{x}, \boldsymbol{IV}))$ non-constant. With this technique, proper IVs can now be found with MILP model rather than time-consuming trial & summations in the offline phase as in [12,22]. According to our experiments, the flag technique has a perfect 100% accuracy for finding proper non-cube IV assignments in most cases. Note that our flag technique has partially proved the availability of the two assumptions since we are able to find proper $\boldsymbol{IV}$'s in all our attacks.

Degree Evaluation for Going Beyond the $|I| + |J| < n$ Restriction. To avoid constructing the whole truth table using cube summations, we introduce a new technique that can upper bound the algebraic degree, denoted as d, of the superpoly using the MILP-aided bit-based division property. With the knowledge of its degree d (and key indices J), the superpoly can be represented with its $\binom{|J|}{\leq d}$ coefficients rather than the whole truth table, where $\binom{|J|}{\leq d}$ is defined as

$$\binom{|J|}{\leq d} := \sum_{i=0}^{d} \binom{|J|}{i}. \tag{1}$$

When $d = |J|$, the complexity by our new method and that by [12] are equal. For $d < |J|$, we know that the coefficients of the monomials with degree higher than d are constantly 0. The complexity of superpoly recovery can be reduced from $2^{|I|+|J|}$ to $2^{|I|} \times \binom{|J|}{\leq d}$. In fact, for some lightweight ciphers, the algebraic degrees of their round functions are quite low. Therefore, the degrees d are often much smaller than the number of involved key bits $|J|$, especially when high-dimensional cubes are used. Since $d \ll |J|$ for all previous attacks, we can improve the complexities of previous results and use larger cubes mounting to more rounds even if $|I| + |J| \geq n$.

Precise Term Enumeration for Further Lowering Complexities. Since the superpolies are generated through iterations, the number of higher-degree monomials in the superpoly is usually much smaller than its low-degree counterpart. For example, when the degree of the superpoly is $d < |J|$, the number of d-degree monomials are usually much smaller than the upper bound $\binom{|J|}{d}$. We propose a MILP model technique for enumerating all t-degree ($t = 1, \ldots, d$) monomials that may appear in the superpoly, so that the complexities of several attacks are further reduced.

Relaxed Term Enumeration. For some primitives (such as 750-round ACORN), our MILP model can only enumerate the d-degree monomials since the number of lower-degree monomials are too large to be exhausted. Alternately, for $t = 1, \ldots, d - 1$, we can find a set of key indices $JR_t \subseteq J$ s.t. all t-degree monomials in the superpoly are composed of x_j, $j \in JR_t$. As long as $|JR_t| < |J|$ for some $t = 1, \ldots, d - 1$, we can still reduce the complexities of superpoly recovery.

Combining the flag technique and the degree evaluation above, we are able to lower the complexities of the previous best cube attacks in [6,12,22]. Particularly, we can further provide key recovery results on 839-round TRIVIUM[2], 891-round Kreyvium, 184-round Grain-128a, and 750-round ACORN. Furthermore, the precise & relaxed term enumeration techniques allow us to lower the complexities of 833-round TRIVIUM, 849-round Kreyvium, 184-round Grain-128a and 750-round ACORN. Our concrete results are summarized in Table 1.[3] In [26], Todo et al. revisit the fast correlation attack and analyze the key-stream generator (rather than the initialization) of the Grain family (Grain-128a, Grain-128, and Grain-v1). As a result, the key-stream generators of the Grain family are insecure. In other words, they can recover the internal state after initialization more efficiently than by exhaustive search. And the secret key is recovered from the internal state because the initialization is a public permutation. To the best of our knowledge, all our results of Kreyvium, Grain-128a, and ACORN are the current best key recovery attacks on the initialization of the targeted ciphers. However, none of our results seems to threaten the security of the ciphers.

Clique View of the Superpoly Recovery. In order to lower the complexity of the superpoly recovery, the term enumeration technique has to execute many MILP instances, which is difficult for some applications. We represent the resultant superpoly as a graph, so that we can utilize the clique concept from the

---

[2] While this paper was under submission, Fu et al. released a paper on ePrint [24] and claimed that 855 rounds initialization of TRIVIUM can be attacked.

[3] Because of the page limitation, we put part of detailed applications about Kreyvium, Grain-128a and ACORN in the full version [25].

graph theory to upper bound the complexity of the superpoly recovery phase, without requiring MILP solver as highly as the term enumeration technique.

Organizations. Section. 2 provides the background of cube attacks, division property, MILP model etc. Section 3 introduces our flag technique for identifying proper assignments to non-cube IVs. Section 4 details the degree evaluation technique upper bounding the algebraic degree of the superpoly. Combining the flag technique and degree evaluation, we give improved key recovery cube attacks on 4 targeted ciphers in Sect. 5. The precise & relaxed term enumeration as well as their applications are given in Sect. 6. We revisit the term enumeration technique from the clique overview in Sect. 7. Finally, we conclude in Sect. 8.

Table 1. Summary of our cube attack results

Applications	#Full rounds	#Rounds	Cube size	$|J|$	Complexity	Reference
Trivium	1152	799	32†	–	Practical	[4]
		832	72	5	2^{77}	[12,22]
		833	**73**	**7**	$2^{76.91}$	Sect. 6.1
		835	37/36*	–	2^{75}	[6]
		836	**78**	**1**	2^{79}	Sect. 5.1
		839	**78**	**1**	2^{79}	Sect. 5.1
Kreyvium	1152	849	61	23	2^{84}	[22]
		849	**61**	**23**	$2^{81.7}$	**Full version [25]**
		849	**61**	**23**	$2^{73.41}$	Sect. 6.2
		872	85	39	2^{124}	[22]
		872	**85**	**39**	$2^{94.61}$	**Full version [25]**
		891	**113**	**20**	$2^{120.73}$	**Full version [25]**
Grain-128a	256	177	33	–	Practical	[27]
		182	88	18	2^{106}	[12,22]
		182	**88**	**14**	2^{102}	**Full version [25]**
		183	92	16	2^{108}	[12,22]
		183	**92**	**16**	$2^{108} - 2^{96.08}$	**Full version [25]**
		184	**95**	**21**	$2^{109.61}$	Sect. 6.3
ACORN	1792	503	5‡	-	Practical‡	[5]
		704	64	58	2^{122}	[12,22]
		704	**64**	**63**	$2^{77.88}$	Sect. 6.4
		750	**101**	**81**	$2^{125.71}$	**Full version [25]**
		750	**101**	**81**	$2^{120.92}$	Sect. 6.4

†18 cubes whose size is from 32 to 37 are used, where the most efficient cube is shown to recover one bit of the secret key.

*28 cubes of sizes 36 and 37 are used, following the correlation cube attack scenario. It requires an additional 2^{51} complexity for preprocessing.

‡The attack against 477 rounds is mainly described for the practical attack in [5]. However, when the goal is the superpoly recovery and to recover one bit of the secret key, 503 rounds are attacked.

2 Preliminaries

2.1 Mixed Integer Linear Programming

MILP is an optimization or feasibility program whose variables are restricted to integers. A MILP model $\mathcal{M}$ consists of variables $\mathcal{M}.var$, constraints $\mathcal{M}.con$, and an objective function $\mathcal{M}.obj$. MILP models can be solved by solvers like Gurobi [17]. If there is no feasible solution at all, the solver simply returns *infeasible*. If no objective function is assigned, the MILP solver only evaluates the feasibility of the model. The application of MILP model to cryptanalysis dates back to the year 2011 [28], and has been widely used for searching characteristics corresponding to various methods such as differential [29,30], linear [30], impossible differential [31,32], zero-correlation linear [31], and integral characteristics with division property [16]. We will detail the MILP model of [16] later in this section.

2.2 Cube Attack

Considering a stream cipher with n secret key bits $\boldsymbol{x} = (x_1, x_2, \ldots, x_n)$ and m public initialization vector (IV) bits $\boldsymbol{v} = (v_1, v_2, \ldots, v_m)$. Then, the first output keystream bit can be regarded as a polynomial of $\boldsymbol{x}$ and $\boldsymbol{v}$ referred as $f(\boldsymbol{x}, \boldsymbol{v})$. For a set of indices $I = \{i_1, i_2, \ldots, i_{|I|}\} \subset \{1, 2, \ldots, n\}$, which is referred as cube indices and denote by t_I the monomial as $t_I = v_{i_1} \cdots v_{i_{|I|}}$, the algebraic normal form (ANF) of $f(\boldsymbol{x}, \boldsymbol{v})$ can be uniquely decomposed as

$$f(\boldsymbol{x}, \boldsymbol{v}) = t_I \cdot p(\boldsymbol{x}, \boldsymbol{v}) + q(\boldsymbol{x}, \boldsymbol{v}),$$

where the monomials of $q(\boldsymbol{x}, \boldsymbol{v})$ miss at least one variable from $\{v_{i_1}, v_{i_2}, \ldots, v_{i_{|I|}}\}$. Furthermore, $p(\boldsymbol{x}, \boldsymbol{v})$, referred as the superpoly in [1], is irrelevant to $\{v_{i_1}, v_{i_2}, \ldots, v_{i_{|I|}}\}$. The value of $p(\boldsymbol{x}, \boldsymbol{v})$ can only be affected by the secret key bits $\boldsymbol{x}$ and the assignment to the non-cube IV bits v_s ($s \notin I$). For a secret key $\boldsymbol{x}$ and an assignment to the non-cube IVs $\boldsymbol{IV} \in \mathbb{F}_2^m$, we can define a structure called cube, denoted as $C_I(\boldsymbol{IV})$, consisting of $2^{|I|}$ 0-1 vectors as follows:

$$C_I(\boldsymbol{IV}) := \{\boldsymbol{v} \in \mathbb{F}_2^m : \boldsymbol{v}[i] = 0/1, i \in I \bigwedge \boldsymbol{v}[s] = \boldsymbol{IV}[s], s \notin I\}. \tag{2}$$

It has been proved by Dinur and Shamir [1] that the value of superpoly p corresponding to the key $\boldsymbol{x}$ and the non-cube IV assignment $\boldsymbol{IV}$ can be computed by summing over the cube $C_I(\boldsymbol{IV})$ as follows:

$$p(\boldsymbol{x}, \boldsymbol{IV}) = \bigoplus_{\boldsymbol{v} \in C_I(\boldsymbol{IV})} f(\boldsymbol{x}, \boldsymbol{v}). \tag{3}$$

In the remainder of this paper, we refer to the value of the superpoly corresponding to the assignment $\boldsymbol{IV}$ in Eq. (3) as $p_{\boldsymbol{IV}}(\boldsymbol{x})$ for short. We use C_I as the cube corresponding to arbitrary $\boldsymbol{IV}$ setting in Eq. (2). Since C_I is defined according to I, we may also refer I as the "cube" without causing ambiguities. The size of I, denoted as $|I|$, is also referred as the dimension of the cube.

Note: since the superpoly p is irrelevant to cube IVs $v_i, i \in I$, the value of $\boldsymbol{IV}[i], i \in I$ cannot affect the result of the summation in Eq. (3) at all. Therefore in Sect. 5, our $\boldsymbol{IV}[i]$'s ($i \in I$) are just assigned randomly to 0-1 values.

2.3   Bit-Based Division Property and Its MILP Representation

At 2015, the division property, a generalization of the integral property, was proposed in [13] with which better integral characteristics for word-oriented cryptographic primitives have been detected. Later, the bit-based division property was introduced in [15] so that the propagation of integral characteristics can be described in a more precise manner. The definition of the bit-based division property is as follows:

Definition 1 ((Bit-Based) Division Property). *Let $\mathbb{X}$ be a multiset whose elements take a value of $\mathbb{F}_2^n$. Let $\mathbb{K}$ be a set whose elements take an n-dimensional bit vector. When the multiset $\mathbb{X}$ has the division property $\mathcal{D}_{\mathbb{K}}^{1^n}$, it fulfills the following conditions:*

$$\bigoplus_{x \in \mathbb{X}} x^u = \begin{cases} \text{unknown} & \text{if there exist } k \in \mathbb{K} \text{ s.t. } u \succeq k, \\ 0 & \text{otherwise,} \end{cases}$$

where $u \succeq k$ if $u_i \geq k_i$ for all i, and $x^u = \prod_{i=1}^n x_i^{u_i}$.

When the basic bitwise operations COPY, XOR, AND are applied to the elements in $\mathbb{X}$, transformations of the division property should also be made following the propagation corresponding rules copy, xor, and proved in [13,15]. Since round functions of cryptographic primitives are combinations of bitwise operations, we only need to determine the division property of the chosen plaintexts, denoted by $\mathcal{D}_{\mathbb{K}_0}^{1^n}$. Then, after r-round encryption, the division property of the output ciphertexts, denoted by $\mathcal{D}_{\mathbb{K}_r}^{1^n}$, can be deduced according to the round function and the propagation rules. More specifically, when the plaintext bits at index positions $I = \{i_1, i_2, \ldots, i_{|I|}\} \subset \{1, 2, \ldots, n\}$ are active (the active bits traverse all $2^{|I|}$ possible combinations while other bits are assigned to static $0/1$ values), the division property of such chosen plaintexts is $\mathcal{D}_{\boldsymbol{k}}^{1^n}$, where $k_i = 1$ if $i \in I$ and $k_i = 0$ otherwise. Then, the propagation of the division property from $\mathcal{D}_{\boldsymbol{k}}^{1^n}$ is evaluated as

$$\{\boldsymbol{k}\} := \mathbb{K}_0 \to \mathbb{K}_1 \to \mathbb{K}_2 \to \cdots \to \mathbb{K}_r,$$

where $\mathcal{D}_{\mathbb{K}_i}$ is the division property after i-round propagation. If the division property $\mathbb{K}_r$ does not have an unit vector e_i whose only ith element is 1, the ith bit of r-round ciphertexts is balanced.

However, when round r gets bigger, the size of $\mathbb{K}_r$ expands exponentially towards $O(2^n)$ requiring huge memory resources. So the bit-based division property has only been applied to block ciphers with tiny block sizes, such as SIMON32 and Simeck32 [15]. This memory-crisis has been solved by Xiang et al. using the MILP modeling method.

Propagation of Division Property with MILP. At ASIACRYPT 2016, Xiang et al. first introduced a new concept *division trail* defined as follows:

Definition 2 (Division Trail [16]). *Let us consider the propagation of the division property* $\{k\} \stackrel{\text{def}}{=} \mathbb{K}_0 \to \mathbb{K}_1 \to \mathbb{K}_2 \to \cdots \to \mathbb{K}_r$. *Moreover, for any vector* $k_{i+1}^* \in \mathbb{K}_{i+1}$, *there must exist a vector* $k_i^* \in \mathbb{K}_i$ *such that* k_i^* *can propagate to* k_{i+1}^* *by the propagation rule of the division property. Furthermore, for* $(k_0, k_1, \ldots, k_r) \in (\mathbb{K}_0 \times \mathbb{K}_1 \times \cdots \times \mathbb{K}_r)$ *if* k_i *can propagate to* k_{i+1} *for all* $i \in \{0, 1, \ldots, r-1\}$, *we call* $(k_0 \to k_1 \to \cdots \to k_r)$ *an* r-round division trail.

Let E_k be the target r-round iterated cipher. Then, if there is a division trail $k_0 \xrightarrow{E_k} k_r = e_j$ $(j = 1, \ldots, n)$, the summation of jth bit of the ciphertexts is unknown; otherwise, if there is no division trial s.t. $k_0 \xrightarrow{E_k} k_r = e_j$, we know the ith bit of the ciphertext is balanced (the summation of the ith bit is constant 0). Therefore, we have to evaluate all possible division trails to verify whether each bit of ciphertexts is balanced or not. Xiang et al. proved that the basic propagation rules copy, xor, and of the division property can be translated as some variables and constraints of an MILP model. With this method, all possible division trials can be covered with an MILP model $\mathcal{M}$ and the division property of particular output bits can be acquired by analyzing the solutions of the $\mathcal{M}$. After Xiang et al.'s work, some simplifications have been made to the MILP descriptions of copy, xor, and in [12,18]. We present the current simplest MILP-based copy, xor, and as follows:

Proposition 1 (MILP Model for COPY [18]). *Let* $a \xrightarrow{COPY} (b_1, b_2, \ldots, b_m)$ *be a division trail of COPY. The following inequalities are sufficient to describe the propagation of the division property for copy.*

$$\begin{cases} \mathcal{M}.var \leftarrow a, b_1, b_2, \ldots, b_m \text{ as binary.} \\ \mathcal{M}.con \leftarrow a = b_1 + b_2 + \cdots + b_m \end{cases}$$

Proposition 2 (MILP Model for XOR [18]). *Let* $(a_1, a_2, \ldots, a_m) \xrightarrow{XOR} b$ *be a division trail of XOR. The following inequalities are sufficient to describe the propagation of the division property for xor.*

$$\begin{cases} \mathcal{M}.var \leftarrow a_1, a_2, \ldots, a_m, b \text{ as binary.} \\ \mathcal{M}.con \leftarrow a_1 + a_2 + \cdots + a_m = b \end{cases}$$

Proposition 3 (MILP Model for AND [12]). *Let* $(a_1, a_2, \ldots, a_m) \xrightarrow{AND} b$ *be a division trail of AND. The following inequalities are sufficient to describe the propagation of the division property for and.*

$$\begin{cases} \mathcal{M}.var \leftarrow a_1, a_2, \ldots, a_m, b \text{ as binary.} \\ \mathcal{M}.con \leftarrow b \geq a_i \text{ for all } i \in \{1, 2, \ldots, m\} \end{cases}$$

Note: Proposition 3 includes redundant propagations of the division property, but they do not affect preciseness of the obtained characteristics [12].

2.4   The Bit-Based Division Property for Cube Attack

When the number of initialization rounds is not large enough for a thorough diffusion, the superpoly $p(\boldsymbol{x}, \boldsymbol{v})$ defined in Eq. (2) may not be related to all key bits $x_1, \ldots, x_n$ corresponding to some high-dimensional cube I. Instead, there is a set of key indices $J \subseteq \{1, \ldots, n\}$ s.t. for arbitrary $\boldsymbol{v} \in \mathbb{F}_2^m$, $p(\boldsymbol{x}, \boldsymbol{v})$ can only be related to x_j's $(j \in J)$. In CRYPTO 2017, Todo et al. proposed a method for determining such a set J using the bit-based division property [12]. They further showed that, with the knowledge of such J, cube attacks can be launched to recover some information related to the secret key bits. More specifically, they proved the following Lemma 1 and Proposition 4.

Lemma 1. *Let $f(\boldsymbol{x})$ be a polynomial from $\mathbb{F}_2^n$ to $\mathbb{F}_2$ and $a_{\boldsymbol{u}}^f \in \mathbb{F}_2$ $(\boldsymbol{u} \in \mathbb{F}_2^n)$ be the ANF coefficients of $f(x)$. Let $\boldsymbol{k}$ be an n-dimensional bit vector. Assuming there is no division trail such that $\boldsymbol{k} \xrightarrow{f} 1$, then $a_{\boldsymbol{u}}^f$ is always 0 for $\boldsymbol{u} \succeq \boldsymbol{k}$.*

Proposition 4. *Let $f(\boldsymbol{x}, \boldsymbol{v})$ be a polynomial, where $\boldsymbol{x}$ and $\boldsymbol{v}$ denote the secret and public variables, respectively. For a set of indices $I = \{i_1, i_2, \ldots, i_{|I|}\} \subset \{1, 2, \ldots, m\}$, let C_I be a set of $2^{|I|}$ values where the variables in $\{v_{i_1}, v_{i_2}, \ldots, v_{i_{|I|}}\}$ are taking all possible combinations of values. Let $\boldsymbol{k}_I$ be an m-dimensional bit vector such that $\boldsymbol{v}^{\boldsymbol{k}_I} = t_I = v_{i_1} v_{i_2} \cdots v_{i_{|I|}}$, i.e. $k_i = 1$ if $i \in I$ and $k_i = 0$ otherwise. Assuming there is no division trail such that $(\boldsymbol{e}_\lambda, \boldsymbol{k}_I) \xrightarrow{f} 1$, x_λ is not involved in the superpoly of the cube C_I.*

When f represents the first output bit after the initialization iterations, we can identify J by checking whether there is a division trial $(\boldsymbol{e}_\lambda, \boldsymbol{k}_I) \xrightarrow{f} 1$ for $\lambda = 1, \ldots, n$ using the MILP modeling method introduced in Sect. 2.3. If the division trial $(\boldsymbol{e}_\lambda, \boldsymbol{k}_I) \xrightarrow{f} 1$ exists, we have $\lambda \in J$; otherwise, $\lambda \notin J$.

When J is determined, we know that for some assignment to the non-cube $\boldsymbol{IV} \in \mathbb{F}_2^m$, the corresponding superpoly $p_{\boldsymbol{IV}}(\boldsymbol{x})$ is not constant 0, and it is a polynomial of $x_j, j \in J$. With the knowledge of J, we recover offline the superpoly $p_{\boldsymbol{IV}}(\boldsymbol{x})$ by constructing its truth table using cube summations defined as Eq. (3). As long as $p_{\boldsymbol{IV}}(\boldsymbol{x})$ is not constant, we can go to the online phase where we sum over the cube $C_I(\boldsymbol{IV})$ to get the exact value of $p_{\boldsymbol{IV}}(\boldsymbol{x})$ and refer to the precomputed truth table to identify the $x_j, j \in J$ assignment candidates. We summarize the whole process as follows:

1. **Offline Phase: Superpoly Recovery.** Randomly pick an $\boldsymbol{IV} \in \mathbb{F}_2^m$ and prepare the cube $C_I(\boldsymbol{IV})$ defined as Eq. (2). For $\boldsymbol{x} \in \mathbb{F}_2^n$ whose $x_j, j \in J$ traverse all $2^{|J|}$ 0-1 combinations, we compute and store the value of the superpoly $p_{\boldsymbol{IV}}(\boldsymbol{x})$ as Eq. (3). The $2^{|J|}$ values compose the truth table of $p_{\boldsymbol{IV}}(\boldsymbol{x})$ and the ANF of the superpoly is determined accordingly. If $p_{\boldsymbol{IV}}(\boldsymbol{x})$ is constant, we pick another $\boldsymbol{IV}$ and repeat the steps above until we find an appropriate one s.t. $p_{\boldsymbol{v}}(\boldsymbol{x})$ is not constant.

2. **Online Phase: Partial Key Recovery.** Query the cube $C_I(\boldsymbol{IV})$ to encryption oracle and get the summation of the $2^{|I|}$ output bits. We denoted the

summation by $\lambda \in \mathbb{F}_2$ and we know $p_{IV}(x) = \lambda$ according to Eq. (3). So we look up the truth table of the superpoly and only reserve the $x_j, j \in J$ s.t. $p_{IV}(x) = \lambda$.

3. **Brute-Force Search.** Guess the remaining secret variables to recover the entire value in secret variables.

Phase 1 dominates the time complexity since it takes $2^{|I|+|J|}$ encryptions to construct the truth table of size $2^{|J|}$. It is also possible that $p_{IV}(x)$ is constant so we have to run several different IV's to find the one we need. The attack can only be meaningful when (1) $|I| + |J| < n$; (2) appropriate IV's are easy to be found. The former requires the adversary to use "good" cube I's with small J while the latter is the exact reason why Assumptions 1 and 2 are proposed [12,22].

3   Modeling the Constant Bits to Improve the Preciseness of the MILP Model

In the initial state of stream ciphers, there are secret key bits, public modifiable IV bits and constant 0/1 bits. In the previous MILP model, the initial bit-based division properties of the cube IVs are set to 1, while those of the non-cube IVs, constant state bits or even secret key bits are all set to 0.

Obviously, when constant 0 bits are involved in multiplication operations, it always results in an constant 0 output. But, as is pointed out in [22], such a phenomenon cannot be reflected in previous MILP model method. In the previous MILP model, the widely used COPY+AND operation:

$$\text{COPY+AND} : (s_1, s_2) \rightarrow (s_1, s_2, s_1 \wedge s_2). \tag{4}$$

can result in division trials $(x_1, x_2) \xrightarrow{COPY+AND} (y_1, y_2, a)$ as follows:

$$(1,0) \xrightarrow{COPY+AND} (0,0,1),$$
$$(0,1) \xrightarrow{COPY+AND} (0,0,1).$$

Assuming that either s_1 or s_2 of Eq. (4) is a constant 0 bit, $(s_1 \wedge s_2)$ is always 0. In this occasion, the division property of $(s_1 \wedge s_2)$ must be 0 which is overlooked by the previous MILP model. To prohibit the propagation above, an additional constraint $\mathcal{M}.con \leftarrow a = 0$ should be added when either s_1 or s_2 is constant 0.

In [22], the authors only consider the constant 0 bits. They thought the model can be precise enough when all the state bits initialized to constant 0 bits are handled. But in fact, although constant 1 bits do not affect the division property propagation, we should still be aware because 0 bits might be generated when even number of constant 1 bits are XORed during the updating process. This is shown in Example 2 for Kreyvium in Appendix A [25].

Therefore, for all variables in the MILP $v \in \mathcal{M}.var$, we give them an additional flag $v.F \in \{1_c, 0_c, \delta\}$ where 1_c means the bit is constant 1, 0_c means constant 0 and δ means variable. Apparently, when $v.F = 0_c/1_c$, there is always a constraint $v = 0 \in \mathcal{M}.con$. We define $=$, $\oplus$ and $\times$ operations for the elements of set $\{1_c, 0_c, \delta\}$. The $=$ operation tests whether two elements are equal(naturally $1_c = 1_c$, $0_c = 0_c$ and $\delta = \delta$). The $\oplus$ operation follows the rules:

$$\begin{cases} 1_c \oplus 1_c = 0_c \\ 0_c \oplus x = x \oplus 0_c = x \text{ for arbitrary } x \in \{1_c, 0_c, \delta\} \\ \delta \oplus x = x \oplus \delta = \delta \end{cases} \quad (5)$$

The $\times$ operation follows the rules:

$$\begin{cases} 1_c \times x = x \times 1_c = x \\ 0_c \times x = x \times 0_c = 0_c \text{ for arbitrary } x \in \{1_c, 0_c, \delta\} \\ \delta \times \delta = \delta \end{cases} \quad (6)$$

Therefore, in the remainder of this paper, the MILP models for COPY, XOR and AND should also consider the effects of flags. So the previous copy, xor, and and should now add the assignment to flags. We denote the modified versions as copyf, xorf, and andf and define them as Propositions 5, 6 and 7 as follows.

Proposition 5 (MILP Model for COPY with Flag). *Let a* $\xrightarrow{COPY}$ *$(b_1, b_2, \ldots, b_m)$ be a division trail of COPY. The following inequalities are sufficient to describe the propagation of the division property for copyf.*

$$\begin{cases} \mathcal{M}.var \leftarrow a, b_1, b_2, \ldots, b_m \text{ as binary.} \\ \mathcal{M}.con \leftarrow a = b_1 + b_2 + \cdots + b_m \\ a.F = b_1.F = \ldots = b_m.F \end{cases}$$

We denote this process as $(\mathcal{M}, b_1, \ldots, b_m) \leftarrow \text{copyf}(\mathcal{M}, a, m)$.

Proposition 6 (MILP Model for XOR with Flag). *Let $(a_1, a_2, \ldots, a_m)$* $\xrightarrow{XOR} b$ *be a division trail of XOR. The following inequalities are sufficient to describe the propagation of the division property for xorf.*

$$\begin{cases} \mathcal{M}.var \leftarrow a_1, a_2, \ldots, a_m, b \text{ as binary.} \\ \mathcal{M}.con \leftarrow a_1 + a_2 + \cdots + a_m = b \\ b.F = a_1.F \oplus a_2.F \oplus \cdots \oplus a_m.F \end{cases}$$

We denote this process as $(\mathcal{M}, b) \leftarrow \text{xorf}(\mathcal{M}, a_1, \ldots, a_m)$.

Proposition 7 (MILP Model for AND with Flag). *Let $(a_1, a_2, \ldots, a_m)$* $\xrightarrow{AND} b$ *be a division trail of AND. The following inequalities are sufficient to describe the propagation of the division property for andf.*

$$\begin{cases} \mathcal{M}.var \leftarrow a_1, a_2, \ldots, a_m, b \text{ as binary.} \\ \mathcal{M}.con \leftarrow b \geq a_i \text{ for all } i \in \{1, 2, \ldots, m\} \\ b.F = a_1.F \times a_2.F \times \cdots a_m.F \\ \mathcal{M}.con \leftarrow b = 0 \quad \text{if } b.F = 0_c \end{cases}$$

We denote this process as $(\mathcal{M}, b) \leftarrow \text{andf}(\mathcal{M}, a_1, \ldots, a_m)$.

Algorithm 1. Evaluate secret variables by MILP with Flags

1: **procedure** `attackFramework`(Cube indices I, specific assignment to non-cube IVs IV or $IV = $ NULL)
2:    Declare an empty MILP model $\mathcal{M}$
3:    Declare x as n MILP variables of $\mathcal{M}$ corresponding to secret variables.
4:    Declare v as m MILP variables of $\mathcal{M}$ corresponding to public variables.
5:    $\mathcal{M}.con \leftarrow v_i = 1$ and assign $v_i.F = \delta$ for all $i \in I$
6:    $\mathcal{M}.con \leftarrow v_i = 0$ for all $i \in (\{1, 2, \ldots, n\} - I)$
7:    $\mathcal{M}.con \leftarrow \sum_{i=1}^{n} x_i = 1$ and assign $x_i.F = \delta$ for all $i \in \{1, \ldots, n\}$
8:    **if** $IV = $ NULL **then**
9:        $v_i.F = \delta$ for all $i \in (\{1, 2, \ldots, m\} - I)$
10:   **else**
11:       Assign the flags of v_i, $i \in (\{1, 2, \ldots, m\} - I)$ as:

$$v_i.F = \begin{cases} 1_c & \text{if } IV[i] = 1 \\ 0_c & \text{if } IV[i] = 0 \end{cases}$$

12:   **end if**
13:   Update $\mathcal{M}$ according to round functions and output functions
14:   **do**
15:       solve MILP model $\mathcal{M}$
16:       **if** $\mathcal{M}$ is feasible **then**
17:           pick index $j \in \{1, 2, \ldots, n\}$ s.t. $x_j = 1$
18:           $J = J \cup \{j\}$
19:           $\mathcal{M}.con \leftarrow x_j = 0$
20:       **end if**
21:   **while** $\mathcal{M}$ is feasible
22:   **return** J
23: **end procedure**

With these modifications, we are able to improve the preciseness of the MILP model. The improved attack framework can be written as Algorithm 1. It enables us to identify the involved keys when the non-cube IVs are set to specific constant 0/1 values by imposing corresponding flags to the non-cube MILP binary variables. With this method, we can determine an $IV \in \mathbb{F}_2^m$ s.t. the corresponding superpoly $p_{IV}(x) \neq 0$.

4   Upper Bounding the Degree of the Superpoly

For an $IV \in \mathbb{F}_2^m$ s.t. $p_{IV}(x) \neq 0$, the ANF of $p_{IV}(x)$ can be represented as

$$p_{IV}(x) = \sum_{u \in \mathbb{F}_2^n} a_u x^u \tag{7}$$

where a_u is determined by the values of the non-cube IVs. If the degree of the superpoly is upper bounded by d, then for all u's with Hamming weight satisfying $hw(u) > d$, we constantly have $a_u = 0$. In this case, we no longer have to build the whole truth table to recover the superpoly. Instead, we only need to determine the coefficients a_u for $hw(u) \leq d$. Therefore, we select $\sum_{i=0}^{d} \binom{|J|}{i}$ different x's and construct a linear system with $\left(\sum_{i=0}^{d} \binom{|J|}{i}\right)$ variables and the coefficients as well as the whole ANF of $p_{IV}(x)$ can be recovered by solving such a linear system. So the complexity of Phase 1 can be reduced from $2^{|I|+|J|}$ to $2^{|I|} \times \sum_{i=0}^{d} \binom{|J|}{i}$. For the simplicity of notations, we denote the summation $\sum_{i=0}^{d} \binom{|J|}{i}$ as $\binom{|J|}{\leq d}$ in the remainder of this paper. With the knowledge of the involved key indices $J = \{j_1, j_2, \ldots, j_{|J|}\}$ and the degree of the superpoly $d = \deg p_{IV}(x)$, the attack procedure can be adapted as follows:

1. **Offline Phase: Superpoly Recovery.** For all $\binom{|J|}{\leq d}$ x's satisfying $hw(x) \leq d$ and $\bigoplus_{j \in J} e_j \succeq x$, compute the values of the superpolys as $p_{IV}(x)$ by summing over the cube $C_I(IV)$ as Eq. (3) and generate a linear system of the $\binom{|J|}{\leq d}$ coefficients a_u $(hw(u) \leq d)$. Solve the linear system, determine the coefficient a_u of the $\binom{|J|}{\leq d}$ terms and store them in a lookup table T. The ANF of the $p_{IV}(x)$ can be determined with the lookup table.
2. **Online Phase: Partial Key Recovery.** Query the encryption oracle and sum over the cube $C_I(IV)$ as Eq. (3) and acquire the exact value of $p_{IV}(x)$. For each of the $2^{|J|}$ possible values of $\{x_{j_1}, \ldots, x_{j_{|J|}}\}$, compute the values of the superpoly as Eq. (7) (the coefficient a_u acquired by looking up the precomputed table T) and identify the correct key candidates.
3. **Brute-force search phase.** Attackers guess the remaining secret variables to recover the entire value in secret variables.

The complexity of Phase 1 becomes $2^{|I|} \times \binom{|J|}{\leq d}$. Phase 2 now requires $2^{|I|}$ encryptions and $2^{|J|} \times \binom{|J|}{\leq d}$ table lookups, so the complexity can be regarded as $2^{|I|} + 2^{|J|} \times \binom{|J|}{\leq d}$. The complexity of Phase 3 remains 2^{n-1}. Therefore, the number of encryptions a feasible attack requires is

$$\max\left\{2^{|I|} \times \binom{|J|}{\leq d}, 2^{|I|} + 2^{|J|} \times \binom{|J|}{\leq d}\right\} < 2^n. \tag{8}$$

The previous limitation of $|I| + |J| < n$ is removed.

The knowledge of the algebraic degree of superpolys can largely benefit the efficiency of the cube attack. Therefore, we show how to estimate the algebraic degree of superpolys using the division property. Before the introduction of the method, we generalize Proposition 4 as follows.

Proposition 8. *Let $f(x, v)$ be a polynomial, where x and v denote the secret and public variables, respectively. For a set of indices $I = \{i_1, i_2, \ldots, i_{|I|}\} \subset \{1, 2, \ldots, m\}$, let C_I be a set of $2^{|I|}$ values where the variables in $\{v_{i_1}, v_{i_2}, \ldots, v_{i_{|I|}}\}$ are taking all possible combinations of values. Let k_I be an m-dimensional bit vector such that $v^{k_I} = t_I = v_{i_1} v_{i_2} \cdots v_{i_{|I|}}$. Let k_Λ be an n-dimensional bit vector. Assuming there is no division trail such that $(k_\Lambda || k_I) \xrightarrow{f} 1$, the monomial x^{k_Λ} is not involved in the superpoly of the cube C_I.*

Proof. The ANF of $f(\boldsymbol{x}, \boldsymbol{v})$ is represented as follows

$$f(\boldsymbol{x}, \boldsymbol{v}) = \bigoplus_{\boldsymbol{u} \in \mathbb{F}_2^{n+m}} a_{\boldsymbol{u}}^f \cdot (\boldsymbol{x} \| \boldsymbol{v})^{\boldsymbol{u}},$$

where $a_{\boldsymbol{u}}^f \in \mathbb{F}_2$ denotes the ANF coefficients. The polynomial $f(\boldsymbol{x}, \boldsymbol{v})$ is decomposed into

$$f(\boldsymbol{x}, \boldsymbol{v}) = \bigoplus_{\boldsymbol{u} \in \mathbb{F}_2^{n+m} | \boldsymbol{u} \succeq (\boldsymbol{0} \| \boldsymbol{k}_I)} a_{\boldsymbol{u}}^f \cdot (\boldsymbol{x} \| \boldsymbol{v})^{\boldsymbol{u}} \oplus \bigoplus_{\boldsymbol{u} \in \mathbb{F}_2^{n+m} | \boldsymbol{u} \not\succeq (\boldsymbol{0} \| \boldsymbol{k}_I)} a_{\boldsymbol{u}}^f \cdot (\boldsymbol{x} \| \boldsymbol{v})^{\boldsymbol{u}}$$

$$= t_I \cdot \bigoplus_{\boldsymbol{u} \in \mathbb{F}_2^{n+m} | \boldsymbol{u} \succeq (\boldsymbol{0} \| \boldsymbol{k}_I)} a_{\boldsymbol{u}}^f \cdot (\boldsymbol{x} \| \boldsymbol{v})^{\boldsymbol{u} \oplus (\boldsymbol{0} \| \boldsymbol{k}_I)} \oplus \bigoplus_{\boldsymbol{u} \in \mathbb{F}_2^{n+m} | \boldsymbol{u} \not\succeq (\boldsymbol{0} \| \boldsymbol{k}_I)} a_{\boldsymbol{u}}^f \cdot (\boldsymbol{x} \| \boldsymbol{v})^{(\boldsymbol{0} \| \boldsymbol{u})}$$

$$= t_I \cdot p(\boldsymbol{x}, \boldsymbol{v}) \oplus q(\boldsymbol{x}, \boldsymbol{v}).$$

Therefore, the superpoly $p(\boldsymbol{x}, \boldsymbol{v})$ is represented as

$$p(\boldsymbol{x}, \boldsymbol{v}) = \bigoplus_{\boldsymbol{u} \in \mathbb{F}_2^{n+m} | \boldsymbol{u} \succeq (\boldsymbol{0} \| \boldsymbol{k}_I)} a_{\boldsymbol{u}}^f \cdot (\boldsymbol{x} \| \boldsymbol{v})^{\boldsymbol{u} \oplus (\boldsymbol{0} \| \boldsymbol{k}_I)}.$$

Since there is no division trail $(\boldsymbol{k}_\Lambda \| \boldsymbol{k}_I) \xrightarrow{f} 1$, $a_{\boldsymbol{u}}^f = 0$ for $\boldsymbol{u} \succeq (\boldsymbol{k}_\Lambda \| \boldsymbol{k}_I)$ because of Lemma 1. Therefore,

$$p(\boldsymbol{x}, \boldsymbol{v}) = \bigoplus_{\boldsymbol{u} \in \mathbb{F}_2^{n+m} | \boldsymbol{u} \succeq (\boldsymbol{0} \| \boldsymbol{k}_I), \boldsymbol{u}^{\boldsymbol{k}_\Lambda \| \boldsymbol{0}} = 0} a_{\boldsymbol{u}}^f \cdot (\boldsymbol{x} \| \boldsymbol{v})^{\boldsymbol{u} \oplus (\boldsymbol{0} \| \boldsymbol{k}_I)}.$$

This superpoly is independent of the monomial $\boldsymbol{x}^{\boldsymbol{k}_\Lambda}$ since $\boldsymbol{u}^{\boldsymbol{k}_\Lambda \| \boldsymbol{0}}$ is always 0. □

According to Proposition 8, the existence of the division trial $(\boldsymbol{k}_\Lambda \| \boldsymbol{k}_I) \xrightarrow{f} 1$ is in accordance with the existence of the monomial $\boldsymbol{x}^{\boldsymbol{k}_\Lambda}$ in the superpoly of the cube C_I.

If there is $d \geq 0$ s.t. for all $\boldsymbol{k}_\Lambda$ of hamming weight $hw(\boldsymbol{k}_\Lambda) > d$, the division trail $\boldsymbol{x}^{\boldsymbol{k}_\Lambda}$ does not exist, then we know that the algebraic degree of the superpoly is bounded by d. Using MILP, this d can be naturally modeled as the maximum of the objective function $\sum_{j=1}^{n} x_j$. With the MILP model $\mathcal{M}$ and the cube indices I, we can bound the degree of the superpoly using Algorithm 2. Same with Algorithm 1, we can also consider the degree of the superpoly for specific assignment to the non-cube IVs. So we also add the input $\boldsymbol{IV}$ that can either be a specific assignment or a NULL referring to arbitrary assignment. The solution $\mathcal{M}.obj = d$ is the upper bound of the superpoly's algebraic degree. Furthermore, corresponding to $\mathcal{M}.obj = d$ and according to the definition of $\mathcal{M}.obj$, there should also be a set of indices $\{l_1, \ldots, l_d\}$ s.t. the variables representing the initially declared $\boldsymbol{x}$ (representing the division property of the key bits) satisfy the constraints $x_{l_1} = \ldots = x_{l_d} = 1$. We can also enumerate all t-degree ($1 \leq t \leq d$) monomials involved in the superpoly using a similar technique which we will detail later in Sect. 6.

Algorithm 2. Evaluate upper bound of algebraic degree on the superpoly

---

1: **procedure** DegEval(Cube indices I, specific assignment to non-cube IVs IV or IV = NULL)
2:     Declare an empty MILP model $\mathcal{M}$.
3:     Declare x be n MILP variables of $\mathcal{M}$ corresponding to secret variables.
4:     Declare v be m MILP variables of $\mathcal{M}$ corresponding to public variables.
5:     $\mathcal{M}.con \leftarrow v_i = 1$ and assign the flags $v_i.F = \delta$ for all $i \in I$
6:     $\mathcal{M}.con \leftarrow v_i = 0$ for $i \in (\{1, \ldots, n\} - I)$
7:     **if** IV = NULL **then**
8:         Assign the flags $v_i.F = \delta$ for $i \in (\{1, \ldots, n\} - I)$
9:     **else**
10:         Assign the flags of v_i, $i \in (\{1, 2, \ldots, n\} - I)$ as:

$$v_i.F = \begin{cases} 1_c & \text{if } IV[i] = 1 \\ 0_c & \text{if } IV[i] = 0 \end{cases}$$

11:     **end if**
12:     Set the objective function $\mathcal{M}.obj \leftarrow \sum_{i=1}^{n} x_i$
13:     Update $\mathcal{M}$ according to round functions and output functions
14:     Solve MILP model $\mathcal{M}$
15:     **return** The solution of $\mathcal{M}$.
16: **end procedure**

---

5   Applications of Flag Technique and Degree Evaluation

We apply our method to 4 NLFSR-based ciphers namely TRIVIUM, Kreyvium, Grain-128a and ACORN. Among them, TRIVIUM, Grain-128a and ACORN are also targets of [12]. Using our new techniques, we can both lower the complexities of previous attacks and give new cubes that mount to more rounds. We give details of the application to TRIVIUM in this section, and the applications to Kreyvium, Grain-128a and ACORN in our full version [25].

5.1   Specification of Trivium

TRIVIUM is an NLFSR-based stream cipher, and the internal state is represented by 288-bit state $(s_1, s_2, \ldots, s_{288})$. Figure 1 shows the state update function of TRIVIUM. The 80-bit key is loaded to the first register, and the 80-bit IV is loaded to the second register. The other state bits are set to 0 except the least three bits in the third register. Namely, the initial state bits are represented as

$$(s_1, s_2, \ldots, s_{93}) = (K_1, K_2, \ldots, K_{80}, 0, \ldots, 0),$$
$$(s_{94}, s_{95}, \ldots, s_{177}) = (IV_1, IV_2, \ldots, IV_{80}, 0, \ldots, 0),$$
$$(s_{178}, s_{279}, \ldots, s_{288}) = (0, 0, \ldots, 0, 1, 1, 1).$$

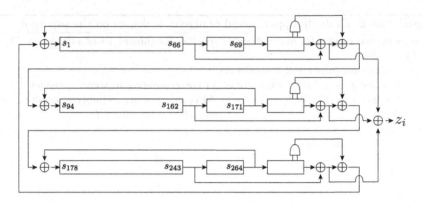

Figure 1. Structure of TRIVIUM

The pseudo code of the update function is given as follows.

$$t_1 \leftarrow s_{66} \oplus s_{93}$$
$$t_2 \leftarrow s_{162} \oplus s_{177}$$
$$t_3 \leftarrow s_{243} \oplus s_{288}$$
$$z \leftarrow t_1 \oplus t_2 \oplus t_3$$
$$t_1 \leftarrow t_1 \oplus s_{91} \cdot s_{92} \oplus s_{171}$$
$$t_2 \leftarrow t_2 \oplus s_{175} \cdot s_{176} \oplus s_{264}$$
$$t_3 \leftarrow t_3 \oplus s_{286} \cdot s_{287} \oplus s_{69}$$
$$(s_1, s_2, \ldots, s_{93}) \leftarrow (t_3, s_1, \ldots, s_{92})$$
$$(s_{94}, s_{95}, \ldots, s_{177}) \leftarrow (t_1, s_{94}, \ldots, s_{176})$$
$$(s_{178}, s_{279}, \ldots, s_{288}) \leftarrow (t_2, s_{178}, \ldots, s_{287})$$

Here z denotes the 1-bit key stream. First, in the key initialization, the state is updated $4 \times 288 = 1152$ times without producing an output. After the key initialization, one bit key stream is produced by every update function.

5.2   MILP Model of Trivium

The only non-linear component of TRIVIUM is a 2-degree core function denoted as f_{core} that takes as input a 288-bit state s and 5 indices $i_1, \ldots, i_5$, and outputs a new 288-bit state $s' \leftarrow f_{core}(s, i_1, \ldots, i_5)$ where

$$s_i' = \begin{cases} s_{i_1} s_{i_2} + s_{i_3} + s_{i_4} + s_{i_5}, & i = i_5 \\ s_i, & \text{otherwise} \end{cases} \quad (9)$$

The division property propagation for the core function can be represented as Algorithm 3. The input of Algorithm 3 consists of $\mathcal{M}$ as the current MILP model, a vector of 288 binary variables x describing the current division property of the

288-bit NFSR state, and 5 indices i_1, i_2, i_3, i_4, i_5 corresponding to the input bits. Then Algorithm 3 outputs the updated model $\mathcal{M}$, and a 288-entry vector y describing the division property after f_{core}.

Algorithm 3. MILP model of division property for the core function (Eq. (9))

1: **procedure** Core$(\mathcal{M}, x, i_1, i_2, i_3, i_4, i_5)$
2:    $(\mathcal{M}, y_{i_1}, z_1) \leftarrow$ copyf$(\mathcal{M}, x_{i_1})$
3:    $(\mathcal{M}, y_{i_2}, z_2) \leftarrow$ copyf$(\mathcal{M}, x_{i_2})$
4:    $(\mathcal{M}, y_{i_3}, z_3) \leftarrow$ copyf$(\mathcal{M}, x_{i_3})$
5:    $(\mathcal{M}, y_{i_4}, z_4) \leftarrow$ copyf$(\mathcal{M}, x_{i_4})$
6:    $(\mathcal{M}, a) \leftarrow$ andf$(\mathcal{M}, z_1, z_2)$
7:    $(\mathcal{M}, y_{i_5}) \leftarrow$ xorf$(\mathcal{M}, a, z_2, z_3, z_4, x_{i_5})$
8:    **for all** $i \in \{1, 2, \ldots, 288\}$ w/o i_1, i_2, i_3, i_4, i_5 **do**
9:      $y_i = x_i$
10:   **end for**
11:   **return** $(\mathcal{M}, y)$
12: **end procedure**

With the definition of Core, the MILP model of R-round TRIVIUM can be described as Algorithm 4. This algorithm is a subroutine of Algorithm 1 for generating the MILP model $\mathcal{M}$, and the model $\mathcal{M}$ can evaluate all division trails for TRIVIUM whose initialization rounds are reduced to R. Note that constraints to the input division property are imposed by Algorithm 1.

5.3 Experimental Verification

Identical to [12], we use the cube $I = \{1, 11, 21, 31, 41, 51, 61, 71\}$ to verify our attack and implementation. The experimental verification includes: the degree evaluation using Algorithm 2, specifying involved key bits using Algorithm 1 with IV = NULL or specific non-cube IV settings.

Example 1 (Verification of Our Attack against 591-round TRIVIUM). With IV = NULL using Algorithm 1, we are able to identify $J = \{23, 24, 25, 66, 67\}$. We know that with some assignment to the non-cube IV bits, the superpoly can be a polynomial of secret key bits $x_{23}, x_{24}, x_{25}, x_{66}, x_{67}$. These are the same with [12]. Then, we set IV to random values and acquire the degree through Algorithm 2, and verify the correctness of the degree by practically recovering the corresponding superpoly.

– When we set IV = 0xcc2e487b, 0x78f99a93, 0xbeae, and run Algorithm 2, we get the degree 3. The practically recovered superpoly is also of degree 3:

$$p_v(x) = x_{66}x_{23}x_{24} + x_{66}x_{25} + x_{66}x_{67} + x_{66},$$

which is in accordance with the deduction by Algorithm 2 through MILP model.

Algorithm 4. MILP model of division property for TRIVIUM

1: **procedure** TriviumEval(round R)
2:     Prepare empty MILP Model $\mathcal{M}$
3:     $\mathcal{M}.var \leftarrow v_i$ for $i \in \{1, 2, \ldots, 80\}$.   ▷ Declare Public Modifiable IVs
4:     $\mathcal{M}.var \leftarrow x_i$ for $i \in \{1, 2, \ldots, 80\}$.   ▷ Declare Secret Keys
5:     $\mathcal{M}.var \leftarrow s_i^0$ for $i \in \{1, 2, \ldots, 288\}$
6:     $s_i^0 = x_i$, $s_{i+93}^0 = v_i$ for $i = 1, \ldots, 80$.
7:     $\mathcal{M}.con \leftarrow s_i^0 = 0$ for $i = 81, \ldots, 93, 174, \ldots, 288$.
8:     $s_i^0.F = 0_c$ for $i = 81, \ldots, 285$ and $s_j^0.F = 1_c$ for $j = 286, 287, 288$.   ▷ Assign the flags for constant state bits
9:     **for** $r = 1$ to R **do**
10:         $(\mathcal{M}, x) = \text{Core}(\mathcal{M}, s^{r-1}, 66, 171, 91, 92, 93)$
11:         $(\mathcal{M}, y) = \text{Core}(\mathcal{M}, x, 162, 264, 175, 176, 177)$
12:         $(\mathcal{M}, z) = \text{Core}(\mathcal{M}, y, 243, 69, 286, 287, 288)$
13:         $s^r = z \ggg 1$
14:     **end for**
15:     **for all** $i \in \{1, 2, \ldots, 288\}$ w/o $66, 93, 162, 177, 243, 288$ **do**
16:         $\mathcal{M}.con \leftarrow s_i^R = 0$
17:     **end for**
18:     $\mathcal{M}.con \leftarrow (s_{66}^R + s_{93}^R + s_{162}^R + s_{177}^R + s_{243}^R + s_{288}^R) = 1$
19:     **return** $\mathcal{M}$
20: **end procedure**

- When we set $\boldsymbol{IV} = \text{0x61fbe5da}, \text{0x19f5972c}, \text{0x65c1}$, the degree evaluation of Algorithm 2 is 2. The practically recovered superpoly is also of degree 2:

$$p_v(\boldsymbol{x}) = x_{23}x_{24} + x_{25} + x_{67} + 1.$$

- When we set $\boldsymbol{IV} = \text{0x5b942db1}, \text{0x83ce1016}, \text{0x6ce}$, the degree is 0 and the superpoly recovered is also constant 0.

On the Accuracy of MILP Model with Flag Technique. As a comparison, we use the cube above and conduct practical experiments on different rounds namely 576, 577, 587, 590, 591 (selected from Table 2 of [22]). We try 10000 randomly chosen $\boldsymbol{IV}$'s. For each of them, we use the MILP method to evaluate the degree d, in comparison with the practically recovered ANF of the superpoly $p_{IV}(\boldsymbol{x})$. For 576, 577, 587 and 590 rounds, the accuracy is 100%. In fact, such 100% accuracy is testified for most of our applied ciphers, which is shown in [25]. For 591-round, the accuracies are distributed as:

1. When the MILP model gives degree evaluation $d = 0$, the accuracy is 100% that the superpoly is constant 0.
2. When the MILP model gives degree evaluation $d = 3$, there is an accuracy 49% that the superpoly is a 3-degree polynomial. For the rest, the superpoly is constant 0.
3. When the MILP model gives degree evaluation $d = 2$, there is accuracy 43% that the superpoly is a 2-degree polynomial. For the rest, the superpoly is constant 0.

The ratios of error can easily be understood: for example, in some case, one key bit may multiply with constant 1 in one step $x_i \cdot 1$ and be canceled by XORing with itself in the next round, this results in a newly generated constant 0 bit $((x_i \cdot 1) \oplus x_i = 0)$. However, by the flag technique, this newly generated bit has flag value $\delta = (\delta \times 1_c) + \delta$. In our attacks, the size of cubes tends to be large, which means most of the IV bits become active, the above situation of $(x_i \cdot 1) \oplus x_i = 0$ will now become $(x_i \cdot v_j) \oplus x_i$. Therefore, when larger cubes are used, fewer constant $0/1$ flags are employed, and the MILP models are becoming closer to those of $\boldsymbol{IV} = NULL$. It is predictable that the accuracy of the flag technique tends to increase when larger cubes are used. To verify this statement, we construct a 10-dimensional cube $I = \{5, 13, 18, 22, 30, 57, 60, 65, 72, 79\}$ for 591-round TRIVIUM. When $\boldsymbol{IV} = NULL$, we acquire the same upper bound of the degree $d = 3$. Then, we tried thousands of random IVs, and get an overall accuracy 80.9%. From above, we can conclude that the flag technique has high preciseness and can definitely improve the efficiency of the division property based cube attacks.

5.4 Theoretical Results

The best result in [12] mounts to 832-round TRIVIUM with cube dimension $|I| = 72$ and the superpoly involves $|J| = 5$ key bits. The complexity is 2^{77} in [12]. Using Algorithm 2, we further acquire that the degree of such a superpoly is 3. So the complexity for superpoly recovery is $2^{72} \times \binom{5}{\leq 3} = 2^{76.7}$ and the complexity for recovering the partial key is $2^{72} + 2^3 \times \binom{5}{3}$. Therefore, according to Eq. (8), the complexity of this attack is $2^{76.7}$.

We further construct a 77-dimensional cube, $I = \{1, \ldots, 80\} \setminus \{5, 51, 65\}$. Its superpoly after 835 rounds of initialization only involves 1 key bit $J = \{57\}$. So the complexity of the attack is 2^{78}. Since there are only 3 non-cube IVs, we let $\boldsymbol{IV}$ be all 2^3 possible non-cube IV assignments and run Algorithm 1. We find that x_{57} is involved in all of the 2^3 superpolys. So the attack is available for any of the 2^3 non-cube IV assignments. This can also be regarded as a support to the rationality of Assumption 1.

According previous results, TRIVIUM has many cubes whose superpolys only contain 1 key bit. These cubes are of great value for our key recovery attacks. Firstly, the truth table of such superpoly is balanced and the Partial Key Recovery phase can definitely recover 1 bit of secret information. Secondly, the Superpoly Recovery phase only requires $2^{|I|+1}$ and the online Partial Key Recovery only requires $2^{|I|}$ encryptions. Such an attack can be meaningful as long as $|I| + 1 < 80$, so we can try cubes having dimension as large as 78. Therefore, we investigate 78-dimensional cubes and find the best cube attack on TRIVIUM is 839 rounds. By running Algorithm 1 with $2^2 = 4$ different assignments to non-cube IVs, we know that the key bit x_{61} is involved in the superpoly for $\boldsymbol{IV} = \texttt{0x0, 0x4000, 0x0}$ or $\boldsymbol{IV} = \texttt{0x0, 0x4002, 0x0}$. In other words, the 47-th IV bit must be assigned to constant 1. The summary of our new results about TRIVIUM is in Table 2.

Table 2. Summary of theoretical cube attacks on TRIVIUM. The time complexity in this table shows the time complexity of Superpoly Recovery (Phase 1) and Partial Key Recovery (Phase 2).

#Rounds	$|I|$	Degree	Involved keys J	Time complexity
832	72†	3	34, 58, 59, 60, 61 ($|J| = 5$)	$2^{76.7}$
833	73‡	3	49, 58, 60, 74, 75, 76 ($|J| = 7$)	2^{79}
833	74∗	1	60 ($|J| = 1$)	2^{75}
835	77⋆	1	57 ($|J| = 1$)	2^{78}
836	78○	1	57 ($|J| = 1$)	2^{79}
839	78●	1	61 ($|J| = 1$)	2^{79}

†: $I = \{1, 2, ..., 65, 67, 69, ..., 79\}$
‡: $I = \{1, 2, ..., 67, 69, 71, ..., 79\}$
∗: $I = \{1, 2, ..., 69, 71, 73, ..., 79\}$
⋆: $I = \{1, 2, 3, 4, 6, 7, ..., 50, 52, 53, ..., 64, 66, 67, ..., 80\}$
○: $I = \{1, ..., 11, 13, ..., 42, 44, ..., 80\}$
●: $I = \{1, ..., 33, 35, ..., 46, 48, ..., 80\}$ and $IV[47] = 1$

6   Lower Complexity with Term Enumeration

In this section, we show how to further lower the complexity of recovering the superpoly (Phase 1) in Sect. 4.

With cube indices I, key bits J and degree d, the complexity of the current superpoly recovery is $2^I \times \binom{|J|}{\leq d}$, where $\binom{|J|}{\leq d}$ corresponds to all 0-, 1-..., d-degree monomials. When $d \leq |J|/2$ (which is true in most of our applications), we constantly have $\binom{|J|}{0} \leq ... \leq \binom{|J|}{d}$. But in practice, high-degree terms are generated in later iterations and the high-degree monomials should be fewer than their low-degree counterparts. Therefore, for all $\binom{|J|}{i}$ monomials, only very few of them may appear in the superpoly. Similar to Algorithm 1 that decides all key bits appear in the superpoly, we propose Algorithm 5 that enumerates all t-degree monomials that may appear in the superpoly. Apparently, when we use $t = 1$, we can get $J_1 = J$, the same output as Algorithm 1 containing all involved keys. If we use $t = 2, 3, ..., d$, we get $J_2, ..., J_d$ that contains all possible monomials of degrees $2, 3, ..., d$. Therefore, we only need to determine $1 + |J_1| + |J_2| + ... + |J_d|$ coefficients in order to recover the superpoly and apparently, $|J_t| \leq \binom{|J|}{t}$ for $t = 1, ... d$. With the knowledge of $J_t, t = 1, ..., d$, the complexity for Superpoly Recovery (Phase 1) has now become

$$2^{|I|} \times \left(1 + \sum_{t=1}^{d} |J_t|\right) \leq 2^{|I|} \times \binom{|J|}{\leq d}. \tag{10}$$

And the size of the lookup table has also reduced to $(1 + \sum_{t=1}^{d} |J_t|)$. So the complexity of the attack is now

$$\max\{2^{|I|} \times (1 + \sum_{t=1}^{d} |J_t|), 2^{|I|} + 2^{|J|} \times (1 + \sum_{t=1}^{d} |J_t|)\}. \tag{11}$$

Furthermore, since high-degree monomials are harder to be generated through iterations than low-degree ones, we can often find $|J_i| < \binom{|J|}{i}$ when i approaches d. So the complexity for superpoly recovery has been reduced.

Note: J_t's $(t = 1, \ldots, d)$ can be generated by `TermEnum` of Algorithm 5 and they satisfy the following Property 1. This property is equivalent to the "Embed Property" given in [19].

Property 1. For $t = 2, \ldots, d$, if there is $T = (i_1, i_2, \ldots, i_t) \in J_t$ and $T' = (i_{s_1}, \ldots, i_{s_l})$ $(l < t)$ is a subsequence of T $(1 \le s_1 < \ldots < s_l \le t)$. Then, we constantly have $T' \in J_l$.

Before proving Property 1, we first prove the following Lemma 2.

Lemma 2. *If $k \succeq k'$ and there is division trial $k \xrightarrow{f} l$, then there is also division trial $k' \xrightarrow{f} l'$ s.t. $l \succeq l'$.*

Proof. Since f is a combination of COPY, AND and XOR operations, and the proofs when f equals to each of them are similar, we only give a proof of the case when f equals to COPY. Let $f : (*, \ldots, *, x) \xrightarrow{COPY} (*, \ldots, *, x, x)$.

First assume the input division property be $k = (k_1, 0)$, since $k \succeq k'$, there must be $k' = (k_1', 0)$ and $k_1 \succeq k_1'$. We have $l = k$, $l' = k'$, thus the property holds.

When the input division property is $k = (k_1, 1)$, we know that the output division property can be $l \in \{(k_1, 0, 1), (k_1, 1, 0)\}$. Since $k \succeq k'$, we know $k' = (k_1', 1)$ or $k' = (k_1', 0)$, and $k_1 \succeq k_1'$. When $k' = (k_1', 0)$, then $l' = k' = (k_1', 0)$, the relation holds. When $k' = (k_1', 1)$, we know $l' \in \{(k_1', 0, 1), (k_1', 1, 0)\}$, the relation still holds. $\square$

Now we are ready to prove Property 1.

Proof. Let $k, k \in \mathbb{F}_2^n$ satisfy $k_i = 1$ for $i \in T$ and $k_i = 0$ otherwise; $k_i' = 1$ for $i \in T'$ and $k_i' = 0$ otherwise. Since $T \in J_t$, we know that there is division trial $(k, k_I) \xrightarrow{R-Rounds} (0, 1)$ Since $k \succeq k'$, we have $(k, k_I) \succeq (k', k_I)$ and according to Lemma 2, there is division trial s.t. $(k', k_I) \xrightarrow{R-Rounds} (0^{m+n}, s)$ where $(0^{m+n}, 1) \succeq (0^{m+n}, s)$. Since the hamming weight of (k', k_I) is larger than 0 and there is no combination of COPY, AND and XOR that makes non-zero division property to all-zero division property. So we have $s = 1$ and there exist division trial $(k', k_I) \xrightarrow{R-Rounds} (0, 1)$. $\square$

Property 1 reveals a limitation of Algorithm 5. Assume the superpoly is

$$p_v(x_1, x_2, x_3, x_4) = x_1 x_2 x_3 + x_1 x_4.$$

We can acquire $J_3 = \{(1,2,3)\}$ by running TermEnum of Algorithm 5. But, if we run TermEnum with $t = 2$, we will not acquire just $J_2 = \{(1,4)\}$ but $J_2 = \{(1,4), (1,2), (1,3), (2,3)\}$ due to $(1,2,3) \in J_3$ and $(1,2)$, $(1,3)$, $(2,3)$ are its subsequences. Although there are still redundant terms, the reduction from $\binom{|J|}{d}$ to $|J_d|$ is usually huge enough to improve the existing cube attack results.

Applying such term enumeration technique, we are able to lower complexities of many existing attacks namely: 832-, 833-round TRIVIUM, 849-round Kreyvium, 184-round Grain-128a and 704-round ACORN. The attack on 750-round ACORN can also be improved using a relaxed version of TermEnum which is presented as RTermEnum on the righthand side of Algorithm 5. In the relaxed algorithm, RTermEnum is acquired from TermEnum by replacing some states which are marked in red in Algorithm 5, and we state details later in Sect. 6.4.

6.1   Application to Trivium

As can be seen in Table 2, the attack on 832-round TRIVIUM has $J = J_1 = 5$ and degree $d = 3$, so we have $\binom{5}{\leq 3} = 26$ using previous technique. But by running Algorithm 5, we find that $|J_2| = 5$, $|J_3| = 1$, so we have $1 + \sum_{t=1}^{3} |J_t| = 12 < \binom{5}{\leq 3} = 26$. Therefore, the complexity has now been reduced from $2^{76.7}$ to $2^{75.8}$. Similar technique can also be applied to the 73 dimensional cube of Table 2. Details are shown in Table 3.

Table 3. Results of TRIVIUM with Precise Term Enumeration

#Rounds	$|I|$	$|J_1|$	$|J_2|$	$|J_3|$	$|J_4|$	$|J_5|$	$|J_t|, t \geq 6$	$1 + \sum_{t=1}^{d} |J_t|$	Previous	Improved
832	72	5	5	1	0	0	0	$12 \approx 2^{3.58}$	$2^{76.7}$	$2^{75.58}$
833	73	7	6	1	0	0	0	$15 \approx 2^{3.91}$	2^{79}	$2^{76.91}$

6.2   Applications to Kreyvium

We revisit the 61-dimensional cube first given in [23] and transformed to a key recovery attack on 849-round Kreyvium in [22]. The degree of the superpoly is 9, so the complexity is given as $2^{81.7}$ in Appendix A of [25]. Since $J = J_1$ is of size 23, we enumerate all the terms of degree 2–9 and acquire the sets $J_2, \ldots, J_9$. $1 + \sum_{t=1}^{d} |J_t| = 5452 \approx 2^{12.41}$. So the complexity is now lowered to $2^{73.41}$. The details are listed in Table 4.

Table 4. Results of Kreyvium with Precise Term Enumeration

#Rounds	$|I|$	$|J_1|$	$|J_2|$	$|J_3|$	$|J_4|$	$|J_5|$	$|J_6|$	$|J_7|$	$|J_8|$	$|J_9|$	$1 + \sum_{t=1}^{d} |J_t|$	Previous	Improved
849	61	23	158	555	1162	1518	1235	618	156	26	$5452 \approx 2^{12.41}$	$2^{81.7}$	$2^{73.41}$

Algorithm 5. Enumerate all the terms of degree t

1: **procedure** TermEnum(Cube indices I, specific assignment to non-cube IVs IV or IV = NULL, targeted degree t)
2:    Declare an empty MILP model $\mathcal{M}$ and an empty set $J_t = \phi \subseteq \{1, \ldots, n\}^n$
3:    Declare $\boldsymbol{x}$ as n MILP variables of $\mathcal{M}$ corresponding to secret variables.
4:    Declare $\boldsymbol{v}$ as m MILP variables of $\mathcal{M}$ corresponding to public variables.
5:    $\mathcal{M}.con \leftarrow v_i = 1$ and assign $v_i.F = \delta$ for all $i \in I$
6:    $\mathcal{M}.con \leftarrow v_i = 0$ for all $i \in (\{1, 2, \ldots, n\} - I)$
7:    $\mathcal{M}.con \leftarrow \sum_{i=1}^{n} x_i = t$ and assign $x_i.F = \delta$ for all $i \in \{1, \ldots, n\}$
8:    **if** IV = NULL **then**
9:        $v_i.F = \delta$ for all $i \in (\{1, 2, \ldots, n\} - I)$
10:   **else**
11:       Assign the flags of v_i, $i \in (\{1, 2, \ldots, n\} - I)$ as:

$$v_i.F = \begin{cases} 1_c & \text{if } IV[i] = 1 \\ 0_c & \text{if } IV[i] = 0 \end{cases}$$

12:   **end if**
13:   Update $\mathcal{M}$ according to round functions and output functions
14:   **do**
15:       solve MILP model $\mathcal{M}$
16:       **if** $\mathcal{M}$ is feasible **then**
17:           pick index sequence $(j_1, \ldots, j_t) \subseteq \{1, \ldots, n\}^t$ s.t. $x_{j_1} = \ldots = x_{j_t} = 1$
18:           $J_t = J_t \cup \{(j_1, \ldots, j_t)\}$
19:           $\mathcal{M}.con \leftarrow \sum_{i=1}^{t} x_{j_i} \leq t - 1$
20:       **end if**
21:   **while** $\mathcal{M}$ is feasible
22:   **return** J_t
23: **end procedure**

1: **procedure** RTermEnum(Cube indices I, specific assignment to non-cube IVs IV or IV = NULL, targeted degree t)
2:    Declare an empty MILP model $\mathcal{M}$ and an empty set $JR_t = \phi \subseteq \{1, \ldots, n\}$
3:    Declare $\boldsymbol{x}$ as n MILP variables of $\mathcal{M}$ corresponding to secret variables.
4:    Declare $\boldsymbol{v}$ as m MILP variables of $\mathcal{M}$ corresponding to public variables.
5:    $\mathcal{M}.con \leftarrow v_i = 1$ and assign $v_i.F = \delta$ for all $i \in I$
6:    $\mathcal{M}.con \leftarrow v_i = 0$ for all $i \in (\{1, 2, \ldots, n\} - I)$
7:    $\mathcal{M}.con \leftarrow \sum_{i=1}^{n} x_i \geq t$ and assign $x_i.F = \delta$ for all $i \in \{1, \ldots, n\}$
8:    **if** IV = NULL **then**
9:        $v_i.F = \delta$ for all $i \in (\{1, 2, \ldots, n\} - I)$
10:   **else**
11:       Assign the flags of v_i, $i \in (\{1, 2, \ldots, n\} - I)$ as:

$$v_i.F = \begin{cases} 1_c & \text{if } IV[i] = 1 \\ 0_c & \text{if } IV[i] = 0 \end{cases}$$

12:   **end if**
13:   Update $\mathcal{M}$ according to round functions and output functions
14:   **do**
15:       solve MILP model $\mathcal{M}$
16:       **if** $\mathcal{M}$ is feasible **then**
17:           pick index set $\{j_1, \ldots, j_{t'}\} \subseteq \{1, \ldots, n\}$ s.t. $t' \geq t$ and $x_{j_1} = \ldots = x_{j_{t'}} = 1$
18:           $JR_t = JR_t \cup \{j_1, \ldots, j_{t'}\}$
19:           $\mathcal{M}.con \leftarrow \sum_{i \notin JR_t} x_i \geq 1$
20:       **end if**
21:   **while** $\mathcal{M}$ is feasible
22:   **return** JR_t
23: **end procedure**

6.3   Applications to Grain-128a

For the attack on 184-round Grain-128a, the superpoly has degree $d = 14$, the number of involved key bits is $|J| = |J_1| = 21$ and we are able to enumerate all terms of degree 1–14 as Table 5.

Table 5. Results of Grain-128a with term Enumeration

#Rounds	$|I|$	$|J_1|$	$|J_i|$ ($2 \leq i \leq 14$)	$1 + \sum_{t=1}^{d} |J_t|$	Previous	Improved
184	95	21	157, 651, 1765, 3394, 4838, 5231, 4326, 2627, 1288, 442, 104, 15, 1	$2^{14.61}$	$2^{115.95}$	$2^{109.61}$

6.4   Applications to ACORN

For the attack on 704-round ACORN, with the cube dimension 64, the number of involved key bits in the superpoly is 72, and the degree is 7. We enumerate all the terms of degree from 2 to 7 as in Table 6, therefore we manage to improve the complexity of our cube attack in the previous section.

Table 6. Results of ACORN with Precise Term Enumeration

#Rounds	$|I|$	$|J_1|$	$|J_2|$	$|J_3|$	$|J_4|$	$|J_5|$	$|J_6|$	$|J_7|$	$1 + \sum_{t=1}^{d} |J_t|$	Previous	Improved
704	64	72	1598	4911	5755	2556	179	3	$2^{13.88}$	$2^{93.23}$	$2^{77.88}$

Relaxed Algorithm 5. For the attack on 750-round ACORN (the superpoly is of degree $d = 5$), The left part of Algorithm 5 can only be carried out for the 5-degree terms $|J_5| = 46$. For $t = 2, 3, 4$, the sizes of J_t are too large to be enumerated. We settle for the index set JR_t containing the key indices that composing all the t-degree terms. For example, when $J_3 = \{(1,2,3), (1,2,4)\}$, we have $JR_3 = \{1, 2, 3, 4\}$. The relationship between J_t and JR_t is $|J_t| \leq \binom{|JR_t|}{t}$ and $J_1 = JR_1$. The searching space for J_t in Algorithm 5 is $\binom{|J_1|}{t}$ while that of the relaxed algorithm is only $\binom{|JR_t|}{t}$. So it is much easier to enumerate JR_t, therefore the complexity can still be improved (in comparison with Eq. (8)) as long as $|JR_t| < |J_1|$. The complexity of this relaxed version can be written as

$$\max\{2^{|I|} \times (1 + \sum_{t=1}^{d-1} \binom{|JR_t|}{t} + J_d), 2^{|I|} + 2^{|J|} \times (1 + \sum_{t=1}^{d-1} \binom{|JR_t|}{t} + J_d)\} \quad (12)$$

For 750-round ACORN, we enumerate J_5 and $JR_1, \ldots, JR_4$ whose sizes are listed in Table 7. The improved complexity, according to Eq. (12), is $2^{120.92}$, lower than the original $2^{125.71}$ given in Appendix A in [25].

Table 7. Results of ACORN with Relaxed Term Enumeration

#Rounds	$|I|$	$|JR_1|$	$|JR_2|$	$|JR_3|$	$|JR_4|$	$|J_5|$	$1 + \sum_{t=1}^{d-1} \binom{|JR_t|}{t} + |J_d|$	Previous	Improved
750	101	81	81	77	70	46	$2^{19.92}$	$2^{125.71}$	$2^{120.92}$

7 A Clique View of the Superpoly Recovery

The precise & relaxed term enumeration technique introduced in Sect. 6 have to execute many MILP instances, which is difficult for some applications. In this section, we represent the resultant superpoly as a graph, which is called *superpoly graph*, so that we can utilize the clique concept from the graph theory to upper bound the complexity of the superpoly recovery phase in our attacks, without requiring MILP solver as highly as the term enumeration technique.

Definition 3 (Clique [33]). *In a graph $G = (V, E)$, where V is the set of vertices and E is the set of edges, a subset $C \subseteq V$, s.t. each pair of vertices in C is connected by an edge is called a clique.*

A *i-clique* is defined as a clique consists of i vertices, and i is called the *clique number*. A 1-clique is a vertex, a 2-clique is just an edge, and a 3-clique is called a triangle.

Given a cube C_I, by running Algorithm 5 for degree i, we determine J_i, which is the set of all the degree-i terms that might appear in the superpoly $p(\boldsymbol{x}, \boldsymbol{v})$ (see Sect. 6). Then we represent $p(\boldsymbol{x}, \boldsymbol{v})$ as a graph $G = (J_1, J_2)$, where the vertices in J_1 correspond to the involved secret key bits in $p(\boldsymbol{x}, \boldsymbol{v})$, the edges between any pairs of the vertices reveal the quadratic terms involved in $p(\boldsymbol{x}, \boldsymbol{v})$, We call the graph $G = (J_1, J_2)$ the *superpoly graph* of the cube C_I. The set of i-cliques in the superpoly graph is denoted as $\mathcal{K}_i$. Note that there is a natural one-to-one correspondence between the sets J_i and $\mathcal{K}_i$ for $i = 1, 2$.

It follows from the definition of a clique that any i-clique in $\mathcal{K}_i$ ($i \geq 2$) represents a monomial of degree i whose all divisors of degree 2 belong to J_2. On the other hand, due to the "embed" Property 1 in Sect. 6, we have that all its quadratic divisors must be in J_2. Then any monomial in J_i can be represented by an i-clique in $\mathcal{K}_i$. Hence for all $i \geq 2$, J_i corresponds to a subset of $\mathcal{K}_i$. Denote the number of i-cliques as $|\mathcal{K}_i|$, then $|J_i| \leq |\mathcal{K}_i|$. Apparently, $|\mathcal{K}_i| \leq \binom{|J|}{i}$ for all $1 \leq i \leq d$.

Now we show a simple algorithm for constructing $\mathcal{K}_i$ from J_1 and J_2 for $i \geq 3$. For instance, when constructing $\mathcal{K}_3$, we take the union operation of all possible combinations of three elements from J_2, and only keep the elements of degree 3. Similarly, we construct $\mathcal{K}_i$ for $3 < i \leq d$, where d is the degree of the superpoly. Therefore, all the i-cliques ($3 \leq i \leq d$) are found by the simple algorithm, i.e. the number of i-cliques $|\mathcal{K}_i|$ in $G(J_1, J_2)$ is determined. We therefore can upper bound the complexity of the offline phase as

$$2^{|I|} \times \left(1 + \sum_{i=1}^{d} |\mathcal{K}_i|\right). \tag{13}$$

Note that we have $|J_i| \leq |\mathcal{K}_i| \leq \binom{|J_1|}{i}$. It indicates that the upper bound of the superpoly recovery given by clique theory in Eq. (13) is better than the one provided by our degree evaluation in Eq. (8), while it is weaker than the one presented by our term enumeration techniques in Eq. (10). However, it is unclear if there exists a specific relation between $|\mathcal{K}_i|$ and $\binom{|JR_i|}{i}$ in the relaxed terms enumeration technique.

Advantage over the Terms Enumeration Techniques. In Sect. 6 when calculating J_i ($i \geq 3$) by Algorithm 5, we set the target degree as i and solve the newly generated MILP to obtain J_i, regardless of the knowledge of J_{i-1} we already hold. On the other hand, as is known in some cases, the MILP solver might take long time before providing J_i as desired. However, by using clique theory, we first acquire J_1 and J_2, which are essential for the term enumeration method as well. According to the "embed" property, we then make full use of the knowledge of J_1 and J_2, to construct $\mathcal{K}_i$ for $i \geq 3$ by an algorithm which is actually just performing simple operations (like union operations among elements, or removal of repeated elements, etc) in sets. So hardly any cost is required to find all the $\mathcal{K}_i$ ($3 \leq i \leq d$) we want. This significantly saves the computation costs since solving MILP is usually very time-consuming.

8   Conclusion

Algebraic properties of the resultant superpoly of the cube attacks were further studied. We developed a division property based framework of cube attacks enhanced by the flag technique for identifying proper non-cube IV assignments. The relevance of our framework is three-fold: For the first time, it can identify proper non-cube IV assignments of a cube leading to a non-constant superpoly, rather than randomizing trails & summations in the offline phase. Moreover, our model derived the upper bound of the superpoly degree, which can break the $|I| + |J| < n$ barrier and enable us to explore even larger cubes or mount to attacks on more rounds. Furthermore, our accurate term enumeration techniques further reduced the complexities of the superpoly recovery, which brought us the current best key recovery attacks on ciphers namely TRIVIUM, Kreyvium, Grain-128a and ACORN.

Besides, when term enumeration cannot be carried out, we represent the resultant superpoly as a graph. By constructing all the cliques of our super-poly graph, an upper bound of the complexity of the superpoly recovery can be obtained.

Acknowledgements. We would like to thank Christian Rechberger, Elmar Tischhauser, Lorenzo Grassi and Liang Zhong for their fruitful discussions, and the anonymous reviewers for their valuable comments. This work is supported by University of Luxembourg project - FDISC, National Key Research and Development Program of China (Grant No. 2018YFA0306404), National Natural Science Foundation of China (No. 61472250, No. 61672347), Program of Shanghai Academic/Technology

Research Leader (No. 16XD1401300), the Research Council KU Leuven: C16/15/058, OT/13/071, the Flemish Government through FWO projects and by European Union's Horizon 2020 research and innovation programme under grant agreement No. H2020-MSCA-ITN-2014-643161 ECRYPT-NET.

References

1. Dinur, I., Shamir, A.: Cube attacks on tweakable black box polynomials. In: Joux, A. (ed.) EUROCRYPT 2009. LNCS, vol. 5479, pp. 278–299. Springer, Heidelberg (2009)
2. Aumasson, J.-P., Dinur, I., Meier, W., Shamir, A.: Cube testers and key recovery attacks on reduced-round MD6 and Trivium. In: Dunkelman, O. (ed.) FSE 2009. LNCS, vol. 5665, pp. 1–22. Springer, Heidelberg (2009)
3. Dinur, I., Shamir, A.: Breaking Grain-128 with dynamic cube attacks. In: Joux, A. (ed.) FSE 2011. LNCS, vol. 6733, pp. 167–187. Springer, Heidelberg (2011)
4. Fouque, P.-A., Vannet, T.: Improving key recovery to 784 and 799 rounds of Trivium using optimized cube attacks. In: Moriai, S. (ed.) FSE 2013. LNCS, vol. 8424, pp. 502–517. Springer, Heidelberg (2014)
5. Salam, M.I., Bartlett, H., Dawson, E., Pieprzyk, J., Simpson, L., Wong, K.K.-H.: Investigating cube attacks on the authenticated encryption stream cipher ACORN. In: Batten, L., Li, G. (eds.) ATIS 2016. CCIS, vol. 651, pp. 15–26. Springer, Singapore (2016)
6. Liu, M., Yang, J., Wang, W., Lin, D.: Correlation cube attacks: from weak-key distinguisher to key recovery. In: Nielsen, J.B., Rijmen, V. (eds.) EUROCRYPT 2018, Part II. LNCS, vol. 10821, pp. 715–744. Springer, Cham (2018)
7. Dinur, I., Morawiecki, P., Pieprzyk, J., Srebrny, M., Straus, M.: Cube attacks and cube-attack-like cryptanalysis on the round-reduced Keccak sponge function. In: Oswald, E., Fischlin, M. (eds.) EUROCRYPT 2015, Part I. LNCS, vol. 9056, pp. 733–761. Springer, Heidelberg (2015)
8. Huang, S., Wang, X., Xu, G., Wang, M., Zhao, J.: Conditional cube attack on reduced-round Keccak sponge function. In: Coron, J.-S., Nielsen, J.B. (eds.) EUROCRYPT 2017, Part II. LNCS, vol. 10211, pp. 259–288. Springer, Cham (2017)
9. Li, Z., Bi, W., Dong, X., Wang, X.: Improved conditional cube attacks on Keccak keyed modes with MILP method. In: Takagi, T., Peyrin, T. (eds.) ASIACRYPT 2017, Part I. LNCS, vol. 10624, pp. 99–127. Springer, Cham (2017)
10. Li, Z., Dong, X., Wang, X.: Conditional cube attack on round-reduced ASCON. IACR Trans. Symmetric Cryptol. **2017**(1), 175–202 (2017)
11. Dong, X., Li, Z., Wang, X., Qin, L.: Cube-like attack on round-reduced initialization of Ketje Sr. IACR Trans. Symmetric Cryptol. **2017**(1), 259–280 (2017)
12. Todo, Y., Isobe, T., Hao, Y., Meier, W.: Cube attacks on non-blackbox polynomials based on division property. In: Katz, J., Shacham, H. (eds.) CRYPTO 2017, Part III. LNCS, vol. 10403, pp. 250–279. Springer, Cham (2017)

13. Todo, Y.: Structural evaluation by generalized integral property. In: Oswald, E., Fischlin, M. (eds.) EUROCRYPT 2015, Part I. LNCS, vol. 9056, pp. 287–314. Springer, Heidelberg (2015)

14. Todo, Y.: Integral cryptanalysis on full MISTY1. In: Gennaro, R., Robshaw, M. (eds.) CRYPTO 2015, Part I. LNCS, vol. 9215, pp. 413–432. Springer, Heidelberg (2015)

15. Todo, Y., Morii, M.: Bit-based division property and application to SIMON family. In: Peyrin, T. (ed.) FSE 2016. LNCS, vol. 9783, pp. 357–377. Springer, Heidelberg (2016)

16. Xiang, Z., Zhang, W., Bao, Z., Lin, D.: Applying MILP method to searching integral distinguishers based on division property for 6 lightweight block ciphers. In: Cheon, J.H., Takagi, T. (eds.) ASIACRYPT 2016, Part I. LNCS, vol. 10031, pp. 648–678. Springer, Heidelberg (2016)

17. Gu, Z., Rothberg, E., Bixby, R.: Gurobi optimizer. http://www.gurobi.com/

18. Sun, L., Wang, W., Wang, M.: MILP-aided bit-based division property for primitives with non-bit-permutation linear layers. Cryptology ePrint Archive, Report 2016/811 (2016). https://eprint.iacr.org/2016/811

19. Sun, L., Wang, W., Wang, M.: Automatic search of bit-based division property for ARX ciphers and word-based division property. In: Takagi, T., Peyrin, T. (eds.) ASIACRYPT 2017, Part I. LNCS, vol. 10624, pp. 128–157. Springer, Cham (2017)

20. Funabiki, Y., Todo, Y., Isobe, T., Morii, M.: Improved integral attack on HIGHT. In: Pieprzyk, J., Suriadi, S. (eds.) ACISP 2017, Part I. LNCS, vol. 10342, pp. 363–383. Springer, Cham (2017)

21. Wang, Q., Grassi, L., Rechberger, C.: Zero-sum partitions of PHOTON permutations. In: Smart, N.P. (ed.) CT-RSA 2018. LNCS, vol. 10808, pp. 279–299. Springer, Cham (2018)

22. Todo, Y., Isobe, T., Hao, Y., Meier, W.: Cube attacks on non-blackbox polynomials based on division property (full version). Cryptology ePrint Archive, Report 2017/306 (2017). https://eprint.iacr.org/2017/306

23. Liu, M.: Degree evaluation of NFSR-based cryptosystems. In: Katz, J., Shacham, H. (eds.) CRYPTO 2017, Part III. LNCS, vol. 10403, pp. 227–249. Springer, Cham (2017)

24. Fu, X., Wang, X., Dong, X., Meier, W.: A key-recovery attack on 855-round Trivium. Cryptology ePrint Archive, Report 2018/198 (2018). https://eprint.iacr.org/2018/198

25. Wang, Q., Hao, Y., Todo, Y., Li, C., Isobe, T., Meier, W.: Improved division property based cube attacks exploiting algebraic properties of superpoly (full version). Cryptology ePrint Archive, Report 2017/1063 (2017). https://eprint.iacr.org/2017/1063

26. Todo, Y., Isobe, T., Meier, W., Aoki, K., Zhang, B.: Fast correlation attack revisited-cryptanalysis on full Grain-128a, Grain-128, and Grain-v1. In: Shacham, H., Boldyreva, A. (eds.) CRYPTO 2018. LNCS, vol. 10991, pp. 129–159. Springer, Cham (2018)

27. Lehmann, M., Meier, W.: Conditional differential cryptanalysis of Grain-128a. In: Pieprzyk, J., Sadeghi, A.-R., Manulis, M. (eds.) CANS 2012. LNCS, vol. 7712, pp. 1–11. Springer, Heidelberg (2012)

28. Mouha, N., Wang, Q., Gu, D., Preneel, B.: Differential and linear cryptanalysis using mixed-integer linear programming. In: Wu, C.-K., Yung, M., Lin, D. (eds.) Inscrypt 2011. LNCS, vol. 7537, pp. 57–76. Springer, Heidelberg (2012)

29. Sun, S., Hu, L., Wang, P., Qiao, K., Ma, X., Song, L.: Automatic security evaluation and (related-key) differential characteristic search: application to SIMON, PRESENT, LBlock, DES(L) and other bit-oriented block ciphers. In: Sarkar, P., Iwata, T. (eds.) ASIACRYPT 2014, Part I. LNCS, vol. 8873, pp. 158–178. Springer, Heidelberg (2014)

30. Sun, S., Hu, L., Wang, M., Wang, P., Qiao, K., Ma, X., Shi, D., Song, L., Fu, K.: Towards finding the best characteristics of some bit-oriented block ciphers and automatic enumeration of (related-key) differential and linear characteristics with predefined properties. Cryptology ePrint Archive, Report 2014/747 (2014). https://eprint.iacr.org/2014/747

31. Cui, T., Jia, K., Fu, K., Chen, S., Wang, M.: New automatic search tool for impossible differentials and zero-correlation linear approximations. Cryptology ePrint Archive, Report 2016/689 (2016). https://eprint.iacr.org/2016/689

32. Sasaki, Y., Todo, Y.: New impossible differential search tool from design and cryptanalysis aspects. In: Coron, J.-S., Nielsen, J.B. (eds.) EUROCRYPT 2017, Part III. LNCS, vol. 10212, pp. 185–215. Springer, Cham (2017)

33. Bondy, J.A., Murty, U.S.R.: Graph Theory with Applications, vol. 290. Macmillan, London (1976)

Generic Attacks Against
Beyond-Birthday-Bound MACs

Gaëtan Leurent[1(✉)], Mridul Nandi[2(✉)], and Ferdinand Sibleyras[1(✉)]

[1] Inria, Paris, France
{gaetan.leurent,ferdinand.sibleyras}@inria.fr
[2] Indian Statistical Institute, Kolkata, India
mridul.nandi@gmail.com

Abstract. In this work, we study the security of several recent MAC constructions with provable security beyond the birthday bound. We consider block-cipher based constructions with a double-block internal state, such as SUM-ECBC, PMAC+, 3kf9, GCM-SIV2, and some variants (LightMAC+, 1kPMAC+). All these MACs have a security proof up to $2^{2n/3}$ queries, but there are no known attacks with less than 2^n queries.

We describe a new cryptanalysis technique for double-block MACs based on finding quadruples of messages with four pairwise collisions in halves of the state. We show how to detect such quadruples in SUM-ECBC, PMAC+, 3kf9, GCM-SIV2 and their variants with $\mathcal{O}(2^{3n/4})$ queries, and how to build a forgery attack with the same query complexity. The time complexity of these attacks is above 2^n, but it shows that the schemes do not reach full security in the information theoretic model. Surprisingly, our attack on LightMAC+ also invalidates a recent security proof by Naito.

Moreover, we give a variant of the attack against SUM-ECBC and GCM-SIV2 with time and data complexity $\tilde{\mathcal{O}}(2^{6n/7})$. As far as we know, this is the first attack with complexity below 2^n against a deterministic beyond-birthday-bound secure MAC.

As a side result, we also give a birthday attack against 1kf9, a single-key variant of 3kf9 that was withdrawn due to issues with the proof.

Keywords: Modes of operation · Cryptanalysis
Message authentication codes · Beyond-birthday-bound security

1 Introduction

Message authentication codes (or MACs) ensure the authenticity of messages in the secret-key setting. They are a core element of real-world security protocols such as TLS, SSH, or IPSEC. A MAC takes a message (and optionally a nonce) and a secret key to generate a tag that is sent with the message. Traditionally, they are classified into three types: deterministic, nonce-based, and probabilistic.

Deterministic MAC designs are the most popular, with widely used constructions based on block-cipher (CBC-MAC [4,13], OMAC [18], PMAC [5],

© International Association for Cryptologic Research 2018
H. Shacham and A. Boldyreva (Eds.): CRYPTO 2018, LNCS 10991, pp. 306–336, 2018.
https://doi.org/10.1007/978-3-319-96884-1_11

LightMAC [29], ...) and hash functions (HMAC [2], NMAC [2], NI-MAC [1], ...). However, there is a generic forgery attack against all deterministic iterated MACs, using collisions in the internal state, due to Preneel and van Oorschot [37]. Therefore, these MACs only achieve security up to the birthday bound, *i.e.* when the number of queries by the adversary is bounded by $2^{n/2}$, with n the state size. This is equivalently called $n/2$-bit security.

One way to increase the security is to use a *nonce*, a unique value provided by the user (in practice, the nonce is usually a counter). This approach has been pioneered by Wegman and Carter [41] based on an earlier work by Gilbert *et al.* [15]. Later a few follow ups like EDM and EWCDM [7], and Dual EDM [30] have been proposed to achieve beyond birthday security.

Alternatively, a probabilistic MAC uses a random coin for the extra value, which is usually called a *salt*, and must be transmitted with the MAC. Probabilistic MACs have the advantage that they can stay secure when called with the same input twice, and don't require a state to keep the nonce unique. Some popular probabilistic MAC constructions are XMACR [3], RMAC [22] and EHtM [31]. In particular, RMAC and EHtM have security beyond the birthday bound.

However, deterministic MACs are easier to use in practice, and there has been an important research effort to build deterministic MAC with security beyond the birthday bound, using an internal state larger than the primitive size. In particular, several constructions use a $2n$-bit internal state so that collisions in the state are only expected after 2^n queries. Yasuda first proposed SUM-ECBC [42], a beyond birthday bound (BBB) secure deterministic MAC that achieves $2n/3$-bit security. However, this construction has rate $1/2$ and later Yasuda himself proposed one of the most popular BBB secure MAC PMAC+ [43] achieving rate 1. Later several other constructions like 3kf9 [44], LightMAC+ [33], GCM-SIV2 [20], and single key PMAC+ [9] have been proposed. Interestingly, all the above designs share a common structure: a double-block universal hash function outputs a $2n$-bit hash value (seen as two n-bit halves), and a finalization function generates the tag by XORing encrypted values of the two n-bit hash values. This structure has been called double-block-hash-then-sum, and it will be the focus of our paper.

More recently, variants of PMAC+ based on tweakable block-cipher have also been proposed, such as PMAC_TBC [32], PMACx [27], ZMAC [21], and ZMAC+ [28].

Our results. We focus on the security of deterministic block-cipher based MACs with security beyond the birthday bound and double-block hash construction. Several previous works have been focused on security proofs, showing that they are secure up to $2^{2n/3}$ queries [9,20,33,42–44]. For most of these constructions, the advantage of an adversary making q short queries is bounded by $\mathcal{O}(q^3/2^{2n})$. Recently, Naito [34] gave an improved security proof for LighMAC+, with advantage at most $\mathcal{O}(q_t^2 q_v/2^{2n})$, with q_t MAC queries and q_v verification queries. In particular, this would prove security up to 2^n when the adversary can only do a single verification query.

In this work, we take the opposite approach and look for generic attacks against these modes. We use a cryptanalysis technique that can be seen as a generalisation of the collision attack of Preneel and van Oorschot [37]. Instead of looking for a pair of messages so that the full state collides, we look for a quadruple of messages, which can be seen either as two pairs colliding on the first half of the state, or two pairs colliding on the second half. Since the finalization function combines the halves with a sum, we can detect such a quadruple because the corresponding MACs sum to zero, and can usually amplify this filtering. Moreover, when the message are well constructed, the relations defining the four collisions create a linear system of rank only three, so that we expect one good quadruple out of 2^{3n}. Therefore, we only need four lists of $2^{3n/4}$ queries, and we expect one good quadruple out of the 2^{3n} choices in the four lists.

Table 1. Summary of the security for studied modes and our main results. q is the number of queries, ℓ is maximum size of a query, σ is total number of processed blocks. The expected lower bound and attack complexity is in number of constant length queries ($\ell = \mathcal{O}(1)$). We use "U" for universal forgeries, and "E" for existential forgeries.

Mode	Provable security bounds		Attacks (this work)		
	Advantage	Queries	Queries	Time	Type
SUM-ECBC [42]	$\mathcal{O}(\frac{q^3\ell^3}{2^{2n}})$	$\Omega(2^{2n/3})$	$\mathcal{O}(2^{3n/4})$	$\tilde{\mathcal{O}}(2^{3n/2})$	U
			$\mathcal{O}(2^{6n/7})$	$\tilde{\mathcal{O}}(2^{6n/7})$	U
GCM-SIV2 [20]	$\mathcal{O}(\frac{q^3\ell^2}{2^{2n}})$	$\Omega(2^{2n/3})$	$\mathcal{O}(2^{3n/4})$	$\tilde{\mathcal{O}}(2^{3n/2})$	U
			$\mathcal{O}(2^{6n/7})$	$\tilde{\mathcal{O}}(2^{6n/7})$	U
PMAC+ [43]	$\mathcal{O}(\frac{q^3\ell^3}{2^{2n}})$	$\Omega(2^{2n/3})$	$\mathcal{O}(2^{3n/4})$	$\tilde{\mathcal{O}}(2^{3n/2})$	E
LightMAC+ [33]	$\mathcal{O}(\frac{q^3}{2^{2n}})$	$\Omega(2^{2n/3})$	$\mathcal{O}(2^{3n/4})$	$\tilde{\mathcal{O}}(2^{3n/2})$	E
1kPMAC+ [9]	$\mathcal{O}(\frac{\sigma}{2^n} + \frac{q\sigma^2}{2^{2n}})$	$\Omega(2^{2n/3})$	$\mathcal{O}(2^{3n/4})$	$\tilde{\mathcal{O}}(2^{3n/2})$	E
3kf9 [44]	$\mathcal{O}(\frac{q^3\ell^3}{2^{2n}} + \frac{q\ell}{2^n})$	$\Omega(2^{2n/3})$	$\mathcal{O}(\sqrt[4]{n}\cdot 2^{3n/4})$	$\tilde{\mathcal{O}}(2^{5n/4})$	U
1kf9 [8]	$\mathcal{O}(\frac{q\ell^2}{2^n} + \frac{q^3\ell^4}{2^{2n}} + \frac{q^4\ell^4}{2^{3n}} + \frac{q^4\ell^6}{2^{4n}})$	$\Omega(2^{2n/3})$	$\mathcal{O}(2^{n/2})$	$\tilde{\mathcal{O}}(2^{n/2})$	U

Table 1 shows a summary of our main results and how they compare with their respective provable security claims. In particular, we have forgeries attacks with $\mathcal{O}(2^{3n/4})$ MAC queries against SUM-ECBC, GCM-SIV2, PMAC+, LightMAC+, 1kPMAC+, and 3kf9. As far as we know, these are the first attacks with less than 2^n queries against these constructions. Our attack against LighMAC+ contradicts the recent security bound for LighMAC+ [34], because we have an attack with $\mathcal{O}(2^{3n/4})$ MAC queries, and a single verification query. The other attacks do not contradict the security proofs, but they make an important step towards understanding the actually security of these modes: we now have a lower bound of $2^{2n/3}$ queries from the proofs, and an upper bound of $2^{3n/4}$ from our attacks.

The attacks have a complexity of $2^{3n/4}$ in the information theoretic model (the model used for most MAC security proofs), but we note that an attacker needs more than 2^n operations to create a forgery. However, we have found a variant of our attack against SUM-ECBC and GCM-SIV2 with total complexity below 2^n, using $\mathcal{O}(2^{6n/7})$ queries and $\tilde{\mathcal{O}}(2^{6n/7})$ operations.

We have also found an attack with only $\mathcal{O}(2^{n/2})$ queries and $\tilde{\mathcal{O}}(2^{n/2})$ operations against 1kf9 [8], a single key variant of 3kf9 with claimed security up to $2^{2n/3}$ queries. 1kf9 has been withdrawn due to issues with its security proof, but no attack was known previously.

Related works. There has been extensive work on security proofs for modes of operations, with a recent focus on security beyond the birthday bound. An interesting example is the encryption mode CENC by Iwata [17]: the initial proof was only up to $2^{2n/3}$ queries, but a later proof showed that it actually remains secure close to 2^n queries [19]. Our results show that in the case of double-block-hash-then-sum MACs, the security is lower than n-bit security.

Similarly, the initial proof of the randomized MAC EHtM only gave security up to $2^{2n/3}$, but a later proof showed security up to $2^{3n/4}$ [11]. This result also includes a matching attack, using a technique similar to ours based on looking for quadruples. However in the case of EHtM the attacker can observe part of the state, which allows him to find a right quadruple in $\mathcal{O}(2^{3n/4})$ time and memory. In our case we can't observe the internal state at all, thus we need to use different tricks tailored to each construction in order to amplify the filtering and avoid the many false-positives. In particular, this significantly increases the time and memory complexity.

There has also been intensive work on generic attacks to complement the security proof results. After the generic collision attack of Preneel and van Oorschot [37], more advanced attacks against MACs have been described, with stronger outcomes than existential forgeries, starting with a key-recovery attack against the envelop MAC by the same authors [38]. In particular, a series of attacks against hash-based MACs [10,16,26,36] led to universal forgery attacks against long challenges, and key-recovery attacks when the hash function has an internal checksum (like the GOST family). Against PMAC, Lee et al. showed a universal forgery attack in 2006 [25]. Later, Fuhr, Leurent and Suder gave a key-recovery attack against the PMAC variant used in AEZv3 [14]. Issues with GCM authentication with truncated tags were also pointed out by Ferguson [12]. These attacks don't contradict the security proofs of the schemes, but they are important results to understand the security degradation after the birthday bound.

Organization of the paper. We first explain our attack technique using quadruples of messages in Sect. 2, and give three concrete attacks using this technique: an attack against SUM-ECBC and GCM-SIV2 in Sect. 3, an attack against PMAC+ and related constructions in Sect. 4, and an attack against 3kf9 in Sect. 5. Finally, we show a variant of the technique using special properties of the single-key constructions of [8,9] in Sect. 6.

Notations. We denote the concatenation of messages blocks x and y as $x \parallel y$. When x and y fit together in one block, we use $x|y$ to denote their concatenation. We use $L[i]$ to denote element i of list L, $x_{[i]}$ to denote bit i of x, and $x_{[i:j]}$ to denote bits i to $j - 1$. Finally, we use a curly brace for systems of equations.

2   Generic Attack Against Double-Block-Hash MACs

We first explain our attacks in a generic way, and leave the specific details to later sections focused on concrete MAC constructions.

We consider MACs where the $2n$-bit internal state is divided in two n-bit parts, that we denote Σ and Θ, and the final MAC is computed as:

$$\mathrm{MAC}(M) = E\big(\Sigma(M)\big) \oplus E'\big(\Theta(M)\big),$$

where E and E' denote the block cipher with potentially different keys. The functions Σ and Θ can be seen as two n-bit universal hash functions computed on the message, hence the name double-block-hash-then-sum MAC.

Our attacks exploit the fact that the two halves are combined with a sum, where one side depends only on Σ, and the other side depends only on Θ. They do not seem applicable to constructions with more intricate finalization functions, such as LightMAC+2 [33], or the tweakable block-cipher based constructions PMAC_TBC [32], PMACx [27], ZMAC [21], or ZMAC+ [28].

2.1   Using Quadruples

Our strategy consists in looking for a quadruple of messages (X, Y, Z, T) such that pairs of values collide for one half of the state. More precisely, we look for quadruples satisfying a relation $\mathcal{R}(X, Y, Z, T)$ defined as:

$$\mathcal{R}(X, Y, Z, T) := \begin{cases} \Sigma(X) = \Sigma(Y) \\ \Theta(Y) = \Theta(Z) \\ \Sigma(Z) = \Sigma(T) \\ \Theta(T) = \Theta(X) \end{cases}$$

In particular, since the MAC is computed as $\mathrm{MAC}(M) = E\big(\Sigma(M)\big) \oplus E'\big(\Theta(M)\big)$, it follows that:

$$\mathcal{R}(X, Y, Z, T) \implies \mathrm{MAC}(X) \oplus \mathrm{MAC}(Y) \oplus \mathrm{MAC}(Z) \oplus \mathrm{MAC}(T) = 0. \quad (1)$$

In addition, if the messages X, Y, Z, T are well constructed, the relation $\mathcal{R}$ reduces to a linear system of rank only three, i.e.

$$[\Sigma(X) = \Sigma(Y) \text{ and } \Theta(Y) = \Theta(Z) \text{ and } \Sigma(Z) = \Sigma(T)] \implies \Theta(T) = \Theta(X).$$

Therefore, we expect to find one quadruple satisfying the relation out of 2^{3n}, and we can construct 2^{3n} quadruples with just $4 \times 2^{3n/4}$ queries. This gives an attack with data complexity $\mathcal{O}(2^{3n/4})$.

In practice, we consider lists of $2^{3n/4}$ messages, generated with two message injection functions ϕ and ψ. These functions are different in every attack, but they mostly correspond to adding two distinct prefixes, as in the following example:

$$\phi(i) = 0 \parallel i \qquad\qquad \psi(i) = 1 \parallel i$$
$$X = \phi(x) = 0 \parallel x \qquad\qquad Y = \psi(y) = 1 \parallel y$$
$$Z = \phi(z) = 0 \parallel z \qquad\qquad T = \psi(t) = 1 \parallel t,$$

In particular, the pairs (X, Y), (Y, Z), (Z, T) and (T, X) that we consider always contain a message built with ϕ and message built with ψ. Therefore, we will have the required collisions in Σ or Θ if the difference introduced in the half-state by the second block cancels the difference found after processing the first block.

This type of attack has some similarities with a higher order differential attack. Indeed, in the easiest case (e. g. our attack against SUM-ECBC), the relation $\mathcal{R}$ can be written as $\mathcal{R}(x, y, z, t) \iff \big[x \oplus y = z \oplus t = \Delta_1$ and $x \oplus t = y \oplus z = \Delta_3 \big]$ for some secret values Δ_1 and Δ_3. This idea of looking for quadruples is also very similar to the attack on EHtM [11], but the full attack will turn out quite different. Indeed, in the case of EHtM, the attacker can observe the salt R which represent half of the $2n$-bit internal state. Here this would be the equivalent of observing $\Sigma(m)$ for all processed messages m. This is clearly not possible for the studied constructions and we need something more to discriminate and find a good quadruple that satisfies $\mathcal{R}$.

2.2 Detecting Quadruples: Generalized Birthday Algorithms

To finish the attack we usually need to locate one good quadruple. The relation $\mathrm{MAC}(X) \oplus \mathrm{MAC}(Y) \oplus \mathrm{MAC}(Z) \oplus \mathrm{MAC}(T) = 0$ in itself is too weak because we expect one quadruple out of 2^n to satisfy it randomly, but we can usually amplify the filtering using related quadruples that satisfy $\mathcal{R}$ simultaneously (the exact details depend on the MAC construction).

In most of our attacks, we can express the search for a quadruple as an instance of the 4-sum problem, and solve it using variants of Wagner's generalized birthday algorithm [40]. This reduces the time complexity of the attacks (compared to a naive search), and provides trade-offs between the query, memory and time complexities.

More precisely, our problem can be stated as follow:

Definition 1 (4-sum problem). *Given four lists L_1, L_2, L_3, L_4 of 2^s elements, with on average 2^p quadruples $(x, y, z, t) \in L_1 \times L_2 \times L_3 \times L_4$ such that $x \oplus y \oplus z \oplus t = 0$, find one of them.*

Note that if the lists contain random n-bit words, we expect to have $2^p = 2^{4s-n}$ solutions, but in some of our instances there are more solutions because of the structure of the lists.

We denote the join operator as $\bowtie$; it computes the pairwise sum of two lists, and keeps the initial values attached to the sum. In addition, the join operator with filtering $\bowtie_t^\alpha$ only keeps values such that the t least significant bits of the sum agree with the value α:

$$A \bowtie B = \{(a \oplus b, a, b) \; : \; (a, b) \in A \times B\}$$
$$A \bowtie_t^\alpha B = \{(a \oplus b, a, b) \; : \; (a, b) \in A \times B, \; a_{[0:t]} \oplus b_{[0:t]} = \alpha\}$$

In particular, we have $\bowtie \; = \bowtie_0^0$. We also denote as $\bowtie_\infty$ the joint operator with filtering over the full input values. The filtered joint operator is the basis of Wagner's algorithm, and it can be computed in almost linear time by sorting the two input lists, and stepping through them simultaneously.

Direct algorithm. While a naive algorithm for our 4-sum instances would take time 2^{4s} to examine all quadruples, there is a simple improvement with time and memory $\tilde{O}(2^{2s})$. First, the attacker builds $L_{12} = L_1 \bowtie L_2$ and $L_{34} = L_3 \bowtie L_4$. Then, he looks for a collision between the first component of L_{12} and L_{34}. A collision directly yields a solution. This always finds a solution if it exists in $\tilde{O}(2^{2s})$ operations but it also takes $O(2^{2s})$ memory.

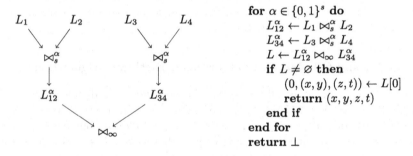

```
for α ∈ {0,1}ˢ do
    L₁₂ᵅ ← L₁ ⋈ˢᵅ L₂
    L₃₄ᵅ ← L₃ ⋈ˢᵅ L₄
    L ← L₁₂ᵅ ⋈∞ L₃₄ᵅ
    if L ≠ ∅ then
        (0, (x, y), (z, t)) ← L[0]
        return (x, y, z, t)
    end if
end for
return ⊥
```

Fig. 1. Generalized Birthday algorithm to find good quadruples.

Memory efficient algorithm. We can reduce the memory complexity of the algorithm if we avoid constructing the full lists L_{12} and L_{34}. An algorithm with low memory complexity was first described by Chose *et al.* [6], but we use the description given by Wagner in the full version of [40].

Instead of building the full lists L_{12} and L_{34}, we filter values such that s least significant bits differ by some fixed value α. This reduces the expected size of the lists to only 2^s: $E[|L_{34}^\alpha|] = E[|L_{12}^\alpha|] = |L_1| \cdot |L_2|/2^s = 2^s$. If this algorithm is repeated for every s-bit value α, it will eventually find all solutions.

Actually, one run of the algorithm detects the solutions whose least significant bits of $x \oplus y$ are equal to α. If there are 2^p solutions in total, there is one such solution with probability 2^{p-s}, and this algorithm will find the first solution after trying 2^{s-p} values of α on average. Therefore, the expected time complexity of the algorithm given by Fig. 1 is only $\tilde{O}(2^{2s-p})$.

Related work. In a 2016 work, Nikolic and Sasaki [35] investigate the 4-sum where we need to find 4 different inputs x, y, z, t to a function f such that $f(x) \oplus f(y) \oplus f(z) \oplus f(t) = 0$. They also mention that their algorithm is adaptable to pairwise identical functions, *i. e.* $f(x) \oplus g(y) \oplus f(z) \oplus g(t) = 0$.

Most of our attacks can be written in this way; concretely, they are equivalent to instances of random functions with $3n$-bit outputs. In this setting our algorithm takes time $\tilde{\mathcal{O}}(2^{3n/2})$ and memory $\mathcal{O}(2^{3n/4})$, while Nikolic and Sasaki's work can reach $\tilde{\mathcal{O}}(2^{9n/8})$ time and $\mathcal{O}(2^{3n/4})$ memory. Unfortunately, their algorithm requires $\tilde{\mathcal{O}}(2^{9n/8})$ queries to the functions; this would translate to $\tilde{\mathcal{O}}(2^{9n/8})$ queries to the MAC, which is not interesting in our context.

3 Attacking SUM-ECBC-like constructions

We start with attacks against SUM-ECBC [42] and GCM-SIV2 [20]; while the constructions are quite different, they have a similar structure and the same attacks can be used in both cases. We give a universal forgery attack with $\mathcal{O}(2^{3n/4})$ queries and $\tilde{\mathcal{O}}(2^{3n/2})$ operations (using memory $\mathcal{O}(2^{3n/4})$), and a variant with total complexity below 2^n, with $\mathcal{O}(2^{6n/7})$ queries and $\tilde{\mathcal{O}}(2^{6n/7})$ operations.

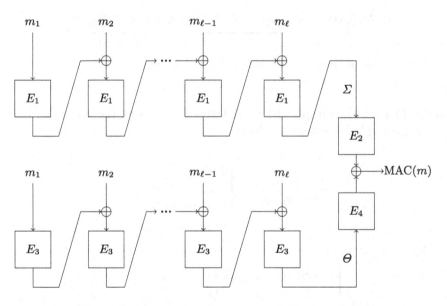

Fig. 2. Diagram for SUM-ECBC with a $\ell-$block message.

3.1 Attacking SUM-ECBC

SUM-ECBC was designed by Yasuda in 2010 [42], inspired by MAC constructions summing two CBC-MACs in the ISO 9797-1 standard. The scheme uses a block cipher keyed with four independent keys, denoted as E_1, E_2, E_3, E_4. The message M is first padded with 10^* padding, and divided into n-bit blocks. In the following we ignore the padding and consider the padded message as the input: this makes our description easier, and any padded message whose last block is non-zero can be "un-padded" to generate a valid input message. The construction is defined as follows (see also Fig. 2):

$$\Sigma(M) = \sigma_\ell \qquad\qquad \sigma_0 = 0 \qquad \sigma_i = E_1(\sigma_{i-1} \oplus m_i)$$
$$\Theta(M) = \theta_\ell \qquad\qquad \theta_0 = 0 \qquad \theta_i = E_3(\theta_{i-1} \oplus m_i)$$
$$\mathrm{MAC}(M) = E_2(\Sigma(M)) \oplus E_4(\Theta(M))$$

Attack. Following the framework of Sect. 2, we consider quadruple of messages, built with two message injection functions:

$$\phi(i) = 0 \,\|\, i \qquad\qquad\qquad \psi(i) = 1 \,\|\, i$$

In particular, we have

$$\mathrm{MAC}(\phi(i)) = E_2\Big(\underbrace{E_1\big(i \oplus E_1(0)\big)}_{\Sigma_0(i)}\Big) \oplus E_4\Big(\underbrace{E_3\big(i \oplus E_3(0)\big)}_{\Theta_0(i)}\Big)$$

$$\mathrm{MAC}(\psi(i)) = E_2\Big(\underbrace{E_1\big(i \oplus E_1(1)\big)}_{\Sigma_1(i)}\Big) \oplus E_4\Big(\underbrace{E_3\big(i \oplus E_3(1)\big)}_{\Theta_1(i)}\Big)$$

Next, we build quadruples of messages X, Y, Z, T with

$$X = \phi(x) \qquad Y = \psi(y) \qquad Z = \phi(z) \qquad T = \psi(t),$$

and we look for a quadruple with partial state collisions for the underlying pairs, *i. e.* a quadruple following the relation:

$$\mathcal{R}(x, y, z, t) := \begin{cases} \Sigma_0(x) = \Sigma_1(y) \\ \Sigma_0(z) = \Sigma_1(t) \\ \Theta_0(z) = \Theta_1(y) \\ \Theta_0(x) = \Theta_1(t). \end{cases}$$

We have

$$\mathcal{R}(x, y, z, t) \Leftrightarrow \begin{cases} x \oplus E_1(0) = y \oplus E_1(1) \\ z \oplus E_3(0) = y \oplus E_3(1) \\ z \oplus E_1(0) = t \oplus E_1(1) \\ x \oplus E_3(0) = t \oplus E_3(1) \end{cases} \Leftrightarrow \begin{cases} x \oplus y \oplus z \oplus t = 0 \\ x \oplus y = E_1(0) \oplus E_1(1) \\ x \oplus t = E_3(0) \oplus E_3(1) \end{cases}$$

As promised in Sect. 2, $\mathcal{R}$ defines a $3n$−bit relation. We can easily observe when $x \oplus y \oplus z \oplus t = 0$, and we can also detect the relation on the sum of the MACs following Eq. (1):

$$\mathcal{R}(x, y, z, t) \Rightarrow \mathrm{MAC}(\phi(x)) \oplus \mathrm{MAC}(\psi(y)) \oplus \mathrm{MAC}(\phi(z)) \oplus \mathrm{MAC}(\psi(t)) = 0$$

Moreover, we observe that $\mathcal{R}(x, y, z, t)$ is satisfied if and only if $\mathcal{R}(x \oplus c, y \oplus c, z \oplus c, t \oplus c)$ is satisfied for any constant c. We use this relation to build several quadruples that satisfy $\mathcal{R}$ simultaneously:

$$\mathcal{R}(x, y, z, t) \iff \forall c, \mathcal{R}(x \oplus c, y \oplus c, z \oplus c, t \oplus c) \tag{2}$$

This leads to an attack with $\mathcal{O}(2^{3n/4})$ queries: we consider four sets $\mathcal{X}, \mathcal{Y}, \mathcal{Z}, \mathcal{T}$ of $2^{3n/4}$ values, and we look for a quadruple $(x, y, z, t) \in \mathcal{X} \times \mathcal{Y} \times \mathcal{Z} \times \mathcal{T}$ with:

$$\begin{cases} x \oplus y \oplus z \oplus t = 0 \\ \mathrm{MAC}(\phi(x)) \oplus \mathrm{MAC}(\psi(y)) \oplus \mathrm{MAC}(\phi(z)) \oplus \mathrm{MAC}(\psi(t)) = 0 \\ \mathrm{MAC}(\phi(x \oplus 1)) \oplus \mathrm{MAC}(\psi(y \oplus 1)) \oplus \mathrm{MAC}(\phi(z \oplus 1)) \oplus \mathrm{MAC}(\psi(t \oplus 1)) = 0. \end{cases} \tag{3}$$

Because we need a fair distribution of values $x \oplus y$ and $x \oplus t$ to find the good quadruple we build the sets as:

$$\mathcal{X} = \{x \in \{0,1\}^n : x_{[0:n/4]} = 0\} \qquad \mathcal{Y} = \{x \in \{0,1\}^n : x_{[n/4:n/2]} = 0\}$$
$$\mathcal{Z} = \{x \in \{0,1\}^n : x_{[n/2:3n/4]} = 0\} \qquad \mathcal{T} = \{x \in \{0,1\}^n : x_{[3n/4:n]} = 0\}$$

With this construction, there is exactly one quadruple $(x, y, z, t) \in \mathcal{X} \times \mathcal{Y} \times \mathcal{Z} \times \mathcal{T}$ that respects $\mathcal{R}$, given by:

$$x = v_1|w_2|u_3|0 \qquad y = w_1|v_2|0|u_4 \qquad z = u_1|0|v_3|w_4 \qquad t = 0|u_2|w_3|v_4,$$

where:

$$E_1(0) \oplus E_1(1) =: u_1|u_2|u_3|u_4$$
$$E_3(0) \oplus E_3(1) =: v_1|v_2|v_3|v_4$$
$$E_1(0) \oplus E_1(1) \oplus E_3(0) \oplus E_3(1) =: w_1|w_2|w_3|w_4.$$

We expect on average one random quadruple satisfying (3) (with 2^{3n} potential quadruples, and a $3n$-bit filtering), in addition to the quadruple satisfying $\mathcal{R}$. The correct quadruple can easily be checked with a few extra queries.

In practice, we use the generalized birthday algorithms of Sect. 2.2 in order to optimize the complexity of the attack. We consider four lists:

$$L_1 = \{x \| \mathrm{MAC}(\phi(x)) \| \mathrm{MAC}(\phi(x \oplus 1)) : x \in \mathcal{X}\}$$
$$L_2 = \{y \| \mathrm{MAC}(\psi(y)) \| \mathrm{MAC}(\psi(y \oplus 1)) : y \in \mathcal{Y}\}$$
$$L_3 = \{z \| \mathrm{MAC}(\phi(z)) \| \mathrm{MAC}(\phi(z \oplus 1)) : z \in \mathcal{Z}\}$$
$$L_4 = \{t \| \mathrm{MAC}(\psi(t)) \| \mathrm{MAC}(\psi(t \oplus 1)) : t \in \mathcal{T}\}$$

Notice that we can build those lists with $5 \cdot 2^{3n/4}$ queries as, by construction, for any element i of $\mathcal{Y}, \mathcal{Z}, \mathcal{T}$ the element $(i \oplus 1)$ also belongs to $\mathcal{Y}, \mathcal{Z}, \mathcal{T}$, respectively. We use the algorithm of Sect. 2.2 to find $(x, y, z, t) \in \mathcal{X} \times \mathcal{Y} \times \mathcal{Z} \times \mathcal{T}$ such that $L_1[x] \oplus L_2[y] \oplus L_3[z] \oplus L_4[t] = 0$ with $\tilde{\mathcal{O}}(2^{3n/2})$ operations, using a memory of size $\mathcal{O}(2^{3n/4})$. After finding a collision, we verify that it is not a false positive by testing the relation for another value c. As there are on average $\mathcal{O}(1)$ random quadruples the attack is indeed using a total of $5 \cdot 2^{3n/4} + \mathcal{O}(1) = \mathcal{O}(2^{3n/4})$ queries.

Universal Forgeries. This attack can be extended to a universal forgery. Indeed, the fixed prefix 0 and 1 can be replaced by v and $v \oplus 1$ for any block v, and when we identify a right quadruple (x, y, z, t) we deduce the value $\Delta_1 = E_1(v) \oplus E_1(v \oplus 1)$ and $\Delta_3 = E_3(v) \oplus E_3(v \oplus 1)$. There is also a length extension property: if (x, y, z, t) is a right quadruple, then $\mathrm{MAC}(v \,\|\, x \,\|\, s) \oplus \mathrm{MAC}(v \oplus 1 \,\|\, y \,\|\, s) \oplus \mathrm{MAC}(v \,\|\, z \,\|\, s) \oplus \mathrm{MAC}(v \oplus 1 \,\|\, t \,\|\, s) = 0$ for any suffix s.

Therefore if we want to forge a MAC for any message m of size $\ell \geq 2$ blocks we parse it as $m = v \,\|\, w \,\|\, s$ (where s has zero, one, or several blocks) and perform the attack to recover Δ_1 and Δ_3. Then we can forge using the previous relation, and Eq. (2):

$$\mathrm{MAC}(v \,\|\, w \,\|\, s) = \mathrm{MAC}(v \oplus 1 \,\|\, w \oplus \Delta_1 \,\|\, s) \oplus \mathrm{MAC}(v \,\|\, w \oplus \Delta_3 \,\|\, s)$$
$$\oplus \mathrm{MAC}(v \oplus 1 \,\|\, w \oplus \Delta_1 \oplus \Delta_3 \,\|\, s)$$

Optimizing the time complexity. Equation (2) can also be used to reduce the time complexity below 2^n, at the cost of more oracle queries. Indeed, if we consider a subset $\mathcal{C}$ of $\{0,1\}^n$, we have:

$$\mathcal{R}(x, y, z, t) \Leftrightarrow \forall c \in \mathcal{C}, \mathcal{R}(x \oplus c, y \oplus c, z \oplus c, t \oplus c)$$
$$\Rightarrow \forall c \in \mathcal{C}, \mathrm{MAC}(\phi(x \oplus c)) \oplus \mathrm{MAC}(\psi(y \oplus c))$$
$$\oplus \mathrm{MAC}(\phi(z \oplus c)) \oplus \mathrm{MAC}(\psi(t \oplus c)) = 0$$
$$\Rightarrow \bigoplus_{c \in \mathcal{C}} \mathrm{MAC}(\phi(x \oplus c)) \oplus \bigoplus_{c \in \mathcal{C}} \mathrm{MAC}(\psi(y \oplus c))$$
$$\oplus \bigoplus_{c \in \mathcal{C}} \mathrm{MAC}(\phi(z \oplus c)) \oplus \bigoplus_{c \in \mathcal{C}} \mathrm{MAC}(\psi(t \oplus c)) = 0 \quad (4)$$

If we select $\mathcal{C}$ as a linear subspace, then the last expression does not depend on the full (x, y, z, t), but only on their projection on the orthogonal of $\mathcal{C}$. Concretely, we use $\mathcal{C} = \{x : x_{[3n/7:n]} = 0\} = \{x : x < 2^{3n/7}\}$, so that the value $\bigoplus_{c \in \mathcal{C}} \mathrm{MAC}(\phi(x \oplus c))$ is independent of bits 0 to $3n/7 - 1$ of x.

Therefore, we consider the rewritten MAC function

$$\mathrm{MAC}'(v \,\|\, w) = \bigoplus_{c \in \mathcal{C}} \mathrm{MAC}(v \,\|\, w \oplus c),$$

the following message injections, with a $4n/7$-bit input

$$\phi'(i) = 0 \,\|\, i|0 \qquad\qquad \psi'(i) = 1 \,\|\, i|0,$$

and a reduced relation over $4n/7$-bit values:

$$\mathcal{R}'(x, y, z, t) := \begin{cases} x \oplus y = (E_1(0) \oplus E_1(1))_{[3n/7:n]} \\ y \oplus z = (E_3(0) \oplus E_3(1))_{[3n/7:n]} \\ z \oplus t = (E_1(0) \oplus E_1(1))_{[3n/7:n]} \\ t \oplus x = (E_3(0) \oplus E_3(1))_{[3n/7:n]} \end{cases}$$

$$\Leftrightarrow \begin{cases} x \oplus y \oplus z \oplus t = 0 \\ x \oplus y = (E_1(0) \oplus E_1(1))_{[3n/7:n]} \\ x \oplus t = (E_3(0) \oplus E_3(1))_{[3n/7:n]} \end{cases}$$

Thanks to Eq. 4, we still have:

$$\mathcal{R}'(x, y, z, t) \Rightarrow \text{MAC}'(\phi'(x)) \oplus \text{MAC}'(\psi'(y)) \oplus \text{MAC}'(\phi'(z)) \oplus \text{MAC}'(\psi'(t)) = 0$$

Since the relation $\mathcal{R}'$ is now only a $12n/7$-bit condition, we can use shorter lists than before, with just $2^{3n/7}$ elements. We can also increase the filtering using the same trick as previously, considering the following lists:

$$L_1' = \left\{ x \parallel \text{MAC}'(\phi'(x)) \parallel \text{MAC}'(\phi'(x \oplus 1)) \; : \; x \in \{0,1\}^{4n/7}, \; x_{[0:n/7]} = 0 \right\}$$

$$L_2' = \left\{ y \parallel \text{MAC}'(\psi'(y)) \parallel \text{MAC}'(\psi'(y \oplus 1)) \; : \; y \in \{0,1\}^{4n/7}, \; y_{[n/7:2n/7]} = 0 \right\}$$

$$L_3' = \left\{ z \parallel \text{MAC}'(\phi'(z)) \parallel \text{MAC}'(\phi'(z \oplus 1)) \; : \; z \in \{0,1\}^{4n/7}, \; z_{[2n/7:3n/7]} = 0 \right\}$$

$$L_4' = \left\{ t \parallel \text{MAC}'(\psi'(t)) \parallel \text{MAC}'(\psi'(t \oplus 1)) \; : \; t \in \{0,1\}^{4n/7}, \; t_{[3n/7:4n/7]} = 0 \right\}$$

Finally, using the algorithm of Sect. 2.2 with $s = 3n/7$ and $p = 0$, we can locate a right quadruple using $\tilde{\mathcal{O}}(2^{6n/7})$ queries, $\tilde{\mathcal{O}}(2^{6n/7})$ operations, and $\mathcal{O}(2^{3n/7})$ memory. This recovers only $4n/7$ bits of $E_1(0) \oplus E_1(1)$ and $E_3(0) \oplus E_3(1)$, but we can easily recover the remaining bits, either by brute force, or by repeating the attack with a different set $\mathcal{C}$.

3.2 Attacking GCM-SIV2

GCM-SIV2 is an authenticated encryption mode designed by Iwata and Minematsu [20] as a double-block-hash version of GCM-SIV (in the following, we consider GCM-SIV2 with GHASH as the underlying universal hash function). For simplicity, we focus on the authentication part of GCM-SIV2, using inputs with a non-empty associated data, and an empty message. In this case, GCM-SIV2 becomes a nonce-based MAC. The message M (considered as associated data for the mode) is zero-padded, divided into n-bit blocks, and the length is appended in an extra block. Then the construction is defined as follows, with $\odot$ a finite field multiplication (see also Fig. 3):

$$\Sigma(N, M) = N \oplus \ell \odot H_1 \oplus \bigoplus_{i=1}^{\ell} m_i \odot H_1^{\ell+2-i}$$

$$\Theta(N, M) = N \oplus \ell \odot H_2 \oplus \bigoplus_{i=1}^{\ell} m_i \odot H_2^{\ell+2-i}$$

$$\text{MAC}(N, M) = E_1(\Sigma(M)) \oplus E_2(\Theta(M)) \; \big\| \; E_3(\Sigma(M)) \oplus E_4(\Theta(M))$$

Attack. The structure of the authentication part of GCM-SIV2 is essentially the same as the structure of SUM-ECBC, where the block cipher calls E_1 and E_3 are replaced by multiplication by H_1 and H_2. The finalization function has a $2n$-bit output $\text{MAC}^1, \text{MAC}^2$, but quadruples following $\mathcal{R}$ will collide on both outputs. Thus, we can essentially repeat the SUM-ECBC attack, but there is an important difference: GCM-SIV2 is a nonce-based MAC, rather than a deterministic one.

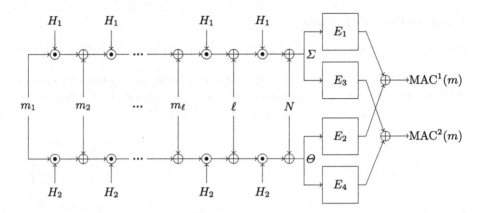

Fig. 3. Diagram for authentication in GCM-SIV2 using GHASH with a ℓ-block message, a nonce N, hash keys H_1 and H_2.

Therefore, all queries must include a nonce N, and we should not query two different messages with the same nonce. We adapt the previous attack using message injection functions that output both a nonce and a message, so that we use two fixed messages, 0 and 1, with variable nonces:

$$\phi(i) = (i, 0) \qquad\qquad \psi(i) = (i, 1)$$

$$\mathrm{MAC}(\phi(i)) = \quad E_1\underbrace{\left(i \oplus H_1\right)}_{\Sigma_0(i)} \oplus E_2\underbrace{\left(i \oplus H_2\right)}_{\Theta_0(i)} \;\Big\|\; E_3\big(\Sigma_0(i)\big) \oplus E_4\big(\Theta_0(i)\big)$$

$$\mathrm{MAC}(\psi(i)) = E_1\underbrace{\left(i \oplus H_1 \oplus H_1^2\right)}_{\Sigma_1(i)} \oplus E_2\underbrace{\left(i \oplus H_2 \oplus H_2^2\right)}_{\Theta_1(i)} \;\Big\|\; E_3\big(\Sigma_1(i)\big) \oplus E_4\big(\Theta_1(i)\big).$$

We consider quadruples of nonce/messages X, Y, Z, T with

$$X = \phi(x) \qquad Y = \psi(y) \qquad Z = \phi(z) \qquad T = \psi(t),$$

and we have the same kind of relations as in the previous attack:

$$\mathcal{R}(x, y, z, t) := \begin{cases} \Sigma_0(x) = \Sigma_1(y) \\ \Sigma_0(z) = \Sigma_1(t) \\ \Theta_0(z) = \Theta_1(y) \\ \Theta_0(x) = \Theta_1(t). \end{cases} \Leftrightarrow \begin{cases} x \oplus y \oplus z \oplus t = 0 \\ x \oplus y = H_1^2 \\ x \oplus t = H_2^2 \end{cases}$$

$$\Rightarrow \mathrm{MAC}(\phi(x)) \oplus \mathrm{MAC}(\psi(y)) \oplus \mathrm{MAC}(\phi(z)) \oplus \mathrm{MAC}(\psi(t)) = 0$$

Since the MAC output is $2n$-bit long, we can directly build an attack with $\mathcal{O}(2^{3n/4})$ queries: we consider four distinct sets $\mathcal{X}, \mathcal{Y}, \mathcal{Z}, \mathcal{T}$ of $2^{3n/4}$ values, and we look for a quadruple $(x, y, z, t) \in \mathcal{X} \times \mathcal{Y} \times \mathcal{Z} \times \mathcal{T}$, such that

$$\begin{cases} x \oplus y \oplus z \oplus t = 0 \\ \mathrm{MAC}(\phi(x)) \oplus \mathrm{MAC}(\psi(y)) \oplus \mathrm{MAC}(\phi(z)) \oplus \mathrm{MAC}(\psi(t)) = 0 \end{cases} \tag{5}$$

we expect to find one good quadruple that respects $\mathcal{R}$ along with $\mathcal{O}(1)$ quadruples that randomly satisfy the observable filter (5). This leads to an attack with $\mathcal{O}(2^{3n/4})$ queries and time $\tilde{O}(2^{3n/2})$. Since we recover H_1 and H_2 (from $H_1^2 = x \oplus y$ and $H_2^2 = x \oplus t$), we can do universal forgeries. In addition, we can also easily adapt the attack with $\mathcal{O}(2^{6n/7})$ queries and time $\tilde{O}(2^{6n/7})$.

4   Attacking PMAC-like Constructions

We now describe attacks against PMAC+ [43] and related constructions: 1kMAC+ [9], and LightMAC+ [33]. We have an existential forgery attack with $\mathcal{O}(2^{3n/4})$ queries and $\tilde{O}(2^{3n/2})$ operations (using memory $\mathcal{O}(2^{3n/4})$), with a range of time-memory trade-offs with $\mathcal{O}(2^t)$ queries, with $3n/4 < t < n$, and $\tilde{O}(2^{3n-2t})$ operations (using memory $\mathcal{O}(2^t)$).

4.1   Attacking PMAC+

PMAC+ was designed by Yasuda in 2011 [43], as a variant of PMAC [5] with a larger internal state. The scheme internally uses a tweakable block cipher construction inspired by the XE construction [39], that we denote as $\tilde{E}_i$. The message M is first padded with 10* padding, and divided into n-bit blocks, but for simplicity we ignore the padding in our description. The construction is shown in Fig. 4[1]:

$$\Sigma(M) = \bigoplus_{i=1}^{\ell} \tilde{E}_i(m_i) \qquad \tilde{E}_i(x) = E_1(x \oplus 2^i \odot \Delta_0 \oplus 2^{2i} \odot \Delta_1)$$
$$\Theta(M) = \bigoplus_{i=1}^{\ell} 2^{\ell-i} \odot \tilde{E}_i(m_i) \qquad \Delta_0 = E_1(0) \quad \Delta_1 = E_1(1)$$
$$\mathrm{MAC}(M) = E_2(\Sigma(M)) \oplus E_3(\Theta(M))$$

Attack. As in the previous attack, we use message injection functions with two different prefixes, but we include an extra block u to define related quadruples:

$$\phi_u(i) = u \parallel 0 \parallel i \qquad\qquad \psi_u(i) = u \parallel 1 \parallel i$$

$$\mathrm{MAC}(\phi_u(i)) = E_2\Big(\underbrace{\tilde{E}_1(u) \oplus \tilde{E}_2(0) \oplus \tilde{E}_3(i)}_{\Sigma_{u,0}(i)}\Big) \oplus E_3\Big(\underbrace{4\tilde{E}_1(u) \oplus 2\tilde{E}_2(0) \oplus \tilde{E}_3(i)}_{\Theta_{u,0}(i)}\Big)$$

$$\mathrm{MAC}(\psi_u(i)) = E_2\Big(\underbrace{\tilde{E}_1(u) \oplus \tilde{E}_2(1) \oplus \tilde{E}_3(i)}_{\Sigma_{u,1}(i)}\Big) \oplus E_3\Big(\underbrace{4\tilde{E}_1(u) \oplus 2\tilde{E}_2(1) \oplus \tilde{E}_3(i)}_{\Theta_{u,1}(i)}\Big).$$

Next, we build quadruples of messages X, Y, Z, T with

$$X = \phi_u(x) \qquad Y = \psi_u(y) \qquad Z = \phi_u(z) \qquad T = \psi_u(t),$$

---

[1] The algorithm and the figure given in [43] differ in the coefficients used to compute Θ. We use the algorithmic description because it matches later PMAC+ variants, but the attack can easily be adapted to the other case.

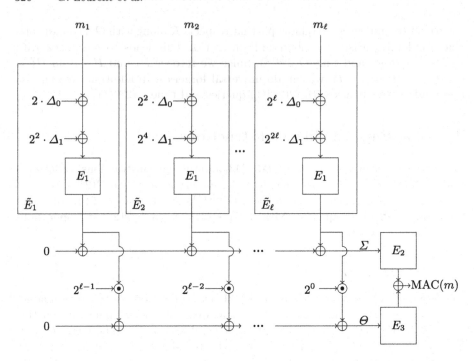

Fig. 4. Diagram for PMAC+ with a ℓ-block message where $\Delta_0 = E_1(0)$ and $\Delta_1 = E_1(1)$.

and we look for a quadruple with partial state collisions for the underlying pairs, *i.e.* a quadruple following the relation:

$$\mathcal{R}(x, y, z, t) := \begin{cases} \Sigma_{u,0}(x) = \Sigma_{u,1}(y) \\ \Sigma_{u,0}(z) = \Sigma_{u,1}(t) \\ \Theta_{u,0}(z) = \Theta_{u,1}(y) \\ \Theta_{u,0}(x) = \Theta_{u,1}(t). \end{cases}$$

We have

$$\mathcal{R}(x, y, z, t) \Leftrightarrow \begin{cases} \tilde{E}_3(x) \oplus \tilde{E}_2(0) = \tilde{E}_3(y) \oplus \tilde{E}_2(1) \\ \tilde{E}_3(z) \oplus \tilde{E}_2(0) = \tilde{E}_3(t) \oplus \tilde{E}_2(1) \\ \tilde{E}_3(y) \oplus 2\tilde{E}_2(1) = \tilde{E}_3(z) \oplus 2\tilde{E}_2(0) \\ \tilde{E}_3(t) \oplus 2\tilde{E}_2(1) = \tilde{E}_3(x) \oplus 2\tilde{E}_2(0) \end{cases}$$

$$\Leftrightarrow \begin{cases} \tilde{E}_3(x) \oplus \tilde{E}_3(y) \oplus \tilde{E}_3(z) \oplus \tilde{E}_3(t) = 0 \\ \tilde{E}_3(x) \oplus \tilde{E}_3(y) = \tilde{E}_2(0) \oplus \tilde{E}_2(1) \\ \tilde{E}_3(t) \oplus \tilde{E}_3(x) = 2\tilde{E}_2(0) \oplus 2\tilde{E}_2(1) \end{cases}$$

Again, $\mathcal{R}$ defines a $3n$−bit relation, and we can detect it through the sum of the MACs following Eq. (1):

$$\mathcal{R}(x, y, z, t) \Rightarrow \text{MAC}(\phi_u(x)) \oplus \text{MAC}(\psi_u(y)) \oplus \text{MAC}(\phi_u(z)) \oplus \text{MAC}(\psi_u(t)) = 0$$

In addition, the relation $\mathcal{R}$ is independent of the value u, so that we can easily build several quadruples that satisfy $\mathcal{R}$ simultaneously. This leads to an attack with $\mathcal{O}(2^{3n/4})$ queries: we consider four sets $\mathcal{X}, \mathcal{Y}, \mathcal{Z}, \mathcal{T}$ of $2^{3n/4}$ random values, and we look for a quadruple $(x, y, z, t) \in \mathcal{X} \times \mathcal{Y} \times \mathcal{Z} \times \mathcal{T}$, such that

$$\forall u \in \{0, 1, 2\}, \text{MAC}(\phi_u(x)) \oplus \text{MAC}(\psi_u(y)) \oplus \text{MAC}(\phi_u(z)) \oplus \text{MAC}(\psi_u(t)) = 0$$

We expect on average one random quadruple (with 2^{3n} potential quadruples, and a $3n$-bit filtering), and one quadruple satisfying $\mathcal{R}$ (also a $3n$-bit condition). The correct quadruple can easily be checked with a few extra queries.

In practice, we use the generalized birthday algorithms of Sect. 2.2 in order to optimize the complexity of the attack. We consider four lists:

$$L_1 = \{\text{MAC}(\phi_0(x)) \parallel \text{MAC}(\phi_1(x)) \parallel \text{MAC}(\phi_2(x)) \; : \; x \in \mathcal{X}\}$$
$$L_2 = \{\text{MAC}(\psi_0(y)) \parallel \text{MAC}(\psi_1(y)) \parallel \text{MAC}(\psi_2(y)) \; : \; y \in \mathcal{Y}\}$$
$$L_3 = \{\text{MAC}(\phi_0(z)) \parallel \text{MAC}(\phi_1(z)) \parallel \text{MAC}(\phi_2(z)) \; : \; z \in \mathcal{Z}\}$$
$$L_4 = \{\text{MAC}(\psi_0(t)) \parallel \text{MAC}(\psi_1(t)) \parallel \text{MAC}(\psi_2(t)) \; : \; t \in \mathcal{T}\}$$

and we look for a quadruple $(x, y, z, t) \in \mathcal{X} \times \mathcal{Y} \times \mathcal{Z} \times \mathcal{T}$ such that $L_1[x] \oplus L_2[y] \oplus L_3[z] \oplus L_4[t] = 0$. This can be done with $\tilde{\mathcal{O}}(2^{3n/2})$ operations, using a memory of size $\mathcal{O}(2^{3n/4})$. Finally, once a quadruple (x, y, z, t) satisfying $\mathcal{R}(x, y, z, t)$ has been detected, it can be used to generate forgeries. Indeed, we can predict the MAC of a new message by making three new queries using Eq. (1):

$$\forall u, \text{MAC}(\phi_u(x)) = \text{MAC}(\psi_u(y)) \oplus \text{MAC}(\psi_u(z)) \oplus \text{MAC}(\phi_u(t))$$

Time-Query Trade-offs. As opposed to the SUM-ECBC attack, we don't have an analogue to Eq. (2) that can be used to reduce the time complexity. However, the time complexity of the algorithm can be slightly reduced when using more than $\mathcal{O}(2^{3n/4})$ queries. If we consider sets $\mathcal{X}, \mathcal{Y}, \mathcal{Z}, \mathcal{T}$ of size 2^t with $3n/4 < t < n$, the resulting 4-sum is slightly easier, because there are 2^{4t-3n} expected solutions. Using the algorithm of Sect. 2.2, this can be solved in time $\tilde{\mathcal{O}}(2^{3n-2t})$, using a memory of size $\mathcal{O}(2^t)$.

4.2 Attacking LightMAC+

LightMAC+ was designed by Naito [33] using ideas from PMAC+ [43] and LightMAC [29]. If we consider it as based on a tweakable block cipher $\tilde{E}$, it follows the same structure as PMAC+ (see Fig. 5), but $\tilde{E}$ takes a message block smaller than n bits:

$$\Sigma(M) = \bigoplus_{i=1}^{\ell} \tilde{E}_i(m_i) \qquad\qquad \tilde{E}_i(x) = E_1(i\|x)$$
$$\Theta(M) = \bigoplus_{i=1}^{\ell} 2^{\ell-i} \odot \tilde{E}_i(m_i)$$
$$\text{MAC}(M) = E_2(\Sigma(M)) \oplus E_3(\Theta(M))$$

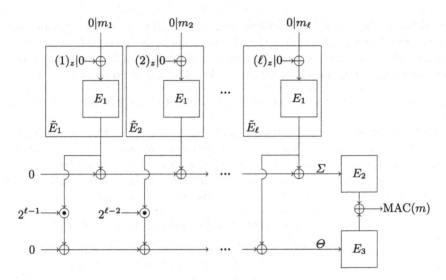

Fig. 5. Diagram for LightMAC+ with $(n-z)$-bit blocks of a ℓ-block message where $(v)_z$ is the value v written over z bits.

Since the structure of LightMAC+ is the same as the structure of PMAC+, we can use the same attack. The only difference from our point of view is that the message blocks are shorter than the block-size. As long as one message block is big enough to fit $2^{3n/4}$ different values, our attack will succeed.

This attack violates the improved security proof recently published at CT-RSA [34], with a security bound of $\mathcal{O}(q_t^2 q_v/2^{2n})$ (with q_t MAC queries and q_v verification queries). Indeed, our attack reaches a constant success probability with $q_t = \mathcal{O}(2^{3n/4})$ and $q_v = 1$. We have shared our attack with Naito and he agreed that his proof is flawed.

4.3 Attacking 1kPMAC+

1kPMAC+ is a single-key variant of PMAC+ [43] designed by Datta et al. [9], shown in Fig. 8.

Since the structure of 1kPMAC+ is the same as the structure of PMAC+, we can use the same attack. Alternatively, we can take advantage of the fix functions to mount a more straightforward attack, as shown in Sect. 6.

5   Attacking f9-like Constructions

Our third attack is applicable to 3kf9 [44] and similar constructions. We have a universal forgery attack with $\mathcal{O}(2^{3n/4})$ queries and $\tilde{\mathcal{O}}(2^{5n/4})$ operations using memory $\mathcal{O}(2^n)$, with a possible time-memory trade-offs.

5.1   Attacking 3kf9

3kf9 [44], designed by Xhang, Wu, Sui and Wang, is a three-key variant of the f9 mode used in 3G telephony. While the original f9 does not have security beyond the birthday bound [24], 3kf9 is secure up to $2^{2n/3}$ queries. We describe 3kf9 in Fig. 6:

$$\Sigma(M) = \sigma_\ell \qquad\qquad \sigma_0 = 0 \qquad \sigma_i = E_1(\sigma_{i-1} \oplus m_i)$$
$$\Theta(M) = \bigoplus_{i=1}^{\ell} \sigma_i$$
$$\mathrm{MAC}(M) = E_2(\Sigma(M)) \oplus E_3(\Theta(M))$$

Attack. Our attack follows the same structure as the previous attacks. We start with messages of the form:

$$\phi(i) = 0 \parallel i \qquad\qquad \psi(i) = 1 \parallel i,$$

and the corresponding MACs:

$$\mathrm{MAC}(\phi(i)) = E_2\Big(\underbrace{E_1\big(x \oplus E_1(0)\big)}_{\Sigma_0(x)} \Big) \oplus E_3\Big(\underbrace{E_1\big(x \oplus E_1(0)\big) \oplus E_1(0)}_{\Theta_0(x)} \Big)$$
$$\mathrm{MAC}(\psi(i)) = E_2\Big(\underbrace{E_1\big(x \oplus E_1(1)\big)}_{\Sigma_1(x)} \Big) \oplus E_3\Big(\underbrace{E_1\big(x \oplus E_1(1)\big) \oplus E_1(1)}_{\Theta_1(x)} \Big).$$

We use quadruples of messages X, Y, Z, T with

$$X = \phi(x) \qquad Y = \psi(y) \qquad Z = \phi(z) \qquad T = \psi(t),$$

and we look for a quadruple with partial state collisions for the underlying pairs, i. e. a quadruple following the relation:

$$\mathcal{R}(x, y, z, t) := \begin{cases} \Sigma_0(x) = \Sigma_1(y) \\ \Sigma_0(z) = \Sigma_1(t) \\ \Theta_0(z) = \Theta_1(y) \\ \Theta_0(x) = \Theta_1(t). \end{cases}$$

$$\Leftrightarrow \begin{cases} x \oplus E_1(0) = y \oplus E_1(1) \\ z \oplus E_1(0) = t \oplus E_1(1) \\ E_1(z \oplus E_1(0)) \oplus E_1(0) = E_1(y \oplus E_1(1)) \oplus E_1(1) \\ E_1(x \oplus E_1(0)) \oplus E_1(0) = E_1(t \oplus E_1(1)) \oplus E_1(1) \end{cases}$$

$$\Leftrightarrow \begin{cases} x \oplus y \oplus z \oplus t = 0 \\ x \oplus y = E_1(0) \oplus E_1(1) \\ E_1(x \oplus E_1(0)) \oplus E_1(t \oplus E_1(1)) = E_1(0) \oplus E_1(1) \end{cases}$$

$$\Rightarrow \mathrm{MAC}(\phi(x)) \oplus \mathrm{MAC}(\psi(y)) \oplus \mathrm{MAC}(\phi(z)) \oplus \mathrm{MAC}(\psi(t)) = 0.$$

As in the previous attacks, $\mathcal{R}$ defines a $3n-$bit relation. Moreover, we can easily observe when $x \oplus y \oplus z \oplus t = 0$, and the relation $x \oplus y = E_1(0) \oplus E_1(1)$ can be verified across several quadruples. We don't have related quadruples satisfying $\mathcal{R}$ simultaneously as in the previous attacks, but we can use those properties to detect right quadruples. This leads to an attack with $\tilde{\mathcal{O}}(2^{3n/4})$ queries: we consider four sets $\mathcal{X}, \mathcal{Y}, \mathcal{Z}, \mathcal{T}$ of $\sqrt[4]{n} \times 2^{3n/4}$ random values, and we look for quadruples $(x, y, z, t) \in \mathcal{X} \times \mathcal{Y} \times \mathcal{Z} \times \mathcal{T}$, such that:

$$\begin{cases} x \oplus y \oplus z \oplus t = 0 \\ \mathrm{MAC}(\phi(x)) \oplus \mathrm{MAC}(\psi(y)) \oplus \mathrm{MAC}(\phi(z)) \oplus \mathrm{MAC}(\psi(t)) = 0. \end{cases} \tag{6}$$

Since this a $2n$-bit condition, we expect on average $n \cdot 2^n$ quadruples (x, y, z, t) satisfying (6). In order to filter out the right ones, we look at the value $x \oplus y$ for all these quadruples. While the wrong quadruples should have a random $x \oplus y$, the right ones have $x \oplus y = E_1(0) \oplus E_1(1)$. Therefore, with high probability, the most frequent value for $x \oplus y$ is equal to $E_1(0) \oplus E_1(1)$, and quadruples satisfying this extra relation are right quadruples with probability $1/2$. More precisely, we expect on average n wrong quadruples for each value of $x \oplus y$, and n right quadruples with $x \oplus y = E_1(0) \oplus E_1(1)$.

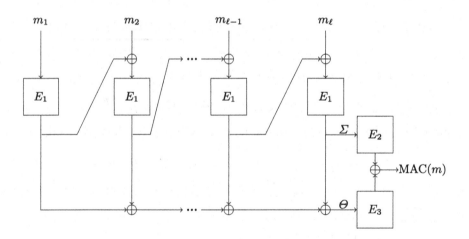

Fig. 6. Diagram for 3kf9 with a $\ell-$block message.

Optimizing the time complexity. While the algorithm of Sect. 2.2 would take time $\tilde{\mathcal{O}}(2^{3n/2})$ with $\tilde{\mathcal{O}}(2^{3n/4})$ queries, we can reduce the time complexity using sets $\mathcal{X}, \mathcal{Y}, \mathcal{Z}, \mathcal{T}$ with some structure. More precisely, we use:

$$\mathcal{X} = \mathcal{Z} = \left\{ x \in \{0, 1\}^n : x_{[0:n/4]} = 0 \right\}$$
$$\mathcal{Y} = \mathcal{T} = \left\{ x \in \{0, 1\}^n : x_{[n/4:n/2]} = 0 \right\}$$

so that quadruples can be written as

$$x =: x_3|x_2|x_1|0 \in \mathcal{X} \qquad\qquad y =: y_3|y_2|0|y_0 \in \mathcal{Y}$$
$$z =: z_3|z_2|z_1|0 \in \mathcal{Z} \qquad\qquad t =: t_3|t_2|0|t_0 \in \mathcal{T}.$$

In particular, right quadruples satisfy $x \oplus y \oplus z \oplus t = 0$, therefore $x_1 = z_1$, $y_0 = t_0$, and $x_3|x_2 \oplus z_3|z_2 = y_3|y_2 \oplus t_3|t_2$. We use these properties to adapt the algorithm of Sect. 2.2 and locate the quadruples efficiently. First we guess the $n/2$-bit value $\alpha_3|\alpha_2 := x_3|x_2 \oplus z_3|z_2 = y_3|y_2 \oplus t_3|t_3$. Then, for each $x = x_3|x_2|x_1|0$, there is a single candidate $z = (x_3 \oplus \alpha_3)|(x_2 \oplus \alpha_2)|x_1|0$ that could be part of a right quadruple. Similarly, every $y = y_3|y_2|0|y_0$ can be paired with a single $t = (y_3 \oplus \alpha_3)|(y_2 \oplus \alpha_2)|0|y_0$. Therefore, we consider the two following lists:

$$L_1 = \{\mathrm{MAC}(\phi(x_3|x_2|x_1|0)) \oplus \mathrm{MAC}((x_3 \oplus \alpha_3)|(x_2 \oplus \alpha_2)|x_1|0) \, : \, x_3|x_2|x_1|0 \in \mathcal{X}\}$$
$$L_2 = \{\mathrm{MAC}(\phi(y_3|y_2|0|y_0)) \oplus \mathrm{MAC}((y_3 \oplus \alpha_3)|(y_2 \oplus \alpha_2)|0|y_0) \, : \, y_3|y_2|0|y_0 \in \mathcal{Y}\}$$

After sorting the lists, we look for matches, and the corresponding quadruples x, y, z, t are exactly the quadruples satisfying

$$\begin{cases} x \oplus y \oplus z \oplus t = 0 \\ (x \oplus z)_{[n/2:n]} = \alpha_3|\alpha_2 \\ \mathrm{MAC}(\phi(x)) \oplus \mathrm{MAC}(\psi(y)) \oplus \mathrm{MAC}(\phi(z)) \oplus \mathrm{MAC}(\psi(t)) = 0. \end{cases} \qquad (7)$$

More precisely, a match $L_1[x] = L_2[y]$ suggests $z = x \oplus \alpha_3|\alpha_2|0|0$ and $t = y \oplus \alpha_3|\alpha_2|0|0$, but there are four corresponding quadruples: (x, y, z, t), (z, y, x, t), (x, t, z, y), (z, t, x, y), and two candidate values for $E_1(0) \oplus E_1(1)$: $x \oplus y$ and $x \oplus y \oplus \alpha_3|\alpha_2|0|0$.

We need $\tilde{\mathcal{O}}(2^{3n/4})$ operations to generate those quadruples. We repeat this $2^{n/2}$ times to exhaust all $n/2$-bit values $\alpha_3|\alpha_2$ and generate all quadruples satisfying (6). Finally, we use an array to count the number of occurrences of each possible value of $x \oplus y$. Each counter receives an average two values, but the counter corresponding to $E_1(0) \oplus E_1(1)$ will receive three values on average. After repeating all the operations $\mathcal{O}(n)$ times, with some arbitrary constants in place of the zero bits, the highest counter corresponds to $E_1(0) \oplus E_1(1)$ with high probability, as proved in Sect. 5.2. This gives an attack with $\tilde{\mathcal{O}}(2^{3n/4})$ queries, $\tilde{\mathcal{O}}(2^{5n/4})$ operations, and $\mathcal{O}(2^n)$ memory[2].

Time-Memory Trade-offs. We can reduce the memory usage if we store only a subset of the counters, and repeat the whole algorithm until the whole set has been covered. Concretely, we store only the counters with a fixed value for bits $[0 : n/8]$ and $[n/4 : 3n/8]$ of $x \oplus y$. Because of the way the lists L_1 and L_2 are constructed, we have actually fixed $n/8$ bits of y_0 and x_1, and we can reduce the lists to size $2^{5n/8}$. Therefore we evaluate $2^{3n/4}$ counters in time $\tilde{\mathcal{O}}(2^{n/2} \cdot 2^{5n/8})$,

---

[2] We can actually reduce the polynomial factors by fixing only $(n - \log_2(n))/4$ bits to zero, in order to have sets of size $\sqrt[4]{n} \cdot 2^{3n/4}$.

using only $\mathcal{O}(2^{3n/4})$ memory. We repeat iteratively over the full counter set, so we need time $\tilde{\mathcal{O}}(2^{n/4} \cdot 2^{n/2} \cdot 2^{5n/8}) = \tilde{\mathcal{O}}(2^{11n/8})$. More generally, we have a time-memory trade-off with time $\tilde{\mathcal{O}}(2^{5n/4+t/2})$ and memory $\mathcal{O}(2^{n-t})$ for $0 < t < n/4$.

Forgeries. Once we found a quadruple (x, y, z, t) that respects $\mathcal{R}(x, y, z, t)$ we know that after processing message $\phi(x) = 0 \| x$ and $\psi(t) = 1 \| t$, there is no difference in the Θ part of the state ($\Theta_0(x) = \Theta_1(t)$). Moreover we have $\Theta_0(x) = \Sigma_0(x) \oplus E_1(0)$ and $\Theta_1(t) = \Sigma_1(x) \oplus E_1(1)$; this implies that there is a difference $E_1(0) \oplus E_1(1) = x \oplus y$ in the Σ part of the state. Therefore, we can build a full state collision with message $0 \| x \| 0$ and $1 \| t \| x \oplus y$. In particular, the following relation can be used to create forgeries with an arbitrary message m (of any length):

$$\text{MAC}(0 \| x \| 0 \| m) = \text{MAC}(1 \| t \| x \oplus y \| m).$$

Universal Forgeries. We can even forge the tag of an arbitrary message of length at least $(2n + 2)$ blocks with complexity only $n + 1$ times the complexity of the simple forgery attack. The technique is more advanced and inspired by the multi-collision attack described by Joux [23]. For ease of notation we'll show how to forge the signature for a message starting with $2n + 2$ blocks of zero, but this can be trivially adapted for any message.

First, we find a quadruple (x_1, y_1, z_1, t_1) as before. Then we consider messages $0\|0$ and $1\|x_1 \oplus y_1$. Since $x_1 \oplus y_1 = E_1(0) \oplus E_1(1)$, we have $\Sigma(0\|0) = \Sigma(1\|x_1 \oplus y_1)$, i. e. the Σ part of the state collides. Moreover, we know the difference in the Θ part: $\Theta(0 \| 0) \oplus \Theta(1 \| x_1 \oplus y_1) = x_1 \oplus y_1$.

More generally, at step i we use message injection functions

$$\phi_i(x) = \underbrace{0 \| 0 \| \ldots \| 0}_{\times 2(i-1)} \| 0 \| x \qquad \psi_i(x) = \underbrace{0 \| 0 \| \ldots \| 0}_{\times 2(i-1)} \| 1 \| x,$$

to look for a quadruple of messages

$$X_i = \phi_i(x_i) \qquad Y_i = \psi_i(y_i) \qquad Z_i = \phi_i(z_i) \qquad T_i = \psi_i(t_i).$$

When a right quadruple (x_i, y_i, z_i, t_i) has been identified, we can deduce that the MACs for $0 \| 0 \| \ldots \| 0 \| 0 \| 0$ and $0 \| 0 \| \ldots \| 0 \| 1 \| x_i \oplus y_i$ will match on the Σ branch and differ by $x_i \oplus y_i$ in their Θ branch.

After several iterations, we have actually built a multi-collision: all the messages $h_1 \| h_2 \| \ldots \| h_n \| h_{n+1}$ with $h_i \in \{(1 \| x_i \oplus y_i), (0 \| 0)\}$ collide on the Σ branch. In addition, we also know the difference in the Θ branch for those messages: it is equal to $\bigoplus_{\{i \,:\, h_i \neq 0\|0\}}(x_i \oplus y_i)$.

After at most $n + 1$ steps, we can find a non empty subset $\mathcal{I} \subseteq [1 : n + 1]$ such that $\bigoplus_{i \in \mathcal{I}}(x_i \oplus y_i) = 0$ by simple linear algebra[3]. This gives a collision on the full state, using messages $m_0 = 0 \| 0 \| \ldots \| 0$ (with $2(n + 1)$ blocks) and

---

[3] We construct the kernel of the linear function $\lambda_i \mapsto \bigoplus_i \lambda_i(x_i \oplus y_i)$.

$h = h_1 \parallel h_2 \parallel \ldots \parallel h_n \parallel h_{n+1}$ with $h_i = 1 \parallel x_i \oplus y_i$ if $i \in \mathcal{I}$, $h_i = 0 \parallel 0$ otherwise. Since the full state collides, we have for any message m (of any length):

$$\mathrm{MAC}(h \parallel m) = \mathrm{MAC}(m_0 \parallel m).$$

5.2 Detailed Complexity Analysis

We want to prove the claim that one will need to find $\mathcal{O}(n \cdot 2^n)$ quadruples in order to finish the attack on 3kf9 described in Sect. 5.1. We say the attack finishes when we recover the target value $T = E(0) \oplus E(1)$.

Assuming that each quadruple we find respects $\mathcal{R}$ with probability $1/2^n$, we fill a list of counters for every suspected values of T; a random quadruple gives two random values and a right one gives one value equal to T and one random value. Therefore we sum up the distribution of an observable value x as:

$$x \begin{cases} \xleftarrow{\$} \{0,1\}^n & \text{with probability } 1 - 1/2^{n+1} \\ \longleftarrow T & \text{with probability } 1/2^{n+1} \end{cases}$$

Let N be the number of observed values, and X_i^c represents the indicator that the ith value equals c (following a Bernoulli distribution), so that the counter corresponding to c is $X^c = \sum_{i=1}^{N} X_i^c$. Now we have to discriminate between the distributions of X^c with $c \neq T$, and the distribution of X^T:

$$\Pr(X_i^T = 1) = \Pr(x = T) = (1 - 1/2^{n+1})/2^n + 1/2^{n+1} = (3/2 - 1/2^{n+1})/2^n$$
$$\implies \mathbf{E}[X^T] = N(3/2 - 1/2^{n+1})/2^n$$
$$\Pr(X_i^c = 1) = \Pr(x = c) = (1 - 1/2^{n+1})/2^n$$
$$\implies \mathbf{E}[X^c] = N(1 - 1/2^{n+1})/2^n$$
$$\implies \mathbf{E}[X^T] \geq 3/2 \cdot \mathbf{E}[X^c]$$

We use the Chernoff bound to get a lower bound on the probability that a given counter is higher than the average value of X^T:

$$\Pr(X^c \geq \mathbf{E}[X^T]) \leq \Pr(X^c \geq 3/2 \cdot \mathbf{E}[X^c]) \leq e^{-N(1-1/2^{n+1})/2^{n+1}}$$

and assuming the counters are independent:

$$Pr(X^c < \mathbf{E}[X^T]) \geq 1 - e^{-N(1-1/2^{n+1})/2^{n+1}}$$
$$Pr(\forall c \neq T : X^c < \mathbf{E}[X^T]) \geq \left(1 - e^{-N(1-1/2^{n+1})/2^{n+1}}\right)^{2^n}$$

This expression will asymptotically converge to a strictly positive constant when $e^{-N(1-1/2^{n+1})/2^{n+1}} \simeq 2^{-n}$. Therefore, we use

$$N \simeq n \ln(2) \cdot \frac{2^{n+1}}{(1 - 1/2^{n+1})} = \mathcal{O}(n \cdot 2^n).$$

Since we observe 2 values per quadruples, this makes $\mathcal{O}(n \cdot 2^n)$ quadruples. Moreover, the event '$X^T \geq \mathbf{E}[X^T]$' has a probability close to 0.5, therefore after

$\mathcal{O}(n \cdot 2^n)$ quadruples, we indeed have a $\Omega(1)$ probability that X^T is greater than all of the other counters, which allows to recover the value T. Performing the attack until the end with probability $\Omega(1)$ also requires $\mathcal{O}(n \cdot 2^n)$ quadruples.

To get to this result some assumptions have been made, like the independence of the counters, but they all tend to be either conservative or asymptotically true.

5.3   Attacking 1kf9

1kf9 is a single-key variant of 3kf9 suggested in [8], and later withdrawn. Since the structure of 1kf9 is the same as the structure of 3kf9, we can use the same attack. However, in the next section, we give an attack with birthday complexity using properties of the fix functions.

6   Attacks Using Collision in fix Functions

Finally, we show attacks against single key variant of beyond-birthday-bound MACs based on fix functions, as defined by Datta et al. [8,9]. The fix functions just fix the least significant bit an n-bit value to zero or one, and are used for domain separation:

$$\texttt{fix0} : x \mapsto x_{[1:n]} \| 0 \qquad\qquad \texttt{fix1} : x \mapsto x_{[1:n]} \| 1$$

Datta *et al.* used those function to build a single-key variant of PMAC+ called 1kPMAC+ [9], and a single-key variant of 3kf9 called 1kf9 [8], both with security up to $2^{2n/3}$ queries. However, 1kf9 has been withdrawn because of issues in its security proof. In this section, we exploit trivial collisions in the fix functions to build colliding pairs or quadruples more easily:

$$\texttt{fix0}(x) = \texttt{fix0}(x \oplus 1) \qquad\qquad \texttt{fix1}(x) = \texttt{fix1}(x \oplus 1)$$

This allows a more straightforward attack against 1kPMAC+ with the same complexity as the attacks in Sect. 4, and an attack against 1kf9 [8] with birthday complexity, violating its security claims.

6.1   Attacking 1kf9

The 1kf9 mode uses the fix function for domain separation to build a single-key variant of 3kf9, as shown in Fig. 7:

$$\sigma_0 = 0 \qquad\qquad \sigma_i = E(\sigma_{i-1} \oplus m_i)$$
$$\Sigma'(M) = \sigma_\ell \qquad\qquad \Sigma(M) = 2 \odot \texttt{fix0}(\Sigma'(M))$$
$$\Theta'(M) = \bigoplus_{i=1}^{\ell} \sigma_i \qquad\qquad \Theta(M) = 2 \odot \texttt{fix1}(\Theta'(M))$$
$$\text{MAC}(M) = E(\Sigma(M)) \oplus E(\Theta(M))$$

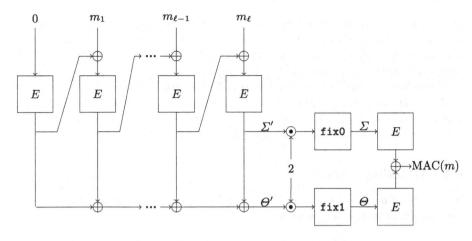

Fig. 7. Diagram for `1kf9` with a ℓ−block message.

Attack. Because of a mistake in the proof of `1kf9`, we can use pairs of messages instead of quadruples. More precisely, instead of looking for a quadruple with pairwise collisions in Σ and Θ, we look for a pair of message X, Y colliding on Σ', and with a difference in Θ' that will be absorbed by the `fix1` function. Therefore, we define the relation $\mathcal{R}$ as:

$$\mathcal{R}(X, Y) := \begin{cases} \Sigma'(X) = \Sigma'(Y) \\ 2\Theta'(X) = 2\Theta'(Y) \oplus 1 \end{cases}$$
$$\Rightarrow \mathrm{MAC}(X) = \mathrm{MAC}(Y).$$

We build the messages with different postfixes, parametrized by u:

$$X = \phi_u(x) = x \parallel u \qquad\qquad Y = \psi_u(y) = y \parallel u \oplus d,$$

where d is the inverse of 2 in the finite field. With this construction, we have

$$\Sigma'(\phi_u(x)) = E\big(u \oplus E(x \oplus E(0))\big)$$
$$\Theta'(\phi_u(x)) = E\big(u \oplus E(x \oplus E(0))\big) \oplus E\big(x \oplus E(0)\big) \oplus E\big(0\big)$$
$$\Sigma'(\psi_u(y)) = E\big(u \oplus d \oplus E(y \oplus E(0))\big)$$
$$\Theta'(\psi_u(y)) = E\big(u \oplus d \oplus E(y \oplus E(0))\big) \oplus E\big(y \oplus E(0)\big) \oplus E\big(0\big)$$

In particular, we observe

$$E(x \oplus E(0)) \oplus E(y \oplus E(0)) = d \Leftrightarrow \Sigma'(\phi_u(x)) = \Sigma'(\psi_u(y))$$
$$\Rightarrow \Theta'(\phi_u(x)) \oplus \Theta'(\psi_u(y)) = d$$
$$\Rightarrow \mathrm{MAC}(\phi_u(x)) = \mathrm{MAC}(\psi_u(y)). \qquad (8)$$

From this observation, we construct a birthday attack against `1kf9`. We build two lists:

$$L_0 = \Big\{\mathrm{MAC}(\phi_0(x)) \ : \ x < 2^{n/2}\Big\} \qquad L_1 = \Big\{\mathrm{MAC}(\psi_0(y)) \ : \ y < 2^{n/2}\Big\},$$

and we look for a match between the lists. We expect on average one pair to match randomly, and one pair to match because of (8). Moreover, when we have a collision candidate $L_0[x], L_1[y]$, we can verify whether it is a right pair by comparing MAC($x \parallel 1$) and MAC($y \parallel d \oplus 1$).

Therefore, we find a pair satisfying $\mathcal{R}(X, Y)$ with complexity $2^{n/2}$, and this leads to simple forgeries using (8). This contradicts the security proof of 1kf9 given in [8]. Note that this attack is still valid if we use different multiplications for the two branches in the finalization function.

6.2   Attacking 1kPMAC+

The 1kPMAC+ mode uses the fix function for domain separation to build a single-key variant of PMAC+, as shown in Fig. 8.

$$\Sigma'(M) = \bigoplus_{i=1}^{\ell} \tilde{E}_i(m_i) \qquad\qquad \Sigma(M) = \texttt{fix0}(\Sigma'(M))$$
$$\Theta'(M) = \bigoplus_{i=1}^{\ell} 2^{\ell+1-i} \odot \tilde{E}_i(m_i) \qquad \Theta(M) = \texttt{fix1}(\Theta'(M))$$
$$\text{MAC}(M) = E(\Sigma(M)) \oplus E(\Theta(M))$$

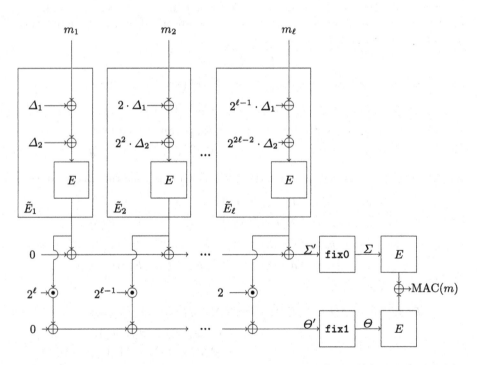

Fig. 8. Diagram for 1kPMAC+ with a ℓ-block message where $\Delta_1 = E(1)$ and $\Delta_2 = E(2)$.

Attack. Since the `fix` functions used in the finalization have collisions, we can build a variant of the attacks from Sect. 4 using differences in Σ' and/or Θ' that are absorbed by the `fix` functions. More precisely, we use the following relation $\mathcal{R}$ on quadruple of messages:

$$\mathcal{R}(X, Y, Z, T) := \begin{cases} \Sigma'(X) = \Sigma(Y)' \oplus 1 \\ \Theta'(Y) = \Theta(Z)' \oplus 1 \\ \Sigma'(Z) = \Sigma(T)' \oplus 1 \\ \Theta'(T) = \Theta(X)' \oplus 1 \end{cases}$$

$$\Rightarrow \mathrm{MAC}(X) \oplus \mathrm{MAC}(Y) \oplus \mathrm{MAC}(Z) \oplus \mathrm{MAC}(T) = 0.$$

We can find quadruple of messages satisfying $\mathcal{R}$ using a single message injection function:

$$\phi_u(i) = u \parallel i$$
$$X = \phi_u(x) = u \parallel x \quad Y = \psi_u(y) = u \parallel y \quad Z = \phi_u(z) = u \parallel z \quad T = \psi_u(t) = u \parallel t$$

Indeed we have

$$\mathrm{MAC}(\phi_u(i)) = E\Big(\mathtt{fix0}\big(\underbrace{\tilde{E}_1(u) \oplus \tilde{E}_2(x)}_{\Sigma'_u(i)}\big)\Big) \oplus E\Big(\mathtt{fix1}\big(\underbrace{4\tilde{E}_1(u) \oplus 2\tilde{E}_2(x)}_{\Theta'_u(i)}\big)\Big).$$

We observe that:

$$\mathcal{R}(x, y, z, t) \Leftrightarrow \begin{cases} \tilde{E}_2(x) = \tilde{E}_2(y) \oplus 1 \\ \tilde{E}_2(z) = \tilde{E}_2(t) \oplus 1 \\ 2\tilde{E}_2(x) = 2\tilde{E}_2(z) \oplus 1 \\ 2\tilde{E}_2(y) = 2\tilde{E}_2(t) \oplus 1 \end{cases}$$

$$\Leftrightarrow \begin{cases} \tilde{E}_2(x) \oplus \tilde{E}_2(y) \oplus \tilde{E}_2(z) \oplus \tilde{E}_2(t) = 0 \\ \tilde{E}_2(x) = \tilde{E}_2(y) \oplus 1 \\ \tilde{E}_2(x) = \tilde{E}_2(z) \oplus d \end{cases}$$

Therefore, $\mathcal{R}$ defines a $3n-$bit relation that is independent of the value u. This can be used for attacks in the same way as in the previous sections, using a single list

$$L = \Big\{ \mathrm{MAC}(\phi_0(x)) \parallel \mathrm{MAC}(\phi_1(x)) \parallel \mathrm{MAC}(\phi_2(x)) \ : \ x < 2^{3n/4} \Big\}$$

We can find a quadruple of four distinct values (x, y, z, t) such that $L[x] \oplus L[y] \oplus L[z] \oplus L[t] = 0$ with $\tilde{\mathcal{O}}(2^{3n/2})$ operations, using a memory of size $\mathcal{O}(2^{3n/4})$, and this easily leads to forgeries.

7 Conclusion

In this paper we have introduced a cryptanalysis technique to attack double-block-hash MACs using quadruples of messages. We show three variants of the technique, with attacks with $\mathcal{O}(2^{3n/4})$ queries against SUM-ECBC, GCM-SIV2, PMAC+, LightMAC+, 1kPMAC+ and 3kf9. All these modes have a security proof up to $2^{2n/3}$ queries, but no attacks with fewer than 2^n queries were known before our work.

Our main attacks are in the information theoretic model, and an attacker would need more than 2^n operations to perform a forgery. On the other hand, we also have a variant of the attack against SUM-ECBC and GCM-SIV2 with time complexity $\tilde{O}(2^{6n/7})$. This opens the path for attack with total complexity below 2^n for other double-block-hash MACs.

We believe that studying generic attacks is important in order to understand the security of these MACs, and is needed in addition to security proofs. In particular our results show that they do not reach full security, and we invalidate a recent proof for LightMAC+. However, there is still a gap between the $2^{2n/3}$ bound of the proofs, and our attacks with $\mathcal{O}(2^{3n/4})$ queries. Further work is needed to determine whether the attacks can be improved, or whether better proofs are possible.

Acknowledgement. Mridul Nandi is supported by R.C.Bose Centre for Cryptology and Security.

Part of this work was supported by the French DGA.

A   SageMath Implementation

In order to verify that the algorithm is correct, we have implemented the attack against SUM-ECBC with complexity $\tilde{O}(2^{6n/7})$ given in Sect. 3.1 with SageMath:

```
xor  = lambda x, y: x.__xor__(y)
txor = lambda a,b: tuple(xor(u,v) for u,v in zip(a,b) )
def random_perm(n):
  pp = Permutations(n).random_element()
  return lambda x: pp(x+1)-1

def CBC(E,M):
  x = 0
  for m in M:
    x = E(x.__xor__(m))
  return x
def SUMECBC(E1,E2,E3,E4,M):
  a = E2(CBC(E1,M))
  b = E4(CBC(E3,M))
  return a.__xor__(b)

E1, E2, E3, E4 = (random_perm(2^21) for _ in range(4))
MAC = lambda x: SUMECBC(E1,E2,E3,E4,x)
print "Values to recover       | {0:06x} {1:06x}".format(
  xor(E1(0),E1(1)), xor(E3(0),E3(1)))
```

```
print "Generating data..."
L1,L2,L3,L4 = [], [], [], []
for i in range(2^12):
  if (i&0b000000000111 == 0): L1.append(i)
  if (i&0b000000111000 == 0): L2.append(i)
  if (i&0b000111000000 == 0): L3.append(i)
  if (i&0b111000000000 == 0): L4.append(i)
def macs(u,i):
  x = (0,0)
  for j in range(i,2^21,2^12):
    x = txor(x,(MAC([u,j]), MAC([u, xor(1,j)])))
  return (i,x)
L1 = [ macs(0,i) for i in L1 ]
L2 = [ macs(0,i) for i in L2 ]
L3 = [ macs(1,i) for i in L3 ]
L4 = [ macs(1,i) for i in L4 ]

print "Looking for quadruples..."
L13 = sorted((txor(a[1],b[1]),a[0],b[0]) for a in L1 for b in L3)
L24 = sorted((txor(a[1],b[1]),a[0],b[0]) for a in L2 for b in L4)

i,j = 0,0
while i<len(L13) and j<len(L24):
  if L13[i][0] == L24[j][0]:
    if L13[i] != L24[j]:
      print "{:06x} {:06x} {:06x} {:06x} | {:06x} {:06x}".format(
        L13[i][1], L13[i][2], L24[j][1], L24[j][2],
        xor(L13[i][1],L13[i][2]), xor(L13[i][1],L24[j][2]))
    if L13[i] < L24[j]:
      i+=1
    else:
      j+=1
  elif L13[i][0] < L24[j][0]:
    i+=1
  else:
    j+=1
```

334   G. Leurent et al.

References

1. An, J.H., Bellare, M.: Constructing VIL-MACs from FIL-MACs: message authentication under weakened assumptions. In: Wiener, M. (ed.) CRYPTO 1999. LNCS, vol. 1666, pp. 252–269. Springer, Heidelberg (1999). https://doi.org/10.1007/3-540-48405-1_16

2. Bellare, M., Canetti, R., Krawczyk, H.: Keying hash functions for message authentication. In: Koblitz, N. (ed.) CRYPTO 1996. LNCS, vol. 1109, pp. 1–15. Springer, Heidelberg (1996). https://doi.org/10.1007/3-540-68697-5_1

3. Bellare, M., Guérin, R., Rogaway, P.: XOR MACs: new methods for message authentication using finite pseudorandom functions. In: Coppersmith, D. (ed.) CRYPTO 1995. LNCS, vol. 963, pp. 15–28. Springer, Heidelberg (1995). https://doi.org/10.1007/3-540-44750-4_2

4. Bellare, M., Kilian, J., Rogaway, P.: The security of the cipher block chaining message authentication code. J. Comput. Syst. Sci. **61**(3), 362–399 (2000)

5. Black, J., Rogaway, P.: A block-cipher mode of operation for parallelizable message authentication. In: Knudsen, L.R. (ed.) EUROCRYPT 2002. LNCS, vol. 2332, pp. 384–397. Springer, Heidelberg (2002). https://doi.org/10.1007/3-540-46035-7_25

6. Chose, P., Joux, A., Mitton, M.: Fast correlation attacks: an algorithmic point of view. In: Knudsen, L.R. (ed.) EUROCRYPT 2002. LNCS, vol. 2332, pp. 209–221. Springer, Heidelberg (2002). https://doi.org/10.1007/3-540-46035-7_14

7. Cogliati, B., Seurin, Y.: EWCDM: an efficient, beyond-birthday secure, nonce-misuse resistant MAC. In: Robshaw, M., Katz, J. (eds.) CRYPTO 2016, Part I. LNCS, vol. 9814, pp. 121–149. Springer, Heidelberg (2016). https://doi.org/10.1007/978-3-662-53018-4_5

8. Datta, N., Dutta, A., Nandi, M., Paul, G., Zhang, L.: Building single-key beyond birthday bound message authentication code. Cryptology ePrint Archive, Report 2015/958 (2015). http://eprint.iacr.org/2015/958

9. Datta, N., Dutta, A., Nandi, M., Paul, G., Zhang, L.: Single key variant of PMAC_Plus. IACR Trans. Symm. Cryptol. **2017**(4), 268–305 (2017)

10. Dinur, I., Leurent, G.: Improved generic attacks against hash-based MACs and HAIFA. In: Garay, J.A., Gennaro, R. (eds.) CRYPTO 2014, Part I. LNCS, vol. 8616, pp. 149–168. Springer, Heidelberg (2014). https://doi.org/10.1007/978-3-662-44371-2_9

11. Dutta, A., Jha, A., Nandi, M.: Tight security analysis of EHtM MAC. IACR Trans. Symm. Cryptol. **2017**(3), 130–150 (2017)

12. Ferguson, N.: Authentication weaknesses in GCM. Comment to NIST (2005). http://csrc.nist.gov/groups/ST/toolkit/BCM/documents/comments/CWC-GCM/Ferguson2.pdf

13. Computer data authentication: National Bureau of Standards, NIST FIPS PUB 113. U.S, Department of Commerce (1985)

14. Fuhr, T., Leurent, G., Suder, V.: Collision attacks against CAESAR candidates. In: Iwata, T., Cheon, J.H. (eds.) ASIACRYPT 2015, Part II. LNCS, vol. 9453, pp. 510–532. Springer, Heidelberg (2015). https://doi.org/10.1007/978-3-662-48800-3_21

15. Gilbert, E.N., MacWilliams, F.J., Sloane, N.J.: Codes which detect deception. Bell Labs Tech. J. **53**(3), 405–424 (1974)

16. Guo, J., Peyrin, T., Sasaki, Y., Wang, L.: Updates on generic attacks against HMAC and NMAC. In: Garay, J.A., Gennaro, R. (eds.) CRYPTO 2014, Part I. LNCS, vol. 8616, pp. 131–148. Springer, Heidelberg (2014). https://doi.org/10.1007/978-3-662-44371-2_8

17. Iwata, T.: New blockcipher modes of operation with beyond the birthday bound security. In: Robshaw, M. (ed.) FSE 2006. LNCS, vol. 4047, pp. 310–327. Springer, Heidelberg (2006). https://doi.org/10.1007/11799313_20

18. Iwata, T., Kurosawa, K.: OMAC: one-key CBC MAC. In: Johansson, T. (ed.) FSE 2003. LNCS, vol. 2887, pp. 129–153. Springer, Heidelberg (2003). https://doi.org/10.1007/978-3-540-39887-5_11

19. Iwata, T., Mennink, B., Vizár, D.: CENC is optimally secure. Cryptology ePrint Archive, Report 2016/1087 (2016). http://eprint.iacr.org/2016/1087

20. Iwata, T., Minematsu, K.: Stronger security variants of GCM-SIV. IACR Trans. Symm. Cryptol. **2016**(1), 134–157 (2016). http://tosc.iacr.org/index.php/ToSC/article/view/539

21. Iwata, T., Minematsu, K., Peyrin, T., Seurin, Y.: ZMAC: a fast tweakable block cipher mode for highly secure message authentication. In: Katz, J., Shacham, H. (eds.) CRYPTO 2017, Part III. LNCS, vol. 10403, pp. 34–65. Springer, Cham (2017). https://doi.org/10.1007/978-3-319-63697-9_2

22. Jaulmes, É., Joux, A., Valette, F.: On the security of randomized CBC-MAC beyond the birthday paradox limit a new construction. In: Daemen, J., Rijmen, V. (eds.) FSE 2002. LNCS, vol. 2365, pp. 237–251. Springer, Heidelberg (2002). https://doi.org/10.1007/3-540-45661-9_19

23. Joux, A.: Multicollisions in iterated hash functions. Application to cascaded constructions. In: Franklin, M. (ed.) CRYPTO 2004. LNCS, vol. 3152, pp. 306–316. Springer, Heidelberg (2004). https://doi.org/10.1007/978-3-540-28628-8_19

24. Knudsen, L.R., Mitchell, C.J.: Analysis of 3GPP-MAC and two-key 3GPP-MAC. Discrete Appl. Math. **128**(1), 181–191 (2003). http://www.sciencedirect.com/science/article/pii/S0166218X02004444. International Workshop on Coding and Cryptography (WCC 2001)

25. Lee, C., Kim, J., Sung, J., Hong, S., Lee, S.: Forgery and key recovery attacks on PMAC and Mitchell's TMAC variant. In: Batten, L.M., Safavi-Naini, R. (eds.) ACISP 2006. LNCS, vol. 4058, pp. 421–431. Springer, Heidelberg (2006). https://doi.org/10.1007/11780656_35

26. Leurent, G., Peyrin, T., Wang, L.: New generic attacks against hash-based MACs. In: Sako, K., Sarkar, P. (eds.) ASIACRYPT 2013, Part II. LNCS, vol. 8270, pp. 1–20. Springer, Heidelberg (2013). https://doi.org/10.1007/978-3-642-42045-0_1

27. List, E., Nandi, M.: Revisiting full-PRF-secure PMAC and using it for beyond-birthday authenticated encryption. In: Handschuh, H. (ed.) CT-RSA 2017. LNCS, vol. 10159, pp. 258–274. Springer, Cham (2017). https://doi.org/10.1007/978-3-319-52153-4_15

28. List, E., Nandi, M.: ZMAC$^+$ - an efficient variable-output-length variant of ZMAC. IACR Trans. Symm. Cryptol. **2017**(4), 306–325 (2017)

29. Luykx, A., Preneel, B., Tischhauser, E., Yasuda, K.: A MAC mode for lightweight block ciphers. In: Peyrin, T. (ed.) FSE 2016. LNCS, vol. 9783, pp. 43–59. Springer, Heidelberg (2016). https://doi.org/10.1007/978-3-662-52993-5_3

30. Mennink, B., Neves, S.: Encrypted Davies-Meyer and its dual: towards optimal security using mirror theory. In: Katz, J., Shacham, H. (eds.) CRYPTO 2017, Part III. LNCS, vol. 10403, pp. 556–583. Springer, Cham (2017). https://doi.org/10.1007/978-3-319-63697-9_19

31. Minematsu, K.: How to Thwart birthday attacks against MACs via small randomness. In: Hong, S., Iwata, T. (eds.) FSE 2010. LNCS, vol. 6147, pp. 230–249. Springer, Heidelberg (2010). https://doi.org/10.1007/978-3-642-13858-4_13

32. Naito, Y.: Full PRF-secure message authentication code based on tweakable block cipher. In: Au, M.-H., Miyaji, A. (eds.) ProvSec 2015. LNCS, vol. 9451, pp. 167–182. Springer, Cham (2015). https://doi.org/10.1007/978-3-319-26059-4_9

33. Naito, Y.: Blockcipher-based MACs: beyond the birthday bound without message length. In: Takagi, T., Peyrin, T. (eds.) ASIACRYPT 2017, Part III. LNCS, vol. 10626, pp. 446–470. Springer, Cham (2017). https://doi.org/10.1007/978-3-319-70700-6_16

34. Naito, Y.: Improved security bound of LightMAC_Plus and its single-key variant. In: Smart, N.P. (ed.) CT-RSA 2018. LNCS, vol. 10808, pp. 300–318. Springer, Cham (2018). https://doi.org/10.1007/978-3-319-76953-0_16

35. Nikolić, I., Sasaki, Y.: Refinements of the k-tree algorithm for the generalized birthday problem. In: Iwata, T., Cheon, J.H. (eds.) ASIACRYPT 2015, Part II. LNCS, vol. 9453, pp. 683–703. Springer, Heidelberg (2015). https://doi.org/10.1007/978-3-662-48800-3_28

36. Peyrin, T., Wang, L.: Generic universal forgery attack on iterative hash-based MACs. In: Nguyen, P.Q., Oswald, E. (eds.) EUROCRYPT 2014. LNCS, vol. 8441, pp. 147–164. Springer, Heidelberg (2014). https://doi.org/10.1007/978-3-642-55220-5_9

37. Preneel, B., van Oorschot, P.C.: MDx-MAC and building fast MACs from hash functions. In: Coppersmith, D. (ed.) CRYPTO 1995. LNCS, vol. 963, pp. 1–14. Springer, Heidelberg (1995). https://doi.org/10.1007/3-540-44750-4_1

38. Preneel, B., van Oorschot, P.C.: On the security of two MAC algorithms. In: Maurer, U. (ed.) EUROCRYPT 1996. LNCS, vol. 1070, pp. 19–32. Springer, Heidelberg (1996). https://doi.org/10.1007/3-540-68339-9_3

39. Rogaway, P.: Efficient instantiations of tweakable blockciphers and refinements to modes OCB and PMAC. In: Lee, P.J. (ed.) ASIACRYPT 2004. LNCS, vol. 3329, pp. 16–31. Springer, Heidelberg (2004). https://doi.org/10.1007/978-3-540-30539-2_2

40. Wagner, D.: A generalized birthday problem. In: Yung, M. (ed.) CRYPTO 2002. LNCS, vol. 2442, pp. 288–304. Springer, Heidelberg (2002). https://doi.org/10.1007/3-540-45708-9_19

41. Wegman, M.N., Carter, L.: New hash functions and their use in authentication and set equality. J. Comput. Syst. Sci. **22**, 265–279 (1981)

42. Yasuda, K.: The sum of CBC MACs is a secure PRF. In: Pieprzyk, J. (ed.) CT-RSA 2010. LNCS, vol. 5985, pp. 366–381. Springer, Heidelberg (2010). https://doi.org/10.1007/978-3-642-11925-5_25

43. Yasuda, K.: A new variant of PMAC: beyond the birthday bound. In: Rogaway, P. (ed.) CRYPTO 2011. LNCS, vol. 6841, pp. 596–609. Springer, Heidelberg (2011). https://doi.org/10.1007/978-3-642-22792-9_34

44. Zhang, L., Wu, W., Sui, H., Wang, P.: 3kf9: enhancing 3GPP-MAC beyond the birthday bound. In: Wang, X., Sako, K. (eds.) ASIACRYPT 2012. LNCS, vol. 7658, pp. 296–312. Springer, Heidelberg (2012). https://doi.org/10.1007/978-3-642-34961-4_19

Searchable Encryption and Differential Privacy

Structured Encryption and Leakage Suppression

Seny Kamara[1]([✉]), Tarik Moataz[1], and Olya Ohrimenko[2]

[1] Brown University, Providence, USA
{seny,tarik_moataz}@brown.edu
[2] Microsoft Research, Cambridge, UK
oohrim@microsoft.com

Abstract. Structured encryption (STE) schemes encrypt data structures in such a way that they can be privately queried. One aspect of STE that is still poorly understood is its leakage. In this work, we describe a general framework to design STE schemes that do not leak the query/search pattern (i.e., if and when a query was previously made).

Our framework consists of two compilers. The first can be used to make any dynamic STE scheme *rebuildable* in the sense that the encrypted structures it produces can be rebuilt efficiently using only $O(1)$ client storage. The second transforms any rebuildable scheme that leaks the query/search pattern into a new scheme that does not. Our second compiler is a generalization of Goldreich and Ostrovsky's square root oblivious RAM (ORAM) solution but does not make use of black-box ORAM simulation. We show that our framework produces STE schemes with query complexity that is asymptotically better than ORAM simulation in certain (natural) settings and comparable to special-purpose oblivious data structures.

We use our framework to design a new STE scheme that is "almost" zero-leakage in the sense that it reveals an, intuitively-speaking, small amount of information. We also show how the scheme can be used to achieve zero-leakage queries when one can tolerate a probabilistic guarantee of correctness. This construction results from applying our compilers to a new STE scheme we design called the *piggyback scheme*. This scheme is a general-purpose STE construction (in the sense that it can encrypt any data structure) that leaks the search/query pattern but hides the response length on non-repeating queries.

1 Introduction

A structured encryption (STE) scheme encrypts data in such a way that it can be privately queried. An STE scheme is secure if it does not reveal any partial information about the data or query beyond a given leakage profile. Special cases of STE include searchable symmetric encryption (SSE) [8,13,17,25,26,36,37] and graph encryption [10,29]. STE has received attention due to its applications to the design of secure cloud services, secure databases, lawful surveillance

© International Association for Cryptologic Research 2018
H. Shacham and A. Boldyreva (Eds.): CRYPTO 2018, LNCS 10991, pp. 339–370, 2018.
https://doi.org/10.1007/978-3-319-96884-1_12

[22] and network provenance [41]. In recent years, a lot of progress has been made on improving various characteristics of STE including its efficiency [13], its dynamism [7,25,26,31], its parallelism and locality [3,7,9,14,25], its security [5,13,37] and its expressiveness [8,10,15,24,33].

One aspect that is still poorly understood, however, is its leakage. In the context of SSE, we currently know of four attacks. All of these attacks are query-recovery attacks in the sense that they aim to recover information about the queries. The IKK attack [20] exploits co-occurrence leakage (i.e., how often each pair of queries occur together in a document) assuming knowledge of the client's data collection. The Count attack [6] exploits co-occurrence and response length leakage (i.e., how many documents contain the query) assuming knowledge of the client's data collection and of a subset of its queries.[1] The LZWT attack [28] exploits search pattern leakage. File injection attacks [40] are query-recovery attacks where the adversary needs the ability to inject documents/files.

Oblivious RAM (ORAM). One approach that is often suggested for handling leakage is to avoid STE completely and use one of two ORAM-based approaches. The first, which we refer to as ORAM simulation, is to store the data (represented as an array) in an ORAM and query it by simulating every read and write operation of the query algorithm with an ORAM access. Note that this approach is general-purpose. The second approach is to design a custom oblivious data structure and query it with a dedicated oblivious query algorithm. We briefly note that while ORAM simulation is often cited as a zero-leakage (ZL) solution,[2] its exact query leakage actually depends on the data structure being managed. More precisely, ORAM simulation is only ZL for structures with constant query complexity. For structures that do not satisfy this constraint (e.g., inverted indexes) some form of padding must be applied which increases both the storage and query complexity of the solution.

Leakage suppression. Another direction, which we initiate here, is to focus on designing general tools and techniques to *suppress* the leakage of existing schemes. We focus mainly on two kinds of techniques: *compilers*, which take schemes with a given leakage profile and produce new schemes with an improved profile; and *transforms*, which modify queries and/or data in such a way that they can be safely used with schemes that have a certain leakage profile. Our goal is to find compilers and transforms for as wide a class of schemes as possible and that incur the smallest overhead possible. In this work, we propose a leakage suppression framework (i.e., a set of compilers and transforms) for *query equality* leakage. The query equality, which is typically referred to as the search/query pattern in the encrypted search literature, reveals if and when a

---

[1] It was shown experimentally in [6] that the IKK and Count attacks need to know at least 90% and 75% of the client's data, respectively. In addition, the Count attack also needs to know at least 5% of the client's queries whenever it knows less than 100% of the client's data.

[2] In this work, a solution is ZL if its leakage reveals only information that is derived from the security parameter or other public parameters.

query has occurred in the past. Interestingly, our main compiler is a generalization of Goldreich and Ostrovsky's square-root ORAM solution [18] but uses STE to avoid ORAM simulation.

Our leakage suppression framework—which combines both STE and ORAM—can result in STE schemes that are asymptotically more efficient than ORAM simulation under certain assumptions on the data and queries which we make precise in Sect. 8.[3] We also find that these schemes can achieve the same asymptotic efficiency as custom oblivious data structures (specifically, we compare to the case of oblivious trees). While we focus here on query equality, suppression frameworks for other common leakage patterns would be of interest.

1.1 Our Contributions and Techniques

In this work, we consider leakage suppression techniques focusing on query equality. We make several contributions which we summarize below.

Modeling leakage. Because the terminology and formalism used in previous work is sometimes inconsistent and contradictory, we extend the definitional approach of [10,13] with a more intuitive nomenclature and precise descriptions. The details are in Sect. 5.1 but, as an example, we mention that the search/query pattern is referred to as the *query equality* pattern in our framework and is modeled as a function $\mathsf{qeq} : \mathbb{D} \times \mathbb{Q}^t \to \{0,1\}^{t \times t}$, where $\mathbb{D}$ is a space of data objects, $\mathbb{Q}^t$ is a sequence of queries from a query space $\mathbb{Q}$, and $\{0,1\}^{t \times t}$ is the set of binary $t \times t$ matrices. The function qeq takes a data object and a query sequence and outputs a binary matrix with a 1 at location (i,j) if the ith and j^{th} queries in the sequence are equal and 0 otherwise. We also identify and formalize the notion of *sub-pattern* leakage which captures the behavior of a leakage pattern on a specified subset of query sequences. As we will see, sub-patterns are important in understanding and analyzing our suppression techniques.

Reinterpreting the square-root solution. Our main suppression compiler is based on the seminal square-root ORAM solution of Goldreich and Ostrovsky [18] which works as follows. Items are encrypted and stored in a main memory together with encrypted dummy items after being randomly shuffled. In addition, a cache is maintained in which encrypted items are moved after being accessed. The ORAM structure consists of the main memory and the cache. Reading from the ORAM requires accessing the entire cache to look for the item and retrieving from main memory either a dummy item if the item was found in the cache, or the real item if the item was not found in the cache.

We observe that the square-root solution can be reinterpreted through the lens of STE as follows: the main memory is an encrypted array that leaks the query equality pattern (since reading the same location twice requires sending the same randomly permuted address to the server) and the cache is an encrypted dictionary with no query leakage. The access protocol can then be understood

---

[3] When these assumptions do not apply, the schemes are comparable in efficiency to ORAM simulation.

as a mechanism that leverages the ZL queries of the cache to suppress the query equality leakage of the encrypted array.

The cache-based compiler. As we show, the ideas that underlie the square-root solution are not only applicable to encrypted arrays but can be generalized to more complex constructions like encrypted multi-maps and dictionaries. In other words, instead of using a ZL cache to suppress the query equality leakage of an encrypted array (i.e., the main memory) we want to use the cache to suppress the query equality leakage of complex encrypted structures. Though there are technical subtleties that must be addressed when moving to more complex structures we describe and analyze this generalization of the square-root solution which we refer to as the *cache-based compiler* (CBC).

The main advantage of using the CBC to suppress query equality leakage is that we can avoid ORAM simulation; that is, we do not have to represent our data structure as an array and simulate every read and write instruction of the query algorithm with an ORAM access. As we show in Sect. 8, our framework induces an additive overhead over the optimal query complexity. This is in contrast to ORAM simulation which induces a multiplicative overhead. Comparing the efficiency of the two approaches over arbitrary data and queries, however, is not possible so we show that under certain natural conditions (e.g., known to occur in the keyword search setting), our framework results in schemes that are asymptotically faster than ORAM simulation and comparable to dedicated oblivious data structure constructions (here, we consider the case of oblivious trees). While the CBC allows us to avoid ORAM simulation, our framework can still benefit from improvements in ORAM design. The reason is that while ORAM is not used to manage the main data structure, it can (and should) be used to implement the cache. Note also that the CBC yields a static scheme even though it requires a dynamic ZL dictionary. Designing a dynamic variant of the CBC is left as an important open problem.

Non-repeating sub-patterns. In analyzing the security of the schemes that result from the CBC, we find that their query leakage is a *sub-pattern* of the base scheme's query leakage. Specifically, it is what we refer to as the *non-repeating* sub-pattern which is the leakage that occurs on sequences of non-repeating queries. This suggests that a future goal in STE design might be to focus on schemes with low non-repeating sub-pattern leakage as opposed to focusing on schemes with low query leakage directly.

Safe extensions. As mentioned above, there are several technicalities that must be handled when adapting the square-root solution to more complex structures. The first is that the structure must be extendable in the sense that it must be able to hold and query dummy items. We formalize this process as an *extension* scheme, which takes as input a data structure and outputs a new one with the same items plus a given number of dummy items. While, a-priori, this might seem straightforward, one has to handle dummy items with care because the leakage of the scheme (which was not originally designed to handle dummies) could reveal information that enables the adversary to distinguish between real and dummy items. In addition, the way in which dummy items are handled could be correlated with the real items and this could be revealed to the adversary through the leakage of the scheme. In Sect. 6.1, we formally define the security properties that extension schemes must satisfy in order to be safely used with the CBC.

The rebuild compiler (RBC). Another challenge is that the CBC requires the base scheme to be efficiently rebuildable, i.e., equipped with an efficient protocol that can reconstruct the structure with new randomness. Most STE schemes were not designed with this in mind so we describe a general-purpose protocol that can be used to rebuild any dynamic STE scheme. If the base scheme has $O(\log^2 n)$ update complexity,[4] where n is the number of items stored in the structure, then our protocol has computation and communication complexity $O(n \log^2 n)$. In addition, our rebuild protocol does not affect the latency of the scheme in the sense that queries can still be made and answered while a rebuild is taking place. Note, however, that the output of the RBC is a *static* rebuildable scheme, therefore losing the dynamism of the base construction. The question of designing a variant of the RBC that preserves dynamism is left open.

The piggyback scheme (PBS). As discussed, the CBC results in new constructions that leak the non-repeating sub-pattern of their base scheme. Our goal, therefore is reduced to designing schemes with low non-repeating sub-pattern leakage. In the setting of encrypted arrays, this is relatively straightforward because the base scheme that implicitly underlies the square-root solution (i.e., encrypt and randomly shuffle the items, and fetch by reading the permuted location) does not reveal anything when queried on non-repeating sequences. This is not the case, however, for standard encrypted multi-map or dictionary constructions which reveal the response identity (i.e., the plaintext result of the query) if they are response revealing; or the response length if they are response hiding. In particular, this means that these leakages may persist even after applying the CBC.

To address this we design a new scheme called PBS with low non-repeating sub-pattern leakage. There are two variants of the scheme: one that reveals the total sequence response length (i.e., the sum of the response lengths over all queries in the sequence) and another that reveals nothing. The former achieves standard correctness whereas the latter is correct with only a certain probability. At a high level, PBS results from applying a transform to the data and queries so that they can be safely used with an encrypted multi-map that leaks the response length. Our approach is to modify the data in such a way that, at query time, the client can retrieve a fixed number of words per query (which we refer to as a *batch*) no matter how large the response is. To maintain correctness, incoming queries are queued and processed at the next available time. This introduces a delay in the querying process but by carefully tuning the batch size we can ensure the entire response is retrieved in a reasonable amount of time. PBS is

---

[4] As far as we know, all dynamic SSE schemes have update complexity ranging from constant to $O(\log^2 n)$.

general-purpose in the sense that it encrypts *any* data structure. As far as we know, this is the first general-purpose STE scheme and may be of independent interest.

New constructions. Our framework results in several new schemes. First, by applying our compilers to PBS, we get a new general-purpose STE scheme called AZL that is "almost" ZL. Specifically, when used on a sequence of t queries $(q_1, \ldots, q_t)$, its query leakage reveals nothing on queries $(q_1, \ldots, q_{t-1})$ and then reveals the sum of the sequence's response lengths on query q_t. We then show that by applying our compilers to a variant of PBS, we can get a "fully" ZL construction at the cost of achieving a weaker notion of correctness. As discussed above, the query complexities of both AZL and its fully ZL variant, FZL, are asymptotically smaller than ORAM simulation under natural assumptions.

Of course, our compilers can also be applied to other constructions with query equality leakage, including schemes for single-keyword and boolean SSE [2,5,7, 8,10,13,24], encrypted relational databases [23] or encrypted graphs [10,29]. We stress, however, that the resulting schemes may not be ZL (or even almost ZL) since, as discussed above, our framework suppresses query equality leakage but still reveals the base scheme's non-repeating sub-pattern.

2   Related Work

Structured and searchable encryption. Searchable encryption was first considered explicitly by Song, Wagner and Perrig in [36]. In [13], Curtmola et al. introduced adaptive security and proposed the first schemes with optimal search complexity $O(\#\mathsf{DB}[w])$, where $\#\mathsf{DB}[w]$ is the number of documents that contain the keyword w. The notion of structured encryption was introduced by Chase and Kamara [10] as a generalization of SSE that supports queries on arbitrarily-structured data. Subsequent works have considered the problems of dynamic [5,7,17,25,26,37], I/O-efficient [7,30], local [3,9,14], more secure [5,16,37], expressive [8,10,15,24,33], and multi-user [13,21] SSE.

Recently, Garg et al. [16] presented a dynamic SSE construction that hides the query equality pattern by leveraging ORAM and garbled RAM techniques. Their construction has non-optimal search complexity $\widetilde{O}(\#\mathsf{DB}[w] \cdot \log N + \log^3 N)$, where $N = \sum_{w \in \mathbb{W}} \mathsf{DB}[w]$. We note that while this scheme does not reveal the query equality explicitly, it still leaks the response length which is often correlated with the query equality. Our AZL and FZL constructions, on the other hand, hide the query equality and reveal only the *sequence* response length and $\perp$, respectively. In addition, they achieve this without the multiplicative $\log N$ overhead and with less computation on the client side.

Oblivious RAM. The seminal work of Goldreich and Ostrovsky [18] introduced the notion of ORAM and described the Square-Root and Hierarchical solutions. Many subsequent constructions improved ORAM upon several dimensions including communication complexity, number of rounds, client storage and storage overhead [16,19,27,32,35,38,39].

3   Preliminaries and Notation

Notation. The set of all binary strings of length n is denoted as $\{0,1\}^n$, and the set of all finite binary strings as $\{0,1\}^*$. $[n]$ is the set of integers $\{1,\ldots,n\}$, and $2^{[n]}$ is the corresponding power set. We write $x \leftarrow \chi$ to represent an element x being sampled from a distribution χ, and $x \overset{\$}{\leftarrow} X$ to represent an element x being sampled uniformly at random from a set X. The output x of an algorithm $\mathcal{A}$ is denoted by $x \leftarrow \mathcal{A}$. Given a sequence $\mathbf{v}$ of n elements, we refer to its ith element as v_i or $\mathbf{v}[i]$. If S is a set then $\#S$ refers to its cardinality. If s is a string then $|s|_2$ refers to its bit length.

Sorting networks. A sorting network is a circuit of comparison-and-swap gates. A sorting network for n elements takes as input a collection of n elements $(a_1,\ldots,a_n)$ and outputs them in increasing order. Each gate g in an n-element network SN_n specifies two input locations $i,j \in [n]$ and, given a_i and a_j, returns the pair (a_i, a_j) if $a_i < a_j$ and (a_j, a_i) otherwise. Sorting networks can be instantiated with the asymptotically-optimal Ajtai-Komlos-Szemeredi network [1] which has size $O(n \log n)$ or Batcher's more practical network [4] with size $O(n \log^2 n)$ but with small constants.

The word RAM. Our model of computation is the word RAM. In this model, we assume memory holds an infinite number of w-bit words and that arithmetic, logic, read and write operations can all be done in $O(1)$ time. We denote by $|x|_w$ the word-length of an item x; that is, $|x|_w = |x|_2/w$. Here, we assume that $w = \Omega(\log k)$.

Abstract data types. An *abstract data type* specifies the functionality of a data structure. It is a collection of data objects together with a set of operations defined on those objects. Examples include sets, dictionaries (also known as key-value stores or associative arrays) and graphs. The operations associated with an abstract data type fall into one of two categories: query operations, which return information about the objects; and update operations, which modify the objects. If the abstract data type supports only query operations it is *static*, otherwise it is *dynamic*. For simplicity we define data types as having a single operation and note that the definitions can be extended to capture multiple operations in the natural way. We model a dynamic data type $\mathbf{T}$ as a collection of four spaces $\mathbb{D} = \{\mathbb{D}_k\}_{k\in\mathbb{N}}$, $\mathbb{Q} = \{\mathbb{Q}_k\}_{k\in\mathbb{N}}$, $\mathbb{R} = \{\mathbb{R}_k\}_{k\in\mathbb{N}}$ and $\mathbb{U} = \{\mathbb{U}_k\}_{k\in\mathbb{N}}$ and two maps $\mathsf{qu} : \mathbb{D} \times \mathbb{Q} \to \mathbb{R}$ and $\mathsf{up} : \mathbb{D} \times \mathbb{U} \to \mathbb{D}$, where $\mathbb{D}, \mathbb{Q}, \mathbb{R}$ and $\mathbb{U}$ are, respectively, $\mathbf{T}$'s object, query, response and update spaces. In Sect. 9, we make the additional assumption that $\mathbb{U} = \mathbb{Q} \times \mathbb{R}$, i.e., an update can be written as a pair composed of a query and its response. When specifying a data type $\mathbf{T}$ we will often just describe its maps $(\mathsf{qu}, \mathsf{up})$ from which the object, query, response and update spaces can be deduced. The spaces are ensembles of finite sets of finite strings indexed by the security parameter. We assume that $\mathbb{R}$ includes a special element $\bot$ and that $\mathbb{D}$ includes an empty object d_0 such that for all $q \in \mathbb{Q}$, $\mathsf{qu}(d_0, q) = \bot$.

Data structures. A type-$\mathbf{T}$ *data structure* is a representation of data objects in $\mathbb{D}$ in some computational model (as mentioned, here it is the word RAM).

Typically, the representation is optimized to support qu as efficiently as possible; that is, such that there exists an efficient algorithm Query that computes the function qu. For data types that support multiple queries, the representation is often optimized to efficiently support as many queries as possible. As a concrete example, the dictionary type can be represented using various data structures depending on which queries one wants to support efficiently. Hash tables support Get and Put in expected $O(1)$ time whereas balanced binary search trees support both operations in worst-case $\log(n)$ time.

Definition 1 (Structuring scheme). *Let* $\mathbf{T} = (\mathsf{qu} : \mathbb{D} \times \mathbb{Q} \to \mathbb{R}, \mathsf{up} : \mathbb{D} \times \mathbb{U} \to \mathbb{D})$ *be a dynamic type. A type-$\mathbf{T}$ structuring scheme* $\mathsf{SS} = (\mathsf{Setup}, \mathsf{Query}, \mathsf{Update})$ *is composed of three polynomial-time algorithms that work as follows:*

- *$\mathsf{DS} \leftarrow \mathsf{Setup}(d)$: is a possibly probabilistic algorithm that takes as input a data object $d \in \mathbb{D}$ and outputs a data structure DS. Note that d can be represented in any arbitrary manner as long as its bit length is polynomial in k. Unlike DS, its representation does not need to be optimized for any particular query.*
- *$r \leftarrow \mathsf{Query}(\mathsf{DS}, q)$: is an algorithm that takes as input a data structure DS and a query $q \in \mathbb{Q}$ and outputs a response $r \in \mathbb{R}$.*
- *$\mathsf{DS} \leftarrow \mathsf{Update}(\mathsf{DS}, u)$: is a possibly probabilistic algorithm that takes as input a data structure DS and an update $u \in \mathbb{U}$ and outputs a new data structure DS.*

Here, we allow Setup and Update to be probabilistic but not Query. This captures most data structures but the definition can be extended to include structuring schemes with probabilistic query algorithms. We say that a data structure DS *instantiates* a data object $d \in \mathbb{D}$ if for all $q \in \mathbb{Q}$, $\mathsf{Query}(\mathsf{DS}, q) = \mathsf{qu}(d, q)$. We denote this by $\mathsf{DS} \equiv d$. We denote the set of queries supported by a structure DS as $\mathbb{Q}_{\mathsf{DS}}$; that is,

$$\mathbb{Q}_{\mathsf{DS}} \overset{def}{=} \left\{ q \in \mathbb{Q} : \mathsf{Query}(\mathsf{DS}, q) \neq \perp \right\}.$$

Similarly, the set of responses supported by a structure DS is denoted $\mathbb{R}_{\mathsf{DS}}$.

Definition 2 (Correctness). *Let* $\mathbf{T} = (\mathsf{qu} : \mathbb{D} \times \mathbb{Q} \to \mathbb{R}, \mathsf{up} : \mathbb{D} \times \mathbb{U} \to \mathbb{D})$ *be a dynamic type. A type-$\mathbf{T}$ structuring scheme* $\mathsf{SS} = (\mathsf{Setup}, \mathsf{Query}, \mathsf{Update})$ *is perfectly correct if it satisfies the following properties:*

1. *(static correctness) for all $d \in \mathbb{D}$,*

$$\Pr\left[\, \mathsf{DS} \equiv d : \mathsf{DS} \leftarrow \mathsf{Setup}(d) \,\right] = 1,$$

   where the probability is over the coins of Setup.
2. *(dynamic correctness) for all $d \in \mathbb{D}$ and $u \in \mathbb{U}$, for all $\mathsf{DS} \equiv d$,*

$$\Pr\left[\, \mathsf{Update}(\mathsf{DS}, u) \equiv \mathsf{up}(d, u) \,\right] = 1,$$

   where the probability is over the coins of Update.

Note that the second condition guarantees the correctness of an updated structure whether the original structure was generated by a setup operation or a previous update operation. Weaker notions of correctness (e.g., for data structures like Bloom filters) can be derived from Definition 2.

Basic data structures. We use structures for several basic data types including arrays, dictionaries and multi-maps which we recall here. Throughout, we will make black-box use of these data types which means that they can be instantiated with any appropriate data structure. To highlight this black-box usage, we refer to the data structure by its type's name. For example, we will write RAM, DX and MM to refer to some arbitrary array, dictionary and multi-map[5] data structures.

An array RAM of capacity n stores n items at locations 1 through n and supports read and write operations. We write $v := \mathsf{RAM}[i]$ to denote reading the item at location i and $\mathsf{RAM}[i] := v$ the operation of storing an item at location i. A dictionary structure DX of capacity n holds a collection of n label/value pairs $\{(\ell_i, v_i)\}_{i \le n}$ and supports get and put operations. We write $v_i := \mathsf{DX}[\ell_i]$ to denote getting the value associated with label ℓ_i and $\mathsf{DX}[\ell_i] := v_i$ to denote the operation of associating the value v_i in DX with label ℓ_i. A multi-map structure MM with capacity n is a collection of n label/tuple pairs $\{(\ell_i, \mathbf{v}_i)_i\}_{i \le n}$ that supports get and put operations. Similarly to dictionaries, we write $\mathbf{v}_i := \mathsf{MM}[\ell_i]$ to denote getting the tuple associated with label ℓ_i and $\mathsf{MM}[\ell_i] := \mathbf{v}_i$ to denote operation of associating the tuple $\mathbf{v}_i$ to label ℓ_i.

Data structure logs. Given a structure DS that instantiates an object d, we will be interested in the shortest sequence of update operations needed to create a new structure DS' that also instantiates d. We refer to this as the *update log* of DS and assume the existence of an efficient algorithm Log that takes as input DS and outputs a sequence $(u_1, \dots, u_n)$ such that adding $u_1, \dots, u_n$ to an empty structure results in some $\mathsf{DS}' \equiv d$.

Extensions. An important property we will need from a data structure is that it be *extendable* in the sense that, given a structure DS one can create another structure $\overline{\mathsf{DS}} \neq \mathsf{DS}$ that is functionally equivalent to DS but that also supports a number of *dummy* queries. We say that a structure is efficiently λ-extendable, for $\lambda \geq 1$, if there exists a query set $\overline{\mathbb{Q}} \supset \mathbb{Q}$ of size $\#\mathbb{Q} + \lambda$ and a probabilistic polynomial time algorithm $\mathsf{Ext_T}$ that takes as input DS and λ and returns a new structure $\overline{\mathsf{DS}}$ of the same type $\mathbf{T}$ such that: (1) $\overline{\mathsf{DS}} \equiv d$; and (2) for all $q \in \overline{\mathbb{Q}} \setminus \mathbb{Q}$, $\mathsf{Query}(\overline{\mathsf{DS}}, q) = \bot$.[6] We say that $\overline{\mathsf{DS}}$ is an extension of DS and that DS is a sub-structure of $\overline{\mathsf{DS}}$.

Cryptographic protocols. We denote by $(\mathsf{out}_A, \mathsf{out}_B) \leftarrow \Pi_{A,B}(X, Y)$ the execution of a two-party protocol Π between parties A and B, where X and Y are

---

[5] Multi-maps are the abstract data type instantiated by an inverted index. In the encrypted search literature multi-maps are sometimes referred to as indexes, databases or tuple-sets (T-sets).

[6] Note that we make the implicit assumption that adding dummy queries to the query space of some data type does not change the type.

the inputs provided by A and B, respectively; and out_A and out_B are the outputs returned to A and B, respectively. We sometimes write $\Pi_{A,\mathcal{A}}$ to denote an execution of Π where the first party follows the protocol and the second party is some adversary $\mathcal{A}$. Similarly we sometimes write $\Pi_{\mathcal{S},\mathcal{A}}$ to denote an execution of Π between a simulator $\mathcal{S}$ and an adversary $\mathcal{A}$. We quantify the round complexity of a protocol in either *moves* (i.e., messages sent between the parties) or *rounds* (i.e., pairs of messages exchanged between the parties).

4   Re-defining Structured Encryption

An STE scheme can be roughly viewed as a data structuring scheme that works over encrypted data. Several types of STE schemes were described in [11] (the full version of [10]) but here we consider structure-only schemes. This variant only encrypts objects as opposed to standard schemes which encrypt both a data structure and data items (e.g., documents, emails, user profiles). At a high-level, the formulation proposed in [10] works as follows. During a setup phase, the client constructs an encrypted data structure EDS under a key K. The client then sends EDS to the server. During the query phase, the client constructs and sends a token tk generated from its query q and secret key K. The server then uses the token tk to query EDS and recover a response r. Below, we formally describe our notion of STE. Our definition generalizes that of [10] in several respects.

Interaction. In the standard variant of STE, the query phase is *non-interactive*; that is, it requires only a single round that consists of the client sending a token and the server returning an encrypted data item. All the constructions proposed in [10] are non-interactive and many SSE constructions are as well. There are, however, several constructions that are interactive including [8,25,34]. The use of interaction in STE provides a lot of power and most interactive constructions are able to improve on the leakage of non-interactive schemes. For example [25] uses interaction during the update phase to leak less than [8,26] uses interaction to leak less than the naive boolean SSE construction which consists of the server taking intersections and unions of results.

Rebuilding. Since previous notions of STE did not consider rebuilding, the standard security notions of [10,13] have to be augmented appropriately. In particular, the definition has to properly capture the effect of rebuilding operations on the security of the scheme. Functionally, the result of rebuilding an encrypted structure EDS should be equivalent to re-running the scheme's Setup algorithm (with new coins) on the structure underlying EDS. From a security perspective, the purpose of rebuilding is to reduce the scheme's leakage.

4.1   Syntax and Correctness

In Definition 3 below we extend the syntax of STE to include interactive operations and rebuilding. We do this by adding an additional protocol for rebuilding operations. When using data structures, it is sometimes convenient to build a structure with a Setup operation that takes as input a data object. Other times, it is more convenient to build an empty structure with an Init operation and add items subsequently. Here, we only define a Setup algorithm but capture Init operations by inputting an empty structure $DS_0 \equiv d_0$.

Definition 3 (Structured encryption). *A type-***T** *interactive structured encryption scheme* STE $=$ (Setup, Query$_{C,S}$, Update$_{C,S}$, Rebuild$_{C,S}$) *consists of an algorithm and three two-party protocols that work as follows:*

- $(K, st, \text{EDS}) \leftarrow$ Setup$(1^k, \lambda, \text{DS})$: *is a probabilistic polynomial-time algorithm that takes as input a security parameter* 1^k, *a query capacity* $\lambda \geq 1$ *and a type-***T** *structure* DS. *It outputs a secret key* K, *a state st and an encrypted structure* EDS. *If* DS $\equiv d_0$, *it outputs an empty* EDS. *We sometimes write this as* Setup$(1^k, \lambda, \perp)$.
- $((st, r), \perp) \leftarrow$ Query$_{C,S}((K, st, q), \text{EDS})$: *is a two-party protocol executed between a client and a server where the client inputs a secret key* K, *a state st and a query q and the server inputs an encrypted data structure* EDS. *The client receives as output an updated state st and a response r while the server receives* $\perp$.
- $(st', \text{EDS}') \leftarrow$ Update$_{C,S}((K, st, u), \text{EDS})$: *is a two-party protocol executed between a client and server where the client inputs a secret key* K, *a state st and an update u and the server inputs an encrypted data structure* EDS. *The client receives a new state st' as output and the server receives* EDS'.
- $((st', K'), \text{EDS}') \leftarrow$ Rebuild$_{C,S}((K, st), \text{EDS})$: *is a two-party protocol executed between the client and server where the client inputs a secret key* K *and a state st. The server inputs an encrypted data structure* EDS. *The client receives an updated state st' and a new key* K' *as output while the server receives a new structure* EDS'.

For visual clarity, we sometimes omit the subscripts of the protocols when the parties involved are clear from the context.

We say that a type-**T** encrypted structure EDS instantiates a data object $d \in \mathbb{D}$ if for all $q \in \mathbb{Q}$, Query$((K, st, q), \text{EDS})$ outputs $((st, r), \perp))$ such that $r = \text{qu}(d, q)$, where K and st are the key and state of EDS. We write this as EDS $\equiv d$ and sometimes write EDS $\equiv$ DS to mean that EDS and DS instantiate the same data object.

Definition 4 (Correctness). *A type-***T** *structured encryption scheme* STE $=$ (Setup, Query$_{C,S}$, Update$_{C,S}$, Rebuild$_{C,S}$) *is correct if it satisfies the following properties:*

- *(static correctness) for all* $k \in \mathbb{N}$, *for all* $d \in \mathbb{D}$, *for all* DS *that instantiate d, for all* $\lambda \geq 1$,

$$\Pr\left[\text{EDS} \equiv \text{DS} : (K, st, \text{EDS}) \leftarrow \text{Setup}(1^k, \lambda, \text{DS}) \right] \geq 1 - \text{negl}(k),$$

where the probability is over the coins of Setup *and of* Query.

- *(dynamic correctness) for all $k \in \mathbb{N}$, for all $d \in \mathbb{D}$, for all EDS that instantiate d, for all $u \in \mathbb{U}$, for all $\lambda \geq 1$,*

$$\Pr\left[\, \mathsf{EDS}' \equiv \mathsf{up}(d, u) : (st, \mathsf{EDS}') \leftarrow \mathsf{Update}\big((K, st, u), \mathsf{EDS}\big) \,\right] \geq 1 - \mathsf{negl}(k),$$

where K and st are the key and state of EDS and the probability is over the coins of Update.

- *(rebuild correctness) for all $k \in \mathbb{N}$, for all $d \in \mathbb{D}$, for all EDS that instantiate d,*

$$\Pr\left[\, \mathsf{EDS}' \equiv d : \big((st, K'), \mathsf{EDS}'\big) \leftarrow \mathsf{Rebuild}\big((K, st), \mathsf{EDS}\big) \,\right] \geq 1 - \mathsf{negl}(k),$$

where K and st are the key and state of EDS and the probability is over the coins of Rebuild.

Structured encryption variants. The syntax of the different variants of STE can all be recovered from Definition 3. Stateless schemes can be recovered by omitting the state from the inputs and outputs of the algorithms. Schemes with non-interactive queries and/or updates can be recovered by requiring that the $\mathsf{Query}_{\mathsf{C,S}}$ or $\mathsf{Update}_{\mathsf{C,S}}$ protocols have only one message referred to as the search and update tokens, respectively. Response-revealing schemes have the following query syntax $(st, r) \leftarrow \mathsf{Query}_{\mathsf{C,S}}\big((K, st, q), \mathsf{EDS}\big)$.

5 Defining Security

As discussed in the previous Section, the standard notion of security for STE guarantees that an encrypted structure reveals no information about its underlying structure beyond the setup leakage $\mathcal{L}_{\mathsf{St}}$, that the query protocol reveals no information about the structure and queries beyond the query leakage $\mathcal{L}_{\mathsf{Qr}}$, and that the update protocol reveals no information about the structure and updates beyond the update leakage $\mathcal{L}_{\mathsf{Up}}$. If this holds for non-adaptively chosen operations then this is referred to as non-adaptive semantic security. If, on the other hand, the operations are chosen adaptively, this leads to the stronger notion of adaptive semantic security [13]. This notion of security was first proposed and formalized by Curtmola *et al.* in the context of SSE [13] and later generalized to STE in [10].

5.1 Modeling Leakage

We use the approach of [10,13] to capture leakage in STE. Every STE operation is associated with leakage which itself can be composed of multiple *leakage patterns*. The collection of all of these leakage functions is the scheme's *leakage profile*. Leakage patterns are (families of) functions over the various spaces associated with the underlying data type.

Leakage patterns. For concreteness, we describe several well-known leakage patterns. Because the terminology used in previous work to describe leakage is very inconsistent, we propose new terminology and nomenclature. Our goal here is to provide a nomenclature for leakage patterns that gives names that are precise, concise, unique and intuitive. We refer to any leakage pattern that reveals an item completely as an *identity pattern*, any leakage pattern that reveals whether two items are equal as an *equality pattern*, any leakage pattern that reveals the size of a set as a *size pattern* and any leakage pattern that reveals the length of an item as a *length pattern*. Let $\mathbf{T} = (\mathsf{qu} : \mathbb{D} \times \mathbb{Q} \to \mathbb{R}, \mathsf{up} : \mathbb{D} \times \mathbb{U} \to \mathbb{D})$ be a dynamic data type and consider the following leakage patterns:

- the *query equality pattern* is the function family $\mathsf{qeq} = \{\mathsf{qeq}_{k,t}\}_{k,t \in \mathbb{N}}$ with $\mathsf{qeq}_{k,t} : \mathbb{D}_k \times \mathbb{Q}_k^t \to \{0,1\}^{t \times t}$ such that $\mathsf{qeq}_{k,t}(d, q_1, \ldots, q_t) = M$, where M is a binary $t \times t$ matrix such that $M[i,j] = 1$ if $q_i = q_j$ and $M[i,j] = 0$ if $q_i \neq q_j$. The query equality pattern is referred to as the search pattern in the SSE literature;
- the *response identity pattern* is the function family $\mathsf{rid} = \{\mathsf{rid}_{k,t}\}_{k,t \in \mathbb{N}}$ with $\mathsf{rid}_{k,t} : \mathbb{D}_k \times \mathbb{Q}_k^t \to \mathbb{R}_k$ such that $\mathsf{rid}_{k,t}(d, q_1, \ldots, q_t) = \big(\mathsf{qu}(d, q_1), \cdots, \mathsf{qu}(d, q_t)\big)$. The response identity pattern is referred to as the access pattern in the SSE literature;
- the *data identity pattern* is the function family $\mathsf{did} = \{\mathsf{did}_k\}_{k \in \mathbb{N}}$ with $\mathsf{did}_k : \mathbb{D}_k \to \mathbb{D}_k$ such that $\mathsf{did}_k(d) = d$.
- the *response equality pattern* is the function family $\mathsf{req} = \{\mathsf{req}_{k,t}\}_{k,t \in \mathbb{N}}$ with $\mathsf{req}_{k,t} : \mathbb{D}_k \times \mathbb{Q}_k^t \to \{0,1\}^{t \times t}$ such that $\mathsf{req}_{k,t}(d, q_1, \ldots, q_t) = M$, where M is a binary $t \times t$ matrix such that $M[i,j] = 1$ if $\mathsf{qu}(d, q_i) = \mathsf{qu}(d, q_j)$.

Note that the patterns described above can be defined over any data type. Some leakage patterns, however, can only be defined over data types with spaces that have additional structure. As examples, consider the following patterns where we assume that the underlying type is defined over data, query and response spaces that are equipped with "length functions" $| \cdot |_{\mathbb{D}} : \mathbb{D} \to \mathbb{N}$, $| \cdot |_{\mathbb{Q}} : \mathbb{Q} \to \mathbb{N}$ and $| \cdot |_{\mathbb{R}} : \mathbb{R} \to \mathbb{N}$ (we drop the subscripts for visual clarity since the space is clear from the context):

- the *query length pattern* is the function family $\mathsf{qlen} = \{\mathsf{qlen}_{k,t}\}_{k,t \in \mathbb{N}}$ with $\mathsf{qlen}_{k,t} : \mathbb{D}_k \times \mathbb{Q}_k^t \to \mathbb{N}$ such that $\mathsf{qlen}_{k,t}(d, q_1, \ldots, q_t) = \big(|q_1|, \cdots, |q_t|\big)$;
- the *response length pattern* is the function family $\mathsf{rlen} = \{\mathsf{rlen}_{k,t}\}_{k,t \in \mathbb{N}}$ with $\mathsf{rlen}_{k,t} : \mathbb{D}_k \times \mathbb{Q}_k^t \to \mathbb{N}$ such that $\mathsf{rlen}_{k,t}(d, q_1, \ldots, q_t) = \big(|\mathsf{qu}(d, q_t)|, \cdots, |\mathsf{qu}(d, q_t)|\big)$;
- the *maximum query length pattern* is the function family $\mathsf{mqlen} = \{\mathsf{mqlen}_{k,t}\}_{k,t \in \mathbb{N}}$ with $\mathsf{mqlen}_{k,t} : \mathbb{D}_k \times \mathbb{Q}_k^t \to \mathbb{N}$ such that $\mathsf{mqlen}_{k,t}(d, q_1, \ldots, q_t) = \max_{q \in \mathbb{Q}_k} |q|$;
- the *maximum response length pattern* is the function family $\mathsf{mrlen} = \{\mathsf{mrlen}_{k,t}\}_{k,t \in \mathbb{N}}$ with $\mathsf{mrlen}_{k,t} : \mathbb{D}_k \times \mathbb{Q}_k^t \to \mathbb{N}$ such that $\mathsf{mrlen}_{k,t}(d, q_1, \ldots, q_t) = \max_{q \in \mathbb{Q}_k} |\mathsf{qu}(d, q)|$;
- the *total response length pattern* is the function family $\mathsf{trlen} = \{\mathsf{trlen}_k\}_{k \in \mathbb{N}}$ with $\mathsf{trlen}_k : \mathbb{D}_k \to \mathbb{N}$ such that $\mathsf{trlen}_k(d) = \sum_{q \in \mathbb{Q}_k} |\mathsf{qu}(d, q)|$;
- the *data size pattern* is the function family $\mathsf{dsize} = \{\mathsf{dsize}_k\}_{k \in \mathbb{N}}$ with $\mathsf{dsize}_k : \mathbb{D}_k \to \mathbb{N}$ such that $\mathsf{dsize}_k(d) = |d|$.

We say that a pattern is ZL if it depends only on the security parameter and other public parameters. Note that this does not imply that no leakage occurred but rather that whatever leakage did occur is not useful since it could have been derived solely from the public parameters. For example, the maximum query length is a ZL pattern since it can be derived from the security parameter. Given some query leakage pattern $\mathsf{patt} : \mathbb{D} \times \mathbb{Q}^t \to \mathbb{X}$, we will often abuse notation and write $\mathsf{patt}(\mathsf{DS}, q_1, \ldots, q_t)$ to mean $\mathsf{patt}(d, q_1, \ldots, q_t)$ where $d \equiv \mathsf{DS}$. Similarly, for some setup leakage pattern $\mathsf{patt} : \mathbb{D} \to \mathbb{X}$, we sometimes write $\mathsf{patt}(\mathsf{DS})$ to mean $\mathsf{patt}(d)$ where $d \equiv \mathsf{DS}$. We use the same notation for update and rebuild leakage patterns.

Leakage sub-patterns. Given a leakage pattern patt we can decompose it into sub-patterns that capture its behavior on restricted classes of query sequences. In this work, we are particularly interested in how certain schemes behave when used on non-repeating query sequences—as opposed to arbitrary sequences. We refer to this as patt's *non-repeating sub-pattern*.

Definition 5 (Non-repeating sub-patterns). *Let* $\mathbf{T} = (\mathsf{qu} : \mathbb{D} \times \mathbb{Q} \to \mathbb{R})$ *be a static data type and* $\mathsf{patt} : \mathbb{D} \times \mathbb{Q}^t \to \mathbb{X}$ *be a query leakage pattern. We say that* $\mathsf{nrp} : \mathbb{D} \times \mathbb{Q}^t \to \mathbb{X}$ *is* patt's *non-repeating sub-pattern if there exists some function* $\mathsf{other} : \mathbb{D} \times \mathbb{Q}^t \to \mathbb{X}$ *such that for all* DS *of type* $\mathbf{T}$ *and all sequences* $(q_1, \ldots, q_t) \in \mathbb{Q}^t$,

$$\mathsf{patt}(\mathsf{DS}, q_1, \ldots, q_t) = \begin{cases} \mathsf{nrp}(\mathsf{DS}, q_1, \ldots, q_t) & \text{if } q_i \neq q_j \text{ for all } i, j \in [t], \\ \mathsf{other}(\mathsf{DS}, q_1, \ldots, q_t) & \text{otherwise.} \end{cases}$$

Definition 5 can be extended to any other operation in the natural way.

Operational leakage. Each operation of an STE scheme (e.g., setup, query, update) generates some leakage which is the *direct product* of one or more leakage patterns. As an example, consider the setup and query leakage of typical static SSE schemes (e.g., [7,13]). The setup leakage is $\mathcal{L}_{\mathsf{St}} = \mathsf{trlen}$ and the query leakage is $\mathcal{L}_{\mathsf{Qr}} = (\mathsf{qeq}, \mathsf{rid}) = \mathsf{qeq} \times \mathsf{rid}$. Note that during the Ideal experiment used to formalize SSE security, the simulator will receive $\mathsf{trlen}(\mathsf{DB}) = \sum_{w \in \mathbb{W}} \#\mathsf{DB}(w)$ in order to simulate EDB and

$$\mathsf{qeq} \times \mathsf{rid}(\mathsf{DB}, w_1, \ldots, w_t) = \big(\mathsf{qeq}(\mathsf{DB}, w_1, \ldots, w_t), \mathsf{rid}(\mathsf{DB}, w_1, \ldots, w_t)\big),$$

in order to simulate the tth search token. We say that an operation is ZL if its leakage includes only ZL patterns.

Leakage profiles. A leakage profile is a collection of leakages for a set of operations. For example, the standard leakage profile for static response-revealing SSE schemes like [7,13] is

$$\Lambda_{\mathsf{RR}} = (\mathcal{L}_{\mathsf{St}}, \mathcal{L}_{\mathsf{Qr}}) = \left(\mathsf{trlen}, \left(\mathsf{qeq}, \mathsf{rid}\right)\right).$$

The response-hiding variants of these constructions, however, have leakage profile

$$\Lambda_{\mathsf{RH}} = (\mathcal{L}_{\mathsf{St}}, \mathcal{L}_{\mathsf{Qr}}) = \left(\mathsf{trlen}, \left(\mathsf{qeq}, \mathsf{rlen} \right) \right).$$

Leakage upper bounds. Another useful notion for our purposes is that of a leakage upper bound which allows us to argue that some leakage pattern reveals nothing beyond some other operational leakage.

Definition 6. *Let* patt_1 *and* patt_2 *be two query leakage patterns. We say that* patt_1 *leaks at most* patt_2 *if there exists a probabilistic polynomial time simulator* $\mathcal{S}$ *such that for all probabilistic polynomial time distinguishers* $\mathcal{D}$, *for all* $d \in \mathbb{D}$, *for all* $\mathsf{DS} \equiv d$, *for all* $t \in \mathbb{N}$, *for all sequences* $(q_1, \ldots, q_t) \in \mathbb{Q}^t$, *the following expression is negligible in* k,

$$\left| \Pr\left[\mathcal{D}\left(\mathsf{patt}_1(\mathsf{DS}, q_1, \ldots, q_t) \right) = 1 \right] - \Pr\left[\mathcal{D}\left(\mathcal{S}\left(\mathsf{patt}_2(\mathsf{DS}, q_1, \ldots, q_t) \right) \right) = 1 \right] \right|.$$

We write this as $\mathsf{patt}_1 \leq \mathsf{patt}_2$.

Similar notions can be defined for Setup, Rebuild and Update operations in the natural way.

5.2   Adaptive Semantic Security

In this Section, we extend the notion of adaptive semantic security for STE from [10,13]. Obviously, since we consider interactive Query and Update protocols, we require that the entire interaction between the adversary and the challenger be simulatable (with appropriate leakage) as opposed to just the tokens as is the case in the non-interactive definitions. Also, to capture the effect of rebuilding, the adversary is allowed to execute rebuild operations.

Definition 7 (Adaptive semantic security). *Let* $\mathsf{STE} = (\mathsf{Setup}, \mathsf{Query}_{\mathbf{C},\mathbf{S}},$ $\mathsf{Update}_{\mathbf{C},\mathbf{S}}, \mathsf{Rebuild}_{\mathbf{C},\mathbf{S}})$ *be a type-***T** *structured encryption scheme and consider the following probabilistic experiments where* $\mathcal{C}$ *is a stateful challenger,* $\mathcal{A}$ *is a stateful adversary,* $\mathcal{S}$ *is a stateful simulator,* $\Lambda = (\mathsf{patt}_{\mathsf{St}}, \mathsf{patt}_{\mathsf{Qr}}, \mathsf{patt}_{\mathsf{Up}}, \mathsf{patt}_{\mathsf{Rb}})$ *is a leakage profile,* $\lambda \geq 1$ *is a query capacity and* $z \in \{0,1\}^*$:

$\mathbf{Real}_{\mathsf{STE},\mathcal{C},\mathcal{A}}(k)$: *given* z *and* λ *the adversary* $\mathcal{A}$ *outputs a structure* DS *of type* **T** *and receives* EDS *from the challenger, where* $(K, st, \mathsf{EDS}) \leftarrow \mathsf{Setup}(1^k, \lambda, \mathsf{DS})$. $\mathcal{A}$ *then adaptively chooses a polynomial-size sequence of operations* $(\mathsf{op}_1, \ldots \mathsf{op}_m)$. *For all* $t \in [m]$ *the challenger and adversary do the following:*

1. *if* op_t *is a query operation* $q \in \mathbb{Q}$, *they execute* $\mathsf{Query}_{\mathcal{C},\mathcal{A}}\big((K, st, q), \mathsf{EDS}\big)$;
2. *if* op_t *is an update operation* $u \in \mathbb{U}$, *they execute* $\mathsf{Update}_{\mathcal{C},\mathcal{A}}\big((K, st, u), \mathsf{EDS}\big)$;
3. *if* op_t *is a rebuild operation, they execute* $\mathsf{Rebuild}_{\mathcal{C},\mathcal{A}}\big((K, st), \mathsf{EDS}\big)$.

Finally, $\mathcal{A}$ outputs a bit b that is output by the experiment.

Ideal$_{\mathsf{STE},\mathcal{A},\mathcal{S}}(k)$: *given z and λ the adversary $\mathcal{A}$ outputs a structure* DS *of type* **T**. *Given* $\mathsf{patt}_{\mathsf{St}}(\mathsf{DS})$, *the simulator returns an encrypted structure* EDS *to $\mathcal{A}$. $\mathcal{A}$ then adaptively chooses a polynomial-size sequence of operations* $(\mathsf{op}_1, \ldots, \mathsf{op}_m)$. *For all $t \in [m]$, the challenger, simulator and adversary do the following:*

1. *if* op_t *is a query operation* $q \in \mathbb{Q}$, *they execute* $\mathsf{Query}_{\mathcal{S},\mathcal{A}}\big(\mathsf{patt}_{\mathsf{Qr}}(\mathsf{DS}, q), \mathsf{EDS}\big)$;

2. *if* op_t *is an update operation* $u \in \mathbb{U}$, *they execute* $\mathsf{Update}_{\mathcal{S},\mathcal{A}}\big(\mathsf{patt}_{\mathsf{Up}}(\mathsf{DS}, u), \mathsf{EDS}\big)$;

3. *if* op_t *is a rebuild operation, they execute* $\mathsf{Rebuild}_{\mathcal{S},\mathcal{A}}\big(\mathsf{patt}_{\mathsf{Rb}}(\mathsf{DS}), \mathsf{EDS}\big)$.

Finally, $\mathcal{A}$ outputs a bit b that is output by the experiment.

We say that STE *is adaptively Λ-semantically secure if there exists a probabilistic polynomial time simulator $\mathcal{S}$ such that for all probabilistic polynomial time adversaries $\mathcal{A}$, all $\lambda \geq 1$, and all $z \in \{0,1\}^*$,*

$$\big|\Pr\big[\mathbf{Real}_{\mathsf{STE},\mathcal{A}}(k) = 1\big] - \Pr\big[\mathbf{Ideal}_{\mathsf{STE},\mathcal{A},\mathcal{S}}(k) = 1\big]\big| \leq \mathsf{negl}(k).$$

Connection to ORAM and PIR. STE captures other primitives like ORAM and PIR. In particular, the syntax and security definitions of both primitives can be recovered from Definition 7 as follows. ORAM can be viewed as an adaptively Λ_{ORAM}-secure array encryption scheme with

$$\Lambda_{\mathsf{ORAM}} = \big(\mathcal{L}_{\mathsf{St}}, \mathcal{L}_{\mathsf{Rd}}, \mathcal{L}_{\mathsf{Wr}}\big) = \big(\mathsf{dsize}, \bot, \bot\big),$$

where $\mathcal{L}_{\mathsf{St}}$, $\mathcal{L}_{\mathsf{Rd}}$ and $\mathcal{L}_{\mathsf{Wr}}$ are the setup, read and write leakages. Similarly, PIR can be viewed as an adaptively Λ_{PIR}-secure array encryption scheme where

$$\Lambda_{\mathsf{PIR}} = \big(\mathcal{L}_{\mathsf{St}}, \mathcal{L}_{\mathsf{Rd}}\big) = \big(\mathsf{did}, \bot\big).$$

where $\mathcal{L}_{\mathsf{St}}$ and $\mathcal{L}_{\mathsf{Rd}}$ are the setup and read leakages.

6   The Cache-Based Compiler

STE provides a natural way to understand the square-root solution of Goldreich and Ostrovsky [18]. More precisely, the construction consists of two components: a main memory in which the encrypted data and dummy items are stored and a cache in which items are moved after being accessed. Access to this ORAM structure requires constantly accessing the cache to look for the desired item and either retrieving a dummy item (in case the item was in the cache) or the real item from main memory (in case the item was not in the cache).

We observe that this ORAM structure can be viewed through the lens of structured encryption as follows: the main memory is an encrypted array that leaks the query equality pattern and the cache is a ZL encrypted dictionary. The access protocol can then be understood as a mechanism that leverages the ZL property of the cache to suppress the query equality leakage of the encrypted array. We now describe this view in more detail.

A structured view of the square-root solution. We assume familiarity with the square-root solution and refer the reader to [18] for a detailed exposition. Given an array RAM of N items the square-root solution produces a structure ORAM = ($\overline{\text{ERAM}}$, EDX) which consists of an encrypted array $\overline{\text{ERAM}}$ and an encrypted dictionary EDX. $\overline{\text{ERAM}}$ is an encryption of a $\sqrt{N}$-extension $\overline{\text{RAM}}$ of RAM, where $\sqrt{N}$ is the capacity with which RAM has been extended. Concretely, it consists of encryptions of the data items in RAM and of $\sqrt{N}$ dummy items all permuted at random.[7] We refer to an item's location in RAM as its virtual address and to its location in $\overline{\text{ERAM}}$ as its real address. To allow for space-efficient rebuilding, the permutation is instantiated by sorting on random tags that are associated to each item and that are generated by evaluating a PRF on the item's virtual address. To access the item with virtual address i, one executes a Read protocol which re-computes the item's random tag and performs a binary search to find it. Since the tags are deterministic the locations accessed by the binary search are also deterministic and, therefore, the Get protocol reveals the query equality (but nothing else since the labels are pseudo-random). The cache simply consists of encryptions of elements of the form $\langle i, v \rangle$, where i is the virtual address of item v. To retrieve the item with virtual address i, one executes a protocol Get which retrieves and decrypts each element of the cache and returns to the client the one with prefix i. The purpose of concatenating virtual addresses i to items v is to allow for retrievals based on virtual address as opposed to based on location in the cache. More abstractly, it instantiates a dictionary with pairs that consist of data items labeled with their virtual address. Finally, by retrieving the entire cache every time a query is made to EDX, we ensure that the Get protocol for EDX is ZL and that nothing is revealed about the query or response.

So we have two structures: $\overline{\text{ERAM}}$, which holds $N + \sqrt{N}$ items (i.e., the real items plus the dummy items) and has query leakage qeq; and EDX, which holds $\sqrt{N}$ items and has query leakage $\perp$. Clearly, accessing $\overline{\text{ERAM}}$ directly more than once leaks information so the goal is to leverage the obliviousness of EDX to *suppress* the leakage of $\overline{\text{ERAM}}$. At a high-level, Goldreich and Ostrovsky's idea is as follows. To retrieve the item at virtual address i, the client executes a Get(i) operation on EDX to check if the item is in the cache. If so, the client executes a Read(j) operation on $\overline{\text{ERAM}}$, where j is the virtual address of a dummy item, followed by a Put operation on EDX to store the dummy item in the cache. If the ith item was not in EDX, then the client executes a Read(i) operation on $\overline{\text{ERAM}}$ followed by a Put operation on EDX to store the item just retrieved from $\overline{\text{ERAM}}$. This protocol has several properties: (1) the client always retrieves the desired item; (2) for any two virtual addresses accessed, the view of the server is identically distributed; and (3) $\overline{\text{ERAM}}$ is never queried more than once on the same address. The first property guarantees correctness. The second guarantees that no partial information is revealed about the address queried. The third

---

[7] Note that after $\sqrt{N}$ queries, the entire ORAM needs to be rebuilt.

property guarantees queries cannot be linked; effectively suppressing the leakage of $\overline{\text{ERAM}}$.

Overview of the CBC. As argued above, the square-root solution can be seen as an instantiation of a more general approach that consists of using a ZL encrypted dictionary to suppress the query equality pattern of an encrypted RAM. We observe that this approach is not only applicable to encrypted RAMs (as in the case of the square-root solution) but to a larger class of encrypted structures. We formalize this by abstracting and generalizing this approach. The result is a compiler that, given a structured encryption scheme STE_{EDS} with query leakage qeq × patt and a dictionary encryption scheme STE_{EDX}, with query leakage $\bot$, yields a new structured encryption scheme STE_{SDS} with query leakage nrp, where nrp is the non-repeating sub-pattern of patt. If nrp = $\bot$, then the resulting scheme has ZL queries.

The CBC works as follows. Given a data structure DS of type **T** and a query capacity $\lambda \geq 1$, it creates a new structure SDS = $(\overline{\text{EDS}}, \text{EDX})$ which consists of: (1) an encryption $\overline{\text{EDS}}$ of a λ-extension of DS; and (2) an encrypted dictionary EDX with capacity λ. To perform a query q on SDS, the client executes a Get on the cache EDX for q. If this results in $\bot$ (i.e., there is no value in the cache with label q) the client queries the main structure $\overline{\text{EDS}}$ with q and updates EDX with the pair (q, r), where r is the result of the query. If, on the other hand, the initial EDX query resulted in a value $v \neq \bot$, the client queries the main structure $\overline{\text{EDS}}$ with an unused dummy. It then updates EDX with the pair (q, v). Rebuilding is handled by creating a new encrypted dictionary EDX and executing the Rebuild protocol of STE_{EDS}. Due to space limitations, we defer a more detailed/pseudo-code description to the full version of this work.

Correctness is easy to verify and, intuitively, one can see that EDS will not leak the query equality because it will be queried with any q at most once. There are, however, some subtleties that come up when trying to apply the CBC to structures other than encrypted RAMs. We discuss some of these challenges below.

6.1 Safe Extensions

As highlighted above, the CBC relies on the ability to query the main encrypted structure EDS on dummy values. In other words, EDS must be an encryption of an *extension* $\overline{\text{DS}}$ of the underlying structure DS. In particular, this means that the setup and query leakage of STE_{EDS} will be on the extension $\overline{\text{DS}}$ as opposed to the original structure DS. This creates some technical problems that have to be treated carefully.

Extension leakage. The first difficulty is that leakage on $\overline{\text{DS}}$ could reveal useful information about its sub-structure DS. As a concrete example, consider an array encryption scheme with the setup leakage $\mathcal{L}_{\text{St}} = \text{dsize}$ which, in this case, reveals the size of the array. Let $\lambda \geq 1$, $s = \text{dsize}(\text{RAM})$ and consider an extended array $\overline{\text{RAM}}$ with size $2 \cdot (s + \lambda)$ if the first element of the sub-array is even and size

$2 \cdot (s + \lambda) + 1$ otherwise. Clearly, the size (i.e., the setup leakage) of the extension $\overline{\mathsf{RAM}}$ reveals a bit of information about the first element of its sub-array.

Definition 8 (Safe extensions). *Let* $\Lambda = (\mathsf{patt}_{\mathsf{St}}, \mathsf{patt}_{\mathsf{Qr}}, \mathsf{patt}_{\mathsf{Rb}})$ *be a type-*$\mathbf{T}$ *leakage profile. We say that an extension* Ext *is* Λ-*safe if for all* $k \in \mathbb{N}$, *for all* $d \in \mathbb{D}_k$, *for all* $\mathsf{DS} \equiv d$, *for all* $\lambda \geq 1$, *for all* $\overline{\mathsf{DS}}$ *output by* $\mathsf{Ext}(\mathsf{DS}, \lambda)$, *for all* $t \in \mathbb{N}$, *for all* $(q_1, \ldots, q_t) \in \mathbb{Q}_k^t$, $\mathsf{patt}_{\mathsf{St}}(\overline{\mathsf{DS}}) \leq \mathsf{patt}_{\mathsf{St}}(\mathsf{DS})$, $\mathsf{patt}_{\mathsf{Qr}}(\overline{\mathsf{DS}}, q_1, \ldots, q_t) \leq \mathsf{patt}_{\mathsf{Qr}}(\mathsf{DS}, q_1, \ldots, q_t)$, *and* $\mathsf{patt}_{\mathsf{Rb}}(\overline{\mathsf{DS}}) \leq \mathsf{patt}_{\mathsf{Rb}}(\mathsf{DS})$.

6.2   Security of the Cache-Based Compiler

We are now ready to analyze the security of the CBC. In Theorem 1 below, we precisely describe the leakage of the supressed scheme as a function of the leakage of the base scheme, of the extension and of the underlying cache.

Theorem 1. *If* $\mathsf{STE}_{\mathsf{EDS}}$ *is a static and rebuildable* $(\mathsf{patt}_{\mathsf{St}}, \mathsf{qeq} \times \mathsf{patt}, \mathsf{patt}_{\mathsf{Rb}})$-*secure scheme of type* $\mathbf{T}$, *if* Ext *is an* $(\mathsf{patt}_{\mathsf{St}}, \mathsf{nrp}, \mathsf{patt}_{\mathsf{Rb}})$-*safe extension scheme, and if* $\mathsf{STE}_{\mathsf{EDX}}$ *is a* $(\mathsf{patt}'_{\mathsf{St}}, \bot, \bot)$-*secure dictionary encryption scheme, then* $\mathsf{STE}_{\mathsf{SDS}}$ *is a*

$$\left(\mathsf{patt}_{\mathsf{St}}, \mathsf{nrp}, \mathsf{patt}_{\mathsf{Rb}} \right) \text{-secure}$$

scheme of type $\mathbf{T}$, *where* nrp *is the non-repeating sub-pattern of* patt.

The proof of Theorem 1 is deferred to the full version of the paper.

7   The Rebuild Compiler

In this section, we describe a compiler that turns any dynamic STE scheme into a rebuildable *static* STE scheme. Recall that for most applications of STE, the client outsources its data to the server. The client, therefore, does not have a local copy of the data from which it can build a new encrypted structure. One possible solution is to have the client retrieve the encrypted structure, "extract" the underlying data structure and set up a new one. This naive approach, however, does not always work as there are many STE schemes that do not support a form of extraction in the sense above. This is the case, for example, for the SSE constructions of Goh [17] and the ZMF construction of Kamara and Moataz [24].[8] Another issue occurs if the client does not have enough local storage to store the encrypted structure. In such a case, the rebuild protocol has to be space-efficient for the client and, preferably, make use of only $O(1)$ space.

Overview of the RBC. There are three main challenges in making an encrypted structure rebuildable. The first is that our approach needs to be general-purpose; that is, it should work for any dynamic encrypted structure.

---

[8] Technically, this is also true for the schemes in [7,8,10,13,26] but they can be easily modified to achieve this property.

Second, the rebuild operation should be time-efficient for the server and client, and space-efficient for the client. The third is that the rebuild operation's leakage should be minimal.

At a high-level, our approach works as follows. When the client constructs an encrypted structure EDS from a plaintext structure DS, it also builds what we refer to as an "encrypted log" RAM. This log is an array that holds encryptions of all the add operations necessary to build the structure DS. The log is stored at the server with EDS. To rebuild EDS, the client will use a sorting network to randomly shuffle the encrypted log at the server. The client and server will then initialize a new (empty) encrypted structure EDS_N. The client then retrieves each ciphertext in the log, decrypts it to recover an update u and executes with the server an add operation on EDS_N for u. After processing every element of the log, EDS_N becomes the new structure. We note that our approach works for both response-hiding and response-revealing constructions.

Detailed description. Let $STE_{EDS} = (Setup, Query, Add)$ be a dynamic type-**T** STE scheme. Our compiler converts STE_{EDS} into a new *static* rebuildable scheme $RSTE_{EDS} = (Setup, Query, Rebuild)$.

Setup takes as input a static data structure DS and encrypts it using STE_{EDS}.Setup. This results in a key K_M and an encrypted structure EDS_M. It then creates an array RAM that stores encryptions of the updates needed to build DS. That is, it computes $(u_1, \ldots, u_m) := Log(DS)$ and, for all $i \in [m]$, it sets

$$RAM[i] := SKE.Enc(K_L, u_i),$$

where K_L is a symmetric key. Setup outputs $EDS = (EDS_M, RAM)$, the keys K_M and K_L for EDS_M and RAM, respectively, and state that includes the state of EDS_M and a counter cnt that will be used to keep track of the number of queries executed.

Query takes as input the secret key, the state and a query from the client, and the encrypted structure from the server. It uses the counter cnt to check if the number of queries since the last Rebuild has not exceeded λ. If so it executes the query protocol of STE_{EDS} and increments cnt. If not, it aborts.

The $Rebuild_{C,S}$ protocol takes as inputs the secret key from the client and the encrypted structure $EDS = (EDS_M, RAM)$ from the server. First, the server creates a copy RAM' of RAM. The client and server then obliviously permute RAM'. To do this, the client samples a random permutation π over $[m]$ and the client and server choose a sorting network for $[m]$ items. For each gate $g = (i, j)$ of the network, the server sends the ciphertexts $ct_i = RAM'[i]$ and $ct_j = RAM'[j]$ to the client. The client decrypts them and swaps them as follows: if $\pi(i) < \pi(j)$, then it returns the pair (ct'_i, ct'_j) otherwise it returns the pair (ct'_j, ct'_i), where ct'_i and ct'_j are re-encryptions of ct_i and ct_j under the same key K_L. The server then stores the first element of the pair at RAM'[i] and the second at RAM'[j]. At the end of this phase, RAM' holds a set of randomly permuted and re-encrypted ciphertexts. Next, the client and server initialize a new encrypted structure $((K_N, st_N), EDS_N) \leftarrow STE_{EDS}.Setup(1^k, \bot)$. The client sequentially retrieves and decrypts all the elements in RAM' and uses the result to update EDS_N. More precisely, for all retrieved ciphertexts ct_i, it computes $u_i :=$ $\mathsf{Dec}(K_L, \mathrm{ct}_i)$ and executes $(st_N, \mathsf{EDS}_N) \leftarrow \mathsf{Add}((K_N, st_N, u_i), \mathsf{EDS}_N)$. Finally, it sets the counter cnt back to 0 and sets EDS_M to be the new encrypted structure EDS_N. Due to space constraints, we provide a more detailed description in the full version of this work.

Remark on amortization and latency. The encrypted structures that result from our rebuild compiler are "amortized" in the sense that an entire Rebuild operation needs to be executed after every λ queries. We note, however, that Rebuild executions do not affect the latency of Query executions because the two operate on different structures: namely, Rebuild works on RAM and EDS_N whereas Query works on EDS_M.

Security. We prove the security of our compiler in Theorem 2 below. We give a black-box leakage analysis and later discuss specific instantiations. We show that the resulting scheme is adaptively-secure with a slightly augmented setup leakage, the same query leakage, and rebuild leakage that depends on the underlying scheme's add leakage.

Theorem 2. *If* $\mathsf{STE}_{\mathsf{EDS}}$ *is a dynamic and non-rebuildable* $(\mathsf{patt}_{\mathsf{St}}, \mathsf{patt}_{\mathsf{Qr}}, \mathsf{patt}_{\mathsf{Ad}})$-*secure scheme of type* $\mathbf{T}$, *then* $\mathsf{RSTE}_{\mathsf{EDS}}$ *is a static and rebuildable* $(\mathsf{patt}_{\mathsf{St}} \times \mathsf{lsize} \times \mathsf{mllen}, \mathsf{patt}_{\mathsf{Qr}}, \mathsf{patt}_{\mathsf{Rb}})$-*secure scheme of type* $\mathbf{T}$ *where,*

$$\mathsf{patt}_{\mathsf{Rb}}(\mathsf{DS}) = \left(\mathsf{patt}_{\mathsf{Ad}}(\mathsf{DS}, u) \right)_{u \in \mathsf{Log}(\mathsf{DS})}.$$

Due to space limitation, the proof of Theorem 2 is deferred to the full version of the paper.

Efficiency. The resulting scheme produces encrypted structures of size $O(\mathsf{S}^{\mathsf{eds}}(\mathsf{DS}) + |\mathsf{Log}(\mathsf{DS})|_w)$, where $\mathsf{S}^{\mathsf{eds}}(\mathsf{DS})$ is the space complexity of the underlying STE scheme. The query complexity is the same as the underlying scheme's. The complexity of the rebuild operation depends on the sorting network used and the amount of local storage at the client. Using Batcher's bitonic sort [4] with $O(1)$ client local storage, Rebuild has communication complexity

$$O\left(\sum_{r \in \mathbb{R}_{\mathsf{DS}}} |r|_w \cdot \log^2 \#\mathbb{Q}_{\mathsf{DS}} + \#\mathbb{Q}_{\mathsf{DS}} \cdot \max_{u \in \mathbb{U}} \mathrm{T}_{\mathsf{Ad}}^{\mathsf{eds}}(u) \right)$$

where $\mathrm{T}_{\mathsf{Ad}}^{\mathsf{eds}}(u)$ is the add complexity of $\mathsf{STE}_{\mathsf{EDS}}$ and $r = \mathsf{qu}(q)$. Note that if $\max_{u \in \mathbb{U}} \mathrm{T}_{\mathsf{Ad}}^{\mathsf{eds}}(u) = O\left(\log^2 \#\mathbb{Q}_{\mathsf{DS}} \right)$, then the rebuild communication complexity is

$$O\left(\sum_{r \in \mathbb{R}_{\mathsf{DS}}} |r|_w \cdot \log^2 \#\mathbb{Q}_{\mathsf{DS}} \right)$$

The round complexity of Rebuild is $O\big(\#\mathbb{Q}_{\mathsf{DS}} \cdot \log^2 \#\mathbb{Q}_{\mathsf{DS}} + \#\mathbb{Q}_{\mathsf{DS}} \cdot \max_{u \in \mathbb{U}} \mathrm{T}_{\mathsf{Ad}}^{\mathsf{eds}}(u)\big)$.

8 Efficiency of the Cache-Based Compiler

In this section, we give the asymptotic overhead of the constructions that result from both the CBC and ORAM simulation (when using tree-based ORAM) and provide a comparison of the two. We defer a more detailed analysis and additional comparisons (e.g., to ORAM simulation with the square-root solution and to custom oblivious data structures) to the full version of this work.

Recall that $\mathsf{STE_{SDS}}$.Query executes: (1) $\mathsf{STE_{EDS}}$.Query in order to query the main structure EDS; (2) $\mathsf{STE_{EDX}}$.Get to query the cache EDX; and (3) $\mathsf{STE_{SDS}}$.Rebuild to rebuild the cache when the counter reaches capacity λ. The *un-amortized* query complexity of the suppressed structure over a query sequence $(q_1, \ldots, q_\lambda)$ is therefore

$$T_{Qr}^{sds}(q_1, \ldots, q_\lambda) = \sum_{i=1}^{\lambda} T_{Qr}^{eds}(q_i) + \sum_{i=1}^{\lambda} T_{Qr}^{edx}(q_i) + T_{Rb}^{eds}(\lambda) + T_{Rb}^{edx}(\lambda), \qquad (1)$$

where $T_{Qr}^{sds}(q_1, \ldots, q_\lambda)$ is the query complexity of SDS, $T_{Qr}^{eds}(q_i)$ is the query complexity of EDS, $T_{Qr}^{edx}(q_i)$ is the query complexity of the cache EDX, and $T_{Rb}^{eds}(\lambda)$ and $T_{Rb}^{edx}(\lambda)$ are the rebuild complexities of EDS and EDX, respectively.

CBC with a tree-based cache. If the CBC is instantiated with tree-based cache, then we have $T_{Qr}^{edx}(q_1, \ldots, q_i) = O\left(\max_{j \in [i]} |r_j|_w \cdot \log^2 i\right)$, where $r_j = \mathsf{qu}(DS, q_j)$. Replacing the rebuild cost in Eq. (1) with the cost of the RBC, we have

$$T_{Qr}^{sds}(q_1, \ldots, q_\lambda) = \sum_{i=1}^{\lambda} T_{Qr}^{eds}(q_i) + O\left(\lambda \cdot \max_{q \in \mathbf{q}} |r|_w \cdot \log^2 \lambda\right) + O\left(\sum_{r \in \mathbb{R}_{DS}} |r|_w \cdot \log^2 \#\mathbb{Q}_{DS}\right)$$

where $\mathbf{q} = (q_1, \ldots, q_\lambda)$.

ORAM simulation with the tree-based ORAM. ORAM simulation of a structure DS using tree-based ORAM has the following complexity.

$$T_{Qr}^{tree}(q_1, \ldots, q_\lambda) = \sum_{i=1}^{\lambda} B(q_i) \cdot O\left(\log^2 \frac{|DS|_2}{B}\right) \cdot \frac{B}{w},$$

where B is the block-size of the ORAM and $B(q_i)$ denotes the number of blocks that need to be read to answer query q_i. Setting $B = \max_{r \in \mathbb{R}_{DS}} |r|_2$, we have

$$T_{Qr}^{tree}(q_1, \ldots, q_\lambda) = \sum_{i=1}^{\lambda} B(q_i) \cdot O\left(\log^2 \frac{|DS|_2}{B}\right) \cdot \max_{r \in \mathbb{R}_{DS}} |r|_w.$$

CBC vs. ORAM simulation. In the following proposition, we compare the efficiency of the CBC with the efficiency of ORAM simulation.

Proposition 1. *Let* DS *be a type-***T** *data structure and* $\mathbf{q} = (q_1, \ldots, q_\lambda)$ *be a query sequence with* $1 \leq \lambda \leq \#\mathbb{Q}_{\mathsf{DS}}$. *If*

$$\sum_{r \in \mathbb{R}_{\mathsf{DS}}} |r|_w = o\left(\sum_{i=1}^{\lambda} \mathrm{B}(q_i) \cdot \max_{r \in \mathbb{R}_{\mathsf{DS}}} |r|_w\right) \quad and \quad \lambda \cdot \max_{q \in \mathbf{q}} |\mathsf{qu}(\mathsf{DS}, q)|_w = O\left(\sum_{r \in \mathbb{R}_{\mathsf{DS}}} |r|_w\right)$$

then

$$\mathrm{T}^{\mathsf{sds}}_{\mathsf{Qr}}(q_1, \ldots, q_\lambda) = o\left(\mathrm{T}^{\mathsf{tree}}_{\mathsf{Qr}}(q_1, \ldots, q_\lambda)\right).$$

Note that for structures with constant-time queries, $\mathrm{B}(q_i) = 1$, our approach improves asymptotically over ORAM simulation whenever

$$\sum_{r \in \mathbb{R}_{\mathsf{DS}}} |r|_w = o\left(\lambda \cdot \max_{r \in \mathbb{R}_{\mathsf{DS}}} |r|_w\right).$$

However, for structures with non-constant query complexity (e.g. search trees, graphs), our approach improves over ORAM simulation whenever

$$\sum_{r \in \mathbb{R}_{\mathsf{DS}}} |r|_w = o\left(\omega(1) \cdot \lambda \cdot \max_{r \in \mathbb{R}_{\mathsf{DS}}} |r|_w\right).$$

A note on our assumptions. We note that the assumptions in Proposition 1 are natural and are satisfied in many scenarios of interest. For example, if the response lengths of DS are distributed according to a power law (a common assumption in the context of keyword search), there always exists a λ for which the first assumption holds. Furthermore, the second assumption follows whenever queries with small responses are more likely than queries with large responses. Again, this is a common assumption in keyword search where users are more likely to search for keywords contained in smaller numbers of documents than keywords that are stored in large number of documents.

9   PBS: The Piggyback Scheme

We describe a general-purpose STE scheme that reveals the query equality and response length on arbitrary query sequences, but only the total response length on sequences of distinct queries. As we will see in Sect. 10, this construction, combined with the RBC and the CBC, results in a scheme that only leaks the total response length on arbitrary sequences. The main idea behind the scheme is to trade-off query latency for leakage.[9] At a high level, our approach is to hide response lengths by ensuring the client retrieves a fixed number of words per query (a batch); no matter what the response length. To maintain correctness, incoming queries are queued and processed at the next available time. Naturally,

---

[9] This approach was first suggested in [23] but never described or analyzed formally.

this introduces a delay/latency in the querying process but by carefully tuning the batch size we can ensure that the entire response is retrieved in a reasonable amount of time. For ease of exposition, we describe a slightly simpler variant of our construction which achieves correctness under some assumptions (which we describe below).

Overview. Our scheme makes black-box use of a dynamic dictionary encryption scheme $\mathsf{STE_{EDX}} = (\mathsf{Setup}, \mathsf{Get}, \mathsf{Put})$. Given a batch size $\alpha \geq 1$ and a data structure DS, if DS $\equiv d_0$, the Setup initializes an empty encrypted dictionary EDX. Otherwise, for every query $q \in \mathbb{Q}_{DS}$, it does the following. It divides q's response r into N words $(r_1, \ldots, r_N)$ and pads it with enough $\bot$ symbols to make it a multiple of the batch size α. It then adds the pairs $((q\|1, r_1), \ldots, (q\|N + p, r_{N+p}))$ to DX, where p is the number of $\bot$ symbols. It also keeps track of q's batch size $(N + p)/\alpha$ in a dictionary DX_{st}. After processing every query in $\mathbb{Q}_{DS}$, it encrypts DX with $\mathsf{STE_{EDX}}$. The output includes a key K_D, the encrypted dictionary EDX and a state st_D. The state of scheme st includes the batch size α, a timeout parameter θ assumed to be larger than the maximum time gap between updates, the encrypted dictionary state st_D, the dictionary DX_{st} and two empty queues Q_s and Q_u. Note that the reason we pad is to guarantee the ability to retrieve α words even when the queue contains a single query. For example, if we did not pad and the first query's response consisted of less than α words, the server would clearly learn the response length of that query.

Query is a two-party protocol between the client and the server. It takes as input from the client a key K, a state st and a query q and from the server EDS = EDX. The client starts by adding the pair $(q, 0)$ to Q_s. It then peeks at Q_s to recover the pair (q', c) and retrieves α words by executing $\mathsf{STE_{EDX}}.\mathsf{Get}$ on labels $q'\|c \cdot \alpha + 1, \ldots, q'\|c \cdot \alpha + \alpha$. It uses DX_{st} to check if this was the last batch of words for q' and if so it removes (q', c) from Q_s. If not, it updates c to $c + 1$.

Add is a two-party protocol between the client and the server. It takes as input from the client a key K, a state st and an update u and from the server EDS = EDX. It checks if the last update occurred more than θ time ago. If so, it flushes Q_u by executing $\mathsf{STE_{EDS}}.\mathsf{Put}$ on all the remaining updates in Q_u and aborts. If not, it parses the update u as a pair composed of the query q and its response r. Similar to Setup, it divides q's response r into N words $(r_1, \ldots, r_N)$ and pads it with enough $\bot$ symbols to make it a multiple of the batch size α. The padded response now has length $c = (N + p)/\alpha$, where p is the number of $\bot$ symbols added. It also keeps track of q's batch size $(N + p)/\alpha$ in a dictionary DX_{st}. The client then adds the pair $((q, r), c - 1)$ to the queue Q_u. It then peeks at Q_u to recover the pair $((q', r'), c')$ and updates EDX by executing $\mathsf{STE_{EDX}}.\mathsf{Put}$ on the update sequence $(q'\|c' \cdot \alpha + 1, r'_1), \ldots, (q'\|c' \cdot \alpha + \alpha, r'_\alpha)$. It removes all r'_i from r', for $i \in [\alpha]$. Finally, if the counter c' is equal to 0, then it removes the pair (u', c) from Q_u, otherwise, it updates c' by $c' - 1$.

Note that, as described, the construction will be correct as long as: (1) the updates $u = (q, r)$ are only for new queries; and that (2) we never query on queries that are still being updated (i.e., that are still in Q_u). In the full version of this work, we show how to lift these restrictions and provide a detailed description of the scheme.

9.1 Security of PBS

In this Section, we analyze the security of PBS. Even though the scheme makes black-box use of an encrypted dictionary we find that here a black-box leakage analysis is not as informative as a concrete leakage analysis. Therefore, in Theorem 3 below we consider the security of PBS instantiated with any response-hiding dynamic encrypted dictionary that has the following leakage profile

$$\Lambda_{\mathsf{EDX}} = \big(\mathcal{L}_{\mathsf{St}}, \mathcal{L}_{\mathsf{Gt}}, \mathcal{L}_{\mathsf{Pt}}\big) = \Big(\mathsf{trlen}, \mathsf{qeq}, \perp\Big).$$

This profile can be achieved by extending well-known constructions [7,13]. We give such an example in the full version of the paper.

Setup leakage. The setup leakage of PBS is the *total batched response length* which reveals the total number of padded word batches needed to store the responses in the structure. More precisely, this is defined as $\mathsf{tbrlen} = \{\mathsf{tbrlen}_{k,\alpha}\}_{k,\alpha \in \mathbb{N}}$, where $\mathsf{tbrlen}_{k,\alpha} : \mathbb{D}_k \rightarrow \mathbb{N}$ with:

$$\mathsf{tbrlen}_{k,\alpha}(\mathsf{DS}) = \sum_{r \in \mathbb{R}_{\mathsf{DS}}} |r| + \alpha - \big(|r| \mod \alpha\big)$$

$$= \mathsf{trlen}(\mathsf{DS}) + \sum_{r \in \mathbb{R}_{\mathsf{DS}}} \alpha - \big(|r| \mod \alpha\big).$$

Query leakage. The query leakage of PBS is the *repeated query equality pattern* $\mathsf{rqeq} = \{\mathsf{rqeq}_{k,m}\}_{k,m \in \mathbb{N}}$, where $\mathsf{rqeq}_{k,m} : \mathbb{D}_k \times \mathbb{Q}_k^t$ is defined as:

$$\mathsf{rqeq}_{k,m}(\mathsf{DS}, q_1, \ldots, q_t) = \begin{cases} \perp & \text{if } t < m \text{ and } q_i \neq q_j \text{ for all } i, j \in [t], \\ \gamma_m & \text{if } t = m \text{ and } q_i \neq q_j \text{ for all } i, j \in [t], \\ \mathsf{qeq} \times \mathsf{rlen}(\mathsf{DS}, q_1, \ldots, q_t) & \text{otherwise,} \end{cases}$$

where

$$\gamma_m \overset{def}{=} \Big(\sum_{i \in [m]} |\mathsf{qu}(\mathsf{DS}, q_i)| + \alpha - \big(|\mathsf{qu}(\mathsf{DS}, q_i)| \mod \alpha\big) \Big) \cdot \alpha^{-1} - (m - 1).$$

Note that the non-repeating sub-pattern of rqeq is

$$\mathsf{nrp}_{k,m}(\mathsf{DS}, q_1, \ldots, q_t) = \begin{cases} \perp & \text{if } t < m \text{ and } q_i \neq q_j \text{ for all } i, j \in [t], \\ \gamma_m & \text{if } t = m \text{ and } q_i \neq q_j \text{ for all } i, j \in [t]. \end{cases}$$

The non-repeating sub-pattern reveals nothing except on the last query where it reveals γ_m, i.e., the total number of batches needed to finish retrieving the entire sequence. For repeated sequences, rqeq reveals the query equality and the response length patterns.

Note that, intuitively speaking, it seems that PBS leaks "less" than rqeq. Specifically, it doesn't leak qeq × rlen on every repeating sequence. Nevertheless, the scheme's leakage on non-repeating patterns is captured precisely by nrp which is ultimately what is relevant for use with the CBC.

Add leakage. The add leakage of PBS is the *add length pattern* alen $=$ $\{\mathsf{alen}_{k,m}\}_{k,m\in\mathbb{N}}$, where $\mathsf{alen}_{k,m} : \mathbb{D}_k \times \mathbb{U}_k^t$ is defined as:

$$\mathsf{alen}_{k,m}(\mathsf{DS}, u_1, \ldots, u_t) = \begin{cases} \bot & \text{if } t < m, \\ \gamma_m & \text{if } t = m, \end{cases}$$

The add length pattern reveals nothing except on the last update where it reveals γ_m, i.e., the total number of batches needed to finish the update sequence.

Theorem 3. *If* $\mathsf{STE}_{\mathsf{EDX}}$ *is* $(\mathsf{trlen}, \mathsf{qeq}, \bot)$-*secure, then* PBS *is* $(\mathsf{tbrlen}, \mathsf{rqeq}, \mathsf{alen})$-*secure.*

The proof of Theorem 3 is deferred to the full version of the paper.

9.2   Latency of PBS

We now analyze the latency of our construction; that is, how long the client has to wait until it receives the entire response for its query. For a query sequence $(q_1, \ldots, q_t)$, the client's waiting time at time t is equal to the number of queries left in the queue at that time. In the worst-case, this is

$$t \cdot \left(\frac{\max_{r \in \mathbb{R}_{\mathsf{DS}}} |r|_w}{\alpha} - 1 \right).$$

Note that if α is set to $\max_{r \in \mathbb{R}_{\mathsf{DS}}} |r|_w$, the scheme does not introduce any latency. This, of course, comes at the cost of a large amount of padding which translates to storage and communication overhead.

The above bound on the latency helps us understand the limitations of the scheme in the worst case but it does not tell us much about its latency in general. Given that, in practice, a client is very unlikely to exclusively search for queries with maximum response length, we are interested in more likely scenarios where client queries and their response lengths follow some known distributions.

The Zipf distribution. To get a more interesting bound on latency, we have to make assumptions on how queries are sampled and how the response lengths are distributed. Here, we will assume queries are sampled from a Zipf distribution $\mathcal{Z}_{n,s}$ with probability mass function $f_{n,s} : [n] \to [0, 1]$, $f_{n,s}(r) = r^{-s}/H_{n,s}$, where r is the rank of the query and $H_{n,s}$ is the harmonic number $H_{n,s} = \sum_{i=1}^n i^{-s}$. Recall that the Zipf distribution is defined over ranks so we assume an implicit ranking function that maps queries to their rank.

We also assume that the lengths of the responses are Zipf distributed by which we mean that the rth response has word length

$$\frac{r^{-s}}{H_{n,s}} \cdot \sum_{r \in \mathbb{R}_{\mathsf{DS}}} |r|_w.$$

Here again, we assume the existence of a ranking function that maps responses to their rank. From our second assumption, it follows that the set of all response lengths is

$$
L = \left\{ \frac{T}{1 \cdot H_{n,s}}, \ldots, \frac{T}{n^s \cdot H_{n,s}} \right\},
$$

where $T \stackrel{def}{=} \sum_{r \in \mathbb{R}_{\mathsf{DS}}} |r|_w$. In our analysis, we will set $s = 1$. All these assumptions are inspired from the information retrieval literature [12,42] where it is common to assume that keyword search queries are sampled from a distribution $\mathcal{Z}_{n,s}$ and that the number of documents in which keywords appear is distributed according to $\mathcal{Z}_{n,s}$. Furthermore, for English language queries and datasets, it common to set $s = 1$.

Before we can finish our analysis, we need to make a third assumption. Specifically, we have to choose a mapping between the rth ranked query and a response. Here, we will assume that high-rank queries have low-rank responses. The intuition is that, in the context of keyword search, we tend to search more often for keywords that appear less frequently in the dataset. Alternatively, we tend to search less for keywords that appear frequently in the data. In our analysis, we will refer to this as the *inverted query hypothesis*.

In the following Theorem we bound the probability that the client will retrieve all of its responses as a function of the number of *additional* query operations it executes, i.e., the number of operations beyond the minimal t.

Theorem 4. *If the rank of the client's queries is sampled i.i.d. from $\mathcal{Z}_{n,1}$ and if the lengths of the responses are distributed according to the $\mathcal{Z}_{n,1}$ distribution then, under the inverted query hypothesis, the client will retrieve all of its responses after an additional $\varepsilon \cdot t$ query operations with probability at least*

$$
1 - \exp\left(- 2t \left(\varepsilon \cdot \frac{\alpha}{\mu} \right)^2 \right),
$$

where $\mu \stackrel{def}{=} \max_{r \in \mathbb{R}_{\mathsf{DS}}} |r|_w$.

The proof of Theorem 4 is deferred to the full version of the paper.

Correctness vs. leakage. As described above, PBS achieves perfect correctness and the client will retrieve the responses for all its queries. For this, however, the client needs to perform additional queries (i.e., more than the t queries in its sequence) in order to empty its queue Q_s.

The scheme, however, can be used differently. Specifically, if the client is willing to trade correctness for lower leakage, it can stop querying after m query operations. Theorem 4 shows that after a sequence of t queries, with probability that is a function of ε, the client needs to perform an additional $\varepsilon \cdot t$ query operations to empty its queue (of course assuming the queries are sampled according to a Zipf distribution). Assuming the client sets the size of the queue to meet its requirements, if it stops querying after m query operations, the leakage profile of PBS becomes

$$
\Lambda_{\mathsf{PBS}} = \left(\mathcal{L}_{\mathsf{St}}, \mathcal{L}_{\mathsf{Qr}}, \mathcal{L}_{\mathsf{Ad}} \right) = \left(\mathsf{tbrlen}, \mathsf{rqeq}', \perp \right),
$$

where

$$\mathsf{rqeq}'_{k,m}(\mathsf{DS}, q_1, \ldots, q_t) = \begin{cases} \perp & \text{if } q_i \neq q_j \text{ for all } i, j \in [t], \\ \mathsf{qeq} \times \mathsf{rlen}(\mathsf{DS}, q_1, \ldots, q_t) & \text{otherwise.} \end{cases}$$

In this case, the scheme's non-repeating sub-pattern leakage is just $\perp$.

10   (Almost) Zero-Leakage Structured Encryption

We now describe an almost zero-leakage STE scheme, AZL, followed by a fully ZL variant we refer to as FZL. AZL results from first applying the RBC to PBS, and then applying CBC to the result. In the following, we describe the leakage profiles of the intermediate constructions that result from this process.

Applying the RBC to PBS. Let RPBS be the scheme that results from applying the RBC to PBS. It follows by Theorem 2 that the concrete leakage profile of this scheme is,

$$\Lambda_{\mathsf{RPBS}} = \left(\mathcal{L}_{\mathsf{St}}, \mathcal{L}_{\mathsf{Qr}}, \mathcal{L}_{\mathsf{Rb}}\right) = \left(\left(\mathsf{tbrlen}, \mathsf{lsize}, \mathsf{mllen}\right), \mathsf{rqeq}, \left(\mathsf{ulen}, \mathsf{lsize}, \mathsf{mllen}\right)\right),$$

where $\mathsf{lsize} = \{\mathsf{lsize}_k\}_{k\in\mathbb{N}}$ is defined as $\mathsf{lsize}_k(\mathsf{DS}) = \#\mathsf{Log}(\mathsf{DS})$, $\mathsf{mllen} = \{\mathsf{mllen}_k\}_{k\in\mathbb{N}}$ is defined as $\mathsf{mllen}_k(\mathsf{DS}) = \max_{u\in\mathsf{Log}(\mathsf{DS})} |u|$, and $\mathsf{ulen} = \{\mathsf{ulen}_{k,m}\}_{k,m\in\mathbb{N}}$ is defined as

$$\mathsf{ulen}_{k,m}(\mathsf{DS}) = \left(\mathsf{alen}_{k,m}(u)\right)_{u\in\mathsf{Log}(\mathsf{DS})}.$$

Safely extending RPBS. We now show how to safely extend RPBS so that it can be used as the base scheme of the CBC. Here, we assume that λ and α are publicly-known parameters and that all queries in the query space $\mathbb{Q}_{\mathsf{DS}}$ have the same bit length. Let $(\tilde{q}_1, \cdots, \tilde{q}_\lambda)$ be dummy queries. For all $i \in [\lambda]$, compute $\overline{\mathsf{DS}} \leftarrow \mathsf{Update}(\overline{\mathsf{DS}}, (\tilde{q}_i, \mathbf{0}))$, where $|\mathbf{0}|_w = \max_{r\in\mathbb{R}_{\mathsf{DS}}} |r|_w$.

Theorem 5. *The extension scheme described above is*

$$\left(\left(\mathsf{tbrlen}, \mathsf{lsize}, \mathsf{mllen}\right), \mathsf{nrp}, \left(\mathsf{ulen}, \mathsf{lsize}, \mathsf{mllen}\right)\right)\text{-}safe.$$

The proof of Theorem 5 is deferred to the full version of the paper.

Applying the CBC. Let AZL be the scheme that results from applying the CBC to RPBS using the extension scheme described above. It follows by Theorem 1 that the concrete leakage profile of AZL is

$$\Lambda_{\mathsf{AZL}} = \left(\mathcal{L}_{\mathsf{St}}, \mathcal{L}_{\mathsf{Qr}}, \mathcal{L}_{\mathsf{Rb}}\right) = \left(\left(\mathsf{tbrlen}, \mathsf{lsize}, \mathsf{mllen}\right), \mathsf{nrp}, \left(\mathsf{ulen}, \mathsf{lsize}, \mathsf{mllen}\right)\right),$$

where

$$\mathsf{nrp}(\mathsf{DS}, q_1, \ldots, q_t) = \begin{cases} \bot & \text{if } t < m \text{ and } q_i \neq q_j \text{ for all } i, j \in [t], \\ \gamma_\lambda & \text{if } t = \lambda \text{ and } q_i \neq q_j \text{ for all } i, j \in [t]. \end{cases}$$

Note that the setup leakage of the cache is mllen which is already included in the setup leakage of RPBS.

Efficiency. AZL has query complexity

$$T_{\mathsf{Qr}}^{\mathsf{SDS}}(q_1, \ldots, q_\lambda) = \sum_{i=1}^{\lambda} T_{\mathsf{Qr}}^{\mathsf{EDS}}(q_i) + O\left(\lambda \cdot \max_{q \in \mathbf{q}} |r|_w \cdot \log^2 \lambda\right) + O\left(\sum_{r \in \mathbb{R}_{\mathsf{DS}}} |r|_w \cdot \log^2 \#\mathbb{Q}_{\mathsf{DS}}\right),$$

and storage complexity

$$O\left(\lambda \cdot (\alpha + \max_{u \in \mathsf{Log}(\mathsf{DS})} |u|_w) + \#\mathbb{Q}_{\mathsf{DS}} \cdot (\alpha + \max_{r \in \mathbb{R}_{\mathsf{DS}}} |r|_w) + (\lambda + \#\mathbb{Q}_{\mathsf{DS}}) \cdot \max_{u \in \mathsf{Log}(\mathsf{DS})} |u|_w\right).$$

If $\max_{u \in \mathsf{Log}(\mathsf{DS})} |u|_w = O(\max_{r \in \mathbb{R}_{\mathsf{DS}}} |r|_w)$, the storage overhead simplifies to

$$O\left((\lambda + \#\mathbb{Q}_{\mathsf{DS}}) \cdot (\alpha + \max_{r \in \mathbb{R}_{\mathsf{DS}}} |r|_w)\right).$$

Achieving zero-leakage. As discussed in Sect. 9, the non-repeating sub-pattern leakage of PBS is $\bot$ if we are willing to tolerate probabilistic correctness. In such a case, applying the RBC and then the CBC results in a scheme FZL with query leakage,

$$\Lambda_{\mathsf{FZL}} = (\mathcal{L}_{\mathsf{St}}, \mathcal{L}_{\mathsf{Qr}}, \mathcal{L}_{\mathsf{Rb}}) = \left(\left(\mathsf{tbrlen}, \mathsf{lsize}, \mathsf{mllen}\right), \bot, \left(\mathsf{ulen}, \mathsf{lsize}, \mathsf{mllen}\right)\right).$$

The efficiency of FZL is the same as AZL.

An improved extension for probabilistic correctness. We briefly note that under probabilistic correctness, we can extend RPBS more efficiently than described above. The extension works as follows. Let $(\tilde{q}_1, \ldots, \tilde{q}_\lambda)$ be dummy queries. For all $i \in [\lambda]$, compute $\overline{\mathsf{DS}} \leftarrow \mathsf{Update}(\overline{\mathsf{DS}}, (\tilde{q}_i, \mathbf{0}))$, where $|\mathbf{0}|_w = \alpha$. Note that the setup and rebuild leakage of this variant are the same as those considered in Theorem 5 so they can be simulated exactly as in the proof of that Theorem. The non-repeating query sub-pattern is $\mathsf{nrp}(\overline{\mathsf{DS}}, q_1, \ldots, q_t) = \mathsf{nrp}(\mathsf{DS}, q_1, \ldots, q_t) = \bot$ which can be simulated trivially.

11 Conclusions and Future Directions

In this work, we introduced a new framework to cope with leakage based on compilers and transformations that suppress the leakage of STE schemes. Our work motivates several interesting directions for future work. The most immediate is the design of a query equality suppression framework for dynamic STE schemes. Another interesting challenge is to design compilers with lower computational overhead. Here, trying to improve the cost of our rebuild compiler—even for restricted classes of encrypted structures—might be a good start. As far as we know, our PBS construction is the first STE scheme to hide the response length without naive padding (i.e., padding to the maximum response length). To achieve this, we used queuing techniques which introduce a delay in the query process. Can the latency of PBS be improved? Can response lengths be suppressed without introducing delays at all? Finally, in this work we focused on suppressing query equality and response length leakage but an important direction for future work is to find suppression techniques and frameworks for other common leakage patterns.

Acknowledgments. We are grateful to Hajar Alturki for useful feedback on the PBS construction and to the anonymous reviewers for helpful suggestions.

References

1. Ajtai, M., Komlós, J., Szemerédi, E.: An o(n log n) sorting network. In: ACM Symposium on Theory of Computing (STOC 1983), pp. 1–9 (1983)
2. Amjad, G., Kamara, S., Moataz, T.: Breach-resistant structured encryption. IACR Cryptology ePrint Archive 2018:195 (2018)
3. Asharov, G., Naor, M., Segev, G., Shahaf, I.: Searchable symmetric encryption: optimal locality in linear space via two-dimensional balanced allocations. In: STOC 2016, pp. 1101–1114. ACM, New York (2016)
4. Batcher, K.: Sorting networks and their applications. In: Proceedings of the Joint Computer Conference, pp. 307–314 (1968)
5. Bost, R.: Sophos - forward secure searchable encryption. In: ACM CCS 2016 (2016)
6. Cash, D., Grubbs, P., Perry, J., Ristenpart, T.: Leakage-abuse attacks against searchable encryption. In: ACM CCS 2015, pp. 668–679. ACM (2015)
7. Cash, D., Jaeger, J., Jarecki, S., Jutla, C., Krawczyk, H., Rosu, M., Steiner, M.: Dynamic searchable encryption in very-large databases: data structures and implementation. In: NDSS 2014 (2014)
8. Cash, D., Jarecki, S., Jutla, C., Krawczyk, H., Roşu, M.-C., Steiner, M.: Highly-scalable searchable symmetric encryption with support for Boolean queries. In: Canetti, R., Garay, J.A. (eds.) CRYPTO 2013. LNCS, vol. 8042, pp. 353–373. Springer, Heidelberg (2013). https://doi.org/10.1007/978-3-642-40041-4_20
9. Cash, D., Tessaro, S.: The locality of searchable symmetric encryption. In: Nguyen, P.Q., Oswald, E. (eds.) EUROCRYPT 2014. LNCS, vol. 8441, pp. 351–368. Springer, Heidelberg (2014). https://doi.org/10.1007/978-3-642-55220-5_20
10. Chase, M., Kamara, S.: Structured encryption and controlled disclosure. In: Abe, M. (ed.) ASIACRYPT 2010. LNCS, vol. 6477, pp. 577–594. Springer, Heidelberg (2010). https://doi.org/10.1007/978-3-642-17373-8_33
11. Chase, M., Kamara, S.: Structured encryption and controlled disclosure. Technical report 2011/010.pdf, IACR Cryptology ePrint Archive (2010)
12. Chaudhuri, S., Church, K.W., König, A.C., Sui, L.: Heavy-tailed distributions and multi-keyword queries. In: ACM SIGIR 2007 (2007)
13. Curtmola, R., Garay, J., Kamara, S., Ostrovsky, R.: Searchable symmetric encryption: improved definitions and efficient constructions. In: CCS 2006 (2006)

14. Demertzis, I., Papamanthou, C.: Fast searchable encryption with tunable locality. In: SIGMOD 2017 (2017)
15. Fisch, B.A., et al.: Malicious-client security in blind seer: a scalable private DBMS. In: IEEE Symposium on Security and Privacy, pp. 395–410. IEEE (2015)
16. Garg, S., Mohassel, P., Papamanthou, C.: TWORAM: efficient oblivious RAM in two rounds with applications to searchable encryption. In: Robshaw, M., Katz, J. (eds.) CRYPTO 2016. LNCS, vol. 9816, pp. 563–592. Springer, Heidelberg (2016). https://doi.org/10.1007/978-3-662-53015-3_20
17. Goh, E.-J.: Secure indexes. Technical report 2003/216, IACR ePrint Cryptography Archive (2003). http://eprint.iacr.org/2003/216
18. Goldreich, O., Ostrovsky, R.: Software protection and simulation on oblivious RAMs. J. ACM **43**(3), 431–473 (1996)
19. Goodrich, M., Mitzenmacher, M., Ohrimenko, O., Tamassia, R.: Oblivious RAM simulation with efficient worst-case access overhead. In: CCSW 2011 (2011)
20. Islam, M.S., Kuzu, M., Kantarcioglu, M.: Access pattern disclosure on searchable encryption: ramification, attack and mitigation. In: NDSS 2012 (2012)
21. Jarecki, S., Jutla, C., Krawczyk, H., Rosu, M., Steiner, M.: Outsourced symmetric private information retrieval. In: ACM CCS 2013 (2013)
22. Kamara, S.: Restructuring the NSA metadata program. In: Böhme, R., Brenner, M., Moore, T., Smith, M. (eds.) FC 2014. LNCS, vol. 8438, pp. 235–247. Springer, Heidelberg (2014). https://doi.org/10.1007/978-3-662-44774-1_19
23. Kamara, S., Moataz, T.: SQL on structurally-encrypted databases. IACR Cryptology ePrint Archive 2016, 453 (2016)
24. Kamara, S., Moataz, T.: Boolean searchable symmetric encryption with worst-case sub-linear complexity. In: Coron, J.-S., Nielsen, J.B. (eds.) EUROCRYPT 2017. LNCS, vol. 10212, pp. 94–124. Springer, Cham (2017). https://doi.org/10.1007/978-3-319-56617-7_4
25. Kamara, S., Papamanthou, C.: Parallel and dynamic searchable symmetric encryption. In: Sadeghi, A.-R. (ed.) FC 2013. LNCS, vol. 7859, pp. 258–274. Springer, Heidelberg (2013). https://doi.org/10.1007/978-3-642-39884-1_22
26. Kamara, S., Papamanthou, C., Roeder, T.: Dynamic searchable symmetric encryption. In: ACM CCS 2012 (2012)
27. Kushilevitz, E., Lu, S., Ostrovsky, R.: On the (in)security of hash-based oblivious RAM and a new balancing scheme. In: SODA 2012 (2012)
28. Liu, C., Zhu, L., Wang, M., Tan, Y.: Search pattern leakage in searchable encryption: attacks and new construction. Inf. Sci. **265**, 176–188 (2014)
29. Meng, X., Kamara, S., Nissim, K., Kollios, G.: GRECS: graph encryption for approximate shortest distance queries. In: CCS 2015 (2015)
30. Miers, I., Mohassel, P.: IO-DSSE: scaling dynamic searchable encryption to millions of indexes by improving locality. Cryptology ePrint Archive, Report 2016/830 (2016). http://eprint.iacr.org/2016/830
31. Naveed, M., Prabhakaran, M., Gunter, C.: Dynamic searchable encryption via blind storage. In: IEEE Symposium on Security and Privacy (S&P 2014) (2014)
32. Ostrovsky, R., Shoup, V.: Private information storage. In: ACM Symposium on Theory of Computing (STOC 1997), pp. 294–303 (1997)
33. Pappas, V., et al.: Blind seer: a scalable private DBMS. In: 2014 IEEE Symposium on Security and Privacy (SP), pp. 359–374. IEEE (2014)
34. Sedghi, S., van Liesdonk, P., Doumen, J.M., Hartel, P.H., Jonker, W.: Adaptively secure computationally efficient searchable symmetric encryption. Technical report TR-CTIT-09-13 (2009)

35. Shi, E., Chan, T.-H.H., Stefanov, E., Li, M.: Oblivious RAM with $O((\log N)^3)$ worst-case cost. In: Lee, D.H., Wang, X. (eds.) ASIACRYPT 2011. LNCS, vol. 7073, pp. 197–214. Springer, Heidelberg (2011). https://doi.org/10.1007/978-3-642-25385-0_11
36. Song, D., Wagner, D., Perrig, A.: Practical techniques for searching on encrypted data. In: IEEE S&P, pp. 44–55. IEEE Computer Society (2000)
37. Stefanov, E., Papamanthou, C., Shi, E.: Practical dynamic searchable encryption with small leakage. In: NDSS 2014 (2014)
38. Stefanov, E., et al.: Path ORAM: an extremely simple oblivious RAM protocol. In: CCS 2013 (2013)
39. Williams, P., Sion, R., Carbunar, B.: Building castles out of mud: practical access pattern privacy and correctness on untrusted storage. In: CCS 2008 (2008)
40. Zhang, Y., Katz, J., Papamanthou, C.: All your queries are belong to us: the power of file-injection attacks on searchable encryption. In: USENIX 2016 (2016)
41. Zhang, Y., O'Neill, A., Sherr, M., Zhou, W.: Privacy-preserving network provenance. Proc. VLDB Endow. **10**(11), 1550–1561 (2017)
42. Zipf, G.K.: The psycho-biology of language (1935)

Searchable Encryption with Optimal Locality: Achieving Sublogarithmic Read Efficiency

Ioannis Demertzis[1(✉)], Dimitrios Papadopoulos[2],
and Charalampos Papamanthou[1]

[1] University of Maryland, College Park, USA
{yannis,cpap}@umd.edu
[2] Hong Kong University of Science and Technology, Kowloon, Hong Kong
dipapado@cse.ust.hk

Abstract. We propose the first linear-space searchable encryption scheme with constant locality and *sublogarithmic* read efficiency, strictly improving the previously best known read efficiency bound (Asharov et al., STOC 2016) from $\Theta(\log N \log \log N)$ to $O(\log^\gamma N)$ where $\gamma = \frac{2}{3} + \delta$ for any fixed $\delta > 0$ and where N is the number of keyword-document pairs. Our scheme employs four different allocation algorithms for storing the keyword lists, depending on the size of the list considered each time. For our construction we develop (i) new probability bounds for the offline two-choice allocation problem; (ii) and a new I/O-efficient oblivious RAM with $\tilde{O}(n^{1/3})$ bandwidth overhead and zero failure probability, both of which can be of independent interest.

1 Introduction

Searchable Encryption (SE), first proposed by Song et al. [30] and then formalized by Curtmola et al. [12], enables a data owner to outsource a private dataset $\mathcal{D}$ to a server, so that the latter can answer keyword queries without learning too much information about the underlying dataset and the posed queries. An alternative to expensive primitives such as oblivious RAM and fully homomorphic encryption, SE schemes are practical at the expense of formally-specified leakage. In typical SE schemes, the data owner prepares a private index which is sent to the server. To perform a query on keyword w, the data owner engages in a protocol with the server such that by the end of the protocol the data owner retrieves the list of document identifiers $\mathcal{D}(w)$ of documents containing w. During this process, the server should learn nothing except for the (number of) retrieved document identifiers—referred to as (size of) *access pattern*—and whether the keyword search query w was repeated in the past or not—referred to as *search pattern*.

To retrieve the document identifiers $\mathcal{D}(w)$ (also referred to as *keyword list* in the rest of the paper), most existing SE schemes require the server access approximately $|\mathcal{D}(w)|$ randomly-assigned memory locations [10,12,14,23,24,30, 31]. While this random allocation is essential for security, it creates a big

© International Association for Cryptologic Research 2018
H. Shacham and A. Boldyreva (Eds.): CRYPTO 2018, LNCS 10991, pp. 371–406, 2018.
https://doi.org/10.1007/978-3-319-96884-1_13

bottleneck when accessing large indexes stored on disk.[1] Therefore the afore-mentioned schemes cannot scale for big data that do not fit in memory due to poor *locality*—the number of non-contiguous memory locations that must be read to retrieve the result.

Locality and Read Efficiency Trade-offs. One trivial way to design an SE scheme that has optimal locality $L = 1$ is to have the client download the whole encrypted index for every query w. Unfortunately, such an approach requires $O(N)$ bandwidth, where N is the total number of keyword-document pairs. Cash et al. [10] were the first to observe this trade-off: To improve the locality of SE, one should expect to read additional entries per query. The ratio of the total number of entries read over the size of the query result was defined as *read effi-ciency*. This trade-off was subsequently formalized by Cash and Tessaro [11] who showed it is impossible[2] to construct an SE scheme with linear space, optimal locality and optimal read efficiency.

In response to this impossibility result, several positive results with vari-ous trade-offs have appeared. Cash and Tessaro [11] presented a scheme with $\Theta(N \log N)$ space, $O(1)$ read efficiency and $O(\log N)$ locality, which was later improved to $O(1)$ by Asharov et al. [6]. Demertzis and Papamanthou [16] pre-sented a scheme with bounded locality $O(N^\epsilon)$, $O(1)$ read efficiency and linear space (for constant $\epsilon < 1$). More recently, Asharov et al. [5] studied the locality in Oblivious RAMs, proposing a construction that, for an arbitrary sequence of accesses (and therefore for SE as well), achieves $O(N)$ space, $O(\log^2 N \log^2 \log N)$ read efficiency, and $O(\log N \log^2 \log N)$ locality. While asymptotically worse than [6], this work has better security as it leaks no access pattern. Finally, significant speedups due to locality in SE implementations have been observed by Miers and Mohassel [25] and Demertzis et al. [15, 16].

Constant Locality with Linear Space. Practical reasons described above have motivated the study of even more asymptotically-efficient SE schemes, and in particular those with *constant locality* and *linear space*. Asharov et al. [6] presented two such SE schemes based on a two-dimensional generalization of the "balls and bins" problem. In particular, the first scheme (A1) uses two-choice allocation, has very low read efficiency $\Theta(\log \log N \log^2 \log \log N)$ but is based on the assumption that all lists $\mathcal{D}(w)$ have size $\leq N^{1-1/\log \log N}$.[3] Recently, Asharov et al. [7] provided a version of A1 with improved read efficiency and a

---

[1] Demertzis and Papamanthou [16] recently showed that low-locality SE may improve practical performance for in-memory data too, due to reduced number of server crypto operations.

[2] The result holds for a setting where lists $\mathcal{D}(w)$ are stored at non-overlapping posi-tions.

[3] We tested this assumption for 4 real datasets: One containing crime records in Chicago since 2001 [1], the Enron email dataset [2], the USPS dataset [4] and the TPC-H dataset [3]. The Enron email dataset does not violate the assumption, which is not the case for the other datasets where almost half of the contained attributes violate it. For the crimes dataset, for example, the assumption was violated in 12 out of 21 attributes for 31% of the keywords on average.

better bound $N/\log^3 N$ for the maximum $\mathcal{D}(w)$ size. The second construction of [6] (A2) has $\tilde{\Theta}(\log N)$ read efficiency, uses one-choice allocation and makes no assumptions about the dataset. To our knowledge, A2 is the best SE scheme with $O(1)$ locality and $O(N)$ space known to-date for *general datasets*.

Our Contribution. Motivated by the above positive results and the impossibility result of [11], we ask whether it is possible to build an SE scheme for general datasets with: (i) *linear space*, (ii) *constant locality*, and (iii) *sublogarithmic read efficiency*. We answer this question in the affirmative by designing the first such SE scheme, strictly improving upon the best known scheme A2 [6]. For the rest of the paper we set $\gamma = 2/3 + \delta$ for $\delta > 0$ arbitrarily small. We show that the read efficiency of our scheme is $O(\log^\gamma N)$ as opposed to that of A2 which is $\Theta(\log N \log \log N)$. Parameter δ above affects the constants in the asymptotic notation which grow with $O(1/\delta)$.

1.1   Summary of Our Techniques

Our techniques (like previous works on low-locality SE) use the notion of an *allocation algorithm*, whose goal is to store the dataset's keyword lists in memory such that each keyword list $\mathcal{D}(w)$ can be efficiently retrieved by accessing memory locations *independent* of the distribution the SE dataset—this is needed for security reasons. Common techniques to achieve this, store keyword lists using a balls-and-bins procedure [6].

Starting Point. We first observe that keyword lists of size less than $N^{1-1/\log^{1-\gamma} N}$, for some $\gamma < 1$, can be allocated using (as a black box) the parameterized version of scheme A1 of Asharov et al. [6]. In particular, we show in Theorem 6 that for $\gamma = 2/3 + \delta$ scheme A1 yields $\Theta(\log^\gamma N)$ read efficiency, as desired. Therefore we only need to focus on allocating the dataset's keyword lists that have size $> N^{1-1/\log^{1-\gamma} N}$.

Our Main Technique: Different Allocation Algorithms for Different Size Ranges. Let $\gamma = 2/3 + \delta$ as defined above. We develop three allocation algorithms for the remaining ranges: Lists with size in $(N^{1-1/\log^{1-\gamma} N}, N/\log^2 N]$, also called *medium*, are allocated using an *offline two-choice allocation* procedure [28] and multiple stashes to handle overflows. Lists with size in $(N/\log^2 N, N/\log^\gamma N]$, also called *large*, are first split into further subranges based on their size and then each subrange is allocated into a separate array using the same algorithm. Finally, for lists with size in $(N/\log^\gamma N, N]$, also called *huge*, there is no special allocation algorithm: We just read the whole dataset. We now provide a summary of our allocation algorithms for medium and large lists.

1.2   Medium Keyword Lists Allocation

Our allocation algorithm for medium keyword lists is using an offline two-choice allocation (OTA),[4] where there are m balls and n bins and for each ball two possible bins are chosen independently and uniformly at random. After *all choices* have been made, one can run a maximum flow algorithm to find the final assignment of balls to bins such that the maximum load is minimized. This strategy yields an almost perfectly balanced allocation (where max-load $\leq \lceil m/n \rceil + 1$) with probability at least $1 - O(1/n)$ [28].

Central Idea: One OTA Per Size and Then Merge. We use one OTA separately for every size s that falls in the range $(N^{1-1/\log^{1-\gamma} N}, N/\log^2 N]$ as follows: Let $\mathbf{A}_s$ be an array of M buckets $\mathbf{A}_s[1], \mathbf{A}_s[2], \ldots, \mathbf{A}_s[M]$, for some appropriately chosen M. One can visualize a bucket $\mathbf{A}_s[i]$ as a vertical structure of unbounded capacity. Let k_s be the number of keyword lists of size s and let $b_s = M/s$ be the number of superbuckets in $\mathbf{A}_s$, where a superbucket is a collection of s consecutive buckets in $\mathbf{A}_s$. We perform an OTA of k_s keyword lists to the b_s superbuckets. From [28], there will be at most $\lceil k_s/b_s \rceil + 1$ lists of size s in each superbucket with probability at least $1 - O(1/b_s)$, meaning the load of each bucket due to lists of size s will be at most $\lceil k_s/b_s \rceil + 1$ with the same probability, given there are s buckets in a superbucket.

Our final allocation merges arrays $\mathbf{A}_s$ for all sizes s corresponding to medium keyword lists into a new array $\mathbf{A}$ of M buckets—see Fig. 5. To bound the final load of each bucket $\mathbf{A}[i]$ in the merged array $\mathbf{A}$ one can compute $\sum_s (\lceil k_s/b_s \rceil + 1)$ which is $O(N/M + \log^\gamma N)$—see Lemma 4. If we set $M = N/\log^\gamma N$, our allocation occupies linear space and each bucket $\mathbf{A}[i]$ has load $O(\log^\gamma N)$—thus to read one list, one reads the two superbuckets initially picked by the OTA yielding read efficiency $O(\log^\gamma N)$.

Handling Bucket Overflows with Additional Stashes. Our analysis above assumes the maximum load of each bucket is at most $\lceil k_s/b_s \rceil + 1$. However, there is a noticeable probability $O(1/b_s)$ of overflowing beyond this bound—this will cause our allocation to fail, leaking information about the dataset. To deal with this problem, for each size s, we place the lists of size s that overflow in a stash $\mathbf{B}_s$ (at the server) that can store up to $O(\log^2 N)$ such overflowing lists. In particular, we prove that when the OTA described previously is performed for medium lists, at most $O(\log^2 N)$ lists of size s overflow with non-negligible probability and thus our stashes $\mathbf{B}_s$ suffice, see Lemma 5. We also stress that we need the condition $s \leq N/\log^2 N$ to keep the space of the stashes linear—see Theorem 7, justifying the pick of $N/\log^2 N$ as endpoint of the range where we apply OTA. Finally, the existence of stashes $\mathbf{B}_s$ differentiates our allocation from those of [6], allowing us to avoid their impossibility result (see discussion in Sect. 7).

---

[4] Deriving the results of this paper using the, more lightweight, online version of the problem is an interesting open problem. Section 7 elaborates on the difficulties that arise in that case.

New Probability Bounds for OTA. Our proof for the $O(\log^2 N)$ stash size extends the analysis of [28] non-trivially—we prove two new results in Sect. 3: First, in Theorem 1 we show that in an OTA, the probability $> \tau$ bins overflow decreases with $(1/\tau)^\tau$. For this proof we show the 0/1 random variables indicating bin overflow are *negatively associated* [17]. Second, in Theorem 2 we show the probability an OTA of m balls to n bins yields a maximum load of $> \lceil m/n \rceil + \tau$ is $\leq O(1/n)^\tau + exp(-n)$.

Accessing Stashes Obliviously. Because keyword lists of size s *might* now live in the stash $\mathbf{B}_s$, retrieving a keyword list $\mathcal{D}(w)$ is a two-step process: First, access the superbuckets that were initially assigned by the OTA and then access a position x in the stash. In case $\mathcal{D}(w)$ is not in the stash (because it was not an overflowing list), x should be still assigned a stash position chosen from the unoccupied ones, if such a position exists. If not, there will be a collision, in which case the adversary can deduce information about the dataset distribution, e.g., that the dataset contains at least $\log^2 N$ lists of size $|\mathcal{D}(w)|$. To avoid such leakage, the stash must be accessed obliviously.

New ORAM with $o(\sqrt{n})$ Bandwidth, $O(1)$ Locality & Zero Failure Probability. Since the stash has only $\log^2 N$ entries of size $|\mathcal{D}(w)|$ each, one can access it obliviously by reading it all. But this increases read efficiency to $\log^2 N$, which is no longer sublogarithmic. Thus, we need an ORAM with (i) $O(1)$ locality, (ii) $o(\sqrt{n})$ bandwidth and (iii) zero failure probability since it will be applied on only $\log^2 N$ indices. In Sect. 4, we devise a new ORAM satisfying the above (with $O(n^{1/3}\log^2 n)$ bandwidth) based on one recursive application of Goldreich's and Ostrovsky's square-root ORAM [18]. This protocol can be of independent interest. To finally ensure our new ORAM has $O(1)$ locality, we use I/O-efficient oblivious sorting by Goodrich and Mitzenmacher [20].

1.3 Large Keyword Lists Allocation

We develop an Algorithm AllocateLarge(min, max) that can allocate lists with sizes in a general range (min, max]. We will be applying this algorithm for lists in the range $(N/\log^2 N, N/\log^\gamma N]$. The algorithm works as follows. Let $\mathbf{A}$ be an array that has $2N$ entries, organized in $N/$max buckets of capacity 2max each. To store a list of size $s \in$ (min, max], a bucket with available size at least s is chosen. To retrieve a list, the entire bucket where the list is stored is accessed using our ORAM construction from Sect. 4—note that ORAM is relatively cheap for this range, since $N/$max is small.

In this way we always pay the cost of accessing lists of size max, even for smaller list sizes $s >$ min. The read efficiency of this approach is clearly at least max/min, which for the specified range above is $\log^2 N/\log^\gamma N = \omega(\log N)$ for $\gamma < 1$. Still, this is not enough for our target, which is sublogarithmic read efficiency. Therefore, we need to further split this range into multiple subranges and apply the algorithm for each subrange independently. The number of subranges depends on the target read efficiency, i.e., it depends on γ (but not on N). For example, for $\gamma < 1$ it suffices to have 3 subranges, whereas setting $\gamma = 0.75$ would require splitting $(N/\log^2 N, N/\log^\gamma N]$ into a fixed number of 11 subranges. In general, as $\delta > 0$ decreases and $\gamma = 2/3 + \delta$ gets closer to $2/3$ the number of subranges will increase. We note that using an ORAM of better worst-case bandwidth (e.g., $O(\log^{1/5} \log^2 N)$ instead of $O(\log^{1/3} \log^2 N)$) would reduce the necessary number of subranges (see discussion in Sect. 7).

2   Notation and Definitions

We use the notation $\langle C', S' \rangle \leftrightarrow \Pi \langle C, S \rangle$ to indicate that protocol Π is executed between a client with input C and a server with input S. After the execution of the protocol the client receives C' and the server receives S'. Server operations are in light gray background. All other operations are performed by the client. The client typically interacts with the server via an `Encrypt-And-Write` *data* operation, with which the client encrypts *data* locally with a CPA-secure encryption scheme and writes the encrypted data *data* remotely to server and via a `Read-And-Decrypt` *data* operation, with which the client reads encrypted data *data* from server and decrypts them locally.

In the following, $\mathcal{D}$ will denote the searchable encryption dataset (SE dataset) which is a set of keywords lists $\mathcal{D}(w_i)$. Each keyword list $\mathcal{D}(w_i)$ is a set of keyword-document pairs (w_i, id), called *elements*, where id is the document identifier containing keyword w_i. We denote with N the size of our dataset, i.e., $N = \sum_{w \in \mathbf{W}} |\mathcal{D}(w)|$, where $\mathbf{W}$ is the set of unique keywords of our dataset $\mathcal{D}$. Without loss of generality, we will assume that all keyword lists $\mathcal{D}(w_i)$ have size $|\mathcal{D}(w_i)|$ that is a power of two. This can always be enforced by padding with dummy elements, and will only increase the space at most by a factor of 2. Finally, a function $f(\kappa)$ is *negligible*, denoted $\mathsf{neg}(\kappa)$, if for sufficiently large κ it is less than $1/p(\kappa)$, for all polynomials $p(\kappa)$.

2.1   Searchable Encryption

Our new SE scheme uses a modification of the square-root ORAM protocol as a black box, which is a two-round protocol. Therefore to model our SE scheme we use the protocol-based definition (SETUP, SEARCH) as proposed by Stefanov et al. [31].

- $\langle st, \mathcal{I} \rangle \leftrightarrow \text{SETUP}\langle (1^\kappa, \mathcal{D}), 1^\kappa \rangle$: SETUP takes as input security parameter κ and SE dataset $\mathcal{D}$ and outputs secret state st (for client), and encrypted index $\mathcal{I}$ (for server).

- $\langle (\mathcal{D}(w), st'), \mathcal{I}' \rangle \leftrightarrow \text{SEARCH}\langle (st, w), \mathcal{I} \rangle$: SEARCH is a protocol between client and server, where the client's input is secret state st and keyword w. Server's input is encrypted index $\mathcal{I}$. Client's output is set of document identifiers $\mathcal{D}(w)$ matching w and updated secret state st' and server's output is updated encrypted index $\mathcal{I}'$.

Just like in previous works [6], the goal of our SE protocols is for the client to retrieve the document identifiers (i.e., the list $\mathcal{D}(w)$) for a specific keyword w. The document themselves can be downloaded from the server in a second

---

$bit \leftarrow \mathbf{Real}^{\mathsf{SE}}(\kappa)$:

1: $D_0 \leftarrow \mathsf{Adv}(1^\kappa)$; $\langle st_0, \mathcal{I}_0 \rangle \leftrightarrow \textsc{Setup}\langle(1^\kappa, \mathcal{D}_0), 1^\kappa\rangle$;
2: **for** $k = 1$ to q **do**
3:     $w_k \leftarrow \mathsf{Adv}(1^k, \mathcal{I}_0, M_1, \ldots, M_{k-1})$;
4:     $\langle(\mathcal{D}(w_k), st_k), \mathcal{I}_k\rangle \leftrightarrow \textsc{Search}\langle(st_{k-1}, w_k), \mathcal{I}_{k-1}\rangle$;
5:     Let M_k be the messages from client to server in the Search protocol above;
6: $bit \leftarrow \mathsf{Adv}(1^k, \mathcal{I}_0, M_1, M_2, \ldots, M_q)$;
7: **return** bit;

$bit \leftarrow \mathbf{Ideal}^{\mathsf{SE}}_{\mathcal{L}_1, \mathcal{L}_2}(\kappa)$:

1: $D_0 \leftarrow \mathsf{Adv}(1^\kappa)$; $(st_{\mathcal{S}}, \mathcal{I}_0) \leftarrow \textsc{SimSetup}(1^\kappa, \mathcal{L}_1(\mathcal{D}_0))$;
2: **for** $k = 1$ to q **do**
3:     $w_k \leftarrow \mathsf{Adv}(1^k, \mathcal{I}_0, M_1, \ldots, M_{k-1})$;
4:     $(st_{\mathcal{S}}, M_k, \mathcal{I}_k) \leftarrow \textsc{SimSearch}(st_{\mathcal{S}}, \mathcal{L}_2(w_k), \mathcal{I}_{k-1})$;
5: $bit \leftarrow \mathsf{Adv}(1^k, \mathcal{I}_0, M_1, M_2, \ldots, M_q)$;
6: **return** bit;

Fig. 1. Real and ideal experiments for the SE scheme.

round, by just providing $\mathcal{D}(w)$. This is orthogonal to our protocols and we do not consider/model it here explicitly. We also note that we focus only on static SE. However, by using generic techniques, e.g., [14], we can extend our schemes to the dynamic setting. The correctness definition of SE is given in the extended version [13]. We now provide the security definition.

Definition 1 (Security of SE). *An SE scheme* (Setup, Search) *is secure in the semi-honest model if for any PPT adversary* Adv*, there exists a stateful PPT simulator* (SimSetup, SimSearch) *such that*

$$|\Pr[\mathbf{Real}^{\mathsf{SE}}(\kappa) = 1] - \Pr[\mathbf{Ideal}^{\mathsf{SE}}_{\mathcal{L}_1, \mathcal{L}_2}(\kappa) = 1]| \leq \mathsf{neg}(\kappa),$$

where experiments $\mathbf{Real}^{\mathsf{SE}}(\kappa)$ *and* $\mathbf{Ideal}^{\mathsf{SE}}_{\mathcal{L}_1, \mathcal{L}_2}(\kappa)$ *are defined in Fig. 1 and where the randomness is taken over the random bits used by the algorithms of the SE scheme, the algorithms of the simulator and* Adv.

Leakage Functions $\mathcal{L}_1$ and $\mathcal{L}_2$. As in prior work [6], $\mathcal{L}_1$ and $\mathcal{L}_2$ are leakage functions such that $\mathcal{L}_1(\mathcal{D}_0) = |\mathcal{D}_0| = N$ and $\mathcal{L}_2(w_i)$ leaks the access pattern *size* $|\mathcal{D}(w_i)|$ and the *search pattern* of w_i. Formally for a keyword w_i searched at time i, $\mathcal{L}_2(w_i)$ is

$$\mathcal{L}_2(w_i) = \begin{cases} (|\mathcal{D}(w_i)|, j) & \text{if } w_i \text{ was searched at time } j < i \\ (|\mathcal{D}(w_i)|, \perp) & \text{if } w_i \text{ was never searched before} \end{cases} \quad (1)$$

2.2 Oblivious RAM

Oblivious RAM (ORAM), introduced by Goldreich and Ostrovsky [18] is a compiler that encodes the memory such that accesses on the compiled memory do

```
(chosen, alternative) ← OfflineTwoChoiceAllocation(m, n)
1: Let {1, ..., m} be a set of balls and {1, ..., n} be a set of bins;
2: Initialize A and B to be empty arrays of m entries;
3: for i = 1, ..., m do
4:     Pick two bins a_i and b_i from {1, ..., n} independently and uniformly at random;
5:     A[i] = a_i; B[i] = b_i;
6: (chosen, alternative) ← MaxFlowSchedule(m, n, A, B);
7: return (chosen, alternative);
```

Fig. 2. Offline two-choice allocation of m balls to n bins.

not reveal access patterns on the original memory. Formal correctness and security definitions of ORAM are given in the Appendix. We give the definition for a read-only ORAM as this is needed in our scheme—the definition naturally extends for writes as well:

- $\langle \sigma, \mathsf{EM} \rangle \leftrightarrow \text{ORAMINITIALIZE}\langle (1^\kappa, \mathsf{M}), 1^\kappa \rangle$: ORAMINITIALIZE takes as input security parameter κ and memory array M of n values $(1, v_1), \ldots, (n, v_n)$ of λ bits each and outputs secret state σ (for client), and encrypted memory EM (for server).

- $\langle (v_i, \sigma'), \mathsf{EM}' \rangle \leftrightarrow \text{ORAMACCESS}\langle (\sigma, i), \mathsf{EM} \rangle$: ORAMACCESS is a protocol between client and server, where the client's input is secret state σ and an index i. Server's input is encrypted memory EM. Client's output is value v_i assigned to i and updated secret state σ'. Server's output is updated encrypted memory EM'.

3   New Bounds for Offline Two-Choice Allocation

As mentioned in the introduction, our medium-list allocation uses a variation of the classic balls-in-bins problem, known as *offline two-choice allocation*—see Fig. 2. Assume m balls and n bins. In the selection phase, for the i-th ball, two bins a_i and b_i are chosen independently and uniformly at random. After selection, in a post-processing phase, the i-th ball is mapped to either bin a_i or b_i such that the maximum load is minimized. This assignment is achieved by a maximum flow algorithm [28] (for completeness, we provide this algorithm in Fig. 13 in the Appendix). The bin that ball i is finally mapped to is stored in an array chosen[i] whereas the other bin that was chosen for ball i is stored in an array alternative[i]. Let $L_{\max}^*$ denote the maximum load across all bins after this allocation process completes. Sanders et al. [28] proved the following.

Lemma 1 (Sanders et al. [28]). *Algorithm* OfflineTwoChoiceAllocation *in Fig. 2 outputs an allocation* chosen *of m balls to n bins such that $L_{\max}^* > \lceil \frac{m}{n} \rceil + 1$*

with probability at most $O(1/n)$.[5] Moreover, the allocation can be performed in time $O(n^3)$.

For our purposes, the bounds derived by Sanders et al. [28] do not suffice. In the following we derive new bounds. In particular:

1. In Sect. 3.1, we derive probability bounds on the *number of overflowing bins*, i.e., the bins that contain more than $\lceil \frac{m}{n} \rceil + 1$ balls after the allocation completes.

2. In Sect. 3.2, we derive probability bounds on the *overflow size*, i.e., the number of balls beyond $\lceil \frac{m}{n} \rceil + 1$ that a bin contains.

3. In Sect. 3.3, we combine these to bound the total number of overflowing balls.

3.1   Bounding the Number of Overflowing Bins

For every bin $\ell \in [n]$, let us define a random 0-1 variable Z_ℓ such that Z_ℓ is 1 if bin ℓ contains more than $\lceil \frac{m}{n} \rceil + 1$ balls after OfflineTwoChoiceAllocation returns and 0 otherwise. What we want is to bound is the random variable $Z = \sum_{\ell=1}^{n} Z_i$, representing the total number of overflowing bins. Unfortunately we cannot use a Chernoff bound directly, since (i) the variables Z_i are not independent; (ii) we do not know the exact expectation $\mathbb{E}[Z]$. However, we observe that if we show that the variables Z_i are *negatively associated* (at a high level negative association indicates that for a set of variables, whenever some of them increase the rest tend to decrease—see the Appendix for a precise definition) and if we get an *upper bound* on $\mathbb{E}[Z]$ we can then derive a Chernoff-like bound for the number of overflowing bins. We begin by proving the following.

Lemma 2. *The set of random variables $Z_1, Z_2, \ldots, Z_n$ is negatively associated.*

Proof. For all $i \in [n]$, $j \in [n]$ and $k \in [m]$ let X_{ijk} be the random variable such that

$$X_{ijk} = \begin{cases} 1 & \text{if OfflineTwoChoiceAllocation chose the two bins } i \text{ and } j \text{ for ball } k \\ 0 & \text{otherwise} \end{cases}.$$

For each k it holds that $\sum_{i,j} X_{ijk} = 1$, since only one pair of bins is chosen for ball k. Therefore, by [17, Proposition 11], it follows that each set $\mathbf{X}_k = \{X_{ijk}\}_{i \in [n], j \in [n]}$ is negatively associated. Moreover, since the sets $\mathbf{X}_k, \mathbf{X}_{k'}$ for $k \neq k'$ consist of mutually independent variables (as the selection of bins is made independently for each ball), it follows from [17, Proposition 7.1] that the set $\mathbf{X} = \{X_{ijk}\}_{i \in [n], j \in [n], k \in [m]}$ is negatively associated. Now consider the disjoint sets U_ℓ for $\ell \in [n]$ defined as $U_\ell = \{X_{ijk} \mid \mathsf{chosen}[k] = \ell \ \wedge \ (\ell = i \vee \ell = j)\}$, where chosen is the array output by OfflineTwoChoiceAllocation. Let us now define $h_\ell(X_{ijk}, X_{ijk} \in U_\ell) = \sum_{X_{ijk} \in U_\ell} X_{ijk}$ for $\ell \in [n]$. Clearly each h_ℓ is a non-decreasing function and therefore by [17, Proposition 7.2] the set of random

---

[5] Sanders et al. [28] gave a better bound $O(1/n)^{\lceil \frac{m}{n} \rceil + 1}$ which is $O(1/n)$ since $\lceil m/n \rceil \geq 0$. Our analysis is simplified when we take this looser bound $O(1/n)$.

variables $\mathbf{Y} = \{Y_\ell\}_{\ell \in [n]}$ where $Y_\ell = h_\ell$ is also negatively associated. We can finally define Z_ℓ for $\ell = 1, \ldots, n$ as

$$Z_\ell = f(Y_\ell) = \begin{cases} 0 & \text{if } Y_\ell \leq \lceil m/n \rceil + 1 \\ 1 & \text{otherwise} \end{cases}.$$

Since f is also a non-decreasing function (as whenever Y_ℓ grows, $Z_\ell = f(Y_\ell)$ may only increase) therefore, again by [17, Proposition 7.2], it follows that the set of random variables $Z_1, Z_2, \ldots, Z_n$ is also negatively associated.   □

Lemma 3. *The expected number of overflowing bins $\mathbb{E}[Z]$ is $O(1)$.*

Proof. For all bins $\ell \in [n]$, it is $\mathbb{E}[Z_\ell] = \Pr[Y_\ell > \lceil m/n \rceil + 1] \leq \Pr[L^*_{\max} > \lceil m/n \rceil + 1] = O(1/n)$, by Lemma 1 (where $L^*_{\max}$ is the maximum load across all bins after allocation). By linearity of expectation and since $Z = \sum Z_i$, it is $\mathbb{E}[Z] = O(1)$.   □

Theorem 1. *Assume OfflineTwoChoiceAllocation from Fig. 2 is used to allocate m balls into n bins. Let Z be the number of bins that receive more than $\lceil m/n \rceil + 1$ balls. Then there exists a fixed positive constant c such that for sufficiently large n[6] and for any $\tau > 1$ we have $\Pr[Z \geq c \cdot \tau] \leq \left(\frac{e}{\tau}\right)^{c \cdot \tau}$.*

Proof. By Lemma 3 we have that there exists a fixed constant c such that $\mathbb{E}[Z] \leq c$ for sufficiently large n. Therefore, by Lemmas 2 and 8 in the Appendix (where we set $\mu_H = c$ since $\mathbb{E}[Z] \leq c$) we have that for any $\delta > 0$

$$\Pr[Z \geq (1 + \delta) \cdot c] \leq \left(\frac{e^\delta}{(1+\delta)^{(1+\delta)}}\right)^c \leq \left(\frac{e^{1+\delta}}{(1+\delta)^{(1+\delta)}}\right)^c.$$

Setting $\delta = \tau - 1$ which is > 0 for $\tau > 1$, we get the desired result.   □

3.2   Bounding the Overflow Size

Next, we turn our attention to the number of balls Y_ℓ that can be assigned to bin ℓ. In particular, we want to derive a probability bound $\Pr[Y_\ell > \lceil m/n \rceil + \tau]$ defined in general for parameter $\tau \geq 2$—Sanders et al. [28] studied only the case where $\tau = 1$. To do that, we will bound the probability that after OfflineTwoChoiceAllocation returns the maximum load $L^*_{\max}$ is larger than $\lceil m/n \rceil + \tau$ for $\tau \geq 2$. We now prove the following.

Theorem 2. *Assume OfflineTwoChoiceAllocation from Fig. 2 is used to allocate m balls into n bins. Let $L^*_{\max}$ be the maximum load across all bins. Then for any $\tau \geq 2$*

$$\Pr\left[L^*_{\max} \geq \left\lceil \frac{m}{n} \right\rceil + \tau\right] \leq O(1/n)^\tau + O(\sqrt{n} \cdot 0.9^n).$$

---

[6] This means that there exists a fixed constant n_0 such that for $n \geq n_0$ the statement holds—we provide an estimate of the constants c and n_0 in the extended version [13].

Proof. Our analysis here closely follows the one of [28]. Without loss of generality, we assume the number of balls m to be a multiple of the number of bins n[7] and we will set $b = m/n$. Let now (a_i, b_i) be the two random choices that OfflineTwoChoiceAllocation makes for ball i where $i = 1, \ldots, m$. For a subset $U \subseteq \{1, \ldots, n\}$ of bins we define the random variables $X_1^U, \ldots, X_m^U$ such that $X_i^U = 1$, if $a_i, b_i \in U$, and 0 otherwise, i.e., X_i^U is 1 only if both selections for the i-th ball are from subset U, which unavoidably leads to this ball being assigned to a bin within subset U. The random variable $L_U = \sum_{i=1}^m X_i^U$ is called the *unavoidable load* of U. Also, for a set U and a parameter τ, let $P_U = \Pr[L_U \geq (b + \tau)|U| + 1]$. Finally, let $L_{\max}^*$ be the *optimal load*, namely the minimum maximum load that can be derived by considering all possible allocations *given* the random choices $(a_1, b_1), \ldots, (a_m, b_m)$. Since MaxFlowSchedule computes an allocation with the optimal load, we must compute the probability $\Pr[L_{\max}^* > b + \tau]$, where $\tau \geq 2$. From [29, Lemma 5] we have $L_{\max}^* = \max_{\emptyset \neq U \subseteq \{1, \ldots, n\}} \{L_U / |U|\}$. Thus,

$$\Pr[L_{\max}^* > b + \tau] = \Pr[\exists U \subseteq [n] : L_U / |U| > b + \tau]$$
$$\leq \sum_{\emptyset \neq U \subseteq [n]} \Pr[L_U \geq (b + \tau)|U| + 1] = \sum_{|U|=1}^n \binom{n}{|U|} P_U,$$

where the inequality follows from a simple union bound and for the last step we used the fact that P_U is the same for all sets U of the same cardinality. This is because for all sets U_1 and U_2 with $|U_1| = |U_2|$ we have that $\Pr[L_{U_1} \geq (b + \tau)|U_1| + 1] = \Pr[L_{U_2} \geq (b + \tau)|U_2| + 1]$ since U_1 and U_2 are identically distributed. Next, we need to bound the sum $\sum_{|U|=1}^n \binom{n}{|U|} P_U$. For this we will split the sum into three separate summands

$$T_1 = \sum_{1 \leq |U| \leq \frac{n}{8}} \binom{n}{|U|} P_U, \quad T_2 = \sum_{\frac{n}{8} < |U| < \frac{nb}{b+\tau}} \binom{n}{|U|} P_U \text{ and } T_3 = \sum_{\frac{nb}{b+\tau} \leq |U| \leq n} \binom{n}{|U|} P_U.$$

We begin with the simple observation that $T_3 = 0$. To see why, note that for $|U| \geq nb/(b + \tau)$ it holds that $P_U = \Pr[L_U \geq (b + \tau)|U| + 1] = \Pr[L_U \geq (b + \tau)nb/(b+\tau) + 1] = \Pr[L_U \geq m + 1] = 0$ as m is a natural upper bound for L_U (i.e., if both selections fall within U for all balls). Regarding T_2, from [28, Lemma 9] we have $\sum_{\frac{n}{8} < |U| < \frac{nb}{b+1}} \binom{n}{|U|} P_U^* = O(\sqrt{n} \cdot 0.9^n)$, where $P_U^* = \Pr[L_U \geq (b+1)|U|+1]$. Clearly, for all U, $P_U \leq P_U^*$. Moreover, $\sum_{\frac{n}{8} < |U| < \frac{nb}{b+\tau}} P_U^* \leq \sum_{\frac{n}{8} < |U| < \frac{nb}{b+1}} P_U^*$ for all $\tau \geq 2$. Putting it all together, we have

$$T_2 \leq \sum_{\frac{n}{8} < |U| < \frac{nb}{b+1}} \binom{n}{|U|} P_U^* = O(\sqrt{n} \cdot 0.9^n).$$

---

[7] If not, we pad to $m = n\lceil m'/n \rceil$ balls, where m' is the original number of balls. Then, to get an allocation for the m' balls, we get an allocation for the m balls and we remove the unnecessary balls. Clearly, if L^* is the optimal maximum load for the m' balls, then $L^* \leq L_{\max}^*$ (if $L^* > L_{\max}^*$ you can get a better allocation for the m' balls by allocating m balls, a contradiction) and therefore whatever probability bounds we derive for $L_{\max}^*$ holds for L^*.

By Lemma 9 in the Appendix, $T_1 = O(1/n)^{b+\tau} = O(1/n)^{b+\tau}$ for all $\tau \geq 2$ hence for all $\tau \geq 2$ it is $\sum_{|U|=1}^{n} \binom{n}{|U|} P_U = O(1/n)^{\tau}$, as $b \geq 0$, which completes the proof. $\qquad\square$

3.3   Bounding the Total Number of Overflowing Balls

Let $T > 0$ be the number of overflowing balls, i.e., $T = \sum_{i=1}^{\ell} Z_i(Y_i - \lceil m/n \rceil - 1)$. Using Theorems 1 and 2, and by a simple application of the law of total probability, we can now prove the following result.

Theorem 3. *Assume* OfflineTwoChoiceAllocation *from Fig. 2 is used to allocate m balls into n bins. Let T be the number of overflowing balls as defined above. Then there exist positive constants c, c_1, c_2 such that for large n and for any $\tau \geq 2$ it is*

$$\Pr[T > c \cdot \tau^2] \leq \left(\frac{e}{\tau}\right)^{c \cdot \tau} + \left(\frac{c_1}{n}\right)^{\tau} + c_2\sqrt{n} \cdot 0.9^n .$$

Proof. Define the events $E : T > c \cdot \tau^2$, $E_1 : Z > \tau$ and $E_2 : L_{\max}^* > \lceil m/n \rceil + \tau$, for some $\tau \geq 2$. There is no way there can be more than τ^2 overflowing balls if both the number of overflowing bins and the maximum overflow per bin is at most τ. This implies that $E \subseteq E_1 \cup E_2$. By a standard union bound and applying Theorems 1 and 2, we have $\Pr[E] \leq \left(\frac{e}{\tau}\right)^{c \cdot \tau} + O(1/n)^{\tau} + O(\sqrt{n} \cdot 0.9^n)$, which completes the proof by taking c_1 and c_2 to be the constants in $O(1/n)$ and $O(\sqrt{n} \cdot 0.9^n)$ respectively. $\qquad\square$

4   New ORAM with $O(1)$ Locality and $o(\sqrt{n})$ Bandwidth

Our constant-locality SE construction uses an ORAM scheme as a black box. In particular, the ORAM scheme that is used must have the following properties:

1. It needs to have constant locality, meaning that for each oblivious access it should only read $O(1)$ non-contiguous locations in the encrypted memory. Existing ORAM constructions with polylogarithmic bandwidth have *logarithmic* locality. For example, a path ORAM access [33] traverses $\log n$ binary tree nodes stored in non-contiguous memory locations—therefore we cannot use it here. This property is required as our underlying SE scheme must have $O(1)$ locality;

2. It needs to have bandwidth cost $o(\sqrt{n} \cdot \lambda)$. This property is required because we would be applying the ORAM scheme on an array of $O(\log^2 N)$ entries, yielding overall bandwidth equal to $o(\log N \cdot \lambda)$, which would imply sublogarithmic read efficiency for the underlying SE scheme.

We note here that an existing scheme that almost satisfies both properties above is the ORAM construction from [27, Theorem 7] by Ohrimenko et al. (where we set $c = 3$). This ORAM has $O(1)$ locality and $O(n^{1/3} \log n \cdot \lambda)$ bandwidth. However we cannot apply it here due to its failure probability which is $\mathsf{neg}(n)$, where n is the size of the memory array. Unfortunately, since our array has $O(\log^2 N)$ entries (N is the size of the SE dataset), this gives a probability of failure $\mathsf{neg}(\log^2 N)$ which is not $\mathsf{neg}(N)$.

Our proposed ORAM construction is a hierarchical application of the square-root ORAM construction of Goldreich and Ostrovsky [18]. Here, we provide a description of the amortized version of our construction (i.e., the read-efficiency and locality bounds we achieve are amortized over n accesses) in Fig. 3. The deamortized version of our ORAM construction is achieved using techniques of Goodrich et al. [21] for deamortizing the square root ORAM, in a straightforward manner (formal description and analysis of the deamortized version can be found in the extended version [13]).

ORAM Setup. Given memory M with n index-value pairs $(1, v_1), \ldots, (n, v_n)$ we allocate three main arrays for storage: A of size $n_a = n + n^{2/3}$, B of size $n_b = n^{2/3} + n^{1/3}$, and C of size $n_c = n^{1/3}$. Initially A stores all elements encrypted with CPA-secure encryption and permuted with a pseudorandom permutation[8] $\pi_a : [n_a] \to [n_a]$ and B and C are empty, containing encryptions of dummy values. We also initialize another pseudorandom permutation $\pi_b : [n_b] \to [n_b]$ used for accessing elements from array B. In particular, if an element $x \in [n]$ is stored in array B, it is located at position $\pi_b[\mathsf{Tab}[x]]$ of B, where Tab is a locally-stored hash table mapping an element $x \in [n]$ to $\mathsf{Tab}[x] \in [n_b]$. Note the hash table is needed to index elements in B as $n_b < n$.

ORAM Access. To access element x, the algorithm always downloads, decrypts and sequentially scans array C. Similarly to the square-root ORAM, we consider two cases:

1. *Element x is in C.* In this case the requested element has been found and the algorithm performs two additional dummy accesses for security reasons: it accesses a random[9] position in array A and a random position in array B.

2. *Element x is not in C.* In this case we distinguish the following subcases.
   - Element x is not in B.[10] In this case x can be retrieved by accessing the random position $\pi_a[x]$ of array A. Like previously, the algorithm also accesses a random position in array B.
   - Element x is in B. In this case x can be retrieved by accessing the random position $\pi_b[\mathsf{Tab}[x]]$ of array B. Like previously, the algorithm also accesses a random position in array A.

After the access above, the retrieved element x is written in the next available position of C, the algorithm computes a fresh encryption of C and writes C back to the server. Just like in square-root ORAM, some oblivious reshuffling must occur: In particular, every $n^{1/3}$ accesses, array C becomes full and both C and the contents of B are obliviously reshuffled into B. Every $n^{2/3}$ accesses, when

---

[8] In practice π_a is implemented with efficient small-domain PRPs (e.g., [22,26,32]).

[9] This position is not entirely random—it is chosen from those that have not been chosen so far.

[10] This can be decided by checking whether $\mathsf{Tab}[x]$ is null or not.

B becomes full, all elements are obliviously reshuffled into A. We describe this reshuffling process next.

Reshuffling, epochs and superepochs. Our algorithm for obliviously accessing an element x described proceeds in *epochs* and *superepochs*. An epoch is defined as a sequence of $n^{1/3}$ accesses. A superepoch is defined as a sequence of $n^{2/3}$ accesses.

At the end of every epoch C becomes full, and all elements in C along with the ones that have been accessed in the current superepoch (and are now stored in B) are obliviously reshuffled into B using a fresh pseudorandom permutation π_b. In our implementation in Fig. 3, we store all the elements that must be reshuffled in an array SCRATCH. After the reshuffling C can be emptied (denoted with $\perp$ Line 30) so that it can be used again in the future. At the end of every superepoch all the elements of the dataset are obliviously reshuffled into array A using a fresh pseudorandom permutation π_a and arrays B, C and SCRATCH are emptied.

Oblivious Sorting with Good Locality. As in previous works, our reshuffling in the ORAM protocol is performed using an oblivious sorting protocol. Since we are using the ORAM scheme in an SE scheme that must have good locality, we must ensure that the oblivious sorting protocol used has good locality as well, i.e., it does not access too many non-contiguous locations. One way to achieve that is to download the whole encrypted array, decrypt it, sort it and encrypt it back. This has excellent locality $L = 1$ but requires linear client space. A standard oblivious sorting protocol such as Batcher's odd-even mergesort [8] does not work either since its locality can be linear.

Fortunately, Goodrich and Mitzenmacher [20] developed an oblivious sorting protocol for an external memory setting that is a perfect fit for our application—see Fig. 16 in the Appendix. The client interacts with the server only by reading and writing b consecutive blocks of memory. We call each b-block access (either for read or write) an *I/O operation*. The performance of their protocol is characterized in the following theorem.

Theorem 4 (Goodrich and Mitzenmacher [20], Goodrich [19]). *Given an array X containing n comparable blocks, we can sort X with a data-oblivious external-memory protocol that uses $O((n/b) \log^2(n/b))$ I/O operations and local memory of $4b$ blocks, where an I/O operation is defined as the read/write of b consecutive blocks of X.*

In the above oblivious sorting protocol, value b (the number of consecutive blocks downloaded/uploaded in one I/O) can be parameterized, affecting the local space accordingly. In our case, we set b to be equal to $n^{1/3} \log^2 n$—see Lines 24 and 29 in Fig. 3, which is enough for achieving constant locality in our SE scheme.

Our final result is as follows (proof can be found in the Appendix).

Theorem 5. *Let n be the size of the memory array and λ be the size of the block. Our ORAM scheme (i) is correct according to Definition 2; (ii) is secure according to Definition 3, assuming pseudorandom permutations and CPA-secure*

Protocol $\langle \sigma, \mathsf{EM} \rangle \leftrightarrow \mathrm{ORAMINITIALIZE}\langle (1^{\kappa}, \mathsf{M}), \perp \rangle$:

1: Parse M as $(1, v_1), (2, v_2), \ldots, (n, v_n)$ where $|i, v_i| = \lambda$ (the values are λ bits long);
2: Let $n_a \leftarrow n + n^{2/3}$, $n_b \leftarrow n^{2/3} + n^{1/3}$, $n_c \leftarrow n^{1/3}$;
3: Let A, B and C be arrays of size n_a, n_b and n_c respectively. Initialize them with **0** entries;
4: Let SCRATCH be an array of size n_b. Initialize it with **0** entries;
5: Let $\pi_a : [n_a] \to [n_a]$ and $\pi_b : [n_b] \to [n_b]$ be pseudorandom permutations;
6: For $i = 1, \ldots, n$, store (i, v_i) at location $\pi_a[i]$ in A;
7: **Encrypt-And-Write** arrays A, B, C and SCRATCH and add them to EM ;
8: Let $\mathsf{count}_a \leftarrow 0$ and $\mathsf{count}_b \leftarrow 0$;
9: Let Tab be an empty hash table;
10: Set $\sigma = (\pi_a, \pi_b, \mathsf{Tab}, \mathsf{count}_a, \mathsf{count}_b)$;
11: **return** $\langle \sigma, \mathsf{EM} \rangle$;

Protocol $\langle (v_i, \sigma'), \mathsf{EM}' \rangle \leftrightarrow \mathrm{ORAMACCESS}\langle (\sigma, i), \mathsf{EM} \rangle$:

1: Parse σ as $(\pi_a, \pi_b, \mathsf{Tab}, \mathsf{count}_a, \mathsf{count}_b)$ and EM as $(A, B, C, \mathsf{SCRATCH})$;
2: Increment count_a and count_b;
3: **Read-And-Decrypt** array C;
4: **if** $(i, v_i) \in C$ **then**   ▷ (i, v_i) was accessed before and is stored in C
5:     $\mathsf{index}_a \leftarrow \pi_a[n + \mathsf{count}_a]$;
6:     $\mathsf{index}_b \leftarrow \pi_b[n^{2/3} + \mathsf{count}_b]$;
7: **else**
8:     **if** $\mathsf{Tab}[i] \neq$ null **then**   ▷ (i, v_i) is stored in $B[\mathsf{index}_b]$
9:         $\mathsf{index}_a \leftarrow \pi_a[n + \mathsf{count}_a]$;
10:         $\mathsf{index}_b \leftarrow \pi_b[\mathsf{Tab}[i]]$;
11:     **else**   ▷ (i, v_i) is stored in $A[\mathsf{index}_a]$
12:         $\mathsf{index}_a \leftarrow \pi_a[i]$;
13:         $\mathsf{index}_b \leftarrow \pi_b[n^{2/3} + \mathsf{count}_b]$;
14: **Read-And-Decrypt** $A[\mathsf{index}_a]$;
15: **Read-And-Decrypt** $B[\mathsf{index}_b]$;
16: Retrieve (i, v_i) from either $A[\mathsf{index}_a]$ or $B[\mathsf{index}_b]$ or C;
17: $C[\mathsf{count}_b] \leftarrow (i, v_i)$;
18: **Encrypt-And-Write** array C;
19: $\mathsf{Tab}[i] \leftarrow \mathsf{count}_a$;
20: **Encrypt-And-Write** element $(\mathsf{Tab}[i], v_i)$ at position count_a of array SCRATCH;
21: **if** $\mathsf{count}_a > n^{2/3}$ **then**   ▷ Transition to a new superepoch
22:     Let π_a and π_b be new pseudorandom permutations;
23:     $\mathsf{count}_a \leftarrow 0$ and $\mathsf{count}_b \leftarrow 0$;
24:     $\langle \perp, A \rangle \leftrightarrow \mathrm{OBLIVIOUSSORTING}\langle (\pi_a, n_a, n^{1/3} \log^2 n), A \rangle$;   ▷ large rebuild
25:     Set $B \leftarrow \perp$; $C \leftarrow \perp$; SCRATCH $\leftarrow \perp$; Set Tab $\leftarrow \perp$;
26: **if** $\mathsf{count}_b > n^{1/3}$ **then**   ▷ Transition to a new epoch
27:     Let π_b be new pseudorandom permutation;
28:     $\mathsf{count}_b \leftarrow 0$;
29:     $\langle \perp, B \rangle \leftrightarrow \mathrm{OBLIVIOUSSORTING}\langle (\pi_b, n_b, n^{1/3} \log^2 n), \mathsf{SCRATCH} \rangle$;   ▷ small rebuild
30:     Set $C \leftarrow \perp$;
31: **return** $\langle (v_i, (\pi_a, \pi_b, \mathsf{Tab}, \mathsf{count}_a, \mathsf{count}_b)), (A, B, C, \mathsf{SCRATCH}) \rangle$;

Fig. 3. Read-only ORAM construction with $O(n^{1/3} \log^2 n \cdot \lambda)$ amortized bandwidth and $O(1)$ amortized locality.

encryption; (ii) has $O(n^{1/3} \log^2 n \cdot \lambda)$ amortized bandwidth and $O(1)$ amortized locality per access and requires client space $O(n^{2/3} \log n + n^{1/3} \log^2 n \cdot \lambda)$.

Standard deamortization techniques from [21] can be applied to make the overheads of our ORAM worst-case as opposed to amortized. A formal treatment of this is presented in the extended version of our paper [13], giving the following result.

Corollary 1. *Let $\lambda = \Omega(n^{1/3})$ bits be the block size. Then our ORAM scheme has $O(n^{1/3} \log^2 n \cdot \lambda)$ worst-case bandwidth per access, $O(1)$ worst-case locality per access and $O(n^{1/3} \log^2 n \cdot \lambda)$ client space.*

5   Allocation Algorithms

As we mentioned in the introduction, to construct our final SE scheme we are going to use a series of *allocation algorithms*. The goal of an allocation algorithm for an SE dataset $\mathcal{D}$ consisting of q keyword lists $\mathcal{D}(w_1), \mathcal{D}(w_2), \ldots, \mathcal{D}(w_q)$ is to store/allocate the elements of all lists into an array **A** (or multiple arrays).

Retrieval Instructions. To be useful, an allocation algorithm should also output a hash table Tab such that Tab[w] contains "instructions" on how to correctly retrieve a keyword list $\mathcal{D}(w)$ after the list is stored. For example, for a keyword list $\mathcal{D}(w)$ that contains four elements stored at positions $5, 16, 26, 27$ of **A** by the allocation algorithm, some valid alternatives for the instructions Tab[w] are: (i) "*access positions 5, 16, 26, 27 of array* **A**"; (ii) "*access all positions from 3 to 28 of array* **A**"; (iii) "*access the whole array* **A**". Clearly, there are different tradeoffs among the above.

Algorithm (**A**, Tab) ← AllocateSmall($\mathcal{D}, N$): (taken from [6])

1: Set $\epsilon \leftarrow 1/\log^{1-\gamma} N$;
2: Let max $\leftarrow N^{1-\epsilon}$, $C = c_s \cdot \log^\gamma N$ and $B \leftarrow N/C^a$;
3: Let **A** be an array of B buckets—each bucket has capacity C;
4: Initialize an empty hash table Tab;
5: **for** sizes $s = $ max, max/2, max/4, ..., 1 **do**
6:     **for** each keyword w such that $|\mathcal{D}(w)| = s$ **do**
7:         Pick α and β from $\{1, \ldots, \frac{B}{s}\}$ independently and uniformly at random;
8:         Let **A**$\{\alpha, s\}$ and **A**$\{\beta, s\}$ be two superbuckets;
9:         Let $x \in \{\alpha, \beta\}$ correspond to the superbucket with the minimum load;
10:        Store $\mathcal{D}(w)$ horizontally into superbucket **A**$\{x, s\}$;
11:        Tab[w] = $(s, \alpha, \beta, \perp)$;
12: **if** there is a bucket **A**[i] that overflows **then return** FAIL;
13: **else**
14:    Pad every bucket **A**[i] to C elements using dummy values;
15:    **return** (**A**, Tab);

---
^{a} Constant c_s can be appropriately chosen in [6].

Fig. 4. Allocation algorithm for small sizes from Asharov et al. [6].

Independence Property. For security purposes, and in particular for simulating the search procedure of the SE scheme, it is important that the instructions Tab[w] output by an allocation algorithm for a keyword list $\mathcal{D}(w)$ are *independent* of the distribution of the rest of the dataset—intuitively this implies that accessing $\mathcal{D}(w)$ does not reveal information about the rest of the data. This independence property is easy to achieve with a "read-all" algorithm, where the whole array is read every time a keyword is accessed, but this is very inefficient. Another way to achieve this property is to store the lists using a random permutation π—this is actually the allocation algorithm used by most existing SE schemes, e.g., [12]. This "permute" approach has however very bad locality since it requires $|\mathcal{D}(w)|$ random jumps in the memory to retrieve $\mathcal{D}(w)$. In the following we present the details of our allocation algorithms for small, medium, large and huge lists. We begin with some terminology.

5.1 Buckets and Superbuckets

Following terminology from [6], our allocation algorithms use fixed-capacity *buckets* for storage. A bucket with capacity C can store up to C *elements*—in our case an *element* is a keyword-document pair (w, id). To simplify notation, we represent a set of B buckets $A_1, A_2, \ldots, A_B$ as an array $\mathbf{A}$ of B buckets, referring to bucket A_i as $\mathbf{A}[i]$. Additionally, a *superbucket* $\mathbf{A}\{k, s\}$ is a set of the following s consecutive buckets

$$\mathbf{A}[(k-1)s+1], \mathbf{A}[(k-1)s+2], \ldots, \mathbf{A}[ks].$$

We say that we store a keyword list $\mathcal{D}(w) = \{(w, id_1), (w, id_2), \ldots, (w, id_s)\}$ *horizontally* into superbucket $\mathbf{A}\{k, s\}$ when each element (w, id_i) is stored in a separate bucket of the superbucket.[11] Finally, the *load* of a bucket or a superbucket is the number of elements stored in each bucket or superbucket.

5.2 Allocating Small Lists with Two-Dimensional Allocation

For small keyword lists we use the two-dimensional allocation algorithm of Asharov et al. [6], by carefully setting the parameters from scratch. For completeness we provide the algorithm in Fig. 4, which we call AllocateSmall. Let $C = c_s \cdot \log^\gamma N$, for some appropriately chosen constant c_s. The algorithm uses $B = N/C$ buckets of capacity C each. It then considers all small keyword lists starting from the largest to the smallest, and depending on the list's size s, it picks two superbuckets from $\{1, 2 \ldots, B/s\}$ uniformly at random, horizontally placing the keyword list into the superbucket with the minimum load. The algorithm records both superbuckets as instructions in a hash table Tab. If, during

---

[11] E.g., consider an array $\mathbf{A}$ consisting of 20 buckets $\mathbf{A}[1], \mathbf{A}[2] \ldots, \mathbf{A}[20]$ where each bucket $\mathbf{A}[i]$ has capacity $C = 5$. Superbucket $\mathbf{A}\{3, 4\}$ contains the buckets $\mathbf{A}[9], \ldots, \mathbf{A}[12]$. Horizontally storing $\{a_1, a_2, \ldots, a_4\}$ into $\mathbf{A}\{3, 4\}$ means storing a_1 into $\mathbf{A}[9]$, a_2 into $\mathbf{A}[10]$, and so on.

this allocation process some bucket overflows, then the algorithm fails. We now have the following result.

Theorem 6. *Algorithm* AllocateSmall *in Fig. 4 outputs* FAIL *with probability* neg(N). *Moreover the output array of buckets* **A** *occupies space* $O(N)$.

Proof.   For the algorithm to fail, the load of some bucket $\mathbf{A}[i]$ (i.e., maximum load) must exceed $O(\log^\gamma N)$. We show this probability is negligible for our choice of $\gamma = 2/3 + \delta$. We recall AllocateSmall allocates all keyword lists using a two-dimensional balanced allocation [6]. For our proof we apply [6, Theorem 3.5] that states: For max $= N^{1-\epsilon}$, $B \geq N/\log N$ and for non-decreasing function $f(N)$ such that $f(N) = \Omega(\log\log N)$, $f(N) = O(\sqrt{\log N})$ and $f(2N) = O(f(N))$ the maximum load of a two-dimensional balanced allocation is $\frac{4N}{B} + O(\log \epsilon^{-1} \cdot f(n))$ with probability at least $1 - O(\log \epsilon^{-1}) \cdot N^{-\Omega(\epsilon \cdot f(N^\epsilon))}$. In our case, it $\epsilon = 1/\log^{1-\gamma} N$ and $B = N/\log^\gamma N$ and we also pick $f(N) = \sqrt{\log N}$. Note that all conditions for f and B and ϵ of [6, Theorem 3.5] are satisfied assuming $1/2 < \gamma < 1$. Also, for this choice of parameters we have that the probability the maximum load is more than $O(\log^\gamma N)$ is at most $O(\log(\log^{1-\gamma} N)) \cdot N^{-\Omega(\ell(N))}$ where $\ell(N)$ is

$$\frac{1}{\log^{1-\gamma} N} \sqrt{\log\left(N^{1/\log^{1-\gamma} N}\right)} = \frac{\sqrt{\log^\gamma N}}{\log^{1-\gamma} N} = \log^{3\gamma/2-1} N.$$

Since our construction uses $\gamma = 2/3 + \delta$ for any small $\delta > 0$ it is always $3\gamma/2 - 1 > 0$ and therefore the above probability is negligible. □

Note now that a list of size s can be read by accessing s consecutive buckets (i.e., a superbucket), therefore the read efficiency for these lists is $O(\log^\gamma N)$.

5.3   Allocating Medium Lists with OTA

The allocation process for medium lists is shown in Fig. 5 and the algorithm is described in Fig. 6. The algorithm uses an array **A** of $B = N/\log^\gamma N$ buckets, where each bucket has capacity $C = 3 \cdot \log^\gamma N$. Just like AllocateSmall, the allocation algorithm for medium sizes stores a list $\mathcal{D}(w)$ of size s horizontally into one of the superbuckets $\mathbf{A}\{1, s\}, \mathbf{A}\{2, s\}, \ldots, \mathbf{A}\{B/s, s\}$.

However, unlike AllocateSmall, the superbucket that is finally chosen to store $\mathcal{D}(w)$ depends only on keyword lists of the *same* size with $\mathcal{D}(w)$ that have already been allocated and not on all other keyword lists encountered so far. In particular, let k_s be the number of keyword lists that have size s. Let also $b_s = B/s$ be the number of superbuckets with respect to size s. To figure out which superbucket to pick for horizontally storing a particular keyword list of size s, the algorithm views the k_s keyword lists as *balls* and the b_s superbuckets as *bins* and performs an offline two-choice allocation of k_s keyword lists (balls) into b_s superbuckets (bins), as described in Sect. 3. When, during this process some superbucket contains $\lceil k_s/b_s \rceil + 1$ keyword lists of size s, any subsequent keyword list of size s meant for this superbucket is instead placed into a stash

$\mathbf{B}_s$ that contains exactly $c \cdot \log^2 N$ buckets of size s each for some fixed constant c derived in Theorem 1. Our algorithm will fail, if

- Some bucket $\mathbf{A}[i]$ overflows (i.e., the number of elements that are eventually stored in $\mathbf{A}[i]$ exceeds its capacity C), which as we show in Lemma 4 never happens; or
- More than $c \cdot \log^2 N$ keyword lists of size s must be stored at some stash $\mathbf{B}_s$, which as we show in Lemma 5 happens with negligible probability.

All the choices that the algorithm makes, such as the two superbuckets originally chosen for every list during the offline two-choice allocation as well as the position in the stash (in case the list was an overflowing one) are recorded in Tab as retrieval instructions. We now prove the following lemma.

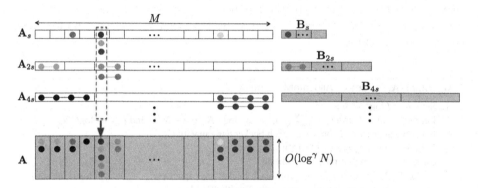

Fig. 5. Allocation of medium lists. Each ball represents a list of size $N^{1-1/\log^{1-\gamma} N}$. Two balls chained together represent a keyword list of double the size and so on. Arrays $\mathbf{A}_i$ show the OTA assignments for all lists of a specific size i. Arrays $\mathbf{A}_i$ are merged into array $\mathbf{A}$ of M buckets of capacity $O(\log^{\gamma} N)$ each. Overflowing lists of size i are placed in the stash $\mathbf{B}_i$. Only light-gray arrays are stored at the server—white arrays are only used for illustrating the intermediate results.

Lemma 4. *During the execution of algorithm* AllocateMedium *in Fig. 6, no bucket* $\mathbf{A}[i]$ *(for all* $i = 1, \ldots, B$*) will ever overflow.*

Proof. For each size $s = 2\min, 4\min, \ldots, \max$, Line 15 of AllocateMedium allows at most $\lceil k_s/b_s \rceil + 1$ keyword lists of size s to be stored in any superbucket $\mathbf{A}\{i, s\}$. Since every keyword list of size s is stored horizontally in a superbucket $\mathbf{A}\{i, s\}$, it follows that every bucket $\mathbf{A}[i]$ within every superbucket $\mathbf{A}\{i, s\}$ will have load, due to keywords lists of size s, at most $s \cdot (\lceil k_s/b_s \rceil + 1)/s = \lceil k_s/b_s \rceil + 1$. Therefore the total load of a bucket $\mathbf{A}[i]$ due to all sizes $s = 2\min, 4\min, \ldots, \max$ is at most $\sum_s \left(\left\lceil \frac{k_s}{b_s} \right\rceil + 1 \right) \leq \sum_s \frac{k_s}{b_s} + \sum_s 2$. We now bound the above sums separately. Since $b_s = B/s$, $\sum_s k_s \cdot s \leq N$ and $B = N/\log^{\gamma} N$ it is $\sum_s \frac{k_s}{b_s} = \frac{1}{B} \sum_s k_s \cdot s \leq \frac{N}{B} = \log^{\gamma} N$. As $\min = 2^{\log N - \log^{\gamma} N + 1}$, $\max = N/\log^2 N = 2^{\log N - 2\log\log N}$ and size s takes only powers of 2, there are at most $\log^{\gamma} N - 2\log\log N$ terms in the sum $\sum_s 2$ and therefore $\sum_s \left(\left\lceil \frac{k_s}{b_s} \right\rceil + 1 \right) \le 3 \cdot \log^\gamma N - 4 \cdot \log \log N \le 3 \cdot \log^\gamma N$, which equals the bucket capacity C in AllocateMedium. Thus no bucket will ever overflow.   □

Lemma 5. *During the execution of algorithm* AllocateMedium *in Fig. 6, no stash* $\mathbf{B}_s$ *(for* $s = 2\min, 4\min, \ldots, \max)$ *will ever overflow, except with probability* $\mathrm{neg}(N)$.

Proof. Recall that for each $s = 2\min, 4\min, \ldots, \max$, placing the k_s keyword lists of size s into the b_s superbuckets of size s is performed via an offline two-choice allocation of k_s balls into b_s bins. Also recall that the lists that end up in the stash $\mathbf{B}_s$ (that has capacity $\log^2 N$) are originally placed by the allocation algorithm in superbuckets containing more than $\lceil k_s/b_s \rceil + 1$ keyword lists of size s, thus they are *overflowing*. Let T_s be the number of these lists. By Theorem 3, where we set $T = T_s$ and $n = b_s$ and $\tau = \log N$, we have that for large b_s and for fixed constants c, c_1 and c_2

Algorithm $(\mathbf{A}, \mathcal{B}, \mathsf{Tab}) \leftarrow$ AllocateMedium$(\mathcal{D}, N)$:

1: Set $\epsilon \leftarrow 1/\log^{1-\gamma} N$;
2: Let $\min \leftarrow N^{1-\epsilon}$, $\max \leftarrow \frac{N}{\log^2 N}$, $C \leftarrow 3 \cdot \log^\gamma N$, $B \leftarrow N/C$ and $\ell = c \cdot \log^2 N$;[a]
3: Let $\mathbf{A}$ be an array of B buckets—each bucket has capacity C;
4: Initialize an empty hash table Tab;
5: **for** sizes $s = 2\min, 4\min, \ldots, \max$ **do**
6:     Let $\mathbf{B}_s$ be an array of ℓ buckets—each bucket has capacity s;   ▷ This is the stash
7:     Let k_s be the number of keywords in $\mathcal{D}$ with $|\mathcal{D}(w)| = s$;
8:     Let $b_s \leftarrow B/s$ be the number of superbuckets with respect to size s;
9:     Let $\mathrm{inStash}_s \leftarrow 0$; $i \leftarrow 0$;
10:     (chosen, alternative) $\leftarrow$ OfflineTwoChoiceAllocation(k_s, b_s);
11:     **for** each keyword w such that $|\mathcal{D}(w)| = s$ **do**
12:         Increment i;
13:         Set $\alpha \leftarrow$ chosen$[i]$; Set $\beta \leftarrow$ alternative$[i]$;
14:         **if** superbucket $\mathbf{A}\{\alpha, s\}$ contains $\le \lceil \frac{k_s}{b_s} \rceil$ keyword lists of size s **then**
15:             Store $\mathcal{D}(w)$ horizontally into superbucket $\mathbf{A}\{\alpha, s\}$;
16:             $\mathsf{Tab}[w] = (s, \alpha, \beta, 1)$;
17:         **else**   ▷ Move to stash
18:             Increment $\mathrm{inStash}_s$;
19:             **if** $\mathrm{inStash}_s > \ell$ **then return** FAIL;   ▷ Stash overflows
20:             Store $\mathcal{D}(w)$ in the bucket $\mathbf{B}_s[\mathrm{inStash}_s]$;
21:             $\mathsf{Tab}[w] = (s, \alpha, \beta, \mathrm{inStash}_s)$;
22: **if** there is a bucket $\mathbf{A}[i]$ that has overflown **then return** FAIL;
23: **else** Pad every bucket $\mathbf{A}[i]$ to C elements using dummy values;
24: **return** $(\mathbf{A}, (\mathbf{B}_{2\min}, \mathbf{B}_{4\min}, \mathbf{B}_{8\min}, \mathbf{B}_{16\min} \ldots, \mathbf{B}_{\max}), \mathsf{Tab})$;

---

[a] Constant c is derived by Theorem 1.

Fig. 6. Allocation algorithm for medium sizes.

$$\Pr[T_s > c \cdot \log^2 N] \le \left(\frac{e}{\log N}\right)^{c \cdot \log N} + \left(\frac{c_1}{b_s}\right)^{\log N} + c_2 \sqrt{b_s} \cdot 0.9^{b_s} = \mathsf{neg}(N),$$

as $b_s = B/s = N/s \log^\gamma N \ge \log^{2-\gamma} N = \omega(\log N)$ as $s \le \mathsf{max} = N/\log^2 N$. □

Theorem 7. *Algorithm* AllocateMedium *in Fig. 6 outputs* FAIL *with probability* $\mathsf{neg}(N)$. *Moreover, the size of the output array $\boldsymbol{A}$ and the stashes $\mathcal{B}$ is $O(N)$.*

Proof. AllocateMedium can fail either because a bucket $\mathbf{A}[i]$ overflows, which by Lemma 4 happens with probability 0, or because some stash $\mathbf{B}_s$ ends up having to store more than $\log^2 N$ elements for some $s = 2\mathsf{min}, 4\mathsf{min}, \dots, \mathsf{max}$, which by Lemma 5 happens with probability $\mathsf{neg}(N)$. For the space complexity, since no bucket $\mathbf{A}[i]$ overflows, array $\mathbf{A}$ occupies space $O(N)$. Also each stash $\mathbf{B}_s$ contains $\log^2 N$ buckets of size s each so the total size required by the stashes is $c \cdot \log^2 N(\mathsf{min} + 2\mathsf{min} + 4\mathsf{min} + \dots + \mathsf{max})$. Since $\mathsf{max} = N/\log^2 N$, the above is $\le 2c \log^2 N\mathsf{max} = O(N)$. □

Algorithm $(A, \mathsf{Tab}) \leftarrow$ AllocateLarge$(\mathcal{D}, N, \mathsf{min}, \mathsf{max})$:
1: Initialize an empty hash table Tab;
2: Let $\mathbf{A}$ be an array of $t = N/\mathsf{max}$ buckets—each bucket has capacity $2\mathsf{max}$;
3: **for** each keyword w such that $\mathsf{min} < |\mathcal{D}(w)| \le \mathsf{max}$ **do**
4:    **if** there exists a bucket $\mathbf{A}[k]$ with at least $|\mathcal{D}(w)|$ available space **then**
5:        Store $\mathcal{D}(w)$ in bucket $\mathbf{A}[k]$;
6:        $\mathsf{Tab}[w] \leftarrow (|\mathcal{D}(w)|, k, \perp, \perp)$;
7:    **else return** FAIL;
8: **return** (A, Tab);

Fig. 7. Allocation algorithm for large sizes.

5.4   Allocating Large Lists

Recall that we call a keyword list large, if its size is in the range $N/\log^2 N$ and $N/\log^\gamma N$ (recall $\gamma = 2/3 + \delta$). Algorithm AllocateLarge in Fig. 7 is used to allocate lists whose size falls within a specific subrange $(\mathsf{min}, \mathsf{max}]$ of the above range. Let step be an appropriately chosen parameter such that $\mathsf{step} < 3\delta/2$ and partition the range $(N/\log^2 N, N/\log^\gamma N]$ into $\frac{2-\gamma}{\mathsf{step}}$ consecutive subranges[12]

$$\left(\frac{N}{\log^2 N}, \frac{N}{\log^{2-\mathsf{step}} N}\right], \left(\frac{N}{\log^{2-\mathsf{step}} N}, \frac{N}{\log^{2-2\cdot\mathsf{step}} N}\right], \dots, \left(\frac{N}{\log^{\gamma-\mathsf{step}} N}, \frac{N}{\log^\gamma N}\right].$$

---

[12] If $\frac{2-\gamma}{\mathsf{step}}$ is not an integer, we round up. Without loss of generality, the last subrange may be of smaller size than the previous ones in order to stop at $N/\log^\gamma N$. Note that, this can only make allocation easier (since it may only reduce the number of lists in the last subrange).

For a given subrange $(\mathsf{min}, \mathsf{max}]$, AllocateLarge stores all keyword lists in an array $\mathbf{A}$ of $t = N/\mathsf{max}$ buckets of capacity $2\mathsf{max}$ each. In particular, for a large keyword list $\mathcal{D}(w)$ of size s, the algorithm places the list in the first bucket that it can find with available space. We later prove that there will always be such a bucket, and therefore no overflow will ever happen. The formal description of the algorithm is shown in Fig. 7.

Theorem 8. *Algorithm* AllocateLarge *in Fig. 7 never outputs* FAIL.

Proof. Assume AllocateLarge fails. This means that at the time some list $\mathcal{D}(w)$ is considered, *all* buckets of $\mathbf{A}$ store at least $2\mathsf{max} - s + 1$ elements each. Therefore the total number of elements considered so far is $\frac{N}{\mathsf{max}}(2\mathsf{max} - s + 1) \geq \frac{N}{\mathsf{max}}(\mathsf{max} + 1) \geq N + \frac{N}{\mathsf{max}} \geq N + \log^{\gamma} N$, since $s \leq \mathsf{max} \leq N/\log^{\gamma} N$. This is a contradiction, however, since the number of entries of our dataset is exactly N.   □

5.5   Allocating Huge Lists with a Read-All Algorithm

Keyword lists that have size greater than $N/\log^{\gamma} N$ up to N are stored directly in an array $\mathbf{A}$ of N entries, one after the other—see Fig. 8. To read a huge list in our actual construction, one would have to read the whole array $\mathbf{A}$—however, due to the huge size of the list, the read efficiency would still be small.

Algorithm $(A, \mathsf{Tab}) \leftarrow$ AllocateHuge$(\mathcal{D}, N)$:
1: Let $\mathsf{min} \leftarrow N/\log^{\gamma} N$;
2: Initialize an empty hash table Tab;
3: Let $\mathbf{A}$ be an array of N entries; $\mathsf{count} \leftarrow 1$;
4: **for** all keywords w such that $|\mathcal{D}(w)| > \mathsf{min}$ **do**
5:     Store $\mathcal{D}(w)$ in positions $\mathsf{count}, \mathsf{count} + 1, \ldots, \mathsf{count} + |\mathcal{D}(w)| - 1$ of array $\mathbf{A}$;
6:     $\mathsf{count} \leftarrow \mathsf{count} + |\mathcal{D}(w)|$;
7:     $\mathsf{Tab}[w] \leftarrow (|\mathcal{D}(w)|, \bot, \bot, \bot)$;
8: **return** $(\mathbf{A}, \mathsf{Tab})$;

Fig. 8. Allocation algorithm for huge sizes.

6   Our SE Construction

We now present our main construction that uses the ORAM scheme presented in Sect. 4 and the allocation algorithms presented in Sect. 5 as black boxes. Our formal protocols are shown in Figs. 9 and 10.

6.1   Setup Protocol of SE Scheme

Our setup algorithm allocates lists depending on whether they are small, medium, large or huge, as defined in Sect. 5. We describe the details below.

Small Keyword Lists. These are allocated to superbuckets using AllocateSmall from Sect. 5.2. The allocation algorithm outputs an array of buckets $\mathbf{S}$ storing the small keyword lists and the instructions hash table Tab_S storing, for each small keyword list $\mathcal{D}(w)$, its size s and the superbuckets α and β assigned for this keyword list by the allocation algorithm. The setup protocol of the SE scheme finally encrypts and writes bucket array $\mathbf{S}$ and stores it remotely—see Line 5 in Fig. 9. It stores Tab_S locally.

Medium Keyword Lists. These are allocated to superbuckets using AllocateMedium from Sect. 5.3. AllocateMedium outputs (i) an array of buckets $\mathbf{M}$; (b) the set of stashes $\{\mathbf{B}_s\}_s$ that handle the overflows, for all sizes s in the range; (iii) the instructions hash table Tab_M storing, for each keyword list $\mathcal{D}(w)$ that falls into this range, its size s, the superbuckets α and β assigned for this keyword list and a stash position x in the stash $\mathbf{B}_s$ where the specific keyword list could have been potentially stored, had it caused an overflow (otherwise a dummy position is stored). The setup protocol finally encrypts and writes $\mathbf{M}$ and stores it remotely—see Line 8 in Fig. 9. It also builds an ORAM per stash $\mathbf{B}_s$—see Line 15 in Fig. 9. Finally, it stores Tab_M locally.

Large Keyword Lists. These are allocated to buckets using AllocateLarge from Sect. 5.4. To keep read efficiency small, we run AllocateLarge for $\frac{2-\gamma}{\mathsf{step}}$ distinct subranges, as we detailed in Sect. 5. For the subrange of $(N/\log^{2-(h-1)\cdot\mathsf{step}} N, N/\log^{2-h\cdot\mathsf{step}} N]$, AllocateLarge outputs an array of buckets $\mathbf{L}_h$ and a hash table $\mathsf{Tab}_{\mathbf{L}_h}$. The setup protocol builds an ORAM for the array $\mathbf{L}_h$ and it stores $\mathsf{Tab}_{\mathbf{L}_h}$ locally.

Huge Keyword Lists. For these lists, we use AllocateHuge from Sect. 5.5. This algorithm outputs an array $\mathbf{H}$ and a hash table Tab_H. Our setup protocol encrypts and writes $\mathbf{H}$ remotely and stores Tab_H locally.

Local State and Using Tokens. For the sake of simplicity and readability of Fig. 9, we assume that the client keeps locally the hash table Tab—see Line 13. This occupies linear space $O(N)$ but can be securely outsourced using standard SE techniques [31], and without affecting the efficiency (read efficiency and locality): For every hash table entry $w \mapsto [s, \alpha, \beta, x]$, store at the server the "encrypted" hash table entry $t_w \to \mathsf{ENC}_{k_w}(s||\alpha||\beta||x)$, where t_w and k_w comprise the *tokens* for keyword w (these are the outputs of a PRF applied on w with two different secret keys that the client stores) and ENC is a CPA-secure encryption scheme. To search for keyword w, the client just needs to send to the server the tokens t_w and k_w and the server can then search the encrypted hash table and retrieve the information $s||\alpha||\beta||x$ by decrypting.

Handling ORAM State and Failures. Our setup protocol does not store locally the ORAM states σ_s and σ_h of the stashes $\mathbf{B}_s$ and the arrays $\mathbf{L}_h$ for which we build an ORAM. Instead, it encrypts and writes them remotely and downloads them when needed—see Line 17 in Fig. 9. Also, our setup algorithm fails whenever any of the allocation algorithms fail. By Theorems 6, 7 and 8 we have the following:

Lemma 6. *Protocol* SETUP *in Fig. 9 fails with probability* $\mathsf{neg}(N)$.

Protocol $\langle st, \mathcal{I} \rangle \leftrightarrow \text{SETUP}\langle (1^\kappa, \mathcal{D}), \bot \rangle$:

1: Let $N \leftarrow \sum_{w \in \mathbf{W}} |\mathcal{D}(w)|$; Set **step** $< 3\delta/2$;
2: Let Tab be an empty hash table of capacity N;
3: $(\mathbf{S}, \text{Tab}_S) \leftarrow \text{AllocateSmall}(\mathcal{D}, N)$;
4: **for** all buckets $\mathbf{S}[i] \in \mathbf{S}$ **do**
5:     **Encrypt-And-Write** bucket $\mathbf{S}[i]$ and add encrypted $\mathbf{S}[i]$ to server index $\mathcal{I}$;
6: $(\mathbf{M}, \mathbf{B}_M, \text{Tab}_M) \leftarrow \text{AllocateMedium}(\mathcal{D}, N)$;
7: **for** all buckets $\mathbf{M}[i] \in \mathbf{M}$ **do**
8:     **Encrypt-And-Write** bucket $\mathbf{M}[i]$ and add encrypted $\mathbf{M}[i]$ to server index $\mathcal{I}$;
9: **for** $h = 1, \ldots, \frac{2-\gamma}{\text{step}}$ **do**
10:     $(\mathbf{L}_h, \text{Tab}_{L_h}) \leftarrow \text{AllocateLarge}(\mathcal{D}, N, N/\log^{2-(h-1)\cdot\text{step}} N, N/\log^{2-h\cdot\text{step}} N)$;
11: $(\mathbf{H}, \text{Tab}_H) \leftarrow \text{AllocateHuge}(\mathcal{D}, N)$;
12: **Encrypt-And-Write** array $\mathbf{H}$ and add encrypted $\mathbf{H}$ to server index $\mathcal{I}$;
13: Set Tab $\leftarrow$ Tab$_S$ $\cup$ Tab$_M$ $\cup$ $\left(\bigcup_{h=1}^{\frac{2-\gamma}{\text{step}}} \text{Tab}_{L_h} \right)$;
14: $st \leftarrow$ Tab;
15: **for** every stash $\mathbf{B}_s \in \mathbf{B}_M$ corresponding to size s **do**
16:     $\langle \sigma_s, \text{EM}_s \rangle \leftrightarrow \text{ORAMINITIALIZE}\langle (1^\kappa, \mathbf{B}_s), \bot \rangle$;
17:     **Encrypt-And-Write** σ_s and add σ_s and EM_s to server index $\mathcal{I}$;
18: **for** $h = 1, \ldots, \frac{2-\gamma}{\text{step}}$ **do**
19:     $\langle \sigma_h, \text{EM}_h \rangle \leftrightarrow \text{ORAMINITIALIZE}\langle (1^\kappa, \mathbf{L}_h), \bot \rangle$;
20:     **Encrypt-And-Write** σ_h and add σ_h and EM_h to server index $\mathcal{I}$;
21: **if** AllocateSmall or AllocateMedium or AllocateLarge called above output FAIL **then**
22:     **return** FAIL;
23: **return** $\langle st, \mathcal{I} \rangle$;

Fig. 9. The setup protocol of our SE construction.

Lemma 7. *Protocol* SETUP *in Fig. 9 outputs an encrypted index* $\mathcal{I}$ *that has* $O(N)$ *size and runs in* $O(N)$ *time.*

Proof. The space complexity follows from Theorems 6 and 7, by the fact that array $\mathbf{H}$ output by AllocateHuge has size $O(N)$, by the fact that we keep a number of arrays for large keyword lists that is independent of N, and by the fact that the ORAM states σ_s and σ_h, being asymptotically less than the ORAM themselves, occupy at most linear space. For the running time, note that AllocateSmall, AllocateLarge, AllocateHuge run in linear time and the ORAM setup algorithms also run in linear time (same analysis with the space can be made). By Lemma 1, AllocateMedium must perform a costly $O(n^3)$ offline allocation (a maximum flow computation) where n is the number of superbuckets defined for every size s in the range. The maximum number of superbuckets M is achieved for the smallest size handled by AllocateMedium and is equal to $M = \frac{N}{N^{1-1/\log^{1-\gamma} N} \cdot \log^\gamma N} = N^{1/\log^{1-\gamma} N}/\log^\gamma N$.

Recall that there are at most $\log^\gamma N$ sizes handled by AllocateMedium and therefore the time required to do the offline allocation is at most $O\left(\log^\gamma N \cdot M^3 \right)$

Protocol $\langle(\mathcal{D}(w), st'), \mathcal{I}'\rangle \leftrightarrow \text{SEARCH}\langle(st, w), \mathcal{I}\rangle$:

1: Parse st as Tab and $\mathcal{I}$ as $(\mathbf{S}, \mathbf{M}, \mathbf{H}, \{\sigma_s, \text{EM}_s\}, \{\sigma_h, \text{EM}_h\})$;
2: Let $(s, \alpha, \beta, x) \leftarrow \text{Tab}[w]$; Set step $< 3\delta/2$;
3: **if** $s > N/\log^\gamma N$ **then**   ▷ Huge sizes
4:      **Read-And-Decrypt** array $\mathbf{H}$;
5:      Retrieve $\mathcal{D}(w)$ from $\mathbf{H}$;
6: **else**
7:      **if** $s \leq N^{1-1/\log^{1-\gamma} N}$ **then**   ▷ Small sizes
8:          **Read-And-Decrypt** superbuckets $\mathbf{S}\{\alpha, s\}$ and $\mathbf{S}\{\beta, s\}$;
9:          Retrieve $\mathcal{D}(w)$ from $\mathbf{S}\{\alpha, s\}$ and $\mathbf{S}\{\beta, s\}$;
10:     **else**
11:         **if** $N^{1-1/\log^{1-\gamma} N} < s \leq N/\log^2 N$ **then**   ▷ Medium sizes
12:             **Read-And-Decrypt** σ_s;
13:             $\langle(v_x, \sigma_s), \text{EM}_s\rangle \leftrightarrow \text{ORAMACCESS}\langle(\sigma_s, x), \text{EM}_s\rangle$;
14:             **Encrypt-And-Write** σ_s;
15:             **Read-And-Decrypt** superbuckets $\mathbf{M}\{\alpha, s\}$ and $\mathbf{M}\{\beta, s\}$;
16:             Retrieve $\mathcal{D}(w)$ from $\mathbf{M}\{\alpha, s\}$ and $\mathbf{M}\{\beta, s\}$ or v_x;
17:         **else**   ▷ Large sizes
18:             Find $h \in \{1, \ldots, \frac{2-\gamma}{\text{step}}\}$ s.t. $N/\log^{2-(h-1)\cdot\text{step}} N < s \leq N/\log^{2-h\cdot\text{step}} N$;
19:             **Read-And-Decrypt** σ_h;
20:             $\langle(v_\alpha, \sigma_h), \text{EM}_h\rangle \leftrightarrow \text{ORAMACCESS}\langle(\sigma_h, \alpha), \text{EM}_h\rangle$;
21:             **Encrypt-And-Write** σ_h;
22:             Retrieve $\mathcal{D}(w)$ from v_α;
23: **return** $\langle(\mathcal{D}(w), st), \mathcal{I}\rangle$;

Fig. 10. The search protocol of our SE construction.

which is equal to $O(N^{3/\log^{1-\gamma} N}/\log^{2\gamma} N) = O(N)$. Therefore, the running time is $O(N)$.   □

6.2   Search Protocol of SE Scheme

Given a keyword w, the client first retrieves information (s, α, β, x) from $\text{Tab}[w]$. Depending on the size s of $\mathcal{D}(w)$ the client takes the following actions (see Fig. 10):

- If the list $\mathcal{D}(w)$ is *small*, the client reads two superbuckets $\mathbf{S}\{\alpha, s\}$ and $\mathbf{S}\{\beta, s\}$ and decrypts them. Since the size of the buckets $\mathbf{S}[i]$ is $\log^\gamma N$ and each superbucket contains s of them, it follows that the read efficiency for small sizes is $\Theta(\log^\gamma N)$. Also, since only two superbuckets are read, the locality for small lists is $O(1)$.

- If the list $\mathcal{D}(w)$ is *medium*, the client reads two superbuckets $\mathbf{M}\{\alpha, s\}$ and $\mathbf{M}\{\beta, s\}$ and decrypts them. Also he performs an ORAM access in the stash $\mathbf{B}_s$ for location x. Since the size of the buckets $\mathbf{M}[i]$ is $O(\log^\gamma N)$ and each superbucket has s of them, the read efficiency for medium sizes due to accessing array $\mathbf{M}$ is $O(\log^\gamma N)$.
  For the ORAM access, note that in our case it is $n = c \cdot \log^2 N$. Therefore, by Corollary 1, and since our block size is at least $N^{1-1/\log\log N}$

which is $\Omega(\log^{2/3} N)$, the bandwidth required is $O(n^{1/3} \log^2 n \cdot s) = O(\log^{2/3} N \log^2 \log N \cdot s)$ and therefore the read efficiency due to the ORAM access is $O(\log^{2/3} N \log^2 \log N) = o(\log^\gamma N)$, since $\gamma = 2/3 + \delta$. Therefore, the overall read efficiency for medium sizes is $O(\log^\gamma N)$. Again, since only two superbuckets are read and the ORAM locality is $O(1)$ (Corollary 1), it follows that the locality for medium lists is $O(1)$.

- Suppose now the list $\mathcal{D}(w)$ is large such that $\mathsf{min} < |\mathcal{D}(w)| \le \mathsf{max}$ where $\mathsf{min} = N/\log^{2-(h-1)\cdot\mathsf{step}} N$ and $\mathsf{max} = N/\log^{2-h\cdot\mathsf{step}} N$ for some $h \in \{1, 2, \ldots, \frac{2-\gamma}{\mathsf{step}}\}$. To retrieve the list, our search algorithm performs our ORAM access on an array on N/max blocks of size $2 \cdot \mathsf{max}$ each. By Corollary 1, we have that the worst-case bandwidth for this access is

$$O\left(\left(\frac{N}{\mathsf{max}}\right)^{1/3} \log^2\left(\frac{N}{\mathsf{max}}\right) \mathsf{max}\right) = O\left(N\left(\log^{2-h\cdot\mathsf{step}} N\right)^{-2/3} \log^2 \log N\right).$$

For read efficiency, note that the client must use this bandwidth to read a keyword list of size $s \ge \mathsf{min} = N/\log^{2-(h-1)\cdot\mathsf{step}} N$. Thus, the read efficiency is at most

$$O\left(\log^{2-(h-1)\cdot\mathsf{step}} N \cdot \left(\log^{2-h\cdot\mathsf{step}} N\right)^{-2/3} \log^2 \log N\right) = O(\log^{\gamma'} N \log^2 \log N),$$

where for all $h \ge 1$ it is $\gamma' \le 2/3 + 2 \cdot \mathsf{step}/3 < 2/3 + \delta = \gamma$ since $\mathsf{step} < 3\delta/2$. Therefore, the above is $o(\log^\gamma N)$ as required.

- For huge sizes, the read efficiency is at most $O(\log^\gamma N)$ and the locality is constant since the whole array $\mathbf{H}$ is read.

Therefore, overall, the locality is $O(1)$, the read efficiency is $O(\log^\gamma N)$ and the space required at the server is $O(N)$.

Rounds of Interaction. Our protocol requires $O(1)$ rounds for interaction for each query. In particular, for small and huge list sizes our construction requires a single round of interaction, as can be easily inferred from Fig. 10. For medium and large sizes, the deamortized version of our protocol which uses the deamortized ORAM from the extended version [13], requires four rounds of interaction.

Client Space. Finally, we measure the storage at the client (assuming, as discussed in Sect. 6.1 that Tab is stored at the server). For small lists, it follows from our above analysis for read efficiency that the storage at the client is $O(\log^\gamma N \cdot s)$. Note that, from Corollary 1, for medium and large list sizes the necessary space at the client due to the ORAM protocol is $O(n^{1/3} \log^2 n \cdot s)$, where n is the number of ORAM indices and s is the result list size (this result uses the deamortized version of our ORAM from the extended version [13]). Since $n \le \log^2 N$, this becomes $O(\log^{2/3} N \log^2 \log N \cdot s)$. Specifically for medium lists, the client also needs to download two superbuckets for total storage $O(\log^\gamma N \cdot s)$. For huge list sizes, recall that the client downloads the entire array $\mathbf{H}$ which results in space $O(N)$. However, note that in this case $s > N/\log^\gamma N$, therefore $N < s \cdot \log^\gamma N$

and the client storage can be written as $O(\log^{\gamma} N \cdot s)$. We stress that any searchable encryption scheme requires $\Omega(s)$ space at the client simply to download the result of a query. Thus, in all cases our scheme imposes just a multiplicative overhead for the client storage that is sub-logarithmic in the database size, compared to the minimum requirement. Moreover, we stress that this storage is *transient*, i.e., it is only necessary when issuing a query; between queries, the client requires $O(1)$ space.

6.3   Security of Our Construction

We now prove the security of our construction. For this, we build a simulator SIMSETUP and SIMSEARCH in Figs. 11 and 12 respectively.

Algorithm $(st_S, \mathcal{I}_0) \leftarrow$ SIMSETUP$(1^{\kappa}, \mathcal{L}_1(\mathcal{D}_0))$:

1: Parse $\mathcal{L}_1(\mathcal{D}_0)$ as N;
2: Let **S** to be an array that contains N dummy elements; **Encrypt-And-Write S**;
3: Let **M** to be an array of N dummy elements; **Encrypt-And-Write M**;
4: **for** $h = 1, 2, \ldots, \frac{2-\gamma}{\mathsf{step}}$ **do**
5:     Set $\mathsf{max} = N / \log^{2-h \cdot \mathsf{step}} N$;
6:     $(st_S^h, \mathsf{EM}_h) \leftarrow$ SIMORAMINITIALIZE$(1^{\kappa}, (N/\mathsf{max}, \mathsf{max}))$;
7:     Parse st_S^h as σ_h; **Encrypt-And-Write** σ_h;
8: Let **H** be an array of N dummy elements; **Encrypt-And-Write H**;
9: Let $\mathsf{min} = N^{1-1/\log^{1-\gamma} N}$ and $\mathsf{max} = N / \log^2 N$
10: **for** $s = 2\mathsf{min}, 4\mathsf{min}, 8\mathsf{min}, \ldots, \mathsf{max}$ **do**
11:     Set $\mathbf{B}_s$ to be an array of $c \cdot \log^2 N$ entries of s dummy elements each;
12:     $(st_S^s, \mathsf{EM}_s) \leftarrow$ SIMORAMINITIALIZE$(1^{\kappa}, (c \cdot \log^2 N, s))$;
13:     Parse st_S^s as σ_s; **Encrypt-And-Write** σ_s;
14: Let messages be an empty hash table;
15: Set $\mathcal{I}_0 = (\mathbf{S}, \mathbf{M}, \mathbf{H}, \{\sigma_s, \mathsf{EM}_s\}, \{\sigma_h, \mathsf{EM}_h\})$;
16: **return** $((N, \mathsf{messages}, \{st_S^s\}, \{st_S^h\}, \mathcal{I}_0);$

Fig. 11. The simulator of the setup protocol of our SE scheme.

Simulation of the Setup Protocol. To simulate the setup protocol, our simulator must output $\mathcal{I}_0$ by just using the leakage $\mathcal{L}_1(\mathcal{D}_0) = N$. Our SIMSETUP algorithm outputs $\mathcal{I}_0$ as CPA-secure encryptions of arrays $(\mathbf{S}, \mathbf{M}, \mathbf{H})$ that contain dummy values and have the same dimensions with the arrays of the actual setup algorithm. Also, it calls the ORAM simulator and also outputs $\{\sigma_s, \mathsf{EM}_s\})$ and $\{\sigma_h, \mathsf{EM}_h\})$. Due to the security of the underlying ORAM scheme and the CPA-security of the underlying encryption scheme, the adversary cannot distinguish between the two outputs.

One potential problem, however, is the fact that SIMSETUP always succeeds while there is a chance that the setup algorithm can fail, which will enable the adversary to distinguish between the two. However, by Lemma 6, this happens with probability $\mathsf{neg}(N) = \mathsf{neg}(\kappa)$, as required by our security definition, Definition 1.

Simulation of the Search Protocol. The simulator of the SEARCH protocol is shown in Fig. 12. For a keyword query w_k, the simulator takes as input the leakage $\mathcal{L}_2(w_k) = (s, b)$, as defined in Relation 1.

If the query on w_k was performed before (thus $b \neq \bot$), the simulator just outputs the previous messages M_b plus the messages that were output by the ORAM simulator.

If the query on w_k was not performed before, then the simulator generates the messages M_k depending on the size s of the list $\mathcal{D}(w_k)$. In particular note that all accesses on $(\mathbf{S}, \mathbf{M}, \mathbf{H}, \mathbf{L}_h)$ are independent of the dataset and therefore can be simulated by repeating the same process with the real execution.

7   Conclusions and Observations

Basing the Entire Scheme on ORAM. Our construction is using ORAM as a black box and therefore one could wonder why not use ORAM from the very beginning and on the whole dataset. While ORAM can provide much better security guarantees, it suffers from high read efficiency. E.g., to the best of our knowledge, there is no ORAM that we could use that yields sublogarithmic read efficiency (irrespective of the locality).

Avoiding the Lower Bound of [6]**.** We note that Proposition 4.6 by Asharov et al. [6] states that one could not expect to construct an allocation algorithm where the square of the locality × the read efficiency is $O(\log N / \log \log N)$. This is the case with our construction! The reason this proposition does not apply to our approach is because our allocation algorithm is using multiple structures for storage, e.g., stashes and multiple arrays, and therefore does not fall into the model used to prove the negative result.

Reducing the ORAM Read Efficiency. Our technique for building our ORAM in Sect. 4 relies on one hierarchical application of the method of square-root ORAM [18]. We believe this approach can be generalized to yield read efficiency $O(n^{1/k} \log^2 n \cdot \lambda)$ for general k. The necessary analysis, while tedious, seems technically non-challenging and we leave it for future work (e.g., we could revisit some ideas from [34]). Such an ORAM could also help us decrease the number of subranges on which we apply our AllocateLarge algorithm.

Using Online Two-Choice Allocation. Our construction uses the offline variant of the two-choice allocation problem. This allows us to achieve low bounds on both the number of overflowing bins and the total overflow size in Sect. 3. However it requires executing a maximum flow algorithm during our construction's setup. A natural question is whether we can use instead the (more efficient) *online* two-choice allocation problem. The best known result [9] for the online version yields a maximum load of $O(\log \log n)$ beyond the expected value m/n, which suffices to bound the maximum number of overflowing bins with our technique. However, deriving a similar bound for the total overflow size would require entirely different techniques and we leave it as an open problem. Still, it seems that even if we could get the same bound for the overflow size as in the offline

---

Algorithm $(st_\mathcal{S}, M_k, \mathcal{I}_k) \leftarrow$ SIMSEARCH$(st_\mathcal{S}, \mathcal{L}_2(w_k), \mathcal{I}_{k-1})$:

1: Parse $st_\mathcal{S}$ as $(N, \mathsf{messages}, \{st_\mathcal{S}^s\}, \{st_\mathcal{S}^h\})$;
2: Parse $\mathcal{I}_{k-1}$ as $(\mathbf{S}, \mathbf{M}, \mathbf{H}, \{\sigma_s, \mathsf{EM}_s\}, \{\sigma_h, \mathsf{EM}_h\})$;
3: Parse $\mathcal{L}_2(w_k)$ as (s, b);
4: Set $m_k = null$; $m_1 = null$; $m_2 = null$;
5: **if** $N/\log^2 N < s \le N/\log^\gamma N$ **then**   ▷ For large sizes, perform a fresh ORAM access
6:     Find $h \in \{1, 2, \ldots, \frac{2-\gamma}{\mathsf{step}}\}$ such that $N/\log^{2-(h-1)\cdot\mathsf{step}} N < s \le N/\log^{2-h\cdot\mathsf{step}} N$;
7:     **Read-And-Decrypt** σ_h. Let m_1 be this message;
8:     $(st_\mathcal{S}^h, \mathsf{EM}_h, m_k) \leftarrow$ SIMORAMACCESS$(st_\mathcal{S}^h, \mathsf{EM}_h)$;
9:     **Encrypt-And-Write** σ_h. Let m_2 be this message;
10:     Set $\mathsf{messages}[k] \leftarrow null$;
11:     Set $st_\mathcal{S} \leftarrow (N, \mathsf{messages}, \{st_\mathcal{S}^s\}, \{st_\mathcal{S}^h\})$;
12:     Set $\mathcal{I}_k \leftarrow (\mathbf{S}, \mathbf{M}, \mathbf{H}, \{\sigma_s, \mathsf{EM}_s\}, \{\sigma_h, \mathsf{EM}_h\})$;
13:     **return** $(st_\mathcal{S}, (m_k, m_1, m_2), \mathcal{I}_k)$;
14: **if** $b \ne \bot$ **then**   ▷ Query has been asked before
15:     **if** $N^{1-1/\log^{1-\gamma} N} < s \le N/\log^2 N$ **then**
16:         **Read-And-Decrypt** σ_h. Let m_1 be this message;
17:         $(st_\mathcal{S}^s, \mathsf{EM}_s, m_k) \leftarrow$ SIMORAMACCESS$(st_\mathcal{S}^s, \mathsf{EM}_s)$;
18:         **Encrypt-And-Write** σ_h. Let m_2 be this message;
19:     Set $st_\mathcal{S} \leftarrow (N, \mathsf{messages}, \{st_\mathcal{S}^s\}, \{st_\mathcal{S}^h\})$;
20:     Set $M_k \leftarrow (\mathsf{messages}[b], (m_k, m_1, m_2))$;
21:     Set $\mathcal{I}_k \leftarrow (\mathbf{S}, \mathbf{M}, \mathbf{H}, \{\sigma_s, \mathsf{EM}_s\}, \{\sigma_h, \mathsf{EM}_h\})$;
22:     **return** $(st_\mathcal{S}, M_k, \mathcal{I}_k)$;
23: **if** $s > N/\log^\gamma N$ **then**   ▷ Huge sizes
24:     **Read-And-Decrypt** array $\mathbf{H}$;
25:     Add the above message to $\mathsf{messages}[k]$;
26: **if** $s \le N^{1-1/\log^{1-\gamma} N}$ **then**   ▷ Small sizes
27:     Set $C \leftarrow c_s \cdot \log^\gamma N^a$ and $B \leftarrow N/C$;
28:     Pick α and β independently and uniformly at random from $\{1, 2, \ldots, \frac{B}{s}\}$;
29:     **Read-And-Decrypt** superbuckets $\mathbf{S}\{\alpha, s\}$ and $\mathbf{S}\{\beta, s\}$;
30:     Add the above message to $\mathsf{messages}[k]$;
31: **if** $N^{1-1/\log^{1-\gamma} N} < s \le N/\log^2 N$ **then**   ▷ Medium sizes
32:     Set $C \leftarrow 3 \cdot \log^\gamma N$ and $B \leftarrow N/C$;
33:     Pick α and β independently and uniformly at random from $\{1, 2, \ldots, \frac{B}{s}\}$;
34:     **Read-And-Decrypt** superbuckets $\mathbf{M}\{\alpha, s\}$ and $\mathbf{M}\{\beta, s\}$;
35:     Add the above message to $\mathsf{messages}[k]$;
36:     **Read-And-Decrypt** σ_h. Let m_1 be this message;
37:     $(st_\mathcal{S}^s, \mathsf{EM}_s, m_k) \leftarrow$ SIMORAMACCESS$(st_\mathcal{S}^s, \mathsf{EM}_s)$;
38:     **Encrypt-And-Write** σ_h. Let m_2 be this message;
39: Set $st_\mathcal{S} \leftarrow (N, \mathsf{messages}, \{st_\mathcal{S}^s\}, \{st_\mathcal{S}^h\})$;
40: Set $M_k \leftarrow (\mathsf{messages}[k], (m_k, m_1, m_2))$;
41: Set $\mathcal{I}_k \leftarrow (\mathbf{S}, \mathbf{M}, \mathbf{H}, \{\sigma_s, \mathsf{EM}_s\}, \{\sigma_h, \mathsf{EM}_h\})$;
42: **return** $(st_\mathcal{S}, M_k, \mathcal{I}_k)$;

---

^{a} Constant c_s is appropriately chosen in [6].

Fig. 12. The simulator of the search protocol of our SE scheme.

case, the read efficiency would be $O(\log^\gamma N \log\log N)$, as opposed to the better $O(\log^\gamma N)$, which is what we achieve here.

Reducing the Read Efficiency for Small Lists. The read efficiency of our scheme for small lists can be strictly improved if instead of using [6], we use the construction of Asharov et al. [7] that was proposed in concurrent work. In this manner, the read efficiency for a small keyword list with size $N^{1-\epsilon}$ would be $\omega(1) \cdot \epsilon^{-1} + O(\log\log\log N)$.

Acknowledgments. We thank Jiaheng Zhang for indicating a tighter analysis for Theorem 6 and for his feedback on the algorithm for allocating large keyword lists, and the reviewers for their comments. Work supported in part by NSF awards #1526950, #1514261 and #1652259, HKUST award IGN16EG16, a Symantec PhD fellowship, and a NIST award.

References

1. Crimes 2001 to present (City of Chicago). https://data.cityofchicago.org/public-safety/crimes-2001-to-present/ijzp-q8t2
2. Enron Email Dataset. https://www.cs.cmu.edu/./enron/
3. TPC-H Dataset. http://www.tpc.org/tpch/
4. USPS Dataset. http://www.app.com
5. Asharov, G., Chan, T.H., Nayak, K., Pass, R., Ren, L., Shi, E.: Oblivious computation with data locality. IACR Cryptology ePrint (2017)
6. Asharov, G., Naor, M., Segev, G., Shahaf, I.: Searchable symmetric encryption: optimal locality in linear space via two-dimensional balanced allocations. In: STOC (2016)
7. Asharov, G., Segev, G., Shahaf, I.: Tight tradeoffs in searchable symmetric encryption. In: Shacham, H., Boldyreva, A. (eds.) CRYPTO 2018. LNCS, vol. 10991, pp. 407–436. Springer, Heidelberg (2018)
8. Batcher, K.E.: Sorting networks and their applications. In: AFIPS (1968)
9. Berenbrink, P., Czumaj, A., Steger, A., Vöcking, B.: Balanced allocations: the heavily loaded case. In: STOC (2000)
10. Cash, D., et al.: Dynamic searchable encryption in very-large databases: data structures and implementation. In: NDSS (2014)
11. Cash, D., Tessaro, S.: The locality of searchable symmetric encryption. In: Nguyen, P.Q., Oswald, E. (eds.) EUROCRYPT 2014. LNCS, vol. 8441, pp. 351–368. Springer, Heidelberg (2014). https://doi.org/10.1007/978-3-642-55220-5_20
12. Curtmola, R., Garay, J.A., Kamara, S., Ostrovsky, R.: Searchable symmetric encryption: improved definitions and efficient constructions. JCS **9**(5), 895–934 (2011)
13. Demertzis, I., Papadopoulos, D., Papamanthou, C.: Searchable encryption with optimal locality: achieving sublogarithmic read efficiency. In: CRYPTO 2018 (2018). https://eprint.iacr.org/2017/749
14. Demertzis, I., Papadopoulos, S., Papapetrou, O., Deligiannakis, A., Garofalakis, M.: Practical private range search revisited. In: SIGMOD (2016)
15. Demertzis, I., Papadopoulos, S., Papapetrou, O., Deligiannakis, A., Garofalakis, M., Papamanthou, C.: Practical private range search in depth. In: TODS (2018)
16. Demertzis, I., Papamanthou, C.: Fast searchable encryption with tunable locality. In: SIGMOD (2017)

17. Dubhashi, D.P., Ranjan, D.: Balls and bins: a study in negative dependence. Random Struct. Algorithms **13**(2), 99–124 (1998)
18. Goldreich, O., Ostrovsky, R.: Software protection and simulation on oblivious rams. J. ACM **43**(3), 431–473 (1996)
19. Goodrich, M.T.: Data-oblivious external-memory algorithms for the compaction, selection, and sorting of outsourced data. In: SPAA (2011)
20. Goodrich, M.T., Mitzenmacher, M.: Privacy-preserving access of outsourced data via oblivious RAM simulation. In: Aceto, L., Henzinger, M., Sgall, J. (eds.) ICALP 2011. LNCS, vol. 6756, pp. 576–587. Springer, Heidelberg (2011). https://doi.org/10.1007/978-3-642-22012-8_46
21. Goodrich, M.T., Mitzenmacher, M., Ohrimenko, O., Tamassia, R.: Oblivious RAM simulation with efficient worst-case access overhead. In: CCSW (2011)
22. Granboulan, L., Pornin, T.: Perfect block ciphers with small blocks. In: Biryukov, A. (ed.) FSE 2007. LNCS, vol. 4593, pp. 452–465. Springer, Heidelberg (2007). https://doi.org/10.1007/978-3-540-74619-5_28
23. Kamara, S., Papamanthou, C.: Parallel and dynamic searchable symmetric encryption. In: Sadeghi, A.-R. (ed.) FC 2013. LNCS, vol. 7859, pp. 258–274. Springer, Heidelberg (2013). https://doi.org/10.1007/978-3-642-39884-1_22
24. Kamara, S., Papamanthou, C., Roeder, T.: Dynamic searchable symmetric encryption. In: CCS (2012)
25. Miers, I., Mohassel, P.: IO-DSSE: scaling dynamic searchable encryption to millions of indexes by improving locality. In: NDSS (2017)
26. Morris, B., Rogaway, P.: Sometimes-recurse shuffle - almost-random permutations in logarithmic expected time. In: Nguyen, P.Q., Oswald, E. (eds.) EUROCRYPT 2014. LNCS, vol. 8441, pp. 311–326. Springer, Heidelberg (2014). https://doi.org/10.1007/978-3-642-55220-5_18
27. Ohrimenko, O., Goodrich, M.T., Tamassia, R., Upfal, E.: The Melbourne shuffle: improving oblivious storage in the cloud. In: Esparza, J., Fraigniaud, P., Husfeldt, T., Koutsoupias, E. (eds.) ICALP 2014. LNCS, vol. 8573, pp. 556–567. Springer, Heidelberg (2014). https://doi.org/10.1007/978-3-662-43951-7_47
28. Sanders, P., Egner, S., Korst, J.H.M.: Fast concurrent access to parallel disks. Algorithmica **35**(1), 21–55 (2003)
29. Schoenmakers, L.A.: A new algorithm for the recognition of series parallel graphs. Technical report, Amsterdam, The Netherlands (1995)
30. Song, D.X., Wagner, D., Perrig, A.: Practical techniques for searches on encrypted data. In: SP (2000)
31. Stefanov, E., Papamanthou, C., Shi, E.: Practical dynamic searchable encryption with small leakage. In: NDSS (2014)
32. Stefanov, E., Shi, E.: FastPRP: fast pseudo-random permutations for small domains. IACR Cryptology ePrint (2012)
33. Stefanov, E., et al.: Path ORAM: an extremely simple oblivious RAM protocol. In: CCS (2013)
34. Zahur, S., et al.: Revisiting square-root ORAM: efficient random access in multiparty computation. In: SP (2016)

Appendix

Definition 2 (Correctness of ORAM). *Let* (ORAMINITIALIZE, ORAMACCESS) *be an ORAM scheme. Let* $\langle\sigma_0, \mathsf{EM}_0\rangle \leftrightarrow$ ORAMINITIALIZE$\langle(1^\kappa, \mathsf{M}_0), 1^\kappa)\rangle$ *for some initial memory* M_0 *of n indexed values* $(1, v_1), (2, v_2), \ldots, (n, v_n)$. *Consider q arbitrary requests* $i_1, \ldots, i_q$. *We say that the ORAM scheme is correct if* $\langle(v_{i_k}, \sigma_k), \mathsf{EM}_k\rangle$ *are the final outputs of the protocol* ORAMACCESS$\langle(\sigma_{k-1}, i_k), \mathsf{EM}_{k-1}\rangle$ *for any* $1 \leq k \leq q$, *where* M_k, EM_k, σ_k *are the memory array, the encrypted memory array and the secret state, respectively, after the k-th access operation, and* ORAMACCESS *is run between an honest client and server.*

Definition 3 (Security of ORAM). *Assume* (ORAMINITIALIZE, ORAMACCESS) *is an ORAM scheme. The ORAM scheme is secure if for any PPT adversary* Adv, *there exists a stateful PPT simulator* (SIMORAMINITIALIZE, SIMORAMACCESS) *such that* $|\Pr[\mathbf{Real}^{\mathsf{ORAM}}(\kappa) = 1] - \Pr[\mathbf{Ideal}^{\mathsf{ORAM}}(\kappa) = 1]| \leq \mathsf{neg}(\kappa)$, *where experiments* $\mathbf{Real}^{\mathsf{ORAM}}(\kappa)$ *and* $\mathbf{Ideal}^{\mathsf{ORAM}}(\kappa)$ *are defined in Fig. 14 and where the randomness is taken over the random bits used by the algorithms of the ORAM scheme, the algorithms of the simulator and* Adv.

Definition 4 (Dubhashi and Ranjan [17]). *A set of random variables* $\{X_1, \ldots, X_n\}$ *is negatively associated if for every two disjoint index sets* $I \in [n]$ *and* $J \subseteq [n]$ *it is*

$$\mathbb{E}[f(X_i, i \in I)g(X_j, j \in J)] \leq \mathbb{E}[f(X_i, i \in I)]\mathbb{E}[g(X_j, j \in J)]$$

(chosen, alternative) ← MaxFlowSchedule$(m, n, \mathsf{A}, \mathsf{B})$

1: Let G be a graph that has n nodes and the following m unit-capacity directed edges
   $\{(\mathsf{A}[1], \mathsf{B}[1]), (\mathsf{A}[2], \mathsf{B}[2]) \ldots, (\mathsf{A}[m], \mathsf{B}[m])\}$;
2: Let s and t be two new nodes added to G serving as the source and the sink;
3: For all $v \in G$ such that $indeg(v) > \lceil m/n \rceil + 1$, add a directed edge (s, v) of capacity
   $indeg(v) - (\lceil m/n \rceil + 1)$;
4: For all $v \in G$ such that $indeg(v) < \lceil m/n \rceil + 1$, add a directed edge (v, t) of capacity
   $(\lceil m/n \rceil + 1) - indeg(v)$;
5: Compute the maximum flow in G from s to t;
6: **if** the maximum flow in G from s to t saturates all the edges having s as origin **then**
7:     Change the direction of all edges $(\mathsf{A}[i], \mathsf{B}[i])$ by calling swap$(\mathsf{A}[i], \mathsf{B}[i])$ that carry flow;
8: Let chosen and alternative be empty arrays of m entries;
9: **for** $i = 1$ to m **do**
10:    Set chosen$[i]$ ← $\mathsf{B}[i]$ and alternative$[i]$ ← $\mathsf{A}[i]$;
11: **return** (chosen, alternative);

Fig. 13. Maximum flow algorithm for finding allocation.

for all $f : \mathbb{R}^{|I|} \to \mathbb{R}$, $g : \mathbb{R}^{|J|} \to \mathbb{R}$ that are both non-increasing or non-decreasing[13].

The following lemmas are used when proving Theorems 1 and 2. Proofs appear in the extended version [13].

Lemma 8. *Let $\{X_1, \ldots, X_n\}$ be negatively associated 0-1 random variables and X be their sum. Let $\mu = \mathbb{E}[X]$ and $\mu_H \in \mathbb{R}$ such that $\mu < \mu_H$. Then, for any $\delta > 0$, the following version of the Chernoff bound holds: $\Pr[X \geq (1 + \delta)\mu_H] \leq \left(\frac{e^\delta}{(1+\delta)^{(1+\delta)}}\right)^{\mu_H}$.*

Lemma 9. *For any set $U \subseteq \{1, \ldots, n\}$ and for any $\tau \geq 2$ it holds that $\sum_{1 \leq |U| \leq \frac{n}{8}} \binom{n}{|U|} P_U \leq \left(\frac{|U|}{n}\right)^{(b+\tau-1)|U|+1} \cdot e^{(b+1)|U|+1} = O(1/n)^{b+\tau}$, where $P_U = \Pr[L_U \geq (b + \tau)|U| + 1]$ and L_U is the unavoidable load of a subset of bins U, where the unavoidable load L_U is defined in Sect. 3.2.*

Correctness proof for our ORAM construction. It is enough to prove that for all indices i, (i, v_i) will be always stored either in C or in $A[\pi_a[i]]$ or in $B[\pi_b[\mathsf{Tab}[i]]]$—these are the values from which we retrieve v_i in Line 16 of our construction in Fig. 3. We consider the following disjoint cases.

1. (*i has been accessed since the last reshuffle*): Then, (i, v_i) can be found in C since it was stored there during the last access to it and C has not been emptied since.

```
bit ← RealORAM(κ):
  1: M₀ ← Adv(1^κ); ⟨σ₀, EM₀⟩ ↔ ORAMINITIALIZE⟨(1^κ, M₀), 1^κ⟩;
  2: for k = 1 to q do
  3:     i_k ← Adv(1^κ, EM₀, m₁, ..., m_{k-1});
  4:     ⟨(v_{i_k}, σ_k), EM_k⟩ ↔ ORAMACCESS⟨(σ_{k-1}, i_k), EM_{k-1}⟩;
  5:     Let m_k be the messages from client to server in the ORAMACCESS protocol above;
  6: bit ← Adv(1^k, EM₀, m₁, m₂, ..., m_q);
  7: return bit;
bit ← IdealORAM(κ):
  1: M₀ ← Adv(1^κ); (st_S, EM₀) ← SIMORAMINITIALIZE(1^κ, |M₀|);
  2: for k = 1 to q do
  3:     (st_S, EM_k, m_k) ← SIMORAMACCESS(st_S, EM_{k-1});
  4: bit ← Adv(1^k, EM₀, m₁, m₂, ..., m_q);
  5: return bit;
```

Fig. 14. Real and ideal experiments for the ORAM scheme.

---

[13] A function $h : \mathbb{R}^k \to \mathbb{R}$ is non-decreasing when $h(\mathbf{x}) \leq h(\mathbf{y})$ whenever $\mathbf{x} \leq \mathbf{y}$ in the component-wise ordering on $\mathbb{R}^k$.

2. (*i* **has not been accessed since the last large reshuffle**): Then, (i, v_i) can be found in $A[\pi[i]]$ since during a large reshuffle all the elements of the dataset are reshuffled into A (and stay there if not accessed afterwards).

3. (*i* **has been accessed since the last large reshuffle but not since the last small reshuffle**): Then, the element can be found in $B[\pi_b[\mathsf{Tab}[i]]]$. This is because, after its first access that occurred after the large reshuffle element i moved to C and after the small reshuffle element i moved to B with a new index $\mathsf{Tab}[i]$ in B and it was stored at location $\pi_b[\mathsf{Tab}[i]]$ during the small reshuffle. Since it was never accessed after the small reshuffle, it remained in B. □

Security proof for our ORAM construction. Our simulator is shown in Fig. 15. Note that all EM_i are trivially indistinguishable from the EM_i output by the real game due to the CPA-security of the encryption scheme that is used— recall that whatever is being written on the server by our protocols is always freshly encrypted. We now argue that the messages $m_1, m_2, \ldots, m_q$ in the real game are indistinguishable from the messages $m_1, m_2, \ldots, m_q$ output by the simulator. This is because for each $1 \le k \le q$, the set of message m_k is entirely independent of the queried value i_k had we used truly random permutations for π_a and π_b. This follows from the following facts:

– When accessing i_k, array C is accessed in its entirety. Also $(\mathsf{Tab}[i_k], v_{i_k})$ is uploaded encrypted at a fixed position count_a in $\mathsf{SCRATCH}$ (see Line 20). So both memory accesses are independent of the index i_k.

– When accessing i_k within a specific superepoch, a location $x = \pi_a[y]$ from array A is accessed for the first and last time within the specific superepoch. Since x is the output of a truly random permutation and is accessed only once within the specific superepoch, x is independent of i_k. The same argument applies for the accesses made to array B. Now if we replace the truly random permutation with the pseudorandom permutation of our construction, the adversary can gain a negligible advantage which is acceptable.

– When accessing i_k at the end of the current superepoch, an oblivious sorting is executed whose memory accesses do not depend on the actual data that are being sorted, but only on the size of the array that is being sorted. The same argument applies for the case when i_k is accessed at the end of an epoch. □

Asymptotic complexity or our ORAM scheme. Over the course of n accesses, each access $1 \le i \le n$ incurs the following:

– $O(n^{1/3} \cdot \lambda)$ bandwidth and $O(1)$ locality due to access of A, B, C and $\mathsf{SCRATCH}$;

– $O(n^{2/3} \log^2 n \cdot \lambda)$ bandwidth and $O(n^{1/3})$ locality due to the small rebuilding which happens only when $i \bmod n^{1/3} = 0$ (i.e., $n^{2/3}$ times);

– $O(n \log^2 n \cdot \lambda)$ bandwidth and $O(n^{2/3})$ locality due to the large rebuilding which happens only when $i \bmod n^{2/3} = 0$ (i.e., $n^{1/3}$ times).

Algorithm $(st_\mathcal{S}, \mathsf{EM}_0) \leftarrow \text{SimOramInitialize}(1^\kappa, |\mathsf{M}_0|)$:

1: Let $(n, \lambda) = |\mathsf{M}_0|$;   ▷ Recall λ is the size of the ORAM block
2: **for** $i = 1$ to n **do**
3:    Set $v_i = \mathbf{0}^\lambda$; $\mathsf{M}_0[i] = (i, v_i)$;
4: $\langle \sigma_0, \mathsf{EM}_0 \rangle \leftrightarrow \text{OramInitialize}\langle (1^\kappa, \mathsf{M}_0), \bot \rangle$;
5: **return** $(\sigma_0, \mathsf{EM}_0)$;

Algorithm $(st_\mathcal{S}, \mathsf{EM}_k, m_k) \leftarrow \text{SimOramAccess}(st_\mathcal{S}, \mathsf{EM}_{k-1})$:

Parse $st_\mathcal{S}$ as σ_{k-1};
Choose $i_k \in [n]$;
$\langle (v_{i_k}, \sigma_k), \mathsf{EM}_k \rangle \leftrightarrow \text{OramAccess}\langle (\sigma_{k-1}, i_k), \mathsf{EM}_{k-1} \rangle$;
Let m_k be the messages sent from client to server during the above OramAccess protocol;
return $(\sigma_k, \mathsf{EM}_k, m_k)$;

Fig. 15. The simulator for the ORAM scheme of Fig. 3

Note that in order to derive the locality of the rebuilding above, we used Theorem 4 for $b = n^{1/3} \log^2 n$. Now, the amortized bandwidth is

$$\lambda \cdot \frac{n \cdot O(n^{1/3}) + n^{2/3} \cdot O(n^{2/3} \log^2 n) + n^{1/3} \cdot O(n \log^2 n)}{n} = O(n^{1/3} \log^2 n \cdot \lambda)$$

and the amortized locality is $\frac{n \cdot O(1) + n^{2/3} \cdot O(n^{1/3}) + n^{1/3} \cdot O(n^{2/3})}{n} = O(1)$. Finally, the client must store Tab locally, that consists of $n^{2/3}$ entries of $\log n$ bits each and also needs to have $O(n^{1/3} \log^2 n \cdot \lambda)$ space locally for the oblivious sorting—see Theorem 4.   □

Protocol $\langle \bot, Y \rangle \leftrightarrow$ OBLIVIOUSSORTING$\langle (\pi, n, b), X \rangle$:

▷ Assume n and b are powers of 2 ▷ Also assume that $X[i]$ also stores the respective index i, so that comparisons using π are possible while elements are being moved around

1: **if** $n \leq b$ **then**
2:     **Read–And–Decrypt** array X. Set Y to be the sorted version of X^a;
3: **else**
4:     $\langle \bot, Y_1 \rangle \leftrightarrow$ OBLIVIOUSSORTING$\langle (\pi, n/2, b), X[1, \ldots, n/2] \rangle$;
5:     $\langle \bot, Y_2 \rangle \leftrightarrow$ OBLIVIOUSSORTING$\langle (\pi, n/2, b), X[n/2 + 1, \ldots, n] \rangle$;
6:     $\langle \bot, Y \rangle \leftrightarrow$ OBLIVIOUSMERGE$\langle (\pi, n, b), (Y_1, Y_2) \rangle$;
7: **Encrypt–And–Write** array Y;
8: **return** $\langle \bot, Y \rangle$;

Protocol $\langle \bot, Y \rangle \leftrightarrow$ OBLIVIOUSMERGE$\langle (\pi, n, b), (Y_1, Y_2) \rangle$:   ▷ Y_1, Y_2 must be sorted

1: **if** $n \leq b$ **then**
2:     **Read–And–Decrypt** array Y_1;
3:     **Read–And–Decrypt** array Y_2;
4:     Set Y to be the merged array of Y_1 and Y_2;
5: **else**
6:     Let D be a $2 \times n/2$ matrix and Y be a length n array stored at the server;
7:     $j = 0$;
8:     **for** $i = 1, 2b + 1, 4b + 1, \ldots, n/2 - 2b + 1$ **do**
9:         Initialize arrays D_1, D_2, D_3, D_4 of size b;
10:        Store $Y_1[i], Y_1[i + 2], \ldots, Y_1[i + 2b - 2]$ at the first available position of D_1;
11:        Store $Y_1[i + 1], Y_1[i + 3], \ldots, Y_1[i + 2b - 1]$ at the first available position of D_3;
12:        Store $Y_2[i], Y_2[i + 2], \ldots, Y_2[i + 2b - 2]$ at the first available position of D_2;
13:        Store $Y_2[i + 1], Y_2[i + 3], \ldots, Y_2[i + 2b - 1]$ at the first available position of D_4;
14:        Store D_1 in D's row 1, from position $1 + j \cdot b$ onwards;
15:        Store D_2 in D's row 1, from position $n/4 + 1 + j \cdot b$ onwards;
16:        Store D_3 in D's row 2, from position $1 + j \cdot b$ onwards;
17:        Store D_4 in D's row 2, from position $n/4 + 1 + j \cdot b$ onwards;
18:        $j \leftarrow j + 1$;
19:     $\langle \bot, D[1, :] \rangle \leftrightarrow$ OBLIVIOUSMERGE$\langle (\pi, n/2, b), (D[1, 1 : n/4], D[1, n/4 + 1 : n/2]) \rangle$;
20:     $\langle \bot, D[2, :] \rangle \leftrightarrow$ OBLIVIOUSMERGE$\langle (\pi, n/2, b), (D[2, 1 : n/4], D[2, n/4 + 1 : n/2]) \rangle$;
21:     Let $Z_1, \ldots, Z_{n/2b}$ be the $2 \times b$ submatrices that result from partitioning D horizontally;
22:     **for** $i = 1$ to $n/2b - 1$ **do**
23:         **Read–And–Decrypt** Z_i;
24:         **Read–And–Decrypt** Z_{i+1};
25:         Sort $Z_i \cup Z_{i+1}$ and let $y_1, \ldots, y_{2b}$ be the smallest resulting elements;
26:         **Encrypt–And–Write** $[y_1, \ldots, y_b]$ starting at the first available position of Y;
27:         **Encrypt–And–Write** $[y_{b+1}, \ldots, y_{2b}]$ starting at the first available position of Y;
28:     Sort $Z_{n/2b}$ and let $y_1, \ldots, y_{2b}$ be the sorted sequence;
29:     **Encrypt–And–Write** $[y_1, \ldots, y_b]$ starting at the first available position of Y;
30:     **Encrypt–And–Write** $[y_{b+1}, \ldots, y_{2b}]$ starting at the first available position of Y;
31: **return** $\langle \bot, Y \rangle$;

---

^{a} We use π to perform comparisons between two elements of X, i.e., $X[i]$ isLessThan $X[j]$ iff $p[i] < p[j]$.

Fig. 16. Data-oblivious and I/O efficient sorting by Goodrich and Mitchenmacher [20].

Tight Tradeoffs in Searchable Symmetric Encryption

Gilad Asharov[1], Gil Segev[2], and Ido Shahaf[2(✉)]

[1] Cornell Tech, New York, NY, USA
asharov@cornell.edu
[2] School of Computer Science and Engineering,
Hebrew University of Jerusalem, 91904 Jerusalem, Israel
{ido.shahaf,segev}@cs.huji.ac.il

Abstract. A searchable symmetric encryption (SSE) scheme enables a client to store data on an untrusted server while supporting keyword searches in a secure manner. Recent experiments have indicated that the practical relevance of such schemes heavily relies on the tradeoff between their *space overhead*, *locality* (the number of non-contiguous memory locations that the server accesses with each query), and *read efficiency* (the ratio between the number of bits the server reads with each query and the actual size of the answer). These experiments motivated Cash and Tessaro (EUROCRYPT '14) and Asharov et al. (STOC '16) to construct SSE schemes offering various such tradeoffs, and to prove lower bounds for natural SSE frameworks. Unfortunately, the best-possible tradeoff has not been identified, and there are substantial gaps between the existing schemes and lower bounds, indicating that a better understanding of SSE is needed.

We establish tight bounds on the tradeoff between the space overhead, locality and read efficiency of SSE schemes within two general frameworks that capture the memory access pattern underlying all existing schemes. First, we introduce the "pad-and-split" framework, refining that of Cash and Tessaro while still capturing the same existing schemes. Within our framework we significantly strengthen their lower bound, proving that any scheme with locality L must use space $\Omega(N \log N / \log L)$ for databases of size N. This is a tight lower bound, matching the tradeoff provided by the scheme of Demertzis and Papamanthou (SIGMOD '17) which is captured by our pad-and-split framework.

Then, within the "statistical-independence" framework of Asharov et al. we show that their lower bound is essentially tight: We construct a scheme whose tradeoff matches their lower bound within an additive $O(\log \log \log N)$ factor in its read efficiency, once again improving

G. Asharov—Supported by a Junior Fellow award from the Simons Foundation.
G. Segev and I. Shahaf—Supported by the European Union's Horizon 2020 Framework Program (H2020) via an ERC Grant (Grant No. 714253), by the Israel Science Foundation (Grant No. 483/13), by the Israeli Centers of Research Excellence (I-CORE) Program (Center No. 4/11), and by the US-Israel Binational Science Foundation (Grant No. 2014632).

H. Shacham and A. Boldyreva (Eds.): CRYPTO 2018, LNCS 10991, pp. 407–436, 2018.
https://doi.org/10.1007/978-3-319-96884-1_14

upon the existing schemes. Our scheme offers optimal space and locality, and nearly-optimal read efficiency that depends on the frequency of the queried keywords: For a keyword that is associated with $n = N^{1-\epsilon(n)}$ document identifiers, the read efficiency is $\omega(1) \cdot \epsilon(n)^{-1} + O(\log \log \log N)$ when retrieving its identifiers (where the $\omega(1)$ term may be arbitrarily small, and $\omega(1) \cdot \epsilon(n)^{-1}$ is the lower bound proved by Asharov et al.). In particular, for any keyword that is associated with at most $N^{1-1/o(\log \log \log N)}$ document identifiers (i.e., for any keyword that is not exceptionally common), we provide read efficiency $O(\log \log \log N)$ when retrieving its identifiers.

1   Introduction

A searchable symmetric encryption (SSE) scheme [11,27] enables a client to store data on an untrusted server and later perform keyword searches: Given a keyword w, the client should be able to retrieve all data items that are associated with w (e.g., all document identifiers that contain w). This typically consists of a two-stage process: First, the client encrypts her database and uploads it to the server, and then the client repeatedly queries the server with various keywords by providing the server with keyword-specific search tokens. Informally, the security requirement of SSE schemes asks that the server does not learn any information about keywords for which the client did not issue any queries.

The practical relevance of SSE schemes. Motivated by the increasingly-growing technological interest in outsourcing data to remote (and thus potentially untrusted) servers, a very fruitful line of research in the cryptography community focused on the design of SSE schemes (e.g., [2,5–12,14,19,20,22,23, 27,29]). Most of the proposed schemes offer strong and meaningful notions of security, and some even extend the basic keyword search functionality to more expressive ones.

Despite these promising developments, Cash et al. [7] showed via experiments with real-world databases that the practical performance of the known schemes is quite disappointing, and scales badly to large databases. Somewhat surprisingly, they observed that performance issues resulting from impractical memory layouts may be significantly more crucial compared to performance issues resulting from the cryptographic processing of the data. More specifically, Cash et al. observed that schemes with poor *locality* (i.e., schemes in which the server has to access a rather large number of *non-contiguous* memory locations with each query) have poor practical performance when dealing with large databases that require the usage of disk-storage mechanisms.

Practical locality, however, is obviously insufficient: Any practically-relevant SSE scheme should (at least) not suffer from either a significant *space overhead* (i.e., encrypted databases should not be much larger than the original databases), or from a poor *read efficiency* (i.e., servers should not read much more data than needed for answering each query)[1].

Efficiency tradeoffs and existing lower bounds. This state of affairs naturally poses the challenge of constructing an SSE scheme that simultaneously enjoys asymptotically-optimal space overhead, locality, and read efficiency – but unfortunately no such scheme is currently known. This has motivated Cash and Tessaro [8] to initiate the study of understanding the tradeoff between these central measures of efficiency. They proved a lower bound showing that, for a large and natural class of SSE schemes, it is in fact impossible to simultaneously enjoy asymptotically-optimal space overhead, locality, and read efficiency. Specifically, they considered the class of SSE schemes with "non-overlapping reads": Schemes in which distinct keywords induce non-overlapping memory regions which the server may access upon their respective queries (we refer the reader to the work of Cash and Tessaro [8] for a formal definition of their notion of non-overlapping reads).

The class of SSE schemes with non-overlapping reads captures the basic techniques underlying all existing SSE schemes other than two schemes proposed by Asharov et al. [2]. These two schemes may have arbitrary overlapping reads, and offer an improved tradeoff between their space overhead, locality, and read efficiency compared to the previously suggested schemes. This tradeoff, however, is still non-optimal, and Asharov et al. showed that this is in fact inherent to their approach. Similarly to Cash and Tessaro, they proved that also for a different class of SSE schemes, it is impossible to simultaneously enjoy asymptotically-optimal space overhead, locality, and read efficiency. Specifically, they considered the class of SSE scheme with "statistically-independent reads": Schemes in which distinct keywords induce statistically-independent memory regions which the server accesses upon their respective queries.

The lower bounds proved by Cash and Tessaro and by Asharov et al. capture all of the existing SSE schemes (except for various schemes with non-standard leakage or functionality that we do not consider in this work). That is, the basic techniques underlying each of the known SSE schemes belong either to the class of "non-overlapping reads" or to the class of "statistically-independent reads". In both cases, however, the existing lower bounds are not tight, as there are still noticeable gaps between the lower bounds and the performance guarantees of the existing schemes (as we detail in the next section). This unsatisfying situation calls for obtaining a better understanding of SSE techniques: Either by strengthening the known lower bounds, or by designing new schemes with better performance guarantees.

---

[1] We consider the notions of locality and read efficiency as formalized by Cash and Tessaro [8]: The locality of a scheme is the number of non-contiguous memory accesses that the server performs with each query, and the read efficiency of a scheme is the ratio between the number of bits the server reads with each query and the actual size of the answer. We refer the reader to Sect. 2.1 for the formal definitions.

1.1 Our Contributions

We prove tight bounds on the tradeoff between the space overhead, locality, and read efficiency of SSE schemes within the following two general frameworks:

The pad-and-split framework: We formalize a framework that refines the non-overlapping reads framework of Cash and Tessaro [8] while still capturing the same existing SSE schemes (i.e., all existing schemes other than those of Asharov et al. [2])[2]. We refer to this framework as the "pad-and-split" framework given the structure of the SSE schemes that it captures.

Within this framework we significantly strengthen the lower bound of Cash and Tessaro: We show that any pad-and-split scheme with locality L must use space $\Omega (N \cdot \log N / \log L)$ for databases of size N. For example, for any constant locality (i.e., $L = O(1)$) and for any logarithmic locality (i.e., $L = O(\log N)$) our lower bound shows that any such scheme must use space $\Omega(N \log N)$ and $\Omega(N \log N / \log \log N)$, respectively, and is thus not likely to be of substantial practical relevance (whereas the lower bound of Cash and Tessaro would only yield space $\omega(N)$ when the locality is constant).

Then, we observe that our lower bound is in fact tight, as it is matched by a recent scheme proposed by Demertzis and Papamanthou [14] that is captured by our framework (i.e., their scheme is an optimal instantiation of our framework). We refer the reader to Sects. 1.2 and 3 for a high-level overview and for a detailed description of this framework, its instantiations, and of our lower bound, respectively.

The statistical-independence framework: We consider the statistical-independence framework of Asharov et al. [2], and show that their lower bound for SSE schemes in this framework is essentially tight: Based on the existence of any one-way function, we construct a scheme whose efficiency guarantees match their lower bound for constant locality within an additive $O(\log \log \log N)$ factor in the read efficiency, and improve upon those of their two schemes.

Specifically, for databases of size N, our scheme offers both optimal space and optimal locality (i.e., space $O(N)$ and locality $O(1)$), and comes very close to offering optimal read efficiency as well. The read efficiency of our scheme when querying for a keyword w depends on the length of the list $\mathsf{DB}(w)$ that is associated with w (that is, the read efficiency depends on the number of identifiers that are associated with w).[3] When querying for a keyword that is associated with $n = N^{1-\epsilon(n)}$ identifiers, the read efficiency of our scheme is $f(N) \cdot \epsilon(n)^{-1} + O(\log \log \log N)$, where $f(N) = \omega(1)$ may be any pre-determined function, and $\omega(1) \cdot \epsilon(n)^{-1}$ is a lower bound as proved by

---

[2] Each of the schemes that are captured by our framework offers other important implementation details, improvements and optimizations that we do not intend to capture, since these are not directly related to the tradeoff between space, locality, and read efficiency.

[3] We emphasize that this does not hurt the security of SSE schemes, and still results in minimal leakage as required.

Asharov et al. [2]. In particular, for any keyword that is associated with at most $N^{1-1/o(\log\log\log N)}$ identifiers (i.e., for any keyword that is not exceptionally common), the read efficiency of our scheme when retrieving its identifiers is $O(\log\log\log N)$. We refer the reader to Sects. 1.2 and 4 for a high-level overview and for a detailed description of this framework and of our new scheme, respectively.

Our results in the pad-and-split and statistical-independence frameworks, which are summarized in Table 1 and presented in more detail in Sect. 1.2, show a significant gap between the performance guarantees that can be offered within these two frameworks. In both frameworks we establish tight bounds that capture the basic techniques underlying all of the existing SSE schemes. Thus, any attempt to further improve upon the tradeoff between the space overhead, locality and read efficiency of our schemes must be based on new techniques that deviate from all known SSE schemes.

Table 1. A summary of our contributions. We denote by N the size of the database. The read efficiency in the lower bound of Asharov et al. [2] and in our statistical-independence scheme (Theorem 1.2) when querying for a keyword w depends on the number $n = N^{1-\epsilon(n)}$ of identifiers that are associated with w. In addition, our statistical-independence scheme is based on the modest assumption that no keyword is associated with more than $N/\log^3 N$ identifiers, whereas the scheme of Asharov et al. [2] is based on the stronger assumption that no keyword is associated with more than $N^{1-1/\log\log N}$ identifiers (thus, the read efficiency of their scheme does not contradict their lower bound, and our scheme has better read efficiency compared to their scheme). Finally, we note that the $\omega(1)$ term in the read efficiency of our scheme can be set to any super-constant function (e.g., $\log\log\log\log N$).

	Space	Locality	Read Efficiency
This work (Theorem 1.1): Pad-and-split lower bound	$\Omega(N\log N/\log L)$	L	$O(1)$
[14]: Pad-and-split scheme	$O(N\log N/\log L)$	L	$O(1)$
[2]: Statistical-independence lower bound	$O(N)$	$O(1)$	$\omega(1)\cdot\epsilon(n)^{-1}$
[2]: Statistical-independence scheme	$O(N)$	$O(1)$	$\tilde{O}(\log\log N)$
This work (Theorem 1.2): Statistical-independence scheme	$O(N)$	$O(1)$	$\omega(1)\cdot\epsilon(n)^{-1} + O(\log\log\log N)$

1.2 Overview of Our Contributions

In this section we provide an overview of the two frameworks that we consider in this work, and present our results within each framework. As standard in the line of research on searchable symmetric encryption, we represent a database as a collection $\mathsf{DB} = \{\mathsf{DB}(w_1),\ldots,\mathsf{DB}(w_{n_{\mathsf{W}}})\}$, where $w_1,\ldots,w_{n_{\mathsf{W}}}$ are distinct keywords, and $\mathsf{DB}(w)$ is the list of all identifiers that are associated with each keyword w. We denote by $N = \sum_{i=1}^{n_{\mathsf{W}}}|\mathsf{DB}(w_i)|$ the size of the database.

Our pad-and-split framework. Our pad-and-split framework considers schemes that are characterized by an algorithm denoted SplitList and consist of two phases. In the first phase, given a database $\mathsf{DB} = \{\mathsf{DB}(w_1), \ldots, \mathsf{DB}(w_{n_\mathsf{W}})\}$ of size N, for each keyword w_i the scheme invokes the SplitList algorithm on the length n_i of its corresponding list $\mathsf{DB}(w_i)$, to obtain a vector $(x_i^{(1)}, \ldots, x_i^{(m)})$ of integers. The scheme then potentially pads the list $\mathsf{DB}(w_i)$ by adding "dummy" elements, and splits the padded list into sublists of lengths $\mathsf{len}^{(1)}, \ldots, \mathsf{len}^{(m)}$, where $x_i^{(j)}$ denotes the number of sublists of each length $\mathsf{len}^{(j)}$. Then, in the second phase, for each possible length $\mathsf{len}^{(j)}$, the scheme groups together all sublists of length $\mathsf{len}^{(j)}$, and independently processes each such group to produce an encrypted database EDB.

We consider any possible instantiation of the SplitList algorithm (satisfying the necessary requirement that no list is longer than the sum of lengths of its sublists), and this enables us to describe a general template for constructing an SSE scheme based on any such algorithm given any one-way function. Our template yields schemes whose space usage and locality are essentially inherited from similar properties of their underlying SplitList algorithm, and whose read efficiency is always constant. We then demonstrate that this template captures the memory access patterns underlying essentially all existing schemes other than those of Asharov et al. [2]. Specifically, we show that each of these schemes can be obtained as an instantiation of our template using a suitable SplitList algorithm.

A tight lower bound for pad-and-split schemes. Equipped with our general notion of pad-and-split schemes, we prove a lower bound on the asymptotic efficiency guarantees of such schemes. Whereas the lower bound of Cash and Tessaro [8] states that SSE schemes with non-overlapping reads cannot simultaneously offer asymptotically-optimal space overhead and locality, we prove the following lower bound (capturing the same existing schemes) stating that the efficiency guarantees of pad-and-split schemes must in fact be very far from optimal:

Theorem 1.1. *Any pad-and-split SSE scheme for databases of size N with locality $L = L(N)$ uses space $\Omega\left(N \log N / \log L\right)$.*

We show that this lower bound is tight, as it matches the tradeoff offered by the scheme of Demertzis and Papamanthou [14] (i.e., their scheme is an optimal instantiation of our framework). We refer the reader to Sect. 3 for a detailed and more formal presentation of our results, including an in-depth discussion of the existing pad-and-split instantiations.

The statistical-independence framework. The statistical-independence framework of Asharov et al. [2] considers symmetric searchable encryption schemes that are characterized by a pair of algorithms, denoted RangesGen and Allocation, and consist of two phases. In the first phase, given a database $\mathsf{DB} = \{\mathsf{DB}(w_1), \ldots, \mathsf{DB}(w_{n_\mathsf{W}})\}$ of size N, for each keyword w_i the scheme invokes the RangesGen algorithm on the length n_i of its corresponding list $\mathsf{DB}(w_i)$, to obtain a set of *possible locations* in which the scheme may place the elements of the list $\mathsf{DB}(w_i)$.[4] Then, in the second phase, given the sets of possible locations for all keywords, the scheme invokes the Allocation algorithm on these sets to obtain the *actual locations* for the corresponding lists. A key property of this framework is that the RangesGen algorithm, which determines the set of possible locations for each list $\mathsf{DB}(w_i)$, is applied separately and independently to the length of each list. Thus, the possible locations of each list are independent of the possible locations of all other lists (in contrast, the actual locations of the lists are naturally correlated).

Asharov et al. referred to a pair (RangesGen, Allocation) of such algorithms as an allocation scheme, and showed that any such allocation scheme can be used to construct an SSE scheme. Then, by constructing two allocation schemes they obtained two SSE schemes with space $O(N)$ and locality $O(1)$. Without making any assumptions on the structure of the database, their first scheme has read efficiency $\tilde{O}(\log N)$, and under the assumption that no keyword is associated with more than $N^{1-1/\log\log N}$ identifiers, their second scheme has read efficiency $\tilde{O}(\log\log N)$.

Our leveled two-choice scheme. Within the statistical-independence framework, as discussed above, we construct a scheme whose tradeoff between space, locality, and read efficiency matches the lower bound proved by Asharov et al. for scheme in this framework to within an additive $O(\log\log\log N)$ factor in its read efficiency (see Sect. 4 for a formal statement of their lower bound).

Specifically, we construct a scheme whose read efficiency when querying for a keyword w depends on the length of the list $\mathsf{DB}(w)$ that is associated with w (that is, the read efficiency depends on the number of identifiers that are associated with w). For any $n \le N$ we denote by $\mathsf{r}(N, n)$ the read efficiency when retrieving a list of length n, and prove the following theorem:

Theorem 1.2. *Assuming the existence of any one-way function, for any function $f(N) = \omega(1)$ there exists an adaptively-secure symmetric searchable encryption scheme for databases of size N in which no keyword is associated with more than $N/\log^3 N$ identifiers, with the following guarantees:*

- *Space $O(N)$.*
- *Locality $O(1)$.*
- *Read efficiency $\mathsf{r}(N, n) = f(N) \cdot \epsilon(n)^{-1} + O(\log\log\log N)$, where $n = N^{1-\epsilon(n)}$.*
- *Token size $O(1)$.*

Our construction applies to databases of size N under the modest assumption that no keyword is associated with more than $N/\log^3 N$ identifiers (note that the construction of Asharov et al. [2] is based on the stronger assumption that no keyword is associated with more than $N^{1-1/\log\log N}$ identifiers). One can always

---

[4] Looking ahead, when supplied with a token corresponding to a keyword w_i, the server will return to the client all data stored in the possible locations of the list $\mathsf{DB}(w_i)$ (the server will not actually know in which of the possible locations the elements of the list are actually placed).

generically deal (in a secure manner) with such extremely-common keywords by first excluding them from the database and applying our proposed scheme, and then applying in addition any other scheme for these extremely-common keywords (e.g., the "one-choice scheme" of Asharov et al. [2] or the recent scheme of Demertzis et al. [13] – see Sect. 1.3 for more details).

When comparing our scheme to the scheme of Asharov et al. (see Table 1), both schemes offer space $O(N)$ and locality $O(1)$, where the read efficiency of our scheme is strictly better than the read efficiency of their scheme – see Fig. 1. In particular, for any keyword that is not exceptionally frequent (specifically, associated with at most $N^{1-1/o(\log \log \log N)}$ identifiers), our scheme provides read efficiency $O(\log \log \log N)$ whereas their scheme provides read efficiency $\tilde{O}(\log \log N)$.

The structure of our scheme. Our scheme is a *leveled* generalization of the "two-choice" scheme of Asharov et al. and consists of three levels for storing the elements of a given database. The first level consists of the two-choice SSE scheme of Asharov et al. but with *an exponentially improved read efficiency*. Our key observation is that when viewing the first level as a collection of "bins", then by allowing a few elements to "overflow" we can reduce the maximal load of each bin from $\tilde{O}(\log \log N)$ (as in [2]) to $O(\log \log \log N)$ and also handle much longer lists (i.e., much more frequent keywords). This then translates into improving the read efficiency in this level from $\tilde{O}(\log \log N)$ to $O(\log \log \log N)$, while still using space $O(N)$ and locality $O(1)$.

At this point, however, we have to store the overflowing elements. We store the vast majority of these elements in our second level, which consists of roughly $\log N$ cuckoo hashing tables [26], where the j hash table is designed to store at most $\hat{N}/2^j$ values each of which of size 2^j. Our specific choice of cuckoo hashing as a static dictionary (i.e., a hash table) is due to its specific properties that guarantee the security of our scheme (see Sect. 2.3 for a discussion of these specific properties). In particular, our third level consists of a cuckoo hashing *stash* for each of the second-level cuckoo hashing tables. The goal of introducing this level is to reduce the failure probably of cuckoo hashing from noticeable to negligible, which is essential for the security of our resulting SSE scheme. We refer the reader to Sect. 4 for a detailed description of our scheme.

1.3   Related Work

The notion of searchable symmetric encryption was put forward by Song et al. [27] who suggested several practical constructions. Formal notions of security and functionality for SSE, as well as the first constructions satisfying them, were later provided by Curtmola et al. [11,12]. Additional work in this line of research developed searchable symmetric encryption schemes with various efficiency properties, support for data updates, authenticity, support for more advanced searches, and more (see [2,5–12,14,16,19,20,22,23,27,29] and the references therein). The two frameworks that we consider in this work capture schemes that satisfy that standard notions of SSE introduced by Curtmola et al. [11,12]. These schemes are

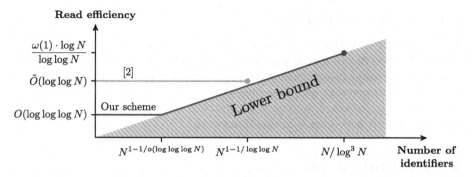

Fig. 1. The read efficiency of our statistical-independence scheme compared to that of Asharov et al. [2] and to the lower bound. The read efficiency of our scheme is depicted by the blue line, and the read efficiency of the scheme of Asharov et al. is depicted by the yellow line (recall that our scheme supports keywords that are associated with up to $N/\log^3 N$ identifiers, whereas the scheme of Asharov et al. only supports keywords that are associated with at most $N^{1-1/\log\log N}$ identifiers). The read efficiency lower bound of Asharov et al. is depicted by the red triangle (note that it coincides with our blue line for keywords that are associated with at least $N^{1-1/o(\log\log\log N)}$ and at most $N/\log^3 N$ identifiers). In all three cases the read efficiency is presented as a function of the number of identifiers that are associated with the queried keyword. (Color figure online)

discussed in Sect. 3.2 as instantiations of our pad-and-split framework, and in Sect. 4.2 as instantiations of the statistical-independence framework of Asharov et al. [2].

Our statistical-independence scheme can be applied to any database in which no keyword is associated with more than $N/\log^3 N$ identifiers. As discussed above, one can always generically deal (in a secure manner) with such extremely-frequent keywords by first excluding them from the database and applying our proposed scheme, and then applying in addition any other scheme for these extremely-common keywords. For example, for these keywords one can apply the "one-choice scheme" of Asharov et al. or the recent scheme of Demertzis, Papadopoulos and Papamanthou [13] that provides a sub-logarithmic read efficiency when searching for extremely frequent keywords[5]. Specifically, Demertzis et al. proposed a scheme that handles such extremely frequent keywords and improves their read efficiency from $\tilde{O}(\log N)$ as guaranteed by the "one-choice scheme" of Asharov et al. to $O(\log^{2/3+\delta} N)$ for any fixed constant $\delta > 0$ (for all other keywords they use the two schemes of Asharov et al., which can now be replaced by our new scheme in its appropriate range of parameters).

---

[5] The scheme of Demertzis et al. [13] is not captured by the two frameworks we consider in this work, as it requires the server to modify its stored data (i.e., the encrypted database) and the user to update her local state whenever a search query is issued.

1.4   Paper Organization

The remainder of this paper is organized as follows. In Sect. 2 we review the standard notion of symmetric searchable encryption schemes, as well as various tools that are used in our constructions. Then, in Sect. 3 we put forward our pad-and-split framework and then present our lower bound and new scheme in this framework. In Sect. 4 we review the statistical-independence framework and then present our new scheme in this framework.

2   Preliminaries

In this section we present the notation, definitions, and basic tools that are used in this work. We denote by $\lambda \in \mathbb{N}$ the security parameter. For a distribution X we denote by $x \leftarrow X$ the process of sampling a value x from the distribution X. Similarly, for a set $\mathcal{X}$ we denote by $x \leftarrow \mathcal{X}$ the process of sampling a value x from the uniform distribution over $\mathcal{X}$. For an integer $n \in \mathbb{N}$ we denote by $[n]$ the set $\{1, \ldots, n\}$. A function $\mathsf{negl} : \mathbb{N} \to \mathbb{R}^+$ is negligible if for every constant $c > 0$ there exists an integer N_c such that $\mathsf{negl}(n) < n^{-c}$ for all $n > N_c$. All logarithms in this paper are to the base of 2.

2.1   Searchable Symmetric Encryption

Let $\mathsf{W} = \{w_1, \ldots, w_{n_\mathsf{W}}\}$ denote a set of keywords, where each keyword w_i is associated with a list $\mathsf{DB}(w_i) = \{\mathsf{id}_1, \ldots, \mathsf{id}_{n_i}\}$ of document identifiers (these may correspond, for example, to documents in which the keyword w_i appears). A database $\mathsf{DB} = \{\mathsf{DB}(w_1), \ldots, \mathsf{DB}(w_{n_\mathsf{W}})\}$ consists of several such lists. We assume that each keyword and document identifier can be represented using a constant number of machine words, each of length $O(\lambda)$ bits, in the unit-cost RAM model[6]. There are various different syntaxes for SSE schemes in the literature, where the main differences are in the flavor of interaction between the server and the client with each query. In this work we consider both a setting where the server decrypts the set of identifiers by itself, and a setting where the server does not decrypt this but rather sends encrypted data back to the client (who can then decrypt and learn the set of identifiers).

Functionality

A searchable symmetric encryption scheme is a 5-tuple (KeyGen, EDBSetup, TokGen, Search, Resolve) of probabilistic polynomial-time algorithms satisfying the following requirements:

- The key-generation algorithm KeyGen takes as input the security parameter $\lambda \in \mathbb{N}$ in unary representation and outputs a secret key K.

---

[6] The unit cost word-RAM model is considered the standard model for analyzing the efficiency of data structures (see, for example, [15,17,18,24,25] and the references therein).

- The database setup EDBSetup algorithm takes as input a secret key K and a database DB, and outputs an encrypted database EDB.
- The token-generation algorithm TokGen takes as input a secret key K and a keyword w, and outputs a token τ and some internal state ρ.
- The search algorithm Search takes as input a token τ and an encrypted database EDB, and outputs a list R of results.
- The resolve algorithm Resolve takes as input a list R of results and an internal state ρ, and outputs a list M of document identifiers.

An SSE scheme for databases of size $N = N(\lambda)$ is correct if for any database DB of size N and for any keyword w, with an overwhelming probability in the security parameter $\lambda \in \mathbb{N}$, it holds that $M = \mathsf{DB}(w)$ at the end of the following experiment:

1. $K \leftarrow \mathsf{KeyGen}(1^\lambda)$.
2. $\mathsf{EDB} \leftarrow \mathsf{EDBSetup}(K, \mathsf{DB})$.
3. $(\tau, \rho) \leftarrow \mathsf{TokGen}(K, w)$.
4. $R \leftarrow \mathsf{Search}(\tau, \mathsf{EDB})$.
5. $M = \mathsf{Resolve}(\rho, R)$.

We note that one can also consider a more adversarially-flavored notion of correctness, where an adversary adaptively interacts with a server with the goal of producing a query that results in an incorrect output. We refer the reader to [2] for more details, and here we only point out that our schemes in this paper satisfy such a notion as well.

Efficiency Measures

Our notions of space usage, locality and read efficiency follow those introduced by Cash and Tessaro [8].

Space. A symmetric searchable encryption scheme (KeyGen, EDBSetup, TokGen, Search, Resolve) uses space $s = s(\lambda, N)$ if for any $\lambda, N \in \mathbb{N}$, for any database DB of size N, and for any key K produced by $\mathsf{KeyGen}(1^\lambda)$, the algorithm $\mathsf{EDBSetup}(K, \mathsf{DB})$ produces encrypted databases that can be represented using s machine words.

Locality. The search procedure of any SSE scheme can be decomposed into a sequence of contiguous reads from the encrypted database EDB, and the locality is defined as the number of such reads. Specifically, locality is defined by viewing the Search algorithm of an SSE scheme as an algorithm that does not obtain as input the actual encrypted database, but rather only obtains oracle access to it. Each query to this oracle consists of an interval $[a_i, b_i]$, and the oracle replies with the machine words that are stored in this interval of EDB. At first, the Search algorithm is invoked on a token τ and queries its oracle with some interval $[a_1, b_1]$. Then, it iteratively continues to compute the next interval to read based on τ and all previously read intervals. We denote these intervals by $\mathsf{ReadPat}(\mathsf{EDB}, \tau)$.

Definition 2.1 (Locality). *An SSE scheme Π is* d-local *(or has locality* d*) if for every* λ, DB *and* $w \in$ W, $K \leftarrow$ KeyGen(1^λ), EDB $\leftarrow$ EDBSetup(K, DB) *and* $\tau \leftarrow$ TokGen(K, w) *we have that* ReadPat(EDB, τ) *consists of at most* d *intervals.*

Read efficiency. The notion of read efficiency compares the overall size of the portion of EDB that is read on each query to the size of the actual answer to the query. For a given DB and w, we let $\|$DB(w)$\|$ denote the number of words in the encoding of DB(w).

Definition 2.2 (Read efficiency). *An SSE scheme Π is* r-read efficient *(or has read efficiency* r*) if for any* λ, DB, *and* $w \in$ W, *we have that* ReadPat(τ, EDB) *consists of intervals of total length at most* r $\cdot \|$DB(w)$\|$ *words.*

Security Notions

The standard security definition for SSE schemes follows the ideal/real simulation paradigm. We consider both static and adaptive security, where the difference is whether the adversary chooses its queries statically (i.e., before seeing any token), or in an adaptive manner (i.e., the next query may be a function of the previous tokens). In both cases, some information is leaked to the server, which is formalized by letting the simulator receive the evaluation of some "leakage function" on the database itself and the real tokens. We start with the static case.

The real execution. The real execution is parameterized by the scheme Π, the adversary $\mathcal{A}$, and the security parameter λ. In the real execution the adversary is invoked on 1^λ, and outputs a database DB and a list of queries $\mathbf{w} = \{w_i\}_i$. Then, the experiment invokes the key-generation algorithm and the database setup algorithms, $K \leftarrow$ KeyGen(1^λ) and EDB $\leftarrow$ EDBSetup(K, DB). Then, for each query $\mathbf{w} = \{w_i\}_i$ that the adversary has outputted, the token generator algorithm is run to obtain $\tau_i =$ TokGen(w_i). The adversary is given the encrypted database EDB and the resulting tokens $\boldsymbol{\tau} = \{\tau_i\}_{w_i \in \mathbf{w}}$, and outputs a bit b.

The ideal execution. The ideal execution is parameterized by the scheme Π, a leakage function $\mathcal{L}$, the adversary $\mathcal{A}$, a simulator $\mathcal{S}$ and the security parameter λ. In this execution, the adversary $\mathcal{A}$ is invoked on 1^λ, and outputs (DB, $\mathbf{w}$) similarly to the real execution. However, this time the simulator $\mathcal{S}$ is given the evaluation of the leakage function on (DB, $\mathbf{w}$) and should output EDB, $\boldsymbol{\tau}$ (i.e., (EDB, $\boldsymbol{\tau}$) $\leftarrow \mathcal{S}(\mathcal{L}($DB, $\mathbf{w})))$. The execution follows by giving (EDB, $\boldsymbol{\tau}$) to the adversary $\mathcal{A}$, which outputs a bit b.

Let SSE-REAL$_{\Pi,\mathcal{A}}(\lambda)$ denote the output of the real execution, and let SSE-IDEAL$_{\Pi,\mathcal{L},\mathcal{A},\mathcal{S}}(\lambda)$ denote the output of the ideal execution, with the adversary $\mathcal{A}$, simulator $\mathcal{S}$ and leakage function $\mathcal{L}$. We now ready to define security of SSE:

Definition 2.3 (Static $\mathcal{L}$-secure SSE). *Let* $\Pi =$ (KeyGen, EDBSetup, TokGen, Search) *be an SSE scheme and let* $\mathcal{L}$ *be a leakage function. We say that* *the scheme Π is* static $\mathcal{L}$-secure searchable encryption *if for every* PPT *adversary* $\mathcal{A}$, *there exists a* PPT *simulator* $\mathcal{S}$ *and a negligible function* $\mathsf{negl}(\cdot)$ *such that*

$$\left| \Pr\left[\text{SSE-REAL}_{\Pi,\mathcal{A}}(\lambda) = 1\right] - \Pr\left[\text{SSE-IDEAL}_{\Pi,\mathcal{L},\mathcal{A},\mathcal{S}}(\lambda) = 1\right]\right| < \mathsf{negl}(\lambda)$$

Adaptive setting. In the adaptive setting, the adversary is not restricted to specifying all of its queries $\mathbf{w}$ in advance, but can instead choose its queries during the execution in an adaptive manner, depending on the encrypted database EDB and on the tokens that it sees. Let SSE-REAL$_{\Pi,\mathcal{A}}^{\mathsf{adapt}}(\lambda)$ denote the output of the real execution in this adaptive setting. In the ideal execution, the simulator $\mathcal{S}$ is now an interactive Turing machine, which interacts with the experiment by responding to queries. First, the simulator $\mathcal{S}$ is invoked on $\mathcal{L}(\mathsf{DB})$ and outputs EDB. Then, for every query w_i that $\mathcal{A}$ may output, the function $\mathcal{L}$ is invoked on DB and all previously queries $\{w_j\}_{j<i}$ and the new query w_i, outputs some new leakage information which is given to the simulator $\mathcal{S}$. The latter outputs some t_i, which is given back to $\mathcal{A}$, who may then issue a new query. At the end of the execution, $\mathcal{A}$ outputs a bit b. Let SSE-IDEAL$_{\Pi,\mathcal{L},\mathcal{A},\mathcal{S}}^{\mathsf{adapt}}(\lambda)$ be the output of the ideal execution. The adaptive security of SSE is defined as follows:

Definition 2.4 (Adaptive $\mathcal{L}$-secure SSE). *Let $\Pi = (\mathsf{KeyGen}, \mathsf{EDBSetup}, \mathsf{TokGen}, \mathsf{Search})$ be an SSE scheme and let $\mathcal{L}$ be a leakage function. We say that the scheme Π is* adaptive $\mathcal{L}$-secure searchable encryption *if for every* PPT *adversary $\mathcal{A}$, there exists a* PPT *simulator $\mathcal{S}$ and a negligible function $\mathsf{negl}(\cdot)$ such that*

$$\left| \Pr\left[\text{SSE-REAL}_{\Pi,\mathcal{A}}^{\mathsf{adapt}}(\lambda) = 1\right] - \Pr\left[\text{SSE-IDEAL}_{\Pi,\mathcal{L},\mathcal{A},\mathcal{S}}^{\mathsf{adapt}}(\lambda) = 1\right]\right| < \mathsf{negl}(\lambda)$$

The leakage function. Following the standard notions of security for SSE we consider the leakage function $\mathcal{L}_{\mathsf{min}}$ for one-round protocols and the leakage function $\mathcal{L}_{\mathsf{sizes}}$ for two-round protocols, where

$$\mathcal{L}_{\mathsf{min}}(\mathsf{DB}, \mathbf{w}) = \left(N, \{\mathsf{DB}(w)\}_{w \in \mathbf{w}}\right),$$
$$\mathcal{L}_{\mathsf{sizes}}(\mathsf{DB}, \mathbf{w}) = \left(N, \{|\mathsf{DB}(w)|\}_{w \in \mathbf{w}}\right),$$

and $N = \sum_{w \in W} |\mathsf{DB}(w)|$ is the size of the database. That is, both functions return the size of the database, and the difference between them is that the function $\mathcal{L}_{\mathsf{min}}$ returns the actual documents that contain each keyword $w \in \mathbf{w}$ that the adversary has queried, whereas the function $\mathcal{L}_{\mathsf{sizes}}$ returns only the number of such documents.

The leakage functions in the adaptive setting are defined analogously. That is, for a database DB, a set of "previous" queries $\{w_j\}_{j<i}$, and a new query w_i, we define

$$\mathcal{L}_{\mathsf{min}}^{\mathsf{adap}}(\mathsf{DB}, \{w_j\}_{j<i}, w_i) = \begin{cases} N & \text{if } (\{w_j\}_{j<i}, w_i) = (\bot, \bot) \\ \mathsf{DB}(w_i) & \text{otherwise} \end{cases}$$

$$\mathcal{L}_{\mathsf{size}}^{\mathsf{adap}}(\mathsf{DB}, \{w_j\}_{j<i}, w_i) = \begin{cases} N & \text{if } (\{w_j\}_{j<i}, w_i) = (\bot, \bot) \\ |\mathsf{DB}(w_i)| & \text{otherwise} \end{cases}.$$

2.2   Static Hash Tables

In our schemes we rely on static hash tables (also known as static dictionaries). These are data structures that given a set S can support lookup operations in constant time in the standard unit-cost word-RAM model. Specifically, a static hash table consists of a pair of algorithms denoted (HTSetup, HTLookup). The algorithm HTSetup gets as input a set $S = \{(\ell_i, d_i)\}_{i=1}^{k}$ of pairs (ℓ_i, d_i) of strings, where $\ell_i \in \{0,1\}^s$ is the label and $d_i \in \{0,1\}^r$ is the data. The output of this algorithm is a hash table $\mathsf{HT}(S)$. The lookup algorithm HTLookup on input $(\mathsf{HT}(S), \ell)$ returns d if $(\ell, d) \in S$, and $\bot$ otherwise.

There exist many constructions of static hash tables that use linear space (i.e., $O(k(r + s))$ bits) and answer lookup queries by reading a constant number of contiguous s-bit blocks and r-bit blocks (see, for example, [1,26], and the many references therein).

2.3   Cuckoo Hashing with a Stash

Cuckoo hashing is an efficient and practical hash table designed by Pagh and Rodler [26], providing worst-case constant lookup time and uses linear space. An important property of cuckoo hashing is that by storing a few elements in a secondary (small) data structure, referred to as a "stash", it is possible to decrease its failure probability from noticeable to negligible [21]. For our purposes in this work, it suffices to consider the following abstraction of cuckoo hashing with a stash:

- The memory is an abstract array $[m]$, where each cell may contain a single element or NULL.
- The potential locations of any element are randomly sampled (instead of being determined by hash functions).

We now summarize the abstract properties of cuckoo hashing with a stash in which we are interested for our construction in Sect. 4:

1. For storing n lists, where each list consists of ℓ elements, an array of size $O(n \cdot \ell)$ is used. The array is partitioned into two segments – a cuckoo hashing segment of size $O(n \cdot \ell)$ and a stash segment of size $s \cdot \ell$.
2. Fetching a list requires accessing two random locations (of size ℓ each) in the cuckoo hashing segment and accessing the entire stash segment.
3. When using a stash of size $s = n^{o(1)}$, the probability that n lists can be successfully stored is $1 - O(n^{s/2})$ [4, Theorem 2].[7]

---

[7] Note that in the original work of Kirsch et al. [21] they considered a constant-sized stash, whereas in this work we are interested in a stash whose size is not necessarily constant, and thus we rely on [4].

3 The Pad-and-Split Framework: A Stronger Lower Bound

In this section we first formalize our pad-and-split framework for the design of symmetric searchable encryption schemes (Sect. 3.1). Then, we show that it captures the memory access patterns underlying essentially all of the existing symmetric searchable encryption schemes other than the schemes of Asharov et al. [2] (Sect. 3.2), and discuss the instantiation of Demertzis and Papamanthou (Sect. 3.3) whose tradeoff matches our lower bound (Sect. 3.4).

3.1 The Pad-and-Split Framework

Our framework considers symmetric searchable encryption schemes that are characterized by a deterministic algorithm denoted SplitList, and consist of the following two phases:

- Given a database $DB = \{DB(w_1), \ldots, DB(w_{n_W})\}$ of size N, for each keyword w_i the scheme invokes the SplitList algorithm on the length n_i of its corresponding list $DB(w_i)$, to obtain a vector $(x_i^{(1)}, \ldots, x_i^{(m)})$ of integers. The scheme then potentially pads the list $DB(w_i)$ by adding "dummy" elements, and splits the padded list into sublists of lengths $\text{len}^{(1)}, \ldots, \text{len}^{(m)}$, where $x_i^{(j)}$ denotes the number of sublists of each length $\text{len}^{(j)}$.
- For each possible length $\text{len}^{(j)}$, the scheme groups together all sublists of length $\text{len}^{(j)}$, and independently processes each such group to produce an encrypted database EDB.

A key property of our framework is that the SplitList algorithm, which determines the number of sublists of each length, does not take as input an actual list $DB(w_i)$ but only its length $n_i = |DB(w_i)|$. This algorithm is parameterized by the possible lengths $\text{len}^{(1)}, \ldots, \text{len}^{(m)}$ of sublists, and also by upper bounds $s^{(1)}, \ldots, s^{(m)}$ on the total number of sublists of lengths $\text{len}^{(1)}, \ldots, \text{len}^{(m)}$, respectively. We allow the parameters m, $\text{len}^{(1)}, \ldots, \text{len}^{(m)}$ and $s^{(1)}, \ldots, s^{(m)}$ to depend on the total length $N = \sum_{i=1}^{n_W} |DB(w_i)|$ of the database, but do not explicitly denote this for ease of notation.

We consider any possible instantiation of the SplitList algorithm subject to satisfying two natural requirements. First, we require that each list $DB(w_i)$ is split into sublists whose total length is at least the length of $DB(w_i)$. Second, we require that for every possible sublist length $\text{len}^{(j)}$ there are at most $s^{(j)}$ sublists of length $\text{len}^{(j)}$ in the worst-case over all possible databases of size N. Formally:

Definition 3.1. *We say that a* SplitList *algorithm, parameterized by* $(\text{len}^{(1)}, \ldots, \text{len}^{(m)})$ *and* $(s_1, \ldots, s_m)$ *is valid if for every integer N and for every vector of lengths $(n_1, \ldots, n_k)$ with $\sum_{i=1}^{k} n_i = N$, it holds that:*

- **Each list is not longer than the sum of lengths of its sublists:** For every n_i it holds that $x_i^{(1)} \cdot \mathsf{len}^{(1)} + \cdots + x_i^{(m)} \cdot \mathsf{len}^{(m)} \geq n_i$, where $(x_i^{(1)}, \ldots, x_i^{(m)}) = \mathsf{SplitList}(N, n_i)$.
- **Each $s^{(j)}$ upper bounds the number of sublists of length $\mathsf{len}^{(j)}$:** For every $j \in [m]$ it holds that $\sum_{i=1}^{k} x_i^{(j)} \leq s^{(j)}$, where $(x_i^{(1)}, \ldots, x_i^{(m)}) = \mathsf{SplitList}(N, n_i)$ for every $i \in [k]$.

In addition, we say that SplitList has locality L if each list $\mathsf{DB}(w_i)$ is split into at most L sublists. Formally:

Definition 3.2. *We say that a* SplitList *algorithm has locality* $L = L(N)$ *if for every* n_i *it holds that* $x_i^{(1)} + \cdots + x_i^{(m)} \leq L$, *where* $(x_i^{(1)}, \ldots, x_i^{(m)}) = \mathsf{SplitList}(N, n_i)$.

Equipped with our notion of a valid SplitList algorithm, we describe a general template (see Construction 3.5) for constructing symmetric searchable encryption schemes given any such algorithm. We rely, in addition, on a pseudorandom function PRF and a private-key encryption scheme (Enc, Dec) with pseudorandom ciphertexts – both of which can be constructed based on any one-way function. This yields the following theorem:

Theorem 3.3. *Given a valid* SplitList *algorithm with parameters* $(\mathsf{len}^{(1)}, \ldots, \mathsf{len}^{(m)})$ *and* $(s^{(1)}, \ldots, s^{(m)})$, *a pseudorandom function* PRF, *and a private-key encryption scheme* (Enc, Dec) *with pseudorandom ciphertexts, Construction 3.5 is a static* $\mathcal{L}_{\min}$ *-secure symmetric searchable encryption scheme for databases of size* N *with the following parameters:*

- *Space* $O\left(\sum_{j=1}^{m} s^{(j)} \cdot \mathsf{len}^{(j)}\right)$.
- *Locality* $O(L(N))$, *where* SplitList *has locality* $L(N)$.
- *Read efficiency* $O(1)$.
- *Token size* $O(1)$.

Moreover, Construction 3.5 is an adaptive $\mathcal{L}_{\min}^{\mathsf{adap}}$ *-secure symmetric searchable encryption scheme in the random-oracle model, when instantiating* PRF *and* (Enc, Dec) *appropriately.*

Note that Theorem 3.3 guarantees that Construction 3.5 is statically-secure in the standard model (although we do prove it can be made adaptively secure in the random-oracle model). The next theorem shows that a simple modification of the scheme (described as Construction 3.6), based on an idea sketched by Stefanov et al. [28], is in fact adaptively secure in the standard model. This comes at the cost of increasing the token size from tokens of size $O(1)$ to tokens of size $O(L)$, where L is the locality of the SplitList algorithm. In this scheme, the client decrypts the results sent by the server (using the Resolve algorithm), and thus the scheme leaks only the size of the results. This is in contrast to the scheme described in Construction 3.5, where the server decrypts the results, and thus the scheme leaks the results themselves[8].

Theorem 3.4. *Given a valid* SplitList *algorithm with parameters* $(\mathsf{len}^{(1)}, \ldots,$ $\mathsf{len}^{(m)})$ *and* $(s^{(1)}, \ldots, s^{(m)})$, *a pseudorandom function* PRF, *and a private-key encryption scheme* (Enc, Dec) *with pseudorandom ciphertexts, Construction 3.6 is an adaptive* $\mathcal{L}_{\mathsf{size}}^{\mathsf{adap}}$-*secure symmetric searchable encryption scheme for databases of size* N *with the following parameters:*

- *Space* $O\left(\sum_{j=1}^{m} s^{(j)} \cdot \mathsf{len}^{(j)}\right)$.
- *Locality* $O(L(N))$, *where* SplitList *has locality* $L(N)$.
- *Read efficiency* $O(1)$.
- *Token size* $O(L(N))$.

In the remainder of this section we provide a high-level overview of these schemes. The proofs of Theorems 3.3 and 3.4 can be found in the full version of this paper [3].

Overview of the schemes. In both schemes each list of document identifiers $\mathsf{DB}(w_i)$ is padded and split as dictated by the output $(x_i^{(1)}, \ldots, x_i^{(m)}) =$ SplitList$(N, |\mathsf{DB}(w_i)|)$. That is, $\mathsf{DB}(w_i)$ is padded to length $x_i^{(1)} \cdot \mathsf{len}^{(1)} + \cdots + x_i^{(m)} \cdot \mathsf{len}^{(m)}$, and split into sublists, where for each $j \in [m]$ there are $x_i^{(j)}$ sublists of length $\mathsf{len}^{(j)}$. Then, we construct an encrypted database which consists of the following hash tables:

- A hash table that stores (in encrypted manner) the lengths of all lists, and is padded to contain exactly N elements.
- For every $j \in [m]$ a hash table that stores (in an encrypted manner) all sublists of length $\mathsf{len}^{(j)}$, and is padded to contain exactly $s^{(j)}$ sublists of this length.

In all hash tables we store the various elements according to pseudorandom labels that are derived from each corresponding keyword w via a pseudorandom function whose key is known only to the client. Intuitively speaking, the scheme is secure for any valid SplitList algorithm due to the following three reasons: (1) The number of padded elements and the number of sublists each list is split into depend only on the length of each list, (2) each hash table consists of encrypted elements with pseudorandom labels, and (3) the size of each hash table depends only on the size of the database.

The main differences between Construction 3.5 (providing static security) and Construction 3.6 (providing adaptive security) are as follows:

1. The lengths of the lists in Construction 3.6 are encrypted using one-time pads. This is required in order to allow "explaining" a random value as the encryption of any particular length on the fly (given that adversaries may be adaptive).

---

[8] Note that any scheme in which the server decrypts the results can be easily transformed into a scheme where only the client decrypts the results by adding an additional encryption layer – but this does not necessarily hold in the other direction.

2. In Construction 3.5, for each searched keyword the server is given keys derived from that keyword, allowing it to compute the labels and decrypt the document identifiers associated with that keyword. In Construction 3.6 the server is given the labels themselves (thus, the token size is $O(L)$), and can only locate the encrypted document identifiers, but not to decrypt them.

3.2   The Generality of the Pad-and-Split Framework

We now demonstrate that our pad-and-split framework captures the memory access patterns underlying the vast majority of existing symmetric searchable encryption schemes for supporting keywords search (i.e., we show that these schemes can be obtained as instantiations of our framework). We note that each of these schemes offers other important implementation details, improvements and optimizations that we do not intend to capture using our framework (since these are not directly related to the tradeoff between space, locality, and read efficiency), and we refer to the relevant papers for further details.

The scheme of Curtmola et al. [11]. This is the most common technique underlying the vast majority of existing schemes (in particular, [7,10,12,20,23, 29]). In this scheme each list is split into single elements (i.e., sublists of length 1), and those are stored in the same hash table. This is captured by our framework when setting $m = 1$, $\mathsf{len}^{(1)} = 1$, $s^{(1)} = N$, and $\mathsf{SplitList}(N, n_i) = (n_i)$. This results in a scheme with space $O(N)$, locality $O(N)$, and read efficiency $O(1)$.

The 2lev scheme of Cash et al. [6]. This scheme can be viewed as a pad-and-split scheme with two possible lengths, b and B, where $b < B$. A list of length at most b is padded to length b, and a list of length greater than b is padded to a length that is a multiple of B and then split into sublists of length B (in order to reduce space overhead, this scheme does not add dummy lists, thus resulting in a non-standard leakage function). This results in a scheme with space $O(N \cdot (b + \frac{B}{b+1}))$, locality $O(N/B)$, and read efficiency $O(1)$.

A simple scheme with $O(N^2)$ space. In this scheme each list is padded to the maximal possible length (i.e., to length N, where $N = \sum_{i=1}^{n_w} |\mathsf{DB}(w_i)|$), and all lists are stored in the same hash table. This is captured by our framework when setting $m = 1$, $\mathsf{len}^{(1)} = N$, $s^{(1)} = N$ and $\mathsf{SplitList}(N, n_i) = (1)$. This results in a scheme with space $O(N^2)$, locality $O(1)$, and read efficiency $O(1)$.

The scheme of Cash and Tessaro [8]. This scheme splits a list of length n_i into at most $\log n_i$ sublists of lengths that are powers of 2 according to the binary representation of n_i. Then, for each possible power of 2, the scheme stores sublists of that length in a separate hash table. This is captured by our framework when setting $m = \lfloor \log N \rfloor + 1$, $\mathsf{len}^{(j)} = 2^{j-1}$, $s^{(j)} = N/2^{j-1}$, and the SplitList algorithm on input n_i outputs a binary vector of length m which corresponds to the binary representation of n_i. This results in a scheme with space $O(\sum_{j=1}^{m} \mathsf{len}^{(j)} s^{(j)}) = O(N \log N)$, locality $O(\log N)$, and read efficiency $O(1)$.

CONSTRUCTION 3.5 (One-Round Pad-and-Split SSE Scheme)

A pad-and-split SSE scheme is parameterized by a SplitList algorithm, and by the following values (all values are functions of the size N of the database):

1. Locality parameter L.
2. Possible lengths $\mathsf{len}^{(1)}, \ldots, \mathsf{len}^{(m)}$ of sublists.
3. Upper bounds $s^{(1)}, \ldots, s^{(m)}$ on the total number of sublists of lengths $\mathsf{len}^{(1)}, \ldots, \mathsf{len}^{(m)}$, respectively.

Key generator. The algorithm KeyGen on input 1^λ samples and outputs a key $K \leftarrow \{0,1\}^\lambda$ for PRF.

Setup. The algorithm EDBSetup on input (K, DB) is defined as follows:

1. Initialize $t+1$ empty sets $T, T_1, \ldots, T_m$, where T will consist of the lengths of the lists, and each set T_j will consist of all sublists of length $\mathsf{len}^{(j)}$.
2. For every keyword $w_i \in \mathsf{W}$ with an associated list $\mathsf{DB}(w_i) = \{\mathsf{id}_1, \ldots, \mathsf{id}_{n_i}\}$:
   (a) Compute $(\mathsf{label}_i, K_i, \widehat{K_i}) = \mathsf{PRF}_K(w_i)$.
   (b) Compute $\widehat{n_i} = \mathsf{Enc}_{K_i}(n_i)$ and add the pair $(\mathsf{label}_i, \widehat{n_i})$ to the set T.
   (c) Compute $(x_i^{(1)}, \ldots, x_i^{(m)}) = \mathsf{SplitList}(N, n_i)$.
   (d) For every $j = 1, \ldots, m$:
       i. For every $x = 1, \ldots, x_i^{(j)}$:
          A. Take the next $\mathsf{len}^{(j)}$ elements from the list $\mathsf{DB}(w_i)$ and create a block $\{\mathsf{id}'_1, \ldots, \mathsf{id}'_{\mathsf{len}(j)}\}$. If there are less than $\mathsf{len}^{(j)}$ elements left in $\mathsf{DB}(w_i)$, then pad with dummy elements.
          B. Compute a label: $\mathsf{label}_{j,x} = \mathsf{PRF}_{\widehat{K_i}}(j, x)$.
          C. Encrypt $d_{j,x} = (\mathsf{Enc}_{K_i}(\mathsf{id}'_1), \ldots, \mathsf{Enc}_{K_i}(\mathsf{id}'_{\mathsf{len}(j)}))$.
          D. Insert the pair $(\mathsf{label}_{j,x}, d_{j,x})$ into the set T_j.
3. Pad the set T to contain exactly N elements by adding dummy elements, and pad each set T_j to contain exactly $s^{(j)}$ elements by adding dummy elements.
4. For each set $T, T_1, \ldots, T_m$, uniformly shuffle the set, and generate a hash table by invoking the HTSetup algorithm for obtaining hash tables $\mathsf{HT}(T), \mathsf{HT}(T_1), \ldots, \mathsf{HT}(T_m)$.
5. Output $\mathsf{EDB} = (\mathsf{HT}(T), (\mathsf{HT}(T_1), \ldots, \mathsf{HT}(T_m)))$.

Token generator. The algorithm TokGen on input (K, w_i) computes and outputs the token $\tau_i = (\mathsf{label}_i, K_i, \widehat{K_i}) = \mathsf{PRF}_K(w_i)$.

Search. The algorithm Search on input (τ_i, EDB), where $\tau_i = (\mathsf{label}_i, K_i, \widehat{K_i})$ and $\mathsf{EDB} = (\mathsf{HT}(T), \mathsf{HT}(T_1), \ldots, \mathsf{HT}(T_m))$, is defined as follows:

1. Initialize a list of document identifiers $R = \emptyset$.
2. Invoke HTLookup on the hash table $\mathsf{HT}(T)$ and label label_i to retrieve $\widehat{n_i} = \mathsf{Enc}_{K_i}(n_i)$. Decrypt $\widehat{n_i}$ using the key K_i, and compute $(x_i^{(1)}, \ldots, x_i^{(m)}) = \mathsf{SplitList}(N, n_i)$.
3. For every $j \in [m]$ and for every $x \in \left[x_i^{(j)}\right]$ compute $\mathsf{label}_{j,x} = \mathsf{PRF}_{\widehat{K_i}}(j, x)$. Invoke HTLookup on the hash table $\mathsf{HT}(T_j)$ for the label $\mathsf{label}_{j,x}$, and obtain the block $d_{j,x}$. Decrypt the block using the key K_i and add the elements to the list R. For a block that contains dummy elements, obtain and decrypt only the part that does not contain dummy elements.

CONSTRUCTION 3.6 (Two-Round Pad-and-Split SSE Scheme)

A pad-and-split SSE scheme is parameterized by a SplitList algorithm, and the following values (all values are functions of the size N of the database):

1. Locality parameter L.
2. Possible lengths $\mathsf{len}^{(1)}, \ldots, \mathsf{len}^{(m)}$ of sublists.
3. Upper bounds $s^{(1)}, \ldots, s^{(m)}$ on the total number of sublists of lengths $\mathsf{len}^{(1)}, \ldots, \mathsf{len}^{(m)}$, respectively.

Key generator. The algorithm KeyGen on input 1^λ samples a key $K \leftarrow \{0,1\}^\lambda$ for PRF, samples a key $\widehat{K} \leftarrow \{0,1\}^\lambda$ for (Enc, Dec), and outputs $(K, \widehat{K})$.

Setup. The algorithm EDBSetup on input $((K, \widehat{K}), \mathsf{DB})$ is defined as follows:

1. Initialize $t+1$ empty sets $T, T_1, \ldots, T_m$, where T will consist of the lengths of the lists, and each set T_j will consist of all sublists of length $\mathsf{len}^{(j)}$.
2. For every keyword $w_i \in \mathsf{W}$ with an associated list $\mathsf{DB}(w_i) = \{\mathsf{id}_1, \ldots, \mathsf{id}_{n_i}\}$:
   (a) Compute $((\mathsf{label}_i, K_i), (\mathsf{label}_{i,1}, \ldots, \mathsf{label}_{i,L})) = \mathsf{PRF}_K(w_i)$.
   (b) Compute $\widehat{n}_i = K_i \oplus n_i$ and add the pair $(\mathsf{label}_i, \widehat{n}_i)$ to the set T.
   (c) Compute $(x_i^{(1)}, \ldots, x_i^{(m)}) = \mathsf{SplitList}(N, n_i)$.
   (d) For every $j = 1, \ldots, m$:
       i. For every $x = 1, \ldots, x_i^{(j)}$:
          A. Take the next $\mathsf{len}^{(j)}$ elements from the list $\mathsf{DB}(w_i)$ and create a block $\{\mathsf{id}_1', \ldots, \mathsf{id}_{\mathsf{len}^{(j)}}'\}$. If there are less than $\mathsf{len}^{(j)}$ elements left in $\mathsf{DB}(w_i)$, then pad with dummy elements.
          B. Let label be the first unused label from $(\mathsf{label}_{i,1}, \ldots, \mathsf{label}_{i,L})$.
          C. Encrypt $d_{j,x} = \left(\mathsf{Enc}_{\widehat{K}}(\mathsf{id}_1'), \ldots, \mathsf{Enc}_{\widehat{K}}(\mathsf{id}_{\mathsf{len}^{(j)}}')\right)$.
          D. Insert the pair $(\mathsf{label}, d_{j,x})$ into the set T_j.
3. Pad the set T to contain exactly N elements by adding dummy elements, and pad each set T_j to contain exactly $s^{(j)}$ elements by adding dummy elements.
4. For each set $T, T_1, \ldots, T_m$, uniformly shuffle the set, and generate a hash table by invoking the HTSetup algorithm for obtaining hash tables $\mathsf{HT}(T), \mathsf{HT}(T_1), \ldots, \mathsf{HT}(T_m)$.
5. Output $\mathsf{EDB} = (\mathsf{HT}(T), (\mathsf{HT}(T_1), \ldots, \mathsf{HT}(T_m)))$.

Token generator. The algorithm TokGen on input $((K, \widehat{K}), w_i)$ computes and outputs the token $\tau_i = ((\mathsf{label}_i, K_i), (\mathsf{label}_{i,1}, \ldots, \mathsf{label}_{i,L})) = \mathsf{PRF}_K(w_i)$.

Search. The algorithm Search on input (τ_i, EDB), where $\tau_i = ((\mathsf{label}_i, K_i), (\mathsf{label}_{i,1}, \ldots, \mathsf{label}_{i,L}))$ and $\mathsf{EDB} = (\mathsf{HT}(T), \mathsf{HT}(T_1), \ldots, \mathsf{HT}(T_m))$, is defined as follows:

1. Initialize a list of results $R = \emptyset$.
2. Invoke HTLookup on the hash table $\mathsf{HT}(T)$ and label label_i to retrieve $\widehat{n}_i = K_i \oplus n_i$. Decrypt $n_i = K_i \oplus \widehat{n}_i$ and compute $(x_i^{(1)}, \ldots, x_i^{(m)}) = \mathsf{SplitList}(N, n_i)$.
3. For every $j \in [m]$ and for every $x \in \left[x_i^{(j)}\right]$, let label be the first unused label from $(\mathsf{label}_{i,1}, \ldots, \mathsf{label}_{i,L})$. Invoke HTLookup on the hash table $\mathsf{HT}(T_j)$ for the label $\mathsf{label}_{j,x}$, obtain the block $d_{j,x}$, and add its elements to the list R. For a block that contains dummy elements, obtain only the part that does not contain dummy elements.

Resolve. The algorithm Resolve on input $((K, \widehat{K}), R)$ computes and outputs the identifiers $M = \{\mathsf{Dec}_{\widehat{K}}(c) : c \in R\}$.

The scheme of Asharov et al. [2, Sect. 5]. This scheme improves the one of Cash and Tessaro [8]. In this scheme, a list of length $2^{p_i-1} < n_i \le 2^{p_i}$ is padded to length 2^{p_i} and stored as a whole. This is captured by our framework when setting $m = \lceil \log N \rceil + 1$, $\mathsf{len}^{(j)} = 2^{j-1}$, $s^{(j)} = 2N/2^{j-1}$, and the SplitList algorithm, on input n_i, outputs a vector of length m where all the entries are zeros except for a one that appears in the location $\lceil \log n_i \rceil + 1$. This results in a scheme with space $O(\sum_{j=1}^m \mathsf{len}^{(j)} s^{(j)}) = O(N \log N)$, locality $O(1)$, and read efficiency $O(1)$.

3.3   An Optimal Instantiation for Any Locality

As discussed in Sect. 1.1, the lower bound that we prove for schemes in the pad-and-split framework matches the tradeoff provided by the scheme of Demertzis and Papamanthou [14] (which is captured by our framework). Specifically, when setting the read efficiency of their scheme to $O(1)$, one obtains a statically-secure scheme with space $O(N \log N / \log L)$, locality L, and read efficiency $O(1)$. It should be noted that their scheme supports also non-constant read efficiency, but in that case it is not captured by our framework as it leaks additional information (in particular, the random choices made by the setup algorithm).

In what follows we describe their instantiation within our above-described template. Their scheme is obtained by splitting each list to sublists of lengths that are a power of the locality L. In our notation, we set $m = \lfloor \log N / \log L \rfloor + 1 = \lfloor \log_L N \rfloor + 1$, $\mathsf{len}^{(j)} = L^{j-1}$, and $s^{(j)} = 2N/\mathsf{len}^{(j)}$ for every $j \in [m]$. As for the splitting algorithm, a list of length $L^{j-1} \le n_i < L^j$ is padded to a length that is a multiple of L^{j-1}, and split into at most L sublists of length L^{j-1}. More formally, SplitList(N, n_i) outputs a vector of length m, where all the entries are zeros except for the entry in the position $j = \lfloor \log_L(n_i) \rfloor + 1$, which is set to the value $\lceil n_i/L^{j-1} \rceil \in \{1, \ldots, L\}$.

This is indeed a valid SplitList algorithm, and its locality is L. Specifically, for each n_i and j it holds that $\lceil n_i/L^{j-1} \rceil \cdot L^{j-1} \ge n_i$, that is, each list is not longer than the sum of the lengths of its sublists. Moreover, for $j = \lfloor \log_L(n_i) \rfloor + 1$ it also holds that $\lceil n_i/L^{j-1} \rceil \le L$ and $\lceil n_i/L^{j-1} \rceil \cdot L^{j-1} < 2 \cdot n_i$. This means that the locality is L, and that the padding at most doubles the length of the list. Therefore, it is suffices to set $s^{(j)} = 2N/\mathsf{len}^{(j)}$, and thus it holds that $\sum_{j=1}^m \mathsf{len}^{(j)} \cdot s^{(j)} = m \cdot 2N = O(N \cdot \log N / \log L)$.

According to Theorems 3.3 and 3.4, the above splitting algorithm results in a searchable symmetric encryption schemes with space $O(N \cdot \log N / \log L)$, locality $O(L)$, and read efficiency $O(1)$. This yields the following corollaries:

Corollary 3.7 ([14]). *Assuming the existence of any one-way function, for any $L = L(N) > c$ (where c is an absolute constant) there exists a static $\mathcal{L}_{\min}$-secure symmetric searchable encryption scheme for databases of size N with the following parameters:*

- *Space $O(N \cdot \log N / \log L)$.*
- *Locality $L(N)$.*

- *Read efficiency $O(1)$.*
- *Token size $O(1)$.*

Moreover, the scheme is adaptively $\mathcal{L}_{\min}^{\mathrm{adap}}$-secure in the random-oracle model, when instantiating its building blocks appropriately.

Corollary 3.8. *Assuming the existence of any one-way function, for any $L = L(N) > c$ (where c is an absolute constant) there exists an adaptive $\mathcal{L}_{\mathrm{size}}^{\mathrm{adap}}$-secure symmetric searchable encryption scheme for databases of size N with the following parameters:*

- *Space $O(N \cdot \log N / \log L)$.*
- *Locality $L(N)$.*
- *Read efficiency $O(1)$.*
- *Token size $O(L(N))$.*

Better efficiency for super-constant sub-polynomial locality. For locality $L(N)$ satisfying $\omega(1) \leq L(N) \leq N^{o(1)}$ we can in fact instantiate our framework in a manner that reduces the expression $\sum_{j=1}^{m} \mathrm{len}^{(j)} s^{(j)}$ to $(1 + o(1))(N \cdot \log N / \log L)$. This matches our lower bound, which is shown to be $(1 - o(1))(N \cdot \log N / \log L)$, to within an additive lower-order term.

This is done as follows. Let $\widehat{L} = \lfloor L / \log L \rfloor$, and for a list of length n_i let j such that $\widehat{L}^j \leq n_i < \widehat{L}^{j+1}$. Represent $n_i = a \cdot \widehat{L}^j + b \cdot \widehat{L}^{j-1} + c$, where $a \in \{1, \ldots, \widehat{L} - 1\}$, $b \in \{0, \ldots, \widehat{L} - 1\}$, and $c \in \{0, \ldots, \widehat{L}^{j-1} - 1\}$. If $a \geq \log L$, then pad and split the list into at most $\widehat{L}$ sublists of length $\widehat{L}^j$. Otherwise, pad and split the list into at most $\widehat{L} \cdot \log L \leq L$ sublists of length $\widehat{L}^{j-1}$. This way, we never pad a list more than $(1 + 1/\log L)$ times its length, so for any j, we can set $s^{(j)} = (1 + 1/\log L) N / \mathrm{len}^{(j)}$, and obtain

$$\sum_{j=1}^{m} \mathrm{len}^{(j)} s^{(j)} = \left(1 + \frac{1}{\log L}\right) N \cdot \left(\left\lfloor \frac{\log N}{\log \widehat{L}} \right\rfloor + 1\right) = (1 + o(1)) N \cdot \frac{\log N}{\log L},$$

where the last equality holds since $\omega(1) \leq L \leq N^{o(1)}$.

3.4   Our Lower Bound for Pad-and-Split Schemes

In this section we present our lower bound on the trade-off between the space and the locality of any pad-and-split scheme. Recall that each such a scheme is characterized by a SplitList algorithm that satisfies a modest validity requirement (recall Definition 3.1), and is associated with the following parameters (all of which may be functions of the size N of the database):

- The possible lengths $\mathrm{len}^{(1)}, \ldots, \mathrm{len}^{(m)}$ of sublists to which the SplitList algorithm splits the list associated with each keyword, as described in Sect. 3.1.
- Upper bounds $s^{(1)}, \ldots, s^{(m)}$ on the total number of sublists of lengths $\mathrm{len}^{(1)}, \ldots, \mathrm{len}^{(m)}$, respectively, that are produced by the SplitList algorithm when processing an entire database.

Equipped with the above parameters, recall from Theorems 3.3 and 3.4 that the space usage of a pad-and-split scheme is $O\left(\sum_{j=1}^{m} s^{(j)} \cdot \mathsf{len}^{(j)}\right)$, and the locality of such a scheme is $O(L)$ where $L = L(N)$ is the locality of its SplitList algorithm (i.e., each list is split into at most L sublists). Thus, proving a lower bound on the trade-off between the space and the locality of pad-and-split schemes translates to proving such a lower bound on the corresponding parameters of their underlying SplitList algorithm. Theorem 1.1 follows as an immediate corollary of the following theorem, which we prove in the full version of this paper [3]:

Theorem 3.9. *Let* SplitList *be a valid splitting algorithm with parameters* $\mathsf{len}^{(1)}, \ldots, \mathsf{len}^{(m)}$ *and* $s^{(1)}, \ldots, s^{(m)}$, *and with locality* $L = L(N)$. *Then, for any* $0 < c < 1$ *it holds that*

$$\sum_{j=1}^{m} \mathsf{len}^{(j)} \cdot s^{(j)} \geq (1-c) \cdot N \cdot \left(\frac{\log N}{\log L - \log c + C_1} - C_2 \right),$$

where C_1 *and* C_2 *are small absolute constants.*

In particular, by setting $c = 1/2$ we obtain the lower bound $\sum_{j=1}^{m} \mathsf{len}^{(j)} \cdot s^{(j)} = \Omega(N \cdot \log N / \log L)$, which implies Theorem 1.1. In addition, if $\omega(1) \leq L(N) \leq N^{o(1)}$ then by setting $c = 1/\log L$ we obtain the tighter lower bound $\sum_{j=1}^{m} \mathsf{len}^{(j)} \cdot s^{(j)} \geq (1 - o(1))N \cdot \log N / \log L$.

4 The Statistical-Independence Framework: A Leveled Two-Choice Scheme

In this section we consider the statistical-independence framework introduced by Asharov et al. [2] for the design of symmetric searchable encryption schemes. As discussed in Sect. 1.2, within this framework we construct a scheme whose read efficiency when querying for a keyword w may depend on the length of the list $\mathsf{DB}(w)$ that is associated with w, and for any $n \leq N$ we denote by $\mathsf{r}(N, n)$ the read efficiency when retrieving a list of length n.[9] We prove the following theorem:

Theorem 4.1. *Assuming the existence of any one-way function, for any function* $f(N) = \omega(1)$ *there exists an adaptive* $\mathcal{L}_{\mathsf{size}}^{\mathsf{adap}}$*-secure symmetric searchable encryption scheme for databases of size* N *in which no keyword is associated with more than* $N/\log^3 N$ *identifiers, with the following parameters:*

- *Space* $O(N)$.
- *Locality* $O(1)$.
- *Read efficiency* $\mathsf{r}(N, n) = f(N) \cdot \epsilon(n)^{-1} + O(\log \log \log N)$, *where* $n = N^{1-\epsilon(n)}$.
- *Token size* $O(1)$.

---

[9] We emphasize that having the read efficiency depend on the length of the retrieved list does not hurt the security of SSE schemes, and our scheme still results in minimal leakage as required.

Comparing the performance of our new scheme with the lower bound of Asharov et al. in the statistical-independence framework, Theorem 4.1 matches their lower bound to within an additive $O(\log\log\log N)$ factor in the read efficiency. Specifically, Asharov et al. proved the following lower bound for schemes in the statistical-independence framework (restated to consider read efficiency $r(N, n)$ that may depend on the length n of each list, and to consider constant locality):

Theorem 4.2 ([2]). *For any searchable symmetric encryption scheme in the statistical-independence framework with space $O(N \log N)$, locality $O(1)$, and read efficiency $r(N, n)$, there exists a function $f(N) = \omega(1)$ such that $r(N, n) = f(N) \cdot \epsilon(n)^{-1}$ for every $1 \le n \le N/\log N$, where $n = N^{1-\epsilon(n)}$.*

In the remainder of this section we first overview the statistical independence framework for the design of symmetric searchable encryption schemes (Sect. 4.1), and then present our new scheme within this framework (Sect. 4.2).

4.1   The Statistical-Independence Framework

The statistical-independence framework of Asharov et al. [2] considers symmetric searchable encryption schemes that are characterized by a pair of algorithms, denoted RangesGen and Allocation, and consist of the following two phases:

- Given a database $DB = \{DB(w_1), \ldots, DB(w_{n_W})\}$ of size N, for each keyword w_i the scheme invokes the RangesGen algorithm on the length n_i of its corresponding list $DB(w_i)$, to obtain a set of *possible locations* in which the scheme may place the elements of the list $DB(w_i)$. This set consists of several intervals and we denote it by $R_i = \{[a_1, b_1], \ldots, [a_d, b_d]\} \leftarrow \mathsf{RangesGen}(N, n_i)$.
  Looking ahead, when supplied with a token corresponding to a keyword w_i, the server will return to the client all data stored in the possible locations of the list $DB(w_i)$ (the server will not actually know in which of the possible locations the elements of the list are actually placed).
- Given the sets of possible locations $R_1, \ldots, R_{n_W}$ of the lists corresponding to all keywords $w_1, \ldots, w_{n_W}$, respectively, the scheme invokes the Allocation algorithm on these sets (and on the respective lengths of the lists) to obtain the *actual locations* for the elements of all lists. We denote the actual locations as an array $\mathsf{map} \leftarrow \mathsf{Allocation}((n_1, R_1), \ldots, (n_{n_W}, R_{n_W}))$, where each of its entries is either a pair (i, j) (representing that this entry is the actual location of the jth element from the list $DB(w_i)$) or NULL (representing an empty entry).

A key property of this framework is that the RangesGen algorithm, which determines the set of possible locations for each list $DB(w_i)$, is applied separately and independently to the length of each list. Thus, the possible locations of each list are independent of the possible locations of all other lists (in contrast, the actual locations of the lists are naturally allowed to be correlated).

Asharov et al. referred to a pair (RangesGen, Allocation) of such algorithms as an allocation scheme, and showed that any such allocation scheme satisfying a natural correctness requirement can be used to construct a searchable symmetric encryption scheme. The correctness requirement asks that for any database, with all but a negligible probability, these algorithms produce an actual allocation map in which each element has exactly one actual placement (where the probability is taken over the internal coin tosses of the algorithms RangesGen and Allocation).

The resulting scheme of Asharov et al. inherits its space, locality and read efficiency from those of its underlying allocation scheme, defined as follows:

Definition 4.3. *A pair* (RangesGen, Allocation) *of algorithms satisfying the above correctness requirement is an* $(\mathsf{s}, \mathsf{d}, \mathsf{r})$*-allocation scheme, for some functions* $\mathsf{s}(\cdot)$, $\mathsf{d}(\cdot)$ *and* $\mathsf{r}(\cdot, \cdot)$, *if the following properties hold:*

- **Space:** *For any input* $(n_1, \ldots, n_k)$, *the array* map $\leftarrow$ Allocation$((n_1, R_1),$ $\ldots, (n_k, R_k))$, *where* $R_i = \{[a_1, b_1], \ldots, [a_\mathsf{d}, b_\mathsf{d}]\} \leftarrow$ RangesGen(N, n_i) *for every* $i \in [k]$, *is of size at most* $\mathsf{s}(N)$, *where* $N = \sum_{i=1}^{k} n_i$.
- **Locality:** *For any input* (N, n_i), *the algorithm* RangesGen *outputs at most* $\mathsf{d}(N)$ *ranges.*
- **Read efficiency:** *For any input* (N, n_i) *for the algorithm* RangesGen *it holds that:*

$$\frac{\sum_{j=1}^{\mathsf{d}} (b_j - a_j + 1)}{n_i} \leq \mathsf{r}(N, n_i) ,$$

where $\{[a_1, b_1], \ldots, [a_\mathsf{d}, b_\mathsf{d}]\} \leftarrow$ RangesGen(N, n_i).

Equipped with the above notation, Asharov et al. proved the following:

Theorem 4.4 ([2]). *Given any* $(\mathsf{s}, \mathsf{d}, \mathsf{r})$*-allocation scheme and any one-way function, there exists an* $\mathcal{L}_{\mathsf{size}}^{\mathsf{adap}}$*-secure searchable symmetric encryption scheme for databases of size* N *with space* $O(\mathsf{s}(N))$, *locality* $O(\mathsf{d}(N))$, *and read efficiency* $O(\mathsf{r}(N, \cdot))$.

From allocation algorithms to SSE schemes. We conclude our high-level description of the statistical-independence framework by briefly overviewing the generic transformation from allocation schemes to SSE scheme. The reader is referred to [2] for the complete formal description of this transformation.

In a nutshell, the client runs the RangesGen and the Allocation procedures as described above to obtain the actual allocation map of all elements. Then, the client encrypts each identifier from each list DB(w) in map with a key that is derived from the keyword w using a pseudorandom function. In addition, any unused entry in the array is filled with a uniform string of the appropriate length.

When issuing a query corresponding to a keyword w, the client asks the server to retrieve the encrypted content of all possible locations of the list DB(w).[10]

---

[10] The details here are quite subtle. The server obtains the pseudorandom key that was used to produce randomness for the relevant invocation of RangesGen. In addition, the server stores the lengths of the lists in an encrypted manner, and can only reveal the lengths of the already-queried lists. Knowing both the pseudorandom key and the list length allows the server to compute the possible locations of the list DB(w).

Since these locations are chosen independently at random, this does not reveal any additional information on the structure of the database except for the length of the queried list. The client then identifies the actual locations and decrypts the data by itself.

4.2 Our Leveled Two-Choice Scheme

In this section we present our new allocation scheme from which Theorem 4.4 provides the searchable symmetric encryption schemes guaranteed by Theorem 4.1. Our scheme consists of the following three levels for storing the elements of any given database DB of size N:

- The first level, named the "two-choice array", consists of the two-choice SSE scheme of Asharov et al. [2] but with *an exponentially improved read efficiency*. In this array, each list $\mathsf{DB}(w_i)$ can be stored in one out of two possible intervals of consecutive locations, in a manner that we describe below as part of our Allocation algorithm. However, unlike the scheme of Asharov et al. we do not store all of the N elements of the database in this array. Instead, the key observation underlying our new scheme is that when viewing this array as a collection of bins, then by allowing a few lists to "overflow" from this level to the second level (overall at most $\widehat{N} = N/\log N$ elements will overflow with all but a negligible probability), we can reduce the maximal load of each bin from $\tilde{O}(\log \log N)$ (as in [2]) to $O(\log \log \log N)$. This then translates into improving the read efficiency in this level from $\tilde{O}(\log \log N)$ to $O(\log \log \log N)$.
- The second level, named the "cuckoo hashing level", stores the vast majority of the elements that overflow from the first level. This level consists of roughly $\log N$ cuckoo hashing tables (see Sect. 2.3), where the j hash table is designed to store at most $\widehat{N}/2^j$ values each of which of size 2^j. These values are the lists that overflow from the first level (the jth table will store overflowing lists of length roughly 2^j).
- The third level, named the "stash level", consists of a cuckoo hashing stash for each of the second-level cuckoo hashing tables. The goal of introducing this level is to reduce the failure probably of cuckoo hashing from noticeable to negligible (see Sect. 2.3), which is essential for the security of the resulting SSE scheme.

This leveled structure of our allocation scheme, and thus of our SSE scheme, guarantees that the possible locations for a list $\mathsf{DB}(w)$ of length n are its two possible intervals in the two-choice array, its two locations in the jth cuckoo hashing table for $j = \log n$, and anywhere in the stash of the jth cuckoo hashing table. In what follows we formally describe our allocation scheme (see Algorithm 4.7), which we prove to have space $O(N)$, locality 5, and read efficiency $\omega(1) \cdot \epsilon(n)^{-1} + O(\log \log \log N)$ when retrieving lists of length $n = N^{1-\epsilon(n)}$.

Theorem 4.5. *For any function* $f(N) = \omega(1)$, *Algorithm 4.7 describes an* $(O(N), 5, \mathsf{r}(N, n))$-*allocation scheme for databases of size* N *in which no keyword is associated with more than* $N/\log^3 N$ *identifiers, where* $\mathsf{r}(N, n) = f(N) \cdot \epsilon(n)^{-1} + O(\log \log \log N)$ *and* $n = N^{1-\epsilon(n)}$.

Proof of Theorem 4.5. We assume without loss of generality that $f(N) = o(\log \log N)$ (since otherwise, we may take $\tilde{f}(N) = \min(f(N), o(\log \log N))$ instead). For the two-choice part of the algorithm, we make use of the following theorem from [2].

Theorem 4.6 ([2] Theorem 3.5 Part 1). *Let* $S \geq n_1$ *be a bound on the maximal length, and let* m *be the number of bins. Consider the two-choice allocation algorithm. Then, with probability* $1 - N^{-\Omega(\log N)}$, *there are at most* $S \log^2 N$ *elements at level greater than* $\frac{4N}{m} + \log \log \frac{N}{S} + 2$, *where the level of an element is the load of its bin right after inserting the element (e.g., the first element that is interested to the bin has level 1).*

In Algorithm 4.7, we set $S = N/\log^3 N$, $m = N/\log \log \log N$, and $\mathsf{BinSize} = O(\log \log \log N)$. Therefore, with an overwhelming probability there are at most $\widehat{N} = N/\log N$ overflowing elements, and in this case, we place at most $\widehat{N}$ elements in the cuckoo hashing tables with the stashes.

Now we analyze the placement of the elements in the hash tables, assuming that the number of elements in LeftOvers is at most $\widehat{N}$. For each $0 \leq j \leq t$, we set the stash size $s_j = f(N) \cdot \epsilon_j^{-1}$ where ϵ_j is chosen such that $2^j = N^{1-\epsilon_j}$. We obtain that the algorithm fails to insert the lists into the cuckoo hash table H_j with its stash with probability at most $O((\widehat{N}/2^j)^{-s_j/2})$ (see Sect. 2.3). Note that $N^{\epsilon_j} \geq \log^3 N$, so it holds that

$$(\widehat{N}/2^j)^{-s_i/2} = (N^{\epsilon_j}/\log N)^{-s_j/2}$$
$$\leq (N^{\frac{2}{3}\epsilon_j})^{-s_j/2}$$
$$= N^{-f(N)/3}.$$

Thus, the insertion of overflowing elements fails with a negligible probability, and we conclude that Algorithm 4.7 fulfills the correctness requirement. Regarding read efficiency, the overhead of the 2-choice is $O(\log \log \log N)$, the overhead of the cuckoo hash table is 2, and the overhead of the stash is $f(N) \cdot \epsilon(n)^{-1}$, where $n = N^{1-\epsilon(n)}$, so in total we get an overhead of $f(N) \cdot \epsilon_i^{-1} + O(\log \log \log N)$ as claimed. Locality of 5 easily follows from the description of SplitList. Regarding the space overhead, the bins require space of $m \cdot \mathsf{BinSize} = O(N)$, each cuckoo hash table with stash requires space of $O(\widehat{N}) = O(N/\log N)$, and there are less than $\log N$ tables. So in total, the space overhead is $O(N)$. ∎

ALGORITHM 4.7 (Our Allocation Scheme (RangesGen, Allocation))

Input: A vector of integers $(n_1, \ldots, n_k)$ representing the lengths of the lists $L_1, \ldots, L_k$ in the database. We let $N = \sum_{i=1}^{k} n_i$, $\widehat{N} = N/\log N$, and assume for concreteness that the n_i's are powers of 2, and that $n_1 \geq n_2 \geq \cdots \geq n_k$.

Parameters:

- A bound $S = N/\log^3 N$ on the length of the longest list in the database.
- The number $m = N/\log \log \log N$ of bins in the two-choice array (it is chosen as a power of 2 and such that $m \geq n_1$).
- A bound $\mathsf{BinSize} = O(\log \log \log N)$ on the size of each bin in the two-choice array.
- Stash sizes $s_0, \ldots, s_t$ where $t = \log S$ and $s_j = f(N) \cdot \epsilon_j$ for every $j \in [t]$, where $2^j = N^{1-\epsilon_j}$ and $\omega(1) \leq f(N) \leq o(\log \log N)$ may be any pre-specified function.

The memory layout. The memory is partitioned into the following segments:

1. m bins $B_0, \ldots, B_{m-1}$, each of size $\mathsf{BinSize}$.
2. Hash tables $H_0, \ldots, H_t$, where each hash table H_j is implemented as a cuckoo hash table for $\widehat{N}/2^j$ data items of size 2^j each with a stash of size s_j.

The RangesGen algorithm. On input N and n_i:

1. Uniformly sample $\alpha_{i,1}, \alpha_{i,2} \leftarrow \{0, \ldots, \frac{m}{n_i} - 1\}$.
   Consider the two super bins $\widetilde{B}_{\alpha_{i,1}} = (B_{n_i \cdot \alpha_1 + j})_{j=0}^{n_i-1}$ and $\widetilde{B}_{\alpha_{i,2}} = (B_{n_i \cdot \alpha_2 + j})_{j=0}^{n_i-1}$.
2. Sample two hash table locations $\beta_{i,1}, \beta_{i,2}$ for the cuckoo hash table $H_{\log n_i}$.
3. The possible ranges R_i are (1) The above two super-bins; (2) The two cells $\beta_{i,1}, \beta_{i,2}$ in the hashtable $H_{\log n_i}$; (3) The stash of the table $H_{\log n_i}$.

The Allocation algorithm.

1. Initialize m empty bins $B_0, \ldots, B_{m-1}$, and an empty set LeftOvers.
2. Initialize hash tables $H_0, \ldots, H_t$, where each hash table H_j is implemented as a cuckoo hash table for $\widehat{N}/2^j$ entries of size 2^j with a stash of size s_j.
3. For every list L_i with size n_i and ranges R_i, reconstruct $(\alpha_{i,1}, \alpha_{i,2})$ and $(\beta_{i,1}, \beta_{i_2})$ from R_i, and place the list L_i as follows:
   (a) Consider the two super bins $\widetilde{B}_{\alpha_{i,1}} = (B_{n_i \cdot \alpha_{i,1} + j})_{j=0}^{n_i-1}$ and $\widetilde{B}_{\alpha_{i,2}} = (B_{n_i \cdot \alpha_{i,2} + j})_{j=0}^{n_i-1}$. Let $\beta \in \{\alpha_{i,1}, \alpha_{i,2}\}$ be the index of the least loaded super bin among $\widetilde{B}_{\alpha_{i,1}}$ and $\widetilde{B}_{\alpha_{i,2}}$, where the load of a super bin is defined as the sum of loads of the bins that constitutes that super bin. If the load of the bins in $\widetilde{B}_\beta$ is $\mathsf{BinSize}$, then add L_i to LeftOvers. Otherwise, place the list L_i in the super bin $\widetilde{B}_\beta$. That is, for every $j = 0, \ldots, n_i - 1$, place the jth element of the list L_i in the bin $B_{n_i \cdot \beta + j}$.
   (b) If the list was not placed, then insert L_i into the cuckoo hash table $H_{\log n_i}$ using the locations $\beta_{i,1}$ and $\beta_{i,2}$. Note that the list might be placed in the stash. If the insertion fails, then output $\perp$ and abort.

References

1. Arbitman, Y., Naor, M., Segev, G.: Backyard cuckoo hashing: constant worst-case operations with a succinct representation. In: Proceedings of the 51st Annual IEEE Symposium on Foundations of Computer Science, pp. 787–796 (2010)
2. Asharov, G., Naor, M., Segev, G., Shahaf, I.: Searchable symmetric encryption: optimal locality in linear space via two-dimensional balanced allocations. In: Proceedings of the 48th Annual ACM Symposium on Theory of Computing, pp. 1101–1114 (2016)
3. Asharov, G., Segev, G., Shahaf, I.: Tight tradeoffs in searchable symmetric encryption. Cryptology ePrint Archive, Report 2018/507 (2018). https://eprint.iacr.org/2018/507
4. Aumüller, M., Dietzfelbinger, M., Woelfel, P.: Explicit and efficient hash families suffice for cuckoo hashing with a stash. Algorithmica 70(3), 428–456 (2014)
5. Cash, D., Grubbs, P., Perry, J., Ristenpart, T.: Leakage-abuse attacks against searchable encryption. In: Proceedings of the 22nd ACM Conference on Computer and Communications Security, pp. 668–679 (2015)
6. Cash, D., Jaeger, J., Jarecki, S., Jutla, C.S., Krawczyk, H., Rosu, M., Steiner, M.: Dynamic searchable encryption in very-large databases: data structures and implementation. In: Proceedings of the 21st Annual Network and Distributed System Security Symposium (2014)
7. Cash, D., Jarecki, S., Jutla, C., Krawczyk, H., Roşu, M.-C., Steiner, M.: Highly-scalable searchable symmetric encryption with support for boolean queries. In: Canetti, R., Garay, J.A. (eds.) CRYPTO 2013. LNCS, vol. 8042, pp. 353–373. Springer, Heidelberg (2013). https://doi.org/10.1007/978-3-642-40041-4_20
8. Cash, D., Tessaro, S.: The locality of searchable symmetric encryption. In: Nguyen, P.Q., Oswald, E. (eds.) EUROCRYPT 2014. LNCS, vol. 8441, pp. 351–368. Springer, Heidelberg (2014). https://doi.org/10.1007/978-3-642-55220-5_20
9. Chang, Y.-C., Mitzenmacher, M.: Privacy preserving keyword searches on remote encrypted data. In: Ioannidis, J., Keromytis, A., Yung, M. (eds.) ACNS 2005. LNCS, vol. 3531, pp. 442–455. Springer, Heidelberg (2005). https://doi.org/10.1007/11496137_30
10. Chase, M., Kamara, S.: Structured encryption and controlled disclosure. In: Abe, M. (ed.) ASIACRYPT 2010. LNCS, vol. 6477, pp. 577–594. Springer, Heidelberg (2010). https://doi.org/10.1007/978-3-642-17373-8_33
11. Curtmola, R., Garay, J.A., Kamara, S., Ostrovsky, R.: Searchable symmetric encryption: improved definitions and efficient constructions. In: Proceedings of the 13th ACM Conference on Computer and Communications Security, pp. 79–88 (2006)
12. Curtmola, R., Garay, J.A., Kamara, S., Ostrovsky, R.: Searchable symmetric encryption: Improved definitions and efficient constructions. J. Comput. Secur. 19(5), 895–934 (2011)
13. Demertzis, I. Papadopoulos, D., Papamanthou, C.: Searchable encryption with optimal locality: achieving sublogarithmic read efficiency. Cryptology ePrint Archive, Report 2017/749 (2017)
14. Demertzis, I., Papamanthou, C.: Fast searchable encryption with tunable locality. In: Proceedings of the 2017 ACM Special Interest Group on Management of Data (SIGMOD) Conference, pp. 1053–1067 (2017)
15. Dietzfelbinger, M., Pagh, R.: Succinct data structures for retrieval and approximate membership. In: Proceedings of the 35th International Colloquium on Automata, Languages and Programming, pp. 385–396 (2008)

16. Goh, E.: Secure indexes. Cryptology ePrint Archive, Report 2003/216 (2003)
17. Hagerup, T.: Sorting and searching on the word RAM. In: Proceedings of the 15th Annual Symposium on Theoretical Aspects of Computer Science, pp. 366–398 (1998)
18. Hagerup, T., Miltersen, P.B., Pagh, R.: Deterministic dictionaries. J. Algorithms **41**(1), 69–85 (2001)
19. Kamara, S., Papamanthou, C.: Parallel and dynamic searchable symmetric encryption. In: Proceedings of the 16th International Conference on Financial Cryptography and Data Security, pp. 258–274 (2013)
20. Kamara, S., Papamanthou, C., Roeder, T.: Dynamic searchable symmetric encryption. In: Proceedings of the 19th ACM Conference on Computer and Communications Security, pp. 965–976 (2012)
21. Kirsch, A., Mitzenmacher, M., Wieder, U.: More robust hashing: cuckoo hashing with a stash. SIAM J. Comput. **39**(4), 1543–1561 (2009)
22. Kurosawa, K., Ohtaki, Y.: UC-secure searchable symmetric encryption. In: Keromytis, A.D. (ed.) FC 2012. LNCS, vol. 7397, pp. 285–298. Springer, Heidelberg (2012). https://doi.org/10.1007/978-3-642-32946-3_21
23. Kurosawa, K., Ohtaki, Y.: How to update documents verifiably in searchable symmetric encryption. In: Proceedings of the 12th International Conference on Cryptology and Network Security, pp. 309–328 (2013)
24. Miltersen, P.B.: Cell probe complexity - a survey. In: Proceedings of the 19th Conference on the Foundations of Software Technology and Theoretical Computer Science, Advances in Data Structures Workshop (1999)
25. Pagh, A., Pagh, R.: Uniform hashing in constant time and optimal space. SIAM J. Comput. **38**(1), 85–96 (2008)
26. Pagh, R., Rodler, F.F.: Cuckoo hashing. J. Algorithms **51**(2), 122–144 (2004)
27. Song, D.X., Wagner, D., Perrig, A.: Practical techniques for searches on encrypted data. In: Proceedings of the 21st Annual IEEE Symposium on Security and Privacy, pp. 44–55 (2000)
28. Stefanov, E., Papamanthou, C., Shi, E.: Practical dynamic searchable encryption with small leakage. In: Proceedings of the 21st Annual Network and Distributed System Security Symposium (2014)
29. van Liesdonk, P., Sedghi, S., Doumen, J., Hartel, P.H., Jonker, W.: Computationally efficient searchable symmetric encryption. In: Proceedings of 7th VLDB Workshop on Secure Data Management, pp. 87–100 (2010)

Hardness of Non-interactive Differential Privacy from One-Way Functions

Lucas Kowalczyk[1], Tal Malkin[1], Jonathan Ullman[2(✉)], and Daniel Wichs[2]

[1] Department of Computer Science, Columbia University, New York, USA
{luke,tal}@cs.columbia.edu
[2] College of Computer and Information Science,
Northeastern University, Boston, USA
{jullman,wichs}@ccs.neu.edu

Abstract. A central challenge in differential privacy is to design computationally efficient non-interactive algorithms that can answer large numbers of *statistical queries* on a sensitive dataset. That is, we would like to design a differentially private algorithm that takes a dataset $D \in X^n$ consisting of some small number of elements n from some large data universe X, and efficiently outputs a summary that allows a user to efficiently obtain an answer to any query in some large family Q.

Ignoring computational constraints, this problem can be solved even when X and Q are exponentially large and n is just a small polynomial; however, all algorithms with remotely similar guarantees run in exponential time. There have been several results showing that, under the strong assumption of indistinguishability obfuscation, no efficient differentially private algorithm exists when X and Q can be exponentially large. However, there are no strong separations between information-theoretic and computationally efficient differentially private algorithms under any standard complexity assumption.

In this work we show that, if one-way functions exist, there is no general purpose differentially private algorithm that works when X and Q are exponentially large, and n is an arbitrary polynomial. In fact, we show that this result holds even if X is just subexponentially large (assuming only polynomially-hard one-way functions). This result solves an open problem posed by Vadhan in his recent survey [52].

1 Introduction

A central challenge in privacy research is to generate rich private *summaries* of a sensitive dataset. Doing so creates a tension between two competing goals. On one hand we would like to ensure *differential privacy* [22]—a strong notion of individual privacy that guarantees no individual's data has a significant influence on the summary. On the other hand, the summary should enable a user to obtain approximate answers to some large set of queries. Since the summary must be generated without knowing which queries the user will need to answer, we would like Q to be very large. This problem is sometimes called *non-interactive query*

© International Association for Cryptologic Research 2018
H. Shacham and A. Boldyreva (Eds.): CRYPTO 2018, LNCS 10991, pp. 437–466, 2018.
https://doi.org/10.1007/978-3-319-96884-1_15

release, in contrast with *interactive query release* where the user is required specify the (much smaller) set of queries that he needs to answer in advance, and the private answers may be tailored to just these queries.

More specifically, there is a sensitive dataset $D = (D_1, \ldots, D_n) \in X^n$ where each element of D is the data of some individual, and comes from some *data universe* X. We are interested in generating a summary that allows the user to answer *statistical queries* on D, which are queries of the form "What fraction of the individuals in the dataset satisfy some property q?" [38]. Given a set of statistical queries Q and a data universe X, we would like to design a differentially private algorithm M that takes a dataset $D \in X^n$ and outputs a summary that can be used to obtain an approximate answer to every query in Q. Since differential privacy requires hiding the information of single individuals, for a fixed (X, Q) generating a private summary becomes easier as n becomes larger. The overarching goal is to find algorithms that are both private and accurate for X and Q as large as possible and n as small as possible.

Since differential privacy is a strong guarantee, *a priori* we might expect differentially private algorithms to be very limited. However, a seminal result of Blum et al. [7] showed how to generate a differentially private summary encoding answers to *exponentially many* queries. After a series of improvements and extensions [23,26,33–35,41,43,49], we know that any set of queries Q over any universe X can be answered given a dataset of size $n \gtrsim \sqrt{\log |X|} \cdot \log |Q|$ [35]. Thus, it is information-theoretically possible to answer huge sets of queries using a small dataset.

Unfortunately, all of these algorithms have running time $\text{poly}(n, |X|, |Q|)$, which can be exponential in the *dimension* of the dataset, and in the *description* of a query. For example if $X = \{0, 1\}^d$, so each individual's data consists of d binary attributes, then the dataset has size nd but the running time will be at least 2^d. Thus, these algorithms are only efficient when both $|X|$ and $|Q|$ have polynomial size. There are computationally efficient algorithms when one of $|Q|$ and $|X|$ is very large, provided that the other is extremely small—at most $n^{2-\Omega(1)}$. Specifically, (1) the classical technique of perturbing the answer to each query with independent noise requires a dataset of size $n \gtrsim \sqrt{|Q|}$ [6,19,22,25], and (2) the folklore *noisy histogram algorithm* (see e.g. [52]) requires a dataset of size $n \gtrsim \sqrt{|X| \cdot \log |Q|}$. Thus there are huge gaps between the power of information-theoretic and computationally efficient differentially private algorithms.

Beginning with the work of Dwork et al. [23], there has been a series of results giving evidence that this gap is inherent using a connection to *traitor-tracing schemes* [16]. The first such result by Dwork et al. [23] showed the first separation between efficient and inefficient differentially private algorithms, proving a polynomial-factor separation in sample complexity between the two cases assuming bilinear cryptography. Subsequently, Boneh and Zhandry [10] proved that, under the much stronger assumption that indistinguishability obfuscation (iO) exists, then for a worst-case family of statistical queries, there is no computationally efficient algorithm with $\text{poly}(\log |Q| + \log |X|)$ sample complexity.

Table 1. Comparison of Hardness Results for Offline Differentially Private Query Release. Each row corresponds to an informal statement of the form "If the assumption holds, then there is no general purpose differentially private algorithm that works when the data universe has size at least $|X|$, the number of queries is at least $|Q|$, and the size of the dataset is at most n." All assumptions are polynomial-time hardness.

Reference	Data Universe $|X_\kappa|$	# of Queries $|Q_\kappa|$	Dataset Size $n(\kappa)$	Assumption
[8, 23]	$\geq \exp(\kappa)$	$\geq \exp(\kappa)$	$\leq \kappa^{2-\Omega(1)}$	Bilinear Maps
[10, 23]	$\geq \exp(\kappa)$	$\geq \exp(\kappa)$	$\leq \mathrm{poly}(\kappa)$	iO + OWF
[40]	$\geq \exp(\kappa)$	$\geq \tilde{O}(n^7)$	$\leq \mathrm{poly}(\kappa)$	iO + OWF
[40]	$\geq \tilde{O}(n^7)$	$\geq \exp(\kappa)$	$\leq \mathrm{poly}(\kappa)$	iO + OWF
This work	$\geq \exp(\kappa^{o(1)})$	$\geq \exp(\kappa)$	$\leq \mathrm{poly}(\kappa)$	OWF

More recently, Kowalczyk et al. [40] strengthened these results to show that the two efficient algorithms mentioned above—independent perturbation and the noisy histogram—are optimal up to polynomial factors, also assuming iO.

These results give a relatively clean picture of the complexity of non-interactive differential privacy, but only if we assume the existence of iO. Recently, in his survey on the foundations of differential privacy [52], Vadhan posed it as an open question to prove hardness of non-interactive differential privacy using standard cryptographic assumptions. In this work, we resolve this open question by proving a strong hardness result for non-interactive differential privacy making only the standard assumption that one-way functions (OWF) exist.

Theorem 1. *There is a sequence of pairs* $\{(X_\kappa, Q_\kappa)\}_{\kappa \in \mathbb{N}}$ *where*

$$|X_\kappa| = 2^{2^{\mathrm{poly}(\log \log \kappa)}} = 2^{\kappa^{o(1)}}, |Q_\kappa| = 2^\kappa$$

such that, assuming the existence of one-way functions, for every polynomial $n = n(\kappa)$, *there is no polynomial time differentially private algorithm that takes a dataset* $D \in X_\kappa^n$ *and outputs an accurate answer to every query in* Q_κ *up to an additive error of* $\pm 1/3$.

We remark that, in addition to removing the assumption of iO, Theorem 1 is actually stronger than that of Boneh and Zhandry [10], since the data universe size can be subexponential in κ, even if we only make standard polynomial-time hardness assumptions. We leave it as an interesting open question to obtain quantitatively optimal hardness results matching (or even improving) those of [40] using standard assumptions. Table 1 summarizes existing hardness results as compared to our work.

Like all of the aforementioned hardness results, the queries constructed in Theorem 1 are somewhat complex, and involve computing some cryptographic functionality. A major research direction in differential privacy has been to construct efficient non-interactive algorithms for specific large families of *simple* queries, or prove that this problem is hard. The main technique for constructing such algorithms has been to leverage efficient *PAC learning* algorithms. Specifically, a series of works [7,32,33,36] have shown that an efficient PAC learning algorithm for a class of concepts related to Q can be used to obtain efficient differentially private algorithms for answering the queries Q. Thus, hardness results for differential privacy imply hardness results for PAC learning. However, it is relatively easy to show the hardness of PAC learning using just OWFs [42], and one can even show the hardness of learning simple concept classes (e.g. DNF formulae [17,18]) by using more structured complexity assumptions. One roadblock to proving hardness results for privately answering simple families of queries is that, prior to our work, even proving hardness results for worst-case families of queries required using extremely powerful cryptographic primitives like iO, leaving little room to utilize more structured complexity assumptions to obtain hardness for simple queries. By proving hardness results for differential privacy using only the assumption of one-way functions, we believe our results are an important step towards proving hardness results for simpler families of queries.

Relationship to [31]. A concurrent and independent work by Goyal, Koppula, and Waters also shows how to prove hardness results for non-interactive differential privacy from weaker assumptions than iO. Specifically, they propose a new primitive called *risky traitor tracing* that has weaker security than standard traitor tracing, but is still strong enough to rule out the existence of computationally efficient differentially private algorithms, and construct such schemes under certain assumptions on composite-order bilinear maps. Unlike our work, their new primitive has applications outside of differential privacy. However, within the context of differential privacy, Theorem 1 is stronger than what they prove in two respects: (1) their bilinear-map assumptions are significantly stronger than our assumption of one-way functions, and (2) their hardness result requires a data universe of size $|X_\kappa| = \exp(\kappa)$, rather than our result, which allows $|X_\kappa| = \exp(\kappa^{o(1)})$.

1.1   Techniques

Differential Privacy and Traitor-Tracing Schemes. Our results build on the connection between differentially private algorithms for answering statistical queries and *traitor-tracing schemes*, which was discovered by Dwork et al. [23]. Traitor-tracing schemes were introduced by Chor et al. [16] for the purpose of identifying pirates who violate copyright restrictions. Roughly speaking, a (fully collusion-resilient) traitor-tracing scheme allows a sender to generate keys for n users so that (1) the sender can broadcast encrypted messages that can be decrypted by any user, and (2) any efficient pirate decoder capable of decrypting messages can be traced to at least one of the users who contributed a key to it, even if an arbitrary coalition of the users combined their keys in an arbitrary efficient manner to construct the decoder.

Dwork et al. show that the existence of traitor-tracing schemes implies hardness results for differential privacy. Very informally, they argue as follows.

Suppose a coalition of users takes their keys and builds a dataset $D \in X^n$ where each element of the dataset contains one of their user keys. The family Q will contain a query q_c for each possible ciphertext c. The query q_c asks "What fraction of the elements (user keys) in D would decrypt the ciphertext c to the message 1?" Every user can decrypt, so if the sender encrypts a message $b \in \{0, 1\}$ as a ciphertext c, then every user will decrypt c to b. Thus, the answer to the statistical query q_c will be b. Now, suppose there were an efficient algorithm that outputs an accurate answer to each query q_c in Q. Then the coalition could use it to efficiently produce a summary of the dataset D that enables one to efficiently compute an approximate answer to every query q_c, which would also allow one to efficiently decrypt the ciphertext. Such a summary can be viewed as an efficient pirate decoder, and thus the tracing algorithm can use the summary to trace one of the users in the coalition. However, if there is a way to identify one of the users in the dataset from the summary, then the summary is not private.

Hardness of Privacy from OWF. In order to instantiate this outline, we need a sufficiently good traitor-tracing scheme. Traitor-tracing schemes can be constructed from any *functional encryption scheme for comparison functions* [8][1] This is a cryptographic scheme in which secret keys are associated with functions f and ciphertexts are associated with a message x, and decrypting the ciphertext with a secret key corresponding to f reveals $f(x)$ and "nothing else." In our application, the functions are of the form f_z where $f_z(x) = 1$ if and only if $x \geq z$ (as integers).

Using techniques from [40] (also closely related to arguments in [14]), we show that, in order to prove hardness results for differentially private algorithms it suffices to have a functional encryption scheme for comparison functions that is non-adaptively secure for just two ciphertexts and n secret keys. That is, if an adversary chooses to receive keys for n functions $f_1, \ldots, f_n$, and ciphertexts for two messages x_1, x_2, then he learns nothing more than $\{f_i(x_1), f_i(x_2)\}_{i \in [n]}$. Moreover, the comparison functions only need to support inputs in $\{0, 1, \ldots, n\}$ (i.e. $\log n$ bits). Lastly, it suffices for us to have a symmetric-key functional encryption scheme where both the encryption and key generation can require a private master secret key.

We then construct this type of functional encryption (FE) using the techniques of Gorbunov et al. [30] who constructed bounded-collusion FE from any public-key encryption. There are two important differences between the type of FE that we need and bounded-collusion FE in [30]: (1) we want a symmetric-key FE based on one-way functions (OWFs), whereas they constructed public-key FE using public-key encryption, (2) we want security for only 2 ciphertexts but many secret keys, whereas they achieved security for many ciphertexts but only a small number of secret keys. It turns out that their construction can be rather easily scaled down from the public-key to the symmetric-key setting by replacing public-key encryption with symmetric-key encryption (as previously observed by, e.g., [11]). Going from many ciphertexts and few secret keys to many secret keys

---

[1] These were called *private linear broadcast encryption schemes* by [8], but we use the more modern terminology of functional encryption.

and few ciphertexts essentially boils down to exchanging the role of secret keys and ciphertexts in their scheme, but this requires care. We give the full description and analysis of this construction. Lastly, we rely on one additional property: for the simple functions we consider with logarithmic input length, we can get a scheme where the ciphertext size is extremely small $\kappa^{o(1)}$, where κ is the security parameter, while being able to rely on standard polynomial hardness of OWFs. To do so, we replace the garbled circuits used in the construction of [30] with information-theoretic randomized encodings for simple functions and leverage the fact that we are in the more restricted nonadaptive secret-key setting. The resulting small ciphertext size allows us to get DP lower bounds even when the data universe is of size $|X| = \exp(\kappa^{o(1)})$.

We remark that Tang and Zhang [47] proved that any black-box construction of a traitor-tracing scheme from a random oracle must have either keys or ciphertexts of length $n^{\Omega(1)}$, provided that the scheme does not make calls to the random oracle when generating the user keys. Our construction uses one-way functions during key generation, and thus circumvents this barrier.

Why Two-Ciphertext Security? In the hardness reduction sketched above, the adversary for the functional encryption scheme will use the efficient differentially private algorithm to output some stateless program (the summary) that correctly decrypts ciphertexts for the functional encryption scheme (by approximately answering statistical queries). The crux of the proof is to use differential privacy to argue that the scheme must violate security of the functional encryption scheme by distinguishing encryptions of the messages x and $x - 1$ even if it does not possess a secret key for the function f_x, which is the only function in the family of comparison functions that would give different output on these two messages, and therefore an adversary without this key should not be able to distinguish between these two messages.

Thus, in order to obtain a hardness result for differential privacy we need a functional encryption scheme with the following non-standard security definition: for every polynomial time adversary that obtains a set of secret keys corresponding to functions other than f_x and outputs some stateless program, with high probability that program has small advantage in distinguishing encryptions of x from $x - 1$. Implicit in the work of Kowalczyk et al. [40] is a lemma that says that this property is satisfied by any functional encryption scheme that satisfies the standard notion of security for two messages. At a high level, security for one-message allow for the possibility that the adversary sometimes outputs a program with large positive advantage and sometimes outputs a program with large negative advantage, whereas two-message security bounds the average *squared advantage*, meaning that the advantage must be small with high probability. This argument is similar to one used by Dodis and Yu [20] in a completely different setting.

1.2   Additional Related Work

(Hardness of) Interactive Differential Privacy. Another area of focus is *interactive* differential privacy, where the mechanism gets the dataset D and a (relatively small) set of queries Q chosen by the analyst and must output answers to each query in Q. Most differentially private algorithms for answering a large number of arbitrary queries actually work in this setting [23,26,34], or even in a more challenging setting where the queries in Q arrive online and may be adaptively chosen. [33,35,43,49]. Ullman [50] showed that, assuming one-way functions exist, there is no polynomial-time differentially private algorithm that takes a dataset $D \in X^n$ and a set of $\tilde{O}(n^2)$ arbitrary statistical queries and outputs an accurate answer to each of these queries. The hardness of interactive differential privacy has also been extended to a seemingly easier model of *interactive data analysis* [37,45], which is closely related to differential privacy [3,21], even though privacy is not an explicit requirement in that model. These results however do not give any specific set of queries Q that can be privately summarized information-theoretically but not by a computationally efficient algorithm, and thus do not solve the problem addressed in thus work.

The Complexity of Simple Statistical Queries. As mentioned above, a major open research direction is to design non-interactive differentially private algorithms for simple families of statistical queries. For example, there are polynomial time differentially private algorithms with polynomial sample complexity for summarizing *point queries* and *threshold queries* [5,12], using an information-theoretically optimal number of samples. Another class of focus has been marginal queries [15,24,32,36,48]. A marginal query is defined on the data universe $\{0,1\}^\kappa$. It is specified by a set of positions $S \subseteq \{1,\ldots,\kappa\}$, and a pattern $t \in \{0,1\}^{|S|}$ and asks "What fraction of elements of the dataset have each coordinate $j \in S$ set to t_j?" Specifically, Thaler et al. [48], building on the work of Hardt et al. [36] gave an efficient differentially private algorithm for answering all marginal queries up to an additive error of $\pm.01$ when the dataset is of size $n \gtrsim 2^{\sqrt{\kappa}}$. If we assume sufficiently hard one-way functions exist, then Theorem 1 would show that these parameters are not achievable for an arbitrary set of queries. It remains a central open problem in differential privacy to either design an optimal computationally efficient algorithm for marginal queries or to give evidence that this problem is hard.

Hardness of Synthetic Data. There have been several other attempts to explain the accuracy vs. computation tradeoff in differential privacy by considering restricted classes of algorithms. For example, Ullman and Vadhan [51] (building on Dwork et al. [23]) show that, assuming one-way functions, no differentially private and computationally efficient algorithm that outputs a *synthetic dataset* can accurately answer even the very simple family of 2-way marginals. A synthetic dataset is a specific type of summary that is interchangeable with the real dataset—it is a set $\hat{D} = (\hat{D}_1,\ldots,\hat{D}_n) \in X^n$ such that the answer to each query on $\hat{D}$ is approximately the same as the answer to the same query on D. 2-way marginals are just the subset of marginal queries above where we only allow $|S| \leq 2$, and these queries capture the mean covariances of the attributes. This result is incomparable to ours, since it applies to a very small and simple family of statistical queries, but only applies to algorithms that output synthetic data.

Information-Theoretic Lower Bounds. A related line of work [1,4,13,29, 46] uses ideas from *fingerprinting codes* [9] to prove information-theoretic lower bounds on the number of queries that can be answered by differentially private algorithms, and also devise realistic attacks against the privacy of algorithms that attempt to answer too many queries [27,28]. Most relevant to this work is the result of [13] which says that if the size of the data universe is 2^{n^2}, then there is a fixed set of n^2 queries that no differentially private algorithm, even a computationally unbounded one, can answer accurately. Although these results are orthogonal to ours, the techniques are quite related, as fingerprinting codes are essentially the information-theoretic analogue of traitor-tracing schemes.

2   Differential Privacy Preliminaries

2.1   Differentially Private Algorithms

A *dataset* $D \in X^n$ is an ordered set of n rows, where each row corresponds to an individual, and each row is an element of some *data universe* X. We write $D = (D_1, \ldots, D_n)$ where D_i is the i-th row of D. We will refer to n as the *size* of the dataset. We say that two datasets $D, D' \in X^*$ are *adjacent* if D' can be obtained from D by the addition, removal, or substitution of a single row, and we denote this relation by $D \sim D'$. In particular, if we remove the i-th row of D then we obtain a new dataset $D_{-i} \sim D$. Informally, an algorithm A is differentially private if it is randomized and for any two adjacent datasets $D \sim D'$, the distributions of $A(D)$ and $A(D')$ are similar.

Definition 1 (Differential Privacy [22]). *Let $A : X^n \to S$ be a randomized algorithm. We say that A is (ε, δ)-differentially private if for every two adjacent datasets $D \sim D'$ and every $E \subseteq S$,*

$$\mathbb{P}\left[A(D) \in E\right] \leq e^\varepsilon \cdot \mathbb{P}\left[A(D') \in E\right] + \delta.$$

In this definition, ε, δ may be functions of n.

2.2   Algorithms for Answering Statistical Queries

In this work we study algorithms that answer *statistical queries* (which are also sometimes called *counting queries*, *predicate queries*, or *linear queries* in the literature). For a data universe X, a statistical query on X is defined by a predicate $q : X \to \{0, 1\}$. Abusing notation, we define the evaluation of a query q on a dataset $D = (D_1, \ldots, D_n) \in X^n$ to be

$$\frac{1}{n} \sum_{i=1}^{n} q(D_i).$$

A single statistical query does not provide much useful information about the dataset. However, a sufficiently large and rich set of statistical queries is sufficient to implement many natural machine learning and data mining algorithms [38], thus we are interested in differentially private algorithms to answer such sets. To this end, let $Q = \{q : X \to \{0, 1\}\}$ be a set of statistical queries on a data universe X.

Informally, we say that a mechanism is accurate for a set Q of statistical queries if it answers every query in the family to within error $\pm\alpha$ for some suitable choice of $\alpha > 0$. Note that $0 \le q(D) \le 1$, so this definition of accuracy is meaningful when $\alpha < 1/2$.

Before we define accuracy, we note that the mechanism may represent its answer in any form. That is, the mechanism outputs may output a *summary* $S \in \mathcal{S}$ that somehow represents the answers to every query in Q. We then require that there is an *evaluator* $Eval : \mathcal{S} \times Q \to [0, 1]$ that takes the summary and a query and outputs an approximate answer to that query. That is, we think of $Eval(S, q)$ as the mechanism's answer to the query q. We will abuse notation and simply write $q(S)$ to mean $Eval(S, q)$.[2]

Definition 2 (Accuracy). *For a family Q of statistical queries on X, a dataset $D \in X^n$ and a summary $S \in \mathcal{S}$, we say that S is α-accurate for Q on D if*

$$\forall q \in Q \qquad |q(D) - q(S)| \le \alpha.$$

For a family of statistical queries Q on X, we say that an algorithm $A : X^n \to \mathcal{S}$ is (α, β)-accurate for Q given a dataset of size n if for every $D \in X^n$,

$$\mathbb{P}\left[A(D) \text{ is } \alpha\text{-accurate for } Q \text{ on } X\right] \ge 1 - \beta.$$

In this work we are typically interested in mechanisms that satisfy the very weak notion of $(1/3, o(1/n))$-accuracy, where the constant $1/3$ could be replaced with any constant $<1/2$. Most differentially private mechanisms satisfy quantitatively much stronger accuracy guarantees. Since we are proving hardness results, this choice of parameters makes our results stronger.

2.3   Computational Efficiency

Since we are interested in asymptotic efficiency, we introduce a computation parameter $\kappa \in \mathbb{N}$. We then consider a sequence of pairs $\{(X_\kappa, Q_\kappa)\}_{\kappa \in \mathbb{N}}$ where Q_κ is a set of statistical queries on X_κ. We consider databases of size n where $n = n(\kappa)$ is a polynomial. We then consider algorithms A that take as input a

---

[2] If we do not restrict the running time of the algorithm, then it is without loss of generality for the algorithm to simply output a list of real-valued answers to each queries by computing $Eval(S, q)$ for every $q \in Q$. However, this transformation makes the running time of the algorithm at least $|Q|$. The additional generality of this framework allows the algorithm to run in time sublinear in $|Q|$. This generality is crucial for our results, which apply to settings where the family of queries is superpolynomially large in the size of the dataset.

dataset X_κ^n and output a summary in S_κ where $\{S_\kappa\}_{\kappa \in \mathbb{N}}$ is a sequence of output ranges. There is an associated evaluator Eval that takes a query $q \in Q_\kappa$ and a summary $s \in S_\kappa$ and outputs a real-valued answer. The definitions of differential privacy and accuracy extend straightforwardly to such sequences.

We say that such an algorithm is *computationally efficient* if the running time of the algorithm and the associated evaluator run in time polynomial in the computation parameter κ. In principle, it could require as many as $|X|$ bits even to specify a statistical query, in which case we cannot hope to answer the query efficiently, even ignoring privacy constraints. Thus, we restrict attention to statistical queries that are specified by a circuit of size polylog$|X|$, and thus can be evaluated in time polylog$|X|$, and so are not the bottleneck in computation. To remind the reader of this fact, we will often say that Q is a family of *efficiently computable statistical queries*.

2.4 Notational Conventions

Given a boolean predicate P, we will write $\mathbb{I}\{P\}$ to denote the value 1 if P is true and 0 if P is false. We also say that a function $\varepsilon = \varepsilon(n)$ is *negligible* if $\varepsilon(n) = O(1/n^c)$ for every constant $c > 0$, and denote this by $\varepsilon(n) = \text{negl}(n)$.

3 Weakly Secure Traitor-Tracing Schemes

In this section we describe a very relaxed notion of traitor-tracing schemes whose existence will imply the hardness of differentially private data release.

3.1 Syntax and Correctness

For a function $n : \mathbb{N} \to \mathbb{N}$ and a sequence $\{K_\kappa, C_\kappa\}_{\kappa \in \mathbb{N}}$, an $(n, \{K_\kappa, C_\kappa\})$-*traitor-tracing scheme* is a tuple of efficient algorithms $\Pi = (\text{Setup}, \text{Enc}, \text{Dec})$ with the following syntax.

- Setup takes as input a security parameter κ, runs in time poly(κ), and outputs $n = n(\kappa)$ secret *user keys* $\text{sk}_1, \ldots, \text{sk}_n \in K_\kappa$ and a secret *master key* msk. We will write $\textbf{sk} = (\text{sk}_1, \ldots, \text{sk}_n, \text{msk})$ to denote the set of keys.
- Enc takes as input a master key msk and an *index* $i \in \{0, 1, \ldots, n\}$, and outputs a ciphertext $c \in C_\kappa$. If $c \leftarrow_R \text{Enc}(j, \text{msk})$ then we say that c is *encrypted to index j*.
- Dec takes as input a ciphertext c and a user key sk_i and outputs a single bit $b \in \{0, 1\}$. We assume for simplicity that Dec is deterministic.

Correctness of the scheme asserts that if $\textbf{sk}$ are generated by Setup, then for any pair i, j, $\text{Dec}(\text{sk}_i, \text{Enc}(\text{msk}, j)) = \mathbb{I}\{i \leq j\}$. For simplicity, we require that this property holds with probability 1 over the coins of Setup and Enc, although it would not affect our results substantively if we required only correctness with high probability.

Definition 3 (Perfect Correctness). *An* $(n, \{K_\kappa, C_\kappa\})$-*traitor-tracing scheme is* perfectly correct *if for every* $\kappa \in \mathbb{N}$, *and every* $i, j \in \{0, 1, \ldots, n\}$

$$\mathop{\mathbb{P}}_{\mathsf{sk}=\mathsf{Setup}(\kappa),\, c=\mathsf{Enc}(\mathsf{msk},j)} [\mathsf{Dec}(\mathsf{sk}_i, c) = \mathbb{I}\{i \leq j\}] = 1.$$

3.2 Index-Hiding Security

Intuitively, the security property we want is that any computationally efficient adversary who is missing one of the user keys sk_{i*} cannot distinguish cipher-texts encrypted with index i^* from index $i^* - 1$, even if that adversary holds all $n - 1$ other keys sk_{-i*}. In other words, an efficient adversary cannot infer anything about the encrypted index beyond what is implied by the correctness of decryption and the set of keys he holds.

More precisely, consider the following two-phase experiment. First the adversary is given every key except for sk_{i*}, and outputs a decryption program S. Then, a challenge ciphertext is encrypted to either i^* or to $i^* - 1$. We say that the traitor-tracing scheme is secure if for every polynomial time adversary, with high probability over the setup and the decryption program chosen by the adversary, the decryption program has small advantage in distinguishing the two possible indices.

Definition 4 (Index Hiding). *A* traitor-tracing scheme Π *satisfies* (weak) index-hiding security *if for every sufficiently large* $\kappa \in \mathbb{N}$, *every* $i^* \in [n(\kappa)]$, *and every* $\mathrm{poly}(\kappa)$-*time adversary* A,

$$\mathop{\mathbb{P}}_{\substack{\mathsf{sk}=\mathsf{Setup}(\kappa)\\ S=A(\mathsf{sk}_{-i*})}} \left[\mathbb{P}\left[S(\mathsf{Enc}(\mathsf{msk}, i^*)) = 1\right] - \mathbb{P}\left[S(\mathsf{Enc}(\mathsf{msk}, i^* - 1)) = 1\right] > \frac{1}{4en} \right] \leq \frac{1}{4en} \tag{1}$$

In the above, the inner probabilities are taken over the coins of Enc *and* S.

Note that in the above definition we have fixed the success probability of the adversary for simplicity. Moreover, we have fixed these probabilities to relatively large ones. Requiring only a polynomially small advantage is crucial to achieving the key and ciphertext lengths we need to obtain our results, while still being sufficient to establish the hardness of differential privacy.

3.3 Index-Hiding Security Implies Hardness for Differential Privacy

It was shown by Kowalczyk et al. [40] (refining similar results from [23,50]) that a traitor-tracing scheme satisfying index-hiding security implies a hardness result for non-interactive differential privacy.

Theorem 2. *Suppose there is an* $(n, \{K_\kappa, C_\kappa\})$-*traitor-tracing scheme that satisfies perfect correctness (Definition 3) and index-hiding security (Definition 4). Then there is a sequence of pairs* $\{X_\kappa, Q_\kappa\}_{\kappa \in \mathbb{N}}$ *where* Q_κ *is a set of statistical queries on* X_κ, $|Q_\kappa| = |C_\kappa|$, *and* $|X_\kappa| = |K_\kappa|$ *such that there is no algorithm* A *that is simultaneously*

1. *computationally efficient,*
2. $(1, 1/4n)$-*differentially private, and*
3. $(1/3, 1/2n)$-*accurate for* Q_κ *on datasets* $D \in X_\kappa^{n(\kappa)}$.

3.4   Two-Index-Hiding-Security

While Definition 4 is the most natural to prove hardness of privacy, it is not consistent with the usual security definition for functional encryption because of the nested "probability-of-probabilities." In order to apply more standard notions of functional encryption, we show that index-hiding security follows from a more natural form of security for two ciphertexts.

First, consider the following IndexHiding game (Fig. 1).

The challenger generates keys $\mathbf{sk} = (\mathsf{sk}_1, \ldots, \mathsf{sk}_n, \mathsf{msk}) \leftarrow_{\mathrm{R}} \mathsf{Setup}(\kappa)$.
The adversary A is given keys sk_{-i^*} and outputs a decryption program S.
The challenger chooses a bit $b \leftarrow_{\mathrm{R}} \{0, 1\}$
The challenger generates an encryption to index $i^* - b$, $c \leftarrow_{\mathrm{R}} \mathsf{Enc}(\mathsf{msk}, i^* - b)$
The adversary makes a guess $b' = S(c)$

Fig. 1. IndexHiding$_{i^*}$

Let IndexHiding$_{I^*, \mathbf{sk}, S}$ be the game IndexHiding$_{I^*}$ where we fix the choices of $\mathbf{sk}$ and S. Also, define

$$\mathrm{Adv}_{i^*, \mathbf{sk}, S} = \underset{\mathsf{IndexHiding}_{i^*, \mathbf{sk}, S}}{\mathbb{P}} [b' = b] - \frac{1}{2}.$$

so that

$$\underset{\mathsf{IndexHiding}_{i^*}}{\mathbb{P}} [b' = b] - \frac{1}{2} = \underset{\substack{\mathbf{sk} = \mathsf{Setup}(\kappa) \\ S = A(\mathsf{sk}_{-i^*})}}{\mathbb{E}} [\mathrm{Adv}_{i^*, \mathbf{sk}, S}]$$

Then the following statement implies (1) in Definition 4:

$$\underset{\mathbf{sk} = \mathsf{Setup}(\kappa), \, S = A(\mathsf{sk}_{-i^*})}{\mathbb{P}} \left[\mathrm{Adv}_{i^*, \mathbf{sk}, S} > \frac{1}{4en} \right] \le \frac{1}{2en} \tag{2}$$

We can define a related two-index-hiding game.

Analogous to what we did with IndexHiding, we can define TwoIndexHiding$_{I^*, \mathbf{sk}, S}$ to be the game TwoIndexHiding$_{i^*}$ where we fix $\mathbf{sk}$ and S, and define

$$\mathrm{TwoAdv}_{i^*} = \underset{\mathsf{TwoIndexHiding}_{i^*}}{\mathbb{P}} [b' = b_0 \oplus b_1] - \frac{1}{2}$$

Kowalczyk et al. [40] proved the following lemma that will be useful to connect our new construction to the type of security definition that implies hardness of differential privacy.

The challenger generates keys $\mathsf{sk} = (\mathsf{sk}_1, \ldots, \mathsf{sk}_n, \mathsf{msk}) \leftarrow_{\mathrm{R}} \mathsf{Setup}$.
The adversary A is given keys sk_{-i^*} and outputs a decryption program S.
Choose $b_0 \leftarrow_{\mathrm{R}} \{0, 1\}$ and $b_1 \leftarrow_{\mathrm{R}} \{0, 1\}$ independently.
Let $c_0 \leftarrow_{\mathrm{R}} \mathsf{Enc}(i^* - b_0; \mathsf{msk})$ and $c_1 \leftarrow_{\mathrm{R}} \mathsf{Enc}(i^* - b_1; \mathsf{msk})$.
Let $b' = S(c_0, c_1)$.

Fig. 2. TwoIndexHiding$_{i^*}$

Lemma 1. *Let Π be a traitor-tracing scheme such that for every efficient adversary A, every $\kappa \in \mathbb{N}$, and index $i^* \in [n(\kappa)]$,*

$$\mathrm{TwoAdv}_{i^*} \leq \frac{1}{300n^3}$$

Then Π satisfies weak index-hiding security.

In the rest of the paper, we will construct a scheme satisfying the assumption of the above lemma with suitable key and ciphertext lengths, which we can immediately plug into Theorem 2 to obtain Theorem 1 in the introduction.

4 Cryptographic Tools

4.1 Decomposable Randomized Encodings

Let $\mathcal{F} = \{f : \{0,1\}^\ell \to \{0,1\}^k\}$ be a family of Boolean functions. An *(information-theoretic) decomposable randomized encoding* for $\mathcal{F}$ is a pair of efficient algorithms (DRE.Encode, DRE.Decode) such that the following hold:

- DRE.Encode takes as input a function $f \in \mathcal{F}$ and randomness R and outputs a randomized encoding consisting of a set of ℓ pairs of labels

$$\tilde{F}(f, R) = \left\{ \begin{matrix} \tilde{F}_1(f, 0, R) \cdots \tilde{F}_\ell(f, 0, R) \\ \tilde{F}_1(f, 1, R) \cdots \tilde{F}_\ell(f, 1, R) \end{matrix} \right\}$$

  where the i-th pair of labels corresponds to the i-th bit of the input x.
- **(Correctness)** DRE.Decode takes as input a set of ℓ labels corresponding to some function f and input x and outputs $f(x)$. Specifically,

$$\forall f \in \mathcal{F}, \ x \in \{0,1\}^\ell \quad \mathsf{DRE.Decode}\big(\tilde{F}_1(f, x_1, R), \ldots, \tilde{F}_\ell(f, x_\ell, R)\big) = f(x)$$

  with probability 1 over the randomness R.
- **(Information-Theoretic Security)** For every function f and input y, the set of labels corresponding to f and y reveal nothing other than $f(y)$. Specifically, there exists a randomized simulator DRE.Sim that depends only on the output $f(x)$ such that

$$\forall f \in \mathcal{F}, \ x \in \{0,1\}^\ell \quad \big\{\tilde{F}_1(f, x_1, R), \ldots, \tilde{F}_\ell(f, x_\ell, R)\big\} \sim \mathsf{DRE.Sim}(f(x))$$

where $\sim$ denotes that the two random variables are identically distributed.

- The *length* of the randomized encoding is the maximum length of $\tilde{F}(f, R)$ over all choices of $f \in \mathcal{F}$ and the randomness R.

We will utilize the fact that functions computable in low depth have small decomposable randomized encodings.

Theorem 3 ([2,39]). *If $\mathcal{F}$ is a family of functions such that a universal function for $\mathcal{F}$, $U(f, x) = f(x)$, can be computed by Boolean formulae of depth d (with fan-in 2, over the basis $\{\wedge, \vee, \neg\}$), then $\mathcal{F}$ has an information-theoretic decomposable randomized encoding of length $O(4^d)$.*

4.2 Private Key Functional Encryption

Let $\mathcal{F} = \{f : \{0, 1\}^\ell \to \{0, 1\}^k\}$ be a family of functions. A *private key functional encryption scheme for $\mathcal{F}$* is a tuple of polynomial-time algorithms $\Pi_{FE} =$ (FE.Setup, FE.KeyGen, FE.Enc, FE.Dec) with the following syntax and properties:

- FE.Setup takes a security parameter 1^κ and outputs a master secret key FE.msk.
- FE.KeyGen takes a master secret key FE.msk and a function $f \in \mathcal{F}$ and outputs a secret key FE.sk$_f$ corresponding to the function f.
- FE.Enc takes the master secret key FE.msk and an input $x \in \{0, 1\}^\ell$ and outputs a ciphertext c corresponding to the input x.
- **(Correctness)** FE.Dec takes a secret key FE.sk corresponding to a function f and a ciphertext c corresponding to an input x and outputs $f(x)$. Specifically, for every FE.msk is in the support of FE.Setup

$$\text{FE.Dec}(\text{FE.KeyGen}(\text{FE.msk}, f), \text{FE.Enc}(\text{FE.msk}, x)) = f(x)$$

- The *key length* is the maximum length of FE.sk over all choices of $f \in \mathcal{F}$ and the randomness of FE.Setup, FE.Enc. The *ciphertext length* is the maximum length of c over all choices of $x \in \{0, 1\}^\ell$ and the randomness of FE.Setup, FE.Enc.
- **(Security)** We will use a *non-adaptive* simulation-based definition of security. In particular, we are interested in security for a large number of keys n and a small number of ciphertexts m. We define security through the pair of games in Fig. 3. We say that Π_{FE} is (n, m, ε)-secure if there exists a polynomial-time simulator FE.Sim such that for every polynomial-time adversary $\mathcal{A}$ and every κ,

$$\left| \mathbb{P}\left[E^{\text{real}}_{\kappa,n,m}(\Pi_{FE}, \mathcal{A}) = 1 \right] - \mathbb{P}\left[E^{\text{ideal}}_{\kappa,n,m}(\Pi_{FE}, \mathcal{A}, \text{FE.Sim}) = 1 \right] \right| \leq \varepsilon(\kappa)$$

Our goal is to construct a functional encryption scheme that is $(n, 2, \frac{1}{300n^3})$-secure and has short ciphertexts and keys, where $n = n(\kappa)$ is a polynomial in the security parameter. Although it is not difficult to see, in Sect. 7 we prove that the definition of security above implies the definition of two-index-hiding security that we use in Lemma 1.

$E^{\text{real}}_{\kappa,n,m}(\Pi_{FE}, \mathcal{A})$:
  $\mathcal{A}$ outputs at most n functions $f_1, \ldots, f_n$ and m inputs $x_1, \ldots, x_m$
  Let FE.msk $\leftarrow_{\text{R}}$ FE.Setup(1^κ) and

$$\forall\, i \in [n] \quad \text{FE.sk}_{f_i} \leftarrow_{\text{R}} \text{FE.KeyGen}(\text{FE.msk}, f_i)$$

$$\forall\, j \in [m] \quad c_j \leftarrow_{\text{R}} \text{FE.Enc}(\text{FE.msk}, x_j)$$

  $\mathcal{A}$ receives $\{\text{FE.sk}_{f_i}\}_{i=1}^n$ and $\{c_j\}_{j=1}^m$ and outputs b

$E^{\text{ideal}}_{\kappa,n,m}(\Pi_{FE}, \mathcal{A}, \text{FE.Sim})$:
  $\mathcal{A}$ outputs at most n functions $f_1, \ldots, f_n$ and m inputs $x_1, \ldots, x_m$

$$\left(\{\text{FE.sk}_{f_i}\}_{i=1}^n , \{c_j\}_{j=1}^m \right) \leftarrow_{\text{R}} \text{FE.Sim}\left(\left\{ f_i, \{f_i(x_j)\}_{j=1}^m \right\}_{i=1}^n \right)$$

  $\mathcal{A}$ receives $\{\text{FE.sk}_{f_i}\}_{i=1}^n$ and $\{c_j\}_{j=1}^m$ and outputs b

Fig. 3. Security of functional encryption

Function-Hiding Functional Encryption. As an ingredient in our construction we also need a notion of **function-hiding security** for a (one-message) functional encryption scheme. Since we will only need this definition for a single message, we will specialize to that case in order to simplify notation. We say that Π_{FE} is function-hiding $(n, 1, \varepsilon)$-secure if there exists a polynomial-time simulator FE.Sim such that for every polynomial-time adversary $\mathcal{A}$ and every κ,

$$\left| \mathbb{P}\left[\bar{E}^{\text{real}}_{\kappa,n,1}(\Pi_{FE}, \mathcal{A}) = 1 \right] - \mathbb{P}\left[\bar{E}^{\text{ideal}}_{\kappa,n,1}(\Pi_{FE}, \mathcal{A}, \text{FE.Sim}) = 1 \right] \right| \leq \varepsilon(\kappa)$$

where $\bar{E}^{\text{real}}_{\kappa,n,1}, \bar{E}^{\text{ideal}}_{\kappa,n,1}$ are the same experiments as $E^{\text{real}}_{\kappa,n,1}, E^{\text{ideal}}_{\kappa,n,1}$ except that the simulator in $\bar{E}^{\text{ideal}}_{\kappa,n,1}$ is not given the functions f_i as input. Namely, in $\bar{E}^{\text{ideal}}_{\kappa,n,1}$:

$$\left(\{\text{FE.sk}_{f_i}\}_{i=1}^n , c \right) \leftarrow_{\text{R}} \text{FE.Sim}\left(\{f_i(x)\}_{i=1}^n \right)$$

A main ingredient in the construction will be a function-hiding functional encryption scheme that is $(n, 1, \text{negl}(\kappa))$-secure. The construction is a small variant of the constructions of Sahai and Seyalioglu [44] and Gorbunov et al. [30]

Theorem 4 (Variant of [30,44]). *Let $\mathcal{F}$ be a family of functions such that a universal function for $\mathcal{F}$ has a decomposable randomized encoding of length L. That is, the function $U(f, x) = f(x)$ has a DRE of length L. If one-way functions exist, then for any polynomial $n = n(\kappa)$ there is an $(n, 1, \text{negl}(\kappa))$-function-hiding-secure functional encryption scheme Π with key length L and ciphertext length $O(\kappa L)$.*

Although this theorem follows in a relatively straightforward way from the techniques of [30,44], we will give a proof of this theorem in Sect. 5. The main novelty in the theorem is to verify that in settings where we have a very short DRE—shorter than the security parameter κ—we can make the secret keys have length proportional to the length of the DRE rather than proportional to the security parameter.

5 One-Message Functional Encryption

We will now construct $\Pi_{OFE} = ($OFE.Setup, OFE.KeyGen, OFE.Enc, OFE.Dec$)$: a function-hiding $(n, 1, \mathrm{negl}(\kappa))$-secure functional encryption scheme for functions with an (information-theoretic) decomposable randomized encoding DRE. The construction is essentially the same as the (public key) variants given by [30, 44] except we consider information theoretic randomized encodings instead of computationally secure ones and instead of encrypting the labels under a public key encryption scheme, we take advantage of the private-key setting to use an encryption method that produces ciphertexts with size equal to the message if the message is smaller than the security parameter. Encrypting the labels of a short randomized encoding, this allows us to argue that keys for our scheme are small. To perform this encryption, we use a PRF evaluated on known indices to mask each short label of DRE.

Let $n = \mathrm{poly}(\kappa)$ denote the number of users for the scheme. We assume for simplicity that $\lg n$ is an integer. Our construction will rely on the following primitives:

- A PRF family $\{\mathrm{PRF}_{\mathrm{sk}} : \{0,1\}^{\lg n} \rightarrow \{0,1\}^{\lg n} \mid s \in \{0,1\}^{\kappa}\}$.
- A DRE of $f_y(x) = \mathbb{I}\{x \geq y\}$ where $x, y \in \{0,1\}^{\log n}$ (Fig. 4).

Setup(1^{κ}) :
    Choose seeds $\mathrm{sk}_{k,b} \leftarrow_{\mathrm{R}} \{0,1\}^{\kappa}$ for $k \in [\lg n], b \in \{0,1\}$.
    Choose randomness R_j for the randomized encoding for each $j \in [n]$.
    Choose $x \leftarrow_{\mathrm{R}} \{0,1\}^{\lg n}$.
    Define
$$\left\{K_{k,b}^{(j)}\right\} := \left\{\mathrm{PRF}_{\mathrm{sk}_{k,b}}(j) \oplus \tilde{F}_k(f_j, x_k \oplus b, R_j)\right\}$$
    where $k \in \{1,..,\lg n\}, b \in \{0,1\}$
    Let each user's secret key be $sk_j = (j, \{K_{k,b}^{(j)}\})$.
    Let the master key be $\mathsf{msk} = \{\mathrm{sk}_{k,b}\}$.

Enc$(i, \mathsf{msk} = \{\mathrm{sk}_{k,b}\})$:
    Output $c_i = (y := x \oplus i, \mathrm{sk}_{1,y_1}, ..., \mathrm{sk}_{\lg n, y_{\lg n}})$

Dec(c_i, sk_j):
    Output $\mathsf{DRE.Decode}(\mathrm{PRF}_{\mathrm{sk}_{1,y_1}}(j) \oplus K_{1,y_1}^{(j)}, ..., \mathrm{PRF}_{\mathrm{sk}_{\lg n, y_{\lg n}}}(j) \oplus K_{\lg n, y_{\lg n}}^{(j)})$

Fig. 4. Our scheme Π_{OFE}.

5.1 Proof of Correctness

$$\mathsf{Dec}(c_i, \mathsf{sk}_j) = \mathsf{DRE.Decode}(\mathsf{PRF}_{\mathsf{sk}_{1,y_1}}(j) \oplus K_{1,y_1}^{(j)}, ..., \mathsf{PRF}_{\mathsf{sk}_{\lg n, y_{\lg n}}}(j) \oplus K_{\lg n, y_{\lg n}}^{(j)})$$
$$= \mathsf{DRE.Decode}(\tilde{F}_1(f_j, y_1 \oplus x_1, R_j), ..., \tilde{F}_{\lg n}(f_j, y_{\lg n} \oplus x_{\lg n}, R_j))$$
$$= \mathsf{DRE.Decode}(\tilde{F}_1(f_j, i_1, R_j), ..., \tilde{F}_{\lg n}(f_j, i_{\lg n}, R_j))$$
$$= f_j(i)$$

Where the last step uses the (perfect) correctness of the randomized encoding scheme. So:

$$\mathop{\mathbb{P}}_{\mathsf{sk}=\mathsf{Setup}(\kappa),\ c_i=\mathsf{Enc}(\mathsf{msk},i)} [\mathsf{Dec}(c_i, \mathsf{sk}_i) = \mathbb{I}\{i \leq j\}]$$
$$= \mathop{\mathbb{P}}_{R} \left[\mathsf{DRE.Decode}(\tilde{F}_1(f_j, i_1, R_j), ..., \tilde{F}_{\lg n}(f_j, i_{\lg n}, R_j)) = f_j(i) \right] = 1$$

5.2 Proof of Security

Lemma 2. $\left| \mathbb{P}\left[E_{\kappa,n,m}^{\mathrm{real}}(\Pi_{OFE}, \mathcal{A}) = 1\right] - \mathbb{P}\left[E_{\kappa,n,m}^{\mathrm{ideal}}(\Pi_{OFE}, \mathcal{A}, \mathsf{FE.Sim}) = 1\right] \right| \leq \varepsilon(\kappa)$

Proof. Consider the hybrid scheme Π_{OFE}^* defined in Fig. 5, which uses a truly random string instead of the output of a PRF for the encrypted labels corresponding to the off-bits of $y = x \oplus i$. Note that this scheme is only useful in the nonadaptive security game, where i is known at time of Setup (since it is needed to compute y). We can easily show that the scheme is indistinguishable from the original scheme in the nonadaptive security game.

Lemma 3.

$$\left| \mathbb{P}\left[E_{\kappa,n,m}^{\mathrm{real}}(\Pi_{OFE}, \mathcal{A}) = 1\right] - \mathbb{P}\left[E_{\kappa,n,m}^{\mathrm{real}}(\Pi_{OFE}^*, \mathcal{A}) = 1\right] \right| \leq \lg n \cdot \mathsf{PRF.Adv}(\kappa) = \varepsilon(\kappa)$$

Proof. Follows easily by the security of the PRF (applied $\lg n$ times in a hybrid for each function in the off-bits of y).

In Fig. 6 we define a simulator for the ideal setting that is indistinguishable from our hybrid scheme Π_{OFE}^*. The simulator uses the simulator for the decomposable randomized encoding scheme to generate the labels to be encrypted using only the knowledge of the output value of the functions on the input.

Lemma 4. $\left| \mathbb{P}\left[E_{\kappa,n,m}^{\mathrm{real}}(\Pi_{OFE}^*, \mathcal{A}) = 1\right] - \mathbb{P}\left[E_{\kappa,n,m}^{\mathrm{ideal}}(\Pi_{OFE}, \mathcal{A}, \mathsf{FE.Sim}) = 1\right] \right| = 0$

Proof. Follows easily by the information-theoretic security of the randomized encoding.

Setup(1^κ) :
  Choose seeds $\mathsf{sk}_k \leftarrow_R \{0,1\}^\kappa$ for $k \in [\lg n]$.
  Choose randomness R_j for the randomized encoding for each $j \in [n]$.
  Choose $x \leftarrow_R \{0,1\}^{\lg n}$.
  Let $y = x \oplus i$.
  Define
  $$\left\{ K^{(j)}_{k,y_k} \right\} := \left\{ \mathsf{PRF}_{\mathsf{sk}_k}(j) \oplus \tilde{F}_k(f_j, i_k, R_j) \right\}$$
  Construct
  $$\left\{ K^{(j)}_{k,\bar{y}_k} \right\} \leftarrow_R \{0,1\}^{\lg n}$$
  where $k \in \{1, .., \lg n\}$
  Let each user's secret key be $sk_j = (j, \{K^{(j)}_{k,b}\})$.
  Let the master key be $\mathsf{msk} = \{\mathsf{sk}_k\}$.

Enc($i, \mathsf{msk} = \{\mathsf{sk}_k\}$) :
  Output $c_i = (y = x \oplus i, \mathsf{sk}_1, ..., \mathsf{sk}_{\lg n})$

Dec(c_i, sk_j):
  Output $\mathsf{DRE.Decode}(\mathsf{PRF}_{\mathsf{sk}_1}(j) \oplus K^{(j)}_{1,y_1}, ..., \mathsf{PRF}_{\mathsf{sk}_{\lg n}}(j) \oplus K^{(j)}_{\lg n, y_{\lg n}})$

Fig. 5. Hybrid scheme Π^*_{OFE}.

Adding the statements of Lemma 3 and Lemma 4 gives us the original statement of security.

So, Π_{OFE} is a function-hiding-secure private-key functional encryption scheme $(n, 1, \mathsf{negl}(\kappa))$ with security based on the hardness of a PRF (which can be instantiated using a one-way function) and the existence of an information-theoretic randomized encoding for the family of comparison functions $\{f_j : \{0,1\}^{\lg n} \to \{0,1\}\}$ of length L. Furthermore, note that the length of ciphertexts is: $\lg n \cdot \kappa + \lg n = O(\kappa L)$ and the length of each key is L, satisfying the conditions of Theorem 4.

6   A Two-Message Functional Encryption Scheme for Comparison

We now use the one-message functional encryption scheme Π_{OFE} described in Sect. 5 to construct a functional encryption scheme Π_{FE} that is $(n, 2, \frac{1}{300n^3})$-secure for the family of comparison functions. For any $y \in \{0,1\}^\ell$, let

$$f_y(x) = \mathbb{I}\{x \geq y\}$$

where the comparison operation treats x, y as numbers in binary. We define the family of functions

$$\mathcal{F}_{\mathrm{comp}} := \left\{ f_y : \{0,1\}^\ell \to \{0,1\} \mid y \in \{0,1\}^\ell \right\}$$

OFE.Sim:

Input: 1^κ, and the evaluation of n (unknown) functions on 1 unknown input: $\{y_i\}_{i=1}^n$

Let DRE.Sim be the simulator for information-theoretic randomized encoding DRE.

// Generate ciphertext

Choose $y \leftarrow_R \{0,1\}^{\lg n}$.

For $k \in [\lg n]$:

  Choose $\mathsf{sk}_k \leftarrow_R \{0,1\}^\kappa$

Let $c = (y, \mathsf{sk}_1, ..., \mathsf{sk}_{\lg n})$

// Generate keys for $j \in [n]$ using DRE.Sim

For $j \in [n]$:

  Generate: $\left(\tilde{F}_1^{(j)}, ..., \tilde{F}_{\lg n}^{(j)} \right) \leftarrow$ DRE.Sim(y_j)

  Let: $K_{k,y_k}^{(j)} = \mathsf{PRF}_{\mathsf{sk}_k}(j) \oplus \tilde{F}_k^{(j)}$ for $k \in [\lg n]$

  Choose: $K_{k,\bar{y}_k}^{(j)} \leftarrow_R \{0,1\}^{\lg n}$ for $k \in [\lg n]$

  Let: OFE.$\mathsf{sk}_j = \left\{ K_{k,b}^{(j)} \right\}_{\substack{k \in [\lg n], \\ b \in \{0,1\}}}$

// Output the n simulated secret keys and simulated ciphertext

Output: $\{\mathsf{OFE.sk}_i\}_{i=1}^n$, c

Fig. 6. The simulator OFE.Sim for Π_{OFE}.

In our application, we need $x, y \in \{0, 1, \ldots, n\}$, so we will set $\ell = \lceil \log_2(n + 1) \rceil = O(\log n)$. One important property of our construction is that the user key length will be fairly small as a function of ℓ, so that when $\ell = O(\log n)$, the overall length of user keys will be $n^{o(1)}$ (in fact, nearly polylogarithmic in n).

6.1 Construction

Our construction will be for a generic family of functions $\mathcal{F}$, and we will only specialize the construction to $\mathcal{F}_{\mathrm{comp}}$ when setting the parameters and bounding the length of the scheme. Before giving the formal construction, let's gather some notation and ingredients. Note that we will introduce some additional parameters that are necessary to specify the scheme, but we will leave many of these parameters to be determined later.

- Let n be a parameter bounding the number of user keys in the scheme, and let $\mathbb{F}$ be a finite field whose size we will determine later.
- Let $\mathcal{F} = \{f : \{0,1\}^\ell \to \{0,1\}\}$ be a family of functions. For each function $f \in \mathcal{F}$ we define an associated polynomial $\tilde{f} : \mathbb{F}^{\ell+1} \to \mathbb{F}$ as follows:
  1. Let $\hat{f} : \mathbb{F}^\ell \to \mathbb{F}$ be a polynomial computing f
  2. Define $\tilde{f} : \mathbb{F}^{\ell+1} \to \mathbb{F}$ to be $\tilde{f}(x_1, \ldots, x_\ell, z) = \hat{f}(x_1, \ldots, x_\ell) + z$

Global Parameters: A family of functions $\mathcal{F} = \{f : \{0,1\}^\ell \to \{0,1\}^\ell\}$ is a family of functions, a finite field $\mathbb{F}$, a bound D on the degree of polynomials $\tilde{f} : \mathbb{F}^{\ell+1} \to \mathbb{F}$ computing a function in $\mathcal{F}$, a parameter $R \in \mathbb{N}$, and parameters $U = RD + 1$, and $T = U^2$.

FE.Setup(1^κ) :
  Generate T independent master keys for OFE.msk$_t \leftarrow_R$ OFE.Setup(1^κ)
  Output FE.msk $= \{$OFE.msk$_t\}_{t=1}^T$

FE.KeyGen(FE.msk, f_i) :
  // To aid in our proofs later, we index invocations of FE.KeyGen with i
  Let $\tilde{f}_i : \mathbb{F}^\ell \to \mathbb{F}$ be a multivariate polynomial computing f_i
  Choose a random polynomial $r_i : \mathbb{F} \to \mathbb{F}$ of degree RD such that $r_i(0) = 0$
  For every $t \in [T]$:
    Define the function $\tilde{f}_{i,r_i}(x, t) = \tilde{f}_i(x) + r_i(t)$
    Let OFE.sk$_{i,t} \leftarrow_R$ OFE.KeyGen(OFE.msk$_t$, $\tilde{f}_{i,r_i}(\cdot, t)$)
  Output FE.sk$_i = \{$OFE.sk$_{i,t}\}_{t=1}^T$

FE.Enc(FE.msk, x) :
  Choose a random polynomial map $q : \mathbb{F} \to \mathbb{F}^\ell$ of degree R so that $q(0) = x$.
  Choose a random $\mathcal{U} \subseteq [T]$ of size U
  For every $t \in \mathcal{U}$: let $c_t \leftarrow_R$ OFE.Enc(OFE.msk$_t$, $q(t)$)
  Output $c = \{c_t\}_{t \in \mathcal{U}}$

FE.Dec(FE.sk$_i$, c) :
  Let FE.sk$_i = \{$OFE.sk$_{i,t}\}_{t=1}^T$, $c = \{c_t\}_{t \in \mathcal{U}}$
  For every $t \in \mathcal{U}$, let $\tilde{p}(t) = $ OFE.Dec(OFE.sk$_{i,t}$, c_t)
  Extend the polynomial $\tilde{p}(t)$ to all of $\mathbb{F}$ by interpolation
  Output $\tilde{p}(0)$

Fig. 7. A functional encryption scheme for 2 messages

Let D and S be such that every for every $f \in \mathcal{F}$, the associated polynomial $\tilde{f}$ has degree at most D and can be computed by an arithmetic circuit of size at most S. These degree and size parameters will depend on $\mathcal{F}$.

- Let $\mathcal{P}_{D',S',\mathbb{F}}$ be the set of all univariate polynomials $p : \mathbb{F} \to \mathbb{F}$ of degree at most D' and size at most S'. Let $\Pi_{OFE} = $ (OFE.Setup, OFE.KeyGen, OFE.Enc, OFE.Dec) be an $(n, 1, \text{negl}(\kappa))$-function-hiding-secure functional encryption scheme (i.e. secure for n keys and one message) for the family of polynomials $\mathcal{P}_{D',S',\mathbb{F}}$.

We're now ready to describe the construction of the two-message functional encryption scheme Π_{FE}. The scheme is specified in Fig. 7.

Correctness of Π_{FE}. Before going on to prove security, we will verify that encryption and decryption are correct for our scheme. Fix any $f_i \in \mathcal{F}$ and let $\tilde{f}_i : \mathbb{F}^\ell \to \mathbb{F}$ be the associated polynomial, and fix any input $x \in \mathbb{F}^\ell$. Let $r_i : \mathbb{F} \to \mathbb{F}$ be the degree RD polynomial chosen by FE.KeyGen on input f_i and let $\tilde{f}_{i,r_i}(\cdot, t)$ be the function used to generate the key OFE.sk$_{i,t}$. Let $q : \mathbb{F} \to \mathbb{F}^\ell$ be the degree R polynomial map chosen by FE.Enc on input x. Observe that, by correctness of Π_{OFE}, when we run FE.Dec we will have

$$\tilde{p}(t) = \text{OFE.Dec}(\text{OFE.sk}_{i,t}, c_t) = \tilde{f}_{i,r_i}(q(t), t) = \tilde{f}_i(q(t)) + r_i(t).$$

Now, consider the polynomial $\tilde{f}_{i,q,r_i} : \mathbb{F} \to \mathbb{F}$ defined by

$$\tilde{f}_{i,q,r_i}(t) = \tilde{f}_i(q(t)) + r_i(t).$$

Since $\tilde{f}_i$ has degree at most D, q has degree at most R, and r_i has degree at most RD, the degree of $\tilde{f}_{i,q,r_i}$ is at most RD. Since $|\mathcal{U}| = RD + 1$, the polynomial $\tilde{p}$ agrees with $\tilde{f}_{i,q,r_i}$ at $RD + 1$ distinct points, and thus $\tilde{p} \equiv \tilde{f}_{i,q,r_i}$. In particular, $\tilde{p}(0) = \tilde{f}_{i,q,r_i}(0)$. Since we chose r_i and q such that $r_i(0) = 0$ and $q(0) = x$, we have $\tilde{p}(0) = \tilde{f}_i(q(0)) + r_i(0) = \tilde{f}_i(x)$. This completes the proof of correctness.

6.2   Security for Two Messages

Theorem 5. *For every polynomial* $n = n(\kappa)$, Π_{FE} *is* $(n, 2, \delta)$-*secure for* $\delta = T \cdot \text{negl}(\kappa) + 2^{-\Omega(R)}$.

First we describe at a high level how to simulate. To do so, it will be useful to first introduce some terminology. Recall that FE.Setup instantiates T independent copies of the one-message scheme T. We refer to each instantiation as a *component*. Thus, when we talk about generating a secret key for a function f_i we will talk about generating each of the T components of that key and similarly when we talk about generating a ciphertext for an input x_b we will talk about generating each of the U components of that ciphertext. Thus the simulator has to generate a total of nT components of keys and $2U$ components of ciphertexts. The simulator will consider several types of components:

- Components $t \in \mathcal{U}_1 \cap \mathcal{U}_2$ where $\mathcal{U}_1, \mathcal{U}_2$ are the random sets of components chosen by the encryption scheme for the two inputs, respectively. The adversary obtains two ciphertexts for these components, so we cannot use the simulator for the one-message scheme. Thus for these components we simply choose uniformly random values for all the keys and ciphertexts and use the real one-message scheme.
- Components $t \in \mathcal{U}_1 \bigtriangleup \mathcal{U}_2$ (where $\bigtriangleup$ is the symmetric difference). For these we want to use the simulator for the one-message scheme to generate both the keys for each function and the ciphertexts for these components (recall that the one-message scheme is function-hiding). To do so, we need to feed the simulator with the evaluation of each of the functions on the chosen input. We show how to generate these outputs by leveraging the random high-degree polynomials r_i included in the keys. These values are then fed into the simulator to produce the appropriate key and ciphertext components.

FE.Sim:
Input: 1^κ, n functions and their evaluations on 2 unknown inputs $\{f_i, y_i^1, y_i^2\}_{i=1}^n$
Let $\tilde{f}_i : \mathbb{F}^{\ell+1} \to \mathbb{F}$ and $\tilde{y}_i^1, \tilde{y}_i^2 \in \mathbb{F}$ be the associated polynomial and field elements
Let OFE.Sim be the simulator for the one-message scheme Π_{OFE}

Choose random $r_1, \ldots, r_n : \mathbb{F} \to \mathbb{F}$ of degree RD s.t. for all i, $r_i(0) = 0$, let
$\tilde{f}_{i,r_i}(\cdot, t) = \tilde{f}_i(\cdot) + r_i(t)$
Choose two random sets $\mathcal{U}^1, \mathcal{U}^2 \subseteq [T]$ of size $U = RD + 1$
Let $\mathcal{I} = \mathcal{U}^1 \cap \mathcal{U}^2$. If $|\mathcal{I}| > R$, halt, output $\perp$ (henceforth we assume $|\mathcal{I}| \leq R$).

// Generate the keys and ciphertexts for components $t \in \mathcal{I}$ using Π_{OFE}
For $t \in \mathcal{I}$:
    For $b \in \{1, 2\}$ choose a random value $\alpha_t^b \in \mathbb{F}^\ell$
    Let OFE.msk$_t \leftarrow_R$ OFE.Setup(1^κ)
    For every i, let OFE.sk$_{i,t} \leftarrow_R$ OFE.KeyGen(OFE.msk$_t$, $\tilde{f}_{i,r_i}(\cdot, t)$)
    For $b \in \{1, 2\}$, let $c_t^b \leftarrow_R$ OFE.Enc(OFE.msk$_t$, α_t^b)

// Interpolate consistent evaluations to give OFE.Sim for $t \in \mathcal{U}^b \setminus \mathcal{I}$
For every i, generate a set of evaluations $\{\tilde{y}_{i,t}^b\}_{t \in \mathcal{U}^b}$ as follows:
    // Choose random values consistent with the choices we made for $t \in \mathcal{I}$
    For every i and every $t \in \mathcal{I}$, set $\tilde{y}_{i,t}^b = \tilde{f}_{i,r_i}(\alpha_t^b)$
    For all except one point $t \in \mathcal{U}^b \setminus \mathcal{I}$, choose uniformly random values for $\tilde{y}_{i,t}^b$
    // Ensure consistency with the final output.
    For all i, interpolate a polynomial $\tilde{p}_i^b$ of degree RD
      such that $\tilde{p}_i^b(0) = \tilde{y}_i^b$, $\tilde{p}_i^b(t) = \tilde{y}_{i,t}^b$
    For the last point $t \in \mathcal{U}^b$, set $\tilde{y}_{i,t}^b = \tilde{p}_i^b(t)$

// Generate keys and ciphertexts for $t \in \mathcal{U}_b \setminus \mathcal{I}$ using OFE.Sim
For $t \in \mathcal{U}_b \setminus \mathcal{I}$: $(\{OFE.sk_{i,t}\}_{i=1}^n, c_t) \leftarrow_R$ OFE.Sim $(\{\tilde{y}_{i,t}^b\}_{i=1}^n)$

// Generate keys (but no ciphertexts) for $t \notin \mathcal{U}_1 \cup \mathcal{U}_2$ obliviously
For $t \in [T] \setminus (\mathcal{U}^1 \cup \mathcal{U}^2)$ and $i \in [n]$, let OFE.sk$_{i,t} \leftarrow_R$ OFE.Sim().

// Output the n simulated secret keys and 2 simulated ciphertexts
Output: $\{\{OFE.sk_{i,t}\}_{t=1}^T\}_{i=1}^n, \{c_t^1\}_{t \in \mathcal{U}_1}, \{c_t^2\}_{t \in \mathcal{U}_2}$

Fig. 8. The simulator FE.Sim for Π_{FE}.

– Components $t \notin \mathcal{U}_1 \cup \mathcal{U}_2$. For these components the real scheme would not generate a ciphertext so the distribution can be simulated by a simulator that takes no inputs.

With this outline in the place, it is not too difficult to construct and analyze the simulator.

Proof (Proof of Theorem 5). We prove security via the simulator described in Fig. 8.

First we make a simple claim showing that there is only a small probability that the simulator has to halt and output $\perp$ because $\mathcal{I}$ is too large.

Claim. $\mathbb{P}\left[\mathsf{FE.Sim} = \perp\right] = 2^{-\Omega(R)}$.

Proof (Proof Sketch for Claim in Sect. 6.2). Recall that $\mathcal{I}$ is defined to be $\mathcal{U}_1 \cap \mathcal{U}_2$. Since $\mathcal{U}_1, \mathcal{U}_2$ are random subsets of $[T]$, each of size U, and we set $T = U^2$, we have $\mathbb{E}\left[|\mathcal{I}|\right] = 1$. Moreover, the intersection of the two sets has a hypergeometric distribution, and by a standard tail bound for the hypergeometric distribution we have $\mathbb{P}\left[\mathsf{FE.Sim} = \perp\right] = \mathbb{P}\left[|\mathcal{I}| > R\right] \leq 2^{-\Omega(R)}$.

In light of the above claim, we will assume for the remainder of the analysis that the simulator does not output $\perp$, and thus $|\mathcal{I}| \leq R$, and this will only cost add $2^{-\Omega(R)}$ to the simulation error. In what follows, we will simplify notation by referring only to components corresponding to keys and one of the ciphertexts, and will drop the superscript b. All of our arguments also applies to the second ciphertext, since this ciphertext is generated in a completely symmetric way.

Components Used in Both Ciphertexts. First, we claim that the simulator produces the correct distribution of the keys and ciphertexts for the components $t \in \mathcal{I}$. Note that the simulator chooses the keys in exactly the same way as the real scheme would: it generates keys for the functions $\tilde{f}_{i,r_i}(\cdot, t)$ where r_i is a random degree RD polynomial with the constant coefficient 0. The ciphertexts in the real scheme would contain the messages $\{q(t)\}_{t \in \mathcal{U}}$ where q is a random degree R polynomial with constant coefficient equal to the (unknown) input x. Since $|\mathcal{I}| \leq R$, this is a uniformly random set of values. Thus, the distribution of $\{\alpha_t\}_{t \in \mathcal{U}}$ is identical to $\{q(t)\}_{t \in \mathcal{U}}$, and therefore the simulated ciphertext components and the real ciphertext components have the same distribution.

Components Used in Exactly One Ciphertext. Next we claim that the simulated keys and ciphertexts for the components $t \in \mathcal{U} \setminus \mathcal{I}$ are computationally indistinguishable from those of the real scheme. Since in these components we only need to generate a single ciphertext, we can rely on the simulator OFE.Sim for the one-message scheme. OFE.Sim takes evaluations of n functions each at a single input and simulates the keys for those n functions and the ciphertext for that single input. In order to apply the indistinguishability guarantee for OFE.Sim, we need to argue that the evaluations that FE.Sim feeds to OFE.Sim are jointly identically distributed to the real scheme.

Recall that in the real scheme, each key corresponds to a function $\tilde{f}_{i,r_i}(\cdot, t)$ and this function gets evaluated on points $q(t)$. Thus for each function i, and each ciphertext component t, the evaluation is $\tilde{f}_{i,q,r_i}(t) = \tilde{f}_i(q(t)) + r_i(t)$. The polynomials $q, r_1, \ldots, r_n$ are chosen so that for every i, $\tilde{f}_{i,q,r_i}(0) = \tilde{f}_i(x)$ where x is the (unknown) input. We need to argue that the set of evaluations $\{\tilde{y}_{i,t}\}$ generated by the simulator have the same distribution as $\{\tilde{f}_{i,q,r_i}(t)\}$. Observe that, since r_i is a random polynomial of degree RD with constant coefficient 0, its evaluation on any set of RD points is jointly uniformly random. Therefore, for every q chosen independently of r, the evaluation of $\tilde{f}_{i,q,r_i}$ on any set of RD points is also jointly uniformly random. On the other hand, the evaluation of $\tilde{f}_{i,q,r_i}$ on any set of $RD+1$ points determines the whole function and thus determines $\tilde{f}_{i,q,r_i}(0)$, therefore conditioned on evaluations at any set of RD points, and the desired value of $\tilde{f}_{i,q,r_i}(0)$, the evaluation at any other point is uniquely determined.

Now, in the simulator, for every i, we choose RD evaluations $\tilde{y}_{i,t}$ uniformly randomly—for the points $t \in \mathcal{I}$ they are uniformly random because the polynomials r_i and the values $\alpha_{i,t}$ were chosen randomly, and then for all but one point in $\mathcal{U} \setminus \mathcal{I}$ we explicitly chose them to be uniformly random. For the remaining point, we chose $\tilde{y}_{i,t}$ to be the unique point such that we obtain the correct evaluation of $\tilde{f}_{i,q,r_i}(0)$, which is the value $\tilde{y}_i$ that was given to the simulator. Thus, we have argued that for any individual i, the distribution of the points $\tilde{y}_{i,t}$ that we give to the simulator is identical to that of the real scheme. The fact that this holds jointly over all i follows immediately by independence of the polynomials $r_1, \ldots, r_n$.

Components Used in Neither Ciphertext. Since the underlying one-message scheme satisfies function-hiding, it must be the case that the distribution of n keys and no messages is computationally indistinguishable from a fixed distribution. That is, it can be simulated given no evaluations. Thus we can simply generate the keys for these unused components in a completely oblivious way.

Since we have argued that all components are simulated correctly, we can complete the proof by taking a hybrid argument over the simulation error for each of the T components, and a union bound over the failure probability corresponding to the case where $|\mathcal{I}| > R$. Thus we argue that FE.Sim and the real scheme are computationally indistinguishable with the claimed parameters.

6.3   Bounding the Scheme Length for Comparison Functions

In the application to differential privacy, we need to instantiate the scheme for the family of comparison functions of the form $f_y(x) = \mathbb{I}\{x \geq y\}$ where $x, y \in \{0,1\}^{\log n}$, and we need to set the parameters to ensure $(n, 2, \frac{1}{300n^3})$-security where $n = n(\kappa)$ is an arbitrary polynomial.

Theorem 6. *For every polynomial $n = n(\kappa)$ there is a $(n, 2, \frac{1}{300n^3})$-secure functional encryption scheme for the family of comparison functions on $O(\log n)$ bits with keys are in K_κ and ciphertexts in C_κ where*

$$|K_\kappa| = 2^{2^{\mathrm{poly}(\log \log n)}} = 2^{n^{o(1)}} \quad and \quad |C_\kappa| = 2^\kappa.$$

Theorem 1 follows by combining Theorem 6 with Theorem 2. Note that Theorem 6 constructs a different scheme for every polynomial $n = n(\kappa)$. However, we can obtain a single scheme that is secure for every polynomial $n(\kappa)$ by instantiating this construction for some $n'(\kappa) = \kappa^{\omega(1)}$.

Proof (Proof of Theorem 6). By Theorem 5, if the underlying one-message scheme Π_{OFE} is $(n, 1, \text{negl}(\kappa))$-function-hiding secure, then the final scheme Π_{FE} will be $(n, 2, \delta)$-secure for $\delta = T \cdot \text{negl}(\kappa) + 2^{-\Omega(R)}$. If we choose an appropriate $R = \Theta(\log n)$ then we will have $\delta = T \cdot \text{negl}(\kappa) + \frac{1}{600n^3}$. As we will see, T will be a polynomial in n, so for sufficiently large values of κ, we will have $\delta \leq \frac{1}{300n^3}$. To complete the proof, we bound the length of the keys and ciphertexts:

The functions constructed in FE.KeyGen *have small DREs.* For the family of comparison functions on $\log n$ bits, there is a universal Boolean formula $u(x, y) : \{0, 1\}^{\log n} \times \{0, 1\}^{\log n} \to \{0, 1\}$ of size $S = O(\log n)$ and depth $d = O(\log \log n)$ that computes $f_y(x)$. Thus, for any field $\mathbb{F}$, the polynomial $\tilde{u}(x, y) : \mathbb{F}^{\log n} \times \mathbb{F}^{\log n} \to \mathbb{F}$ is computable by an arithmetic circuit of size $S = O(\log n)$ and depth $d = O(\log \log n)$, and this polynomial computes $\tilde{f}_y(x)$. For any value $r \in \mathbb{F}$, the polynomial $\tilde{u}_r(x, y) = \tilde{u}(x, y) + r$ is also computable by an arithmetic circuit of size $S + 1 = O(\log n)$ with degree d. Note that this polynomial is a universal evaluation for the polynomials $\tilde{f}_{y,r}(\cdot, t) = \tilde{f}_y(\cdot) + r(t)$ created in FE.KeyGen.

To obtain a DRE, we can write $\tilde{u}_r(x, y)$ as a Boolean formula $u_{r,\mathbb{F}}(x, y) : \{0, 1\}^{(\log n)(\log |\mathbb{F}|)} \times \{0, 1\}^{(\log n)(\log |\mathbb{F}|)} \to \{0, 1\}^{\log |\mathbb{F}|}$ with depth $d' = d \cdot \text{depth}(\mathbb{F})$ and size $S' = S \cdot \text{size}(\mathbb{F})$ where $\text{depth}(\mathbb{F})$ and $\text{size}(\mathbb{F})$ are the depth and size of Boolean formulae computing operations in the field $\mathbb{F}$, respectively. Later we will argue that it suffices to choose a field of size $\text{poly}(\log n)$, and thus $d_{\mathbb{F}}, S_{\mathbb{F}} = \text{poly}(\log \log n)$. Therefore these functions can be computed by formulae of depth $d' = \text{poly}(\log \log n)$ and size $S' = \text{poly}(\log n)$. Finally, by Theorem 3, the universal evaluator for this family has DREs of length $O(4^{d'}) = \exp(\text{poly}(\log \log n))$.

The secret keys and ciphertexts for each component are small. Π_{FE} generates key and ciphertext components for up to T independent instantiations of Π_{OFE}. Each function for Π_{OFE} corresponds to a formula of the form $u_{r,\mathbb{F}}$ defined above. By Theorem 4, we can instantiate Π_{OFE} so that each key component has length $\exp(\text{poly}(\log \log n))$ and each ciphertext component has length $\kappa \cdot \exp(\text{poly}(\log \log n)) = \text{poly}(\kappa)$, where the last inequality is because $n = \text{poly}(\kappa)$.

The number of components T and the size of the field $\mathbb{F}$ is small. In Π_{FE} we take $T = U^2 = (RD + 1)^2$ where $D \leq 2^d$ is the degree of the polynomials computing the comparison function over $\mathbb{F}$. As we argued above, we can take $R = O(\log n)$ and $D = \text{poly}(\log n)$. Therefore we have $T = \text{poly}(\log n)$. We need to ensure that $|\mathbb{F}| \geq T + 1$, since the security analysis relies on the fact that each component $t \in [T]$ corresponds to a different non-zero element of $\mathbb{F}$. Therefore, it suffices to have $|\mathbb{F}| = \text{poly}(\log n)$. In particular, this justifies the calculations above involving the complexity of field operations.

Putting it together. By the above, each component of the secret keys has length $\exp(\text{poly}(\log \log n))$ and there are $\text{poly}(\log n)$ components, so the overall length of the keys for Π_{FE} is $\exp(\text{poly}(\log \log n))$. Each component of the ciphertexts has length $\text{poly}(\kappa)$ and there are $\text{poly}(\log n) = \text{poly}(\log \kappa))$ components, so the overall length of the ciphertexts for Π_{FE} is poly(κ). The theorem statement now follows by rescaling κ and converting the bound on the length of the keys and ciphertexts to a bound on their number.

7   Two-Message Functional Encryption ⇒ Index Hiding

As discussed in Subsect. 3.2, Lemma 1 tells us that if we can show that any adversary's advantage in the TwoIndexHiding game is small, then the game's traitor-tracing scheme satisfies weak index-hiding security and gives us the lower bound of Theorem 2. First, note that one can use a private key functional encryption scheme for comparison functions directly as a traitor-tracing scheme, since they have the same functionality. We will now show that any private key functional encryption scheme that is $(n, 2, \frac{1}{300n^3})$-secure is a secure traitor-tracing scheme in the TwoIndexHiding game.

In Fig. 9, we describe a variant of the TwoIndexHiding game from Fig. 2 that uses the simulator FE.Sim for the functional encryption scheme $\Pi_{FE} =$ (FE.Setup, FE.KeyGen, FE.Enc, FE.Dec) for comparison functions $f_y(x) = \mathbb{I}\{x \geq y\}$ where $x, y \in \{0, 1\}^{\log n}$ that is $(n, 2, \frac{1}{300n^3})$-secure. Note that the challenger can give the simulator inputs that are independent of the game's b_0, b_1 since for all indices $j \neq i^*$, the output values of the comparison function for j on both inputs $i^* - b_0, i^* - b_1$ are always identical: $\mathbb{I}\{j > i^*\}$ (for all $b_0, b_1 \in \{0, 1\}$).

The challenger runs the simulator to produce:
$$\left(\{\mathsf{sk}_j\}_{j \neq i^* \in [n]}, \{c_0, c_1\} \right) \leftarrow_{\mathrm{R}} \mathsf{FE.Sim} \left(\{f_j, \{\mathbb{I}\{j > i^*\}, \mathbb{I}\{j > i^*\}\}\}_{j \neq i^* \in [n]} \right)$$
The adversary A is given keys sk_{-i^*} and outputs a decryption program S.
Choose $b_0 \leftarrow_{\mathrm{R}} \{0, 1\}$ and $b_1 \leftarrow_{\mathrm{R}} \{0, 1\}$ independently.
Let $b' = S(c_0, c_1)$.

Fig. 9. SimTwoIndexHiding[i^*]

Defining:

$$\mathrm{SimTwoAdv}[i^*] = \mathop{\mathbb{P}}_{\mathrm{SimTwoIndexHiding}[i^*]} [b' = b_0 \oplus b_1] - \frac{1}{2}$$

We can then prove the following lemmas:

Lemma 5. *For all p.p.t. adversaries,* $\mathrm{SimTwoAdv}[i^*] = 0$.

Proof. In SimTwoIndexHiding[i^*], b_0, b_1 are chosen uniformly at random and independent of the adversary's view. Therefore, the probability that the adversary outputs $b' = b_0 \oplus b_1$ is exactly $\frac{1}{2}$, and so

$$\mathrm{SimTwoAdv}[i^*] = \mathop{\mathbb{P}}_{\mathrm{SimTwoIndexHiding}[i^*]} [b' = b_0 \oplus b_1] - \frac{1}{2} = 0.$$

Lemma 6. *For all p.p.t. adversaries,* $|\text{TwoAdv}[i^*] - \text{SimTwoAdv}[i^*]| \leq \frac{1}{300n^3}$.

Proof. This follows easily from the simulation security of the 2-message FE scheme.

We can now show that any adversary's advantage in the TwoIndexHiding game is small:

Lemma 7. *Given a Two-Message Functional Encryption scheme for comparison functions* $f_y(x) = \mathbb{I}\{x \geq y\}$ *where* $x, y \in \{0, 1\}^{\log n}$ *that is* $(n, 2, \frac{1}{300n^3})$-*secure,*

$$\Pi_{FE} = (\text{FE.Setup}, \text{FE.KeyGen}, \text{FE.Enc}, \text{FE.Dec})$$

then for all i^*,

$$\text{TwoAdv}[i^*] \leq \frac{1}{300n^3}$$

Proof. Adding the statements of Lemma 5 and Lemma 6 gives us the statement of the lemma: $\text{TwoAdv}[i^*] \leq \frac{1}{300n^3}$ This completes the proof.

Combining Lemma 7 with Lemma 1, the $(n, 2, \frac{1}{300n^3})$-secure Two-Message Functional Encryption scheme from Sect. 6 is therefore a $(n, \{K_\kappa, C_\kappa\})$-traitor tracing scheme with weak index-hiding security. From Theorem 6, we have that

$$|K_\kappa| = 2^{2^{\text{poly}(\log \log n)}} = 2^{n^{o(1)}} \quad \text{and} \quad |C_\kappa| = 2^\kappa.$$

which when combined with Theorem 2 gives us our main Theorem 1.

Acknowledgements. The authors are grateful to Salil Vadhan for many helpful discussions.

The first and second authors are supported in part by the Defense Advanced Research Project Agency (DARPA) and Army Research Office (ARO) under Contract W911NF-15-C-0236, and NSF grants CNS-1445424 and CCF-1423306. Any opinions, findings and conclusions or recommendations expressed are those of the authors and do not necessarily reflect the views of the Defense Advanced Research Projects Agency, Army Research Office, the National Science Foundation, or the U.S. Government. The first author is also supported by NSF grant CNS-1552932 and NSF Graduate Research Fellowship DGE-16-44869.

The third author is supported by NSF CAREER award CCF-1750640, NSF grant CCF-1718088, and a Google Faculty Research Award.

The fourth author is supported by NSF grants CNS-1314722, CNS-1413964.

References

1. Bafna, M., Ullman, J.: The price of selection in differential privacy. In: COLT 2017 - The 30th Annual Conference on Learning Theory (2017)
2. Barrington, D.A.: Bounded-width polynomial-size branching programs recognize exactly those languages in NC^1. In: Proceedings of the 18th ACM Symposium on Theory of Computing (STOC) (1986)

3. Bassily, R., Nissim, K., Smith, A.D., Steinke, T., Stemmer, U., Ullman, J.: Algorithmic stability for adaptive data analysis. In: Proceedings of the 48th Annual ACM on Symposium on Theory of Computing, STOC (2016)
4. Bassily, R., Smith, A., Thakurta, A.: Private empirical risk minimization: efficient algorithms and tight error bounds. In: FOCS, pp. 464–473. IEEE, 18–21 October 2014
5. Beimel, A., Nissim, K., Stemmer, U.: Private learning and sanitization: pure vs. approximate differential privacy. In: Raghavendra, P., Raskhodnikova, S., Jansen, K., Rolim, J.D.P. (eds.) APPROX/RANDOM – 2013. LNCS, vol. 8096, pp. 363–378. Springer, Heidelberg (2013). https://doi.org/10.1007/978-3-642-40328-6_26
6. Blum, A., Dwork, C., McSherry, F., Nissim, K.: Practical privacy: the SuLQ framework. In: Symposium on Principles of Database Systems (PODS) (2005)
7. Blum, A., Ligett, K., Roth, A.: A learning theory approach to noninteractive database privacy. J. ACM **60**(2), 12 (2013)
8. Boneh, D., Sahai, A., Waters, B.: Fully collusion resistant traitor tracing with short ciphertexts and private keys. In: Vaudenay, S. (ed.) EUROCRYPT 2006. LNCS, vol. 4004, pp. 573–592. Springer, Heidelberg (2006). https://doi.org/10.1007/11761679_34
9. Boneh, D., Shaw, J.: Collusion-secure fingerprinting for digital data. IEEE Trans. Inf. Theory **44**(5), 1897–1905 (1998)
10. Boneh, D., Zhandry, M.: Multiparty key exchange, efficient traitor tracing, and more from indistinguishability obfuscation. In: Garay, J.A., Gennaro, R. (eds.) CRYPTO 2014. LNCS, vol. 8616, pp. 480–499. Springer, Heidelberg (2014)
11. Brakerski, Z., Segev, G.: Function-private functional encryption in the private-key setting. J. Cryptol. **31**(1), 202–225 (2018)
12. Bun, M., Nissim, K., Stemmer, U., Vadhan, S.: Differentially private release and learning of threshold functions. In: IEEE Annual Symposium on Foundations of Computer Science (FOCS) (2015)
13. Bun, M., Ullman, J., Vadhan, S.P.: Fingerprinting codes and the price of approximate differential privacy. In: STOC, pp. 1–10. ACM, 31 May–3 June 2014
14. Bun, M., Zhandry, M.: Order-revealing encryption and the hardness of private learning. In: Kushilevitz, E., Malkin, T. (eds.) TCC 2016. LNCS, vol. 9562, pp. 176–206. Springer, Heidelberg (2016). https://doi.org/10.1007/978-3-662-49096-9_8
15. Chandrasekaran, K., Thaler, J., Ullman, J., Wan, A.: Faster private release of marginals on small databases. In: Innovations in Theoretical Computer Science (ITCS) (2014)
16. Chor, B., Fiat, A., Naor, M.: Tracing traitors. In: Desmedt, Y.G. (ed.) CRYPTO 1994. LNCS, vol. 839, pp. 257–270. Springer, Heidelberg (1994). https://doi.org/10.1007/3-540-48658-5_25
17. Daniely, A., Linial, N., Shalev-Shwartz, S.: From average case complexity to improper learning complexity. In: Symposium on Theory of Computing (STOC) (2014)
18. Daniely, A., Shalev-Shwartz, S.: Complexity theoretic limitations on learning DNFs. In: COLT (2016)
19. Dinur, I., Nissim, K.: Revealing information while preserving privacy. In: Principles of Database Systems (PODS). ACM (2003)
20. Dodis, Y., Yu, Y.: Overcoming weak expectations. In: Sahai, A. (ed.) TCC 2013. LNCS, vol. 7785, pp. 1–22. Springer, Heidelberg (2013). https://doi.org/10.1007/978-3-642-36594-2_1
21. Dwork, C., Feldman, V., Hardt, M., Pitassi, T., Reingold, O., Roth, A.: Preserving statistical validity in adaptive data analysis. In: STOC. ACM (2015)

22. Dwork, C., McSherry, F., Nissim, K., Smith, A.: Calibrating noise to sensitivity in private data analysis. In: Halevi, S., Rabin, T. (eds.) TCC 2006. LNCS, vol. 3876, pp. 265–284. Springer, Heidelberg (2006). https://doi.org/10.1007/11681878_14

23. Dwork, C., Naor, M., Reingold, O., Rothblum, G.N., Vadhan, S.P.: On the complexity of differentially private data release: efficient algorithms and hardness results. In: Symposium on Theory of Computing (STOC). ACM (2009)

24. Dwork, C., Nikolov, A., Talwar, K.: Using convex relaxations for efficiently and privately releasing marginals. In: Symposium on Computational Geometry (SOCG) (2014)

25. Dwork, C., Nissim, K.: Privacy-preserving datamining on vertically partitioned databases. In: Franklin, M. (ed.) CRYPTO 2004. LNCS, vol. 3152, pp. 528–544. Springer, Heidelberg (2004). https://doi.org/10.1007/978-3-540-28628-8_32

26. Dwork, C., Rothblum, G.N., Vadhan, S.P.: Boosting and differential privacy. In: Foundations of Computer Science (FOCS). IEEE (2010)

27. Dwork, C., Smith, A., Steinke, T., Ullman, J.: Exposed! a survey of attacks on private data (2017)

28. Dwork, C., Smith, A., Steinke, T., Ullman, J., Vadhan, S.: Robust traceability from trace amounts. In: FOCS. IEEE (2015)

29. Dwork, C., Talwar, K., Thakurta, A., Zhang, L.: Analyze gauss: optimal bounds for privacy-preserving principal component analysis. In: Symposium on Theory of Computing, STOC, pp. 11–20 (2014)

30. Gorbunov, S., Vaikuntanathan, V., Wee, H.: Functional encryption with bounded collusions via multi-party computation. In: Safavi-Naini, R., Canetti, R. (eds.) CRYPTO 2012. LNCS, vol. 7417, pp. 162–179. Springer, Heidelberg (2012). https://doi.org/10.1007/978-3-642-32009-5_11

31. Goyal, R., Koppula, V., Waters, B.: Risky traitor tracing and new differential privacy negative results. Cryptology ePrint Archive, Report 2017/1117 (2017)

32. Gupta, A., Hardt, M., Roth, A., Ullman, J.: Privately releasing conjunctions and the statistical query barrier. SIAM J. Comput. **42**(4), 1494–1520 (2013)

33. Gupta, A., Roth, A., Ullman, J.: Iterative constructions and private data release. In: Cramer, R. (ed.) TCC 2012. LNCS, vol. 7194, pp. 339–356. Springer, Heidelberg (2012). https://doi.org/10.1007/978-3-642-28914-9_19

34. Hardt, M., Ligett, K., McSherry, F.: A simple and practical algorithm for differentially private data release. In: Advances in Neural Information Processing Systems (NIPS) (2012)

35. Hardt, M., Rothblum, G.: A multiplicative weights mechanism for privacy-preserving data analysis. In: Foundations of Computer Science (FOCS) (2014)

36. Hardt, M., Rothblum, G.N., Servedio, R.A.: Private data release via learning thresholds. In: Symposium on Discrete Algorithms (SODA) (2012)

37. Hardt, M., Ullman, J.: Preventing false discovery in interactive data analysis is hard. In: FOCS. IEEE (2014)

38. Kearns, M.J.: Efficient noise-tolerant learning from statistical queries. In: Symposium on Theory of Computing (STOC). ACM (1993)

39. Kilian, J.: Founding cryptography on oblivious transfer. In: Proceedings of the 20th ACM Symposium on Theory of Computing (STOC) (1988)

40. Kowalczyk, L., Malkin, T., Ullman, J., Zhandry, M.: Strong hardness of privacy from weak traitor tracing. In: Hirt, M., Smith, A. (eds.) TCC 2016. LNCS, vol. 9985, pp. 659–689. Springer, Heidelberg (2016). https://doi.org/10.1007/978-3-662-53641-4_25

41. Nikolov, A., Talwar, K., Zhang, L.: The geometry of differential privacy: the small database and approximate cases. SIAM J. Comput. **45**(2), 575–616 (2016). https://doi.org/10.1137/130938943
42. Pitt, L., Valiant, L.G.: Computational limitations on learning from examples. J. ACM (JACM) **35**(4), 965–984 (1988)
43. Roth, A., Roughgarden, T.: Interactive privacy via the median mechanism. In: Symposium on Theory of Computing (STOC). ACM (2010)
44. Sahai, A., Seyalioglu, H.: Worry-free encryption: functional encryption with public keys. In: Conference on Computer and Communications Security (CCS) (2010)
45. Steinke, T., Ullman, J.: Interactive fingerprinting codes and the hardness of preventing false discovery. In: Proceedings of the 28th Conference on Learning Theory, COLT, pp. 1588–1628 (2015)
46. Steinke, T., Ullman, J.: Tight lower bounds for differentially private selection. In: IEEE 58th Annual Symposium on Foundations of Computer Science, FOCS, pp. 634–649 (2017)
47. Tang, B., Zhang, J.: Barriers to black-box constructions of traitor tracing systems. In: Kalai, Y., Reyzin, L. (eds.) TCC 2017. LNCS, vol. 10677, pp. 3–30. Springer, Cham (2017). https://doi.org/10.1007/978-3-319-70500-2_1
48. Thaler, J., Ullman, J., Vadhan, S.: Faster algorithms for privately releasing marginals. In: Czumaj, A., Mehlhorn, K., Pitts, A., Wattenhofer, R. (eds.) ICALP 2012. LNCS, vol. 7391, pp. 810–821. Springer, Heidelberg (2012). https://doi.org/10.1007/978-3-642-31594-7_68
49. Ullman, J.: Private multiplicative weights beyond linear queries. In: PODS. ACM (2015)
50. Ullman, J.: Answering $n^{2+o(1)}$ counting queries with differential privacy is hard. SIAM J. Comput. **45**(2), 473–496 (2016)
51. Ullman, J., Vadhan, S.: PCPs and the hardness of generating private synthetic data. In: Ishai, Y. (ed.) TCC 2011. LNCS, vol. 6597, pp. 400–416. Springer, Heidelberg (2011). https://doi.org/10.1007/978-3-642-19571-6_24
52. Vadhan, S.: The complexity of differential privacy. In: Lindell, Y. (ed.) Tutorials on the Foundations of Cryptography. ISC, pp. 347–450. Springer, Cham (2017). https://doi.org/10.1007/978-3-319-57048-8_7

Risky Traitor Tracing and New
Differential Privacy Negative Results

Rishab Goyal[(✉)], Venkata Koppula, Andrew Russell, and Brent Waters

University of Texas at Austin, Austin, USA
{rgoyal,kvenkata,ahr,bwaters}@cs.utexas.edu

Abstract. In this work we seek to construct collusion-resistant traitor tracing systems with small ciphertexts from standard assumptions that also move toward practical efficiency. In our approach we will hold steadfast to the principle of collusion resistance, but relax the requirement on catching a traitor from a successful decoding algorithm. We define a f-risky traitor tracing system as one where the probability of identifying a traitor is $f(\lambda, n)$ times the probability a successful box is produced. We then go on to show how to build such systems from prime order bilinear groups with assumptions close to those used in prior works. Our core system achieves, for any $k > 0$, $f(\lambda, n) \approx \frac{k}{n+k-1}$ where ciphertexts consists of $(k+4)$ group elements and decryption requires $(k+3)$ pairing operations.

At first glance the utility of such a system might seem questionable since the f we achieve for short ciphertexts is relatively small. Indeed an attacker in such a system can more likely than not get away with producing a decoding box. However, we believe this approach to be viable for four reasons:

1. A risky traitor tracing system will provide deterrence against risk averse attackers. In some settings the consequences of being caught might bear a high cost and an attacker will have to weigh his utility of producing a decryption D box against the expected cost of being caught.

2. Consider a broadcast system where we want to support low overhead broadcast encrypted communications, but will periodically allow for a more expensive key refresh operation. We refer to an adversary produced algorithm that maintains the ability to decrypt across key refreshes as a persistent decoder. We show how if we employ a risky traitor tracing systems in this setting, even for a small f, we can amplify the chances of catching such a "persistent decoder" to be negligibly close to 1.

3. In certain resource constrained settings risky traitor tracing provides a best tracing effort where there are no other collusion-resistant alternatives. For instance, suppose we had to support 100 K users over a radio link that had just 10 KB of additional resources for extra ciphertext overhead. None of the existing $\sqrt{N}$ bilinear map systems can fit in these constraints. On the other hand a risky traitor tracing system provides a spectrum of tracing probability versus overhead tradeoffs and can be configured to at least give some deterrence in this setting.

H. Shacham and A. Boldyreva (Eds.): CRYPTO 2018, LNCS 10991, pp. 467–497, 2018.
https://doi.org/10.1007/978-3-319-96884-1_16

4. Finally, we can capture impossibility results for differential privacy from $\frac{1}{n}$-risky traitor tracing. Since our ciphertexts are short ($O(\lambda)$), we get the negative result which matches what one would get plugging in the obfuscation based tracing system Boneh-Zhandry [9] solution into the prior impossibility result of Dwork et al. [14].

1   Introduction

A traitor tracing [11] system is an encryption system in which a setup algorithm produces a public key pk, master secret key msk and n private keys $sk_1, sk_2, \ldots, sk_n$ that are distributed to n user devices. One can encrypt a message m using the public key to produce a ciphertext ct which can be decrypted using any of the private keys; however, is inaccessible by an attacker that is bereft of any keys. The tracing aspect comes into play if we consider an attacker that corrupts some subset $S \subseteq \{1, \ldots, n\}$ of the devices and produces a decryption algorithm D that decrypts ciphertext with some non-negligible probability $\epsilon(\lambda)$ where λ is the security parameter. An additional Trace algorithm will take as input the master secret key msk and with just oracle access to D will identify at least one user from the corrupted set S (and no one outside it). Importantly, any secure system must be able to handle attackers that will construct D in an arbitrary manner including using techniques such as obfuscation.

While the concept of traitor tracing was originally motivated by the example of catching users that created pirate decoder boxes in broadcast TV systems, there are several applications that go beyond that setting. For example ciphertexts could be encryptions of files stored on cloud storage. Or one might use a broadcast to transmit sensitive information to first responders on an ad-hoc deployed wireless network. In addition, the concepts and techniques of traitor tracing have had broader impacts in cryptography and privacy. Most notably Dwork et al. [14] showed that the existence of traitor tracing schemes leads to certain impossibility results in the area of differential privacy [13]. Briefly, they consider the problem of constructing a "sanitizer" $\mathcal{A}$ that takes in a database $x_1, \ldots, x_n$ of entries and wishes to efficiently produce a sanitized summary of database that can evaluate a set of predicate queries on the database. The sanitized database should both support giving an average of answers without too much error and the database should be differentially private in that no one entry should greatly impact the output of the sanitization process. The authors show that an efficient solution to such a problem is impossible to achieve (for certain parameters) assuming the existence of a (collusion resistant) traitor tracing system. The strength of their negative results is directly correlated with the size of ciphertexts in the traitor tracing system.

A primary obstacle in building traitor tracing systems is achieving (full) collusion resistance. There have been several proposals [4–6,19,20,25,29] for building systems that are k-collusion resistant where the size of the ciphertexts grows as some polynomial function of k. These systems are secure as long as the number of corrupted keys $|S| \leq k$; however, if the size of the corrupted set exceeds k the attacker will be able to produce a decryption box that is untraceable. Moreover, the collusion bound of k is fixed at system setup so an attacker will know how many keys he needs to exceed to beat the system. In addition, the impossibility results of Dwork et al. [14] only apply for fully collusion resistant encryption systems. For these reasons we will focus on collusion resistant systems in the rest of the paper.

The existing approaches for achieving collusion resistant broadcast encryption can be fit in the framework of Private Linear Broadcast Encryption (PLBE) introduced by Boneh et al. [7]. In a PLBE system the setup algorithm takes as input a security parameter λ and the number of users n. Like a traitor tracing system it output a public key pk, master secret key msk and n private keys $sk_1, sk_2, \ldots, sk_n$ where a user with index j is given key sk_j. Any of the private keys is capable of decrypting a ciphertext ct created using pk. However, there is an additional TrEncrypt algorithm that takes in the master secret key, a message and an index i. This produces a ciphertext that only users with index $j \geq i$ can decrypt. Moreover, any adversary produced decryption box D that was created with a set of S where $i \notin S$ would not be able to distinguish between encryption to index i or $i + 1$. These properties lead to a tracing system where the tracer measures for each index the probability that D decrypts a ciphertext encrypted (using TrEncrypt) for that index and reports all indices i where there is a significant discrepancy between i and $i + 1$. These properties imply that such a PLBE based traitor tracing system will catch at least one user in S with all but negligible probability and not falsely accuse anyone in S.

The primary difficulty in achieving collusion resistant traitor tracing is to do so with short ciphertext size. There are relatively few approaches for achieving this goal. First, one can achieve PLBE in a very simple way from public key encryption. Simply create n independent public and private key pairs from the PKE system and lump all the individual public keys together as the PLBE public key. To encrypt one just encrypts to each sub public key in turn. The downside of this method is that the ciphertext size grows as $O(n \cdot \lambda)$ as each of the n users need their own slot in the PLBE ciphertext. If one plugs this into the Dwork et al. [14] impossibility result it rules out systems with a query set $\mathcal{Q}$ of size $2^{O(n \cdot \lambda)}$ or larger. Boneh et al. [7] showed how ciphertexts in a PLBE system can be compressed to $O(\sqrt{n} \cdot \lambda)$ using bilinear maps of composite order. Future variants [15,17] moved this to the decision linear assumption in prime order groups. While this was an improvement and worked under standard assumptions, there was still a large gap between this and the ideal case where ciphertext size has only polylogarithmic dependence on n.

To achieve really short ciphertexts one needs to leverage heavier tools such as collusion resistant functional encryption or indistinguishability obfuscation [2,16]. For instance, a simple observation shows that one can make a PLBE scheme directly from a collusion resistant FE scheme such as the [16]. Boneh and Zhandry [9] gave a construction of PLBE from indistinguishability obfuscation. These two approaches get ciphertexts that grow proportionally to $\log n$ and thus leading to differential privacy impossibility results with smaller query sets of size $n \cdot 2^{O(\lambda)}$. However, general functional encryption and indistinguishability obfuscation candidates currently rely on multilinear map candidates, many of which have been broken and the security of which is not yet well understood. In addition, the actual decryption time resulting from using obfuscation is highly impractical.

Our Results. In this work we seek to construct collusion resistant traitor tracing systems with small ciphertexts from standard assumptions geared towards practical efficiency. In our approach we will hold steadfast to the principle of collusion resistance, but relax the requirement on catching a traitor from a successful decoding algorithm. We define a f-risky traitor tracing system as one where the probability of identifying a traitor is $f(\lambda, n)$ times the probability a successful box is produced. We then go on to show how to build such systems from prime order bilinear groups. Our core system achieves $f(\lambda, n) \approx \frac{k}{n+k-1}$ where ciphertexts consist of $(k + 4)$ group elements and decryption requires $(k + 3)$ pairing operations, where $k > 0$ is a system parameter fixed at setup time. For the basic setting, i.e. $k = 1$, this gives us a success probability of $\frac{1}{n}$, ciphertext consisting of 5 group elements, and decryption requiring just 4 pairing operations in primer order groups.[1] In addition, we show a generic way to increase f by approximately a factor of c at the cost of increasing the size of the ciphertext and decryption time also by a factor of c.

Finally, we show that the argument of Dwork et al. applies to $\frac{1}{n}$-risky traitor tracing. Interestingly, when we structure our argument carefully we achieve the same negative results as when it is applied to a standard traitor tracing system. Since our ciphertexts are short $(O(\lambda))$, we get the negative result which matches what one would get plugging in the obfuscation based tracing system Boneh-Zhandry [9] solution into the prior impossibility result of Dwork et al. [14].

1.1 Technical Overview

In this section, we give a brief overview of our technical approach. We start by discussing the definitional work. That is, we discuss existing traitor tracing definitions, mention their limitations and propose a stronger (and possibly more useful) definition, and finally introduce a weaker notion of traitor tracing which we call *risky* traitor tracing. Next, we describe our construction for risky traitor tracing from bilinear maps. Lastly, we discuss the differential privacy negative results implied by existence of risky traitor tracing schemes.

Definitional Work. A traitor tracing system consists of four poly-time algorithms — Setup, Enc, Dec, and Trace. The setup algorithm takes as input security parameter λ, and number of users n and generates a public key pk, a master secret key msk, and n private keys $sk_1, \ldots, sk_n$. The encrypt algorithm encrypts

---

[1] In addition to our construction from prime-order bilinear groups, we also provide a construction from composite order bilinear groups where ciphertexts consist of three group elements and decryption requires two pairing operations only.

messages using pk and the decrypt algorithm decrypts a ciphertext using any one of the private keys sk_i. The tracing algorithm takes msk as input and is given a black-box oracle access to a pirate decoder D. It either outputs a special failure symbol $\perp$, or an index $i \in \{1, \dots, n\}$ signalling that the key sk_i was used to create the pirate decoder.

Traditionally, a traitor tracing scheme is required to satisfy two security properties. First, it must be IND-CPA secure, i.e. any PPT adversary, when given no private keys, should not be able to distinguish between encryptions of two different messages. Second, it is required that if an adversary, given private keys $\{\mathsf{sk}_i\}_{i \in S}$ for any set S of its choice, builds a good pirate decoding box D (that is, a decoding box that can decrypt encryptions of random messages with non-negligible probability), then the trace algorithm should be able to catch one of the private keys used to build the pirate decoding box. Additionally, the trace algorithm should not falsely accuse any user with non-negligible probability. This property is referred to as secure traitor tracing.

Now a limitation of the traitor tracing property as traditionally defined is that a pirate box is labeled as a *good decoder* only if it extracts the entire message from a non-negligible fraction of ciphertexts.[2] In numerous practical scenarios such a definition could be useless and problematic. For instance, consider a pirate box that can always decrypt encryptions of messages which lie in a certain smaller set but does not work on others. If the size of this special set is negligible, then it won't be a good decoder as per existing definitions, but might still be adversarially useful in practice. There are also other reasons why the previous definitions of traitor tracing are problematic (see Sect. 3.2 for more details). To this end, we use an indistinguishability-based secure-tracing definition, similar to that used in [26], in which a pirate decoder is labeled to a good decoder if it can distinguish between encryptions of messages chosen by the adversary itself. We discuss this in more detail in Sect. 3.2.

In this work, we introduce a weaker notion of traitor tracing called f-*risky* traitor tracing, where f is a function that takes the security parameter λ and number of users n as inputs. The syntax as well as IND-CPA security requirement is identical to that of standard traitor tracing schemes. The difference is in the way security of tracing traitors is defined. In an f-risky system, we only require that the trace algorithm must catch a traitor with probability at least $f(\lambda, n)$ whenever the adversary outputs a good decoder. This property is referred to as f-risky secure traitor tracing. Note that a 1-risky traitor tracing scheme is simply a standard traitor tracing scheme, and as f decreases, this progressively becomes weaker.

Constructing Risky Traitor Tracing from Bilinear Maps. As mentioned before, our main construction is based on prime order bilinear groups, and leads to a $\frac{k}{n+k-1}$-risky traitor tracing where k is chosen at setup time. However, for ease of technical exposition we start with a simpler construction that uses composite order bilinear groups and leads to $\frac{1}{n}$-risky traitor tracing scheme. This scheme conveys the basic idea and will serve as a basis for our prime order construction.

---

[2] The tracing algorithm only needs to work when the pirate box is a good decoder.

Let $\mathbb{G}, \mathbb{G}_T$ be groups of order $N = p_1 p_2 p_3 p_4$ such that there exists a bilinear mapping $e : \mathbb{G} \times \mathbb{G} \to \mathbb{G}_T$ (that is, a mapping which maps (g^a, g^b) to $e(g, g)^{a \cdot b}$ for all $a, b \in \mathbb{Z}_N$). Since these groups are of composite order, $\mathbb{G}$ has subgroups $\mathbb{G}_1, \mathbb{G}_2, \mathbb{G}_3, \mathbb{G}_4$ of prime order p_1, p_2, p_3 and p_4 respectively. Moreover, pairing any element in $\mathbb{G}_i$ with an element in $\mathbb{G}_j$ (for $i \neq j$) results in the identity element (we will say that elements in $\mathbb{G}_i$ and $\mathbb{G}_j$ are orthogonal to each other).

At a high level, our construction works as follows. There are three key-generation algorithms: 'less-than' key-generation, 'equal' key-generation and 'greater-than' key-generation. Similarly, we have three encryption algorithms : 'standard' encryption, 'less-than' encryption and 'less-than-equal' encryption. Out of these encryption algorithms, the 'less-than' and 'less-than-equal' encryptions require the master secret key, and are only used for tracing traitors. The decryption functionality can be summarized by Table 1.

Table 1. Decryption functionality for different encryption/key-generation algorithms. The symbol ✓ denotes that decryption works correctly, while ✗ denotes that decryption fails.

	'less-than' keygen	'equal' keygen	'greater-than' keygen
standard enc	✓	✓	✓
'less-than' enc	✗	✓	✓
'less-than-equal' enc	✗	✗	✓

The master secret key consists of a 'cutoff' index i chosen uniformly at random from $\{1, \ldots, n\}$. For any index $j < i$, it uses the 'less-than' key-generation algorithm to generate keys. For $j > i$, it uses the 'greater-than' key-generation algorithm, and for $j = i$, it uses the 'equal' key-generation algorithm. The ciphertext for a message m is a 'standard' encryption of m. From Table 1, it is clear that decryption works. The trace algorithm tries to identify if the cutoff index i is used by the pirate box D. It first checks if D can decrypt 'less-than' encryptions. If so, then it checks if D can decrypt 'less-than-equal' encryptions. If D works in the 'less-than' case, but not in the 'less-than-equal' case, then the trace algorithm identifies index i as one of the traitors.

Let us now look at how the encryption/key generation algorithms work at a high level. The public key in our scheme consists of $g_1 \in \mathbb{G}_1$ and $e(g_1, g_1)^\alpha$, while the master secret key has the cut-off index i, element α, as well as generators for all subgroups of $\mathbb{G}$. The 'less-than' keys are set to be $g_1^\alpha \cdot w_3 \cdot w_4$, where w_3, w_4 are random group elements from $\mathbb{G}_3, \mathbb{G}_4$ respectively. The 'equal' key is $g_1^\alpha \cdot w_2 \cdot w_4$, where $w_2 \leftarrow \mathbb{G}_2, w_4 \leftarrow \mathbb{G}_4$. Finally, the 'greater-than' key has no $\mathbb{G}_2$ or $\mathbb{G}_3$ terms, and is set to be $g_1^\alpha \cdot w_4$.

The 'standard' encryption of message m is simply $(m \cdot e(g_1, g_1)^{\alpha \cdot s}, g_1^s)$. In the 'less-than' and 'less-than-equal' ciphertexts, the first component is computed similarly but the second component is modified. For 'less-than' encryptions, the ciphertext is $(m \cdot e(g_1, g_1)^{\alpha \cdot s}, g_1^s \cdot h_3)$, where h_3 is a uniformly random group element in $\mathbb{G}_3$. For 'less-than-equal' encryptions the ciphertext is $(m \cdot e(g_1, g_1)^{\alpha \cdot s}, g_1^s \cdot h_2 \cdot h_3)$, where h_2 and h_3 are uniformly random group elements in $\mathbb{G}_2$ and $\mathbb{G}_3$ respectively.

To decrypt a ciphertext $\mathsf{ct} = (\mathsf{ct}_1, \mathsf{ct}_2)$ using a key K, one must compute $\mathsf{ct}_1 / e(\mathsf{ct}_2, K)$. It is easy to verify that the keys and encryptions follow the decryption behavior described in Table 1. For instance, an 'equal' key $K = g_1^{\alpha} \cdot w_2 \cdot w_4$ can decrypt a 'less-than' encryption $(m \cdot e(g_1, g_1)^{\alpha \cdot s}, g_1^s \cdot h_3)$ because $e(\mathsf{ct}_2, K) = e(g_1, g_1)^{\alpha \cdot s}$. However, an 'equal' key cannot decrypt a 'less-than-equal' ciphertext $\mathsf{ct} = (m \cdot e(g_1, g_1)^{\alpha \cdot s}, g_1^s \cdot h_2)$ because $e(\mathsf{ct}_2, K) = e(g_1, g_1)^{\alpha \cdot s} \cdot e(h_2, w_2)$.

Given this construction, we need to prove two claims. First, we need to show that no honest party is implicated by our trace algorithm; that is, if an adversary does not receive key for index i, then the trace algorithm must not output index i. We show that if an adversary does not have key for index i, then the pirate decoding box must not be able to distinguish between 'less-than' and 'less-than-equal' encryptions (otherwise we can break the subgroup-decision assumption on composite order bilinear groups). Next, we show that if an adversary outputs a pirate decoding box that works with probability ρ, then we can identify a traitor with probability ρ/n. To prove this, we show that if ρ_i denotes the probability that the adversary outputs a ρ-functional box and i is the cutoff-index, then the sum of all these ρ_i quantities is close to ρ. The above scheme is formally described in the full version along with a detailed security proof. Next we move on to our risky traitor tracing construction from prime order bilinear groups.

Moving to Prime Order Bilinear Maps and $\frac{k}{n+k-1}$-Risky. The starting point for building $\frac{k}{n+k-1}$-risky traitor tracing scheme from prime order bilinear groups is the aforementioned scheme. Now to increase the success probability of the tracing algorithm by a factor k, we increase the types of secret keys and ciphertexts from 3 to $k+2$ such that the decryptability of ciphertexts w.r.t. secret keys can again be described as an upper-triangular matrix of dimension $k+2$ as follows (Table 2).

Table 2. New decryption functionality.

	'$< w$' keygen	'$= w$' keygen	'$= w+1$' keygen	$\cdots$	'$= w+k-1$' keygen	'$\geq w+k$' keygen
standard enc	✓	✓	✓	$\cdots$	✓	✓
'$< w$' enc	✗	✓	✓	$\cdots$	✓	✓
'$< w+1$' enc	✗	✗	✓	$\cdots$	✓	✓
$\vdots$	$\vdots$	$\vdots$	$\ddots$	$\ddots$	$\vdots$	$\vdots$
'$< w+k-1$' enc	✗	✗	✗	$\cdots$	✓	✓
'$< w+k$' enc	✗	✗	✗	$\cdots$	✗	✓

The basic idea will similar to the one used previously, except now we choose a cutoff window $W = \{w, w + 1, \ldots, w + k - 1\}$ of size k uniformly at random. (Earlier the window had size 1, that is we choose a single index.) The first $w - 1$ users are given '$< w$' keys. For $w \leq j < w + k$, the jth user gets '$= j$' key, and rest of the users get the '$\geq w + k$' keys. The remaining idea is similar to what we used which is that the tracer estimates the successful decryption probability for a decoder D on all the special index encryptions (i.e., '$< j$' encryptions), and outputs the indices of all those users where there is a gap in decoding probability while moving from type '$< j$' to '$< j + 1$'.

Now instead of directly building a scheme that induces such a decryption functionality, we provide a general framework for building risky traitor tracing schemes. In this work, we introduce a new intermediate primitive called Mixed Bit Matching Encryption (mBME) and show that it is sufficient to build risky traitor tracing schemes. In a mBME system, the secret keys and ciphertexts are associated with bit vectors $\mathbf{x}, \mathbf{y} \in \{0, 1\}^{\ell}$ (respectively) for some ℓ. And decryption works whenever $f(\mathbf{x}, \mathbf{y}) = 1$ where f computes an 'AND-of-ORs' over vectors $\mathbf{x}, \mathbf{y}$ (i.e., for every $i \leq \ell$, either $\mathbf{x}_i = 1$ or $\mathbf{y}_i = 1$). Using the public parameters, one could encrypt to the 'all-ones' vector, and using the master secret key one could sample a ciphertext (or secret key) for any vector. For security, we require that the ciphertexts and the secret keys should not reveal non-trivial information about their associated vectors. In other words, the only information an adversary learns about these vectors is by running the decryption algorithm. In the sequel, we provide a generic construction of risky traitor tracing from a mBME scheme, and also give a construction of mBME scheme using prime order bilinear groups. They are described in detail later in Sects. 5 and 6.

Finally, we also provide a performance evaluation of our risky traitor tracing scheme in Sect. 7.

Relation to BSW Traitor Tracing Scheme. Boneh et al. [7] constructed a (fully) collusion-resistant traitor tracing scheme with $O(\sqrt{n} \cdot \lambda)$ size ciphertexts. The BSW construction introduced the *private linear broadcast encryption* (PLBE) abstraction, showed how to build traitor tracing using PLBE, and finally gave a PLBE construction using composite-order bilinear groups.

Our framework deviates from the PLBE abstraction in that we support encryptions to only $k+1$ adjacent indices (that is, if w is starting index of the cutoff window, then we support encryptions to either $w, \ldots, w + k$) and index 0. As a result, the trace algorithm can only trace an index in the window $w, \ldots, w + k$. The main difficulty in our proof argument is that encrypting to index j is not defined for indices outside the cutoff window, i.e. $j \notin \{0, w, w + 1, \ldots, w + k\}$. As a result, we need to come up with a new way to link success probabilities across different setups and weave these into an argument.

Negative Results for Differential Privacy. Given a database $D = (x_1, x_2, \ldots, x_n) \in \mathcal{X}^n$, in which each row represents a single record of some sensitive information contributed by an individual and each record is an element in the data universe $\mathcal{X}$, the problem of privacy-preserving data analysis is to allow statistical analyses of D while protecting the privacy of individual contributors. The problem is formally defined in the literature by representing the database with a *sanitized* data structure s that can be used to answer all queries q in some query class $\mathcal{Q}$ with reasonable accuracy, with the restriction that the sanitization of any two databases D, D' which differ at only a single position are indistinguishable. In this work, we will focus on counting (or statistical) queries. Informally, a counting query q on a database D tells what fraction of records in D satisfy the property associated with q.

Dwork et al. [14] first showed that secure traitor tracing schemes can be used to show hardness results for efficient differentially private sanitization. In their hardness result, the data universe is the private key space of traitor tracing scheme and the query space is the ciphertext space. A database consists of n private keys and each query is associated with either an encryption of 0 or 1. Formally, for a ciphertext ct, the corresponding query q_{ct} on input a private key sk outputs the decryption of ct using sk. They show that if the underlying traitor tracing scheme is secure, then there can not exist sanitizers that are simultaneously accurate, differentially private, and efficient. At a very high level, the idea is as follows. Suppose there exists an efficient sanitizer A that, on input $D = (sk_1, \ldots, sk_n)$ outputs a sanitization s. The main idea is to use sanitizer A to build a pirate decoding box such that the tracing algorithm falsely accuses a user with non-negligible probability, thereby breaking secure traitor property. Concretely, let $\mathcal{B}$ be an attacker on the secure tracing property that works as follows — $\mathcal{B}$ queries for private keys of all but i^{th} party, and then uses sanitizer A to generate sanitization s of the database containing all the queried private keys, and finally it outputs the pirate decoding box as the sanitization evaluation algorithm which has s hardwired inside and on input a ciphertext outputs its evaluation given sanitization s.[3]

To prove that the tracing algorithm outputs i (with non-negligible probability) given such a decoding box, Dwork et al. crucially rely on the fact that A is differentially private. First, they show that if an adversary uses all private keys to construct the decoding box, then the tracing algorithm always outputs an index and never aborts.[4] Then, they argue that there must exist an index i such that tracing algorithm outputs i with probability $p \geq 1/n$. Finally, to complete the claim they show that even if i^{th} key is removed from the database, the tracing algorithm will output i with non-negligible probability since the sanitizer is differentially private with parameters $\epsilon = O(1)$ and $\delta = o(1/n)$.

In this work, we show that their analysis can be adapted to risky traitor tracing as well. Concretely, we show that f-risky secure traitor tracing schemes can be used to show hardness results for efficient differentially private sanitization, where f directly relates to the differential privacy parameters. At a high level,

---

[3] Technically, the decoding box must round the output of evaluation algorithm in order to remove evaluation error.

[4] In the full proof, one could only argue that tracing algorithm outputs an index with probability at least $1 - \beta$ where β is the accuracy parameter of sanitizer A.

the proof strategy is similar, i.e. we also show that an efficient sanitizer could be used to build a good pirate decoding box. The main difference is that now we can only claim that if an adversary uses all private keys to construct the decoding box, then (given oracle access to the box) the tracing algorithm outputs an index with probability at least f, i.e. the trace algorithm could potentially abort with non-negligible probability. Next, we can argue that there must exist an index i such that tracing algorithm outputs i with probability $p \geq f/n$. Finally, using differential privacy of A we can complete the argument. An important caveat in the proof is that since the lower bounds in the probability terms have an additional multiplicative factor of f, thus f-risky traitor tracing could only be used to argue hardness of differential privacy with slightly lower values of parameter δ, i.e. $\delta = o(f/n)$.

However, we observe that if the risky traitor tracing scheme additionally satisfies what we call "singular trace" property, then we could avoid the $1/n$ loss. Informally, a risky scheme is said to satisfy the singular trace property if the trace algorithm always outputs either a fixed index or the empty set. One could visualize the fixed index to be tied to the master secret and public keys. Concretely, we show that f-risky traitor tracing with singular trace property implies hardness of differential privacy for $\delta = o(f)$, thereby matching that achieved by previous obfuscation based result of [9]. We describe our hardness result in detail in Sect. 8.2.

Amplifying the Probability of Tracing — Catching Persistent Decoders. While an f-risky traitor tracing system by itself gives a small probability of catching a traitor, there can be ways to deploy it that increase this dramatically. We discuss one such way informally here.

Consider a broadcast system where we want to support low overhead broadcast encrypted communications, but will periodically allow for a more expensive key refresh operation. Suppose that we generate the secret keys $\mathsf{sk}_1, \mathsf{sk}_2, \ldots, \mathsf{sk}_n$ for a risky traitor tracing system and in addition generate standard secret keys $\mathsf{SK}_1, \ldots, \mathsf{SK}_n$. In this system an encryptor can use the traitor tracing public key pk to compute a ciphertext. A user i will use secret key sk_i to decrypt. The system will allow this to continue for a certain window of time. (Note during the window different ciphertexts may be created by different users.) Then at some point in time the window will close and a new risky tracing key pk' and secret keys $\mathsf{sk}_1', \mathsf{sk}_2', \ldots, \mathsf{sk}_n'$ will be generated. The tracing secret keys will be distributed by encrypting each sk_i' under the respective permanent secret key SK_i. And the encryptors will be instructed to only encrypt using the new public key pk'. This can continue for an arbitrary number of windows followed by key refreshes. Note that each key refresh requires $O(n\lambda)$ size communication.

Consider an attacker that wishes to disperse a stateless decoder D that is capable of continuing to work through multiple refresh cycles. Such a "persistent decoder" can be traced with very high probability negligibly close to 1. The tracing algorithm must simply give it multiple key refreshes followed by calls to the Trace algorithm and by the risky property it will eventually pick one that can trace one of the contributors.

We emphasize that care must be taken when choosing the refresh size window. If the window is too small the cost of key refreshes will dominate communication — in one extreme if a refresh happens at the frequency that ciphertexts are created then the communication is as bad as the trivial PLBE system. In addition, dispersing new public keys very frequently can be an issue. On the other hand if a refresh window is very long, then an attacker might decide there is value in producing a decoding box that works only for the given window and we are back to having only an $f(\lambda, n)$ chance of catching him.

1.2   Additional Related Work

Our traitor tracing system allows for public key encryption, but requires a master secret key to trace users as do most works. However, there exists exceptions [8–10,21,27,28,31] where the tracing can be done using a public key. In a different line of exploration, Kiayias and Yung [20] argue that a traitor tracing system with higher overhead can be made "constant rate" with long enough messages. Another interesting point in the space of collusion resistant systems is that of Boneh and Naor [6]. They show how to achieve short ciphertext size, but require private keys that grow quadratically in the number of users as $O(n^2\lambda)$. In addition, this is only achievable assuming a perfect decoder. If the decoder D works with probability δ then the secret key grows to $O(n^2\lambda/\delta^2)$. Furthermore, the system must be configured a-priori with a specific δ value and once it is set one will not necessarily be able to identify a traitor from a box D that works with smaller probability. Such systems have been called threshold traitor tracing systems [12,24]. Both [12,24] provide combinatorial and probabilistic constructions in which the tracing algorithm is guaranteed to work with high probability, and to trace t traitors they get private keys of size $O(t \cdot \log n)$. In contrast we can capture any traitor strategy that produces boxes that work with any non-negligible function $\epsilon(\lambda)$. Chor et al. [12] also considered a setting for traitor tracing in which the tracing algorithm only needs to correctly trace with probability $1 - p$, where p could the scheme parameter. However, this notion has not been formally defined or explored since then.

Dwork et al. [14] first showed that existence of collusion resistant traitor tracing schemes implies hardness results for efficient differentially private sanitization. In their hardness result, the database consists of n secret keys and each query is associated with an encryption of 0/1. Thus, the size of query space depends on the size of ciphertexts. Instantiating the result of Dwork et al. with the traitor tracing scheme of Boneh et al. [7], we get that under assumptions on bilinear groups, there exist a distribution on databases of size n and a query space of size $O(2^{\sqrt{n}\cdot\lambda})$ such that it is not possible to efficiently sanitize the database in a differentially private manner.

Now the result of Dwork et al. gives hardness of *one-shot* sanitization. A one-shot sanitizer is supposed to produce a summary of an entire database from which approximate answers to any query in the query set could be computed. A weaker setting could be where we consider interactive sanitization, in which the queries are fixed and given to the sanitizer as an additional input and the sanitizer only needs to output approximate answers to all those queries instead of a complete summary. Ullman [30] showed that, under the assumption that one-way functions exist, there is no algorithm that takes as input a database of n records along with an arbitrary set of about $O(n^2)$ queries, and approximately answers each query in polynomial time while preserving differential privacy. Ullman's result differs from the result of Dwork et al. in that it applies to algorithms answering any arbitrary set of $O(n^2)$ queries, whereas Dwork et al. show that it is impossible to sanitize a database with respect to a fixed set of $O(2^{\sqrt{n} \cdot \lambda})$ queries.

Recently a few works [9,23] have improved the size of query space for which (one-shot) sanitization is impossible from $O(2^{\sqrt{n} \cdot \lambda})$ to $n \cdot O(2^{\lambda})$ to $\mathsf{poly}(n)$.[5] [9] showed the impossibility by first constructing a fully collusion resistant scheme with short ciphertexts, and later simply applying the Dwork et al. result. On the other hand, [23] first construct a weakly secure traitor tracing scheme by building on top of PLBE abstraction, and later adapt the Dwork et al. impossibility result for this weaker variant. These works however assume existence of a stronger cryptographic primitive called *indistinguishability-obfuscator* ($i\mathcal{O}$) [2,16]. Currently we do not know of any construction of $i\mathcal{O}$ from a standard cryptographic assumption. In this work, we are interested in improving the state-of-the-art hardness results in differential privacy based on standard assumptions.

More recent related work. In an independent and concurrent work, Kowalczyk et al. [22] gave similar differential privacy negative results from one way functions. The negative results they achieve are similar to ours, but also apply for slightly smaller database sizes. However, the paths taken in our and their work diverge significantly. Our approach has been to focus on weaker notion of traitor tracing that suffices for DP impossibility while still being useful as a standalone primitive. On the other hand, KMUW instead closely follow the approach taken in [23], and they build special purpose functional encryption scheme for comparisons that supports 2 ciphertexts and bounded number of secret keys (succinctly) and achieves a very weak notion of IND-based security. Thus, the focus of their work is on negative results for differential privacy, whereas our risky tracing framework has both positive applications as well as lead to differential privacy impossibility results.

Subsequent to our work, Goyal et al. [18] gave a collusion resistant tracing system from the Learning with Errors assumption where the ciphertext size grows polynomially in $\lambda, \lg(N)$. Their result could be directly plugged into the Dwork et al. [14] differential privacy result as is, however they do not develop paths for weakening TT.

2   Preliminaries

Notations. For any set $\mathcal{X}$, let $x \leftarrow \mathcal{X}$ denote a uniformly random element drawn from the set $\mathcal{X}$. Given a PPT algorithm D, let A^D denote an algorithm A that uses D as an oracle (that is, A sends queries to D, and for each query x, it receives $D(x)$). Throughout this paper, we use PPT to denote probabilistic

---

[5] In this work, we only focus on the size of query space.

polynomial-time. We will use lowercase bold letters for vectors (e.g. $\mathbf{v}$), and we will sometimes represent bit vectors $\mathbf{v} \in \{0,1\}^\ell$ as bit-strings of appropriate length.

2.1 Assumptions

In this work, we will be using bilinear groups. Let Grp-Gen be a PPT algorithm that takes as input security parameter λ (in unary), and outputs a λ-bit prime p, an efficient description of groups $\mathbb{G}_1, \mathbb{G}_2, \mathbb{G}_T$ of order p, generators $g_1 \in \mathbb{G}_1$, $g_2 \in \mathbb{G}_2$ and an efficient non-degenerate bilinear mapping $e : \mathbb{G}_1 \times \mathbb{G}_2 \to \mathbb{G}_T$ (that is, $e(g_1, g_2) \neq 1_{\mathbb{G}_T}$, and for all $a, b \in \mathbb{Z}_p$, $e(g_1^a, g_2^b) = e(g_1, g_2)^{a \cdot b}$).

We will be using the following assumptions in this work.

Assumption 1. *For every PPT adversary $\mathcal{A}$, there exists a negligible function $negl(\cdot)$ s.t. for all $\lambda \in \mathbb{N}$,*

$$\Pr\left[b \leftarrow \mathcal{A}\left(\begin{array}{c} \text{params,} \\ g_1^x, g_1^y, g_1^{y \cdot z}, g_2^y, g_2^z, T_b \end{array} \right) : \begin{array}{l} \text{params} = (p, \mathbb{G}_1, \mathbb{G}_2, \mathbb{G}_T, g_1, g_2, e(\cdot, \cdot)) \leftarrow \text{Grp-Gen}(1^\lambda); \\ x, y, z, r \leftarrow \mathbb{Z}_p, T_0 = g_1^{x \cdot y \cdot z}, T_1 = g_1^{x \cdot y \cdot z + r}, b \leftarrow \{0, 1\} \end{array} \right] \leq 1/2 + negl(\lambda).$$

Assumption 2. *For every PPT adversary $\mathcal{A}$, there exists a negligible function $negl(\cdot)$ s.t. for all $\lambda \in \mathbb{N}$,*

$$\Pr\left[b \leftarrow \mathcal{A}\left(\begin{array}{c} \text{params,} \\ g_1^y, g_1^z, g_2^x, g_2^y, T_b \end{array} \right) : \begin{array}{l} \text{params} = (p, \mathbb{G}_1, \mathbb{G}_2, \mathbb{G}_T, g_1, g_2, e(\cdot, \cdot)) \leftarrow \text{Grp-Gen}(1^\lambda); \\ x, y, z, r \leftarrow \mathbb{Z}_p, T_0 = g_2^{x \cdot y \cdot z}, T_1 = g_2^{x \cdot y \cdot z + r}, b \leftarrow \{0, 1\} \end{array} \right] \leq 1/2 + negl(\lambda).$$

3 Risky Traitor Tracing

In this section, we will first introduce the traditional definition of traitor tracing based on that given by Boneh et al. [7]. We provide a "public key" version of the definition in which the encryption algorithm is public, but the tracing procedure will require a master secret key. Our definition will by default capture full collusion resistance.

A limitation of this definition is that the tracing algorithm is only guaranteed to work on decoders that entirely decrypt encryptions of randomly selected messages with non-negligible probability. We will discuss why this definition can be problematic and then provide an *indistinguishability* based definition for secure tracing.

Finally, we will present our new notion of *risky* traitor tracing which captures the concept of a trace algorithm that will identify a traitor from a working pirate box with probability close to $f(\lambda, n)$. Our main definition for risky traitor tracing will be a public key one using the indistinguishability; however we will also consider some weaker variants that will be sufficient for obtaining our negative results in differential privacy.

3.1 Public Key Traitor Tracing

A traitor tracing scheme with message space $\mathcal{M}$ consists of four PPT algorithms Setup, Enc, Dec and Trace with the following syntax:

$(\mathsf{msk}, \mathsf{pk}, (\mathsf{sk}_1, \ldots, \mathsf{sk}_n)) \leftarrow \mathsf{Setup}(1^\lambda, 1^n)$: The setup algorithm takes as input the security parameter λ, number of users n, and outputs a master secret key msk, a public key pk and n secret keys $\mathsf{sk}_1, \mathsf{sk}_2, \ldots, \mathsf{sk}_n$.

$\mathsf{ct} \leftarrow \mathsf{Enc}(\mathsf{pk}, m \in \mathcal{M})$: The encryption algorithm takes as input a public key pk, message $m \in \mathcal{M}$ and outputs a ciphertext ct.

$y \leftarrow \mathsf{Dec}(\mathsf{sk}, \mathsf{ct})$: The decryption algorithm takes as input a secret key sk, ciphertext ct and outputs $y \in \mathcal{M} \cup \{\bot\}$.

$S \leftarrow \mathsf{Trace}^D(\mathsf{msk}, 1^y)$: The tracing algorithm takes a parameter $y \in \mathbb{N}$ (in unary) as input, has black box access to an algorithm D, and outputs a set $S \subseteq \{1, 2, \ldots, n\}$.

Correctness. For correctness, we require that if ct is an encryption of message m, then decryption of ct using one of the valid secret keys must output m. More formally, we require that for all $\lambda \in \mathbb{N}$, $n \in \mathbb{N}$, $(\mathsf{msk}, \mathsf{pk}, (\mathsf{sk}_1, \ldots, \mathsf{sk}_n)) \leftarrow \mathsf{Setup}(1^\lambda, 1^n)$, $m \in \mathcal{M}$, $\mathsf{ct} \leftarrow \mathsf{Enc}(\mathsf{pk}, m)$ and $i \in \{1, 2, \ldots, n\}$, $\mathsf{Dec}(\mathsf{sk}_i, \mathsf{ct}) = m$.

Security. A secure traitor tracing scheme must satisfy two security properties. First, the scheme must be IND-CPA secure (that is, any PPT adversary, when given no secret keys, cannot distinguish between encryptions of m_0, m_1). Next, we require that if an adversary, using some secret keys, can build a pirate decoding box, then the trace algorithm should be able to catch at least one of the secret keys used to build the pirate decoding box. In this standard definition, the trace algorithm identifies a traitor if the pirate decoding box works with non-negligible probability in extracting the entire message from an encryption of a random message.

Definition 1 (IND-CPA security). *A traitor tracing scheme $\mathcal{T} = (\mathsf{Setup}, \mathsf{Enc}, \mathsf{Dec}, \mathsf{Trace})$ is IND-CPA secure if for any PPT adversary $\mathcal{A} = (\mathcal{A}_1, \mathcal{A}_2)$, polynomial $n(\cdot)$, there exists a negligible function $\mathsf{negl}(\cdot)$ such that for all $\lambda \in \mathbb{N}$, $|\Pr[1 \leftarrow \mathsf{Expt\text{-}IND\text{-}CPA}_{\mathcal{A}}^{\mathcal{T}}(1^\lambda, 1^n)] - 1/2| \leq \mathsf{negl}(\lambda)$, where $\mathsf{Expt\text{-}IND\text{-}CPA}_{\mathcal{T}, \mathcal{A}}$ is defined below.*

- $(\mathsf{msk}, \mathsf{pk}, (\mathsf{sk}_1, \ldots, \mathsf{sk}_n)) \leftarrow \mathsf{Setup}(1^\lambda, 1^{n(\lambda)})$
- $(m_0, m_1, \sigma) \leftarrow \mathcal{A}_1(\mathsf{pk})$
- $b \leftarrow \{0, 1\}$, $\mathsf{ct} \leftarrow \mathsf{Enc}(\mathsf{pk}, m_b)$
- $b' \leftarrow \mathcal{A}_2(\sigma, \mathsf{ct})$. *Experiment outputs 1 iff $b = b'$.*

Definition 2 (Secure traitor tracing). *Let $\mathcal{T} = (\mathsf{Setup}, \mathsf{Enc}, \mathsf{Dec}, \mathsf{Trace})$ a traitor tracing scheme. For any polynomial $n(\cdot)$, non-negligible function $\epsilon(\cdot)$ and PPT adversary $\mathcal{A}$, consider the following experiment $\mathsf{Expt}_{\mathcal{A}, n, \epsilon}^{\mathcal{T}}(\lambda)$:*

- $(\mathsf{msk}, \mathsf{pk}, (\mathsf{sk}_1, \ldots, \mathsf{sk}_{n(\lambda)})) \leftarrow \mathsf{Setup}(1^\lambda, 1^{n(\lambda)})$.
- $D \leftarrow A^{O(\cdot)}(\mathsf{pk})$

- $S_D \leftarrow \mathsf{Trace}^D(\mathsf{msk}, 1^{1/\epsilon(\lambda)})$.

Here, $O(\cdot)$ is an oracle that has $\{\mathsf{sk}_1, \mathsf{sk}_2, \ldots, \mathsf{sk}_{n(\lambda)}\}$ hardwired, takes as input an index $i \in \{1, 2, \ldots, n(\lambda)\}$ and outputs sk_i. Let S be the set of indices queried by $\mathcal{A}$. Based on this experiment, we will now define the following (probabilistic) events and the corresponding probabilities (which is a function of λ, parameterized by $\mathcal{A}, n, \epsilon$):

- Good-Decoder : $\Pr[D(\mathsf{ct}) = m \ : \ m \leftarrow \mathcal{M}, \mathsf{ct} \leftarrow \mathsf{Enc}(\mathsf{pk}, m)] \geq \epsilon(\lambda)$
  $\Pr\text{-G-D}_{\mathcal{A},n,\epsilon}(\lambda) = \Pr[\mathsf{Good\text{-}Decoder}]$.
- Cor-Tr : $S_D \subseteq S \ \wedge \ S_D \neq \emptyset$
  $\Pr\text{-Cor-Tr}_{\mathcal{A},n,\epsilon}(\lambda) = \Pr[\mathsf{Cor\text{-}Tr}]$.
- Fal-Tr : $S_D \setminus S \neq \emptyset$
  $\Pr\text{-Fal-Tr}_{\mathcal{A},n,\epsilon}(\lambda) = \Pr[\mathsf{Fal\text{-}Tr}]$.

A traitor tracing scheme $\mathcal{T}$ is said to be secure if for every PPT adversary $\mathcal{A}$, polynomials $n(\cdot)$, $p(\cdot)$ and non-negligible function $\epsilon(\cdot)$, there exists negligible functions $\mathsf{negl}_1(\cdot)$, $\mathsf{negl}_2(\cdot)$ such that for all $\lambda \in \mathbb{N}$ such that $\epsilon(\lambda) > 1/p(\lambda)$, $\Pr\text{-Fal-Tr}_{\mathcal{A},n,\epsilon}(\lambda) \leq \mathsf{negl}_1(\lambda)$ and $\Pr\text{-Cor-Tr}_{\mathcal{A},n,\epsilon}(\lambda) \geq \Pr\text{-G-D}_{\mathcal{A},n,\epsilon}(\lambda) - \mathsf{negl}_2(\lambda)$.

3.2 Indistinguishability Security Definition for Traitor Tracing Schemes

A limitation of the previous definition is that the tracing algorithm is only guaranteed to work on decoders that *entirely* decrypt a *randomly* selected message with non-negligible probability. This definition can be problematic for the following reasons.

- First, there could be pirate boxes which do not extract the entire message from a ciphertext, but can extract some information about the message underlying a ciphertext. For example, a box could paraphrase English sentences or further compress an image. Such boxes could be very useful to own in practice yet the tracing definition would give no guarantees on the ability to trace them.
- Second, a pirate decoder may not be very successful in decrypting random ciphertexts, but can decrypt encryptions of messages from a smaller set. In practice the set of useful or typical messages might indeed fall in a smaller set.
- Finally, if the message space is small (that is, of polynomial size), then one can always construct a pirate decoder which succeeds with non-negligible probability and can not get caught (the pirate decoder box simply outputs a random message for each decryption query. If $\mathcal{M}$ is the message space, then decryption will be successful with probability $1/|\mathcal{M}|$). Since such a strategy does not use any private keys, it cannot be traced. Therefore the above definition is only sensible for superpolynomial sized message spaces.

To address these issues, we provide a stronger definition, similar to that used in [26], in which a pirate decoder is successful if it can distinguish between encryptions of messages chosen by the decoder itself. For this notion, we also need to modify the syntax of the Trace algorithm. Our security notion is similar to the one above except that an attacker will output a box D along with two messages (m_0, m_1). If the box D is able to distinguish between encryptions of these two messages with non-negligible probability then the tracing algorithm can identify a corroborating user.

Trace^D(msk, $1^y, m_0, m_1$): The trace algorithm has oracle access to a program D, it takes as input a master secret key msk, y (in unary) and two messages m_0, m_1. It outputs a set $S \subseteq \{1, 2, \ldots, n\}$.

Definition 3 (Ind-secure traitor tracing). *Let* $\mathcal{T} = $ (Setup, Enc, Dec, Trace) *be a traitor tracing scheme. For any polynomial* $n(\cdot)$, *non-negligible function* $\epsilon(\cdot)$ *and PPT adversary* $\mathcal{A}$, *consider the experiment* Expt-TT$^{\mathcal{T}}_{\mathcal{A},n,\epsilon}(\lambda)$ *defined in Fig. 1. Based on this experiment, we will now define the following (probabilistic) events and the corresponding probabilities (which is a function of* λ, *parameterized by* $\mathcal{A}, n, \epsilon$):

- Good-Decoder : $\Pr[D(\text{ct}) = b \; : \; b \leftarrow \{0, 1\}, \text{ct} \leftarrow \text{Enc}(\text{pk}, m_b)] \geq 1/2 + \epsilon(\lambda)$
  Pr -G-D$_{\mathcal{A},n,\epsilon}(\lambda) = \Pr[\text{Good-Decoder}]$.
- Cor-Tr : $S_D \subseteq S \; \wedge \; S_D \neq \emptyset$
  Pr -Cor-Tr$_{\mathcal{A},n,\epsilon}(\lambda) = \Pr[\text{Cor-Tr}]$.
- Fal-Tr : $S_D \setminus S \neq \emptyset$
  Pr -Fal-Tr$_{\mathcal{A},n,\epsilon}(\lambda) = \Pr[\text{Fal-Tr}]$.

A traitor tracing scheme $\mathcal{T}$ *is said to be ind-secure if for every PPT adversary* $\mathcal{A}$, *polynomials* $n(\cdot)$, $p(\cdot)$ *and non-negligible function* $\epsilon(\cdot)$, *there exists negligible functions* $negl_1(\cdot)$, $negl_2(\cdot)$ *such that for all* $\lambda \in \mathbb{N}$ *satisfying* $\epsilon(\lambda) > 1/p(\lambda)$, Pr -Fal-Tr$_{\mathcal{A},n,\epsilon}(\lambda) \leq negl_1(\lambda)$ *and* Pr -Cor-Tr$_{\mathcal{A},n,\epsilon}(\lambda) \geq$ Pr -G-D$_{\mathcal{A},n,\epsilon}(\lambda) - negl_2(\lambda)$.

Experiment Expt-TT$^{\mathcal{T}}_{\mathcal{A},n,\epsilon}(\lambda)$

- $(\text{msk}, \text{pk}, (\text{sk}_1, \ldots, \text{sk}_{n(\lambda)})) \leftarrow \text{Setup}(1^\lambda, 1^{n(\lambda)})$.
- $(D, m_0, m_1) \leftarrow \mathcal{A}^{O(\cdot)}(\text{pk})$
- $S_D \leftarrow \text{Trace}^D(\text{msk}, 1^{1/\epsilon(\lambda)}, m_0, m_1)$.

Here, $O(\cdot)$ is an oracle that has $\{\text{sk}_1, \text{sk}_2, \ldots, \text{sk}_{n(\lambda)}\}$ hardwired, takes as input an index $i \in \{1, 2, \ldots, n(\lambda)\}$ and outputs sk_i. Let S be the set of indices queried by $\mathcal{A}$.

Fig. 1. Experiment Expt-TT

3.3 Risky Traitor Tracing

In this section, we will introduce the notion of risky traitor tracing. The syntax is same as that of ind-secure traitor tracing. However, for security, if the adversary outputs a good decoder, then the trace algorithm will catch a traitor with probability f where f is a function of λ and the number of users.

Definition 4 (f-risky secure traitor tracing). *Let $f : \mathbb{N} \times \mathbb{N} \to [0,1]$ be a function and $\mathcal{T} = (\mathsf{Setup}, \mathsf{Enc}, \mathsf{Dec}, \mathsf{Trace})$ a traitor tracing scheme. For any polynomial $n(\cdot)$, non-negligible function $\epsilon(\cdot)$ and PPT adversary $\mathcal{A}$, consider the experiment $\mathsf{Expt\text{-}TT}^{\mathcal{T}}_{\mathcal{A},n,\epsilon}(\lambda)$ (defined in Fig. 1). Based on this experiment, we will now define the following (probabilistic) events and the corresponding probabilities (which are functions of λ, parameterized by $\mathcal{A}, n, \epsilon$):*

- Good-Decoder : $\Pr[D(\mathsf{ct}) = b \ : \ b \leftarrow \{0,1\}, \mathsf{ct} \leftarrow \mathsf{Enc}(\mathsf{pk}, m_b)] \geq 1/2 + \epsilon(\lambda)$
  $\Pr\text{-}\mathsf{G\text{-}D}_{\mathcal{A},n,\epsilon}(\lambda) = \Pr[\mathsf{Good\text{-}Decoder}]$.
- Cor-Tr : $S_D \subseteq S \ \wedge \ S_D \neq \emptyset$
  $\Pr\text{-}\mathsf{Cor\text{-}Tr}_{\mathcal{A},n,\epsilon}(\lambda) = \Pr[\mathsf{Cor\text{-}Tr}]$.
- Fal-Tr : $S_D \setminus S \neq \emptyset$
  $\Pr\text{-}\mathsf{Fal\text{-}Tr}_{\mathcal{A},n,\epsilon}(\lambda) = \Pr[\mathsf{Fal\text{-}Tr}]$.

A traitor tracing scheme $\mathcal{T}$ is said to be f-risky secure if for every PPT adversary $\mathcal{A}$, polynomials $n(\cdot)$, $p(\cdot)$ and non-negligible function $\epsilon(\cdot)$, there exists negligible functions $\mathsf{negl}_1(\cdot)$, $\mathsf{negl}_2(\cdot)$ such that for all $\lambda \in \mathbb{N}$ satisfying $\epsilon(\lambda) > 1/p(\lambda)$, $\Pr\text{-}\mathsf{Fal\text{-}Tr}_{\mathcal{A},n,\epsilon}(\lambda) \leq \mathsf{negl}_1(\lambda)$ and $\Pr\text{-}\mathsf{Cor\text{-}Tr}_{\mathcal{A},n,\epsilon}(\lambda) \geq \Pr\text{-}\mathsf{G\text{-}D}_{\mathcal{A},n,\epsilon}(\lambda) \cdot f(\lambda, n(\lambda)) - \mathsf{negl}_2(\lambda)$.

We also define another interesting property for traitor tracing schemes which we call "singular" trace. Informally, a scheme satisfies it if the trace algorithm always outputs either a fixed index or the reject symbol. The fixed index could depend on the master secret and public keys. Below we define it formally.

Definition 5 (Singular Trace). *A traitor tracing scheme $\mathcal{T} = (\mathsf{Setup}, \mathsf{Enc}, \mathsf{Dec}, \mathsf{Trace})$ is said to satisfy singular trace property if for every polynomial $n(\cdot)$, $\lambda \in \mathbb{N}$, keys $(\mathsf{msk}, \mathsf{pk}, (\mathsf{sk}_1, \ldots, \mathsf{sk}_n)) \leftarrow \mathsf{Setup}(1^\lambda, 1^n)$, there exists an index $i^* \in \{1, \ldots, n\}$ such that for every poly-time algorithm D, parameter $y \in \mathbb{N}$, any two messages m_0, m_1,*

$$\Pr[\mathsf{Trace}^D(\mathsf{msk}, 1^y, m_0, m_1) \in \{\{i^*\}, \emptyset\}] = 1,$$

where the probability is taken over random coins of Trace.

One can analogously define the notion of private key risky traitor tracing, which suffices for our differential privacy lower bound. We present this notion in the full version.

4   A New Abstraction for Constructing Risky Traitor Tracing

Let $\{\mathcal{M}_\lambda\}_\lambda$ denote the message space. A mixed bit matching encryption scheme for $\mathcal{M}$ consists of five algorithms with the following syntax.

Setup$(1^\lambda, 1^\ell) \to (\mathsf{pk}, \mathsf{msk})$: The setup algorithm takes as input security parameter λ, a parameter ℓ and outputs a public key pk and master secret key msk.

KeyGen$(\mathsf{msk}, \mathbf{x} \in \{0,1\}^\ell) \to \mathsf{sk}$: The key generation algorithm takes as input the master secret key msk and a vector $\mathbf{x} \in \{0,1\}^\ell$. It outputs a secret key sk corresponding to $\mathbf{x}$.

Enc-PK$(\mathsf{pk}, m \in \mathcal{M}) \to \mathsf{ct}$: The public-key encryption algorithm takes as input a public key pk and a message m, and outputs a ciphertext ct.

Enc-SK$(\mathsf{msk}, m \in \mathcal{M}, \mathbf{y} \in \{0,1\}^\ell) \to \mathsf{ct}$: The secret-key encryption algorithm takes as input master secret key msk, message m, and an attribute vector $\mathbf{y} \in \{0,1\}^\ell$. It outputs a ciphertext ct.

Dec$(\mathsf{sk}, \mathsf{ct}) \to z$: The decryption algorithm takes as input a ciphertext ct, a secret key sk and outputs $z \in \mathcal{M} \cup \{\bot\}$.

Permissions. Define $f : \{0,1\}^\ell \times \{0,1\}^\ell \to \{0,1\}$ by the following:

$$f(\mathbf{x}, \mathbf{y}) = \bigwedge_{i=1}^{\ell} x_i \vee y_i$$

We will use this function to determine when secret keys with attribute vectors $\mathbf{x}$ are "permitted" to decrypt ciphertexts with attribute vectors $\mathbf{y}$.

Correctness. We require the following properties for correctness:

- For every $\lambda \in \mathbb{N}, \ell \in \mathbb{N}$, $(\mathsf{pk}, \mathsf{msk}) \leftarrow \mathsf{Setup}(1^\lambda, 1^\ell)$, $\mathbf{x} \in \{0,1\}^\ell$, $\mathsf{sk} \leftarrow \mathsf{KeyGen}(\mathsf{msk}, \mathbf{x})$, message $m \in \mathcal{M}_\lambda$ and $\mathsf{ct} \leftarrow \mathsf{Enc\text{-}PK}(\mathsf{pk}, m)$, $\mathsf{Dec}(\mathsf{sk}, \mathsf{ct}) = m$.
- For every $\lambda \in \mathbb{N}, \ell \in \mathbb{N}$, $(\mathsf{pk}, \mathsf{msk}) \leftarrow \mathsf{Setup}(1^\lambda, 1^\ell)$, $\mathbf{x} \in \{0,1\}^\ell$, $\mathsf{sk} \leftarrow \mathsf{KeyGen}(\mathsf{msk}, \mathbf{x})$, message $m \in \mathcal{M}_\lambda$, $\mathbf{y} \in \{0,1\}^\ell$ and $\mathsf{ct} \leftarrow \mathsf{Enc\text{-}SK}(\mathsf{msk}, m, \mathbf{y})$, if $f(\mathbf{x}, \mathbf{y}) = 1$ then $\mathsf{Dec}(\mathsf{sk}, \mathsf{ct}) = m$.

4.1   Security

Oracles. To begin, we define two oracles we use to enable the adversary to query for ciphertexts and secret keys. Let m be a message, and $\mathbf{x}, \mathbf{y}, \in \{0,1\}^\ell$.

- $\mathcal{O}_{\mathsf{msk}}^{\mathsf{sk}}(\mathbf{x}) \leftarrow \mathsf{KeyGen}(\mathsf{msk}, \mathbf{x})$.
- $\mathcal{O}_{\mathsf{msk}}^{\mathsf{ct}}(m, \mathbf{y}) \leftarrow \mathsf{Enc\text{-}SK}(\mathsf{msk}, m, \mathbf{y})$.

Experiments. We will now define three security properties that a mixed bit matching encryption scheme must satisfy. These definitions are similar to the indistinguishability-based data/function privacy definitions for attribute based encryption. For each of these experiments we restrict the adversary's queries to the ciphertext and secret key oracles to prevent trivial distinguishing strategies. Also, we will be considering selective definitions, since our constructions achieve selective security, and selective security suffices for our risky traitor tracing application. One could also consider full (adaptive) versions of these security definitions.

Definition 6. *A mixed bit matching encryption scheme* mBME = (Setup, KeyGen, Enc-PK, Enc-SK, Dec) *is said to satisfy pk-sk ciphertext indistinguishability if for any polynomial $\ell(\cdot)$ and stateful PPT adversary $\mathcal{A}$, there exists a negligible function negl$(\cdot)$ such that for all security parameters $\lambda \in \mathbb{N}$,* $\Pr[1 \leftarrow$ Expt-pk-sk-ct$^{\mathsf{mBME}}_{\ell(\lambda),\mathcal{A}}(1^\lambda)] \leq 1/2 + negl(\lambda)$, *where* Expt-pk-sk-ct *is defined in Fig. 2.*

Experiment Expt-pk-sk-ct$^{\mathsf{mBME}}_{\ell(\lambda),\mathcal{A}}(1^\lambda)$

- $(\mathsf{pk}, \mathsf{msk}) \leftarrow \mathsf{Setup}(1^\lambda, 1^{\ell(\lambda)})$
- $m \leftarrow \mathcal{A}^{\mathcal{O}^{\mathsf{sk}}_{\mathsf{msk}}, \mathcal{O}^{\mathsf{ct}}_{\mathsf{msk}}}(\mathsf{pk})$.
- $\mathsf{ct}_0 \leftarrow \mathsf{Enc}\text{-}\mathsf{SK}(\mathsf{msk}, m, 1^{\ell(\lambda)})$, $\mathsf{ct}_1 \leftarrow \mathsf{Enc}\text{-}\mathsf{PK}(\mathsf{pk}, m)$, $b \leftarrow \{0,1\}$
- $b' \leftarrow \mathcal{A}^{\mathcal{O}^{\mathsf{sk}}_{\mathsf{msk}}, \mathcal{O}^{\mathsf{ct}}_{\mathsf{msk}}}(\mathsf{ct}_b)$.
- Output 1 if $b = b'$, and 0 otherwise.

Fig. 2. Public-key vs secret-key ciphertext indistinguishability experiment

Definition 7. *A mixed bit matching encryption scheme* mBME = (Setup, KeyGen, Enc-PK, Enc-SK, Dec) *is said to satisfy selective ciphertext hiding if for any polynomial $\ell(\cdot)$ and stateful PPT adversary $\mathcal{A}$, there exists a negligible function negl$(\cdot)$ such that for all security parameters $\lambda \in \mathbb{N}$,* $\Pr[1 \leftarrow$ Expt-ct-ind$^{\mathsf{mBME}}_{\ell(\lambda),\mathcal{A}}(1^\lambda)] \leq 1/2 + negl(\lambda)$, *where* Expt-ct-ind *is defined in Fig. 3.*

Definition 8. *A mixed bit matching encryption scheme* mBME = (Setup, KeyGen, Enc-PK, Enc-SK, Dec) *is said to satisfy selective key hiding if for any polynomial $\ell(\cdot)$ and stateful PPT adversary $\mathcal{A}$, there exists a negligible function negl$(\cdot)$ such that for all security parameters $\lambda \in \mathbb{N}$,* $\Pr[1 \leftarrow$ Expt-key-ind$^{\mathsf{mBME}}_{\ell(\lambda),\mathcal{A}}(1^\lambda)] \leq 1/2 + negl(\lambda)$, *where* Expt-key-ind *is defined in Fig. 4.*

4.2 Simplified Ciphertext Hiding

As a tool for proving mixed bit matching encryption constructions secure, we define two simplified ciphertext hiding experiments, and then show that they imply the original (selective) ciphertext hiding security game.

Experiment Expt-ct-ind$_{\ell(\lambda),\mathcal{A}}^{\mathsf{mBME}}(1^\lambda)$

- $(\mathbf{y}_0, \mathbf{y}_1) \leftarrow \mathcal{A}(1^\lambda)$.
- $(\mathsf{pk}, \mathsf{msk}) \leftarrow \mathsf{Setup}(1^\lambda, 1^{\ell(\lambda)})$
- $(m_0, m_1) \leftarrow \mathcal{A}^{\mathcal{O}_{\mathsf{msk}}^{\mathsf{sk}}, \mathcal{O}_{\mathsf{msk}}^{\mathsf{ct}}}(\mathsf{pk})$
- $b \leftarrow \{0,1\}$, $\mathsf{ct}_b \leftarrow \mathsf{Enc\text{-}SK}(\mathsf{msk}, m_b, \mathbf{y}_b)$
- $b' \leftarrow \mathcal{A}^{\mathcal{O}_{\mathsf{msk}}^{\mathsf{sk}}, \mathcal{O}_{\mathsf{msk}}^{\mathsf{ct}}}(\mathsf{ct}_b)$
- Output 1 if $b = b'$, and 0 otherwise.

Adversarial Restrictions: For all queries $\mathbf{x}$ made by $\mathcal{A}$ to $\mathcal{O}_{\mathsf{msk}}^{\mathsf{sk}}$ the following conditions must hold:

- If $m_0 = m_1$, then $f(\mathbf{x}, \mathbf{y}_0) = f(\mathbf{x}, \mathbf{y}_1)$.
- If $m_0 \neq m_1$, then $f(\mathbf{x}, \mathbf{y}_0) = f(\mathbf{x}, \mathbf{y}_1) = 0$.

Fig. 3. Ciphertext hiding experiment

Experiment Expt-key-ind$_{\ell(\lambda),\mathcal{A}}^{\mathsf{mBME}}(1^\lambda)$

- $(\mathbf{x}_0, \mathbf{x}_1) \leftarrow \mathcal{A}(1^\lambda)$
- $(\mathsf{pk}, \mathsf{msk}) \leftarrow \mathsf{Setup}(1^\lambda, 1^{\ell(\lambda)})$
- $b \leftarrow \{0,1\}$, $\mathsf{sk}_b \leftarrow \mathsf{KeyGen}(\mathsf{msk}, \mathbf{x}_b)$
- $b' \leftarrow \mathcal{A}^{\mathcal{O}_{\mathsf{msk}}^{\mathsf{sk}}, \mathcal{O}_{\mathsf{msk}}^{\mathsf{ct}}}(\mathsf{pk}, \mathsf{sk}_b)$
- Output 1 if $b = b'$, and 0 otherwise.

Adversarial Restrictions: For all queries $(m, \mathbf{y})$ made by $\mathcal{A}$ to $\mathcal{O}_{\mathsf{msk}}^{\mathsf{ct}}$ the following equality must hold: $f(\mathbf{x}_0, \mathbf{y}) = f(\mathbf{x}_1, \mathbf{y})$.

Fig. 4. Key hiding experiment

Definition 9. *A mixed bit matching encryption scheme* $\mathsf{mBME} = (\mathsf{Setup}, \mathsf{KeyGen}, \mathsf{Enc\text{-}PK}, \mathsf{Enc\text{-}SK}, \mathsf{Dec})$ *is said to satisfy* selective 1-attribute ciphertext hiding *if for any polynomial* $\ell(\cdot)$ *and stateful PPT adversary* $\mathcal{A}$, *there exists a negligible function* $negl(\cdot)$ *such that for all security parameters* $\lambda \in \mathbb{N}$, $\Pr[1 \leftarrow \mathsf{Expt\text{-}1\text{-}attr\text{-}ct\text{-}ind}_{\ell(\lambda),\mathcal{A}}^{\mathsf{mBME}}(1^\lambda)] \leq 1/2 + negl(\lambda)$, *where* $\mathsf{Expt\text{-}1\text{-}attr\text{-}ct\text{-}ind}$ *is defined in Fig. 5.*

Definition 10. *A mixed bit matching encryption scheme* $\mathsf{mBME} = (\mathsf{Setup}, \mathsf{KeyGen}, \mathsf{Enc\text{-}PK}, \mathsf{Enc\text{-}SK}, \mathsf{Dec})$ *is said to satisfy* selective ciphertext indistinguishability under chosen attributes *if for any polynomial* $\ell(\cdot)$ *and stateful PPT adversary* $\mathcal{A}$, *there exists a negligible function* $negl(\cdot)$ *such that for all security parameters* $\lambda \in \mathbb{N}$, $\Pr[1 \leftarrow \mathsf{Expt\text{-}IND\text{-}CA}_{\ell(\lambda),\mathcal{A}}^{\mathsf{mBME}}(1^\lambda)] \leq 1/2 + negl(\lambda)$, *where* $\mathsf{Expt\text{-}IND\text{-}CA}$ *is defined in Fig. 6.*

Experiment Expt-1-attr-ct-ind$_{\ell(\lambda),\mathcal{A}}^{\text{mBME}}(1^\lambda)$

- $(\mathbf{y}_0, \mathbf{y}_1) \leftarrow \mathcal{A}(1^\lambda)$ where $y_{0,i} \neq y_{1,i}$ for at most one $i \in \{1, \ldots, \ell(\lambda)\}$.
- $(\text{pk}, \text{msk}) \leftarrow \text{Setup}(1^\lambda, 1^{\ell(\lambda)})$
- $m \leftarrow \mathcal{A}^{\mathcal{O}_{\text{msk}}^{\text{sk}}, \mathcal{O}_{\text{msk}}^{\text{ct}}}(\text{pk})$
- $b \leftarrow \{0, 1\}$, $\text{ct}_b \leftarrow \text{Enc-SK}(\text{msk}, m_b, \mathbf{y}_b)$.
- $b' \leftarrow \mathcal{A}^{\mathcal{O}_{\text{msk}}^{\text{sk}}, \mathcal{O}_{\text{msk}}^{\text{ct}}}(\text{pk}, \text{ct}_b)$
- Output 1 if $b = b'$, and 0 otherwise.

Adversarial Restrictions: For all queries $\mathbf{x}$ made by $\mathcal{A}$ to $\mathcal{O}_{\text{msk}}^{\text{sk}}$ the following equality must hold: $f(\mathbf{x}, \mathbf{y}_0) = f(\mathbf{x}, \mathbf{y}_1)$.

Fig. 5. 1-Attribute ciphertext hiding experiment

Experiment Expt-IND-CA$_{\ell(\lambda),\mathcal{A}}^{\text{mBME}}(1^\lambda)$

- $\mathbf{y} \leftarrow \mathcal{A}(1^\lambda)$.
- $(\text{pk}, \text{msk}) \leftarrow \text{Setup}(1^\lambda, 1^{\ell(\lambda)})$
- $(m_0, m_1) \leftarrow \mathcal{A}^{\mathcal{O}_{\text{msk}}^{\text{sk}}, \mathcal{O}_{\text{msk}}^{\text{ct}}}(\text{pk})$
- $b \leftarrow \{0, 1\}$, $\text{ct}_b \leftarrow \text{Enc-SK}(\text{msk}, m_b, \mathbf{y})$.
- $b' \leftarrow \mathcal{A}^{\mathcal{O}_{\text{msk}}^{\text{sk}}, \mathcal{O}_{\text{msk}}^{\text{ct}}}(\text{pk}, \text{ct}_b)$
- Output 1 if $b = b'$, and 0 otherwise.

Adversarial Restrictions: If $m_0 \neq m_1$, then for all queries $\mathbf{x}$ made by $\mathcal{A}$ to $\mathcal{O}_{\text{msk}}^{\text{sk}}$ the following equality must hold: $f(\mathbf{x}, \mathbf{y}) = 0$.

Fig. 6. Ciphertext indistinguishability under chosen attributes experiment

Theorem 1. *If a mixed bit matching encryption scheme* mBME = (Setup, KeyGen, Enc-PK, Enc-SK, Dec) *satisfies selective 1-attribute ciphertext hiding (Definition 9) and selective ciphertext indistinguishability under chosen attributes (Definition 10), then it also satisfies selective ciphertext hiding (Definition 7).*

The proof of above theorem is provided in the full version.

4.3 Simplified Key Hiding

We also define a similar simplified experiment for the key hiding security property.

Definition 11. *A mixed bit matching encryption scheme* mBME = (Setup, KeyGen, Enc-PK, Enc-SK, Dec) *is said to satisfy* selective 1-attribute key hiding *if for any polynomial ℓ and stateful PPT adversary $\mathcal{A}$, there exists a negligible function* negl($\cdot$) *such that for all security parameters* $\lambda \in \mathbb{N}$, Pr[1 $\leftarrow$ Expt-1-attr-key-ind$_{\ell(\lambda),\mathcal{A}}^{\text{mBME}}(1^\lambda)$] $\leq$ 1/2 + negl(λ), *where* Expt-1-attr-key-ind *is defined in Fig. 7.*

Theorem 2. *If a mixed bit matching encryption scheme* mBME = (Setup, KeyGen, Enc-PK, Enc-SK, Dec) *satisfies 1-attribute key hiding (Definition 11) then it satisfies key hiding (Definition 8).*

The proof of above theorem is provided in the full version.

Experiment Expt-1-attr-key-ind$_{\ell(\lambda),\mathcal{A}}^{\text{mBME}}(1^\lambda)$

- $(\mathbf{x}_0, \mathbf{x}_1) \leftarrow \mathcal{A}(1^\lambda)$ where $x_{0,i} \neq x_{1,i}$ for at most one i.
- $(\text{pk}, \text{msk}) \leftarrow$ Setup$(1^\lambda, 1^{\ell(\lambda)})$.
- $b \leftarrow \{0,1\}$, sk$_b \leftarrow$ KeyGen(msk, $\mathbf{x}_b$).
- $b' \leftarrow \mathcal{A}^{\mathcal{O}_{\text{msk}}^{\text{sk}}, \mathcal{O}_{\text{msk}}^{\text{ct}}}(\text{pk}, \text{sk}_b)$.
- Output 1 if $b = b'$.

Adversarial Restrictions: For all queries $(m, \mathbf{y})$ made by $\mathcal{A}$ to $\mathcal{O}_{\text{msk}}^{\text{ct}}$ the following equality must hold: $f(\mathbf{x}_0, \mathbf{y}) = f(\mathbf{x}_1, \mathbf{y})$.

Fig. 7. 1-Attribute key hiding experiment

5 Building Risky Traitor Tracing Using Mixed Bit Matching Encryption

In this section, we provide a generic construction for risky traitor tracing schemes from any mixed bit matching encryption scheme. Our transformation leads to a risky traitor tracing scheme with secret-key tracing. The *risky-ness* of the scheme will be $f = \frac{k}{n+k-1} - O\left(\frac{k(k-1)}{n^2}\right)$,[6] where k can be thought of as a scheme parameter fixed during setup, and the size of ciphertext will grow with k.

5.1 Construction

- Setup($1^\lambda, 1^n$): The setup algorithm chooses a key pair for mixed bit matching encryption system as (mbme.pk, mbme.msk) $\leftarrow$ mBME.Setup($1^\lambda, 1^{k+1}$). Next, it samples an index w as $w \leftarrow \{-k+2, -k+3, \ldots, n-1, n\}$, and sets vectors $\mathbf{x}_i$ for $i \in [n]$ as

$$\mathbf{x}_i = \begin{cases} 0^{k+1} & \text{if } i < w, \\ 0^{k-i+w}1^{i-w+1} & \text{if } w \leq i < w+k, \\ 1^{k+1} & \text{otherwise.} \end{cases}$$

---

[6] We want to point out that for $k = 1$ we get the tight *risky-ness*, i.e. prove that our scheme is $\frac{1}{n}$-risky secure.

It sets the master secret key as msk $=$ (mbme.msk, w), public key as pk $=$ mbme.pk, and computes the n user secret keys as $\mathsf{sk}_i \leftarrow$ mBME.KeyGen (mbme.msk, $\mathbf{x}_i$) for $i \in [n]$.

- Enc(pk, m): The encryption algorithm outputs the ciphertext ct as ct $\leftarrow$ mBME.Enc-PK(pk, m).
- Dec(sk, m): The decryption algorithm outputs the message m as $m = $ mBME.Dec(sk, ct).
- Trace^D(msk, 1^y, m_0, m_1): Let msk $=$ (mbme.msk, w). To define the trace algorithm, we first define a special index encryption algorithm Enc-ind which takes as input a master secret key msk, message m, and an index $i \in [k+1]$.

Enc-ind(msk, m, i): The index encryption algorithm outputs ct $\leftarrow$ mBME.Enc-SK(msk, m, $1^{k+1-i}0^i$).

Next, consider the Subtrace algorithm defined in Fig. 8. The sub-tracing algorithm simply tests whether the decoder box uses the key for user $i + w - 1$ where i is one of the inputs provided to Subtrace. Now the tracing algorithm simply runs the Subtrace algorithm for all indices $i \in [k]$, and for each index i where the Subtrace algorithm outputs 1, the tracing algorithm adds index $i + w - 1$ to the set of traitors. Concretely, the algorithm runs as follows:

- Let $S = \emptyset$. For $i = 1$ to k:
  * Compute $b \leftarrow$ Subtrace(mbme.msk, 1^y, m_0, m_1, i).
  * If $b = 1$, set $S := S \cup \{i + w - 1\}$.
- Output S.

Algorithm Subtrace(msk, 1^y, m_0, m_1, i)

Inputs: Key msk, parameter y, messages m_0, m_1, index i
Output: 0/1
Let $\epsilon = \lfloor 1/y \rfloor$. It sets $T = \lambda \cdot n/\epsilon$, and $\mathsf{count}_1 = \mathsf{count}_2 = 0$. For $j = 1$ to T, it computes the following:

1. It chooses $b_j \leftarrow \{0,1\}$ and computes $\mathsf{ct}_{j,1} \leftarrow$ Enc-ind(msk, m_{b_j}, i) and sends $\mathsf{ct}_{j,1}$ to D. If D outputs b_j, set $\mathsf{count}_1 = \mathsf{count}_1 + 1$, else set $\mathsf{count}_1 = \mathsf{count}_1 - 1$.
2. It chooses $c_j \leftarrow \{0,1\}$ and computes $\mathsf{ct}_{j,2} \leftarrow$ Enc-ind(msk, $m_{c_j}, i+ 1$) and sends $\mathsf{ct}_{j,2}$ to D. If D outputs c_j, set $\mathsf{count}_2 = \mathsf{count}_2 + 1$, else set $\mathsf{count}_2 = \mathsf{count}_2 - 1$.

If $\mathsf{count}_1 - \mathsf{count}_2 > T \cdot (\epsilon/4n)$, output 1, else output 0.

Fig. 8. Subtrace

Correctness. Since the encryption algorithm simply runs the public-key encryption algorithm for mixed bit matching encryption the correctness of above scheme follows directly from the correctness of the mixed bit matching encryption scheme.

Singular Trace Property. Note that if k is fixed to be 1, then our scheme satisfies the singular trace property as defined in Definition 5. This is because the trace algorithm will either output the fixed index w (chosen during setup), or output an empty set.

Due to space constraints, the proof of security is provided in the full version.

6 Construction: Mixed Bit Matching Encryption Scheme

Let Grp-Gen be an algorithm that takes as input security parameter 1^λ and outputs params $= (p, \mathbb{G}_1, \mathbb{G}_2, \mathbb{G}_T, e(\cdot, \cdot), g_1, g_2)$ where p is a λ bit prime, $\mathbb{G}_1, \mathbb{G}_2, \mathbb{G}_T$ are groups of order p, $e : \mathbb{G}_1 \times \mathbb{G}_2 \to \mathbb{G}_T$ is an efficiently computable non-degenerate bilinear map and g_1, g_2 are generators of $\mathbb{G}_1, \mathbb{G}_2$ respectively.

$(\mathsf{pk}, \mathsf{msk}) \leftarrow \mathsf{mBME.Setup}(1^\lambda, 1^\ell)$: The setup algorithm first chooses params $= (p, \mathbb{G}_1, \mathbb{G}_2, \mathbb{G}_T, e(\cdot, \cdot), g_1, g_2) \leftarrow \mathsf{Grp\text{-}Gen}(1^\lambda)$. It chooses $\alpha \leftarrow \mathbb{Z}_p$, $a_i \leftarrow \mathbb{Z}_p, b_i \leftarrow \mathbb{Z}_p, c_i \leftarrow \mathbb{Z}_p$ for each $i \in [\ell]$. The public key consists of params, $e(g_1, g_2)^\alpha$, $\prod_{i \in [\ell]} g_1^{a_i \cdot b_i + c_i}$ and $\{g_1^{a_i}\}_{i \in [\ell]}$, while the master secret key consists of $\left(\mathsf{params}, \alpha, \{a_i, b_i, c_i\}_{i \in \ell}\right)$.

$\mathsf{sk} \leftarrow \mathsf{mBME.KeyGen}(\mathbf{x}, \mathsf{msk})$: Let $\mathsf{msk} = \left(\mathsf{params}, \alpha, \{a_i, b_i, c_i\}_{i \in \ell}\right)$. The key generation algorithm first chooses $t \leftarrow \mathbb{Z}_p$ and $u_i \leftarrow \mathbb{Z}_p$ for each $i \in [\ell]$. It computes $K_0 = g_2^\alpha \cdot \left(\prod_{i \in [\ell]} g_2^{-t \cdot c_i}\right) \cdot \left(\prod_{i : x_i = 0} g_2^{-u_i \cdot a_i}\right)$. Next, it sets $K_1 = g_2^t$, and for each $i \in [\ell]$, $K_{2,i} = g^{-t \cdot b_i}$ if $x_i = 1$, else $K_{2,i} = g_2^{-t \cdot b_i + u_i}$. The key is $\left(K_0, K_1, \{K_{2,i}\}_{i \in [\ell]}\right)$.

$\mathsf{ct} \leftarrow \mathsf{mBME.Enc\text{-}SK}(m, \mathbf{y}, \mathsf{msk})$: Let $\mathsf{msk} = \left(\mathsf{params}, \alpha, \{a_i, b_i, c_i\}_{i \in \ell}\right)$. The secret key encryption algorithm first chooses $s \leftarrow \mathbb{Z}_p$, and for each $i \in [\ell]$ such that $y_i = 0$, it chooses $r_i \leftarrow \mathbb{Z}_p$. It sets $C = m \cdot e(g_1, g_2)^{\alpha \cdot s}$, $C_0 = g_1^s$, $C_1 = \left(\prod_{i : y_i = 1} g_1^{s \cdot (a_i \cdot b_i + c_i)}\right) \cdot \left(\prod_{i : y_i = 0} g_1^{s \cdot c_i + a_i \cdot b_i \cdot r_i}\right)$. For each $i \in [\ell]$, it sets $C_{2,i} = g_1^{a_i \cdot s}$ if $y_i = 1$, else $C_{2,i} = g_1^{a_i \cdot r_i}$ if $y_i = 0$. The ciphertext is $\left(C, C_0, C_1, \{C_{2,i}\}_{i \in [\ell]}\right)$.

$\mathsf{ct} \leftarrow \mathsf{mBME.Enc\text{-}PK}(m, \mathsf{pk})$: Let $\mathsf{pk} = (\mathsf{params}, e(g_1, g_2)^\alpha, \prod_{i \in \ell} g_1^{a_i \cdot b_i + c_i}, \{g_1^{a_i}\}_{i \in [\ell]})$. The public key encryption algorithm is identical to the secret key encryption algorithm. It first chooses $s \leftarrow \mathbb{Z}_p$. It sets $C = m \cdot e(g_1, g_2)^{\alpha \cdot s}$, $C_0 = g_1^s$, $C_1 = \left(\prod_{i \in \ell} g_1^{a_i \cdot b_i + c_i}\right)^s$. For each $i \in [\ell]$, it sets $C_{2,i} = (g_1^{a_i})^s$. The ciphertext is $\left(C, C_0, C_1, \{C_{2,i}\}_{i \in [\ell]}\right)$.

$z \leftarrow \mathsf{mBME.Dec}(\mathsf{ct}, \mathsf{sk})$: Let $\mathsf{ct} = \left(C, C_0, C_1, \{C_{2,i}\}_{i \in [\ell]}\right)$ and $\mathsf{sk} = \left(K_0, K_1, \{K_{2,i}\}_{i \in [\ell]}\right)$. The decryption algorithm outputs

$$\frac{C}{e(C_0, K_0) \cdot e(C_1, K_1) \cdot \prod_{i \in [\ell]} e(C_{2,i}, K_{2,i})}.$$

6.1 Correctness

Fix any security parameter λ, message m, vectors $\mathbf{x}, \mathbf{y}$ such that $f(\mathbf{x}, \mathbf{y}) = 1$ and public key $\mathsf{pk} = (\mathsf{params}, e(g_1, g_2)^\alpha, \prod_{i\in[\ell]} g_1^{a_i \cdot b_i + c_i}, \{g_1^{a_i}\}_{i\in[\ell]})$. Let $(s, \{r_i\}_{i:y_i=0})$ be the randomness used during encryption, $(t, \{u_i\}_{i:x_i=0})$ the randomness used during key generation, ciphertext $\mathsf{ct} = (C, C_0, C_1, \{C_{2,i}\}_{i\in[\ell]})$ and key $\mathsf{sk} = (K_0, K_1, \{K_{2,i}\}_{i\in[\ell]})$. To show that decryption works correctly, it suffices to show that $e(C_0, K_0) \cdot e(C_1, K_1) \cdot \left(\prod_{i\in[\ell]} e(C_{2,i}, K_{2,i})\right) = e(g_1, g_2)^{\alpha \cdot s}$.

$$
e(C_0, K_0) \cdot e(C_1, K_1) \cdot \left(\prod_{i\in[\ell]} e(C_{2,i}, K_{2,i})\right)
$$
$$
= \left(e(g_1, g_2)^{\alpha \cdot s - (\sum_i s \cdot t \cdot c_i) - (\sum_{i:x_i=0} s \cdot u_i \cdot a_i)}\right)
$$
$$
\cdot \left(e(g_1, g_2)^{(\sum_i s \cdot t \cdot c_i) + (\sum_{i:y_i=1} s \cdot t \cdot a_i \cdot b_i) + (\sum_{i:y_i=0} t \cdot a_i \cdot b_i \cdot r_i)}\right)
$$
$$
\cdot \left(e(g_1, g_2)^{-(\sum_{i:y_i=1} t \cdot s \cdot a_i \cdot b_i) - (\sum_{i:y_i=0} t \cdot a_i \cdot b_i \cdot r_i) + (\sum_{i:x_i=0} a_i \cdot s \cdot u_i)}\right)
$$

In the second step, we use the fact that since $f(\mathbf{x}, \mathbf{y}) = 1$, whenever $x_i = 0$, $y_i = 1$ (if this was not the case, then we would have, for all i such that $x_i = y_i = 0$, $e(g_1, g_2)^{u_i \cdot a_i \cdot r_i}$ terms in the product). Simplifying the expression, we get the desired product $e(g_1, g_2)^{\alpha \cdot s}$.

Due to space constraints, the proof of security is provided in the full version.

7 Performance Evaluation

We provide the performance evaluation of our risky traitor tracing scheme obtained by combining the mixed bit matching encryption scheme and the transformation to risky TT provided in Sects. 6 and 5, respectively. Our performance evaluation is based on concrete measurements made using the RELIC library [1] written in the C language.

We use the BN254 curve for pairings. It provides 126-bit security level [3]. All running times below were measured on a server with 2.93 GHz Intel Xeon CPU and 40 GB RAM. Averaged over 10000 iterations, the time taken to perform an exponentiation in the groups $\mathbb{G}_1$, $\mathbb{G}_2$ and $\mathbb{G}_T$ is approximately 0.28 ms, 1.60 ms and 0.90 ms, respectively. The time for perform a pairing operation is around 2.22 ms. The size of elements in group $\mathbb{G}_1$ is 96 bytes.

Based on the above measurements, for risky traitor tracing with parameter k we get the ciphertext size as $(96 \cdot k + 288)$ bytes, encryption time $(0.28 \cdot k + 1.74)$ ms, and decryption time $(2.226 \cdot k + 6.66)$ ms.[7] We point out in the above evaluations we consider the KEM version of our risky traitor tracing in which the message is encrypted using a symmetric key encryption with the hash of the

---

[7] In these estimations, we ignore the time to evaluate the hash function on the element in the target group $\mathbb{G}_T$ since it has an insiginicant effect on the running time.

first component of ciphertext $e(g_1, g_2)^{\alpha \cdot s}$ is used as the secret key. That is, the hashed value could be used as an AES key to perform message encryptions. For the basic setting of risky traitor tracing, i.e. $k = 1$, we get the ciphertext size, encryption time, and decryption time to be around 384 bytes, 2.16 ms, 8.89 ms (respectively).

8 Hardness of Differentially Private Sanitization

In this section, we show that the Dwork et al. [14] result works even if the traitor tracing scheme is f-risky secure. This, together with our risky TT constructions, results in a hardness result with query set size $2^{O(\lambda)}$ and based on assumptions over bilinear groups. First, we introduce some differential privacy related preliminaries following the notations from [23]. Next, we describe our hardness result.

8.1 Definitions

Differentially Private Algorithms. A database $D \in \mathcal{X}^n$ is a collection of n rows $x_1, \ldots, x_n$, where each row is an element of the date universe $\mathcal{X}$. We say that two databases $D, D' \in \mathcal{X}^*$ are adjacent, denoted by $D \sim D'$, if D' can be obtained from D by the addition, removal, or substitution of a single row (i.e., they differ only on a single row). Also, for any database $D \in \mathcal{X}^n$ and index $i \in \{1, 2, \ldots, n\}$, we use D_{-i} to denote a database where the ith element/row in D is set removed. At a very high level, an algorithm is said to be differentially private if its behavior on all adjacent databases is similar. The formal definition is provided below.

Definition 12 (Differential Privacy [13]). *Let $A : \mathcal{X}^n \to \mathcal{S}_n$ be a randomized algorithm that takes a database as input and outputs a summary. A is (ϵ, δ)-differentially private if for every pair of adjacent databases $D, D' \in \mathcal{X}^n$ and every subset $T \subseteq \mathcal{S}_n$,*

$$\Pr[A(D) \in T] \leq e^{\epsilon} \Pr[A(D') \in T] + \delta.$$

Here parameters ϵ and δ could be functions in n, the size of the database.

Accuracy of Sanitizers. Note that any algorithm A that always outputs a fixed symbol, say $\perp$, already satisfies Definition 12. Clearly such a summary will never be useful as the summary does not contain any information about the underlying database. Thus, we also need to specify what it means for the sanitizer to be useful. As described before, in this work we study the notion of differentially private sanitizers that give accurate answers to *statistical* queries.[8] A statistical query on data universe $\mathcal{X}$ is defined by a binary predicate $q : \mathcal{X} \to \{0, 1\}$. Let $\mathcal{Q} = \{q : \mathcal{X} \to [0, 1]\}$ be a set of statistical queries on the data universe $\mathcal{X}$. Given any $n \in \mathbb{N}$, database $D \in \mathcal{X}^n$ and query $q \in \mathcal{Q}$, let $q(D) = \dfrac{\sum_{x \in D} q(x)}{n}$.

---

[8] Statistical queries are also referred as counting queries, predicate queries, or linear queries in the literature.

Before we define accuracy, we would like to point out that the algorithm A might represent the summary s of a database D is any arbitrary form. Thus, to extract the answer to each query q from summary s, we require that there exists an evaluator $\mathsf{Eval} : S \times Q \to [0,1]$ that takes the summary and a query, and outputs an approximate answer to that query. As in prior works, we will abuse notation and simply write $q(s)$ to denote $\mathsf{Eval}(s,q)$, i.e. the algorithm's answer to query q. At a high level, an algorithm is said to be accurate if it answers every query to within some bounded error. The formal definition follows.

Definition 13 (Accuracy). *For a set Q of statistical queries on X, a database $D \in X^n$ and a summary $s \in S$, we say that s is α-accurate for Q on D if*

$$\forall q \in Q, |q(D) - q(s)| \leq \alpha.$$

A randomized algorithm $A : X^n \to S$ is said to be an (α, β)-accurate sanitizer if for every database $D \in X^n$,

$$\Pr_{A's\ coins}[A(D)\ is\ \alpha\text{-}accurate\ for\ Q\ on\ D] \geq 1 - \beta.$$

The parameters α and β could be functions in n, the size of the database.

Efficiency of Sanitizers. In this work, we are interested in asymptotic efficiency, thus we introduce a computation parameter $\lambda \in \mathbb{N}$. The data universe and query space, both will be parameterized by λ; that is, for every $\lambda \in \mathbb{N}$, we have a data universe X_λ and a query space Q_λ. The size of databases will be bounded by $n = n(\lambda)$, where $n(\cdot)$ is a polynomial. Now the algorithm A takes as input a database X_λ^n and output a summary in S_λ, where $\{S_\lambda\}_{\lambda \in \mathbb{N}}$ is a sequence of output ranges. And, there is an associated evaluator Eval that takes a query $q \in Q_\lambda$ and a summary $S \in S_\lambda$ and outputs a real-valued answer. The definitions of differential privacy and accuracy readily extend to such sequences.

Definition 14 (Efficiency). *A sanitizer A is efficient if, on input a database $D \in X_\lambda^n$, A runs in time $\mathsf{poly}(\lambda, \log(|X_\lambda|), \log(|Q_\lambda|))$, as well as on input a summary $s \in S_\lambda$ and query $q \in Q_\lambda$, the associated evaluator Eval runs in time $\mathsf{poly}(\lambda, \log(|X_\lambda|), \log(|Q_\lambda|))$.*

8.2 Hardness of Efficient Differentially Private Sanitization from Risky Traitor Tracing

In this section, we prove hardness of efficient differentially private sanitization from risky traitor tracing schemes. The proof is an adaptation of the proofs in [14, 23, 30] to this restricted notion. At a high level, the idea is to set the data universe to the secret key space and each query will be associated with a ciphertext such that answer to a query on any secret key will correspond to the output of decryption of associated ciphertext using the secret key. Now to show hardness of sanitization we will prove by contradiction. The main idea is that if there exists an efficient (accurate) sanitizer, then that could be successfully used as a pirate box in the traitor tracing scheme. Next, assuming that the sanitizer satisfies differential privacy, we can argue that the sanitizer could still be a useful pirate box even if one of keys in the database is deleted, however the tracing algorithm will still output the missing key as a traitor with non-negligible probability, thereby contradicting the property that the tracing algorithm incorrectly traces with only negligible probability.

Below we state the formal theorem. The proof of this theorem can be found in the full version. Later we also show to get a stronger hardness result if the underlying risky traitor tracing schemes also satisfies "singular trace" property (Definition 5).

Hardness from Risky Traitor Tracing

Theorem 3. *If there exists a f-risky secure private-key no-query traitor tracing scheme $\mathcal{T} = (\mathsf{Setup}, \mathsf{Enc}, \mathsf{Dec}, \mathsf{Trace})$, then there exists a data universe and query family $\{\mathcal{X}_\lambda, \mathcal{Q}_\lambda\}_\lambda$ such that there does not any sanitizer $A : \mathcal{X}_\lambda^n \to \mathcal{S}_\lambda$ that is simultaneously — (1) (ϵ, δ)-differentially private, (2) (α, β)-accurate for query space $\mathcal{Q}_\lambda$ on $\mathcal{X}_\lambda^n$, and (3) computationally efficient — for any $\epsilon = O(\log \lambda), \alpha < 1/2, \beta = o(1)$ and $\delta \leq f \cdot (1 - \beta)/4n$.*

Theorem 4. *If there exists a f-risky secure private-key no-query traitor tracing scheme $\mathcal{T} = (\mathsf{Setup}, \mathsf{Enc}, \mathsf{Dec}, \mathsf{Trace})$ satisfying singular trace property (Definition 5), then there exists a data universe and query family $\{\mathcal{X}_\lambda, \mathcal{Q}_\lambda\}_\lambda$ such that there does not any sanitizer $A : \mathcal{X}_\lambda^n \to \mathcal{S}_\lambda$ that is simultaneously — (1) (ϵ, δ)-differentially private, (2) (α, β)-accurate for query space $\mathcal{Q}_\lambda$ on $\mathcal{X}_\lambda^n$, and (3) computationally efficient — for any $\epsilon = O(\log \lambda), \alpha < 1/2, \beta = o(1)$ and $\delta \leq f \cdot (1 - \beta)/4$.*

Hardness from Assumptions over Bilinear Groups. Combining Theorem 4 with our risky TT scheme over prime order bilinear groups, we get the following corollary.

Corollary 1. *If Assumption 1 and Assumption 2 hold, then there exists a data universe and query family $\{\mathcal{X}_\lambda, \mathcal{Q}_\lambda\}_\lambda$ such that there does not any sanitizer $A : \mathcal{X}_\lambda^n \to \mathcal{S}_\lambda$ that is simultaneously — (1) (ϵ, δ)-differentially private, (2) (α, β)-accurate for query space $\mathcal{Q}_\lambda$ on $\mathcal{X}_\lambda^n$, and (3) computationally efficient — for any $\epsilon = O(\log \lambda), \alpha < 1/2, \beta = o(1)$ and $\delta \leq (1 - \beta)/4n$.*

Similary, combining Theorem 4 with our risky TT scheme over composite order bilinear groups, we get the following corollary.

Corollary 2. *Assuming subgroup decision and subgroup hiding in target group assumptions, there exists a data universe and query family $\{\mathcal{X}_\lambda, \mathcal{Q}_\lambda\}_\lambda$ such that there does not any sanitizer $A : \mathcal{X}_\lambda^n \to \mathcal{S}_\lambda$ that is simultaneously — (1) (ϵ, δ)-differentially private, (2) (α, β)-accurate for query space $\mathcal{Q}_\lambda$ on $\mathcal{X}_\lambda^n$, and (3) computationally efficient — for any $\epsilon = O(\log \lambda), \alpha < 1/2, \beta = o(1)$ and $\delta \leq (1 - \beta)/4n$.*

Acknowledgements. The fourth author is supported by NSF CNS-1414082, DARPA SafeWare, Microsoft Faculty Fellowship, and Packard Foundation Fellowship.

References

1. Relic toolkit. https://github.com/relic-toolkit/relic
2. Barak, B., et al.: On the (Im)possibility of obfuscating programs. In: Kilian, J. (ed.) CRYPTO 2001. LNCS, vol. 2139, pp. 1–18. Springer, Heidelberg (2001). https://doi.org/10.1007/3-540-44647-8_1
3. Beuchat, J.-L., González-Díaz, J.E., Mitsunari, S., Okamoto, E., Rodríguez-Henríquez, F., Teruya, T.: High-speed software implementation of the optimal ate pairing over Barreto–Naehrig curves. In: Joye, M., Miyaji, A., Otsuka, A. (eds.) Pairing 2010. LNCS, vol. 6487, pp. 21–39. Springer, Heidelberg (2010). https://doi.org/10.1007/978-3-642-17455-1_2
4. Billet, O., Phan, D.H.: Efficient traitor tracing from collusion secure codes. In: Safavi-Naini, R. (ed.) ICITS 2008. LNCS, vol. 5155, pp. 171–182. Springer, Heidelberg (2008). https://doi.org/10.1007/978-3-540-85093-9_17
5. Boneh, D., Franklin, M.: An efficient public key traitor tracing scheme. In: Wiener, M. (ed.) CRYPTO 1999. LNCS, vol. 1666, pp. 338–353. Springer, Heidelberg (1999). https://doi.org/10.1007/3-540-48405-1_22
6. Boneh, D., Naor, M.: Traitor tracing with constant size ciphertext. In: Proceedings of the 2008 ACM Conference on Computer and Communications Security, CCS 2008, Alexandria, Virginia, USA, 27–31 October 2008, pp. 501–510 (2008)
7. Boneh, D., Sahai, A., Waters, B.: Fully collusion resistant traitor tracing with short ciphertexts and private keys. In: Vaudenay, S. (ed.) EUROCRYPT 2006. LNCS, vol. 4004, pp. 573–592. Springer, Heidelberg (2006). https://doi.org/10.1007/11761679_34
8. Boneh, D., Waters, B.: A fully collusion resistant broadcast, trace, and revoke system. In: Proceedings of the 13th ACM Conference on Computer and Communications Security, pp. 211–220. ACM (2006)
9. Boneh, D., Zhandry, M.: Multiparty key exchange, efficient traitor tracing, and more from indistinguishability obfuscation. In: Garay, J.A., Gennaro, R. (eds.) CRYPTO 2014, Part I. LNCS, vol. 8616, pp. 480–499. Springer, Heidelberg (2014). https://doi.org/10.1007/978-3-662-44371-2_27
10. Chabanne, H., Phan, D.H., Pointcheval, D.: Public traceability in traitor tracing schemes. In: Cramer, R. (ed.) EUROCRYPT 2005. LNCS, vol. 3494, pp. 542–558. Springer, Heidelberg (2005). https://doi.org/10.1007/11426639_32
11. Chor, B., Fiat, A., Naor, M.: Tracing traitors. In: Desmedt, Y.G. (ed.) CRYPTO 1994. LNCS, vol. 839, pp. 257–270. Springer, Heidelberg (1994). https://doi.org/10.1007/3-540-48658-5_25
12. Chor, B., Fiat, A., Naor, M., Pinkas, B.: Tracing traitors. IEEE Trans. Inf. Theor. **46**(3), 893–910 (2000)
13. Dwork, C., McSherry, F., Nissim, K., Smith, A.: Calibrating noise to sensitivity in private data analysis. In: Halevi, S., Rabin, T. (eds.) TCC 2006. LNCS, vol. 3876, pp. 265–284. Springer, Heidelberg (2006). https://doi.org/10.1007/11681878_14
14. Dwork, C., Naor, M., Reingold, O., Rothblum, G.N., Vadhan, S.: On the complexity of differentially private data release: efficient algorithms and hardness results. In: Proceedings of the Forty-first Annual ACM Symposium on Theory of Computing, STOC 2009, pp. 381–390. ACM, New York (2009)

15. Freeman, D.M.: Converting pairing-based cryptosystems from composite-order groups to prime-order groups. In: Gilbert, H. (ed.) EUROCRYPT 2010. LNCS, vol. 6110, pp. 44–61. Springer, Heidelberg (2010). https://doi.org/10.1007/978-3-642-13190-5_3

16. Garg, S., Gentry, C., Halevi, S., Raykova, M., Sahai, A., Waters, B.: Candidate indistinguishability obfuscation and functional encryption for all circuits. In: FOCS (2013)

17. Garg, S., Kumarasubramanian, A., Sahai, A., Waters, B.: Building efficient fully collusion-resilient traitor tracing and revocation schemes. In: Proceedings of the 17th ACM Conference on Computer and Communications Security, CCS 2010, Chicago, Illinois, USA, 4–8 October 2010, pp. 121–130 (2010)

18. Goyal, R., Koppula, V., Waters, B.: Collusion Resistant Traitor Tracing from Learning with Errors (2018)

19. Kiayias, A., Pehlivanoglu, S.: Encryption for Digital Content. Advances in Information Security, vol. 52. Springer, Boston (2010). https://doi.org/10.1007/978-1-4419-0044-9

20. Kiayias, A., Yung, M.: Traitor tracing with constant transmission rate. In: Knudsen, L.R. (ed.) EUROCRYPT 2002. LNCS, vol. 2332, pp. 450–465. Springer, Heidelberg (2002). https://doi.org/10.1007/3-540-46035-7_30

21. Kiayias, A., Yung, M.: Breaking and repairing asymmetric public-key traitor tracing. In: Digital Rights Management: ACM CCS-9 Workshop, DRM 2002, Washington, DC, USA, 18 November 2002, Revised Papers (2003)

22. Kowalczyk, L., Malkin, T., Ullman, J., Wichs, D.: Hardness of non-interactive differential privacy from one-way functions. Cryptology ePrint Archive, Report 2017/1107 (2017). https://eprint.iacr.org/2017/1107

23. Kowalczyk, L., Malkin, T., Ullman, J., Zhandry, M.: Strong hardness of privacy from weak traitor tracing. In: Hirt, M., Smith, A. (eds.) TCC 2016. LNCS, vol. 9985, pp. 659–689. Springer, Heidelberg (2016). https://doi.org/10.1007/978-3-662-53641-4_25

24. Naor, M., Pinkas, B.: Threshold traitor tracing. In: Krawczyk, H. (ed.) CRYPTO 1998. LNCS, vol. 1462, pp. 502–517. Springer, Heidelberg (1998). https://doi.org/10.1007/BFb0055750

25. Naor, M., Pinkas, B.: Efficient trace and revoke schemes. In: Frankel, Y. (ed.) FC 2000. LNCS, vol. 1962, pp. 1–20. Springer, Heidelberg (2001). https://doi.org/10.1007/3-540-45472-1_1

26. Nishimaki, R., Wichs, D., Zhandry, M.: Anonymous traitor tracing: how to embed arbitrary information in a key. In: Fischlin, M., Coron, J.-S. (eds.) EUROCRYPT 2016, Part II. LNCS, vol. 9666, pp. 388–419. Springer, Heidelberg (2016). https://doi.org/10.1007/978-3-662-49896-5_14

27. Pfitzmann, B.: Trials of traced traitors. In: Anderson, R. (ed.) IH 1996. LNCS, vol. 1174, pp. 49–64. Springer, Heidelberg (1996). https://doi.org/10.1007/3-540-61996-8_31

28. Pfitzmann, B., Waidner, M.: Asymmetric fingerprinting for larger collusions. In: Proceedings of the 4th ACM Conference on Computer and Communications Security, pp. 151–160. ACM (1997)

29. Sirvent, T.: Traitor tracing scheme with constant ciphertext rate against powerful pirates. Cryptology ePrint Archive, Report 2006/383 (2006). http://eprint.iacr.org/2006/383

30. Ullman, J.: Answering $n_{\{2+o(1)\}}$ counting queries with differential privacy is hard. In: Symposium on Theory of Computing Conference, STOC 2013, Palo Alto, CA, USA, 1–4 June 2013, pp. 361–370 (2013)
31. Watanabe, Y., Hanaoka, G., Imai, H.: Efficient asymmetric public-key traitor tracing without trusted agents. In: Naccache, D. (ed.) CT-RSA 2001. LNCS, vol. 2020, pp. 392–407. Springer, Heidelberg (2001). https://doi.org/10.1007/3-540-45353-9_29

Secret Sharing

Non-malleable Secret Sharing for General Access Structures

Vipul Goyal[1] and Ashutosh Kumar[2]($\boxtimes$)

[1] CMU, Mount Pleasant, USA
goyal@cs.cmu.edu
[2] UCLA, Los Angeles, USA
a@ashutoshk.com

Abstract. Goyal and Kumar (STOC'18) recently introduced the notion of *non-malleable secret sharing*. Very roughly, the guarantee they seek is the following: the adversary may potentially tamper with all of the shares, and still, *either* the reconstruction procedure outputs the original secret, *or*, the original secret is "destroyed" and the reconstruction outputs a string which is completely "unrelated" to the original secret. Prior works on non-malleable codes in the 2 split-state model imply constructions which can be seen as 2-out-of-2 non-malleable secret sharing (NMSS) schemes. Goyal and Kumar proposed constructions of t-out-of-n NMSS schemes. These constructions have already been shown to have a number of applications in cryptography.

We continue this line of research and construct NMSS for more general access structures. We give a generic compiler that converts any statistical (resp. computational) secret sharing scheme realizing any access structure into another statistical (resp. computational) secret sharing scheme that not only realizes the same access structure but also ensures statistical non-malleability against a computationally unbounded adversary who tampers each of the shares arbitrarily and independently. Instantiating with known schemes we get unconditional NMMS schemes that realize any access structures generated by polynomial size monotone span programs. Similarly, we also obtain conditional NMMS schemes realizing access structure in **monotone P** (resp. **monotone NP**) assuming one-way functions (resp. witness encryption).

Towards considering more general tampering models, we also propose a construction of n-out-of-n NMSS. Our construction is secure even if the adversary could divide the shares into any two (possibly overlapping) subsets and then arbitrarily tamper the shares in each subset. Our construction is based on a property of inner product and an observation that the inner-product based construction of Aggarwal, Dodis and Lovett (STOC'14) is in fact secure against a tampering class that is stronger than 2 split-states. We also show applications of our construction to the problem of non-malleable message transmission.

© International Association for Cryptologic Research 2018
H. Shacham and A. Boldyreva (Eds.): CRYPTO 2018, LNCS 10991, pp. 501–530, 2018.
https://doi.org/10.1007/978-3-319-96884-1_17

1   Introduction

Secret sharing is a fundamental primitive in cryptography which allows a dealer to distribute shares of a secret among several parties, such that only authorized subsets of parties can recover the secret; the secret is "hidden" from all the unauthorized set of parties. Shamir [Sha79] and Blakley [Bla79] initiated the study of secret sharing by constructing threshold secret sharing schemes that only allows at least t-out-of-n parties to reconstruct the secret. A rich line of works have studied the construction of secret sharing schemes for more advanced access structures [KW93, Bei, Bei11, KNY14].

A number of works have studied the setting where the primary goal of the adversary is to instead *tamper* with the secret. This relates to the line of works on error detecting codes such as algebraic manipulation detection(AMD) codes [CDF+08], and, verifiable secret sharing [RBO89]. A more detailed overview of the related works can be found later in this section.

Non-malleable Secret Sharing. Very recently, Goyal and Kumar [GK18] initiated a systematic study of what they call *non-malleable secret sharing*. Very roughly, the guarantee is the following: the adversary may potentially tamper with all of the shares, and still, *either* the reconstruction procedure outputs the original secret, *or*, the original secret is "destroyed" and the reconstruction outputs a string which is completely "unrelated" to the original secret. This is a natural guarantee which is inspired by applications in cryptography.

As noted by [GK18], 2-out-of-2 non-malleable secret sharing (NMSS) is equivalent to non-malleable codes in the 2 split-state model. Constructing such split state non-malleable codes has proven to be surprisingly hard. Though a brilliant line of works [DPW10, LL12, DKO13, ADL14, CGL16, Li17], such 2-split-state codes have been constructed. However such an implication does not hold if the number of shares is more than 2. To see this, consider a (contrived) example of a 3 split-state non-malleable code where the encoding functions encodes the message using a 2 split-state non-malleable code to obtain the first two states and outputs the message (in the clear) in the third state. The decoding function simply ignores the third state and uses the first two states to decode the message. Such a construction is a valid 3 split-state non-malleable code that is not a 3-out-of-3 secret sharing scheme (in fact, it has no secrecy at all). Towards that end, Goyal and Kumar proposed a construction of t-out-of-n NMSS scheme where reconstruction could be done given *any* t shares, any set of less than t shares has no information about the original secret, and, non-malleability is guaranteed even if an adversary may tamper with each share.

Even though a relatively new primitive, non-malleable coding in the split state model (or 2-out-of-2 NMSS) has already found a number of applications in cryptography including in tamper-resilient cryptography [DPW10], designing multi-prover interactive proof systems [GJK15] and obtaining efficient encryption schemes [CDTV16]. Very recently, non-malleable codes in the split-state model were used as 2-out-of-2 non-malleable secret sharing scheme to obtain 3-round protocol for non-malleable commitments [GPR16].

Our Question. We study the following natural question in this work:

Can we get non-malleable secret sharing schemes for access structures beyond threshold?

As noted before, known results on split state non-malleable codes provide 2-out-of-2 NMSS. Goyal and Kumar [GK18] recently took a significant step forward by constructing t-out-of-n NMSS schemes. However to our knowledge, NMSS are not known access structures beyond threshold. For example, can we get NMSS schemes for access structures which can be represented using log depth circuits or polynomial sized boolean formulas? Can we get a NMSS for all of **monotone P**? Or even better, can we get a NMSS for all of **monotone NP**?

Existing Secret-Sharing Schemes. As noted by Goyal and Kumar, most of the secret sharing schemes known are linear [Bei, Chap. 4] and have nice algebraic and geometric properties, which are harnessed to obtain efficient sharing and reconstruction procedures. *Non-malleable secret sharing schemes on the other hand cannot be linear.* As the secret is a linear combination of the shares in a linear secret sharing scheme, the adversary can perform local operations on each of the shares and encode any linear function of the secret. Indeed, the malleability of linear secret sharing schemes, such as polynomials based Shamir's secret sharing scheme [Sha79], forms the basis of secure multi-party computation protocols [BOGW88]. For the purpose of constructing NMMS, any such alteration is an "attack" and the goal is to build secret sharing schemes that necessarily prohibit any such attacks.

1.1   Our Results

Generic Compiler for Individual Tampering. Recall that an access structure $\mathcal{A}$ is a monotone collection of subsets of parties (such that every subsets of parties in this set are authorized to reconstruct the secret; other subset of parties are unauthorized). Our first main result is the following:

Theorem 1 *(informal). For any access structure $\mathcal{A}$ that does not contain singletons[1], if there exists an efficient statistical (resp. computational) secret sharing scheme realizing access structure $\mathcal{A}$, then there exists an efficient statistical (resp. computational) secret sharing scheme realizing $\mathcal{A}$ that is statistically non-malleable against an adversary who tampers each of the shares arbitrarily and independently.*

Karchmer and Wigderson [KW93] gave an efficient[2] secret sharing scheme for access structures that can be described by a polynomial-size monotone *span program*. This is a general class for which efficient secret sharing schemes are known

---

[1] We note that this is a necessary assumption, as otherwise the notion of non malleability becomes meaningless. A single authorized party can recover the message and trivially encode any related message.

[2] A statistical secret sharing scheme is efficient if the sharing and reconstruction functions run in $\mathbf{poly}\big(n, k, \log(1/\epsilon)\big)$ time where k is the size of the message and $\epsilon > 0$ is the statistical error.

and includes undirected connectivity in a graph. Instantiating our compiler with their scheme, we obtain the following corollary.

Corollary 1 *(informal). For any access structure that can be described by a polynomial-size monotone span program and does not contain a singleton, there exists an efficient statistical secret sharing scheme that is statistically non-malleable against an adversary who arbitrarily tampers each of the shares independently.*

In an unpublished work (mentioned in [Bei11,KNY14]), Yao constructed an efficient computational secret-sharing scheme for access structures whose characteristic function are computable by monotone circuit of polynomial-size (assuming just one-way functions). Using this scheme, we get,

Corollary 2 *(informal). If one-way functions exist, then for any access structure $\mathcal{A}$ that does not contain singletons and is computable by monotone boolean circuits of polynomial size, there exists an efficient computational secret sharing scheme that realizes $\mathcal{A}$ and is statistically non-malleable against an adversary who arbitrarily tampers each of the shares independently.*

Observe that the secret sharing scheme resulting from the above theorem has *statistical* non-malleability (even though the secrecy is computational). Furthermore, Komargodski et al. [KNY14], constructed efficient computational secret sharing scheme for every **monotone NP** access structure assuming one way functions and witness-encryption for **NP** [GGSW13]. This gives us the following:

Corollary 3 *(informal). If one-way functions and witness-encryption for **NP** exist, then for every **monotone NP** access structure $\mathcal{A}$ that does not contain singletons and supports efficient membership queries, there exists an efficient computational secret sharing scheme that realizes $\mathcal{A}$ and is statistically non-malleable against an adversary who arbitrarily tampers each of the shares independently.*

We say that an access structure supports efficient membership queries, if it is possible to efficiently decide whether a given subset of parties is authorized or not. For t-out-of-n, this is trivial. Similarly, for access structures based on polynomial sized monotone boolean circuits, one can execute the corresponding circuit to decide whether the input subset is authorized or not.

Towards Stronger Tampering Models. In addition to the individual tampering model, Goyal and Kumar [GK18] also considered *joint* tampering where an adversary may divide the set of shares into two *disjoint* sets and may tamper with the shares in each set jointly. They additionally required the two subsets to have different cardinalities (i.e., both of them must not have equal number of shares). This holds even for the basic case of n-out-of-n secret sharing. We present a new construction of n-out-of-n NMSS against a significantly more general class of tampering functions. In particular, the adversary may partition the shares into any two (possibly overlapping) sets having up to $n-1$ shares. For example, the adversary may use the first $n-1$ shares to produce the tampered version of first $\frac{n}{2}$ shares, and uses the last $n-1$ shares to produce the last $\frac{n}{2}$ shares.

Theorem 2 *(informal). For any integer $n \geq 2$, there exists an efficient statistical secret sharing scheme that encodes a secret into n shares, allows for reconstruction of the secret only when all the n shares are available, and is also statistically non-malleable against an adversary who partitions the n shares into any two (possibly overlapping) non-empty subsets of its choice having up to $n-1$ shares each, and then, arbitrarily tampers the shares in each of the subsets (independently of the shares in the other subset).*

Our techniques in fact extend to allow the tampering of each share to depend on *all* the n shares in a limited way (see Sect. 4 for more details).

Ito et al. [ISN89] showed that every access structure has a (possibly inefficient) secret sharing scheme. In a manner similar to their construction, we can use the above n-out-of-n NMSS scheme for every minimal authorized set and obtain the following existential result.

Corollary 4. *For any access structure $\mathcal{A}$ that does not contain singletons, there exists a (possibly inefficient) statistical secret sharing scheme that realizes $\mathcal{A}$ and is statistically non-malleable against an adversary who chooses any minimal authorized set, partitions it into two subset and arbitrarily tampers shares in each of the subsets independently.*

Interesting Corollaries of Our Techniques. We observe that the inner-product construction of non-malleable codes of Aggarwal et al. [ADL14] can in fact withstand tampering which is stronger than 2 split state tampering.

Corollary 5 *(informal). The 2 split-state non-malleable code of Aggarwal et al. [ADL14] encodes a message as two vectors L and R of length λ over prime field Z_p. This scheme is even secure against an adversary*

$$\widetilde{L} \leftarrow f_1(L) \odot g_1(R)$$

$$\widetilde{R} \leftarrow f_2(L) \odot g_2(R)$$

where (f_1, f_2, g_1, g_2) are arbitrary tampering functions and $\odot$ represents coordinate-wise multiplication of two vectors (that is $L \odot R = (L_1 \times R_1, L_2 \times R_2, \ldots, L_\lambda \times R_\lambda))$.

Compared to leakage-resilient non-malleable codes where the tampering of the left share can depend on a bounded amount of information about the right share, in the above, *the tampered left share can be exactly equal to the right share.*

As an application of NMSS, [GK18] initiated the study of *non-malleable message transmission.* This guarantees that the receiver either receives the original message, or, the original message is essentially destroyed and the receiver receives an "unrelated" message, when the network is under the influence of an adversary who can execute arbitrary protocol on each of the nodes in the network (apart from the sender and the receiver). The adversary is even allowed to add a bounded number of arbitrary hidden links which it can use in addition to the original links for communicating amongst corrupt nodes.

Our techniques allow us to obtain a *strict improvement* over the results in [GK18]. In fact, our result is tight. We first informally define the notion of non-malleable paths. For a network represented by an undirected graph G, let G' be the induced subgraph of G with sender S and R removed. We define a collection of paths from S to R to be non-malleable if in the induced subgraph G' any node is reachable by nodes present on at most one of these paths.

Corollary 6. *In any network, with a designated sender S and receiver R, if there exists a collection of n non-malleable paths from S to R, then non-malleable secure message transmission protocol is possible with respect to an adversary which adds at most $n - 2$ arbitrary hidden links in the network and byzantinely corrupts all nodes other than S and R. Moreover, the bound of $n - 2$ is tight.*

1.2 Our Techniques

First we briefly recall the construction of t-out-of-n NMSS secure against an adversary which tampers each share independently [GK18].

Construction of [GK18]. Assume $t \geq 3$. First they encode the secret m using a 2 split-state non-malleable code to obtain $l, r \leftarrow \mathbf{NMEnc}(m)$. Then they share l using any t-out-of-n secret-sharing scheme to obtain $l_1, \ldots, l_n$, and, encode r using a 2-out-of-n *leakage-resilient* secret-sharing scheme to obtain $r_1, \ldots, r_n$. Final shares are of the form $share_i = (l_i, r_i)$. Given an adversary A who tampers with each share $share_i$ arbitrarily and independently, we would like to construct a split state adversary (f, g) against the underlying non-malleable code. A (somewhat oversimplified) high level structure of their proof is as follows:

1. Fix shares $l_1, \ldots, l_{t-1}$ independent of the secret m. This can be done since l is shared using a t-out-of-n secret-sharing and $t \geq 3$. Shares $l_1, \ldots, l_{t-1}$ are hardcoded in the description of f and g.
2. The function g gets r as input and must output $\tilde{r}$, the tampered version of r. Given r, g samples r_1, r_2 and hence now has $share_1 = (l_1, r_1)$ and $share_2 = (l_2, r_2)$ (since l_1 and l_2 are hardcoded). Use adversary A to compute $\widetilde{share_1}$ and $\widetilde{share_2}$, and hence, $\tilde{r}_1$ and $\tilde{r}_2$. Reconstruct $\tilde{r}$ using $\tilde{r}_1$ and $\tilde{r}_2$ (recall r was shared using a 2-out-of-n scheme) and output it.
3. The function f gets l as input and must output $\tilde{l}$. As the first step, f uses l to sample l_t *which is consistent with the fixed shares* $l_1, \ldots, l_{t-1}$. Next, f must run adversary A to compute tampered shares $\widetilde{share_1}, \ldots, \widetilde{share_t}$ which would allow for recovery of $\tilde{l}_1, \ldots, \tilde{l}_t$ and hence $\tilde{l}$. However note that f does not have $(r_1, \ldots, r_t)$ and therefore cannot even compute $share_1$. In fact, it cannot have any two shares of r, as the tampering function f needs to be

independent of r. Towards that end, [GK18] rely on the leakage resilience of the secret sharing scheme to compute $\tilde{l}_1, \ldots, \tilde{l}_t$.

Note that the above proof structure does not work when $t = 2$. For this case, they device a (completely separate) 2-out-of-n NMSS scheme by giving every pair an independent non-malleable encoding of the secret m.

Getting NMSS for General Access Structures. The natural starting point would be to replace the t-out-of-n secret sharing used to share l by the given secret sharing for the access structure in question. Instantiating this with various computational and information theoretic secret sharing schemes would presumably lead to NMSS for a variety of access structures including **monotone P**. However this idea fails because of the following two basic issues.

Firstly, we have to deal with authorized sets of size two ('pairs') in the given access structure (in case there are any). In case of [GK18], this was achieved by simply giving an entirely different construction (with a separate proof) for the case of $t = 2$. However in the setting of general access structures, the authorized set of size two may coexist with authorized sets of larger size. We solve this issue by efficiently constructing another access structure that has all authorized sets that contain an authorized subset of size two, in addition to the original access structure. Our hope would be to run NMSS for both these access structures in "parallel" for the same message. However this leads to additional difficulties in the proof of security related to composition: any authorized *pair* of parties will now have the same message encoded under two different schemes, and the split-state reduction to non-malleable codes fails.

Secondly, the construction in [GK18] heavily makes use of the fact that one can sample some of the shares without having knowledge of the secret at all. Then once the secret is available, you can "adjust" the remaining shares such that the resulting set of shares altogether is sampled from the correct distribution. As an example, see how the share l_t is sampled in step 3 (see the summary of [GK18] construction above). Indeed, such sampling is not just done once but at multiple steps in the [GK18] construction. In the computational case however, such an approach inherently breaks down. Since each share may have complete information about the secret (the secret may only be computationally hidden), one may not be able to sample a few shares independently of the secret and then "adjust" the rest so that overall, they come from the correct distribution. One could try to argue that even if the shares are sampled incorrectly, since the tampering function does not get all of them as input, it may anyway be indistinguishable to the tampering functions. However, such a guarantee is not sufficient for non-malleability. The tampering functions individually may not be able to distinguish correct shares from incorrect ones, and yet, the distribution of their joint output might change completely.

To solve these issues, we use two additional ideas to make our construction work.

1. Introduce "limited" information theoretic secrecy: We first compile the underlying statistical (resp. computational) secret sharing scheme into

another which additionally guarantees that any two shares hide the secret *information theoretically* (even if the secret sharing scheme was computational to begin with). This not only solves the first issue, but also paves a way to the solution of the second issue. For the first issue, this approach allows us to use non-malleable codes in a black-box way, as opposed to an alternative approach, where we could have strengthened the underlying split-state code to ensure non-malleability against "parallel" tamperings. For the second issue, we are now allowed to fix up to two shares of l even for computational schemes.

2. We use a secret sharing scheme with stronger leakage resilience properties: For any two secrets, suppose an adversary is given some valid shares of each of the secrets (potentially enough even to reconstruct the secret). Additionally, the adversary is given individual leakage from the rest of the shares of one secret. It should be statistically impossible for the adversary to identify whether the leakage corresponds to the first or the second secret. This property is significantly stronger than the one needed by Goyal and Kumar [GK18]. Unlike the proof of [GK18], this allows our reduction to generate t shares that are statistically quite far from any valid set of t shares, and still achieve statistical non-malleability.

Towards Stronger Tampering Models. Let us try to construct n-out-of-n secret sharing schemes that are non-malleable against an adversary that arbitrarily partitions the n shares into two non-empty subsets and jointly tampers the shares in each of the these subsets independently.

First Attempt. Let us try to use a 2 split-state non-malleable code that encodes the message into two parts, say l and r. We let l be the first share, and obtain the last $n - 1$ shares by secret sharing r using a traditional (n-1)-out-of-(n-1) secret sharing scheme. However, if the adversary tampers the first and last shares together, the tampered versions of last share (in particular r) may depend of the first share l and we will be not be able to obtain a split-state reduction to the underlying non-malleable code.

Second Attempt. What about a tree-based construction? Consider, for example, a complete binary tree with 2^k leaves corresponding to 2^k parties. To share a secret, we put the secret at the root of this tree, and encode it using a non-malleable code to obtain the value of nodes at level 1 (children of root). We can recursively apply this process using several non-malleable codes to obtain the value of all the 2^k leaves, and these values correspond to the shares of 2^k parties. While this seems like a promising approach, the share size increases exponentially with the depth of the tree (as constant rate statistical split-state non-malleable codes are not yet constructed). Even more fundamentally, it is not clear how to prove that such a construction is secure against arbitrary joint tampering. As a concrete example, consider a simple depth 2 tree having 4 leaves. Suppose adversary tampers the first and the last leaf together, and independently tampers the second and the third leaf. It seems that stronger notions of non-malleable codes (while maintaining constant rate) are needed. Moreover, it appears that different properties might be needed for different choices of partitioning.

Third Attempt. Can we extend the techniques of [GK18]? Unfortunately, when the two subsets are of equal cardinality, their technique of using different degree polynomials no longer seems to work.

Our Construction: We take a step back and construct n-out-of-n scheme in a manner similar to the first attempt described above. Recall that we were struck while trying to obtain a split-state reduction to the underlying non-malleable code. Nevertheless, we observe an underlying 'multiplicative structure' present in the code of Aggarwal et al. [ADL14] (hereby refered to as ADL construction) to achieve split-state reduction avoiding the problem mentioned in the first attempt.

We begin by recalling the elegant inner-product based ADL construction. They prove an amazing property of inner product, which roughly states that any independent tampering of left and right vector can be translated to an affine tampering of the output of inner product. This observation, reduces the problem to creating non-malleable codes against split-state arbitrary tampering functions to creating non-malleable codes against an affine function. To this end, they introduce affine evasive function, which ensures non-malleability against tampering by affine functions. Their proof relies on the linearity property satisfied by inner-product and is highly non-trivial relying on new results proved in additive combinatorics.

Given two equal length vectors over some finite field, the decoder of ADL computes inner-product and then applies the affine-evasive function to the output. Instead of viewing the first step as inner-product, we take a more fine-grained approach, and consider coordinate-wise multiplication of vectors to be the first step, followed by an addition of the coordinates. Our main observation is that the set of equal length vectors containing non-zero coordinates forms a finite abelian group under the operation of coordinate-wise multiplication of vectors. Next, we recall that Karnin et al. [KGH83] have shown how to use any abelian group to construct a n-out-of-n secret sharing scheme. The resulting scheme is quite simple, the reconstruction function will perform coordinate-wise multiplication of all the n vectors to obtain the secret vector, and we can proceed as in ADL, by computing sum of coordinates and then applying the affine evasive function to the sum.

We elaborated our scheme in the above fashion, instead of directly stating that we will use generalized inner-product instead of inner-product, because it is more insightful in conveying our proof ideas. In particular, we essentially use the associativity and commutativity of the mentioned abelian group (formed by coordinate-wise multiplication of non-zero field elements) to handle arbitrary partitions. Given any partitioning of n vectors into two subsets, we can use the commutativity of the abelian group to collect all the vectors of the first subset, and independently collect all the vectors of the second subset together. After which we can use the associativity of the same group to coordinate-wise multiply all the vectors in the first subset together, and independently coordinate-wise multiply all the vectors of the second subset. Notice, that now we are left with exactly two vectors corresponding to each of the two subsets, and we might be able to utilize the non-malleability of the ADL construction which works for two vectors. If we did not rely on this structure, we would have had to generalize the entire additive-combinatorics based proof of the ADL construction.

Paper Organization. We define various primitives in Sect. 2. We give our generic compiler in Sect. 3. We give the construction of n-out-of-n schemes supporting joint-tampering in Sect. 4.

Related Works. A number of works in the literature ensure that the correct secret is recovered even when some number of shares are arbitrarily corrupted. Concepts from error correcting codes have been useful in obtaining such schemes [Sha79, MS81]. In a seminal work [RBO89], Rabin and Ben-Or introduced verifiable secret sharing, which allowed the adversary to tamper almost half the shares, and still ensured that the adversary cannot cause the reconstruction procedure to output an incorrect message (except with exponentially small error probability). Cramer et al. [CDF+08], in a beautiful work introduced algebraic manipulation detection(AMD) codes and gave almost optimal constructions for them. These codes allow the adversary to "blindly" add any value to the codeword, and ensure that any such algebraic tampering will be detected with high probability. They used such codes to construct robust secret sharing schemes, which allowed adversary to tamper with any unauthorized subset of shares.

   As already noted, 2 split state non-malleable codes can be seen as 2-out-of-2 non-malleable secret sharing schemes in which both the shares can be independently tampered. Though a brilliant line of works, such split-state non-malleable codes have been constructed [DPW10, LL12, DKO13, ADL14, CGL16, Li17]. [GK18] construct t-out-of-n non-malleable secret sharing schemes.

2   Definitions

We use capital letters to denote distributions and their support, and corresponding small letters to denote a sample from the distribution. Let $[m]$ denote the set $\{1, 2, \ldots, m\}$, and U_r denote the uniform distribution over $\{0, 1\}^r$. Unless otherwise stated, $\mathbb{F}_p$ is a finite field of order prime (power) p. For any set $B \in [n]$, let $\otimes_{i \in B} S_i$ denote the Cartesian product $S_{i_1} \times S_{i_2} \times \ldots \times S_{i_{|B|}}$, where $i_1, i_2 \ldots i_{|B|}$ are ordered elements of B, such that $i_j < i_{j+1}$.

Definition 1 (min-entropy). *The min-entropy of a source X is defined as*

$$H_\infty(X) = \min_{x \in Support(X)} \left\{ \frac{1}{log(Pr[X = x])} \right\}$$

A (n, k)-source is a distribution on $\{0, 1\}^n$ with min-entropy k. A distribution D is flat if it is uniform over a set S.

**Definition 2 *(Statistical Distance).* ** *Let D_1 and D_2 be two distributions on a set S. The statistical distance between D_1 and D_2 is defined to be:*

$$|D_1 - D_2| = \max_{T \subseteq S} |D_1(T) - D_2(T)| = \frac{1}{2} \sum_{s \in S} |Pr[D_1 = s] - Pr[D_2 = s]|$$

We say D_1 is ϵ-close to D_2 if $|D_1 - D_2| \leq \epsilon$. Sometimes we represent the same using $D_1 \approx_\epsilon D_2$.

2.1 Non-malleable Codes

Definition 3 *(Coding Schemes)* ([ADL14]). *A coding scheme consists of two functions: an encoding function (possibly randomized) $Enc : \mathcal{M} \to \mathcal{C}$, and a deterministic decoding function $Dec : \mathcal{C} \to \mathcal{M} \cup \{\bot\}$ such that, for each $m \in \mathcal{M}$, $Pr(Dec(Enc(m)) = m) = 1$ (over the randomness of the encoding function).*

Definition 4 *(Non-Malleable Codes)* ([ADL14]). *Let $\mathcal{F}$ be some family of tampering functions. For each $f \in \mathcal{F}$, and $m \in \mathcal{M}$, define the tampering experiment*

$$\mathbf{Tamper}_m^f = \begin{Bmatrix} c \leftarrow Enc(m) \\ \tilde{c} \leftarrow f(c) \\ \tilde{m} \leftarrow Dec(\tilde{c}) \\ Output : \tilde{m} \end{Bmatrix}$$

*which is random variable over the randomness of the encoding function Enc. We say a coding scheme (Enc, Dec) is ϵ-**non-malleable** w.r.t $\mathcal{F}$ if for each $f \in \mathcal{F}$, there exists a distribution D^f (corresponding to the simulator) over $\mathcal{M} \cup \{same^*, \bot\}$ such that, for all $m \in \mathcal{M}$, we have that the statistical distance between $Tamper_m^f$ and*

$$\mathbf{Sim}_m^f = \begin{Bmatrix} \tilde{m} \leftarrow D^f \\ Output : m \, if \, \tilde{m} = same^*, or \, \tilde{m}, otherwise \end{Bmatrix}$$

is at most ϵ. Additionally, D^f should be efficiently samplable given oracle access to $f(.)$.

2.2 Secret Sharing Schemes

The following definition is inspired from the survey [Bei11].

Definition 5 *(Access Structure and Sharing function).* *A collection $\mathcal{A}$ is called monotone if $B \in \mathcal{A}$ and $B \subseteq C$, then $C \in \mathcal{A}$. Let $[n] = \{1, 2, \ldots, n\}$ be a set of identities of n parties. An **access structure** is a monotone collection $\mathcal{A} \subseteq 2^{\{1, \ldots, n\}}$ of non-empty subsets of $[n]$. Sets in $\mathcal{A}$ are called **authorized**, and sets not in $\mathcal{A}$ are called **unauthorized**.*

*For any access structure $\mathcal{A}$, we define **minimal basis access structure** of $\mathcal{A}$, denoted by $\mathcal{A}^{min}$, as the minimal subcollection of $\mathcal{A}$, such that for all* *authorized set $T \in \mathcal{A}$, there exists an authorized subset $B \subseteq T$ which is an element of $\mathcal{A}^{min}$.*

Let $\mathcal{M}$ be the domain of secrets. A **sharing function** Share is a randomized mapping from $\mathcal{M}$ to $S_1 \times S_2 \times \ldots \times S_n$, where S_i is called the domain of shares of party with identity j. A dealer distributes a secret $m \in \mathcal{M}$ by computing the vector $Share(m) = (s_1, \ldots, s_n)$, and privately communicating each share s_j to the party j. For a set $S \subseteq \{p_1, \ldots, p_n\}$, we denote $Share(m)_S$ to be a restriction of $Share(m)$ to its S entries.

Definition 6 *(Secret Sharing Scheme [Bei11]). Let $\mathcal{M}$ be a finite set of secrets, where $|\mathcal{M}| \geq 2$. A sharing function Share with domain of secrets $\mathcal{M}$ is a (n, ϵ)-**Secret Sharing Scheme** realizing an access structure $\mathcal{A}$ if the following two properties hold:*

1. **Correctness.** *The secret can be reconstructed by any authorized set of parties. That is, for any set $B \in \mathcal{A}$, where $B = \{i_1, \ldots, i_{|B|}\}$, there exists a deterministic reconstruction function $Rec^B : \otimes_{i \in B} S_i \to \mathcal{M}$ such that for every $m \in \mathcal{M}$,*

$$Pr[\mathbf{Rec}^B(\mathbf{Share}(m)_B) = m] = 1$$

   (over the randomness of the Sharing function)

2. **Statistical Privacy.** *Any collusion of unauthorized parties should have "almost" no information about the underlying secret. More formally, for any unauthorized set $T \notin \mathcal{A}$, and for every pair of secrets $a, b \in \mathcal{M}$, for any distinguisher D with output in $\{0, 1\}$, the following holds:*

$$|Pr_{shares \leftarrow Share(a)}[D(shares_T) = 1] - Pr_{shares \leftarrow Share(b)}[D(shares_T) = 1]| \leq \epsilon$$

   The special case of $\epsilon = 0$, is known as Perfect Privacy.

We use the definition of leakage-resilience from [GK18].

Definition 7 *(Leakage-Resilient Secret Sharing Schemes). Let $\mathcal{L}$ be some family of leakage functions. We say that the (n, ϵ)-secret sharing scheme, (Share, Rec), realizing access structure $\mathcal{A}$ is ϵ'-**leakage-resilient** w.r.t $\mathcal{L}$ if for each $f \in \mathcal{L}$, and for any two messages $a, b \in \mathcal{M}$, any distinguisher D with output in $\{0, 1\}$, the following holds:*

$$|Pr_{shares \leftarrow Share(a)}[D(f(shares)) = 1] - Pr_{shares \leftarrow Share(b)}[D(f(shares)) = 1]| \leq \epsilon'$$

We generalize the definition of non-malleable secret sharing schemes of [GK18] to general access structures.

Definition 8 *(Non-Malleable Secret Sharing Schemes). Let $\mathcal{A}$ be some access structure. Let $\mathcal{A}^{min}$ be its corresponding minimal basis access structure. Let $\mathcal{F}$ be some family of tampering functions. For each $f \in \mathcal{F}$, $m \in \mathcal{M}$ and authorized $T \in \mathcal{A}^{min}$, define the tampering experiment*

$$\mathbf{STamper}_m^{f,T} = \begin{Bmatrix} shares \leftarrow Share(m) \\ \widetilde{shares} \leftarrow f(shares) \\ \tilde{m} \leftarrow Rec(\widetilde{shares}_T) \\ Output : \tilde{m} \end{Bmatrix}$$

*which is a random variable over the randomness of the sharing function Share. We say that the (n, ϵ)-secret sharing scheme, (Share, Rec), realizing access structure $\mathcal{A}$ is ϵ'-**non-malleable** w.r.t $\mathcal{F}$ if for each $f \in \mathcal{F}$ and authorized $T \in \mathcal{A}^{min}$, there exists a distribution $SD^{f,T}$ (corresponding to the simulator) over $\mathcal{M} \cup \{same^*, \perp\}$ such that, for all $m \in \mathcal{M}$ and all authorized $T \in \mathcal{A}^{min}$, we have that the statistical distance between $STamper_m^{f,T}$ and*

$$\mathbf{SSim_m^{f,T}} = \left\{ \begin{array}{c} \tilde{m} \leftarrow SD^{f,T} \\ Output : m \ if \tilde{m} = same^*, or \ \tilde{m}, otherwise \end{array} \right\}$$

is at most ϵ'.

2.3 Threshold Access Structure $\mathcal{A}_n^t$

Apart from general access structure we will be interested in a special access structure which allows any t-out-of-n parties to pool their secret and reconstruct the secret. This threshold access structure can be formally represented as $\mathcal{A}_n^t = \{B \subseteq [n] : |B| \geq t\}$. We use the notation of (t, n, ϵ)-secret sharing sharing scheme for denoting (n, ϵ)-secret sharing scheme realizing access structure $\mathcal{A}_n^t$.

3 Non-malleable Secret Sharing Against Individual Tampering

In this section we show how to convert any secret sharing scheme into a non-malleable one against an adversary who arbitrarily tampers each of the shares independently. We begin recalling the tampering family from [GK18]:

Split-State Tampering Family $\mathcal{F}_n^{split}$
Let **Share** be a sharing function that takes as input a message $m \in \mathcal{M}$ and outputs a shares $shares \in \otimes_{i \in [n]} \mathcal{S}_i$. Parse the output $shares$ into n blocks, namely $share_1, share_2, \ldots, share_n$ where each $share_i \in \mathcal{S}_i$. For each $i \in [n]$, let $\mathbf{f_i} : \mathcal{S}_i \rightarrow \mathcal{S}_i$ be an arbitrary tampering function, that takes as input $share_i$, the ith share. Let $\mathcal{F}_n^{split}$ be a family of such n functions $(\mathbf{f_1}, \mathbf{f_2}, \ldots, \mathbf{f_n})$.

Note that above definition is written with respect to a sharing function. It is just for ease of presentation, we can use this family of tampering functions with respect to a coding scheme, by treating the encoding procedure as a sharing function. We also recall a lemma, which can be used to show that every 2 split-state non-malleable code is a 2-out-of-2 non-malleable secret sharing scheme.

Lemma 1 ([ADKO15]). *Let* **Enc** $: \mathcal{M} \rightarrow \mathcal{C}^2$ *be the encoding function, and* **Dec** $: \mathcal{C}^2 \rightarrow \mathcal{M} \cup \{\perp\}$ *be a deterministic decoding function. If a coding scheme* (**Enc**, **Dec**) *is ϵ-non-malleable w.r.t $\mathcal{F}_2^{split}$ then* (**Enc**, **Dec**) *is also a $(2, 2\epsilon)$-secret sharing scheme that is ϵ-non-malleable w.r.t $\mathcal{F}_2^{split}$, where* **Enc** *acts as a sharing function.*

Access Structures Based Definitions. As our building blocks, we will use secret-sharing schemes that allow any authorized "pair" to reconstruct the secret. We formally define such "paired" access structures below, and construct these schemes in the full version.

Definition 9 *(Paired Access Structures)*. *An access structure $\mathcal{A}$ is called a* ***paired access structure****, if each authorized set contains an authorized subset of size two. Formally, for all $B \in \mathcal{A}$, there exists a subset $C \subseteq B$ such that C is authorized and has cardinality two.*

Notice that, if $\mathcal{A}$ is a paired access structure then its corresponding minimal basis access structure $\mathcal{A}^{min}$ will only contain authorized sets of size two.

Definition 10 *(Authorized Paired Access Structures)*. *For any access structure $\mathcal{A}$, we call a paired access structure $\mathcal{A}_{pairs}$ an* ***authorized paired access structure*** *corresponding to $\mathcal{A}$ if $\mathcal{A}_{pairs}$ is the maximal subcollection of $\mathcal{A}$. Formally,*

$$\mathcal{A}_{pairs} = \{B \in \mathcal{A} : \exists C \subseteq B, (C \in \mathcal{A}) \wedge (|C| = 2)\}$$

Notice that $\mathcal{A}_{pairs}^{min}$ will be equal to the set of all the authorized sets of size two in $\mathcal{A}$.

Leakage Family. We also use a 2-out-of-n leakage-resilient secret sharing scheme. While in [GK18] split state family of leakage-resilience was needed, we require leakage-resilience against the following stronger leakage family.

Leakage Family $\mathcal{L}_{\mu}^{pair}$
Let $(\textbf{LRShare}, \textbf{LRRec})$ be any $(2, n, \epsilon)$-secret sharing scheme with message space $\mathcal{M}$. For any $i, j \in [n]$, for each $k \in [n] \setminus \{i, j\}$, let f_k be an arbitrary function that takes $share_i$ as input and outputs μ bits of information about its input. For any collection of such functions, any pair of message $a^0, a^1 \in \mathcal{M}$, any independently chosen bit $b \in \{0, 1\}$, we define the leakage experiment as,

$$\textbf{Leak}_b^{a^0, a^1} = \left\{ \begin{array}{c} a_1^0, \ldots, a_n^0 \leftarrow \textbf{LRShare}(a^0) \\ a_1^1, \ldots, a_n^1 \leftarrow \textbf{LRShare}(a^1) \\ Output : a_i^0, a_j^0, a_i^1, a_j^1, \otimes_{k \in [n] \setminus \{i,j\}} f_k(a_k^b) \end{array} \right\}$$

We say that the scheme $(\textbf{LRShare}, \textbf{LRRec})$ is ϵ-leakage-resilient w.r.t. $\mathcal{L}_{\mu}^{pair}$ if for every pair of message $a^0, a^1 \in \mathcal{M}$, we have that

$$\textbf{Leak}_0^{a^0, a^1} \approx_\epsilon \textbf{Leak}_1^{a^0, a^1}$$

In full version, we prove that the construction of [GK18] is in fact leakage-resilient against $\mathcal{L}_{\mu}^{pair}$.

Building Blocks. In our constructions for general access structure, we need a method to find a minimal authorized set, when given any authorized set. For any access structure $\mathcal{A}$ not containing singletons, we define a deterministic procedure **FindMinSet** : $\mathcal{A} \to \mathcal{A}^{min}$, which takes an authorized set and outputs a minimal authorized set contained in that set. The description follows:

Procedure *FindMinSet*$^{\mathcal{A}}(S)$

On input an authorized set S for an access structure $\mathcal{A}$, if there exists an $i \in S$ and $j \in S$ such that $i \neq j$ and $\{i, j\} \in \mathcal{A}$, then return the lexicographical smallest pair $\{i, j\}$ satisfying these conditions, otherwise initialize $T \gets D$ and execute the following loop: let T be an ordered set of t elements $i_1, i_2, \ldots, i_t$. For $j \in [t]$, check if $T \setminus \{i_j\}$ belongs to $\mathcal{A}$, in which case set $T \gets T \setminus \{i_j\}$ and go the beginning of the loop. If no such j exists, then break from the loop and output T.

The runtime of the above procedure is $O(n^2)$, because in each step of the loop it removes one element from the set T, whose size is upper bounded by the number of parties n. Note that, we assumed a membership query oracle, which decides whether the given set is authorized or not.

Pruning Compiler. As a building block towards our generic compiler, we need another compiler that given any statistical (resp. computational) secret sharing scheme realizing any access structure, outputs another secret sharing scheme that deauthorizes all authorized pairs while preserving the underlying statistical/computational secrecy. That is, it additionally guarantees that any two shares perfectly hide the secret.

Lemma 2. *For any efficient statistical (resp. computational) secret sharing scheme* (**AShare**, **ARec**) *realizing access structure $\mathcal{A}$ that does not contain singletons, there exists another efficient statistical (resp. computational) secret sharing scheme* (**APShare**, **APRec**) *which satisfies the following properties.*

1. (**APShare**, **APRec**) *realizes the access structure $\mathcal{A}$ with authorized pairs removed. The statistical error remains the same if the input is a statistical scheme.*
2. (**APShare**, **APRec**) *ensures that given any two shares, the secret is perfectly hidden.*

Proof. We give the construction of (**APShare**, **APRec**), deferring the proof to full version. Let n be the number of parties, and $\mathbb{F}$ be the secret space. Let **AShare** share an element of $\mathbb{F}$ into n elements of field $\mathbb{F}_1$. Let (**TShare**$_n^3$, **TRec**$_n^3$) and (**TShare**$_2^2$, **TRec**$_2^2$) be two threshold secret sharing scheme instantiated with Shamir's Secret Sharing scheme [Sha79] mapping an element of $\mathbb{F}_1$ into shares in $\mathbb{F}_1$ having threshold 2 and 3 respectively.

- **Sharing function APShare.** On input $m \in F$, share m using **AShare** to obtain $m_1, \ldots, m_n \gets$ **AShare**(m). For each $i \in [n]$, share m_i using **TShare**$_2^2$ to obtain $l_i, r_i \gets$ **TShare**$_2^2(m_i)$ and share r_i using **TShare**$_n^3$ to obtain $r_i^1, \ldots, r_i^n \gets$ **TShare**$_n^3(r_i)$. For each $i \in [n]$ construct $share_i$ as $l_i, r_1^i, \ldots, r_n^i$. Output $share_1, \ldots, share_n$.

- **Reconstruction Function APRec.** On input the shares $\otimes_{i \in T} share_i$ corresponding to authorized set $T \in \mathcal{A}$ with $|T| \geq 3$, for each $i \in T$, parse $share_i$ as $l_i, r_1^i, \ldots, r_n^i$. For each $i \in [n]$, reconstruct $r_i \leftarrow \mathbf{TRec_n^3}(\otimes_{i \in T} r_i)$. For each $i \in T$, reconstruct $m_i \leftarrow \mathbf{TRec_2^2}(l_i, r_i)$. Reconstruct $m \leftarrow \mathbf{ARec}(\otimes_{i \in T} m_i)$. Output m.

As our compiler also works with computational schemes, we first define them. Please refer to the book by Goldreich [Gol07] for definition of computational indistinguishability.

Definition 11 *(Computational Secret Sharing). Let $\mathcal{M}$ be a finite set of secrets, where $|\mathcal{M}| \geq 2$. An efficient sharing function Share with domain of secrets $\mathcal{M}$ is a **Computational Secret Sharing Scheme** realizing an access structure $\mathcal{A}$ if the following two properties hold:*

1. **Correctness.** *The secret m can be reconstructed by any authorized set of parties. That is, for any set $B \in \mathcal{A}$(where $B = \{p_{i_1}, \ldots, p_{i_{|B|}}\}$), there exists an efficient deterministic reconstruction function $Reconstruct^B : S_{i_1} \times S_{i_2} \times \ldots S_{i_{|B|}} \rightarrow \mathcal{M}$ such that for every $m \in \mathcal{M}$,*

$$Pr[Reconstruct^B(Share(m)_B) = m] = 1$$

   (over the randomness of the Sharing function)

2. **Computational Privacy.** *An unauthorized set of parties should be unable to distinguish whether the hidden secret is m_0 or m_1 for all $m_0, m_1 \in \mathcal{M}$. More formally, for any set $T \notin \mathcal{A}$, for every two secrets $a, b \in \mathcal{M}$, any PPT adversary should not be able to distinguish between,*

$$Share(a)_T \approx Share(b)_T$$

where the two distributions are computationally indistinguishable.

Main Result for General Access Structures. We are now in position to give our main result.

Theorem 3. *For any number of parties n, and any access structure $\mathcal{A}$ that does not contain singletons. If we have the following primitives:*

1. *For any $\epsilon_1 \geq 0$, let $(\mathbf{NMEnc}, \mathbf{NMDec})$ be any coding scheme that is ϵ_1-non-malleable wrt $\mathcal{F}_2^{split}$, which encodes an element of the set $\mathbb{F}_0$ into two elements of the field $\mathbb{F}_1$.*
2. *For any $\epsilon_2 \geq 0$, let $(\mathbf{AShare}, \mathbf{ARec})$ be any (n, ϵ_2)-secret sharing scheme (resp. computational) realizing access structure $\mathcal{A}$, which shares an element of field $\mathbb{F}_1$ into n elements of the field $\mathbb{F}_2$.*
3. *Let $\mu \leftarrow \log|\mathbb{F}_2|$. For any $\epsilon_3 \geq 0$, let $(\mathbf{LRShare}, \mathbf{LRRec})$, be any $(2, n, \epsilon_3)$-secret sharing scheme that is ϵ_3-leakage-resilient w.r.t. $\mathcal{L}_\mu^{pair}$, which shares an element of the field $\mathbb{F}_1$ into n elements of the field $\mathbb{F}_3$.*

4. *For any $\epsilon_4 \geq 0$, let* (**PNMShare, PNMRec**), *be any* (n, ϵ_4)-*secret sharing scheme realizing the authorized paired access structure* $\mathcal{A}_{pairs}$ *that is* ϵ_4-*non-malleable wrt* $\mathcal{F}_n^{split}$, *which shares an element of the set* $\mathbb{F}_0$ *into* n *elements of the field* $\mathbb{F}_4$.

then there exists $(n, 2\epsilon_1 + \epsilon_2 + \epsilon_4)$-*secret sharing scheme (resp. computational) realizing access structure* $\mathcal{A}$ *that is* $(2\epsilon_1 + \epsilon_2 + \epsilon_3 + \epsilon_4)$-*non-malleable w.r.t* $\mathcal{F}_n^{split}$. *The resulting scheme,* (**NMShare, NMRec**), *shares an element of the set* $\mathbb{F}_0$ *into* n *shares where each share is an element of* $(\mathbb{F}_2 \times \mathbb{F}_3 \times \mathbb{F}_4)$. *Further, if the four primitives have efficient construction (polynomial time sharing and reconstruction functions), then the constructed scheme is also efficient.*

Proof. We begin with the construction of the desired non-malleable secret sharing scheme. Apply Lemma 2 to the computational secret sharing scheme (**AShare, ARec**) to obtain a pruned secret sharing scheme (**APShare, APRec**).

- **Sharing function NMShare**: Encode the secret input $m \in \mathbb{F}_1$ using the encoding function of the non-malleable code. Let $l, r \leftarrow$ **NMEnc**(m). Share l using a **APShare** to obtain $l_1, \ldots, l_n \leftarrow$ **APShare**(l). Share r using a 2-out-of-n leakage-resilient secret sharing scheme. Let $r_1, \ldots, r_n \leftarrow$ **LRRec**(r). Use the sharing procedure **PNMShare** to share m. Let $(p_1, \ldots, p_n) \leftarrow$ **PNMShare**$(\mathbf{m})$. Then for each $i \in [n]$, construct $share_i$ as l_i, r_i, p_i.
- **Reconstruction function NMRec**: On input the shares $\otimes_{i \in D} share_i$ corresponding to authorized set D, for each $i \in D$, parse $share_i$ as (l_i, r_i, p_i). Find the minimal authorized set $T \in \mathcal{A}^{min}$ by running the procedure **findMinSet** with input D. Let T be a set containing t indices $\{i_1, i_2, \ldots, i_t\}$ such that $i_j < i_{j+1}$ for each $j \in [t-1]$. If $D \in \mathcal{A}_{pairs}$, use the decoding procedure **PNMRec**$^{\{i_1, i_2\}}$ to obtain the hidden secret $m \leftarrow$ **PNMRec**$^{\{i_1, i_2\}}(p_{i_1}, p_{i_2})$. Otherwise, run the reconstruction procedure **APRec** on t shares of l, to obtain $l \leftarrow$ **APRec**$(\otimes_{i \in T} l_i)$. Run the reconstruction procedure of the leakage-resilient secret sharing scheme on the first 2 shares of r, to obtain $r \leftarrow$ **LRRec**$^{\{i_1, i_2\}}(r_{i_1}, r_{i_2})$. Decode l and r using decoding process of underlying non-malleable code to obtain: $m \leftarrow$ **NMDec**$(\mathbf{l}, \mathbf{r})$. Output m.

Correctness and Efficiency: Trivially follows from the construction.

Statistical (resp. Computational Privacy): We prove statistical privacy using hybrid argument. For ease of understanding, let $share_i$ be of the form al_i, ar_i, ap_i when the secret a is encoded by the sharing procedure **NMShare**. Similarly, let $share_i$ be of the form bl_i, br_i, bp_i when the secret b is encoded. Let T be an unauthorized set containing t indices $\{i_1, i_2, \ldots, i_t\}$ such that $i_j < i_{j+1}$ for each $j \in [t-1]$. We describe the hybrids below:

1. **Hybrid$_1$**: for each $i \in T$, $share_i$ is of the form al_i, ar_i, ap_i. The distribution of these t shares is identical to distribution obtained on running the **NMShare** on input a. Output $\otimes_{i \in T} share_i$.

2. **Hybrid$_2$**: Sample the shares as in **Hybrid$_1$**, the previous hybrid. For each $i \in T$, replace al_i with bl_i to obtain share of the form bl_i, ar_i, ap_i. Output $\otimes_{i \in T} share_i$.
3. **Hybrid$_3$**: Sample the shares as in **Hybrid$_2$**, the previous hybrid. For each $i \in T$, replace ar_i with br_i to obtain share of the form bl_i, br_i, ap_i. Output $\otimes_{i \in T} share_i$.
4. **Hybrid$_4$**: Sample the shares as in **Hybrid$_3$**, the previous hybrid. For each $i \in T$, replace ap_i with bp_i to obtain share of the form bl_i, br_i, bp_i. Output $\otimes_{i \in T} share_i$. The distribution of these t shares is identical to distribution obtained on running the **NMShare** on input b. Output $\otimes_{i \in T} share_i$.

<u>Claim:</u> For any pair of secrets $a, b \in F_0$, any unauthorized $T \notin \mathcal{A}$, the statistical distance between **Hybrid$_1$** and **Hybrid$_2$** is at most ϵ_2 (resp. **Hybrid$_1$** and **Hybrid$_2$** are computationally indistinguishable).
<u>Proof:</u> The two hybrids only differ in the shares of l. As T is unauthorized in $\mathcal{A}$, the claim follows from the statistical (resp. computational) privacy of the secret scheme (**AShare, ARec**). ∎

<u>Claim:</u> For any pair of secrets $a, b \in F_0$, any unauthorized $T \notin \mathcal{A}$, the statistical distance between **Hybrid$_2$** and **Hybrid$_3$** is at most $2\epsilon_1$.
<u>Proof:</u> As in [GK18], the two hybrids are statistically indistinguishable by the $(2, 2\epsilon_1)$-secrecy satisfied by the non-malleable code (**NMEnc, NMDec**) (as in Lemma 1), by utilizing that fact knowing only r reveals nothing about the underlying message m. ∎

<u>Claim:</u> For any pair of secrets $a, b \in F_0$, any unauthorized $T \notin \mathcal{A}$, the statistical distance between **Hybrid$_3$** and **Hybrid$_4$** is at most ϵ_4.
<u>Proof:</u> $T \notin \mathcal{A}$, implies that $T \notin \mathcal{A}_{pairs}$. The two hybrids only differ in the shares corresponding to output of **PNMShare**. The claim follows from the statistical privacy of (**PNMShare, PNMRec**). ∎

By repeated application of triangle inequality, we get that for any $a, b \in F_0$, any unauthorized $T \notin \mathcal{A}$, the statistical distance between **Hybrid$_1$** and **Hybrid$_4$** is at most $2\epsilon_1 + \epsilon_2 + \epsilon_4$ (resp. the hybrids **Hybrid$_1$** and **Hybrid$_4$** are computationally indistinguishable). This proves the statistical (resp. computational) privacy of our scheme.

Statistical Non Malleability: To prove non-malleability of the current secret sharing scheme, we give a simulator for every admissible tampering attack on our scheme by using the simulator of the underlying non-malleable code after we have given an equivalent split-state tampering attack.

Let us begin with the intuition for the procedure **FindMinSet**. Notice that for general access structures, it is possible that the given authorized set has an authorized subset of size two, and another disjoint (minimal) authorized set of size three. Moreover, in our construction different schemes are being used to encode for these subsets. In case our output depends on all these five shares, we cannot hope to achieve a reduction to the underlying non-malleable code (because by definition, non-malleability holds only when the adversary is given one encoding of the message, and it tampers to produce only one encoding. In the present case it gets two encodings of the same message). We solve such an issue by giving the procedure **FindMinSet** in Subsect. 3, which prunes the given authorized set efficiently and ensures that no proper subset of the output (minimal) authorized set is authorized. It is easy to see that this procedure needs to be deterministic for us to be able to argue that share reconstructed in real experiment is equal to the one in reduction. Given this observation, without loss of generality we can assume that adversary chooses an authorized set $T \in \mathcal{A}^{min}$ to be used for reconstruction of the secret, as otherwise we can use the function **FindMinSet** to compute $T \in \mathcal{A}^{min}$ from any $D \in \mathcal{A}$. As the adversary belongs to $\mathcal{F}_n^{split}$, it also specifies a set of n tampering functions $\{\mathbf{f_i} : i \in [n]\}$. All these functions act on their respective shares independently of the other shares, i.e. every $\mathbf{f_i}$ takes $share_i$ as input and outputs the tampered $\widetilde{share}_i$. We can also assume without loss of generality that all these tampering functions are deterministic, as the computationally unbounded adversary can compute the optimal randomness. Unlike [GK18], depending on the cardinality of T, we use these tampering functions to create explicit split-state function to tamper with either non-malleable code or paired non-malleable secret-sharing.

CASE 1 ($|T| = 2$)
Let i_1 and i_2 be the two indices of T such that $i_1 < i_2$. In this case, we use the tampering functions $\mathbf{f}_{i_1}$ and $\mathbf{f}_{i_2}$ for the scheme (**NMShare, NMRec**) to create explicit tampering functions $\mathbf{F}_{i_1}$ and $\mathbf{F}_{i_2}$ for the underlying scheme (**PNMShare, PNMRec**). The reduction is described below:

1. (**Initial Setup**): Randomly choose a message $m_\$ \in \mathcal{M}$, and run the sharing function **NMShare** with input $m_\$$ to obtain temporary shares. That is, $(tShare_1, \ldots, tShare_n) \leftarrow \mathbf{NMShare}(m_\$)$. For each $i \in [n]$, parse $tShare_i$ as tl_i, tr_i, tp_i.
2. The **tampering function** $\mathbf{F}_{i_1}$ is defined as follows: On input $p_{i_1} \in \mathbb{F}_4$, replace tp_{i_1} by p_{i_1} in $tShare_{i_1}$ to obtain $share_{i_1}$. Run $\mathbf{f}_{i_1}$ on $share_{i_1}$ to obtain $\widetilde{share}_{i_1}$. Parse $\widetilde{share}_{i_1}$ as $\widetilde{l}_{i_1}, \widetilde{r}_{i_1}, \widetilde{p}_{i_1}$. Output $\widetilde{p}_{i_1}$.
3. The **tampering function** $\mathbf{F}_{i_2}$ is defined as follows: On input $p_{i_2} \in \mathbb{F}_4$, replace tp_{i_2} by p_{i_2} in $tShare_{i_2}$ to obtain $share_{i_2}$. Run $\mathbf{f}_{i_2}$ on $share_{i_2}$ to obtain $\widetilde{share}_{i_2}$. Parse $\widetilde{share}_{i_2}$ as $\widetilde{l}_{i_2}, \widetilde{r}_{i_2}, \widetilde{p}_{i_2}$. Output $\widetilde{p}_{i_2}$.

The functions $\mathbf{F}_{i_1}$ and $\mathbf{F}_{i_2}$ have been defined in this way to ensure that the secret hidden by the shares l_{i_1} and l_{i_2} of the scheme (**PNMShare, PNMRec**) is the same as the secret hidden by $share_{i_1}$ and $share_{i_2}$ of the scheme (**NMShare, NMRec**). We also need to argue that the reduction generates $share_{i_1}$ and $share_2$ from the right distribution, as otherwise the functions $\mathbf{f}_{i_1}$ and $\mathbf{f}_{i_2}$ may detect the change in distribution and stop working. Similar to the proof of statistical privacy, we can use hybrid argument to show that, for any p_{i_1} and p_{i_2} encoding message $m \leftarrow \mathbf{PNMRec}^{\{i_1, i_2\}}(p_{i_1}, p_{i_2})$, the statistical distance between the distribution of $share_{i_1}, share_{i_2}$ generated while executing **NMShare**(m) and the two shares generated by the reduction is at most $2\epsilon_1$. We rely on 2-out-of-2 secrecy property satisfied by non-malleable codes to show that even after learning r from the two shares, we learn nothing about the underlying secret. We also relied on the fact that two shares of l reveal nothing about l by the property of the pruning compiler (as in Lemma 2). Note that here we relied on the pruning compiler to ensure that any authorized pair will only get the encoding of the message under the pair-wise scheme (**PNMShare, PNMRec**) and not the other scheme.

For all $i \in [n] \setminus \{i_1, i_2\}$, let $\mathbf{F}_i$ be the identity function. The created set of functions $\{F_i : i \in [n]\}$ belongs to $\mathcal{F}_n^{split}$. Therefore, the tampering experiments of the two non-malleable secret-sharing scheme (see Definition 8) are statistically indistinguishable, specifically,

$$\mathbf{STamper}_{\mathbf{m}}^{\mathbf{f,T}} \approx_{2\epsilon_1} \mathbf{STamper}_{\mathbf{m}}^{\mathbf{F,T}}$$

By the ϵ_4-non malleability of the scheme (**PNMShare, PNMRec**), there exists a simulator $\mathbf{SSim}_{\mathbf{m}}^{\mathbf{F,T}}$ such that $\mathbf{STamper}_{\mathbf{m}}^{\mathbf{F,T}} \approx_{\epsilon_4} \mathbf{SSim}_{\mathbf{m}}^{\mathbf{F,T}}$. We use the underlying simulator as our simulator, and let $\mathbf{SSim}_{\mathbf{m}}^{\mathbf{f,T}} \equiv \mathbf{SSim}_{\mathbf{m}}^{\mathbf{F,T}}$. Applying triangle inequality to the above relations we prove the statistical non malleability for this case.

$$\mathbf{STamper}_{\mathbf{m}}^{\mathbf{f,T}} \approx_{2\epsilon_1 + \epsilon_4} \mathbf{SSim}_{\mathbf{m}}^{\mathbf{f,T}}$$

CASE 2 ($|T| \geq 3$)

Let $T = \{i_1, i_2 \ldots i_t\}$ be an ordered set of t indices, such that $i_j < i_{j+1}$. In this case, we use the tampering functions $\{\mathbf{f_i} : i \in T\}$ that tamper the shares of the scheme (**NMShare, NMRec**) to create explicit tampering functions $\mathbf{F}$ and $\mathbf{G}$ which tamper the two parts of non-malleable code. Note that as $\mathcal{F}_2^{split}$ allows arbitrary computation, the functions $\mathbf{F}$ and $\mathbf{G}$ are allowed to brute force over any finite subset. The reduction giving explicit $(\mathbf{F}, \mathbf{G}) \in \mathcal{F}_2^{split}$ is described below.

1. **(Initial Setup):** Fix an arbitrary $m_\$$ and let $l_\$, r_\$ \leftarrow \mathbf{NMEnc}(m_\$)$. Run the sharing function **APShare** with input $l_\$$ to obtain $\otimes_{i \in [n]} tl_i$. Run the sharing function $\mathbf{LRShare}_n^2(r_\$)$ to obtain $\otimes_{i \in [n]} tr_i$. Run the sharing function **PNMShare**$(m_\$)$ to obtain $\otimes_{i \in [n]} tp_i$. For each $i \in [n]$, create $tshare_i$ as tl_i, tr_i, tp_i. For all $i \in T$, **fix** $p_i \leftarrow tp_i$. For each $i \in \{i_1, i_2\}$, run f_i on $tShare_i$ to obtain $\widetilde{tShare_i} \leftarrow f_i(tShare_i)$. Parse $\widetilde{tshare_i}$ as $\widetilde{tl_i}, \widetilde{tr_i}, \widetilde{tp_i}$. **Fix** $l_i \leftarrow tl_i$ and $\widetilde{l_i} \leftarrow \widetilde{tl_i}$. For $i \in \{i_3, \ldots, i_t\}$, fix $r_i \leftarrow tr_i$. (Note that, here we rely on our pruning compiler for a different purpose: fixing l_{i_1}, l_{i_2} is allowed by property 2 of lemma 2. We would not have been able to do the same with a computational secret sharing directly. Also note that we depart significantly from initial step of [GK18], where $t - 1$ shares of l and only the last share of r was fixed. This was allowed because any $t - 1$ shares (resp. one share) does not reveal anything about the underlying l (resp. r). We on the other hand have fixed $t - 2$ shares of r, which encode a random value of $r_\$$).

2. The **tampering function F** is defined as follows: On input l, sample the value of $l_{i_3}, \ldots, l_{i_t}$ such that the shares $\{l_i : i \in T\}$ hide the secret l under (**APShare, APRec**) and the distribution of sampled l_{i_t} is identical to the distribution produced on running **APShare** with input l conditioned on fixing $\{l_i : i \in \{i_1, i_2\}\}$. In case such a sampling is not possible, then abort. Otherwise, for each $i \in T \setminus \{i_1, i_2\}$, construct $share_i$ as l_i, r_i, p_i using the fixed values of r_i and p_i. Run the tampering function f_i on $share_i$ to obtain tampered $\widetilde{share_i}$. Parse $\widetilde{share_i}$ as $\widetilde{l}_i, \widetilde{r}_i, \widetilde{p}_i$. Run the reconstruction function **APRec** with input $\otimes_{i \in T} \widetilde{l}_i$ to obtain $\widetilde{l}$. Output $\widetilde{l}$. (Note that unlike [GK18] we invoked the tampering functions with 'incorrect' shares of r).

3. The **tampering function G** is defined as follows: On input r, sample the values of first two shares of r, namely $\{r_{i_1}, r_{i_2}\}$ satisfying the following constraints:

   – The two shares $\{r_{i_1}, r_{i_2}\}$ encode the secret r under the (**LRShare,LRRec**). Moreover, the two shares should be distributed according to the output distribution of scheme (**LRShare,LRRec**).

   – For each $i \in \{i_1, i_2\}$, let $Share_i$ be l_i, r_i, p_i, run f_i on $share_i$ to obtain $\widetilde{share_i}$. Parse $\widetilde{share_i}$ as $\widetilde{nl}_i, \widetilde{nr}_i, \widetilde{np}_i$. The value of $\widetilde{nl}_i$ should be equal to $\widetilde{l}_i$ (the value that was fixed in the initial step of reduction). This can be achieved via brute force over the all the possibilities.

   In case such a sampling is not possible, then abort. Otherwise, run the reconstruction procedure of the leakage-resilient scheme to obtain $\widetilde{r}$, using the tampered values of first 2 shares of r. That is $\widetilde{r} \leftarrow \mathbf{LRRec}^{\{\mathbf{i_1, i_2}\}}(\widetilde{nr_{i_1}}, \widetilde{nr_{i_2}})$. Output $\widetilde{r}$. (Unlike [GK18], we now only ensure that the first two shares are from the correct distribution.)

The reduction given above creates t shares corresponding to indices in T. Unlike the proof of [GK18], here the distribution of the t shares is not close to the distribution of the t shares during actual sharing (in fact statistically it is quite far). Nevertheless, we show that an adversary cannot notice this change without violating the leakage resilience of the (**LRShare, LRRec**).

We achieve this using hybrid argument, however, instead of outputting t shares $\otimes_{i \in T} share_i$ as in [GK18], we output $\mathbf{NMRec}(\otimes_{i \in T} f_i(share_i))$, the output of the tampering experiment. For ease of understanding, let $share_i$ be of the form al_i, ar_i, ap_i when the shares are produced by the reduction on input l and r, with the fixing of $l_\$$ and $r_\$$. Similarly, let $share_i$ be of the form bl_i, br_i, bp_i when the secret m is encoded by the sharing procedure **NMShare** conditioned on output of **NMEnc**(m) being l, r.

1. **Hybrid$_1$**: for each $i \in T$, $share_i$ is of the form al_i, ar_i, ap_i. The distribution of these t shares is identical to distribution of the shares produced by the reduction on input l and r, with the fixing of $l_\$$ and $r_\$$. Output $\mathbf{NMRec}(\otimes_{i \in T} f_i(share_i))$.

2. **Hybrid$_2$**: In the initial setup phase of the reduction, for each $i \in T$, fix bp_i instead of ap_i. Proceed with the reduction to create t shares of the form al_i, ar_i, bp_i. Output $\mathbf{NMRec}(\otimes_{i \in T} f_i(share_i))$.

3. **Hybrid$_3$**: Fix $l_\$ \leftarrow l$ in the initial setup phase. Fix shares of p like **Hybrid$_2$**. Output **NMRec**($\otimes_{i \in T} f_i(share_i)$).
4. **Hybrid$_4$**: Fix $l_\$ \leftarrow l$ and fix $r_\$ \leftarrow r$ in the initial setup phase. Fix the shares of p as in previous hybrid **Hybrid$_3$**. Proceed with the reduction to create the t shares. Output **NMRec**($\otimes_{i \in T} f_i(share_i)$).
5. **Hybrid$_5$**: For each $i \in [n]$, let $share_i$ be of the form bl_i, br_i, bp_i. The distribution of these t shares is identical to distribution obtained on running the **NMShare** conditioned on output of **NMEnc**(m) being l, r. Output **NMRec**($\otimes_{i \in T} f_i(share_i)$).

<u>Claim:</u> For any authorized $T \in \mathcal{A}^{min}$ with cardinality greater than 2, the statistical distance between **Hybrid$_1$** and **Hybrid$_2$** is at most ϵ_4.
<u>Proof:</u> As $|T| \geq 3$, T does not belong to $\mathcal{A}_{pairs}$. The two hybrids only differ in the shares corresponding to output of **PNMShare**. The claim follows from the statistical privacy of (**PNMShare**, **PNMRec**).   ∎

<u>Claim:</u> For any $l, l_\$$, any authorized $T \in \mathcal{A}^{min}$, **Hybrid$_2$** is identical to **Hybrid$_3$**.
<u>Proof:</u> The two hybrids differ in the initial setup phase. In **Hybrid$_2$**, 2 shares of $l_\$$ are fixed, while in **Hybrid$_3$** 2 shares of l are fixed. Lemma 2 ensures that the secret is perfectly hidden even when two shares of **APShare** are revealed.   ∎
    The above also shows that the function **F** in the reduction never aborts.
<u>Claim:</u> For any $r, r_\$$, any authorized $T \in \mathcal{A}^{min}$ with cardinality greater than 2, the statistical distance between **Hybrid$_3$** and **Hybrid$_4$** is at most ϵ_3.
<u>Proof:</u> Assume towards contradiction that there exists $r, r_\$ \in \mathbb{F}_1$, $T \in \mathcal{A}^{min}$ and a distinguisher D that is successful in distinguishing **Hybrid$_3$** and **Hybrid$_4$** with probability greater than ϵ_3. We use distinguisher D to construct another distinguisher D_1 and a leak function $g \in \mathcal{L}_\mu^{pair}$ which violates the property of leakage-resilience satisfied by the scheme (**LRShare$_n^2$**, **LRRec$_n^2$**) for the secrets $r, r_\$$. The reduction is described below:

1. (**Initial Setup**): Run the sharing function **APRec** with input l to obtain $\otimes_{i \in [n]} tl_i$. Run the sharing function **PNMShare**(m) to obtain $\otimes_{i \in [n]} tp_i$. For all $i \in T$, fix $p_i \leftarrow tp_i$ and $l_i \leftarrow tl_i$.
    Give $r, r_\$$ to the adversary, who then specifies $r_1, r_2, r_1^\$, r_2^\$$. Use l_1, r_1, p_1 and l_2, r_2, p_2 to create the first two shares $share_1$ and $share_2$. Tamper the shares using f_1 and f_2 to obtain $\tilde{l}_1, \tilde{r}_1, \tilde{p}_1$ and $\tilde{l}_2, \tilde{r}_2, \tilde{p}_2$. Compute $\tilde{r} \leftarrow$ **LRRec**($\tilde{r}_1, \tilde{r}_2$). Fix $\tilde{l}_1, \tilde{l}_2$.
2. (**Leak function g**): We define a specific leakage function $g = \{g_i : i \in T \setminus \{i_1, i_2\}\}$ which leaks μ bits independently from each of the $t - 2$ shares.
    – For each $i \in T \setminus \{i_1, i_2\}$, define g_i as the following function which takes r_i as input. Create $tShare_i$ as l_i, r_i, p_i. Run f_i on $tShare_i$ to obtain $\widetilde{tShare_i} \leftarrow f_i(tShare_i)$. Parse $\widetilde{tshare_i}$ as $\tilde{tl}_i, \tilde{tr}_i, \tilde{tp}_i$. Output $\tilde{tl}_i$.
    As $\tilde{tl}_i$ is an element of $\mathbb{F}_2$, it can be represented by at most $\log |\mathbb{F}_2|$ bits, which is equal to μ. This shows that the above leak function g belongs to the class $\mathcal{L}_\mu^{pair}$.

3. (**Distinguisher** D_1): The distinguisher D_1 is defined as follows: On input $g(r_3, \ldots, r_t)$, parse it as $\widetilde{tl_{i_3}}, \ldots, \widetilde{tl_{i_t}}$. Compute $\tilde{l} \leftarrow \mathbf{APRec}(\tilde{l}_1, \ldots, \tilde{l}_t)$. Compute $\tilde{m} \leftarrow \mathbf{NMDec}(\tilde{l}, \tilde{r})$. Invoke the distinguisher D with $\tilde{m}$ and output its output.

Notice, in the case the secret hidden by the leakage-resilient scheme was $r_\$$, D will be invoked with input distributed according to **Hybrid$_2$**. In the other case, in which r was hidden, D will be invoked with distributed according to **Hybrid$_3$**. Therefore the success probability of D_1 will be equal to the advantage of D in distinguishing these two hybrids, which is greater than ϵ_3 by assumption. Hence, we have arrived at a contradiction to statistical leakage-resilience property of the scheme (**LRShare, LRRec**). ∎

The above also shows that the function **G** in the reduction aborts with probability less than ϵ_3.

Claim: For any l, r, **Hybrid$_4$** is identical to **Hybrid$_5$**.

Proof: In **Hybrid$_4$**, the shares of $r_\$$ (resp. $l_\$$) that are sampled in the initial setup already encode the value r (resp. l). Therefore, all the t shares created in **Hybrid$_4$** will be identically distributed to the ones produced while executing **NMShare** with the output of **NMEnc** being (l, r). ∎

By repeated application of triangle inequality, we get that for any $a, b \in \mathbb{F}_0$, the statistical distance between **Hybrid$_1$** and **Hybrid$_5$** is at most $\epsilon_2 + \epsilon_3 + \epsilon_4$. This proves that the set of shares created by our reduction is statistically close the set of shares created during the real sharing by the scheme, and thus the tampering functions $\mathbf{f} = \{\mathbf{f_i} : i \in T\}$ can be successfully invoked.

From our construction of **F** and **G**, it is clear that for any l and r, if the reduction is successful in creating the t shares, then the secret hidden is these t shares is the same as the message encoded by l and r (under non-malleable code). That is,

$$\mathbf{NMRec}(\{share_i : i \in T\}) = \mathbf{NMDec}(l, r)$$

Similarly, we can say that the secret hidden is the t tampered shares is the same as the message encoded by tampered $\tilde{l}$ and tampered $\tilde{r}$. That is,

$$\mathbf{NMRec}(\{\mathbf{f_i}(share_i) : i \in T\}) = \mathbf{NMDec}(\mathbf{F}(l), \mathbf{G}(r))$$

Therefore, the tampering experiments of non-malleable codes (see Definition 4) and non-malleable secret-sharing schemes (see Definition 8) are statistically indistinguishable, specifically,

$$\mathbf{STamper}_m^{f,T} \approx_{\epsilon_2 + \epsilon_3 + \epsilon_4} \mathbf{Tamper}_m^{F,G}$$

By the ϵ_1-non malleability of the scheme (**NMEnc, NMDec**), there exists a simulator $\mathbf{Sim}_m^{F,G}$ such that $\mathbf{Tamper}_m^{F,G} \approx_{\epsilon_1} \mathbf{Sim}_m^{F,G}$. We use the underlying simulator as our simulator and let $\mathbf{SSim}_m^{f,T} \equiv \mathbf{Sim}_m^{F,G}$. Applying triangle inequality to the above relations we prove the statistical non malleability.

$$\mathbf{STamper}_m^{f,T} \approx_{\epsilon_1 + \epsilon_2 + \epsilon_3 + \epsilon_4} \mathbf{SSim}_m^{f,T}$$

As the statistical distances between $\mathbf{STamper}_m^{f,T}$ and $\mathbf{SSim}_m^{f,T}$ in the two cases are $(2\epsilon_1 + \epsilon_4)$ and $(\epsilon_1 + \epsilon_2 + \epsilon_3 + \epsilon_4)$, we take $(2\epsilon_1 + \epsilon_2 + \epsilon_3 + \epsilon_4)$ as the worst case statistical error of our scheme $(\mathbf{NMShare}, \mathbf{NMRec})$.   □

4   n-out-of-n NMSS Against Joint Tampering

Tampering Family. We now formally define the supported tampering family, in which, we allow the tampered value of each share to depend on all the n shares in a restricted fashion.

Tampering Family $\mathcal{F}_n^{general}$

Assume that input shares are of equal length vectors over some finite field of prime order. The adversary specifies four subsets of $[n]$, namely $B_f^{in}, B_f^{out}, B_g^{in}, B_g^{out}$ and also specifies four arbitrary tampering functions f_1, g_1, f_2, g_2 such that

$$f_1 : \{share_i : i \in B_f^{in}\} \to \{\widetilde{fshare}_i : i \in B_f^{out}\}$$

$$g_1 : \{share_i : i \in B_g^{in}\} \to \{\widetilde{gshare}_i : i \in B_f^{out}\}$$

$$f_2 : \{share_i : i \in B_f^{in}\} \to \{\widetilde{fshare}_i : i \in B_g^{out}\}$$

$$g_2 : \{share_i : i \in B_g^{in}\} \to \{\widetilde{gshare}_i : i \in B_g^{out}\}$$

such that for all $i \in [n]$, the final tampered share is of the form

$$\widetilde{share}_i \leftarrow \widetilde{fshare}_i \odot \widetilde{gshare}_i$$

where $\odot$ represents element wise multiplication of the two vectors over the given finite field. Here $B_f^{in} \subset [n]$ denotes the set of identities of parties whose shares are available as input to function f_1 and f_2. Similarly, B_f^{out} denotes the set of identities of parties whose tampered shares are produced by functions f_1 and g_1. B_g^{in} and B_g^{out} are analogous. The four subsets can be arbitrarily chosen by the adversary as long as they satisfy the following natural constraints:

- The input to tampering function f_1 contains atleast one share, which does not occur as the input of the tampering function g_1 and vice versa. That is, $|B_f^{in} \setminus B_g^{in}| \geq 1$ and $|B_g^{in} \setminus B_f^{in}| \geq 1$.
- The output sets B_f^{out} and B_g^{out} are disjoint. For the sake of simplicity, we further assume w.l.o.g that $B_f^{out} \cup B_g^{out} = [n]$.

Construction of [ADL14]. As we use the construction of Aggarwal et al. [ADL14] in a non-black-box way, we recall it for convenience:

Definition 12 ([ADL14]). *Affine Evasive Function: A surjective function* $h : F_p \to \mathcal{M} \cup \{\bot\}$ *is called* (γ, δ)-*affine-evasive if for any* $a, b \in F_p$ *such that* $a \neq 0$, *and* $(a, b) \neq (1, 0)$, *and for any* $m \in \mathcal{M}$,

– $Pr(h(aU + b) \neq \perp) \leq \gamma$
– $Pr(h(aU + b) \neq \perp | h(U) = m) \leq \delta$
– A uniformly random X such that $h(X) = m$ is efficiently samplable.

Using these affine evasive functions, they arrive at the construction of split-state non-malleable codes by composing it with inner product. For $L, R \in F_p^\lambda$, let $\langle L, R \rangle$ represent the inner product $\langle L, R \rangle = \sum_{i=1}^\lambda L[i] \times R[i]$. Their scheme is as follows:

– The **decoding function ADLDec** : $F_p^\lambda \times F_p^\lambda \to \mathcal{M} \cup \{\perp\}$ is defined using affine evasive function h as follows: $\mathbf{ADLDec(L, R)} := \mathbf{h}(\langle \mathbf{L}, \mathbf{R} \rangle)$
– The **encoding function ADLEnc** : $\mathcal{M} \to F_p^\lambda \times F_p^\lambda$ is defined as $\mathbf{ADLEnc}(m) = (L, R)$ where L, R are chosen uniformly at random from $F_p^\lambda \times F_p^\lambda$ conditioned on the fact that $\mathbf{ADLDec}(L, R) = m$.

Theorem 4 ([ADL14]). *Let $\mathcal{M} = \{1, 2, \ldots, K\}$ and let $p \geq (\frac{4K}{\epsilon})^{\rho \log \log(4K/\epsilon)}$ be a prime. Let λ be $(\lceil \frac{2 \log p}{c} \rceil)^6$. Let $\mathbf{ADLEnc} : \mathcal{M} \to F_p^\lambda \times F_p^\lambda$, $\mathbf{ADLDec} : F_p^\lambda \times F_p^\lambda \to \mathcal{M} \cup \{\perp\}$ be as defined above. Then the scheme $(\mathbf{ADLEnc}, \mathbf{ADLDec})$ is ϵ-non-malleable w.r.t $\mathcal{F}_2^{split}$.*

Multiplicative Secret Sharing Scheme of [KGH83]. We recall the result of Karnin et al. [KGH83], in which they construct $(n, 0)$-secret sharing scheme realizing access structure $\mathcal{A}_n^n$ over arbitrary Abelian group. Let $(\mathbb{F}_p, +, \times)$ be a finite field. Let $\mathbb{F}_p^*$ be the set of non zero elements of the field $\mathbb{F}_p$, and this set along with the operation $\times$ forms an abelian group.

– **MultShare$_n$**: Let $MultShare_n : \mathbb{F}_p^* \to \otimes_{i \in [n]} \mathbb{F}_p^*$ be a randomized sharing function. On input a secret $s \in \mathbb{F}_p^*$, sample the first $n - 1$ shares, namely $s_1, s_2, \ldots, s_{n-1}$, randomly from $\mathbb{F}_p^*$. Compute the last share using the secret s and the sampled shares as $s_n \leftarrow x / \prod_{i=1}^{n-1} s_i$ Output $s_1, \ldots, s_n$.
– **MultRec$_n$**: Let $MultRec_n : \otimes_{i \in [n]} \mathbb{F}_p^* \to \mathbb{F}_p^*$ be a deterministic function for reconstruction. On input n shares, namely $s_1, s_2, \ldots, s_n$, compute $s \leftarrow \prod_{i=1}^n s_i$ and output the result s.

Theorem 5 ([KGH83]). **MultShare$_n$**, **MultRec$_n$** *is an $(n, 0)$-secret sharing scheme realizing access structure $\mathcal{A}_n^n$.*

Our n-out-of-n Non-malleable Secret Sharing Scheme

Theorem 6. *Let the message space $\mathcal{M}$, prime p and vector length λ be as in the construction of $(\mathbf{ADLEnc}, \mathbf{ADLDec})$, the coding scheme of Aggarwal et al. [ADL14] that is ϵ-non-malleable against $\mathcal{F}_n^{split}$. Then for any number of parties $n \geq 2$, there exists an efficient construction of $(n, n, 2\epsilon + \frac{2\lambda}{p})$-secret sharing scheme that is $(\epsilon + \frac{2\lambda}{p})$-non-malleable w.r.t $\mathcal{F}_n^{general}$.*

Corollary 7. *The coding scheme of Aggarwal et al. [ADL14] is also statistically-non-malleable w.r.t. $\mathcal{F}_2^{general}$. (which allows the tampering of left share to partially depend on the right share).*

Proof. (of theorem)

We begin with the description of our secret sharing scheme:

- The **reconstruction function JNMRec$_n$** : $F_p^{\lambda \times n} \rightarrow \mathcal{M} \cup \{\bot\}$ is defined using affine evasive function h as follows:

$$\mathbf{JNMRec_n(sh_1, sh_2 \ldots sh_n)} := \mathbf{h}(\langle \mathbf{sh_1, sh_2 \ldots sh_n} \rangle)$$

where $\langle a_1, a_2 \ldots a_n \rangle = \sum_{i=1}^{\lambda} \prod_{j=1}^{n} a_j[i]$ is the generalized inner product function.

- The **sharing function JNMShare$_n$** : $\mathcal{M} \rightarrow F_p^{\lambda \times n}$ is defined as follows. On input m, output $(sh_1, sh_2 \ldots sh_n)$ where $sh_1, sh_2 \ldots sh_n$ are chosen uniformly at random from $F_p^{\lambda \times n}$ conditioned on the fact that $\mathbf{JNMRec_n}(sh_1, sh_2 \ldots sh_n) = m$.

Correctness, Efficiency, and Statistical Privacy: Correctness and efficiency trivially follows from the construction. Statistical privacy follows from the non-malleability proved below (in a manner similar to Lemma 1).

Statistical Non Malleability: We transform an attack on our scheme to an attack on the underlying split-state non-malleable code. Let the adversary choose a four tampering functions (f_1, g_1, f_2, g_2) and corresponding four subsets $(B_f^{in}, B_f^{out}, B_g^{out}, B_g^{in})$ from the allowed tampering class $\mathcal{F}_n^{general}$. Let $n_f \leftarrow |B_f^{in}|$ and $n_g \leftarrow |B_g^{in}|$ denote the cardinality of input set of indices of function f and g respectively. Similarly, let $n_f^{out} \leftarrow |B_f^{out}|$ and $n_g^{out} \leftarrow |B_g^{out}|$ denote the cardinality of output set of indices of function f and g respectively. Using these tampering functions, we give explicit pair of tampering function $(F, G) \in \mathcal{F}_2^{split}$. The description of the reduction follows:

- **(Initial Setup)**: Start with fixing the shares which occurs as input of both the tampering functions. Let $B_{fix} \leftarrow B_f^{in} \cap B_g^{in}$. Let $n_{fix} \leftarrow |B_{fix}|$ denote the cardinality of this common set. For each $i \in B_{fix}$, fix $share_i \leftarrow a_1^i, a_2^i, \ldots, a_\lambda^i$ randomly such that each $a_i^j \in \mathbb{F}_p^*$.

- The **tampering function F(l)**
  - On input a vector $l \in F_p^{\lambda}$, parse it as $l_1, l_2 \ldots l_\lambda$ such that $l_i \in F_p$ for each $i \in [\lambda]$. If there exists an $i \in [\lambda]$ such that $l_i = 0$, then abort. Otherwise, for each $i \in [\lambda]$, calculate $prod_i \leftarrow (l_i / (\prod_{j \in B_f^{out} \cap B_{fix}} a_i^j))$ and use multiplicative sharing to share $prod_i$ into $n_f - n_{fix}$ shares. That is, let $\{a_i^j : j \in B_f^{in} \setminus B_{fix}\} \leftarrow \mathbf{MultShare_{n_f - n_{fix}}}(prod_i)$. Construct $share_i$ as $a_1^i, a_2^i, \ldots, a_\lambda^i$ for each $i \in B_f^{in} \setminus B_{fix}$. (We are excluding shares in B_{fix} as they have already been fixed earlier)
  - Tamper the shares by executing the adversary specified function f_1.

$$\{\widetilde{fshare_j} : j \in B_f^{out}\} \leftarrow \mathbf{f_1}(\{share_j : j \in B_f^{in}\})$$

Similarly, compute the tampered shares using f_2.

$$\{\widetilde{fshare_j} : j \in B_g^{out}\} \leftarrow \mathbf{f_2}(\{share_j : j \in B_f^{in}\})$$

- Parse the tampered shares as $\widetilde{fshare}_i = (\widetilde{a}_1^i, \widetilde{a}_2^i \ldots \widetilde{a}_\lambda^i)$ for each $i \in [n]$ (recall $[n] = B_f^{in} \cup B_f^{out}$ by assumption). Reconstruct the tampered value of l_i for each $i \in [\lambda]$ using the reconstruction function of multiplicative sharing. Let $\widetilde{l}_i \leftarrow \mathbf{MultRec}_{n_f^{out}}(\{\widetilde{a}_i^j : j \in [n]\})$.
- Then construct the tampered vector $\widetilde{l}$ as $(\widetilde{l}_1, \widetilde{l}_2, \ldots, \widetilde{l}_\lambda)$ and output $\widetilde{l}$.
- The **tampering function G(r)**
  - On input a vector $r \in F_p^\lambda$, parse it as $r_1, r_2 \ldots r_\lambda$ such that $r_i \in F_p$ for each $i \in [\lambda]$. If there exists an $i \in [\lambda]$ such that $r_i = 0$, then **abort**. Otherwise, for each $i \in [\lambda]$, calculate $prod_i \leftarrow (r_i/(\prod_{j \in B_g^{out} \cap B_{fix}} a_i^j))$ and use multiplicative sharing to share $prod_i$ into $n_g - n_{fix}$ shares. That is, let $\{a_i^j : j \in B_g^{in} \setminus B_{fix}\} \leftarrow \mathbf{MultShare}_{n_g - n_{fix}}(prod_i)$. Construct $share_i$ as $a_1^i, a_2^i, \ldots, a_\lambda^i$ for each $i \in B_g^{in} \setminus B_{fix}$. (We are excluding shares in B_{fix} as they have already been fixed earlier)
  - Tamper the shares by executing the adversary specified function g_1.

$$\{\widetilde{gshare}_j : j \in B_g^{out}\} \leftarrow \mathbf{g_1}(\{share_j : j \in B_g^{in}\})$$

Similarly, compute the tampered shares using g_2.

$$\{\widetilde{gshare}_j : j \in B_f^{out}\} \leftarrow \mathbf{g_2}(\{share_j : j \in B_g^{in}\})$$

  - Parse the tampered shares as $\widetilde{gshare}_i = \widetilde{b}_1^i, \widetilde{b}_2^i \ldots \widetilde{b}_\lambda^i$ for each $i \in [n]$. Reconstruct the tampered value of r_i for each $i \in [\lambda]$ using the reconstruction function of multiplicative sharing. Let $\widetilde{r}_i \leftarrow \mathbf{MultRec}_{n_g^{out}}(\{\widetilde{b}_i^j : j \in [n]\})$.
  - Then construct the tampered vector $\widetilde{r}$ as $(\widetilde{r}_1, \widetilde{r}_2, \ldots, \widetilde{r}_\lambda)$ and output $\widetilde{r}$.

It is easy to see that the reduction does not terminate with probability at least $\left(1 - \frac{2\lambda}{p}\right)$.

<u>Claim:</u> For any l and r, if the reduction is successful in creating the n shares, then the secret hidden is these n shares is the same as the message encoded by l and r.

<u>Proof:</u> The reduction constructs an instance of the secret sharing scheme using l and r in a split-state manner. Basically, for all $i \in [\lambda]$, it creates parts of shares such that $\left(\prod_{j \in B_f^{out}} a_i^j\right) = l_i$ and $\left(\prod_{j \in B_g^{out}} a_i^j\right) = r_i$. In this way, it is ensured that the secret hidden by n shares is the same as the message encoded by challenge shares l and r of the underlying non-malleable code. This can be seen by the following calculation:

$$\mathbf{JNMRec_n}(\{share_i : i \in [n]\}) = h\Big(\sum_{i=1}^{\lambda} \prod_{j=1}^{n} a_i^j\Big)$$

$$= h\Big(\sum_{i=1}^{\lambda} \big(\prod_{j \in B_f^{out}} a_i^j\big) \times \big(\prod_{j \in B_g^{out}} a_i^j\big)\Big)$$

$$= h\Big(\sum_{i=1}^{\lambda} l_i \times r_i\Big) = h(\langle l, r \rangle)$$

$$= \mathbf{ADLDec}(l, r)$$

∎

Claim: For any l and r, if the reduction is successful in creating the t shares, then the secret hidden is the t tampered shares is the same as the message encoded by the tampered l and the tampered r.

Proof: Let $\{\widetilde{share}_i : i \in [n]\}$ be the disjoint union of outputs of two tampering functions $\mathbf{f}(\{share_i : i \in B_f^{in}\})$ and $\mathbf{g}(\{share_i : i \in B_g^{in}\})$. Now the reduction transforms the tampered shares back to two tampered parts of non-malleable code. Let $(\mathbf{F}, \mathbf{G})$ be as defined in the reduction.

$$\mathbf{ADLDec}\big(\mathbf{F}(l), \mathbf{G}(r)\big) = h\big(\langle \mathbf{F}(l), \mathbf{G}(r) \rangle\big)$$

$$= h\big(\langle \widetilde{l}, \widetilde{r} \rangle\big)$$

$$= h\Big(\sum_{i=1}^{\lambda} \widetilde{l}_i \times \widetilde{r}_i\Big)$$

$$= h\Bigg(\sum_{i=1}^{\lambda} \Big(\prod_{j \in [n]} \widetilde{a}_i^j\Big) \times \Big(\prod_{j \in [n]} \widetilde{b}_i^j\Big)\Bigg)$$

$$= h\Big(\sum_{i=1}^{\lambda} \prod_{j=1}^{n} (\widetilde{a}_i^j \times \widetilde{b}_i^j)\Big)$$

$$= h\Big(\sum_{i=1}^{\lambda} \prod_{j=1}^{n} (\widetilde{share}_j[i])\Big)$$

$$= \mathbf{JNMRec_n}\big(\{\widetilde{share}_j : j \in [n]\}\big)$$

∎

By design the tampering functions $\mathbf{F}$ and $\mathbf{G}$ belongs to $\mathcal{F}_2^{split}$. By the ϵ-non malleability of the scheme $(\mathbf{ADLEnc}, \mathbf{ADLDec})$, we know that there exists a distribution $D^{F,G}$ such that

$$\mathbf{Sim_m^{F,G}} \approx_\epsilon \mathbf{Tamper_m^{F,G}}$$

Using the observation about the equivalence of tampering, and assuming that the adversary succeeds in case the reduction terminates by executing abort, we get that

$$\mathbf{STamper_m^{f,T}} \approx_{\epsilon + \frac{2\lambda}{p}} \mathbf{SSim_m^{f,T}}$$

This proves the non malleability of our scheme.   □

Acknowledgments. We thank the anonymous reviewers, as their detailed and insightful reviews significantly helped in improving the presentation of this article.

The first author is supported by a grant from Northrop Grumman.
A part of this work was done while the second author was at Microsoft Research, India. Work done at UCLA is supported in part from NSF grant 1619348, NSF frontier award 1413955, US-Israel BSF grants 2012366, 2012378, and by the Defense Advanced Research Projects Agency (DAPRA) SAFEWARE program through the ARL under Contract W911NF-15-C-0205 and through a subcontract with Galois, inc. The views expressed are those of the authors and do not reflect the official policy or position of the Department of Defense, the National Science Foundation, or the U.S. Government.

References

[ADKO15] Dodis, Y., Nielsen, J.B. (eds.): TCC 2015. LNCS, vol. 9014. Springer, Heidelberg (2015). https://doi.org/10.1007/978-3-662-46494-6

[ADL14] Aggarwal, D., Dodis, Y., Lovett, S.: Non-malleable codes from additive combinatorics. In: Proceedings of the 46th Annual ACM Symposium on Theory of Computing, pp. 774–783. ACM (2014)

[Bei] Beimel, A.: Secure schemes for secret sharing and key distribution. Ph.D. thesis (1996)

[Bei11] Beimel, A.: Secret-sharing schemes: a survey. In: Chee, Y.M., et al. (eds.) IWCC 2011. LNCS, vol. 6639, pp. 11–46. Springer, Heidelberg (2011). https://doi.org/10.1007/978-3-642-20901-7_2

[Bla79] Blakley, G.R.: Safeguarding cryptographic keys. In: AFIPS National Computer Conference (NCC 1979), pp. 313–317. IEEE Computer Society, Los Alamitos (1979)

[BOGW88] Ben-Or, M., Goldwasser, S., Wigderson, A.: Completeness theorems for non-cryptographic fault-tolerant distributed computation. In: Proceedings of the Twentieth Annual ACM Symposium on Theory of Computing, pp. 1–10. ACM (1988)

[CDF+08] Cramer, R., Dodis, Y., Fehr, S., Padró, C., Wichs, D.: Detection of algebraic manipulation with applications to robust secret sharing and fuzzy extractors. In: Smart, N. (ed.) EUROCRYPT 2008. LNCS, vol. 4965, pp. 471–488. Springer, Heidelberg (2008). https://doi.org/10.1007/978-3-540-78967-3_27

[CDTV16] Coretti, S., Dodis, Y., Tackmann, B., Venturi, D.: Non-malleable encryption: simpler, shorter, stronger. In: Kushilevitz, E., Malkin, T. (eds.) TCC 2016. LNCS, vol. 9562, pp. 306–335. Springer, Heidelberg (2016). https://doi.org/10.1007/978-3-662-49096-9_13

[CGL16] Chattopadhyay, E., Goyal, V., Li, X.: Non-malleable extractors and codes, with their many tampered extensions. In: STOC (2016)

[DKO13] Dziembowski, S., Kazana, T., Obremski, M.: Non-malleable codes from two-source extractors. In: Canetti, R., Garay, J.A. (eds.) CRYPTO 2013. LNCS, vol. 8043, pp. 239–257. Springer, Heidelberg (2013). https://doi.org/10.1007/978-3-642-40084-1_14

[DPW10] Dziembowski, S., Pietrzak, K., Wichs, D.: Non-malleable codes. In: Innovations in Computer Science - ICS 2010, Tsinghua University, Beijing, China, 5–7 January 2010, Proceedings, pp. 434–452 (2010)

[GGSW13] Garg, S., Gentry, C., Sahai, A., Waters, B.: Witness encryption and its applications. In: Proceedings of the Forty-Fifth Annual ACM Symposium on Theory of Computing, pp. 467–476. ACM (2013)

[GJK15] Goyal, V., Jain, A., Khurana, D.: Non-malleable multi-prover interactive proofs and witness signatures. Cryptology ePrint Archive, Report 2015/1095 (2015). http://eprint.iacr.org/2015/1095

[GK18] Goyal, V., Kumar, A.: Non-malleable secret sharing. In: Proceedings of the Fiftieth ACM STOC. ACM (2018, to appear)

[Gol07] Goldreich, O.: Foundations of Cryptography: Volume 1, Basic Tools. Cambridge University Press, Cambridge (2007)

[GPR16] Goyal, V., Pandey, O., Richelson, S.: Textbook non-malleable commitments. In: Proceedings of the 48th Annual ACM SIGACT Symposium on Theory of Computing, STOC 2016, Cambridge, MA, USA, 18–21 June 2016, pp. 1128–1141 (2016)

[ISN89] Ito, M., Saito, A., Nishizeki, T.: Secret sharing scheme realizing general access structure. Electron. Commun. Jpn. (Part III Fundam. Electron. Sci.) $72(9)$, 56–64 (1989)

[KGH83] Karnin, E., Greene, J., Hellman, M.: On secret sharing systems. IEEE Trans. Inf. Theory $29(1)$, 35–41 (1983)

[KNY14] Komargodski, I., Naor, M., Yogev, E.: Secret-sharing for NP. In: Sarkar, P., Iwata, T. (eds.) ASIACRYPT 2014. LNCS, vol. 8874, pp. 254–273. Springer, Heidelberg (2014). https://doi.org/10.1007/978-3-662-45608-8_14

[KW93] Karchmer, M., Wigderson, A.: On span programs. In: 1993, Proceedings of the Eighth Annual Structure in Complexity Theory Conference, pp. 102–111. IEEE (1993)

[Li17] Li, X.: Improved non-malleable extractors, non-malleable codes and independent source extractors. In: STOC. ACM (2017)

[LL12] Liu, F.-H., Lysyanskaya, A.: Tamper and leakage resilience in the split-state model. In: Safavi-Naini, R., Canetti, R. (eds.) CRYPTO 2012. LNCS, vol. 7417, pp. 517–532. Springer, Heidelberg (2012). https://doi.org/10.1007/978-3-642-32009-5_30

[MS81] McEliece, R.J., Sarwate, D.V.: On sharing secrets and Reed-Solomon codes. Commun. ACM $24(9)$, 583–584 (1981)

[RBO89] Rabin, T., Ben-Or, M.: Verifiable secret sharing and multiparty protocols with honest majority. In: STOC 1989, pp. 73–85. ACM, New York (1989)

[Sha79] Shamir, A.: How to share a secret. Commun. ACM $22(11)$, 612–613 (1979)

On the Local Leakage Resilience of Linear Secret Sharing Schemes

Fabrice Benhamouda[1], Akshay Degwekar[2($\boxtimes$)], Yuval Ishai[3], and Tal Rabin[1]

[1] IBM Research, Yorktown Heights, NY, USA
fabrice.benhamouda@normalesup.org, talr@us.ibm.com
[2] MIT, Cambridge, MA, USA
akshayd@mit.edu
[3] Technion, Haifa, Israel
yuvali@cs.technion.ac.il

Abstract. We consider the following basic question: to what extent are standard secret sharing schemes and protocols for secure multiparty computation that build on them resilient to leakage? We focus on a simple *local leakage* model, where the adversary can apply an arbitrary function of a bounded output length to the secret state of each party, but cannot otherwise learn joint information about the states.

We show that additive secret sharing schemes and high-threshold instances of Shamir's secret sharing scheme are secure under local leakage attacks when the underlying field is of a large prime order and the number of parties is sufficiently large. This should be contrasted with the fact that any linear secret sharing scheme over a small characteristic field is clearly insecure under local leakage attacks, regardless of the number of parties. Our results are obtained via tools from Fourier analysis and additive combinatorics.

We present two types of applications of the above results and techniques. As a positive application, we show that the "GMW protocol" for honest-but-curious parties, when implemented using shared products of random field elements (so-called "Beaver Triples"), is resilient in the local leakage model for sufficiently many parties and over certain fields. This holds even when the adversary has *full access* to a constant fraction of the views. As a negative application, we rule out multi-party variants of the share conversion scheme used in the 2-party homomorphic secret sharing scheme of Boyle et al. (Crypto 2016).

1 Introduction

The recent attacks of Meltdown and Spectre [38, 41] have brought back to the forefront the question of side-channel leakage and its effects. Starting with the early works of Kocher et al. [39, 40], side-channel attacks have demonstrated vulnerabilities in cryptographic primitives. Moreover, there are often inherent tradeoffs between efficiency and leakage resilience, where optimizations increase the susceptibility to side-channel attacks.

© International Association for Cryptologic Research 2018
H. Shacham and A. Boldyreva (Eds.): CRYPTO 2018, LNCS 10991, pp. 531–561, 2018.
https://doi.org/10.1007/978-3-319-96884-1_18

A large body of work on the theory of *leakage resilient cryptography* (cf. [1,21,42]) studies the possibility of constructing cryptographic schemes that remain secure in the presence of partial leakage of the internal state. One prominent direction of investigation has been designing leakage resilient cryptographic protocols for general computations [19,22,26,31,35].

The starting point for most of these works is the observation that some standard cryptographic schemes are vulnerable to very simple types of leakage. Moreover, analyzing the leakage resilience of others seems difficult. This motivates the design of new cryptographic schemes that deliver strong provable leakage resilience guarantees.

In this work, we forgo designing *special-purpose* leakage resilient schemes and focus on studying the properties of existing common designs. We want to understand:

To what extent are standard *cryptographic schemes leakage resilient?*

We restrict our attention to *linear* secret sharing schemes and secure multiparty computation (MPC) protocols that build on them. In particular, we would like to understand the leakage resilience properties of the most commonly used secret sharing schemes, like additive secret sharing and Shamir's scheme, as well as simple MPC protocols that rely on them.

Analyzing existing schemes has a big advantage, as it can potentially allow us to enjoy their design benefits while at the same time enjoying a strong leakage-resilience guarantee. Indeed, classical secret sharing schemes and MPC protocols have useful properties which the specially designed leakage-resilient schemes are not known to achieve. For instance, linear secret sharing schemes can be manipulated via additive (and sometimes multiplicative) homomorphism, and standard MPC protocols can offer resilience to faults and a large number of *fully corrupted* parties. Finally, classical schemes are typically more efficient than special-purpose leakage-resilient schemes.

Local Leakage. We study leakage resilience under a simple and natural model of *local leakage* attacks. The local leakage model has the following three properties: (1) The attacker can leak information about each server's state *locally*, independently of the other servers; this is justified by physical separation. (2) Only a *few bits of information* can be leaked about the internal state of each server; this is justified by the limited precision of measurements of physical quantities such as time or power. (3) The leakage is *adversarial*, in the sense that the adversary can decide what function of the secret state to leak. This is due to the fact that the adversary may have permission to legally execute programs on the server or have other forms of influence that can somewhat control the environment.

The local leakage model we consider is closely related to other models that were considered in the literature under the names "only computation leaks" (OCL) [7,15,26,42], "intrusion resilience" [20], or "bounded communication leakage" [31]. These alternative models are typically more general in that they allow the leakage to be *adaptive*, or computable by an interactive protocol, whereas the leakage model we consider is non-adaptive.

Despite its apparent simplicity, our local leakage model can be quite powerful and enable very damaging attacks. In particular, in any linear secret sharing scheme over a field $\mathbb{F}_{2^k}$ of characteristic 2, an adversary can learn a bit of the secret by leaking just one bit from each share. Surprisingly, in the case of Shamir's scheme, full recovery of a multi-bit secret is possible by leaking only one bit from each share [34]. Some of the most efficient implementations of MPC protocols (such as the ones in [2,16,36]) are based on secret sharing schemes over $\mathbb{F}_{2^k}$ and are thus susceptible to such an attack.

As mentioned earlier, most prior works on leakage-resilient cryptography (see Sect. 1.2 below) design *special-purpose* leakage-resilient schemes. These works have left open the question of analyzing (variants of) standard schemes and protocols. Such an analysis is motivated by the hope to obtain better efficiency and additional security features.

1.1   Our Results

We obtain three kinds of results. First, we analyze the local leakage resilience of linear secret sharing schemes. Then, we apply these results to prove the leakage resilience of some natural MPC protocols. Finally, we present a somewhat unexpected application of these techniques to rule out the existence of certain *local share conversion* schemes. Our results are based on Fourier analytic techniques developed in the context of additive combinatorics. See Sect. 1.2 for details. We now give a more detailed overview of these results.

Leakage resilience of linear secret sharing schemes. In a linear secret sharing scheme over a finite field $\mathbb{F}$, the secret is an element $s \in \mathbb{F}$ and the share obtained by each party consists of one or more linear combinations of s and ℓ random field elements. We consider a scenario where n parties hold a linear secret sharing of either s_0 or s_1 specified by the adversary $\mathcal{A}$. (Due to linearity, we can assume without loss of generality that $s_0 = 0$ and $s_1 = 1$.) The adversary can also specify arbitrary leakage functions $\tau^{(1)}, \tau^{(2)}, \ldots, \tau^{(n)}$ such that each function $\tau^{(j)}$ outputs m bits of leakage from the share held by the j-th party. The adversary's goal is to determine if the shared secret is s_0 or s_1. In this setting we provide the following theorems.

Theorem 1.1 (Informally, Additive Secret Sharing). *Let p be a prime. There exists a constant $c_p < 1$ such that, for sufficiently large n, the additive secret sharing scheme over $\mathbb{F}_p$ is local leakage resilient when $\lfloor (\log p)/4 \rfloor$ bits are leaked from every share.*

In more detail, for any $\lfloor (\log p)/4 \rfloor$-bit output leakage functions $\tau^{(1)}, \ldots \tau^{(n)}$ and any two secrets s_0, s_1, the statistical distance between the leakage distributions $\boldsymbol{\tau}(\mathbf{x})$ and $\boldsymbol{\tau}(\mathbf{y})$ is at most pc_p^n, where $\boldsymbol{\tau}(\mathbf{x}) = (\tau^{(1)}(x^{(1)}), \ldots \tau^{(n)}(x^{(n)}))$ is obtained by applying the leakage functions to a random share $\mathbf{x} = (x^{(1)}, \ldots, x^{(n)})$ of s_0 and, similarly, $\boldsymbol{\tau}(\mathbf{y})$ is obtained by leaking from random shares of s_1.

For a more precise statement see Corollaries 4.6 and 4.7.

In contrast to the theorem above, if the additive secret sharing were over $\mathbb{F}_{2^k}$, the adversary could distinguish between the two secrets by just leaking the least significant bit of each share and adding those up to reveal the least significant bit of the secret. We show the following result for Shamir's secret sharing.

Theorem 1.2 (Informally, Shamir Secret Sharing). *For large enough n, for primes $p \approx n$, the (n, t)-Shamir secret sharing[1] over $\mathbb{F}_p$ is local leakage resilient for $t = n - o(\log n)$ when $(\log p)/4$ bits are leaked from every share.*

Shamir's secret sharing is typically used with threshold $t = cn$ for some constant $c > 0$, in which case the above result is not applicable. While we cannot prove local leakage resilience, we do not know of attacks in this parameter regime. We conjecture the following:

Conjecture 1.3 (Shamir Secret Sharing). Let $c > 0$ be a constant. For large enough n, $(n, t = cn)$-Shamir Secret Sharing is 1-bit local leakage resilient. That is, for any family of functions $\tau^{(1)}, \ldots, \tau^{(n)}$ with 1-bit output,

$$\mathrm{SD}(\boldsymbol{\tau}(\mathbf{x}), \boldsymbol{\tau}(\mathbf{u})) < \mathsf{negl}(n)$$

where $\boldsymbol{\tau}(\mathbf{x}) = \left(\tau^{(1)}(x^{(1)}), \ldots \tau^{(n)}(x^{(n)})\right)$ where $\mathbf{x} \leftarrow \mathsf{ShaSh}_{n,cn}(0)$ and $\mathbf{u} \leftarrow \mathbb{F}_p^n$.

Classical MPC protocols like the BGW protocol use Shamir secret sharing with $c = 1/3$ or $1/2$. Observe that proving the conjecture for a specific constant c immediately implies the conjecture for any constant $c' > c$. This follows from the fact that (n, cn)-Shamir Shares can be locally converted to random $(n, c'n)$-Shamir Shares for $c' > c$.[2]

Application to leakage-resilient MPC. We use the leakage resilience of linear secret sharing schemes to show that the *honest-but-curious* variant of the GMW [25] protocol with a "Beaver Triples" setup [3] (that we call GMW with shared product preprocessing) is local leakage resilient.

For the MPC setting, we modify the leakage model as follows to allow for a stronger adversary. The adversary $\mathcal{A}$ is allowed to corrupt a fraction of the parties, see their shares and views of the entire protocol execution. In addition, $\mathcal{A}$ specifies local leakage functions for the non-corrupted parties and receives the corresponding leakage on their individual views.

The honest-but-curious GMW protocol with shared product preprocessing works as follows. The parties wish to evaluate an arithmetic circuit C on an input x. The parties receive random shares of the input x under a linear secret sharing scheme and random shares of Beaver triples under the same scheme.[3]

---

[1] In the whole paper, a (n, t)-Shamir secret sharing scheme or Shamir secret sharing scheme with threshold t uses polynomials of degree t, so that the secret cannot be recovered from a collusion of up to t parties. The secret can be recovered from $t + 1$ parties.

[2] This can be done by locally adding shares of a random $(n, c'n)$-Shamir share of 0 to the given (n, cn)- Shamir shares.

[3] A Beaver triple consists of (a, b, ab) where a, b are randomly chosen field elements.

The protocol proceeds gate by gate where the parties maintain a secret sharing of the value at each gate. For input, addition and inverse (-1) gates, parties locally manipulate their existing shares to generate the shares for these gates. For multiplication gate, where we multiply z_1 and z_2 to get z, the parties first construct $z_1 - a$ and $z_2 - b$ by subtracting the shares of the inputs and Beaver triples (a, b, ab) and broadcasting these values. Then the parties can locally construct a secret sharing of $z = z_1 \cdot z_2$ by using the following relation:

$$z = (z_1 - a)(z_2 - b) + a(z_2 - b) + b(z_1 - a) + ab .$$

We show that when the underlying protocol is local leakage resilient, this protocol can also tolerate local leakage. We can prove leakage resilience in a simulation-based definition. See Sect. 5 for details. Informally, when the additive secret sharing scheme is used, we show the following.

Theorem 1.4 (Informally, Leakage Resilience of GMW). *For large enough n, for any prime p, the GMW protocol with shared product preprocessing and additive secret sharing over $\mathbb{F}_p$ is local leakage resilient where the adversary can corrupt $n/2$ parties, learn their entire state and, then locally leak $(\log p)/4$ bits each from all the uncorrupted parties.*

On the impossibility of local share conversion. In the problem of local share conversion [4,14], n parties hold a share of a secret s under a secret sharing scheme $\mathcal{L}$. Their goal is to *locally*, without interaction, convert their shares to shares of a related secret s' under a different secret sharing scheme $\mathcal{L}'$ such that (s, s') satisfy a pre-specified relation R. We assume R is not trivial in the sense that it is permissible to map shares of every secret s to shares of a fixed constant. Local share conversion has been used to design protocols for Private Information Retrieval [4]. More recently, different kinds of local share conversion were used to construct Homomorphic Secret Sharing (HSS) schemes [11,17,23]. Using techniques similar to the ones for leakage resilience, we rule out certain nontrivial instances of local share conversion. We first state our results and then discuss their relevance to constructions of HSS schemes.

Theorem 1.5 (Informally, Impossibility of Local Share Conversion). *Three-party additive secret sharing over $\mathbb{F}_p$, for any prime $p > 2$, cannot be converted to additive secret sharing over $\mathbb{F}_2$, with constant success probability $(> 5/6)$, for any non-trivial relation R on the secrets.*

The proof of this result uses a Fourier analytic technique similar to the analysis of the Blum-Luby-Rubinfeld linearity test [9]. We also show a similar impossibility result for Shamir secret sharing. See the full version for the precise general statement. This result relies crucially on a technique of Green and Tao [33]. We elaborate more in Sect. 2.

Relevance to HSS Schemes. At the heart of the DDH-based 2-party HSS scheme of Boyle et al. [11] and its Paillier-based variant of Fazio et al. [23] is an efficient local share conversion algorithm of the following special form. The two parties hold shares g^x and g^y respectively of $b \in \{0, 1\}$, such that $g^b = g^x \cdot g^y$. The conversion algorithm enables them to *locally* compute additive shares of the bit b over the *integers* $\mathbb{Z}$, with small (inverse polynomial) failure probability. Note that this implies similar conversion to additive sharing over $\mathbb{Z}_2$. One approach to constructing 3-party HSS schemes would be to generalize this local share conversion scheme to 3 parties, i.e., servers holding random g^x, g^y and g^z respectively, such that $g^b = g^x \cdot g^y \cdot g^z$, can locally convert these shares to additive shares of the bit b over integers. We rule out this approach by showing that even when given the exponents x, y and z in the clear (i.e. $x + y + z = b$ over $\mathbb{Z}_p$), locally computing additive shares of b over $\mathbb{Z}_2$ (or the integers) is impossible. A similar share conversion from (noisy) additive sharing over $\mathbb{Z}_p$ to additive sharing over $\mathbb{Z}_2$ was used by Dodis et al. [17] to obtain an LWE-based construction of 2-party HSS and spooky encryption. However, in this case there is an alternative route of reducing the multi-party case to the 2-party case that avoids our impossibility result.

1.2   Related Work

Our work was inspired by the surprising result of Guruswami and Wootters [34] mentioned above. This work turned attention to the fact that some natural linear secret sharing schemes miserably fail to offer local leakage resilience over fields of characteristic 2, in that leaking only one bit from each share is sufficient to fully recover a multi-bit secret.

The traditional "leakage" model considered in multi-party cryptography allows the adversary to fully corrupt up to t parties and learn their entire secret state. This t-bounded leakage model motivated secret sharing schemes designed to protect *information* [8,43] and secure multiparty computation (MPC) protocols designed to protect *computation* [6,13,25,45]. The same leakage model was also considered at the hardware level, where parties are replaced by atomic gates [35]. The t-bounded leakage considered in all these works is quite different from the local leakage model we consider: we allow *partial* leakage from *every* secret state, whereas the t-bounded model allows *full* leakage from up to t secret states. While resilience to t-bounded leakage was shown to imply resilience to certain kinds of "noisy leakage" [18,22] or "low-complexity leakage" [10], it clearly does not imply local leakage resilience in general. Indeed, additive secret sharing over $\mathbb{F}_{2^k}$ is highly secure in the t-bounded model and yet is totally insecure in the local leakage model.

The literature on leakage resilient cryptography is extensive, thus we discuss a few of the most relevant works. A closely related work by Dziembowski and Pietrzak [21] is one of those works that design new constructions to withstand leakage. Their secret sharing scheme uses artificially long shares that are hard to retrieve in full, as the model bounds the amount of bits that can be leaked.

The length of the shares of course impacts the performance of the protocol. The reconstruction of the secret is an interactive process.

Boyle et al. [12] consider the problem of leakage-resilient coin-tossing and reduce it to a certain kind of leakage-resilient verifiable secret sharing. Here too, a new construction of (nonlinear) secret sharing is developed in order to achieve these results.

Goldwasser and Rothblum [26] give a general transformation that takes any algorithm and creates a related algorithm that computes the same function and can tolerate leakage. This approach can be viewed as a special-purpose MPC protocol for a constant number of parties that offers local leakage resilience (and beyond) [7]. However, this construction is quite involved and offers poor concrete leakage resilience and efficiency overhead.

Most relevant to our MPC-related results is the recent work of Goyal et al. [31] on leakage resilient secure two-party computation (see also [24]). This work analyzes the resilience of a GMW-style protocol under a similar (in fact, more general) type of leakage to the local leakage model we consider. One key difference is that the protocol from [31] modifies the underlying circuit (incurring a considerable overhead) whereas we apply the GMW protocol to the original circuit. Also, our approach applies to a large number of parties of which a large fraction can be entirely corrupted, whereas the construction in [31] is restricted to the two-party setting.

Our results use techniques developed in the context of additive combinatorics. See Tao and Vu [44] for an exposition on Fourier analytic methods used in additive combinatorics. The works most relevant to ours are works by Green and Tao [33] and follow-ups by Gowers and Wolf [28–30]. The relation of these works and their techniques to ours is discussed in Sect. 2.4.

2 Overview of the Techniques

2.1 Leakage Resilience of Secret Sharing Schemes

Very simple local leakage attacks exist for linear secret sharing schemes over small characteristic fields. These attacks stem from the abundance of additive subgroups in these fields. This gives rise to the hope that linear schemes over fields of prime order, that lack such subgroups, are leakage resilient. We start by considering the simpler case of additive secret sharing.

Additive secret sharing. We define $\mathsf{AddSh}(s)$ to be a function that outputs random shares $x^{(1)}, ..., x^{(n)}$ such that $\sum x^{(i)} = s$. Let $\boldsymbol{\tau} = \tau^{(1)}, \tau^{(2)}, \ldots, \tau^{(n)}$ be some leakage functions. We want to show that for all secrets $s_0, s_1 \in \mathbb{F}$, the leakage distributions are statistically close. That is,

$$\left\{ \boldsymbol{\tau}(\mathbf{x}) : \mathbf{x} \leftarrow \mathsf{AddSh}(s_0) \right\} \approx \left\{ \boldsymbol{\tau}(\mathbf{x}) : \mathbf{x} \leftarrow \mathsf{AddSh}(s_1) \right\}$$

where $\boldsymbol{\tau}(\mathbf{x}) = \tau^{(1)}(x^{(1)}), \ldots, \tau^{(n)}(x^{(n)})$ is the total leakage the adversary sees on the shares $\mathbf{x} = x^{(1)}, x^{(2)}, \ldots, x^{(n)}$.

We know that there is a local leakage attack on $\mathbb{F}_{2^k}$: simply leak the least significant bit (lsb) from all the parties and add the outputs to reconstruct the lsb of the secret. What enables the attack on $\mathbb{F}_{2^k}$ while $\mathbb{F}_p$ is unaffected?

To understand this difference, it is instructive to start with an example. Let us consider additive secret sharing over $\mathbb{F}_{2^k}$ for 3 parties. We know that,

$$\mathsf{lsb}(x) = \mathsf{lsb}(x^{(1)}) + \mathsf{lsb}(x^{(2)}) + \mathsf{lsb}(x^{(3)}).$$

This attack works because $\mathbb{F}_{2^k}$ has many subgroups that are closed under addition. Let $A_0 = \mathsf{lsb}^{-1}(0)$ and $A_1 = \mathsf{lsb}^{-1}(1)$. The set A_0 is an additive subgroup of $\mathbb{F}_{2^k}$ and A_1 is a coset of A_0. Furthermore, the lsb function is a homomorphism from $\mathbb{F}_{2^k}$ to the quotient group[4] $\mathbb{F}_{2^k}/A_0$. The lsb leakage tells us which coset each share $x^{(j)}$ is in. Then by adding these leakages, we can infer if $x \in A_0$ or $x \in A_1$ (i.e., to which coset it belongs).

Let us consider the analogous situation over $\mathbb{F}_p$ for a prime p. The group $\mathbb{F}_p$ does not have any subgroups. In fact, it has an opposite kind of expansion property: that adding *any* two sets results in a larger set.

Theorem 2.1 (Cauchy-Davenport Inequality). *Let $A, B \subset \mathbb{F}_p$. Let $A+B = \{a + b : a \in A \text{ and } b \in B\}$. Then,*

$$|A + B| \geq \min(p, |A| + |B| - 1).$$

So, if we secret shared a random secret over $\mathbb{F}_p$ and got back leakage output indicating that $x^{(1)} \in B_1$, $x^{(2)} \in B_2$, and $x^{(3)} \in B_3$, we can infer that $x \in B_1 + B_2 + B_3$. But because of this expansion property, the set $B_1 + B_2 + B_3$ is a lot larger than the sets B_i's individually. This is in contrast to the $\mathbb{F}_{2^k}$ case where e.g. $A_0 + A_1$ was the same size as A_0.

This gives an idea of why the lsb attack does not work. Some information is lost because of expansion. This is not sufficient for us though. What we need to show is stronger. We want to show that even given the leakage, the secret is *almost completely hidden*. This is a more "distributional" statement.

We model it as follows: Let us say that we have n parties where party j holds the share $x^{(j)}$. The adversary $\mathcal{A}$ has specified leakage functions $\tau^{(j)} : \mathbb{F}_p \to \{0, 1\}^m$ and received back the leakage $\boldsymbol{\ell} = \ell_1, \ell_2, \ldots, \ell_n$ where $\ell_j = \tau^{(j)}(x^{(j)})$: the leakage on the j-th share. We want to show that even conditioned on this leakage, the probability that the secret was s_0 vs s_1 is close to a half. That is, we want to show the following:

$$\Pr_{\mathbf{x} \leftarrow \mathsf{AddSh}(s_0)} [\boldsymbol{\tau}(\mathbf{x}) = \boldsymbol{\ell}] \approx \Pr_{\mathbf{x} \leftarrow \mathsf{AddSh}(s_1)} [\boldsymbol{\tau}(\mathbf{x}) = \boldsymbol{\ell}] . \tag{1}$$

---

[4] To recall, in the quotient group $\mathbb{F}_{2^k}/A_0$, the elements are the cosets A_0, A_1. The sum of two cosets is the coset formed by the sum of elements of the first cosets with elements of the second coset. Because of the structure, $A_0 + A_0 = A_0$, $A_0 + A_1 = A_1$ and so on.

Below, we will sketch an argument showing that leaking from the additive shares of 0 is statistically close to leaking from a uniformly random element

$$\Pr_{\mathbf{x} \leftarrow \mathsf{AddSh}(0)} [\tau(\mathbf{x}) = \ell] \approx \Pr_{\mathbf{u} \leftarrow U} [\tau(\mathbf{u}) = \ell] . \tag{2}$$

From Eq. (2), Eq. (1) follows by a simple hybrid argument as shares of any other secret s are simply shares of 0 with the secret s added to the first party's share. That is, let $\mathbf{e}_1 = (1, 0, 0, \dots, 0)$,

$$\{\mathbf{x} + s \cdot \mathbf{e}_1 : \mathbf{x} \leftarrow \mathsf{AddSh}(0)\} \equiv \{\mathbf{y} : \mathbf{y} \leftarrow \mathsf{AddSh}(s)\} .$$

To understand this probability better, let us consider the following operator:

$$\Lambda(f_1, f_2, \dots, f_n) = \mathbb{E}_{\mathbf{x} \leftarrow \mathsf{AddSh}(0)} \left[f_1(x^{(1)}) \cdot f_2(x^{(2)}) \cdots f_n(x^{(n)}) \right] .$$

By picking the functions f_j's appropriately, we can model the probability. Define $1_{\ell_j} : \mathbb{F}_p \rightarrow \{0, 1\}$ as follows: $1_{\ell_j}(x) = 1$ if the output of the leakage function $\tau^{(j)}$ on input x is ℓ_j, i.e., $\tau^{(j)}(x) = \ell_j$ and, 0 otherwise. Notice that we can write the probability of leakage output being ℓ in terms of the operator Λ as follows,

$$\Pr_{\mathbf{x} \leftarrow \mathsf{AddSh}(0)} [\tau(\mathbf{x}) = \ell] = \Lambda(1_{\ell_1}, 1_{\ell_2}, \dots, 1_{\ell_n}) .$$

The probability of the leakage being ℓ on the uniform distribution is simply a product of the expectations:

$$\Pr_{\mathbf{u} \leftarrow U} [\tau(\mathbf{u}) = \ell] = \mathbb{E}_{\mathbf{u} \leftarrow U} [1_\ell(\mathbf{u})] = \mathbb{E}_{\mathbf{u} \leftarrow U} [1_{\ell_1}(u_1) \cdot 1_{\ell_2}(u_2) \cdots 1_{\ell_n}(u_n)]$$

where $1_\ell(\mathbf{u}) = 1_{\ell_1}(u^{(1)}) \cdot 1_{\ell_2}(u^{(2)}) \cdots 1_{\ell_n}(u^{(n)})$. So, we want to show:

$$\Lambda(1_{\ell_1}, 1_{\ell_2}, \dots, 1_{\ell_n}) = \mathbb{E}_{\mathbf{u} \leftarrow U} [1_\ell(\mathbf{u})] + \varepsilon .$$

The tool we use to bound the difference $|\Lambda(1_\ell) - \mathbb{E}_{\mathbf{u} \leftarrow U} [1_\ell(\mathbf{u})]|$ is Fourier analysis. At the heart of this is a nice property of the Λ operator: the Fourier spectrum of Λ is very similar to the standard form as follows. For Λ defined over a linear code C:

$$\Lambda(f_1, f_2, \dots, f_n) = \mathbb{E}_{\mathbf{x} \leftarrow C} \left[f_1(x^{(1)}) \cdot f_2(x^{(2)}) \cdots f_n(x^{(n)}) \right] ,$$

Λ can be equivalently represented on the dual code $C^\perp$ (see Lemma 4.9),

$$= \sum_{\vec{\alpha} \in C^\perp} \widehat{f_1}(\alpha_1) \cdot \widehat{f_2}(\alpha_2) \cdots \widehat{f_n}(\alpha_n)$$

with the 'Fourier coefficients' $\widehat{f}(a) = \mathbb{E}_{y \leftarrow \mathbb{F}_p} [f(x) \cdot \omega^{\alpha x}]$ where $\omega = \exp(2\pi i / p)$ is a root of unity. Observe that as $\widehat{1_\ell}(0) = \mathbb{E}_x [1_\ell(x)]$. So, $\mathbb{E}_{\mathbf{u} \leftarrow U} [1_\ell(\mathbf{u})] = \widehat{1_{\ell_j}}(0) \cdot$

$\widehat{1_{\ell_j}}(0) \cdots \widehat{1_{\ell_n}}(0)$, the term corresponding to the all-zeros codeword in the dual code. Hence, the error term we have to bound is the following:

$$\Lambda(1_\ell) - \mathop{\mathbb{E}}_{\mathbf{u} \leftarrow U}[1_\ell(\mathbf{u})] = \sum_{\vec{a} \in C^\perp \setminus \{0\}} \widehat{1_{\ell_1}}(\alpha_1) \cdot \widehat{1_{\ell_2}}(\alpha_2) \cdots \widehat{1_{\ell_n}}(\alpha_n) \,.$$

Note that, at this point, it is interesting to observe how the presence of subgroups (over $\mathbb{F}_{2^k}$) and the lack thereof (over $\mathbb{F}_p$) manifests itself. Over $\mathbb{F}_{2^k}$ because of the non-trivial subgroups, these non-zero Fourier coefficients can be large and hence the error term is not small. On the other hand, over $\mathbb{F}_p$, we can show that each non-zero Fourier coefficient is *strictly smaller* than the zero-th coefficient and measurably so. This lets us bound the error term. First we elaborate on the large Fourier coefficient over $\mathbb{F}_{2^k}$ and then we state results for $\mathbb{F}_p$.

Large coefficients over $\mathbb{F}_{2^k}$. Each Fourier basis function over $\mathbb{F}_{2^k}$ is indexed by a vector $\vec{a} \in \{0,1\}^k$ and the Fourier coefficient for $\vec{a}$ is given by $\hat{f}(\vec{a}) = \mathbb{E}_{\vec{x} \leftarrow \mathbb{F}_{2^k}}[f(\vec{x})(-1)^{\langle \vec{a}, \vec{x} \rangle}]$. Over $\mathbb{F}_{2^k}$, non-zero Fourier coefficients can be as large as the zero-th coefficient, which is always the largest for binary valued functions.

To use the running example, in the case of the lsb function, let $\tau^{(j)} = $ lsb and consider the $1_{\mathsf{lsb}=1}$ to be the function which returns 1 if the lsb is 1 and 0 otherwise. So, $1_{\mathsf{lsb}=1}$ is 1 on the set A_1 and 0 on A_0. The non-zero Fourier coefficient indexed by $\vec{e}_k = (0,0,\ldots 0,1) \in \{0,1\}^k$ is as large as the zero-th Fourier coefficient since: $\widehat{1_{\mathsf{lsb}=1}}(\vec{0}) = \mathbb{E}_{\vec{x}}[1_{\mathsf{lsb}=1}(\vec{x})] = 0.5$ as half of the inputs satisfy lsb $= 1$, and also, $\widehat{1_{\mathsf{lsb}=1}}(\vec{e}_k) = \mathbb{E}_{\vec{x}}[1_{\mathsf{lsb}=1}(\vec{x}) \cdot (-1)^{x_k}] = \mathbb{E}_{\vec{x}}[1_{\mathsf{lsb}=1}(\vec{x}) \cdot (-1)] = -0.5$ because when $1_{\mathsf{lsb}=1}(x) = 1$, then $x_k = 1$ and $1_{\mathsf{lsb}=1}(\vec{x}) \cdot (-1)^{x_k} = -1$. So, these two Fourier coefficients are equally large in magnitude. Hence the error term can be quite large.

Bounds on $\mathbb{F}_p$. Bounding $\widehat{1_{\ell_j}}(\alpha)$ for non-zero $\alpha \in \mathbb{F}_p$, we prove prove the following result:

$$\mathrm{SD}\left(\tau(C), \tau(U)\right) \leq \frac{1}{2} \cdot |C^\perp| \cdot \left(\frac{2^m \sin(\pi/2^m)}{p \sin(\pi/p)}\right)^t,$$

where SD denotes the statistical distance between the two distributions, $\tau(C) = \{\tau(\mathbf{x}) : \mathbf{x} \leftarrow C\}$, $\tau(U) = \{\tau(\mathbf{x}) : \mathbf{x} \leftarrow U\}$ (with U being the uniform distribution over $\mathbb{F}_p^n$), and t is the minimum distance of the dual code $C^\perp$. We prove this formally in Sect. 4.3. When applied to the code $C = \mathsf{AddSh}(0)$, we have $|C^\perp| = p$ and this implies that additive secret sharing is leakage resilient, proving Theorem 1.1. We strengthen the result in Corollary 4.7 to avoid this dependence on $|C^\perp| = p$.

Applying the result to Reed-Solomon Codes, the codes underlying (t,n)-Shamir Secret Sharing, gives us Theorem 1.2. Also note that in the case of Shamir secret sharing, $|C^\perp| = p^{n-t}$ and hence this proof works only when $n-t$ is small.

2.2   Application to Leakage Resilience of MPC Protocols

Given the leakage resilience of additive secret sharing over $\mathbb{F}_p$, we can show that the following *honest-but-curious* variant of the GMW protocol [25] (GMW with

---

GMW Protocol with Shared Product Preprocessing

Setup: Given an arithmetic circuit C over field $\mathbb{F}$ computing f. C has gates from the basis $\mathbb{B} = \{+, \times, -1\}$ where the -1 gate negates the input. For convenience, we have input gates that read a field element from the input.

Input Encoding: On input $\vec{x}$, randomly secret share $\vec{x}$ using additive secret sharing. i.e., $\vec{x}^{(1)}, \vec{x}^{(2)}, \ldots, \vec{x}^{(n)} \leftarrow \mathsf{AddSh}(\vec{x})$. Party j gets $\vec{x}^{(j)}$.

Randomness: Let $G_\times$ be the set of multiplication gates in C. For each multiplication gate g in $G_\times$, generate a Beaver triple: $\mathbf{a}_g \leftarrow \mathsf{AddSh}(a_g)$, $\mathbf{b}_g \leftarrow \mathsf{AddSh}(b_g)$ and $(\mathbf{ab})_g \leftarrow \mathsf{AddSh}(a_g \cdot b_g)$ for $a_g, b_g \leftarrow \mathbb{F}$.

Protocol Π: Party j receives an input $\vec{x}^{(j)}$ and randomness $(a_g^{(j)}, b_g^{(j)}, (ab)_g^{(j)})_{g \in G_\times}$. The parties traverse the gates in the circuit C in a predetermined order where every gate is traversed only after its input gates. Let $\mathbf{z}_g$ denote the secret sharing of the value z_g at gate g. For each gate they do the following:

1. If gate g is not a multiplication gate, the parties *locally* generate:

$$\mathbf{z}_g = \begin{cases} \mathbf{x}_i & \text{if } g \text{ is an input gate reading } x_i \\ -\mathbf{z}_{g_1} & \text{if } g \text{ is a } -1 \text{ gate with input } g_1 \\ \mathbf{z}_{g_1} + \mathbf{z}_{g_2} & \text{if } g \text{ is a } + \text{ gate with input } g_1 \text{ and } g_2 \end{cases}$$

2. If g is a $\times$ gate, with input g_1 and g_2, then the parties do the following:
   (a) Locally compute $\mathbf{a}'_g = \mathbf{z}_{g_1} - \mathbf{a}_g$ and $\mathbf{b}'_g = \mathbf{z}_{g_2} - \mathbf{b}_g$ and broadcast these values.
   (b) Receive the corresponding values from other parties.
   (c) Compute $z_{g_1} - a_g$ and $z_{g_2} - b_g$ by adding all the values received.
   (d) Compute $\mathbf{z}_g = (z_{g_1} - a_g)(z_{g_2} - b_g) \cdot \mathbf{1} + (z_{g_1} - a_g) \cdot \mathbf{b}_g + \mathbf{a}_g \cdot (z_{g_2} - b_g) + (\mathbf{ab})_g$ where $\mathbf{1}$ a fixed secret sharing of the value 1.

Fig. 1. GMW Protocol with Shared Product Preprocessing

shared product preprocessing) using Beaver Triples [3] is leakage resilient. The protocol is described in Fig. 1. Recall that in our leakage model, the adversary $\mathcal{A}$ is allowed to corrupt a fraction of the parties, see their views of the entire protocol execution and then specify leakage functions $\tau^{(j)}$ for the non-corrupted parties and receive this leakage on their individual views.

We consider two settings, the first being with *private outputs* where the adversary does not see the output of the non-corrupted parties and the second with *public outputs* where the parties broadcast their output shares at the end to reconstruct the final output and the adversary sees them.

In both models, we show that the adversary's view (i.e., the views of the corrupted parties and the leakage on all the uncorrupted parties' views) can be simulated by a simulator which gets nothing (in the private-outputs setting) and gets all the shares of the output (in the public-outputs setting).

To prove the result, we need two ingredients: (a) the leakage resilience of additive secret sharing over $\mathbb{F}_p$ and, (b) a lemma formalizing the following intuition: *In the GMW protocol, each party learns a share of a secret sharing of the value at each gate in the circuit and nothing more.* The first ingredient we have shown

above, and we now describe the second. In Lemmas 5.8 and 5.9, we formally state
and prove this intuition in both the private-outputs and public-outputs setting
and here we provide an informal statement.

Lemma 2.2 (Informal). *On an input $\vec{x}$, let $(z_g)_{g \in G_\times}$ denote the value at multiplication gate g. The joint view of any subset Θ of the parties, $\mathsf{view}^{(\Theta)}$, can be
simulated given their shares of the inputs and of the values at each multiplication
gate.*

$$\mathsf{view}^{(\Theta)}(x) \equiv \mathsf{Sim}(\vec{x}^{(\Theta)}, (z_g^{(\Theta)})_{g \in G_\times}).$$

Given the lemma, proving local leakage resilience in the private-outputs setting is a hybrid argument. Because of the lemma, the adversary can leak from
party j a function of $\vec{x}^{(j)}$ and $(z_g^{(j)})_{g \in G_\times}$. The simulator LeakSim, not knowing
the input $\vec{x}$, picks random values $\vec{x}', (z_g')_{g \in G_\times}$ instead, secret shares them and
then leaks from these values according to the leakage functions $\tau^{(j)}$ specified
by $\mathcal{A}$.

Then via a hybrid, we show that these two distributions are close to each
other. If the local leakage can distinguish between the two distributions, then
we can use them to construct leakage functions that violate the local leakage
resilience of a single instance of the underlying secret sharing scheme.

The proof in the public-outputs setting has a subtlety that the adversary
sees not only the local leakage from the uncorrupted parties, but also their
final outputs. In this case, we first observe that the final output is a fixed linear
function of the circuit values z_g of the multiplication gates and of the input values
x_i. Using this observation, the simulator picks the shares of the multiplication
gates conditioned on the output values seen. And we can show a similar reduction
to the local leakage resilience of the underlying secret sharing scheme. This proves
Theorem 1.4.

2.3   On Local Share Conversion

In this section, we sketch the techniques used to show Theorem 1.5: that three-
party additive secret sharing over $\mathbb{F}_p$, for any prime $p > 2$, cannot be converted
to additive secret sharing over $\mathbb{Z}_2$, even with a small error, for any non-trival
relation R on the secrets.

Our results on impossibility of local share conversion are derived by viewing
the output of the share conversion schemes as leakage on the original shares and
that the adversary instead of being able to do arbitrary computation, can only
add the leakage outputs over $\mathbb{Z}_2$.

Impossibility of Share Conversion of Additive Secret Sharing from $\mathbb{F}_p$ to $\mathbb{Z}_2$. We
start with the impossibility of local share conversion of additive secret sharings

from $\mathbb{F}_p$ to $\mathbb{Z}_2$ for any non-trivial relation R on the secrets.[5] The analysis is inspired by Fourier analytic reinterpretations of linearity testing [9] and group homomorphism testing [5].

Assume that $g_1, g_2, g_3 : \mathbb{F}_p \to \mathbb{Z}_2$ form a 3-party local share conversion scheme for additive secret sharing for some relation R where shares of 0 in $\mathbb{F}_p$ have to be mapped to shares of 0 in $\mathbb{Z}_2$ and shares of 1 in $\mathbb{F}_p$ have to be mapped to shares of 1 in $\mathbb{Z}_2$ (with high probability, say 99%).[6] That is, if $x_1 + x_2 + x_3 = b$, then $g_1(x_1) + g_2(x_2) + g_3(x_3) = b$ for $b \in \{0, 1\}$. It is convenient for us to define the real-valued analogues $G_i(x) = (-1)^{g_i(x)}$. At the heart of this proof is the following operator:

$$\Lambda(G_1, G_2, G_3) = \underset{\mathbf{x} \leftarrow \mathsf{AddSh}(0)}{\mathbb{E}} [G_1(x_1) \cdot G_2(x_2) \cdot G_3(x_3)] .$$

The first observation is that if shares of 0 over $\mathbb{F}_p$ are mapped to shares of 0 over $\mathbb{Z}_2$ with high probability (say 99%), then the value of this operator is quite high as,

$$\Lambda(G_1, G_2, G_3) = 1 - 2 \cdot \underset{\mathbf{x} \leftarrow \mathsf{AddSh}(0)}{\Pr} [g_1(x_1) + g_2(x_2) + g_3(x_3) \neq 0] \geq 0.98 .$$

The crux of the argument is an 'inverse theorem' style lemma which characterizes functions G_1's that result in a large value for Λ. This lemma shows that if $\Lambda(G_1, G_2, G_3)$ is high, then each of the functions G_1, G_2 and G_3 are 'almost' constant functions, i.e., for most x's, $G_i(x)$ is the same fixed value. Given this lemma, the impossibility result follows. Because the functions G_i's (and hence g_i's) are almost always constant, even given secret shares of 1 as input, they would still output shares of 0 as output.

To complete the proof, we need to argue that G_1 is an almost constant function. This proof has two parts: the first part which is generic to any field $\mathbb{F}$ is to show that if Λ is large, then G_1 has a large Fourier coefficient. In the second part, we show that if G_1 has a large Fourier coefficient, then G_1 is an almost constant function. This part is specific to $\mathbb{F}_p$.

To show the first part, we rewrite $\Lambda(G_1, G_2, G_3)$ over the Fourier basis (using Lemma 4.9) to get

$$\Lambda(G_1, G_2, G_3) = \sum_{a \in \mathbb{F}_p} \widehat{G}_1(a) \cdot \widehat{G}_2(a) \cdot \widehat{G}_3(a)$$

this follows from Lemma 4.9 as the dual code of additive shares of 0 is the code generated by the all-ones vector. We can now use Cauchy-Schwarz inequality

---

[5] A relation is trivial if no matter what secret is shared, a constant output by the conversion scheme would satisfy correctness. Or put another way, in a non-trivial relation R, there exist s_0 and s_1 such that s_0 has to be mapped to 0 and s_1 has to be mapped to 1 in the relation R.

[6] We consider more general case in the full version which also tolerates a higher error probability of $1/6$.

with the fact that $\sum_a |\widehat{G}_i(a)|^2 = 1$ to get that,

$$\leq \left|\widehat{G}_1\right|_\infty \cdot \left(\sum_a |\widehat{G}_2(a)|^2\right) \cdot \left(\sum_a |\widehat{G}_3(a)|^2\right) \leq \left|\widehat{G}_1\right|_\infty .$$

This implies that $|\widehat{G}_1|_\infty$ is large. Now we show the second part, which is specific to $\mathbb{F}_p$. We need to show that g_1 is almost constant function. We want to show that if some Fourier coefficient of G_1 is large (larger than $\frac{2}{3}$), then it has to be the zero-th coefficient. The zero-th coefficient measures the bias of G_1: if the coefficient is small, then G_1 is close to balanced, and if this coefficient is large, then G_1 is an almost constant function. Although proving this for all primes is somewhat tedious, the intuition is easy to grasp. Let $p = 3$ and $\omega = \exp(2\pi i/3)$ be a root of unity. A non-zero Fourier coefficient of G_1 takes the following form: $\widehat{G}_1(a) = \mathbb{E}_{x \in \mathbb{Z}_3} [G_1(x) \cdot \omega^{ax}]$ for $a \neq 0$. Because G_1 takes values in $\{-1, 1\}$ and ω^{ax} takes all values $\{1, \omega, \omega^2\}$, these two functions cannot be too correlated. And hence the Fourier coefficient cannot be too large: $|\widehat{G}_1(a)| \leq 2/3$. This completes the proof.

The Impossibility of Share Conversion from Shamir Secret Sharing from $\mathbb{F}_p$ to Additive Sharing on $\mathbb{F}_2$. We now briefly discuss the techniques used to prove the result on local conversion of (n, t)-Shamir secret sharing over $\mathbb{F}_p$, for $n = 2t - 1$. Again consider a relation R where Shamir shares of 0 over $\mathbb{F}_p$ have to be mapped to additive shares of 0 over $\mathbb{F}_2$ and Shamir shares of 1 have to be mapped to additive shares of 1 over $\mathbb{F}_2$. Let $g_1, g_2, \ldots, g_5$ be the local share conversion functions used. We want to follow a similar strategy: first show that the corresponding function $G_i = (-1)^{g_i}$ has a large Fourier coefficient. Then, similar to the additive secret sharing proof, show that if G_i has a large Fourier coefficient, then G_i is 'almost constant' and hence derive a contradiction.

In the first part, we want to use the fact that Shamir shares of 0 over $\mathbb{F}_p$ are converted to additive shares of 0 over $\mathbb{F}_2$ to infer that G_1 (say) has a large Fourier coefficient. The proof is a specialized case of the work of Green and Tao [33]. In the proof, the value of an appropriately defined operator Λ:

$$\Lambda(G_1, G_2, \ldots, G_n) = \mathbb{E}_{\mathbf{x} \leftarrow C} [G_1(x_1) \cdot G_2(x_2) \cdots G_n(x_n)]$$

is bound by the "Gowers' Uniformity Norm" (the U^2 norm) of the function G_1. Then using a connection between the U^2 norm and Fourier bias, we can derive that G_1 has a large Fourier coefficient. For details see the full version.

2.4 Additive Combinatorics Context

We provide some context for these techniques. Such Λ style operators have been studied quite a bit in number theory. They can be used to represent many fascinating questions about the distribution of prime numbers. To give some examples, *What is the density of three-term arithmetic progressions in primes?* is a question about the operator $\Lambda = \mathbb{E}_{x,d} [1_P(x)1_P(x + d)1_P(x + 2d)]$ where 1_P is 1 if x is a prime and 0 otherwise. Also, the *twin primes conjecture* can be framed in terms of the operator $\Lambda = \mathbb{E}_x \left[1_P(x) \cdot 1_P(x+2) \right]$. Green and Tao [33] and subsequent works by Wolf and Gowers [28–30] tried to understand the following question: let $L_1, L_2, \ldots, L_m$ be linear equations from $\mathbb{F}^n$ to $\mathbb{F}$. Can we bound the following expectation:

$$\Lambda(f_1, f_2, \ldots f_m) = \mathop{\mathbb{E}}_{\vec{x} \leftarrow \mathbb{F}^n} \left[f_1(L_1(\vec{x})) \cdot f_2(L_2(\vec{x})) \cdots f_m(L_m(\vec{x})) \right] ?$$

This is a very general question. And roughly speaking, they give the following answer. These works define two measures of complexity (termed as Cauchy-Shwarz Complexity and True Complexity respectively) and show that if a system of linear equations has complexity k, then,[7]

$$\Lambda(f_1, f_2, \ldots, f_m) < C \cdot \min_i \ \|f_i\|_{U^k} \ ,$$

where $\|f_i\|_{U^k}$ is the k-th order Gowers' Uniformity Norm [27]. This method of bounding Λ by the Gowers' norm has been very influential in number theory. This method is what we use to prove the results on Shamir secret sharing. We first bound an appropriately defined operator Λ by the Gowers' U^2 norm and then exploit a connection between the U^2 and Fourier analysis. Such a technique does not suffice to give desired results in the case of leakage resilience of $(n, t = cn)$-Shamir secret sharing for two reasons (for some constant $c > 0$). The first reason is that constant C derived from this method is often extremely large and has an exponential dependence on the number of equations m. Also the second reason is that in our setting, the functions f_i's are chosen by the adversary. So, showing that $\|f_i\|_{U^k}$ is small is either very challenging or just not true for some adversarially chosen functions f_i's. On the other hand, we do not know how to translate this in to an local leakage attack on Shamir secret sharing either and hence a strong win-win result eludes us.

3 Preliminaries

We denote by $\mathbb{C}$ the field of complex numbers, by SD the statistical distance (or total variation distance), and by $\equiv$ the equality of distributions.

3.1 Linear Codes

Secret sharing schemes are closely related to linear codes, that we define next.

Definition 3.1 (Linear Code). *A subset $C \subseteq \mathbb{F}^n$ is an $[n, k, d]$-linear code over field $\mathbb{F}$ if C is a subspace of $\mathbb{F}^n$ of dimension k such that: for all $\vec{x} \in$*

---

[7] Both complexity measures do not assign complexity to all possible linear forms. To give an example, the linear form $(L_1(x) = x, L_2(x) = x + 2)$, which corresponds to the twin primes conjecture, is not assigned a complexity value and the twin primes conjecture is still open.

$C \setminus \{\vec{0}\}$, HammingDistance$(\vec{x}) \geq d$ (i.e., the minimum distance between two elements of the code is at least d). A code is called **Maximum Distance Separable (MDS)** if $n - k + 1 = d$. The **dual code** of the code C is defined as $C^{\perp} = \{\vec{y} \in \mathbb{F}^n : \forall \vec{x} \in C, \langle \vec{x}, \vec{y} \rangle = 0\}$.

Proposition 3.2. *The dual code $C^{\perp}$ of an $[n, k, d]$ MDS code C is itself an MDS code with parameters $[n, n - k, k + 1]$.*

Example 3.3 (Reed-Solomon Code). The $[n, k, n-k+1]$-Reed-Solomon code over $\mathbb{F}$ such that $|\mathbb{F}| > n$ interprets a message $\vec{m} \in \mathbb{F}^k$ as $p(x) = m_1 + m_2 x + \cdots + m_k x^{k-1}$ and encodes it as $(p(\alpha_1), p(\alpha_2), \ldots, p(\alpha_n))$ where $A = \{\alpha_1, \alpha_2 \ldots \alpha_n\} \subseteq \mathbb{F}$ is a fixed set of evaluation points. Reed-Solomon code is an MDS code.

3.2 Linear Secret Sharing Schemes

We recall the definition of (threshold) secret sharing schemes.

Definition 3.4 (Secret Sharing Scheme). *An (n, t)-secret sharing scheme over field $\mathbb{F}$ is defined by a pair (Share, Rec) where Share is a randomized mapping of an input $s \in \mathbb{F}$ to shares for each party $\mathbf{s} = (s^{(1)}, s^{(2)}, \ldots, s^{(n)})$ and the reconstruction algorithm Rec is a function mapping a set A and the corresponding shares $\mathbf{s}^{(A)} = (s^{(j)})_{j \in A}$ to a secret $s \in \mathbb{F}$, such that the following properties hold:*

1. *Reconstruction.* Rec$(A, \mathbf{s}^{(A)})$ *outputs the secret s for all A where $|A| > t$.*
2. *Security. For any set A such that $|A| \leq t$, the joint distribution of shares received by the subset of parties A, $\mathbf{s}^{(A)} = (s^{(j)})_{j \in A}$ where $\mathbf{s} \leftarrow$ Share(s), is independent of the secret s.*

When we use these schemes to encode vectors, it should be interpreted as sharing each element of the vector under the underlying scheme.

An important particular case of secret sharing scheme are linear secret sharing schemes. Actually all the schemes we consider in this paper are linear.

Definition 3.5. *An (n, t)-SSS (Share, Rec) over $\mathbb{F}$ is linear if*

1. *the codomain of Share is the vector space $(\mathbb{F}^\ell)^n$, for some positive integer ℓ (i.e., each share is a vector of ℓ field elements),*
2. *for any $s \in \mathbb{F}$, Share(s) is uniformly distributed over an affine subspace of $(\mathbb{F}^\ell)^n$,*
3. *for any $\lambda_0, \lambda_1, s_0, s_1 \in \mathbb{F}$:*

$$\left\{ \lambda_0 \mathbf{s}_0 + \lambda_1 \mathbf{s}_1 : \begin{array}{l} \mathbf{s}_0 \leftarrow \text{Share}(s_0) \\ \mathbf{s}_1 \leftarrow \text{Share}(s_1) \end{array} \right\} \equiv \text{Share}(\lambda_0 s_0 + \lambda_1 s_1).$$

Let us now recall the two classical linear secret sharing schemes we are using.

Example 3.6 (Additive Secret Sharing $(\mathsf{AddSh}_n, \mathsf{AddRec}_n))$*.* The additive secret sharing scheme $(\mathsf{AddSh}_n, \mathsf{AddRec}_n)$ for n parties over a field $\mathbb{F}$ is a linear $(n, n-1)$-secret sharing scheme defined as follows. Shares $\mathsf{AddSh}_n(s) = \mathbf{s}$ of a secret $s \in \mathbb{F}$ are generated as follows: $(s^{(1)}, \ldots, s^{(n-1)}) \leftarrow \mathbb{F}^{n-1}$, and $s^{(n)} = s - (s^{(1)} + \cdots + s^{(n-1)})$. The reconstruction of s from $\mathbf{s}$ is done as follows: $\mathsf{AddRec}_n(\mathbf{s}) = s^{(1)} + \cdots + s^{(n)}$.

Example 3.7 (Shamir Secret Sharing $(\mathsf{ShaSh}_{n,t}, \mathsf{ShaRec}_{n,t}))$*.* The Shamir secret sharing scheme $(\mathsf{ShaSh}_{n,t}, \mathsf{ShaRec}_{n,t})$ of degree t for n parties over a field $\mathbb{F}$ (with $|\mathbb{F}| > n$) is a linear (n, t)-secret sharing scheme defined as follows. Let $\alpha_1, \ldots, \alpha_n \in \mathbb{F}^*$ be n distinct arbitrary non-zero field elements. Shares $\mathsf{ShaSh}_{n,t}(s) = \mathbf{s}$ of a secret $s \in \mathbb{F}$ are generated as follows: generate a uniformly random polynomial P of degree at most t over $\mathbb{F}$ with constant coefficient s (i.e., $P(0) = s$), the share $s^{(j)}$ is $s^{(j)} = P(\alpha_j)$. Given shares $s^{(A)}$ with $A \subseteq [n]$ and $|A| > t$, the reconstruction works as follows: it computes the Lagrange coefficients $\lambda_j = \prod_{i \in A \setminus \{j\}} (\alpha_i / (\alpha_i - \alpha_j))$ and output $\mathsf{ShaRec}_{n,t}(A, s^{(A)}) = \sum_{j \in A} \lambda_j s^{(j)} \in \mathbb{F}$.

3.3 Fourier Analysis

In this section, we present the notion of Fourier coefficients of a function and some of its properties. Most of the calculations needed about Fourier coefficients are deferred to the corresponding sections for the ease of readability. For an excellent survey on how Fourier Analytic methods are used in Additive Combinatorics, see [32].

Let $\mathbb{G}$ be any finite abelian group. A **character** is a homomorphism $\chi : \mathbb{G} \to \mathbb{C}$ from the group $\mathbb{G}$ to $\mathbb{C}$, i.e., $\chi(a + b) = \chi(a) \cdot \chi(b)$ for all $a, b \in \mathbb{G}$. For any finite abelian group $\mathbb{G}$, the set of characters $\widehat{\mathbb{G}}$ is a group (under the operation point-wise product) isomorphic to $\mathbb{G}$. We will use $\widehat{F}(\alpha)$ to denote the Fourier coefficient corresponding to χ_α. The reader should note that while we define Fourier coefficients in generality, we would be primarily use Fourier analysis on the groups $\mathbb{F}_p$ for some prime p.

Definition 3.8 (Fourier Coefficients). *For functions* $f : \mathbb{G} \to \mathbb{C}$*, the* Fourier basis *is composed of the group* $\widehat{\mathbb{G}}$ *of characters* $\chi : \mathbb{G} \to \mathbb{C}$*. We define the* Fourier coefficient $\widehat{f}(\chi)$ *corresponding to a character* χ *as*

$$\widehat{f}(\chi) = \mathop{\mathbb{E}}_{x \leftarrow \mathbb{G}} [f(x) \cdot \chi(x)] \in \mathbb{C}.$$

As we would use Fourier analysis on the additive group $\mathbb{F}_p = \mathbb{F}_p$. We describe the Fourier characters over $\mathbb{F}_p$. Let $\omega = \exp(2\pi i/p)$ be a primitive p-th root of unity. Then, the characters for $\mathbb{F}_p$ are given by $\chi_\alpha(x) = \omega^{\alpha \cdot x}$ where $\alpha \in \mathbb{F}_p$. We sometimes abuse notation and write $\widehat{f}(\alpha)$ instead of $\widehat{f}(\chi_\alpha)$.

We follow the "standard" notation in additive combinatorics. In this notation, when working on the group $\mathbb{G}$, the *Haar measure* is used which assigns the weight $|\mathbb{G}|^{-1}$ to every $x \in \mathbb{G}$ and when working on $\widehat{\mathbb{G}}$, the *counting measure* is used which assigns the weight 1 to every $\alpha \in \widehat{\mathbb{G}}$. Using these measures generally eliminates the need for normalization. So, when we talk about norms, these will always be taken with respect to the underlying measure. That is,

$$\|f\|_1 = \mathop{\mathbb{E}}_x\left[|f(x)|\right] \quad \text{whereas} \quad \|\widehat{f}\|_2 = \left(\sum_\alpha |\widehat{f}(\alpha)|^2\right)^{1/2}.$$

We note that the Fourier Transform has the following properties. These follow easily from the orthogonality relation on the characters: $\sum_{x\in\mathbb{F}_p} \omega^{a\cdot x}$ is p when $a = 0$ and 0 otherwise.

Theorem 3.9. *Let $f, g : \mathbb{G} \to \mathbb{C}$ be two functions. Let $\widehat{\mathbb{G}}$ denote the group of characters of $\mathbb{G}$. The following hold:*

(a) *(Parseval's identity) We have,*

$$\mathop{\mathbb{E}}_{x\leftarrow\mathbb{G}}\left[f(x)\cdot\overline{g(x)}\right] = \sum_{\chi\in\widehat{\mathbb{G}}} \widehat{f}(\chi)\cdot\overline{\widehat{g}(\chi)}$$

In particular, $\|f\|_2 = \|\widehat{f}\|_2$ where $\|f\|_2 = \mathbb{E}_{x\leftarrow\mathbb{G}}\left[f(x)^2\right]$ and $\|\widehat{f}\|_2 = \sum_{\chi\in\widehat{\mathbb{G}}} \widehat{f}(\chi)^2$.

(b) *(Fourier Inversion Formula) For any $x \in \mathbb{G}$, $f(x) = \sum_{\chi\in\widehat{\mathbb{G}}} \widehat{f}(\chi)\cdot\overline{\chi(x)}$.*

Finally, we introduce the notion of bias. A function is biased if it is highly correlated with some Fourier character.

Definition 3.10 (Bias). *For a function $f : \mathbb{G} \to \mathbb{C}$, the bias of f is defined as,*

$$bias(f) = \|\widehat{f}\|_\infty = \max_{\chi\in\widehat{\mathbb{G}}} \widehat{f}(\chi).$$

We need a calculation on certain sums of roots of unity. Let A be a subset of $\mathbb{Z}_k$. And let $\gamma = e^{i\cdot 2\pi/k}$. We want to bound sums of the form $\gamma^A = \sum_{x\in A} \gamma^x$. We state and prove the Lemma below. We will use the lemma to show that non-trivial Fourier coefficients of certain functions have to be smaller than the trivial one.

Lemma 3.11. *Let k be a positive integer. Let $\zeta_k : [0, k] \to \mathbb{R}_{\geq 0}$ be defined as $\zeta_k(x) = \frac{\sin(x\pi/k)}{\sin(\pi/k)}$ with $\zeta(0) = 0$. Let $A \subset \mathbb{Z}_k$ of size t. Let $A^* = \{0, 1, \ldots t-1\}$. Then*

$$\left|\gamma^A\right| \leq \left|\gamma^{A^*}\right| = \frac{\sin(\pi t/k)}{\sin(\pi/k)} = \zeta_k(t).$$

We will show that the sum is maximized when A is an interval. The proof of the claim is an extremal argument. If an element does not lie in the direction of the sum, we can remove it and add something in the direction to increase the norm.

Proof. Pick the A that maximizes this sum. If possible, let A not be an interval. Let $\zeta = \sum_{a\in A} \omega^a$. If $|\zeta| = 0$, then the lemma holds as the sum over an interval of the same size would be higher. Hence, let $|\zeta| > 0$, consider the subset A' of size t consisting of all the roots of unity most 'aligned' with ζ. That is, for all $a \in A'$ and $b \in \{0, 1 \ldots k-1\} \setminus A'$, $\omega^a \circ \zeta \geq \omega^b \circ \zeta$ where $\circ$ is the complex dot product.[8] If $A = A'$, then we are done.

Otherwise, pick $a \in A' \setminus A$ and $b \in A \setminus A'$. Consider the set $A'' = (A \setminus \{b\}) \cup \{a\}$. We claim that it has a bigger sum. That is, $|\gamma^{A''}| \geq |\gamma^A| = |\zeta|$. Observe that $\gamma^{A''} = \zeta - b + a$. And as $\zeta \circ a \geq \zeta \circ b$, $\zeta \circ (a - b) \geq 0$. Hence, $\cos \theta \geq 0$ where θ is the angle between ζ and $(a - b)$. This implies that $\theta \in [-\pi/2, \pi/2]$ and hence $|\zeta - b + a| = |\zeta + (a - b)| \geq |\zeta|$. This yields a contradiction if A is not an interval.

The fact that $\gamma^{A^*} = \zeta_k(t)$ is derived using a basic trigonometry calculation:

$$\left|\gamma^{A^*}\right| = \left|\sum_{i=0}^{t-1} \gamma^i\right| = \frac{|\gamma^t - 1|}{|\gamma - 1|} = \frac{2\sin(\pi t/k)}{2\sin \pi/k}$$

where the last equality follows from the fact that the angle between γ^t and -1 is $(\pi - 2t\pi/k)$ and hence, $|\gamma^t - 1| = 2\cos((\pi - 2t\pi/k)/2) = 2\sin(\pi t/k)$. And the result follows. $\qquad \square$

4 On Leakage Resilience of Secret Sharing Schemes

4.1 Definitions and Basic Properties

We consider a model of leakage where the adversary can first choose a subset of $\Theta \subseteq [n]$ parties and get their full shares and then leak m bits each from all the shares of all the (other) parties. Formally, what is learned by the adversary on a sharing $\mathbf{s}$ is the following:

$$\mathsf{Leak}_{\Theta, \tau} = (\mathbf{s}^{(\Theta)}, (\tau^{(i)}(\mathbf{s}^{(\Theta)}, s^{(i)}))_{i \in [n]}) \tag{3}$$

where $\tau = (\tau^{(1)}, \tau^{(2)} \ldots \tau^{(n)})$ is a family of n leakage functions that output m bits and $\mathbf{s}^{(\Theta)} = (s^{(j)})_{j \in \Theta}$ are the complete shares of the parties corrupted. The adversary can choose the functions $\vec{\tau}$ arbitrarily.

Definition 4.1 (Local Leakage Resilient). *Let Θ be a subset of $[n]$. A secret sharing scheme $(\mathsf{Share}, \mathsf{Rec})$ is said to be (Θ, m, ε)-local leakage resilient (or (Θ, m, ε)-LL resilient for short) if for every leakage function family $\tau = (\tau^{(1)}, \tau^{(2)} \ldots \tau^{(n)})$ where $\tau^{(j)}$ has an m-bit output, and for every pair of secrets s_0, s_1,*

$$SD\left(\{\mathsf{Leak}_{\Theta, \tau}(\mathbf{s}) \; : \; \mathbf{s} \leftarrow \mathsf{Share}(s_0)\}, \{\mathsf{Leak}_{\Theta, \tau}(\mathbf{s}) \; : \; \mathbf{s} \leftarrow \mathsf{Share}(s_1)\}\right) \leq \varepsilon.$$

A secret sharing scheme $(\mathsf{Share}, \mathsf{Rec})$ is said to be (θ, m, ε)-LL resilient if it is (Θ, m, ε)-LL resilient for any subset $\Theta \subseteq [n]$ of size at most θ.

---

[8] $z_1 \circ z_2 = x_1 x_2 + y_1 y_2$ where $z_b = x_b + i \cdot y_b$ is the dot product of z_1 and z_2. Equivalently, $z_1 \circ z_2 = |z_1||z_2| \cos \theta$ where θ is the angle between z_1 and z_2.

Remark 4.2. We remark that we can consider an equivalent definition where for each distribution $\mathcal{D}$ of leakage function family $\tau = (\tau^{(1)}, \tau^{(2)} \ldots \tau^{(n)})$:

$$\mathsf{SD}\left(\left\{\mathsf{Leak}_{\Theta, \tau}(\mathbf{s}) : \begin{matrix} \mathbf{s} \leftarrow \mathsf{Share}(s_0) \\ \tau \leftarrow \mathcal{D} \end{matrix}\right\}, \left\{\mathsf{Leak}_{\Theta, \tau}(\mathbf{s}) : \begin{matrix} \mathbf{s} \leftarrow \mathsf{Share}(s_1) \\ \tau \leftarrow \mathcal{D} \end{matrix}\right\}\right) \leq \varepsilon.$$

Observe that an standard notion of (n, t)-secret sharing scheme corresponds to $(t, 0, 0)$-Local Leakage resilient: that is, complete access to the shares of t parties and no information about the others.

Note that in the leakage model, the adversary is not allowed to adaptively choose the leakage functions. As discussed in the introduction, this is a very meaningful and well-motivated leakage model. Next, demonstrate some attacks in this model. We formalize the observation that linear secret sharing schemes over small characteristic fields are not local leakage resilient.

Example 4.3 (Attack on Schemes Over Small Characteristic Fields). Over fields of small characteristic like $\mathbb{F}_{2^k}$ that have many additive subgroups, secret sharing schemes with linear reconstruction are not local leakage resilient. We give some examples of such attacks. They are not hard to generalize. Let $x \in \mathbb{F}_{2^k}$ be the secret that is shared among n-parties as shares $(x^{(1)}, x^{(2)} \ldots x^{(n)})$. Consider the following attacks:

- *Additive Secret Sharing.* The adversary can locally leak the least significant bit of each share $x^{(j)}$. Adding them up, the adversary can reconstruct the least significant bit of x.
- *Shamir Secret Sharing.* For a similar attack, observe that $x = \lambda_1 x^{(1)} + \lambda_2 x^{(2)} + \cdots + \lambda_n x^{(n)}$ where λ_j's are *fixed* Lagrange coefficients. So to attack the scheme, the adversary locally multiplies the share $x^{(j)}$ with λ_j and leaks the least significant bit. This again reveals the least significant bit of x. The recent work of Guruswami and Wootters [34] shows how such leakage can be used to even *completely reconstruct* x.

Example 4.4 (Attack on Few Parties). If the number of parties n is a constant, then the additive secret sharing over $\mathbb{F}_p$ is not LL-resilient. The adversary can distinguish between secrets $< p/2$ and $> p/2$ by local leakage. The adversary locally leaks $\tau^{(j)}(x^{(j)}) = 1$ if the share $x^{(j)} < p/(2n)$ (seeing the share as integer in $\{0, \ldots, p-1\}$). If all the leakages output 1, the adversary can conclude that the secret $x = x^{(1)} + \cdots + x^{(n)} < p/2$. On the other hand, if the secret is larger than $p/2$, then all the leakage outputs will never be 1 simultaneously. In the $< p/2$ case, the probability of all the secrets being $< p/2n$ is about $(1/2n)^n$, a constant. Similar attacks can also be performed on Shamir secret sharing. We stress that this is not the most effective attack, but it is an attack nonetheless. This attack is similar to the one in [37, Footnote 8].

4.2 Leakage Resilience of Additive and Shamir Secret Sharings

We are now in a position to state the main technical result of this section. That, if no family of local leakage functions can distinguish between shares picked uniformly at random and shares picked from a 'good' linear code. We will then prove a slightly better parameters in the case of additive-secret sharing.

Theorem 4.5. *Let* $C \subset \mathbb{F}_p^n$ *be any linear* $[n, t, n - t]$ *code. Let* $\tau = (\tau^{(1)}, \tau^{(2)}, \ldots, \tau^{(n)})$ *be any family of leakage functions where* $\tau^{(j)} : \mathbb{F}_p \to \{0, 1\}^m$. *Let* $c_m = \frac{2^m \sin(\pi/2^m)}{p \sin(\pi/p)} < 1$ *(when* $2^m < p$*). Then,*

$$SD(\tau(C), \tau(U_n)) \le \tfrac{1}{2} \cdot p^{n-t} \cdot c_m^t$$

where U_n *is the uniform distribution on* $\mathbb{F}_p^n$ *and:*

$$\tau(C) = \left\{ \left(\tau^{(i)}(x_i) \right)_{i \in [n]} : \vec{x} \leftarrow C \right\}, \quad \tau(U_n) = \left\{ \left(\tau^{(i)}(x_i) \right)_{i \in [n]} : \vec{x} \leftarrow U_n \right\}.$$

We observe that Theorem 4.5 yields the following two corollaries for additive secret sharing and Shamir secret sharing. We can strengthen the result slightly for additive secret sharing. We state it in Corollary 4.7. We first prove the corollaries assuming Theorem 4.5 and then prove Theorem 4.5.

Corollary 4.6 (Leakage Resilience of Additive Secret Sharing). *The additive secret sharing* AddSh_n *for* n *parties is* (θ, m, ε)-*LL resilient where:*

$$\varepsilon = p \cdot c_m^{n-1-\theta} \quad and \quad c_m = \frac{2^m \sin(\pi/2^m)}{p \sin(\pi/p)} < 1 \quad (when\ 2^m < p).$$

Proof. This corollary follows from the following claim after remarking that, when θ parties reveal their share, an additive secret sharing with n parties becomes an additive secret sharing with $n - \theta$ parties.

Claim 4.6.1. *Let* $\tau = (\tau^{(1)}, \tau^{(2)}, \ldots, \tau^{(n)})$ *be any family of* m-*bit output leakage functions. Let* $c_m = \frac{2^m \sin(\pi/2^m)}{p \sin(\pi/p)} < 1$ *(when* $2^m < p$*). Then for all secrets* $s_0, s_1 \in \mathbb{F}_p$,

$$SD\left(\tau(\mathsf{AddSh}_n(s_0)), \ \tau(\mathsf{AddSh}_n(s_1)) \right) \le p \cdot c_m^{n-1}$$

Proof. Let C be the support of $\mathsf{AddSh}(0)$. Note that C is an $[n, n-1, 1]$ linear code and $\mathsf{AddSh}(0)$ is uniformly distributed on C. Also note that the distribution $\mathsf{AddSh}(s)$ is a coset of $\mathsf{AddSh}(0)$, i.e., $\mathsf{AddSh}(s)$ can be obtained by first sampling $\mathbf{x} \leftarrow \mathsf{AddSh}(0)$ and then adding a fixed vector $s \cdot \mathbf{e} = (s, 0, 0, \ldots 0)$ to $\mathbf{x}$. So, for any secret s,

$$SD\left(\tau(\mathsf{AddSh}(s)), \ \tau(U_n) \right) = SD\left(\tau(\mathsf{AddSh}(0) + se), \ \tau(U_n) \right)$$
$$= SD\left(\tau'(\mathsf{AddSh}(0)), \ \tau'(U_n - se) \right)$$

where $\tau'^{(1)}(x) = \tau^{(1)}(x + s)$ and $\tau'^{(j)} = \tau^{(j)}$ for $j > 1$.

$$= SD\left(\tau'(\mathsf{AddSh}(0)), \ \tau'(U_n) \right)$$
$$\le \tfrac{1}{2} \cdot p \cdot c_m^{n-1}.$$

Using triangle inequality, we can complete the proof:

$$\text{SD}\left(\boldsymbol{\tau}(\text{AddSh}(s_0)),\ \boldsymbol{\tau}(\text{AddSh}(s_1))\right)$$
$$\leq \text{SD}\left(\boldsymbol{\tau}(\text{AddSh}(s_0)),\ \boldsymbol{\tau}(U_n)\right)\ +\ \text{SD}\left(\boldsymbol{\tau}(U_n),\ \boldsymbol{\tau}(\text{AddSh}(s_1))\right)$$
$$\leq p \cdot c_m^{n-1}$$

☐
☐

In the case of additive secret sharing, we can strength the result slightly to show the following:

Corollary 4.7 (Leakage Resilience of Additive Secret Sharing). *The additive secret sharing* AddSh_n *for* n *parties is* (θ, m, ε)-*LL resilient where:*

$$\varepsilon = 2^m \cdot c_m^{n-2-\theta} \quad and \quad c_m = \frac{2^m \sin(\pi/2^m)}{p \sin(\pi/p)} < 1 \quad (when\ 2^m < p).$$

Note that the difference between Corollary 4.6 and Corollary 4.7 is that in the stronger claim, the bound on the adversary's advantage does not degrade as the prime increases. Corollary 4.7 is proved in Sect. 4.3. In the case of Shamir secret sharing, we can prove the following result.

Corollary 4.8 (Leakage Resilience of Shamir Secret Sharing). *The Shamir secret sharing* ShaSh_n *for* n *parties is* (θ, m, ε)-*LL resilient where:*

$$\varepsilon = p^{n-t} \cdot c_m^{t-\theta} \quad and \quad c_m = \frac{2^m \sin(\pi/2^m)}{p \sin(\pi/p)} < 1 \quad (when\ 2^m < p).$$

We defer the proofs of both corollaries to the full version.

4.3   Proof of Theorem 4.5

In this section, we prove Theorem 4.5. The proof is divided into three parts. In Lemma 4.9, we show that the operator Λ has a good representation in the Fourier Basis. In Lemma 4.10, we show bounds on the Fourier coefficients and then prove Theorem 4.5 using these two lemmas. Due to the lack of space, we prove Lemmas 4.9 and 4.10 in the full version.

On Fourier Expansion of Λ. First, we show that expectation of product of functions over a code can be represented as a sum of products over the dual code.

Lemma 4.9 (Poisson Summation Formula). *Let* $p > 2$ *be a prime and* $\omega = \exp(\frac{2\pi i}{p})$. *Let* $C \subset \mathbb{F}_p^n$ *be a linear code with dual code is* $C^\perp$. *Let* $f_1, f_2 \ldots f_n : \mathbb{F}_p \to \mathbb{C}$ *be functions. Let* Λ *be defined as follows:*

$$\Lambda(f_1, f_2, \ldots, f_n) = \mathop{\mathbb{E}}_{\vec{x} \leftarrow C} [f_1(x_1) \cdot f_2(x_2) \cdots f_n(x_n)] \ ,$$

where $\vec{x} = (x_1, x_2, \ldots, x_n)$. Then, the following holds:

$$\Lambda(f_1, f_2, \ldots, f_n) = \sum_{\vec{\alpha} \in C^\perp} \widehat{f_1}(\alpha_1) \cdot \widehat{f_2}(\alpha_2) \cdots \widehat{f_n}(\alpha_n)$$

where $\vec{\alpha} = (\alpha_1, \alpha_2, \ldots, \alpha_n) \in \mathbb{F}_p^n$.

Bounds on Fourier Coefficients. We want to bound terms of the form $\widehat{1_A}(\alpha)$ for sets A. We now show bounds on such terms over $\mathbb{F}_p$. We state the lemma required in the proof of Theorem 4.5 for such bounds.

Lemma 4.10. *Let* $\omega = e^{i \cdot 2\pi / p}$. *For any set* A, *let* $\omega^A = \sum_{x \in A} \omega^x$. *For any partition* $A_1, A_2 \ldots A_{2^m}$ *of* $\mathbb{F}_p$, *let* $c_m = \frac{2^m \sin(\pi/2^m)}{p \sin(\pi/p)}$. *Then,*

$$\sum_{i=1}^{2^m} \left| \omega^{A_i} \right| \leq p \cdot c_m.$$

Completing the Proof. At this point, given Lemmas 4.9 and 4.10 we can complete the proof of Theorem 4.5.

Proof (Proof of Theorem 4.5). For the sake of simplicity, we abuse notation and define $1_{\ell_j}(x) = 1$ if $\tau^{(j)}(x) = \ell_j$ and 0 otherwise. We can express the statistical distance as follows:

$$\begin{aligned}
SD(\tau(C), \tau(U_n)) &= \frac{1}{2} \sum_{\vec{\ell}} \left| \mathop{\mathbb{E}}_{\vec{x} \leftarrow C} \left[\prod_j 1_{\ell_j}(x_j) \right] - \mathop{\mathbb{E}}_{\vec{x} \leftarrow U_n} \left[\prod_j 1_{\ell_j}(x_j) \right] \right| \\
&= \frac{1}{2} \sum_{\vec{\ell}} \left| \sum_{\vec{\alpha} \in C^\perp} \prod_j \widehat{1_{\ell_j}}(\alpha_j) - \mathop{\mathbb{E}}_{\vec{x} \leftarrow U_n} \left[\prod_j 1_{\ell_j}(x_j) \right] \right| \\
&= \frac{1}{2} \sum_{\vec{\ell}} \left| \sum_{\vec{\alpha} \in C^\perp \setminus \{0\}} \prod_j \widehat{1_{\ell_j}}(\alpha_j) \right| \\
&\leq \frac{1}{2} \sum_{\vec{\ell}} \sum_{\vec{\alpha} \in C^\perp \setminus \{0\}} \prod_j \left| \widehat{1_{\ell_j}}(\alpha_j) \right| = \frac{1}{2} \sum_{\vec{\alpha} \in C^\perp \setminus \{0\}} \prod_j \left(\sum_{\ell_j} \left| \widehat{1_{\ell_j}}(\alpha_j) \right| \right)
\end{aligned}$$

where the second equality follows from Lemma 4.9, the third equality follows from the fact that $\mathbb{E}_{\vec{x} \leftarrow U_n} \left[\prod_j 1_{\ell_j}(x_j) \right] = \prod_j \left(|(\tau^{(j)})^{-1}(\ell_j)|/p \right) = \prod_j \widehat{1_{\ell_j}}(0)$. The first inequality follows from triangle inequality. To complete the proof, we bound each individual term. Before that, we need the following claim:

Claim 4.10.1. *For any* $\alpha \neq 0$ *and a leakage function* $\tau : \mathbb{F}_p \to \{0,1\}^m$, *let* $1_\ell(x) = 1$ *if and only if* $\tau(x) = \ell$. *Then,* $\sum_{\ell \in \{0,1\}^m} \left| \widehat{1_\ell}(\alpha) \right| < c_m$.

Proof. Observe that $\widehat{1_\ell}(\alpha) = \mathbb{E}_x \left[1_\ell(x) \cdot \omega^{\alpha x} \right] = p^{-1} \cdot \omega^{\alpha \tau^{-1}(\ell)}$. As $\alpha \neq 0$, as the sets $(\tau^{-1}(\ell))_\ell$ partition $\mathbb{F}_p$, so do the sets $(A_\ell = \alpha \tau^{-1}(\ell))_\ell$. Thus, using Lemma 4.10, we get that,

$$\sum_{\ell_j \in \{0,1\}^m} \left| \widehat{1_{\ell_j}}(\alpha_j) \right| = \sum_{\ell_j \in \{0,1\}^m} \left| \omega^{A_\ell} \right| / p \leq c_m$$

This completes the proof. $\qquad\square$

$$\sum_{\vec{\ell} \in (\{0,1\}^m)^n} \prod_j \left| \widehat{1_{\ell_j}}(\alpha_j) \right| = \prod_{j \in [n]} \left(\sum_{\ell_j \in \{0,1\}^m} \left| \widehat{1_{\ell_j}}(\alpha_j) \right| \right)$$

We observe that $\widehat{1_{\ell_j}}(\alpha_j) = \mathbb{E}_x \left[1_{\ell_j}(x) \cdot \omega^{\alpha_j x} \right] = p^{-1} \cdot \omega^{\alpha_j \tau_j^{-1}(\ell_j)}$. If $\alpha_i = 0$, then the sum $\sum_{\ell_j \in \{0,1\}^m} \left| \widehat{1_{\ell_j}}(\alpha_j) \right| = 1$. On the other hand, if $\alpha_j \neq 0$, as the sets $(\tau_j^{-1}(\ell_j))_{\ell_j}$ partition $\mathbb{F}_p$, so do the sets $(A_{\ell_j} = \alpha_j \tau_j^{-1}(\ell_j))_{\ell_j}$. Thus, if $\alpha_j \neq 0$, using Lemma 4.10, we get that, $\sum_{\ell_j \in \{0,1\}^m} \left| \widehat{1_{\ell_j}}(\alpha_j) \right| = \sum_{\ell_j \in \{0,1\}^m} \left| \omega^{A_{\ell_j}} \right| / p \leq c_m$. Hence, we get that,

$$\leq c_m^{\mathsf{HW}(\vec{\alpha})} \leq c_m^t$$

where $\mathsf{HW}(\cdot)$ denotes the Hamming weight. The last inequality follows from the fact that the dual code $C^\perp$ has minimum distance t, as C is a $[n, t, n-t]$ MDS linear code. $\qquad\square$

Now we add up the contributions from each $\vec{\alpha} \in C^\perp$ to complete the proof.

5   Leakage Resilience of GMW with Preprocessing

In this section, we describe an application of the results on leakage resilience of secret sharing to MPC protocols. We show that a variant of the GMW protocol with preprocessing is leakage resilient.

We consider arithmetic circuits over a field $\mathbb{F}$ over a basis $\mathbb{B} = \{+, \times, -1\}$ where the -1 gate negates the input. For convenience, we have input gates that read a field element from the input. The following definition of an MPC protocol is adapted from [31] (Definition 3).

Definition 5.1 (n-party protocol with encoded input and output). *An n-party protocol for* $f : \mathbb{F}^{n_{\text{in}}} \to \mathbb{F}^{n_{\text{out}}}$ *is defined by* $\Pi = (I, \mathbf{R}, \mathbf{M}, O)$, *where:*

- Input Encoder. $I : \mathbb{F}^{n_{\text{in}}} \to (\mathbb{F}^{\hat{n}_{\text{in}}})^n$ *is a randomized* input encoder *circuit, which maps an input* $\vec{x}$ *for* f *to a tuple of protocol inputs* $\mathbf{x} = (\vec{x}^{(1)}, \vec{x}^{(2)}, \ldots, \vec{x}^{(n)})$ *one for each party.*
- Randomness. $\mathbf{R} = (R^{(1)}, R^{(2)}, \ldots, R^{(n)})$ *are distributions over* $\mathbb{F}^{n_r}$ *that capture the random inputs of the parties. They are assumed to be correlated due to preprocessing.*

- $\mathbf{M} = (M^{(1)}, M^{(2)}, \ldots, M^{(n)})$ are deterministic next message functions where $M^{(j)}$ determines the next message sent by party j as a function of its input $\vec{x}^{(j)}$, random input $r^{(j)}$, and the sequence of messages received in the previous rounds. Messages are sent in rounds where each party sends a message to possibly every other party. After a predetermined number of rounds, the function $M^{(j)}$ returns a local output $\vec{y}^{(j)} \in \mathbb{F}^{\hat{n}_{\text{out}}}$ for party j.
- $O : (\mathbb{F}^{\hat{n}_{\text{out}}})^n \to \mathbb{F}^{n_{\text{out}}}$ is a deterministic output decoder circuit, which maps a tuple of protocol outputs $\vec{\mathbf{y}} = (\vec{y}^{(1)}, \ldots, \vec{y}^{(n)})$ to an output $\vec{y}$ of f.

For $\vec{x} \in \mathbb{F}^{\hat{n}_{\text{in}}}$, we denote by $\Pi(\vec{x})$ the output of Π on input $\vec{x}$, namely the result of applying the input encoder I to $\vec{x}$, interacting as specified by $\mathbf{R}, \mathbf{M}$, and applying the output decoder O to the vector of protocol outputs. We say that Π is correctly computes $f : \mathbb{F}^{n_{\text{in}}} \to \mathbb{F}^{n_{\text{out}}}$ if for every input $\vec{x} \in \mathbb{F}^{n_{\text{in}}}$, we have $\Pr[\Pi(\vec{x}) = f(\vec{x})] = 1$.

We denote by $\mathbf{view}(\vec{x})$ the joint distribution $(\text{view}^{(1)}(\vec{x}), \ldots, \text{view}^{(n)}(\vec{x}))$ obtained by running Π on input $\vec{x}$, where $\text{view}^{(j)}$ includes the encoded input $\vec{x}^{(j)}$, the random input $r^{(j)}$ (sampled from $R^{(j)}$), and the sequence of messages received by party j. (The messages sent by party j as well as its output $\vec{y}^{(j)}$ are uniquely determined by $\text{view}^{(j)}$.)

We denote by $\mathbf{out}(\vec{x})$ the joint distribution of the outputs $\vec{\mathbf{y}}$.

5.1 Security Definitions

The definition we consider uses the simulation paradigm. We only consider an *honest-but-curious* definition, albeit one where the adversary can leak information from the views of the uncorrupted parties. We consider two security notions: private-output local leakage resilience and public-outputs local leakage resilience.

In the private-output case, the adversary does not learn the local outputs $\vec{y}^{(j)}$ of non-corrupted parties nor the output $\vec{y} = \Pi(\vec{x})$. This would model the setting where a client wants to delegate some computation $f(\vec{x})$ to some leaky servers: the client secret shares $\vec{x}$ into $\vec{\mathbf{x}}$, sends each share $\vec{x}^{(i)}$ to the server i, the servers run the protocol Π, and each server i sends back its output share $\vec{y}^{(i)}$ to the client.

In the public-outputs case, the adversary learns all the local outputs $\vec{\mathbf{y}}$ of all the parties (and in particular learns the output $\vec{y} = O(\vec{\mathbf{y}}) = \Pi(\vec{x})$). This models a setting where at the end of the computation, the parties would broadcast their local outputs $\vec{y}^{(j)}$ to jointly reconstruct the output $\vec{y}$.

Definition 5.2 (Private-Output Local Leakage Resilient Protocol). *We say that Π is (Θ, m, ε)-private-output local leakage resilient for f (or (Θ, m, ε)-priv-LL-resilient for short) if Π correctly computes f, and the following security requirement holds. For any family of local leakage functions $\tau = (\tau^{(1)}, \tau^{(2)}, \ldots, \tau^{(n)})$ where $\tau^{(j)}$ is a function that outputs m bits, there exists a simulator $\mathsf{LeakSim}_{\Theta, \vec{\tau}}$ such that, and for any input $\vec{x} \in \mathbb{F}^{n_{\text{in}}}$, we have*

$$SD\left(\mathsf{Leak}_{\Theta, \vec{\tau}}(\mathbf{view}(\vec{x})), \mathsf{LeakSim}_{\Theta, \vec{\tau}}()\right) \le \varepsilon.$$

We say that Π is (θ, m, ε)-priv-LL-resilient if Π is (Θ, m, ε)-LL-resilient for all subsets $\Theta \subseteq [n]$ of at most size θ.

Definition 5.3 (Public-Outputs Local Leakage Resilient Protocol). *We say that Π is (Θ, m, ε)-public-outputs local leakage resilient for f (or (Θ, m, ε)-pub-LL-resilient for short) if Π correctly computes f, and the following security requirement holds. For any family of local leakage functions $\tau = (\tau^{(1)}, \tau^{(2)}, \ldots, \tau^{(n)})$ where $\tau^{(j)}$ is a function that outputs m bits, there exists a simulator $\mathsf{LeakSim}_{\Theta, \vec{\tau}}$ such that, and for any input $\vec{x} \in \mathbb{F}^{n_{\mathrm{in}}}$, we have*

$$SD\left((\mathbf{out}(\vec{x}), \mathsf{Leak}_{\Theta, \vec{\tau}}(\mathbf{view}(\vec{x}))), \ \mathsf{LeakSim}_{\Theta, \vec{\tau}}(f(\vec{x}))\right) \leq \varepsilon.$$

We say that Π is (θ, m, ε)-pub-LL-resilient if Π is (Θ, m, ε)-pub-LL-resilient for all subsets $\Theta \subseteq [n]$ of at most size θ.

Both definitions model a protocol executed in the presence of a real-world adversary $\mathcal{A}$ that may corrupt a subset Θ of the parties. The adversary learns the entire view of corrupted parties (and in the second case, also the output of all parties). The adversary also leaks independently m bits from each party. As we consider *semi-honest* corruptions, the adversary can only observe their views but does not modify the messages they send.

Note that the classical notion of security against semi-honest adversaries corrupting θ parties is equivalent to $(\theta, 0, \varepsilon)$-priv-LL-resilient.

5.2 GMW with Shared Product Preprocessing

Notation. Let f be a function computed by a given circuit C. Let G be the set of all gates in C and $G_\times$ be the set of multiplication gates in C. For any input $\vec{x}$, let z_g denote the value at gate $g \in G$ in the circuit C when the input is $\vec{x}$.

In Fig. 2, we describe a variant of the GMW [25] protocol based on the ideas of Beaver triples [3] that we call GMW with shared product preprocessing. The protocol works with any linear secret sharing. We show that if the underlying linear secret sharing is local leakage resilient, then the protocol is pub-LL-resilient and priv-LL-resilient.

Let us first prove correctness.

Proposition 5.4 (Correctness). *The protocol Π in Fig. 2 on any input $\vec{x}$ correctly computes $f(\vec{x})$.*

Proof. To prove correctness, we show that at every gate g, the parties maintain a linear secret sharing of the value z_g. This is easy to verify for the addition, -1 and input gates. We will only do the verification for the multiplication case.

Consider any multiplication gate g with input gates g_1, g_2. Assume that the parties have a valid secret sharing $\mathbf{z}_{g_1}$ and $\mathbf{z}_{g_2}$ of values z_{g_1} and z_{g_2} respectively. Pick any valid Beaver triple $(\mathbf{a}_g, \mathbf{b}_g, (\mathbf{ab})_g)$. We need to show that $\mathbf{z}_g$ as computed is a valid secret sharing of $z_g = z_{g_1} z_{g_2}$. We remark that:

$$\mathbf{z}_g = (z_{g_1} - a_g)(z_{g_2} - b_g)\mathbf{1} + (z_{g_1} - a_g) \cdot \mathbf{b}_g + \mathbf{a}_g \cdot (z_{g_2} - b_g) + (\mathbf{ab})_g.$$

By linearity $\mathbf{z}_g$ is a secret sharing of:

$$(z_{g_1} - a_g)(z_{g_2} - b_g) \cdot 1 + (z_{g_1} - a_g) \cdot b_g + a_g \cdot (z_{g_2} - b_g) + a_g b_g = z_{g_1} z_{g_2}. \quad (4)$$

This concludes the proof. $\qquad\qquad\qquad\qquad\qquad\qquad\qquad\qquad\qquad\qquad\qquad$ $\square$

We have the following security theorems.

Theorem 5.5. *If the linear secret sharing scheme* (Share, Rec) *is* (Θ, m, ε)-*LL-resilient then the protocol* Π *in Fig. 2 is* (Θ, m, ε)-*priv-LL-resilient.*

GMW with Shared Product Preprocessing for computing f with circuit C on field $\mathbb{F}$
Parameters: n the number of parties. (Share, Rec) a secret sharing scheme for n parties. **1** an arbitary sharing of 1.

Input Encoder $I(\vec{x})$:
1. Sample $\vec{x} \leftarrow$ Share($\vec{x}$).
2. Output $\vec{x}$.

Output Decoder $I(\vec{y})$:
1. Output $\vec{y} = $ Rec($\vec{y}$)

Randomness $R(C)$:
1. For each multiplication gate g in C,
   (a) Generate $a_g, b_g \leftarrow \mathbb{F}$.
   (b) Generate $\mathbf{a}_g \leftarrow$ Share(a_g), $\mathbf{b}_g \leftarrow$ Share(b_g), and $(\mathbf{ab})_g \leftarrow$ Share($a_g \cdot b_g$).
   (c) Append to $r^{(j)}$ the tuple $(a_g^{(j)}, b_g^{(j)}, (ab)_g^{(j)})$.
2. Output $\mathbf{r} = (r^{(1)}, r^{(2)}, \ldots, r^{(n)})$.

Protocol run by Party j (defining $M^{(j)}$)
1. Set state$^{(j)} = (n, C, \vec{x}^{(j)})$.
2. Iterate over gates in C fixed order such that for every gate, its input gates are visited before the gate. And run the subprotocol "Process Gate" below.
3. Output $z_{g_{\text{out}}}^{(j)}$: the share of the output gate g_{out}

Process Gate g:
1. If gate g is (a) an input gate with input x_i, or, (b) a (-1) gate with input from gate g', or, (c) a $+$ gate with inputs g_1, g_2, then, set $z_g^{(j)}$ as follows:

$$z_g^{(j)} = \begin{cases} x_i^{(j)} & \text{if } g \text{ is an input gate} \\ -z_{g'}^{(j)} & \text{if } g \text{ is a } -1 \text{ gate} \\ z_{g_1}^{(j)} + z_{g_2}^{(j)} & \text{if } g \text{ is a } + \text{ gate} \end{cases}$$

and append $z_g^{(j)}$ to the list state$^{(j)}$.
2. If g is a $\times$ gate, with input gates g_1 and g_2, then do the following:
   (a) Compute $a_g'^{(j)} = z_{g_1}^{(j)} - a_g^{(j)}$ and $b_g'^{(j)} = z_{g_2}^{(j)} - b_g^{(j)}$ and broadcast these values.
   (b) Receive the corresponding values from other parties.
   (c) Compute $z_{g_1} - a_g$ and $z_{g_2} - b_g$ by adding all the values received.
   (d) Compute $z_g^{(j)} = (z_{g_1} - a_g)(z_{g_2} - b_g) \cdot 1^{(j)} + (z_{g_1} - a_g) \cdot b_g^{(j)} + a_g^{(j)} \cdot (z_{g_2} - b_g) + (ab)_g^{(j)}$
   (e) Append $z_g^{(j)}$ and $(a_g^{(j)}, b_g^{(j)}, (ab)_g^{(j)})$ to state$^{(j)}$.

Fig. 2. GMW Protocol with Shared Product Preprocessing

Theorem 5.6. *If the linear secret sharing scheme* (Share, Rec) *is* (Θ, m, ε)-*LL-resilient then the protocol* Π *in Fig. 2 is* (Θ, m, ε)-*pub-LL-resilient.*

Since an (n, t)-secret sharing scheme is $(t, 0, 0)$-LL-resilient, when instantiated with an (n, t)-secret sharing scheme, the protocol is $(t, 0, 0)$-priv-LL resilient and thus secure against a semi-honest adversaries corrupting up to t parties.

Before we prove Theorems 5.5 and 5.6, let us state the following lemma that is proven in the full version.

Lemma 5.7 (Parallel Composition of LL-Resilience). *If* (Share, Rec) *is a* (Θ, m, ε)-*LL-resilient linear secret sharing scheme, then for any leakage function family* $\tau = (\tau^{(1)}, \tau^{(2)} \ldots \tau^{(n)})$ *where* $\tau^{(j)}$ *has an m-bit output, and for any* $\vec{y}, \vec{y'} \in \mathbb{F}^k$:

$$SD\left(\left\{\mathsf{Leak}_{\Theta, \tau}(\vec{y}) \ : \ \vec{y} \leftarrow \mathsf{Share}(\vec{y})\right\}, \left\{\mathsf{Leak}_{\Theta, \tau}(\vec{y'}) \ : \ \vec{y'} \leftarrow \mathsf{Share}(\vec{y'})\right\}\right) \leq \varepsilon.$$

Note that the bound on statistical distance does not degrade with the size of the vectors.

5.3   Proof of Private-Output Local Leakage Resilience (Theorem 5.5)

To prove the private-output local leakage resilience (Theorem 5.5), we first start with a lemma that characterizes what information the parties see, both individually and jointly. Informally, we show that, when the protocol evaluates the circuit C on input $\vec{x}$, the view of each party (or any subset of parties) can be simulated given a set of common random values and additive shares given to the party (or parties) of the value in each gate. Then, the leakage resilience of the secret sharing scheme allows us to replace the secret sharings used by the simulator by secret sharings of any arbitrary value.

Lemma 5.8. *There exists simulator* S *such that for every input* $\vec{x}$, *the following two distributions are identical.*

$$\mathbf{view}(\vec{x}) \equiv \left\{ \left(\mathsf{S}(j, \vec{x}^{(j)}, (z_g^{(j)}, \mathbf{a}_g', \mathbf{b}_g')_{g \in G_\times})\right)_{j \in [n]} \ : \ \begin{array}{l} \vec{\mathbf{x}} \leftarrow \mathsf{Share}(\vec{x}) \\ (\mathbf{z}_g \leftarrow \mathsf{Share}(z_g))_{g \in G_\times} \\ (a_g', b_g' \leftarrow \mathbb{F})_{g \in G_\times} \\ (\mathbf{a}_g' \leftarrow \mathsf{Share}(a_g'))_{g \in G_\times} \\ (\mathbf{b}_g' \leftarrow \mathsf{Share}(b_g'))_{g \in G_\times} \end{array} \right\}.$$

Assuming Lemma 5.8, the proof of Theorem 5.5 (private-output-LL-resilience of Π) is immediate. Formal proofs are provided in the full version.

5.4   Proof of Public-Outputs Local Leakage Resilience (Theorem 5.6)

To prove the public-outputs local leakage resilience (Theorem 5.6), we extend Lemma 5.8 to take into account the output shares.

Lemma 5.9. *There exists a simulator* S′ *such that for every input* $\vec{x}$, *the following two distributions are identical.*

$(\mathbf{out}(\vec{x}), \mathbf{view}(\vec{x}))$

$$\equiv \left\{ \left(\vec{\mathbf{y}}, \, \mathsf{S}'(j, \vec{\mathbf{y}}, \vec{x}^{(j)}, (z_g^{(j)}, \mathbf{a}_g', \mathbf{b}_g')_{g \in G_\times}) \right)_{j \in [n]} \; : \; \begin{array}{c} \vec{\mathbf{x}} \leftarrow \mathsf{Share}(\vec{x}) \\ (\mathbf{z}_g \leftarrow \mathsf{Share}(z_g))_{g \in G_\times} \\ (a_g', b_g' \leftarrow \mathbb{F})_{g \in G_\times} \\ (\mathbf{a}_g' \leftarrow \mathsf{Share}(a_g'))_{g \in G_\times} \\ (\mathbf{b}_g' \leftarrow \mathsf{Share}(b_g'))_{g \in G_\times} \end{array} \right\}.$$

Assuming Lemma 5.9, the proof of Theorem 5.5 (pub-LL-resilience of Π) is immediate. Formal proofs are provided in the full version.

Acknowledgements. We thank the Crypto reviewers for helpful comments.

The first and fourth authors were supported by the Defense Advanced Research Projects Agency (DARPA) and U.S. Army Research Office (ARO) under Contract No. W911NF-15-C-0236. The second author did some of the work when he was a summer intern at IBM Research. He was supported in part by NSF Grants CNS-1413920 and CNS-1350619, and by the Defense Advanced Research Projects Agency (DARPA) and the U.S. Army Research Office (ARO) under contracts W911NF-15-C-0226 and W911NF-15-C-0236. The third author was supported in part by ERC grant 742754, ISF grant 1709/14, NSF-BSF grant 2015782, and a grant from the Ministry of Science and Technology, Israel and Department of Science and Technology, Government of India.

References

1. Akavia, A., Goldwasser, S., Vaikuntanathan, V.: Simultaneous hardcore bits and cryptography against memory attacks. In: Reingold, O. (ed.) TCC 2009. LNCS, vol. 5444, pp. 474–495. Springer, Heidelberg (2009). https://doi.org/10.1007/978-3-642-00457-5_28

2. Araki, T., Furukawa, J., Lindell, Y., Nof, A., Ohara, K.: High-throughput semi-honest secure three-party computation with an honest majority. In: CCS (2016)

3. Beaver, D.: Efficient multiparty protocols using circuit randomization. In: Feigenbaum, J. (ed.) CRYPTO 1991. LNCS, vol. 576, pp. 420–432. Springer, Heidelberg (1992). https://doi.org/10.1007/3-540-46766-1_34

4. Beimel, A., Ishai, Y., Kushilevitz, E., Orlov, I.: Share conversion and private information retrieval. In: CCC (2012)

5. Ben-Or, M., Coppersmith, D., Luby, M., Rubinfeld, R.: Non-abelian homomorphism testing, and distributions close to their self-convolutions. Random Struct. Algorithms (2008)

6. Ben-Or, M., Goldwasser, S., Wigderson, A.: Completeness theorems for non-cryptographic fault-tolerant distributed computation (extended abstract). In: STOC (1988)

7. Bitansky, N., Dachman-Soled, D., Lin, H.: Leakage-tolerant computation with input-independent preprocessing. In: Garay, J.A., Gennaro, R. (eds.) CRYPTO 2014. LNCS, vol. 8617, pp. 146–163. Springer, Heidelberg (2014). https://doi.org/10.1007/978-3-662-44381-1_9

8. Blakley, G.: Safeguarding cryptographic keys. In: AFIPS National Computer Conference (1979)
9. Blum, M., Luby, M., Rubinfeld, R.: Self-testing/correcting with applications to numerical problems. J. Comput. Syst. Sci. **47**, 549–595 (1993)
10. Bogdanov, A., Ishai, Y., Viola, E., Williamson, C.: Bounded indistinguishability and the complexity of recovering secrets. In: Robshaw, M., Katz, J. (eds.) CRYPTO 2016, Part III. LNCS, vol. 9816, pp. 593–618. Springer, Heidelberg (2016). https://doi.org/10.1007/978-3-662-53015-3_21
11. Boyle, E., Gilboa, N., Ishai, Y.: Breaking the circuit size barrier for secure computation under DDH. In: Robshaw, M., Katz, J. (eds.) CRYPTO 2016. LNCS, vol. 9814, pp. 509–539. Springer, Heidelberg (2016). https://doi.org/10.1007/978-3-662-53018-4_19
12. Boyle, E., Goldwasser, S., Kalai, Y.T.: Leakage-resilient coin tossing. In: Peleg, D. (ed.) DISC 2011. LNCS, vol. 6950, pp. 181–196. Springer, Heidelberg (2011). https://doi.org/10.1007/978-3-642-24100-0_16
13. Chaum, D., Crépeau, C., Damgård, I.: Multiparty unconditionally secure protocols (extended abstract). In: Pomerance, C. (ed.) CRYPTO 1987. LNCS, vol. 293, pp. 462–462. Springer, Heidelberg (1988). https://doi.org/10.1007/3-540-48184-2_43
14. Cramer, R., Damgård, I., Ishai, Y.: Share conversion, pseudorandom secret-sharing and applications to secure computation. In: Kilian, J. (ed.) TCC 2005. LNCS, vol. 3378, pp. 342–362. Springer, Heidelberg (2005). https://doi.org/10.1007/978-3-540-30576-7_19
15. Dachman-Soled, D., Liu, F., Zhou, H.: Leakage-resilient circuits revisited – optimal number of computing components without leak-free hardware. In: Oswald, E., Fischlin, M. (eds.) EUROCRYPT 2015. LNCS, vol. 9057, pp. 131–158. Springer, Heidelberg (2015). https://doi.org/10.1007/978-3-662-46803-6_5
16. Damgård, I., Pastro, V., Smart, N.P., Zakarias, S.: Multiparty computation from somewhat homomorphic encryption. In: Safavi-Naini, R., Canetti, R. (eds.) CRYPTO 2012. LNCS, vol. 7417, pp. 643–662. Springer, Heidelberg (2012). https://doi.org/10.1007/978-3-642-32009-5_38
17. Dodis, Y., Halevi, S., Rothblum, R.D., Wichs, D.: Spooky encryption and its applications. In: Robshaw, M., Katz, J. (eds.) CRYPTO 2016, Part III. LNCS, vol. 9816, pp. 93–122. Springer, Heidelberg (2016). https://doi.org/10.1007/978-3-662-53015-3_4
18. Duc, A., Dziembowski, S., Faust, S.: Unifying leakage models: from probing attacks to noisy leakage. In: Nguyen, P.Q., Oswald, E. (eds.) EUROCRYPT 2014. LNCS, vol. 8441, pp. 423–440. Springer, Heidelberg (2014). https://doi.org/10.1007/978-3-642-55220-5_24
19. Dziembowski, S., Faust, S.: Leakage-resilient circuits without computational assumptions. In: Cramer, R. (ed.) TCC 2012. LNCS, vol. 7194, pp. 230–247. Springer, Heidelberg (2012). https://doi.org/10.1007/978-3-642-28914-9_13
20. Dziembowski, S., Pietrzak, K.: Intrusion-resilient secret sharing. In: FOCS (2007)
21. Dziembowski, S., Pietrzak, K.: Leakage-resilient cryptography. In: FOCS (2008)
22. Faust, S., Rabin, T., Reyzin, L., Tromer, E., Vaikuntanathan, V.: Protecting circuits from leakage: the computationally-bounded and noisy cases. In: Gilbert, H. (ed.) EUROCRYPT 2010. LNCS, vol. 6110, pp. 135–156. Springer, Heidelberg (2010). https://doi.org/10.1007/978-3-642-13190-5_7
23. Fazio, N., Gennaro, R., Jafarikhah, T., Skeith III, W.E.: Homomorphic secret sharing from paillier encryption. In: Okamoto, T., Yu, Y., Au, M.H., Li, Y. (eds.) ProvSec 2017. LNCS, vol. 10592, pp. 381–399. Springer, Cham (2017). https://doi.org/10.1007/978-3-319-68637-0_23

24. Genkin, D., Ishai, Y., Weiss, M.: How to construct a leakage-resilient (stateless) trusted party. In: Kalai, Y., Reyzin, L. (eds.) TCC 2017. LNCS, vol. 10678, pp. 209–244. Springer, Cham (2017). https://doi.org/10.1007/978-3-319-70503-3_7

25. Goldreich, O., Micali, S., Wigderson, A.: How to play any mental game or a completeness theorem for protocols with honest majority. In: STOC 1987 (1987)

26. Goldwasser, S., Rothblum, G.N.: How to compute in the presence of leakage. SICOMP (2015). https://doi.org/10.1137/130931461

27. Gowers, W.T.: A new proof of Szemerédi's theorem. Geom. Funct. Anal. **11**, 465–588 (2001)

28. Gowers, W.T., Wolf, J.: The true complexity of a system of linear equations. Proc. London Math. Soc. **100**, 155–176 (2010)

29. Gowers, W.T., Wolf, J.: Linear forms and higher-degree uniformity for functions on $\mathbb{F}_n^p$. Geom. Funct. Anal. **21**, 36–39 (2011)

30. Gowers, W.T., Wolf, J.: Linear forms and quadratic uniformity for functions on $\mathbb{F}_n^p$. Mathematika **57**, 215–237 (2011)

31. Goyal, V., Ishai, Y., Maji, H.K., Sahai, A., Sherstov, A.A.: Bounded-communication leakage resilience via parity-resilient circuits. In: FOCS (2016)

32. Green, B.: Montréal notes on quadratic Fourier analysis. Add. Comb. **43**, 69–102 (2007)

33. Green, B., Tao, T.: Linear equations in primes. Ann. Math. **171**, 1753–1850 (2010)

34. Guruswami, V., Wootters, M.: Repairing reed-solomon codes. IEEE Trans. Inf. Theory **63**, 5684–5698 (2017)

35. Ishai, Y., Sahai, A., Wagner, D.A.: Private circuits: securing hardware against probing attacks. In: Boneh, D. (ed.) CRYPTO 2003. LNCS, vol. 2729, pp. 463–481. Springer, Heidelberg (2003). https://doi.org/10.1007/978-3-540-45146-4_27

36. Keller, M., Orsini, E., Scholl, P.: MASCOT: faster malicious arithmetic secure computation with oblivious transfer. In: CCS (2016)

37. Kiltz, E., Pietrzak, K.: Leakage resilient elgamal encryption. In: Abe, M. (ed.) ASIACRYPT 2010. LNCS, vol. 6477, pp. 595–612. Springer, Heidelberg (2010). https://doi.org/10.1007/978-3-642-17373-8_34

38. Kocher, P., Genkin, D., Gruss, D., Haas, W., Hamburg, M., Lipp, M., Mangard, S., Prescher, T., Schwarz, M., Yarom, Y.: Spectre attacks: exploiting speculative execution. ArXiv e-prints, January 2018

39. Kocher, P.C.: Timing attacks on implementations of Diffie-Hellman, RSA, DSS, and other systems. In: Koblitz, N. (ed.) CRYPTO 1996. LNCS, vol. 1109, pp. 104–113. Springer, Heidelberg (1996). https://doi.org/10.1007/3-540-68697-5_9

40. Kocher, P., Jaffe, J., Jun, B.: Differential power analysis. In: Wiener, M. (ed.) CRYPTO 1999. LNCS, vol. 1666, pp. 388–397. Springer, Heidelberg (1999). https://doi.org/10.1007/3-540-48405-1_25

41. Lipp, M., Schwarz, M., Gruss, D., Prescher, T., Haas, W., Mangard, S., Kocher, P., Genkin, D., Yarom, Y., Hamburg, M.: Meltdown. ArXiv e-prints

42. Micali, S., Reyzin, L.: Physically observable cryptography. In: Naor, M. (ed.) TCC 2004. LNCS, vol. 2951, pp. 278–296. Springer, Heidelberg (2004). https://doi.org/10.1007/978-3-540-24638-1_16

43. Shamir, A.: How to share a secret. Commun. ACM **22**, 612–613 (1979)

44. Tao, T., Vu, V.H.: Additive Combinatorics. Cambridge University Press, Cambridge (2006)

45. Yao, A.C.: How to generate and exchange secrets (extended abstract). In: FOCS (1986)

Encryption

Threshold Cryptosystems from Threshold Fully Homomorphic Encryption

Dan Boneh[1], Rosario Gennaro[2], Steven Goldfeder[3], Aayush Jain[4],
Sam Kim[1(✉)], Peter M. R. Rasmussen[4], and Amit Sahai[4]

[1] Stanford University, Stanford, USA
skim13@cs.stanford.edu
[2] City College of New York, New York, USA
[3] Princeton University, Princeton, USA
[4] Center for Encrypted Functionalities, UCLA, Los Angeles, USA

Abstract. We develop a general approach to adding a threshold functionality to a large class of (non-threshold) cryptographic schemes. A threshold functionality enables a secret key to be split into a number of shares, so that only a threshold of parties can use the key, without reconstructing the key. We begin by constructing a *threshold* fully-homomorphic encryption scheme (ThFHE) from the learning with errors (LWE) problem. We next introduce a new concept, called a *universal thresholdizer*, from which many threshold systems are possible. We show how to construct a universal thresholdizer from our ThFHE. A universal thresholdizer can be used to add threshold functionality to many systems, such as CCA-secure public-key encryption (PKE), signature schemes, pseudorandom functions, and others primitives. In particular, by applying this paradigm to a (non-threshold) lattice signature system, we obtain the first single-round threshold signature scheme from LWE.

1 Introduction

Threshold cryptography [25,26,28] is a general technique used to protect a cryptographic secret by splitting it into N shares and storing each share on a different server. Any subset of t servers can use the secret without re-constructing it. However, an adversary that compromises $t-1$ servers should not be able to recover or use the secret. Two examples of threshold tasks are:

- **Threshold signatures:** distribute the signing key of a signature system among N servers, so that any t servers can generate a signature. The scheme must provide *anonymity* and *succinctness*. Anonymity means that the same signature is produced, no matter which subset of t servers is used. Succinctness means that the signature size can depend on the security parameter, but must be independent of N and t.

The full version of this paper is available at [12].

H. Shacham and A. Boldyreva (Eds.): CRYPTO 2018, LNCS 10991, pp. 565–596, 2018.
https://doi.org/10.1007/978-3-319-96884-1_19

- **Threshold decryption:** distribute the decryption key of a CCA-secure public-key encryption scheme among N servers, so that any t servers can decrypt. The scheme must be succinct, meaning that ciphertext size must be independent of N and t.

Moreover, the time to verify signatures or encrypt messages should be independent of N and t. Other threshold tasks include threshold (H)IBE key generation, threshold ABE key generation, threshold pseudorandom functions, and many others (see Sect. 1.2). All have similar anonymity, succinctness, and efficiency requirements.

A common goal for threshold systems is to minimize the amount of interaction in the system, and in particular, construct *one-round* schemes. For example, in the case of signatures, an entity called a *combiner* wishes to sign a message m. The combiner sends m to all N servers, and some t of them reply. The combiner combines the t replies, and obtains the signature. No other interaction is allowed. In particular, the servers may not communicate with one another, or interact further with the combiner. Similarly, for threshold decryption, the combiner sends the ciphertext to all N servers, some t servers reply, and the combiner combines the replies to obtain the plaintext. No other interaction is allowed. We will often refer to the servers as *partial signers* or *partial decryptors*.

Many signature and encryption schemes have been thresholdized. For example, RSA signatures and encryption [25,28,31,50], Schnorr signatures [52], (EC)DSA signatures [29,30], BLS signatures [9,14], Cramer-Shoup encryption [21], Regev encryption [7], and many more [10,27,51]. Despite this great success, thresholdizing many basic lattice-based cryptographic primitives has been challenging. For example, it is still an open problem to construct lattice-based *one-round* threshold signatures or CCA-secure threshold PKE satisfying strong succinctness properties, as discussed in related work (Sect. 1.2). Thresholdizing more advanced lattice-based primitives, such as fully homomorphic encryption or functional encryption, has been largely unexplored.

1.1   Our Contributions

Our main contributions are twofold. First, we define the notion of *threshold fully homomorphic encryption* (ThFHE) and construct it from the learning with errors assumption (LWE). As in a threshold PKE, a threshold FHE scheme allows the decryption key to be split into shares such that any t-out-of-N partial decryptions can be combined into a complete decryption of a given ciphertext in a single round. Furthermore, an evaluated ciphertext should be compact meaning that its size is independent of the original message and the number of decryptors N (Sect. 5.1). Second, we present a general framework for universally thresholdizing many (non-threshold) cryptographic schemes using threshold-FHE. This framework lets us resolve the long-standing problem of one-round threshold signatures from lattices.

Threshold FHE. A general obstacle to constructing succinct threshold cryptosystems from lattice assumptions is the noise blow up that results from a multiplication by a large Lagrange coefficient, which prevents correct reconstruction of the message by the combiner.[1] We handle this difficulty in two different ways. Our first method relies on linear secret sharing schemes where the reconstruction coefficients are always binary. We show that this class of secret sharing schemes is compatible with the decryption operation of a fully homomorphic encryption scheme and is also expressive enough to contain threshold access structures, along with other more general structures. In our second method, to achieve better efficiency, we focus on using the standard t-out-of-N Shamir secret sharing scheme, but we modify the noise distribution such that a multiplication by a Lagrange coefficient does not blow up the noise too much. By combining our methods with a suitable FHE scheme, we obtain a secure ThFHE (Sects. 5.2 and 5.3) with strong compactness properties.

A universal thresholdizer. Our second contribution is a general framework for universally thresholdizing many (non-threshold) cryptographic schemes using a ThFHE. For this, we define a new primitive called a *universal thresholdizer*. We show how to construct a universal thresholdizer with strong compactness properties from our ThFHE scheme (Sect. 7). A universal thresholdizer takes in a cryptographic key and produces a number of key shares that can be used to individually evaluate a cryptographic function. Each of these individual evaluation shares can then be combined to result in the final evaluation of the function. We require that the scheme guarantees *privacy* meaning that no $t - 1$ key shares or their evaluation shares reveal any information about the original key. Furthermore, we require that the scheme satisfies *robustness*, meaning that a maliciously generated evaluation share can always be detected.

With these guarantees, a universal thresholdizer scheme can be used to thresholdize many different types of systems. For example, we can take *any* (non-threshold) signature scheme as a black box and construct from it a one-round threshold signature scheme (Sect. 8.1). Since a universal thresholdizer can be proven secure based on LWE, and because there are known (non-threshold) signature schemes based on LWE [15,32,41], we obtain the first one-round threshold signature scheme based on LWE that is both succinct and anonymous. This resolves a long-standing open problem in lattice-based cryptography.

Beyond signatures, a universal thresholdizer can be composed with an existing CCA-secure PKE scheme [1,32,42,45,47] to obtain the first lattice-based (one-round) threshold CCA-secure PKE where the public key size and encryption time are independent of the number of servers. Similarly, composing universal thresholdizer with a functional encryption scheme gives functional encryption with threshold key generation. A universal thresholdizer, on its own, gives a function secret sharing scheme [16,17] that can support threshold access structures. We provide the details of these constructions in [12].

---

[1] In the weaker model where the set of t servers that will respond to the combiner is known ahead of time, there exist simple methods to preserve correctness. In this work, we work in the traditional model of threshold cryptography where the combiner *does not* know the set of t servers ahead of time.

Decentralized threshold FHE. Our basic ThFHE scheme requires a trusted setup procedure to split the secret FHE key into shares. In Sect. 6, we define a *decentralized threshold fully homomorphic encryption* (dThFHE). In a dThFHE scheme, each decryption server generates its own public/secret key pair. At encryption time, the encryptor specifies a set of public keys and a threshold t to produce a ciphertext that can only be decrypted by combining t partial decryptions corresponding to the specific public keys. We construct a dThFHE in Sect. 6. However, while it achieves the added flexibility of decentralized key generation, our dThFHE scheme does not satisfy as strong compactness properties as our ThFHE and universal thresholdizer. We leave improving the compactness of our dThFHE as an important open problem following from our work.

1.2 Related Work on Threshold Lattice Systems

Before describing our results, we first survey the existing work on threshold lattice cryptosystems.

Non-compact systems. We begin with threshold systems that have public key and ciphertext/signature sizes that are linear in N. Bendlin and Damgård [7] gave a threshold version of Regev's CPA-secure encryption scheme [48], and Myers et al. [44] applied the technique to fully homomorphic encryption. Xie et al. [54] gave a threshold CCA secure PKE scheme from lossy trapdoor functions, which can be instantiated from LWE [47]. In all these schemes, both the size of the public key and the ciphertext scales at least linearly in the number of decryptors. For signatures, Cayrel et al. [22] gave a lattice-based threshold ring signature scheme in which at least t signers are needed to create an anonymous signature. In this system, each signer has its own public key, and the verification time of a signature grows linearly with the number of signers.

Online/offline systems. The threshold Gaussian sampling protocol of Bendlin et al. [8] as well as the commitment and zero-knowledge protocols of Baum et al. [5] provide compact threshold cryptosystems. However, the limitation of these systems is that the servers can only perform an *a priori* bounded number of *online* non-interactive decryption/signing operations before they must perform an *offline* interactive step.

MPC systems. Recent advances in low-round MPC from LWE [3,19,23,36, 43,46] give t-out-of-N threshold cryptosystems for a restricted set of t. In these constructions, each party that is involved in the protocol encrypts an input to a joint function using an FHE scheme and broadcasts the ciphertexts to other parties. Then, each party homomorphically evaluates on the ciphertexts that it receives and participates in a single round distributed decryption protocol. If a trusted authority thresholdizes an FHE key at setup and distributes the keys to each servers, then the single round distributed decryption protocol can be used to construct a threshold FHE scheme. A crucial limitation of these single-round distributed decryption protocols, however, is that in order for decryption to be successful, *all* parties must participate in the decryption protocol.

Even if only a single party is corrupt or is absent from the protocol, the rest of the parties cannot recover the encrypted message, which results in only an N-out-of-N threshold FHE scheme.

We note that a t-out-of-N threshold FHE for small thresholds can be constructed from a N-out-of-N scheme generically. For each $\binom{N}{t}$ sets of size t, a trusted authority can thresholdize an FHE key using a t-out-of-t scheme and provide the corresponding key shares to each parties that are contained in the set. To decrypt a ciphertext, each party can provide the partial decryptions using each of the key shares it has, which a combiner can re-combine only when at least t users provide a correct partial decryption. We note, however, that such extension clearly does not work for larger values of t.

Fully-thresholdized systems. Very few existing lattice cryptosystems overcome all the limitations described above. One exception is threshold distributed PRFs [13] built from key-homomorphic PRFs [4,13,20].

Decentralized key generation. Generally, in threshold cryptosystems, the key shares for each parties must be generated either by a trusted authority or via a highly interactive multiparty computation. In this work, we study threshold cryptosystems with *non-interactive* decentralized key generation. In the setting of functional encryption, a non-interactive decentralized key generation for threshold access structures was considered in [18].

2 Overview of the Main Construction

In this section, we provide an overview of the main threshold fully homomorphic encryption (ThFHE) construction and its applications to thresholdizing cryptographic systems through a universal thresholdizer. We provide the full ThFHE construction in Sects. 5.2 and 5.3. We define and construct a universal thresholdizer scheme in Sect. 7 and discuss its applications in greater depth in Sect. 8.

2.1 Distributing FHE Decryption

Our starting point is a standard LWE based fully homomorphic encryption schemes such as GSW [33]. Recall that a ciphertext ct is a matrix in $\mathbb{Z}_q^{n \times m}$ and a secret key sk is a vector in $\mathbb{Z}_q^n$ for appropriately chosen LWE parameters n, m, q. To decrypt a ciphertext ct, the decryptor takes a specific column ct_m of the ciphertext matrix and computes its inner product with the secret key sk. That is, the decryptor computes $\langle \mathsf{ct}_m, \mathsf{sk} \rangle \in \mathbb{Z}_q$. If the resulting value is small, the underlying plaintext is interpreted as 0; otherwise, it is interpreted as 1.

Since inner product is linear, one might try to thresholdize FHE decryption by applying Shamir t-out-of-N secret sharing to sk. This will produce N keys $\mathsf{sk}_1, \ldots, \mathsf{sk}_N$, one for each user. Then to decrypt a ciphertext ct, each user can compute the inner product $\langle \mathsf{ct}_m, \mathsf{sk}_i \rangle$ as its partial decryption. The combiner can then compute the Lagrange coefficients $\lambda_i^{(S)}$ for some subset $S \subseteq \{1, \ldots, N\}$ of size t and recombine the shares as

$$\sum_{i \in S} \lambda_i^{(S)} \cdot \left\langle \mathsf{ct}_m, \mathsf{sk}_i \right\rangle = \left\langle \mathsf{ct}_m, \sum_{i \in S} \lambda_i^{(S)} \cdot \mathsf{sk}_i \right\rangle = \left\langle \mathsf{ct}_m, \mathsf{sk} \right\rangle.$$

Unfortunately, this construction is insecure. For $i \in \{1, \ldots, N\}$, every time decryptor i computes a partial decryption, it leaks information about its secret share sk_i by publishing the inner product of sk_i with a public vector ct_m.

One way to resolve this issue is for decryptor i to add small additive noise to the inner product

$$\mathsf{p}_i = \langle \mathsf{ct}_m, \mathsf{sk}_i \rangle + \mathsf{noise}.$$

However, for a t-out-of-N threshold scheme, this additive error prevents correct reconstruction of the key. The Lagrange coefficients, when interpreted as elements in $\mathbb{Z}_q$, are large and therefore blow up the noise when multiplied to the partial decryptions.

Problem with Bit Decomposition. A typical solution to the problem of noise blow-up is called bit decomposition. For example, every decryptor can compute the inner product and scale it by powers of two as

$$\mathsf{p}_i^{(j)} = 2^j \cdot \langle \mathsf{ct}_m, \mathsf{sk}_i \rangle + \mathsf{noise}^{(j)}, \qquad \text{for } j = 1, \ldots, \log_2 q.$$

It sends $\{\mathsf{p}_i^{(j)}\}_j$ to the combiner. Using the binary representation of the Lagrange coefficients, the combiner can recombine the shares without blowing up the noise. However, such bit decomposition introduces a problem in the security proof. In order to prove that the partial decryption shares $\{\mathsf{p}_i^{(j)}\}_j$ do not leak information about the underlying key sk_i, it should be possible to statistically simulate them given only $t-1$ secret keys of the other parties and the plaintext m. Interpreting the $t-1$ partial decryptions of the other parties and the message m as t shares of decryptor i's partial decryption, we can simulate the inner product $\lambda_i^{(S)} \cdot \langle \mathsf{ct}_m, \mathsf{sk}_i \rangle + \mathsf{noise}$, but not the individually scaled inner products $\{\mathsf{p}_i^{(j)}\}_j$.

The difficulty with the simulation is, in fact, warranted as there are direct attacks that can recover information about the key share sk_i from the scaled partial decryptions. One such attack can be described as follows. Consider an adversary that has access to a number of partial decryptions produced by decryptor i on ciphertexts $\mathsf{ct}_1, \ldots, \mathsf{ct}_n \in \mathbb{Z}_q^n$ where the ciphertexts form a linearly independent set as vectors over $\mathbb{Z}_q$. Then, since the matrix $\mathbf{C} = [\mathsf{ct}_1 | \ldots | \mathsf{ct}_n] \in \mathbb{Z}_q^{n \times n}$ is full rank, there exists a vector $\mathbf{u} \in \mathbb{Z}_q^n$ such that $\mathbf{C} \cdot \mathbf{u} = \mathbf{e}_1$ for the elementary basis vector $\mathbf{e}_1 = (1, 0, \ldots, 0)$. To recover the first vector position of $\mathsf{sk}_i \in \mathbb{Z}_q^n$, the adversary can bit-decompose the vector $\mathbf{u}$ and linearly combine the noisy inner products

$$\{\langle 2^j \cdot \mathsf{ct}_1, \mathsf{sk}_i \rangle + \mathsf{noise}^{(j)}\}_{j \in \log q}$$
$$\{\langle 2^j \cdot \mathsf{ct}_2, \mathsf{sk}_i \rangle + \mathsf{noise}^{(j)}\}_{j \in \log q}$$
$$\vdots$$
$$\{\langle 2^j \cdot \mathsf{ct}_n, \mathsf{sk}_i \rangle + \mathsf{noise}^{(j)}\}_{j \in \log q}$$

to form $\langle \mathbf{e}_1, \mathsf{sk}_i \rangle$ + noise. Since $\mathbf{u}$ is bit-decomposed, noise remains small and hence, the high-ordered bits of the first vector position of sk_i is leaked. The adversary can then repeat the attack (with different elementary basis vectors $\mathbf{e}_i$) to recover the high-ordered bits of the rest of the components of sk_i, which can completely compromise security.

As such attack demonstrates, managing the noise blow up to achieve both correct decryption and security is difficult. In this work, we show how to handle this noise blow up in two different ways. We provide the high level overview of the two techniques below.

Using $\{0,1\}$-LSSS. In our first method, instead of coping with the Lagrange coefficients directly in the construction itself, we abstract it out by using a different secret sharing scheme. Specifically, we first define a class of access structures, denoted $\{0,1\}$-LSSS, that consists of the set of access structures that can be supported by a linear secret sharing scheme where the reconstruction coefficients are always binary (Definition 4.4). More precisely, for a fixed access structure, such secret sharing scheme divides a secret sk into a set of shares $\mathsf{sk}_1, \ldots, \mathsf{sk}_N$ such that each sk_i itself consists of a set of individual shares in $\mathbb{Z}_q$, $\mathsf{sk}_i = \{\mathsf{s}_{i,j}\}_{j \in [\ell]}$ for some fixed bound ℓ. Then, we require that for any set $S \subseteq [N]$ that satisfies the access structure, there exists a subset of the individual shares $S' \subseteq \bigcup_{i \in S} \mathsf{sk}_i$ such that $\sum_{S'} \mathsf{s}_{i,j} = \mathsf{sk}$. With such requirement, it is straightforward to construct a correct threshold FHE scheme for $\{0,1\}$-LSSS following the approach outlined above. For simulatability, we use the fact that for a linear secret sharing scheme for $\{0,1\}$-LSSS, there exists a *maximal invalid share set* $S^* \subseteq \bigcup_{i \in [N]} \mathsf{sk}_i$ such that S^* itself does not reveal any information about the secret sk or any other individual shares $\mathsf{s}_{i,j} \in \bigcup_{i \in [N]} \mathsf{sk}_i$, but any set that strictly contains S^* completely determines the secret sk as well as all *individual* shares $\mathsf{s}_{i,j} \in \bigcup_{i \in [N]} \mathsf{sk}_i$. With careful analysis (Sect. 5.2), we show that this property allows the simulation of each partial decryptions.

The remaining question is how expressive is the class $\{0,1\}$-LSSS? Two obvious access structures which are contained in $\{0,1\}$-LSSS are undirected s-t-connectivity and N-out-of-N, but other than these, it is not clear what other useful access structures are contained in $\{0,1\}$-LSSS. However, we show that, in fact, the class is fairly large and contains the set of access structures defined by *monotone Boolean formulas* [40]. A classic result of Valiant [34,53] shows that every threshold function can be expressed as a polynomial size monotone formula. Therefore, a $\{0,1\}$-LSSS contains the set of threshold access structures that we need.

We start with the observation that the set of access structures defined by monotone Boolean formulas with input fan-out 1 (*special* monotone Boolean formula) belongs to $\{0,1\}$-LSSS through a folklore algorithm [6,37] Let $C : \{0,1\}^N \to \{0,1\}$ be a special monotone Boolean formula with an associated tree T whose internal nodes are assigned either AND or OR, and the N leaf nodes are INPUT gates that are assigned x_i. Then, we can define a linear secret sharing scheme for $\mathsf{s} \in \mathbb{Z}_q$ described as follows.

1. Assign the root r of T with the secret to be shared s.
2. If r is an INPUT gate, then simply return. Otherwise:
   - If r is an AND gate, then additively secret share k by sampling $\alpha \overset{R}{\leftarrow} \mathbb{Z}_q$ and define two shares $s_\ell = \alpha$ and $s_r = k - \alpha$.
   - If r is an OR gate, then duplicate k into shares by setting $s_\ell = k$ and $s_r = k$.
3. For each child node v_ℓ and v_r, let T_ℓ and T_r be the sub-trees having v_ℓ and v_r as roots respectively. Then, recurse on the sub-trees T_ℓ and T_r with secrets s_ℓ and s_r respectively.

At the end of the recursive process, each leaf node that is assigned x_i is assigned with a secret share s_i. It is not difficult to see that for $x \in \{0,1\}^N$, the secret s can be reconstructed from the set of shares $\{s_i\}_{x_i=1}$ if and only if $C(x) = 1$. Furthermore, given $\{s_i\}_{x_i=1}$ for $C(x) = 1$, the reconstruction procedure consists of simply identifying a subset of the shares $S \subseteq \{s_i\}_{x_i=1}$ (for the OR gates), and summing up the shares $s = \sum_{i \in S} s_i$ (for the AND gates). In Sect. 4.1, we prove that this construction indeed yields a correct and secure secret sharing scheme for special monotone Boolean formulas.

Our next observation is that the secret sharing mechanism above can also be used for *regular* monotone Boolean formulas, which have multiple input fan-out. Consider a monotone Boolean formula $C : \{0,1\}^N \to \{0,1\}$ with multiple input fan-out that is bounded by ℓ. Then, we can derive a new special monotone Boolean formula $\tilde{C} : \{0,1\}^{\ell N} \to \{0,1\}$ by letting every fan-out of an input gate of C to be a separate input. Now, applying the secret sharing mechanism above to $\tilde{C}$ yields a set of shares $\{s_i\}_{i \in [\ell N]}$ shares. Partitioning this set into the corresponding input x_i in C, we get a set of N shares $\{s_{i,j}\}$ that still abides to the syntax of a linear secret sharing scheme required for $\{0,1\}$-LSSS. Furthermore, since the circuit C can still be evaluated from $\tilde{C}$, the secret s can be reconstructed from the union of the set of shares $\bigcup_{i \in S}\{s_{i,j}\}$ for any satisfying set S.

It remains to prove that this secret sharing scheme is secure under collusion. If so, then we obtain a *secure* ThFHE scheme. We indeed show that this is the case.

Clearing out denominators. Although the use of $\{0,1\}$-LSSS to achieve threshold decryption results in a clean construction that does not require any significant modification to the existing fully homomorphic construction, the use of monotone Boolean formulas to express threshold access structure introduces significant overhead to the resulting ThFHE construction. In particular, the size of the key shares sk_i for $i = 1, \ldots, N$ is at least $\Omega(N^4)$, introducing significant space overhead. In Sect. 5.3, we introduce another approach where the share sizes are quasilinear $\tilde{O}(N)$.[2]

The high level idea of our second method is to use the technique of "clearing out the denominators" [2,50]. The observation is that since the Lagrange coefficients are rational numbers, we can scale them to be integers. In particular, for a t-out-of-N secret sharing, for any set S of size t and $i \in S$, the term $(N!)^2 \cdot \lambda_i^{(S)}$

---

[2] We describe the precise trade-offs between the two methods in [12].

is an integer. This means that even when interpreted as an element in $\mathbb{Z}_q$, the term $(N!)^2 \cdot \lambda_i^{(S)}$ is bounded by a fixed positive integer. Hence, by modifying the construction so that every signer first scales the noise that it adds by $(N!)^2$, and sufficiently increasing the modulus of the scheme to support its additional noise growth, we can preserve correct reconstruction. In fact, with careful analysis (Sect. 5.3), such a ThFHE construction can be made secure.

An evident limitation of the method above is the increase in the modulus and hence, an increase in the size of the ciphertext. Since the size of elements in $\mathbb{Z}_q$ increases by $\log N! = O(N \log N)$ bits, the size of the ciphertext depends linearly on the number of servers N, violating our compactness requirement (Definition 5.2). However, we show that any non-compact ThFHE can be boosted to a compact one by combining it with any compact (non-threshold) FHE. The idea is to first construct the notion of universal thresholdizer (Definition 7.1) via a (non-compact) ThFHE scheme and then use the thresholdizer to thresholdize a compact FHE. We provide a high level description of the universal thresholdizer in Sect. 2.2 and provide the formal details of this boosting step in the full version [12].

2.2   Universal Thresholdizer: A General Tool

We next put our new ThFHE to use. We define a new primitive called a *universal thresholdizer* (UT) that can be used to thresholdize many existing systems including signatures (Sect. 8.1). The resulting systems are secure one-round threshold systems that also provide robustness guarantees against malicious key share holders. Our universal thresholdizer abstraction provides a modular design for threshold systems and also simplifies the proof of security.

A UT scheme consists of a setup algorithm, an evaluation algorithm, and a combining algorithm. The setup of a UT scheme takes in a secret message x and divides it into a set of shares $s_1, \ldots, s_N$, which are distributed to N users. On input a circuit C, each user can independently compute an evaluation share y_i of $C(x)$ using their shares s_i. For a set $S = \{y_i\}$ for which $|S| \geq t$, the evaluation shares can be combined to produce $y = C(x)$. For robustness, we define an extra verification algorithm that given C and y_i, checks whether y_i was computed correctly.

The privacy guarantee of a UT scheme states that the shares $s_1, \ldots, s_N$ as well as the evaluation shares y_i can be simulated only given access to the circuit C and $C(x)$. The robustness guarantee of a UT scheme simply states that it is hard for an adversary to produce an improperly computed evaluation share y_i for a circuit C such that the verification algorithm accepts.

With these security guarantees, it is easy thresholdize existing cryptographic functions. To demonstrate the idea, consider the case of distributed PRF where a key k can be divided into a number of key shares such that independent PRF evaluations using these key shares can be combined into a final PRF evaluation. To construct a distributed PRF $\tilde{F}$ from a regular PRF $F : \mathcal{K} \times \mathcal{X} \to \mathcal{Y}$, we sample a key $k \xleftarrow{\text{R}} \mathcal{K}$ and invoke UT setup with k to generate the key shares

$s_1, \ldots, s_N$. Then, to evaluate $\tilde{F}$ on an input $x \in \mathcal{X}$, each party generates the evaluation share y_i for the circuit $C_x(k) = F(k, x)$. The evaluation share can then be combined in a threshold manner to produce the final PRF evaluation $y = F(k, x)$.

The robustness of $\tilde{F}$ follows from the robustness condition of UT straightforwardly. To prove pseudorandomness of $\tilde{F}$, we simply erase the original PRF key k from the security experiment by invoking the privacy simulator of UT. This allows us to reduce pseudorandomness directly to the underlying PRF security game of F.

In Sect. 7.2, we construct a robust universal thresholdizer using non-interactive zero knowledge proofs (NIZK). We note that constructing NIZKs from lattices is still an open problem. However, our setting allows the use of NIZK with preprocessing [24,39], which can be constructed from lattices [38]. A better way to ensure robustness is using *homomorphic signatures* (Sect. 7.3). Because homomorphic signatures [11,35] give more compact proofs than NIZKs, we can get partial evaluations y_i whose size is independent of the original secret message x and the size of circuit C that is used for the evaluation. Unlike NIZK, homomorphic signatures can be constructed from the SIS problem [35].

3   Preliminaries

Basic Notations. For an integer n, we write $[n]$ to denote the set $\{1, \ldots, n\}$. We use bold lowercase letters (*e.g.* $\mathbf{v}, \mathbf{w}$) to denote vectors and bold uppercase letters (*e.g.* $\mathbf{A}, \mathbf{B}$) to denote matrices. Throughout this work, we will always use infinity norm for vectors. This means that for a vector $\mathbf{x}$, the norm $\|\mathbf{x}\|$ is the maximal absolute value of an element in $\mathbf{x}$. For any set X, we denote $\mathcal{P}(X)$ as the power set of X. For any $Y, Z \in \{0,1\}^n$, we say that $Y \subseteq Z$ if for each index $i \in [n]$ such that $Y_i = 1$, we have $Z_i = 1$.

We write λ for the security parameter. We say that a function $\epsilon(\lambda)$ is negligible in λ if $\epsilon(\lambda) = o(1/\lambda^c)$ for every $c \in \mathbb{N}$, and we write $\mathsf{negl}(\lambda)$ to denote a negligible function in λ. For a distribution X over a finite domain Ω, we write $\omega \leftarrow X$ to denote that ω is sampled at random according to distribution X. For a uniform distribution, we simply write $\omega \xleftarrow{\text{R}} \Omega$. For a distribution ensemble $\chi = \chi(\lambda)$ over the integers, and an integer bound $B = B(\lambda)$, we say that χ is B-bounded if $\Pr_{x \leftarrow \chi(\lambda)}[|x| \leq B(\lambda)] = 1$. In the full version of this paper [12], we provide additional preliminaries in statistical distance, lattice cryptography, as well as definitions of basic cryptographic primitives.

4   Secret Sharing for Threshold Access Structures

In this section, we provide general results on secret sharing that we use throughout this work. We provide additional background on basic notations and terms that we use in the full version [12]. In Sect. 4.1, we define threshold access structures and recall Shamir secret sharing. In Sect. 4.2, we define a special class of access structures that we call $\{0,1\}$-LSSS and show that it contains the class of threshold access structures.

4.1   Threshold Access Structures

In this section, we define the class of threshold access structures TAS and describe Shamir secret sharing [49].

Definition 4.1 (TAS). *Let $P = \{P_1, \ldots, P_N\}$ be a set of parties. An access structure $\mathbb{A}_t$ is called a threshold access structure if for every set of parties $S \subseteq P$, we have $S \in \mathbb{A}_t$ if and only if $|S| \geq t$. We define TAS to be the class of all access structures $\mathbb{A}_t$ for all $t \in \mathbb{N}$.*

Instead of defining the algorithms of Shamir secret sharing formally, we just describe the properties of the scheme that we need.

Theorem 4.2 (Shamir Secret Sharing). *Let $P = \{P_1, \ldots, P_N\}$ be a set of parties and let TAS be the class of threshold access structures on P. Then, there exists a linear secret sharing scheme SS with secret space $\mathcal{K} = \mathbb{Z}_p$ for some prime p satisfying the following properties:*

- *For any secret $\mathsf{k} \in \mathbb{Z}_p$ and $\mathbb{A}_t \in$ TAS, each share for party P_i consists of a single element $w_i \in \mathbb{Z}_p$. For convenience of notation, we denote $w_0 = \mathsf{k}$.*
- *For every $i, j \in [N] \cup \{0\}$ and set $S \subset [N] \cup \{0\}$ of size t, there exists an efficiently computable Lagrange coefficients $\lambda_{i,j}^S \in \mathbb{Z}_q$ such that*

$$w_j = \sum_{i \in S} \lambda_{i,j}^S \cdot w_i.$$

For our purposes, we want the Lagrange coefficients to be "low-norm" values. However, a regular Lagrange coefficient have no bound on its norm. Therefore, for our construction, we take advantage of the fact that the Lagrange coefficients can be defined to be rational numbers and therefore, we can "clear out their denominators" [2,50].

Lemma 4.3 ([2]). *Let $P = \{P_1, \ldots, P_N\}$ be a set of parties, TAS the class of threshold access structures on P, and SS a Shamir secret sharing scheme with secret space $\mathbb{Z}_p$ for some prime p with $(N!)^3 \leq p$. Then, for any set $S \subset [N] \cup \{0\}$ of size t, and for any $i, j \in [N]$, the product $(N!)^2 \cdot \lambda_{i,j}^S$ is an integer and is bounded*

$$\left| (N!)^2 \cdot \lambda_{i,j}^S \right| \leq (N!)^3.$$

4.2   Access Structures {0, 1}-LSSS

In this section, we define a special class of access structures that we denote by {0, 1}-LSSS that is contained in LSSS. This is the class of access structures that can be supported by a linear secret sharing scheme where the recovery coefficients are always binary. We show that the class {0, 1}-LSSS contains the class of threshold access structures. In Sect. 5.2, we construct a threshold fully homomorphic encryption scheme for these classes of access structures.

Definition 4.4 ({0,1}-LSSS). *Let* $P = \{P_1, \ldots, P_N\}$ *be a set of parties. The class of access structure* {0,1}-$LSSS_N$ *is the collection of access structures* $\mathbb{A} \in LSSS_N$ *for which there exists an efficient linear secret sharing scheme* SS = (SS.Share, SS.Combine) *over the secret space* $\mathcal{K} = \mathbb{Z}_p$ *satisfying the following property:*

- *Let* k *be a shared secret and* $\{w_j\}_{j \in T_i}$ *be the share of party* P_i *for* $i \in [N]$. *Then, for every set* $S \in \mathbb{A}$, *there exists a subset* $T \subseteq \bigcup_{i \in S} T_i$ *such that* $\mathsf{k} = \sum_{j \in T} w_j$.

We call a linear secret sharing scheme that satisfies the properties above as a special linear secret sharing scheme.

We note that for any special linear secret sharing scheme SS, and for any minimal valid share set $T \subseteq [\ell]$, we have that $\sum_{j \in T} w_j = \mathsf{k}$.

Now, the fact that every access structure $\mathbb{A} \in \{0,1\}$-LSSS is efficient follows directly from the efficiency of the LSSS class. However, it is less clear that the set T of the definition above can be computed efficiently given any $S \subseteq \mathbb{A}$. We show that this is indeed the case in the following lemma. We provide the proof in the full version [12].

Lemma 4.5. *Let* $P = \{P_1, \ldots, P_N\}$ *be a set of parties, and* SS *a special linear secret sharing scheme for* {0,1}-LSSS. *Then, for any access structure* $\mathbb{A} \in \{0,1\}$-LSSS, *and* $S \in \mathbb{A}$, *the set* $T \subseteq S$ *as specified in Definition 4.4 can be computed efficiently.*

We now state the main theorem of this section.

Theorem 4.6. TAS $\subseteq \{0,1\}$-LSSS.

To prove the lemma, we first define the class of access structures induced by monotone Boolean formulas.

Definition 4.7 (Monotone Boolean Formula). *A monotone Boolean formula* $C : \{0,1\}^N \to \{0,1\}$ *is a Boolean circuit with the following properties:*

- *There is a single output gate.*
- *Every gate is one of AND or OR gate with fan-in 2 and fan-out 1.*
- *The input wires can have multiple fan-out.*

Definition 4.8 (MBF). *Let* $P = \{P_1, \ldots, P_N\}$ *be a set of parties and* $C : \{0,1\}^N \to \{0,1\}$ *a monotone Boolean formula. An access structure* $\mathbb{A}_C$ *is called a monotone boolean formula access structure if for every set of parties* $S \subseteq P$, *we have* $S \in \mathbb{A}$ *if and only if* $C(\mathbf{x}) = 1$. *We define* MBF *to be the class of all access structures* $\mathbb{A}_C$ *for all monotone Boolean formula* C.

Now, Theorem 4.6 is implied by the following.

Theorem 4.9 ([34,53]). TAS $\subseteq$ MBF.

Theorem 4.10 ([40]). MBF $\subseteq \{0,1\}$-LSSS.

Although Theorem 4.10 is folklore, we provide the formal proof in the full version [12].

5 Threshold Fully Homomorphic Encryption

In this section, we present the definition of threshold fully homomorphic encryption (ThFHE) for any class of access structures. Then, in Sects. 5.2 and 5.3, we construct ThFHE for the class of threshold access structure TAS. In the full version of this work [12], we provide the performance comparisons of the two constructions.

5.1 Definitions

Definition 5.1 (Threshold Fully Homomorphic Encryption (ThFHE)).
Let $P = \{P_1, \ldots, P_N\}$ be a set of parties and let $\mathbb{S}$ be a class of efficient access structures on P. A threshold fully homomorphic encryption scheme for $\mathbb{S}$ is a tuple of PPT algorithms ThFHE = (ThFHE.Setup, ThFHE.Encrypt, ThFHE.Eval, ThFHE.PartDec, ThFHE.FinDec) with the following properties:

- *ThFHE.Setup($1^\lambda, 1^d, \mathbb{A}$) → (pk, sk$_1$, ..., sk$_N$): On input the security parameter λ, a depth bound d, and an access structure $\mathbb{A}$, the setup algorithm outputs a public key pk, and a set of secret key shares sk$_1$, ..., sk$_N$.*
- *ThFHE.Encrypt(pk, μ) → ct: On input a public key pk, and a single bit plaintext $\mu \in \{0, 1\}$, the encryption algorithm outputs a ciphertext ct.*
- *ThFHE.Eval(pk, C, ct$_1$, ... ct$_k$) → ĉt: On input a public key pk, circuit C : $\{0, 1\}^k \to \{0, 1\}$ of depth at most d, and a set of ciphertexts ct$_1$, ..., ct$_k$, the evaluation algorithm outputs a ciphertext ĉt.*
- *ThFHE.PartDec(pk, ct, sk$_i$) → p$_i$: On input a public key pk, a ciphertext ct, and a secret key share sk$_i$, the partial decryption algorithm outputs a partial decryption p$_i$ related to the party P_i.*
- *ThFHE.FinDec(pk, B) → $\hat{\mu}$: On input a public key pk, and a set $B = \{p_i\}_{i \in S}$ for some $S \subseteq \{P_1, \ldots, P_N\}$, the final decryption algorithm outputs a plaintext $\hat{\mu} \in \{0, 1, \bot\}$.*

As in a standard FHE scheme, we require that a ThFHE scheme satisfies compactness, correctness, and security.

Definition 5.2 (Compactness). *We say that a ThFHE scheme is compact if there exists polynomials $\mathsf{poly}_1(\cdot)$ and $\mathsf{poly}_2(\cdot)$ such that for all λ, depth bound d, circuit C : $\{0, 1\}^k \to \{0, 1\}$ of depth at most d, and $\mu \in \{0, 1\}$, the following holds. For (pk, sk$_1$, ..., sk$_N$) ← ThFHE.Setup($1^\lambda, 1^d, \mathbb{A}$), ct$_i$ ← ThFHE.Encrypt(pk, μ_i) for $i \in [k]$, ĉt ← ThFHE.Eval(pk, C, ct$_1$, ..., ct$_k$), p$_j$ ← ThFHE.PartDec(pk, ct, sk$_j$) for any $j \in [N]$, $|\hat{ct}| \leq \mathsf{poly}(\lambda, d)$ and $|p_j| \leq \mathsf{poly}(\lambda, d, N)$.*

Definition 5.3 (Evaluation Correctness). *We say that a ThFHE scheme satisfies evaluation correctness if for all λ, depth bound d, access structure $\mathbb{A}$, circuit $C : \{0, 1\}^k \to \{0, 1\}$ of depth at most d, $S \in \mathbb{A}$, and $\mu_i \in \{0, 1\}$ for $i \in [k]$, the following condition holds. For (pk, sk$_1$, ..., sk$_N$) ← ThFHE.Setup($1^\lambda, 1^d, \mathbb{A}$), ct$_i$ ← ThFHE.Encrypt(pk, μ_i) for $i \in [k]$, ĉt ← ThFHE.Eval(pk, C, ct$_1$, ..., ct$_k$), $\Pr[\mathsf{ThFHE.FinDec}(\mathsf{pk}, \{\mathsf{ThFHE.PartDec}(\mathsf{pk}, \mathsf{ct}, \mathsf{sk}_i)\}_{i \in S}) = C(\mu_1, \ldots, \mu_k)] = 1 - \mathsf{negl}(\lambda)$.*

Definition 5.4 (Semantic Security). *We say that a* ThFHE *scheme satisfies semantic security if for all* λ, *and depth bound* d, *the following holds. For any PPT adversary* $\mathcal{A}$, *the following experiment* $\mathsf{Expt}_{\mathcal{A},\mathsf{ThFHE},\mathsf{sem}}(1^\lambda, 1^d)$ *outputs 1 with negligible probability:*

$\mathsf{Expt}_{\mathcal{A},\mathsf{ThFHE},\mathsf{sem}}(1^\lambda, 1^d)$:

1. *On input the security parameter* 1^λ *and a circuit depth* 1^d, *the adversary* $\mathcal{A}$ *outputs* $\mathbb{A} \in \mathbb{S}$.
2. *The challenger runs* $(\mathsf{pk}, \mathsf{sk}_1, \ldots, \mathsf{sk}_N) \leftarrow \mathsf{ThFHE}.\mathsf{Setup}(1^\lambda, 1^d, \mathbb{A})$ *and provides* pk *to* $\mathcal{A}$.
3. $\mathcal{A}$ *outputs a set* $S \subseteq \{P_1, \ldots, P_N\}$ *such that* $S \notin \mathbb{A}$.
4. *The challenger provides* $\{\mathsf{sk}_i\}_{i \in S}$ *along with* $\mathsf{ThFHE}.\mathsf{Encrypt}(\mathsf{pk}, b)$ *for* $b \xleftarrow{\text{R}} \{0,1\}$ *to* $\mathcal{A}$.
5. $\mathcal{A}$ *outputs a guess* b'. *The experiment outputs 1 if* $b = b'$.

In addition to the standard semantic security notion for ThFHE, we require a ThFHE scheme to satisfy simulation security. Semantic security guarantees that the *ciphertexts* of a ThFHE scheme does not reveal any information to an adversary with an unqualified set of partial decryption keys. For most use cases for ThFHE, the adversary additionally gets access to valid *partial decryptions* of ciphertexts. Simulation security guarantees that the partial decryptions also do not leak information to the adversary.

Definition 5.5 (Simulation Security). *We say that a* ThFHE *scheme satisfies simulation security if for all* λ, *depth bound* d, *and access structure* $\mathbb{A}$, *the following holds. There exists a stateful PPT algorithm* $\mathcal{S} = (\mathcal{S}_1, \mathcal{S}_2)$ *such that for any PPT adversary* $\mathcal{A}$, *the following experiments* $\mathsf{Expt}_{\mathcal{A},\mathsf{Real}}(1^\lambda, 1^d)$ *and* $\mathsf{Expt}_{\mathcal{A},\mathsf{Ideal}}(1^\lambda, 1^d)$ *are indistinguishable:*

$\mathsf{Expt}_{\mathcal{A},\mathsf{Real}}(1^\lambda, 1^d)$:

1. *On input the security parameter* 1^λ *and a circuit depth* 1^d, *the adversary* $\mathcal{A}$ *outputs* $\mathbb{A} \in \mathbb{S}$.
2. *The challenger runs* $(\mathsf{pk}, \mathsf{sk}_1, \ldots, \mathsf{sk}_N) \leftarrow \mathsf{ThFHE}.\mathsf{Setup}(1^\lambda, 1^d, \mathbb{A})$ *and provides* pk *to* $\mathcal{A}$.
3. $\mathcal{A}$ *outputs a maximal invalid party set* $S^* \subseteq \{P_1, \ldots, P_N\}$ *and messages* $\mu_1, \ldots, \mu_k \in \{0,1\}$.
4. *The challenger provides the keys* $\{\mathsf{sk}_i\}_{i \in S^*}$ *and* $\{\mathsf{ThFHE}.\mathsf{Encrypt}(\mathsf{pk}, \mu_i)\}_{i \in [k]}$ *to* $\mathcal{A}$.
5. $\mathcal{A}$ *issues a polynomial number of adaptive queries of the form* $(S \subseteq \{P_1, \ldots, P_N\}, C)$ *for circuits* $C : \{0,1\}^k \rightarrow \{0,1\}$ *of depth at most* d. *For each query, the challenger computes* $\hat{\mathsf{ct}} \leftarrow \mathsf{ThFHE}.\mathsf{Eval}(\mathsf{pk}, C, \mathsf{ct}_1, \ldots, \mathsf{ct}_k)$ *and provides* $\{\mathsf{ThFHE}.\mathsf{PartDec}(\mathsf{pk}, \hat{\mathsf{ct}}, \mathsf{sk}_i)\}_{i \in S}$ *to* $\mathcal{A}$.
6. *At the end of the experiment,* $\mathcal{A}$ *outputs a distinguishing bit* b.

$\mathsf{Expt}_{\mathcal{A},\mathsf{Ideal}}(1^\lambda, 1^d)$:

1. *On input the security parameter* 1^λ *and a circuit depth* 1^d, *the adversary* $\mathcal{A}$ *outputs* $\mathbb{A} \in \mathbb{S}$.

2. *The challenger runs* $(\mathsf{pk}, \mathsf{sk}_1, \ldots \mathsf{sk}_N, \mathsf{st}) \leftarrow \mathcal{S}_1(1^\lambda, 1^d, \mathbb{A})$ *and provides* pk *to* $\mathcal{A}$.

3. $\mathcal{A}$ *outputs a maximal invalid party set* $S^* \subseteq \{P_1, \ldots, P_N\}$ *and messages* $\mu_1, \ldots, \mu_k \in \{0, 1\}$.

4. *The challenger provides the keys* $\{\mathsf{sk}_i\}_{i \in S^*}$ *and* $\{\mathsf{ThFHE.Encrypt}(\mathsf{pk}, \mu_i)\}_{i \in [k]}$ *to* $\mathcal{A}$.

5. $\mathcal{A}$ *issues a polynomial number of adaptive queries of the form* $(S \subseteq \{P_1, \ldots, P_N\}, C)$ *for circuits* $C : \{0, 1\}^k \to \{0, 1\}$ *of depth at most* d. *For each query, the challenger runs the simulator* $\{\mathsf{p}_i\}_{i \in S} \leftarrow \mathcal{S}_2(C, \{\mathsf{ct}_1, \ldots, \mathsf{ct}_k\}, C(\mu_1, \ldots, \mu_k), S, \mathsf{st})$ *and sends* $\{\mathsf{p}_i\}_{i \in S}$ *to* $\mathcal{A}$.

6. *At the end of the experiment,* $\mathcal{A}$ *outputs a distinguishing bit* b.

We note that it is possible to unify the two security definitions, and we do so for our definition of universal thresholdizers (Sect. 7). However, for ThFHE, we present the two definitions separately for more intuition and modularity in the security proof.

5.2 ThFHE Using $\{0, 1\}$-LSSS

In this section, we present our construction of ThFHE for the class of access structures $\{0, 1\}$-LSSS. We note that by Theorem 4.6, this gives a ThFHE scheme for the class of threshold access structures TAS.

Construction 5.6. *Let* $P = \{P_1, \ldots, P_N\}$ *be a set of parties. Our* ThFHE *construction relies on the following primitives:*

- *Let* FHE $= (\mathsf{FHE.Setup}, \mathsf{FHE.Encrypt}, \mathsf{FHE.Eval}, \mathsf{FHE.Decrypt})$ *be a special fully homomorphic encryption scheme with noise bound* $B = B(\lambda, d, q)$ *and multiplicative constant 1 [12, Definition 3.9].*
- *Let* SS $= (\mathsf{SS.Share}, \mathsf{SS.Combine})$ *be a special linear secret sharing scheme (Definition 4.4). We use* T_i *to denote a partition of the share matrix and use* $\{\mathsf{s}_j\}_{j \in T_i}$ *to denote a share associated with* P_i *consisting of elements in* $\mathbb{Z}_q$. *We also use* $\ell = \ell(\lambda, N)$ *to denote a fixed polynomial bound on the size of the share:* $|T_i| \le \ell$ *for all* $i \in [N]$.

We also fix a parameter B_{sm} that specifies the bound on the smudging noise (see Sect. 5.2.1). We construct ThFHE $= (\mathsf{ThFHE.Setup}, \mathsf{ThFHE.Encrypt}, \mathsf{ThFHE.Eval}, \mathsf{ThFHE.PartDec}, \mathsf{ThFHE.FinDec})$ as follows:

- $\mathsf{ThFHE.Setup}(1^\lambda, 1^d, \mathbb{A})$: *On input the security parameter* λ, *depth bound* d, *and an access structure* $\mathbb{A}$, *the setup algorithm generates the* FHE *keys* $(\mathsf{fhepk}, \mathsf{fhesk}) \leftarrow \mathsf{FHE.Setup}(1^\lambda, 1^d)$. *Then, it divides the key* fhesk *into shares* $(\mathsf{fhesk}_1, \ldots, \mathsf{fhesk}_N) \leftarrow \mathsf{SS.Share}(\mathsf{fhesk}, \mathbb{A})$. *It sets* $\mathsf{pk} = \mathsf{fhepk}$ *and* $\mathsf{sk}_i = \mathsf{fhesk}_i$ *for* $i = 1, \ldots, N$.
- $\mathsf{ThFHE.Encrypt}(\mathsf{pk}, \mu)$: *On input the public key* pk, *and a message* $\mu \in \{0, 1\}$, *the encryption algorithm computes* $\mathsf{ct} \leftarrow \mathsf{FHE.Encrypt}(\mathsf{pk}, \mu)$ *and outputs* ct.

- ThFHE.Eval(pk, C, ct$_1$, ..., ct$_k$): *On input a public key* pk, *a circuit* C, *and a set of ciphertexts* ct$_1$, ..., ct$_k$ *the evaluation algorithm computes* ĉt $\leftarrow$ FHE.Eval(C, ct$_1$, ..., ct$_k$) *and outputs* ĉt.
- ThFHE.PartDec(pk, ct, sk$_i$): *On input a public key* pk, *a ciphertext* ct, *and a decryption key share* sk$_i$ = {s$_j$}$_{j \in T_i}$ *for each* s$_j \in \mathbb{Z}_q^n$, *the partial decryption algorithm samples a smudging error* $e_j \xleftarrow{\text{R}} [-B_{\text{sm}}, B_{\text{sm}}]$ *and computes* $\tilde{\mathbf{p}}_j$ = FHE.Decode$_0$(s$_j$, ct) $+ e_j \in \mathbb{Z}_q$ *for* $j \in T_i$. *It outputs the set* p$_i$ = {$\tilde{\mathbf{p}}_j$}$_{j \in T_i}$ *as its partial decryption.*
- ThFHE.FinDec(pk, B): *On input a public key* pk *and a set of partial decryption shares* {p$_i$}$_{i \in S}$, *it first checks if* $S \in \mathbb{A}$. *If this is not the case, then it outputs* $\perp$. *Otherwise, it computes a minimal valid share set* $T \subseteq \bigcup_{i \in S} T_i$ *and computes* $\mu \leftarrow$ FHE.Decode$_1$$\left(\sum_{j \in T} \tilde{\mathbf{p}}_j \right)$. *It outputs* μ.

We now state the compactness, correctness, and security theorems for Construction 5.6.

Theorem 5.7. *Suppose* FHE *is a compact fully homomorphic encryption scheme [12, Definition 3.6]. Then, the* ThFHE *scheme from Construction 5.6 satisfies compactness (Definition 5.2).*

Theorem 5.8. *Suppose* FHE *is a special fully homomorphic encryption scheme that satisfies correctness [12, Definition 3.7] with noise bound* B *and* SS *is a secret sharing scheme that satisfies correctness [12, Definition 4.5]. Then, the* ThFHE *scheme from Construction 5.6 with parameter* B_{sm} *such that* $B + \ell \cdot B_{\text{sm}} \leq \lfloor \frac{q}{4} \rceil$ *satisfies evaluation correctness (Definition 5.3).*

Theorem 5.9. *Suppose* FHE *is a fully homomorphic encryption scheme that satisfies security [12, Definition 3.8]. Then, the* ThFHE *scheme from Construction 5.6 satisfies semantic security (Definition 5.4).*

Theorem 5.10. *Suppose* FHE *is a fully homomorphic encryption scheme that satisfies security [12, Definition 3.8] and* SS *is a secret sharing scheme that satisfies security [12, Definition 4.6]. Then, the* ThFHE *scheme from Construction 5.6 with parameter* B_{sm} *such that* $B/B_{\text{sm}} = $ negl(λ) *satisfies simulation security (Definition 5.5).*

The compactness and semantic security of Construction 5.6 (Theorems 5.7 and 5.9) follow from the compactness and security of the underlying FHE and SS schemes in a straightforward way. We provide the formal proofs of evaluation correctness and simulation security (Theorems 5.8 and 5.10) in the full version [12].

5.2.1   Parameter Instantiation

For correctness and security, we require the parameters to satisfy:

- $B + \ell \cdot B_{\text{sm}} \leq \frac{q}{4}$ (Theorem 5.8).
- $B/B_{\text{sm}} = $ negl(λ) (Theorem 5.10).

For a depth bound d, there exists a special FHE scheme with an associated noise bound $B = 2^{\tilde{O}(d)}$ assuming the hardness of $\mathsf{LWE}(n, m, q, \chi)$ for $B = \mathsf{poly}(\lambda)$ and $q = 2^{\tilde{O}(d)+\omega(\log n)}$. Then, if we set $B_{\mathsf{sm}} = 2^{\tilde{O}(d)+\omega(\log n)}$, the two conditions above are satisfied. In particular, this translates to approximating worst-case lattice problems with sub-exponential approximation factors.

5.3 ThFHE from Shamir Secret Sharing

In this section, we present our construction of ThFHE using a standard Shamir secret sharing scheme. This construction does not satisfy our notion of compactness 5.2. However, in the full version [12], we show how to transform a non-compact ThFHE scheme to a compact one generically using UT.

Construction 5.11. *Let* $P = \{P_1, \ldots, P_N\}$ *be a set of parties. Our* ThFHE *construction relies on the following primitives:*

- *Let* FHE $=$ (FHE.Setup, FHE.Encrypt, FHE.Eval, FHE.Decrypt) *be a special fully homomorphic encryption scheme with noise bound* $B = B(\lambda, d, q)$ *and multiplicative constant* $(N!)^2$ *([12, Definition 3.9]).*
- *Let* SS $=$ (SS.Share, SS.Combine) *be a Shamir secret sharing scheme (Theorem 4.2).*

We also fix a parameter B_{sm} *that specifies the bound on the smudging noise (see Sect. 5.3.1). We construct* ThFHE $=$ (ThFHE.Setup, ThFHE.Encrypt, ThFHE.Eval, ThFHE.PartDec, ThFHE.FinDec) *as follows:*

- ThFHE.Setup($1^\lambda, 1^d, \mathbb{A}_t$): *On input the security parameter* λ, *depth bound* d, *and an access structure* $\mathbb{A}_t \in \mathsf{TAS}$, *the setup algorithm generates the* FHE *keys* (fhepk, fhesk) $\leftarrow$ FHE.Setup($1^\lambda, 1^d$). *Then, it divides the key* fhesk *into shares using Shamir secret sharing* (fhesk$_1, \ldots,$ fhesk$_N$) $\leftarrow$ SS.Share(fhesk, $\mathbb{A}_t$). *It sets* pk $=$ fhepk *and* sk$_i =$ fhesk$_i \in \mathbb{Z}_q^n$ *for* $i = 1, \ldots, N$.
- ThFHE.Encrypt(pk, μ): *On input the public key* pk, *and a message* $\mu \in \{0, 1\}$, *the encryption algorithm computes* ct $\leftarrow$ FHE.Encrypt(pk, μ) *and outputs* ct.
- ThFHE.Eval(pk, C, ct$_1, \ldots,$ ct$_k$): *On input a public key* pk, *a circuit* C, *and a set of ciphertexts* ct$_1, \ldots,$ ct$_k$ *the evaluation algorithm computes* $\hat{\mathsf{ct}} \leftarrow$ FHE.Eval(C, ct$_1, \ldots,$ ct$_k$) *and outputs* $\hat{\mathsf{ct}}$.
- ThFHE.PartDec(pk, ct, sk$_i$): *On input a public key* pk, *a ciphertext* ct, *and a decryption key share* sk$_i \in \mathbb{Z}_q^n$, *the partial decryption algorithm samples a smudging error* $e \xleftarrow{\mathsf{R}} [-B_{\mathsf{sm}}, B_{\mathsf{sm}}]$ *and computes* p$_i =$ FHE.Decode$_0$(sk$_i$, ct) $+$ $(N!)^2 \cdot e \in \mathbb{Z}_q$. *It outputs* p$_i$.
- ThFHE.FinDec(pk, B): *On input a public key* pk *and a set of partial decryption shares* $\{\mathsf{p}_i\}_{i \in S}$, *it first checks if* $S \in \mathbb{A}$. *If this is not the case, then it output* $\bot$. *Otherwise, it arbitrary chooses a satisfying set* $S' \subseteq S$ *of size* t *and computes the Lagrange coefficients* $\lambda_{i,0}^{S'}$ *for all* $i \in S'$. *Then, it computes* $\mu \leftarrow$ FHE.Decode$_1\left(\sum_{i \in S'} \lambda_{i,0}^{S'} \cdot \mathsf{p}_i\right)$, *and outputs* μ.

We now state the correctness and security theorems for Construction 5.11.

Theorem 5.12. *Suppose* FHE *is a compact fully homomorphic encryption scheme ([12, Definition 3.7]) with noise bound B and* SS *is a Shamir secret sharing scheme that satisfies correctness (Theorem 4.2). Then, the* ThFHE *scheme from Construction 5.11 with parameter $B + (N!)^3 \cdot N \cdot B_{sm} \leq \frac{q}{4}$ satisfies evaluation correctness (Definition 5.3).*

Theorem 5.13. *Suppose* FHE *is a fully homomorphic encryption scheme that satisfies security ([12, Definition 3.8]). Then, the* ThFHE *scheme from Construction 5.6 satisfies semantic security (Definition 5.4).*

Theorem 5.14. *Suppose* FHE *is a fully homomorphic encryption scheme that satisfies security ([12, Definition 3.8]) and* SS *is a secret sharing scheme that satisfies security ([12, Definition 4.6]). Then, the* ThFHE *scheme from Construction 5.11 with parameter B_{sm} such that $B/B_{sm} = \mathsf{negl}(\lambda)$ satisfies simulation security (Definition 5.5).*

The semantic security of Construction 5.11 (Theorem 5.13) follows from the semantic security of the underlying FHE in a straightforward way. We provide the formal proofs of evaluation correctness and simulation security (Theorems 5.8 and 5.10) in the full version [12].

5.3.1   Parameter Instantiation

For correctness and security, we require the parameters to satisfy:

- $B + (N!)^3 \cdot N \cdot B_{sm} \leq \frac{q}{4}$ (Theorem 5.12).
- $B/B_{sm} = \mathsf{negl}(\lambda)$ (Theorem 5.14).

For a depth bound d, there exists a special FHE scheme with an associated noise bound $B = 2^{\tilde{O}(d)}$ assuming the hardness of $\mathsf{LWE}(n, m, q, \chi)$ for $B = \mathsf{poly}(\lambda)$ and $q = 2^{\tilde{O}(d)+\omega(\log n)}$. Then, if we set $B_{sm} = 2^{\tilde{O}(d)+\omega(\log n)}/(N!)^3$, the two conditions above are satisfied. In particular, this translates to approximating worst-case lattice problems with sub-exponential approximation factors.

6   Decentralized ThFHE

In Sect. 5, we defined the notion of a threshold fully homomorphic encryption scheme to have a central setup. Namely, the setup algorithm takes in an access structure $\mathbb{A}$ for a fixed set of parties as input and produces a set of decryption key shares $\mathsf{sk}_1, \ldots, \mathsf{sk}_N$ for the servers. In practice, the set of parties that participate in the decryption protocol can always change and the access structure updated. When using a standard ThFHE scheme in this dynamic setting, a trusted setup algorithm must be run each time a new decryption server enters or leaves a protocol.

In this section, we define and construct an extension to the notion of ThFHE that we name *decentralized threshold fully homomorphic encryption* (dThFHE).

In a dThFHE scheme, there is no setup algorithm. Rather, each party can generate its own $(\mathsf{pk}_i, \mathsf{sk}_i)$ key pair from a public key encryption scheme of its choice. The encryption algorithm then takes in a set of public keys $\{\mathsf{pk}_i\}_{i \in [N]}$ and an access structure $\mathbb{A}$ to encrypt to a message x. A ciphertext that is generated in this way can only be decrypted with a set of keys $\{\mathsf{sk}_i\}_{i \in S}$ for a satisfying set $S \in \mathbb{A}$. Due to space limitations, we provide the formal definition of dThFHE in Sect. 6.1 and provide the construction in the full version [12].

6.1 Definition

In this subsection, we define our notion of decentralized fully homomorphic encryption. To capture the fact that a party can use any general public key encryption scheme, we allow the dThFHE encryption algorithm to take in the actual PKE encryption algorithms of party P_i denoted Enc_i. We assume that Enc_i consists of the description of the PKE encryption algorithm as well as a hardcoded public key pk_i. We denote a decryption algorithm by Dec_i similarly.

Definition 6.1. *A decentralized threshold fully homomorphic encryption scheme for a class of access structures* $\mathbb{S}$ *is a tuple of PPT algorithms* dThFHE = (dThFHE.Encrypt, dThFHE.Eval, dThFHE.PartDec, dThFHE.FinDec) *with the following properties:*

- dThFHE.Encrypt$(1^\lambda, 1^d, \mathsf{Enc}_1, \ldots, \mathsf{Enc}_N, \mathbb{A}, x) \rightarrow \mathsf{ct}$: *On input the security parameter* λ, *a depth bound* d, *a set of encryption algorithms* $\mathsf{Enc}_1, \ldots, \mathsf{Enc}_N$, *an access structure* $\mathbb{A}$ *on* $\{P_1, \ldots, P_N\}$, *and a message* $x \in \{0,1\}^k$, *the encryption algorithm outputs a ciphertext* ct.
- dThFHE.Eval$(C, \mathsf{ct}) \rightarrow \hat{\mathsf{ct}}$: *On input a circuit* $C : \{0,1\}^k \rightarrow \{0,1\}$, *and a ciphertext* ct, *the evaluation algorithm outputs an evaluated ciphertext* $\hat{\mathsf{ct}}$.
- dThFHE.PartDec$(\hat{\mathsf{ct}}, \mathsf{Dec}_i)$: *On input a ciphertext* $\hat{\mathsf{ct}}$, *and a secret key* sk_i, *the partial decryption algorithm outputs a partial decryption* p_i *associated with party* P_i.
- dThFHE.FinDec(B): *On input a set of partial decryptions* $\{\mathsf{p}_i\}_{i \in S}$, *the final decryption algorithm outputs a message* x'.

We require a dThFHE scheme to satisfy the following compactness, correctness, and security properties. We note that our compactness notion for dThFHE is weaker than Definition 5.2 as we allow the size of an evaluated ciphertext to depend on N.

Definition 6.2 (Weak Compactness). *We say that a* dThFHE *scheme for* $\mathbb{S}$ *is compact if there exists a polynomial* $\mathsf{poly}(\cdot)$ *such that for all* λ, *depth bound* d, *circuit* $C : \{0,1\}^k \rightarrow \{0,1\}$ *of depth at most* d, *encryption algorithms* Enc_i *for* $i \in [N]$, *access structure* $\mathbb{A} \in \mathbb{S}$, *and* $x \in \{0,1\}^k$, *the following holds. For* $\mathsf{ct} \leftarrow$ dThFHE.Encrypt$(1^\lambda, 1^d, \mathsf{Enc}_1, \ldots, \mathsf{Enc}_N, \mathbb{A}, x)$, $\hat{\mathsf{ct}} \leftarrow$ dThFHE.Eval(C, ct), *and* $\mathsf{p}_i \leftarrow$ dThFHE.PartDec$(\hat{\mathsf{ct}}, \mathsf{Dec}_i)$ *for* $i \in [N]$, *we have* $|\hat{\mathsf{ct}}|, |\mathsf{p}_i| \leq \mathsf{poly}(\lambda, d, N)$.

Definition 6.3 (Evaluation Correctness). *We say that a* dThFHE *scheme for* $\mathbb{S}$ *satisfies evaluation correctness if for all* λ, *depth bound* d, *circuit* C : $\{0,1\}^k \rightarrow \{0,1\}$ *of depth at most* d, *correct encryption and decryption algorithms* $(\mathsf{Enc}_i, \mathsf{Dec}_i)$ *for* $i \in [N]$, *access structure* $\mathbb{A} \in \mathbb{S}$, *and* $x \in \{0,1\}^k$, *the following holds. For* $\mathsf{ct} \leftarrow \mathsf{dThFHE.Encrypt}(1^\lambda, 1^d, \mathsf{Enc}_1, \ldots, \mathsf{Enc}_N, \mathbb{A}, x)$, *and* $\hat{\mathsf{ct}}, \leftarrow \mathsf{dThFHE.Eval}(C, \mathsf{ct})$, *we have*

$$\Pr[\mathsf{dThFHE.FinDec}(\{\mathsf{dThFHE.PartDec}(\hat{\mathsf{ct}}, \mathsf{Dec}_i)\}_{i \in S}) = C(x)] = 1 - \mathsf{negl}(\lambda).$$

Definition 6.4 (Semantic Security). *We say that a* dThFHE *scheme for* $\mathbb{S}$ *satisfies semantic security if for all* λ, *depth bound* d, *and secure encryption algorithms* Enc_i *for* $i \in [N]$, *the following holds. For any PPT adversary* $\mathcal{A}$, *the following experiment* $\mathsf{Expt}_{\mathcal{A},\mathsf{dThFHE},\mathsf{sem}}(1^\lambda, 1^d, \{\mathsf{Enc}_i\}_{i \in [N]})$ *outputs 1 with negligible probability:*

$\mathsf{Expt}_{\mathcal{A},\mathsf{dThFHE},\mathsf{sem}}(1^\lambda, 1^d, \{\mathsf{Enc}_i\}_{i \in [N]})$:
1. *On input the security parameter* 1^λ, *depth bound* 1^d, *and encryption algorithms* $\{\mathsf{Enc}_i\}_{i \in [N]}$, *the challenger provides* $\mathsf{Enc}_1, \ldots, \mathsf{Enc}_N$ *to* $\mathcal{A}$.
2. $\mathcal{A}$ *outputs an access structure* $\mathbb{A} \in \mathbb{S}$, *a pair of messages* $x_0, x_1 \in \{0,1\}^k$, *and an unsatisfying set* $S \subseteq \{P_1, \ldots, P_N\}$.
3. *The challenger encrypts* $\mathsf{ct}_b \leftarrow \mathsf{dThFHE.Encrypt}(1^\lambda, 1^d, \mathsf{Enc}_1, \ldots, \mathsf{Enc}_N, \mathbb{A}, x_b)$ *for* $b \xleftarrow{\mathrm{R}} \{0,1\}$ *and sends it to* $\mathcal{A}$ *along with* $\{\mathsf{Dec}_i\}_{i \in S}$.
4. $\mathcal{A}$ *outputs its guess* b'. *The experiment outputs 1 if* $b' = b$.

Definition 6.5 (Simulation Security). *We say that a* dThFHE *scheme for* $\mathbb{S}$ *satisfies simulation security if for all* λ, *depth bound* d, *and secure encryption and decryption algorithms* $(\mathsf{Enc}_i, \mathsf{Dec}_i)$ *for* $i \in [N]$, *the following holds. There exists a stateful simulator* $\mathcal{S} = (\mathcal{S}_1, \mathcal{S}_2)$ *such that for any PPT adversary* $\mathcal{A}$, *the following two experiments* $\mathsf{Expt}_{\mathcal{A},\mathsf{dThFHE},\mathsf{Real}}(1^\lambda, 1^d, \{(\mathsf{Enc}_i, \mathsf{Dec}_i)\}_{i \in [N]})$ *and* $\mathsf{Expt}_{\mathcal{A},\mathsf{dThFHE},\mathsf{Ideal}}(1^\lambda, 1^d, \{(\mathsf{Enc}_i, \mathsf{Dec}_i)\}_{i \in [N]})$ *are computationally indistinguishable:*

$\mathsf{Expt}_{\mathcal{A},\mathsf{dThFHE},\mathsf{Real}}(1^\lambda, 1^d, \{(\mathsf{Enc}_i, \mathsf{Dec}_i)\}_{i \in [N]})$:
1. *On input the security parameter* 1^λ, *depth bound* 1^d, *and a set of algorithms* $\{(\mathsf{Enc}_i, \mathsf{Dec}_i)\}_{i \in [N]}$, *the challenger provides* $\mathsf{Enc}_1, \ldots, \mathsf{Enc}_N$ *to* $\mathcal{A}$.
2. $\mathcal{A}$ *outputs an access structure* $\mathbb{A}$, *a message* $x \in \{0,1\}^k$, *and a maximal invalid party set* $S^* \subseteq \{P_1, \ldots, P_N\}$.
3. *The challenger encrypts* $\mathsf{ct} \leftarrow \mathsf{dThFHE.Encrypt}(1^\lambda, 1^d, \mathsf{Enc}_1, \ldots, \mathsf{Enc}_N, \mathbb{A}, x)$ *and provides* $(\mathsf{ct}, \{\mathsf{Dec}_i\}_{i \in S^*})$ *to* $\mathcal{A}$.
4. $\mathcal{A}$ *issues a polynomial number of adaptive queries of the form* $(S \subseteq \{P_1, \ldots, P_N\}, C)$ *for circuits* $C : \{0,1\}^k \rightarrow \{0,1\}$ *of depth at most* d. *For each query, the challenger computes* $\hat{\mathsf{ct}} \leftarrow \mathsf{dThFHE.Eval}(C, \mathsf{ct})$ *and provides* $\{\mathsf{dThFHE.PartDec}(\hat{\mathsf{ct}}, \mathsf{Dec}_i)\}_{i \in S}$ *to* $\mathcal{A}$.
5. *At the end of the experiment,* $\mathcal{A}$ *outputs a distinguishing bit* b.

$\mathsf{Expt}_{\mathcal{A},\mathsf{dThFHE},\mathsf{Ideal}}(1^\lambda, 1^d, \{(\mathsf{Enc}_i, \mathsf{Dec}_i)\}_{i \in [N]})$:

1. *On input the security parameter* 1^λ, *depth bound* 1^d, *and a set of algorithms* $\{(\mathsf{Enc}_i, \mathsf{Dec}_i)\}_{i \in [N]}$, *the challenger provides* $\mathsf{Enc}_1, \ldots, \mathsf{Enc}_N$ *to* $\mathcal{A}$.
2. $\mathcal{A}$ *outputs an access structure* $\mathbb{A}$, *a message* $X \in \{0,1\}^k$, *and a maximal invalid party set* $S^* \subseteq \{P_1, \ldots, P_N\}$.
3. *The challenger computes* $(\mathsf{ct}, \mathsf{st}) \leftarrow \mathcal{S}_1(1^\lambda, 1^d, \{\mathsf{Enc}_i\}_{i \in [N]}, \{\mathsf{Dec}_i\}_{i \in S^*}, \mathbb{A})$ *and provides* $(\mathsf{ct}, \{\mathsf{Dec}_i\}_{i \in S^*})$ *to* $\mathcal{A}$.
4. $\mathcal{A}$ *issues a polynomial number of adaptive queries of the form* $(S \subseteq \{P_1, \ldots, P_N\}, C)$ *for circuits* $C : \{0,1\}^k \to \{0,1\}$ *of depth at most* d. *For each query, the challenger runs the simulator* $\{\mathsf{p}_i\}_{i \in S} \leftarrow \mathcal{S}_2(C, C(x), \mathsf{st})$ *and sends* $\{\mathsf{p}_i\}_{i \in S}$ *to* $\mathcal{A}$.
5. *At the end of the experiment,* $\mathcal{A}$ *outputs a distinguishing bit* b.

7   Universal Thresholdizer

The notion of ThFHE is a natural generalization of a standard fully homomorphic encryption scheme that has numerous applications in threshold cryptography. Specifically, it can be used to *generically* construct a *thresholdized* variant of any basic cryptographic function. For these type of applications, it is natural to view the notion of ThFHE as a *thresholdizer* mechanism. In these settings, we do not require the full generality of the ThFHE syntax. Furthermore, for ThFHE to be useful as a thresholdizer tool, we require it to be robust, meaning that there exists an efficient public mechanism to verify whether a partial decryption was done correctly. Therefore, we define a natural notion of *universal thresholdizer* (UT) that captures these properties. We use universal thresholdizers for our applications in Sect. 8.

7.1   Definition

Informally, the setup and the encryption algorithms for ThFHE are merged into a single UT setup algorithm, and the evaluation and partial decryption algorithms for ThFHE is merged into a single UT evaluation algorithm. Furthermore, semantic security (Definition 5.4) and simulation security (Definition 5.5) is merged into a single definition for simplicity. Finally, there is an additional verification algorithm that checks whether an evaluation was done correctly.

Definition 7.1 (Universal Thresholdizer). *Let* $P = \{P_1, \ldots, P_N\}$ *be a set of parties and let* $\mathbb{S}$ *be a class of efficient access structures on* P. *A universal thresholdizer scheme for* $\mathbb{S}$ *and* $\mathcal{M}$ *is a tuple of PPT algorithms* $\mathsf{UT} = (\mathsf{UT.Setup}, \mathsf{UT.Eval}, \mathsf{UT.Verify}, \mathsf{UT.Combine})$ *with the following properties:*

- $\mathsf{UT.Setup}(1^\lambda, 1^d, \mathbb{A}, x) \to (\mathsf{pp}, \mathsf{s}_1, \ldots, \mathsf{s}_N)$: *On input the security parameter* λ, *a depth bound* d, *an access structure* $\mathbb{A}$, *and a message* $x \in \{0,1\}^k$, *the setup algorithm outputs the public parameters* pp, *and a set of shares* $\mathsf{s}_1, \ldots, \mathsf{s}_N$.

- UT.Eval(pp, s_i, C) → y_i: *On input the public parameters* pp, *a share* s_i, *and a circuit* $C : \{0,1\}^k → \{0,1\}$ *of depth at most* d, *the evaluation algorithm outputs a partial evaluation* y_i.
- UT.Verify(pp, y_i, C) → $\{0,1\}$: *On input the public parameters* pp, *a partial evaluation* y_i, *and a circuit* $C : \{0,1\}^k → \{0,1\}$, *the verification algorithm accepts or rejects.*
- UT.Combine(pp, B) → y: *On input the public parameters* pp, *a set of partial evaluations* $B = \{y_i\}_{i \in S}$, *the combining algorithm outputs the final evaluation* y.

We require a UT scheme satisfy the following compactness, correctness, and security properties. The compactness and evaluation correctness definitions are natural analogues of the ThFHE definitions. The security requirement of a ThFHE scheme combines the semantic and simulation security definitions of ThFHE. Verification correctness and robustness are additions to the definition to capture verifiable evaluation.

Definition 7.2 (Compactness). *We say that a* UT *scheme is compact if there exists a polynomial* poly($\cdot$) *such that for all* λ, *depth bound* d, *circuit* $C : \{0,1\}^k → \{0,1\}$ *of depth at most* d, *and* $\mu \in \{0,1\}$, *the following holds. For* (pk, $sk_1, \ldots, sk_N$) ← UT.Setup($1^\lambda, 1^d, \mathbb{A}, x$), y_i ← UT.Eval(pp, s_i, C) *for any* $i \in [N]$, *we have* $|y_i| \leq$ poly(λ, d, N).

Definition 7.3 (Evaluation Correctness). *We say that a* UT *scheme satisfies evaluation correctness if for all* λ, *depth bound* d, *access structure* $\mathbb{A}$, *message* $x \in \{0,1\}^k$, *circuit* $C : \{0,1\}^k → \{0,1\}$ *of depth at most* d, *and* $S \in \mathbb{A}$, *the following condition holds. For* (pp, $s_1, \ldots, s_N$) ← UT.Setup($1^\lambda, 1^d, \mathbb{A}, x$),

$$\Pr[\text{UT.Combine(pp, } \{\text{UT.Eval(pp, } s_i, C)\}_{i \in S}) = C(x)] = 1 - \mathsf{negl}(\lambda).$$

Definition 7.4 (Verification Correctness). *We say that a* UT *scheme satisfies verification correctness if for all* λ, *depth bound* d, *access structure* $\mathbb{A}$, *message* $x \in \{0,1\}^k$, *and circuit* $C : \{0,1\}^k → \{0,1\}$ *of depth at most* d, *the following holds. For* (pp, $s_1, \ldots, s_N$) ← UT.Setup($1^\lambda, 1^d, \mathbb{A}, x$), y_i ← UT.Eval(pp, s_i, C) *for any* $i \in [N]$, *we have that*

$$\Pr[\text{UT.Verify(pp, } y_i, C) = 1] = 1.$$

Definition 7.5 (Security). *We say that a* UT *scheme satisfies security if for all* λ, *and depth bound* d, *the following holds. There exists a stateful PPT algorithm* $\mathcal{S} = (\mathcal{S}_1, \mathcal{S}_2)$ *such that for any PPT adversary* $\mathcal{A}$, *we have that the following experiments* $\mathsf{Expt}_{\mathcal{A},\text{UT,Real}}(1^\lambda, 1^d)$ *and* $\mathsf{Expt}_{\mathcal{A},\text{UT,Ideal}}(1^\lambda, 1^d)$ *are computationally indistinguishable:*

$\mathsf{Expt}_{\mathcal{A},\text{UT,Real}}(1^\lambda, 1^d)$:
1. *On input the security parameter* 1^λ, *and circuit depth* 1^d, *the adversary* $\mathcal{A}$ *outputs an access structure* $\mathbb{A} \in \mathbb{S}$, *and a message* $x \in \{0,1\}^k$.

2. *The challenger runs* $(pp, s_1, \ldots, s_N) \leftarrow UT.Setup(1^\lambda, 1^d, \mathbb{A}, \underline{x})$ *and provides* pp *to* $\mathcal{A}$.
3. $\mathcal{A}$ *outputs a maximal invalid party set* $S^* \subseteq \{P_1, \ldots, P_N\}$ *for* $\mathbb{A}$.
4. *The challenger provides the shares* $\{s_i\}_{i \in S^*}$ *to* $\mathcal{A}$.
5. $\mathcal{A}$ *issues a polynomial number of adaptive queries of the form* $(S \subseteq \{P_1, \ldots, P_N\}, C)$ *for circuits* $C : \{0, 1\}^k \to \{0, 1\}$ *of depth at most* d. *For each query, the challenger provides* $\{y_i \leftarrow UT.Eval(pp, s_i, C)\}_{i \in S}$ *to* $\mathcal{A}$.
6. *At the end of the experiment,* $\mathcal{A}$ *outputs a distinguishing bit* b.

$Expt_{\mathcal{A}, UT, Ideal}(1^\lambda, 1^d)$:

1. *On input the security parameter* 1^λ *and a circuit depth* 1^d, *the adversary* $\mathcal{A}$ *outputs an access structure* $\mathbb{A} \in \mathbb{S}$, *and a message* $x \in \{0, 1\}^k$.
2. *The challenger runs* $(pp, s_1, \ldots, s_N, st) \leftarrow \mathcal{S}_1(1^\lambda, 1^d, \mathbb{A})$ *and provides* pp *to* $\mathcal{A}$.
3. $\mathcal{A}$ *outputs a maximal invalid party set* $S^* \subseteq \{P_1, \ldots, P_N\}$ *for* $\mathbb{A}$.
4. *The challenger provides the shares* $\{s_i\}_{i \in S^*}$ *to* $\mathcal{A}$.
5. $\mathcal{A}$ *issues a polynomial number of adaptive queries of the form* $(S \subseteq \{P_1, \ldots, P_N\}, C)$ *for circuits* $C : \{0, 1\}^k \to \{0, 1\}$ *of depth at most* d. *For each query, the challenger runs the simulator* $\{y_i\}_{i \in S} \leftarrow \mathcal{S}_2(pp, C, C(x), S, st)$ *and sends* $\{y_i\}_{i \in S}$ *to* $\mathcal{A}$.
6. *At the end of the experiment,* $\mathcal{A}$ *outputs a distinguishing bit* b.

Definition 7.6 (Robustness). *We say that a* UT *scheme satisfies robustness if for all* λ, *and depth bound* d, *the following holds. For any PPT adversary* $\mathcal{A}$, *the following experiment* $Expt_{\mathcal{A}, Robust}(1^\lambda, 1^d)$ *outputs 1 with negligible probability:*

– $Expt_{\mathcal{A}, UT, rb}(1^\lambda, 1^d)$:

1. *On input the security parameter* 1^λ *and circuit depth* 1^d, *the adversary* $\mathcal{A}$ *outputs a message* $x \in \{0, 1\}^k$ *and* $\mathbb{A} \in \mathbb{S}$.
2. *The challenger runs* $(pp, s_1, \ldots, s_N) \leftarrow UT.Setup(1^\lambda, 1^d, \mathbb{A}, x)$ *and provides* $(pp, s_1, \ldots, s_N)$ *to* $\mathcal{A}$.
3. $\mathcal{A}$ *outputs a fake partial evaluation* y_i^*.
4. *The challenger returns 1 if* $y_i^* \neq UT.Eval(pp, s_i, C)$ *and* $UT.Verify(pp, y_i^*, C) = 1$.

7.2 Universal Thresholdizer from **ThFHE** and **PZK**

In this section, we construct a universal thresholdizer generically from threshold fully homomorphic encryption (Sect. 5) and NIZK with pre-processing (see [12]).

Construction 7.7. *Our universal thresholdizer construction relies on the following primitives:*

– *Let* ThFHE = (ThFHE.Setup, ThFHE.Encrypt, ThFHE.Eval, ThFHE.PartDec, ThFHE.FinDec) *be a threshold fully homomorphic encryption scheme.*
– *Let* PZK = (PZK.Pre, PZK.Prove, PZK.Verify) *be a NIZK with pre-processing scheme.*
– *Let* C = (C.Com) *be a non-interactive commitment scheme.*

We construct a universal thresholdizer scheme $\mathsf{UT} = (\mathsf{UT.Setup}, \mathsf{UT.Eval},$ $\mathsf{UT.Verify}, \mathsf{UT.Combine})$ *as follows:*

- $\mathsf{UT.Setup}(1^\lambda, 1^d, \mathbb{A}, x)$: *On input the security parameter* λ, *depth bound* d, *access structure* $\mathbb{A}$, *and message* $x \in \{0,1\}^k$, *the setup algorithm first generates the* ThFHE *keys* $(\mathsf{tfhepk}, \mathsf{tfhesk}_1, \ldots, \mathsf{tfhesk}_N) \leftarrow \mathsf{ThFHE.Setup}(1^\lambda, 1^d, \mathbb{A})$ *and ciphertexts* $\mathsf{ct}_i \leftarrow \mathsf{ThFHE.Encrypt}(\mathsf{tfhepk}, x_i)$ *for* $i = 1, \ldots k$. *Then, it generates reference strings* $(\sigma_{V,i}, \sigma_{P,i}) \leftarrow \mathsf{PZK.Pre}(1^\lambda)$, *commitment randomness* $r_i \xleftarrow{\mathrm{R}} \{0,1\}^\lambda$, *and commitments* $\mathsf{com}_i \leftarrow \mathsf{C.Com}(\mathsf{tfhesk}_i; r_i)$ *for* $i = 1, \ldots N$. *It sets*

$$\mathsf{pp} = \left(\mathsf{tfhepk}, \{\mathsf{ct}_i\}_{i \in [k]}, \{\sigma_{V,i}\}_{i \in [N]}, \{\mathsf{com}_i\}_{i \in [N]}\right) \qquad \mathsf{s}_i = \left(\mathsf{tfhesk}_i, \sigma_{P,i}, r_i\right).$$

- $\mathsf{UT.Eval}(\mathsf{pp}, \mathsf{s}_i, C)$: *On input the public parameters* pp, *a share* s_i, *and a circuit* C, *the evaluation algorithm first computes the evaluated ciphertext* $\hat{\mathsf{ct}} \leftarrow \mathsf{ThFHE.Eval}(\mathsf{tfhepk}, C, \mathsf{ct}_1, \ldots, \mathsf{ct}_k)$ *and partial decryption* $\mathsf{p}_i \leftarrow \mathsf{ThFHE.PartDec}(\mathsf{tfhepk}, \hat{\mathsf{ct}}, \mathsf{tfhesk}_i)$. *Then, it constructs the statement* $\Psi_i = \Psi_i(\mathsf{com}_i, \hat{\mathsf{ct}}, \mathsf{p}_i)$ *asserting that the value* p_i *is consistent with the committed secret key* tfhesk_i:

$$\exists\, (\mathsf{tfhesk}_i, r_i) : \mathsf{com}_i = \mathsf{C.Com}(\mathsf{tfhesk}_i; r_i) \wedge \mathsf{p}_i = \mathsf{ThFHE.PartDec}(\mathsf{pp}, \hat{\mathsf{ct}}, \mathsf{tfhesk}_i).$$

It generates a NIZK proof $\pi_i \leftarrow \mathsf{PZK.Prove}(\sigma_{P,i}, \Psi_i, (\mathsf{tfhesk}_i, r_i))$ *and returns* $\mathsf{y}_i = (\mathsf{p}_i, \pi_i)$.
- $\mathsf{UT.Verify}(\mathsf{pp}, \mathsf{y}_i, C)$: *On input the public parameters* pp, *a partial evaluation* y_i, *and a circuit* C, *the verification algorithm first computes the evaluated ciphertext* $\hat{\mathsf{ct}} \leftarrow \mathsf{ThFHE.Eval}(\mathsf{pp}, C, \mathsf{ct}_1, \ldots, \mathsf{ct}_k)$ *and constructs the statement* $\Psi_i = \Psi_i(\mathsf{com}_i, \hat{\mathsf{ct}}, \mathsf{p}_i)$. *It then parses* $\mathsf{y}_i = (\mathsf{p}_i, \pi_i)$ *and returns the result of* $\mathsf{PZK.Verify}(\sigma_{V,i}, \Psi_i, \pi_i)$.
- $\mathsf{UT.Combine}(\mathsf{pp}, B)$: *On input the public parameters* pp, *and a set of partial evaluations* $B = \{\mathsf{y}_i\}_{i \in S}$ *for some* $S \subseteq \{P_1, \ldots, P_N\}$, *the combining algorithm first parses* $\mathsf{y}_i = (\mathsf{p}_i, \pi_i)$ *for* $i \in S$ *and outputs* $\mathsf{ThFHE.FinDec}(\mathsf{tfhepk}, \{\mathsf{p}_i\}_{i \in S})$.

We now state the compactness, correctness, and security theorems for Construction 7.7.

Theorem 7.8. *Suppose* ThFHE *is a compact threshold fully homomorphic encryption scheme (Definition 5.2). Then, the universal thresholdizer scheme from Construction 7.7 satisfies compactness (Definition 7.2).*

Theorem 7.9. *Suppose* ThFHE *is a threshold fully homomorphic encryption scheme that satisfies evaluation correctness (Definition 5.3). Then, the universal thresholdizer scheme from Construction 7.7 satisfies evaluation correctness (Definition 5.3).*

Theorem 7.10. *Suppose* PZK *is a complete zero knowledge proof system with pre-processing ([12, Definition 3.4]). Then, the universal thresholdizer scheme from Construction 7.7 satisfies verification correctness (Definition 7.4).*

Theorem 7.11. *Suppose* ThFHE *satisfies semantic security (Definition 5.4) and simulation security (Definition 5.5),* PZK *is a zero knowledge proof system with pre-processing that satisfies zero-knowledge ([12, Definition 3.4]), and* C *is a non-interactive commitment scheme that satisfies computational hiding ([12, Definition A.1]). Then, the universal thresholdizer scheme from Construction 7.7 satisfies security (Definition 7.5).*

Theorem 7.12. *Suppose* PZK *is a zero knowledge proof system with pre-processing that satisfies soundness ([12, Definition 3.4]) and* C *is a non-interactive commitment scheme that satisfies perfect binding ([12, Definition A.1]). Then, the universal thresholdizer scheme from Construction 7.7 satisfies robustness 7.6.*

We provide the formal proofs of the theorems above in the full version [12].

7.3 Robustness from Homomorphic Signatures

In Sect. 7.2, we used NIZK with pre-processing to enforce robustness. Another way to enforce robustness is to use homomorphic signatures [11,35]. A homomorphic signature scheme is like a regular signature scheme, but it additionally allows a signature σ_x of a message x to be homomorphically evaluated with a circuit C. The resulting signature $\sigma_{C(x)}$ is compact in that its size depends only on the depth of C and $|C(x)|$; and it certifies that a value $y = C(x)$ is indeed the output of C evaluated on the original message x. Furthermore, the signature $\sigma_{C(x)}$ itself does not leak any information about the original message x other than what can be inferred from C and $C(x)$.

To enforce robustness for the construction in Sect. 7.2, the setup algorithm can simply use a homomorphic signature to sign each decryption key share of a ThFHE scheme and include it as part of each party's share. Then, to evaluate on the shares, each user can homomorphically compute on the ThFHE ciphertexts and compute the partial decryption as before, but at the same time homomorphically evaluate on the signatures to derive a new signature that certifies correct partial decryption. The unforgeability property of the homomorphic signature scheme guarantees that no cheating adversary can generate a falsen signature on a value $y \neq C(x)$.

The benefit of using a homomorphic signature is that the proof size depends only on the depth of the circuit C to be computed and the evaluation share y. Using NIZK's, on the other hand, the proof size grows in the secret size $|x|$ and size of the circuit $|C|$. For applications that require long secret x, homomorphic signatures can give significant savings in the size of the evaluation shares. Since homomorphic signatures for circuits can be constructed from LWE [35], its use does not introduce any new assumption to our construction. We provide the formal construction from homomorphic signatures in the full version [12].

8 Applications

In this section, we describe our applications of a universal thresholdizer scheme. Due to space constraints, we defer some of our applications in the full version [12]. In [12], we show that a universal thresholdizer scheme for a class of access structures immediately give rise to a function secret sharing scheme for the same class of access structures. We also show that a universal thresholdizer scheme for the class of threshold access structures can be combined with existing cryptographic primitives to produce their thresholdized variants. As discussed in Sect. 1.1, these give rise to *threshold signatures, CCA threshold PKE, distributed PRFs,* and even *functional encryption* with thresholdized key generation. In this work, we provide just two of these applications: threshold signatures (Sect. 8.1) and CCA threshold PKE ([12]). These two notions demonstrate how to use a universal thresholdizer as a general tool. The methods that we develop in this section can be applied to a wide range of other applications in a straightforward way. In [12], we also show that a non-compact universal threholsidzer scheme can be used to thresholdize a compact fully homomorphic encryption scheme to construct a compact ThFHE scheme to a compact one.

For full generality, we define the notions of functional secret sharing, threshold signatures, and CCA threshold PKE with respect to general access structures. By Theorem 4.6, all applications in this section can be instantiated for the class of threshold access structure TAS (Definition 4.1).

8.1   Threshold Signatures

In this section, we construct a threshold signatures scheme from universal thresholdizers. In a threshold signature scheme, the signing key of a signer is divided into a number of key shares and are distributed to multiple signers. When sigining a message, each of the signers creates a partial signature with its own share of the signing key. Then, a combining algorithm combines the partial signatures into a full signature. For generality, we present the definition of threshold signatures with respect to a general class of access structures.

We provide the full definition of threshold signatures in the full version [12].

8.1.1   Construction
We construct threshold signature scheme from a universal thresholdizer (Sect. 7) and a signature scheme.

Construction 8.1. *Our threshold signature construction relies on the following primitives:*

- *Let* UT = (UT.Setup, UT.Eval, UT.Verify, UT.Combine) *be a universal thresholdizer scheme for the class of access structures* S.
- *Let* S = (S.KeyGen, S.Sign, S.Verify) *be a signature scheme. For our construction, we assume that the signing algorithm* S.Sign *is a deterministic algorithm. This is without loss of generality since any randomized signature scheme can be derandomized (i.e. using PRFs).*

Now, we construct a threshold signature scheme TS = (TS.Setup, TS.PartSign, TS.PartSignVerify, TS.Combine, TS.Verify) *for* S *as follows:*

- TS.Setup(1^λ, $\mathbb{A}$): *On input the security parameter λ, and an access structure $\mathbb{A}$, the setup algorithm first generates the keys for the signature scheme* (ssk, svk) $\leftarrow$ S.KeyGen(1^λ). *Then it instantiates the universal thresholdizer scheme* (utpp, uts$_1$, ..., uts$_N$) $\leftarrow$ UT.Setup(1^λ, 1^d, $\mathbb{A}$, ssk) *where d is the depth of the signing algorithm* S.Sign. *Then, it sets*

$$\mathsf{pp} = \mathsf{utpp}, \quad \mathsf{vk} = \mathsf{svk}, \quad \mathsf{sk}_i = \mathsf{uts}_i \quad \forall i \in [N].$$

- TS.PartSign(pp, sk$_i$, m): *On input the public parameters* pp $=$ utpp, *a partial signing key* sk$_i$ $=$ uts$_i$, *and a message $m \in \{0,1\}^*$, the partial signing algorithm outputs $\sigma_i \leftarrow$* UT.Eval(utpp, uts$_i$, C_m) *where the circuit C_m is defined as*

$$C_m(\mathsf{ssk}) = \mathsf{S.Sign}(\mathsf{ssk}, m).$$

- TS.PartSignVerify(pp, m, σ_i): *On input the public parameters* pp, *message $m \in \{0,1\}^*$, and a partial signature σ_i, the partial signature verification algorithm outputs* UT.Verify(utpp, σ_i, C_m).
- TS.Combine(pp, B): *On input the public parameters* pp, *and a set of partial signatures $B = \{\sigma_i\}_{i \in S}$, the signature combining algorithm outputs* UT.Combine(utpp, B).
- TS.Verify(vk, m, σ): *On input the signature verification key* vk $=$ svk, *a message $m \in \{0,1\}^*$, and a signature σ, the verification algorithm outputs* S.Verify(vk, m, σ).

We now state the compactness, correctness, and security theorems for Construction 8.1.

Theorem 8.2. *Suppose* UT *is a universal thresholdizer scheme that satisfies evaluation correctness (Definition 7.3). Then, the threshold signature scheme from Construction 8.1 satisfies compactness ([12, Definition 8.10]).*

Theorem 8.3. *Suppose* UT *is a universal thresholdizer scheme that satisfies evaluation correctness (Definition 7.3) and* S *is a signature scheme that satisfies correctness ([12, Definition A.4]). Then, the threshold signature scheme from Construction 8.1 satisfies evaluation correctness ([12, Definition 8.11]).*

Theorem 8.4. *Suppose* UT *is a universal thresholdizer scheme that satisfies evaluation verification correctness (Definition 7.4). Then, the threshold signature scheme from Construction 8.1 satisfies partial verification correctness ([12, Definition 8.12]).*

Theorem 8.5. *Suppose* UT *is a universal thresholdizer scheme that satisfies security (Definition 7.5) and* S *is a signature scheme that satisfies unforgeability ([12, Definition A.5]). Then, the threshold signature scheme from Construction 8.1 satisfies unforgeability ([12, Definition 8.13]).*

Theorem 8.6. *Suppose* UT *is a universal thresholdizer scheme that satisfiesd robustness (Definition 7.6). Then, the threshold signature scheme from Construction 8.1 satisfies robustness ([12, Definition 8.14]).*

Theorem 8.7. *Suppose* UT *is a universal thresholdizer scheme that satisfies evaluation correctness (Definition 7.3). Then, the threshold signature scheme from Construction 8.1 satisfies anonymity ([12, Definition 8.15]).*

We provide formal proofs of the theorems above in the full version [12].

9   Conclusion and Open Problems

In this work, we proposed a general framework for constructing various threshold cryptosystems from standard lattice assumptions. We first defined the notion of *threshold fully homomorphic encryption* (ThFHE) and constructed it from LWE. Then, we showed that ThFHE can be used to instantiate a new abstraction called *universal thresholdizers*, which can be combined with existing cryptographic primitives like digital signatures and CCA-secure PKEs to form new *threshold signatures* and *CCA-secure threshold PKEs* from LWE.

Our work gives rise to many new open problems in threshold cryptography. A universal thresholdizer can be used as a tool to construct a variety of different primitives in threshold cryptography. Can universal thresholdizers be realized from other standard assumptions such as DDH or assumptions on bilinear maps? Are there more efficient constructions of universal thresholdizers from LWE?

On the theoretical side, we show how to construct threshold signatures or threshold PKEs via a generic, but primitive dependent transformation using universal thresholdizers. Is it possible to formalize what it means to thresholdize *any* cryptographic function?

Acknowledgements. We thank the anonymous Crypto reviewers for their helpful comments. D. Boneh and S. Kim are supported by NSF, DARPA, the Simons foundation, and a grant from ONR. R. Gennaro is supported by NSF grant 1545759. S. Goldfeder is supported by NSF Graduate Research Fellowship under grant number DGE 1148900. A. Jain, P. Rasmussen, and A. Sahai are supported by a DARPA/ARL SAFEWARE award, NSF Frontier Award 1413955, NSF grants 1619348, 1228984, 1136174, and 1065276, a Xerox Faculty Research Award, a Google Faculty Re- search Award, an equipment grant from Intel, and an Okawa Foundation Research Grant. This material is based upon work supported by the Defense Advanced Research Projects Agency through the ARL under Contract W911NF-15-C-0205. The views expressed are those of the author and do not reflect the official policy or position of the Department of Defense, the National Science Foundation, or the U.S. Government.

References

1. Agrawal, S., Boneh, D., Boyen, X.: Efficient lattice (H)IBE in the standard model. In: Gilbert, H. (ed.) EUROCRYPT 2010. LNCS, vol. 6110, pp. 553–572. Springer, Heidelberg (2010). https://doi.org/10.1007/978-3-642-13190-5_28
2. Agrawal, S., Boyen, X., Vaikuntanathan, V., Voulgaris, P., Wee, H.: Functional encryption for threshold functions (or fuzzy IBE) from lattices. In: Fischlin, M., Buchmann, J., Manulis, M. (eds.) PKC 2012. LNCS, vol. 7293, pp. 280–297. Springer, Heidelberg (2012). https://doi.org/10.1007/978-3-642-30057-8_17

3. Asharov, G., Jain, A., López-Alt, A., Tromer, E., Vaikuntanathan, V., Wichs, D.: Multiparty computation with low communication, computation and interaction via threshold FHE. In: Pointcheval, D., Johansson, T. (eds.) EUROCRYPT 2012. LNCS, vol. 7237, pp. 483–501. Springer, Heidelberg (2012). https://doi.org/10. 1007/978-3-642-29011-4_29

4. Banerjee, A., Peikert, C.: New and improved key-homomorphic pseudorandom functions. In: Garay, J.A., Gennaro, R. (eds.) CRYPTO 2014. LNCS, vol. 8616, pp. 353–370. Springer, Heidelberg (2014). https://doi.org/10.1007/978-3-662-44371-2_20

5. Baum, C., Damgård, I., Oechsner, S., Peikert, C.: Efficient commitments and zero-knowledge protocols from Ring-SIS with applications to lattice-based threshold cryptosystems. IACR Cryptology ePrint Archive, 2016:997 (2016)

6. Beimel, A.: Ph.D. thesis. Israel Institute of Technology, Technion, Haifa, Israel (1996)

7. Bendlin, R., Damgård, I.: Threshold decryption and zero-knowledge proofs for lattice-based cryptosystems. In: Micciancio, D. (ed.) TCC 2010. LNCS, vol. 5978, pp. 201–218. Springer, Heidelberg (2010). https://doi.org/10.1007/978-3-642-11799-2_13

8. Bendlin, R., Krehbiel, S., Peikert, C.: How to share a lattice trapdoor: threshold protocols for signatures and (H)IBE. In: Jacobson, M., Locasto, M., Mohassel, P., Safavi-Naini, R. (eds.) ACNS 2013. LNCS, vol. 7954, pp. 218–236. Springer, Heidelberg (2013). https://doi.org/10.1007/978-3-642-38980-1_14

9. Boldyreva, A.: Threshold signatures, multisignatures and blind signatures based on the Gap-Diffie-Hellman-group signature scheme. In: Desmedt, Y.G. (ed.) PKC 2003. LNCS, vol. 2567, pp. 31–46. Springer, Heidelberg (2003). https://doi.org/10. 1007/3-540-36288-6_3

10. Boneh, D., Boyen, X., Halevi, S.: Chosen ciphertext secure public key threshold encryption without random oracles. In: Pointcheval, D. (ed.) CT-RSA 2006. LNCS, vol. 3860, pp. 226–243. Springer, Heidelberg (2006). https://doi.org/10. 1007/11605805_15

11. Boneh, D., Freeman, D.M.: Homomorphic signatures for polynomial functions. In: Paterson, K.G. (ed.) EUROCRYPT 2011. LNCS, vol. 6632, pp. 149–168. Springer, Heidelberg (2011). https://doi.org/10.1007/978-3-642-20465-4_10

12. Boneh, D., et al.: Threshold cryptosystems from threshold fully homomorphic encryption. Cryptology ePrint Archive, Report 2017/956 (2017). https://eprint. iacr.org/2017/956

13. Boneh, D., Lewi, K., Montgomery, H., Raghunathan, A.: Key homomorphic PRFs and their applications. In: Canetti, R., Garay, J.A. (eds.) CRYPTO 2013. LNCS, vol. 8042, pp. 410–428. Springer, Heidelberg (2013). https://doi.org/10.1007/978-3-642-40041-4_23

14. Boneh, D., Lynn, B., Shacham, H.: Short signatures from the Weil pairing. J. Cryptol. 17(4), 297–319 (2004)

15. Boyen, X.: Lattice mixing and vanishing trapdoors: a framework for fully secure short signatures and more. In: Nguyen, P.Q., Pointcheval, D. (eds.) PKC 2010. LNCS, vol. 6056, pp. 499–517. Springer, Heidelberg (2010). https://doi.org/10. 1007/978-3-642-13013-7_29

16. Boyle, E., Gilboa, N., Ishai, Y.: Function secret sharing. In: Oswald, E., Fischlin, M. (eds.) EUROCRYPT 2015. LNCS, vol. 9057, pp. 337–367. Springer, Heidelberg (2015). https://doi.org/10.1007/978-3-662-46803-6_12

17. Boyle, E., Gilboa, N., Ishai, Y.: Breaking the circuit size barrier for secure computation under DDH. In: Robshaw, M., Katz, J. (eds.) CRYPTO 2016. LNCS, vol. 9814, pp. 509–539. Springer, Heidelberg (2016). https://doi.org/10.1007/978-3-662-53018-4_19

18. Brakerski, Z., Chandran, N., Goyal, V., Jain, A., Sahai, A., Segev, G.: Hierarchical functional encryption. In: ITCS (2016)

19. Brakerski, Z., Perlman, R.: Lattice-based fully dynamic multi-key FHE with short ciphertexts. In: Robshaw, M., Katz, J. (eds.) CRYPTO 2016. LNCS, vol. 9814, pp. 190–213. Springer, Heidelberg (2016). https://doi.org/10.1007/978-3-662-53018-4_8

20. Brakerski, Z., Vaikuntanathan, V.: Constrained key-homomorphic PRFs from standard lattice assumptions - or: how to secretly embed a circuit in your PRF. In: Dodis, Y., Nielsen, J.B. (eds.) TCC 2015. LNCS, vol. 9015, pp. 1–30. Springer, Heidelberg (2015). https://doi.org/10.1007/978-3-662-46497-7_1

21. Canetti, R., Goldwasser, S.: An efficient *threshold* public key cryptosystem secure against adaptive chosen ciphertext attack (extended abstract). In: Stern, J. (ed.) EUROCRYPT 1999. LNCS, vol. 1592, pp. 90–106. Springer, Heidelberg (1999). https://doi.org/10.1007/3-540-48910-X_7

22. Cayrel, P.-L., Lindner, R., Rückert, M., Silva, R.: A lattice-based threshold ring signature scheme. In: Abdalla, M., Barreto, P.S.L.M. (eds.) LATINCRYPT 2010. LNCS, vol. 6212, pp. 255–272. Springer, Heidelberg (2010). https://doi.org/10.1007/978-3-642-14712-8_16

23. Choudhury, A., Loftus, J., Orsini, E., Patra, A., Smart, N.P.: Between a rock and a hard place: interpolating between MPC and FHE. In: Sako, K., Sarkar, P. (eds.) ASIACRYPT 2013. LNCS, vol. 8270, pp. 221–240. Springer, Heidelberg (2013). https://doi.org/10.1007/978-3-642-42045-0_12

24. De Santis, A., Micali, S., Persiano, G.: Non-interactive zero-knowledge with preprocessing. In: Goldwasser, S. (ed.) CRYPTO 1988. LNCS, vol. 403, pp. 269–282. Springer, New York (1990). https://doi.org/10.1007/0-387-34799-2_21

25. DeSantis, A., Desmedt, Y., Frankel, Y., Yung, M.: How to share a function securely. In: STOC (1994)

26. Desmedt, Y., Frankel, Y.: Threshold cryptosystems. In: Brassard, G. (ed.) CRYPTO 1989. LNCS, vol. 435, pp. 307–315. Springer, New York (1990). https://doi.org/10.1007/0-387-34805-0_28

27. Dodis, Y., Katz, J.: Chosen-ciphertext security of multiple encryption. In: Kilian, J. (ed.) TCC 2005. LNCS, vol. 3378, pp. 188–209. Springer, Heidelberg (2005). https://doi.org/10.1007/978-3-540-30576-7_11

28. Frankel, Y.: A practical protocol for large group oriented networks. In: Quisquater, J.-J., Vandewalle, J. (eds.) EUROCRYPT 1989. LNCS, vol. 434, pp. 56–61. Springer, Heidelberg (1990). https://doi.org/10.1007/3-540-46885-4_8

29. Gennaro, R., Goldfeder, S., Narayanan, A.: Threshold-optimal DSA/ECDSA signatures and an application to bitcoin wallet security. In: Manulis, M., Sadeghi, A.-R., Schneider, S. (eds.) ACNS 2016. LNCS, vol. 9696, pp. 156–174. Springer, Cham (2016). https://doi.org/10.1007/978-3-319-39555-5_9

30. Gennaro, R., Jarecki, S., Krawczyk, H., Rabin, T.: Robust threshold DSS signatures. Inf. Comput. **164**(1), 54–84 (2001)

31. Gennaro, R., Rabin, T., Jarecki, S., Krawczyk, H.: Robust and efficient sharing of RSA functions. J. Cryptol. **20**(3), 393 (2007)

32. Gentry, C., Peikert, C., Vaikuntanathan, V.: Trapdoors for hard lattices and new cryptographic constructions. In: STOC (2008)

33. Gentry, C., Sahai, A., Waters, B.: Homomorphic encryption from learning with errors: conceptually-simpler, asymptotically-faster, attribute-based. In: Canetti, R., Garay, J.A. (eds.) CRYPTO 2013. LNCS, vol. 8042, pp. 75–92. Springer, Heidelberg (2013). https://doi.org/10.1007/978-3-642-40041-4_5
34. Goldreich, O.: Valiant's polynomial-size monotone formula for majority (2014)
35. Gorbunov, S., Vaikuntanathan, V., Wichs, D.: Leveled fully homomorphic signatures from standard lattices. In: STOC (2015)
36. Dov Gordon, S., Liu, F.-H., Shi, E.: Constant-round MPC with fairness and guarantee of output delivery. In: Gennaro, R., Robshaw, M. (eds.) CRYPTO 2015. LNCS, vol. 9216, pp. 63–82. Springer, Heidelberg (2015). https://doi.org/10.1007/978-3-662-48000-7_4
37. Goyal, V., Pandey, O., Sahai, A., Waters, B.: Attribute-based encryption for fine-grained access control of encrypted data. In: Proceedings of the 13th ACM Conference on Computer and Communications Security (2006)
38. Kim, S., Wu, D.J.: Multi-theorem preprocessing NIZK from Lattices. In: Shacham, S., Boldyreva, A. (eds.) CRYPTO 2018, Part II. LNCS, vol. 10992, pp. 733–765. Springer, Cham (2018)
39. Lapidot, D., Shamir, A.: Publicly verifiable non-interactive zero-knowledge proofs. In: Menezes, A.J., Vanstone, S.A. (eds.) CRYPTO 1990. LNCS, vol. 537, pp. 353–365. Springer, Heidelberg (1991). https://doi.org/10.1007/3-540-38424-3_26
40. Lewko, A., Waters, B.: Decentralizing attribute-based encryption. In: Paterson, K.G. (ed.) EUROCRYPT 2011. LNCS, vol. 6632, pp. 568–588. Springer, Heidelberg (2011). https://doi.org/10.1007/978-3-642-20465-4_31
41. Lyubashevsky, V.: Lattice signatures without trapdoors. In: Pointcheval, D., Johansson, T. (eds.) EUROCRYPT 2012. LNCS, vol. 7237, pp. 738–755. Springer, Heidelberg (2012). https://doi.org/10.1007/978-3-642-29011-4_43
42. Micciancio, D., Peikert, C.: Trapdoors for lattices: simpler, tighter, faster, smaller. In: Pointcheval, D., Johansson, T. (eds.) EUROCRYPT 2012. LNCS, vol. 7237, pp. 700–718. Springer, Heidelberg (2012). https://doi.org/10.1007/978-3-642-29011-4_41
43. Mukherjee, P., Wichs, D.: Two round multiparty computation via multi-key FHE. In: Fischlin, M., Coron, J.-S. (eds.) EUROCRYPT 2016. LNCS, vol. 9666, pp. 735–763. Springer, Heidelberg (2016). https://doi.org/10.1007/978-3-662-49896-5_26
44. Myers, S., Sergi, M., Shelat, A.: Threshold fully homomorphic encryption and secure computation. IACR Cryptology ePrint Archive, 2011:454 (2011)
45. Peikert, C.: Public-key cryptosystems from the worst-case shortest vector problem. In: STOC (2009)
46. Peikert, C., Shiehian, S.: Multi-key FHE from LWE, revisited. In: Hirt, M., Smith, A. (eds.) TCC 2016. LNCS, vol. 9986, pp. 217–238. Springer, Heidelberg (2016). https://doi.org/10.1007/978-3-662-53644-5_9
47. Peikert, C., Waters, B.: Lossy trapdoor functions and their applications. SIAM J. Comput. 40(6), 1803–1844 (2011)
48. Regev, O.: On lattices, learning with errors, random linear codes, and cryptography. J. ACM (JACM) 56(6), 34 (2009)
49. Shamir, A.: How to share a secret. Commun. ACM 22(11), 612–613 (1979)
50. Shoup, V.: Practical threshold signatures. In: Preneel, B. (ed.) EUROCRYPT 2000. LNCS, vol. 1807, pp. 207–220. Springer, Heidelberg (2000). https://doi.org/10.1007/3-540-45539-6_15
51. Shoup, V., Gennaro, R.: Securing threshold cryptosystems against chosen ciphertext attack. J. Cryptol. 15(2), 75–96 (2002)

52. Stinson, D.R., Strobl, R.: Provably secure distributed schnorr signatures and a (t, n) threshold scheme for implicit certificates. In: Varadharajan, V., Mu, Y. (eds.) ACISP 2001. LNCS, vol. 2119, pp. 417–434. Springer, Heidelberg (2001). https://doi.org/10.1007/3-540-47719-5_33

53. Valiant, L.G.: Short monotone formulae for the majority function. J. Algorithms **5**, 363–366 (1984)

54. Xie, X., Xue, R., Zhang, R.: Efficient threshold encryption from lossy trapdoor functions. In: Yang, B.-Y. (ed.) PQCrypto 2011. LNCS, vol. 7071, pp. 163–178. Springer, Heidelberg (2011). https://doi.org/10.1007/978-3-642-25405-5_11

Multi-Input Functional Encryption for Inner Products: Function-Hiding Realizations and Constructions Without Pairings

Michel Abdalla[1,2]([✉]) [iD], Dario Catalano[3], Dario Fiore[4] [iD], Romain Gay[1,2], and Bogdan Ursu[5]

[1] Département informatique de l'ENS, École normale supérieure, CNRS, PSL University, 75005 Paris, France
{michel.abdalla,romain.gay}@ens.fr
[2] INRIA, Paris, France
[3] Dipartimento di Matematica e Informatica, Università di Catania, Catania, Italy
catalano@dmi.unict.it
[4] IMDEA Software Institute, Madrid, Spain
dario.fiore@imdea.org
[5] KIT, Karlsruhe, Germany
bogdan.ursu@kit.edu

Abstract. We present new constructions of multi-input functional encryption (MIFE) schemes for the inner-product functionality that improve the state of the art solution of Abdalla *et al.* (Eurocrypt 2017) in two main directions.

First, we put forward a novel methodology to convert single-input functional encryption for inner products into multi-input schemes for the same functionality. Our transformation is surprisingly simple, general and efficient. In particular, it does not require pairings and it can be instantiated with *all* known single-input schemes. This leads to two main advances. First, we enlarge the set of assumptions this primitive can be based on, notably, obtaining new MIFEs for inner products from plain DDH, LWE, and Decisional Composite Residuosity. Second, we obtain the first MIFE schemes from standard assumptions where decryption works efficiently even for messages of super-polynomial size.

Our second main contribution is the first function-hiding MIFE scheme for inner products based on standard assumptions. To this end, we show how to extend the original, pairing-based, MIFE by Abdalla *et al.* in order to make it function hiding, thus obtaining a function-hiding MIFE from the MDDH assumption.

1 Introduction

Functional Encryption (FE) [8,15,16] is an emerging cryptographic paradigm that allows fine-grained access control over encrypted data. Functional encryption schemes come equipped with a key generation mechanism that allows the

© International Association for Cryptologic Research 2018
H. Shacham and A. Boldyreva (Eds.): CRYPTO 2018, LNCS 10991, pp. 597–627, 2018.
https://doi.org/10.1007/978-3-319-96884-1_20

owner of a master secret key to generate decryption keys that have a some-how restricted capability. Namely, each decryption key sk_f is associated with a function f and using sk_f to decrypt a ciphertext $\mathsf{Enc}(x)$ allows for recovering $f(x)$, with the guarantee that no more information about x is revealed. The basic notion of functional encryption considers functionalities where all the inputs are provided and encrypted by a single party. The more general case of multi-input functionalities is captured by the notion of multi-input functional encryption (MIFE, for short) [12]. Informally, this notion can be thought of as an FE scheme where n encryption slots are explicitly given, in the sense that a user who is assigned the i-th slot can, independently, create a ciphertext $\mathsf{Enc}(x_i)$ from his own plaintext x_i. Given ciphertexts $\mathsf{Enc}(x_1), \ldots, \mathsf{Enc}(x_n)$, one can use a secret key sk_f to retrieve $f(x_1, \ldots, x_n)$, similarly to the basic FE notion. This multi-input capability makes MIFE particularly well suited for many real life scenarios (such as data mining over encrypted data or multi-client delegation of computation) where the (encrypted) data may come from different and unrelated sources.

The security requirement for both FE and MIFE imposes that decryption keys should be collusion resistant. This means that a group of users, holding different decryption keys, should not be able to gain information about the encrypted messages, beyond the union of what they can individually learn. More precisely, the standard notion of security for functional encryption is *indistinguishability*. Informally, this states that an adversary that obtains the secret keys corresponding to functions $f_1, \ldots, f_n$ should not be able to decide which of the challenge messages x_0, x_1 was encrypted, as long as $f_i(x_0) = f_i(x_1)$ for all i. This indistinguishability notion has been put forward in [8,15] and it has been shown inadequate for certain cases (see [8,15] for details). They also proposed an alternative simulation-based security notion which is also problematic as, for instance, it cannot be satisfied in general.

As an additional security property, functional encryption schemes might also be required to guarantee so-called *function hiding*. Intuitively, this means that a secret key sk_f should not reveal information about the function f it encodes, beyond what is implicitly leaked by $f(x)$. Slightly more in detail, in the indistinguishability setting, this is formalized by imposing that the adversary should not be able to decide for which of the challenge functions $f_i^{(0)}, f_i^{(1)}$ it is holding secret keys, as long as as $f_i^{(0)}(x_0) = f_i^{(1)}(x_1)$ for all i. Over the last few years, functional encryption has attracted a lot of interest, both in its basic and in its multi-input incarnations. Known results can be broadly categorized as focusing on (1) feasibility results for general functionalities, and on (2) concrete, efficient realizations for restricted functionalities of practical interest.

For the specific case of MIFE, which is the focus of this paper, constructions of the first type [6,7,9,12] all rely on quite unstable assumptions, such as indistinguishability obfuscation or multilinear maps[1]. The only known construction of the second category has been recently proposed by Abdalla *et al.* in [4].

---

[1] Here we only consider schemes where *unbounded* collusions are allowed. See [9] and references therein for the bounded collusions case.

There, they propose a (secret-key) MIFE scheme for the inner product functionality that relies on the standard k-linear assumption in (prime-order) bilinear groups[2]. Remarkably, their scheme allows for unbounded collusions and supports any (polynomially bounded) number of encryption slots. On the negative side, as in previous discrete-log-based constructions of functional inner-product encryption schemes, it employs an inefficient decryption procedure that requires to extract discrete logarithms and thus imposes serious restrictions on the size of supported messages. Moreover, the scheme is not function hiding as decryption requires the function f to be provided explicitly in the clear.

1.1 Our Contributions

In this paper we propose new constructions of multi-input functional encryption schemes for the inner product functionality that address the aforementioned shortcomings of the state-of-the-art solution of Abdalla et al. [4].

MIFE for inner products without pairings. Our first contribution consists of (secret-key) MIFE schemes for inner products based on a variety of assumptions, notably *without the need of bilinear maps*, and where *decryption works efficiently*, even for messages of super-polynomial size. We achieve this result by proposing a generic construction of MIFE from any single-input FE (for inner products) in which the encryption algorithm is linearly-homomorphic. Our transformation is surprisingly simple, general and efficient. In particular, it does not require pairings (as in the case of [4]), and it can be instantiated with *all* known single-input functional encryption schemes (e.g., [1,2,5]). This allows us to obtain new MIFE for inner products from plain DDH, composite residuosity, and LWE. Beyond the obvious advantage of enlarging the set of assumptions on which MIFE can be based, this result yields schemes that can be used with a much larger message space. Indeed, dropping the bilinear groups requirement allows us to employ schemes where the decryption time is polynomial, rather than exponential, in the message bit size. From a more theoretical perspective, our results also show that, contrary to what was previously conjectured [4], MIFE for inner product does not need any (qualitatively) stronger assumption than their single-input counterpart.

OUR SOLUTION, IN MORE DETAIL. To better describe our solution, let us first explain the basic ideas behind Abdalla *et al.*'s scheme [4]. Informally, the latter builds upon a clever two-step decryption blueprint. The ciphertexts $\mathsf{ct}_1 = \mathsf{Enc}(x_1), \ldots, \mathsf{ct}_n = \mathsf{Enc}(x_n)$ (corresponding to slots $1, \ldots, n$) are all created using different instances of a single-input FE. Decryption is performed in two stages. One first decrypts each single ct_i separately using the secret key sk_{y_i} of the underlying single-input FE, and then the outputs of these decryptions are added up to get the final result.

---

[2] As discussed in detail in [4], we stress that in the public key setting, MIFE for inner products is both easy to achieve (from its single-input counterpart) and of very limited interest, because of its inherent leakage.

The main technical challenge of this approach is that the stage one of the above decryption algorithm leaks information on each partial inner product $\langle \boldsymbol{x}_i, \boldsymbol{y}_i \rangle$. To avoid this leakage, their idea is to let source i encrypt its plaintext vector $\boldsymbol{x}_i$ augmented with some fixed (random) value u_i, which is part of the secret key. Moreover, $\mathsf{sk}_{\boldsymbol{y}_i}$ are built by running the single-input FE key generation algorithm on input $\boldsymbol{y}_i \| r$, i.e., the vector $\boldsymbol{y}_i$ augmented with fresh randomness r.

By these modifications, and skipping many technical details, stage-one decryption then consists of using pairings to compute, in $\mathbb{G}_T$, the values[3] $[\langle \boldsymbol{x}_i, \boldsymbol{y}_i \rangle + u_i r]_T$ for every slot i. From these quantities, the result $[\langle \boldsymbol{x}, \boldsymbol{y} \rangle]_T$ is obtained as

$$\prod_{i=1}^{n} [\langle \boldsymbol{x}_i, \boldsymbol{y}_i \rangle + u_i r]_T \cdot [-(\sum_{i=1}^{n} u_i) r]_T$$

which can be easily computed if $[-(\sum_{i=1}^{n} u_i) r]_T$ is included in the secret key.

Intuitively, the scheme is secure as the quantities $[u_i r]_T$ are all pseudo random (under the DDH assumption) and thus hide all the partial information $[\langle \boldsymbol{x}_i, \boldsymbol{y}_i \rangle + u_i r]_T$ may leak. Notice that, in order for this argument to go through, it is crucial that the quantities $[\langle \boldsymbol{x}_i, \boldsymbol{y}_i \rangle + u_i r]_T$ are all encoded in the exponent, and thus decoding is possible only for small norm exponents. Furthermore, this technique seems to inherently require pairings, as both u_i and r have to remain hidden while allowing to compute an encoding of their product at decryption time. This is why the possibility of a scheme without pairings was considered as "quite surprising" in [4].

We overcome these difficulties via a new FE to MIFE transform, which manages to avoid leakage in a much simpler and efficient way. Our transformation works in two steps. First, we consider a simplified scheme where only one ciphertext query is allowed and messages live in the ring $\mathbb{Z}_L$, for some integer L. In this setting, we build the following multi-input scheme. For each slot i the (master) secret key for slot i consists of one random vector $\boldsymbol{u}_i \in \mathbb{Z}_L^m$. Encrypting $\boldsymbol{x}_i$ merely consists in computing $\boldsymbol{c}_i = \boldsymbol{x}_i + \boldsymbol{u}_i \bmod L$. The secret key for function $\boldsymbol{y} = (\boldsymbol{y}_1, \ldots, \boldsymbol{y}_n)$, is just $z_{\boldsymbol{y}} = \sum_{i=1}^{n} \langle \boldsymbol{u}_i, \boldsymbol{y}_i \rangle \bmod L$. To decrypt, one computes

$$\langle \boldsymbol{x}, \boldsymbol{y} \rangle \bmod L = \langle (\boldsymbol{c}_1, \ldots, \boldsymbol{c}_n), \boldsymbol{y} \rangle - z_{\boldsymbol{y}} \bmod L$$

Security comes from the fact that, if only one ciphertext query is allowed, the above can be seen as the functional encryption equivalent of the one-time pad[4].

Next, to guarantee security in the more challenging setting where many ciphertext queries are allowed, we just add a layer of (functional) encryption on top of the above one-time encryption. More specifically, we encrypt each $\boldsymbol{c}_i$ using a FE (supporting inner products) that is both linearly homomorphic and whose message space is compatible with L. So, given ciphertexts $\{\mathsf{ct}_i = \mathsf{Enc}(\boldsymbol{c}_i)\}$

---

[3] Here we implicitly adopt the, by now standard, bracket notation from [11].

[4] We remark that a similar information theoretic construction was put forward by Wee in [17], as a warm-up scheme towards an FE for inner products achieving simulation security.

and secret key $\mathsf{sk}_{\boldsymbol{y}} = (\{\mathsf{sk}_{\boldsymbol{y}_i}\}_i, z_{\boldsymbol{y}})$, one can first obtain $\{\langle \boldsymbol{c}_i, \boldsymbol{y}_i \rangle = \mathsf{Dec}(\mathsf{ct}_i, \mathsf{sk}_{\boldsymbol{y}_i})\}$, and then extract the result as $\langle \boldsymbol{x}, \boldsymbol{y} \rangle = \sum_{i=1}^{n} \langle \boldsymbol{c}_i, \boldsymbol{y}_i \rangle - \langle \boldsymbol{u}, \boldsymbol{y} \rangle$.

Our transformation actually comes in two flavors: the first one addresses the case where the underlying FE computes inner products over some finite ring $\mathbb{Z}_L$; the second one instead considers FE schemes that compute bounded-norm inner products over the integers. In both cases the transformations are generic enough to be instantiated with known single-input FE schemes for inner products. This gives us new MIFE relying on plain DDH [1], LWE [5] and Composite residuosity [2,5]. Moreover, the proposed transform is security-preserving in the sense that, if the underlying FE achieves adaptive security, so does our resulting MIFE.

Function-Hiding MIFE for inner products. Our second contribution are new MIFE schemes for inner products that achieve function hiding. Our constructions build on the pairing-based solution from [4] and, as such, they also rely on pairings. More precisely, we propose transformations that, starting from the MIFE from [4], build function hiding MIFEs using single input FE for inner products as additional building block. Ours transforms are generic with respect to this latter component, in the sense that they can be instantiated using any single input FE satisfying some natural additional requirements (details of which are given in Sect. 4).

Our methods build from the two-layer encryption technique recently developed by Lin [13] to generically achieve function hiding in the context of (single input) FE for inner products. Intuitively, Lin's idea consists in doing similar operations both at encryption and at key derivation time. Starting from two independent instances of the underlying FE, an "inner" one and an "outer" one, the idea is to encrypt the plaintext $\boldsymbol{x}$ in two steps. One first uses the "inner" FE to compute $\mathsf{ct}_1 = \mathsf{Enc}(\mathsf{msk}_1, \boldsymbol{x})$ and then "extracts" the key corresponding to ct_1, i.e., $\mathsf{ct}_2 = \mathsf{KeyGen}(\mathsf{msk}_2, \mathsf{ct}_1)$. Key derivation is done similarly, one first computes $\mathsf{sk}_1 = \mathsf{KeyGen}(\mathsf{msk}_1, \boldsymbol{y})$ and then encrypts sk_1 using the outer scheme, i.e., $\mathsf{sk}_2 = \mathsf{Enc}(\mathsf{msk}_2, \mathsf{sk}_1)$.

If one encodes ciphertexts in $\mathbb{G}_1$ and secret keys in $\mathbb{G}_2$, then one can use pairings to compute an encoding, in $\mathbb{G}_T$, of $[\langle \mathsf{ct}_2, \mathsf{sk}_2 \rangle]_T$. Since decryption essentially performs inner product, the latter computation actually decrypts also the inner ct_1 component using secret key sk_1, thus yielding an encoding of $\langle \boldsymbol{x}, \boldsymbol{y} \rangle$. Moreover, since now $\boldsymbol{y}$ is encrypted, the FE security also provides function hiding[5].

An obvious drawback of Lin's transformation is that, when applied generically, it would induce an extra-level of multilinearity in the process. This means that, starting from a pairing-free FE for inner products, one ends up with a scheme that is function hiding but also pairing-based.

We propose similar two-layer encryption techniques that *do not*, inherently, induce extra levels of multi-linearity with respect to those of the underlying primitives. Our transforms achieve this by using the MIFE from [4] as inner scheme and, several instances of, a single input FE, one for each encryption slot, as outer

---

[5] Actually the transform sketched here only manages to guarantee a weaker form of function hiding. However this can be generically turned into standard function hiding [14], as described in the full version of the paper [3].

schemes. In particular, by carefully exploiting the specific algebraic properties of the MIFE, we manage to achieve function hiding from the Matrix Decisional Diffie Hellman assumption over *standard* bilinear groups (i.e., without resorting to multi-linear maps). Specifically, our schemes come in two flavors: a simpler one for selective security and a more convoluted one achieving adaptive security. A high level overview of our technique appears in Sect. 4. The MIFE schemes from Lin [13] are selectively secure and function-hiding, but are based on multi-linear maps ($d - 1$ slots require a multilinear map of degree d). In comparison, our schemes support a polynomial number of inputs and achieve adaptive-security, while using only pairings and while being based only on standard assumptions.

GENERALITY OF OUR APPROACH. As mentioned above, our function-hiding transforms are not entirely generic as they impose restrictions on the underlying MIFE. These restrictions, while compatible with the pairing-based realization from [4], do not cope well with our newly constructed MIFEs without pairings. Very informally, this is due to the fact that our transform relies on the two-step decryption blueprint in which one learns $[\langle \boldsymbol{x}_i, \boldsymbol{y}_i \rangle + z_i]$, and each z_i is "sufficiently" random to guarantee security in the MIFE security experiment. Specifically, in Abdalla et al.'s scheme $z_i = u_i r$ whereas in our new scheme $z_i = \langle \boldsymbol{u}_i, \boldsymbol{y}_i \rangle$. While the latter value is sufficiently random in the MIFE indistinguishability experiment, this is no longer the case in the function-hiding experiment, where the adversary asks for pairs of keys $(\boldsymbol{y}^0, \boldsymbol{y}^1)$, and $z_i = \langle \boldsymbol{u}_i, \boldsymbol{y}_i^\beta \rangle$ may actually leak information about which of the two keys was chosen (i.e. information about the value of the bit β). With a different interpretation, if one sees $[\langle \boldsymbol{x}_i, \boldsymbol{y}_i \rangle + z_i]$ as a secret sharing of $\langle \boldsymbol{x}, \boldsymbol{y} \rangle$, then in our new scheme this secret sharing depends on the function $\boldsymbol{y}$ whereas in [4] this is function independent and more suitable for function-hiding. We believe that coming up with more powerful transforms, capable of exploiting the potential of our efficient MIFEs, is a very natural and interesting open problem.

CONCURRENT WORK ON FUNCTION-HIDING. Concurrently and independently of our work, Datta et al. [10] proposed a multi-input function-hiding scheme for inner products. Their construction uses the framework of dual pairing vector spaces and require the use of pairings. They achieve slightly shorter ciphertexts and decryption keys (ciphertexts are shorter by 2 group elements, while decryption keys require 2n+1 less group elements). However, this comes at the expense of a larger master secret key, which contains $4n(m^2 - 1)$ more group elements (a quadratic blow-up in m).

Interestingly, Datta et al. [10] also provide a technique based on pseudo-random functions to extend their multi-input function-hiding scheme to an unbounded number of slots. Although their techniques also appear to be applicable to our schemes, hence capable of extending both the pairing-free and the pairing-based constructions to the unbounded setting, we leave it as future work.

2   Preliminaries

Notation. We denote with $\lambda \in \mathbb{N}$ a security parameter. A *probabilistic polynomial time* (PPT) algorithm $\mathcal{A}$ is a randomized algorithm for which there exists a polynomial $p(\cdot)$ such that for every input x the running time of $\mathcal{A}(x)$ is bounded by $p(|x|)$. We say that a function $\varepsilon : \mathbb{N} \to \mathbb{R}^+$ is *negligible* if for every positive polynomial $p(\lambda)$ there exists $\lambda_0 \in \mathbb{N}$ such that for all $\lambda > \lambda_0$: $\varepsilon(\lambda) < 1/p(\lambda)$. If S is a set, $x \leftarrow_{\mathrm{R}} S$ denotes the process of selecting x uniformly at random in S. If $\mathcal{A}$ is a probabilistic algorithm, $y \leftarrow_{\mathrm{R}} \mathcal{A}(\cdot)$ denotes the process of running $\mathcal{A}$ on some appropriate input and assigning its output to y. For a positive integer n, we denote by $[n]$ the set $\{1, \dots, n\}$. We denote vectors $\boldsymbol{x} = (x_i)$ and matrices $\mathbf{A} = (a_{i,j})$ in bold. For a set S (resp. vector $\boldsymbol{x}$) $|S|$ (resp. $|\boldsymbol{x}|$) denotes its cardinality (resp. number of entries). Also, given two vectors $\boldsymbol{x}$ and $\boldsymbol{x}'$ we denote by $\boldsymbol{x}\|\boldsymbol{x}'$ their concatenation. By $\equiv$, we denote the equality of statistical distributions, and for any $\varepsilon > 0$, we denote by $\approx_{\varepsilon}$ the ε-statistical difference of two distributions.

2.1   Definitions for Multi-Input Functional Encryption

In this section we recall the definitions of multi-input functional encryption [12] specialized to the private-key setting, as this is the one relevant for our constructions.

Definition 1 (Multi-input Function Encryption). *Let $\mathcal{F} = \{\mathcal{F}_n\}_{n \in \mathbb{N}}$ be an ensemble where each $\mathcal{F}_n$ is a family of n-ary functions. A function $f \in \mathcal{F}_n$ is defined as follows $f : \mathcal{X}_1 \times \dots \times \mathcal{X}_n \to \mathcal{Y}$. A multi-input functional encryption scheme $\mathcal{MIFE}$ for $\mathcal{F}$ consists of the following algorithms:*

- Setup$(1^\lambda, \mathcal{F}_n)$ *takes as input the security parameter λ and a description of $\mathcal{F}_n \in \mathcal{F}$, and outputs a master public key* mpk[6] *and a master secret key* msk. *The master public key* mpk *is assumed to be part of the input of all the remaining algorithms.*
- Enc$($msk$, i, x_i)$ *takes as input the master secret key* msk, *an index $i \in [n]$, and a message $x_i \in \mathcal{X}_i$, and it outputs a ciphertext* ct. *Each ciphertext is assumed to be associated with an index i denoting for which slot this ciphertext can be used for. When $n = 1$, the input i is omitted.*
- KeyGen$($msk$, f)$ *takes as input the master secret key* msk *and a function $f \in \mathcal{F}_n$, and it outputs a decryption key* sk$_f$.
- Dec$($sk$_f, ct_1, \dots, ct_n)$ *takes as input a decryption key* sk$_f$ *for function f and n ciphertexts, and it outputs a value $y \in \mathcal{Y}$.*

A scheme $\mathcal{MIFE}$ as defined above is correct if for all $n \in \mathbb{N}$, $f \in \mathcal{F}_n$ and all $x_i \in \mathcal{X}_i$ for $1 \leq i \leq n$, we have

$$\Pr\left[\begin{array}{l} (\mathsf{mpk}, \mathsf{msk}) \leftarrow \mathsf{Setup}(1^\lambda, \mathcal{F}_n); \quad \mathsf{sk}_f \leftarrow \mathsf{KeyGen}(\mathsf{msk}, f); \\ \mathsf{Dec}(\mathsf{sk}_f, \mathsf{Enc}(\mathsf{msk}, 1, x_1), \dots, \mathsf{Enc}(\mathsf{msk}, n, x_n)) = f(x_1, \dots, x_n) \end{array}\right] = 1,$$

---

[6] In the private key setting, we think of mpk as some public parameters common to all algorithms.

where the probability is taken over the coins of Setup, KeyGen *and* Enc.

Security notions. Here we recall the definitions of security for multi-input functional encryption. We give both one-time and many-time indistinguishability-based security definitions. Namely, we consider several security notions denoted xx-AD-IND and xx-SEL-IND, where: xx $\in$ {one, many}. We also give simulation-based security definitions in the full version of the paper [3].

Definition 2 (xx-AD-IND-secure MIFE). *For every multi-input functional encryption* $\mathcal{MIFE}$ *for* $\mathcal{F}$, *every stateful adversary* $\mathcal{A}$, *every security parameter* $\lambda \in \mathbb{N}$, *and every* xx $\in$ {one,many}, *we define the following experiments for* $\beta \in \{0,1\}$:

Experiment **xx-AD-IND**$_{\beta}^{\mathcal{MIFE}}(1^{\lambda}, \mathcal{A})$:

$(\mathsf{mpk}, \mathsf{msk}) \leftarrow \mathsf{Setup}(1^{\lambda}, \mathcal{F}_n)$
$\alpha \leftarrow \mathcal{A}^{\mathsf{KeyGen}(\mathsf{msk}, \cdot), \mathsf{Enc}(\cdot, \cdot, \cdot)}(\mathsf{mpk})$
Output: α

---

where Enc *is an oracle that on input* (i, x_i^0, x_i^1) *outputs* Enc(msk, i, x_i^{β}). *Also,* $\mathcal{A}$ *is restricted to only make queries* f *to* KeyGen(msk, $\cdot$) *satisfying*

$$f(x_1^{j_1,0}, \ldots, x_n^{j_n,0}) = f(x_1^{j_1,1}, \ldots, x_n^{j_n,1})$$

for all $j_1, \ldots, j_n \in [Q_1] \times \cdots \times [Q_n]$, *where for all* $i \in [n]$, Q_i *denotes the number of encryption queries for input slot* i. *We denote by* Q_f *the number of key queries. Note that w.l.o.g. (as shown in [4, Lemma 3]), we can assume that for all* $i \in [n]$, $Q_i > 0$. *When* xx = *one, we also require that* $\mathcal{A}$ *queries* Enc($i, \cdot, \cdot$) *once per slot, namely that* $Q_i = 1$, *for all* $i \in [n]$.

A private-key multi-input functional encryption $\mathcal{MIFE}$ *for* $\mathcal{F}$ *is* xx-AD-IND-*secure if every PPT adversary* $\mathcal{A}$ *has advantage negligible in* λ, *where the advantage is defined as:*

A $\mathsf{dv}_{\mathcal{MIFE}}^{xx\text{-}AD\text{-}IND}(\lambda, \mathcal{A}) =$
$\left| \Pr\left[\textbf{xx-AD-IND}_0^{\mathcal{MIFE}}(1^{\lambda}, \mathcal{A}) = 1 \right] - \Pr\left[\textbf{xx-AD-IND}_1^{\mathcal{MIFE}}(1^{\lambda}, \mathcal{A}) = 1 \right] \right|$

Remark 1 (winning condition). The winning condition may not always efficiently checkable because of the combinatorial explosion in the restrictions on the queries.

Definition 3 (xx-SEL-IND-secure MIFE). *For every multi-input functional encryption* $\mathcal{MIFE}$ *for* $\mathcal{F}$, *every stateful adversary* $\mathcal{A}$, *every security parameter* $\lambda \in \mathbb{N}$, *and every* xx $\in$ {one,many}, *we define the following experiments for* $\beta \in \{0,1\}$:

$Experiment$ **xx-SEL-IND**$_\beta^{\mathcal{MIFE}}(1^\lambda, \mathcal{A})$:

$\{x_i^{j,b}\}_{i\in[n], j\in[Q_i], b\in\{0,1\}} \leftarrow \mathcal{A}(1^\lambda, \mathcal{F}_n)$

$(\mathsf{mpk}, \mathsf{msk}) \leftarrow \mathsf{Setup}(1^\lambda, \mathcal{F}_n)$

$\mathsf{ct}_i^j := \mathsf{Enc}(\mathsf{msk}, x_i^{j,\beta})$

$\alpha \leftarrow \mathcal{A}^{\mathsf{KeyGen}(\mathsf{msk}, \cdot)} \left(\mathsf{mpk}, \{\mathsf{ct}_i^j\}_{i\in[n], j\in[Q_i]}\right)$

Output: α

---

where $\mathcal{A}$ is restricted to only make queries f to $\mathsf{KeyGen}(\mathsf{msk}, \cdot)$ satisfying

$$f(x_1^{j_1,0}, \ldots, x_n^{j_n,0}) = f(x_1^{j_1,1}, \ldots, x_n^{j_n,1})$$

for all $j_1, \ldots, j_n \in [Q_1] \times \cdots \times [Q_n]$. When $xx = one$, we also require that $Q_i = 1$, for all $i \in [n]$.

A $\mathcal{MIFE}$ for $\mathcal{F}$ is xx-SEL-IND-secure if every PPT adversary $\mathcal{A}$ has negligible advantage in λ, where the advantage is defined as:

$$\mathsf{Adv}_{\mathcal{MIFE}, \mathcal{A}}^{xx\text{-}SEL\text{-}IND}(\lambda) =$$

$$\left| \Pr\left[\mathbf{xx\text{-}SEL\text{-}IND}_0^{\mathcal{MIFE}}(1^\lambda, \mathcal{A}) = 1\right] - \Pr\left[\mathbf{xx\text{-}SEL\text{-}IND}_1^{\mathcal{MIFE}}(1^\lambda, \mathcal{A}) = 1\right] \right|.$$

Zero vs multiple queries in the private-key setting. A nice feature enjoyed by all the schemes in Sect. 3 is that the owner of a decryption key sk_y associated with the vector $\boldsymbol{y} = \boldsymbol{y}_1\| \cdots \|\boldsymbol{y}_n$ does not need to know a specific value ct_i of the ciphertext vector $\mathsf{ct} = (\mathsf{ct}_1, \ldots, \mathsf{ct}_n)$ in order to decrypt ct if $\boldsymbol{y}_i = \mathbf{0}$. In other words, Q_i can be 0 whenever $\boldsymbol{y}_i = \mathbf{0}$. In this case, the adversary is only allowed to obtain a secret key sk_y for a vector $\boldsymbol{y}$ satisfying the condition

$$\sum_{i\in I} \langle x_i^{j,0}, \boldsymbol{y}_i \rangle = \sum_{i\in I} \langle x_i^{j,1}, \boldsymbol{y}_i \rangle,$$

for all queries $j \in [Q_i]$, where $I \subseteq [n]$ denotes the set of slots for which the adversary made at least one query to Enc, that is, for which $Q_i > 0$. Though we believe this feature can be useful in practice (for instance, if one of the encrypting parties decides to stop collaborating), certain applications may require at least one ciphertext for each encryption slot in order for decryption to be possible. In such cases, one can apply to our schemes the simple generic compiler given in [4, Lemma 3] to ensure that the set $I = [n]$, thus obtaining new schemes which leak no information in the setting where some $Q_i = 0$. For this reason, we assume without loss of generality that $Q_i > 0$ in all our security definitions and proofs.

2.2 Function-Hiding Multi-Input Functional Encryption

For function-hiding, we focus on indistinguishability security notions. This is because even single-input function-hiding inner-product encryption is known to be unrealizable in a simulation sense under standard assumptions.

Definition 4 (xx-SEL-Function-hiding MIFE). *For every multi-input functional encryption $\mathcal{MIFE}$ for $\mathcal{F}$, every security parameter λ, every stateful adversary $\mathcal{A}$, and every $xx \in \{one, many\}$, we define the following experiments for $\beta \in \{0,1\}$:*

Experiment **xx-SEL-FH-IND**$_\beta^{\mathcal{MIFE}}(1^\lambda, \mathcal{A})$:

$\{x_i^{j,b}\}_{i \in [n], j \in [Q_i], b \in \{0,1\}} \leftarrow \mathcal{A}(1^\lambda, \mathcal{F}_n)$

$\{f^{j,b}\}_{j \in [Q_f], b \in \{0,1\}} \leftarrow \mathcal{A}(1^\lambda, \mathcal{F}_n)$

$(\mathsf{mpk}, \mathsf{msk}) \leftarrow \mathsf{Setup}(1^\lambda, \mathcal{F}_n)$

$\mathsf{ct}_i^j \leftarrow \mathsf{Enc}(\mathsf{msk}, i, x_i^{j,\beta}) \; \forall i \in [n], j \in [Q_i]$

$\mathsf{sk}^j \leftarrow \mathsf{KeyGen}(\mathsf{msk}, f^{j,\beta}) \; \forall j \in [Q_f]$

$\alpha \leftarrow \mathcal{A}\left(\mathsf{mpk}, (\mathsf{ct}_i^j)_{i \in [n], j \in [Q_i]}, (\mathsf{sk}^j)_{j \in [Q_f]}\right)$

Output: α

---

where $\mathcal{A}$ only makes Q_i selective queries of plaintext pairs $(x_i^{j_i,0}, x_i^{j_i,1})$ and Q_f selective queries of key pairs $(f^{j_f,0}, f^{j_f,1})$, that must satisfy:

$$f^{j_f,0}(x_1^{j_1,0}, \ldots, x_n^{j_n,0}) = f^{j_f,1}(x_1^{j_1,1}, \ldots, x_n^{j_n,1})$$

for all $j_1, \ldots, j_n \in [Q_1] \times \cdots \times [Q_n]$ and for all $j_f \in [Q_f]$.

A $\mathcal{MIFE}$ is xx-SEL-FH-IND-secure if every PPT adversary $\mathcal{A}$ has negligible advantage in λ, where the advantage is defined as:

$$\mathsf{Adv}_{\mathcal{MIFE}, \mathcal{A}}^{xx\text{-}SEL\text{-}FH\text{-}IND}(\lambda) = \Big| \Pr\big[\textbf{xx-SEL-FH-IND}_0^{\mathcal{MIFE}}(1^\lambda, \mathcal{A}) = 1\big]$$
$$- \Pr\big[\textbf{xx-SEL-FH-IND}_1^{\mathcal{MIFE}}(1^\lambda, \mathcal{A}) = 1\big]\Big|$$

Definition 5 (xx-AD-Function-hiding MIFE). *For every multi-input functional encryption $\mathcal{MIFE} := (\mathsf{Setup}, \mathsf{Enc}, \mathsf{KeyGen}, \mathsf{Dec})$ for $\mathcal{F}$, every security parameter λ, every stateful adversary $\mathcal{A}$, and every $xx \in \{one, many\}$, we define the following experiments for $\beta \in \{0,1\}$:*

Experiment **xx-AD-FH-IND**$_\beta^{\mathcal{MIFE}}(1^\lambda, \mathcal{A})$:

$(\mathsf{mpk}, \mathsf{msk}) \leftarrow \mathsf{Setup}(1^\lambda, \mathcal{F}_n)$

$\beta' \leftarrow \mathcal{A}^{\mathsf{KeyGen}(\mathsf{msk}, \cdot, \cdot), \mathsf{Enc}(\mathsf{msk}, \cdot, \cdot)}(\mathsf{mpk})$

Output: α

---

where Enc is an oracle that on input (i, x_i^0, x_i^1) outputs $\mathsf{Enc}(\mathsf{msk}, i, x_i^\beta)$ and KeyGen is an oracle that on input (f^0, f^1) outputs $\mathsf{KeyGen}(\mathsf{msk}, f^\beta)$. Additionally, $\mathcal{A}$ queries must satisfy:

$$f^{j_f,0}(x_1^{j_1,0}, \ldots, x_n^{j_n,0}) = f^{j_f,1}(x_1^{j_1,1}, \ldots, x_n^{j_n,1})$$

for all $j_1, \ldots, j_n \in [Q_1] \times \cdots \times [Q_n]$ and for all $j_f \in [Q_f]$.

A $\mathcal{MIFE}$ is xx-AD-FH-IND-secure if every PPT adversary has negligible advantage in λ, where the advantage is defined as:

$$\mathsf{Adv}_{\mathcal{MIFE},\mathcal{A}}^{xx\text{-}AD\text{-}FH\text{-}IND}(\lambda) = \Big| \Pr\big[\textbf{xx-AD-FH-IND}_0^{\mathcal{MIFE}}(1^\lambda, \mathcal{A}) = 1\big]$$
$$- \Pr\big[\textbf{xx-AD-FH-IND}_1^{\mathcal{MIFE}}(1^\lambda, \mathcal{A}) = 1\big]\Big|$$

Definition 6 (Weak function hiding MIFE). *Following the approach from [14], we define the notion of weak function hiding (denoted xx-yy-wFH-IND) in the multi-input case, which is as in Definitions 4 and 5, with the exception that the previous constraints on ciphertext and key challenges:*

$$f^{j_f,0}(x_1^{j_1,0}, \ldots, x_n^{j_n,0}) = f^{j_f,1}(x_1^{j_1,1}, \ldots, x_n^{j_n,1}),$$
$$\text{for all } j_1, \ldots, j_n \in [Q_1] \times \cdots \times [Q_n] \text{ and for all } j_f \in [Q_f]$$

are extended with additional constraints to help with our hybrid proof:

$$f^{j_f,0}(x_1^{j_1,0}, \ldots, x_n^{j_n,0}) = f^{j_f,0}(x_1^{j_1,1}, \ldots, x_n^{j_n,1}) = f^{j_f,1}(x_1^{j_1,1}, \ldots, x_n^{j_n,1}),$$
$$\text{for all } j_1, \ldots, j_n \in [Q_1] \times \cdots \times [Q_n] \text{ and for all } j_f \in [Q_f].$$

2.3 Inner-Product Functionality

In this paper we construct multi-input functional encryption schemes that support the following two variants of the multi-input inner product functionality:

Multi-Input Inner Product over $\mathbb{Z}_L$. This is a family of functions that is defined as $\mathcal{F}_{L,n}^m = \{f_{\boldsymbol{y}_1, \ldots, \boldsymbol{y}_n} : (\mathbb{Z}_L^m)^n \to \mathbb{Z}_L, \text{ for } \boldsymbol{y}_i \in \mathbb{Z}_L^m\}$ where

$$f_{\boldsymbol{y}_1, \ldots, \boldsymbol{y}_n}(\boldsymbol{x}_1, \ldots, \boldsymbol{x}_n) = \sum_{i=1}^{n} \langle \boldsymbol{x}_i, \boldsymbol{y}_i \rangle \bmod L.$$

Multi-Input Bounded-Norm Inner Product over $\mathbb{Z}$. This is defined as $\mathcal{F}_n^{m,X,Y} = \{f_{\boldsymbol{y}_1, \ldots, \boldsymbol{y}_n} : (\mathbb{Z}^m)^n \to \mathbb{Z}\}$ where $f_{\boldsymbol{y}_1, \ldots, \boldsymbol{y}_n}(\boldsymbol{x}_1, \ldots, \boldsymbol{x}_n)$ is the same as above except that the result is not reduced mod L, and vectors are required to satisfy the following bounds: $\|\boldsymbol{x}\|_\infty < X$, $\|\boldsymbol{y}\|_\infty < Y$.

2.4 Computational Assumptions

Prime-order groups. Let GGen be a probabilistic polynomial time (PPT) algorithm that on input 1^λ returns a description $\mathcal{G} = (\mathbb{G}, p, g)$ of an cyclic group $\mathcal{G}$ of order p for a 2λ-bit prime p, whose generator is g.

We use implicit representation of group elements as introduced in [11]. For $a \in \mathbb{Z}_p$, define $[a] = g^a \in \mathbb{G}$ as the *implicit representation* of a in $\mathcal{G}$. More generally, for a matrix $\mathbf{A} = (a_{ij}) \in \mathbb{Z}_p^{n \times m}$ we define $[\mathbf{A}]$ as the implicit representation of $\mathbf{A}$ in $\mathcal{G}$:

$$[\mathbf{A}] := \begin{pmatrix} g^{a_{11}} & \ldots & g^{a_{1m}} \\ & & \\ g^{a_{n1}} & \ldots & g^{a_{nm}} \end{pmatrix} \in \mathbb{G}^{n \times m}$$

We will always use this implicit notation of elements in $\mathbb{G}$, i.e., we let $[a] \in \mathbb{G}$ be an element in $\mathbb{G}$. Note that from a random $[a] \in \mathbb{G}$ it is generally hard to compute the value a (discrete logarithm problem in $\mathbb{G}$). Obviously, given $[a], [b] \in \mathbb{G}$ and a scalar $x \in \mathbb{Z}_p$, one can efficiently compute $[ax] \in \mathbb{G}$ and $[a+b] \in \mathbb{G}$.

Matrix Diffie-Hellman Assumption for prime-order groups. We recall the definition of the Matrix Decision Diffie-Hellman (MDDH) Assumption [11].

Definition 7 (Matrix Distribution). *Let $k \in \mathbb{N}$. We call $\mathcal{D}_k$ a matrix distribution if it outputs matrices in $\mathbb{Z}_p^{(k+1) \times k}$ of full rank k in polynomial time.*

W.l.o.g. we assume the first k rows of $\mathbf{A} \leftarrow_\mathrm{R} \mathcal{D}_k$ form an invertible matrix. The $\mathcal{D}_k$-Matrix Diffie-Hellman problem is to distinguish the two distributions $([\mathbf{A}], [\mathbf{A}\boldsymbol{w}])$ and $([\mathbf{A}], [\boldsymbol{u}])$ where $\mathbf{A} \leftarrow_\mathrm{R} \mathcal{D}_k$, $\boldsymbol{w} \leftarrow_\mathrm{R} \mathbb{Z}_p^k$ and $\boldsymbol{u} \leftarrow_\mathrm{R} \mathbb{Z}_p^\ell$.

Definition 8 ($\mathcal{D}_k$-Matrix Diffie-Hellman ($\mathcal{D}_k$-MDDH) assumption in prime-order groups). *Let $\mathcal{D}_k$ be a matrix distribution. The $\mathcal{D}_k$-Matrix Diffie-Hellman ($\mathcal{D}_k$-MDDH) assumption holds relative to GGen if for all PPT adversaries $\mathcal{A}$,*

$$\mathbf{Adv}_{\mathsf{GGen}, \mathcal{A}}^{\mathcal{D}_k\text{-mddh}}(\lambda) := |\Pr[\mathcal{A}(\mathcal{G}, [\mathbf{A}], [\mathbf{A}\boldsymbol{w}]) = 1] - \Pr[\mathcal{A}(\mathcal{G}, [\mathbf{A}], [\boldsymbol{u}]) = 1]| = \mathsf{negl}(\lambda),$$

where probabilities are over $\mathcal{G} \leftarrow_\mathrm{R} \mathsf{GGen}(1^\lambda)$, $\mathbf{A} \leftarrow_\mathrm{R} \mathcal{D}_k$, $\boldsymbol{w} \leftarrow_\mathrm{R} \mathbb{Z}_p^k$, $\boldsymbol{u} \leftarrow_\mathrm{R} \mathbb{Z}_p^{k+1}$.

Pairing groups. Let PGGen be a probabilistic polynomial time (PPT) algorithm that on input 1^λ returns a description $\mathcal{PG} = (\mathbb{G}_1, \mathbb{G}_2, q, g_1, g_2)$ of asymmetric pairing groups where $\mathbb{G}_1$, $\mathbb{G}_2$, $\mathbb{G}_T$ are cyclic group of order p for a 2λ-bit prime p, g_1 and g_2 are generators of $\mathbb{G}_1$ and $\mathbb{G}_2$, respectively, and $e : \mathbb{G}_1 \times \mathbb{G}_2 \to \mathbb{G}_T$ is an efficiently computable (non-degenerate) bilinear map. Define $g_T := e(g_1, g_2)$, which is a generator of $\mathcal{G}_T$. We again use implicit representation of group elements. For $s \in 1, 2, T$ and $a \in \mathbb{Z}_p$, define $[a]_s = g_s^a \in \mathcal{G}_s$ as the implicit representation of a in G_s. Given $[a]_1$, $[a]_2$, one can efficiently compute $[ab]_T$ using the pairing e. For two matrices $\mathbf{A}, \mathbf{B}$ with matching dimensions define $e([\mathbf{A}]_1, [\mathbf{B}]_2) := [\mathbf{AB}]_T \in \mathbb{G}_T$.

We define the $\mathcal{D}_k$-MDDH assumption in pairing groups similarly than in prime-order groups (see Definition 8).

Definition 9 ($\mathcal{D}_k$-MDDH assumption in pairing groups). *Let $\mathcal{D}_k$ be a matrix distribution. The $\mathcal{D}_k$-MDDH assumption holds relative to PGGen in $\mathbb{G}_s$, for $s \in \{1, 2, T\}$, if for all PPT adversaries $\mathcal{A}$, the following is $\mathsf{negl}(\lambda)$:*

$$\mathbf{Adv}_{\mathbb{G}_s, \mathcal{A}}^{\mathcal{D}_k\text{-mddh}}(\lambda) := |\Pr[\mathcal{A}(\mathcal{PG}, [\mathbf{A}]_s, [\mathbf{A}\boldsymbol{w}]_s) = 1] - \Pr[\mathcal{A}(\mathcal{PG}, [\mathbf{A}]_s, [\boldsymbol{u}]_s) = 1]|$$

where probabilities are over $\mathcal{PG} \leftarrow_\mathrm{R} \mathsf{PGGen}(1^\lambda)$, $\mathbf{A} \leftarrow_\mathrm{R} \mathcal{D}_k$, $\boldsymbol{w} \leftarrow_\mathrm{R} \mathbb{Z}_p^k$, $\boldsymbol{u} \leftarrow_\mathrm{R} \mathbb{Z}_p^{k+1}$.

Next, we recall a result on the uniform distribution over full-rank matrices:

Definition 10 (Uniform distribution). *Let $\ell, k \in \mathbb{N}$, with $\ell > k$. We denote by $\mathcal{U}_{\ell,k}$ the uniform distribution over all full-rank $\ell \times k$ matrices over $\mathbb{Z}_p$.*

Among all possible matrix distributions $\mathcal{D}_k$, the uniform matrix distribution $\mathcal{U}_{\ell,k}$ is the hardest possible instance, so in particular k-Lin $\Rightarrow \mathcal{U}_k$-MDDH, as stated in Lemma 1.

Lemma 1 ($\mathcal{D}_k$-MDDH $\Rightarrow \mathcal{U}_{\ell,k}$-MDDH, [11]). *Let $\ell, k \in \mathbb{N}$ and $\mathcal{D}_k$ a matrix distribution. For any PPT adversary $\mathcal{A}$, there exists a PPT $\mathcal{B}$ such that*

$$\mathbf{Adv}_{\mathbb{G}_s,\mathcal{A}}^{\mathcal{U}_{\ell,k}\text{-mddh}}(\lambda) \leq \mathbf{Adv}_{\mathbb{G}_s,\mathcal{B}}^{\mathcal{D}_k\text{-mddh}}(\lambda).$$

3 From Single to Multi-Input FE for Inner Product

In this section, we give a generic construction of MIFE for inner product from any single-input FE (Setup, Enc, KeyGen, Dec) for the same functionality. More precisely, we show two transformations: the first one addresses FE schemes that compute the inner product functionality over a finite ring $\mathbb{Z}_L$ for some integer L, while the second transformation addresses FE schemes for bounded-norm inner product. The two transformations are almost the same, and the only difference is that in the case of bounded-norm inner product, we require additional structural properties on the single-input FE. Yet we stress that these properties are satisfied by all existing constructions. Both our constructions rely on a simple MIFE scheme that is one-AD-IND secure unconditionally. In particular, our constructions show how to use single-input FE in order to bootstrap the information-theoretic MIFE from one-time to many-time security.

Setup^{ot}($1^\lambda, \mathcal{F}_{L,n}^m$):	**KeyGen^{ot}($u, y_1 \| \cdots \| y_n$):**
For all $i \in [n]$, $u_i \leftarrow_{\text{R}} \mathbb{Z}_L^m$	Return $z := \sum_{i \in [n]} \langle u_i, y_i \rangle \bmod L$
Return $u = \{u_i\}_{i \in [n]}$	
	Dec^{ot}($z, \text{ct}_1, \ldots, \text{ct}_n$):
Enc^{ot}(u, i, x_i):	Return $\sum_{i=1}^{n} \langle \text{ct}_i, y_i \rangle - z \bmod L$
Return $x_i + u_i \bmod L$	

Fig. 1. Private-key, information theoretically secure, multi-input FE scheme $\mathcal{MIFE}^{\text{ot}} = (\text{Setup}^{\text{ot}}, \text{Enc}^{\text{ot}}, \text{KeyGen}^{\text{ot}}, \text{Dec}^{\text{ot}})$ for the class $\mathcal{F}_{L,n}^m$.

3.1 Information-Theoretic MIFE with One-Time Security

Here we present the multi-input scheme $\mathcal{MIFE}^{\text{ot}}$ for the class $\mathcal{F}_{L,n}^m$, and we prove its one-AD-IND security. The scheme is described in Fig. 1.

Theorem 1. *The MIFE described in Fig. 1 is one-AD-IND secure. Namely, for any adversary $\mathcal{A}$, $\mathrm{Adv}_{\mathcal{MIFE},\mathcal{A}}^{one\text{-}AD\text{-}IND}(\lambda) = 0$.*

Proof overview. The proof of Theorem 1 has two main steps. First, we use the fact that any adaptive distinguisher against $\mathcal{MIFE}^{ot}$ with advantage ε can be transformed into a selective distinguisher with advantage $\varepsilon/|X|^2$ by randomly guessing the two challenge input vectors, where $|X|$ is the size of the input space ($|X| = L^{nm}$ in our case). Then, in a second step, we show that any selective distinguisher against $\mathcal{MIFE}^{ot}$ has advantage 0 since $\mathcal{MIFE}^{ot}$ behaves as the FE equivalent of the one-time pad. Hence, it follows that any adaptive distinguisher must also have advantage 0.

Proof. Let $\mathcal{A}$ be an adversary against the one-AD-IND security of the MIFE. First, we use a complexity leveraging argument to build an adversary $\mathcal{B}$ such that:

$$\mathsf{Adv}^{one\text{-}AD\text{-}IND}_{\mathcal{MIFE},\mathcal{A}}(\lambda) \leq L^{-2nm} \cdot \mathsf{Adv}^{one\text{-}SEL\text{-}IND}_{\mathcal{MIFE},\mathcal{B}}(\lambda).$$

The adversary $\mathcal{B}$ simply guesses the challenge $\{x_i^b\}_{i\in[n],b\in\{0,1\}}$ in advance, then simulates $\mathcal{A}$'s experiment using its own selective experiment. When $\mathcal{B}$ receives $\mathcal{A}$'s challenge, it checks if the guess was successful (call E that event): if it was, it continues simulating $\mathcal{A}$'s experiment, otherwise, it returns 0. When the guess is successful, $\mathcal{B}$ perfectly simulate $\mathcal{A}$'s view. Since event E happens with probability exactly L^{-2nm}, and is independent of the adversary $\mathcal{A}$'s view, we obtain $\mathsf{Adv}^{one\text{-}AD\text{-}IND}_{\mathcal{MIFE},\mathcal{A}}(\lambda) \leq L^{-2nm} \cdot \mathsf{Adv}^{one\text{-}SEL\text{-}IND}_{\mathcal{MIFE},\mathcal{B}}(\lambda)$.

It remains to prove that the MIFE presented in Fig. 1 satisfies perfect one-SEL-IND security, namely, for any adversary $\mathcal{B}$, $\mathsf{Adv}^{one\text{-}SEL\text{-}IND}_{\mathcal{MIFE},\mathcal{B}}(\lambda) = 0$. To do so, we introduce hybrid games $\mathsf{H}_\beta(1^\lambda, \mathcal{B})$ described in Fig. 2. We prove that for all $\beta \in \{0,1\}$, $\mathsf{H}_\beta(1^\lambda, \mathcal{B})$ is identical to the experiment one-SEL-IND$_\beta^{\mathcal{MIFE}}(1^\lambda, \mathcal{B})$. This can be seen using the fact that for all $\{x_i^\beta \in \mathbb{Z}^m\}_{i\in[n]}$, the following distributions are identical: $\{u_i \bmod L\}_{i\in[n]}$ and $\{u_i - x_i^\beta \bmod L\}_{i\in[n]}$, with $u_i \leftarrow_R \mathbb{Z}_L^m$. Recall that here $i \in [n]$ is an index for input slots. Note that the independence of the x_i^β from the u_i is only true in the selective security game. Finally, we show that $\mathcal{B}$'s view in $\mathsf{H}_\beta(1^\lambda, \mathcal{B})$ is independent of β. Indeed, the only information about β that leaks in this experiment is $\sum_i \langle x_i^\beta, y_i \rangle$, which is independent of β by definition of the security game. $\qquad\square$

HYB$_\beta(1^\lambda, \mathcal{B})$:	$\mathcal{O}_H(y)$:
$\{x_i^b\}_{i\in[n],b\in\{0,1\}} \leftarrow \mathcal{B}(1^\lambda, \mathcal{F}^m_{L,n})$	For all $i \in [n]$,
For all $i \in [n]$,	$\qquad z := \sum_{i\in[n]}\langle u_i, y_i \rangle - \langle x_i^\beta, y_i \rangle \bmod L$
$\qquad u_i \leftarrow_R \mathbb{Z}_L^m$; $\mathsf{ct}_i \leftarrow u_i$	Return z
$\alpha \leftarrow \mathcal{B}^{\mathcal{O}_H(\cdot)}(\{\mathsf{ct}_i\}_{i\in[\ell]})$	
Output α	

Fig. 2. Experiments for the proof of Theorem 1.

Remark 2 (one-SEL-SIM security). As a result of independent interest, in the full version of the paper [3] we show that the MIFE presented in Fig. 1 satisfies perfect one-SEL-SIM security, which implies perfect one-SEL-IND (which itself implies perfect one-AD-IND security via complexity leveraging, as shown in the proof above).

Remark 3 (Linear homomorphism). We use the fact that $\mathsf{Enc}^{\mathsf{ot}}$ is linearly homomorphic, that is, for all input slots $i \in [n]$, $\boldsymbol{x}_i, \boldsymbol{x}_i' \in \mathbb{Z}_p^m$, $\boldsymbol{u} \leftarrow \mathsf{Setup}^{\mathsf{ot}}(1^\lambda, \mathcal{F}_{L,n}^m)$, $\mathsf{Enc}^{\mathsf{ot}}(\boldsymbol{u}, i, \boldsymbol{x}_i) + \boldsymbol{x}_i' \bmod L = \mathsf{Enc}^{\mathsf{ot}}(\boldsymbol{u}, i, \boldsymbol{x}_i + \boldsymbol{x}_i')$. This property will be used when using the one-time scheme $\mathcal{MIFE}^{\mathsf{ot}}$ from Fig. 1 as a building block to obtain a full-fledged many-AD-IND MIFE.

3.2 Our Transformation for Inner Product over $\mathbb{Z}_L$

We present our multi-input scheme $\mathcal{MIFE}$ for the class $\mathcal{F}_{L,n}^m$ in Fig. 3. The construction relies on the one-time scheme $\mathcal{MIFE}^{\mathsf{ot}}$ of Fig. 1, and any single-input FE for the class $\mathcal{F}_{L,1}^m$.

$\mathsf{Setup}'(1^\lambda, \mathcal{F}_{L,n}^m)$:

$\boldsymbol{u} \leftarrow \mathsf{Setup}^{\mathsf{ot}}(1^\lambda, \mathcal{F}_{L,n}^m)$, for all $i \in [n]$, $(\mathsf{mpk}_i, \mathsf{msk}_i) \leftarrow \mathsf{Setup}(1^\lambda, \mathcal{F}_{L,1}^m)$
$(\mathsf{mpk}, \mathsf{msk}) := \big(\{\mathsf{mpk}_i\}_{i \in [n]}, (\{\mathsf{msk}_i,\}_{i \in [n]}, \boldsymbol{u})\big)$
Return $(\mathsf{mpk}, \mathsf{msk})$

$\mathsf{Enc}'(\mathsf{msk}, i, \boldsymbol{x}_i)$:

$\boldsymbol{w}_i := \mathsf{Enc}^{\mathsf{ot}}(\boldsymbol{u}, i, \boldsymbol{x}_i)$
Return $\mathsf{Enc}(\mathsf{msk}_i, \boldsymbol{w}_i)$

$\mathsf{KeyGen}'(\mathsf{msk}, \boldsymbol{y}_1 \| \cdots \| \boldsymbol{y}_n)$:

For all $i \in [n]$, $\mathsf{sk}_i \leftarrow \mathsf{KeyGen}(\mathsf{msk}_i, \boldsymbol{y}_i)$, $z := \mathsf{KeyGen}^{\mathsf{ot}}(\boldsymbol{u}, \boldsymbol{y}_1 \| \cdots \| \boldsymbol{y}_n)$
$\mathsf{sk}_{\boldsymbol{y}_1 \| \cdots \| \boldsymbol{y}_n} := \big(\{\mathsf{sk}_i\}_{i \in [n]}, z\big)$
Return $\mathsf{sk}_{\boldsymbol{y}_1 \| \cdots \| \boldsymbol{y}_n}$

$\mathsf{Dec}'\big((\{\mathsf{sk}_i\}_{i \in [n]}, z), \mathsf{ct}_1, \ldots, \mathsf{ct}_n\big)$:

For all $i \in [n]$, $D_i \leftarrow \mathsf{Dec}(\mathsf{sk}_i, \mathsf{ct}_i)$
Return $\sum_{i \in [n]} D_i - z \bmod L$

Fig. 3. Private-key multi-input FE scheme $\mathcal{MIFE} := (\mathsf{Setup}', \mathsf{Enc}', \mathsf{KeyGen}', \mathsf{Dec}')$ for the class $\mathcal{F}_{L,n}^m$ from a public-key single-input FE $\mathcal{FE} := (\mathsf{Setup}, \mathsf{Enc}, \mathsf{KeyGen}, \mathsf{Dec})$ for the class $\mathcal{F}_{L,1}^m$, and one-time multi-input FE $\mathcal{MIFE}^{\mathsf{ot}} = (\mathsf{Setup}^{\mathsf{ot}}, \mathsf{Enc}^{\mathsf{ot}}, \mathsf{KeyGen}^{\mathsf{ot}}, \mathsf{Dec}^{\mathsf{ot}})$ for the class $\mathcal{F}_{L,n}^m$.

The correctness of $\mathcal{MIFE}$ follows from the correctness properties of the single-input scheme $\mathcal{FE}$ and the multi-input scheme $\mathcal{MIFE}^{\mathsf{ot}}$. Indeed, correctness of the former implies that, for all input slots $i \in [n]$, $D_i = \langle \boldsymbol{w}_i, \boldsymbol{y}_i \rangle$

mod L, while correctness of $\mathcal{MIFE}^{ot}$ implies that $\sum_{i \in [n]} D_i - z =$ $\mathsf{Dec}^{ot}(z, \boldsymbol{w}_1, \ldots, \boldsymbol{w}_n) = \sum_{i \in [n]} \langle \boldsymbol{x}_i, \boldsymbol{y}_i \rangle \mod L$.

For the security we state the following theorem:

Theorem 2. *If the single-input FE, $\mathcal{FE}$ is many-AD-IND-secure, and the multi-input scheme $\mathcal{MIFE}^{ot}$ is one-AD-IND-secure, then the multi-input FE, $\mathcal{MIFE}$, described in Fig. 3, is many-AD-IND-secure.*

Since the proof of the above theorem is almost the same as the one for the case of bounded-norm inner product, we only provide an overview here, and defer to the proof of Theorem 3 for further details.

Proof overview. Here, for any input slot $i \in [n]$, we denote by $(\boldsymbol{x}_i^{j,0}, \boldsymbol{x}_i^{j,1})$ the j'th query to $\mathsf{Enc}(i, \cdot, \cdot)$, for any $j \in [Q_i]$, where Q_i is the total number of queries to $\mathsf{Enc}(i, \cdot, \cdot)$.

The proof is in two main steps. First, we switch encryptions of $\boldsymbol{x}_1^{1,0}, \ldots, \boldsymbol{x}_n^{1,0}$ to those of $\boldsymbol{x}_1^{1,1}, \ldots, \boldsymbol{x}_n^{1,1}$, using the one-AD-IND security of $\mathcal{MIFE}^{ot}$. For the remaining ciphertexts, we switch from an encryption of $\boldsymbol{x}_i^{j,0} = (\boldsymbol{x}_i^{j,0} - \boldsymbol{x}_i^{1,0}) + \boldsymbol{x}_i^{1,0}$ to that of $(\boldsymbol{x}_i^{j,0} - \boldsymbol{x}_i^{1,0}) + \boldsymbol{x}_i^{1,1}$. In this step we use the fact that one can compute an encryption of $\mathsf{Enc}^{ot}(\boldsymbol{u}, i, (\boldsymbol{x}_i^{j,0} - \boldsymbol{x}_i^{1,0}) + \boldsymbol{x}_i^{1,0})$ from an encryption $\mathsf{Enc}^{ot}(\boldsymbol{u}, i, \boldsymbol{x}_i^{1,0})$, because the encryption algorithm Enc^{ot} of $\mathcal{MIFE}^{ot}$ is linearly homomorphic (see Remark 3). Finally, we apply a hybrid argument across the slots to switch from encryptions of

$$(\boldsymbol{x}_i^{2,0} - \boldsymbol{x}_i^{1,0}) + \boldsymbol{x}_i^{1,1}, \ldots, (\boldsymbol{x}_i^{Q_i,0} - \boldsymbol{x}_i^{1,0}) + \boldsymbol{x}_i^{1,1}$$

to those of

$$(\boldsymbol{x}_i^{2,1} - \boldsymbol{x}_i^{1,1}) + \boldsymbol{x}_i^{1,1}, \ldots, (\boldsymbol{x}_i^{Q_i,1} - \boldsymbol{x}_i^{1,1}) + \boldsymbol{x}_i^{1,1},$$

using the many-AD-IND security of $\mathcal{FE}$.

Instantiations. The construction in Fig. 3 can be instantiated using the single-input FE schemes of Agrawal et al. [5] that are many-AD-IND-secure and allow for computing inner products over a finite ring. Specifically, we obtain:

- A MIFE for inner product over $\mathbb{Z}_p$ for a prime p, based on the LWE assumption. This is obtained by using the LWE-based scheme of Agrawal et al. [5, Sect. 4.2].
- A MIFE for inner product over $\mathbb{Z}_N$ where N is an RSA modulus, based on the Composite Residuosity assumption. This is obtained by using the Paillier-based scheme of Agrawal et al. [5, Sect. 5.2].

We note that since both these schemes in [5] have a stateful key generation, our MIFE inherits this stateful property. Stateless MIFE instantiations are obtained from the transformation in the next section.

3.3   Our Transformation for Inner Product over $\mathbb{Z}$

Here we present our transformation for the case of bounded-norm inner product. In particular, in Fig. 4 we present a multi-input scheme $\mathcal{MIFE}$ for the class $\mathcal{F}_n^{m,X,Y}$ from the one-time scheme $\mathcal{MIFE}^{\mathsf{ot}}$ of Fig. 1, and a (single-input) scheme $\mathcal{FE}$ for the class $\mathcal{F}_1^{m,3X,Y}$.[7] For our transformation to work, we require $\mathcal{FE}$ to satisfy two properties. The first one, that we call *two-step decryption*, intuitively says that the $\mathcal{FE}$ decryption algorithm works in two steps: the first step uses the secret key to output an encoding of the result, while the second step returns the actual result $\langle x, y \rangle$ provided that the bounds $\|x\|_\infty < X$, $\|y\|_\infty < Y$ hold. The second property informally says that the $\mathcal{FE}$ encryption algorithm is additively homomorphic.

We note that the two-step property also says that the encryption algorithm accepts inputs x such that $\|x\|_\infty > X$, yet correctness is guaranteed as long as the encrypted inputs are within the bound at the moment of invoking the second step of decryption.

Two-step decryption is formally defined as follows.

Property 1 (Two-step decryption). An FE scheme $\mathcal{FE}$ = (Setup, Enc, KeyGen, Dec) satisfies *two-step decryption* if it admits PPT algorithms Setup*, $\mathsf{Dec}_1, \mathsf{Dec}_2$ and an encoding function $\mathcal{E}$ such that:

1. For all $\lambda, m, n, X, Y \in \mathbb{N}$, $\mathsf{Setup}^\star(1^\lambda, \mathcal{F}_1^{m,X,Y}, 1^n)$ outputs (msk, mpk) where mpk includes a bound $B \in \mathbb{N}$, and the description of a group $\mathbb{G}$ (with group law $\circ$) of order $L > n \cdot m \cdot X \cdot Y$, which defines the encoding function $\mathcal{E} : \mathbb{Z}_L \times \mathbb{Z} \to \mathbb{G}$.
2. For all (msk, mpk) $\leftarrow \mathsf{Setup}^\star(1^\lambda, \mathcal{F}_1^{m,X,Y}, 1^n)$, $x \in \mathbb{Z}^m$, ct $\leftarrow \mathsf{Enc}(\mathsf{msk}, x)$, $y \in \mathbb{Z}^m$, and sk $\leftarrow \mathsf{KeyGen}(\mathsf{msk}, y)$, we have

$$\mathsf{Dec}_1(\mathsf{ct}, \mathsf{sk}) = \mathcal{E}(\langle x, y \rangle \bmod L, \mathsf{noise}),$$

   for some noise $\in \mathbb{N}$ that depends on ct and sk. Furthermore, it holds that for all $x, y \in \mathbb{Z}^m$, $\Pr[\mathsf{noise} < B] = 1 - \mathsf{negl}(\lambda)$, where the probability is taken over the random coins of Setup*, Enc and KeyGen. Note that there is no restriction on the norm of $\langle x, y \rangle$ here, and that we are assuming that Enc accepts inputs x whose norm may be larger than the bound.
3. Given any $\gamma \in \mathbb{Z}_L$, and mpk, one can efficiently compute $\mathcal{E}(\gamma, 0)$.
4. The encoding $\mathcal{E}$ is linear, that is: for all $\gamma, \gamma' \in \mathbb{Z}_L,$, noise, noise' $\in \mathbb{Z}$, we have

$$\mathcal{E}(\gamma, \mathsf{noise}) \circ \mathcal{E}(\gamma', \mathsf{noise}') = \mathcal{E}(\gamma + \gamma' \bmod L, \mathsf{noise} + \mathsf{noise}').$$

5. For all $\gamma < n \cdot m \cdot X \cdot Y$, and noise $< n \cdot B$, $\mathsf{Dec}_2(\mathcal{E}(\gamma, \mathsf{noise})) = \gamma$.

The second property is as follows.

Property 2 (Linear encryption). For any FE scheme $\mathcal{FE}$ = (Setup, Enc, KeyGen, Dec) satisfying the two-step property, we define the following additional property.

---

[7] The reason why we need $3X$ instead of X is due to maintain a correct distribution of the inputs in the security proof.

There exists a deterministic algorithm Add that takes as input a ciphertext and a message, such that for all $\boldsymbol{x}, \boldsymbol{x}' \in \mathbb{Z}^m$, the following are identically distributed:

$$\mathsf{Add}(\mathsf{Enc}(\mathsf{msk}, \boldsymbol{x}), \boldsymbol{x}'), \quad \text{and} \quad \mathsf{Enc}\big(\mathsf{msk}, (\boldsymbol{x} + \boldsymbol{x}' \bmod L)\big).$$

Note that the value $L \in \mathbb{N}$ is defined as part of the output of the algorithm Setup^* (see the two-step property above). We later use a single input FE with this property as a building block for a multi-input FE (see Fig. 4); this property however is only used in the security proof of our transformation.

Instantiations. It is not hard to check that these two properties are satisfied by known functional encryption schemes for (bounded-norm) inner product. In particular, in the full version of the paper we show that this is satisfied by the many-AD-IND secure FE schemes of Agrawal et al. [5].[8] This allows us to obtain MIFE schemes for bounded-norm inner product based on a variety of assumptions such as plain DDH, Decisional Composite Residuosity, and LWE. In addition to obtaining the first schemes without the need of pairing groups, we also obtain schemes where decryption works efficiently even for large outputs. This stands in contrast to the previous result [4], where decryption requires to extract discrete logarithms.

Correctness. The correctness of the scheme $\mathcal{MIFE}$ follows from (i) the correctness and Property 1 (two-step decryption) of the single-input scheme, and (ii) from the correctness of $\mathcal{MIFE}^{\mathsf{ot}}$ and the linear property of its decryption algorithm $\mathsf{Dec}^{\mathsf{ot}}$.

More precisely, consider any vector $\boldsymbol{x} := (\boldsymbol{x}_1 \| \cdots \| \boldsymbol{x}_n) \in (\mathbb{Z}^m)^n$, $\boldsymbol{y} \in \mathbb{Z}^{mn}$, such that $\|\boldsymbol{x}\|_\infty < X$, $\|\boldsymbol{y}\|_\infty < Y$, and let $(\mathsf{mpk}, \mathsf{msk}) \leftarrow \mathsf{Setup}'(1^\lambda, \mathcal{F}_n^{m,X,Y})$, $\mathsf{sk}_{\boldsymbol{y}} \leftarrow \mathsf{KeyGen}'(\mathsf{msk}, \boldsymbol{y})$, and $\mathsf{ct}_i \leftarrow \mathsf{Enc}'(\mathsf{msk}, i, \boldsymbol{x}_i)$ for all $i \in [n]$.

By (2) of Property 1, the decryption algorithm $\mathsf{Dec}'(\mathsf{sk}_{\boldsymbol{y}}, \mathsf{ct}_1, \ldots, \mathsf{ct}_n)$ computes $\mathcal{E}(\langle \boldsymbol{w}_i, \boldsymbol{y}_i \rangle \bmod L, \mathsf{noise}_i) \leftarrow \mathsf{Dec}_1(\mathsf{sk}_i, \mathsf{ct}_i)$ where for all $i \in [n]$, $\mathsf{noise}_i < B$, with probability $1 - \mathsf{negl}(\lambda)$.

By (4) of Property 1 (linearity of $\mathcal{E}$), and the correctness of $\mathcal{MIFE}^{\mathsf{ot}}$ we have:

$$\mathcal{E}(\langle \boldsymbol{w}_1, \boldsymbol{y}_1 \rangle \bmod L, \mathsf{noise}_1) \circ \cdots \circ \mathcal{E}(\langle \boldsymbol{w}_n, \boldsymbol{y}_n \rangle \bmod L, \mathsf{noise}_n) \circ \mathcal{E}(-z, 0)$$

$$= \mathcal{E}\left(\mathsf{Dec}^{\mathsf{ot}}(z, \boldsymbol{w}_1, \ldots, \boldsymbol{w}_n), \sum_{i \in [n]} \mathsf{noise}_i\right) = \mathcal{E}\left(\langle \boldsymbol{x}, \boldsymbol{y} \rangle \bmod L, \sum_{i \in [n]} \mathsf{noise}_i\right).$$

Since $\langle \boldsymbol{x}, \boldsymbol{y} \rangle < n \cdot m \cdot X \cdot Y < L$ and $\sum_{i \in [n]} \mathsf{noise}_i < n \cdot B$, we have

$$\mathsf{Dec}_2\big(\mathcal{E}(\langle \boldsymbol{x}, \boldsymbol{y} \rangle \bmod L, \sum_{i \in [n]} \mathsf{noise}_i)\big) = \langle \boldsymbol{x}, \boldsymbol{y} \rangle,$$

by (5) of Property 1.

---

[8] While in [5] the FE schemes are proven only one-AD-IND secure (i.e., for adversaries making a single encryption query), note that these are *public-key* schemes and thus many-AD-IND security can be obtained via a standard hybrid argument from one-AD-IND security.

$\mathsf{Setup}'(1^\lambda, \mathcal{F}_n^{m,X,Y})$:

$\boldsymbol{u} \leftarrow \mathsf{Setup}^{ot}(1^\lambda, \mathcal{F}_{L,n}^m)$, for all $i \in [n]$, $(\mathsf{mpk}_i, \mathsf{msk}_i) \leftarrow \mathsf{Setup}^\star(1^\lambda, \mathcal{F}_1^{m,3X,Y}, 1^n)$
$(\mathsf{mpk}, \mathsf{msk}) := \big(\{\mathsf{mpk}_i\}_{i\in[n]}, (\{\mathsf{msk}_i, \}_{i\in[n]}, \boldsymbol{u})\big)$
Return $(\mathsf{mpk}, \mathsf{msk})$

$\mathsf{Enc}'(\mathsf{msk}, i, \boldsymbol{x}_i)$:

$\boldsymbol{w}_i := \mathsf{Enc}^{ot}(\boldsymbol{u}, i, \boldsymbol{x}_i)$
Return $\mathsf{Enc}(\mathsf{msk}_i, \boldsymbol{w}_i)$

$\mathsf{KeyGen}'(\mathsf{msk}, \boldsymbol{y}_1\|\cdots\|\boldsymbol{y}_n)$:

For all $i \in [n]$, $\mathsf{sk}_i \leftarrow \mathsf{KeyGen}(\mathsf{msk}_i, \boldsymbol{y}_i)$, $z \leftarrow \mathsf{KeyGen}^{ot}(\boldsymbol{u}, \boldsymbol{y}_1\|\cdots\|\boldsymbol{y}_n)$
$\mathsf{sk}_{\boldsymbol{y}_1\|\cdots\|\boldsymbol{y}_n} := \big(\{\mathsf{sk}_i\}_{i\in[n]}, z\big)$
Return $\mathsf{sk}_{\boldsymbol{y}_1\|\cdots\|\boldsymbol{y}_n}$

$\mathsf{Dec}'\big((\{\mathsf{sk}_i\}_{i\in[n]}, z), \mathsf{ct}_1, \ldots, \mathsf{ct}_n\big)$:

For all $i \in [n]$, $\mathcal{E}(\langle \boldsymbol{x}_i + \boldsymbol{u}_i, \boldsymbol{y}_i \rangle \bmod L, \mathsf{noise}_i) \leftarrow \mathsf{Dec}_1(\mathsf{sk}_i, \mathsf{ct}_i)$
Return $\mathsf{Dec}_2\big(\mathcal{E}(\langle \boldsymbol{x}_1 + \boldsymbol{u}_1, \boldsymbol{y}_1 \rangle \bmod L, \mathsf{noise}_1) \circ \cdots \circ \mathcal{E}(\langle \boldsymbol{x}_n + \boldsymbol{u}_n, \boldsymbol{y}_n \rangle \bmod L, \mathsf{noise}_n) \circ \mathcal{E}(-z, 0)\big)$

Fig. 4. Private-key multi-input FE scheme $\mathcal{MIFE} = (\mathsf{Setup}', \mathsf{Enc}', \mathsf{KeyGen}', \mathsf{Dec}')$ for the class $\mathcal{F}_n^{m,X,Y}$ from public-key single-input FE scheme $\mathcal{FE} = (\mathsf{Setup}, \mathsf{Enc}, \mathsf{KeyGen}, \mathsf{Dec})$ for the class $\mathcal{F}_1^{m,X,Y}$ and one-time multi-input FE $\mathcal{MIFE}^{ot} = (\mathsf{Setup}^{ot}, \mathsf{Enc}^{ot}, \mathsf{KeyGen}^{ot}, \mathsf{Dec}^{ot})$.

Proof of Security. In the following theorem we show that our construction is a many-AD-IND-secure MIFE, assuming that the underlying single-input FE scheme is many-AD-IND-secure, and the scheme $\mathcal{MIFE}^{ot}$ is one-AD-IND secure.

Theorem 3. *Assume that the single-input FE is many-AD-IND-secure and the multi-input FE $\mathcal{MIFE}^{ot}$ is one-AD-IND-secure. Then the multi-input FE $\mathcal{MIFE}$ in Fig. 4 is many-AD-IND-secure. Namely, for any PPT adversary $\mathcal{A}$, there exist PPT adversaries $\mathcal{B}$ and $\mathcal{B}'$ such that*

$$\mathsf{Adv}_{\mathcal{MIFE}, \mathcal{A}}^{many\text{-}AD\text{-}IND}(\lambda) \le \mathsf{Adv}_{\mathcal{MIFE}^{ot}, \mathcal{B}}^{one\text{-}AD\text{-}IND}(\lambda) + n \cdot \mathsf{Adv}_{\mathcal{FE}, \mathcal{B}'}^{many\text{-}AD\text{-}IND}(\lambda).$$

Proof of Theorem 3. The proof proceeds by a sequence of games where G_0 is the **many-AD-IND**$_0^{\mathcal{MIFE}}(1^\lambda, \mathcal{A})$ experiment. A formal description of all the experiments used in this proof is given in Fig. 6, and a high-level summary is provided in Fig. 5. For any game G_i, we denote by $\mathbf{Adv}_i(\mathcal{A})$ the advantage of $\mathcal{A}$ in G_i, that is, $\Pr[G_i(1^\lambda, \mathcal{A}) = 1]$, where the probability is taken over the random coins of G_i and $\mathcal{A}$. In what follows we adopt the same notation from [4] for queried plaintexts, namely $(\boldsymbol{x}_i^{j,0}, \boldsymbol{x}_i^{j,1})$ denotes the j-th encryption query on the i-th slot.

Game G_1: Here we change the way the challenge ciphertexts are created. In particular, for all slots and all queries simultaneously, we switch from $\mathsf{Enc}'(\mathsf{msk}, i, \boldsymbol{x}_i^{j,0} - \boldsymbol{x}_i^{1,0} + \boldsymbol{x}_i^{1,0})$ to $\mathsf{Enc}'(\mathsf{msk}, i, \boldsymbol{x}_i^{j,0} - \boldsymbol{x}_i^{1,0} + \boldsymbol{x}_i^{1,1})$.

Game	ct_i^j	justification/remark
G_0	$\mathsf{Enc}'(\mathsf{msk}', i, \boldsymbol{x}_i^{j,0} - \boldsymbol{x}_i^{1,0} + \boldsymbol{x}_i^{1,0})$	
G_1	$\mathsf{Enc}'(\mathsf{msk}', i, \boldsymbol{x}_i^{j,0} - \boldsymbol{x}_i^{1,0} + \boxed{\boldsymbol{x}_i^{1,1}})$	one-AD-IND of $\mathcal{MIFE}^{\mathsf{ot}}$, Lemma 2
$G_{1.\ell}$	$\mathsf{Enc}'(\mathsf{msk}', i, \boldsymbol{x}_i^{j,1} - \boldsymbol{x}_i^{1,1} + \boldsymbol{x}_i^{1,1})$, for $i \le \ell$ $\mathsf{Enc}'(\mathsf{msk}', i, \boldsymbol{x}_i^{j,0} - \boldsymbol{x}_i^{1,0} + \boldsymbol{x}_i^{1,1})$, for $i > \ell$	many-AD-IND of $\mathcal{FE}$, Lemma 3
G_2	$\mathsf{Enc}'(\mathsf{msk}', i, \boxed{\boldsymbol{x}_i^{j,1}})$	$G_2 = G_{1.n}$

Fig. 5. An overview of the games used in the proof of Theorem 3.

G_1 can be proved indistinguishable from G_0 by relying on the one-time security of the multi-input scheme. More formally,

Lemma 2. *There exists a PPT adversary $\mathcal{B}_1$ against the one-AD-IND security of $\mathcal{MIFE}^{\mathsf{ot}}$ scheme such that*

$$|\mathbf{Adv}_0(\mathcal{A}) - \mathbf{Adv}_1(\mathcal{A})| \le \mathsf{Adv}_{\mathcal{MIFE}^{\mathsf{ot}}, \mathcal{B}_1}^{one\text{-}AD\text{-}IND}(\lambda).$$

Proof. Here we replace encryptions of $\boldsymbol{x}_i^{j,0} - \boldsymbol{x}_i^{1,0} + \boldsymbol{x}_i^{1,0}$ with encryptions of $\boldsymbol{x}_i^{j,0} - \boldsymbol{x}_i^{1,0} + \boldsymbol{x}_i^{1,1}$ in all slots simultaneously. Recall that here, j is the index of the encryption query while i is the index for the slot. The argument relies on the one-AD-IND security of the multi-input scheme $\mathcal{MIFE}^{\mathsf{ot}}$ and on the fact that ciphertexts produced by the latter can be used as plaintext for the underlying single input FE scheme $\mathcal{FE}$ that we are using as additional basic building block.

More in details, we build the adversary $\mathcal{B}_1$ so that it simulates G_β to $\mathcal{A}$ when interacting with experiment **one-AD-IND**$_\beta^{\mathcal{MIFE}}$.

Initially $\mathcal{B}_1$ does not receive anything, since the one-AD-IND information-theoretically secure MIFE does not have any public key. For all $i \in [n]$ it runs $(\mathsf{mpk}_i, \mathsf{msk}_i) \leftarrow \mathsf{Setup}^\star(1^\lambda, \mathcal{F}_1^{m,3X,Y}, 1^n)$, and hands the public parameters to $\mathcal{A}$. Also, whenever $\mathcal{A}$ queries a secret key, $\mathcal{B}_1$ first queries its own oracle (on the same input) to get a corresponding key z. Next, for all $i \in [n]$, it sets $\mathsf{sk}_i \leftarrow \mathsf{KeyGen}(\mathsf{msk}_i, \boldsymbol{y}_i)$ and gives back to $\mathcal{A}$ the secret key $\mathsf{sk}_{\boldsymbol{y}_1\|\cdots\|\boldsymbol{y}_n} := (\{\mathsf{sk}_i\}_{i \in [n]}, z)$.

When $\mathcal{A}$ asks encryption queries, $\mathcal{B}_1$ proceeds as follows. For each slot i, when receiving the first query $(i, \boldsymbol{x}_i^{1,0}, \boldsymbol{x}_i^{1,1})$, it computes the challenge ciphertext, for slot i, by invoking its own encryption oracle on the same input. Calling $\boldsymbol{w}_i^1 := \mathsf{Enc}^{\mathsf{ot}}(\boldsymbol{u}, i, \boldsymbol{x}_i^{1,\beta})$ the received ciphertext, $\mathcal{B}_1$ computes $\mathsf{ct}_i^1 = \mathsf{Enc}(\mathsf{msk}_i, \boldsymbol{w}_i^1) = \mathsf{Enc}'(\mathsf{msk}, i, \boldsymbol{x}_i^{1,\beta})$.

Subsequent queries, on slot i, are answered as follows. $\mathcal{B}_1$ produces ct_i^j (for $j > 1$) by encrypting $\boldsymbol{x}_i^{j,0} - \boldsymbol{x}_i^{1,0} + \boldsymbol{w}_i^1 \bmod L$, using msk_i. Note that $\mathsf{Enc}^{\mathsf{ot}}$ is linearly homomorphic (see Remark 3), thus, $\boldsymbol{x}_i^{j,0} - \boldsymbol{x}_i^{1,0} + \boldsymbol{w}_i^1 \bmod L = \mathsf{Enc}^{\mathsf{ot}}(\boldsymbol{u}, i, \boldsymbol{x}_i^{1,\beta} + \boldsymbol{x}_i^{j,0} - \boldsymbol{x}_i^{1,0})$.

Finally, $\mathcal{B}_1$ outputs 1 iff $\mathcal{A}$ outputs 1. One can see that $\mathcal{B}_1$ provides a perfect simulation to $\mathcal{A}$ and thus:

$$|\mathbf{Adv}_0(\mathcal{A}) - \mathbf{Adv}_1(\mathcal{A})| \leq \mathsf{Adv}_{\mathcal{MIFE}, \mathcal{B}_1}^{one\text{-}AD\text{-}IND}(\lambda).$$

□

$G_0(1^\lambda, \mathcal{A})$, $\boxed{G_1(1^\lambda, \mathcal{A})}$, $\overline{G_2(1^\lambda, \mathcal{A})}$:	$G_{1.\ell}(1^\lambda, \mathcal{A})$:
$(\mathsf{mpk}, \mathsf{msk}) \leftarrow \mathsf{Setup}'(1^\lambda, \mathcal{F}_n^{m,X,Y})$	$(\mathsf{mpk}, \mathsf{msk}) \leftarrow \mathsf{Setup}'(1^\lambda, \mathcal{F}_n^{m,X,Y})$
$\beta' \leftarrow \mathcal{A}^{\mathsf{KeyGen}'(\mathsf{msk},\cdot), \mathsf{EncO}'(\cdot,\cdot,\cdot)}(\mathsf{mpk})$	$\beta' \leftarrow \mathcal{A}^{\mathsf{KeyGen}'(\mathsf{msk},\cdot), \mathsf{EncO}'(\cdot,\cdot,\cdot)}(\mathsf{mpk})$
return β'	return β'
$\mathsf{EncO}'(i, \boldsymbol{x}_i^{j,0}, \boldsymbol{x}_i^{j,1})$	
$\quad \mathsf{ct}_i^j := \mathsf{Enc}'(\mathsf{msk}, i, \boldsymbol{x}_i^{j,0} - \boldsymbol{x}_i^{1,0} + \boldsymbol{x}_i^{1,0})$	$\mathsf{EncO}'(i, \boldsymbol{x}_i^{j,0}, \boldsymbol{x}_i^{j,1})$
$\quad \boxed{\mathsf{ct}_i^j := \mathsf{Enc}'(\mathsf{msk}, i, \boldsymbol{x}_i^{j,0} - \boldsymbol{x}_i^{1,0} + \boldsymbol{x}_i^{1,1})}$	$\quad$ If $i \leq \ell$ return
$\quad \overline{\mathsf{ct}_i^j := \mathsf{Enc}'(\mathsf{msk}, i, \boldsymbol{x}_i^{j,1} - \boldsymbol{x}_i^{1,1} + \boldsymbol{x}_i^{1,1})}$	$\quad\quad \mathsf{Enc}'(\mathsf{msk}, i, \boldsymbol{x}_i^{j,1} - \boldsymbol{x}_i^{1,1} + \boldsymbol{x}_i^{1,1})$
$\quad$ return ct_i^j	$\quad$ If $i > \ell$ return
	$\quad\quad \mathsf{Enc}'(\mathsf{msk}, i, \boldsymbol{x}_i^{j,0} - \boldsymbol{x}_i^{1,0} + \boldsymbol{x}_i^{1,1})$
$\mathsf{KeyGen}'(\mathsf{msk}, \boldsymbol{y})$	$\mathsf{KeyGen}'(\mathsf{msk}, \boldsymbol{y})$
$\quad$ return $\mathsf{sk}_{\boldsymbol{y}}$	$\quad$ return $\mathsf{sk}_{\boldsymbol{y}}$

Fig. 6. Experiments for the proof of Theorem 3.

Game G_2: Here we change again the way the challenge ciphertexts are created. In particular, for all slots i and all queries j, we switch ct_i^j from $\mathsf{Enc}'(\mathsf{msk}, i, \boldsymbol{x}_i^{j,0} - \boldsymbol{x}_i^{1,0} + \boldsymbol{x}_i^{1,1})$ to $\mathsf{Enc}'(\mathsf{msk}, i, \boldsymbol{x}_i^{j,1} - \boldsymbol{x}_i^{1,1} + \boldsymbol{x}_i^{1,1})$.

G_2 can be proved indistinguishable from G_1 via an hybrid argument over the n slots, relying on the security of the underlying single-input scheme.

By looking at the games defined in Fig. 6, one can see that

$$|\mathbf{Adv}_1(\mathcal{A}) - \mathbf{Adv}_2(\mathcal{A})| = \sum_{\ell=1}^{n} |\mathbf{Adv}_{1,\ell-1}(\mathcal{A}) - \mathbf{Adv}_{1,\ell}(\mathcal{A})|$$

since G_1 corresponds to game $G_{1.0}$ and whereas G_2 is identical to game $G_{1.n}$.

Therefore, for every ℓ we bound the difference between each consecutive pair of games in the following lemma:

Lemma 3. *For every $\ell \in [n]$, there exists a PPT adversary $\mathcal{B}_{1.\ell}$ against the many-AD-IND security of the single-input scheme $\mathcal{FE}$ such that*

$$|\mathbf{Adv}_{1,\ell-1}(\mathcal{A}) - \mathbf{Adv}_{1,\ell}(\mathcal{A})| \leq \mathsf{Adv}_{\mathcal{FE}, \mathcal{B}_{1.\ell}}^{many\text{-}AD\text{-}IND}(\lambda).$$

Proof. Here, we replace encryptions of $\boldsymbol{x}_i^{j,0} - \boldsymbol{x}_i^{1,0} + \boldsymbol{x}_i^{1,1}$ with encryptions of $\boldsymbol{x}_i^{j,1} - \boldsymbol{x}_i^{1,1} + \boldsymbol{x}_i^{1,1}$ in all slots. Let us recall that j is the index of the encryption query while i is the index for the slot. The argument relies on (1) the many-AD-IND security of the underlying single input scheme $\mathcal{FE} := (\mathsf{Setup}, \mathsf{KeyGen}, \mathsf{Enc}, \mathsf{Dec})$, (2) the fact that Enc satisfies Property 2 (linear encryption), and (3) the restrictions imposed by the security game (see [4]). As for this latter point we notice that, indeed, the security experiment restriction in the case of the inner product functionality imposes that $\langle \boldsymbol{x}_i^{j,0} - \boldsymbol{x}_i^{1,0}, \boldsymbol{y}_i \rangle = \langle \boldsymbol{x}_i^{j,1} - \boldsymbol{x}_i^{1,1}, \boldsymbol{y}_i \rangle$, for all slots $i \in [n]$. In our scheme this becomes $\langle \boldsymbol{x}_i^{j,0} - \boldsymbol{x}_i^{1,0}, \boldsymbol{y}_i \rangle \bmod L = \langle \boldsymbol{x}_i^{j,1} - \boldsymbol{x}_i^{1,1}, \boldsymbol{y}_i \rangle \bmod L$, which in turn is equivalent to

$$\langle \boldsymbol{x}_i^{j,0} - \boldsymbol{x}_i^{1,0} + \boldsymbol{x}_i^{1,1} + \boldsymbol{u}_i, \boldsymbol{y}_i \rangle \bmod L = \langle \boldsymbol{x}_i^{j,1} - \boldsymbol{x}_i^{1,1} + \boldsymbol{x}_i^{1,1} + \boldsymbol{u}_i, \boldsymbol{y}_i \rangle \bmod L.$$

More formally, we build an adversary $\mathcal{B}_{1.\ell}$ that simulates $\mathsf{G}_{1.\ell-1+\beta}$ to $\mathcal{A}$ when interacting with the experiment $\mathbf{many\text{-}AD\text{-}IND}_\beta^{\mathcal{FE}}$.

$\mathcal{B}_{1.\ell}$ starts by receiving a public key for the scheme $\mathcal{FE}$, which is set to be the key mpk_ℓ for the ℓ-th instance of $\mathcal{FE}$. Next, it runs $\boldsymbol{u} \leftarrow \mathsf{Setup}^{\mathsf{ot}}$, and for all $i \neq \ell$, it runs $\mathsf{Setup}^\star$ to get $(\mathsf{mpk}_i, \mathsf{msk}_i)$. It gives $(\mathsf{mpk}_1, \ldots, \mathsf{mpk}_n)$ to $\mathcal{A}$.

$\mathcal{B}_{1.\ell}$ answers secret key queries $\boldsymbol{y} = \boldsymbol{y}_1 || \ldots || \boldsymbol{y}_n$ by first running $\mathsf{sk}_i \leftarrow \mathsf{KeyGen}(\mathsf{msk}_i, \boldsymbol{y}_i)$ for $i \neq \ell$. Also it invokes its own key generation oracle on $\boldsymbol{y}_\ell$, to get sk_ℓ. Finally, it computes $z \leftarrow \mathsf{KeyGen}^{\mathsf{ot}}(\boldsymbol{u}, \boldsymbol{y}_1 || \ldots || \boldsymbol{y}_n)$ (recall that $\mathcal{B}_{1.\ell}$ knows $\boldsymbol{u}$). This key material is then sent to $\mathcal{A}$.

$\mathcal{B}_{1.\ell}$ answers encryption queries $(i, \boldsymbol{x}_i^{j,0}, \boldsymbol{x}_i^{j,1})$ to Enc' as follows.

If $i < \ell$, it computes $\mathsf{Enc}(\mathsf{msk}_i, \mathsf{Enc}^{\mathsf{ot}}(\boldsymbol{u}, i, \boldsymbol{x}_i^{j,1}))$.

If $i > \ell$, it computes $\mathsf{Enc}(\mathsf{msk}_i, \mathsf{Enc}^{\mathsf{ot}}(\boldsymbol{u}, i, \boldsymbol{x}_i^{j,0} - \boldsymbol{x}_i^{1,0} + \boldsymbol{x}_i^{1,1}))$.

If $i = \ell$, at the j-th encryption query on slot ℓ, $\mathcal{B}_{1.\ell}$ queries its own oracle on input $(\boldsymbol{x}_\ell^{j,0} - \boldsymbol{x}_\ell^{1,0} + \boldsymbol{x}_\ell^{1,1}, \boldsymbol{x}_\ell^{j,1} - \boldsymbol{x}_\ell^{1,1} + \boldsymbol{x}_\ell^{1,1})$ (note that these vectors have norm less than $3X$, and as such, are valid input to the encryption oracle), to get back $\mathsf{ct}_*^j := \mathsf{Enc}(\mathsf{msk}_\ell, \boldsymbol{x}_\ell^{j,\beta} - \boldsymbol{x}_\ell^{1,\beta} + \boldsymbol{x}_\ell^{1,1})$ from the experiment $\mathbf{many\text{-}AD\text{-}IND}_\beta^{\mathcal{FE}}$. Then, $\mathcal{B}_{1.\ell}$ computes $\mathsf{ct}_\ell^j := \mathsf{Add}(\mathsf{ct}_*^j, \boldsymbol{u}_\ell)$, and sends it to $\mathcal{A}$.

Note that by Property 2 ct_ℓ^j is identically distributed to $\mathsf{Enc}(\mathsf{msk}_\ell, \boldsymbol{x}_\ell^{j,\beta} - \boldsymbol{x}_\ell^{1,\beta} + \boldsymbol{x}_\ell^{1,1} + \boldsymbol{u}_\ell \bmod L)$, the latter being equal to $\mathsf{Enc}(\mathsf{msk}_\ell, \mathsf{Enc}^{\mathsf{ot}}(\boldsymbol{x}_\ell^{j,\beta} - \boldsymbol{x}_\ell^{1,\beta} + \boldsymbol{x}_\ell^{1,1}))$. Also, we remark that because $\mathcal{B}_{1.\ell}$ plays in the many-AD-IND security game, it can make several queries to its encryption oracle, which means that every ct_*^j obtained from the oracle is encrypted under fresh randomness r_j, i.e., $\mathsf{ct}_*^j := \mathsf{Enc}(\mathsf{msk}_\ell, \boldsymbol{x}_\ell^{j,\beta} - \boldsymbol{x}_\ell^{1,\beta} + \boldsymbol{x}_\ell^{1,1}; r_j)$. Therefore, the simulated ciphertext ct_ℓ^j uses randomness r_j which is independent of the randomness $r_{j'}$ used in $\mathsf{ct}_\ell^{j'}$, for all $j \neq j'$. This means ct_ℓ^j is distributed as in game $\mathsf{G}_{1.\ell-1+\beta}$.

Finally, $\mathcal{B}_{1.\ell}$ outputs the same bit β' returned by $\mathcal{A}$. Thus:

$$|\mathbf{Adv}_{1.\ell-1}(\mathcal{A}) - \mathbf{Adv}_{1.\ell}(\mathcal{A})| \leq \mathsf{Adv}_{\mathcal{FE}, \mathcal{B}_{1.\ell}}^{many\text{-}AD\text{-}IND}(\lambda).$$

$\square$

The proof of Theorem 3 follows by combining the bounds obtained in the previous lemmas. $\square$

4   Function-Hiding Multi-Input FE for Inner Product

In this section, we give a function-hiding MIFE. We transform the MIFE for inner product proposed by Abdalla et al. in [4] into a function-hiding scheme, using a double layered encryption approach, similar to the one of Lin [13]. Namely, in Sect. 4.1, we give a generic construction that use any single-input FE on top of the MIFE from [4], which can prove selectively secure. Unlike the results in Sect. 3 that can be instantiated without pairings, for function-hiding we rely on pairing groups. Finally, in Sect. 4.2, we prove adaptive security, considering a specific instantiation of our construction.

Our construction. We present our function-hiding scheme $\mathcal{MIFE}$ in Fig. 8. The construction relies on the multi-input scheme $\mathcal{MIFE}'$ of Abdalla et al. [4] (recalled in Fig. 7), used together with any one-SEL-SIM secure, single-input FE for the functionality

$$\mathcal{F}^{\ell}_{\mathbb{G}_1, \mathbb{G}_2, \mathbb{G}_T} = \{f_{[\boldsymbol{y}]_1} : \mathbb{G}_2^{\ell} \to \mathbb{G}_T \text{ for } [\boldsymbol{y}]_1 \in \mathbb{G}_1^{\ell}\},$$

where

$$f_{[\boldsymbol{y}]_1}([\boldsymbol{x}]_2) := [\langle \boldsymbol{x}, \boldsymbol{y} \rangle]_T,$$

$\mathcal{PG} := (\mathbb{G}_1, \mathbb{G}_2, p, g_1, g_2)$ is a pairing group, and ℓ is the size of the ciphertext and secret keys in $\mathcal{MIFE}'$.

Concretely, we use the single-input FE from [5], generalized to the MDDH assumption, whose one-SEL-SIM security is proven in [4,17], and whose description is recalled in the full version of the paper. Note that this single-input FE happens to be public-key, but this is not a property that we need for our overall MIFE.

Outline of the construction. Our starting point is the MIFE scheme for inner-products from [4], denoted by $\mathcal{MIFE}' := (\mathsf{Setup}', \mathsf{Enc}', \mathsf{KeyGen}', \mathsf{Dec}')$ and recalled in Fig. 7. This scheme is clearly not function-hiding, as the vector $\boldsymbol{y}$ is given in the clear as part of functional secret key, in order to make decryption possible. In order to avoid the leakage of $\boldsymbol{y}$, we employ an approach similar to the one proposed in [13], which intuitively consists into adding a layer of encryption on top of the MIFE keys and ciphertexts; this is done by using a *single-input* inner product encryption scheme $\mathcal{FE}$. Slightly more in detail, using the $\mathcal{FE}$ and $\mathcal{MIFE}'$ schemes, we design our new function-hiding multi-input scheme $\mathcal{MIFE}$ as follows.

We generate master keys $(\mathsf{mpk}_i, \mathsf{msk}_i) \leftarrow \mathcal{FE}.\mathsf{Setup}(1^{\lambda}, \mathcal{F}^{\ell}_{\mathbb{G}_1, \mathbb{G}_2, \mathbb{G}_T})$ for computing inner products on vectors of dimension ℓ, where ℓ is the size of the ciphertexts and secret keys of $\mathcal{MIFE}'$. To encrypt $\boldsymbol{x}_i \in \mathbb{Z}_p^m$ for each slot $i \in [n]$, we first compute $[\mathsf{ct}_i^{\mathsf{in}}]_1$ using $\mathcal{MIFE}'$, and then we compute $\mathsf{ct}_i^{\mathsf{out}} := \mathcal{FE}.\mathsf{KeyGen}(\mathsf{msk}_i, [\mathsf{ct}_i^{\mathsf{in}}]_1)$. To generate a key for $\boldsymbol{y} := (\boldsymbol{y}_1 \| \cdots \| \boldsymbol{y}_n) \in \mathbb{Z}_p^{nm}$, we first compute the keys $\mathsf{sk}^{\mathsf{in}}$ from $\mathcal{MIFE}'$, and then we would like to encrypt these keys using $\mathcal{FE}$ in order to hide information about $\boldsymbol{y}$. A generic way to do it would be to set our secret key to be $\mathsf{Enc}(\mathsf{msk}_i, \mathsf{sk}^{\mathsf{in}})$, for all possible $i \in [n]$, so that we can compute the inner product of $[\mathsf{ct}_i^{\mathsf{in}}]_1$ with $\mathsf{sk}^{\mathsf{in}}$ for all $i \in [n]$. But that would yield keys of size $O(n^2m)$, since the key $\mathsf{sk}^{\mathsf{in}}$ itself is of size $O(nm)$. We can do better, however. If we consider the specific $\mathcal{MIFE}'$ scheme from [4], a secret key $\mathsf{sk}^{\mathsf{in}}$ for $\boldsymbol{y}$ consists of the components $([\mathsf{sk}_1^{\mathsf{in}}\|\ldots\|\mathsf{sk}_n^{\mathsf{in}}]_2, [z]_T)$, where each $[\mathsf{sk}_i^{\mathsf{in}}]_2$ only depends on $\boldsymbol{y}_i$ and is of size $O(m)$, while $[z]_T \in \mathbb{G}_T$ does not depend on $\boldsymbol{y}$ at all. Hence, we encrypt each vectors $[\mathsf{sk}_i^{\mathsf{in}}]_2$ to obtain $\mathsf{sk}_i^{\mathsf{out}} := \mathcal{FE}.\mathsf{Enc}(\mathsf{mpk}_i, [\mathsf{sk}_i^{\mathsf{in}}]_2)$, which gives us a secret key $\mathsf{sk}^{\mathsf{out}} := \left(\{\mathsf{sk}_i^{\mathsf{out}}\}_{i\in[n]}, [z]_T\right)$ of total size $O(nm)$.

This way, decrypting the outer layer as $\mathcal{FE}.\mathsf{Dec}(\mathsf{sk}_i^{\mathsf{out}}, \mathsf{ct}_i^{\mathsf{out}})$ yields $[\langle\mathsf{sk}_i^{\mathsf{in}}, \mathsf{ct}_i^{\mathsf{in}}\rangle]_T$, which is what needs to be computed in the $\mathcal{MIFE}'$ decryption algorithm Dec'. More precisely, correctness of $\mathcal{MIFE}$ follows from the correctness of $\mathcal{MIFE}'$, and the structural requirement of $\mathcal{FE}.\mathsf{Dec}$ that is used in the $\mathcal{MIFE}'$ decryption algorithm, namely:

$$\mathcal{MIFE}.\mathsf{Dec}(\{\mathsf{sk}_i^{\mathsf{out}}\}_{i\in[n]}, [z]_T, \{\mathsf{ct}_i^{\mathsf{out}}\}_{i\in[n]})$$

$$= \prod_{i=1}^{n} \mathcal{FE}.\mathsf{Dec}(\mathsf{ct}_i^{\mathsf{out}}, \mathsf{sk}_i^{\mathsf{out}})/[z]_T = \prod_{i=1}^{n} [\langle\mathsf{sk}_i^{\mathsf{in}}, \mathsf{ct}_i^{\mathsf{in}}\rangle]_T/[z]_T$$

$$= \mathcal{MIFE}'.\mathsf{Dec}(\{[\mathsf{sk}_i^{\mathsf{in}}]_2\}_{i\in[n]}, [z]_T, \{[\mathsf{ct}_i^{\mathsf{in}}]_1\}_{i\in[n]}).$$

Definition 11 (one-SEL-SIM-secure FE). *A single-input functional encryption $\mathcal{FE}$ for the functionality $\mathcal{F}_{\mathbb{G}_1,\mathbb{G}_2,\mathbb{G}_T}^{\ell}$ is one-SEL-SIM-secure if there exist PPT simulator algorithms $(\widetilde{\mathsf{Setup}}, \widetilde{\mathsf{Enc}}, \widetilde{\mathsf{KeyGen}})$ such that for every PPT (stateful) adversary $\mathcal{A}$ and every $\lambda \in \mathbb{N}$, the following two distributions are computationally indistinguishable:*

Experiment $\mathbf{REAL}_{\mathsf{SEL}}^{\mathcal{MIFE}}(1^\lambda, \mathcal{A})$:	*Experiment* $\mathbf{IDEAL}_{\mathsf{SEL}}^{\mathcal{MIFE}}(1^\lambda, \mathcal{A})$:
$\boldsymbol{x} \leftarrow \mathcal{A}(1^\lambda, \mathcal{F}_{\mathbb{G}_1,\mathbb{G}_2,\mathbb{G}_T}^{\ell})$	$\boldsymbol{x} \leftarrow \mathcal{A}(1^\lambda, \mathcal{F}_{\mathbb{G}_1,\mathbb{G}_2,\mathbb{G}_T}^{\ell})$
$(\mathsf{mpk}, \mathsf{msk}) \leftarrow \mathsf{Setup}(1^\lambda, \mathcal{F}_{\mathbb{G}_1,\mathbb{G}_2,\mathbb{G}_T}^{\ell})$	$(\widetilde{\mathsf{mpk}}, \widetilde{\mathsf{msk}}) \leftarrow \widetilde{\mathsf{Setup}}(1^\lambda, \mathcal{F}_{\mathbb{G}_1,\mathbb{G}_2,\mathbb{G}_T}^{\ell})$
$\mathsf{ct} \leftarrow \mathsf{Enc}(\mathsf{msk}, \boldsymbol{x})$	$\mathsf{ct} \leftarrow \widetilde{\mathsf{Enc}}(\widetilde{\mathsf{msk}})$
$\alpha \leftarrow \mathcal{A}^{\mathsf{KeyGen}(\mathsf{msk},\cdot)}(\mathsf{mpk}, \mathsf{ct})$	$\alpha \leftarrow \mathcal{A}^{\mathcal{O}(\cdot)}(\widetilde{\mathsf{mpk}}, \mathsf{ct})$
Output: α	***Output:*** α

The oracle $\mathcal{O}(\cdot)$ in the ideal experiment above is given access to another oracle that, given $[\boldsymbol{y}]_1 \in \mathcal{F}_{\mathbb{G}_1,\mathbb{G}_2,\mathbb{G}_T}^{\ell}$, returns $[\langle\boldsymbol{x}, \boldsymbol{y}\rangle]_1$, and then $\mathcal{O}(\cdot)$ returns $\widetilde{\mathsf{KeyGen}}\left(\widetilde{\mathsf{msk}}, [\boldsymbol{y}]_1, [\langle\boldsymbol{x}, \boldsymbol{y}\rangle]_1\right)$.

For every stateful adversary $\mathcal{A}$, we define its advantage as

$$\mathsf{Adv}_{\mathcal{FE},\mathcal{A}}^{one\text{-}SEL\text{-}SIM}(\lambda)$$
$$= \left|\Pr\left[\mathbf{REAL}_{\mathsf{SEL}}^{\mathcal{FE}}(1^\lambda, \mathcal{A}) = 1\right] - \Pr\left[\mathbf{IDEAL}_{\mathsf{SEL}}^{\mathcal{FE}}(1^\lambda, \mathcal{A}) = 1\right]\right|,$$

and we require that for every PPT $\mathcal{A}$, there exists a negligible function negl such that for all $\lambda \in \mathbb{N}$, $\mathsf{Adv}_{\mathcal{FE},\mathcal{A}}^{one\text{-}SEL\text{-}SIM}(\lambda) = \mathsf{negl}(\lambda)$.

Multi-input scheme $\mathcal{MIFE'}[3]$	
$\mathsf{Setup'}(1^\lambda, \mathcal{F}_n^{m,X,Y})$: $\forall i \in [n]$: $\quad (\mathsf{mpk}_i', \mathsf{msk}_i') \leftarrow \mathcal{FE'}.\mathsf{Setup}(1^\lambda, \mathcal{F}_1^{m+k,X,Y})$ $\forall i \in [n] : z_i \leftarrow_{\mathrm{R}} \mathbb{Z}_q^k$ $\mathsf{mpk'} := (\{\mathsf{mpk}_i'\}_{i \in [n]})$ $\mathsf{msk'} := (\{\mathsf{msk}_i', z_i\}_{i \in [n]})$ $\mathrm{return}\ (\mathsf{mpk'}, \mathsf{msk'})$ $\underline{\mathsf{Dec'}\left((\{[\mathsf{sk}_i^{\mathsf{in}}]_2\}_{i \in [n]}, [z]_T), \{[\mathsf{ct}_i^{\mathsf{in}}]_1\}_{i \in [n]} \right)}$: $\forall i \in [n] : [a_i]_T \leftarrow \mathcal{FE'}.\mathsf{Dec}([\mathsf{sk}_i^{\mathsf{in}}]_2, [\mathsf{ct}_i^{\mathsf{in}}]_1)$ return the discrete log of $\left(\prod_{i=1}^{n}[a_i]_T\right)/[z]_T$	$\underline{\mathsf{Enc'}(\mathsf{msk}, i, \boldsymbol{x}_i)}$: $[\mathsf{ct}_i^{\mathsf{in}}]_1 := \mathcal{FE'}.\mathsf{Enc}(\mathsf{mpk}_i', \boldsymbol{x}_i\|z_i)$ $\quad \mathrm{return}\ [\mathsf{ct}_i^{\mathsf{in}}]_1$ $\underline{\mathsf{KeyGen'}(\mathsf{msk}, \boldsymbol{y}_1\|\cdots\|\boldsymbol{y}_n)}$: $r \leftarrow_{\mathrm{R}} \mathbb{Z}_q^k$ $\forall i \in [n] :$ $\quad \mathsf{sk}_i^{\mathsf{in}} \leftarrow \mathcal{FE'}.\mathsf{KeyGen}(\mathsf{msk}_i', \boldsymbol{y}_i\|r)$ $z := \langle z_1 + \cdots + z_n, r \rangle$ $\mathsf{sk}^{\mathsf{in}} := (\{[\mathsf{sk}_i^{\mathsf{in}}]_2\}_{i \in [n]}, [z]_T)$ $\mathrm{return}\ \mathsf{sk}^{\mathsf{in}}$

Fig. 7. Multi-input, FE for $\mathcal{F}_n^{m,X,Y}$ from [4], whose many-SEL-IND relies on the $\mathcal{D}_k$-MDDH assumption. Here $\mathcal{FE'} := (\mathcal{FE'}.\mathsf{Setup}, \mathcal{FE'}.\mathsf{Enc}, \mathcal{FE'}.\mathsf{KeyGen}, \mathcal{FE'}.\mathsf{Dec})$ is a one-SEL-SIM secure, public-key, single-input FE for $\mathcal{F}_1^{m+k,X,Y}$, where k is the parameter used by the $\mathcal{D}_k$-MDDH assumption (concretely, $k = 1$ for SXDH, $k = 2$ for DLIN).

4.1 Proof of Selective Security

In the following theorem we state the selective security of our scheme $\mathcal{MIFE}$. Precisely, the theorem proves that our scheme is weakly function-hiding. We stress that this does not entail any limitation in the final result, as full-fledged function-hiding can be achieved in a generic way via a simple transformation, proposed in [14] (for single-input FE). The main idea is to work with slightly larger vectors where both input vectors $\boldsymbol{x}$ and secret-key vectors $\boldsymbol{y}$ are padded with zeros. In the full version of the paper we show how to do this transformation in the multi-input setting.

Theorem 4 (many-SEL-wFH-IND security). *Let* $\mathcal{MIFE'}$ *be the many-SEL-IND secure multi-input FE from Fig. 7. Suppose the single-input* $\mathcal{FE} := (\mathcal{FE}.\mathsf{Setup}, \mathcal{FE}.\mathsf{Enc}, \mathcal{FE}.\mathsf{KeyGen}, \mathcal{FE}.\mathsf{Dec})$ *is one-SEL-SIM-secure. Then the multi-input scheme* $\mathcal{MIFE} := (\mathsf{Setup}, \mathsf{Enc}, \mathsf{KeyGen}, \mathsf{Dec})$ *in Fig. 8 is many-SEL-wFH-IND-secure.*

Proof Overview. The proof is done via a hybrid argument that consists of two main phases: we first switch the ciphertexts from encryptions of $\boldsymbol{x}_i^{j_i,0}$ to encryptions of $\boldsymbol{x}_i^{j_i,1}$ for all slots $i \in [n]$, and ciphertext queries $j_i \in [Q_i]$, where Q_i denotes the number of ciphertext query on the i'th slot. This change is justified by the many-SEL-IND security of the underlying $\mathcal{MIFE'}$ in a black box manner. In addition, this change relies on the weak-function-hiding property that imposes the constraints $\sum_{i=1}^{n}\langle \boldsymbol{x}_i^{j_i,0}, \boldsymbol{y}_i^{j_f,0} \rangle = \sum_{i=1}^{n}\langle \boldsymbol{x}_i^{j_i,1}, \boldsymbol{y}_i^{j_f,0} \rangle$, for all secret key queries $j_f \in [Q_f]$, where Q_f denotes the number of secret key queries, which thus disallow the adversary from trivially distinguishing the two games.

The second main change in the proof is to switch the decryption keys from keys corresponding to $\boldsymbol{y}_1^{j,0}\|\cdots\|\boldsymbol{y}_n^{j,0}$ to keys corresponding to $\boldsymbol{y}_1^{j,1}\|\cdots\|\boldsymbol{y}_n^{j,1}$ for

---

New function-hiding scheme $\mathcal{MIFE}$

$\mathsf{Setup}(1^\lambda, \mathcal{F}_n^{m,X,Y})$:
$(\mathsf{mpk}', \mathsf{msk}') \leftarrow \mathsf{Setup}'(1^\lambda, \mathcal{F}_n^{m,X,Y})$
$\forall i \in [n] : (\mathsf{mpk}_i, \mathsf{msk}_i) \leftarrow \mathcal{FE}.\mathsf{Setup}(1^\lambda, \mathcal{F}_1^{\ell,X,Y})$
$\mathsf{mpk} := \big(\{\mathsf{mpk}_i\}_{i\in[n]}, \mathsf{mpk}'\big),\ \mathsf{msk} := \big(\{\mathsf{msk}_i\}_{i\in[n]}, \mathsf{msk}'\big)$
return $(\mathsf{mpk}, \mathsf{msk})$

$\mathsf{Enc}(\mathsf{msk}, i, \boldsymbol{x}_i)$:
$[\mathsf{ct}_i^{\mathsf{in}}]_1 := \mathsf{Enc}'(\mathsf{msk}', i, \boldsymbol{x}_i)$
$\mathsf{ct}_i^{\mathsf{out}} := \mathcal{FE}.\mathsf{KeyGen}(\mathsf{msk}_i, [\mathsf{ct}_i^{\mathsf{in}}]_1)$
return $\mathsf{ct}_i^{\mathsf{out}}$

$\mathsf{KeyGen}(\mathsf{msk}, \boldsymbol{y}_1 \| \cdots \| \boldsymbol{y}_n)$:
$\big(\{[\mathsf{sk}_i^{\mathsf{in}}]_2\}_{i\in[n]}, [z]_T\big) \leftarrow \mathsf{KeyGen}'(\mathsf{msk}', \boldsymbol{y}_1 \| \cdots \| \boldsymbol{y}_n)$
$\forall i \in [n] : \mathsf{sk}_i^{\mathsf{out}} \leftarrow \mathcal{FE}.\mathsf{Enc}(\mathsf{msk}_i, [\mathsf{sk}_i^{\mathsf{in}}]_2)$
$\mathsf{sk}_{\boldsymbol{y}_1 \| \cdots \| \boldsymbol{y}_n} := \big(\{\mathsf{sk}_i^{\mathsf{out}}\}_{i\in[n]}, [z]_T\big)$
return $\mathsf{sk}_{\boldsymbol{y}_1 \| \cdots \| \boldsymbol{y}_n}$

$\mathsf{Dec}\big(\big(\{\mathsf{sk}_i^{\mathsf{out}}\}_{i\in[n]}, [z]_T\big), \{\mathsf{ct}_i^{\mathsf{out}}\}_{i\in[n]}\big)$:
$\forall i \in [n] : [a_i]_T \leftarrow \mathcal{FE}.\mathsf{Dec}(\mathsf{ct}_i^{\mathsf{out}}, \mathsf{sk}_i^{\mathsf{out}})$
return the discrete log of $\big(\prod_{i=1}^n [a_i]_T\big) / [z]_T$

Fig. 8. Many-SEL-wFH-IND secure, private-key, multi-input, FE for the class $\mathcal{F}_n^{m,X,Y}$. Here $\mathcal{FE} := (\mathcal{FE}.\mathsf{Setup}, \mathcal{FE}.\mathsf{Enc}, \mathcal{FE}.\mathsf{KeyGen}, \mathcal{FE}.\mathsf{Dec})$ is a one-SEL-SIM secure, single-input FE for $\mathcal{F}_1^{\ell,X,Y}$, where by ℓ we denote the output size of Enc' and KeyGen', and $\mathcal{MIFE}' := (\mathsf{Setup}', \mathsf{Enc}', \mathsf{KeyGen}', \mathsf{Dec}')$ is the many-AD-IND secure, multi-input FE from Fig. 7.

every $j \in [Q_f]$. This in turn requires a hybrid argument over all decryption keys, changing one key at a time. To switch the ρ'th key, we use the selective simulation security of the underlying $\mathcal{FE}$ to embed the value $\langle \boldsymbol{x}_i^{j,1}, \boldsymbol{y}_i^{\rho,\beta} \rangle + \langle \boldsymbol{r}^\rho, \boldsymbol{z}_i \rangle$ in the ciphertexts ct_i^j, for all slots $i \in [n]$ and all $j \in [Q_i]$. Next, we use the $\mathcal{D}_k$-MDDH

Game	$[\mathsf{ct}_i^{\mathsf{in},k}]_1$	$[\mathsf{sk}_i^{\mathsf{in},j}]_2$	justification/remark
G_0	$\mathsf{Enc}'(\mathsf{msk}', i, \boldsymbol{x}_i^{k,0})$	$\mathsf{KeyGen}'(\mathsf{msk}', \boldsymbol{y}_1^{j,0} \| \cdots \| \boldsymbol{y}_n^{j,0})$	many-SEL-wFH-IND$_0$ security game
G_1	$\mathsf{Enc}'(\mathsf{msk}', i, \boldsymbol{x}_i^{k,1})$	$\mathsf{KeyGen}'(\mathsf{msk}', \boldsymbol{y}_1^{j,0} \| \cdots \| \boldsymbol{y}_n^{j,0})$	many-SEL-IND of $\mathcal{MIFE}'$
$G_{1.\rho}$	$\mathsf{Enc}'(\mathsf{msk}', i, \boldsymbol{x}_i^{k,1})$	$\mathsf{KeyGen}'(\mathsf{msk}', \boldsymbol{y}_1^{j,1} \| \cdots \| \boldsymbol{y}_n^{j,1})$, for $j < \rho$ $\mathsf{KeyGen}'(\mathsf{msk}', \boldsymbol{y}_1^{j,0} \| \cdots \| \boldsymbol{y}_n^{j,0})$, for $j \geq \rho$	Lemma 5

Fig. 9. An overview of the games used in the proof of Theorem 4. By $[\mathsf{ct}_i^{\mathsf{in},k}]_1$ and $[\mathsf{sk}_i^{\mathsf{in},j}]_2$ we denote the kth ciphertext and the jth decryption key of the inner scheme $\mathcal{MIFE}'$.

assumption to argue that $[\langle r^\rho, z_i \rangle]_T$ is indistinguishable from a uniform random value and thus perfectly hides $\langle x_i^{1,1}, y_i^{\rho,\beta} \rangle$ for the first ciphertext of each slot: ct_i^1. For all the other remaining $\langle x_i^{j,1}, y_i^{\rho,\beta} \rangle$, for $j \in [Q_i]$, $j > 1$, we use the fact that $\langle x_i^{j,1} - x_i^{1,1}, y_i^{\rho,0} \rangle = \langle x_i^{j,1} - x_i^{1,1}, y_i^{\rho,1} \rangle$, as implied by the game's restrictions.

Proof of Theorem 4. We proceed via a series of Games $\mathsf{G}_0, \mathsf{G}_1, \mathsf{G}_{1,\rho}$, for $\rho \in [Q_f + 1]$, described in Fig. 10. An overview is provided in Fig. 9. Let $\mathcal{A}$ be a PPT adversary, and $\lambda \in \mathbb{N}$ be the security parameter. We denote by $\mathsf{Adv}_{\mathsf{G}_i}(\mathcal{A})$ the advantage of $\mathcal{A}$ in game G_i.

G_0: is the experiment **many-SEL-wFH-IND**$_0^{\mathcal{MIFE}}$ (see Definition 6).
G_1: we replace the inner encryption of $x_i^{j,0}$ by encryptions of $x_i^{j,1}$, for all $i \in [n], j \in [Q_i]$, using the many-SEL-IND security of $\mathcal{MIFE}'$. This is possible due to the weak function-hiding constraint, which states in particular that $\sum_{i=1}^n \langle x_i^{j_i,0}, y_i^{j_f,0} \rangle = \sum_{i=1}^n \langle x_i^{j_i,1}, y_i^{j_f,0} \rangle$, for all indices $j_i \in [Q_i], j_f \in [Q_f]$.
$\mathsf{G}_{1,\rho}$: for the first $\rho - 1$ queries to KeyGen, we replace inner secret key $\mathsf{KeyGen}'(\mathsf{msk}', y_1^0 \| \cdots \| y_n^0)$, by $\mathsf{KeyGen}'(\mathsf{msk}', y_1^1 \| \cdots \| y_n^1)$. Note that G_1 is the same as $\mathsf{G}_{1.1}$, and $\mathsf{G}_{1.Q_f+1}$ is the same as **many-SEL-wFH-IND**$_1^{\mathcal{MIFE}}$.

We prove $\mathsf{G}_0 \approx_c \mathsf{G}_1$ in Lemma 4, and $\mathsf{G}_{1,\rho} \approx_c \mathsf{G}_{1,\rho+1}$ for all $\rho \in [Q_f]$ in Lemma 5. $\qquad\qquad\square$

$\mathsf{G}_0, \boxed{\mathsf{G}_1, \underline{\mathsf{G}_{1,\rho}}}$, for $\rho \in [Q_f + 1]$:

$\{x_i^{j,\beta}\}_{i\in[n], j\in[Q_i], \beta\in\{0,1\}}, \{y_i^{j,\beta}\}_{i\in[n], j\in[Q_f], \beta\in\{0,1\}} \leftarrow \mathcal{A}(1^\lambda, \mathcal{F}_n^{m,X,Y})$
$(\mathsf{mpk}', \mathsf{msk}') \leftarrow \mathsf{Setup}'(1^\lambda, \mathcal{F}_n^{m,X,Y})$
$\forall i \in [n] : (\mathsf{mpk}_i, \mathsf{msk}_i) \leftarrow \mathcal{FE}.\mathsf{Setup}(1^\lambda, \mathcal{F}_1^{\ell,X,Y})$
$\mathsf{mpk} := (\{\mathsf{mpk}_i\}_{i\in[n]}, \mathsf{mpk}'), \mathsf{msk} := (\{\mathsf{msk}_i\}_{i\in[n]}, \mathsf{msk}')$
$\forall i \in [n], j \in [Q_i]$:
$\quad [\mathsf{ct}_i^{\mathsf{in},j}]_1 := \mathsf{Enc}'(\mathsf{msk}', i, x_i^{j,0}), \quad \boxed{[\mathsf{ct}_i^{\mathsf{in},j}]_1 := \mathsf{Enc}'(\mathsf{msk}', i, x_i^{j,1})}$
$\quad \mathsf{ct}_i^{\mathsf{out},j} := \mathcal{FE}.\mathsf{KeyGen}(\mathsf{msk}_i, [\mathsf{ct}_i^{\mathsf{in},j}]_1)$
$\alpha \leftarrow \mathcal{A}^{\mathsf{KeyGen}(\mathsf{msk},\cdot)}(\mathsf{mpk}, \{\mathsf{ct}_i^{\mathsf{out},j}\}_{i\in[n], j\in[Q_i]})$
Output: α.

$\mathsf{KeyGen}\big(\mathsf{msk}, (y_1^{j,\beta} \| \cdots \| y_n^{j,\beta})_{\beta\in\{0,1\}}\big)$:

$\quad \big(\{[\mathsf{sk}_i^{\mathsf{in}}]_2\}_{i\in[n]}, [z]_T\big) \leftarrow \mathsf{KeyGen}'(\mathsf{msk}', y_1^0 \| \cdots \| y_n^0)$
$\quad \underline{\text{If } j < \rho: \big(\{[\mathsf{sk}_i^{\mathsf{in}}]_2\}_{i\in[n]}, [z]_T\big) \leftarrow \mathsf{KeyGen}'(\mathsf{msk}', y_1^1 \| \cdots \| y_n^1)}$
$\quad \mathsf{sk}_i^{\mathsf{out}} \leftarrow \mathcal{FE}.\mathsf{Enc}(\mathsf{msk}_i, [\mathsf{sk}_i^{\mathsf{in}}]_2)$
$\quad \mathsf{sk}_{y_1 \| \cdots \| y_n} := \big(\{\mathsf{sk}_i^{\mathsf{out}}\}_{i\in[n]}, [z]_T\big)$
$\quad \text{return } \mathsf{sk}_{y_1 \| \cdots \| y_n}$

Fig. 10. Games for the proof of Theorem 4. In each procedure, the components inside a solid (dotted) frame are only present in the games marked by a solid (dotted) frame.

Lemma 4 (G_0 to G_1). *There exists a PPT adversary $\mathcal{B}_1$ such that*

$$\mathsf{Adv}_{G_0}(\mathcal{A}) - \mathsf{Adv}_{G_1}(\mathcal{A}) \leq \mathsf{Adv}_{\mathcal{MIFE}',\mathcal{B}_1}^{many\text{-}SEL\text{-}IND}(\lambda).$$

Proof. In order to show that we can switch $x_i^{j,0}$ to $x_i^{j,1}$, we rely on the security of the underlying $\mathcal{MIFE}'$ scheme. Intuitively, adding an additional layer of encryption on the decryption keys $\mathsf{sk}_i^{\mathsf{in}}$ cannot invalidate the security of the underlying $\mathcal{MIFE}'$.

More formally, we design an adversary $\mathcal{B}_1$ against the many-SEL-IND security of $\mathcal{MIFE}'$. Adversary $\mathcal{B}_1$ draws public and secret keys for the outer encryption layer and then uses its own experiment to simulate either G_0 or G_1. We describe adversary $\mathcal{B}_1$ in the full version of the paper and give a textual description here.

Simulation of master public key mpk. Since the game is selective, the adversary $\mathcal{B}_1$ first gets the challenges $\{x_i^{j,b}\}_{i \in [n], j \in [Q_i], b \in \{0,1\}}$ from $\mathcal{A}$, and it sends them to its experiment many-SEL-IND$_\beta^{\mathcal{MIFE}'}$. Then, $\mathcal{B}_1$ receives the public key mpk' of the $\mathcal{MIFE}'$ scheme. To construct the full public key, it draws $(\mathsf{mpk}_i, \mathsf{msk}_i) \leftarrow \mathcal{FE}.\mathsf{Setup}(1^\kappa, \mathcal{F}_1^{\ell,X,Y})$, for all slots $i \in [n]$, independently. It then sets $\mathsf{mpk} := \{\mathsf{mpk}_i\}_{i \in [n]} \cup \{\mathsf{mpk}'\}$ and returns mpk to adversary $\mathcal{A}$.

Simulation of the challenge ciphertexts. The adversary $\mathcal{B}_1$ receives $[\mathsf{ct}_i^{\mathsf{in},j}]_1$ from the encryption oracle of the experiment many-SEL-IND$_\beta^{\mathcal{MIFE}'}$, for all $i \in [n]$. This corresponds to encryptions of either $x_i^{j,\beta}$, for $\beta = 0$ or 1. Since it knows msk_i, it computes $\mathsf{ct}_i^{\mathsf{out},j} := \mathcal{FE}.\mathsf{KeyGen}(\mathsf{msk}_i, [\mathsf{ct}_i^{\mathsf{in},j}]_1)$ for all $i \in [n]$ and returns $\{\mathsf{ct}_i^{\mathsf{out},j}\}_{i \in [n]}$ to $\mathcal{A}$.

Simulation of $\mathsf{KeyGen}(\mathsf{msk}, \cdot)$. On every secret key query $(\boldsymbol{y}_1^b \| \dots \| \boldsymbol{y}_n^b)_{b \in \{0,1\}}$, adversary $\mathcal{B}_1$ queries the KeyGen' oracle of the experiment many-SEL-IND$_\beta^{\mathcal{MIFE}'}$ on $\boldsymbol{y}_1^0 \| \dots \| \boldsymbol{y}_n^0$. It obtains $\{[\mathsf{sk}_i^{\mathsf{in}}]_2\}_{i \in [n]}, [z]_T$. Finally, it computes $\mathsf{sk}_i^{\mathsf{out}} := \mathcal{FE}.\mathsf{Enc}(\mathsf{mpk}_i, [\mathsf{sk}_i^{\mathsf{in}}]_2)$ and returns $(\{\mathsf{sk}_i^{\mathsf{out}}\}_{i \in [n]}, [z]_T)$ to $\mathcal{A}$. $\qquad\square$

Lemma 5 ($G_{1.\rho}$ to $G_{1.\rho+1}$). *For all $\rho \in [Q_f]$, there exist PPT adversaries $\mathcal{B}_\rho$ and $\mathcal{B}_\rho'$ such that*

$$\mathsf{Adv}_{G_{1.\rho}}(\mathcal{A}) - \mathsf{Adv}_{G_{1.\rho+1}}(\mathcal{A}) \leq 2n \cdot \mathsf{Adv}_{\mathcal{FE},\mathcal{B}_\rho}^{one\text{-}SEL\text{-}SIM}(\lambda,) + 2 \cdot \mathbf{Adv}_{G_1,\mathcal{B}_\rho'}^{\mathcal{D}_k\text{-}mddh}(\lambda) + \frac{2k}{p}.$$

For lack of space the proof of Lemma 5 appears in the full version of the paper [3].

4.2 Adaptively-Secure Multi-Input Function-Hiding FE for Inner Product

In this section, we prove that if we instantiate the construction described in Fig. 8 (Sect. 4), with the many-AD-IND-secure, single-input FE from [5], we obtain an adaptively secure function-hiding MIFE. Specifically, we consider the generalized version of single-input FE, as described in [4] (recalled in the full version of the paper [3]). For completeness, we present this new MIFE instantiation in Fig. 11. Proving adaptive security for our construction in a generic way would require the underlying $\mathcal{FE}$ to achieve strong security notions, such as one-AD-SIM (which is not achieved by any known scheme). We overcome this issue, managing to prove adaptive security of our concrete MIFE in Fig. 8, using non-generic techniques inspired by [4].

Theorem 5 (many-AD-IND-wFH security). *If the $\mathcal{D}_k$-MDDH assumption holds in $\mathbb{G}_1$ and $\mathbb{G}_2$, then the multi-input FE for $\mathcal{F}_n^{m,X,Y}$ described in Fig. 11 is many-AD-IND-wFH-secure.*

Proof overview. Similarly to the selective-security proof presented in Sect. 4.1, we prove weakly-function-hiding. This is sufficient, since it can be transformed generically into a fully function-hiding MIFE by using techniques from [14] (see in the full version of the paper [3] for more details).

To prove weak function-hiding we proceed in two stages. First, we switch from $\mathsf{Enc}(\mathsf{msk}, i, \boldsymbol{x}_i^{j,0})$ to $\mathsf{Enc}(\mathsf{msk}, i, \boldsymbol{x}_i^{j,1})$ for all slots $i \in [n]$ and all queries $j \in [Q_i]$ simultaneously, using the many-AD-IND security of $\mathcal{MIFE}'$ (the underlying

---

$\mathsf{Setup}(1^\lambda, \mathcal{F}_n^{m,X,Y})$:

$\quad \mathcal{PG} \leftarrow_{\mathrm{R}} \mathsf{GGen}(1^\lambda), \mathbf{A}_1, \mathbf{B}_1, \ldots, \mathbf{A}_n, \mathbf{B}_n \leftarrow_{\mathrm{R}} \mathcal{D}_k, \mathbf{U}_1, \ldots, \mathbf{U}_n \leftarrow_{\mathrm{R}} \mathbb{Z}_p^{(k+m) \times (k+1)}$

$\quad \mathbf{V}_1, \ldots, \mathbf{V}_n \leftarrow_{\mathrm{R}} \mathbb{Z}_p^{(2k+m+1) \times (k+1)}, \; z_1, \ldots, z_n \leftarrow_{\mathrm{R}} \mathbb{Z}_p^k$

$\quad \mathsf{mpk} := \mathcal{PG}, \quad \mathsf{msk} := \{\mathbf{A}_i, \mathbf{B}_i, \mathbf{U}_i, \mathbf{V}_i, z_i\}_{i \in [n]}$

$\quad$ return $(\mathsf{mpk}, \mathsf{msk})$

$\mathsf{Enc}(\mathsf{msk}, i, \boldsymbol{x}_i \in \mathbb{Z}_p^m)$:

$\quad \boldsymbol{s}_i \leftarrow_{\mathrm{R}} \mathbb{Z}_p^k, \; \boldsymbol{c}_i := \mathbf{A}_i \boldsymbol{s}_i, \; \boldsymbol{c}_i' := \begin{pmatrix} \boldsymbol{x}_i \\ \boldsymbol{z}_i \end{pmatrix} + \mathbf{U}_i \mathbf{A}_i \boldsymbol{s}_i, \; \boldsymbol{c}_i'' := \begin{pmatrix} \boldsymbol{c}_i \\ \boldsymbol{c}_i' \end{pmatrix}^\top \mathbf{V}_i$

$\quad$ return $([\boldsymbol{c}_i]_1, [\boldsymbol{c}_i']_1, [\boldsymbol{c}_i'']_1)$

$\mathsf{KeyGen}(\mathsf{msk}, \boldsymbol{y}_1 \| \cdots \| \boldsymbol{y}_n \in (\mathbb{Z}_p^m)^n)$:

$\quad \boldsymbol{r} \leftarrow_{\mathrm{R}} \mathbb{Z}_p^k, \; z := \langle z_1 + \cdots + z_n, \boldsymbol{r} \rangle$

$\quad \forall i \in [n] : \boldsymbol{t}_i \leftarrow_{\mathrm{R}} \mathbb{Z}_p^k, \; \boldsymbol{d}_i := \mathbf{B}_i \boldsymbol{t}_i, \; \boldsymbol{d}_i' := \begin{pmatrix} -\mathbf{U}_i^\top \begin{pmatrix} \boldsymbol{y}_i \\ \boldsymbol{r} \end{pmatrix} \\ \begin{pmatrix} \boldsymbol{y}_i \\ \boldsymbol{r} \end{pmatrix} \end{pmatrix} + \mathbf{V}_i \mathbf{B}_i \boldsymbol{t}_i$

$\quad$ return $(\{[\boldsymbol{d}_i]_2, [\boldsymbol{d}_i']_2\}_{i \in [n]}, [z]_T)$

$\mathsf{Dec}\Big((\{[\boldsymbol{d}_i]_2, [\boldsymbol{d}_i']_2\}_{i \in [n]}, [z]_T), \{[\boldsymbol{c}_i]_1, [\boldsymbol{c}_i']_1, [\boldsymbol{c}_i'']_1\}_{i \in [n]} \Big)$:

$\quad out \leftarrow \left(\prod_i \left(e\left(\begin{bmatrix} \boldsymbol{c}_i \\ \boldsymbol{c}_i' \end{bmatrix}_1^\top, [\boldsymbol{d}_i']_2 \right) / e\left([\boldsymbol{c}''_i]_1, [\boldsymbol{d}_i]_2 \right) \right) \right) / [z]_T$

$\quad$ return discrete log of out

Fig. 11. Many-AD-IND-wFH secure, multi-input FE scheme for the class $\mathcal{F}_n^{m,X,Y}$ (self-contained description).

MIFE from [4]). For completeness, we also give a concrete description of $\mathcal{MIFE}'$ in the full version of the paper.

Secondly, we use a hybrid argument over all Q_f queried keys, switching them one by one from $\mathsf{KeyGen}(\mathsf{msk}, \boldsymbol{y}_1^0\| \cdots \|\boldsymbol{y}_n^0)$ to $\mathsf{KeyGen}(\mathsf{msk}, \boldsymbol{y}_1^1\| \cdots \|\boldsymbol{y}_n^1)$. To switch the ρ'th key, we use the security of $\mathcal{FE}$ in a non-generic way. Structurally, we do a proof similar to the selective one of the previous section. In order to apply complexity leveraging, we first do all the computational steps. Afterwards, only at some particular transition in the proof (transition from $\mathsf{H}''_{\rho.0}$ to $\mathsf{H}''_{\rho.1}$ the full version of the paper), we use complexity leveraging, and we simulate the selective proof arguments. This multiplies the security loss by an exponential factor. We can do so here because this particular transition is perfect: the exponential term is multiplied by a zero advantage.

Although this proof strategy shares similarities with the adaptive security proof the MIFE in [4], our proof has some crucial differences: mainly, the role of the keys and ciphertexts in our proof is switched. Since the multi-input model is asymmetric with respect to the ciphertexts and decryption keys (only ciphertexts can be mixed-and-matched), this results in a different proof strategy.

For lack of space, the full proof appears in the full version of the paper.

Acknowledgments. Michel Abdalla was supported in part by SAFEcrypto (H2020 ICT-644729) and by the European Union's Horizon 2020 Research and Innovation Programme under grant agreement 780108 (FENTEC). Dario Fiore was partially supported by the Spanish Ministry of Economy under project references TIN2015-70713-R (DEDETIS), RTC-2016-4930-7 (DataMantium), and by the Madrid Regional Government under project N-Greens (ref. S2013/ICE-2731). Romain Gay was partially supported by a Google PhD Fellowship in Privacy and Security and by the ERC Project aSCEND (H2020 639554). Bogdan Ursu was partially supported by ANR-14-CE28-0003 (Project EnBiD) and by the ERC Project PREP-CRYPTO (H2020 724307).

References

1. Abdalla, M., Bourse, F., De Caro, A., Pointcheval, D.: Simple functional encryption schemes for inner products. In: Katz, J. (ed.) PKC 2015. LNCS, vol. 9020, pp. 733–751. Springer, Heidelberg (2015). https://doi.org/10.1007/978-3-662-46447-2_33
2. Abdalla, M., Bourse, F., De Caro, A., Pointcheval, D.: Better security for functional encryption for inner product evaluations. Cryptology ePrint Archive, Report 2016/011 (2016). http://eprint.iacr.org/2016/011
3. Abdalla, M., Catalano, D., Fiore, D., Gay, R., Ursu, B.: Multi-Input Functional Encryption for Inner Products: Function-Hiding Realizations and Constructions without Pairings. Cryptology ePrint Archive, Report 2017/972 (2017). http://eprint.iacr.org/2017/972
4. Abdalla, M., Gay, R., Raykova, M., Wee, H.: Multi-input inner-product functional encryption from pairings. In: Coron, J.-S., Nielsen, J.B. (eds.) EUROCRYPT 2017, Part I. LNCS, vol. 10210, pp. 601–626. Springer, Cham (2017). https://doi.org/10.1007/978-3-319-56620-7_21

5. Agrawal, S., Libert, B., Stehlé, D.: Fully secure functional encryption for inner products, from standard assumptions. In: Robshaw, M., Katz, J. (eds.) CRYPTO 2016, Part III. LNCS, vol. 9816, pp. 333–362. Springer, Heidelberg (2016). https://doi.org/10.1007/978-3-662-53015-3_12

6. Ananth, P., Jain, A.: Indistinguishability obfuscation from compact functional encryption. In: Gennaro, R., Robshaw, M. (eds.) CRYPTO 2015, Part I. LNCS, vol. 9215, pp. 308–326. Springer, Heidelberg (2015). https://doi.org/10.1007/978-3-662-47989-6_15

7. Badrinarayanan, S., Gupta, D., Jain, A., Sahai, A.: Multi-input functional encryption for unbounded arity functions. In: Iwata, T., Cheon, J.H. (eds.) ASIACRYPT 2015, Part I. LNCS, vol. 9452, pp. 27–51. Springer, Heidelberg (2015). https://doi.org/10.1007/978-3-662-48797-6_2

8. Boneh, D., Sahai, A., Waters, B.: Functional encryption: definitions and challenges. In: Ishai, Y. (ed.) TCC 2011. LNCS, vol. 6597, pp. 253–273. Springer, Heidelberg (2011). https://doi.org/10.1007/978-3-642-19571-6_16

9. Brakerski, Z., Komargodski, I., Segev, G.: Multi-input functional encryption in the private-key setting: stronger security from weaker assumptions. In: Fischlin, M., Coron, J.-S. (eds.) EUROCRYPT 2016, Part II. LNCS, vol. 9666, pp. 852–880. Springer, Heidelberg (2016). https://doi.org/10.1007/978-3-662-49896-5_30

10. Datta, P., Okamoto, T., Tomida, J.: Full-hiding (unbounded) multi-input inner product functional encryption from the k-linear assumption. In: Abdalla, M., Dahab, R. (eds.) PKC 2018, Part II. LNCS, vol. 10770, pp. 245–277. Springer, Cham (2018). https://doi.org/10.1007/978-3-319-76581-5_9

11. Escala, A., Herold, G., Kiltz, E., Ràfols, C., Villar, J.: An algebraic framework for Diffie-Hellman assumptions. In: Canetti, R., Garay, J.A. (eds.) CRYPTO 2013, Part II. LNCS, vol. 8043, pp. 129–147. Springer, Heidelberg (2013). https://doi.org/10.1007/978-3-642-40084-1_8

12. Goldwasser, S., et al.: Multi-input functional encryption. In: Nguyen, P.Q., Oswald, E. (eds.) EUROCRYPT 2014. LNCS, vol. 8441, pp. 578–602. Springer, Heidelberg (2014). https://doi.org/10.1007/978-3-642-55220-5_32

13. Lin, H.: Indistinguishability obfuscation from SXDH on 5-linear maps and locality-5 PRGs. In: Katz, J., Shacham, H. (eds.) CRYPTO 2017, Part I. LNCS, vol. 10401, pp. 599–629. Springer, Cham (2017). https://doi.org/10.1007/978-3-319-63688-7_20

14. Lin, H., Vaikuntanathan, V.: Indistinguishability obfuscation from DDH-like assumptions on constant-degree graded encodings. In: Dinur, I. (ed.) 57th FOCS, pp. 11–20. IEEE Computer Society Press, October 2016

15. O'Neill, A.: Definitional issues in functional encryption. Cryptology ePrint Archive, Report 2010/556 (2010). http://eprint.iacr.org/2010/556

16. Sahai, A., Waters, B.: Fuzzy identity-based encryption. In: Cramer, R. (ed.) EUROCRYPT 2005. LNCS, vol. 3494, pp. 457–473. Springer, Heidelberg (2005). https://doi.org/10.1007/11426639_27

17. Wee, H.: Attribute-hiding predicate encryption in bilinear groups, revisited. In: Kalai, Y., Reyzin, L. (eds.) TCC 2017, Part I. LNCS, vol. 10677, pp. 206–233. Springer, Cham (2017). https://doi.org/10.1007/978-3-319-70500-2_8

Symmetric Cryptography

Encrypt or Decrypt? To Make a Single-Key Beyond Birthday Secure Nonce-Based MAC

Nilanjan Datta[1], Avijit Dutta[2(✉)], Mridul Nandi[2], and Kan Yasuda[3]

[1] Indian Institute of Technology, Kharagpur, Kharagpur, India
nilanjan_isi_jrf@yahoo.com
[2] Indian Statistical Institute, Kolkata, India
avirocks.dutta13@gmail.com, mridul.nandi@gmail.com
[3] NTT Secure Platform Laboratories, NTT Corporation, Tokyo, Japan
yasuda.kan@lab.ntt.co.jp

Abstract. At CRYPTO 2016, Cogliati and Seurin have proposed a highly secure nonce-based MAC called Encrypted Wegman-Carter with Davies-Meyer (EWCDM) construction, as $\mathsf{E}_{K_2}\big(\mathsf{E}_{K_1}(N) \oplus N \oplus \mathsf{H}_{K_h}(M)\big)$ for a nonce N and a message M. This construction achieves roughly $2^{2n/3}$ bit MAC security with the assumption that E is a PRP secure n-bit block cipher and H is an almost xor universal n-bit hash function. In this paper we propose Decrypted Wegman-Carter with Davies-Meyer (DWCDM) construction, which is structurally very similar to its predecessor EWCDM except that the outer encryption call is replaced by decryption. The biggest advantage of DWCDM is that we can make a truly single key MAC: the two block cipher calls can use the same block cipher key $K = K_1 = K_2$. Moreover, we can derive the hash key as $K_h = \mathsf{E}_K(1)$, as long as $|K_h| = n$. Whether we use encryption or decryption in the outer layer makes a huge difference; using the decryption instead enables us to apply an extended version of the mirror theory by Patarin to the security analysis of the construction. DWCDM is secure beyond the birthday bound, roughly up to $2^{2n/3}$ MAC queries and 2^n verification queries against nonce-respecting adversaries. DWCDM remains secure up to $2^{n/2}$ MAC queries and 2^n verification queries against nonce-misusing adversaries.

Keywords: EDM · EWCDM · Mirror theory
Extended mirror theory · H-Coefficient

1 Introduction

Pseudo-Random Functions or in short PRF is an important tool for studying almost all symmetric-key cryptographic systems that use secret keys, including encryption, authentication and authenticated-encryption. But unfortunately, very few PRFs are actually available in practice, and it is not easy to construct

© International Association for Cryptologic Research 2018
H. Shacham and A. Boldyreva (Eds.): CRYPTO 2018, LNCS 10991, pp. 631–661, 2018.
https://doi.org/10.1007/978-3-319-96884-1_21

a sufficiently secure PRF. As a result, Pseudo-Random Permutations or in short PRPs or block ciphers, which are available in plenty [9,10,15,20], replace the PRF and are deployed as building blocks for almost every cryptographic systems.

Although various available block ciphers [9,10,15,20] can be assumed to be PRFs, but such an assumption comes at the cost of quadratic security degradation due to the PRF-PRP switch [5], which is often called the "birthday bound security degradation". This loss of security is sometimes acceptable in practice if the block size of the cipher is large enough (e.g. AES-128). But with lightweight block ciphers with relatively small block sizes (e.g. 64-bit), whose number has grown tremendously in recent years (e.g. [1,2,9,10,20]), this security loss severely limits their applicability, and as a result it seems to be challenging to use these small ciphers in modern-day lightweight cryptography (e.g. Smart Card, RFID etc.).

In order to save these ciphers from obsolescence, various PRP-to-PRF constructions have been proposed in recent years that guarantee higher security than the usual birthday bound security. Such constructions are often called BBB (Beyond Birthday Bound)—i.e., security against more than $2^{n/2}$ queries where n is the block size of the underlying cipher. A popular BBB construction is the XOR of permutations [3,6,23,28].

XOR OF PERMUTATIONS. Bellare et al. [6] suggested a way to construct a PRF from PRPs by taking the xor (more generally sum) of two independent PRPs,

$$\mathsf{XOR}_{\mathsf{E}_{K_1},\mathsf{E}_{K_2}}(x) = \mathsf{E}_{K_1}(x) \oplus \mathsf{E}_{K_2}(x).$$

This construction was later analyzed by Lucks [23] who proved its security up to $2^{2n/3}$ queries. Bellare and Impagliazzo have shown a BBB security $O(nq/2^n)$ of single-keyed variant of this construction [3]. However, their proof was sketchy and hard to verify. Subsequently, a lot of efforts have been invested towards improving the bound of XOR construction and its single-keyed variant (even proving up to n-bit security) by Patarin [28,31,32], but the proof contains serious gaps. Later Cogliati et al. generalized this result to the xor of three or more independent PRPs [12]. Recently, Dai et al. [16] have provided a verifiable n-bit security proof of the XOR construction using the *chi-squared* method. Although, the original proof contained a glitch, as pointed out by Bhattacharya and Nandi [8], it was later fixed in the full version of [16].

The XOR construction provides a solution for encryption by combining itself with the counter (CTR) mode of encryption, resulting in a BBB secure nonce-based encryption mode, called CENC, proposed by Iwata [21], who showed its security upto $O(2^{2n/3})$ queries against all nonce-respecting adversaries. Later, Iwata et al. [22] provided its optimal security bound based on the mirror theory technique [32]. Recently, Bhattacharya and Nandi [8] have given its optimal security bound by analysing the PRF security of variable output length xor of permutations using chi-squared method.

Though useful for encryption, the XOR construction does not seem to be directly usable for authentication as we have to extend the domain size, so that the construction can authenticate long messages. This can be done by hashing the message, but with the XOR construction it seems that we need some subtle combination with a double-block hash function, as employed in PMAC_Plus [33], 1K-PMAC_Plus [17] and LightMAC_Plus [26].

ENCRYPTED DAVIES-MEYER. The above problem with the XOR construction in authentication was solved by Cogliati and Seurin [13], who proposed a PRP-to-PRF conversion method, called Encrypted Davies-Meyer (EDM). The EDM construction is defined as follows:

$$\mathsf{EDM}_{\mathsf{E}_{K_1},\mathsf{E}_{K_2}}(x) = \mathsf{E}_{K_2}(\mathsf{E}_{K_1}(x) \oplus x).$$

EDM uses two independent block-cipher keys and achieves $O(q^3/2^{2n})$ security [13]. Soon after, Dai et al. [16] improved its bound to $O(q^4/2^{3n})$ by applying chi-squared method. Concurrently, Mennink and Neves [24] proved its almost optimal security, i.e. $O(2^n/67n)$, using mirror theory technique. Recently, Cogliati and Seurin have proved a BBB security $O(q/2^{2n/3})$ of single-keyed EDM [14], as originally conjectured by themselves [13].

ENCRYPTED WEGMAN-CARTER WITH DAVIES-MEYER. Following the construction of EDM, Cogliati and Seurin extended the idea to construct EWCDM, a nonce-based BBB secure MAC, which is defined as follows:

$$\mathsf{EWCDM}_{\mathsf{E}_{K_1},\mathsf{E}_{K_2},\mathsf{H}_{K_h}}(N, M) = \mathsf{E}_{K_2}\big(\mathsf{E}_{K_1}(N) \oplus N \oplus \mathsf{H}_K(M)\big),$$

where N is the nonce and M is the message to be authenticated. Note that, EWCDM uses two independent block-cipher keys, K_1 and K_2, and also another independent hash-key K_h for the AXU hash function.[1] In this way, EDM obviated the necessity of using double-block hash function that existed with the XOR construction. It has been proved that EWCDM is secure against all nonce-respecting MAC adversaries[2] that make at most $2^{2n/3}$ MAC queries and 2^n verification queries. Cogliati and Seurin also proved $O(2^{n/2})$ security of the construction against nonce-misusing adversaries. Later, Mennink and Neves [24] proved its n-bit PRF security using mirror theory in the nonce respecting setting and mentioned that the analysis straightforwardly generalizes to the analysis for unforgeability or for the nonce-misusing setting of the construction. The trick involved in proving the optimal security of EWCDM is by replacing the last block cipher call with its inverse. This subtle change does not make any difference in the output distribution and as a bonus, it trivially allows one to express the output of the construction as a sum of two random permutations (or in general a bi-variate affine equation[3]). It is only this feature which is captured by the *mirror theory* to derive the security bound of the construction.

---

[1] An AXU hash function is a keyed hash function such that for any two distinct messages, the probability, over a random draw of a hash key, of the hash differential being equal to a specific output is small.

[2] Adversaries who never repeat the same value of N in their MAC queries.

[3] For two variables, P, Q and $\lambda \in \mathrm{GF}(2^n)$ we call an equation of the form $P \oplus Q = \lambda$, a bi-variate affine equation.

MOTIVATION BEHIND THIS WORK. As evident from the definition of the construction, EWCDM requires three keys; two block cipher keys K_1 and K_2 and one hash key K_h. Constructions with multiple keys necessarily demand larger storage space for storing the secret keys, which is sometimes infeasible for lightweight crypto devices. All popular MACs, including CMAC [27] and HMAC [4], require only a single secret key. But most of the time reducing the number of keys without compromising the security is not a trivial task.

Cogliati and Seurin [13] believed that BBB security should hold for single-keyed EWCDM (with $K_1 = K_2$) but be likely cumbersome to prove. As mentioned earlier, Cogliati and Seurin recently proved that single-keyed EDM (not EWCDM) is BBB secure, but the proof is highly complicated. Moreover, it is not clear at all how to build on this result to prove the MAC security of EWCDM construction with $K_1 = K_2$. In fact, Cogliati and Seurin, in their proof of single-keyed EDM [14], state that

"For now, we have been unable to extend the current (already cumbersome) counting used for the proof of the single-permutation EDM construction to the more complicated case of single-key EWCDM."

Thus, we expect that proving the MAC security of single-keyed EWCDM should be a notably hard task and very likely require heavy mathematical tools like *Sum Capture Lemma* as already used for single-keyed EDM. This motivates us to design an another single-keyed, nonce-based MAC built from block ciphers (and a hash function) with BBB security that can be proven by a simpler approach.

Our Contribution. Our contribution in this paper is fourfold which we outline as follows:

- DWCDM: NEW NONCE-BASED MAC. We propose *Decrypted Wegman-Carter with Davies-Meyer*, in short DWCDM, a nonce-based BBB secure MAC. The design philosophy of DWCDM is inspired from the trick used in [24] while proving the optimal security of EWCDM. Recall that, in [24], authors replace the last block cipher call with its inverse so that the output of EWCDM can be expressed as a sum of *two independent* PRPs. But the same trick does not work at the time of using the same block cipher key in the construction. This phenomenon triggers us to design a nonce based MAC, very similar to EWCDM, in which instead of using the encryption algorithm in the last block-cipher call, we use its **decryption** algorithm so that the output of the construction can be expressed as a sum of two *identical* PRPs and hence the name **Decrypted** Wegman-Carter with Davies-Meyer. The construction is *single-keyed* in the sense that the *same block cipher key* is used for the two cipher calls. Schematic diagram of DWCDM is shown in Fig. 1 where the last $n/3$ bits of the nonce N is zero, i.e. $N = N^* \| 0^{n/3}$. We would like to mention here that one cannot use the full n-bit nonce in DWCDM as that would end up with a birthday bound MAC attack which is described in Sect. 4.1. We show that DWCDM is secure up to $2^{2n/3}$ MAC queries and 2^n verification queries against nonce-respecting adversaries. We also show that DWCDM is secure up to $2^{n/2}$ MAC queries and 2^n verification queries in the nonce-misuse setting, where the bound is tight. As a concrete example of DWCDM, we present an instantiation of DWCDM with the AXU hash function being realized via PolyHash [25]. We show that nPolyMAC achieves $2^{2n/3}$-bit MAC security in the nonce-respecting setting.

- EXTENDED MIRROR THEORY. Since, our study of interest is the MAC security of the construction, we require to analyze the number of solutions of a system of affine bi-variate equations along with *affine uni-variate and bi-variate non-equations*[4]. Such a general treatment of analysing system of affine equations with non-equations was only mentioned in [32] without giving any formal analysis. To the best of our knowledge, this is the first time we analyse such a generic system of equations with non-equations, which we regard to as *extended mirror theory* and our MAC security proofs of DWCDM and 1K-DWCDM are crucially based on this new result.

- 1K-DWCDM: "PURE" SINGLE-KEYED VARIANT OF DWCDM. Moreover, we exhibit a truly single-keyed nonce-based MAC construction, 1K-DWCDM. Under the condition that the length of the hash key is equal to the block size as $|K_h| = n$, we can even derive the hash key as $K_h = \mathsf{E}_K(0^{n-1}\|1)$, which results in the construction 1K-DWCDM. We prove that 1K-DWCDM is essentially as secure as DWCDM.

- POTENTIALITY OF ACHIEVING HIGHER SECURITY. Finally, we show how one can boost the security for DWCDM type constructions using extended generalized version of Mirror Theory.

PROOF APPROACH. Our MAC security proof of DWCDM and 1K-DWCDM is fundamentally relied on Patarin's H-coefficient technique [29]. Similar to the technique of [13,19], we cast the unforgeability game of MAC to an equivalent indistinguishability game, with some suitable choice of ideal world, that allows us to apply the H-coefficient technique for bounding the distinguishing advantage of the construction of our concern.

As mentioned earlier that one can express the output of DWCDM as a sum of two identical permutations. Thus, q many such evaluations of DWCDM gives us a system of q many affine bi-variate equations

$$
\begin{cases}
\mathsf{E}_K(N_1) \oplus \mathsf{E}_K(T_1) = N_1 \oplus H_{K_h}(M_1) \\
\mathsf{E}_K(N_2) \oplus \mathsf{E}_K(T_2) = N_2 \oplus H_{K_h}(M_2) \\
\quad\quad\quad\vdots \\
\mathsf{E}_K(N_q) \oplus \mathsf{E}_K(T_q) = N_q \oplus H_{K_h}(M_q)
\end{cases}
$$

Along with this, we also need to ensure that the verification attempt of the adversary should fail (as a part of the good transcript), i.e. for a verification query (N', M', T'), chosen by the adversary, we should always have

$$
\mathsf{E}_K^{-1}(\mathsf{E}_K(N') \oplus N' \oplus H_{K_h}(M')) \neq T'.
$$

---

[4] For two variables, P, Q and $\lambda \in \mathrm{GF}(2^n) \setminus 0^n$ we call $P \oplus Q \neq \lambda$, an affine bi-variate non-equation and $P \neq \lambda$ is an affine uni-variate non-equation.

Hence, it tells us that we also need to incorporate *affine non-equations* along with the system of bi-variate affine equations. This leads us to *extend* the mirror theory technique (extension as in incorporating affine non-equations along with affine bi-variate equations). We require the result of extended mirror theory while lower bounding the real interpolation probability for a good transcript.

Remark 1. We would like to point out that a possible alternative approach is to use the *chi-square method*, a recently discovered technique which has been reported in [7,8,16]. It is interesting to observe that in some settings chi-square outperforms H-coefficient technique in terms of guaranteeing security with quadratic improvement on the number of queries that adversary can make [16]. However, it is difficult to apply this technique in our construction. The reason behind this is the lack of sufficient entropy of the conditional distribution when we condition on the hash key. The same holds true for the analysis of EWCDM as well. In fact, this negative phenomenon motivates us to consider DWCDM so that we can represent the construction as a sum of permutations and eventually apply extended mirror theory.

2   Preliminaries

SYMBOLS AND NOTATIONS. For a set $\mathcal{X}$, $X \xleftarrow{\$} \mathcal{X}$ denotes that X is sampled uniformly at random from $\mathcal{X}$ and independent to all random variables defined so far. $\{0,1\}^n$ denotes the set of all binary strings of length n. The set of all functions from $\mathcal{X}$ to $\mathcal{Y}$ is denoted as $\mathsf{Func}(\mathcal{X},\mathcal{Y})$ and the set of all permutations over $\mathcal{X}$ is denoted as $\mathsf{Perm}(\mathcal{X})$. $\mathsf{Func}_{\mathcal{X}}$ denotes the set of all functions from $\mathcal{X}$ to $\{0,1\}^n$ and Perm denotes the set of all permutations over $\{0,1\}^n$. We often write Func instead of $\mathsf{Func}_{\mathcal{X}}$ when the domain of the functions is understood from the context. We write $[q]$ to refer to the set $\{1,\ldots,q\}$.

For any binary string x, $|x|$ denotes the length i.e. the number of bits in x. For $x,y \in \{0,1\}^n$, we write $z = x \oplus y$ to denote the modulo 2 addition of x and y. We write $\mathbf{0}$ to denote the zero element of the field $\{0,1\}^n$ (i.e. 0^n) and $\mathbf{1}$ to denote $0^{n-1}\|1$. For integers $1 \leq b \leq a$, we write $(a)_b$ to denote $a(a-1)\ldots(a-b+1)$, where $(a)_0 = 1$ by convention.

2.1   Security Definitions

PRF AND PRP AND SPRP. A keyed function with key space $\mathcal{K}$, domain $\mathcal{X}$ and range $\mathcal{Y}$ is a function $\mathsf{F} : \mathcal{K} \times \mathcal{X} \to \mathcal{Y}$ and we denote $\mathsf{F}(K,X)$ by $\mathsf{F}_K(X)$. Similarly, a keyed permutation with key space $\mathcal{K}$ and domain $\mathcal{X}$ is a mapping $\mathsf{E} : \mathcal{K} \times \mathcal{X} \to \mathcal{X}$ such that for all key $K \in \mathcal{K}$, $X \mapsto \mathsf{E}(K,X)$ is a permutation over $\mathcal{X}$ and we denote $\mathsf{E}_K(X)$ for $\mathsf{E}(K,X)$.

PRF. Given an oracle algorithm A with oracle access to a function from $\mathcal{X}$ to $\mathcal{Y}$, making at most q queries, running time is at most t and outputting a single bit. We define the prf-advantage of A against the family of keyed functions F as

$$\mathbf{Adv}_{\mathsf{F}}^{\mathrm{PRF}}(\mathsf{A}) := |\Pr[K \xleftarrow{\$} \mathcal{K} : \mathsf{A}^{\mathsf{F}_K} = 1] - \Pr[\mathsf{RF} \xleftarrow{\$} \mathsf{Func}(\mathcal{X},\mathcal{Y}) : \mathsf{A}^{\mathsf{RF}} = 1]|.$$

We say that F is a (q, t, ϵ) secure PRF, if $\mathbf{Adv}_{\mathsf{F}}^{\mathrm{PRF}}(q, t) := \max_{\mathsf{A}} \mathbf{Adv}_{\mathsf{F}}^{\mathrm{PRF}}(\mathsf{A}) \leq \epsilon$, where the maximum is taken over all adversaries A that makes q many queries and running time is at most t.

PRP. Given an oracle algorithm A with oracle access to a permutation of $\mathcal{X}$, making at most q queries, running time is at most t and outputting a single bit. We define the prp-advantage of A against the family of keyed permutations E as

$$\mathbf{Adv}_{\mathsf{E}}^{\mathrm{PRP}}(\mathsf{A}) := |\Pr[K \leftarrow_{\$} \mathcal{K} : \mathsf{A}^{\mathsf{E}_K} = 1] - \Pr[\Pi \leftarrow_{\$} \mathsf{Perm}(\mathcal{X}) : \mathsf{A}^{\Pi} = 1]|.$$

We say that E is a (q, t, ϵ) secure PRP, if $\mathbf{Adv}_{\mathsf{E}}^{\mathrm{PRP}}(q, t) := \max_{\mathsf{A}} \mathbf{Adv}_{\mathsf{E}}^{\mathrm{PRP}}(\mathsf{A}) \leq \epsilon$, where the maximum is taken over all adversaries A that makes q many queries and running time is at most t.

SPRP. Given an oracle algorithm A with oracle access to a permutation and its inverse over $\mathcal{X}$, making at most q^{+} queries to permutation and q^{-} queries to inverse permutation, running time is at most t and outputting a single bit. We define the sprp-advantage of A against the family of keyed permutations E as

$$\mathbf{Adv}_{\mathsf{E}}^{\mathrm{SPRP}}(\mathsf{A}) := |\Pr[K \leftarrow_{\$} \mathcal{K} : \mathsf{A}^{\mathsf{E}_K, \mathsf{E}_K^{-1}} = 1] - \Pr[\Pi \leftarrow_{\$} \mathsf{Perm}(\mathcal{X}) : \mathsf{A}^{\Pi, \Pi^{-1}} = 1]|.$$

We say that E is a (q, t, ϵ) secure SPRP, if $\mathbf{Adv}_{\mathsf{E}}^{\mathrm{SPRP}}(q, t) := \max_{\mathsf{A}} \mathbf{Adv}_{\mathsf{E}}^{\mathrm{SPRP}}(\mathsf{A}) \leq \epsilon$, where the maximum is taken over all adversaries A that makes q many encryption and decryption queries altogether and running time is at most t.

MACs. Given four non-empty finite sets $\mathcal{K}, \mathcal{N}, \mathcal{M}$ and $\mathcal{T}$, a nonce based keyed function with key space $\mathcal{K}$, nonce space $\mathcal{N}$, message space $\mathcal{M}$ and range $\mathcal{T}$ is a keyed function whose domain is $\mathcal{N} \times \mathcal{M}$ and range is $\mathcal{T}$ and we write $\mathsf{F}(K, N, M)$ as $\mathsf{F}_K(N, M)$.

Definition 1 (Nonce Based MAC). *Let $\mathcal{K}, \mathcal{N}, \mathcal{M}$ and $\mathcal{T}$ be four non-empty finite sets and $\mathsf{F} : \mathcal{K} \times \mathcal{N} \times \mathcal{M} \to \mathcal{T}$ be a nonce based keyed function. For $K \in \mathcal{K}$, let Ver_K be the verification oracle that takes as input $(N, M, T) \in \mathcal{N} \times \mathcal{M} \times \mathcal{T}$ and outputs 1 if $\mathsf{F}_K(N, M) = T$, otherwise outputs 0. A (q_m, q_v, t) adversary against the MAC security of F is an adversary A with access to two oracles F_K and Ver_K for $K \in \mathcal{K}$ such that it makes at most q_m many MAC queries to first oracle and q_v many verification queries to second oracle. We say that A forges F if any of its queries to Ver_K returns 1. The advantage of A against the MAC security of F is defined as*

$$\mathbf{Adv}_{\mathsf{F}}^{\mathrm{MAC}}(\mathsf{A}) := \Pr[K \leftarrow_{\$} \mathcal{K} : \mathsf{A}^{\mathsf{F}_K, \mathsf{Ver}_K} \text{ forges }],$$

where the probability is taken over the randomness of the underlying key and the random coin of adversary A (if any). We assume that A does not make any verification query (N, M, T) to Ver_K if T is obtained in previous MAC query with input (N, M) and it does not repeat any query. We call such an adversary as "non-trivial" adversary. The adversary is said to be "nonce respecting" if it does not repeat nonces in its queries to the MAC oracle[5].

REGULAR AND AXU HASH FUNCTION. Let $\mathcal{K}_h, \mathcal{X}, \mathcal{Y}$ be three non-empty finite sets and H be a keyed function $\mathsf{H} : \mathcal{K}_h \times \mathcal{X} \to \mathcal{Y}$. Then,

(1) H is said to be an ϵ regular hash function, if for any $X \in \mathcal{X}$ and any $Y \in \mathcal{Y}$,

$$\Pr[K_h \leftarrow_\$ \mathcal{K}_h : \mathsf{H}_{K_h}(X) = Y] \le \epsilon. \tag{1}$$

(2) H is said to be an ϵ almost xor universal (AXU) hash function if for any distinct $X, X' \in \mathcal{X}$ and for any $Y \in \mathcal{Y}$,

$$\Pr[K_h \leftarrow_\$ \mathcal{K}_h : \mathsf{H}_{K_h}(X) \oplus \mathsf{H}_{K_h}(X') = Y] \le \epsilon. \tag{2}$$

(3) H is said to be an ϵ 3-way regular hash function if for any distinct $X_1, X_2, , X_3 \in \mathcal{X}$ and for any *non-zero* $Y \in \mathcal{Y}$,

$$\Pr[K_h \leftarrow_\$ \mathcal{K}_h : \mathsf{H}_{K_h}(X_1) \oplus \mathsf{H}_{K_h}(X_2) \oplus \mathsf{H}_{K_h}(X_3) = Y] \le \epsilon. \tag{3}$$

In the following, we state that PolyHash [25] is one of the examples of algebraic hash function which is $\ell/2^n$ regular, AXU as well as 3-way regular hash function.

Proposition 1. *Let* Poly $: \{0,1\}^n \times (\{0,1\}^n)^* \to \{0,1\}^n$ *be a hash function defined as follows: For a fixed key* $K_h \in \{0,1\}^n$ *and for a fixed message* M, *we first apply an injective padding such as* 10^* *i.e., pad 1 followed by minimum number of zeros so that the total number of bits in the padded message becomes multiple of n. Let the padded message be* $M^* = M_1 \| M_2 \| \dots \| M_l$ *where for each* $i, |M_i| = n$. *Then we define*

$$\mathsf{Poly}_{K_h}(M) = M_l \cdot K_h \oplus M_{l-1} \cdot K_h^2 \oplus \dots \oplus M_1 \cdot K_h^l, \tag{4}$$

where l is the number of n-bit blocks. Then, Poly *is* $\ell/2^n$ *regular, AXU and 3-way regular hash function, where* ℓ *denotes the maximum number of message blocks of size n-bits.*

The proof of the result lies around in finding the number of roots of a non-zero polynomial over the hash key K_h with message blocks being the coefficients of the polynomial. The details of the proof of can be found in [18].

3   Patarin's Mirror Theory

Mirror theory, as defined in [32] is the theory of evaluating the number of solutions of affine system of equalities and non-equalities in a finite group. Patarin,

---

[5] Similar to nonce respecting adversary, we say that an adversary is nonce misusing if the adversary is not restricted to make queries to the MAC oracle with distinct nonces.

who coined this theory, has given a lower bound on the number of solutions of a finite system of affine bi-variate equations using an inductive proof when the variables in the equations are wor samples [30]. The proof is tractable upto the order of $2^{2n/3}$ security bound, but the proof becomes highly complex and too difficult to verify in the case of deriving the optimal security bound. In specific, once the first-order recursion is considered, one needs to consider a second-order recursion, and so on, until the n-th recursion. For the i-th order recursion, there are $O(2^i)$ many cases and Patarin's proof only addresses the first (and perhaps the second) order recursion by a tedious analysis, but the cases of the higher-order ones are quite different, and it's not at all clear how to bridge the gap, given an exponential number of cases that one has to consider. Moreover, to the best of our knowledge, the proof did not consider any affine non-equation as well.

In this section we extend the Mirror theory in the context of our MAC security to incorporate the affine non-equations (that includes uni-variate and bi-variate non-equations) along with a system of affine bi-variate equations. In the following, we prove that when the number of affine bi-variate equations is $q \leq 2^{2n/3}$ and the number of non-equations is $v \leq 2^n$ (v is the total number of affine uni-variate and bi-variate non equations), then the number of solutions becomes at least $(2^n)_{3q/2}/2^{nq}$. For the sake of presentation and interoperability with the results in the remainder of the paper, we use different parameterization and naming convention.

3.1 General Setting of Mirror Theory

Given a bi-variate affine equation $P \oplus Q = \lambda$ over $\mathrm{GF}(2^n)$, the associated linear equation of this affine equation is $P \oplus Q = \mathbf{0}$. Now, given $\lambda_1, \ldots, \lambda_q \in \mathrm{GF}(2^n) \setminus \mathbf{0}$ which we write as $\Lambda = (\lambda_1, \ldots, \lambda_q)$, let us consider a system of q many bi-variate affine equations over $\mathrm{GF}(2^n)$:

$$\mathcal{E}_\Lambda = \{P_{n_1} \oplus P_{t_1} = \lambda_1, P_{n_2} \oplus P_{t_2} = \lambda_2, \ldots, P_{n_q} \oplus P_{t_q} = \lambda_q\}.$$

Given a function $\phi : \{n_1, t_1, \ldots, n_q, t_q\} \to \mathcal{I}$, called *index mapping function*, we associate another system of bi-variate affine equations:

$$\mathcal{E}_{\Lambda,\phi} = \{P_{\phi(n_1)} \oplus P_{\phi(t_1)} = \lambda_1, P_{\phi(n_2)} \oplus P_{\phi(t_2)} = \lambda_2, \ldots, P_{\phi(n_q)} \oplus P_{\phi(t_q)} = \lambda_q\}.$$

Let α denotes the cardinality of the image set of ϕ. Then, $\mathcal{E}_{\Lambda,\phi}$ is a system of bi-variate affine equations over α variables. In our paper, a specific choice of $\mathcal{I}$ would be $\{0,1\}^n$.

Example. Consider a system of equations:

$$\{P_1 \oplus P_2 = \lambda_1, P_1 \oplus P_3 = \lambda_2, P_2 \oplus P_4 = \lambda_3\}.$$

Then, the index mapping function for the above system of equations is $\phi(n_1) = 1, \phi(t_1) = 2, \phi(n_2) = 1, \phi(t_2) = 3, \phi(n_3) = 2, \phi(t_3) = 4$. For this system of equations $\alpha = 4$.

EQUATION-DEPENDENT GRAPH. For index mapping function $\phi : \{n_1, t_1, \ldots, n_q, t_q\} \to \mathcal{I}$, we associate a undirected graph $G_\phi = ([q], \mathcal{S})$ where $\{i, j\} \in \mathcal{S}$ if

$$|\{\phi(n_i), \phi(t_i)\} \cap \{\phi(n_j), \phi(t_j)\} \mid \geq 1$$

or if $i = j$ and $\phi(n_i) = \phi(t_i)$. We call such an edge a *self-loop*. In other words, we introduce an edge between two equations (node represents the equation number) in the equation-dependent graph if the corresponding equations have at least one common unknown variable. Note that the set $\{\phi(n_i), \phi(t_i)\}$ can be a multi-set.

For a subset $\{i_1, \ldots, i_c\} \subseteq [q]$, let

$$\{P_{\phi(n_{i_1})} \oplus P_{\phi(t_{i_1})} = \mathbf{0}, P_{\phi(n_{i_2})} \oplus P_{\phi(t_{i_2})} = \mathbf{0}, \ldots, P_{\phi(n_{i_c})} \oplus P_{\phi(t_{i_c})} = \mathbf{0}\}$$

be the sub-system of associated linear equations. We say this sub-system of associated linear equations is linearly dependent if $\{i_1, \ldots, i_c\}$ is the minimal set and all variables P_x, which appeared in the above sub-system, appears exactly twice. Depending on the value of c (for the minimal linearly dependent sub-system), we have the following three cases;

(i) $c = 1$: SELF-LOOP. If there exists i such that $\phi(n_i) = \phi(t_i)$.
(ii) $c = 2$: PARALLEL-EDGE. If there exists $i \neq j$ such that either:

(a) $\phi(n_i) = \phi(n_j)$ and $\phi(t_i) = \phi(t_j)$ or (b) $\phi(n_i) = \phi(t_j)$ and $\phi(t_i) = \phi(n_j)$.

(iii) $c \geq 3$: ALTERNATING-CYCLE. If there exists distinct $i_1, i_2, \ldots, i_c$ such that for every $j \in [c]$ either
  - $\phi(n_{i_j}) \in \{\phi(n_{i_{j+1}}), \phi(t_{i_{j+1}})\}$ and $\phi(t_{i_j}) \in \{\phi(n_{i_{j-1}}), \phi(t_{i_{j-1}})\}$ or
  - $\phi(t_{i_j}) \in \{\phi(n_{i_{j+1}}), \phi(t_{i_{j+1}})\}$ and $\phi(n_{i_j}) \in \{\phi(n_{i_{j-1}}), \phi(t_{i_{j-1}})\}$.

When $i = 1$, $i - 1$ is considered as c and when $i = c$, $i + 1$ is considered as 1. We say that ϕ is *dependent* if any one of the above condition holds. Otherwise, we call it *independent*. Given an independent ϕ, the graph G_ϕ becomes a simple graph and $\mathcal{E}_{\Lambda,\phi}$ becomes linearly independent. In this case, the number of variables present in a connected component $C = \{i_1, \ldots, i_c\}$ of G_ϕ (i.e., the size of the set $\{\phi(n_{i_1}), \phi(t_{i_1}), \ldots, \phi(n_{i_c}), \phi(t_{i_c})\}$) is exactly $c + 1$. We call the set $\{\phi(n_{i_1}), \phi(t_{i_1}), \ldots, \phi(n_{i_c}), \phi(t_{i_c})\}$ a *block*. The *block maximality*, denoted by $\zeta_{\max}$, of an independent ϕ is defined as $\zeta_{\max} + 1$ where $\zeta_{\max}$ is the size of the maximum connected components of G_ϕ (Note that, a block with p many elements introduces $p - 1$ many affine equations.).

3.2   Extended Mirror Theory

In this section, we introduce the extended Mirror theory technique by incorporating two types of non-equations with a finite number of bi-variate affine equations. We consider (i) uni-variate affine non-equation of the form $X_i \neq c$ and (ii) bi-variate affine non-equation of the form $X_i \oplus Y_i \neq c$, where c is a non-zero constant. In particular, we lower bound the number of solutions of a finite number of affine equations[6] and uni(bi-) variate affine non-equations. To begin with, let us investigate what happens when we introduce a single uni(bi-) variate affine non-equation with a finite number of affine equations. Let $\mathcal{E}^=$ be a system of q many affine equations of the form

$$\mathcal{E}^= = \{P_{n_1} \oplus P_{t_1} = \lambda_1, \ldots, P_{n_q} \oplus P_{t_q} = \lambda_q\}. \tag{5}$$

Let ϕ be an index mapping function that maps from $\{n_1, t_1, \ldots, n_q, t_q\} \to \mathcal{I}$. Let $\Lambda_= = (\lambda_1, \lambda_2, \ldots, \lambda_q)$, where each $\lambda_i \in \mathrm{GF}(2^n) \setminus \mathbf{0}$. Now, for an independent choice of ϕ, $\mathcal{E}^=_{\phi, \Lambda_=}$ is a linearly independent set of q many affine equations. Let $\mathcal{E}^{\neq}$ be a system of r many bi-variate affine non-equations and $v - r$ many uni-variate affine non-equations of the form

$$\mathcal{E}^{\neq} = \{P_{n_{q+1}} \oplus P_{t_{q+1}} \neq \lambda_1', \ldots, P_{n_{(q+r)}} \oplus P_{t_{(q+r)}} \neq \lambda_r'\}$$
$$\bigcup \{P_{n_{q+r+1}} \neq \lambda_{r+1}', \ldots, P_{n_{(q+v)}} \neq \lambda_v'\}.$$

We denote $\Lambda_{\neq} = (\lambda_1', \lambda_2', \ldots, \lambda_v')$, where each $\lambda_i' \in \mathrm{GF}(2^n) \setminus \mathbf{0}$, and $\Lambda' = (\lambda_1, \lambda_2, \ldots, \lambda_q, \lambda_1', \lambda_2', \ldots, \lambda_v')$. Now, for the system of affine equations and non-equations $\mathcal{E} := \mathcal{E}^= \cup \mathcal{E}^{\neq}$, we consider the index mapping function

$$\phi' : \{n_1, t_1, \ldots, n_q, t_q, n_{q+1}, t_{q+1}, \ldots, n_{q+v}, t_{q+v}\} \to \mathcal{I}.$$

Moreover, we denote $\phi := \phi'_{|q}$ to be the index mapping function that maps $\{n_1, t_1, \ldots, n_q, t_q\} \to \mathcal{I}$ and $\Lambda_= := \Lambda'_{|q}$ to be $(\lambda_1, \lambda_2, \ldots, \lambda_q)$.

CHARACTERIZING GOOD (ϕ', Λ'). We say that a pair (ϕ', Λ') is good if

- (C1) ϕ is independent and for all $x \neq y$, $P_{\phi(x)} = P_{\phi(y)}$ cannot be generated from the system of equations $\mathcal{E}^=_{\phi, \Lambda_=}$.

- (C2) for all $j \in [v]$ and $i_1, \ldots, i_c \in [q]$, $c \geq 0$, such that $\{i_1, \ldots, i_c, q+j\}$ is dependent system then $\lambda_{i_1} \oplus \cdots \oplus \lambda_{i_c} \oplus \lambda_j' \neq \mathbf{0}$.

In words, a good (ϕ', Λ') says that: (i) the system of equation $\mathcal{E}_{\phi, \Lambda_=}$ is linearly independent system of equations and one cannot generate an equation of the form $P_{\phi(x)} = P_{\phi(y)}$ by linearly combining the equation of $\mathcal{E}_{\phi, \Lambda_=}$. Moreover, (ii) by linearly combining the equation of $\mathcal{E}_{\phi, \Lambda_=}$, one cannot generate an equation of the form $P_x \oplus P_y = \lambda_{x,y}$ such that $P_x \oplus P_y \neq \lambda_{x,y}$ already exist in $\mathcal{E}^{\neq}_{\phi', \Lambda'}$. Summarizing above, we state and prove the following main theorem, which we call as *Extended Mirror Theorem for $\xi_{\max} = 3$*. For the notational simplicity we assume the index set $\mathcal{I} = [\alpha]$.

Theorem 1. *Let $(\mathcal{E}^= \cup \mathcal{E}^{\neq}, \phi', \Lambda')$ be a system of q many affine equations and v many uni(bi-) variate affine non-equations associated with index mapping function ϕ' over $\mathrm{GF}(2^n)$ which are of the form*

$$(a) P_{\phi(n_i)} \oplus P_{\phi(t_i)} = \lambda_i(\neq \mathbf{0}), \ \forall i \in [q]$$
$$(b) P_{\phi(n_j)} \oplus P_{\phi(t_j)} \neq \lambda_j'(\neq \mathbf{0}), \ \forall j \in [q+1, q+r]$$
$$(c) P_{\phi(n_j)} \neq \lambda_j''(\neq \mathbf{0}), \ \forall j \in [q+r+1, q+v]$$

---

[6] When we consider affine equation, we actually refer to the bi-variate affine equation.

over the set of α many unknown variables $\mathcal{P} = \{P_1, \ldots, P_\alpha\}$ such that P_a may be equals to some $P_{\phi(n_i)}$ or $P_{\phi(t_i)}$, where $a \in \{\phi(n_j), \phi(t_j)\}, j \in [q+1, q+v]$. Now, if

- (i) (ϕ', Λ') is good and
- (ii) $\xi_{\max} = 3$

then the number of solutions for $\mathcal{P}$, denoted by $h_{\frac{3q}{2}}$ such that $P_i \neq P_j$ for all distinct $i, j \in \{1, \ldots, \alpha\}$ is

$$h_{\frac{3q}{2}} \geq \frac{(2^n)_{\frac{3q}{2}}}{2^{nq}} \left(1 - \frac{5q^3}{2^{2n}} - \frac{v}{2^n}\right). \tag{6}$$

Proof. As mentioned, our proof is an inductive proof based on the number of blocks u. Our first observation is that as (ϕ', Λ') is good, ϕ is independent and thus $\xi_{\max} = \zeta_{\max} + 1$ and hence, the maximum number of variables P_i that can reside in the same block is 3. For the simplicity of the proof, assume that we have exactly 3 variables at each blocks. Now, it is easy to see that Eq. (6) holds when $u = 1$.

As the next step of the proof, let h_{3u} be the solutions for first $2u$ many affine equations, which we denote as $\mathcal{E}_{2u}^=$. Now as soon as we add the $(u+1)^{th}$ block, we consider the following bi-variate affine equations $P_{3u+1} \oplus P_{3u+2} = \lambda_{2u+1}, P_{3u+1} \oplus P_{3u+3} = \lambda_{2u+2}$ and those bi-variate affine non-equations which are of the form $P_{\sigma_i} \oplus P_{\delta_i} \neq \lambda_i'$, where $\sigma_i \in \{1, \ldots, 3u+3\}, \delta_i \in \{3u+1, 3u+2, 3u+3\}$ and also those uni-variate affine non-equations of the form $P_{\delta_i} \neq \lambda_i''$, where $\delta_i \in \{3u+1, 3u+2, 3u+3\}$. Let v' and v'' be the number of such bi-variate and uni-variate affine non-equations. Now, note that each such bi-variate affine non-equation of the form $P_{\sigma_i} \oplus P_{\delta_i} \neq \lambda_i'$ where $\sigma_i \in \{1, \ldots, 3u+3\}, \delta_i \in \{3u+1, 3u+2, 3u+3\}$ can be written as $P_{3u+1} \neq P_{\sigma_i} \oplus \lambda_i^\star$, where $\sigma_i \in \{1, \ldots, 3u+3\}$ and $\lambda_i^\star \in \{\lambda_i', \lambda_i' \oplus \lambda_{2u+1}, \lambda_i' \oplus \lambda_{2u+2}\}$. Moreover, each such uni-variate affine non-equation of the form $P_{\delta_i} \neq \lambda_i''$ where $\delta_i \in \{3u+1, 3u+2, 3u+3\}$ can be written as $P_{3u+1} \neq \lambda_i^{\star\star}$, where $\lambda_i^{\star\star} \in \{\lambda_i'', \lambda_i'' \oplus \lambda_{2u+1}, \lambda_i'' \oplus \lambda_{2u+2}\}$.

Now h_{3u+3} counts for the number of solutions to $\{P_1, \ldots, P_{3u}, P_{3u+1}, P_{3u+2}, P_{3u+3}\}$ such that

- $\{P_1, \ldots, P_{3u}\}$ is a valid solution of $\mathcal{E}_{2u}^=$.
- $P_{3u+1} \oplus P_{3u+2} = \lambda_{2u+1}, P_{3u+1} \oplus P_{3u+3} = \lambda_{2u+2}$.
- $P_{3u+1} \notin \{P_1, \ldots, P_{3u}, P_1 \oplus \lambda_{2u+1}, \ldots, P_{3u} \oplus \lambda_{2u+1}, P_1 \oplus \lambda_{2u+2}, \ldots, P_{3u} \oplus \lambda_{2u+2}\}$.
- $P_{3u+1} \notin \{P_{\sigma_1} \oplus \lambda_1^\star, \ldots, P_{\sigma_{v'}} \oplus \lambda_{v'}^\star\}$.
- $P_{3u+1} \notin \{\lambda_1^{\star\star}, \ldots, \lambda_{v''}^{\star\star}\}$.

Let $V_1 = \{P_1, \ldots, P_{3u}\}, V_2 = \{P_1 \oplus \lambda_{2u+1}, \ldots, P_{3u} \oplus \lambda_{2u+1}\}, V_3 = \{P_1 \oplus \lambda_{2u+2}, \ldots, P_{3u} \oplus \lambda_{2u+2}\}, V_4 = \{P_{\sigma_1} \oplus \lambda_1^\star, \ldots, P_{\sigma_{v'}} \oplus \lambda_{v'}^\star\}$ and $V_5 = \{\lambda_1^{\star\star}, \ldots, \lambda_{v''}^{\star\star}\}$. Note that, $|V_i| = 3u, i = 1, 2, 3$ and $|V_4| = v', |V_5| = v''$. Therefore, we can write

$$h_{3u+3} = h_{3u}(2^n - |V_1 \cup V_2 \cup V_3 \cup V_4 \cup V_5|) \geq h_{3u}(2^n - |V_1| - |V_2| - |V_3| - |V_4| - |V_5|)$$
$$\geq h_{3u}(2^n - 9u - v' - v'').$$

By applying repeated induction, we obtain

$$h_{\frac{3q}{2}} \geq \left(2^n - 9(\frac{q}{2}-1) - v' - v''\right) h_{3(\frac{q}{2}-1)} \geq \ldots \geq \prod_{u=0}^{q/2-1} (2^n - 9u - v' - v'')$$

for which we have,

$$\frac{h_{\frac{3q}{2}} 2^{nq}}{(2^n)_{\frac{3q}{2}}} \geq \prod_{u=0}^{q/2-1} \frac{2^{2n}(2^n - 9u - v' - v'')}{(2^n - 3u)(2^n - 3u - 1)(2^n - 3u - 2)}$$

$$\geq \prod_{u=0}^{q/2-1} \frac{2^{2n}(2^n - 9u - v' - v'')}{2^{3n} - (9u+3)2^{2n} + (27u^2 + 18u + 2)2^n}$$

$$\overset{[1]}{\geq} \prod_{u=0}^{q/2-1} \left(1 + \frac{3}{2^n} - \frac{27u^2 + 18u + 2}{2^{2n}} - \frac{v' + v''}{2^n}\right)$$

$$\overset{[2]}{\geq} \prod_{u=0}^{q/2-1} \left(1 - \frac{27u^2}{2^{2n}} - \frac{9u^2}{2^{2n}} - \frac{v' + v''}{2^n}\right) \geq \prod_{u=0}^{q/2-1} \left(1 - \frac{36u^2}{2^{2n}} - \frac{v' + v''}{2^n}\right)$$

$$\geq \left(1 - \sum_{u=0}^{q/2-1} \frac{36u^2}{2^{2n}} - \sum_{u=0}^{q/2-1} \frac{v' + v''}{2^n}\right) \overset{[3]}{\geq} \left(1 - \frac{5q^3}{2^{2n}} - \frac{v}{2^n}\right)$$

where [1] follows from the assumptions $u \leq 2^n/9$, [2] follows as $\frac{9u^2}{2^{2n}} \geq \frac{(18u+3)}{2^{2n}} - \frac{3}{2^n}$
and [3] follows as $\sum_{u=0}^{q/2-1} (v' + v'') \leq v$. $\qquad\square$

4 DWCDM and Its Security Result

In this section, we discuss our proposed construction DWCDM and state its security in nonce respecting and nonce misuse setting. Let us recall the DWCDM construction $\mathsf{DWCDM}[\mathsf{E}, \mathsf{E}^{-1}, \mathsf{H}](N, M) := \mathsf{E}_K^{-1}(\mathsf{E}_K(N) \oplus N \oplus \mathsf{H}_{K_h}(M))$ where $N = N^* \| 0^{n/3}$. E_K is a n-bit block cipher and H_{K_h} is an ϵ_1-regular, ϵ_2-AXU and ϵ_3-3-way regular n-bit keyed hash function. A schematic diagram of DWCDM is shown in Fig. 1. Note that, DWCDM is structurally similar to EWCDM, but unlike EWCDM, our construction uses the same block cipher key and the last block cipher call of EWCDM is replaced by its decryption function. Moreover, DWCDM cannot exploit the full nonce space like EWCDM, otherwise its beyond birthday security will be compromised as explained below.

4.1 Why DWCDM Cannot Accommodate Full n-bit Nonce

As mentioned above, for DWCDM we need to reduce the nonce space to $2n/$ 3-bits. If it uses the full nonce space then using a nonce respecting adversary A who set the tags as nonce repeatedly, can mount a birthday bound forging attack on DWCDM as follows:

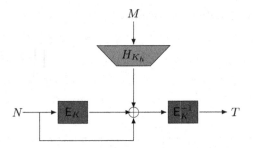

Fig. 1. Decrypted Wegman-Carter with Davies-Meyer construction.

Suppose, an adversary starts with query (N, M) and then makes a chain of queries of the form (T_{i-1}, M) where (T_{i-1}, M) is the i-th query and T_{i-1} is the response of the previous $(i - 1)$-th query, until the first time collision occurs (i.e. a response matches with one of the previous responses). If the adversary makes upto $q \approx 2^{n/2}$ queries, it gets a collision $T_i = T_j$ with high probability. Interestingly, if $(j - i)^7$ is even (which holds with probability $1/2$), then

$$T_j = T_i \text{ iff } (T_i + T_{i+1} + \cdots + T_{j-1} = \mathbf{0}).$$

Now, this property can be easily used A to predict $T_i = T_j$ if it finds $T_i + T_{i+1} + \cdots + T_{j-1} = \mathbf{0}$ for some i, j such that $(j - i)$ is even.

However, if we restrict the nonce space to $2n/3$ bits, then this attack doesn't work because now using the tag as a valid nonce is a probabilistic event. Probability that a tag is a valid nonce is $2^{-n/3}$. This restricts the adversary from forming a chain as used in the attack. In fact, if adversary makes $2^{2n/3}$ many MAC queries then the expected number of tags whose last $n/3$ bits are all zeros is $2^{n/3}$. Now, if adversary uses these $2^{n/3}$ tags as the nonces, then the expected number of tags whose last $n/3$ bits are zeros is 1 and then adversary cannot proceed further. This phenomenon effectively invalidates the above attack to happen.

4.2   Nonce Respecting Security of DWCDM

In this section, we state that DWCDM is secure up to $2^{2n/3}$ MAC queries and 2^n verification queries against nonce respecting adversaries. Formally, the following result bounds the MAC advantage of DWCDM against nonce respecting adversaries.

Theorem 2. *Let* $\mathcal{M}, \mathcal{K}$ *and* $\mathcal{K}_h$ *be finite and non-empty sets. Let* $\mathsf{E} : \mathcal{K} \times \{0, 1\}^n \rightarrow \{0, 1\}^n$ *be a block cipher and* $\mathsf{H} : \mathcal{K}_h \times \mathcal{M} \rightarrow \{0, 1\}^n$ *be an* ϵ_1 *regular,* ϵ_2 *AXU and* ϵ_3 *3-way regular hash function. Then, the MAC advantage for any* (q_m, q_v, t) *nonce respecting adversary against* $\mathsf{DWCDM}[\mathsf{E}, \mathsf{E}^{-1}, \mathsf{H}]$ *is given by,*

---

[7] We assume $j > i$.

$$\mathbf{Adv}^{\mathrm{MAC}}_{\mathrm{DWCDM}[E,E^{-1},H]}(q_m,q_v,t) \leq \mathbf{Adv}^{\mathrm{SPRP}}_{E}(q_m+q_v,t') + \frac{2q_m}{2^{2n/3}} + q_m\epsilon_1 + \frac{2q_m\epsilon_2}{2^{n/3}}$$

$$+ \max\{q_v\epsilon_1, 2q_v\epsilon_2, 2q_v\epsilon_3, \frac{q_m}{2^{2n/3}}\} + \frac{(q_m+q_v)}{2^n} + \frac{5q_m^3}{2^{2n}},$$

where $t' = O(t + (q_m + q_v)t_H)$, t_H *be the time for computing the hash function. By assuming* $\epsilon_1, \epsilon_2, \epsilon_3 \approx 2^{-n}$ *and* $q_m \leq 2^{2n/3}$, DWCDM *is secured up to roughly* $q_m \approx 2^{2n/3}$ *MAC queries and* $q_v \approx 2^n$ *verification queries.*

4.3   Nonce Misuse Security of DWCDM

Similar to EWCDM [13], one can prove that DWCDM$[E, E^{-1}, H]$ is birthday bound secure MAC against nonce misuse adversaries. In particular, DWCDM is secure up to $2^{n/2}$ MAC queries and 2^n verification queries against nonce misuse adversaries and that the security bound is essentially tight. More formally, we have the following MAC security result of DWCDM in nonce misuse setting.

Theorem 3. *Let* $\mathcal{M}, \mathcal{K}$ *and* $\mathcal{K}_h$ *be finite and non-empty sets,* $E : \mathcal{K} \times \{0,1\}^n \rightarrow \{0,1\}^n$ *be a block cipher and* $H : \mathcal{K}_h \times \mathcal{M} \rightarrow \{0,1\}^n$ *be an* ϵ_1 *regular and* ϵ_2 *AXU hash function. Then, the MAC security of* DWCDM$[E, E^{-1}, H]$ *in nonce misuse setting is given by*

$$\mathbf{Adv}^{\mathrm{MAC}}_{\mathrm{DWCDM}}(q_m,q_v,t) \leq \mathbf{Adv}^{\mathrm{SPRP}}_{E}(q_m+q_v,t') + q_m^2\epsilon_2 + \frac{4q_m^2}{2^n} + q_m\epsilon_1 + \frac{(q_m+q_v)}{2^n},$$

where $t' = O(t + q(q_m + q_v)t_H)$, t_H *be the time for computing hash function.*

By assuming $\epsilon_1 \approx 2^{-n}$ and $\epsilon_2 \approx 2^{-n}$, DWCDM is secure up to roughly $q_m \approx 2^{n/2}$ MAC queries and $q_v \approx 2^n$ verification queries. The proof of this theorem can be found in the full version [18].

Tightness of the Bound

We show that the above bound of DWCDM is tight by demonstrating a forging attack which shows thats roughly $2^{n/2}$ MAC queries are enough to break the MAC security of DWCDM when an adversary is allowed to repeat nonce only for once. The attack is as follows:

1. Adversary A makes q many MAC queries (N_i, M_i) with distinct nonces where a collision in the response, i.e. $T_i = T_j$ for some $i < j$ occurs.
2. Make a MAC query (N_j, M_i). Let T_{q+1} be the response.
3. Forge with (N_i, M_j, T_{q+1}).

As $\Pi(T_{q+1}) = \Pi(N_i) \oplus N_i \oplus H_{K_h}(M_j)$, (N_i, M_j, T_{q+1}) is a valid forgery. If we make $q = 2^{n/2}$ many queries, with very high probability, we will get a collision in step 1, and mount the attack. Note that, the attack does not exploit any specific properties of the hash function and a single time repetition of nonce makes the construction vulnerable above birthday bound security.

4.4   nPolyMAC: An Instantiation of DWCDM

In this section, we propose nPolyMAC, an algebraic hash function based instantiation of DWCDM, as defined in Eq. (4), as the underlying hash function of DWCDM construction.

PolyHash [25] is one of the popular examples of algebraic hash function. For a hash key K_h and a for a fixed message M, we first apply an injective padding such as 10^* i.e., pad 1 followed by minimum number of zeros so that the total number of bits in the padded message becomes multiple of n. Let the padded message be $M^* = M_1 \| M_2 \| \ldots \| M_l$ where for each i, $|M_i| = n$. Then we define

$$\mathsf{Poly}_{K_h}(M) = M_l \cdot K_h \oplus M_{l-1} \cdot K_h^2 \oplus \ldots \oplus M_1 \cdot K_h^l.$$

It has already been shown in Proposition 1 that Poly is a $\ell/2^n$ regular, AXU and 3-way regular hash function. Following these results, we show in the following that nPolyMAC[Poly, E, E^{-1}] is secure up to $2^{2n/3}$ MAC and 2^n verification queries against nonce respecting adversaries.

Theorem 4. *Let $\mathcal{K}, \mathcal{K}_h$ and $\mathcal{M}$ be three non-empty finite sets. Let $\mathsf{E} : \mathcal{K} \times \{0,1\}^n \rightarrow \{0,1\}^n$ be a block cipher. Then, the MAC security of nPolyMAC in nonce respecting setting is given by*

$$\mathbf{Adv}_{\mathsf{nPolyMAC}}^{\mathrm{MAC}}(q_m, q_v, t) \leq \mathbf{Adv}_{\mathsf{E}}^{\mathrm{SPRP}}(q_m + q_v, t') + \frac{11 q_m \ell}{2^{2n/3}} + \frac{3 q_v \ell}{2^n},$$

where $t' = O(t + (q_m + q_v)\ell)$, ℓ be the maximum number of message blocks among all q queries.

The proof of the theorem directly follows from Proposition 1 and Theorem 2 with the assumption $q_m \leq 2^{2n/3}$.

5   Proof of Theorem 2

In this section, we prove Theorem 2. We would like to note that we will often refer to the construction DWCDM[E, E^{-1}, H] as simply DWCDM where the underlying primitives are assumed to be understood.

The first step of the proof is the standard switch from the computational setting to the information theoretic one by replacing E_K and E_K^{-1} with an n-bit uniform random permutation Π and Π^{-1} at the cost of $\mathbf{Adv}_{\mathsf{E}}^{\mathrm{SPRP}}(q_m + q_v, t')$ and denote the construction as DWCDM$^*[\Pi, \Pi^{-1}, \mathsf{H}]$. Hence,

$$\mathbf{Adv}_{\mathsf{DWCDM}}^{\mathrm{MAC}}(q_m, q_v, t) \leq \mathbf{Adv}_{\mathsf{E}}^{\mathrm{SPRP}}(q_m + q_v, t') + \underbrace{\mathbf{Adv}_{\mathsf{DWCDM}^*}^{\mathrm{MAC}}(q_m, q_v, t)}_{\delta^*}. \quad (7)$$

To upper bound δ^*, we consider that Rand be a perfect random oracle that on input (N, M) returns T, sampled uniformly at random from $\{0,1\}^n$, whereas Rej be an oracle with inputs (N, M, T), returns always $\bot$ (i.e. rejects). Now, due to [13,19] we write

$$\delta^* := \max_{D} \Pr[D^{\mathsf{TG}[\Pi,\Pi^{-1},H_{K_h}],\mathsf{VF}[\Pi,\Pi^{-1},H_{K_h}]} = 1] - \Pr[D^{\mathsf{Rand},\mathsf{Rej}} = 1],$$

where the maximum is taken over all non-trivial distinguishers D. This formulation allows us to apply the H-Coefficient Technique [29], as we explain in more detail below, to prove

$$\delta^* \leq \frac{2q_m}{2^{2n/3}} + q_m\epsilon_1 + \frac{2q_m\epsilon_2}{2^{n/3}} + \max\{q_v\epsilon_1, 2q_v\epsilon_2, 2q_v\epsilon_3, \frac{q_m}{2^{2n/3}}\} + \frac{(q_m + q_v)}{2^n} + \frac{5q_m^3}{2^{2n}}.$$
$$(8)$$

H-COEFFICIENT TECHNIQUE. From now on, we fix a non-trivial distinguisher D that interacts with either (1) the real oracle ($\mathsf{TG}[\Pi, \Pi^{-1}, H_{K_h}], \mathsf{VF}[\Pi, \Pi^{-1}, H_{K_h}]$) for a random permutation Π, its inverse Π^{-1} and a random hashing key K_h or (2) the ideal oracle ($\mathsf{Rand}, \mathsf{Rej}$) making at most q_m queries to its left (MAC) oracle and at most q_v queries to its right (verification) oracle, and outputting a single bit. We let

$$\mathbf{Adv}(D) = \Pr[D^{\mathsf{TG}[\Pi,\Pi^{-1},H_{K_h}],\mathsf{VF}[\Pi,\Pi^{-1},H_{K_h}]} = 1] - \Pr[D^{\mathsf{Rand},\mathsf{Rej}} = 1].$$

We assume that D is computationally unbounded and hence wlog deterministic and that it never repeats a query. Let

$$\tau_m := \{(N_1, M_1, T_1), (N_2, M_2, T_2), \ldots, (N_{q_m}, M_{q_m}, T_{q_m})\}$$

be the list of MAC queries of D and its corresponding responses. Note that, as D is nonce respecting, there cannot be any repetition of triplet in τ_m. Let also

$$\tau_v := \{(N_1', M_1', T_1', b_1'), (N_2', M_2', T_2', b_2'), \ldots, (N_{q_v}', M_{q_v}', T_{q_v}', b_{q_v}')\}$$

be the list of verification queries of D and its corresponding responses, where for all j, $b_j' \in \{\top, \bot\}$ denotes the accept ($b_j' = \top$) or reject ($b_j' = \bot$). The pair (τ_m, τ_v) constitutes the query transcript of the attack. For convenience, we slightly modify the experiment where we reveal to the distinguisher (after it made all its queries and obtains corresponding responses but before it output its decision) the hashing key K_h, if we are in the real world, or a uniformly random dummy key K_h if we are in the ideal world. All in all, the transcript of the attack is $\tau = (\tau_m, \tau_v, K_h)$ where τ_m and τ_v is the tuple of MAC and verification queries respectively. We will often simply name a tuple $(N, M, T) \in \tau_m$ a *MAC query*, and a tuple $(N', M', T', b) \in \tau_v$ a *verification query*.

A transcript τ is said to be an *attainable* (with respect to D) transcript if the probability to realize this transcript in ideal world is non-zero. For an attainable transcript $\tau = (\tau_m, \tau_v, K_h)$, any verification query $(N_i', M_i', T_i', b_i') \in \tau_v$ is such that $b_i' = \bot$. We denote Θ to be the set of all attainable transcripts and X_{re} and X_{id} denotes the probability distribution of transcript τ induced by the real world and ideal world respectively. In the following we state the main lemma of the H-coefficient technique (see e.g. [11] for the proof).

Lemma 1. *Let* D *be a fixed deterministic distinguisher and* $\Theta = \Theta_g \sqcup \Theta_b$ *(disjoint union) be some partition of the set of all attainable transcripts. Suppose there exists* $\epsilon_{ratio} \geq 0$ *such that for any* $\tau \in \Theta_g$,

$$\frac{\Pr[X_{re} = \tau]}{\Pr[X_{id} = \tau]} \geq 1 - \epsilon_{ratio},$$

and there exists $\epsilon_{bad} \geq 0$ *such that* $\Pr[X_{id} \in \Theta_b] \leq \epsilon_{bad}$. *Then,* $\mathbf{Adv}(D) \leq \epsilon_{ratio} + \epsilon_{bad}$.

The remaining of the proof of Theorem 2 is structured as follows: in Sect. 5.1 we define the transcript graph; in Sect. 5.2 we define bad transcripts and upper bound their probability in the ideal world; in Sect. 5.3, we analyze good transcripts and prove that they are almost as likely in the real and the ideal world. Theorem 2 follows easily by combining Lemma 1, Eqs. (7) and (8) above, and Lemmas 3 and 4 proven below.

5.1   Transcript Graph

Given a transcript $\tau = (\tau_m, \tau_v, K_h)$, we define the following two types of graphs: (a) MAC Graph and (b) Verification Graph.

MAC GRAPH. Given a transcript $\tau = (\tau_m, \tau_v, K_h)$, we define the *MAC graph*, denoted as G_τ^m as follows:

$$G_\tau^m = ([q_m], E^m) \text{ where } E^m = \{(i,j) \in [q_m] \times [q_m] : N_i = T_j \vee N_j = T_i \vee T_i = T_j\}.$$

For the sake of convenience, we denote the edge (i,j) as a dotted line when $T_i = T_j$, else we denote it as a continuous line. Thus, the edge set of G_τ^m consists of two different types of edges as depicted in Fig. 2(a) and (b). Note that, for a MAC graph we cannot have edges of type (c).

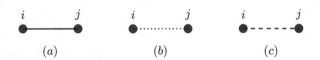

Fig. 2. Different types of edges of MAC and Verification Graphs. $(a) : N_i = T_j/T_i = N_j$, $(b) : T_i = T_j$, $(c) : N_i = N_j$.

Given such a MAC graph, we can partition the set of vertices in the following way: if vertex i and j are connected by an edge then they belong to the same partition. Each partition is called a component of the graph and the number of vertices in the component is called its size, which we denote as ζ.

VERIFICATION GRAPH. Given a MAC graph G_τ^m, we define *Verification graph*, denoted as G_τ^v, by extending G_τ^m with adding one more vertex and at most two edges for incorporating a verification query as follows: For convenience, we reorder the set of MAC queries and verification queries so that all verification queries appears after all MAC queries. Therefore, after such a reordering, j-th verification query becomes $(q_m + j)$-th verification query. Let $(q_m + j)$-th verification query be $(N'_{q_m+j}, M'_{q_m+j}, T'_{q_m+j}, b'_{q_m+j}) \in \tau_v$ and G^m_τ be the MAC graph corresponding to $\tau = (\tau_m, \tau_v, K_h)$. Then we define $G^v_\tau = ([q_m] \cup \{q_m + j\}, E^v)$ where E^v is defined as follows:

$$E^v = E^m \cup \{(q_m + j, r), (q_m + j, s) : r \neq s \in [q_m] \text{ such that either of } (1)-(4) \text{ holds}\}.$$

$$\begin{cases} (1) \ N'_{q_m+j} = N_r \wedge T'_{q_m+j} = N_s \\ (2) \ N'_{q_m+j} = N_r \wedge T'_{q_m+j} = T_s \\ (3) \ N'_{q_m+j} = T_r \wedge T'_{q_m+j} = N_s \\ (4) \ N'_{q_m+j} = T_r \wedge T'_{q_m+j} = T_s \end{cases}$$

Definition 2 (Valid Cycle). *A cycle $C = (i_1, i_2, \ldots, i_p)$ of length p in the MAC graph G^m_τ is said to be valid if the imposed equality pattern of (N, T), generated out of C, derives*

$$0 = \bigoplus_{i \in C} \left(N_i \oplus H_{K_h}(M_i) \right)$$

equation from the given system of equations.

Similar to the definition of valid cycle of MAC graph, one can define the valid cycle for the Verification graph also. Note that, the definition of valid cycle in MAC graph or verification graph actually resembles to the alternating cycle as stated in Sect. 3.1. Now, we make an important observations about the MAC queries (in ideal oracle) as follows:

Lemma 2. *For two MAC queries i, j, we have*

$$(a) \ if \ i < j, \ \Pr[T_j = N_i] = \frac{1}{2^n}; \quad (b) \ if \ i > j, \ \Pr[T_j = N_i] = \frac{1}{2^{n/3}}.$$

Proof. Proof of the first result holds due to the randomness of T_j, i.e. a randomly sampled value T_j is equal to a fixed nonce value N_i holds with probability 2^{-n}. For the later one, condition $i > j$ ensures that one can set the nonce value N_i to a previously sampled tag value T_j. But this would be valid only when the last $n/3$ bits of T_i are all zero, probability of which is $2^{-n/3}$. □

5.2 Definition and Probability of Bad Transcripts

In this section, we define and bound the probability of bad transcript in ideal world. But, before that we first briefly justify the reason about our identified bad events and there after we define the bad transcript accordingly.

Let $\tau = (\tau_m, \tau_v, K_h)$ be an attainable transcript. Then, for all MAC queries (N_i, M_i, T_i) in real oracle, we have

$$i \in \{1, \ldots, q_m\}, \Pi(N_i) \oplus \Pi(T_i) = N_i \oplus \mathsf{H}_{K_h}(M_i).$$

Moreover, for all verification queries (N_a', M_a', T_a', b_a) in real oracle, we have

$$a \in \{1, \ldots, q_v\}, \Pi(N_a') \oplus \Pi(T_a') \neq N_a' \oplus \mathsf{H}_{K_h}(M_a').$$

We refer to the system of equations as "MAC Equations" which involve only the MAC queries. Similarly, we refer to the system of non-equations as "Verification non-equations" which involve only the verification queries.

Therefore, from a given attainable transcript τ, one can write exactly q_m many affine equations and q_v many non-equations. Now, as one needs to lower bound the number of solutions of this system of equations and non-equations (for analyzing the real interpolation probability), it essentially leads us to the model of extended Mirror theory where the equivalence of two set up is established as follows:

$$\begin{cases} \phi'(n_i) = N_i, \ \phi'(t_i) = T_i, \ \lambda_i = N_i \oplus \mathsf{H}_{K_h}(M_i), \ i \in \{1, \ldots, q_m\} \\ \phi'(n_a) = N_a', \ \phi'(t_a) = T_a', \ \lambda_a' = N_a' \oplus \mathsf{H}_{K_h}(M_a'), \ a \in \{1, \ldots, q_v\} \end{cases}$$

Recall that, (ϕ', Λ') where $\Lambda' = (\lambda_1, \ldots, \lambda_{q_m}, \lambda_1', \ldots, \lambda_{q_v}')$, was characterized to be bad if either of the following holds:

(i) $\phi(n_i) = \phi(t_i)$.
(ii)   - $\phi(n_i) = \phi(n_j)$ and $\phi(t_i) = \phi(t_j)$
       - $\phi(n_i) = \phi(t_j)$ and $\phi(t_i) = \phi(n_j)$ for $i \neq j \in [q_m]$.
(iii) there is an alternating cycle.
(iv) for all $j \in [q_v]$ and $i_1, \ldots, i_c \in [q_m]$, $c \geq 0$, such that $\{i_1, \ldots, i_c, q_m + j\}$ is dependent system then $\lambda_{i_1} \oplus \cdots \oplus \lambda_{i_c} \oplus \lambda_j' = \mathbf{0}$.

where $\phi = \phi'_{|q_m}$. Therefore, with the help of equivalence of two set up as established above, we justify our identified bad events:

- $(i) \Rightarrow N_i = T_i$
- $(ii) \Rightarrow$ existence of a valid cycle in the MAC graph G_τ^{m}.
- $(iii) \Rightarrow N_i \oplus \mathsf{H}_{K_h}(M_i) = N_j \oplus \mathsf{H}_{K_h}(M_j), T_i = T_j$ or $N_i = T_j, N_i \oplus \mathsf{H}_{K_h}(M_i) = N_j \oplus \mathsf{H}_{K_h}(M_j)$ such that $i \neq j \in [q_m]$.

Moreover, recall that while considering the non-equation then we considered that any of q_v non-equations can be determined from a subset of q_m many affine equations with their corresponding sum of λ constant becomes zero, which is to say that

- the verification graph G_τ^{v} contains any valid cycle.

Summarizing above, we now define the bad transcript.

Definition 3. *A transcript $\tau = (\tau_m, \tau_v, K_h)$ is said to be* bad *if the associated MAC graph G_τ^m and the Verification graph G_τ^v satisfies the either of the following properties:*

- B0 : $\exists i \in [q_m]$ *such that* $T_i = \mathbf{0}$.
- B1 : G_τ^m *has a component of size 3 or more.*
- B2 : G_τ^m *contains a valid cycle of any arbitrary length that also includes the self loop (that implicitly takes care of the condition $N_i = T_i$).*
- B3 : G_τ^v *contains a valid cycle of any arbitrary length that involves the verification query.*

Moreover, τ is also said to be bad *if*

- B4 : $\exists i \neq j \in [q_m]$ *such that* $N_i \oplus \mathsf{H}_{K_h}(M_i) = N_j \oplus \mathsf{H}_{K_h}(M_j), T_i = T_j$.
- B5: $\exists i \neq j \in [q_m]$ *such that* $N_i = T_j, N_i \oplus \mathsf{H}_{K_h}(M_i) = N_j \oplus \mathsf{H}_{K_h}(M_j)$.
- B6 : $\exists i \in [q_m]$ *such that* $\mathsf{H}_{K_h}(M_i) = N_i$.

Condition B1 actually imposes a restriction on the block maximality as we do not allow to have a larger component size for a good transcript. Condition B6 ensures that for a good transcript, all the elements of the tuple $\left(N_1 \oplus \mathsf{H}_{K_h}(M_1), \ldots, N_{q_m} \oplus \mathsf{H}_{K_h}(M_{q_m})\right)$ are non-zero. Note that, if we do not consider the condition B6, then for a good attainable transcript the real interpolation probability would become zero.

We denote $\Theta_b \subseteq \Theta$ be the set of all attainable bad transcripts and the event B denotes B := B0 ∨ B1 ∨ B2 ∨ B3 ∨ B4 ∨ B5 ∨ B6. We bound the probability of event B in the following lemma, proof of which is deffered to Sect. 5.4.

Lemma 3. *Let X_{id} and Θ_b be defined as above. If $q_m \leq 2^{2n/3}$ and $q_v \leq 2^n$, then*

$$\Pr[X_{\mathrm{id}} \in \Theta_b] \leq \epsilon_{\mathsf{bad}} = \frac{2q_m}{2^{2n/3}} + \frac{q_m}{2^n} + q_m\epsilon_1 + \frac{2q_m\epsilon_2}{2^{n/3}} + \max\left\{q_v\epsilon_1, 2q_v\epsilon_2, 2q_v\epsilon_3, \frac{q_m}{2^{2n/3}}\right\}.$$

5.3 Analysis of Good Transcripts

In this section, we show that for a good transcript τ, realizing τ is almost as likely in the real world as in the ideal world. Formally, we prove the following lemma.

Lemma 4. *Let $\tau = (\tau_m, \tau_v, K_h)$ be a good transcript. Then*

$$\frac{\mathsf{p_{re}}(\tau)}{\mathsf{p_{id}}(\tau)} := \frac{\Pr[X_{\mathrm{re}} = \tau]}{\Pr[X_{\mathrm{id}} = \tau]} \geq (1 - \epsilon_{\mathsf{ratio}}) = \left(1 - \frac{5q_m^3}{2^{2n}} - \frac{q_v}{2^n}\right).$$

Proof. Consider the good transcript $\tau = (\tau_m, \tau_v, K_h)$. Since in the ideal world the MAC oracle is perfectly random and the verification always rejects, one simply has

$$\mathsf{p_{id}} := \Pr[X_{\mathrm{id}} = \tau] = \frac{1}{|\mathcal{K}_h|} \cdot \frac{1}{2^{nq_m}}. \tag{9}$$

We must now lower bound the probability of getting τ in real world. We say that a permutation Π is compatible with τ_m if $\forall i \in [q_m]$, (i) happens and Π is compatible with τ_v if $\forall a \in [q_v]$, (ii) happens

$$(i)\ \Pi(N_i) \oplus \Pi(T_i) = \underbrace{N_i \oplus H_{K_h}(M_i)}_{\lambda_i}, \quad (ii)\ \Pi(N_a') \oplus \Pi(T_a') \neq \underbrace{N_a' \oplus H_{K_h}(M_a')}_{\lambda_a'}.$$

We simply say that Π is compatible with τ if it is compatible with τ_m and τ_v. We denote $\mathsf{Comp}(\tau)$ the set of permutations that are compatible with τ. Therefore,

$$\mathsf{p_{re}}(\tau) = \frac{1}{|\mathcal{K}_h|} \cdot \Pr[\Pi \leftarrow_\$ \mathsf{Perm} : \Pi \in \mathsf{Comp}(\tau)]$$

$$= \frac{1}{|\mathcal{K}_h|} \cdot \underbrace{\Pr[\Pi(N_i) \oplus \Pi(T_i) = \lambda_i, \forall i \in [q_m], \Pi(N_a') \oplus \Pi(T_a') \neq \lambda_a', \forall a \in [q_v]]}_{\mathsf{P}_{mv}}.$$

LOWER BOUNDING P_{mv}: Observe that lower bounding P_{mv} implies lower bounding the probability of the number of solutions to the following system of q_m many equations of the form $\Pi(N_i) \oplus \Pi(T_i) = \lambda_i$ and q_v many non-equations of the form $\Pi(N_a') \oplus \Pi(T_a') \neq \lambda_a'$.

Let us assume the distinct number of random variables in the above set of equations is α. As the transcript τ is good, we have the following properties:

- (i) all λ_i values are non-zero (otherwise condition B6 is satisfied).
- (ii) (ϕ', Λ') is good.
- (iii) Finally, block maximality $\xi_{\max}$ is 3.

Above properties enable us directly to apply Theorem 1 to lower bound P_{mv} as follows:

$$\mathsf{P}_{mv} \geq \frac{1}{2^{nq_m}} \left(1 - \frac{5q_m^3}{2^{2n}} - \frac{q_v}{2^n} \right). \tag{10}$$

Therefore, from Eq. (10), we have

$$\mathsf{p_{re}}(\tau) \geq \frac{1}{|\mathcal{K}_h|} \cdot \frac{1}{2^{nq_m}} \cdot \left(1 - \frac{5q_m^3}{2^{2n}} - \frac{q_v}{2^n} \right). \tag{11}$$

Finally, taking the ratio of Eqs. (11) to (9), the result follows.   □

5.4   Proof of Lemma 3

In this section, we prove Lemma 3. A more detailed version of this proof can be found in the full version of this paper [18]. In order to bound $\Pr[X_{\mathrm{id}} \in \Theta_b]$, it is enough to bound $\Pr[\mathsf{B}]$. Therefore, we write

$$\Pr[\mathsf{B}] \leq \sum_{v \in \{0,1,4,5,6\}} \Pr[\mathsf{Bv}] + \Pr[\mathsf{B2} \mid \overline{\mathsf{B1}}] + \Pr[\mathsf{B3} \mid \overline{\mathsf{B0}} \wedge \overline{\mathsf{B1}} \wedge \overline{\mathsf{B2}}]. \tag{12}$$

In the following, we bound the probabilities of all the bad events individually.

Bounding B0. As the responses are sampled uniformly and independently to all other sampled random variables, $\Pr[\mathsf{B0}] \leq \frac{q_m}{2^n}$.

Bounding B1. Event B1 occurs if there exists a component of size at least 3 in G_τ^{m}, i.e. there exist a chain of two edges. Depending on whether the edges are dotted (Dot) or continuous (Con), there are three possible choices of components: (Dot-Dot), (Dot-Con) and (Con-Con), as depicted in Fig. 3.

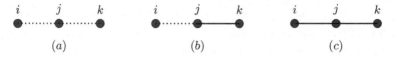

Fig. 3. Different components of size of three. (a) $T_i = T_j = T_k$, (b) $T_i = T_j = N_k$ or $T_i = T_j, N_j = T_k$ and (c) $N_i = T_j, N_j = T_k$ or $T_i = N_j, T_j = N_k$.

Using Lemma 2 and the fact that each T_i is sampled uniformly at random from $\{0,1\}^n$, one can show that having any such component has a probability of $\frac{q_m}{2^{2n/3}}$ and therefore, we have $\Pr[\mathsf{B1}] \leq \frac{q_m}{2^{2n/3}}$.

Bounding B2 $|\overline{\mathsf{B1}}$. Here we bound the existence of a cycle of length one (self loop) and two (parallel edges), as depicted in Fig. 4(a) and (b). Again using Lemma 2 and the fact that each T_i is sampled uniformly at random from $\{0,1\}^n$, one can show that the probability of having a self loop or parallel edges can be bounded by $\frac{q_m}{2^{2n/3}}$ and therefore $\Pr[\mathsf{B2} \mid \overline{\mathsf{B1}}] \leq \frac{q_m}{2^{2n/3}}$.

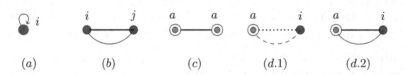

Fig. 4. (a) Self Loop in G_τ^{m}: when $N_i = T_i$, (b) Parallel Edges in G_τ^{m}: $N_i = T_j, N_j = T_i$, (c) Self Loop in G_τ^{v}: when $N_a' = T_a$, (d) Parallel Edges in G_τ^{v}: (d.1) $N_a' = N_i, T_a' = T_i$, (d.2) $N_a' = T_i, T_a' = N_i$. Node with concentric circle denotes the verification query node.

Bounding B3 $\mid \overline{\mathsf{B0}} \wedge \overline{\mathsf{B1}} \wedge \overline{\mathsf{B2}}$. Recall that event B3 holds if there exists any cycle in G_τ^{v} and the sum of the corresponding $N \oplus \mathsf{H}_{K_h}(M)$ is zero. But, as we conditioned on $\overline{\mathsf{B0}} \wedge \overline{\mathsf{B1}} \wedge \overline{\mathsf{B2}}$, it is enough to bound the existence of a cycle of length one (self loop), two (parallel edges) and three (closed triangle).

SELF LOOP. As the hash function is ϵ_1 regular, the probability of having a self loop can be bounded by $q_v \epsilon_1$.

PARALLEL EDGES. A parallel edge or cycle of length 2 in G_τ^{v} implies that the edges would be (i) one dotted and one dashed (Dot-Dash) or (ii) both continuous

(Con-Con), as depicted in Fig. 4(d.1) and (d.2). Using Lemma 2 and the fact that the hash function is ϵ_2 AXU, one can show that the probability of having parallel edges can be bounded by $2q_v\epsilon_2$.

CLOSED TRIANGLE. A closed triangle or cycle of length 3 in G_τ^v essentially implies that the triangle must have been form having edges of the form (Con-Dash-Dot), (Con-Con-Con) and (Dot-Dash-Con), as depicted in Fig. 5. Again using Lemma 2 and the fact that the hash function is ϵ_3 3-way-regular, one can show that the probability of having edges of the above form in G_τ^v is $\max\{2q_v\epsilon_3, \frac{q_m}{2^{2n/3}}\}$. Therefore, combining everything together, $\Pr[\mathrm{B3} \mid \overline{\mathrm{B0}} \wedge \overline{\mathrm{B1}} \wedge \overline{\mathrm{B2}}] \leq \max\{2q_v\epsilon_3, 2q_v\epsilon_2, q_v\epsilon_1, \frac{q_m}{2^{2n/3}}\}$.

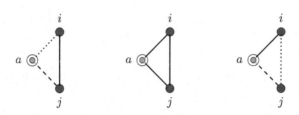

Fig. 5. Cycles of length 3 including the verification query which is denoted by the concentric circle node.

Bounding B4: Since, in the ideal oracle the hash key is sampled independent to all previously sampled MAC responses T_i, we have $\Pr[\mathrm{B4}] \leq \frac{q_m^2 \cdot \epsilon_2}{2^n}$.

Bounding B5: It is easy to see that for fixed i and j, $N_i \oplus \mathsf{H}_{K_h}(M_i) = N_j \oplus \mathsf{H}_{K_h}(M_j)$ holds with probability ϵ_2. Now summing over all possible choices of i and j, using Lemma 2 and assuming $q_m \leq 2^{2n/3}$, we obtain $\Pr[\mathrm{B5}] \leq \frac{q_m\epsilon_2}{2^{n/3}}$.

Bounding B6: For any fixed i the event $N_i = \mathsf{H}_{K_h}(M_i)$ occurs with probability ϵ_1, due to the regular property of the hash function. Summing over all choices of i, we have $\Pr[\mathrm{B6}] \leq q_m\epsilon_1$.

Finally, by assuming $q_m \leq 2^{2n/3}$, Lemma 3 follows from all the above bounds.

6   1K-DWCDM: A Single Keyed DWCDM

Recall that, our proposed construction DWCDM is instantiated with a hash function and a block cipher where the hash key is independent to block cipher keys, leading to have a two-keyed (counting hash key separately from block cipher keys) nonce based MAC. In this section, we transform the DWCDM construction to a purely single keyed construction by setting the underlying hash key K_h to the encryption of $\mathbf{1}$ (i.e. $K_h := \mathsf{E}_K(\mathbf{1})$) and argue that the modified construction (that we call as 1K-DWCDM) is secure.

Now, we state and prove that 1K-DWCDM is secure up to $2^{2n/3}$ MAC queries and 2^n verification queries against all nonce respecting adversaries. We mainly focus on the nonce respecting security of the construction, as its nonce misuse security is very similar to that of DWCDM and hence we skip it.

Theorem 5. *Let $\mathcal{M}$ and $\mathcal{K}$ be finite and non-empty sets. Let $\mathsf{E} : \mathcal{K} \times \{0,1\}^n \to \{0,1\}^n$ be a block cipher and $\mathsf{H} : \mathsf{E}_K(1) \times \mathcal{M} \to \{0,1\}^n$ be an ϵ_1 regular, ϵ_2 AXU and ϵ_3 3-way regular hash function. Then, the MAC advantage of* 1K-DWCDM *is given by:*

$$\mathbf{Adv}^{\mathrm{MAC}}_{\mathrm{1K\text{-}DWCDM}[\mathsf{E},\mathsf{E}^{-1},\mathsf{H}]}(q_m, q_v, t) \leq \mathbf{Adv}^{\mathrm{SPRP}}_{\mathsf{E}}(q_m + q_v, t') + \frac{3q_m}{2^{2n/3}} + \frac{q_m^2 \epsilon_2}{2^n} + \frac{q_v}{2^n - 1}$$

$$+ \max\{q_v \epsilon_1, 2q_v \epsilon_2, 2q_v \epsilon_3, \frac{q_m}{2^{2n/3}}\} + q_v \epsilon_1 + \frac{q_m}{2^n} + \frac{5q_m^3}{2^{2n}},$$

where $t' = O(t + (q_m + q_v)t_H)$, t_H being the time for computing hash function. Assuming ϵ_1, ϵ_2 and $\epsilon_3 \approx 2^{-n}$ and $q_m \leq 2^{2n/3}$, 1K-DWCDM$[\mathsf{E}, \mathsf{E}^{-1}, \mathsf{H}]$ construction is secured up to roughly $2^{2n/3}$ MAC and 2^n verification queries.

Proof. The proof approach is similar to the one used in Theorem 2. Using standard argument, we can replace E_K and E_K^{-1} with an n-bit uniform random permutation Π and its inverse Π^{-1}, denote the construction as 1K-DWCDM$^*[\Pi, \mathsf{E}^{-1}, \mathsf{H}]$ and bound $\mathbf{Adv}^{\mathrm{MAC}}_{\mathrm{1K\text{-}DWCDM}^*[\Pi,\Pi^{-1},\mathsf{H}]}(A)$:

For this, we first define the ideal oracle which works as follows: for each MAC query (N, M), it samples the response T from $\{0,1\}^n$ uniformly at random and returns it to the distinguisher and for each verification query it returns $\perp$. As before, we reveal the hashing key K_h to the distinguisher after it made all it's queries and before the final decision. Note that, the hash key is $\mathsf{E}_K(1)$ in the real world and a uniformly random dummy key K_h, sampled uniformly at random from $\{0,1\}^n$ in the ideal world. Let the transcript of the attack is $\tau = (\tau_m, \tau_v, K_h)$ where τ_m and τ_v is the tuple of MAC and verification queries respectively.

Bad Transcript. The definition of bad transcript is similar to that of defined in Sect. 5.2 and therefore, we have the following result:
Let X_{id} and Θ_b be defined as above. If $q_m \leq 2^{2n/3}$ and $q_v \leq 2^n$, then

$$\Pr[X_{\mathrm{id}} \in \Theta_b] \leq \frac{3q_m}{2^{2n/3}} + \frac{q_m^2 \epsilon_2}{2^n} + \max\left\{q_v \epsilon_1, 2q_v \epsilon_2, 2q_v \epsilon_3, \frac{q_m}{2^{2n/3}}\right\} + q_v \epsilon_1 + \frac{q_m}{2^n}. \quad (13)$$

Analysis of Good Transcripts. Similar to Lemma 4, we prove that for any good transcript τ, realizing τ is almost as likely as real and in the ideal world. As the transcript τ is good, each sampled T_i value is non-zero. Since, in the ideal world the MAC oracle is perfectly random and the verification always rejects, one simply has

$$\mathsf{p}_{\mathrm{id}} := \Pr[X_{\mathrm{id}} = \tau] = \frac{1}{2^n} \cdot \frac{1}{(2^n - 1)^{q_m}}. \quad (14)$$

Now, for the real interpolation probability, we have

$$\Pr[\Pi(N_i) \oplus \Pi(T_i) = \lambda_i, \forall i \in [q_m] \text{ and } \Pi(N_a') \oplus \Pi(T_a') \neq \lambda_a', \forall a \in [q_v]].$$

Additionally, if the adversary makes any verification query (N_a', M_a', T_a') with tag T_a' set to $\mathbf{1}$, then we need to ensure that

$$\Pi(N_a') \neq \underbrace{\Pi(\mathbf{1}) \oplus N_a' \oplus \mathsf{H}_{\Pi(\mathbf{1})}(M_a')}_{\lambda_a''}, \forall a \in [q_v]]. \tag{15}$$

Since, the hash key, i.e., $\Pi(\mathbf{1})$, is revealed to the adversary after the interaction is over, the right hand side of the non-Eq. (15) becomes a constant, which makes it a uni-variate affine non-equation and then it is satisfied by condition (c) of Theorem 1. Therefore, we have

$$\mathsf{p}_{\mathrm{re}}(\tau) = \frac{1}{2^n} \cdot \Pr[\Pi(N_i) \oplus \Pi(T_i) = \lambda_i, \forall i \in [q_m], \Pi(N_a') \oplus \Pi(T_a') \neq \lambda_a',$$

$$\Pi(N_a') \neq \lambda_a'', \forall a \in [q_v]]$$

$$\geq \frac{1}{2^n} \cdot \frac{1}{(2^n - 1)^{q_m}} \cdot \left(1 - \frac{5q_m^3}{2^{2n}} - \frac{q_v}{2^n - 1}\right). \tag{16}$$

The last inequality follows using similar to the proof of Lemma 4 and Eq. (10). Finally, from Eqs. (14) and (16), we compute the ratio as follows:

$$\frac{\mathsf{p}_{\mathrm{re}}(\tau)}{\mathsf{p}_{\mathrm{id}}(\tau)} \geq \left(1 - \frac{5q_m^3}{2^{2n}} - \frac{q_v}{2^n - 1}\right). \tag{17}$$

Finally, Theorem 5 follows from Eqs. (13) and (17).   □

7 Towards Higher Security of DWCDM

In this section, we briefly describe how to boost the security of DWCDM upto $(k-1)/k$-bit for a general k. The underlying construction remains as it is, however the nonce space is increased to $(k-1)n/k$-bits i.e., $\mathsf{DWCDM\_k}[\mathsf{E}, \mathsf{H}](N, M) := \mathsf{E}_K^{-1}(\mathsf{E}_K(N) \oplus N \oplus \mathsf{H}_{K_h}(M))$ but here we consider $N = N^* \| 0^{n/k}$ where N^* is a $(k-1)n/k$ bit nonce. For this, we first state the following conjecture on Mirror theory, which is a generalized version of extended Mirror theorem as introduced in Sect. 3.2.

Conjecture 1 (Extended Mirror Theorem for $\xi_{\max} = k$). Let $(\mathcal{E}^= \cup \mathcal{E}^{\neq}, \phi', \Lambda')$ be a system of q many affine equations and v many affine non-equations associated with index mapping function ϕ' over $\mathrm{GF}(2^n)$ which are of the form $P_{\phi(n_i)} \oplus P_{\phi(t_i)} = \lambda_i$ for $i \in [q]$ and $P_{\phi(n_j)} \oplus P_{\phi(t_j)} \neq \lambda_j' (\neq 0)$ for $j \in [q+1, q+v]$ over the set of α many unknown variables $\mathcal{P} = \{P_1, \ldots, P_\alpha\}$ such that P_a may be equals to some $P_{\phi(n_i)}$ or $P_{\phi(b_j)}$, where $a \in \{\phi(n_j), \phi(t_j)\}, j \in [q, q+v]$. Now, if

- (i) (ϕ', Λ') is good and
- (ii) $\xi_{\max} = k$

then the number of solutions for $\mathcal{P}$, denoted by h_β (where $\beta = \frac{kq}{k-1}$) such that $P_i \neq P_j$ for all distinct $i, j \in \{1, \ldots, \alpha\}$ is

$$h_\beta \geq \frac{(2^n)_\beta}{2^{nq}} \left(1 - O\left(\frac{q^k}{2^{(k-1)n}} + \frac{v}{2^n}\right)\right). \tag{18}$$

Assuming this conjecture holds, we have the following result on the MAC advantage of DWCDM_k:

Theorem 6. *Let* E *be a block cipher and* H *be an* ϵ_1 *regular,* ϵ_2 *AXU and* ϵ_j *j-way regular hash function,*[8] *for all* $3 \leq j \leq k$ *(e.g., PolyHash). Then, the MAC advantage for any* (q_m, q_v, t) *nonce-respecting adversary against* DWCDM_k *is given by,*

$$\mathbf{Adv}^{MAC}_{DWCDM\_k}(q_m, q_v, t) \leq \mathbf{Adv}^{SPRP}_E(q_m + q_v, t') + O(q_m^k/2^{n(k-1)} + q_v.\epsilon),$$

where $q_v = \max\{\epsilon_1, \epsilon_2, \epsilon_j\}$ *and* $t' = O(t + (q_m + q_v)t_H)$.

The proof will be similar to the proof of Theorem 2. We first define the transcript, associated MAC and the verification graph as before.

Now, we call a transcript $\tau = (\tau_m, \tau_v, K_h)$ to be bad if the associated MAC graph G^m_τ and the Verification graph G^v_τ satisfies the either of the following properties:

- B1' : G^m_τ has a component of size k or more.
- B2' : G^m_τ contains a valid cycle of length less than k.
- B3' : G^v_τ contains a valid cycle of length less than or equals to k that involves the verification query.

Moreover, τ is also said to be bad if it satisfies B0, B4, B5, B6 (as defined in Definition 3).

Here we will mainly consider bounding B1', B2' and B3', as the remaining ones are already done. Here we provide a sketch for bounding each of this event:

Bounding B1'. Event B1' occurs if there exists a component of size at least k in G^m_τ. This essentially implies there is a chain of $(k-1)$ edges. Let there are c_1 number of edges are of the form $T_i = N_j$ with $i < j$. Here we claim that

$$\Pr[\text{B1}'] \leq q_m \cdot \left(\frac{q_m}{2^n}\right)^{k-c_1} \cdot \left(\frac{1}{2^{k/n}}\right)^{c_1}.$$

As $k \geq 4$, the above bound is $O(q_m^k/2^{n(k-1)})$.

Bounding B2'. Event B2' occurs if there exists a cycle of size less than k in G^m_τ. Let us bound a cycle of length $c < (k-1)$. Again, assume there are c_1 number of edges of the form $T_i = N_j$ with $i < j$. Using similar argument as above,

$$\Pr[\text{B2}'] \leq \left(\frac{q_m}{2^n}\right)^{c-c_1} \cdot \left(\frac{1}{2^{k/n}}\right)^{c_1}.$$

It is easy to see that for any c, the above bound is $O(q_m/2^n)$.

---

[8] A Hash function H is said to be a ϵ j-way regular hash function if for all distinct $(X_1, \ldots, X_j)$ and for any non-zero Y, $\Pr[H(X_1) \oplus \ldots \oplus H(X_j) = Y] \leq \epsilon$.

Bounding B3'. Event B3' occurs if there exists a cycle of size less than or equals to k in G_τ^v. Extending similar arguments used in Lemma 3 to bound the event B3, one can show that if H is ϵ j-way regular for all $j \leq k$ then

$$\Pr[\text{B3'}] \approx O\left(\frac{q_v.\epsilon.q_m^c}{2^{nc}}\right), c \geq 0.$$

Combining everything together, we have

$$\Pr[\text{B}] \approx O(q_m^k/2^{n(k-1)} + q_v.\epsilon).$$

Next, we fix a good transcript τ. Now, to obtain the lower bound of the probability of getting τ in real world, we need a lower bound on the probability of the number of solutions to a system of q_m many equations and q_v many non-equations. Again, we can do that using an extended Mirror theory result with maximal block size $\xi_{max} = k$. From Conjecture 1, we have

$$\mathsf{p_{re}}(\tau) \geq \frac{1}{|\mathcal{K}_h|} \cdot \frac{1}{2^{nq_m}} \cdot \left(1 - O\left(\frac{q_m^k}{2^{(k-1)n}} + \frac{q_v}{2^n}\right)\right). \tag{19}$$

The theorem follows by applying Patarin's H-Coefficient Technique.   □

Remark 2. We would like to clarify that increasing the nonce space does not have any relation with the increase in security. We have restricted the nonce space of DWCDM to $2n/3$-bit (note that this is minimum as we must allow $2^{2n/3}$ many MAC queries with distinct nonces) purely because of the simplicity of the extended mirror theory analysis. One can of course increase the nonce space to $(k-1)n/k$-bit for any $k \leq n$, but that increases the block maximality (ξ_{max}) to k and hence the analysis of the extended mirror theory would become tedious and involved.

8   Conclusion

In this paper we have proposed DWCDM, a single keyed nonce based MAC, which is structurally identical to EWCDM except that the outer encryption call is replaced by the decryption call and same key is used for both the block cipher calls. Using an extended mirror theory results, we have shown that DWCDM is secure roughly up to $2n/3$-bit against nonce-respecting adversaries and $n/2$-bit against nonce-misuse adversaries. We have also provided an intuition on how to boost the nonce-respecting security of DWCDM upto $(k-1)/k$-bit for a general k.

Acknowledgments. Initial part of this work was done in NTT Lab, Japan when Avijit Dutta was visiting there. Mridul Nandi is supported by R.C.Bose Centre for Cryptology and Security. The authors would like to thank all the anonymous reviewers of CRYPTO 2018 for their invaluable comments and suggestions and also to Eik List and Yaobin Shen for pointing out some minor issues in the paper.

References

1. Beaulieu, R., Shors, D., Smith, J., Treatman-Clark, S., Weeks, B., Wingers, L.: The SIMON and SPECK families of lightweight block ciphers. Cryptology ePrint Archive, Report 2013/404 (2013). http://eprint.iacr.org/2013/404

2. Beierle, C., et al.: The SKINNY family of block ciphers and its low-latency variant MANTIS. In: Robshaw, M., Katz, J. (eds.) CRYPTO 2016, Part II. LNCS, vol. 9815, pp. 123–153. Springer, Heidelberg (2016). https://doi.org/10.1007/978-3-662-53008-5_5

3. Bellare, M., Impagliazzo, R.: A tool for obtaining tighter security analyses of pseudorandom function based constructions, with applications to PRP to PRF conversion. Cryptology ePrint Archive, Report 1999/024 (1999). http://eprint.iacr.org/1999/024

4. Bellare, M., Canetti, R., Krawczyk, H.: Keying hash functions for message authentication. In: Koblitz, N. (ed.) CRYPTO 1996. LNCS, vol. 1109, pp. 1–15. Springer, Heidelberg (1996). https://doi.org/10.1007/3-540-68697-5_1

5. Bellare, M., Kilian, J., Rogaway, P.: The security of cipher block chaining. In: Desmedt, Y.G. (ed.) CRYPTO 1994. LNCS, vol. 839, pp. 341–358. Springer, Heidelberg (1994). https://doi.org/10.1007/3-540-48658-5_32

6. Bellare, M., Krovetz, T., Rogaway, P.: Luby-Rackoff backwards: increasing security by making block ciphers non-invertible. In: Nyberg, K. (ed.) EUROCRYPT 1998. LNCS, vol. 1403, pp. 266–280. Springer, Heidelberg (1998). https://doi.org/10.1007/BFb0054132

7. Bhattacharya, S., Nandi, M.: Full indifferentiable security of the Xor of two or more random permutations using the χ^2 method. In: Nielsen, J.B., Rijmen, V. (eds.) EUROCRYPT 2018, Part I. LNCS, vol. 10820, pp. 387–412. Springer, Cham (2018). https://doi.org/10.1007/978-3-319-78381-9_15

8. Bhattacharya, S., Nandi, M.: Revisiting variable output length XOR pseudorandom function. IACR Trans. Symmetric Cryptol. 2018(1), 314–335 (2018)

9. Bogdanov, A., et al.: PRESENT: an ultra-lightweight block cipher. In: Paillier, P., Verbauwhede, I. (eds.) CHES 2007. LNCS, vol. 4727, pp. 450–466. Springer, Heidelberg (2007). https://doi.org/10.1007/978-3-540-74735-2_31

10. De Cannière, C., Dunkelman, O., Knežević, M.: KATAN and KTANTAN — a family of small and efficient hardware-oriented block ciphers. In: Clavier, C., Gaj, K. (eds.) CHES 2009. LNCS, vol. 5747, pp. 272–288. Springer, Heidelberg (2009). https://doi.org/10.1007/978-3-642-04138-9_20

11. Chen, S., Lampe, R., Lee, J., Seurin, Y., Steinberger, J.: Minimizing the two-round Even-Mansour cipher. In: Garay, J.A., Gennaro, R. (eds.) CRYPTO 2014, Part I. LNCS, vol. 8616, pp. 39–56. Springer, Heidelberg (2014). https://doi.org/10.1007/978-3-662-44371-2_3

12. Cogliati, B., Lampe, R., Patarin, J.: The indistinguishability of the XOR of k permutations. In: Cid, C., Rechberger, C. (eds.) FSE 2014. LNCS, vol. 8540, pp. 285–302. Springer, Heidelberg (2015). https://doi.org/10.1007/978-3-662-46706-0_15

13. Cogliati, B., Seurin, Y.: EWCDM: an efficient, beyond-birthday secure, nonce-misuse resistant MAC. In: Robshaw, M., Katz, J. (eds.) CRYPTO 2016, Part I. LNCS, vol. 9814, pp. 121–149. Springer, Heidelberg (2016). https://doi.org/10.1007/978-3-662-53018-4_5

14. Cogliati, B., Seurin, Y.: Analysis of the single-permutation encrypted Davies-Meyer construction. Des. Codes Cryptogr. (2018, to appear)

15. Daemen, J., Rijmen, V.: Rijndael for AES. In: AES Candidate Conference, pp. 343–348 (2000)

16. Dai, W., Hoang, V.T., Tessaro, S.: Information-theoretic indistinguishability via the chi-squared method. In: Katz, J., Shacham, H. (eds.) CRYPTO 2017, Part III. LNCS, vol. 10403, pp. 497–523. Springer, Cham (2017). https://doi.org/10.1007/978-3-319-63697-9_17

17. Datta, N., Dutta, A., Nandi, M., Paul, G., Zhang, L.: Single key variant of PMAC_plus. IACR Trans. Symmetric Cryptol. **2017**(4), 268–305 (2017)

18. Datta, N., Dutta, A., Nandi, M., Yasuda, K.: Encrypt or decrypt? To make a single-key beyond birthday secure nonce-based MAC. Cryptology ePrint Archive, Report 2018/500 (2018)

19. Dutta, A., Jha, A., Nandi, M.: Tight security analysis of EHtM MAC. IACR Trans. Symmetric Cryptol. **2017**(3), 130–150 (2017)

20. Guo, J., Peyrin, T., Poschmann, A., Robshaw, M.J.B.: The LED block cipher. IACR Cryptology ePrint Archive, 2012:600 (2012)

21. Iwata, T.: New blockcipher modes of operation with beyond the birthday bound security. In: Robshaw, M. (ed.) FSE 2006. LNCS, vol. 4047, pp. 310–327. Springer, Heidelberg (2006). https://doi.org/10.1007/11799313_20

22. Iwata, T., Mennink, B., Vizár, D.: CENC is optimally secure. IACR Cryptology ePrint Archive, 2016:1087 (2016)

23. Lucks, S.: The sum of PRPs is a secure PRF. In: Preneel, B. (ed.) EUROCRYPT 2000. LNCS, vol. 1807, pp. 470–484. Springer, Heidelberg (2000). https://doi.org/10.1007/3-540-45539-6_34

24. Mennink, B., Neves, S.: Encrypted Davies-Meyer and its dual: towards optimal security using mirror theory. In: Katz, J., Shacham, H. (eds.) CRYPTO 2017, Part III. LNCS, vol. 10403, pp. 556–583. Springer, Cham (2017). https://doi.org/10.1007/978-3-319-63697-9_19

25. Minematsu, K., Iwata, T.: Building blockcipher from tweakable blockcipher: extending FSE 2009 proposal. In: Chen, L. (ed.) IMACC 2011. LNCS, vol. 7089, pp. 391–412. Springer, Heidelberg (2011). https://doi.org/10.1007/978-3-642-25516-8_24

26. Naito, Y.: Blockcipher-based MACs: beyond the birthday bound without message length. In: Takagi, T., Peyrin, T. (eds.) ASIACRYPT 2017, Part III. LNCS, vol. 10626, pp. 446–470. Springer, Cham (2017). https://doi.org/10.1007/978-3-319-70700-6_16

27. NIST: Recommendation for block cipher modes of operation: The CMAC mode for authentication. SP 800–38B (2005)

28. Patarin, J.: A proof of security in $O(2^n)$ for the Xor of two random permutations. In: Safavi-Naini, R. (ed.) ICITS 2008. LNCS, vol. 5155, pp. 232–248. Springer, Heidelberg (2008). https://doi.org/10.1007/978-3-540-85093-9_22

29. Patarin, J.: The "Coefficients H" technique. In: Avanzi, R.M., Keliher, L., Sica, F. (eds.) SAC 2008. LNCS, vol. 5381, pp. 328–345. Springer, Heidelberg (2009). https://doi.org/10.1007/978-3-642-04159-4_21

30. Patarin, J.: Introduction to mirror theory: analysis of systems of linear equalities and linear non equalities for cryptography. IACR Cryptology ePrint Archive, 2010:287 (2010)

31. Patarin, J.: Security in $o(2^n)$ for the Xor of two random permutations - proof with the standard H technique. IACR Cryptology ePrint Archive, 2013:368 (2013)
32. Patarin, J.: Mirror theory and cryptography. Appl. Algebra Eng. Commun. Comput. **28**(4), 321–338 (2017)
33. Yasuda, K.: A new variant of PMAC: beyond the birthday bound. In: Rogaway, P. (ed.) CRYPTO 2011. LNCS, vol. 6841, pp. 596–609. Springer, Heidelberg (2011). https://doi.org/10.1007/978-3-642-22792-9_34

Rasta: A Cipher with Low ANDdepth and Few ANDs per Bit

Christoph Dobraunig[1]([⊠]), Maria Eichlseder[1], Lorenzo Grassi[1],
Virginie Lallemand[2], Gregor Leander[2], Eik List[3], Florian Mendel[4],
and Christian Rechberger[1]

[1] Graz University of Technology, Graz, Austria
{christoph.dobraunig,maria.eichlseder,lorenzo.grassi,
christian.rechberger}@iaik.tugraz.at
[2] Horst Görtz Institute for IT Security,
Ruhr-Universität Bochum, Bochum, Germany
{virginie.lallemand,gregor.leander}@rub.de
[3] Bauhaus-Universität Weimar, Weimar, Germany
eik.list@uni-weimar.de
[4] Infineon Technologies AG, Neubiberg, Germany
florian.mendel@infineon.com

Abstract. Recent developments in multi party computation (MPC) and fully homomorphic encryption (FHE) promoted the design and analysis of symmetric cryptographic schemes that minimize multiplications in one way or another. In this paper, we propose with Rasta a design strategy for symmetric encryption that has ANDdepth d and at the same time only needs d ANDs per encrypted bit. Even for very low values of d between 2 and 6 we can give strong evidence that attacks may not exist. This contributes to a better understanding of the limits of what concrete symmetric-key constructions can theoretically achieve with respect to AND-related metrics, and is to the best of our knowledge the first attempt that minimizes both metrics simultaneously. Furthermore, we can give evidence that for choices of d between 4 and 6 the resulting implementation properties may well be competitive by testing our construction in the use-case of removing the large ciphertext-expansion when using the BGV scheme.

Keywords: Symmetric encryption · ASASA
Homomorphic encryption · Multiplicative complexity
Multiplicative depth

1 Introduction

In this paper we study symmetric encryption primitives with few AND gates. This firstly feeds on the curiosity about how few AND gates a cryptographic primitive can have for which we do not know attacks or are otherwise able to

© International Association for Cryptologic Research 2018
H. Shacham and A. Boldyreva (Eds.): CRYPTO 2018, LNCS 10991, pp. 662–692, 2018.
https://doi.org/10.1007/978-3-319-96884-1_22

argue about its security. But secondly, this is motivated by various new developments in applied and theoretical cryptography where AND-related properties are of great interest: Encryption schemes with few AND gates were shown to positively affect the cost of countermeasures against side-channel attacks [35], throughput and latency of various applications of secure multiparty-computation protocols [4, 34], verification time of SNARKs [2], the cost to avoid ciphertext-expansion in homomorphic encryption schemes [4, 17, 44], or reducing the signature size of signature schemes based on Sigma-protocols [18].

In general, we may be interested in three different metrics. One metric refers to what is commonly called multiplicative complexity (MC), which is simply the number of multiplications (in our case AND gates) in a circuit, see e.g. [14]. A natural variant in the context of encryption schemes is the number of AND gates per encrypted bit (MC/bit). The third metric refers to the multiplicative depth of the circuit, which we will subsequently call ANDdepth.

1.1 Motivating Applications

There are many examples where only the number of multiplications matters (perhaps together with the size of the ring in which they operate). SNARKs, protocols for secure multiparty communication based on Yao's garbled circuits, or Sigma-protocol signature schemes come to mind. However, there are also a number of applications where both the ANDdepth and the number of multiplications matter simultaneously, such as the following two important examples.

Preventing ciphertext expansion in homomorphic encryption schemes.
All known fully/somewhat homomorphic encryption schemes come with significant, often prohibitive ciphertext expansion. To prevent the thousand-fold to million-fold ciphertext expansion in (F)HE schemes, a decryption circuit of a symmetric encryption scheme has to be homomorphically evaluated in addition to the actual computations on the ciphertext. The downside of this approach is that application-specific operations on the ciphertext become more costly, as the decryption circuit of the cipher always needs to be evaluated as well.

To prevent bootstrapping, we need to choose the FHE parameters generously enough to accommodate all additional noise from the decryption circuit. This is linked to the homomorphic capacity of a concrete instantiation of an FHE scheme, i.e., the number of operations on the ciphertext before an expensive bootstrapping operation is needed. All known candidates for FHE schemes are using noise-based cryptography. Each operation on the homomorphically encrypted ciphertext incurs an increase in the noise. In many schemes, the noise level grows fast with the multiplicative depth of the circuit [16, 19]. Hence, symmetric encryption scheme proposals aiming for these types of applications minimize first of all the ANDdepth.

While the cost of the application-specific homomorphic operations only depends on the ANDdepth of the cipher, the cost of evaluating the additional decryption circuit itself primarily depends on the number of multiplications. Thus, the number of AND computations is also a relevant metric.

Applications of secure multiparty computation protocols. There are various classes of practically efficient secure multiparty computation (MPC) protocols for securely evaluating Boolean circuits where XOR gates are considerably cheaper (no communication, less computation) than AND gates. There are also many MPC protocols where each AND gate of the evaluated circuit requires interaction and so the performance depends on both the multiplicative complexity (MC) and ANDdepth of the circuit. Examples are the semi-honest secure version of the GMW protocol [33], and tiny-OT [47] with security against malicious adversaries. Applications of symmetric encryption schemes in these protocols include privacy-preserving keyword search based on Oblivious Pseudorandom Functions (OPRFs) [29], set intersection [38] and secure database join [42]. More details to motivate the use of symmetric encryption in MPC are given in [3].

1.2   The Design Strategy Rasta and Its Background

In this paper, we propose a design strategy called Rasta[1] for symmetric encryption that simultaneously achieves very low values in two of the three considered metrics: Symmetric encryption that has ANDdepth d and at the same time only needs d ANDs per encrypted bit. The main result is that even for very low $d = 2, \ldots, 6$, we can give some evidence that attacks may not exist.

We achieve this by putting so-called ASASA-like permutation constructions into a new setting. Generic substitution-permutation designs which interleave a key-dependent affine layer with key-dependent S-boxes have been studied since SASAS [12]. Follow-up work refined and extended this line of inquiry, and put it also into use for the purpose of white-box implementations of symmetric ciphers, and for instantiating schemes with public-key-like properties [10].

Our *new twist* to ASASA-like constructions is to consider a setting where the substitution layer is suitably chosen, public and fixed, but the affine layers are *derived from a public nonce and a counter* such that no affine layer is likely to be ever re-used under a single key (see Fig. 1). This approach prevents the attacks [24,32,45] that broke the proposals of [10], as an adversary will never be able to query the same ASASA-like permutation more than once in Rasta. Since the setup of each instance via the extendable-output function (XOF) [46] depends only on public information (N, i), it does not contribute to the (homomorphic) circuit evaluation cost in applications like FHE. In addition to the number of rounds, we also consider key sizes (and so block sizes of the used permutation) that are bigger than the required security level as tunable security parameter to provide protection against certain attack vectors.

---

[1] The name Rasta originates from the use of randomly looking affine layers A and the repetition of affine and S-box layer (AS)* followed by a last affine layer A. In short R(AS)*A. A C++ reference implementation is available at: https://github.com/iaikkrypto/rasta.

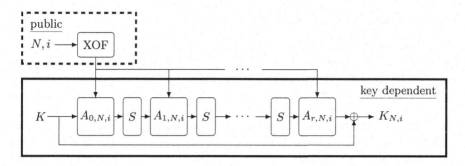

Fig. 1. The r-round Rasta construction to generate the keystream $K_{N,i}$ for block i under nonce N with affine layers $A_{j,N,i}$.

Variants. The practical downside of Rasta with a very low d is that for a fixed security level, the required key size and the number of additions needed for its evaluation grows very fast, see also the comparison in Table 1. Hence we consider such parameters as non-practical and at times use gray coloring in tables or figures for it. Throughout the paper we describe ways to bound various classes of attacks and use them to derive key and block sizes for any depth d. However, we do not have attacks matching these bounds. As we can see in Sect. 3, the attacks we have are rather far away from these bounds. In order to also explore the limits of what this design approach might achieve and to further encourage more cryptanalysis we also propose a variant of Rasta called Agrasta where the key and block size equals the security level (plus one to get odd numbers) and basing the number of rounds on what we can attack plus a security margin. Figure 2 brings the area where we know attacks in relation to the instances of Rasta and Agrasta having 80-bit security. Note that the area is mostly defined by cases, where the maximal number of different monomials becomes so low, that the equation system can be solved by a trivial linearization.

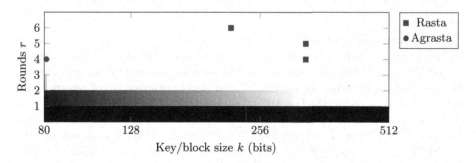

Fig. 2. Security margin of Rasta and Agrasta instances having 80-bit security. White area cannot be attacked with a complexity less than 2^{80}. Black area can be attacked with complexity below 2^{50}.

1.3   Related Work and Comparison

Recently, a number of new primitives were proposed that aim to minimize metrics related to the computation of AND gates. For a conjectured security of 128 bits, LowMC and Kreyvium require an ANDdepth of 11 or more, whereas FLIP manages to have a much lower ANDdepth of 4. The total number of AND computations is however much larger in FLIP (1072 per bit, 112 from the quadratic function, and 960 from the triangular function) than in LowMC or Kreyvium (which can be as low as 3 to 4 ANDs per bit). We give a more detailed discussion and comparison of them in the following. MiMC [2] is a very different design that shows excellent properties in a broad range of use-cases incl. MPC and SNARKs. It's main feature is that it can operate on elements in GF(p) natively and as such is very different to all the other designs we consider in our comparison.

High-level approach. What we do in this approach is to make a significant part of the computations independent of the key. This high-level approach was (perhaps for the first time) used in the FLIP design [44]. While in FLIP it is only the key bits that are permuted in a nonce-dependent way and the rest of the construction is fixed, in our design many essential parts of the construction are nonce-dependent: The derivation of a suitable affine layer, for every block, based on nonce and counter inputs using an extendable-output function (XOF) [46].

The advantage of this idea is that operations which do not depend on the key are in various settings of interest much less costly than operations that do depend on the key. In our experimental validation of the proposed approach, we include in the runtime the construction of each affine layer.

This results in a nice advantage of our approach as it allows for security arguments in the case of outputting many more than a single bit, hence drastically improving the number of ANDs per bit. Note that FLIP mitigates this property by focusing on a class of homomorphic encryption schemes where error growth is quasi-additive when considering a multiplicative chain and hence the large number of AND computations per encrypted bit are less of an issue.

New cryptanalytic insights. As a side-effect of these novel designs, new and interesting cryptanalytic insights continue to emerge. Attacks on earlier versions of LowMC [4] led to new insights on how higher-order properties can get extended because of non-full S-box layers [25, 27] and novel optimization of interpolation attacks [25]. As a result, the LowMC v2 parameters are larger: For 80-bit security at least 12 instead of 11 rounds are needed, and for 128-bit security at least 14 instead of 12 rounds are needed. The 12-round version was shown to offer less than 128-bit of security. In this paper we consider both versions, because comparisons in the past have been done with v1. Another example are attacks on FLIP which showed that guess-and-determine attacks [28] force designers to choose more conservative parameters for their novel design.

Comparison with respect to AND-related metrics. For a security level of 128 bits, LowMCv2 has a depth of at least 14, Kreyvium [17] has depth of at least 12 and the most recent proposal FLIP [44] only needs a depth of 4. The comparison among these three custom designs is however more complicated than these numbers might suggest. Whereas for LowMC the depth remains constant for the encryption of at least 256 bits, for Kreyvium the depth starts to grow after 67 bits already. The very low depth of FLIP comes at the cost of a much larger number of AND computations per encrypted bit: At a security level of 128 bits it is 1072 ANDs/bit for FLIP compared to values as low as 3 to 4 for Kreyvium and LowMC.

Table 1 and Fig. 3 illustrate a comparison of our design with these three earlier designs. They overlap partially in the content they convey, but also complement each other. For simplicity, in Fig. 3, we only show the figure for the security level of 128 bits, also because only for this particular security level, instantiation proposals are available for all design options.

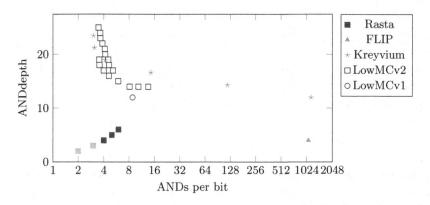

Fig. 3. Comparison with respect to the two most important metrics. All are for a security level of 128 bits. The different points for Kreyvium are derived from varying the number of output bits generated per initialization. For LowMCv2 the given round formula is used to explore various possible trade-offs in the design space.

Figure 3 does not only give single data points for LowMC and Kreyvium, but explores a wider range of usable options. As the stream cipher Kreyvium needs an initialization phase, the number of ANDs/bit is very high if only a small number of keystream bits are generated. The more keystream bits are generated, the more this initialization phase is amortized and hence reduces the number of ANDs/bit, but on the other hand the ANDdepth is growing.

LowMC does not refer to a particular block cipher geometry, but allows to generate instantiations for a wide range of block sizes, security levels, and number of S-boxes per round. For Fig. 3, we fixed the security level to 128-bit, but tried a number of block sizes in the range between 256 and 1024 bits for two different choices of number of S-boxes per round (33 and 63), simply to get an impression of the trade-space between the two metrics.

Table 1. Comparison of Rasta with related designs. ℓ is the number of encrypted bits. For Rasta and LowMC, ℓ needs to be a multiple of the block size.

Name	security	key size	block size	ANDdepth	minAND	ANDs per bit
LowMCv1 [4]	80	80	256	11	1617	6.3
LowMCv2 [3]	80	80	256	12	1764	6.89
LowMCv2 [3]	80	80	128	12	1116	8.72
Trivium [23]	80	80	1	$12 + \log(\ell)$	$1152 + 3 \cdot \ell$	3–1152
FLIP [44]	80	530	1	4	352	352
Rasta	80	219	219	6	$6 \cdot \ell$	6
Rasta	80	327	327	5	$5 \cdot \ell$	5
Rasta	80	327	327	4	$4 \cdot \ell$	4
Rasta	80	3939	3939	3	$3 \cdot \ell$	3
Rasta	80	$\approx 2^{21}$	$\approx 2^{21}$	2	$2 \cdot \ell$	2
LowMCv1 [4]	118	128	256	12	2268	8.86
LowMCv2 [3]	128	128	256	14	2646	10.34
LowMCv2 [3]	128	128	192	14	2592	13.50
Kreyvium [17]	128	128	1	$12 + \log(\ell)$	$1152 + 3 \cdot \ell$	3–1152
FLIP [44]	128	1394	1	4	1072	1072
Rasta	128	351	351	6	$\mathbf{6 \cdot \ell}$	**6**
Rasta	128	525	525	5	$\mathbf{5 \cdot \ell}$	**5**
Rasta	128	1877	1877	4	$\mathbf{4 \cdot \ell}$	**4**
Rasta	128	$\approx 2^{18}$	$\approx 2^{18}$	3	$\mathbf{3 \cdot \ell}$	**3**
Rasta	128	$\approx 2^{33}$	$\approx 2^{33}$	2	$\mathbf{2 \cdot \ell}$	**2**
LowMCv2 [3]	256	256	512	18	3564	6.96
Rasta	256	703	703	6	$6 \cdot 1$	6
Rasta	256	3545	3545	5	$5 \cdot \ell$	5
Rasta	256	$\approx 2^{19}$	$\approx 2^{19}$	4	$4 \cdot \ell$	4
Rasta	256	$\approx 2^{34}$	$\approx 2^{34}$	3	$3 \cdot \ell$	3
Rasta	256	$\approx 2^{65}$	$\approx 2^{65}$	2	$2 \cdot \ell$	2

What is however not visible in Fig. 3 are two other properties these schemes have: key size and block size. This is shown, together with all other metrics in Table 1, for 80-bit, 128-bit and 256-bit security levels. Whereas Kreyvium and LowMC allow for keys as short as the security level in bits, both FLIP and Rasta require longer keys. Rasta, similarly to LowMC, requires a much larger number of XOR operations than FLIP or Kreyvium. Whereas traditionally the cost of linear operations is considered almost negligible compared to non-linear operations, the number of XOR operations is so extreme in the setting we are interested in that it is no longer negligible. Hence as we will see in the validation of the practical usefulness in Sect. 4, more moderate choices of d between 4 and 6 seem more useful.

Implementation comparisons. Even though the main contribution of this work is the analysis of a scheme that has at the same time very low ANDdepth and ANDs/bit, we also aim to validate the design approach by means of actual implementations. As discussed already, the practical downside is that for a fixed security level and for a very low d (meaning very low ANDdepth and ANDs/bit), the required key size and especially the number of additions needed for its evaluation grows very fast. Hence the question is: can variants of Rasta be useful in practice?

It turns out that implementations of our scheme with too low d are not possible. For some parameters this is already obvious from the huge required key and block size. As we will show in Sect. 4 also more moderate sizes can be prohibitive in the FHE setting we use as a test-case. Nevertheless, we also give some evidence that for more moderate choices of d between 4 and 6, the resulting cost of homomorphically evaluating the circuit of our new construction may well be competitive.

1.4 External Cryptanlysis

We are already aware of cryptanalysis attempts outside the design team. This includes work by Raddum on a dedicated attack [48] as well work by Bile et al. [9] on algebraic and Gröbner bases approaches.

1.5 Organization of the Paper

In Sect. 2, we give the detailed specification of the new design approach together with concrete parameters for instantiations of Rasta at various conjectured security levels. In Sect. 3, we thoroughly analyze the security of Rasta, considering both standard symmetric cryptanalysis techniques and some novel, design-specific attack angles. In Sect. 4, we discuss concrete implementations and provide a performance comparison with other designs. Finally, we conclude and discuss open problems in Sect. 5.

2 Specification

2.1 Encryption Mode

Rasta is a family of stream ciphers. To produce the keystream, it applies a permutation with feed-forward, as shown in Fig. 4. The input of the permutation is the secret key K, where the key size k matches the block size n of the underlying family of permutations $P_{N,i}$. The keystream is generated by using different permutations $P_{N,i}$ per encrypted block, which are parametrized by the nonce N and a block counter i. To ensure confidentiality, a new and unique nonce N is required for every encryption.

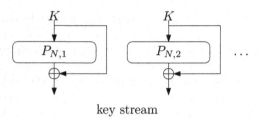

key stream

Fig. 4. Encryption mode of Rasta.

2.2 Permutation $P_{N,i}$

Rasta's family of permutations $P_{N,i}$ applies r rounds of different affine layers $A_{j,N,i}$ and non-linear layers S. After r rounds, a final affine layer is applied:

$$P_{N,i} = A_{r,N,i} \circ S \circ A_{r-1,N,i} \circ \cdots \circ S \circ A_{1,N,i} \circ S \circ A_{0,N,i}$$

Each affine layer is different and depends on the nonce N and block counter i. The family of permutations is parameterizable in the number of rounds r and permutation block size in bits n, where n is an odd number.

Non-linear layer S. For the non-linear layer, we apply the χ-transformation [22, Sect. 6.6.3], previously used in KECCAK [6] and ASASA [10], to the entire block. This transformation is invertible for any odd number of bits n [22]. For input bits x_i and output bits y_i, $0 \leq i < n$, the non-linear layer is defined in (1), with all indices modulo n:

$$y_\ell = x_\ell \oplus x_{\ell+2} \oplus x_{\ell+1}x_{\ell+2} . \tag{1}$$

Affine layers $A_{j,N,i}$. The affine layers are a simple binary multiplication of a binary $n \times n$ matrix $M_{j,N,i}$ to the input vector x, followed by the addition of a round constant $c_{j,N,i}$:

$$y = M_{j,N,i} \cdot x \oplus c_{j,N,i} .$$

Generation of matrices $M_{j,N,i}$ and round constants $c_{j,N,i}$. We propose to generate the different matrices and round constants with the help of an extendable-output function (XOF) [46] that is seeded with the number of rounds r, block size n, nonce N, and i. Hereby, the output of the XOF is first used to generate $M_{0,N,i}$, a unstructured nonsingular Matrix. Several ways are possible to generate such matrices with the help of an XOF output. The first one is to add rows and check if the matrix is invertible. As pointed out by Randall [49], on average it takes three tries before we end up with a nonsingular matrix. In the same paper, Randall [49] gives an algorithm to generate random nonsingular $n \times n$ matrices needing less than $n^2 + 3$ random bits inserted in a clever way in two $n \times n$ matrices, which are multiplied in the end. The choice of the algorithm to generate nonsingular matrices should not influence the security, just the execution time. However, communicating instances have to use the same algorithm.

After the generation of the first matrix, the output of the XOF is used to create $c_{0,N,i}$, $M_{1,N,i}$, and so on. To ensure that the permutation is secure, we expect the XOF to behave like a random oracle up to a certain security level. For instance, it should not be feasible for an attacker to find inputs to the XOF for outputs of the attacker's choice except by repeatedly querying the XOF for different inputs. Furthermore, the internal state of the XOF should be large enough so that internal collisions within its state are prohibited. A suitable choice for an XOF would be for most instances SHAKE256 [46].

2.3   Design Rationale

The essential idea of Rasta is to reduce the ANDdepth as much as possible by creating a moving target, which is only evaluated once per key. Hence, we have decided to use a permutation with feed-forward, where the secret occupies the whole input and the keystream is generated by always evaluating a different permutation. Those permutations are obtained by choosing new matrices and round constants for each new permutation call, based on an XOF seeded with a public nonce and block counter. This technique allows us to treat matrices and round constants during our analysis as if they where randomly created for each different permutation (new nonce and counter pair), with the restriction that same nonce and counter pairs give us always the same permutation and hence, the same matrices and round constants. Since the XOF uses no secret, it can be publicly evaluated and thus, similar as for FLIP [44], the XOF does not influence the AND related metrics.

Choosing the matrices this way minimizes structural similarities between and within the permutation instances. The round constants remove obvious fixed-points, such as the fixed-point 0 that all instances would have in common otherwise. Furthermore, these round constants add an additional layer of protection against attacks between single instances of the permutation. The affine layer is required to be a permutation rather than a function to prevent reduction of the state space. Otherwise, if for example the final affine layer did not have full rank, then the final key feed-forward would allow an attacker to derive information about a linear subspace containing the key with each query.

Instead of smaller S-boxes for the non-linear layer, we use one large χ-transformation [22] across the entire state. Since we only need to evaluate it in forward direction, we benefit from the fact that it is very efficient to evaluate in forward direction with a degree of only 2, while its inverse has a very high degree of $(t + 1)$ for size $(2t + 1)$ [10].

2.4   Instances

Based on the security analysis done in Sect. 3, we propose block sizes for 4 to 6 rounds of Rasta aiming at 80, 128, or 256 bits of security in Table 2. These block sizes are derived from the results of our security analysis shown in Tables 5 and 6 in Sect. 3. Since part of our analysis relies on good diffusion properties, we only recommend instances of Rasta with at least 4 rounds for use. However, from a theoretical perspective, smaller instances are also of interest and hence, we add parameters for 2 and 3 rounds (in gray) to Table 2 as well.

Table 2. Minimal block sizes for 4 to 6 rounds of Rasta aiming to provide 80, 128, or 256 bits of security (2 and 3 rounds in gray).

Security level	Rounds				
	2	3	4	5	6
80-bit	2 320 961	3 939	327	327	219
128-bit	9 506 325 433	246 831	1 877	525	351
256-bit	40 829 356 287 426 864 861	16 167 762 975	445 939	3 545	703

Data limit and related-key attacks. Our goal is to derive instances that have both a small block size and a small ANDdepth. To make this feasible, we limit the data complexity per key to $\sqrt{2^s}/n$ blocks, where s is the targeted security level in bits and n is the block size. Furthermore, to ensure low ANDdepth with our construction in general, related-key attacks are out of scope.

Agrasta. As already mentioned, we have chosen the block sizes for Rasta in a conservative manner based on our security evaluation in Sect. 3. The block sizes of Rasta are mostly determined by the bounds we get on the maximal number of monomials and the bounds on the probability that good linear approximations exists. However, this does not mean that we have matching attacks for these bounds, in fact, as can be seen in Sect. 3, what we actually can attack is far away from these bounds. Hence, we propose Agrasta shown in Table 3.

Table 3. Instances of Agrasta.

Security level	Rounds	Blocksize
80-bit	4	81
128-bit	4	129
256-bit	5	257

Agrasta has a block size that matches the security level (or block size plus 1 for even security levels). For the number of rounds, we consider a minimal number of 4 rounds for the same reasons as for Rasta and add rounds until we cannot attack the construction anymore. As a consequence, Agrasta has a block size of 81-bit for 80-bit security having 4 rounds, 129-bit for 128-bit security having 4 rounds and 257-bits for 256-bit security having 5 rounds (in this case trivial linearization would work for 4 rounds).

Toy version. To encourage cryptanalysis of Rasta, we specify toy versions of Rasta in Table 4. Those toy versions aim at achieving 24-bit security.

Table 4. Toy versions of Rasta.

Security level	Rounds				
	2	3	4	5	6
24-bit	193	193	97	97	65

3   Security Analysis

In this section, we discuss various cryptanalytic approaches and argue the security of the recommended parameter configurations of Rasta. We focus on key recovery attacks and assume that the attacker can obtain the keystream $K \oplus P_{N,i}(K)$ for arbitrary choices of (N, i). In particular, assume the attacker requires instances with a particular property of the affine layers that occurs with probability p. This chosen-nonce setting allows the attacker to obtain keystream for such an instance with about p^{-1} XOF queries (in key-independent precomputation) and 1 Rasta query, instead of p^{-1} Rasta queries. We start out with algebraic attacks via linearization and Gröbner bases, then we describe a potential attack vector via linear approximations and argue why various classical attacks such as differential, integral, or higher order attacks are ruled out or unlikely. In Sect. 3.4 we then briefly describe a dedicated attack on variants that are very close to Agrasta but reduced to three rounds. Last but not least we describe various experiments on toy versions of Rasta, incl. SAT solver, Gröberner-bases, and an analysis of linear properties and monomial counts.

3.1   Algebraic Attacks

We first consider algebraic attacks that aim to recover the secret key K by solving a system of non-linear Boolean equations. These equations are collected by observing the keystream for different nonces, and setting up the algebraic normal form (ANF) for each observed instance of the permutation. All these ANF polynomials share the bits of K as unknowns. In the following, we consider different approaches and trade-offs to solve these equations.

Trivial linearization. In this attack, the resulting non-linear system of equations is solved by substituting the non-linear terms with new variables. Consider a cipher with algebraic degree ϕ and a key of k bits. For such a cipher, the number of unknowns in our equations is upper-bounded by k^ϕ. Thus, after collecting at most k^ϕ linearly independent equations, the system can definitely be solved and the key recovered. The maximum number of different monomials we can get is:

$$U = \sum_{i=0}^{\phi} \binom{k}{i}. \tag{2}$$

For instance, considering Rasta with one S-layer of degree 2 and a key of 1024 bits, we need at most 2^{19} equations. For degree 16 and a 1024-bit key, we already get up to $2^{115.6}$ monomials, and for degree 16 with a 2048-bit key, $2^{131.68}$ monomials. Since the ANDdepth d of the cipher is bound to the degree ϕ ($d = \log_2 \phi$), we have an immediate impact on reasonable ANDdepths.

Key-guessing to reduce monomials. Guessing g out of k key bits reduces the number U of possible monomials and hence the number of variables we need to introduce to linearize and solve the system to

$$U = \sum_{i=0}^{\phi} \binom{k-g}{i}. \tag{3}$$

Guessing g bits of the key reduces the data needed to perform a linearization attack. However, the linear system has to be solved 2^g times for every key guess, giving a maximum number of bits corresponding to the security level of the scheme that can be guessed.

On the number of monomials. While the total number of monomials we can get gives us insight when the system of equations can definitely be solved, the number of monomials we get per output equation plays an important role in algebraic attacks. For instance, getting too many sparse equations might lead to a sub-system that can be solved or might enable other attacks similar to algebraic or fast algebraic attacks on stream ciphers [20,21]. Therefore, we study the average number of monomials after r rounds of Rasta, and we compare it with the worst-case number. We conclude that the number of monomials in the average case is well approximated by the worst case.

We first consider the worst case, which was already studied in the previous section. Since the S-layer has degree 2, the degree of the scheme after r rounds is 2^r, so the number of monomials is upper-bounded by $\sum_{i=0}^{2^r} \binom{k}{i}$.

To analyze the average case, recall that the matrices A of the linear layers are uniformly distributed. Consider one round $S \circ A(x)$ of Rasta with input $x = (x_0, \ldots, x_{k-1})$, and let $A(x) = A \cdot x + c$:

$$S \circ A(x)_i = \bigoplus_{j=0}^{k-1} \bigoplus_{l=j+1}^{k-1} a_{j,l}^i \cdot x_j \cdot x_l \oplus \bigoplus_{j=0}^{k-1} b_j^i \cdot x_j \oplus g^i,$$

where remember that $x^2 = x$ for each $x \in \mathbb{F}_2$ and where

$$a_{j,l}^i = A_{i+1,j} \cdot A_{i+2,l} \oplus A_{i+2,j} \cdot A_{i+1,l},$$
$$b_j^i = A_{i,j} \oplus c_{i+2} \cdot A_{i+1,j} \oplus (1 \oplus c_{i+1}) \cdot A_{i+2,j},$$
$$g^i = c_i \oplus c_{i+2} \oplus c_{i+1} \cdot c_{i+2}.$$

First, we focus on the monomials of degree 2. The probability that a coefficient $a^i_{j,l}$ is equal to 0 is 5/8:

$$\mathbb{P}[a^i_{j,l}=0]=\mathbb{P}[A_{i+1,j}A_{i+2,l}=A_{i+2,j}A_{i+1,l}=0]+\mathbb{P}[A_{i+1,j}A_{i+2,l}=A_{i+2,j}A_{i+1,l}=1]$$

$$=\left(\frac{3}{4}\right)^2+\left(\frac{1}{4}\right)^2=\frac{5}{8}.$$

Thus, the probability that all the coefficients of the variable $x_j \cdot x_l$ with $l > j$ are equal to 0 is

$$\mathbb{P}[a^i_{j,l} = 0 \quad \forall i = 0, \ldots, k - 1] = \left(\frac{5}{8}\right)^k,$$

or equivalently, at least one of these coefficients is different from 0 with probability $1 - (5/8)^k$. Since this is true for each i, we expect an average number of monomials of degree 2 equals to

$$\binom{k}{2} \cdot \left[1 - \left(\frac{5}{8}\right)^k\right] \simeq \binom{k}{2},$$

where the approximation holds for $k \gg 1$. In a similar way, we expect an average number of monomials of degree 1 equal to $k \cdot (1 - 2^{-k}) \simeq k$. It follows that the average number of monomials is well approximated by the worst one[2]. Our experiments confirm this prediction: we observed that all the monomials of degree 1 and 2 are present in the system made by the equations of the output bits of $S \circ A(x)$. Since the same argumentation holds for the following rounds, this justifies our claim.

Another consideration that strengthens our claim is the following. Suppose that the average number v of variables is much lower than the number w of variables in the worst case, that is, $v \ll w$. Thus, given one encryption, the number of variables that one has to consider is only v. However, in order to find a solution for the given system of linear equations, one must consider other encryptions. Due to the previous hypothesis, for each one of these new encryptions, one expects an average number of v variables. On the other hand, since the linear layers change for each encryption, it is not possible to claim that the variables of these texts are always the same. In other words, the variables of a second encryption are in general different from the ones of the first encryption. This implies that if one uses a second encryption to find the solutions of the linear equations, the number of variables that one has to consider is on average not v, but greater than v (and lower than $2 \cdot v$). Due to the same argumentation, using r texts, this number is on average (much) greater than v but lower than $r \cdot v$. Intuitively, if r is big enough, even if for each single text/encryption the

---

[2] Note here that an attacker could rename the variables obtained after the linear layer, making the number of quadratic terms drops to only k. However, this renaming would only be valid for one message, while many messages are necessary to solve the system, which makes this process unlikely to have any impact.

number of variables v is lower than the total one (i.e. w), using r texts to find the real values of these variables implies that this number is closer to w than to v. As a consequence, this provides another intuitive reason while one has to consider the worst number of variables in order to evaluate the security of this proposed encryption scheme.

Gröbner basis computation. Instead of linearization, one could also try to solve the non-linear system by computing a Gröbner basis for the system of equations. However, it is highly unlikely that this attack vector threatens our construction. The main point is that the number of unknowns is very large (starting from 219-bit blocks and keys) and the algebraic degree is not that small (starting from 2^4). This has to be compared with the best results to date, that to the best of our knowledge allow to solve a system with up to 148 *quadratic* equations in 74 variables [39]. Actually, a recent technical report [9] provides further evidence that solving the corresponding system of equations for Rasta becomes quickly infeasible.

Discussion. As we have seen in this section, Rasta gives us a system of non-linear Boolean equations of a certain degree ϕ with dense equations on average. This system of equations depends on the two parameters of Rasta, the block size and the ANDdepth. We have plotted the maximum number of different monomials under guessing 80-bit, 128-bit, 256-bit key information for various block sizes and ANDdepths in Fig. 5.

This is particularly interesting, since Fig. 5 gives us insight in the data an attacker has to acquire before the system can be definitely solved by using trivial linearization. At the moment, the costs of solving a linear system of U linear

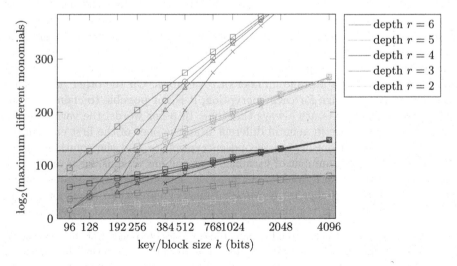

Fig. 5. Trivial linearization attack. □/○/△/× denote 0/80/128/256-bit key guess.

Table 5. Block sizes such that the number of monomials is greater than 2^s.

Security level s	Depth				
	2	3	4	5	6
80-bit	2 320 961	3 938	305	167	161
128-bit	9 506 325 433	246 831	1 876	348	258
256-bit	40 829 356 287 426 864 861	16 167 762 975	445 938	3 545	682

equations in U binary variables is about $\mathcal{O}(U^\omega)$ bit operations, where ω is bigger than 2. Nevertheless, the key can still be recovered with k equations in $\mathcal{O}(2^k)$ using brute-force.

Estimating the required time complexity for solving a non-linear system with $k + \epsilon$ equations is an open research question in the case of Rasta. Since we want to limit the possibilities an attacker has to solve the system of non-linear equations, we set a limit on the number of equations acquirable by the attacker. Therefore, we apply the two following restrictions for a security level of s bits, which will influence the block and key size of Rasta:

- Choose key size k and degree ϕ such that $\sum_{i=0}^{\phi} \binom{k-g}{i} > 2^s$ (Table 5).
- Data complexity limit of $\sqrt{2^s}/k$ blocks.

3.2   Attacks Based on Linear Approximations

In this section, we deal with classes of attacks that exploit linear approximations having a high bias δ.[3] Although classical linear cryptanalysis [43] seems not to be applicable to Rasta in a straight forward manner, since it would require the repeated evaluation of the same Rasta permutation with different inputs, other attacks based on linear approximations can still be a threat. For instance, in the case of Rasta, an attacker can collect single evaluations of different linear approximations that are valid with a certain probability. Here, the attacker faces a problem similar to the LPN-problem and thus, algorithms similar to existing ones designed for solving the LPN-problem [13] might be applicable.

However, attacks utilizing linear approximations have in common that the data-complexity required to perform these attacks is bound to the bias δ of the used linear approximations and usually lies in regions of δ^{-2}. Therefore, attacks exploiting linear approximations are not applicable if linear approximations that have a good bias do not exist. Hence, we want to bound the probability that a certain choice of the nonce N gives us matrices that allow linear approximations with a good bias δ.

Bounds for 2 S-layers (depth 2). First, we restrict our observations on two S-layers with one matrix M_1 in between. If we restrict our observations just to

---
[3] We do not extend this analysis to linear hulls as even for more suitable designs doing this analytically is beyond the state of the art.

the matrix M_1, we can make the following statements about the quality for a certain linear characteristic, where a_0 bits are active at the input of M_1 and a_1 bits are active at the output.

As already shown by Daemen [22], the correlation weight w_c of a linear approximation for the χ-transformation only depends on the active bits of the output of the χ-transformation and its correlation is either zero or 2^{-w_c}. As noted in [5], the correlation weight is equal to half the number of active bits plus a positive number that depends on the position of the active bits. If all bits of the output are active, then the correlation weight is half the block length minus 1. Since in our case the block length is always at least the security level, we can ignore this special case and can upper-bound the bias just relying on half the number of active bits at the output of the S-layers.

However, as mentioned in the beginning of the section, we want to make statements on a linear characteristic just using the information that a_0 bits are active at the output of the first S-layer and a_1 bits are active on the input a second one. Thus, we have to make assumptions on the minimum number of active output bits, dependent on the number of active input bits. Let us assume that just one bit on the output of an S-layer is active. How many bits at the input can be active, so that we get a correlation, or bias different from zero? As we can see from (1), in order to determine the value of the active bit, just 3 bits of the input are needed. Hence, just these 3 bits can be used in a linear approximation of the output bit, while trying to include any other bit in a linear approximation definitely results in a bias of zero. So we know that if we have 3 active bits at the input of the S-layer, at least one output bit has to be active, while for 4 or more active input bits definitely 2, or more outputs have to be active. Considering two active output bits, at most 6 bits are used in their calculations, which in turn can be used in a linear approximation and so on. Therefore, a linear approximation using x active bits at the output might use less than $3x$ input bits, but never more.

Thus, we can upper-bound the correlation of a resulting linear approximation with $2^{-(a_0 + \lceil a_1/3 \rceil + 1)/2}$ and hence, the bias with $2^{-(a_0 + \lceil a_1/3 \rceil)/2 - 1}$. So, our goal is to make statements about the probability that a randomly generated matrix M_1 allows linear characteristics with a_0 active bits at the input and a_1 active bits at the output, so that $(a_0 + \lceil a_1/3 \rceil)$ is smaller than a certain threshold. Consider that in Sect. 3.1, we already introduce a limit on the data complexity, we assume that linear approximations with a bias of $2^{-s/4}$ cannot be exploited in attacks, where s is the security level in bits. The number of invertible matrices that allow a certain fixed linear characteristic is $\prod_{i=1}^{n-1}(2^n - 2^i)$. This number is independent of the concrete pattern. By counting the number of suitable linear characteristics, we get:

$$\mathbb{P}_2\left[-\log_2(\delta) \leq \frac{s}{2}\right] \leq \sum_{\substack{0 < a_0, a_1 \\ (a_0 + \lceil a_1/3 \rceil) \leq s/2}} \binom{n}{a_0}\binom{n}{a_1} \frac{\prod_{i=1}^{n-1}(2^n - 2^i)}{\prod_{i=0}^{n-1}(2^n - 2^i)}$$

$$\leq (2^n - 1)^{-1} \sum_{\substack{0 < a_0, a_1 \\ (a_0 + \lceil a_1/3 \rceil) \leq s/2}} \binom{n}{a_0} \binom{n}{a_1}$$

$$\leq (2^n - 1)^{-1} \sum_{\substack{0 \leq a_0, a_1 \\ (a_0 + a_1) \leq \frac{3s}{2}}} \binom{n}{a_0} \binom{n}{a_1} \leq (2^n - 1)^{-1} \sum_{0 \leq a \leq \frac{3s}{2}} \binom{2n}{a}.$$

$$M_0 \longrightarrow \underbrace{S_0 \longrightarrow M_1 \longrightarrow S_1 \longrightarrow M_2}_{P_1} \longrightarrow \underbrace{S_2 \longrightarrow M_3 \longrightarrow S_3}_{P_2} \longrightarrow M_4$$

Fig. 6. Schematic of the Rasta permutations having 4 S-layers.

Bounds for 4 S-layers (depth 4). Figure 6 shows the Rasta permutation having 4 S-layers. We will show bounds on the probability that randomly generated matrices allow linear characteristics having a bias higher than some threshold. As indicated in Fig. 6, we will calculate the probabilities that randomly generated matrices allowing a certain number of active bits independently for M_1 and M_3, assuming that every randomly generated matrix M_2 is able to connect the resulting linear characteristics.

As already mentioned, we do not want to have linear characteristics with a bias higher than $2^{-s/4}$. Hence, the probability that a matrix M_1 allows transitions with a_0 and a_1 active bits and that a matrix M_3 allows transitions with a_2 and a_3 active bits, where $(a_0 + \lceil a_1/3 \rceil + a_2 + \lceil a_3/3 \rceil) \leq s/2$, should be small. We can calculate this probability by summing up the probabilities for fixed a_0, a_1, a_2, and a_3 up to this bound:

$$\mathbb{P}_4 \left[-\log_2(\delta) \leq \frac{s}{2} \right] \leq (2^n - 1)^{-2} \times \sum_{\substack{0 < a_0, a_1, a_2, a_3 \\ (a_0 + \lceil a_1/3 \rceil + a_2 + \lceil a_3/3 \rceil) \leq \frac{s}{2}}} \binom{n}{a_0} \binom{n}{a_1} \binom{n}{a_2} \binom{n}{a_3}$$

$$\leq (2^n - 1)^{-2} \times \sum_{0 \leq a \leq \frac{3s}{2}} \binom{4n}{a} \tag{4}$$

Bounds for r S-layers (depth r). Finally, we can extend the concept we have used for bounding the probability on 4 rounds to bounding the probability for an arbitrary even number r of rounds of Rasta:

$$\mathbb{P}_r \left[-\log_2(\delta) \leq \frac{s}{2} \right] \leq (2^n - 1)^{-r/2} \times \sum_{0 \leq a \leq \frac{3s}{2}} \binom{r \cdot n}{a} \tag{5}$$

Discussion. We visualize the effects of different block sizes and depths on the probability that a linear approximation for such an instance of Rasta exists in

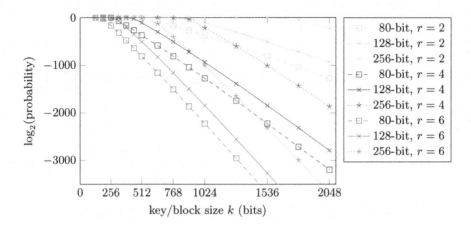

Fig. 7. Probability that a linear trail/approximation with bias above for a certain security level exists.

Table 6. Block sizes where $-\log_2(\mathbb{P})$ matches s.

Security level s	$r = 2$	$r = 4$	$r = 6$
80-bit	654	327	218
128-bit	1050	525	350
256-bit	2106	1053	702

Fig. 7. If we compare 80-bit security for 2 rounds of Rasta with 80-bit security for 4 rounds in Fig. 7, we see that we get similar probabilities for 4 rounds of Rasta, requiring just half the block size compared to two rounds. This gets already clear from Eq. (5), where we see that—at least for our bounds—doubling the depth has similar effects as doubling the block size.

To give a better overview about the influence of linear approximations on the block size, we summarize reasonable block sizes for different security levels and depths in Table 6. Concretely, we do not want to have linear approximations where the squared bias δ^2 matches our data limit. Hence, the block sizes for this table have been chosen so that the probability that a linear approximation with a bias $\geq 2^{-s/4}$ exists (for one random choice of the linear layer) exceeds 2^{-s}, where s is the security level. Therefore, for block sizes exceeding the ones given in Table 6, the probability that one useable bias exists gets smaller and smaller.

3.3 Classical Attacks Exploiting the Structure

In this section, we discuss some classical cryptographic attacks that exploit the "structure" of cryptographic primitives. This includes, among others, differential attacks [8], higher-order differential attacks [41], cube attacks [26] and integral attacks [40]. All these attacks have in common that they rely on the fact that an adversary is able to evaluate the cryptographic primitive (equation system) more than once with different inputs. Since Rasta uses different nonce-based matrices and constants for encryption/decryption, these attacks are thwarted.

Differential attacks. Two prerequisites are needed for performing a differential attack. First of all, an attacker has to find a suitable differential characteristic and in the second step, this attacker has to be able to inject the needed differences. For Rasta such a classical differential attack on the permutation is precluded, since each permutation can be at most evaluated once.

Furthermore, even if we consider related-key attacks, where the attacker is able to use different keys under the same nonce and hence is able to evaluate the permutation of Rasta multiple times, it is unlikely that a good exploitable differential characteristic exists following similar arguments as used in Sect. 3.2.

Higher-order differential and cube attacks. Higher-order differential and cube attacks exploit the algebraic degree of the output functions of a cryptographic primitive. Considering the fact that Rasta aims to have a small AND-depth, the output functions of Rasta only have a low degree. For instance, Rasta with depth 4 has an algebraic degree of 16. Having such a low degree would be a major threat for classical permutation based designs, where a single permutation can be evaluated multiple times.

For instance, in a related-key scenario, an attacker could evaluate the same permutation of depth-4 Rasta 2^{16} times to mount a cube attack or exploit the resulting non-random properties in another way. Hence, Rasta is clearly not secure in such a related-key scenario. However, for single-key attacks, each individual output function can only be evaluated once. Therefore, higher-order differential attacks and cube attacks, which rely on evaluating these functions multiple times are effectively hindered.

Integral attacks. Integral attacks exploit the structure of the linear layer in combination with properties of the used S-boxes. In addition to the fact that the used linear layer of Rasta is highly unstructured, it changes for every application of the nonce. Moreover, we only get one evaluation per permutation instance. Both facts make the application of integral attacks on Rasta unlikely.

3.4 On the Relative Size of Block and Security Parameter for Depth 3

In this section we show a guess-and-determine attack to three-round (toy) variants of Rasta where the block size matches the security parameter ($n = k = s$), or the block size slightly exceeds the security parameter. In particular, the attack is applicable to round-reduced variants of Agrasta. This attack requires knowledge of only a single keystream block K_n. As an additional prerequisite, it needs a weakness in the matrix M_2 of the affine layer $A_{2,N,j}$. The adversary can query the XOF in an off-line phase to identify a nonce-counter pair (N, i) that yields

Table 7. Best found parameters for the guess-and-determine attack on three- round Agrasta. Encs. = encryptions; p_s = Probability of a weak affine layer.

n (bits)	g_1 (bits)	g_2 (bits)	g_3 (bits)	x_r (bits)	y_r (bits)	$\log_2(t)$ (encs.)	$\log_2(p)$
81	41	37	1	4	74	78.19	−18.71
129	65	59	2	6	118	125.76	−55.46
219	110	103	2	6	206	215.39	−66.45
257	129	122	2	6	244	253.58	−66.22
327	164	157	2	6	314	323.86	−65.86

such a weak matrix. In the following, we briefly describe the details of the attack, including an estimation on the probability of a weak matrix to occur.

We enumerate by X^i the states after the $(i-1)$-th affine layer (e.g., $X^1 = A_{0,N,j}(K)$) and by $Y^i = S(X^i)$ the states after the subsequent S-layer. The attack starts at X^1. We guess $g_1 = \lceil n/2 \rceil$ bits of X^1 at consecutive even bit indices. This allows to formulate all bits of Y^1, i.e., the state after the S layer, as equations linear in the $(n - g_1)$ unknowns from X^1. Through $A_{1,N,j}$, we can derive the full state X^2 also as linear equations in unknowns from X^1. In backward direction, we can formulate K, and from it $X^4 = K \oplus K_N$, and $Y^3 = A_{3,N,j}^{-1}(X^4)$ also as a linear equation system in the unknowns from X^1.

Next, from Y^1, we can guess $g_2 < n/2$ bits of X^2 at consecutive even bit indices. Therefore, we obtain g_2 further equations for the $n - g_1$ remaining unknowns. Moreover, we can formulate $2 \cdot g_2$ bits of Y^2 as linear equations in the unknowns from X^1. We have to choose the $2g_2$ bits in a way such that we obtain a subset of $y_r = g_2 - 2$ consecutive bits of X^3 to depend only on the $2g_2$ bits of Y^2 through $A_{2,N,j}$, the "weak" linear layer.

Let Mat_n denote the set of all non-singular matrices from $\mathbb{F}_2^{n \times n}$. Let $x, y \in \mathbb{F}_2^n$ be Boolean vectors with $x = M \cdot y$. Let $u, v \in \mathbb{F}_2^n$ denote further Boolean vectors that represent masks; we call a bit x_i relevant iff $u_i = 1$ and y_i relevant iff $v_i = 1$. We call a matrix $M \in \mathsf{Mat}_n$ weak iff for all $i, j \in \{0, \ldots, n-1\}$ with $v_i = 1$ and $u_j = 0$, it holds that $M_{i,j} = 0$. So, if M is weak, all relevant bits of x depend on only relevant bits of y.

Naturally, the probability for a linear layer to be weak depends on the numbers of relevant bits x_r in x, relevant bits y_r in y, and the options to distribute them. In M, we need $(n - y_r) \cdot x_r$ zero entries and to have the relevant bits in both Y^2 and X^3 to be located at consecutive positions. Hence, there exist $(n - y_r + 1)$ consecutive bit positions in Y^2 and $(n - x_r + 1)$ positions in X^3. So, we can approximate

$$\mathbb{P}\left[M \leftarrow \mathsf{Mat}_n : M \text{ is weak}\right] \geq \frac{(n - x_r + 1)(n - y_r + 1)}{2^{(n - y_r) \cdot x_r}}.$$

We search for a region of x_r consecutive relevant bits in X^3 such they depend only on the y_r relevant bits in Y^2. Since those bits are linear in unknowns from

X^1, we can formulate also the x_r bits of X^3 as those. Since they are consecutive, we can guess $g_3 = \lfloor (x_r - 2)/2 \rfloor$ bits at consecutive even indices among them. This yields again g_3 equations for the unknowns form X^1. Moreover, the guessed bits allow to formulate $2g_3$ further linear equations for $2g_3$ consecutive bits of Y^3 through the S-layer since we already formulated Y^3 as equations from the backward direction before.

We require that the x_r consecutive bits in X^3 depend only on the $y_r = 2g_2$ bits of Y^2. We guess $g = g_1 + g_2 + g_3 = \lceil n/2 \rceil + g_2 + \lfloor (x_r - 2)/2 \rfloor$ bits at X^1, X^2, and X^3. We obtain $g_2 + g_3 + 2 \cdot g_3$ equations for the $n - \lceil n/2 \rceil$ unknowns of X^1. The equation system can be solved with computational complexity

$$t = 2^{g_1 + g_2 + g_3} \cdot (n - \lceil n/2 \rceil)^{2.8} \text{ binary operations.}$$

Rasta with r rounds consists of $(r + 1)(n^2 + n)$ XORs in the linear layers, $2rn$ XORs in the S-layers, n XORs in the feed-forward as well as $(r+1)n^2$ multiplications by constants in the linear layers, plus rn multiplications in the S-layers. For $r = 3$, this sums to $8n^2 + 14n$ binary operations. We use this figure to compare the effort in terms of three- round Rasta encryptions and optimize the choice of g_2, g_3, and x_r with respect to lowest computational effort while remaining a practical probability that a weak matrix occurs of at least 2^{-80}. The results of the optimal values for the interesting state sizes are provided in Table 7. We implemented a proof of concept of the attack for small state sizes of $n \in \{11, 21\}$, tested it with 100 random keys and weak affine layers each.

3.5 Experiments on Small Block Size Variants of Rasta

Number of Monomials. To confirm the analysis made in Sect. 3.1, we looked at a reduced version with $n = 21$ bits. We start with a state where each bit is a variable and compute one round, that is $S \circ A(x)$. As expected, we found out that all the monomials of degree less or equal to 2 are present in the resulting state. Similarly, we start with a formal state x and compute $S^{-1} \circ A^{-1}(x)$. Since $n = 21$, the degree of S^{-1} is 11. On average on 20 instances (i.e. 20 different linear layers), we obtained that the resulting state contains 1394774 monomials out of the 1401292 monomials of degree less or equal to 11, that is 99.5%.

Practical Gröbner basis experiments. As mentioned above, [9] provided some experiments on computing a Gröbner basis using the algorithm F4 as implemented in Magma. Those experiments, limited to two and three rounds and a block size of at most 101 (for two rounds) and 49 (for three rounds) confirmed our intuition that the security of Rasta is not threatened by this attack vector.

Practical SAT solver experiments. To practically evaluate the performance of automatic solvers for Rasta, we solved small-scale key-recovery challenges with a state-of-the-art SAT solver. Each challenge consists of finding a k-bit key input that generates a randomly generated keystream, for a randomly sampled

nonce. Such a challenge is expected to have a valid solution with a probability
of roughly 63 %, which was confirmed by our experiments. We translated 8 challenges per depth $d \in \{1, 2, 3\}$ and block size $k \in \{11, 13, \ldots, 41\}$ (if $d = 1$) or
$k \in \{11, 13, \ldots, 31\}$ (if $d > 1$) to a 3-SAT instance in conjunctive normal form
(CNF) and solved with parallelized SAT solver `plingeling` [7]. The minimum,
average, and maximum runtime out of the 8 challenges per parameter configuration is illustrated in Fig. 8. Only challenges that were (a) valid and (b) solved
within a time limit of 1 hour are included, so there are no results for $d = 3$
and $k > 25$, and only one result for $d = 2$ and $k > 27$. All runs were evaluated
on Intel Xeon E5-2630v3 CPUs (8 cores max. 3.2 GHz, 92 GB DDR4 RAM).
Surprisingly, the solvers performed worse than exhaustive search for all depths
$d > 1$. It appears the large, dense linear layers do not interact well with the SAT
solver's decision heuristics.

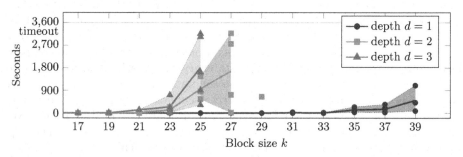

Fig. 8. SAT solver runtime for successful key-recovery on toy parameter sizes: Average,
minimum, and maximum runtime for up to 8 challenges per block size.

Linear properties. To test the linear properties of the cipher, we consider
versions with small block sizes and we compute the linear approximation table
of the full cipher. Recall that $LAT[a, b] = (1/2) \times W_{a,b}$ where $W_{a,b}$ is the Walsh
coefficient. In our case, we have: $W_{a,b} = \sum_{K \in \mathbb{F}_2^n} (-1)^{a \cdot K \oplus b \cdot K_{N,i}}$, where K is the
key and $K_{N,i}$ is the keystream for block i of nonce N.

   As explained previously, to measure the resistance of these reduced versions
we should not only look at the value of the higher coefficients of their LAT but
also see if it is likely that the same linear approximation holds with high probability for various instances. To apprehend these questions, we run the following
experiments for variants on 9 and 11 bits and for different number of rounds:

- Pick 50 linear layers at random. For each of them, look for the maximum
  (in absolute value) of the LAT coefficients. Report the maximum and the
  minimum over the 50 maximums, together with their average (first three
  lines of Table 8).
- Pick 50 linear layers at random. Compute the average LAT (that is a LAT
  where each coefficient is the average of the corresponding (absolute) values
  of the 50 instances). Look for the maximum of this LAT (line 4 in Table 8).

Table 8. Experiments on small scaled variants of Rasta. We look at 50 instances of each variant, and for each we note the maximum absolute coefficient of its LAT. We report below the maximum, the minimum and the average of these maximums. We also report the maximum of the average of the LAT.

$n = 9$	2 rounds	3 rounds	4 rounds	5 rounds	random
max	128	64	60	60	60
min	60	50	46	48	50
average	72.72	53.88	53.40	52.96	52.92
max in av. LAT	13.68	13.76	14.00	14.12	14.36

$n = 11$	2 rounds	3 rounds	4 rounds	5 rounds	random
max	384	160	132	130	126
min	184	112	112	110	112
average	256.64	121.92	119.44	117.64	118.40
max in av. LAT	29.44	30.72	29.32	28.96	29.20

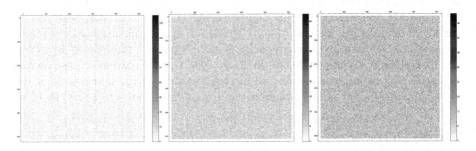

Fig. 9. Pollock representation of the LAT for 1, 2 and 3-round versions of one instance of the cipher on 9 bits.

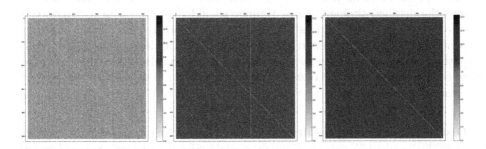

Fig. 10. Pollock representation of the average LAT for 1, 2 and 3-round versions over 50 instances of the cipher on 9 bits.

The obtained results in Table 8 show that the linear properties of variants of 4 and more rounds are close to the one of random permutations.

To visualize this graphically, we use the Pollock representation of the LAT as introduced in [11]. We start by looking at the LAT of specific instances of the small scaled ciphers (Fig. 9). On these, we can observe patterns for variants of 1 and 2 rounds, but these disappear for 2 rounds when looking at the average LAT for 50 random instances (see Fig. 10). Together with Table 8, this seems to indicate that it is really unlikely that several instances have a bias for the same input and output masks.

4   Validation of Design Approach Through Benchmarking of FHE Use-Case

To test the feasibility of our proposed design approach for various choices for the ANDdepth, we implemented Rasta using Helib [36,37], which implements the BGV homomorphic encryption scheme [15] and which was also used to evaluate AES-128 [30,31] and earlier custom designs that minimize the number of multiplications. Our implementation represents each plaintext, ciphertext and key bits as individual HE ciphertexts on which XOR and AND operations are performed. In the HE setting the number of AND gates is not the main determinant of complexity. Instead, the ANDdepth of the circuit largely determines the cost of XOR and AND, where AND is more expensive than XOR. However, due to the high number of XORs in Rasta, the cost of the linear layer is not negligible. In our implementation we use the "method of the four Russians" [1] to reduce the number of HE ciphertext additions from $\mathcal{O}(n^2)$ to $\mathcal{O}(n^2/\log(n))$.

Caveats. All earlier works of custom constructions use Helib for comparative timings. However things are vague in this part of the literature as security levels are not given (automatically determined by Helib from within a wide range of possibilities). Also the number of slots (i.e. blocks processed in parallel) is not under the direct control of the user (and as can be seen from the comparison tables in the literature). Hence we caution the reader to not interpret too much into the detailed timings. These timings should rather be seen as supporting evidence for the practical feasibility of a design approach. In contrast to earlier benchmarks we give the concrete (conjectured) security level of the instantiation of the underlying BGV scheme. We also try to get that BGV security level close to the security level of the cipher. Earlier comparisons were always done with at most 80-bit of security.

Discussion of timing results. A detailed overview of our benchmarks is given in Tables 9 and 10. We use the publicly available Helib implementation of LowMC and compare it with our implementation of Rasta in various settings. We try parameters for 80-bit, 128-bit, and 256-bit security. In addition to the case of a pure cipher evaluation (Table 9), we also consider the case where the BGV

Table 9. Performance comparison of Rasta on Intel(R) Core(TM) i7-4650U CPU @ 1.70GHz CPU with 8GB of RAM. BVG parameters chosen to allow homomorphic evaluation of cipher only. n is the block size or the number of encrypted bits, d is the multiplicative depth, # slots is the number of slots used in HElib, t_{total} is the total running time of the decryption in seconds and includes the time to generate all nonce-dependant computations. Trivium and Kreyvium estimated based on [17], FLIP estimated based on [44], both using LowMCv1 numbers as a point of reference and linear extrapolation.

Cipher	n	d	t_{total}	BGV slots	BGV levels	BGV security
80-bit cipher security						
LowMC v1	128	11	884.7	600	13	64.24
LowMC v2 (low latency)	128	12	546.0	600	14	75.82
LowMC v2 (throughput)	256	12	733.6	600	14	62.83
Trivium	57	12	~500.0	504	–	–
Trivium	136	13	~1500.0	682	–	–
FLIP	1	4	~0.4	378	5	–
Rasta	327	4	110.5	378	5	50.72
Rasta	327	5	206.2	150	7	78.60
Rasta	219	6	159.0	150	7	78.60
Agrasta	81	4	20.0	378	5	50.72
128-bit cipher security						
LowMC v1	256	12	2807.6	720	15	132.26
LowMC v1	256	12	2298.8	720	15	105.72
LowMC v2	256	14	2981.5	720	17	88.01
Kreyvium	46	12	~1090.0	504	–	–
Kreyvium	125	13	~2530.0	682	–	–
FLIP	1	4	~3.3	630	6	–
Rasta	525	5	464.5	330	7	103.97
Rasta	351	6	815.0	600	9	86.37
Agrasta	129	4	50.3	150	5	117.38
256-bit cipher security						
LowMC v2	512	18	8142.7	1285	20	107.67
Kreyvium	Not specified for this security level					
FLIP	Not specified for this security level					
Rasta	703	6	1345.5	720	9	149.43
Agrasta	257	5	1141.1	720	9	212.90

parameters are chosen such that there is enough "room" for more noise coming from operations that constitute the actual reason (Table 10). For comparability with work in [17,44] we also chose to allow for 7 additional levels for this purpose. Note that we also include the time spent on parts that are not depending on the

Table 10. Performance comparison of Rasta, like Table 9, but BGV parameters are chosen to allow homomorphic evaluation of 7 more levels on top of the cipher evaluation.

Cipher	n	d	t_{total}	BGV slots	BGV levels	BGV security
80-bit cipher security						
LowMC v1	128	11	2011.9	720	20	74.05
LowMC v2 (throughput)	256	12	1721.3	600	21	62.83
Trivium	57	12	~1560.0	504	–	–
Trivium	136	13	~4050.0	682	–	–
FLIP	1	4	~3.5	600	12	–
Rasta	327	4	397.8	224	12	89.57
Rasta	327	4	609.6	600	13	62.83
Rasta	327	5	766.7	600	14	62.83
Rasta	219	6	610.6	600	14	62.83
Agrasta	81	4	98.9	600	12	81.41
128-bit cipher security						
LowMC v1	256	12	3785.2	480	21	106.31
Kreyvium	12	42	~1760.0	504	–	–
Kreyvium	13	124	~4430.0	682	–	–
FLIP	1	4	~39.0	720	13	–
Rasta	525	5	912.1	682	14	90.39
Rasta	351	6	2018.6	720	15	110.74
Agrasta	129	4	217.4	682	12	127.50
256-bit cipher security						
LowMCv2	Too big to run					
Kreyvium	Not specified for this security level					
FLIP	Not specified for this security level					
Rasta	703	6	5543.2	720	16	89.93
Agrasta	257	5	1763.8	1800	15	210.68

key, i.e., to generate all the affine layers. For simplicity, we re-use the approach that is used in LowMC to generate random invertible matrices. Only for larger block sizes this is not completely negligible, but it always amounts for less than 10 % of the overall time. This confirms our model to focus on AND-related metrics of those parts of the algorithm that depend on the key.

We cannot directly compare Trivium, Kreyvium and FLIP on our machine as no HElib implementation for them is public. Trivium and Kreyvium numbers are estimates based on [17], FLIP numbers are estimates based on [44], both using LowMCv1 numbers as a point of reference and linear extrapolation.

As can be seen in the tables, even the very conservative Rasta can in several cases offer significantly lower latency than LowMC or Kreyvium/Trivium.

As FLIP is special in this table with the ability to take advantage of producing only a single keystream bit with minimal latency, the latency there is even lower. However, taking the number of encrypted bits into account the comparison with Rasta seems to be in favour of Rasta. One interesting thing to note here is that in some cases we were not able to successfully complete the cipher evaluation, despite seemingly moderate parameter sizes. At the 256-bit security level LowMC instances could only be computed purely, but not when 7 additional levels of homomorphic operations are still allowable. With Rasta it was possible. Other designs do not offer parameter-sets for such high security levels.

5 Conclusion and Future Work

Summary. We studied Rasta constructions where the substitution layer is chosen to be of low ANDdepth and public and fixed, but each affine layer is different, derived from a public nonce and various counters. Our conclusion is that they are interesting candidates for schemes that try to offer *simultaneously* a very low ANDdepth and a very low number of AND computations per encrypted bit. This contributes to a better understanding of the limits of what concrete symmetric-key constructions can achieve.

Implementations and Applications. Applications for symmetric schemes that minimize metrics like those we consider in this paper are currently investigated in various lines of work [2,4,17,34,44]. To test the applicability of our theoretical work, we chose the FHE setting (using HElib) and benchmarked actual implementations of Rasta, concluding that balanced choices of instantiations appear to be in the same ballpark as other specialized approaches. Our more aggressively parameterized variants termed 'Agrasta' result in our HElib experiments in an improvement of around one order of magnitude. It would be interesting to test applications of Rasta in various other settings, like secure multiparty communication protocols. Due to its low depth it will be especially beneficial for all those protocols where the round-complexity is linear in the ANDdepth of the evaluated circuit (e.g. GMW, tiny-OT).

Cryptanalysis. To better understand the security offered by Rasta, we explored various attack vectors including algebraic attacks, linear approximations and statistical attacks and choose parameters for the instantiations of Rasta to rule them out in a *conservative* way. While we conclude that known attacks do not threaten our design, we encourage further cryptanalysis and also proposed concrete toy versions to that end.

Improving the affine layer. As we have shown, the huge amount of XORs influences performance in targeted applications, and even more so considerably slows down a "plain" implementation of Rasta. New ideas for linear-layer design are needed which impose structure in one way or another which on one hand allows for significantly more efficient implementations while at the same time still resist attacks and allows for arguments against such attacks.

Acknowledgments. This research was supported by H2020 project PRISMACLOUD, grant agreement n° 644962 and by the Austrian Science Fund (project P26494-N15).

References

1. Albrecht, M.R., Bard, G.V., Hart, W.: Algorithm 898: efficient multiplication of dense matrices over GF(2). ACM Trans. Math. Softw. **37**(1), 9:1–9:14 (2010)
2. Albrecht, M., Grassi, L., Rechberger, C., Roy, A., Tiessen, T.: MiMC: efficient encryption and cryptographic hashing with minimal multiplicative complexity. In: Cheon, J.H., Takagi, T. (eds.) ASIACRYPT 2016. LNCS, vol. 10031, pp. 191–219. Springer, Heidelberg (2016). https://doi.org/10.1007/978-3-662-53887-6_7
3. Albrecht, M.R., Rechberger, C., Schneider, T., Tiessen, T., Zohner, M.: Ciphers for MPC and FHE. Cryptology ePrint Archive, Report 2016/687 (2016)
4. Albrecht, M.R., Rechberger, C., Schneider, T., Tiessen, T., Zohner, M.: Ciphers for MPC and FHE. In: Oswald, E., Fischlin, M. (eds.) EUROCRYPT 2015. LNCS, vol. 9056, pp. 430–454. Springer, Heidelberg (2015). https://doi.org/10.1007/978-3-662-46800-5_17
5. Bertoni, G., Daemen, J., Peeters, M., Van Assche, G.: The Keccak reference (version 3.0) (2011). http://keccak.noekeon.org
6. Bertoni, G., Daemen, J., Peeters, M., Van Assche, G.: Keccak specifications. Submission to NIST (Round 3) (2011). http://keccak.noekeon.org
7. Biere, A.: Lingeling, plingeling and treengeling entering the SAT Competition 2013. In: Balint, A., Belov, A., Heule, M., Järvisalo, M. (eds.) SAT Competition 2013, vol. B-2013-1, pp. 51–52 (2013). http://fmv.jku.at/lingeling/
8. Biham, E., Shamir, A.: Differential cryptanalysis of DES-like cryptosystems. In: Menezes, A.J., Vanstone, S.A. (eds.) CRYPTO 1990. LNCS, vol. 537, pp. 2–21. Springer, Heidelberg (1991). https://doi.org/10.1007/3-540-38424-3_1
9. Bile, C., Perret, L., Faugère, J.C.: Algebraic cryptanalysis of RASTA. Technical report (2017)
10. Biryukov, A., Bouillaguet, C., Khovratovich, D.: Cryptographic schemes based on the ASASA structure: black-box, white-box, and public-key (extended abstract). In: Sarkar, P., Iwata, T. (eds.) ASIACRYPT 2014. LNCS, vol. 8873, pp. 63–84. Springer, Heidelberg (2014). https://doi.org/10.1007/978-3-662-45611-8_4
11. Biryukov, A., Perrin, L.: On reverse-engineering S-Boxes with hidden design criteria or structure. In: Gennaro, R., Robshaw, M. (eds.) CRYPTO 2015. LNCS, vol. 9215, pp. 116–140. Springer, Heidelberg (2015). https://doi.org/10.1007/978-3-662-47989-6_6
12. Biryukov, A., Shamir, A.: Structural cryptanalysis of SASAS. J. Cryptol. **23**(4), 505–518 (2010)
13. Blum, A., Kalai, A., Wasserman, H.: Noise-tolerant learning, the parity problem, and the statistical query model. J. ACM **50**(4), 506–519 (2003)
14. Boyar, J., Peralta, R., Pochuev, D.: On the multiplicative complexity of Boolean functions over the basis (cap, +, 1). Theor. Comput. Sci. **235**(1), 43–57 (2000)
15. Brakerski, Z., Gentry, C., Vaikuntanathan, V.: Fully homomorphic encryption without bootstrapping. In: ECCC, vol. 18, p. 111 (2011)
16. Brakerski, Z., Gentry, C., Vaikuntanathan, V.: (Leveled) fully homomorphic encryption without bootstrapping. In: ITCS, pp. 309–325. ACM (2012)

17. Canteaut, A., Carpov, S., Fontaine, C., Lepoint, T., Naya-Plasencia, M., Paillier, P., Sirdey, R.: Stream ciphers: a practical solution for efficient homomorphic-ciphertext compression. In: Peyrin, T. (ed.) FSE 2016. LNCS, vol. 9783, pp. 313–333. Springer, Heidelberg (2016). https://doi.org/10.1007/978-3-662-52993-5_16
18. Chase, M., Derler, D., Goldfeder, S., Orlandi, C., Ramacher, S., Rechberger, C., Slamanig, D., Zaverucha, G.: Post-quantum zero-knowledge and signatures from symmetric-key primitives. In: CCS, pp. 1825–1842. ACM (2017)
19. Coron, J.-S., Lepoint, T., Tibouchi, M.: Scale-invariant fully homomorphic encryption over the integers. In: Krawczyk, H. (ed.) PKC 2014. LNCS, vol. 8383, pp. 311–328. Springer, Heidelberg (2014). https://doi.org/10.1007/978-3-642-54631-0_18
20. Courtois, N.T.: Fast algebraic attacks on stream ciphers with linear feedback. In: Boneh, D. (ed.) CRYPTO 2003. LNCS, vol. 2729, pp. 176–194. Springer, Heidelberg (2003). https://doi.org/10.1007/978-3-540-45146-4_11
21. Courtois, N.T., Meier, W.: Algebraic attacks on stream ciphers with linear feedback. In: Biham, E. (ed.) EUROCRYPT 2003. LNCS, vol. 2656, pp. 345–359. Springer, Heidelberg (2003). https://doi.org/10.1007/3-540-39200-9_21
22. Daemen, J.: Cipher and hash function design - strategies based on linear and differential cryptanalysis. Ph.D. thesis, Katholieke Universiteit Leuven (1995)
23. Cannière, C.: TRIVIUM: a stream cipher construction inspired by block cipher design principles. In: Katsikas, S.K., López, J., Backes, M., Gritzalis, S., Preneel, B. (eds.) ISC 2006. LNCS, vol. 4176, pp. 171–186. Springer, Heidelberg (2006). https://doi.org/10.1007/11836810_13
24. Dinur, I., Dunkelman, O., Kranz, T., Leander, G.: Decomposing the ASASA block cipher construction. Cryptology ePrint Archive, Report 2015/507 (2015)
25. Dinur, I., Liu, Y., Meier, W., Wang, Q.: Optimized interpolation attacks on LowMC. In: Iwata, T., Cheon, J.H. (eds.) ASIACRYPT 2015. LNCS, vol. 9453, pp. 535–560. Springer, Heidelberg (2015). https://doi.org/10.1007/978-3-662-48800-3_22
26. Dinur, I., Shamir, A.: Cube attacks on tweakable black box polynomials. In: Joux, A. (ed.) EUROCRYPT 2009. LNCS, vol. 5479, pp. 278–299. Springer, Heidelberg (2009). https://doi.org/10.1007/978-3-642-01001-9_16
27. Dobraunig, C., Eichlseder, M., Mendel, F.: Higher-order cryptanalysis of LowMC. In: Kwon, S., Yun, A. (eds.) ICISC 2015. LNCS, vol. 9558, pp. 87–101. Springer, Cham (2016). https://doi.org/10.1007/978-3-319-30840-1_6
28. Duval, S., Lallemand, V., Rotella, Y.: Cryptanalysis of the FLIP family of stream ciphers. In: Robshaw, M., Katz, J. (eds.) CRYPTO 2016. LNCS, vol. 9814, pp. 457–475. Springer, Heidelberg (2016). https://doi.org/10.1007/978-3-662-53018-4_17
29. Freedman, M.J., Ishai, Y., Pinkas, B., Reingold, O.: Keyword search and oblivious pseudorandom functions. In: Kilian, J. (ed.) TCC 2005. LNCS, vol. 3378, pp. 303–324. Springer, Heidelberg (2005). https://doi.org/10.1007/978-3-540-30576-7_17
30. Gentry, C., Halevi, S., Smart, N.P.: Homomorphic evaluation of the AES circuit. Cryptology ePrint Archive, Report 2012/099
31. Gentry, C., Halevi, S., Smart, N.P.: Homomorphic evaluation of the AES circuit. In: Safavi-Naini, R., Canetti, R. (eds.) CRYPTO 2012. LNCS, vol. 7417, pp. 850–867. Springer, Heidelberg (2012). https://doi.org/10.1007/978-3-642-32009-5_49
32. Gilbert, H., Plût, J., Treger, J.: Key-recovery attack on the ASASA cryptosystem with expanding S-Boxes. In: Gennaro, R., Robshaw, M. (eds.) CRYPTO 2015. LNCS, vol. 9215, pp. 475–490. Springer, Heidelberg (2015). https://doi.org/10.1007/978-3-662-47989-6_23

33. Goldreich, O., Micali, S., Wigderson, A.: How to play any mental game or a completeness theorem for protocols with honest majority. In: STOC, pp. 218–229. ACM (1987)

34. Grassi, L., Rechberger, C., Rotaru, D., Scholl, P., Smart, N.P.: MPC-friendly symmetric key primitives. In: CCS, pp. 430–443. ACM (2016)

35. Grosso, V., Leurent, G., Standaert, F.-X., Varıcı, K.: LS-Designs: bitslice encryption for efficient masked software implementations. In: Cid, C., Rechberger, C. (eds.) FSE 2014. LNCS, vol. 8540, pp. 18–37. Springer, Heidelberg (2015). https://doi.org/10.1007/978-3-662-46706-0_2

36. Halevi, S., Shoup, V.: Design and implementation of a homomorphic-encryption library (2013). https://github.com/shaih/HElib/

37. Halevi, S., Shoup, V.: Algorithms in HElib. In: Garay, J.A., Gennaro, R. (eds.) CRYPTO 2014. LNCS, vol. 8616, pp. 554–571. Springer, Heidelberg (2014). https://doi.org/10.1007/978-3-662-44371-2_31

38. Hazay, C., Lindell, Y.: Efficient protocols for set intersection and pattern matching with security against malicious and covert adversaries. In: Canetti, R. (ed.) TCC 2008. LNCS, vol. 4948, pp. 155–175. Springer, Heidelberg (2008). https://doi.org/10.1007/978-3-540-78524-8_10

39. Joux, A., Vitse, V.: A crossbred algorithm for solving Boolean polynomial systems. Cryptology ePrint Archive, Report 2017/372

40. Knudsen, L., Wagner, D.: Integral cryptanalysis. In: Daemen, J., Rijmen, V. (eds.) FSE 2002. LNCS, vol. 2365, pp. 112–127. Springer, Heidelberg (2002). https://doi.org/10.1007/3-540-45661-9_9

41. Lai, X.: Higher order derivatives and differential cryptanalysis. In: Communications and Cryptography: Two Sides of One Tapestry, pp. 227–233. Kluwer Academic Publishers (1994)

42. Laur, S., Talviste, R., Willemson, J.: From oblivious AES to efficient and secure database join in the multiparty setting. In: Jacobson, M., Locasto, M., Mohassel, P., Safavi-Naini, R. (eds.) ACNS 2013. LNCS, vol. 7954, pp. 84–101. Springer, Heidelberg (2013). https://doi.org/10.1007/978-3-642-38980-1_6

43. Matsui, M.: Linear cryptanalysis method for DES cipher. In: Helleseth, T. (ed.) EUROCRYPT 1993. LNCS, vol. 765, pp. 386–397. Springer, Heidelberg (1994). https://doi.org/10.1007/3-540-48285-7_33

44. Méaux, P., Journault, A., Standaert, F.-X., Carlet, C.: Towards stream ciphers for efficient FHE with low-noise ciphertexts. In: Fischlin, M., Coron, J.-S. (eds.) EUROCRYPT 2016. LNCS, vol. 9665, pp. 311–343. Springer, Heidelberg (2016). https://doi.org/10.1007/978-3-662-49890-3_13

45. Minaud, B., Derbez, P., Fouque, P.-A., Karpman, P.: Key-recovery attacks on ASASA. In: Iwata, T., Cheon, J.H. (eds.) ASIACRYPT 2015. LNCS, vol. 9453, pp. 3–27. Springer, Heidelberg (2015). https://doi.org/10.1007/978-3-662-48800-3_1

46. National Institute of Standards and Technology: FIPS PUB 202: SHA-3 Standard: Permutation-Based Hash and Extendable-Output Functions. U.S. Department of Commerce, August 2015

47. Nielsen, J.B., Nordholt, P.S., Orlandi, C., Burra, S.S.: A new approach to practical active-secure two-party computation. In: Safavi-Naini, R., Canetti, R. (eds.) CRYPTO 2012. LNCS, vol. 7417, pp. 681–700. Springer, Heidelberg (2012). https://doi.org/10.1007/978-3-642-32009-5_40

48. Raddum, H.: Personal Communication (2017)

49. Randall, D.: Efficient generation of random nonsingular matrices. Random Struct. Algorithms 4(1), 111–118 (1993)

Non-Uniform Bounds
in the Random-Permutation,
Ideal-Cipher, and Generic-Group Models

Sandro Coretti[1]($\boxtimes$), Yevgeniy Dodis[1], and Siyao Guo[2]

[1] New York University, New York, USA
{corettis,dodis}@nyu.edu
[2] Northeastern University, Boston, USA
s.guo@neu.edu

Abstract. The random-permutation model (RPM) and the ideal-cipher model (ICM) are idealized models that offer a simple and intuitive way to assess the conjectured standard-model security of many important symmetric-key and hash-function constructions. Similarly, the generic-group model (GGM) captures generic algorithms against assumptions in cyclic groups by modeling encodings of group elements as random injections and allows to derive simple bounds on the advantage of such algorithms.

Unfortunately, both well-known attacks, e.g., based on rainbow tables (Hellman, IEEE Transactions on Information Theory '80), and more recent ones, e.g., against the discrete-logarithm problem (Corrigan-Gibbs and Kogan, EUROCRYPT '18), suggest that the concrete security bounds one obtains from such idealized proofs are often *completely inaccurate* if one considers *non-uniform* or *preprocessing* attacks in the standard model. To remedy this situation, this work

– defines the auxiliary-input (AI) RPM/ICM/GGM, which capture both non-uniform and preprocessing attacks by allowing an attacker to leak an arbitrary (bounded-output) function of the oracle's function table;

– derives the *first* non-uniform bounds for a number of important practical applications in the AI-RPM/ICM, including constructions based on the Merkle-Damgård and sponge paradigms, which underlie the SHA hashing standards, and for AI-RPM/ICM applications with computational security; and

– using simpler proofs, recovers the AI-GGM security bounds obtained by Corrigan-Gibbs and Kogan against preprocessing attackers, for a number of assumptions related to cyclic groups, such as discrete

S. Coretti—Supported by NSF grants 1314568 and 1319051.

Y. Dodis—Partially supported by gifts from VMware Labs and Google, and NSF grants 1619158, 1319051, 1314568.

S. Guo—Supported by NSF grants CNS-1314722 and CNS-1413964; Part of this work done while the author was visiting the Simons Institute for the Theory of Computing at UC Berkeley.

H. Shacham and A. Boldyreva (Eds.): CRYPTO 2018, LNCS 10991, pp. 693–721, 2018.
https://doi.org/10.1007/978-3-319-96884-1_23

logarithms and Diffie-Hellman problems, and provides new bounds
for two assumptions.

An important step in obtaining these results is to port the tools used in
recent work by Coretti et al. (EUROCRYPT '18) from the ROM to the
RPM/ICM/GGM, resulting in very powerful and easy-to-use tools for
proving security bounds against non-uniform and preprocessing attacks.

1   Introduction

The random-permutation and ideal-cipher models. The random-permutation
model (RPM) and the ideal-cipher model (ICM) are idealized models that offer a
simple and intuitive way to prove the (conjectured) security of many important
applications. This holds especially true in the realms of symmetric cryptography and hash-function design since most constructions of block ciphers and hash
functions currently do not have solid theoretical foundations from the perspective
of provable security. In fact, the *exact security bounds* obtained in such idealized models are often viewed as guidance for both designers and cryptanalysts
in terms of the *best possible* security level that can be achieved by the corresponding construct in the standard model. By and large, this method has been
quite successful in practice, as most separations between the standard model and
various idealized models [3,8,10,11,28,35] are somewhat contrived and artificial
and are not believed to affect the security of widely used applications. In fact,
the following *RPM/ICM methodology* appears to be a good way for practitioners
to assess the best possible security level of a given (natural) application.

> **RPM/ICM methodology.** *For "natural" applications of hash functions
> and block ciphers, the concrete security proven in the RPM/ICM is the
> right bound even in the standard model, assuming the "best possible"
> instantiation for the idealized component (permutation or block cipher)
> is chosen.*

Both the RPM and the ICM have numerous very important practical applications. In fact, most practical constructions in symmetric-key cryptography and
hash-function design are naturally defined in the RPM/ICM. The following are
a few representative examples:

- The famous AES cipher is an example of key-alternating cipher, which can
  be abstractly described and analyzed in the RPM [2,12], generalizing the
  Even-Mansour [21,22] cipher $\mathsf{EM}_{\pi,s}(x) = \pi(x \oplus s) \oplus s$, where π is a public
  permutation, s is the secret key, and x is the message.
- The compression function of the SHA-1/2 [38,43] and MD5 [40] hash functions, as well as the popular HMAC scheme [4], is implemented via the Davies-
  Meyer (DM) hash function $\mathsf{DM}_E(x,y) = E_x(y) \oplus y$, for a block cipher E. But
  its collision-resistance can only be analyzed in the ICM [48].
- The round permutation of SHA-3 [37]—as part of the *sponge mode of operation* [6]—can be defined in the RPM: given old n-bit state s and new r-bit

block message x (where $r < n$), the new state is $s' = \pi(s \oplus (x \| 0^{n-r}))$, where π is a public permutation. The sponge mode is useful for building CRHFs, message authentication codes (MACs), pseudorandom functions (PRFs) [7], and key derivation functions [24], among others.

- The round function of MD6 [41] can be written as $f_Q(x) = \mathsf{trunc}_r(\pi(x \| Q))$, where Q is a constant, trunc_r is the truncation to r bits, and π is a public permutation. This construction was shown indifferentiable from a random oracle in the RPM [20].
- Many other candidate collision-resistant hash functions can be described using either ideal ciphers (e.g., the large PGV family [9]) or random permutations (e.g., [6,20,42,45]).

The generic group model. Another well-known idealized model is the so-called generic-group model (GGM), which serves the purpose of proving lower bounds on the complexity of generic attacks against common computational problems in cyclic groups used in public-key cryptography, such as the discrete-logarithm problem (DL), the computational and decisional Diffie-Hellman problems (CDH and DDH), and many more. Generic attacks are algorithms that do not exploit the specific representation of the elements of a group. This property is modeled by considering generic encoding captured by a random injection $\sigma : \mathbb{Z}_N \to [M]$ and allowing the algorithm access to a group-operation oracle, which, given a pair of encodings $(\sigma(x), \sigma(y))$, returns $\sigma(x + y)$.

The justification for the GGM is rooted in the fact that there are no unconditional hardness proofs for important group-related problems, and that there are some groups based on elliptic curves for which no better algorithms than the generic ones are known. Hence, results in the GGM provide at least some indication as to how sensible particular assumptions are. There are a plethora of security bounds proven in the GGM, e.g., lower bounds on the complexity of generic algorithms against DL or CDH/DDH by Shoup [44] or the knowledge-of-exponent assumption by Abe and Fehr [1] and Dent [17].

Non-uniformity and preprocessing. Unfortunately, a closer look reveals that the rosy picture above can only be true if one considers *uniform* attacks (as explained below). In contrast, most works (at least in theoretical cryptography) consider attackers in the *non-uniform* setting, where the attacker is allowed to obtain some arbitrary (but bounded) *advice* before attacking the system. The main rationale for this modeling comes from the realization that a determined attacker will know the parameters of a target system in advance and might be able to invest a significant amount of preprocessing to do something to speed up the actual attack, or to break many instances at once (therefore amortizing the one-time preprocessing cost). Perhaps the best known example of such attacks are *rainbow tables* [30,36] (see also [32, Sect. 5.4.3]) for inverting arbitrary functions; the idea is to use one-time preprocessing to initialize a clever data structure in order to dramatically speed up brute-force inversion attacks. Thus, restricting to uniform attackers might not accurately model realistic preprocessing attacks one would like to protect against.

There are also other, more technical, reasons why the choice to consider non-uniform attackers is convenient (see [14] for details), the most important of which is security under composition. A well-known example are zero-knowledge proofs [26,27], which are *not* closed under (even sequential) composition unless one allows non-uniform attackers and simulators. Of course, being a special case of general protocol composition, this means that any work that uses zero-knowledge proofs as a subroutine must consider security against *non-uniform* attackers in order for the composition to work. Hence, it is widely believed by the theoretical community that *non-uniformity is the right cryptographic modeling of attackers*, despite being overly conservative and including potentially unrealistic attackers—due to the potentially unbounded pre-computation allowed to generate the advice.

Idealized models vs. non-uniformity and preprocessing. When considering non-uniform attackers, it turns out that the RPM/ICM methodology above is blatantly false: once non-uniformity or preprocessing is allowed, the separations between the idealized models and the standard model are *no longer* contrived and artificial, but rather lead to *impossibly good* exact security of most *widely deployed* applications. To see this, consider the following examples:

- *One-way permutations:* Hellman [30] showed that there is a preprocessing attack that takes S bits of advice and makes T queries to a permutation $\pi : [N] \to N$ and inverts a random element of $[N]$ with probability roughly ST/N. Hence, a permutation cannot be one-way against attackers of size beyond $T = S = N^{1/2}$. However, in the RPM, a random permutation is easily shown to be invertible with probability at most T/N, therefore suggesting security against attackers of size up to N.
- *Even-Mansour cipher:* In a more recent publication, Fouque *et al.* [23] showed a non-uniform $N^{1/3}$ attack against the Even-Mansour cipher that succeeds with constant probability. As with OWPs, the analysis in the RPM model suggests an incorrect security level, namely, up to the birthday bound since one easily derives an upper bound of T^2/N on the distinguishing advantage of any attacker in RPM.

Similar examples also exist in the GGM:

- *Discrete logarithms:* A generic preprocessing attack by Mihalcik [34] and Bernstein and Lange [5] (and a recent variant by Corrigan-Gibbs and Kogan [15]) solves the DL problem with advantage ST^2/N in a group of order N, whereas the security of DL in the GGM is known to be T^2/N [44].
- *Square DDH:* A generic preprocessing attack by Corrigan-Gibbs and Kogan [15] breaks the so-called *square DDH (sqDDH)* problem—distinguishing (g^x, g^{x^2}) from (g^x, g^y) in a cyclic group $G = \langle g \rangle$ of order N—with advantage $\sqrt{ST^2/N}$, whereas the security of sqDDH in the GGM can be shown to be T^2/N.

Table 1. Asymptotic upper and lower bounds on the security of applications in the AI-ICM/AI-RPM and in the standard model (SM) against (S,T)-attackers.

	AI Security	SM Security	Best Attack
OWP	$\frac{ST}{N}$	$\frac{T}{N}$	$\frac{ST}{N}$ [30]
EM	$\left(\frac{ST^2}{N}\right)^{1/2} + \frac{T^2}{N}$	$\frac{T^2}{N}$	$\left(\frac{S}{N}\right)^{1/2}$ [16]
BC-IC	$\left(\frac{ST}{K}\right)^{1/2} + \frac{T}{K}$	$\frac{T}{K}$	$\left(\frac{S}{K}\right)^{1/2}$ [16]
PRF-DM	$\left(\frac{ST}{N}\right)^{1/2} + \frac{T}{N}$	$\frac{T}{N}$	$\left(\frac{S}{N}\right)^{1/2}$ [16]
CRHF-DM	$\frac{(ST)^2}{N}$	$\frac{T^2}{N}$	not known
CRHF-S	$\frac{ST^2}{2^c} + \frac{T^2}{2^r}$	$\frac{T^2}{2^c} + \frac{T^2}{2^r}$	$\frac{ST^2}{N}$ [14]
PRF-S	$\left(\frac{ST^2}{2^c}\right)^{1/2}$	$\frac{T^2}{2^c}$	$\left(\frac{S}{N}\right)^{1/2}$ [16]
MAC-S	$\frac{ST^2}{2^c} + \frac{T}{2^r}$	$\frac{T^2}{2^c} + \frac{T}{2^r}$	$\min\left\{\frac{ST}{N}, \left(\frac{S^2T}{N^2}\right)^{1/3}\right\} + \frac{T}{N}$ [30]
CRHF-MD	$\frac{ST^2}{N}$	$\frac{T^2}{N}$	$\frac{ST^2}{N}$ [14]
PRF-MD-N	$\left(\frac{ST^3}{N}\right)^{1/2} + \frac{T^3}{N}$	$\frac{T^3}{N}$	$\left(\frac{S}{N}\right)^{1/2}$ [16]
NMAC/HMAC	$\frac{ST^3}{N}$	$\frac{T^3}{N}$	$\min\left\{\frac{ST}{N}, \left(\frac{S^2T}{N^2}\right)^{1/3}\right\} + \frac{T}{N}$ [30]

1.1 Contributions: Non-Uniform Bounds in the RPM/ICM/GGM

Given the above failure of the idealized-models methodology, this paper revisits security bounds derived in the RPM, ICM, and GGM and re-analyzes a number of applications highly relevant in practice w.r.t. their security against non-uniform attackers or preprocessing. To that end, following the seminal work of Unruh [47] as well as follow-up papers by Dodis *et al.* [18] and Coretti *et al.* [14], the idealized models are replaced by weaker counterparts that adequately capture non-uniformity and preprocessing by allowing the attacker to obtain *oracle-dependent* advice. The resulting models, called the *auxiliary-input* RPM, ICM, and GGM, are parameterized by S ("space") and T ("time") and work as follows: The attacker $\mathcal{A}$ in the AI model consists of two entities $\mathcal{A}_1$ and $\mathcal{A}_2$. The first-stage attacker $\mathcal{A}_1$ is computationally unbounded, gets full access to the idealized primitive $\mathcal{O}$, and computes some advice z of size at most S. This advice is then passed to the second-stage attacker $\mathcal{A}_2$, who may make up to T queries to oracle $\mathcal{O}$ (and, unlike $\mathcal{A}_1$, may have additional application-specific restrictions, such as bounded running time, etc.). The oracle-dependent advice naturally maps to non-uniform advice when the random oracle is instantiated, and, indeed, none of the concerns expressed in the above examples remain valid in the AI-RPM/ICM/GGM.

Symmetric primitives. In the AI-RPM and AI-ICM, this work analyzes and derives non-uniform security bounds for (cf. Table 1 and Sect. 4):

- *basic applications* such as inverting a random permutation (OWP), the Even-Mansour cipher (EM), using the ideal cipher as a block cipher directly (BC-IC), the PRF security of Davies-Meyer (PRF-DM), the collision resistance of a salted version of the Davies-Meyer compression function (CRHF-DM);

Table 2. Asymptotic upper and lower bounds on the security of applications in the generic-group model against (S,T)-attackers in the AI-ROM; new bounds are in a bold-face font. The value t for the one-more DL problem stands for the number of challenges requested by the attacker. The attack against MDL succeeds with constant probability and requires that $ST^2/t + T^2 = \Theta(tN)$.

	AI-GGM Security	GGM Security	Best Attack
DL/CDH	$\frac{ST^2}{N} + \frac{T^2}{N}$	$\frac{T^2}{N}$	$\frac{ST^2}{N}$ [5, 15, 34]
t-fold MDL	$\left(\frac{S(T+t)^2}{tN} + \frac{(T+t)^2}{tN}\right)^t$	$\left(\frac{(T+t)^2}{tN}\right)^t$	see caption [15]
DDH	$\left(\frac{ST^2}{N}\right)^{1/2} + \frac{T^2}{N}$	$\frac{T^2}{N}$	$\frac{ST^2}{N}$ [5, 15, 34]
sqDDH	$\left(\frac{ST^2}{N}\right)^{1/2} + \frac{T^2}{N}$	$\frac{T^2}{N}$	$\left(\frac{ST^2}{N}\right)^{1/2}$ [15]
OM-DL	$\left(\frac{S(T+t)^2}{N}\right) + \frac{(T+t)^2}{N}$	$\frac{T^2}{N}$	$\frac{ST^2}{N}$ [5, 15, 34]
KEA	$\frac{ST^2}{N}$	$\frac{T^2}{N}$	not known

- the collision-resistance, the PRF security, and the MAC security of the *sponge construction*, which underlies the SHA-3 hashing standard;
- the collision-resistance of the *Merkle-Damgård construction with Davies-Meyer (MD-DM)*, which underlies the SHA-1/2 hashing standards, and PRF/MAC security of NMAC and HMAC.

Surprisingly, except for OWPs [16], no non-uniform bounds were known for any of the above applications; not even for applications as fundamental as BC-IC, Even-Mansour, or HMAC.

The bounds derived for OWP and the collision-resistance (CR) of Sponges and MD-DM are tight, i.e., there exist matching attacks by Hellman [30] (for OWPs) and by Coretti *et al.* [14] (for CR). For the remaining primitives significant gaps remain between the derived security bounds and the best known attacks. Closing these gaps is left as an interesting (and important) open problem.

Generic groups. In the AI-GGM, the following applications are analyzed w.r.t. their security against preprocessing (cf. Table 2 and Sect. 5): the discrete-logarithm problem (DL), the multiple-discrete-logarithms problem (MDL), the computational Diffie-Hellman problem (CDH), the decisional Diffie-Hellman problem (DDH), the square decisional Diffie-Hellman problem (sqDDH), the one-more discrete-logarithm problem (OM-DL), and the knowledge-of-exponent assumption (KEA).

- For DL, MDL, CDH, DDH, and sqDDH, the derived bounds match those obtained in recent work by Corrigan-Gibbs and Kogan [15]. As highlighted below, however, the techniques used in this paper allow for much simpler proofs than the one based on incompressibility arguments in [15]. All of these bounds are tight, except those for DDH, for which closing the gap remains an open problem.
- The bounds for OM-DL and KEA are new and may be non-trivial to derive using compression techniques.

Computational security. Idealized models such as the ROM, RPM, and ICM are also often used in conjunction with computational hardness assumptions such as one-way functions, hardness of factoring, etc. Therefore, this paper also analyzes the security of public-key encryption based on trapdoor functions (cf. Sect. 6) in the AI-RPM, specifically, of a scheme put forth by Phan and Pointcheval [39]. Other schemes in the AI-RPM/ICM, e.g., [29,31], can be analyzed similarly.

1.2 Methodology: Pre-Sampling

Bit-fixing oracles and pre-sampling. Unfortunately, while solving the issue of not capturing non-uniformity and preprocessing, the AI models are considerably more difficult to analyze than the traditional idealized models. From a technical point, the key difficulty is the following: *conditioned on the leaked value z, which can depend on the entire function table of $\mathcal{O}$, many of the individual values $\mathcal{O}(x)$ are no longer random to the attacker, which ruins many of the key techniques utilized in the traditional idealized models, such as lazy sampling programmability, etc.

One way of solving the above issues is to use incompressibility arguments, as introduced by Gennaro and Trevisan [25] and successfully applied to OWPs by De *et al.* [16], to the random-oracle model by Dodis *et al.* [14], and to the GGM by Corrigan-Gibbs and Kogan [15]. Compression-based proofs generally lead to tight bounds, but are usually quite involved and, moreover, seem inapplicable to computationally secure applications. Hence, this paper, adopts the much simpler and more powerful pre-sampling approach taken recently by Coretti *et al.* [14] and dating back to Unruh [47]. The pre-sampling technique can be viewed as a general reduction from the auxiliary-input model to the so-called *bit-fixing* (BF) model, where the oracle can be arbitrarily fixed on some P coordinates, for some parameter P, but the remaining coordinates are chosen at random and independently of the fixed coordinates. Moreover, the non-uniform S-bit advice of the attacker in this model can only depend on the P fixed points, but *not* on the remaining truly random points. This makes dealing with the BF model much easier than with the AI model, as many of the traditional proof techniques can again be used, provided that one avoids the fixed coordinates.

Bit-fixing vs. auxiliary input. In order for the BF model to be useful, this work shows that any (S,T)-attack in the AI-RPM/ICM/GGM model will have similar advantage in the P-BF-RPM/ICM/GGM model for an appropriately chosen P, up to an *additive* loss of $\delta(S,T,P) \approx ST/P$. Moreover, for the special case of unpredictability applications (e.g., CRHFs, OWFs, etc.), one can set P to be (roughly) ST, and achieve a *multiplicative* loss of 2 in the exact security. This gives a general recipe for dealing with the AI models as follows: (a) prove security $\varepsilon(S,T,P)$ of the given application in the P-BF model; (b) for unpredictability applications, set $P \approx ST$, and obtain final AI security roughly $2 \cdot \varepsilon(S,T,ST)$; (c) for general applications, choose P to minimize $\varepsilon(S,T,P) + \delta(S,T,P)$.

The proof of the above connection is based on a similar connection between the AI-ROM and BF-ROM shown by [14] (improving a weaker original bound of Unruh [47]). While borrowing a lot of tools from [14], the key difficulty is ensuring that the P-bit-fixing cipher, which "approximates" the ideal cipher conditioned on the auxiliary input z, is actually a valid cipher: the values at fixed points cannot repeat, and the remaining values are chosen at random from the "unused" values (similar issues arise for generic groups). Indeed, the proof in this paper is more involved and the resulting bounds are slightly worse than those in [14].

Using the power of pre-sampling to analyze the applications presented above, the technical bulk consists of showing the security of these applications in the easy-to-handle BF-RPM/ICM/GGM, and then using Theorem 1 to translate the resulting bound to the AI-RPM/ICM/GGM. Most of BF proofs are remarkably straightforward extensions of the traditional proofs (without auxiliary input), which is a great advantage of the pre-sampling methodology over other approaches, such as compression-based proofs.

Computational security. Note that, unlike compression-based techniques [15,18], pre-sampling can be applied to computational reductions, by "hardwiring" the pre-sampling set of size P into the attacker breaking the computational assumption. However, this means that P cannot be made larger than the maximum allowed running time t of such an attacker. Since standard pre-sampling incurs additive cost $\Omega(ST/P)$, one cannot achieve final security better that ST/t, irrespective of the value of ε in the (t, ε)-security of the corresponding computational assumption.

Fortunately, the multiplicative variant of pre-sampling for unpredictability applications sets the list size to be roughly $P \approx ST$, which is polynomial for polynomial S and T and can be made smaller than the complexity t of the standard-model attacker for the computational assumption used. Furthermore, even though the security of public-key encryption is *not* an unpredictability application, the analysis in Sect. 6 shows a way to use multiplicative pre-sampling for the part that involves the reduction to a computational assumption.

1.3   Related Work

Tessaro [46] also adapted the presampling technique by Unruh to the random-*permutation* model; the corresponding bound is suboptimal, however. De *et al.* [16] study the effect of salting for inverting a *permutation* as well as for a specific pseudorandom generator based on one-way permutations.

Corrigan-Gibbs and Kogan [15], investigate the power of preprocessing in the GGM. Besides deriving security bounds for a number of important GGM applications, they also provide new attacks for DL (based on [5,34]), MDL, and sqDDH.

The most relevant papers in the AI-ROM are those by Unruh [47], Dodis *et al.* [18], and Coretti *et al.* [14]. Chung *et al.* [13] study the effects of salting in the design of collision-resistant hash functions, and used Unruh's pre-sampling technique to argue that salting defeats pre-processing in this important case. However, they did not focus on the exact security and obtained suboptimal bounds (compared to the expected "birthday" bound obtained by [18]). Using salting to obtain non-uniform security was also advocated by Mahmoody and Mohammed [33], who used this technique for obtaining non-uniform black-box separation results.

The realization that multiplicative error is enough for unpredictability applications and can lead to non-trivial savings, is related to the work of Dodis *et al.* [19] in the context of improved entropy loss of key derivation schemes.

2 Capturing the Models

This section explains how the various idealized models considered in this paper—the ideal-cipher-model (ICM), the random-permutation model (RPM), and the generic-group model (GGM)—are captured. Attackers in these models are modeled as two-stage attackers $\mathcal{A} = (\mathcal{A}_1, \mathcal{A}_2)$, and applications as (single-stage) challengers C. Both $\mathcal{A}$ and C are given access to an oracle $\mathcal{O}$. Oracles $\mathcal{O}$ have two interfaces pre and main, where pre is accessible only to $\mathcal{A}_1$, which may pass auxiliary information to $\mathcal{A}_2$, and both $\mathcal{A}_2$ and C may access main. In certain scenarios it is also useful to consider an additional interface main-c that is only available to the challenger C.

Notation. Throughout this paper, P, K, N, and M are natural numbers and $[x] = \{0, \ldots, x-1\}$ for $x \in \mathbb{N}$. For applications in the generic-group model, $[N]$ is identified with the cyclic group $\mathbb{Z}_N$. Furthermore, denote by $\mathcal{P}_N$ the set of permutations $\pi : [N] \to [N]$ and by $\mathcal{I}_{N,M}$ the set of injections $f : [N] \to [M]$.

Oracles. An oracle $\mathcal{O}$ has two interfaces $\mathcal{O}$.pre and $\mathcal{O}$.main, where $\mathcal{O}$.pre is accessible only once before any calls to $\mathcal{O}$.main are made. Some oracles may also have an additional interface $\mathcal{O}$.main-c. Oracles used in this work are:

- *Auxiliary-input ideal cipher* AI-IC(K, N): Samples a random permutation $\pi_k \leftarrow \mathcal{P}_N$ for each $k \in [K]$; outputs all π_k at $\mathcal{O}$.pre; answers both forward and backward queries $(k, x) \in [N]$ at $\mathcal{O}$.main by the corresponding value $\pi_k(x) \in [N]$ or $\pi_k^{-1}(x) \in [N]$, respectively.
- *Bit-fixing ideal cipher* BF-IC(P, K, N): Takes a list at $\mathcal{O}$.pre of at most P query/answer pairs (without collisions for each k); samples a random permutation $\pi_k \leftarrow \mathcal{P}_N$ consistent with said list for each k; the other interfaces behave as with AI-IC.
- *Auxiliary-input random permutation* AI-RP(N): Special case of an auxiliary-input ideal cipher with $K = 1$.
- *Bit-fixing random permutation* BF-RP(P, N): Special case of a bit-fixing ideal cipher with $K = 1$.
- *Auxiliary-input generic group* AI-GG(N, M): Samples a random injection $\sigma \leftarrow \mathcal{I}_{N,M}$; outputs all of σ at $\mathcal{O}$.pre; answers *forward* queries $x \in [N]$ at $\mathcal{O}$.main by the corresponding value $\sigma(x) \in [N]$; answers *group-operation* queries (s, s') at $\mathcal{O}$.main as follows: if $s = \sigma(x)$ and $s' = \sigma(y)$ for some x, y, the oracle replies by $\sigma(x + y)$ and by $\perp$ otherwise; answers *inverse* queries s at interface $\mathcal{O}$.main-c by returning $\sigma^{-1}(s)$ if s is in the range of F and by $\perp$ otherwise.

- *Bit-fixing generic group* BF-GG(P, N, M): Samples a random size-N subset $\mathcal{Y}$ of $[M]$ and outputs $\mathcal{Y}$ at $\mathcal{O}$.pre; takes a list at $\mathcal{O}$.pre of at most P query/answer pairs without collisions and all answers in $\mathcal{Y}$; samples a random injection $\sigma \leftarrow \mathcal{I}_{N,M}$ with range $\mathcal{Y}$ and consistent with said list; the other interfaces behave as with AI-GG.
- *Standard model:* None of the interfaces offer any functionality.

The parameters P, K, N, and M are occasionally omitted in contexts where they are of no relevance. Similarly, whenever evident from the context, explicitly specifying which interface is queried is omitted. Note that the non-auxiliary-input versions of the above oracles can be defined by not offering any functionality at $\mathcal{O}$.pre. However, they are not used in this paper.

Attackers with oracle-dependent advice. Attackers $\mathcal{A} = (\mathcal{A}_1, \mathcal{A}_2)$ consist of a preprocessing procedure $\mathcal{A}_1$ and a main algorithm $\mathcal{A}_2$, which carries out the actual attack using the output of the preprocessing. Correspondingly, in the presence of an oracle $\mathcal{O}$, $\mathcal{A}_1$ interacts with $\mathcal{O}$.pre and $\mathcal{A}_2$ with $\mathcal{O}$.main.

Definition 1. *An (S, T)-attacker $\mathcal{A} = (\mathcal{A}_1, \mathcal{A}_2)$ in the $\mathcal{O}$-model consists of two procedures*

- *$\mathcal{A}_1$, which is computationally unbounded, interacts with $\mathcal{O}$.pre, and outputs an S-bit string, and*
- *$\mathcal{A}_2$, which takes an S-bit auxiliary input and makes at most T queries to $\mathcal{O}$.main.*

In certain contexts, if additional restrictions, captured by some parameters p, are imposed on $\mathcal{A}_2$ (e.g., time and space requirements of $\mathcal{A}_2$ or a limit on the number of queries of a particular type that $\mathcal{A}_2$ makes to a challenger it interacts with), $\mathcal{A}$ is referred to as (S, T, p)-attacker.

Applications. Let $\mathcal{O}$ be an arbitrary oracle. An application G in the $\mathcal{O}$-model is defined by specifying a challenger C, which is an oracle algorithm that has access to $\mathcal{O}$.main as well as possibly to $\mathcal{O}$.main-c, interacts with an attacker $\mathcal{A} = (\mathcal{A}_1, \mathcal{A}_2)$, and outputs a bit at the end of the interaction. The *success* of $\mathcal{A}$ on G in the $\mathcal{O}$-model is defined as

$$\mathrm{Succ}_{G,\mathcal{O}}(\mathcal{A}) := \mathsf{P}\big[\mathcal{A}_2^{\mathcal{O}.\mathsf{main}}(\mathcal{A}_1^{\mathcal{O}.\mathsf{pre}}) \leftrightarrow \mathsf{C}^{\mathcal{O}.\mathsf{main},\mathcal{O}.\mathsf{main\text{-}c}} = 1\big],$$

where $\mathcal{A}_2^{\mathcal{O}.\mathsf{main}}(\mathcal{A}_1^{\mathcal{O}.\mathsf{pre}}) \leftrightarrow \mathsf{C}^{\mathcal{O}.\mathsf{main},\mathcal{O}.\mathsf{main\text{-}c}}$ denotes the bit output by C after its interaction with the attacker. This work considers two types of applications, captured by the next definition.

Definition 2. *For an* indistinguishability *application G in the $\mathcal{O}$-model, the advantage of an attacker $\mathcal{A}$ is defined as*

$$\mathrm{Adv}_{G,\mathcal{O}}(\mathcal{A}) := 2\left|\mathrm{Succ}_{G,\mathcal{O}}(\mathcal{A}) - \frac{1}{2}\right|.$$

For an unpredictability *application G, the advantage is defined as*

$$\text{Adv}_{G,\mathcal{O}}(\mathcal{A}) := \text{Succ}_{G,\mathcal{O}}(\mathcal{A}).$$

An application G is said to be $((S, T, p), \varepsilon)$-*secure in the* $\mathcal{O}$-*model if for every* (S, T, p)-*attacker* $\mathcal{A}$,

$$\text{Adv}_{G,\mathcal{O}}(\mathcal{A}) \leq \varepsilon.$$

Combined query complexity. In order to state and prove Theorem 1 in Sect. 3, the interaction of some attacker $\mathcal{A} = (\mathcal{A}_1, \mathcal{A}_2)$ with a challenger C in the $\mathcal{O}$-model must be "merged" into a single entity $\mathcal{D} = (\mathcal{D}_1, \mathcal{D}_2)$ that interacts with oracle $\mathcal{O}$. That is, $\mathcal{D}_1^{(\cdot)} := \mathcal{A}_1^{(\cdot)}$ and $\mathcal{D}_2^{(\cdot)}(z) := \mathcal{A}_2^{(\cdot)}(z) \leftrightarrow \mathsf{C}^{(\cdot)}$ for $z \in \{0,1\}^S$. $\mathcal{D}$ is called the *combination of* $\mathcal{A}$ *and* C, and the number of queries it makes to its oracle is referred to as *the combined query complexity of* $\mathcal{A}$ *and* C. For all applications in this work, there exists an upper bound $T_G^{\text{comb}} = T_G^{\text{comb}}(S, T, p)$ on the combined query complexity of any attacker and the challenger.

3   Auxiliary Input vs. Bit Fixing

Since dealing with idealized models with *auxiliary input (AI)* directly is difficult, this section establishes useful connections between AI models and their *bit-fixing (BF)* counterparts, which are much less cumbersome to analyze. Specifically, for ideal ciphers, random permutations (as special cases of ideal ciphers), and generic groups, Theorem 1 below relates the advantage of attackers in a BF model to that in the corresponding AI model, allowing to translate the security of (1) *any* application at an *additive* security loss and of (2) *unpredictability* applications at a *multiplicative* security loss from the BF setting to the AI setting.

Theorem 1. *Let* $P, K, N, M \in \mathbb{N}$, $N \geq 16$, *and* $\gamma > 0$. *Moreover, let* $(\mathsf{AI}, \mathsf{BF}) \in \{(\mathsf{AI}\text{-}\mathsf{IC}(K, N), \mathsf{BF}\text{-}\mathsf{IC}(P, K, N)), (\mathsf{AI}\text{-}\mathsf{GG}(N, M), \mathsf{BF}\text{-}\mathsf{GG}(P, N, M))\}$. *Then,*

1. *if an application G is* $((S, T, p), \varepsilon')$-*secure in the* BF-*model, it is* $((S, T, p), \varepsilon)$-*secure in the* AI-*model, where*

$$\varepsilon \leq \varepsilon' + \frac{6(S + \log \gamma^{-1}) \cdot T_G^{\text{comb}}}{P} + \gamma;$$

2. *if an* unpredictability *application G is* $((S, T, p), \varepsilon')$-*secure in the* BF-*model for*

$$P \geq 6(S + \log \gamma^{-1}) \cdot T_G^{\text{comb}},$$

it is $((S, T, p), \varepsilon)$-*secure in the* AI-*model for*

$$\varepsilon \leq 2\varepsilon' + \gamma,$$

where T_G^{comb} *is the combined query complexity corresponding to G.*

Proof Outline

This section contains a brief outline of the proof of Theorem 1. The full proof of Theorem 1 is provided in the full version of this paper; it follows the high-level structure of the proof in [14], where a similar theorem is shown for the random-oracle model.

1. *Leaky sources vs. dense sources:* A (K, N)-cipher source X is the random variable corresponding to the function table of a cipher $F : [K] \times [N] \to [M]$. It turns out that if X has min-entropy $H_\infty(X) = K \log N! - S$ for some S, it can be replaced by a convex combination of so-called *dense* sources, which are fixed on a subset of the coordinates and have almost full min-entropy everywhere else:

Definition 3. *A (K, N)-cipher source X is called $(\bar{P}, 1 - \delta)$-dense for $\bar{P} = (P_1, \ldots, P_K) \in [N]^K$ if it is fixed on at most P_k coordinates $(k, \cdot)$ for each $k \in [K]$ and if for all families $I = \{I_k\}_{k \in [K]}$ of subsets I_k of non-fixed coordinates $(k, \cdot)$,*

$$H_\infty(X_I) \geq (1 - \delta) \sum_{k=1}^{K} \log(N - P_k)^{\underline{|I_k|}},$$

where $a^{\underline{b}} := a!/(a-b)!$ and X_I is X restricted to the coordinates in I. X is called $(1 - \delta)$-dense if it is $(0, 1 - \delta)$-dense, and $\bar{P}$-fixed if it is $(\bar{P}, 1)$-dense.

More concretely, one can prove that a cipher source X as above is close to a convex combination of finitely many $(\bar{P}', 1 - \delta)$-dense sources for some $\bar{P} = (P_1, \ldots, P_K)$ satisfying $\sum_{k=1}^{K} P_k \approx \frac{S}{\delta}$. The proof is an adaptation of the proof of the corresponding lemma for random functions in [14], the difference being that the version here handles cipher sources.

2. *Dense sources vs. bit-fixing sources:* Any dense source has a corresponding bit-fixing source, which is simply a function table chosen uniformly at random from all those that agree with the P fixed positions. It turns out that a T-query distinguisher's
   - *advantage* at telling a dense source and its corresponding bit-fixing source apart can be upper bounded by approximately $T\delta$, and that its
   - *probability of outputting 1* is at most a factor of approximately $2^{T\delta}$ larger when interacting with the bit-fixing as compared to the dense source.

   Compared to the case of random functions [14], some additional care is needed to properly handle *inverse* queries. Given the above, by setting $\delta \approx S/P$, one obtains additive and multiplicative errors of roughly ST/P and $2^{ST/P}$, respectively.

3. *From bit fixing to auxiliary input:* The above almost immediately implies that an application that is $((S, T), \varepsilon)$-secure in the BF-ICM is $((S, T), \varepsilon')$-secure in the AI-ICM for

$$\varepsilon' \approx \varepsilon + \frac{ST}{P}$$

and even

$$\varepsilon' \approx 2\varepsilon$$

if it is an unpredictability application, by setting $P \approx ST$. Observe that for the additive case, the final security bound in the AI-ICM is obtained by choosing P in a way that minimizes $\varepsilon(P) + ST/P$.

For the generic-group model, the proof proceeds similarly, with two important observations:

(a) once the range is fixed, a random injection behaves like a random permutation, which is covered by ideal ciphers as a special case;

(b) the group-operation oracle can be implemented by three (two inverse and one forward) calls to the injection.

4   Non-Uniform Bounds for Hash Functions and Symmetric Primitives

This section derives non-uniform security bounds for a number of primitives commonly analyzed in either the random-permutation model (RPM) or the ideal-cipher model (ICM). The primitives in question can be grouped into *basic*, *sponge-based*, and *Merkle-Damgård-based* applications.

In the following, for primitives in the RPM, $\pi, \pi^{-1} : [N] \to [N]$ denote the permutation and its inverse to which AI-RP(N) and BF-RP(P, N) offer access at interface main. Similarly, for primitives in the ICM $E, E^{-1} : [K] \times [N] \to [N]$ denote the ideal cipher and its inverse to which AI-IC(K, N) and BF-IC(P, K, N) offer access at interface main (cf. Sect. 2).

Basic applications. The security of the following basic applications in the RPM resp. ICM is considered:

- *One-way permutation inversion (OWP):* Given $\pi(x)$ for an $x \in [N]$ chosen uniformly at random, an attacker has to find x.
- *Even-Mansour cipher (EM):* The PRF security of the Even-Mansour cipher

$$\mathsf{EM}_{\pi,s}(m) := \pi(m \oplus s_2) \oplus s_1$$

with key $s = (s_1, s_2)$.
- *Ideal cipher as block cipher (ICM):* The PRF security of the ideal cipher used as a block cipher directly.
- *PRF security of Davies-Meyer (PRF-DM):* The PRF security of the Davies-Meyer (DM) compression function DM_E

$$\mathsf{DM}_E(h, m) := E(m, h) \oplus h$$

when h is used as the key.
- *A collision-resistant variant of Davies-Meyer (CRHF-DM):* The collision-resistance of a salted variant

$$\mathsf{DM}_{E,a,b}(h, m) := E(m, h) + am + bh$$

of the DM compression function, where the first-stage attacker $\mathcal{A}_1$ is unaware of the public random salt value (a, b).

Sponge-based constructions. The sponge construction is a popular hash-function design paradigm and underlies the SHA-3 hash-function standard. For $N = 2^n$, $r \leq n$, $c = n - r$, it hashes a message $m = m_1 \cdots m_\ell$ consisting of r-bit blocks m_i to $y := \mathsf{Sponge}_{\pi, \mathsf{IV}}(m)$ as follows, where $\mathsf{IV} \in \{0,1\}^c$ is a c-bit initialization vector (IV):[1]

1. Set $s_0 \leftarrow 0^r \| \mathsf{IV}$.
2. For $i = 1, \ldots, \ell$: set $s_i \leftarrow \pi(m_i \oplus s_{i-1}^{(1)} \| s_{i-1}^{(2)})$, where $s_{i-1}^{(1)}$ denotes the first r bits of s_{i-1} and $s_{i-1}^{(2)}$ the remaining c bits.
3. Output $y := s_\ell^{(1)}$.

This work considers the following applications based on the sponge paradigm:

- *Collision-resistance:* The collision resistance of the sponge construction for a randomly chosen public IV unknown to the first-stage attacker $\mathcal{A}_1$.
- *PRF security:* The PRF security of the sponge construction with the IV serving as the key.
- *MAC security:* The MAC security of the sponge construction with the IV serving as the key.

Merkle-Damgård constructions with Davies-Meyer: Another widely used approach to the design of hash functions is the well-known Merkle-Damgård paradigm. For a compression function $f : [N] \times [K] \rightarrow [N]$ and an IV $\mathsf{IV} \in [N]$, a message $\overline{m} = m_1 \cdots m_\ell$ consisting of ℓ blocks $m_i \in [K]$, is hashed to $y := \mathsf{MD}_{f, \mathsf{IV}}(\overline{m})$ as follows:[2]

1. Set $h_0 \leftarrow \mathsf{IV}$.
2. For $i = 1, \ldots, \ell$: set $h_i \leftarrow f(h_{i-1}, m_i)$.
3. Output $y := h_\ell$.

This work considers the Merkle-Damgård construction with f instantiated by the Davies-Meyer compression function

$$\mathsf{DM}_E(h, m) := E(m, h) \oplus h,$$

resulting in the *Merkle-Damgård-with-Davies-Meyer* function (MD-DM)

$$\mathsf{MD\text{-}DM}_{E, \mathsf{IV}}(\overline{m}) := \mathsf{MD}_{\mathsf{DM}_E, \mathsf{IV}}(\overline{m}),$$

which underlies the SHA-2 hashing standard. This work considers the following applications based on the MD-DM hash function:

- *Collision-resistance:* The collision resistance of the MD-DM construction for a randomly chosen public IV unknown to the first-stage attacker $\mathcal{A}_1$.
- *PRF security:* The PRF security of the NMAC/HMAC variants

$$\mathsf{NMAC}_{E, k}(\overline{m}) := \mathsf{DM}_E(k_1, \mathsf{MD\text{-}DM}_{E, k_2}(\overline{m}))$$

of the MD-DM construction with key $k = (k_1, k_2)$.
- *MAC security:* The MAC security of the NMAC/HMAC variant of the MD-DM construction.

---

[1] To keep things simple, no padding is considered here.
[2] As with the sponge construction, for simplicity no padding is considered here.

Discussion. The asymptotic security bounds derived for the applications listed above are summarized in Table 1. No non-uniform bounds were previously known for any of these primitives, except for OWPs, for which the same bound was derived by De *et al.* [16] using an involved, compression-based proof.

As can be seen from Table 1, a matching attack, derived by Hellman *et al.* [30], is known for OWPs. Moreover, for CRHFs based on sponges and Merkle-Damgård with Davies-Meyer, a variant of a recent attack by Coretti *et al.* [14] closely matches the derived bounds.[3] For the remaining applications, significant gaps remain: For indistinguishability applications such as BI-IC and PRFs, adapting an attack on PRGs by De *et al.* [16] results in an advantage of roughly $\sqrt{S/N}$. For the MAC applications, the best attacks are based on rainbow tables for inverting functions [30].

All security bounds are derived by following the bit-fixing approach: the security of a particular application is assessed in the *bit-fixing (BF)* RPM/ICM, and then Theorem 1 is invoked to obtain a corresponding bound in the *auxiliary-input (AI)* RPM/ICM and similarly for the random-permutation model. Deriving security bounds in the BF-ICM/RPM turns out to be quite straightforward, and all of the proofs closely follow the corresponding proofs in the ICM/RPM without auxiliary input; intuitively, the only difference is that one needs to take the list $\mathcal{L}$ of the at most P input/output pairs where $\mathcal{A}_1$ fixes the random permutation or the ideal cipher.

The security proofs for one-way permutations, the ideal cipher as block cipher, the collision-resistant variant of Davies-Meyer, collision-resistance of the sponge construction, and the PRF and MAC security of NMAC/HMAC with Davies-Meyer are provided after the brief overview below. The precise definitions of the remaining applications as well as the corresponding theorems and proofs can be found in the full version of this paper.

4.1   One-Way Permutations

The one-way-permutation inversion application G^{OWP} is defined via the challenger $\mathsf{C}^{\mathsf{OWP}}$ that randomly and uniformly picks an $x \in N$, passes $y := \pi(x)$ to the attacker, and outputs 1 if and only if the attacker returns x.

Theorem 2 below provides an upper bound on the success probability of any attacker in inverting π in the AI-RP$'$-model, which is defined as the AI-RP-model, except that no queries to π^{-1} are allowed. The bound matches known attacks (up to logarithmic factors) and are also shown by De *et al.* [16] via a more involved compression argument.

Theorem 2. *The application* G^{OWP} *is* $\left((S,T),\ \tilde{O}\left(\frac{ST}{N}\right)\right)$*-secure in the* AI-RP$'(N)$*-model for* $N \geq 16$.

Proof. It suffices to show that G^{OWP} is $\left((S,T), O\left(\frac{P+T}{N}\right)\right)$-secure in the BF-RP$'(P,N)$-model. Then, by observing that $T^{\mathsf{comb}}_{G^{\mathsf{OWP}}} = T + 1$, setting $\gamma := 1/N$

---

[3] The original attack by [14] was devised for Merkle-Damgård with a random compression function.

and $P = 2(S + \log N)(T + 1) = \tilde{O}(ST)$, and applying Theorem 1, the desired conclusion follows.

Assume $P + T < N/2$ since, otherwise, the bound of $O((P + T)/N)$ holds trivially. Let $\mathcal{A} = (\mathcal{A}_1, \mathcal{A}_2)$ be an (S, T)-attacker. Without loss of generality, assume $\mathcal{A}$ is deterministic and $\mathcal{A}_2$ makes distinct queries and always queries its output. Let $\mathcal{L} = \{(x_1', y_1'), \ldots, (x_P', y_P')\}$ be the list submitted by $\mathcal{A}_1$. Recall that the challenger uniformly and randomly picks an x from $[N]$ and outputs $y := \pi(x)$. Let $x_1, \ldots, x_T$ denote the queries made by $\mathcal{A}_2$ and let $y_i := \pi(x_i)$ for $i \in [T]$ be the corresponding answers. Let $\mathcal{E}$ be the event that y appears in $\mathcal{L}$ namely $x = x_i'$ for some $i \in [P]$. Note that

$$\mathrm{Succ}_{G,\mathsf{BF-RP}}(\mathcal{A}) \leq \mathsf{P}[\mathcal{E}] + \mathsf{P}[\exists i \in [T], x_i = x | \neg \mathcal{E}]$$

$$\leq \mathsf{P}[\mathcal{E}] + \sum_{i=1}^{T} \mathsf{P}[x_i = x | \neg \mathcal{E}, x_1 \neq x, \ldots, x_{i-1} \neq x] .$$

Observe that $\mathsf{P}[\mathcal{E}] \leq P/N$. Moreover, conditioned on $y \notin \mathcal{L}$ and any fixed choice of $(x_1, y_1), \ldots, (x_{i-1}, y_{i-1})$, x_i is a deterministic value while x is uniformly distributed over $[N] \setminus \{x_1, \ldots, x_{i-1}, x_1', \ldots, x_P'\}$. Thus,

$$\mathsf{P}[x_i = x | \neg \mathcal{E}, x_1 \neq x, \ldots, x_{i-1} \neq x] \leq 1/(N - P - T) \leq 2/N ,$$

where the second inequality uses $P + T < N/2$. Therefore, $\mathrm{Succ}_{G,\mathsf{BF-RP}}(\mathcal{A}) \leq \frac{P}{N} + \frac{2T}{N} = O(\frac{P+T}{N})$. □

4.2   The Ideal Cipher as a Block Cipher

The ideal cipher can be directly used as a block cipher even in the presence of leakage. The corresponding application $G^{\mathsf{BC\text{-}IC}}$ is defined via the following challenger $\mathsf{C}^{\mathsf{BC\text{-}IC}}$: it initially chooses random bit $b \leftarrow \{0,1\}$; if $b = 0$, it picks a key $k^* \leftarrow [K]$ uniformly at random, and answers forward queries $m \in [N]$ made by $\mathcal{A}_2$ by the value $E(k^*, m)$ and inverse queries $c \in [N]$ by $E^{-1}(k^*, c)$; if $b = 1$, forward queries m are answered by $f(m)$ and inverse queries c by $f^{-1}(c)$, where f is an independently chosen uniform random permutation; the attacker wins if and only if he correctly guesses b.

Theorem 3. *Application $G^{\mathsf{BC\text{-}IC}}$ is $\left((S, T, q), \tilde{O}\left(\frac{T}{K} + \sqrt{S(T + q)/K}\right)\right)$-secure in the* $\mathsf{AI\text{-}IC}(K, N)$*-model for $N \geq 16$.*

Proof. It suffices to show that $G^{\mathsf{BC\text{-}IC}}$ is $((S, T, q), O((T + P)/K))$-secure in the $\mathsf{BF\text{-}IC}(P, K, N)$-model since then the theorem follows by observing that $T_{G^{\mathsf{BC\text{-}IC}}}^{\mathsf{comb}} = T + q$, setting $\gamma := 1/N$ and

$$P := \sqrt{(S + \log N)(T + q)K} = \tilde{\Theta}\left(\sqrt{S(T + q)K}\right),$$

and applying Theorem 1.

Clearly, $\mathcal{A}_2$ only has non-zero advantage in guessing bit b if it makes a (forward or inverse) query involving the key k^* chosen by the challenger or if k^* appears in one of the prefixed query/answer pairs. The latter occurs with probability at most P/K, whereas the former occurs with probability at most $T/(K - (T + P)) \leq 2T/K$, using that $T + P \leq K/2$, an assumption one can always make since, otherwise, $G^{\text{BC-IC}}$ is trivially $O\left((T + P)/K\right)$-secure.   □

4.3   A Collision-Resistant Variant of Davies-Meyer

The plain Davies-Meyer (DM) compression function cannot be collision-resistant against non-uniform attackers, which begs the question of if and how it can be salted to withstand non-uniform attacks. To that end, let $N = K = 2^\kappa$ for some $\kappa \in \mathbb{N}$ and interpret $[N]$ as a finite field of size N. For two values $a, b \in [N]$, let

$$\text{DM}_{E,a,b}(h, m) := E(m, h) + am + bh .$$

Note that for $a = 0$ and $b = 1$, $\text{DM}_{E,a,b}$ is the usual DM compression function.

The application $G^{\text{CRHF-DM}}$ of collision-resistance of the salted DM function is defined via the following challenger $C^{\text{CRHF-DM}}$: it picks two random values $a, b \in [N]$ and passes them to the attacker; the attacker wins if and only if it returns two pairs $(h, m) \neq (h', m')$ such that $\text{DM}_{E,a,b}(h, m) = \text{DM}_{E,a,b}(h', m')$.

Theorem 4. $G^{\text{CRHF-DM}}$ is $\left((S, T), \tilde{O}\left(\frac{(ST)^2}{N}\right)\right)$-secure in the AI-IC$(N, N)$-model for $N \geq 16$.

Proof. At the cost of at most 2 additional queries to E, assume that the pairs (h, m) and (h', m') output by $\mathcal{A}_2$ are such that $\mathcal{A}_2$ has queried its oracle E on all points $\text{DM}_{E,a,b}$ would query E when evaluated on (h, m) and (h', m'). It suffices to show that $G^{\text{CRHF-DM}}$ is $\left((S, T), O\left(\frac{T^2}{N} + \frac{P(P+T)}{N}\right)\right)$-secure in the BF-IC$(P, N, N)$-model. Then, by observing that $T^{\text{comb}}_{G^{\text{CRHF-DM}}} = T + 2$, setting $\gamma := 1/N$ and $P = 2(S + \log N)(T + 2) = \tilde{O}(ST)$, and applying Theorem 1, the desired conclusion follows.

Set $T' := T + 2$ and consider an interaction of $\mathcal{A} = (\mathcal{A}_1, \mathcal{A}_2)$ and $C^{\text{CRHF-DM}}$ in the BF-IC(P, N, N)-model. Denote by $((k_i', x_i'), y_i')$ for $i = 1, \ldots, P$ the query/answer pairs prefixed by $\mathcal{A}_1$ and by $((k_i, x_i), y_i)$ for $i = 1, \ldots, T'$ the queries $\mathcal{A}_2$ makes to E. Let $\mathcal{E}$ be the event that there exists *no* collision among the prefixed values, i.e., there exist no $i \neq j$ such that

$$E(k_i', x_i) + ak_i' + bh_j' = E(k_j', x_j) + ak_j' + bh_j' \tag{1}$$

and that $b \neq 0$. For any fixed $i \neq j$, consider two cases:

1. $k_i \neq k_j$: in this case, the two pairs cause a collision if and only if

$$a = \frac{(y_j' - y_i') - b(x_i' - x_j')}{k_i' - k_j'} ,$$

   which happens with probability at most $1/N$.

2. $k_i = k_j$: in this case, $x_i' \neq x_j'$, and the two pairs cause a collision if and only if $b = (y_j' - y_i')/(x_i' - x_j')$, which happens with probability at most $1/N$ as well.

Summarizing, $\mathsf{P}[\neg \mathcal{E}] \leq (P^2 + 1)/N = O\left(P^2/N\right)$.

Moving to queries made by $\mathcal{A}_2$, let $\mathcal{E}_i'$ be the event that after the ith query made by $\mathcal{A}_2$, there exists no collision between any query pair and a prefixed pair or among the query pairs themselves; the corresponding conditions are analogous to (1). Consider the probability $\mathsf{P}[\neg \mathcal{E}_i' | \mathcal{E}_{i-1}', \mathcal{E}]$. If the ith query is a forward query, then a collision occurs only if $y_i = a(k_i - k_j) + b(x_i - x_j) + y_j$ for some $j < i$ or if the analogous condition holds for a collision with a prefixed pair and some $j \in \{1, \ldots, P\}$; if the ith query is a backward query, then a collision occurs only if

$$x_i = \frac{a(k_i - k_j) - (y_j - y_i)}{b}$$

for some $j < i$ or if the analogous condition holds for a collision with a prefixed pair and some $j \in \{1, \ldots, P\}$ (using that $b \neq 0$). In either case,

$$\mathsf{P}[\neg \mathcal{E}_i' | \mathcal{E}_{i-1}', \mathcal{E}] \leq \frac{(i-1) + P}{N - (T' + P)} \leq \frac{2((i-1) + P)}{N},$$

using that $T' + P \leq N/2$, an assumption on may always make since, otherwise, the desired bound holds trivially. Summarizing, setting $\mathcal{E}' := \mathcal{E}_{T'}'$,

$$\mathsf{P}[\neg \mathcal{E}' | \mathcal{E}] = \mathsf{P}[\neg \mathcal{E}_{T'}' | \mathcal{E}] \leq \sum_{i=1}^{T'} \mathsf{P}[\neg \mathcal{E}_i' | \mathcal{E}_{i-1}', \mathcal{E}]$$

$$\leq \sum_{i=1}^{T'} \frac{2((i-1) + P)}{N}$$

$$= O\left(\frac{T^2}{N} + \frac{TP}{N}\right).$$

Clearly, $\mathcal{A}_2$ only wins if $\mathcal{E}$ or $\mathcal{E}'$ occurs, and hence the overall security in the BF-IC(P, N, N)-model is $O\left(\frac{T^2}{N} + \frac{P(P+T)}{N}\right)$.   □

4.4   CRHFs from Unkeyed Sponges

The application $G^{\mathsf{CRHF\text{-}S}}$ of collision resistance for the sponge construction is defined via the following challenger $\mathsf{C}^{\mathsf{CRHF\text{-}S}}$: it picks an initialization vector $\mathsf{IV} \leftarrow \{0,1\}^c$ uniformly at random, passes it to the attacker, and outputs 1 if and only if the attacker returns two messages $m \neq m'$ such that $\mathsf{Sponge}_{\pi, \mathsf{IV}}(m) = \mathsf{Sponge}_{\pi, \mathsf{IV}}(m')$.

The following theorem provides an upper bound on the probability that an (S, T, ℓ)-attacker finds a collision of the sponge construction in the AI-RPM, where ℓ is an upper bound on the lengths of the messages m and m' the attacker submits to the challenger. The proof follows the approach by Bertoni et al. [6].

Theorem 5. *Application* $G^{\mathsf{CRHF\text{-}S}}$ *is* $\left((S, T, \ell), \tilde{O}\left(\frac{S(T+\ell)^2}{2^c} + \frac{(T+\ell)^2}{2^r}\right)\right)$-*secure in the* AI-RP$(N)$-*model, for* $N = 2^n = 2^{r+c} \geq 16$.

Node graphs. A useful formalism for security proofs of sponge-based constructions is that of node and supernode graphs, as introduced by Bertoni *et al.* [6]. For a permutation $\pi : \{0,1\}^n \rightarrow \{0,1\}^n$, consider the following (directed) *node graph* $G_\pi = (V,E)$ with $V = \{0,1\}^r \times \{0,1\}^c = \{0,1\}^n$ and $E = \{(s,t) \mid \pi(s) = t\}$. Moreover, let $G'_\pi = (V',E')$ be the (directed) *supernode graph*, with $V' = \{0,1\}^c$ and $(s^{(2)}, t^{(2)}) \in E'$ iff $((s^{(1)}, s^{(2)}), (t^{(1)}, t^{(2)})) \in E$ for some $s^{(1)}, t^{(1)} \in \{0,1\}^r$. Observe that the value of $\mathsf{Sponge}_{\pi,\mathsf{IV}}(m)$ for an ℓ-block message $m = m_1 \cdots m_\ell$ is obtained by starting at $s_0 := (0^r, \mathsf{IV}) \in \{0,1\}^n$ in G_π, moving to $s_i \leftarrow \pi(m_i \oplus s_{i-1}^{(1)} \| s_{i-1}^{(2)})$ for $i = 1, \ldots, \ell$, and outputting $s_\ell^{(1)}$. In other words, in the supernode graph, m corresponds to a path of length ℓ starting at node IV and ending at $s_\ell^{(2)}$, and $s_1^{(1)}, \ldots, s_\ell^{(1)} \in \{0,1\}^r$ are the values that *appear* on that path.

Proof. At the cost of at most 2ℓ additional queries to π, assume that the messages m and m' output by $\mathcal{A}_2$ are such that $\mathcal{A}_2$ has queried its oracle π on all points $\mathsf{Sponge}_{\pi,\mathsf{IV}}(\cdot)$ would query π when evaluated on m and m'.

It suffices to show that $G^{\mathsf{CRHF\text{-}S}}$ is $\left((S,T,\ell), O\left(\frac{(T+\ell)^2}{2^r} + \frac{(T+\ell)^2 + (T+\ell)P)}{2^c}\right)\right)$-secure in the $\mathsf{BF\text{-}RP}(P,N)$-model. Then, by observing that $T^{\mathsf{comb}}_{G^{\mathsf{CRHF\text{-}S}}} = T + 2\ell$, setting $\gamma := 1/N$ and $P := 2(S + \log N)(T + \ell) = \tilde{O}\left(S(T+\ell)\right)$, and applying Theorem 1, the desired conclusion follows.

Consider now an interaction of $\mathcal{A}_2$ with $\mathsf{C}^{\mathsf{CRHF\text{-}S}}$ and incrementally build the node and supernode graphs (as defined above), adding edges when $\mathcal{A}_2$ makes the corresponding (forward or inverse) query to π, and starting with the edges that correspond to the at most P prefixed query/answer pairs.

Let $\mathcal{E}_{\mathsf{coll}}$ be the event that a (valid) collision occurs. Clearly, this happens if and only if there exists a value $s^{(1)} \in \{0,1\}^r$ that appears as the last value on two different paths from IV. Let $\mathcal{E}_{\mathsf{path},i}$ be the event that after the ith query to π, there is a unique path from IV to any node in the supernode graph and that no prefixed supernode is reachable from IV.

Observe that when $\mathcal{E}_{\mathsf{path}} := \mathcal{E}_{\mathsf{path},T+2\ell}$ occurs, the values that appear on these paths are uniformly random and independent since every node inside a supernode has the same probability of being chosen. Hence,

$$\mathsf{P}[\mathcal{E}_{\mathsf{coll}} | \mathcal{E}_{\mathsf{path}}] \leq \binom{T + 2\ell}{2} \cdot 2^{-r} = O\left(\frac{(T+\ell)^2}{2^r}\right).$$

Moreover,

$$\mathsf{P}[\neg\mathcal{E}_{\mathsf{path},i} | \mathcal{E}_{\mathsf{path},i-1}] \leq \frac{(i+P) \cdot 2^r}{2^{r+c} - (i - 1 + P)} \leq \frac{i + P}{2^c - (T + 2\ell + P)/2^r} \leq \frac{i + P}{2^{c-1}}$$

if the ith query is a forward query, and

$$\mathsf{P}[\neg\mathcal{E}_{\mathsf{path},i} | \mathcal{E}_{\mathsf{path},i-1}] \leq \frac{i \cdot 2^r}{2^{r+c} - (i - 1 + P)} \leq \frac{i}{2^{c-1}}$$

if the ith query is an inverse query, using that $T + 2\ell + P \leq N/2$, an assumption one may always make since, otherwise, the lemma holds trivially. Letting $T' := T + 2\ell$,

$$
\begin{aligned}
\mathsf{P}[\neg\mathcal{E}_{\mathsf{path}}] = \mathsf{P}[\neg\mathcal{E}_{\mathsf{path},T'}] &\leq \mathsf{P}[\neg\mathcal{E}_{\mathsf{path},T'}|\mathcal{E}_{\mathsf{path},T'-1}] + \mathsf{P}[\neg\mathcal{E}_{\mathsf{path},T'-1}] \\
&\leq \sum_{i=1}^{T'} \mathsf{P}[\neg\mathcal{E}_{\mathsf{path},i}|\mathcal{E}_{\mathsf{path},i-1}] + \mathsf{P}[\mathcal{E}_{\mathsf{path},0}] \\
&\leq \sum_{i=0}^{T'} \frac{(i+P)}{2^{c-1}} = O\left(\frac{(T+\ell)(T+\ell+P)}{2^{-c}}\right),
\end{aligned}
$$

observing that $\mathsf{P}[\mathcal{E}_{\mathsf{path},0}] \leq \frac{P}{2^c}$, the probability that a node inside supernode IV is prefixed. □

4.5   PRFs via NMAC with Davies-Meyer

For simplicity, let $K = N$. Recall that the NMAC construction using the Davies-Meyer compression function is defined as

$$
\mathsf{NMAC}_{E,k}(\overline{m}) := \mathsf{DM}_E(k_1, \mathsf{MD\text{-}DM}_{E,k_2}(\overline{m}))
$$

where $k = (k_1, k_2)$.

The application $G^{\mathsf{PRF\text{-}MD\text{-}N}}$ of PRF security for NMAC is defined via the following challenger $\mathsf{C}^{\mathsf{PRF\text{-}MD\text{-}N}}$: it picks a random bit $b \leftarrow \{0,1\}$ and a key $k \leftarrow [N]$; when the attacker queries a message $m = m_1 \cdots m_\ell$ consisting of blocks m_i, if $b = 0$, the challenger answers by $\mathsf{NMAC}_{E,k}(\overline{m})$, and, if $b = 1$, the challenger answers by a value chosen uniformly at random for each $\overline{m}$. The attacker wins, if and only if he correctly guesses the bit b.

The following theorem provides an upper bound on the advantage of an (S, T, q, ℓ)-attacker in distinguishing the sponge construction from a random function in the AI-ICM, where q is an upper bound on the number of messages $\overline{m}$ the attacker submits to the challenger and ℓ is an upper bound on the length of those messages.

Theorem 6. $G^{\mathsf{PRF\text{-}MD\text{-}N}}$ is $\left((S, T, q, \ell), \tilde{O}\left(\frac{Tq^2\ell}{N} + \sqrt{\frac{S(T+\ell q)q^2\ell}{N}}\right)\right)$-secure in the AI-IC$(N, N)$-model, for $N \geq 16$.

Proof. Follows from Lemma 1 by observing that $T^{\mathsf{comb}}_{G^{\mathsf{PRF\text{-}MD\text{-}N}}} = T + q\ell$, setting $\gamma := 1/N$ and $P := \sqrt{\frac{S(T+q\ell)N}{q^2\ell}}$, and applying Theorem 1. □

Lemma 1. *For any* $P, N \in \mathbb{N}$, $G^{\mathsf{PRF\text{-}MD\text{-}N}}$ *is* $\left((S, T, q, \ell), O\left(q^2\ell\frac{T+P}{N}\right)\right)$*-secure in the* BF-IC$(P, N, N)$*-model.*

The proof of Lemma 1 uses the fact the Merkle-Damgård construction with the DM function is *almost-universal* in the BF-ICM; this property is captured by the application $G^{\text{AU-MD}}$ defined by the following challenger $C^{\text{AU-MD}}$: It expects $\mathcal{A}_2$ to submit two messages $\overline{m}$ and $\overline{m}'$. Then, it picks a random key k. The attacker wins if $\text{MD-DM}_{E,k}(\overline{m}) = \text{MD-DM}_{E,k}(\overline{m}')$.

The proof of almost-universality uses the fact that the DM function is a PRF when keyed by h (cf. full version of this paper).

Lemma 2. *For any* $P, N \in \mathbb{N}$, $G^{\text{AU-MD}}$ *is* $\left((S, T, q, \ell), O\left(\ell \frac{T+P}{N}\right)\right)$-*secure in the* BF-IC$(P, N, N)$-*model.*

Proof (sketch). Consider a sequence of ℓ hybrid experiments, where in the ith hybrid, instead of evaluating $\text{MD-DM}_{E,k}(\overline{m})$ for $\overline{m} = m_1 \cdots m_\ell$, the challenger computes $\text{MD-DM}_{E,k'}(m_{i+1} \cdots m_\ell)$, where $k' \leftarrow f(m_1 \cdots m_i)$ for a uniformly random function $f : [N]^i \to N$. By the PRF security of the Davies-Meyer function, the distance between successive hybrids is at most $8(T + P)/N$. Moreover, in the last hybrid, the success probability of $\mathcal{A}_2$ is at most $1/N$. □

Proof (of Lemma 1, sketch). Using the PRF security of the Davies-Meyer (DM) function, it suffices to show security in the hybrid experiment in which the outer DM evaluation is replaced by a uniform random function f. In this hybrid experiment, $\mathcal{A}_2$ only has non-zero advantage in guessing bit b if two of its q queries to the challenger cause a collision right before f. Let ε be the probability that this event occurs.

Consider the following attacker $\mathcal{A}' := (\mathcal{A}_1, \mathcal{A}'_2)$ against the $C^{\text{AU-MD}}$: $\mathcal{A}'_2$ runs $\mathcal{A}_2$ internally, forwarding its oracle queries to and back from its own oracle, and answering every query $\mathcal{A}_2$ would make to its challenger by a fresh uniformly random value. Once $\mathcal{A}_2$ terminates, $\mathcal{A}'_2$ picks a pair of queries made by $\mathcal{A}_2$ uniformly at random and submits it to its own challenger. It is easily seen that the advantage of $\mathcal{A}'_2$ is at least ε/q^2. Therefore, the final PRF security of NMAC is $q^2\ell(T + P)/N$. □

4.6 MACs via NMAC with Davies-Meyer

The application $G^{\text{MAC-MD-N}}$ of MAC security of the NMAC construction is defined via the following challenger $C^{\text{MAC-MD-N}}$: it initially picks a random key $k \leftarrow [N]$; when the attacker queries a message $\overline{m} = m_1 \cdots m_\ell$ consisting of blocks m_i, the challenger answers by $\text{MD-DM}_{\mathcal{O},k}(\overline{m})$. The attacker wins if he submits a pair $(\overline{m}, y)$ with $\text{MD-DM}_{\mathcal{O},k}(\overline{m}) = y$ for a previously unqueried $\overline{m}$.

Theorem 7. $G^{\text{MAC-MD-N}}$ *is* $\left((S, T, q, \ell), \tilde{O}\left(q^2\ell\frac{S(T+q\ell)}{N}\right)\right)$-*secure in the* AI-IC$(N)$-*model, for* $N \geq 16$.

Proof. It suffices to show that $G^{\text{MAC-MD-N}}$ is $\left((S, T, q, \ell), O\left(q^2\ell\frac{T+P}{N}\right)\right)$-secure in the BF-IC$(P, N)$-model. Then, by observing that $T^{\text{comb}}_{G^{\text{MAC-MD-N}}} = T + q\ell$, setting $\gamma := 1/N$ and $P = 2(S + \log N)(T + q\ell) = \tilde{O}\left(S(T + q\ell)\right)$ and applying Theorem 1, the desired conclusion follows.

The bound in the BF-IC(P, N)-model follows immediately from Lemma 1 and the fact that with a truly random function, the adversary's success probability at breaking the MAC is at most q/N. □

4.7 Extensions to HMAC

Recall that, for simplicity, $K = N$. The HMAC construction using the Davies-Meyer compression function is defined as

$$\mathsf{HMAC}_{E,k}(\overline{m}) := \mathsf{MD\text{-}DM}_{E,\mathsf{IV}}(k \oplus \mathsf{opad}, \mathsf{MD\text{-}DM}_{E,\mathsf{IV}}(k \oplus \mathsf{ipad}, \overline{m})) ,$$

where $\mathsf{IV} \in [N]$ is some fixed initialization vector. As usual, results for NMAC carry over to HMAC, even in the presence of leakage about the ideal cipher. More precisely, the HMAC construction can be seen as a special case of the NMAC by observing that

$$\mathsf{HMAC}_{E,k}(\overline{m}) = \mathsf{NMAC}_{E,k_1,k_2}(\overline{m})$$

for $k_1 = E(k \oplus \mathsf{ipad}, \mathsf{IV}) \oplus \mathsf{IV}$ and $k_2 = E(k \oplus \mathsf{opad}, \mathsf{IV}) \oplus \mathsf{IV}$. Hence, in the BF-IC-model, unless $(k \oplus \mathsf{opad}, \mathsf{IV})$ or $(k \oplus \mathsf{ipad}, \mathsf{IV})$ are prefixed by $\mathcal{A}_1$ or queried by $\mathcal{A}_2$, which happens with probability $O((T + P)/N)$, the NMAC analysis applies.

5   The Generic-Group Model with Preprocessing

This section analyzes the hardness of various problems in the generic-group model (GGM) with preprocessing. Specifically, the following applications are considered, where $N \in \mathbb{N}$ is an arbitrary prime and σ the random injection used in the GGM:

- *Discrete-logarithm problem (DL):* Given $\sigma(x)$ for a uniformly random $x \in [N]$, find x.
- *Multiple-discrete-logarithms problem (MDL):* Given $(\sigma(x_1), \ldots, \sigma(x_t))$ for uniformly random and independent $x_i \in [N]$, find $(x_1, \ldots, x_t)$.
- *Computational Diffie-Hellman problem (CDH):* Given $(\sigma(x), \sigma(y))$ for uniformly random and independent $x, y \in [N]$, find xy.
- *Decisional Diffie-Hellman problem (DDH):* Distinguish $(\sigma(x), \sigma(y), \sigma(xy))$ from $(\sigma(x), \sigma(y), \sigma(z))$ for uniformly random and independent $x, y, z \in [N]$.
- *Square decisional Diffie-Hellman problem (sqDDH):* Distinguish $(\sigma(x), \sigma(x^2))$ from $(\sigma(x), \sigma(y))$ for uniformly random and independent $x, y \in [N]$.
- *One-more-discrete-logarithm problem (OM-DL):* Given access to an oracle creating DL challenges $\sigma(x_i)$, for uniformly random and independent $x_i \in [N]$, as well as a DL oracle, make t queries to the challenge oracle and at most $t - 1$ queries to the DL oracle, and solve *all* t challenges, i.e., find $(x_1, \ldots, x_t)$.
- *Knowledge-of-exponent assumption (KEA):* The KEA assumption states that if an attacker $\mathcal{A}$ is given $\sigma(x)$, for $x \in [N]$ chosen uniformly at random, and outputs A and $\hat{A}$ with $A = \sigma(a)$ and $\hat{A} = \sigma(ax)$, then it must know discrete logarithm a of A. This is formalized by requiring that for every $\mathcal{A}$ there exist an *extractor* $\mathcal{X}_{\mathcal{A}}$ that is run on the same random coins as $\mathcal{A}$ and must output the value a.

The asymptotic security bounds derived for the above applications are summarized in Table 2. The bounds for DL, MDL, CDH, DDH, and sqDDH match previously known bounds from [5,15,34]; they are tight in that there is a matching attack, except for the DDH problem, for which, remarkably, closing the gap remains an open problem. The bounds for OM-DL and KEA are new.

Note that all bounds with preprocessing are considerably worse than those without. For example, in the classical GGM, DL is secure up to roughly $N^{1/2}$ queries, whereas it becomes insecure for $S = T = N^{1/3}$ in the AI-GGM.

All security bounds are derived by following the bit-fixing approach: the security of a particular application is assessed in the *bit-fixing (BF)* GGM, and then Theorem 1 is invoked to obtain a corresponding bound in the *auxiliary-input (AI)* GGM. This approach features great simplicity since deriving security bounds in the BF-GGM turns out to be remarkably straightforward, and all of the proofs closely follow the original proofs in the classical GGM without preprocessing; the only difference is that one needs to take the list $\mathcal{L}$ of the at most P input/output pairs where $\mathcal{A}_1$ fixes σ into account.

Besides simplicity, another advantage of the bit-fixing methodology is applicability: using bit-fixing, in addition to recovering all of the bounds obtained in [15] via much more involved compression-based proofs, one also easily derives bounds for applications that may be challenging to derive using compression-based proofs, such as, e.g., the knowledge-of-exponent assumption.

As representative examples, the proofs for the DL problem and the KEA are provided below. Readers familiar with the original proofs by Shoup [44] for DL and by Abe and Fehr [1] and Dent [17] for the KEA may immediately observe the similarity. The precise definitions of the remaining applications as well as the corresponding theorems and proofs can be found in the full version of this paper.

5.1 Discrete Logarithms

The discrete-logarithm application G^{DL} is defined via the challenger C^{DL} that randomly and uniformly picks an $x \in [N]$, passes $\sigma(x)$ to the attacker, and outputs 1 if and only if the attacker returns x.

Theorem 8 below provides an upper bound on the success probability of any attacker at computing discrete logarithms in the AI-GGM. The bound is matched by the attack of Mihalcik [34] and Bernstein and Lange [5]; a variation of said attack has recently also been presented by Corrigan-Gibbs and Kogan [15].

Theorem 8. G^{DL} *is* $((S,T),\varepsilon)$-*secure in the* $\mathsf{AI\text{-}GG}(N,M)$-*model for any prime* $N \geq 16$ *and*

$$\varepsilon = \tilde{O}\left(\frac{ST^2}{N} + \frac{T^2}{N}\right).$$

Proof. It suffices to show that the application G^{DL} is $\left((S,T), O\left(\frac{TP+T^2}{N}\right)\right)$-secure in the $\mathsf{BF\text{-}GG}(P,N,M)$-model. Then, by observing that $T_{G^{\mathsf{DL}}}^{\mathsf{comb}} = T + 1$, setting $\gamma := 1/N$ and $P = 6(S + \log N)(T + 1) = \tilde{O}(ST)$, and applying the second part of Theorem 1, the desired conclusion follows.

Consider now the interaction of $\mathcal{A} = (\mathcal{A}_1, \mathcal{A}_2)$ with C^{DL} in the BF-GG-model. Recall that the BF-GG-oracle outputs the range $\mathcal{Y}$ of the underlying random injection σ to $\mathcal{A}_1$ via interface pre. Condition on a particular realization of this set for the remainder of the proof.

Define the following hybrid experiment involving $\mathcal{A}_1$ and $\mathcal{A}_2$:

- For each of the at most P query/answer pairs (a', s') where $\mathcal{A}_1$ fixes σ, define a (constant) polynomial $v(X) := a'$ and store the pair (v, s').
- To create the challenge, choose a value s^* uniformly at random from all unused values in $\mathcal{Y}$, define the polynomial $u^*(X) := X$, and store (u^*, s^*).
- A forward query a by $\mathcal{A}_2$ to BF-GG is answered as follows: define the (constant) polynomial $u(X) := a$, choose a value s uniformly at random from all unused values in $\mathcal{Y}$, store the pair (u, s), and return s.
- A group-operation query (s_1, s_2) by $\mathcal{A}_2$ is answered as follows:
  • If s_1 or s_2 is not in $\mathcal{Y}$, return $\bot$.
  • If s_1 has not been recorded, choose a random unused $a \in [N]$, define the (constant) polynomial $u(X) := a$, and store the pair (u, a). Proceed similarly if s_2 has not been recorded. Go to the next item.
  • Let u_1 and u_2 be the polynomials recorded with s_1 and s_2, respectively. If, for $u' := u_1 + u_2$, a pair (u', s') has been recorded, return s'. Otherwise, choose a value s' uniformly at random from all unused values in $\mathcal{Y}$, store the pair (u', s'), and return s'.
- When $\mathcal{A}_2$ outputs a value x', pick a value $x \in [N]$ uniformly at random and output 1 if and only if $x' = x$.

Observe that the hybrid experiment only differs from the original one if for a group-operation query (s_1, s_2), $u'(x) = v(x)$ for some recorded v or $u'(x) = u(x)$ for some recorded u—and similarly for the polynomial u^* corresponding to the challenge. Since in the hybrid experiment, x is chosen uniformly at random *at the end* of the execution, the probability of this event is at most $((T + 1)P + (T + 1)^2)/N$ by the Schwartz-Zippel Lemma and a union bound. Moreover, in the hybrid experiment, the probability that $x' = x$ is $1/N$. The theorem follows. □

5.2   Knowledge-of-Exponent Assumption

Informally, the knowledge-of-exponent assumption (KEA) states that if an attacker $\mathcal{A}$ is given (h, h^x), for a generator h of a cyclic group of order N and $x \in [N]$ chosen uniformly at random, and outputs group elements A and $\hat{A}$ with $\hat{A} = A^x$, then it must know discrete logarithm a of A. This is formalized by requiring that for every $\mathcal{A}$ there exist an *extractor* $\mathcal{X}_\mathcal{A}$ that is run on the same random coins as $\mathcal{A}$ and must output the value a.

The above is captured in the GGM by considering the following experiment $\text{Exp}^{\mathcal{O}}_{\mathcal{A}, \mathcal{X}_\mathcal{A}}$ parameterized by an attacker $\mathcal{A} = (\mathcal{A}_1, \mathcal{A}_2)$, an extractor $\mathcal{X}_\mathcal{A}$, and an oracle $\mathcal{O} \in \{\text{AI-GG}(N, M), \text{BF-GG}(N, M)\}$:

1. Run $\mathcal{A}_1$ to obtain $z \leftarrow \mathcal{A}_1^{\mathcal{O}}$.
2. Choose $x \in [N]$ uniformly at random, let $y \leftarrow \sigma(x)$, pick random coins ρ, and run
   (a) $\mathcal{A}_2$ to get $(A, \hat{A}) \leftarrow \mathcal{A}_2(z, y; \rho)$, and
   (b) $\mathcal{X}_\mathcal{A}$ to get $a \leftarrow \mathcal{X}_\mathcal{A}(z, y; \rho)$.
3. Output 1 if and only if $A = \sigma(a')$ and $\hat{A} = \sigma(a'x)$ for some a', but $a \neq a'$.

The KEA says that for every attacker $\mathcal{A}$ there exists an extractor $\mathcal{X}_\mathcal{A}$ such that the probability of the above experiment outputting 1 is negligible. The following theorem is equivalent to saying that the KEA holds in the AI-GGM.

Theorem 9. *For every attacker $\mathcal{A} = (\mathcal{A}_1, \mathcal{A}_2)$, there exists an extractor $\mathcal{X}_\mathcal{A}$ such that*

$$\mathsf{P}[\mathrm{Exp}_{\mathcal{A}, \mathcal{X}_\mathcal{A}}^{\mathcal{O}} = 1] \leq \tilde{O}\left(\frac{ST^2}{N}\right).$$

Proof (Sketch). The extractor $\mathcal{X}_\mathcal{A}$ internally runs $\mathcal{A}_2$ on the inputs received and keeps track of $\mathcal{A}_2$'s oracle queries using polynomials as in the proof of Theorem 8. If at the end the polynomials u_A and $u_{\hat{A}}$ corresponding to $\mathcal{A}_2$'s outputs $(A, \hat{A})$ have the form $u_A(X) = a$ and $u_{\hat{A}}(X) = aX$, then $\mathcal{X}_\mathcal{A}$ outputs a and otherwise $\perp$.

Observe that if the experiment outputs 1, then

- $u_{\hat{A}} \neq X \cdot u_A$ since $\mathcal{A}_2$ only creates polynomials of degree at most 1, but
- $u_{\hat{A}}(x) = x \cdot u_A(x)$ for the challenge x.

Hence, the extractor only fails if at least two of the polynomials involved (including $u_{\hat{A}}$ and $X \cdot u_A$) collide on x, which is already analyzed in the proof of Theorem 8.

The experiment $\mathrm{Exp}_{\mathcal{A}, \mathcal{X}_\mathcal{A}}^{\mathcal{O}}$ defining KEA does not exactly match the syntax of challenger and attacker to which Theorem 1 caters, but it is easily checked that the corresponding proof can be adapted to fit $\mathrm{Exp}_{\mathcal{A}, \mathcal{X}_\mathcal{A}}^{\mathcal{O}}$.   □

6   Computationally Secure Applications

A main advantage of the pre-sampling methodology over other approaches (such as compression) to dealing with auxiliary-input in idealized models is that it also applies to applications that rely on computational hardness assumptions. To illustrate this fact, this section considers a public-key encryption scheme based on trapdoor functions by Phan and Pointcheval [39] in the auxiliary-input random-permutation model (AI-RPM). Other schemes in the AI-RPM/ICM, e.g., [29,31], can be analyzed similarly.

FDP encryption. Let F be a trapdoor family (TDF) generator. Full-domain permutation (FDP) encryption in the random-permutation model with oracle $\mathcal{O}$ is defined as follows:

- *Key generation:* Run the TDF generator to obtain $(f, f^{-1}) \leftarrow F$, where $f, f^{-1} : [N] \rightarrow [N]$. Set the public key $\mathsf{pk} := f$ and the secret key $\mathsf{sk} := f^{-1}$.

– *Encryption:* To encrypt a message m with randomness r and public key $\mathsf{pk} = f$, compute $\tilde{y} \leftarrow f(y)$ for $y \leftarrow \mathcal{O}(m\|r))$ and output $c = \tilde{y}$.
– *Decryption:* To decrypt a ciphertext $c = y$ with secret key $\mathsf{sk} = f^{-1}$, compute $m\|r \leftarrow \mathcal{O}^{-1}(f^{-1}(y))$ and output m.

The following theorem relates to the CPA security of FDP encryption in the AI-RPM.

Theorem 10. *Let Π be FDP encryption with F. If $G^{\mathsf{TDF},F}$ is $((S', *, t', s'), \varepsilon')$-secure, then, for any $T \in \mathbb{N}$, $G^{\mathsf{PKE},\Pi}$ is $((S, T, t, s), \varepsilon)$-secure in the* AI-RP$(N, N)$-*model, where*

$$\varepsilon = \tilde{O}\left(\varepsilon' + \sqrt{\frac{ST}{2^\rho}}\right)$$

and $S = S' - \tilde{O}(ST)$, $t = t' - \tilde{O}(t_{\mathsf{tdf}} \cdot T)$, and $s = s' - \tilde{O}(ST)$, where t_{tdf} is the time required to evaluate the TDF.

The straightforward approach to proving the security of FDP encryption in the AI-RPM would be to analyze the scheme in the BF-RPM with list size P and then use the general part of Theorem 1 to obtain a bound in the AI-RPM. However, such an approach, due to the additive error in the order of ST/P would require a very large list and therefore make the reduction to TDF security extremely loose.

Instead, the actual proof, which is sketched in the full version of this paper, follows the same high-level structure as that of TDF encryption in the AI-ROM, analyzed in [14]:

1. It first considers a hybrid experiment that is only distinguishable from the original CPA experiment if the attacker queries a particular value to the random permutation. To bound the probability of this event occurring, the proof moves to the BF-RPM and the analysis there—which involves the reduction to TDF security—is carried back to the AI-RPM via the *unpredictability* part of Theorem 1. This allows the list size to remain a moderate $P' \approx ST$ and hence for a tight reduction.
2. To analyze the advantage of the attacker in the hybrid experiment, the BF-RPM is used again, but using the general part of Theorem 1, which requires a larger list size P. However, since this second step involves no reduction to TDF security and is purely information-theoretic, this does not pose a problem.

Acknowledgments. The authors thank Dan Boneh, Henry Corrigan-Gibbs, and Dmitry Kogan for valuable discussions on pre-processing in generic-group models.

References

1. Abe, M., Fehr, S.: Perfect NIZK with adaptive soundness. In: Vadhan, S.P. (ed.) TCC 2007. LNCS, vol. 4392, pp. 118–136. Springer, Heidelberg (2007). https://doi.org/10.1007/978-3-540-70936-7_7
2. Andreeva, E., Bogdanov, A., Dodis, Y., Mennink, B., Steinberger, J.P.: On the indifferentiability of key-alternating ciphers. In: Canetti, R., Garay, J.A. (eds.) CRYPTO 2013, Part I. LNCS, vol. 8042, pp. 531–550. Springer, Heidelberg (2013). https://doi.org/10.1007/978-3-642-40041-4_29
3. Bellare, M., Boldyreva, A., Palacio, A.: An uninstantiable random-oracle-model scheme for a hybrid-encryption problem. In: Cachin, C., Camenisch, J.L. (eds.) EUROCRYPT 2004. LNCS, vol. 3027, pp. 171–188. Springer, Heidelberg (2004). https://doi.org/10.1007/978-3-540-24676-3_11
4. Bellare, M., Canetti, R., Krawczyk, H.: Keying hash functions for message authentication. In: Koblitz, N. (ed.) CRYPTO 1996. LNCS, vol. 1109, pp. 1–15. Springer, Heidelberg (1996). https://doi.org/10.1007/3-540-68697-5_1
5. Bernstein, D.J., Lange, T.: Non-uniform cracks in the concrete: the power of free precomputation. In: Sako, K., Sarkar, P. (eds.) ASIACRYPT 2013, Part II. LNCS, vol. 8270, pp. 321–340. Springer, Heidelberg (2013). https://doi.org/10.1007/978-3-642-42045-0_17
6. Bertoni, G., Daemen, J., Peeters, M., Van Assche, G.: On the indifferentiability of the sponge construction. In: Smart, N. (ed.) EUROCRYPT 2008. LNCS, vol. 4965, pp. 181–197. Springer, Heidelberg (2008). https://doi.org/10.1007/978-3-540-78967-3_11
7. Bertoni, G., Daemen, J., Peeters, M., Van Assche, G.: On the security of the keyed sponge construction. In: Symmetric Key Encryption Workshop (SKEW) (2011)
8. Black, J.: The ideal-cipher model, revisited: an uninstantiable blockcipher-based hash function. In: Robshaw, M. (ed.) FSE 2006. LNCS, vol. 4047, pp. 328–340. Springer, Heidelberg (2006). https://doi.org/10.1007/11799313_21
9. Black, J., Rogaway, P., Shrimpton, T.: Black-box analysis of the block-cipher-based hash-function constructions from PGV. In: Yung, M. (ed.) CRYPTO 2002. LNCS, vol. 2442, pp. 320–335. Springer, Heidelberg (2002). https://doi.org/10.1007/3-540-45708-9_21
10. Canetti, R., Goldreich, O., Halevi, S.: On the random-oracle methodology as applied to length-restricted signature schemes. In: Naor, M. (ed.) TCC 2004. LNCS, vol. 2951, pp. 40–57. Springer, Heidelberg (2004). https://doi.org/10.1007/978-3-540-24638-1_3
11. Canetti, R., Goldreich, O., Halevi, S.: The random oracle methodology, revisited. J. ACM **51**(4), 557–594 (2004)
12. Chen, S., Steinberger, J.: Tight security bounds for key-alternating ciphers. In: Nguyen, P.Q., Oswald, E. (eds.) EUROCRYPT 2014. LNCS, vol. 8441, pp. 327–350. Springer, Heidelberg (2014). https://doi.org/10.1007/978-3-642-55220-5_19
13. Chung, K.-M., Lin, H., Mahmoody, M., Pass, R.: On the power of nonuniformity in proofs of security. In: Innovations in Theoretical Computer Science, ITCS 2013, Berkeley, CA, USA, 9–12 January 2013, pp. 389–400 (2013)
14. Coretti, S., Dodis, Y., Guo, S., Steinberger, J.: Random oracles and non-uniformity. In: Nielsen, J.B., Rijmen, V. (eds.) EUROCRYPT 2018. LNCS, vol. 10820, pp. 227–258. Springer, Cham (2018). https://doi.org/10.1007/978-3-319-78381-9_9
15. Corrigan-Gibbs, H., Kogan, D.: The discrete-logarithm problem with preprocessing. In: Nielsen, J.B., Rijmen, V. (eds.) EUROCRYPT 2018. LNCS, vol. 10821, pp. 415–447. Springer, Cham (2018). https://doi.org/10.1007/978-3-319-78375-8_14

16. De, A., Trevisan, L., Tulsiani, M.: Time space tradeoffs for attacks against one-way functions and PRGs. In: Rabin, T. (ed.) CRYPTO 2010. LNCS, vol. 6223, pp. 649–665. Springer, Heidelberg (2010). https://doi.org/10.1007/978-3-642-14623-7_35

17. Dent, A.W.: The hardness of the DHK problem in the generic group model. Cryptology ePrint Archive, Report 2006/156 (2006). https://eprint.iacr.org/2006/156

18. Dodis, Y., Guo, S., Katz, J.: Fixing cracks in the concrete: random oracles with auxiliary input, revisited. In: Coron, J.-S., Nielsen, J.B. (eds.) EUROCRYPT 2017. LNCS, vol. 10211, pp. 473–495. Springer, Cham (2017). https://doi.org/10.1007/978-3-319-56614-6_16

19. Dodis, Y., Pietrzak, K., Wichs, D.: Key derivation without entropy waste. In: Nguyen, P.Q., Oswald, E. (eds.) EUROCRYPT 2014. LNCS, vol. 8441, pp. 93–110. Springer, Heidelberg (2014). https://doi.org/10.1007/978-3-642-55220-5_6

20. Dodis, Y., Reyzin, L., Rivest, R.L., Shen, E.: Indifferentiability of permutation-based compression functions and tree-based modes of operation, with applications to MD6. In: Dunkelman, O. (ed.) FSE 2009. LNCS, vol. 5665, pp. 104–121. Springer, Heidelberg (2009). https://doi.org/10.1007/978-3-642-03317-9_7

21. Even, S., Mansour, Y.: A construction of a cipher from a single pseudorandom permutation. In: Imai, H., Rivest, R.L., Matsumoto, T. (eds.) ASIACRYPT 1991. LNCS, vol. 739, pp. 210–224. Springer, Heidelberg (1993). https://doi.org/10.1007/3-540-57332-1_17

22. Even, S., Mansour, Y.: A construction of a cipher from a single pseudorandom permutation. J. Cryptol. 10(3), 151–162 (1997)

23. Fouque, P.-A., Joux, A., Mavromati, C.: Multi-user collisions: applications to discrete logarithm, even-mansour and PRINCE. In: Sarkar, P., Iwata, T. (eds.) ASIACRYPT 2014, Part I. LNCS, vol. 8873, pp. 420–438. Springer, Heidelberg (2014). https://doi.org/10.1007/978-3-662-45611-8_22

24. Gaži, P., Tessaro, S.: Provably robust sponge-based PRNGs and KDFs. In: Fischlin, M., Coron, J.-S. (eds.) EUROCRYPT 2016, Part I. LNCS, vol. 9665, pp. 87–116. Springer, Heidelberg (2016). https://doi.org/10.1007/978-3-662-49890-3_4

25. Gennaro, R., Trevisan, L.: Lower bounds on the efficiency of generic cryptographic constructions. In: 41st Annual Symposium on Foundations of Computer Science, FOCS 2000, 12–14 November 2000, Redondo Beach, California, USA, pp. 305–313 (2000)

26. Goldreich, O., Krawczyk, H.: On the composition of zero-knowledge proof systems. SIAM J. Comput. 25(1), 169–192 (1996)

27. Goldreich, O., Oren, Y.: Definitions and properties of zero-knowledge proof systems. J. Cryptol. 7(1), 1–32 (1994)

28. Goldwasser, S., Kalai, Y.T.: On the (in)security of the Fiat-Shamir paradigm. In: Proceedings of the 44th Symposium on Foundations of Computer Science (FOCS 2003), 11–14 October 2003, Cambridge, MA, USA, pp. 102–113 (2003)

29. Granboulan, L.: Short signatures in the random oracle model. In: Zheng, Y. (ed.) ASIACRYPT 2002. LNCS, vol. 2501, pp. 364–378. Springer, Heidelberg (2002). https://doi.org/10.1007/3-540-36178-2_23

30. Hellman, M.E.: A cryptanalytic time-memory trade-off. IEEE Trans. Inf. Theory 26(4), 401–406 (1980)

31. Jonsson, J.: An OAEP variant with a tight security proof. IACR Cryptology ePrint Archive, 2002:34 (2002)

32. Katz, J., Lindell, Y.: Introduction to Modern Cryptography. Chapman and Hall/CRC Press, Boca Raton (2007)

33. Mahmoody, M., Mohammed, A.: On the power of hierarchical identity-based encryption. In: Fischlin, M., Coron, J.-S. (eds.) EUROCRYPT 2016, Part II. LNCS, vol. 9666, pp. 243–272. Springer, Heidelberg (2016). https://doi.org/10.1007/978-3-662-49896-5_9

34. Mihalcik, J.P.: An analysis of algorithms for solving discrete logarithms in fixed groups. Master's thesis, Naval Postgraduate School, Monterey, California (2010)

35. Nielsen, J.B.: Separating random oracle proofs from complexity theoretic proofs: the non-committing encryption case. In: Yung, M. (ed.) CRYPTO 2002. LNCS, vol. 2442, pp. 111–126. Springer, Heidelberg (2002). https://doi.org/10.1007/3-540-45708-9_8

36. Oechslin, P.: Making a faster cryptanalytic time-memory trade-off. In: Boneh, D. (ed.) CRYPTO 2003. LNCS, vol. 2729, pp. 617–630. Springer, Heidelberg (2003). https://doi.org/10.1007/978-3-540-45146-4_36

37. National Institute of Standards and Technology (NIST): FIPS 202. SHA-3 standard: permutation-based hash and extendable-output functions. Technical report, US Department of Commerce, April 2014

38. National Institute of Standards and Technology (NIST): FIPS 180-4. Secure hash standard. Technical report, US Department of Commerce, August 2015

39. Phan, D.H., Pointcheval, D.: Chosen-ciphertext security without redundancy. In: Laih, C.-S. (ed.) ASIACRYPT 2003. LNCS, vol. 2894, pp. 1–18. Springer, Heidelberg (2003). https://doi.org/10.1007/978-3-540-40061-5_1

40. Rivest, R.L.: The MD5 Message-Digest algorithm (RFC 1321). http://www.ietf.org/rfc/rfc1321.txt?number=1321

41. Rivest, R.L., et al.: The MD6 hash function: a proposal to NIST for SHA-3 (2008)

42. Rogaway, P., Steinberger, J.: Constructing cryptographic hash functions from fixed-key blockciphers. In: Wagner, D. (ed.) CRYPTO 2008. LNCS, vol. 5157, pp. 433–450. Springer, Heidelberg (2008). https://doi.org/10.1007/978-3-540-85174-5_24

43. National Technical Information Service: FIPS 180-1. Secure hash standard. Technical report, US Department of Commerce, April 1995

44. Shoup, V.: Lower bounds for discrete logarithms and related problems. In: Fumy, W. (ed.) EUROCRYPT 1997. LNCS, vol. 1233, pp. 256–266. Springer, Heidelberg (1997). https://doi.org/10.1007/3-540-69053-0_18

45. Shrimpton, T., Stam, M.: Building a collision-resistant compression function from non-compressing primitives. In: Aceto, L., Damgård, I., Goldberg, L.A., Halldórsson, M.M., Ingólfsdóttir, A., Walukiewicz, I. (eds.) ICALP 2008, Part II. LNCS, vol. 5126, pp. 643–654. Springer, Heidelberg (2008). https://doi.org/10.1007/978-3-540-70583-3_52

46. Tessaro, S.: Security amplification for the cascade of arbitrarily weak PRPs: tight bounds via the interactive hardcore lemma. In: Ishai, Y. (ed.) TCC 2011. LNCS, vol. 6597, pp. 37–54. Springer, Heidelberg (2011). https://doi.org/10.1007/978-3-642-19571-6_3

47. Unruh, D.: Random oracles and auxiliary input. In: Menezes, A. (ed.) CRYPTO 2007. LNCS, vol. 4622, pp. 205–223. Springer, Heidelberg (2007). https://doi.org/10.1007/978-3-540-74143-5_12

48. Winternitz, R.S.: A secure one-way hash function built from DES. In: Proceedings of the 1984 IEEE Symposium on Security and Privacy, Oakland, California, USA, 29 April–2 May 1984, pp. 88–90 (1984)

Provable Security of (Tweakable)
Block Ciphers Based
on Substitution-Permutation Networks

Benoît Cogliati[1], Yevgeniy Dodis[2], Jonathan Katz[3], Jooyoung Lee[4(✉)],
John Steinberger[6], Aishwarya Thiruvengadam[5], and Zhe Zhang[6]

[1] University of Luxembourg, Esch-sur-Alzette, Luxembourg
benoitcogliati@hotmail.fr
[2] New York University, New York, USA
dodis@cs.nyu.edu
[3] University of Maryland, College Park, USA
jkatz@cs.umd.edu
[4] KAIST, Daejeon, Korea
hicalf@kaist.ac.kr
[5] University of California, Santa Barbara, USA
aish@cs.ucsb.edu
[6] Tsinghua University, Beijing, China
jpstein@gmail.com, leomwzz@gmail.com

Abstract. *Substitution-Permutation Networks* (SPNs) refer to a family of constructions which build a wn-bit block cipher from n-bit public permutations (often called S-boxes), which alternate keyless and "local" substitution steps utilizing such S-boxes, with keyed and "global" permutation steps which are non-cryptographic. Many widely deployed block ciphers are constructed based on the SPNs, but there are essentially no provable-security results about SPNs.

In this work, we initiate a comprehensive study of the provable security of SPNs as (possibly tweakable) wn-bit block ciphers, when the underlying n-bit permutation is modeled as a public random permutation. When the permutation step is *linear* (which is the case for most existing designs), we show that 3 SPN rounds are necessary and sufficient for security. On the other hand, even 1-round SPNs can be secure when non-linearity is allowed. Moreover, 2-round non-linear SPNs can achieve "beyond-birthday" (up to $2^{2n/3}$ adversarial queries) security, and, as the number of non-linear rounds increases, our bounds are meaningful for the number of queries approaching 2^n. Finally, our non-linear SPNs can be made *tweakable* by incorporating the tweak into the permutation layer, and provide good multi-user security.

As an application, our construction can turn two public n-bit permutations (or fixed-key block ciphers) into a tweakable block cipher working on wn-bit inputs, $6n$-bit key and an n-bit tweak (for any $w \geq 2$); the tweakable block cipher provides security up to $2^{2n/3}$ adversarial queries in the random permutation model, while only requiring w calls to each permutation, and $3w$ field multiplications for each wn-bit input.

© International Association for Cryptologic Research 2018
H. Shacham and A. Boldyreva (Eds.): CRYPTO 2018, LNCS 10991, pp. 722–753, 2018.
https://doi.org/10.1007/978-3-319-96884-1_24

Keywords: Substitution-permutation networks
Tweakable block ciphers · Domain extension of block ciphers
Beyond-birthday-bound security

1   Introduction

SUBSTITUTION-PERMUTATION NETWORKS. Modern block ciphers are generally constructed using two main paradigms [KL15]: Feistel networks [Fei73] or substitution-permutation networks (SPNs) [Sha49,Fei73]. Examples of block ciphers based on Feistel networks include DES, FEAL, MISTY and KASUMI; block ciphers based on SPNs include AES, Serpent, and PRESENT. These two approaches share the same goal: namely, to extend a "pseudorandom object" on a small domain to a (keyed) pseudorandom permutation on a larger domain by repeating a few, relatively simple operations several times across multiple rounds. Simplifying somewhat, Feistel networks begin with a keyed pseudorandom function on n-bit inputs and extend this to give a keyed pseudorandom permutation on $2n$-bit inputs. On the other hand, SPNs start with one or more public "random permutations" on n-bit inputs (called S-boxes) and extend them to give a keyed pseudorandom permutation on wn-bit inputs for some w, by iterating the following steps:

1. *Substitution step*: break down the wn-bit state into w disjoint n-bit blocks, and compute an S-box on each n-bit block;
2. *Permutation step*: apply a non-cryptographic, keyed permutation to the whole wn-bit state (which is also applied to the plaintext before the first round).

Proving the security of a concrete block cipher unconditionally is currently beyond our capabilities. Thus, the usual approach is to prove that the high-level structure is sound in a relevant security model. For Feistel networks, a substantial line of work, starting with Luby and Rackoff's seminal work [LR88], and culminating with Patarin's results [Pat03,Pat04], proves optimal security with a sufficient number of rounds. Numerous other articles [Pat10,HR10,HKT11,Tes14,CHK+16] study the security of (variants of) Feistel networks in various security models. In contrast, it is somewhat surprising that there are almost no results about provable security of SPNs (see below.) Here, we address this gap and explore conditions under which SPNs can be proven secure.

DOMAIN EXTENSION OF BLOCK CIPHERS. Block ciphers following the SPNs typically rely on very small S-boxes (e.g. AES uses an 8-bit S-box). However, it is also possible to use a larger domain block cipher with a fixed key (which has non-trivial efficiency gains and avoid related key attacks) as "S-box" in order to extend the domain of the underlying block cipher, or to use a larger dedicated permutation (e.g., Keccak permutation [BDPA09] or with Gimli [BKL+17]), in order to directly obtain a "wide" block cipher. From this point of view, the substitution-permutation networks can also be viewed as enciphering modes of operation (of a fixed input length), in which the length n of the S-box is not necessarily small. Such enciphering modes of operations have applications to disk encryption that protects the confidentiality of data stored on a sector-addressable device, such as a hard disk. In this scenario, the disk is divided into several sectors, and each sector, viewed as a wide block, should be encrypted and decrypted independently of each other. Non-linear 1-round SPNs with *secret* S-boxes have already been used to provide domain extension for block ciphers [CS06,Hal07]. These constructions provide the birthday bound security, while this level of security might not be desirable for an environment where stronger security is required. One of our results will address this limitation.

1.1 Our Contribution

We analyze SPNs in the standard sense as a strong pseudorandom permutation [LR88] (i.e., against adaptive chosen-plaintext and chosen-ciphertext attacks).

LINEAR SPNS. We first characterize the security of *linear* SPNs, where the permutation layer is a linear function (over $GF(2^n)$, where n is the size of the S-box) of the current wn-bit round key and the current wn-bit state. Indeed, most current SPN-based block ciphers (e.g., AES, Serpent, PRESENT, etc.) use linear permutation step, which involves a simple key-mixing step followed by an invertible linear transformation. For this widely used setting we give a general against any 2-round linear SPNs with $w \geq 2$.[1] Complementing this attack, we show that a 3-round linear SPNs *are* secure, for any w, if the keyed linear permutations satisfy some very mild technical requirements. This result critically uses the H-coefficients technique [Pat08,CS14].

NON-LINEAR SPNS. In an effort to reduce the number of rounds (and get other benefits we explain below), we then turn our attention to non-linear SPNs, where the permutation step does not have to be linear (although must remain efficient and "non-cryptographic"). Here we show that even a *1-round SPN can be secure*, if appropriate keyed permutations are used. We identify a combinatorial property on the permutations — which we term *blockwise universality* — that suffices for security in this case, and then study the efficiency of constructing permutations satisfying this property. Specifically, we show a construction of a satisfactory permutation with n-bit keys (but having high degree), and another construction with longer keys but having degree 3.

We then show that, by using such blockwise independent permutations, the security of resulting SPNs increases when we increase the number of rounds: while 1 round already achieves "birthday security", as our main technical result we show that 2-round non-linear SPNs (with independent S-boxes and keys in different rounds) achieves "beyond-birthday" security (for up to $2^{2n/3}$ queries). This result uses the refinement of the H-coefficient technique due to [HT16]. We

---

[1] Even a 1-round linear SPN can be secure if $w = 1$, since this corresponds to the famous Even-Mansour cipher [EM97].

also give an asymptotic analysis of non-linear SPNs built from blockwise universal permutations using the coupling technique of [MRS09,HR10]. In particular, for $r = 2s$ we prove that r-round SPNs are secure as long as the number of adversarial queries is well below $2^{sn/(s+1)}$. Thus, as r grows, our bounds tend towards optimal 2^n security.

As an additional benefit of this setting, we show that the blockwise universal permutations can be efficiently tweaked, meaning that our non-linear SPN constructions yield *tweakable block ciphers* [LRW11], which is important for some settings. Finally, we analyze our non-linear SPNs in the multi-user setting using the point-wise proximity technique of [HT16].

APPLICATION TO WIDE TWEAKABLE BLOCK CIPHERS. Besides providing theoretical insights on SPN-based block ciphers, our results also have a practical interest in the context of domain extension for block ciphers and permutation-based cryptography. For example, if our construction is instantiated with two n-bit permutations and a tweakable permutation TBPE in the permutation layer (as defined in Sect. 2.2), then we can build a wide tweakable block cipher with key space $\{0,1\}^{6n}$, tweak space $\{0,1\}^n$ and message space $\{0,1\}^{wn}$ for any integer $w \geq 2$. This tweakable block cipher requires w calls to each permutation and $3w$ field multiplications for each encryption/decryption call. The multi-user advantage of any adversary is shown to be small as long as the number of its queries is well below $2^{2n/3}$. This means that a 192 bit (resp. 384 bit) permutation or block cipher is sufficient to get a provably secure mode of operation as long as the number of adversarial queries is small in front of 2^{128} (resp. 2^{256}). As far as we know, this is the first construction for domain extension of a block cipher/permutation that enjoys beyond birthday-bound security.

Of course, to instantiate this construction we would need a good public permutation with large domain size n. As mentioned earlier, we could either use a larger domain block cipher with a fixed key, or use a larger dedicated permutation on larger domain, such as Keccak permutation [BDPA09] or Gimli [BKL+17].

OPEN PROBLEMS. We conjecture that r-round non-linear SPNs should actually be enough to prove security up to $\mathcal{O}(2^{rn/(r+1)})$ adversarial queries. Proving it using combinatorial techniques seems very challenging and we leave it as an interesting open problem. It is also interesting if we can prove beyond-birthday security bounds for linear SPNs (with 3 or more rounds), as these SPNs appear to be the ones used in practice. More generally, it would be great to prove tight security bounds and matching attacks for r-round linear and non-linear SPNs.

IMPLICATIONS FOR SMALL BLOCK SIZE. While our results are directly meaningful when the length n of public S-boxes in at least security parameter (e.g., for building *wide* tweakable block ciphers), our bounds are too weak for regular SPN-based ciphers, such as AES, which use very low values of n for their S-boxes. This "2^n provable barrier" is inherent using our current modeling, where the S-box of size 2^n is providing the only source of cryptographic hardness. More generally, establishing a sound theory of building block ciphers from small S-boxes is one of the biggest and most important open problems in symmetric-key cryptography. We hope that our structural results for reduced-round SPN ciphers will be useful in establishing such theory, despite not crossing the fundamental "2^n barrier" mentioned above.

1.2   Related Work

There are only a few prior papers looking at provable security of SPNs. The vast majority of such work analyzes the case of secret, key-dependent S-boxes (rather than public S-boxes as we consider here), and so we survey that work first.

SPNs WITH SECRET S-BOXES. Naor and Reingold [NR99] prove security for what can be viewed as a non-linear, 1-round SPN. Their ideas were further developed, in the context of domain extension for block ciphers (see further discussion below), by Chakraborty and Sarkar [CS06] and Halevi [Hal07].

Iwata and Kurosawa [IK00] analyze SPNs in which the linear permutation step is based on the specific permutations used in the block cipher Serpent. They show an attack against 2-round SPNs of this form, and prove security for 3-round SPNs against non-adaptive adversaries. In addition to the fact that we consider public S-boxes, our linear SPN model considers generic linear permutations and we prove security against adaptive attackers.

Miles and Viola [MV15] study SPNs from a complexity-theoretic viewpoint. Two of their results are relevant here. First, they analyze the security of linear SPNs using S-boxes that are not necessarily injective (so the resulting keyed functions are not, in general, invertible). They show that r-round SPNs of this type (for $r \geq 2$) are secure against chosen-plaintext attacks. (In contrast, our results show that 2-round, linear SPNs are not secure against a combination of chosen-plaintext and chosen-ciphertext attacks when $w \geq 2$.) They also analyze SPNs based on a concrete set of S-boxes, but in this case they only show security against linear/differential attacks (a form of chosen-plaintext attack), rather than all possible attacks, and only when the number of rounds is $r = \Theta(\log n)$.

SPNs WITH PUBLIC S-BOXES. A difference between our work and all the work discussed above is that we treat the S-boxes as public. We are aware of only one prior work analyzing the provable security of SPNs in this setting. Dodis et al. [DSSL16] recently studied the *indifferentiability* [MRH04] of confusion-diffusion networks, which can be viewed as *unkeyed* SPNs. One could translate their results to the keyed setting, but that would require using multiple, key-dependent S-boxes (rather than a fixed, public S-box) and so would not imply our results. We remark further that they show positive results only for 5 rounds and above.

As observed earlier, the Even-Mansour construction [EM97] of a (keyed) pseudorandom permutation from a public random permutation can be viewed as a 1-round, linear SPN in the degenerate case where $w = 1$ (i.e., no domain extension) and all round permutations are instantiated using simple key mixing. Security of the 1-round Even-Mansour construction against adaptive chosen-plaintext/ciphertext attacks, using independent keys for the initial and final key mixing, was shown in the original paper [EM97]. Our positive results imply security of the 1-round Even-Mansour construction (with similar concrete security bounds) as a special case. The r-round generalization of the Even-Mansour cipher has seen a lot of interest over the years, culminating with [CS14,HT16] where it was proved that the r-round Even-Mansour construction is secure up to roughly $2^{rn/(r+1)}$ adversarial queries, when the public S-boxes are uniformly random and independent permutations and the round keys are independent. Chen et al. [CLL+14] also proved that several minimized variants of the 2-round Even-Mansour construction are also secure up to roughly $2^{2n/3}$ adversarial queries. None of these results extend to the setting $w > 1$ considered in this work.

CRYPTANALYSIS OF SPNs. Researchers have also explored cryptanalytic attacks on generic SPNs [BS10,BBK14,DDKL,BK]. These works generally consider a model of SPNs in which round permutations are secret, random (invertible) linear transformations, and S-boxes may be secret as well; this makes the attacks stronger but positive results weaker. In many cases the complexities of the attacks are exponential in n (though still faster than a brute-force search for the key), and hence do not rule out asymptotic security results. On the positive side, Biryukov et al. [BBK14] show that 2-round SPNs (of the stronger form just mentioned) are secure against some specific types of attacks, but other attacks on such schemes have recently been identified [DDKL].

ATTACKS. Attacks due to Joux [Jou03] and to Halevi and Rogaway [HR04], originally developed in the afore-mentioned context of block cipher domain extension (or more exactly, in the construction of *tweakable* block ciphers with large domains from standard block ciphers with "small" domains) can be translated to the context of linear SPNs as well. Specifically, these attacks imply that linear 2-round SPNs of width $w \geq 2$ are insecure, as long as the underlying field has characteristic 2.[2]

DOMAIN EXTENSION OF BLOCK CIPHERS. Non-linear, 1-round SPNs with secret S-boxes have been used for domain extension of block ciphers before [CS06, Hal07]. Other approaches for domain extension, not relying on (pure) SPNs, have also been considered [BD99,HR03,HR04,MF07,CDMS10]. To the best of our knowledge, none of these results achieve beyond-birthday security.

RANDOM PERMUTATION BASED TWEAKABLE BLOCK CIPHERS. Our tweakable SPNs can be viewed as tweakable block ciphers based on public random permutations. It is easy to see that $T : (h, t, x) \mapsto x \oplus h(t)$ is (δ, δ')-blockwise universal (as defined in Sect. 2) if h is chosen from a δ'-almost uniform and δ-almost XOR-universal hash family. So with this permutation layer (and with $w = 1$), we obtain the security bound for the Tweakable Even-Mansour constructions [CLS15] in the multi-user setting. In this line of research, a number of efficient constructions have been proposed [GJMN16,Men16].

---

[2] Indeed, a technical difference with the attack presented here is that our attack does not require a finite field of characteristic 2. Because of this difference, our attack ends up having little (if anything) in common with the attacks of Joux and Halevi-Rogaway.

2   Preliminaries

Throughout this work, we fix positive integers w and n; an element x in $\{0,1\}^{wn}$ can be viewed as a concatenation of w blocks, each of which is of length n. The i-th block of this representation will be denoted x_i for $i = 1, \ldots, w$, so we have

$$x = x_1 \| x_2 \| \cdots \| x_w,$$

sometimes written as $x = (x_1, \ldots, x_w)$.

For a set R and an integer $s \geq 1$, R^{*s} denotes the set of all sequences that consists of s pairwise distinct elements of R. For any integer r such that $r \geq s$, we will write $(r)_s = r!/(r-s)!$ If $|R| = r$, then $(r)_s$ becomes the size of R^{*s}. The sets of non-negative integers and non-negative real numbers are denoted $\mathbb{N}$ and $\mathbb{R}^{\geq 0}$, respectively. The following inequality will be used in our security proof.

Lemma 1. *Let m be an integer and let x be a real number such that $m \geq 2$ and $-1 \leq x < \frac{1}{m-1}$. Then one has*

$$(1+x)^m \leq 1 + \frac{mx}{1 - (m-1)x}.$$

2.1   Tweakable Substitution-Permutation Networks

All the notions below are defined for the general tweak set $\mathcal{T}$; however, the standard "non-tweakable" setting is a special case of the definitions below when $|\mathcal{T}| = 1$.

TWEAKABLE PERMUTATIONS. For an integer $m \geq 1$, the set of all permutations on $\{0,1\}^m$ will be denoted $\mathsf{Perm}(m)$. A tweakable permutation with tweak space $\mathcal{T}$ and message space $\mathcal{X}$ is a mapping $\widetilde{P} : \mathcal{T} \times \mathcal{X} \to \mathcal{X}$ such that, for any tweak $t \in \mathcal{T}$,

$$x \mapsto \widetilde{P}(t, x)$$

is a permutation of $\mathcal{X}$. The set of all tweakable permutations with tweak space $\mathcal{T}$ and message space $\{0,1\}^m$ will be denoted $\widetilde{\mathsf{Perm}}(\mathcal{T}, m)$.

A keyed tweakable permutation with key space $\mathcal{K}$, tweak space $\mathcal{T}$ and message space $\mathcal{X}$ is a mapping $T : \mathcal{K} \times \mathcal{T} \times \mathcal{X} \to \mathcal{X}$ such that, for any key $k \in \mathcal{K}$,

$$(t, x) \mapsto T(k, t, x)$$

is a tweakable permutation with tweak space $\mathcal{T}$ and message space $\mathcal{X}$. We will sometimes write $T(k, t, x)$ as $T_k(t, x)$ or $T_{k,t}(x)$. For an integer $s \geq 1$, let $\mathbf{t} = (t_1, \ldots, t_s) \in \mathcal{T}^s$, and let $\mathbf{x} = (x_1, \ldots, x_s) \in (\mathcal{X})^{*s}$. We will write $(T(k, t_i, x_i))_{1 \leq i \leq s}$ as $T_k(\mathbf{t}, \mathbf{x})$ or $T_{k,\mathbf{t}}(\mathbf{x})$.

TWEAKABLE SPNs. For fixed parameters w and n, let

$$T : \mathcal{K} \times \mathcal{T} \times \{0,1\}^{wn} \longrightarrow \{0,1\}^{wn}$$

be a keyed tweakable permutation with key space $\mathcal{K}$, tweak space $\mathcal{T}$ and message space $\{0,1\}^{wn}$.

For a fixed number of rounds r, an r-round substitution-permutation network (SPN) based on T, denoted SP^T, takes as input a set of n-bit permutations $\mathcal{S} = (S_1, \ldots, S_r)$, and defines a keyed tweakable permutation $\mathsf{SP}^T[\mathcal{S}]$ operating on wn-bit blocks with key space $\mathcal{K}^{r+1}$ and tweak space $\mathcal{T}$: on input $x \in \{0,1\}^{wn}$, key $\mathbf{k} = (k_0, k_1, \ldots, k_r) \in \mathcal{K}^{r+1}$ and tweak $t \in \mathcal{T}$, the output of $\mathsf{SP}^T[\mathcal{S}]$ is computed as follows (see also Fig. 1).

$y \leftarrow x$
for $i \leftarrow 1$ to r **do**
   $y \leftarrow T_{k_{i-1},t}(y)$
   Break $y = y_1 || \cdots || y_w$ into n-bit blocks
   $y \leftarrow S_i(y_1) || \cdots || S_i(y_w)$
$y \leftarrow T_{k_r,t}(y)$
return y

Remark 1. Both of the permutation layer T and the entire construction SP^T can be viewed as keyed tweakable permutations. However, T will typically be built upon non-cryptographic operations such as filed multiplications, while SP^T are based on S-boxes which are modeled as public random permutations.

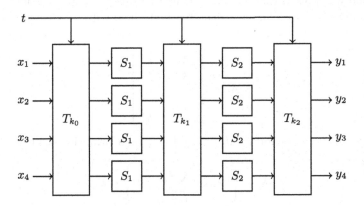

Fig. 1. A 2-round tweakable SPN with $w = 4$. The input and output blocks of the SPN are represented as $x = x_1||x_2||x_3||x_4$ and $y = y_1||y_2||y_3||y_4$, respectively.

BLOCKWISE UNIVERSAL TWEAKABLE PERMUTATIONS. A keyed tweakable permutation

$$T : \mathcal{K} \times \mathcal{T} \times \{0,1\}^{wn} \longrightarrow \{0,1\}^{wn}$$

is called (δ, δ')-*blockwise universal* if the following hold.

1. For all distinct $(t, x, i), (t', x', i') \in \mathcal{T} \times \{0,1\}^{wn} \times \{1, \dots, w\}$, we have

$$\Pr\left[k \xleftarrow{\$} \mathcal{K} : T_{k,t}(x)_i = T_{k,t'}(x')_{i'}\right] \leq \delta.$$

2. For all $(t, x, i, c) \in \mathcal{T} \times \{0,1\}^{wn} \times \{1, \dots, w\} \times \{0,1\}^n$, we have

$$\Pr\left[k \xleftarrow{\$} \mathcal{K} : T_{k,t}(x)_i = c\right] \leq \delta'.$$

Since each pair of key $k \in \mathcal{K}$ and tweak $t \in \mathcal{T}$ defines a permutation $T_{k,t}$ on $\{0,1\}^{wn}$, one can define a keyed tweakable permutation

$$T^{-1} : \mathcal{K} \times \mathcal{T} \times \{0,1\}^{wn} \longrightarrow \{0,1\}^{wn}$$

such that $T^{-1}(k, t, x) = (T_{k,t})^{-1}(x)$. If T and T^{-1} are both (δ, δ')-blockwise universal, then T is called (δ, δ')-*super blockwise universal*.

2.2   An Efficient Super Blockwise Tweakable Universal Permutation

In this section, we show that an efficient xor-blockwise universal construction, dubbed BPE, proposed by Halevi [Hal07] can be made tweakable with a slight modification. Other constructions of (tweakable) blockwise universal permutations can be found in [DKS+17] some of which support tweaks. We present BPE below and will present the remaining constructions in the full version.

Assuming $2^n \geq w + 3$, let $\mathbb{F}$ denote a finite field with 2^n elements. For each $k \in \mathbb{F}$, define a $w \times w$ matrix over $\mathbb{F}$, $M_k =^{\text{def}} A_k + I$, where I is the identity matrix and

$$A_k = \begin{bmatrix} k & k^2 & & k^w \\ k & k^2 & & k^w \\ & & \ddots & \\ k & k^2 & & k^w \end{bmatrix}.$$

Precisely, $(A_k)_{i,j} = k^j$ for $1 \leq i, j \leq w$. Let z be a primitive element of $\mathbb{F}$, and let

$$\mathcal{K} = \left\{ k \in \mathbb{F} : \sum_{i=0}^{w} k^i \neq 0 \right\} \times \mathbb{F}.$$

Then BPE is defined as follows.

$$\mathsf{BPE} : \mathcal{K} \times \{0,1\}^{wn} \longrightarrow \{0,1\}^{wn}$$
$$((k, k'), x) \longmapsto M_k x \oplus a_{k'},$$

where we identify $x \in \{0,1\}^{wn}$ with a w-dimensional column vector over $\mathbb{F}$, and

$$a_{k'} = \begin{bmatrix} k' \\ zk' \\ \vdots \\ z^{w-1}k' \end{bmatrix}.$$

It is easy to check that M_k is invertible if $\sum_{i=0}^{w} k^i \neq 0$; precisely,

$$M_k^{-1} = I \oplus \frac{A_k}{k^*},$$

where $k^* =^{\text{def}} \sum_{i=0}^{w} k^i$. For any $(k, k') \in \mathcal{K}$, $\mathsf{BPE}_{k,k'}$ is also invertible with

$$\mathsf{BPE}_{k,k'}^{-1}(x) = M_k^{-1}(x \oplus a_{k'})$$

for any $x \in \{0,1\}^{wn}$. Halevi [Hal07] also proved that for any pair of distinct $(x, i), (x', i') \in \{0,1\}^{wn} \times \{1, \ldots, w\}$ and $\Delta \in \{0,1\}^n$,

$$\Pr\left[(k,k') \xleftarrow{\$} \mathcal{K} : \mathsf{BPE}_{k,k'}(x)_i \oplus \mathsf{BPE}_{k,k'}(x')_{i'} = \Delta\right] \leq \frac{w}{2^n - w},$$

$$\Pr\left[(k,k') \xleftarrow{\$} \mathcal{K} : \mathsf{BPE}_{k,k'}^{-1}(x)_i \oplus \mathsf{BPE}_{k,k'}^{-1}(x')_{i'} = \Delta\right] \leq \frac{w}{2^n - w}. \tag{1}$$

For a fixed $(x, i, c) \in \{0,1\}^{wn} \times \{1, \ldots, w\} \times \{0,1\}^n$, $\mathsf{BPE}_{k,k'}(x)_i = c$ implies that

$$\sum_{j=1}^{w} x_j k^j \oplus x_i \oplus z^{i-1} k' = c,$$

which holds with probability $\frac{1}{2^n}$ over a random choice of $(k, k') \in \mathcal{K}$. On the other hand, $\mathsf{BPE}_{k,k'}^{-1}(x)_i = c$ implies that

$$\left(z^{i-1} \oplus \frac{1}{k^*}\sum_{j=1}^{w} z^{j-1}k^j\right) k' \oplus \left(c \oplus x_i \oplus \frac{1}{k^*}\sum_{j=1}^{w} x_j k^j\right) = 0.$$

This equation holds with probability at most $\frac{w}{2^n - w} + \frac{1}{2^n}$. To summarize, we have

$$\Pr\left[(k,k') \xleftarrow{\$} \mathcal{K} : \mathsf{BPE}_{k,k'}(x)_i = c\right] \leq \frac{1}{2^n},$$

$$\Pr\left[(k,k') \xleftarrow{\$} \mathcal{K} : \mathsf{BPE}_{k,k'}^{-1}(x)_i = c\right] \leq \frac{w+1}{2^n - w}. \tag{2}$$

Now we define a tweakable variant of BPE, dubbed TBPE (for Tweakable Blockwise Polynomial-Evaluation), with tweak space $\mathcal{T} = \{0,1\}^n$ as follows.

$$\mathsf{TBPE} : \mathcal{K} \times \mathcal{T} \times \{0,1\}^{wn} \longrightarrow \{0,1\}^{wn}$$
$$((k, k'), t, x) \longmapsto M_k(x \oplus b_t) \oplus a_{k'} \oplus b_t,$$

where b_t is the column vector whose entries are all t, namely,

$$b_t = \begin{bmatrix} t \\ t \\ \vdots \\ t \end{bmatrix}.$$

Since each pair of key $(k, k') \in \mathcal{K}$ and tweak $t \in \mathcal{T}$ defines a permutation $\mathsf{TBPE}_{k,k',t}$ on $\{0,1\}^{wn}$, one can define a keyed tweakable permutation

$$\mathsf{TBPE}^{-1} : \mathcal{K} \times \mathcal{T} \times \{0,1\}^{wn} \longrightarrow \{0,1\}^{wn}.$$

Then we can prove the following lemma.

Lemma 2. *Let* TBPE *be the keyed tweakable permutation as defined above, and let* TBPE^{-1} *be its inverse.*

1. *For all distinct* (t, x, i), $(t', x', i') \in \mathcal{T} \times \{0,1\}^{wn} \times \{1, \ldots, w\}$, *we have*

$$\Pr\left[(k,k') \overset{\$}{\leftarrow} \mathcal{K} : \mathsf{TBPE}_{k,k',t}(x)_i = \mathsf{TBPE}_{k,k',t'}(x')_{i'}\right] \leq \frac{w}{2^n - w}.$$

2. *For all* $(t, x, i, c) \in \mathcal{T} \times \{0,1\}^{wn} \times \{1, \ldots, w\} \times \{0,1\}^n$, *we have*

$$\Pr\left[(k,k') \overset{\$}{\leftarrow} \mathcal{K} : \mathsf{TBPE}_{k,k',t}(x)_i = c\right] \leq \frac{1}{2^n}.$$

3. *For all distinct* (t, x, i), $(t', x', i') \in \mathcal{T} \times \{0,1\}^{wn} \times \{1, \ldots, w\}$, *we have*

$$\Pr\left[(k,k') \overset{\$}{\leftarrow} \mathcal{K} : \mathsf{TBPE}^{-1}_{k,k',t}(x)_i = \mathsf{TBPE}^{-1}_{k,k',t'}(x')_{i'}\right] \leq \frac{w}{2^n - w}.$$

4. *For all* $(t, x, i, c) \in \mathcal{T} \times \{0,1\}^{wn} \times \{1, \ldots, w\} \times \{0,1\}^n$, *we have*

$$\Pr\left[(k,k') \overset{\$}{\leftarrow} \mathcal{K} : \mathsf{TBPE}^{-1}_{k,k',t}(x)_i = c\right] \leq \frac{w+1}{2^n - w}.$$

Proof. For distinct (t, x, i) and (t', x', i'), we have

$$\mathsf{TBPE}_{k,k',t}(x)_i \oplus \mathsf{TBPE}_{k,k',t'}(x')_{i'} = \mathsf{BPE}_{k,k'}(x \oplus b_t)_i \oplus \mathsf{BPE}_{k,k'}(x' \oplus b_{t'})_{i'} \oplus t \oplus t'.$$

If $(x \oplus b_t, i) \neq (x' \oplus b_{t'}, i')$, then $\mathsf{BPE}_{k,k'}(x \oplus b_t)_i \oplus \mathsf{BPE}_{k,k'}(x' \oplus b_{t'})_{i'} \oplus t \oplus t' = 0$ with probability at most $\frac{w}{2^n - w}$ by (1). If $(x \oplus b_t, i) = (x' \oplus b_{t'}, i')$, then it implies $t \neq t'$, and hence $\mathsf{BPE}_{k,k'}(x \oplus b_t)_i \oplus \mathsf{BPE}_{k,k'}(x' \oplus b_{t'})_{i'} \oplus t \oplus t' = t \oplus t' \neq 0$.

For a fixed (t, x, i, c), $\mathsf{TBPE}_{k,k',t}(x)_i = c$ if and only if $\mathsf{BPE}_{k,k'}(x \oplus b_t)_i = c \oplus t$, and this equation holds with probability at most $\frac{1}{2^n}$. The remaining properties are proved similarly.   □

From Lemma 2, it follows that TBPE is $\left(\frac{w}{2^n - w}, \frac{w+1}{2^n - w}\right)$-super blockwise universal. Except constant multiplications $z^i k'$, $i = 1, \ldots, w - 1$, (which also can be precomputed), each evaluation of $\mathsf{TBPE}_{k,k',t}(x)$ requires w field multiplications.

2.3   Indistinguishability in the Multi-user Setting

Let $\mathsf{SP}^T[\mathcal{S}]$ be an r-round SPN based on a set of S-boxes $\mathcal{S} = (S_1, \ldots, S_r)$ and a keyed tweakable permutation T with key space $\mathcal{K}$ and tweak space $\mathcal{T}$. So $\mathsf{SP}^T[\mathcal{S}]$

becomes a keyed tweakable permutation on $\{0,1\}^{wn}$ with key space $\mathcal{K}^{r+1}$ and tweak space $\mathcal{T}$.

In the multi-user setting, let ℓ denote the number of users. In the *real* world, ℓ secret keys $\mathbf{k}_1, \ldots \mathbf{k}_\ell \in \mathcal{K}^{r+1}$ are chosen independently at random. A set of independent S-boxes $\mathcal{S} = (S_1, \ldots, S_r)$ is also randomly chosen from $\mathsf{Perm}(n)^r$. A distinguisher $\mathcal{D}$ is given oracle access to $(\mathsf{SP}^T_{\mathbf{k}_1}[\mathcal{S}], \ldots, \mathsf{SP}^T_{\mathbf{k}_\ell}[\mathcal{S}])$ as well as $\mathcal{S} = (S_1, \ldots, S_r)$. In the *ideal* world, $\mathcal{D}$ is given a set of independent random tweakable permutations $\widetilde{\mathcal{P}} = (\widetilde{P}_1, \ldots, \widetilde{P}_\ell) \in \widetilde{\mathsf{Perm}}(\mathcal{T}, wn)^\ell$ instead of $(\mathsf{SP}^T_{\mathbf{k}_1}[\mathcal{S}], \ldots, \mathsf{SP}^T_{\mathbf{k}_\ell}[\mathcal{S}])$. However, oracle access to $\mathcal{S} = (S_1, \ldots, S_r)$ is still allowed in this world.

The adversarial goal is to tell apart the two worlds $(\mathsf{SP}^T_{\mathbf{k}_1}[\mathcal{S}], \ldots, \mathsf{SP}^T_{\mathbf{k}_\ell}[\mathcal{S}], \mathcal{S})$ and $(\widetilde{P}_1, \ldots, \widetilde{P}_\ell, \mathcal{S})$ by adaptively making forward and backward queries to each of the constructions and the S-boxes. Formally, $\mathcal{D}$'s distinguishing advantage is defined by

$$\mathsf{Adv}^{mu}_{\mathsf{SP}^T}(\mathcal{D}) = \Pr\left[\widetilde{P}_1, \ldots, \widetilde{P}_\ell \xleftarrow{\$} \widetilde{\mathsf{Perm}}(\mathcal{T}, wn), \mathcal{S} \xleftarrow{\$} \mathsf{Perm}(n)^r : 1 \leftarrow \mathcal{D}^{\mathcal{S}, \widetilde{P}_1, \ldots, \widetilde{P}_\ell}\right]$$

$$- \Pr\left[\mathbf{k}_1, \ldots, \mathbf{k}_\ell \xleftarrow{\$} \mathcal{K}^{r+1}, \mathcal{S} \xleftarrow{\$} \mathsf{Perm}(n)^r : 1 \leftarrow \mathcal{D}^{\mathcal{S}, \mathsf{SP}^T_{\mathbf{k}_1}[\mathcal{S}], \ldots, \mathsf{SP}^T_{\mathbf{k}_\ell}[\mathcal{S}]}\right].$$

For $p, q > 0$, we define

$$\mathsf{Adv}_{\mathsf{SP}^T}(p, q) = \max_{\mathcal{D}} \mathsf{Adv}_{\mathsf{SP}^T}(\mathcal{D})$$

where the maximum is taken over all adversaries $\mathcal{D}$ making at most p queries to each of the S-boxes and at most q queries to the outer tweakable permutations. In the single-user setting with $\ell = 1$, $\mathsf{Adv}^{mu}_{\mathsf{SP}^T}(\mathcal{D})$ and $\mathsf{Adv}^{mu}_{\mathsf{SP}^T}(p, q)$ will also be written as $\mathsf{Adv}^{su}_{\mathsf{SP}^T}(\mathcal{D})$ and $\mathsf{Adv}^{su}_{\mathsf{SP}^T}(p, q)$, respectively.

H-COEFFICIENT TECHNIQUE. Suppose that a distinguisher $\mathcal{D}$ makes p queries to each of the S-boxes, and total q queries to the construction oracles. The queries made to the j-th construction oracle, denoted C_j, are recorded in a query history

$$\mathcal{Q}_{C_j} = (j, t_{j,i}, x_{j,i}, y_{j,i})_{1 \leq i \leq q_j}$$

for $j = 1, \ldots, \ell$, where q_j is the number of queries made to C_j and $(j, t_{j,i}, x_{j,i}, y_{j,i})$ represents the evaluation obtained by the i-th query to C_j.[3] So according to the instantiation, it implies either $\mathsf{SP}^T_{\mathbf{k}_j}[\mathcal{S}](t_{j,i}, x_{j,i}) = y_{j,i}$ or $\widetilde{P}_j(t_{j,i}, x_{j,i}) = y_{j,i}$. Let

$$\mathcal{Q}_C = \mathcal{Q}_{C_1} \cup \cdots \cup \mathcal{Q}_{C_\ell}.$$

For $j = 1, \ldots, r$, the queries made to S_j are recorded in a query history

$$\mathcal{Q}_{S_j} = (j, u_{j,i}, v_{j,i})_{1 \leq i \leq p},$$

where $(j, u_{j,i}, v_{j,i})$ represents the evaluation $S_j(u_{j,i}) = v_{j,i}$ obtained by the i-th query to S_j. Let

$$\mathcal{Q}_S = \mathcal{Q}_{S_1} \cup \cdots \cup \mathcal{Q}_{S_r}.$$

---

[3] The index j in a construction query can be dropped out in the single-user setting.

Then the pair of query histories

$$\tau = (\mathcal{Q}_C, \mathcal{Q}_S)$$

will be called the *transcript* of the attack: it contains all the information that $\mathcal{D}$ has obtained at the end of the attack. In this work, we will only consider information theoretic distinguishers. Therefore we can assume that a distinguisher is deterministic without making any redundant query, and hence the output of $\mathcal{D}$ can be regarded as a function of τ, denoted $\mathcal{D}(\tau)$ or $\mathcal{D}(\mathcal{Q}_C, \mathcal{Q}_S)$.

Fix a transcript $\tau = (\mathcal{Q}_C, \mathcal{Q}_S)$, a key $\mathbf{k} \in \mathcal{K}^{r+1}$, a tweakable permutation $\widetilde{P} \in \widetilde{\mathsf{Perm}}(\mathcal{T}, wn)$, a set of S-boxes $\mathcal{S} = (S_1, \ldots, S_r) \in \mathsf{Perm}(n)^r$ and $j \in \{1, \ldots, \ell\}$: if $S_j(u_{j,i}) = v_{j,i}$ for every $i = 1, \ldots, p$, then we will write $S_j \vdash \mathcal{Q}_{S_j}$. We will write $\mathcal{S} \vdash \mathcal{Q}_S$ if $S_j \vdash \mathcal{Q}_{S_j}$ for every $j = 1, \ldots, r$. Similarly, if $\mathsf{SP}^T_\mathbf{k}[\mathcal{S}](t_{j,i}, x_{j,i}) = y_{j,i}$ (resp. $\widetilde{P}(t_{j,i}, x_{j,i}) = y_{j,i}$) for every $i = 1, \ldots, q_j$, then we will write $\mathsf{SP}^T_\mathbf{k}[\mathcal{S}] \vdash \mathcal{Q}_{C_j}$ (resp. $\widetilde{P} \vdash \mathcal{Q}_{C_j}$).

Let $\mathbf{k}_1, \ldots, \mathbf{k}_\ell \in \mathcal{K}^{r+1}$ and $\widetilde{P} = (\widetilde{P}_1, \ldots \widetilde{P}_\ell) \in \widetilde{\mathsf{Perm}}(\mathcal{T}, wn)^\ell$. If $\mathsf{SP}^T_{\mathbf{k}_j}[\mathcal{S}] \vdash \mathcal{Q}_{C_j}$ (resp. $\widetilde{P}_j \vdash \mathcal{Q}_{C_j}$) for every $j = 1, \ldots, \ell$, then we will write $(\mathsf{SP}^T_{\mathbf{k}_j}[\mathcal{S}])_{j=1,\ldots,\ell} \vdash \mathcal{Q}_C$ (resp. $\widetilde{P} \vdash \mathcal{Q}_C$).

If there exist $\widetilde{P} \in \widetilde{\mathsf{Perm}}(\mathcal{T}, wn)^\ell$ and $\mathcal{S} \in \mathsf{Perm}(n)^w$ that outputs τ at the end of the interaction with $\mathcal{D}$, then we will call the transcript τ *attainable*. So for any attainable transcript $\tau = (\mathcal{Q}_C, \mathcal{Q}_S)$, there exist $\widetilde{P} \in \widetilde{\mathsf{Perm}}(\mathcal{T}, wn)^\ell$ and $\mathcal{S} \in \mathsf{Perm}(n)^w$ such that $\widetilde{P} \vdash \mathcal{Q}_C$ and $\mathcal{S} \vdash \mathcal{Q}_S$. For an attainable transcript $\tau = (\mathcal{Q}_C, \mathcal{Q}_S)$, let

$$\mathsf{p}_1(\mathcal{Q}_C|\mathcal{Q}_S) = \Pr\left[\widetilde{P} \xleftarrow{\$} \widetilde{\mathsf{Perm}}(\mathcal{T}, wn)^\ell, \mathcal{S} \xleftarrow{\$} \mathsf{Perm}(n)^r : \widetilde{P} \vdash \mathcal{Q}_C \;\middle|\; \mathcal{S} \vdash \mathcal{Q}_S\right],$$

$$\mathsf{p}_2(\mathcal{Q}_C|\mathcal{Q}_S) = \Pr\left[\mathbf{k}_1, \ldots, \mathbf{k}_\ell \xleftarrow{\$} \mathcal{K}^{r+1}, \mathcal{S} \xleftarrow{\$} \mathsf{Perm}(n)^r : (\mathsf{SP}^T_{\mathbf{k}_j}[\mathcal{S}])_j \vdash \mathcal{Q}_C \;\middle|\; \mathcal{S} \vdash \mathcal{Q}_S\right].$$

With these definitions, the following lemma, the core of the H-coefficients technique (without defining "bad" transcripts), will be also used in our security proof.

Lemma 3. *Let $\varepsilon > 0$. Suppose that for any attainable transcript $\tau = (\mathcal{Q}_C, \mathcal{Q}_S)$,*

$$\mathsf{p}_2(\mathcal{Q}_C|\mathcal{Q}_S) \geq (1 - \varepsilon)\mathsf{p}_1(\mathcal{Q}_C|\mathcal{Q}_S). \tag{3}$$

Then one has

$$\mathrm{Adv}^{\mathrm{mu}}_{\mathsf{SP}^T}(\mathcal{D}) \leq \varepsilon.$$

The lower bound (3) is called ε-*point-wise proximity* of the transcript $\tau = (\mathcal{Q}_C, \mathcal{Q}_S)$. The point-wise proximity of a transcript in the multi-user setting is guaranteed by the point-wise proximity of $(\mathcal{Q}_{C_j}, \mathcal{Q}_S)$ for each $j = 1, \ldots, \ell$ in the single-user setting. The following lemma is a restatement of Lemma 3 in [HT16].

Lemma 4. *Let* $\varepsilon : \mathbb{N} \times \mathbb{N} \to \dot{\mathbb{R}}^{\geq 0}$ *be a function such that*

1. $\varepsilon(x,y) + \varepsilon(x,z) \leq \varepsilon(x, y+z)$ *for every* x, y, $z \in \mathbb{N}$,
2. $\varepsilon(\cdot, z)$ *and* $\varepsilon(z, \cdot)$ *are non-decreasing functions on* $\mathbb{N}$ *for every* $z \in \mathbb{N}$.

Suppose that for any distinguisher $\mathcal{D}$ *in the single-user setting that makes* p *primitive queries to each of the underlying S-boxes and makes* q *construction queries, and for any attainable transcript* $\tau = (\mathcal{Q}_C, \mathcal{Q}_S)$ *obtained by* $\mathcal{D}$, *one has*

$$\mathsf{p}_2(\mathcal{Q}_C | \mathcal{Q}_S) \geq (1 - \varepsilon(p, q))\mathsf{p}_1(\mathcal{Q}_C | \mathcal{Q}_S).$$

Then for any distinguisher $\mathcal{D}$ *in the multi-user setting that makes* p *primitive queries to each of the underlying S-boxes and makes total* q *construction queries, and for any attainable transcript* $\tau = (\mathcal{Q}_C, \mathcal{Q}_S)$ *obtained by* $\mathcal{D}$, *one has*

$$\mathsf{p}_2(\mathcal{Q}_C | \mathcal{Q}_S) \geq (1 - \varepsilon(p + wq, q))\mathsf{p}_1(\mathcal{Q}_C | \mathcal{Q}_S).$$

2.4 Coupling Technique

Given a finite event space Ω and two probability distributions μ and ν defined on Ω, the *total variation distance* between μ and ν, denoted $\|\mu - \nu\|$, is defined as

$$\|\mu - \nu\| = \frac{1}{2} \sum_{x \in \Omega} |\mu(x) - \nu(x)|.$$

The following definitions are also all equivalent.

$$\|\mu - \nu\| = \max_{Z \subset \Omega}\{\mu(Z) - \nu(Z)\} = \max_{Z \subset \Omega}\{\nu(Z) - \mu(Z)\} = \max_{Z \subset \Omega}\{|\mu(Z) - \nu(Z)|\}.$$

A *coupling* of μ and ν is a distribution τ on $\Omega \times \Omega$ such that for all $x \in \Omega$, $\sum_{y \in \Omega} \tau(x, y) = \mu(x)$ and for all $y \in \Omega$, $\sum_{x \in \Omega} \tau(x, y) = \nu(x)$. In other words, τ is a joint distribution whose marginal distributions are respectively μ and ν. We will use the following two lemmas in our security proof.

Lemma 5. *Let* μ *and* ν *be probability distributions on a finite event space* Ω, *let* τ *be a coupling of* μ *and* ν, *and let* (X, Y) *be a random variable sampled according to distribution* τ. *Then* $\|\mu - \nu\| \leq \mathsf{Pr}[X \neq Y]$.

Lemma 6. *Let* Ω *be some finite event space and* ν *be the uniform probability distribution on* Ω. *Let* μ *be a probability distribution on* Ω *such that* $\|\mu - \nu\| \leq \varepsilon$. *Then there is a set* $Z \subset \Omega$ *such that*

1. $|Z| \geq (1 - \sqrt{\varepsilon})|\Omega|$,
2. $\mu(x) \geq (1 - \sqrt{\varepsilon})\nu(x)$ *for every* $x \in Z$.

We refer to [LPS12] for the proof of the above two lemmas.

3   Security of Linear SPNs

All the results in this section are for the "non-tweakable" setting ($|\mathcal{T}| = 1$)
Hence, we do not explicitly refer to the tweak in the notation. Further, the
results in this section hold even when a single n-bit permutation S is used, i.e.,
even when $S_1 = \ldots = S_r = S$ and are presented as such. We start by defining
linear (non-tweakable) SPNs.

Definition 1. *Keyed permutation* $T : \mathcal{K} \times \{0,1\}^{wn} \longrightarrow \{0,1\}^{wn}$ *is* linear *if*

$$T(k, x) = (T_k \cdot k) + (T_x \cdot x) + \Delta,$$

where $T_k, T_x \in \mathcal{K} \times \{0,1\}^{wn}$ *are linear transformations,* T_x *is invertible, and*
$\Delta \in \{0,1\}^{wn}$. *An SPN is* linear *if all its round permutations* $\{T_{k_i}\}_{i=0}^r$ *are linear.*

We present an attack showing that 2-round, linear SPNs cannot be secure for
$w \geq 2$. The attack is based on one shown by Halevi and Rogaway [HR04] in a different context (and is a simple application of the boomerang technique [Wag99]);
our contribution here is to observe that the attack is applicable to any 2-round,
linear SPN. The attack relies on the fact that the field $\mathbb{F} = \text{GF}(2^n)$ is of characteristic 2. This attack and an attack that works for fields of arbitrary characteristic
can be found in [DKS+17].

3.1   Security of 3-Round, Linear (non-tweakable) SPNs

We now explore conditions under which 3-round, linear SPNs are secure. Recall
that a 3-round SPN has four round permutations $\{T_i\}_{i=0}^3$, and without loss of
generality we may assume

$$T_i(k_i, x) = \begin{cases} x \oplus k_i & i \in \{0, 3\} \\ T_i' \cdot (x \oplus k_i) & i \in \{1, 2\} \end{cases}, \tag{4}$$

where $T_1', T_2' \in \mathbb{F}^{w \times w}$ are invertible linear transformations. We prove that a
3-round, linear SPN is secure so long as (i) T_1' and $T_2'^{-1}$ contain no zero
entries (Miles and Viola [MV15] show that matrices with maximal branch number [Dae95] satisfy this property), and (ii) round keys k_0 and k_3 are (individually)
uniform. The proof of this theorem can be found in [DKS+17].

Theorem 1. *Assume* $w > 1$. *Let* SP^T *be a 3-round, linear (non-tweakable) SPN
with round permutations as in (4) and with distribution* $\mathcal{K}$ *over keys* k_0, k_1, k_2, k_3.
If k_0 *and* k_3 *are uniformly distributed and the matrices* $T_1', T_2'^{-1}$ *contain no zero
entries, then*

$$\text{Adv}_{\mathsf{SP}^T}^{\text{su}}(p, q) \leq \frac{5w^2q^2 + 4wpq}{2^n - p - 2w} + \frac{q^2}{2^{wn}}.$$

A MINIMAL SECURE (LINEAR) SPN. We proved that a 3-round, linear SPN is secure if the keys k_0 and k_3 are individually uniform and $T_1', T_2'^{-1}$ contain no 0-entries. No assumptions were made about independence of k_0, k_3, nor were any assumptions made about the distributions of k_1, k_2. So the theorem implies security for the following "minimal" 3-round, linear SPN: Let $k_0 = k_3 = k$, where k is uniform, set $k_1 = k_2 = 0^{wn}$, and let $T_1' = T_2'^{-1} = T$ be invertible with no 0-entries. Define keyed permutations

$$\pi_i(k, x) = \begin{cases} x \oplus k & i \in \{0, 3\} \\ T'x & i = 1 \\ T'^{-1}x & i = 2. \end{cases} \tag{5}$$

We have:

Corollary 1. *Assume $w > 1$. Let SP^T be a 3-round, linear SPN with round permutations as in (5) and $\mathcal{K}$ choosing uniform $k_0 = k_3$ and $k_1 = k_2 = 0^{wn}$. Then*

$$\mathrm{Adv}_{\mathsf{SP}^T}^{\mathrm{su}}(p, q) \leq \frac{5w^2q^2 + 4wpq}{2^n - p - 2w} + \frac{q^2}{2^{wn}}.$$

REDUCING KEY-LENGTH. It is in fact sufficient for the wn-bit key k ($= k_0 = k_3$) in Corollary 1 to satisfy the following conditions: informally, for any n-bit constant c and distinct indices i, i', (a) $k[i]$ equals c with negligible probability, and (b) the sum of $k[i]$ and $k[i']$ equals c with negligible probability. This can be achieved by choosing a uniform n-bit key k' and letting $k[i] = a_i \cdot k'$ where a_i are distinct non-zero elements of $\mathbb{F}$. Thus, one can make do with a "master key" of only n bits, while preserving the same security as in Corollary 1.

4   Security of Non-Linear SPNs

In this section, we first show that keyed tweakable blockwise universal permutations help construct (non-linear) tweakable SPNs. As a preliminary step, we show that 1 round is sufficient to obtain this result. However, the security of the SPN is only up to the birthday attack in this case. Towards obtaining a better security bound, we show that 2 rounds suffice to go beyond the birthday bound and in addition, also present multi-user security beyond the birthday bound for the 2-round tweakable SPN. Finally, we show that if T is a super blockwise tweakable universal permutation, then the security of SP^T converges to 2^n as the number of rounds r increases.

4.1   Birthday Security of 1-Round SPNs

We show that a tweakable blockwise-universal permutation is useful in constructing non-linear tweakable SPNs. The proof of the theorem is a straightforward extension of the non-tweakable version found in [DKS+17]. Consider the 1-round SPN SP^T with $T_{k_1} := T_{k_0}^{-1}$ where T is a keyed blockwise universal tweakable permutation.

Theorem 2. *Let T be a (δ, δ')-blockwise universal tweakable permutation. Then for any integers p and q, one has* $\mathbf{Adv}^{\mathrm{su}}_{\mathsf{SPT}}(p, q) \leq q^2 w^2 \delta + pqw\delta'$.

4.2   Beyond-Birthday Security of 2-Round SPNs

In this section, we will prove the following theorem.

Theorem 3. *Let δ, $\delta' > 0$, and let n and w be positive integers such that $w \geq 2$. Let T be a (δ, δ')-super blockwise universal tweakable permutation. Then for any integers p and q such that $wp + 3w^2q < 2^n/2$, one has*

$$\mathbf{Adv}^{\mathrm{su}}_{\mathsf{SPT}}(p, q) \leq w^2 q(\delta'p + \delta wq)(3\delta'p + 3\delta wq + 2\delta'wq) + \frac{q^2}{2^{wn}} + \frac{q(2wp + 6w^2q)^2}{2^{2n}},$$

$$\mathbf{Adv}^{\mathrm{mu}}_{\mathsf{SPT}}(p, q) \leq w^2 q(\delta'p + (\delta + \delta')wq)(3\delta'p + 3\delta wq + 5\delta'wq)$$
$$+ \frac{q^2}{2^{wn}} + \frac{q(2wp + 8w^2q)^2}{2^{2n}}.$$

Remark 2. For the sake of simplicity, we assume that the three keyed layers are actually the same, which is why we require T to be (δ, δ')-super blockwise tweakable universal. However, if one looks closely at the proof, only the middle layer has to be super-blockwise-universal. The first and the last layer only need to be (δ, δ')-blockwise universal.

Remark 3. When the S-boxes are modeled as block ciphers using secret keys, the security bound (in the standard model) is obtained by setting $p = 0$.

The proof of Theorem 3 relies on the following lemma (with the lower bound simplified) and on Lemma 3 and Lemma 4.

Lemma 7. *Let p and q be positive integers such that $wp + 3w^2q < 2^n/2$, and let $\mathcal{D}$ be a distinguisher in the single-user setting that makes p primitive queries to each of S_1 and S_2 and makes q construction queries. Then for any attainable transcript $\tau = (\mathcal{Q}_C, \mathcal{Q}_S)$, one has*

$$\frac{\mathsf{p}_2(\mathcal{Q}_C|\mathcal{Q}_S)}{\mathsf{p}_1(\mathcal{Q}_C|\mathcal{Q}_S)} \geq 1 - w^2 q(\delta'p + \delta wq)(3\delta'p + 3\delta wq + 2\delta'wq) - \frac{q^2}{2^{wn}} - \frac{q(2wp + 6w^2q)^2}{2^{2n}}.$$

OUTLINE OF PROOF OF LEMMA 7. Throughout the proof, we will write a 2-round SP construction as

$$\mathsf{SP}^T[\mathcal{S}]_{\mathbf{k}}(t, x) = T_{k_2, t}\left(S_2^{\|}\left(T_{k_1, t}\left(S_1^{\|}\left(T_{k_0, t}(x)\right)\right)\right)\right),$$

where $\mathcal{S} = (S_1, S_2)$ is a pair of two public random permutations of $\{0, 1\}^n$, $\mathbf{k} = (k_0, k_1, k_2) \in \mathcal{K}^3$ is the key, $x \in \{0, 1\}^{wn}$ is the plaintext, and, for $i = 1, 2$,

$$S_i^{\|} : \{0, 1\}^{wn} \to \{0, 1\}^{wn}$$
$$x = x_1 \| x_2 \| \ldots \| x_w \longmapsto S_i(x_1) \| S_i(x_2) \| \ldots \| S_i(x_w).$$

We also fix a distinguisher $\mathcal{D}$ as described in the statement and fix an attainable transcript $\tau = (\mathcal{Q}_C, \mathcal{Q}_S)$ obtained by $\mathcal{D}$. Let

$$\mathcal{Q}_{S_1}^{(0)} = \{(u,v) \in \{0,1\}^n \times \{0,1\}^n : (1,u,v) \in \mathcal{Q}_S\},$$
$$\mathcal{Q}_{S_2}^{(0)} = \{(u,v) \in \{0,1\}^n \times \{0,1\}^n : (2,u,v) \in \mathcal{Q}_S\}$$

and let

$$U_1^{(0)} = \{u_1 \in \{0,1\}^n : (u_1,v_1) \in \mathcal{Q}_{S_1}^{(0)}\}, \quad V_1^{(0)} = \{v_1 \in \{0,1\}^n : (u_1,v_1) \in \mathcal{Q}_{S_1}^{(0)}\},$$
$$U_2^{(0)} = \{u_2 \in \{0,1\}^n : (u_2,v_2) \in \mathcal{Q}_{S_2}^{(0)}\}, \quad V_2^{(0)} = \{v_2 \in \{0,1\}^n : (u_2,v_2) \in \mathcal{Q}_{S_2}^{(0)}\}$$

denote the domains and ranges of $\mathcal{Q}_{S_1}^{(0)}$ and $\mathcal{Q}_{S_2}^{(0)}$, respectively.

This type of lemma is usually proved by defining a large enough set of "good" keys, and then, for each choice of a good key, lower bounding the probability of observing this transcript, again by lower bounding the number of possible "intermediate" values. A key is usually said to be good if the adversary cannot use the transcript to follow the path of computation of the encryption/decryption of a query up to a contradiction. However, since the S-boxes are used several times in each round, there will not be enough information in the transcript to allow such a naive definition. Therefore, instead of summing over the choice of the key, we will define an extension of the transcript, that will provide the necessary information, and then sum over every possible good extension.

We will first define what we mean by an extension of the transcript τ. Then we will define bad extensions and explain the link between good extended transcripts and the ratio $\frac{\mathsf{p}_2(\mathcal{Q}_C|\mathcal{Q}_S)}{\mathsf{p}_1(\mathcal{Q}_C|\mathcal{Q}_S)}$. Finally, we will show that the number of bad extended transcripts is small enough in Lemma 8, and then show that the probability to obtain any good extension in the real world is sufficiently close to the probability to obtain τ the ideal world in Lemma 9. We stress that extended transcripts are completely virtual and are not disclosed to the adversary. They are just an artificial intermediate step to lower bound the probability to observe transcript τ in the real world.

EXTENSION OF A TRANSCRIPT. We will extend the transcript τ of the attack via a certain randomized process. We begin with choosing a pair of keys $(k_0, k_2) \in \mathcal{K}^2$ uniformly at random. Once these keys have been chosen, some construction queries will become involved in collisions. A *colliding query* is defined as a construction query $(t,x,y) \in \mathcal{Q}_C$ such that one of the following conditions holds:

1. there exist an S-box query $(1,u,v) \in \mathcal{Q}_S$ and an integer $i \in \{1,\ldots,w\}$ such that $T_{k_0,t}(x)_i = u$;
2. there exist an S-box query $(2,u,v) \in \mathcal{Q}_S$ and an integer $i \in \{1,\ldots,w\}$ such that $T_{k_2,t}^{-1}(y)_i = v$;
3. there exist a construction query $(t',x',y') \in \mathcal{Q}_C$ and integers $i,j \in \{1,\ldots,w\}$ such that $(t,x,y,i) \neq (t',x',y',j)$ and $T_{k_0,t}(x)_i = T_{k_0,t'}(x')_j$;
4. there exist a construction query $(t',x',y') \in \mathcal{Q}_C$ and integers $i,j \in \{1,\ldots,w\}$ such that $(t,x,y,i) \neq (t',x',y',j)$ and $T_{k_2,t}^{-1}(y)_i = T_{k_2,t'}^{-1}(y')_j$.

We are now going to build a new set $\mathcal{Q}'_S$ of S-box evaluations that will play the role of an extension of $\mathcal{Q}_S$. For each *colliding* query $(t, x, y) \in \mathcal{Q}_C$, we will add tuples $(1, T_{k_0}(t, x)_i, v')_{1 \leq i \leq w}$ (if (t, x, y) collides at the input of S_1) or $(2, u', T_{k_2, t}^{-1}(y)_i)_{1 \leq i \leq w}$ (if (t, x, y) collides at the output of S_2) by lazy sampling $v' = S_1(T_{k_0, t}(x)_i)$ or $u' = S_2^{-1}(T_{k_2, t}^{-1}(y)_i)$, as long as it has not been determined by any existing query in $\mathcal{Q}_S$. We finally choose a key k_1 uniformly at random. An extended transcript of τ will be defined as a tuple $\tau' = (\mathcal{Q}_C, \mathcal{Q}_S, \mathcal{Q}'_S, \mathbf{k})$ where $\mathbf{k} = (k_0, k_1, k_2)$. For each collision between a construction query and a primitive query, or between two construction queries, the extended transcript will contain enough information to compute a complete round of the evaluation of the SPN. This will be useful to lower bound the probability to get the transcript τ in the real world.

Definition of Bad Transcript Extensions. Let

$$\mathcal{Q}_{S_1}^{(1)} = \{(u, v) \in \{0, 1\}^n \times \{0, 1\}^n : (1, u, v) \in \mathcal{Q}_S \cup \mathcal{Q}'_S\}$$

$$\mathcal{Q}_{S_2}^{(1)} = \{(u, v) \in \{0, 1\}^n \times \{0, 1\}^n : (2, u, v) \in \mathcal{Q}_S \cup \mathcal{Q}'_S\}.$$

In words, $\mathcal{Q}_{S_i}^{(1)}$ summarizes each constraint that is forced on S_i by $\mathcal{Q}_S$ and $\mathcal{Q}'_S$. Let

$$U_1 = \{u_1 \in \{0, 1\}^n : (1, u_1, v_1) \in \mathcal{Q}_{S_1}^{(1)}\}, \quad V_1 = \{v_1 \in \{0, 1\}^n : (1, u_1, v_1) \in \mathcal{Q}_{S_1}^{(1)}\},$$

$$U_2 = \{u_2 \in \{0, 1\}^n : (2, u_2, v_2) \in \mathcal{Q}_{S_2}^{(1)}\}, \quad V_2 = \{v_2 \in \{0, 1\}^n : (2, u_2, v_2) \in \mathcal{Q}_{S_2}^{(1)}\}$$

be the domains and ranges of $\mathcal{Q}_{S_1}^{(1)}$ and $\mathcal{Q}_{S_2}^{(1)}$, respectively. We define two quantities characterizing an extended transcript τ', namely

$$\alpha_1 \stackrel{\text{def}}{=} |\{(x, y) \in \mathcal{Q}_C : T_{k_0}(x)_i \in U_1 \text{ for some } i \in \{1, \ldots, w\}\}|,$$

$$\alpha_2 \stackrel{\text{def}}{=} |\{(x, y) \in \mathcal{Q}_C : T_{k_2}^{-1}(y)_i \in V_2 \text{ for some } i \in \{1, \ldots, w\}\}|.$$

In words, α_1 (resp. α_2) is the number of queries $(t, x, y) \in \mathcal{Q}_C$ which collide with a query $(u_1, v_1) \in \mathcal{Q}_{S_1}^{(1)}$ (resp. which collide with a query $(u_2, v_2) \in \mathcal{Q}_{S_2}^{(1)}$) in the extended transcript. This corresponds to the number of queries $(t, x, y) \in \mathcal{Q}_C$ which collide with either an original query $(u_1, v_1) \in \mathcal{Q}_{S_1}^{(0)}$ (resp. $(u_2, v_2) \in \mathcal{Q}_{S_2}^{(0)}$) or with a query $(t', x', y') \in \mathcal{Q}_C$ at an input of S_1 (resp. at the output of S_2), once the choice of (k_0, k_2) has been made. We will also denote

$$\beta_i = |\mathcal{Q}_{S_i}^{(1)}| - |\mathcal{Q}_{S_i}^{(0)}| = |\mathcal{Q}_{S_i}^{(1)}| - p$$

for $i = 1, 2$, the number of additional queries included in the extended transcript.

We say an extended transcript τ' is *bad* if at least one of the following conditions is fulfilled:

(C-1) there exist $(t, x, y) \in \mathcal{Q}_C$, $i, j \in \{1, \ldots, w\}$, $u_1 \in U_1$, and $v_2 \in V_2$ such that $T_{k_0, t}(x)_i = u_1$ and $T_{k_2, t}^{-1}(y)_j = v_2$;

(C-2) there exist $(t, x, y) \in \mathcal{Q}_C$, $i, j \in \{1, \ldots, w\}$, $u_1 \in U_1$, and $u_2 \in U_2$ such that $T_{k_0,t}(x)_i = u_1$ and $T_{k_1,t} \left(S_1^{\|} (T_{k_0,t}(x)) \right)_j = u_2$[4];

(C-3) there exist $(t, x, y) \in \mathcal{Q}_C$, $i, j \in \{1, \ldots, w\}$, $v_1 \in V_1$, and $v_2 \in V_2$ such that $T_{k_2,t}^{-1}(y)_i = v_2$ and $T_{k_1,t}^{-1} \left((S_2^{-1})^{\|} \left(T_{k_2,t}^{-1}(y) \right) \right)_j = v_1$;

(C-4) there exist $(t, x, y), (t', x', y') \in \mathcal{Q}_C$, $i, i', j, j' \in \{1, \ldots, w\}$ with $(t, x, j) \neq (t', x', j')$, $u_1, u_1' \in U_1$ such that $T_{k_0,t}(x)_i = u_1$, $T_{k_0,t'}(x')_{i'} = u_1'$ and

$$T_{k_1,t} \left(S_1^{\|} (T_{k_0,t}(x)) \right)_j = T_{k_1,t'} \left(S_1^{\|} (T_{k_0,t'}(x')) \right)_{j'};$$

(C-5) there exist $(t, x, y), (t', x', y') \in \mathcal{Q}_C$, $i, i', j, j' \in \{1, \ldots, w\}$ with $(y, j) \neq (y', j')$, $v_2, v_2' \in V_2$ such that $T_{k_2,t}^{-1}(y)_i = v_2$, $T_{k_2,t}^{-1}(y')_{i'} = v_2'$ and

$$T_{k_1,t}^{-1} \left((S_2^{-1})^{\|} \left(T_{k_2,t}^{-1}(y) \right) \right)_j = T_{k_1,t'}^{-1} \left((S_2^{-1})^{\|} \left(T_{k_2,t'}^{-1}(y') \right) \right)_{j'}.$$

Any extended transcript that is not bad will be called *good*. Given an original transcript τ, we denote $\Theta_{\text{good}}(\tau)$ (resp. $\Theta_{\text{bad}}(\tau)$) the set of good (resp. bad) extended transcripts of τ and $\Theta'(\tau)$ the set of all extended transcripts of τ.

FROM ATTAINABLE TRANSCRIPTS TO GOOD EXTENDED TRANSCRIPTS. We are now going to justify the usefulness of extended transcripts. For any extended transcript $\tau' = (\mathcal{Q}_C, \mathcal{Q}_S, \mathcal{Q}_S', \mathbf{k})$, let us denote

$$\mathsf{p}_{\text{re}}(\tau') = \Pr \left[(\mathbf{k}', \mathcal{S}) \xleftarrow{\$} \mathcal{K}^3 \times \mathsf{Perm}(n)^2 : (\mathcal{S} \vdash \mathcal{Q}_S \cup \mathcal{Q}_S') \wedge (\mathsf{SP}_{\mathbf{k}}^T[\mathcal{S}] \vdash \mathcal{Q}_C) \wedge (\mathbf{k}' = \mathbf{k}) \right],$$

$$\mathsf{p}(\tau') = \Pr \left[\mathcal{S} \xleftarrow{\$} \mathsf{Perm}(n)^2 : \mathsf{SP}^T[\mathcal{S}]_{\mathbf{k}} \vdash \mathcal{Q}_C \middle| (S_1 \vdash \mathcal{Q}_{S_1}^{(1)}) \wedge (S_2 \vdash \mathcal{Q}_{S_2}^{(1)}) \right].$$

Note that one has

$$\Pr \left[(\widetilde{P}, \mathcal{S}) \xleftarrow{\$} \widetilde{\mathsf{Perm}}(\mathcal{T}, wn) \times \mathsf{Perm}(n)^2 : (\mathcal{S} \vdash \mathcal{Q}_S) \wedge (\widetilde{P} \vdash \mathcal{Q}_C) \right]$$

$$\leq \frac{1}{(2^{wn})_q (2^n)_p (2^n)_p},$$

$$\Pr \left[(\mathbf{k}, \mathcal{S}) \xleftarrow{\$} \mathcal{K}^3 \times \mathsf{Perm}(n)^2 : (\mathcal{S} \vdash \mathcal{Q}_S) \wedge (\mathsf{SP}_{\mathbf{k}}^T[\mathcal{S}] \vdash \mathcal{Q}_C) \right]$$

$$\geq \sum_{\tau' \in \Theta_{\text{good}}(\tau)} \mathsf{p}_{\text{re}}(\tau') \geq \sum_{\tau' \in \Theta_{\text{good}}(\tau)} \frac{1}{|\mathcal{K}|^3 (2^n)_{p+\beta_1} (2^n)_{p+\beta_2}} \mathsf{p}(\tau'),$$

which gives

$$\mathsf{p}_1(\mathcal{Q}_C | \mathcal{Q}_S) \leq \frac{1}{(2^{wn})_q},$$

$$\mathsf{p}_2(\mathcal{Q}_C | \mathcal{Q}_S) \geq \sum_{\tau' \in \Theta_{\text{good}}(\tau)} \frac{1}{|\mathcal{K}|^3 (2^n - p)_{\beta_1} (2^n - p)_{\beta_2}} \mathsf{p}(\tau').$$

---

[4] Note that the value $S_1^{\|} (T_{k_0,t}(x))$ is well-defined thanks to the additional virtual queries from $\mathcal{Q}_S'$.

Thus one has

$$
\frac{\mathsf{p}_2(\mathcal{Q}_C|\mathcal{Q}_S)}{\mathsf{p}_1(\mathcal{Q}_C|\mathcal{Q}_S)} \geq \sum_{\tau' \in \Theta_{\mathrm{good}}(\tau)} \frac{(2^{wn})_q}{|\mathcal{K}|^3 (2^n - p)_{\beta_1} (2^n - p)_{\beta_2}} \mathsf{p}(\tau')
$$

$$
\geq \min_{\tau' \in \Theta_{\mathrm{good}}(\tau)} ((2^{wn})_q \mathsf{p}(\tau')) \sum_{\tau' \in \Theta_{\mathrm{good}}(\tau)} \frac{1}{|\mathcal{K}|^3 (2^n - p)_{\beta_1} (2^n - p)_{\beta_2}}.
$$

Note that the weighted sum $\sum_{\tau' \in \Theta_{\mathrm{good}}(\tau)} \frac{1}{|\mathcal{K}|^3 (2^n - p)_{\beta_1} (2^n - p)_{\beta_2}}$ corresponds exactly to the probability that a random extended transcript is good when it is sampled as follows:

1. choose keys $k_0, k_2 \in \mathcal{K}$ uniformly and independently at random;
2. choose the partial extension of the S-box queries based on the new collisions $\mathcal{Q}'_S$ uniformly at random (meaning that each possible u or v is chosen uniformly at random in the set of its authorized values);
3. finally choose k_1 uniformly at random, independently from everything else.

Thus, the exact probability of observing the extended transcript τ' is

$$
\frac{1}{|\mathcal{K}|^3 (2^n - p)_{\beta_1} (2^n - p)_{\beta_2}},
$$

and we have

$$
\sum_{\tau' \in \Theta_{\mathrm{good}}(\tau)} \frac{1}{|\mathcal{K}|^3 (2^n - p)_{\beta_1} (2^n - p)_{\beta_2}} = \Pr\left[\tau' \in \Theta_{\mathrm{good}}(\tau)\right].
$$

One finally gets

$$
\frac{\mathsf{p}_2(\mathcal{Q}_C|\mathcal{Q}_S)}{\mathsf{p}_1(\mathcal{Q}_C|\mathcal{Q}_S)} \geq \Pr\left[\tau' \in \Theta_{\mathrm{good}}(\tau)\right] \cdot \min_{\tau' \in \Theta_{\mathrm{good}}(\tau)} ((2^{wn})_q \mathsf{p}(\tau')). \tag{6}
$$

Lemma 8 and Lemma 9 lower bound $\Pr\left[\tau' \in \Theta_{\mathrm{good}}(\tau)\right]$ (by upper bounding $\Pr\left[\tau' \in \Theta_{\mathrm{bad}}(\tau)\right]$) and $\min_{\tau' \in \Theta_{\mathrm{good}}(\tau)} ((2^{wn})_q \mathsf{p}(\tau'))$, respectively. Then combining (6) with Lemma 8 and Lemma 9 will complete the proof of Lemma 7.

Lemma 8. *One has*

$$
\Pr\left[\tau' \in \Theta_{\mathrm{bad}}(\tau)\right] \leq w^2 q (\delta' p + \delta w q)(3\delta' p + 3\delta w q + 2\delta' w q).
$$

Proof. We fix any attainable transcript, denoted $(\mathcal{Q}_C, \mathcal{Q}_{S_1}^{(0)}, \mathcal{Q}_{S_2}^{(0)})$. For any fixed construction query $(t, x, y) \in \mathcal{Q}_C$, define event

$$
\mathsf{Coll}_1(t, x, y) \Leftrightarrow \text{there exist } i \in \{1, \ldots, w\} \text{ and } u_1 \in U_1 \text{ such that } T_{k_0, t}(x)_i = u_1.
$$

This event can be broken down into the following two subevents:

- there exist $i \in \{1, \ldots, w\}$, $j \in \{1, \ldots, p\}$ such that $T_{k_0, t}(x)_i = u_j$,

– there exist $(t', x', y') \in \mathcal{Q}_C$, $i, j \in \{1, \ldots, w\}$ such that $(t, x, y, i) \neq (t', x', y', j)$ and $T_{k_0, t}(x)_i = T_{k_0, t'}(x')_j$.

Note that these events only involve queries from the original transcript, which means that the choice of the key is actually independent from these values. By the blockwise uniformity of T, one has

$$\Pr\left[k_0 \in \mathcal{K} : \mathsf{Coll}_1(t, x, y)\right] \leq \delta' wp + \delta w^2 q. \tag{7}$$

Similarly, let

$\mathsf{Coll}_2(t, x, y) \Leftrightarrow$ there exist $i \in \{1, \ldots, w\}$ and $v_2 \in V_2$ such that $T_{k_2, t}^{-1}(y)_i = v_2$.

Then one has

$$\Pr\left[k_2 \in \mathcal{K} : \mathsf{Coll}_2(x, y)\right] \leq \delta' wp + \delta w^2 q. \tag{8}$$

Also note that one has $|\mathcal{Q}_{S_1}^{(1)}|, |\mathcal{Q}_{S_2}^{(1)}| \leq p + wq$, as additional tuples in $\mathcal{Q}'_S$ come from the completion of partial information about a construction query.

We now upper bound the probabilities of the five conditions in turn. The sets of attainable transcripts fulfilling condition (C-1), (C-2), (C-3), (C-4), (C-5) will be denoted $\Theta_1, \Theta_2, \Theta_3, \Theta_4, \Theta_5$, respectively.

Condition (C-1). One has

$$\Pr\left[\tau' \in \Theta_1\right] \leq \sum_{(t, x, y) \in \mathcal{Q}_C} \Pr\left[\mathsf{Coll}_1(t, x, y) \wedge \mathsf{Coll}_2(t, x, y)\right].$$

Since the random choice of k_0 and k_2 are independent, and by (7) and (8), one has

$$\Pr\left[\tau' \in \Theta_1\right] \leq q(\delta' wp + \delta w^2 q)^2.$$

Condition (C-2) and (C-3). Fix any query $(t, x, y) \in \mathcal{Q}_C$. Since the random choice of k_1 is independent from the queries transcript and from the choice of k_0, the probability, over the random choice of k_1, that there exist $i \in \{1, \ldots, w\}$ and $u_2 \in U_2$ such that $T_{k_1, t}\left(S_1^{\|}\left(T_{k_0, t}(x)\right)\right)_i = u_2$, conditioned on $\mathsf{Coll}_1(t, x, y)$, is upper bounded by $\delta' w(p + wq)$. Thus, by summing over every construction query and using (7), one has

$$\Pr\left[\tau' \in \Theta_2\right] \leq \delta' wq(p + wq)(\delta' wp + \delta w^2 q).$$

Similarly, one has

$$\Pr\left[\tau' \in \Theta_3\right] \leq \delta' wq(p + wq)(\delta' wp + \delta w^2 q).$$

Conditions (C-4), and (C-5). Given two distinct pairs $(i, (t, x, y))$, $(i', (t', x', y')) \in \{1, \ldots, w\} \times \mathcal{Q}_C$ such that (t, x, y) and (t', x', y') are both colliding queries, let us define event

$$\mathsf{Coll}(t, x, y, t', x', y')_{i,i'} \Leftrightarrow T_{k_1,t}\left(S_1^{\|}\left(T_{k_0,t}(x)\right)\right)_i = T_{k_1,t'}\left(S_1^{\|}\left(T_{k_0,t'}(x')\right)\right)_{i'}.$$

Then for any distinct pairs $(i, (t, x, y)), (i', (t', x', y')) \in \{1, \ldots, w\} \times \mathcal{Q}_C$, one has

$$\Pr\left[\mathsf{Coll}_1(t, x, y) \wedge \mathsf{Coll}_1(t', x', y') \wedge \mathsf{Coll}(t, x, y, t', x', y')_{i,i'}\right]$$
$$= \Pr\left[\mathsf{Coll}(t, x, y, t', x', y')_{i,i'} \mid \mathsf{Coll}_1(t, x, y) \wedge \mathsf{Coll}_1(t', x', y')\right]$$
$$\times \Pr\left[\mathsf{Coll}_1(t', x', y') \mid \mathsf{Coll}_1(t, x, y)\right]$$
$$\times \Pr\left[\mathsf{Coll}_1(t, x, y)\right] \leq \delta \cdot 1 \cdot (\delta' wp + \delta w^2 q),$$

where, for the last inequality, we used the (δ, δ')-blockwise uniformity of T and the fact that the event $\mathsf{Coll}_1(t, x, y) \wedge \mathsf{Coll}_1(t', x', y')$ only depends on the choice of k_0 whereas $\mathsf{Coll}(t, x, y, t', x', y')_{i,i'}$ involves the choice of k_1. Thus, by summing over every such pair, one obtains

$$\Pr\left[\tau' \in \Theta_4\right] \leq \delta w^2 q^2 (\delta' wp + \delta w^2 q).$$

Similarly, one has

$$\Pr\left[\tau' \in \Theta_5\right] \leq \delta w^2 q^2 (\delta' wp + \delta w^2 q).$$

The lemma follows by taking a union bound over all the conditions.   □

Our next step is to study good extended transcripts.

Lemma 9. *For any good extended transcript τ', one has*

$$(2^{wn})_q \mathsf{p}(\tau') \geq 1 - \frac{q^2}{2^{wn}} - \frac{q(2wp + 6w^2 q)^2}{2^{2n}}.$$

Proof. Fix any good extended transcript $\tau' = (\mathcal{Q}_C, \mathcal{Q}_S, \mathcal{Q}'_S, (k_0, k_1, k_2))$. Let us denote $p_1 = |\mathcal{Q}_{S_1}^{(1)}|$ and $p_2 = |\mathcal{Q}_{S_2}^{(1)}|$.

Our goal is then to prove that $\mathsf{p}(\tau')$ is close enough to $1/(2^{wn})_q$. In order to do so, we are going to group the construction queries according to the type of collision they are involved in:

$$\mathcal{Q}_{U_1} = \{(t, x, y) \in \mathcal{Q}_C : T_{k_0,t}(x)_i \in U_1 \text{ for } i = 1, \ldots, w\}$$
$$\mathcal{Q}_{V_2} = \{(t, x, y) \in \mathcal{Q}_C : T_{k_2,t}^{-1}(y)_i \in V_2 \text{ for } i = 1, \ldots, w\}$$
$$\mathcal{Q}_0 = \mathcal{Q}_C \setminus (\mathcal{Q}_{U_1} \cup \mathcal{Q}_{V_2}).$$

Note that, thanks to the additional queries from $\mathcal{Q}'_S$, there is an equivalence between the events "$T_{k_0,t}(x)_i \in U_1$ for each $i = 1, \ldots, w$" and "there exists $i \in \{1, \ldots, w\}$ such that $T_{k_0,t}(x)_i \in U_1$". Thus, one has by definition $|\mathcal{Q}_{U_1}| = \alpha_1$. Similarly, one has $|\mathcal{Q}_{V_2}| = \alpha_2$. Also note that these sets form a partition of $\mathcal{Q}_C$:

- $\mathcal{Q}_0 \cap \mathcal{Q}_{U_1} = \emptyset$ by definition;
- $\mathcal{Q}_0 \cap \mathcal{Q}_{V_2} = \emptyset$ by definition;

– $\mathcal{Q}_{U_1} \cap \mathcal{Q}_{V_2} = \emptyset$ since otherwise τ' would satisfy (C-1).

If we denote respectively $\mathsf{E}_{U_1}, \mathsf{E}_{V_2}$ and E_0 the event $\mathsf{SP}^T[\mathcal{S}]_{\mathbf{k}} \vdash \mathcal{Q}_{U_1}, \mathcal{Q}_{V_2}, \mathcal{Q}_0$, the event $\mathsf{SP}^T[\mathcal{S}]_{\mathbf{k}} \vdash \mathcal{Q}_C$ is equivalent to $\mathsf{E}_{U_1} \wedge \mathsf{E}_{V_2} \wedge \mathsf{E}_0$. Note that, by definition of $\mathcal{Q}_{U_1}$, each $(t, x, y) \in \mathcal{Q}_{U_1}$ is such that $T_{k_0,t}(x)_i \in U_1$ for each $i = 1, \ldots, w$; this means that the output of S_1 is already fixed by $\mathcal{Q}_{S_1}^{(1)}$ and E_{U_1} actually only involves S_2. A similar reasoning can be made for E_{V_2}. Thus we have

$$
\begin{aligned}
\mathsf{p}(\tau') &= \Pr\left[\mathsf{E}_{U_1} \wedge \mathsf{E}_{V_2} \wedge \mathsf{E}_0 \;\middle|\; S_1 \vdash \mathcal{Q}_{S_1}^{(1)} \wedge S_2 \vdash \mathcal{Q}_{S_2}^{(1)}\right] \\
&= \Pr\left[\mathsf{E}_{U_1} \wedge \mathsf{E}_{V_2} \;\middle|\; S_1 \vdash \mathcal{Q}_{S_1}^{(1)} \wedge S_2 \vdash \mathcal{Q}_{S_2}^{(1)}\right] \\
&\quad \times \Pr\left[\mathsf{E}_0 \;\middle|\; \mathsf{E}_{U_1} \wedge \mathsf{E}_{V_2} \wedge S_1 \vdash \mathcal{Q}_{S_1}^{(1)} \wedge S_2 \vdash \mathcal{Q}_{S_2}^{(1)}\right] \\
&= \Pr\left[\mathsf{E}_{U_1} \;\middle|\; S_2 \vdash \mathcal{Q}_{S_2}^{(1)}\right] \cdot \Pr\left[\mathsf{E}_{V_2} \;\middle|\; S_1 \vdash \mathcal{Q}_{S_1}^{(1)}\right] \\
&\quad \times \Pr\left[\mathsf{E}_0 \;\middle|\; \mathsf{E}_{U_1} \wedge \mathsf{E}_{V_2} \wedge S_1 \vdash \mathcal{Q}_{S_1}^{(1)} \wedge S_2 \vdash \mathcal{Q}_{S_2}^{(1)}\right],
\end{aligned} \tag{9}
$$

where $\Pr\left[\mathsf{E}_{U_1} \middle| S_2 \vdash \mathcal{Q}_{S_2}^{(1)}\right]$ (resp. $\Pr\left[\mathsf{E}_{V_2} \middle| S_1 \vdash \mathcal{Q}_{S_1}^{(1)}\right]$) is the probability, over the random choice of permutation S_2 (resp. permutation S_1), that S_2 (resp. S_1) is compatible with the additional equations implied by $\mathcal{Q}_{U_1}$ (resp. by $\mathcal{Q}_{V_2}$), conditioned on the event $S_2 \vdash \mathcal{Q}_{S_2}^{(1)}$ (resp. $S_1 \vdash \mathcal{Q}_{S_1}^{(1)}$).

In order to evaluate $\Pr\left[\mathsf{E}_{U_1} \middle| S_2 \vdash \mathcal{Q}_{S_2}^{(1)}\right]$ and $\Pr\left[\mathsf{E}_{V_2} \middle| S_1 \vdash \mathcal{Q}_{S_1}^{(1)}\right]$, we first note that, since we condition on the event $S_2 \vdash \mathcal{Q}_{S_2}^{(1)}$, S_2 is already fixed on p_2 values. Second, remark that this event is actually equivalent to the following equations:

$$
S_2\left(T_{k_1,t}\left(S_1^{\|}\left(T_{k_0,t}(x)\right)\right)_i\right) = T_{k_2,t}^{-1}(y)_i
$$

for every $(t, x, y) \in \mathcal{Q}_{U_1}$ and $i \in \{1, \ldots, w\}$. All the values $T_{k_1,t}\left(S_1^{\|}\left(T_{k_0,t}(x)\right)\right)_i$ are actually pairwise distinct and outside U_2 since otherwise (C-2) or (C-4) would be satisfied. Similarly, the values $T_{k_2,t}^{-1}(y)_i$ are pairwise distinct and outside V_2 since otherwise (C-1) would be satisfied. Indeed, if a collision between two values $T_{k_2,t}^{-1}(y)_i$ had occurred, then these values would also appear in V_2. Hence the event E_{U_1} is actually equivalent to $w\alpha_1$ new and distinct equations on S_2, so that

$$
\Pr\left[\mathsf{E}_{U_1} \middle| S_2 \vdash \mathcal{Q}_{S_2}^{(1)}\right] = \frac{1}{(2^n - p_2)_{w\alpha_1}}. \tag{10}
$$

By a similar reasoning, one has

$$
\Pr\left[\mathsf{E}_{V_2} \middle| S_1 \vdash \mathcal{Q}_{S_1}^{(1)}\right] = \frac{1}{(2^n - p_1)_{w\alpha_2}}. \tag{11}
$$

The next step is to lower bound $\Pr\left[\mathsf{E}_0 \middle| \mathsf{E}_{U_1} \wedge \mathsf{E}_{V_2} \wedge S_1 \vdash \mathcal{Q}_{S_1}^{(1)} \wedge S_2 \vdash \mathcal{Q}_{S_2}^{(1)}\right]$. Conditioned on $\mathsf{E}_{U_1} \wedge \mathsf{E}_{V_2} \wedge S_1 \vdash \mathcal{Q}_{S_1}^{(1)} \wedge S_2 \vdash \mathcal{Q}_{S_2}^{(1)}$, S_1 and S_2 are fixed on respectively $p_1 + w\alpha_2$ and $p_2 + w\alpha_1$ values. Let U_1' (resp. U_2') be the set of values on which S_1 (resp. S_2) is already fixed and $V_1' = \{S_1(u) : u \in U_1'\}$ (resp. $V_2' = \{S_2(u) : u \in U_2'\}$). Let also $q_0 = |\mathcal{Q}_0|$. For clarity, we denote

$$\mathcal{Q}_0 = \{(t_1, x_1, y_1), \ldots, (t_{q_0}, x_{q_0}, y_{q_0})\},$$

using an arbitrary ordering of the queries.

Our goal is now to compute a lower bound on the number of possible "intermediate values" such that the event E_0 is equivalent to new and distinct equations on S_1 and S_2. First note that the values $T_{k_0,t}(x)_i$ for each $(t, x, y) \in \mathcal{Q}_0, i \in \{1, \ldots, w\}$ are pairwise distinct and outside U_1'. Indeed, if this were not the case, then at least one query in $\mathcal{Q}_0$ would be a colliding query. By definition of our security experiment, this means that this query would either be in E_{U_1} or E_{V_2}, depending on the type of collision it is involved in. Similarly, the values $T_{k_2,t}^{-1}(y)_i$ for each $(t, x, y) \in \mathcal{Q}_0, i \in \{1, \ldots, w\}$ are pairwise distinct and outside V_2'.

Let N_0 be the number of tuples of distinct values $(v_{1,i,j})_{1 \leq i \leq q_0, 1 \leq j \leq w}$ in $\{0,1\}^n \backslash V_1'$ such that the values $(T_{k_1,t_i}(\|_{k=1}^w v_{1,i,k})_j)_{1 \leq i \leq q_0, 1 \leq j \leq w}$ are also pairwise distinct and outside U_2'. Let $i \in \{1, \ldots, q_0\}$. There are exactly $(2^n - |V_1'| - w(i-1))_w$ possible tuples of distinct values $(v_{1,i,j})_{1 \leq j \leq w}$ in $\{0,1\}^n \backslash V_1'$ that will also be different from the previous values $v_{1,i,j}$ for $i < q_0$ and $j \in \{1, \ldots, w\}$. Similarly, there are exactly $(2^n - |U_2'| - w(i-1))_w$ possible tuples of distinct values for $(T_{k_1,t_i}(\|_{k=1}^w v_{1,i,k}))_{1 \leq j \leq w}$ in $\{0,1\}^n \backslash U_2'$ that will also be different from the previous values $T_{k_1,t_i}(\|_{k=1}^w v_{1,i,k})$ for $i < q_0$ and $j \in \{1, \ldots, w\}$. This removes at most $2^{wn} - (2^n - |U_2'| - w(i-1))_w$ tuples of values for $(T_{k_1,t_i}(\|_{k=1}^w v_{1,i,k}))_{1 \leq j \leq w}$. Since T_{k_1,t_i} is a permutation, we have to remove at most $2^{wn} - (2^n - |U_2'| - w(i-1))_w$ possible tuples of values for $(v_{1,i,j})_{1 \leq j \leq w}$. Thus

$$N_0 \geq \prod_{i=1}^{q_0} \left((2^n - |V_1'| - w(i-1))_w + (2^n - |U_2'| - w(i-1))_w - 2^{wn}\right). \quad (12)$$

For any tuple of values $(v_{1,i,j})$ fulfilling the previous conditions, then, conditioned on S_1 satisfying $S_1(T_{k_0,t_i}(x_i))_j = v_{1,i,j}$, the event E_0 is equivalent to wq_0 distinct and new equations on S_2. Hence, it follows that

$$\Pr\left[\mathsf{E}_0 \mid \mathsf{E}_{U_1} \wedge \mathsf{E}_{V_2} \wedge S_1 \vdash \mathcal{Q}_{S_1}^{(1)} \wedge S_2 \vdash \mathcal{Q}_{S_2}^{(1)}\right]$$

$$\geq \frac{N_0}{(2^n - p_1 - w\alpha_2)_{wq_0}(2^n - p_2 - w\alpha_1)_{wq_0}}. \quad (13)$$

Combining (9), (10), (11), (12) (13), we obtain

$$(2^{wn})_q \mathsf{p}(\tau') \geq \frac{(2^{wn})_q \prod_{i=0}^{q_0-1} \left(\begin{array}{c}(2^n - p_1 - w(\alpha_2 + i))_w \\ +(2^n - p_2 - w(\alpha_1 + i))_w - 2^{wn}\end{array}\right)}{(2^n - p_1)_{wq_0 + w\alpha_2}(2^n - p_2)_{wq_0 + w\alpha_1}}$$

$$= \frac{(2^{wn})_q}{2^{q_0 wn}(2^n - p_1)_{w\alpha_2}(2^n - p_2)_{w\alpha_1}}$$

$$\times \prod_{i=0}^{q_0-1} \frac{2^{wn}\left(\begin{array}{c}(2^n - p_1 - w(\alpha_2 + i))_w \\ +(2^n - p_2 - w(\alpha_1 + i))_w - 2^{wn}\end{array}\right)}{(2^n - p_1 - w\alpha_2 - wi)_w(2^n - p_2 - w\alpha_1 - wi)_w}$$

$$\geq \frac{(2^{wn})_q}{2^{q_0 wn}(2^n - p_1)_{w\alpha_2}(2^n - p_2)_{w\alpha_1}} \cdot \prod_{i=0}^{q_0-1} \Delta_i$$

where

$$\Delta_i = 1 - \left(\frac{2^{wn}}{(2^n - p_2 - w\alpha_1 - wi)_w} - 1\right)\left(\frac{2^{wn}}{(2^n - p_1 - w\alpha_2 - wi)_w} - 1\right)$$

for $i = 0, \ldots, q_0 - 1$. We also have $\alpha_1 \leq q$ and $p_2 \leq p + wq$, which gives

$$\frac{2^{wn}}{(2^n - p_2 - w\alpha_1 - wi)_w} \leq \left(\frac{2^n}{2^n - p - 3wq}\right)^w \leq \left(1 + \frac{p + 3wq}{2^n - p - 3wq}\right)^w.$$

Then, since $wp + 3w^2q < 2^n/2$, we can apply Lemma 1 and we get

$$\frac{2^{wn}}{(2^n - p_2 - w\alpha_1 - wi)_w} \leq 1 + \frac{wp + 3w^2q}{2^n - wp - 3w^2q} \leq 1 + \frac{2wp + 6w^2q}{2^n}.$$

Similarly, one has

$$\frac{2^{wn}}{(2^n - p_1 - w\alpha_2 - wi)_w} \leq 1 + \frac{2wp + 6w^2q}{2^n}.$$

Thus one has

$$\Delta_i \geq 1 - \left(\frac{2wp + 6w^2q}{2^n}\right)^2.$$

Moreover, one has

$$\frac{(2^{wn})_q}{2^{q_0 wn}(2^n - p_1)_{w\alpha_2}(2^n - p_2)_{w\alpha_1}} \geq \frac{(2^{wn} - q)^q}{2^{qwn}} \geq \left(1 - \frac{q}{2^{wn}}\right)^q \geq 1 - \frac{q^2}{2^{wn}}.$$

Finally, we get

$$(2^{wn})_q \mathsf{p}(\tau') \geq \left(1 - \frac{q^2}{2^{wn}}\right)\left(1 - \left(\frac{2wp + 6w^2q}{2^n}\right)^2\right)^{q_0}$$

$$\geq \left(1 - \frac{q^2}{2^{wn}}\right)\left(1 - \frac{q(2wp + 6w^2q)^2}{2^{2n}}\right)$$

$$\geq 1 - \frac{q^2}{2^{wn}} - \frac{q(2wp + 6w^2q)^2}{2^{2n}}.$$

$\square$

4.3   Asymptotically Optimal Security of SPNs

In this section, we will prove that if T is a super blockwise tweakable universal permutation, then the security of SP^T converges to 2^n (in terms of the threshold number of queries) as the number of rounds r increases.

Theorem 4. *For an even integer r, let SP^T be an r-round substitution-permutation network based on a (δ, δ')-super blockwise tweakable universal permutation T. Then one has*

$$\mathrm{Adv}^{\mathrm{mu}}_{\mathsf{SP}^T}(p, q) \leq 4\sqrt{q}\left(2wp\delta' + 2w^2 q(\delta' + \delta) + w^2\delta\right)^{\frac{r}{4}}.$$

Hence, assuming $\delta, \delta' \simeq 2^{-n}$ and $p = q$, an r-round SP^T is secure up to $2^{\frac{rn}{r+2}}$ queries.

PROOF OF THEOREM 4. We assume that $r = 2s$ for a positive integer s. Let $\overline{\mathsf{SP}}^T[\mathcal{S}]$ denote a variant of $\mathsf{SP}^T[\mathcal{S}]$ without the last permutation layer. Then one has

$$\mathsf{SP}^T[\mathcal{S}] = \left(\overline{\mathsf{SP}}^{T^{-1}}[\mathcal{S}^{(2)}]\right)^{-1} \circ T \circ \overline{\mathsf{SP}}^T[\mathcal{S}^{(1)}]$$

for $\mathcal{S}^{(1)} = (S_1, \ldots, S_s)$ and $\mathcal{S}^{(2)} = (S_{2s}^{-1}, \ldots, S_{s+1}^{-1})$. Our proof strategy is to first prove NCPA-security of $\overline{\mathsf{SP}}$ in the multi-user setting and lift it to CCA-security by doubling the number of rounds.

Suppose that a distinguisher $\mathcal{D}$ makes p primitive queries to each of the underlying S-boxes and makes q construction queries in the multi-user setting, obtaining an attainable transcript $\tau = (\mathcal{Q}_C, \mathcal{Q}_S)$. We can partition $\mathcal{Q}_C$ and $\mathcal{Q}_S$ as follows.

$$\mathcal{Q}_C = \mathcal{Q}_{C_1} \cup \cdots \cup \mathcal{Q}_{C_\ell},$$
$$\mathcal{Q}_S = \mathcal{Q}_{S_1} \cup \cdots \cup \mathcal{Q}_{S_s} \cup \mathcal{Q}_{S_{s+1}} \cup \cdots \cup \mathcal{Q}_{S_{2s}},$$

where we will write

$$\mathcal{Q}_S^{(1)} = \mathcal{Q}_{S_1} \cup \cdots \cup \mathcal{Q}_{S_s},$$
$$\mathcal{Q}_S^{(2)} = \mathcal{Q}_{S_{s+1}} \cup \cdots \cup \mathcal{Q}_{S_{2s}}.$$

Throughout the proof, we will write $\mathcal{Q}_{C_j} = (t_{j,i}, x_{j,i}, y_{j,i})_{1 \leq i \leq q_j}$ for $j = 1, \ldots, \ell$. So q_j denotes the number of queries made to the j-th construction oracle C_j, and $(t_{j,i}, x_{j,i}, y_{j,i})$ represents the evaluation obtained by the i-th query to C_j. We will also write $\mathbf{t} = (\mathbf{t}_j)_{1 \leq j \leq \ell}$, $\mathbf{x} = (\mathbf{x}_j)_{1 \leq j \leq \ell}$, $\mathbf{y} = (\mathbf{y}_j)_{1 \leq j \leq \ell}$, where

$$\mathbf{t}_j = (t_{j,1}, \ldots, t_{j,q_j}), \qquad \mathbf{x}_j = (x_{j,1}, \ldots, x_{j,q_j}), \qquad \mathbf{y}_j = (y_{j,1}, \ldots, y_{j,q_j}),$$

for $j = 1, \ldots, \ell$. Without loss of generality, we can assume that the indices (j, i) have been grouped by their tweaks $t_{j,i}$; suppose that $\mathbf{t}_j$ consists of d different tweaks, $t_1^*, \ldots, t_d^* \in \mathcal{T}$. Then by dropping j for simplicity (when it will be clear from the context), we can write

$$\mathbf{x}_j = (\mathbf{x}_1^*, \ldots, \mathbf{x}_d^*),$$

so that $\mathbf{x}_i^* = (x_{i,1}^*, \ldots, x_{i,q_i'}^*)$ corresponds to t_i^* for $i = 1, \ldots, d$, where q_i' is the multiplicity of t_i^* in $\mathbf{t}_j$ (satisfying $q_1' + \ldots + q_d' = q_\beta$). Let

$$\Omega_{\mathbf{t}_j} = \left\{ (u_1, \ldots, u_{q_j}) \in (\{0,1\}^n)^{q_j} : \forall i \neq i', (t_{j,i}, u_i) \neq (t_{j,i'}, u_{i'}) \right\},$$
$$\Omega_{\mathbf{t}} = \Omega_{\mathbf{t}_1} \times \ldots \times \Omega_{\mathbf{t}_\ell}.$$

With these notations, we define probability distributions μ_1 and μ_2 on $\Omega_{\mathbf{t}}$; for each $\mathbf{z} = (\mathbf{z}_1, \ldots, \mathbf{z}_\ell) \in \Omega_{\mathbf{t}}$,

$$\mu_1(\mathbf{z}) \stackrel{\text{def}}{=} \Pr\left[\mathbf{k}_1, \ldots, \mathbf{k}_\ell \stackrel{\$}{\leftarrow} \mathcal{K}^s, \mathcal{S} \stackrel{\$}{\leftarrow} \mathsf{Perm}(n)^s : \forall j, \overline{\mathsf{SP}}_{\mathbf{k}_j}^T[\mathcal{S}] \vdash (t_{j,i}, x_{j,i}, z_{j,i})_{1 \leq i \leq q_j} \,\Big|\, \mathcal{S} \vdash \mathcal{Q}_\mathcal{S}^{(1)} \right],$$

$$\mu_2(\mathbf{z}) \stackrel{\text{def}}{=} \Pr\left[\mathbf{k}_1, \ldots, \mathbf{k}_\ell \stackrel{\$}{\leftarrow} \mathcal{K}^s, \mathcal{S} \stackrel{\$}{\leftarrow} \mathsf{Perm}(n)^s : \forall j, \overline{\mathsf{SP}}_{\mathbf{k}_j}^T[\mathcal{S}] \vdash (t_{j,i}, y_{j,i}, z_{j,i})_{1 \leq i \leq q_j} \,\Big|\, \mathcal{S} \vdash \mathcal{Q}_\mathcal{S}^{(2)} \right],$$

where we write $\mathbf{z}_j = (z_{j,i})_{1 \leq i \leq q_j}$ for $j = 1, \ldots, \ell$. Using the coupling technique, we can upper bound the statistical distance between μ_c and the uniform probability distribution for $c = 1, 2$. The proof of the following lemma can be found in [CL18].

Lemma 10. *For $c = 1, 2$, let μ_c be the probability distribution defined as above, and let ν be the uniform probability distribution on $\Omega_{\mathbf{t}}$. Then for $c = 1, 2$, one has $\|\mu_c - \nu\| \leq \varepsilon$, where*

$$\varepsilon = \varepsilon(p, q) \stackrel{\text{def}}{=} q \left(2wp\delta' + 2w^2 q(\delta' + \delta) + w^2 \delta \right)^s.$$

By Lemma 6 and Lemma 10, we have a subset $Z_1 \subset \Omega_{\mathbf{t}}$ such that $|Z_1| \geq (1 - \sqrt{\varepsilon})|\Omega_{\mathbf{t}}|$ and

$$\mu_1(\mathbf{z}) \geq (1 - \sqrt{\varepsilon})\nu(\mathbf{z}) = \frac{1 - \sqrt{\varepsilon}}{|\Omega_{\mathbf{t}}|}$$

for every $\mathbf{z} \in Z_1$. Similarly, we also have a subset $Z_2 \subset \Omega_{\mathbf{t}}$ such that $|Z_2| \geq (1 - \sqrt{\varepsilon})|\Omega_{\mathbf{t}}|$ and

$$\mu_2(\mathbf{z}) \geq (1 - \sqrt{\varepsilon})\nu(\mathbf{z}) = \frac{1 - \sqrt{\varepsilon}}{|\Omega_{\mathbf{t}}|}$$

for every $\mathbf{z} \in Z_2$. For a fixed key $(k_1, \ldots, k_\ell) \in \mathcal{K}^\ell$, let

$$Z_2' = \{ (T_{k_1, \mathbf{t}_1}^{-1}(\mathbf{z}_1), \ldots, T_{k_\ell, \mathbf{t}_\ell}^{-1}(\mathbf{z}_\ell)) : (\mathbf{z}_1, \ldots, \mathbf{z}_\ell) \in Z_2 \},$$

and let $Z = Z_1 \cap Z_2'$. Then it follows that

$$p_2(\mathcal{Q}_C | \mathcal{Q}_S) = \Pr\left[\forall j, \mathsf{SP}_{\mathbf{k}_j}^T[\mathcal{S}] \vdash \mathcal{Q}_{C_j} \,\middle|\, \mathcal{S} \vdash \mathcal{Q}_S\right]$$

$$\geq \frac{1}{|\mathcal{K}|^\ell} \sum_{\substack{k_1,\dots,k_\ell \in \mathcal{K} \\ \mathbf{z}_1,\dots,\mathbf{z}_\ell \in Z}} \Pr\left[\forall j, \overline{\mathsf{SP}}_{\mathbf{k}_j}^T[\mathcal{S}] \vdash (\mathbf{t}_j, \mathbf{x}_j, \mathbf{z}_j) \,\middle|\, \mathcal{S} \vdash \mathcal{Q}_S^{(1)}\right]$$

$$\times \Pr\left[\forall j, \overline{\mathsf{SP}}_{\mathbf{k}_j}^{T-1}[\mathcal{S}] \vdash (\mathbf{t}_j, \mathbf{y}_j, T_{k_j, t_j}(\mathbf{z}_j)) \,\middle|\, \mathcal{S} \vdash \mathcal{Q}_S^{(2)}\right]$$

$$\geq (1 - 2\sqrt{\varepsilon})|\Omega_\mathbf{t}| \cdot \left(\frac{1 - \sqrt{\varepsilon}}{|\Omega_\mathbf{t}|}\right)^2 \geq (1 - 4\sqrt{\varepsilon})p_1(\mathcal{Q}_C | \mathcal{Q}_S)$$

since $|Z| \geq (1 - 2\sqrt{\varepsilon})|\Omega_\mathbf{t}|$. By Lemma 3, we complete the proof of Theorem 4.

Acknowledgments. The work of Aishwarya Thiruvengadam was done while at the University of Maryland. Benoît Cogliati was partially supported by the European Union's H2020 Programme under grant agreement number ICT-644209. The work of Yevgeniy Dodis was done in part while visiting the University of Maryland, and was supported by gifts from VMware Labs and Google, as well as NSF grants 1619158, 1319051, and 1314568. The work of Jonathan Katz and Aishwarya Thiruvengadam was performed under financial assistance award 70NANB15H328 from the U.S. Department of Commerce, National Institute of Standards and Technology. Jooyoung Lee was supported by a National Research Foundation of Korea (NRF) grant funded by the Korean government (Ministry of Science and ICT), No. NRF-2017R1E1A1A03070248.

References

[BBK14] Biryukov, A., Bouillaguet, C., Khovratovich, D.: Cryptographic schemes based on the ASASA structure: black-box, white-box, and public-key (extended abstract). In: Sarkar, P., Iwata, T. (eds.) ASIACRYPT 2014, Part I. LNCS, vol. 8873, pp. 63–84. Springer, Heidelberg (2014). https://doi.org/10.1007/978-3-662-45611-8_4

[BD99] Bleichenbacher, D., Desai, A.: A construction of a super-pseudorandom cipher, February 1999. Unpublished manuscript

[BDPA09] Bertoni, G., Daemen, J., Peeters, M., Van Assche, G.: Keccak sponge function family main document. Submission to NIST (Round 2) (2009). http://keccak.noekeon.org/Keccak-main-2.0.pdf

[BK] Biryukov, A., Khovratovich, D.: Decomposition attack on SASASASAS. http://eprint.iacr.org/2015/646

[BKL+17] Bernstein, D.J., et al.: GIMLI: a cross-platform permutation. In: Fischer, W., Homma, N. (eds.) CHES 2017. LNCS, vol. 10529, pp. 299–320. Springer, Cham (2017). https://doi.org/10.1007/978-3-319-66787-4_15. http://eprint.iacr.org/2017/630

[BS10] Biryukov, A., Shamir, A.: Structural cryptanalysis of SASAS. J. Cryptol. **23**(4), 505–518 (2010)

[CDMS10] Coron, J.-S., Dodis, Y., Mandal, A., Seurin, Y.: A domain extender for the ideal cipher. In: Micciancio, D. (ed.) TCC 2010. LNCS, vol. 5978, pp. 273–289. Springer, Heidelberg (2010). https://doi.org/10.1007/978-3-642-11799-2_17

[CHK+16] Coron, J.-S., Holenstein, T., Künzler, R., Patarin, J., Seurin, Y., Tessaro, S.: How to build an ideal cipher: the indifferentiability of the Feistel construction. J. Cryptol. 29(1), 61–114 (2016)

[CL18] Cogliati, B., Lee, J.: Wide tweakable block ciphers based on substitution-permutation networks: security beyond the birthday bound. IACR Cryptology ePrint Archive, Report 2018/488 (2018). http://eprint.iacr.org/2018/488

[CLL+14] Chen, S., Lampe, R., Lee, J., Seurin, Y., Steinberger, J.: Minimizing the two-round Even-Mansour cipher. In: Garay, J.A., Gennaro, R. (eds.) CRYPTO 2014, Part I. LNCS, vol. 8616, pp. 39–56. Springer, Heidelberg (2014). https://doi.org/10.1007/978-3-662-44371-2_3

[CLS15] Cogliati, B., Lampe, R., Seurin, Y.: Tweaking Even-Mansour ciphers. In: Gennaro, R., Robshaw, M. (eds.) CRYPTO 2015, Part I. LNCS, vol. 9215, pp. 189–208. Springer, Heidelberg (2015). https://doi.org/10.1007/978-3-662-47989-6_9

[CS06] Chakraborty, D., Sarkar, P.: A new mode of encryption providing a tweakable strong pseudo-random permutation. In: Robshaw, M. (ed.) FSE 2006. LNCS, vol. 4047, pp. 293–309. Springer, Heidelberg (2006). https://doi.org/10.1007/11799313_19

[CS14] Chen, S., Steinberger, J.: Tight security bounds for key-alternating ciphers. In: Nguyen, P.Q., Oswald, E. (eds.) EUROCRYPT 2014. LNCS, vol. 8441, pp. 327–350. Springer, Heidelberg (2014). https://doi.org/10.1007/978-3-642-55220-5_19

[Dae95] Daemen, J.: Cipher and hash function design strategies based on linear and differential cryptanalysis. Ph.D. thesis, Katholieke Universiteit Leuven (1995)

[DDKL] Dinur, I., Dunkelman, O., Kranz, T., Leander, G.: Decomposing the ASASA block cipher construction. http://eprint.iacr.org/2015/507

[DKS+17] Dodis, Y., Katz, J., Steinberger, J.P., Thiruvengadam, A., Zhang, Z.: Provable security of substitution-permutation networks. IACR Cryptology ePrint Archive, Report 2017/016 (2017). http://eprint.iacr.org/2017/016

[DSSL16] Dodis, Y., Stam, M., Steinberger, J., Liu, T.: Indifferentiability of confusion-diffusion networks. In: Fischlin, M., Coron, J.-S. (eds.) EUROCRYPT 2016, Part II. LNCS, vol. 9666, pp. 679–704. Springer, Heidelberg (2016). https://doi.org/10.1007/978-3-662-49896-5_24

[EM97] Even, S., Mansour, Y.: A construction of a cipher from a single pseudorandom permutation. J. Cryptol. 10(3), 151–162 (1997)

[Fei73] Feistel, H.: Cryptography and computer privacy. Sci. Am. 228(5), 15–23 (1973)

[GJMN16] Granger, R., Jovanovic, P., Mennink, B., Neves, S.: Improved masking for tweakable blockciphers with applications to authenticated encryption. In: Fischlin, M., Coron, J.-S. (eds.) EUROCRYPT 2016, Part I. LNCS, vol. 9665, pp. 263–293. Springer, Heidelberg (2016). https://doi.org/10.1007/978-3-662-49890-3_11

[Hal07] Halevi, S.: Invertible universal hashing and the TET encryption mode. In: Menezes, A. (ed.) CRYPTO 2007. LNCS, vol. 4622, pp. 412–429. Springer, Heidelberg (2007). https://doi.org/10.1007/978-3-540-74143-5_23

[HKT11] Holenstein, T., Künzler, R., Tessaro, S.: The equivalence of the random oracle model and the ideal cipher model, revisited. In: Fortnow, L., Vadhan, S.P. (eds.) Symposium on Theory of Computing - STOC 2011, pp. 89–98. ACM (2011)

[HR03] Halevi, S., Rogaway, P.: A tweakable enciphering mode. In: Boneh, D. (ed.) CRYPTO 2003. LNCS, vol. 2729, pp. 482–499. Springer, Heidelberg (2003). https://doi.org/10.1007/978-3-540-45146-4_28

[HR04] Halevi, S., Rogaway, P.: A parallelizable enciphering mode. In: Okamoto, T. (ed.) CT-RSA 2004. LNCS, vol. 2964, pp. 292–304. Springer, Heidelberg (2004). https://doi.org/10.1007/978-3-540-24660-2_23

[HR10] Hoang, V.T., Rogaway, P.: On generalized Feistel networks. In: Rabin, T. (ed.) CRYPTO 2010. LNCS, vol. 6223, pp. 613–630. Springer, Heidelberg (2010). https://doi.org/10.1007/978-3-642-14623-7_33

[HT16] Hoang, V.T., Tessaro, S.: Key-alternating ciphers and key-length extension: exact bounds and multi-user security. In: Robshaw, M., Katz, J. (eds.) CRYPTO 2016, Part I. LNCS, vol. 9814, pp. 3–32. Springer, Heidelberg (2016). https://doi.org/10.1007/978-3-662-53018-4_1

[IK00] Iwata, T., Kurosawa, K.: On the pseudorandomness of the AES finalists - RC6 and serpent. In: Goos, G., Hartmanis, J., van Leeuwen, J., Schneier, B. (eds.) FSE 2000. LNCS, vol. 1978, pp. 231–243. Springer, Heidelberg (2001). https://doi.org/10.1007/3-540-44706-7_16

[Jou03] Joux, A.: Cryptanalysis of the EMD mode of operation. In: Biham, E. (ed.) EUROCRYPT 2003. LNCS, vol. 2656, pp. 1–16. Springer, Heidelberg (2003). https://doi.org/10.1007/3-540-39200-9_1

[KL15] Katz, J., Lindell, Y.: Introduction to Modern Cryptography, 2nd edn. Chapman & Hall/CRC Press, London (2015)

[LPS12] Lampe, R., Patarin, J., Seurin, Y.: An asymptotically tight security analysis of the iterated Even-Mansour cipher. In: Wang, X., Sako, K. (eds.) ASIACRYPT 2012. LNCS, vol. 7658, pp. 278–295. Springer, Heidelberg (2012). https://doi.org/10.1007/978-3-642-34961-4_18

[LR88] Luby, M., Rackoff, C.: How to construct pseudorandom permutations from pseudorandom functions. SIAM J. Comput. 17(2), 373–386 (1988)

[LRW11] Liskov, M., Rivest, R.L., Wagner, D.A.: Tweakable block ciphers. J. Cryptol. 24(3), 588–613 (2011)

[Men16] Mennink, B.: XPX: generalized tweakable Even-Mansour with improved security guarantees. In: Robshaw, M., Katz, J. (eds.) CRYPTO 2016, Part I. LNCS, vol. 9814, pp. 64–94. Springer, Heidelberg (2016). https://doi.org/10.1007/978-3-662-53018-4_3

[MF07] McGrew, D.A., Fluhrer, S.R.: The security of the extended codebook (XCB) mode of operation. In: Adams, C., Miri, A., Wiener, M. (eds.) SAC 2007. LNCS, vol. 4876, pp. 311–327. Springer, Heidelberg (2007). https://doi.org/10.1007/978-3-540-77360-3_20

[MRH04] Maurer, U.M., Renner, R., Holenstein, C.: Indifferentiability, impossibility results on reductions, and applications to the random oracle methodology. In: Naor, M. (ed.) TCC 2004. LNCS, vol. 2951, pp. 21–39. Springer, Heidelberg (2004). https://doi.org/10.1007/978-3-540-24638-1_2

[MRS09] Morris, B., Rogaway, P., Stegers, T.: How to encipher messages on a small domain. In: Halevi, S. (ed.) CRYPTO 2009. LNCS, vol. 5677, pp. 286–302. Springer, Heidelberg (2009). https://doi.org/10.1007/978-3-642-03356-8_17

[MV15] Miles, E., Viola, E.: Substitution-permutation networks, pseudorandom functions, and natural proofs. J. ACM **62**(6), 46 (2015)

[NR99] Naor, M., Reingold, O.: On the construction of pseudorandom permutations: Luby-Rackoff revisited. J. Cryptol. **12**(1), 29–66 (1999)

[Pat03] Patarin, J.: Luby-Rackoff: 7 rounds are enough for $2^{n(1-\epsilon)}$ security. In: Boneh, D. (ed.) CRYPTO 2003. LNCS, vol. 2729, pp. 513–529. Springer, Heidelberg (2003). https://doi.org/10.1007/978-3-540-45146-4_30

[Pat04] Patarin, J.: Security of random Feistel schemes with 5 or more rounds. In: Franklin, M. (ed.) CRYPTO 2004. LNCS, vol. 3152, pp. 106–122. Springer, Heidelberg (2004). https://doi.org/10.1007/978-3-540-28628-8_7

[Pat08] Patarin, J.: The "Coefficients H" technique. In: Avanzi, R.M., Keliher, L., Sica, F. (eds.) SAC 2008. LNCS, vol. 5381, pp. 328–345. Springer, Heidelberg (2009). https://doi.org/10.1007/978-3-642-04159-4_21

[Pat10] Patarin, J.: Security of balanced and unbalanced Feistel schemes with linear non equalities. IACR Cryptology ePrint Archive, Report 2010/293 (2010). http://eprint.iacr.org/2010/293

[Sha49] Shannon, C.: Communication theory of secrecy systems. Bell Syst. Tech. J. **28**(4), 656–715 (1949)

[Tes14] Tessaro, S.: Optimally secure block ciphers from ideal primitives. In: Iwata, T., Cheon, J.H. (eds.) ASIACRYPT 2015. LNCS, vol. 9453, pp. 437–462. Springer, Heidelberg (2015). https://doi.org/10.1007/978-3-662-48800-3_18

[Wag99] Wagner, D.: The boomerang attack. In: Knudsen, L. (ed.) FSE 1999. LNCS, vol. 1636, pp. 156–170. Springer, Heidelberg (1999). https://doi.org/10.1007/3-540-48519-8_12

Proofs of Work and Proofs of Stake

Verifiable Delay Functions

Dan Boneh[1], Joseph Bonneau[2], Benedikt Bünz[1], and Ben Fisch[1(✉)]

[1] Stanford University, Stanford, USA
{dabo,bfisch}@cs.stanford.edu
[2] New York University, New York, USA

Abstract. We study the problem of building a *verifiable delay function* (VDF). A VDF requires a specified number of sequential steps to evaluate, yet produces a unique output that can be efficiently and publicly verified. VDFs have many applications in decentralized systems, including public randomness beacons, leader election in consensus protocols, and proofs of replication. We formalize the requirements for VDFs and present new candidate constructions that are the first to achieve an exponential gap between evaluation and verification time.

1 Introduction

Consider the problem of running a verifiable lottery using a *randomness beacon*, a concept first described by Rabin [62] as an ideal service that regularly publishes random values which no party can predict or manipulate. A classic approach is to apply an extractor function to a public entropy source, such as stock prices [24]. Stock prices are believed to be difficult to predict for a passive observer, but an active adversary could manipulate prices to bias the lottery. For example, a high-frequency trader might slightly alter the closing price of a stock by executing (or not executing) a few transactions immediately before the market closes.

Suppose the extractor takes only a single bit per asset (e.g. whether the stock finished up or down for the day) and suppose the adversary is capable of changing this bit for k different assets using last-second trades. The attacker could read the prices of the assets it cannot control, quickly simulate 2^k potential lottery outcomes based on different combinations of the k outcomes it can control, and then manipulate the market to ensure its preferred lottery outcome occurs.

One solution is to add a delay function after extraction, making it slow to compute the beacon outcome from an input of raw stock prices. With a delay function of say, one hour, by the time the adversary simulates the outcome of any potential manipulation strategy, the market will be closed and prices finalized, making it too late to launch an attack. This suggests the key security property for a delay function: it should be infeasible for an adversary to distinguish the function's output from random in less than a specified amount of wall-clock time, even given a potentially large number of parallel processors.

A trivial delay function can be built by iterating a cryptographic hash function. For example, it is reasonable to assume it is infeasible to compute 2^{40}

© International Association for Cryptologic Research 2018
H. Shacham and A. Boldyreva (Eds.): CRYPTO 2018, LNCS 10991, pp. 757–788, 2018.
https://doi.org/10.1007/978-3-319-96884-1_25

iterations of SHA-256 in a matter of seconds, even using specialized hardware. However, a lottery participant wishing to verify the output of this delay function must repeat the computation in its entirety (which might take many hours on a personal computer). Ideally, we would like to design a delay function which any observer can quickly verify was computed correctly.

Defining delay functions. In this paper we formalize the requirements for a *verifiable delay function* (VDF) and provide the first constructions which meet these requirements. A VDF consists of a triple of algorithms: Setup, Eval, and Verify. Setup(λ, t) takes a security parameter λ and delay parameter t and outputs public parameters **pp** (which fix the domain and range of the VDF and may include other information necessary to compute or verify it). Eval(**pp**, x) takes an input x from the domain and outputs a value y in the range and (optionally) a short proof π. Finally, Verify(**pp**, x, y, π) efficiently verifies that y is the correct output on x. Crucially, for every input x there should be a *unique* output y that will verify. Informally, a VDF scheme should satisfy the following properties:

- *sequential:* honest parties can compute $(y, \pi) \leftarrow$ Eval(**pp**, x) in t sequential steps, while no parallel-machine adversary with a polynomial number of processors can distinguish the output y from random in significantly fewer steps.
- *efficiently verifiable:* We prefer Verify to be as fast as possible for honest parties to compute; we require it to take total time $O(\text{polylog}(t))$.

A VDF should remain secure even in the face of an attacker able to perform a polynomially bounded amount of pre-computation.

Some VDFs may also offer additional useful properties:

- *decodable:* An input x can be recovered uniquely from an output y. If the decoding is efficient then no additional proof is required. For example, an invertible function or permutation that is sequentially slow to compute but efficient to invert could be used to instantiate an efficiently decodable VDF.
- *incremental:* a single set of public parameters **pp** supports multiple hardness parameters t. The number of steps used to compute y is specified in the proof, instead of being fixed during Setup.

Classic slow functions. Time-lock puzzles [64] are similar to VDFs in that they involve computing an inherently sequential function. An elegant solution uses repeated squaring in an RSA group as a time-lock puzzle. However, time-lock puzzles are not required to be universally verifiable and in all known constructions the verifier uses its secret state to prepare each puzzle and verify the results. VDFs, by contrast, may require an initial trusted setup but then must be usable on any randomly chosen input.

Another construction for a slow function dating to Dwork and Naor [31] is extracting modular square roots. Given a challenge $x \in \mathbb{Z}_p^*$ (with $p \equiv 3 \pmod 4$), computing $y = \sqrt{x} = x^{\frac{p+1}{4}} \pmod p$ can be efficiently verified by checking that $y^2 = x \pmod p$. There is no known algorithm for computing modular exponentiation which is sublinear in the bit-length of the exponent. However, the difficulty of puzzles is limited to $t = O(\log p)$ as the exponent can be reduced modulo $p - 1$ before computation, requiring the use of a very large prime p to produce a difficult puzzle. While it was not originally proposed for its sequential nature, it has subsequently been considered as such several times [39,46]. In particular, Lenstra and Wesolowski [46] proposed chaining a series of such puzzles together in a construction called Sloth, with lotteries as a specific motivation. Sloth is best characterized as a *time-asymmetric encoding*, offering a trade-off in practice between computation and inversion (verification), and thus can be viewed as a pseudo-VDF. However, it does not meet our asymptotic definition of a VDF because it does not offer asymptotically efficient verification: the t-bit modular exponentiation can be computed in parallel time t, whereas the output (a t-bit number) requires $\Omega(t)$ time simply to read, and therefore verification cannot run in total time polylog(t). We give a more complete overview of related work in Sect. 8.

Our contributions: In addition to providing the first formal definitions of VDFs, we contribute the following candidate constructions and techniques:

1. A theoretical VDF can be constructed using *incrementally verifiable computation* [66] (IVC), in which a proof of correctness for a computation of length t can be computed in parallel to the computation with only polylog(t) processors. We prove security of this theoretical VDF using IVC as a black box. IVC can be constructed from succinct non-interactive arguments of knowledge (SNARKs) under a suitable extractor complexity assumption [14].
2. We propose a construction based on injective polynomials over algebraic sets that cannot be inverted faster than computing polynomial GCDs. Computing polynomial GCD is sequential in the degree d of the polynomials on machines with fewer than $O(d^2)$ processors. We propose a candidate construction of time-asymmetric encodings from a particular family of *permutation polynomials* over finite fields [37]. This construction is asymptotically a strict improvement on Sloth, and to the best of our knowledge is the first encoding offering an exponential time gap between evaluation and inversion. We call this a decodable *weak VDF* because it requires the honest Eval to use greater than *polylog(t)* parallelism to run in parallel time t (the delay parameter).
3. We describe a practical boost to constructing VDFs from IVC using time-asymmetric encodings as the underlying sequential computation, offering up to a 20,000 fold improvement (in the SNARK efficiency) over naive hash chains. In this construction decodability of the VDF is maintained, however a SNARK proof is used to boost the efficiency of verification.
4. We construct a VDF secure against bounded pre-computation attacks following a generalization of time-lock puzzles based on exponentiation in a group of unknown order.

2   Applications

Before giving precise definitions and describing our constructions, we first informally sketch several important applications of VDFs in decentralized systems.

Randomness beacons. VDFs are useful for constructing randomness beacons from sources such as stock prices [24] or proof-of-work blockchains (e.g. Bitcoin, Ethereum) [12,17,60]. Proof-of-work blockchains include randomly sampled solutions to computational puzzles that network participants (called *miners*) continually find and publish for monetary rewards. Underpinning the security of proof-of-work blockchains is the strong belief that these solutions have high computational min-entropy. However, similar to potential manipulation of asset prices by high-frequency traders, powerful miners could potentially manipulate the beacon result by refusing to post blocks which produce an unfavorable beacon output.

Again, this attack is only feasible if the beacon can be computed quickly, as each block is fixed to a specific predecessor and will become "stale" if not published. If a VDF with a suitably long delay is used to compute the beacon, miners will not be able to determine the beacon output from a given block before it becomes stale. More specifically, given the desired delay parameter t, the public parameters $\mathbf{pp} = (\mathsf{ek}, \mathsf{vk}) \xleftarrow{R} \mathsf{Setup}(\lambda, t)$ are posted on the blockchain, then given a block b the beacon value is determined to be r where $(r, \pi) = \mathsf{Eval}(\mathsf{ek}, b)$, and anyone can verify correctness by running $\mathsf{Verify}(\mathsf{vk}, b, r, \pi)$. The security of this construction, and in particular the length of delay parameter which would be sufficient to prevent attacks, remains an informal conjecture due to the lack of a complete game-theoretic model capturing miner incentives in Nakamoto-style consensus protocols. We refer the reader to [12,17,60] for proposed models for blockchain manipulation. Note that most formal models for Nakamoto-style consensus such as that of Garay et al. [34] do not capture miners with external incentives such as profiting from lottery manipulation.

Another approach for constructing beacons derives randomness from a collection of participants, such as all participants in a lottery [36,46]. The simplest paradigm is "commit-and-reveal" paradigm where n parties submit commitments to random values $r_1, ..., r_n$ in an initial phase and subsequently reveal their commitments, at which point the beacon output is computed as $r = \bigoplus_i r_i$. The problem with this approach is that a malicious adversary (possibly controlling a number of parties) might manipulate the outcome by refusing to open its commitment after seeing the other revealed values, forcing a protocol restart. Lenstra and Wesolowski proposed a solution to this problem (called "Unicorn"[46]) using a delay function: instead of using commitments, each participant posts their r_i directly and $seed = H(r_1, ..., r_n)$ is passed through a VDF. The outcome of Eval is then posted and can be efficiently verified. With a sufficiently long delay parameter (longer than the time period during which values may be submitted), even the last party to publish their r_i cannot predict what its impact will be on the final beacon outcome. The beacon is unpredictable even to an adversary who controls $n - 1$ of the participating parties. It has linear communication complexity and uses only two rounds. This stands in contrast to coin-tossing beacons which use verifiable secret sharing and are at best resistant to an adversary who controls a minority of the nodes [1,23,65]. These beacons also use super-linear communication and require multiple rounds of interaction. In the two party setting there are tight bounds that an r-round coin-flipping protocol can be biased with $O(1/r)$ bias [54]. The "Unicorn" construction circumvents these bounds by assuming semi-synchronous communication, i.e. there exists a bound to how long an adversary can delay messages.

Resource-efficient blockchains. Amid growing concerns over the long-term sustainability of proof-of-work blockchains like Bitcoin, which consume a large (and growing) amount of energy, there has been concerted effort to develop resource-efficient blockchains in which miners invest an upfront capital expenditure which can then be re-used for mining. Examples include proof-of-stake [13,28,43,44,52], proof-of-space [58], and proof-of-storage [2,53]. However, resource-efficient mining suffers from *costless simulation* attacks. Intuitively, since mining is not computationally expensive, miners can attempt to produce many separate forks easily.

One method to counter simulation attacks is to use a randomness beacon to select new leaders at regular intervals, with the probability of becoming a leader biased by the quality of proofs (i.e. amount of stake, space, etc.) submitted by miners. A number of existing blockchains already construct beacons from tools such as *verifiable random functions, verifiable secret sharing,* or *deterministic threshold signatures* [4,23,28,43]. However, the security of these beacons requires a non-colluding honest majority; with a VDF-based lottery as described above this can potentially be improved to participation of any honest party.

A second approach, proposed by Cohen [26], is to combine proofs-of-resources with incremental VDFs and use the product of resources proved and delay induced as a measure of blockchain quality. This requires a proof-of-resource which is costly to initialize (such as certain types of proof-of-space). This is important such that the resources are committed to the blockchain and cannot be used for other purposes. A miner controlling N units of total resources can initialize a proof π demonstrating control over these N units. Further assume that the proof is non-malleable and that in each epoch there is a common random challenge c, e.g. a block found in the previous epoch, and let H be a random oracle available to everyone. In each epoch, the miner finds $\tau = \min_{1 \le i \le N}\{H(c, \pi, i)\}$ and computes a VDF on input c with a delay proportional to τ. The first miner to successfully compute the VDF can broadcast their block successfully. Note that this process mimics the random delay to find a Bitcoin block (weighted by the amount of resources controlled by each miner), but without each miner running a large parallel computation.

Proof of data replication. Another promising application of VDFs is proofs of replication, a special type of proof of storage of data which requires dedicating storage even if the data is publicly available. For instance, this could be used to prove that a number of replicas of the same file are being stored. Classic *proofs of retrievability* [41] are typically defined in a private-key client/server setting, where the server proves to the client that it can retrieve the client's (private) data, which the client verifies using a private key.

Instead, the goal of a *proof of replication* [2,3,6] is to verify that a given server is storing a *unique replica* of some data which may be publicly available. Armknecht et al. [6] proposed a protocol in the private verifier model using RSA time-lock puzzles. Given an efficiently decodable VDF, we can adapt their construction to create proofs-of-replication which are more transparent (i.e. do not rely on a designated verifier). Given a unique replicator identifier id and public parameters $\mathbf{pp} \xleftarrow{\text{R}} \mathsf{Setup}(\lambda, t)$, the replicator computes a unique *slow* encoding of the file that take sequential time t. This encoding is computed by breaking the file into b-bit blocks $B_1, \ldots, B_n$ and storing $y_1, \ldots, y_n$ where $(y_i, \perp) = \mathsf{Eval}(\mathbf{pp}, B_i \oplus H(id\|i))$ where H is a collision-resistant hash function $H : \{0,1\}^* \rightarrow \{0,1\}^b$. To verify that the replicator has stored this unique copy, a verifier can query an encoded block y_i (which must be returned in significantly less time than it is feasible to compute Eval). The verifier can quickly decode this response and check it for correctness, proving that the replicator has stored (or can quickly retrieve from somewhere) an encoding of this block which is unique to the identifier id. If the unique block encoding y_i has not being stored, the VDF ensures that it cannot be re-computed quickly enough to fool the verifier, even given access to B_i. The verifier can query for as many blocks as desired; each query has a $1 - \rho$ chance of exposing a cheating prover that is only storing a fraction ρ of the encoded blocks. Note that in this application it is critical that the VDF is decodable. Otherwise, the encoding of the file isn't a useful replica because it cannot be used to recover the data if all other copies are lost.

Computational timestamping. All known proof-of-stake systems are vulnerable to long-range forks due to post-hoc stakeholder misbehavior [13,43,44,52]. In proof-of-stake protocols, at any given time the current stakeholders in the system are given voting power proportionate to their stake in the system. An honest majority (or supermajority) is assumed because the current stakeholders are incentivized to keep the system running correctly. However, after stakeholders have divested they no longer have this incentive. Once the majority (eq. supermajority) of stakeholders from a point in time in the past are divested, they can collude (or sell their key material to an attacker) in order to create a long alternate history of the system up until the present. Current protocols typically assume this is prevented through an external timestamping mechanism which can prove to users that the genuine history of the system is much older.

Incremental VDFs can provide computational evidence that a given version of the state's system is older (and therefore genuine) by proving that a long-running VDF computation has been performed on the genuine history just after the point of divergence with the fraudulent history. This potentially enables detecting long-range forks without relying on external timestamping mechanisms.

We note however that this application of VDFs is fragile as it requires precise bounds on the attacker's computation speed. For other applications (such as randomness beacons) it may be acceptable if the adversary can speed up VDF evaluation by a factor of 10 using faster hardware; a higher t can be chosen until even the adversary cannot manipulate the beacon even with a hardware speedup. For computational timestamping, a 10-fold speedup would be a serious problem: once the fraudulent history is more than one-tenth as old as the genuine history, an attacker can fool participants into believing the fraudulent history is actually *older* than the genuine one.

3   Model and Definitions

We now define VDFs more precisely. In what follows we say that an algorithm runs in *parallel time* t with p processors if it can be implemented on a PRAM machine with p parallel processors running in time t. We say *total time* (eq. sequential time) to refer to the time needed for computation on a single processor.

Definition 1. *A VDF $V = (Setup, Eval, Verify)$ is a triple of algorithms as follows:*

- *$Setup(\lambda, t) \rightarrow \mathbf{pp} = (ek, vk)$ is a randomized algorithm that takes a security parameter λ and a desired puzzle difficulty t and produces public parameters $\mathbf{pp}$ that consists of an evaluation key ek and a verification key vk. We require Setup to be polynomial-time in λ. By convention, the public parameters specify an input space $\mathcal{X}$ and an output space $\mathcal{Y}$. We assume that $\mathcal{X}$ is efficiently sampleable. Setup might need secret randomness, leading to a scheme requiring a trusted setup. For meaningful security, the puzzle difficulty t is restricted to be sub-exponentially sized in λ.*
- *$Eval(ek, x) \rightarrow (y, \pi)$ takes an input $x \in \mathcal{X}$ and produces an output $y \in \mathcal{Y}$ and a (possibly empty) proof π. Eval may use random bits to generate the proof π but not to compute y. For all $\mathbf{pp}$ generated by $Setup(\lambda, t)$ and all $x \in \mathcal{X}$, algorithm $Eval(ek, x)$ must run in parallel time t with $poly(\log(t), \lambda)$ processors.*
- *$Verify(vk, x, y, \pi) \rightarrow \{Yes, No\}$ is a deterministic algorithm takes an input, output and proof and outputs Yes or No. Algorithm Verify must run in total time polynomial in $\log t$ and λ. Notice that Verify is much faster than Eval.*

Additionally V must satisfy Correctness (Definition 2), Soundness (Definition 3), and Sequentiality (Definition 4).

Correctness and Soundness. Every output of Eval must be accepted by Verify. We guarantee that the output y for an input x is unique because Eval evaluates a deterministic function on $\mathcal{X}$. Note that we do not require the proof π to be unique, but we do require that the proof is sound and that a verifier cannot be convinced that some different output is the correct VDF outcome. More formally,

Definition 2 (Correctness). *A VDF V is correct if for all λ, t, parameters $(ek, vk) \xleftarrow{\text{R}} Setup(\lambda, t)$, and all $x \in \mathcal{X}$, if $(y, \pi) \xleftarrow{\text{R}} Eval(ek, x)$ then $Verify(vk, x, y, \pi) = Yes$.*

We also require that for no input x can an adversary get a verifier to accept an incorrect VDF output.

Definition 3 (Soundness). *A VDF is sound if for all algorithms $\mathcal{A}$ that run in time $O\left(poly(t, \lambda)\right)$*

$$P \left[\begin{array}{c} Verify(vk, x, y, \pi) = Yes \\ y \neq Eval(ek, x) \end{array} \middle| \begin{array}{c} \mathbf{pp} = (ek, vk) \xleftarrow{\text{R}} Setup(\lambda, t) \\ (x, y, \pi) \xleftarrow{\text{R}} \mathcal{A}(\lambda, \mathbf{pp}, t) \end{array} \right] = negl(\lambda)$$

Size restriction on t. Asymptotically t must be subexponential in λ. The reason for this is that the adversary needs to be able to run in time at least t (Eval requires this), and if t is exponential in λ then the adversary might be able to break the underlying computational security assumptions that underpin both the soundness as well as the sequentiality of the VDF, which we will formalize next.

Parallelism in Eval. The practical implication of allowing more parallelism in Eval is that "honest" evaluators may be required to have this much parallelism in order to complete the challenge in time t. The sequentiality security argument will compare an adversary's advantage to this optimal implementation of Eval. Constructions of VDFs that do not require any parallelism to evaluate Eval in the optimal number of sequential steps are obviously superior. However, it is unlikely that such constructions exist (without trusted hardware). Even computing an iterated hash function or modular exponentiation (used for time-lock puzzles) could be computed faster by parallelizing the hash function or modular arithmetic. In fact, for an decodable VDF it is necessary that $|\mathcal{Y}| > \mathsf{poly}(t)$, and thus the challenge inputs to Eval have size $\mathsf{poly}\log(t)$. Therefore, in our definition we allow algorithm Eval up to $\mathsf{poly}\log(t)$ parallelism.

3.1 VDF Security

We call the security property needed for a VDF scheme σ-*sequentiality*. Essentially, we require that no adversary is able to compute an output for Eval on a random challenge in parallel time $\sigma(t) < t$, even with up to "many" parallel processors and after a potentially large amount of pre-computation. It is critical to bound the adversary's allowed parallelism, and we incorporate this into the definition. Note that for an efficiently decodable VDF, an adversary with $|\mathcal{Y}|$ processors can always compute outputs in $o(t)$ parallel time by simultaneously trying all possible outputs in $\mathcal{Y}$. This means that for efficiently decodable VDFs it is necessary that $|\mathcal{Y}| > \mathsf{poly}(t)$, and cannot achieve σ-sequentiality against an adversary with greater than $|\mathcal{Y}|$ processors.

We define the following sequentiality game applied to an adversary $\mathcal{A} := (\mathcal{A}_0, \mathcal{A}_1)$:

$$
\begin{array}{lll}
\mathbf{pp} \xleftarrow{\text{R}} \mathsf{Setup}(\lambda, t) & // & \text{choose a random } \mathbf{pp} \\
L \xleftarrow{\text{R}} \mathcal{A}_0(\lambda, \mathbf{pp}, t) & // & \text{adversary preprocesses } \mathbf{pp} \\
x \xleftarrow{\text{R}} \mathcal{X} & // & \text{choose a random input } x \\
y_A \xleftarrow{\text{R}} \mathcal{A}_1(L, \mathbf{pp}, x) & // & \text{adversary computes an output } y_A
\end{array}
$$

We say that $(\mathcal{A}_0, \mathcal{A}_1)$ wins the game if $y_A = y$ where $(y, \pi) := \mathsf{Eval}(\mathbf{pp}, x)$.

Definition 4 (Sequentiality). *For functions $\sigma(t)$ and $p(t)$, the VDF is (p, σ)-sequential if no pair of randomized algorithms $\mathcal{A}_0$, which runs in total time $O(\mathsf{poly}(t, \lambda))$, and $\mathcal{A}_1$, which runs in parallel time $\sigma(t)$ on at most $p(t)$ processors, can win the sequentiality game with probability greater than $\mathsf{negl}(\lambda)$.*

The definition captures the fact that even after $\mathcal{A}_0$ computes on the parameters **pp** for a (polynomially) long time, the adversary $\mathcal{A}_1$ cannot compute an output from the input x in time $\sigma(t)$ on $p(t)$ parallel processors. If a VDF is (p, σ)-sequential for any polynomial p, then we simply say the VDF is σ-sequential. In the sequentiality game we do not require the online attack algorithm $\mathcal{A}_1$ to output a proof π. The reason is that in many of our applications, for example in a lottery, the adversary can profit simply by learning the output early, even without being able to prove correctness to a third party.

Values of $\sigma(t)$. Clearly any candidate construction trivially satisfies $\sigma(t)$-sequentiality for *some* σ (e.g. $\sigma(t) = 0$). Thus, security becomes more meaningful as $\sigma(t) \to t$. No construction can obtain $\sigma(t) = t$ because by design Eval runs in parallel time t. Ideal security is achieved when $\sigma(t) = t - 1$. This ideal security is in general unrealistic unless, for example, time steps are measured in rounds of queries to an ideal oracle (e.g. random oracle). In practice, if the oracle is instantiated with a concrete program (e.g. a hash function), then differences in hardware/implementation would in general yield small differences in the response time for each query. An almost-perfect VDF would achieve $\sigma(t) = t - o(t)$ sequentiality. Even $\sigma(t) = t - \epsilon t$ sequentiality for small ϵ is sufficient for most applications. Security degrades as $\epsilon \to 1$. The naive VDF construction combining a hash chain with succinct verifiable computation (i.e. producing a SNARG proof of correctness following the hash chain computation) cannot beat $\epsilon = 1/2$, unless it uses at least $\omega(t)$ parallelism to generate the proof in sublinear time (exceeding the allowable parallelism for VDFs, though see a relaxation to "weak" VDFs below).

Unpredictability and min-entropy. Definition 4 captures an unpredictability property for the output of the VDF, similar to a one-way function. However, similar to random oracles, the output of the VDF on a given input is never indistinguishable from random. It is possible that no depth $\sigma(t)$ circuit can distinguish the output on a randomly sampled challenge from random, but only if the VDF proof is not given to the distinguisher. Efficiently decodable VDFs cannot achieve this stronger property.

For the application to random beacons (e.g. for lotteries), it is only necessary that on a random challenge the output is unpredictable and also has sufficient min-entropy[1] conditioned on previous outputs for different challenges. In fact, σ-sequentiality already implies that min-entropy is $\Omega(\log \lambda)$. Otherwise some fixed output y occurs with probability $1/\text{poly}(\lambda)$ for randomly sampled input x; the adversary $\mathcal{A}_0$ can computes $O(\text{poly}(\lambda))$ samples of this distribution in the preprocessing to find such a y' with high probability, and then $\mathcal{A}_1$ could output y' as its guess. Moreover, if σ-sequentiality is achieved for t superpolynomial (sub-exponential) in λ, then the preprocessing adversary is allowed $2^{o(\lambda)}$ samples, implying some $o(\lambda)$ min-entropy of the output must be preserved. By itself, σ-sequentiality does not imply $\Omega(\lambda)$ min-entropy. Stronger min-entropy

---

[1] A randomness extractor can then be applied to the output to map it to a uniform distribution.

preservation can be demonstrated in other ways given additional properties of the VDF, e.g. if it is a permutation or collision-resistant. Under suitable complexity theoretic assumptions (namely the existence of subexponential $2^{o(n)}$ circuit lower bounds) a combination of Nisan-Wigderson type PRGs and extractors can also be used to generate $\mathsf{poly}(\lambda)$ pseudorandom bits from a string with min-entropy $\log \lambda$.

Random "Delay" Oracle. In the random oracle model, *any* unpredictable string (regardless of its min-entropy) can be used to extract an unpredictable λ-bit uniform random string. For the beacon application, a random oracle H would simply be applied to the output of the VDF to generate the beacon value. We can even model this construction as an ideal object itself, a *Random Delay Oracle*, which implements a random function H' and on any given input x it waits for $\sigma(t)$ steps before returning the output $H'(x)$. Demonstrating a construction from a σ-sequential VDF and random oracle H that is provably *indifferentiable* [50] from a Random Delay Oracle is an interesting research question.[2]

Remark: Removing any single property makes VDF construction easy. We note the existence of well-known outputs if any property is removed:

- If Verify is not required to be fast, then simply iterating a one-way function t times yields a trivial solution. Verification is done by re-computing the output, or a set of ℓ intermediate points can be supplied as a proof which can be verified in parallel time $\Theta(t/\ell)$ using ℓ processors, with total verification time remaining $\Theta(t)$.
- If we do not require *uniqueness*, then the construction of Mahmoody et al. [49] using hash functions and depth-robust graphs suffices. This construction was later improved by Cohen and Pietrzak [19]. This construction fails to ensure uniqueness because once an output y is computed it can be easily mangled into many other valid outputs $y' \neq y$, as discussed in Sect. 8.1.
- If we do not require σ-*sequentiality*, many solutions are possible, such as finding the discrete log of a challenge group element with respect to a fixed generator. Note that computing an elliptic curve discrete log can be done in parallel time $o(t)$ using a parallel version of the Pollard rho algorithm [67].

Weaker VDFs. For certain applications it is still interesting to consider a VDF that requires even more than $\mathsf{polylog}(t)$ parallelism in Eval to compute the output in parallel time t. For example, in the randomness beacon application only one party is required to compute the VDF and all other parties can simply verify the output. It would not be unreasonable to give this one party a significant amount of parallel computing power and optimized hardware. This would yield

---

[2] The difficulty in proving indifferentiability arises because the distinguisher can query the VDF/RO construction and the RO itself separately, therefore the simulator must be able to simulate queries to the random oracle H given only access to the Random Delay Oracle. Indifferentiability doesn't require the simulator to respond in exactly the same time, but it is still required to be *efficient*. This becomes an issue if the delay t is superpolynomial.

a secure beacon as long as no adversary could compute the outputs of Eval in faster that t steps given even more parallelism than this party. Moreover, for small values of t it may be practical for anyone to use up to $O(t)$ parallelism (or more). With this in mind, we define a weaker variant of a VDF that allows additional parallelism in Eval.

Definition 5. *We call a system $V = $ (Setup, Eval, Verify) a weak-VDF if it satisfies Definition 1 with the exception that Eval is allowed up to $poly(t, \lambda)$ parallelism.*

Note that (p, σ)-sequentiality can only be meaningful for a weak-VDF if Eval is allowed strictly less that $p(t)$ parallelism, otherwise the honest computation of Eval would require more parallelism than even the adversary is allowed.

4   VDFs from Incrementally Verifiable Computation

VDFs are by definition sequential functions. We therefore require the existence of sequential functions in order to construct any VDF. We begin by defining a sequential function.

Definition 6. $((t, \epsilon)$**-Sequential function).** *$f : X \rightarrow Y$ is a (t, ϵ)-sequential function if for $\lambda = O(\log(|X|))$, if the following conditions hold.*

1. *There exists an algorithm that for all $x \in X$ evaluates f in parallel time t using $poly(\log(t), \lambda)$ processors.*
2. *For all $\mathcal{A}$ that run in parallel time strictly less than $(1 - \epsilon) \cdot t$ with $poly(t, \lambda)$ processors:*

$$P\big[y_A = f(x) \mid y_A \xleftarrow{\text{R}} \mathcal{A}(\lambda, x), \ x \xleftarrow{\text{R}} X\big] < negl(\lambda).$$

In addition we consider iterated sequential functions that are an iterative composition of some other function with the same domain and image. The key property of an iterated sequential function cannot be evaluated more efficiently then through iteration of the round function.

Definition 7. (Iterated Sequential Function). *Let $g : X \rightarrow X$ be a function which satisfies (t, ϵ)-sequentiality. A function $f : \mathbb{N} \times X \rightarrow X$ defined as $f(k, x) = g^{(k)}(x) = \underbrace{g \circ g \circ \cdots \circ g}_{k \ times}$ is called an iterated sequential function (with round function g) if for all $k = 2^{o(\lambda)}$ the function $h : X \rightarrow X$ such that $h(x) = f(k, x)$ is $(k \cdot t, \epsilon)$-sequential as per Definition 6.*

It is widely believed that a chain of a secure hash function (like SHA-256) is an iterated sequential function with $t = O(\lambda)$ and ϵ negligible in λ. The sequentiality of hash chains can be proved in the random oracle model [45,49]. We will use the functions g explicitly and require it to have an arithmetic circuit representation. Modeling g as an oracle, therefore, does not suffice for our construction.

Another candidate for an iterated sequential function is exponentiation in a finite group of unknown order, where the round function is squaring in the group. The fastest known way to compute this is by repeated squaring which is an iterative sequential computation.

Based on these candidates, we can make the following assumption about the existence of iterated sequential functions:

Assumption 1. *For all $\lambda \in \mathbb{N}$ there exists an ϵ, t with $t = poly(\lambda)$ and a function $g_\lambda : X \to X$ s.t. $\log |X| = \lambda$ and X can be sampled in time $poly(\lambda)$ and g_λ is a (t, ϵ)-sequential function, and the function $f : \mathbb{N} \times X \to X$ with round function g_λ is an iterated sequential function.*

An iterated sequential function by itself gives us many of the properties needed of a secure VDF construction. It is sequential by definition and the trivial algorithm (iteratively computing g) uses only $poly(\lambda)$ parallelism. Such a function by itself, however, does not suffice to construct a VDF. The fastest generic verification algorithm simply recomputes the function. While this ensures soundness it does not satisfy the efficient verification requirement of a VDF. The verifier of a VDF needs to be exponentially more efficient than the prover.

SNARGs and SNARKs. A natural idea to improve the verification time is to use verifiable computation. In verifiable computation the prover computes a succinct argument (SNARG) that a computation was done correctly. The argument can be efficiently verified using resources that are independent of the size of the computation. A SNARG is a weaker form of a succinct non-interactive argument of knowledge (SNARK) [35] for membership in an NP language $\mathcal{L}$ with relation R (Definition 8). The additional requirement of a SNARK is that for any algorithm that outputs a valid proof of membership of an instance $x \in \mathcal{L}$ there is also an extractor that "watches" the algorithm and outputs a witness w such that $(x, w) \in R$. In the special case of providing a succinct proof that a (polynomial size) computation F was done correctly, i.e. y is the output of F on x, the NP witness is empty and the NP relation simply consists of pairs $((x, y), \perp)$ such that $F(x) = y$.

Definition 8 (Verifiable Computation/SNARK). *Let $\mathcal{L}$ denote an NP language with relation $R_{\mathcal{L}}$, where $x \in \mathcal{L}$ iff $\exists w \; R_{\mathcal{L}}(x, w) = 1$. A SNARK system for $R_{\mathcal{L}}$ is a triple of polynomial time algorithms (SNKGen, SNKProve, SNKVerify) that satisfy the following properties:*

   Completeness:

$$\forall (x, w) \in R_{\mathcal{L}}, \; Pr\left[SNKVerify(\text{vk}, x, \pi) = 0 \; \middle| \; \begin{matrix} (\text{vk}, ek) \leftarrow SNKGen(1^\lambda) \\ \pi \leftarrow SNKProve(ek, x, w) \end{matrix} \right] = 0$$

   Succinctness: The length of a proof and complexity of SNKVerify is bounded by $poly(\lambda, log(|y| + |w|))$.

Knowledge extraction: *[sub-exponential adversary knowledge extractor] For all adversaries $\mathcal{A}$ running in time $2^{o(\lambda)}$ there exists an extractor $\mathcal{E}_{\mathcal{A}}$ running in time $2^{o(\lambda)}$ such that for all $\lambda \in \mathbb{N}$ and all auxiliary inputs z of size $poly(\lambda)$:*

$$Pr\left[\begin{array}{c} SNKVerify(\mathsf{vk}, x, \pi) = 1 \\ R_{\mathcal{L}}(x, w) \neq 1 \end{array} \middle| \begin{array}{l} (\mathsf{vk}, \mathsf{ek}) \leftarrow SNKGen(1^{\lambda}) \\ (x, \pi) \leftarrow \mathcal{A}(z, \mathsf{ek}) \\ w \leftarrow \mathcal{E}_{\mathcal{A}}(z, \mathsf{ek}) \end{array}\right] < negl(\lambda)$$

Impractical VDF from SNARGs. Consider the following construction for a VDF from a t, ϵ-sequential function f. Let $\mathbf{pp} = (\mathsf{ek}, \mathsf{vk}) = SNKGen(\lambda)$ be the public parameter of a SNARG scheme for proving membership in the language of pairs (x, y) such that $f(x) = y$. On input $x \in X$ the Eval computes $y = f(x)$ and a succinct argument $\pi = SNKProve(\mathsf{ek}, (x, y), \perp)$. The prover outputs $((x, y), \pi)$. On input $((x, y), \pi)$ the verifier checks $y = f(x)$ by checking $SNKVerify(\mathsf{vk}, (x, y), \pi) = 1$.

This construction clearly satisfies fast verification. All known SNARK constructions are quasi-linear in the length of the underlying computation f [11]. Assuming the cost for computing a SNARG for a computation of length t is $k \cdot t \log(t)$ then the SNARG VDF construction achieves $\sigma(t) = \frac{(1-\epsilon) \cdot t}{(k+1) \cdot \log(t)}$ sequentiality. This does not even achieve the notion of $(1 - \epsilon')t$ sequentiality for any adversary. This means that the adversary can compute the output of the VDF in a small fraction of the time that it takes the honest prover to convince an honest verifier. If, however, SNKProve is sufficiently parallelizable then it is possible to partially close the gap between the sequentiality of f and the sequentiality of the VDF. The Eval simply executes SNKProve in parallel to reduce the relative total running time compared to the computation of f. SNARK constructions can run in parallel time $polylog(t)$ on $O(t \cdot polylog(t))$ processors. This shows that a VDF can theoretically be built from verifiable computation.

The construction has, however, two significant downsides: In practice computing a SNARG is more than 100,000 times more expensive than evaluating the underlying computation [68]. This means that to achieve meaningful sequentiality the SNARG computation would require massive parallelism using hundreds thousands of cores. The required parallelism additionally depends on the time t. Secondly, the construction does not come asymptotically close to the sequentiality induced by the underlying computation f. We, therefore, now give a VDF construction with required parallelism independent of t and σ-sequentiality asymptotically close to $(1-\epsilon)t$ where ϵ will be defined by the underlying sequential computation.

Incremental Verifiable Computation (IVC). IVC provides a direction for circumventing the problem mentioned above. IVC was first studied by Valiant [66] in the context of computationally sound proofs [51]. Bitansky et al. [14] generalized IVC to distributed computations and to other proof systems such as SNARKs. IVC requires that the underlying computation can be expressed as an iterative sequence of evaluations of the same Turing machine. An iterated sequential function satisfies this requirement.

The basic idea of IVC is that at every incremental step of the computation, a prover can produce a proof that a certain state is indeed the current state of the computation. This proof is updated after every step of the computation to produce a new proof. Importantly, the complexity of each proof in proof size and verification cost is bounded by $\mathsf{poly}(\lambda)$ for any sub-exponential length computation. Additionally the complexity of updating the proof is independent of the total length of the computation.

Towards VDFs from IVC. Consider a VDF construction that runs a sequential computation and after each step uses IVC to update a proof that both this step and the previous proof were correct. Unfortunately, for IVC that requires knowledge extraction we cannot prove soundness of this construction for $t > O(\lambda)$. The problem is that a recursive extraction yields an extractor that is exponential in the recursion depth [14].

The trick around this is to construct a binary tree of proofs of limited depth [14,66]. The leaf proofs verify computation steps whereas the internal node proofs prove that their children are valid proofs. The verifier only needs to check the root proof against the statement that all computation steps and internal proofs are correct.

We focus on the special case that the function f is an iterated sequential function. The regularity of the iterated function ensures that the statement that the verifier checks is succinct. We impose a strict requirement on our IVC scheme to output both the output of f and a final proof with only an additive constant number of additional steps over evaluating f alone.

We define *tight* IVC for an iterated sequential functions, which captures the required primitive needed for our theoretical VDF. We require that incremental proving is almost overhead free in that the prover can output the proof almost immediately after the computation has finished. The definition is a special case of Valiant's definition [66].

Definition 9 (Tight IVC for iterated sequential functions). *Let $f_\lambda : \mathbb{N} \times \mathcal{X} \to \mathcal{X}$ be an iterated sequential function with round function g_λ having (t, ϵ)-sequentiality. An IVC system for f_λ is a triple of polynomial time algorithms (IVCGen, IVCProve, IVCVerify) that satisfy the following properties:*
   Completeness:

$$\forall x \in \mathcal{X}, \; Pr\left[IVCVerify(\mathsf{vk}, x, y, k, \pi) = Yes \;\middle|\; \begin{array}{l} (\mathsf{vk}, \mathsf{ek}) \xleftarrow{\text{R}} IVCGen(\lambda, f) \\ (y, \pi) \xleftarrow{\text{R}} IVCProve(\mathsf{ek}, k, x) \end{array}\right] = 1$$

   Succinctness: The length of a proof and the complexity of SNKVerify is bounded by $\mathsf{poly}(\lambda, \log(k \cdot t))$.
   Soundness: [sub-exponential soundness] For all algorithms $\mathcal{A}$ running in time $2^{o(\lambda)}$:

$$Pr\left[\begin{array}{l} IVCVerify(\mathsf{vk}, x, y, k, \pi) = Yes \\ f(k, x) \neq y \end{array} \;\middle|\; \begin{array}{l} (\mathsf{vk}, \mathsf{ek}) \xleftarrow{\text{R}} IVCGen(\lambda, f) \\ (x, y, k, \pi) \xleftarrow{\text{R}} \mathcal{A}(\lambda, \mathsf{vk}, \mathsf{ek}) \end{array}\right] < \mathsf{negl}(\lambda)$$

Tight Incremental Proving: *There exists a k' such that for all $k \geq k'$ and $k = 2^{o(\lambda)}$, IVCProve(ek, k, x) runs in parallel time $k \cdot t + O(1)$ using poly(λ, t)-processors.*

Existence of tight IVC. Bitansky et al. [14] showed that any SNARK system such as [59] can be used to construct IVC. Under strong knowledge of exponent assumptions there exists an IVC scheme using a SNARK tree of depth less than λ (Theorem 1 of [14]). In every computation step the prover updates the proof by computing λ new SNARKs each of complexity poly(λ), each verifying another SNARK and one of complexity t which verifies one evaluation of g_λ, the round function of f_λ. Ben Sasson et al. [10] discuss the parallel complexity of the Pinocchio SNARK [59] and show that for a circuit of size m there exists a parallel prover using $O(m \cdot \log(m))$ processors that computes a SNARK in time $O(\log(m))$. Therefore, using these SNARKs we can construct an IVC proof system (IVCGen, IVCProve, IVCVerify) where, for sufficiently large t, IVCProve uses $\tilde{O}(\lambda + t)$ parallelism to produce each incremental IVC output in time $\lambda \cdot \log(t + \lambda) \leq t$. If t is not sufficiently large, i.e. $t > \lambda \cdot \log(t + \lambda)$ then we can construct an IVC proof system that creates proofs for k' evaluations of g_λ. The IVC proof system chooses k' such that $t \leq \lambda \cdot \log(k' \cdot t + \lambda)$. Given this the total parallel runtime of IVCProve on k iterations of an (t, ϵ)-sequential function would thus be $k \cdot t + \lambda \cdot \log(k' \cdot t + \lambda) = k \cdot t + O(1)$. This shows that we can construct tight IVC from existing SNARK constructions.

VDF$_{IVC}$ construction. We now construct a VDF from a tight IVC. By Assumption 1 we are given a family $\{f_\lambda\}$, where each $f_\lambda : \mathbb{N} \times X_\lambda \to X_\lambda$ is defined by $f_\lambda(k, x) = g_\lambda^{(k)}(x)$. Here g_λ is a (s, ϵ)-sequential function on an efficiently sampleable domain of size $O(2^\lambda)$.

Given a tight IVC proof system (IVCGen, IVCProve, IVCVerify) for f we can construct a VDF that satisfies $\sigma(t)$-sequentiality for $\sigma(t) = (1 - \epsilon) \cdot t - O(1)$:

- Setup(λ, t) : Let g_λ be a (t, ϵ)-sequential function and f_λ the corresponding iterated sequential function as described in Assumption 1. Run $(ek, vk) \overset{R}{\leftarrow}$ IVCGen(λ, f_λ). Set k to be the largest integer such that IVCProve(ek, k, x) takes time less than t. Output $\mathbf{pp} = ((ek, k), (vk))$.
- Eval$((ek, k), x)$: Run $(y, \pi) \overset{R}{\leftarrow}$ IVCProve(ek, k, x), output (y, π).
- Verify$(vk, x, (y, \pi))$: Run and output IVCVerify(vk, x, y, k, π).

Note that t is fixed in the public parameters. It is, however, also possible to give t directly to Eval. VDF$_{IVC}$ is, therefore, *incremental*.

Lemma 1. *VDF$_{IVC}$ satisfies soundness (Definition 3)*

Proof. Assume that an poly(t, λ) algorithm $\mathcal{A}$ outputs (with non-negligible probability in λ) a tuple (x, y, π) on input λ, t, and $\mathbf{pp} \overset{R}{\leftarrow}$ Setup(λ, t) such that Verify$(\mathbf{pp}, x, y, \pi) = \text{Yes}$ but $f_\lambda(k, x) \neq y$. We can then construct an adversary $\mathcal{A}'$ that violates IVC soundness. Given $(vk, ek) \overset{R}{\leftarrow}$ IVCGen(λ, f_λ) the adversary $\mathcal{A}'$ runs $\mathcal{A}$ on λ, t, and (vk, ek). Since (vk, ek) is sampled from the same distribution as $\mathbf{pp} \overset{R}{\leftarrow}$ Setup(λ, t) it follows that, with non-negligible probability in λ,

$\mathcal{A}'$ outputs (x, y, π) such that $\mathsf{Verify}(\mathbf{pp}, x, y, \pi) = \mathsf{IVCVerify}(\mathsf{vk}, x, y, k, \pi) = \mathsf{Yes}$ and $f_\lambda(k, x) \neq y$, which directly violates the soundness of IVC.

Theorem 1 (VDF$_{\mathbf{IVC}}$). *VDF$_{IVC}$ is a VDF scheme with $\sigma(t) = (1 - \epsilon)t - O(1)$ sequentiality.*

Proof. First note that the VDF$_{\mathrm{IVC}}$ algorithms satisfy the definition of the VDF algorithms. IVCProve runs in time $(\frac{t}{s} - 1) \cdot s + s = t$ using $\mathsf{poly}(\lambda, s) = \mathsf{poly}(\lambda)$ processors. IVCVerify runs in total time $\mathsf{poly}(\lambda, \log(t))$. Correctness follows from the correctness of the IVC scheme. Soundness was proved in Lemma 1. The scheme is $\sigma(t)$-sequential because IVCProve runs in time $k \cdot s + O(1) < t$. If any algorithm that uses $\mathsf{poly}(t, \lambda)$ processors can produce the VDF output in time less than $(1 - \epsilon)t - O(1)$ he can directly break the t, ϵ-sequentiality of f_λ. Since s is independent of t we can conclude that VDF$_{\mathrm{IVC}}$ has $\sigma(t) = (1 - \epsilon)t - O(1)$ sequentiality.

5   A Weak VDF Based on Injective Rational Maps

In this section we explore a framework for constructing a weak VDF satisfying $(t^2, o(t))$-sequentiality based on the existence of degree t injective rational maps that cannot be inverted faster than computing polynomial greatest common denominators (GCDs) of degree t polynomials, which we conjecture cannot be solved in parallel time less than $t - o(t)$ on fewer than t^2 parallel processors. Our candidate map will be a permutation polynomial over a finite field of degree t. The construction built from it is a weak VDF because the Eval will require $O(t)$ parallelism to run in parallel time t.

5.1   Injective Rational Maps

Rational maps on algebraic sets. An *algebraic rational function* on finite vector spaces is a function $F : \mathbb{F}_q^n \to \mathbb{F}_q^m$ such that $F = (f_1, \ldots, f_m)$ where each $f_i : \mathbb{F}_q^n \to \mathbb{F}_q$ is a rational function in $\mathbb{F}_q(X_1, \ldots, X_n)$, for $i = 1, \ldots, m$. An *algebraic set* $\mathcal{Y} \subseteq \mathbb{F}_q^n$ is the complete set of points on which some set S of polynomials simultaneously vanish, i.e. $\mathcal{Y} = \{x \in \mathbb{F}_q^n | f(x) = 0 \text{ for all } f \in S\}$ for some $S \subset \mathbb{F}_q[X_1, \ldots, X_n]$. An *injective rational map* of algebraic sets $\mathcal{Y} \subseteq \mathbb{F}_q^n$ to $\mathcal{X} \subseteq \mathbb{F}_q^m$ is an algebraic rational function F that is injective on $\mathcal{Y}$, i.e. if $\mathcal{X} := F(\mathcal{Y})$, then for every $\bar{x} \in \mathcal{X}$ there exists a unique $\bar{y} \in \mathcal{Y}$ such that $F(\bar{y}) = \bar{x}$.

Inverting rational maps. Consider the problem of inverting an injective rational map $F = (f_1, \ldots, f_m)$ on algebraic sets $\mathcal{Y} \subseteq \mathbb{F}_q^n$ to $\mathcal{X} \subseteq \mathbb{F}_q^m$. Here $\mathcal{Y} \subseteq \mathbb{F}_q^n$ is the set of vanishing points of some set of polynomials S. For $x \in \mathbb{F}_q^m$, a solution to $F(\bar{y}) = \bar{x}$ is a point $\bar{y} \in \mathbb{F}_q^n$ such that all polynomials in S vanish at $\bar{y}$ and $f_i(\bar{y}) = x_i$ for $i = 1, \ldots, m$. Furthermore, each $f_i(\bar{y}) = g(\bar{y})/h(\bar{y}) = x_i$ for some polynomials g, h, and hence yields a polynomial constraint $z_i(\bar{y}) := g(\bar{y}) - x_i h(\bar{y}) = 0$. In total we are looking for solutions to $|S| + m$ polynomial constraints on $\bar{y}$.

We illustrate two special cases of injective rational maps that can be inverted by a univariate polynomial GCD computation. In general, inverting injective rational maps on $\mathbb{F}_q^d$ for constant d can be reduced to a univariate polynomial GCD computation using resultants.

- *Rational functions on finite fields.* Consider any injective rational function $F(X) = g(X)/h(X)$, for univariate polynomials h, g, on a finite field $\mathbb{F}_q$. A finite field is actually a special case of an algebraic set over itself; it is the set of roots of the polynomial $X^q - X$. Inverting F on a point $c \in \mathbb{F}_q$ can be done by calculating $GCD(X^q - X, g(X) - c \cdot h(X))$, which outputs $X - s$ for the unique s such that $F(s) = c$.

- *Rational maps on elliptic curves.* An elliptic curve $E(\mathbb{F}_q)$ over $\mathbb{F}_q$ is a 2-dimensional algebraic set of vanishing points in $\mathbb{F}_q^2$ of a bivariate polynomial $E(y, x) = y^2 - x^3 - ax - b$. Inverting an injective rational function F on a point in the image of $F(E(\mathbb{F}_q))$ involves computing the GCD of three bivariate polynomials: E, z_1, z_2, where z_1 and z_2 come from the two rational function components of F. The resultant $R = Res_y(z_1, z_2)$ is a univariate polynomial in x of degree $deg(z_1) \cdot deg(z_2)$ such that $R(x) = 0$ iff there exists y such that (x, y) is a root of both z_1 and z_2. Finally, taking the resultant again $R' = Res_y(R, E)$ yields a univariate polynomial such that any root x of R' has a corresponding coordinate y such that (x, y) is a point on E and satisfies constraints z_1 and z_2. Solving for the unique root of R' reduces to a Euclidean GCD computation as above. Then given x, there are two possible points $(x, y) \in E$, so we can try them both and output the unique point that satisfies all the constraints.

Euclidean algorithm for univariate polynomial GCD. Univariate polynomials over a finite field form a Euclidean domain, and therefore the GCD of two polynomials can be found using the Euclidean algorithm. For two polynomials f and g such that $deg(f) > deg(g) = d$, one first reduces f mod g and then computes $GCD(f, g) = GCD(f \bmod g, g)$. In the example $f = X^q - X$, the first step of reducing X^q mod g requires $O(\log(q))$ multiplications of degree $O(deg(g))$ polynomials. Starting with X, we run the sequence of repeated squaring operations to get X^q, reducing the intermediate results mod g after each squaring operation. Then running the Euclidean algorithm to find $GCD(f \bmod g, g)$ involves $O(d)$ sequential steps where in each step we subtract two $O(d)$ degree polynomials. On a sequential machine this computation takes $O(d^2)$ time, but on $O(d)$ parallel processors this can be computed in parallel time $O(d)$.

NC algorithm for univariate polynomial GCD. There is an algorithm for computing the GCD of two univariate polynomials of degree d in $O(\log^2(d))$ parallel time, but requires $O(d^{3.8})$ parallel processors. This algorithm runs d parallel determinant calculations on submatrices of the Sylvester matrix associated with the two polynomials, each of size $O(d^2)$. Each determinant can be computed in parallel time $O(\log^2(d))$ on $M(d) \in O(d^{2.85})$ parallel processors [25]. The parallel advantage of this method over the euclidean GCD method kicks in after

$O(d^{2.85})$ processors. For any $c \leq d/\log^2(d)$, it is possible to compute the GCD in $O(d/c)$ steps on $c\log^2(d)M(d)$ processors.

Sequentiality of univariate polynomial GCD. The GCD can be calculated in parallel time d using d parallel processors via the Euclidean algorithm. The NC algorithm only beats this bound on strictly greater than $d^{2.85}$ processors, but a hybrid of the two methods can gain an $o(d)$ speedup on only d^2 processors. Specifically, we can run the Euclidean method for $d - d^{2/3}$ steps until we are left with two polynomials of degree $d^{2/3}$, then we can run the NC algorithm using $\log^3(d)M(d^{2/3}) < (d^{2/3})^3 = d^2$ processors to compute the GCD of these polynomials in $O(d^{2/3}/\log(d))$ steps, for a total of $d - \epsilon d^{2/3}$ steps. This improvement can be tightened further, but generally results in $d - o(d)$ steps as long as $M(d) \in \omega(d^2)$.

We pose the following assumption on the parallel complexity of calculating polynomials GCDs on fewer that $O(d^2)$ processors. This assumption would be broken if there is an NC algorithm for computing the determinant of a $n \times n$ matrix on $o(n^2)$ processors, but this would require a significant advance in mathematics on a problem that has been studied for a long time.

Assumption 2. *There is no general algorithm for computing the GCD of two univariate polynomials of degree d over a finite field $\mathbb{F}_q$ (where $q > d^3$) in less than parallel time $d - o(d)$ on $O(d^2)$ parallel processors.*

On the other hand, evaluating a polynomial of degree d can be logarithmic in its degree, provided the polynomial can be expressed as a small arithmetic circuit, e.g. $(ax + b)^d$ can be computed with $O(\log(d))$ field operations.

Abstract weak VDF from an injective rational map. Let $F : \mathbb{F}_q^n \to \mathbb{F}_q^m$ be a rational function that is an injective map from $\mathcal{Y}$ to $\mathcal{X} := F(\mathcal{Y})$. We further require that $\mathcal{X}$ is efficiently sampleable and that F can be evaluated efficiently for all $\bar{y} \in \mathcal{Y}$. When using F in a VDF we will require that $|\mathcal{X}| > \lambda t^3$ to prevent brute force attacks, where t and λ are given as input to the Setup algorithm.

We will need a family $\mathcal{F} := \{(q, F, \mathcal{X}, \mathcal{Y})\}_{\lambda,t}$ parameterized by λ and t. Given such a family we can construct a weak VDF as follows:

- Setup(λ, t): choose a $(q, F, \mathcal{X}, \mathcal{Y}) \in \mathcal{F}$ specified by λ and t, and output $\mathbf{pp} := ((q, F), (q, F))$.
- Eval$((q, F), \bar{x})$: for an output $\bar{x} \in \mathcal{X} \subseteq \mathbb{F}_q^m$ compute $\bar{y} \in \mathcal{Y}$ such that $F(\bar{y}) = \bar{x}$; The proof π is empty.
- Verify$((q, F), \bar{x}, \bar{y}, \pi)$ outputs Yes if $F(\bar{y}) = \bar{x}$.

The reason we require that F be injective on $\mathcal{Y}$ is so that the solution $\bar{y}$ be unique.

The construction is a weak $(p(t), \sigma(t))$-VDF for $p(t) = t^2$ and $\sigma(t) = t - o(t)$ assuming that there is no algorithm that can invert of $F \in \mathcal{F}$ on a random value in less than parallel time $d - o(d)$ on $O(d^2)$ processors. Note that this is a stronger assumption than 2 as the inversion reduces to a specific GCD computation rather than a general one.

Candidate rational maps. The question, of course, is how to instantiate the function family $\mathcal{F}$ so that the resulting weak VDF system is secure. There are many examples of rational maps on low dimensional algebraic sets among which we can search for candidates. Here we will focus on the special case of efficiently computable permutation polynomials over $\mathbb{F}_q$, and one particular family of permutation polynomials that may be suitable.

5.2 Univariate Permutation Polynomials

The simplest instantiation of the VDF system above is when $n = m = 1$ and $\mathcal{Y} = \mathbb{F}_q$. In this case, the function F is a univariate polynomial $f : \mathbb{F}_q \to \mathbb{F}_q$. If f implements an injective map on $\mathbb{F}_q$, then it must be a permutation of $\mathbb{F}_q$, which brings us to the study of *univariate permutation polynomials* as VDFs.

The simplest permutation polynomials are the monomials x^e for $e \geq 1$, where $gcd(e, q - 1) = 1$. These polynomials however, can be easily inverted and do not give a secure VDF. Dickson polynomials [47] $D_{n,\alpha} \in \mathbb{F}_p[x]$ are another well known family of polynomials over $\mathbb{F}_p$ that permute $\mathbb{F}_p$. Dickson polynomials are defined by a recurrence relation and can be evaluated efficiently. Dickson polynomials satisfy $D_{t,\alpha^n}(D_{n,\alpha}(x)) = x$ for all n, t, α where $n \cdot t = 1 \bmod p - 1$, hence they are easy to invert over $\mathbb{F}_p$ and again do not give a secure VDF.

A number of other classes of permutation polynomials have been discovered over the last several decades [38]. We need a class of permutation polynomials over a suitably large field that have a tunable degree, are fast to evaluate (i.e. have polylog(d) circuit complexity), and cannot be inverted faster than running the parallelized Euclidean algorithm on O(d) processors.

Candidate permutation polynomial. We consider the following polynomial of Guralnick and Muller [37] over $\mathbb{F}_{p^m}$:

$$\frac{(x^s - ax - a) \cdot (x^s - ax + a)^s + ((x^s - ax + a)^2 + 4a^2 x)^{(s+1)/2}}{2x^s} \quad (5.1)$$

where $s = p^r$ for odd prime p and a is not a $(s - 1)st$ power in $\mathbb{F}_{p^m}$. This polynomial is a degree s^3 permutation on the field $\mathbb{F}_{p^m}$ for all s, m chosen independently.

Below we discuss why instantiating a VDF with nearly all other examples of permutation polynomials would not be secure and why attacks on these other polynomials do not work against this candidate.

Attacks on other families of permutation polynomials. We list here several other families of permutation polynomials that can be evaluated in $O(\text{polylog}(d))$ time, yet would not yield a secure VDF. We explain why each of these attacks do not work against the candidate polynomial.

1. *Sparse permutation polynomials.* Sparse polynomials have a constant number of terms and therefore can be evaluated in time $O(\log(d))$. There exist families of non-monomial sparse permutation polynomials, e.g. $X^{2^{t+1}+1} + X^3 + X \in$

$\mathbb{F}_{2^{2t+1}}[X]$ [38, Theorem 4.12]. The problem is that the degree of this polynomial is larger than the square root of the field size, which allows for brute force parallel attacks. Unfortunately, all known sparse permutation polynomials have this problem. In our candidate the field size can be made arbitrarily large relative to the degree of the polynomial.

2. *Linear algebraic attacks.* A classic example of a sparse permutation polynomial of tunable degree over an arbitrarily large field, due to Mathieu [33], is the family $x^{p^i} - ax$ over $\mathbb{F}_{p^m}$ where a is not a $p-1$st power. Unfortunately, this polynomial is easy to invert because $x \mapsto x^{p^i}$ is a linear operator in characteristic p so the polynomial can be written as a linear equation over an m-dimensional vector space. To prevent linear algebraic attacks the degree of at least one non-linear term in the polynomial cannot be divisible by the field characteristic p. In our candidate there are many such non-linear terms, e.g. of degree $s+1$ where $s = p^r$.

3. *Exceptional polynomials co-prime to characteristic.* An *exceptional polynomial* is a polynomial $f \in \mathbb{F}_q[X]$ which is a permutation on $\mathbb{F}_{q^m}$ for infinitely many m, which allows us to choose sufficiently large m to avoid brute force attacks. However, all exceptional polynomials over $\mathbb{F}_q$. of degree co-prime to q can be written as the composition of Dickson polynomials and linear polynomials, which are easy to invert [56]. In our candidate, the degree s^3 of the polynomial and field size are both powers of p, and are therefore not co-prime.

Additional application: a new family of one-way permutations. We note that a sparse permutation polynomial of sufficiently high degree over a sufficiently large finite field may be a good candidate for a one-way permutation. This may give a secure one-way permutation over a domain of smaller size than what is possible by other methods.

5.3   Comparison to Square Roots mod p

A classic approach to designing a sequentially slow verifiable function, dating back to Dwork and Naor [31], is computing modular square roots. Given a challenge $x \in \mathbb{Z}_p^*$, computing $y = x^{\frac{p+1}{4}} \pmod{p}$ can be efficiently verified by checking that $y^2 = x \pmod{p}$ (for $p \equiv 3 \pmod{4}$). There is no known way to compute this exponentiation in faster than $\log(p)$ sequential field multiplications.

This is a special case of inverting a rational function over a finite field, namely the polynomial $f(y) = y^2$, although this function is not injective and therefore cannot be calculated with GCDs. An injective rational function with nearly the same characteristics is the permutation $f(y) = y^3$. Since the inverse of 3 mod $p-1$ will be $O(\log p)$ bits, this requires $O(\log p)$ squaring operations to invert. Viewed another way, this degree 3 polynomial can be inverted on a point c by computing the $\mathrm{GCD}(y^p - y, y^2 - c)$, where the first step requires reducing $y^p - y \bmod y^3 - c$, involving $O(\log p))$ repeated squarings and reductions mod $y^3 - c$.

While this approach appears to offer a delay parameter of $t = \log(p)$, as t grows asymptotically the evaluator can use $O(t)$ parallel processors to gain a factor t parallel speedup in field multiplications, thus completing the challenge in parallel time equivalent to one squaring operation on a sequential machine. Therefore, there is asymptotically no difference in the parallel time complexity of the evaluation and the total time complexity of the verification, which is why this does not even meet our definition of a weak VDF. Our approach of using higher degree injective rational maps gives a strict (asymptotic) improvement on the modular square/cubes approach, and to the best of our knowledge is the first concrete algebraic candidate to achieve an exponential gap between parallel evaluation complexity and total verification complexity.

6 ´ Practical Improvements on VDFs from IVC

In this section we propose a practical boost to constructing VDFs from IVC (Sect. 4). In an IVC construction the prover constructs a SNARK which verifies a SNARK. Ben-Sasson et al. [11] showed an efficient construction for IVC using "cycles of Elliptic curves". This construction builds on the pairing-based SNARK [59]. This SNARK system operates on arithmetic circuits defined over a finite field $\mathbb{F}_p$. The proof output consists of elements of an elliptic curve group $E/\mathbb{F}_q$ of prime order p (defined over a field $\mathbb{F}_q$). The SNARK verification circuit, which computes a pairing, is therefore an arithmetic circuit over $\mathbb{F}_q$. Since $q \neq p$, the prover cannot construct a new SNARK that directly operates on the verification circuit, as the SNARK operates on circuits defined over $\mathbb{F}_p$. Ben-Sasson et al. propose using two SNARK systems where the curve order of one is equal to the base field of the other, and vice versa. This requires finding a pair of pairing-friendly elliptic curves E_1, E_2 (defined over two different base fields $\mathbb{F}_1$ and $\mathbb{F}_2$) with the property that the order of each curve is equal to the size of the base field of the other.

The main practical consideration in VDF$_{\text{IVC}}$ is that the evaluator needs to be able to update the incremental SNARK proofs at the same rate as computing the underlying sequential function, and without requiring a ridiculous amount of parallelism to do so. Our proposed improvements are based on two ideas:

1. In current SNARK/IVC constructions (including [11,59]) the prover complexity is proportional to the multiplicative arithmetic complexity of the underlying statement over the field $\mathbb{F}_p$ used in the SNARK ($p \approx 2^{128}$). Therefore, as an optimization, we can use a "SNARK friendly" hash function (or permutation) as the iterated sequential function such that the verification of each iteration has a lower multiplicative arithmetic complexity over $\mathbb{F}_p$.
2. We can use the Eval of a weak VDF as the iterated sequential function, and compute a SNARK over the Verify circuit applied to each incremental output instead of the Eval circuit. This should increase the number of sequential steps required to evaluate the iterated sequential function relative to the number of multiplication gates over which the SNARK is computed.

An improvement of type (1) alone could be achieved by simply using a cipher or hash function that has better multiplicative complexity over the SNARK field $\mathbb{F}_q$ than AES or SHA256 (e.g., see MiMC [5], which has 1.6% complexity of AES). We will explain how using square roots in $\mathbb{F}_q$ or a suitable permutation polynomial over $\mathbb{F}_q$ (from Sect. 5) as the iterated function achieve improvements of both types (1) and (2).

6.1   Iterated Square Roots in $\mathbb{F}_q$

Sloth. A recent construction called Sloth [46] proposed a secure way to chain a series of square root computations in $\mathbb{Z}_p$ interleaved with a simple permutation[3] such that the chain must be evaluated sequentially, i.e. is an iterated sequential function (Definition 7). More specifically, Sloth defines two permutations on $\mathbb{F}_p$: a permutation ρ such that $\rho(x)^2 = \pm x$, and a permutation σ such that $\sigma(x) = x\pm1$ depending on the parity of x. The parity of x is defined as the integer parity of the unique $\hat{x} \in \{0, ..., p-1\}$ such that $\hat{x} = x \mod p$. Then Sloth iterates the permutation $\tau = \rho \circ \sigma$.

The verification of each step in the chain requires a single multiplication over $\mathbb{Z}_p$ compared to the $O(\log(p))$ multiplications required for evaluation. Increasing the size of p amplifies this gap, however it also introduces an opportunity for parallelizing multiplication in $\mathbb{Z}_p$ for up to $O(\log(p))$ speedup.

Using Sloth inside $\mathrm{VDF}_{\mathrm{IVC}}$ would only achieve a practical benefit if $p = q$ for the SNARK field $\mathbb{F}_q$, as otherwise implementing multiplication in $\mathbb{Z}_p$ in an arithmetic circuit over $\mathbb{F}_q$ would have $O(\log^2(p))$ complexity. On modern architectures, multiplication of integers modulo a 256-bit prime is near optimal on a single core, whereas multi-core parallelized algorithms only offer speed-ups for larger primes [8]. Computing a single modular square root for a 256-bit prime takes approximately 570 cycles on an Intel Core i7 [46], while computing SHA256 for 256-bit outputs takes approximately 864 cycles[4]. Therefore, to achieve the same wall-clock time-delay as an iterated SHA256 chain, only twice as many iterations of modular square roots are needed.

The best known arithmetic circuit implementation of SHA256 has 27,904 multiplication gates [9]. In stark contrast, the arithmetic circuit over $\mathbb{F}_p$ for verifying a modular square root is a single multiplication gate. Verifying the permutation σ is more complex as it requires a parity check, but this requires at most $O(\log(p))$ complexity.

Sloth++ extension. Replacing SHA256 with Sloth as the iterated function in $\mathrm{VDF}_{\mathrm{IVC}}$ already gives a significant improvement, as detailed above. Here we suggest yet a further optimization, which we call Sloth++. The main arithmetic complexity of verifying a step of Sloth comes from the fact that the permutation σ is not naturally arithmetic over $\mathbb{F}_p$, which was important for preventing attacks

---

[3] If square roots are iterated on a value x without an interleaved permutation then there is a shortcut to the iterated computation that first computes $v = (\frac{p+1}{4})^{\ell}$ mod p and then the single exponentiation x^v.

[4] http://www.ouah.org/ogay/sha2/.

that factor $\tau^\ell(x)$ as a polynomial over $\mathbb{F}_p$. Our idea here is to compute square roots over a degree 2 extension field $\mathbb{F}_{p^2}$ interleaved with a permutation that is arithmetic over $\mathbb{F}_p$ but not over $\mathbb{F}_{p^2}$.

In any degree r extension field $\mathbb{F}_{p^r}$ of $\mathbb{F}_p$ for a prime $p = 3 \mod 4$ a square root of an element $x \in \mathbb{F}_{p^r}$ can be found by computing $x^{(p^r+1)/4}$. This is computed in $O(r\log(p))$ repeated squaring operations in $\mathbb{F}_p^r$. Verifying a square root requires a single multiplication over $\mathbb{F}_{p^r}$. Elements of $\mathbb{F}_{p^r}$ can be represented as length r vectors over $\mathbb{F}_p$, and each multiplication reduces to $O(r^2)$ arithmetic operations over $\mathbb{F}_p$. For $r = 2$ the verification *multiplicative* complexity over $\mathbb{F}_p$ is exactly 4 gates.

In Sloth++ we define the permutation ρ exactly as in Sloth, yet over $\mathbb{F}_{p^2}$. Then we define a simple non-arithmetic permutation σ on $\mathbb{F}_{p^2}$ that swaps the coordinates of elements in their vector representation over $\mathbb{F}_p$ and adds a constant, i.e. maps the element (x, y) to $(y + c_1, x + c_2)$. The arithmetic circuit over $\mathbb{F}_p$ representing the swap is trivial: it simply swaps the values on the input wires. The overall multiplicative complexity of verifying an iteration of Sloth++ is only 4 gates over $\mathbb{F}_p$. Multiplication can be parallelized for a factor 2 speedup, so 4 gates must be verified roughly every 1700 parallel-time evaluation cycles. Thus, for parameters that achieve the same wall-clock delay, the SNARK verification complexity of Sloth++ is a 14,000 fold improvement over that of a SHA256 chain.

Cube roots. The underlying permutation in both Sloth and Sloth++ can be replaced by cube roots over $\mathbb{F}_q$ when $gcd(3, q - 1) = 1$. In this case the slow function is computing $\rho(x) = x^v$ where $3v = 1 \mod q - 1$. The output can be verified as $\rho(x)^3 = x$.

6.2 Iterated Permutation Polynomials

Similar to Sloth+, we can use our candidate permutation polynomial (Eq. 5.1) over $\mathbb{F}_q$ as the iterated function in VDF$_{IVC}$. Recall that $\mathbb{F}_q$ is an extension field chosen independently from the degree of the polynomial. We would choose $q \approx 2^{256}$ and use the same $\mathbb{F}_q$ as the field used for the SNARK system. For each $O(d)$ sequential provers steps required to invert the polynomial on a point, the SNARK only needs to verify the evaluation of the polynomial on the inverse, which has multiplicative complexity $O(\log(d))$ over $\mathbb{F}_q$. Concretely, for each 10^5 parallel-time evaluation cycles a SNARK needs to verify approximately 16 gates. This is yet another factor 15 improvement over Sloth+. The catch is that the evaluator must use 10^5 parallelism[5] to optimize the polynomial GCD computation. We must also assume that an adversary cannot feasibly amass more than 10^{14} parallel processors to implement the NC parallelized algorithm for polynomial GCD.

---

[5] This is reasonable if the evaluator has an NVIDIA Titan V GPU, which can compute up to 10^{14} pipelined arithmetic operations per second (https://www.nvidia.com/en-us/titan/titan-v/).

From a theory standpoint, using permutation polynomials inside VDF_{IVC} reduces it to a weak VDF because the degree of the polynomial must be super-polynomial in λ to prevent an adversary from implementing the NC algorithm on $poly(\lambda)$ processors, and therefore the honest evaluator is also required to use super-polynomial parallelism. However, the combination does yield a better weak VDF, and from a practical standpoint appears quite promising for many applications.

7   Towards VDFs from Exponentiation in a Finite Group

The sequential nature of large exponentiation in a finite group may appear to be a good source for secure VDF systems. This problem has been used extensively in the past for time-based problems such as time-lock puzzles [64], benchmarking [21], timed commitments [16], and client puzzles [31,46]. Very recently, Pietrzak [61] showed how to use this problem to construct a VDF that requires a trusted setup. The trusted setup can be eliminated by instead choosing a sufficiently large random number N so that N has two large prime factors with high probability. However, the large size of N provides the adversary with more opportunity for parallelizing the arithmetic. It also increases the verifier's running time. Alternatively, one can use the class group of an imaginary quadratic order [20], which is an efficient group of unknown order with a public setup [48].

7.1   Exponentiation-Based VDFs with Bounded Pre-computation

Here we suggest a simple exponentiation-based approach to constructing VDFs whose security would rely on the assumption that the adversary cannot run a long pre-computation between the time that the public parameters **pp** are made public and the time when the VDF needs to be evaluated. Therefore, in terms of security this construction is subsumed by the more recent solution of Pietrzak [61], however it yields much shorter proofs. We use the following notation to describe the VDF:

- let $L = \{\ell_1, \ell_2, \ldots, \ell_t\}$ be the first t odd primes, namely $\ell_1 = 3$, $\ell_2 = 5$, etc. Here t is the provided delay parameter.
- let P be the product of the primes in L, namely $P := \ell_1 \cdot \ell_2 \cdots \ell_t$. This P is a large integer with about $t \log t$ bits.

With this notation, the trusted setup procedure works as follows: construct an RSA modulus N, say 4096 bits long, where the prime factors are strong primes. The trusted setup algorithm knows the factorization of N, but no one else will. Let $\mathbb{G} := (\mathbb{Z}/N\mathbb{Z})^*$. We will also need a random hash function $H : \mathbb{Z} \to \mathbb{G}$. Next, for a given preprocessing security parameter B, say $B = 2^{30}$, do:

- for $i = 1, \ldots, B$: compute $h_i \leftarrow H(i) \in \mathbb{G}$ and then compute $g_i := h_i^{1/P} \in \mathbb{G}$.

– output
$$\mathsf{ek} := (\mathbb{G},\ H,\ g_1,\ldots,g_B) \qquad \text{and} \qquad \mathsf{vk} := (\mathbb{G},\ H).$$

Note that the verifier's public parameters are short, but the evaluators parameters are not.

Solving a challenge x: Algorithm $\mathsf{Eval}(\mathbf{pp}_{\mathrm{eval}}, x)$ takes as input the public parameters $\mathbf{pp}_{\mathrm{eval}}$ and a challenge $x \in \mathcal{X}$.

– using a random hash function, map the challenge x to a random subset $L_x \subseteq L$ of size λ, and a random subset S_x of λ values in $\{1,\ldots,B\}$.
– Let P_x be the product of all the primes in L_x, and let g be $g := \prod_{i \in S_x} g_i \in \mathbb{G}$.
– the challenge solution y is simply $y \leftarrow g^{P/P_x} \in \mathbb{G}$, which takes $O(t \log t)$ multiplications in $\mathbb{G}$.

Verifying a solution y: Algorithm $\mathsf{Verify}(\mathbf{pp}_{\mathrm{verify}}, x, y)$ works as follows:

– Compute P_x and S_x as in algorithm $\mathsf{Eval}(\mathbf{pp}_{\mathrm{eval}}, x)$.
– let h be $h := \prod_{i \in S_x} H(i) \in \mathbb{G}$.
– output yes if and only if $y^{P_x} = h$ in $\mathbb{G}$.

Note that exactly one $y \in \mathbb{G}$ will be accepted as a solution for a challenge x. Verification takes only $\tilde{O}(\lambda)$ group operations.

Security. The scheme does not satisfy the definition of a secure VDF, but may still be useful for some of the applications described in Sect. 2. In particular, the system is not secure against an adversary who can run a large pre-computation once the parameters $\mathbf{pp}$ are known. There are several pre-computation attacks possible that require tB group operations in $\mathbb{G}$. Here we describe one such instructive attack. It uses space $O(sB)$, for some $s > 0$, and gives a factor of s speed up for evaluating the VDF.

Consider the following pre-computation, for a given parameter s, say $s = 100$. Let $b = \lfloor P^{1/s} \rfloor$, then the adversary computes and stores a table of size sB:

$$\text{for all } i = 1,\ldots,B: \qquad g_i^b,\ g_i^{(b^2)},\ \ldots,\ g_i^{(b^s)} \quad \in \mathbb{G}. \tag{7.1}$$

Computing these values is comparable to solving B challenges. Once computed, to evaluate the VDF at input x, the adversary uses the precomputed table to quickly compute
$$g^b,\ g^{(b^2)},\ \ldots,\ g^{(b^s)} \in \mathbb{G}.$$

Now, to compute g^{P/P_x}, it can write P/P_x in base b as:
$P/P_x = \alpha_0 + \alpha_1 b + \alpha_2 b^2 + \ldots + \alpha_s b^s$ so that

$$g^{P/P_x} = g^{\alpha_0} \cdot (g^b)^{\alpha_1} \cdot (g^{(b^2)})^{\alpha_2} \cdots (g^{(b^s)})^{\alpha_s}.$$

This expression can be evaluated in parallel and gives a parallel adversary a factor of s speed-up over a sequential solver, which violates the sequentiality property of the VDF.

To mount this attack, the adversary must compute the entire table (7.1) for all $g_1, \ldots, g_B$, otherwise it can only gain a factor of two speed-up with negligible probability in λ. Hence, the scheme is secure for only B challenges, after which new public parameters need to be generated. This may be sufficient for some applications of a VDF.

8   Related Work

Taking a broad perspective, VDFs can be viewed as an example of *moderately hard* cryptographic functions. Moderately hard functions are those whose difficulty to compute is somewhere in between 'easy' (designed to be as efficient as possible) and 'hard' (designed to be so difficult as to be intractable). The use of moderately hard cryptographic functions dates back at least to the use of a deliberately slow DES variant for password hashing in early UNIX systems [55]. Dwork and Naor [31] coined the term *moderately hard* in a classic paper proposing client puzzles or "pricing functions" for the purpose of preventing spam. Juels and Brainard proposed the related notion of a *client puzzle*, in which a TCP server creates a puzzle which must be solved before a client can open a connection [42]. Both concepts have been studied for a variety of applications, including TLS handshake requests [7,29], node creation in peer-to-peer networks [30], creation of digital currency [27,57,63] or censorship resistance [18]. For interactive client puzzles, the most common construction is as follows: the server chooses a random ℓ-bit value x and sends to the client $H(x)$ and $x[\ell - \log_2 t - 1]$. The client must send back the complete value of x. That is, the server sends the client $H(x)$ plus all of the bits of x except the final $\log_2 t + 1$ bits, which the client must recover via brute force.

8.1   Inherently Sequential Puzzles

The simple interactive client puzzle described above is embarrassingly parallel and can be solved in constant time given t processors. In contrast, the very first construction of a client puzzle proposed by Dwork and Naor involved computing modular square roots and is believed to be inherently sequential (although they did not discuss this as a potential advantage).

The first interest in designing puzzles that require an inherently sequential solving algorithm appears to come for the application of hardware benchmarking. Cai et al. [21,22] proposed the use of inherently sequential puzzles to verify claimed hardware performance as follows: a customer creates an inherently-sequential puzzle and sends it to a hardware vendor, who then solves it and returns the solution (which the customer can easily verify) as quickly as possible. Note that this work predated the definition of client puzzles. Their original construction was based on exponentiation modulo an RSA number N, for which the customer has created N and therefore knows $\varphi(N)$. They later proposed solutions based on a number of other computational problems not typically used in cryptography, including Gaussian elimination, fast Fourier transforms, and matrix multiplication.

Time-lock puzzles. Rivest, Shamir, and Wagner [64] constructed a time-lock encryption scheme, also based on the hardness of RSA factoring and the conjectured sequentiality of repeated exponentiation in a group of unknown order. The encryption key K is derived as $K = x^{2^t} \in \mathbb{Z}_N$ for an RSA modulus N and a published starting value x. The encrypting party, knowing $\varphi(N)$, can reduce the exponent $e = 2^t \bmod \varphi(N)$ to quickly derive $K = x^e \bmod N$. The key K can be publicly recovered slowly by 2^t iterated squarings. Boneh and Naor [16] showed that the puzzle creator can publish additional information enabling an efficient and sound proof that K is correct. In the only alternate construction we are aware of, Bitansky et al. [15] show how to construct time-lock puzzles from randomized encodings assuming any inherently-sequential functions exist.

Time-lock puzzles are similar to VDFs in that they involve computing an inherently sequential function. However, time-lock puzzles are defined in a private-key setting where the verifier uses its private key to prepare each puzzle (and possibly a verification proof for the eventual answer). In contrast to VDFs, this trusted setup must be performed per-puzzle and each puzzle takes no unpredictable input.

Proofs of sequential work. Mahmoody et al. [49] proposed publicly verifiable proofs of sequential work (PoSW) which enable proving to any challenger that a given amount of sequential work was performed on a specific challenge. As noted, time-lock puzzles are a type of PoSW, but they are not publicly verifiable. VDFs can be seen as a special case of publicly verifiable proofs of sequential work with the additional guarantee of a unique output (hence the use of the term "function" versus "proof").

Mahmoody et al.'s construction uses a sequential hash function H (modeled as a random oracle) and depth robust directed-acyclic graph G. Their puzzle involves computing a *labeling* of G using H salted by the challenge c. The label on each node is derived as a hash of all the labels on its parent nodes. The labels are committed to in a Merkle tree and the proof involves opening a randomly sampled fraction. Very briefly, the security of this construction is related to graph pebbling games (where a pebble can be placed on a node only if all its parents already have pebbles) and the fact that depth robust graphs remain sequentially hard to pebble even if a constant fraction of the nodes are removed (in this case corresponding to places where the adversary cheats). Mahmoody et al. proved security unconditionally in the random oracle model. Depth robust graphs and parallel pebbling hardness are use similarly to construct memory hard functions [40] and proofs of space [32]. Cohen and Pietrzak [19] constructed a similar PoSW using a simpler non-depth-robust graph based on a Merkle tree.

PoSWs based on graph labeling don't naturally provide a VDF because removing any single edge in the graph will change the output of the proof, yet is unlikely to be detected by random challenges.

Sequentially hard functions. The most popular solution for a slow function which can be viewed as a proto-VDF, dating to Dwork and Naor [31], is computing modular square roots. Given a challenge $x \in \mathbb{Z}_p^*$, computing $y = x^{\frac{p+1}{4}}$ (mod p) can be efficiently verified by checking that $y^2 = x$ (mod p) (for $p \equiv 3$ (mod 4)). There is no known algorithm for computing modular exponentiation which is sublinear in the exponent. However, the difficulty of puzzles is fixed to $t = \log p$ as the exponent can be reduced modulo $p - 1$ before computation, requiring the use of a very large prime p to produce a difficult puzzle.

This puzzle has been considered before for similar applications as our VDFs, in particular randomness beacons [39,46]. Lenstra and Wesolowski [46] proposed creating a more difficult puzzle for a small p by chaining a series of such puzzles together (interleaved with a simple permutation) in a construction called Sloth. We proposed a simple improvement of this puzzle in Sect. 6. Recall that this does not meet our asymptotic definition of a VDF because it does not offer (asymptotically) efficient verification, however we used it as an important building block to construct a more practical VDF based on IVC. Asymptotically, Sloth is comparable to a hash chain of length t with t checkpoints provided as a proof, which also provides $O(\text{polylog}(t))$-time verification (with t processors) and a solution of size $\Theta(t \cdot \lambda)$.

9   Conclusions

Given their large number of interesting applications, we hope this work stimulates new practical uses for VDFs and continued study of theoretical constructions. We still lack a theoretically optimal VDF, consisting of a simple inherently sequential function requiring low parallelism to compute but yet being very fast (e.g. logarithmic) to invert. These requirements motivate the search for new problems which have not traditionally been used in cryptography. Ideally, we want a VDF that is also post-quantum secure.

Acknowledgments. We thank Micheal Zieve for his help with permutation polynomials. We thank the CRYPTO reviewers for their helpful comments. This work was supported by NSF, a grant from ONR, the Simons Foundation, and a Google faculty fellowship.

References

1. RANDAO: A DAO working as RNG of Ethereum. Technical report (2016)
2. Filecoin: A decentralized storage network. Protocol Labs (2017). https://filecoin. io/filecoin.pdf
3. Proof of replication. Protocol Labs (2017). https://filecoin.io/proof-of-replication. pdf
4. Threshold relay. Dfinity (2017). https://dfinity.org/pdfs/viewer.html?file=../ library/threshold-relay-blockchain-stanford.pdf

5. Albrecht, M., Grassi, L., Rechberger, C., Roy, A., Tiessen, T.: MiMC: efficient encryption and cryptographic hashing with minimal multiplicative complexity. In: Cheon, J.H., Takagi, T. (eds.) ASIACRYPT 2016. LNCS, vol. 10031, pp. 191–219. Springer, Heidelberg (2016). https://doi.org/10.1007/978-3-662-53887-6_7

6. Armknecht, F., Barman, L., Bohli, J.-M., Karame, G.O.: Mirror: enabling proofs of data replication and retrievability in the cloud. In: USENIX Security Symposium, pp. 1051–1068 (2016)

7. Aura, T., Nikander, P., Leiwo, J.: DOS-resistant authentication with client puzzles. In: Christianson, B., Malcolm, J.A., Crispo, B., Roe, M. (eds.) Security Protocols 2000. LNCS, vol. 2133, pp. 170–177. Springer, Heidelberg (2001). https://doi.org/10.1007/3-540-44810-1_22

8. Baktir, S., Savas, E.: Highly-parallel montgomery multiplication for multi-core general-purpose microprocessors. In: Gelenbe, E., Lent, R. (eds.) Computer and Information Sciences III, pp. 467–476. Springer, London (2013). https://doi.org/10.1007/978-1-4471-4594-3_48

9. Ben-Sasson, E., et al. Zerocash: decentralized anonymous payments from Bitcoin. In: IEEE Symposium on Security and Privacy (2014)

10. Ben-Sasson, E., Chiesa, A., Genkin, D., Tromer, E., Virza, M.: SNARKs for C: verifying program executions succinctly and in zero knowledge. In: Canetti, R., Garay, J.A. (eds.) CRYPTO 2013. LNCS, vol. 8043, pp. 90–108. Springer, Heidelberg (2013). https://doi.org/10.1007/978-3-642-40084-1_6

11. Ben-Sasson, E., Chiesa, A., Tromer, E., Virza, M.: Scalable zero knowledge via cycles of elliptic curves. Algorithmica **79**, 1102–1160 (2014)

12. Bentov, I., Gabizon, A., Zuckerman, D.: Bitcoin beacon. arXiv preprint arXiv:1605.04559 (2016)

13. Bentov, I., Pass, R., Shi, E.: Snow white: provably secure proofs of stake. IACR Cryptology ePrint Archive, 2016 (2016)

14. Bitansky, N., Canetti, R., Chiesa, A., Tromer, E.: Recursive composition and bootstrapping for SNARKs and proof-carrying data. In: Proceedings of the Forty-Fifth Annual ACM Symposium on Theory of Computing, pp. 111–120. ACM (2013)

15. Bitansky, N., Goldwasser, S., Jain, A., Paneth, O., Vaikuntanathan, V., Waters, B.: Time-lock puzzles from randomized encodings. In: ACM Conference on Innovations in Theoretical Computer Science (2016)

16. Boneh, D., Naor, M.: Timed commitments. In: Bellare, M. (ed.) CRYPTO 2000. LNCS, vol. 1880, pp. 236–254. Springer, Heidelberg (2000). https://doi.org/10.1007/3-540-44598-6_15

17. Bonneau, J., Clark, J., Goldfeder, S.: On bitcoin as a public randomness source (2015). https://eprint.iacr.org/2015/1015.pdf

18. Bonneau, J., Xu, R.: Scrambling for lightweight censorship resistance. In: Christianson, B., Crispo, B., Malcolm, J., Stajano, F. (eds.) Security Protocols 2011. LNCS, vol. 7114, pp. 296–302. Springer, Heidelberg (2011). https://doi.org/10.1007/978-3-642-25867-1_28

19. Cohen, B., Pietrzak, K.: Simple proofs of sequential work. In: Nielsen, J.B., Rijmen, V. (eds.) EUROCRYPT 2018. LNCS, vol. 10821, pp. 451–467. Springer, Cham (2018). https://doi.org/10.1007/978-3-319-78375-8_15

20. Buchmann, J., Williams, H.C.: A key-exchange system based on imaginary quadratic fields. J. Cryptol. **1**(2), 107–118 (1988)

21. Cai, J., Lipton, R.J., Sedgewick, R., Yao, A.C.: Towards uncheatable benchmarks. In: Structure in Complexity Theory (1993)

22. Cai, J.-Y., Nerurkar, A., Wu, M.-Y.: The design of uncheatable benchmarks using complexity theory (1997)

23. Cascudo, I., David, B.: Scrape: scalable randomness attested by public entities. Cryptology ePrint Archive, Report 2017/216 (2017). http://eprint.iacr.org/2017/216

24. Clark, J., Hengartner, U.: On the use of financial data as a random beacon. In: Usenix EVT/WOTE (2010)

25. Codenottia, B., Datta, B.N., Datta, K., Leoncini, M.: Parallel algorithms for certain matrix computations. Theor. Comput. Sci. **180**, 287–308 (1997)

26. Cohen, B.: Proofs of space and time. In: Blockchain Protocol Analysis and Security Engineering (2017). https://cyber.stanford.edu/sites/default/files/bramcohen.pdf

27. Dai, W.: B-money. Consulted **1**, 2012 (1998)

28. David, B., Gaži, P., Kiayias, A., Russell, A.: Ouroboros Praos: an adaptively-secure, semi-synchronous proof-of-stake blockchain. In: Nielsen, J.B., Rijmen, V. (eds.) EUROCRYPT 2018. LNCS, vol. 10821, pp. 66–98. Springer, Cham (2018). https://doi.org/10.1007/978-3-319-78375-8_3

29. Dean, D., Stubblefield, A.: Using client puzzles to protect TLS. In: USENIX Security Symposium, vol. 42 (2001)

30. Douceur, J.R.: The Sybil attack. In: Druschel, P., Kaashoek, F., Rowstron, A. (eds.) IPTPS 2002. LNCS, vol. 2429, pp. 251–260. Springer, Heidelberg (2002). https://doi.org/10.1007/3-540-45748-8_24

31. Dwork, C., Naor, M.: Pricing via processing or combatting junk mail. In: Brickell, E.F. (ed.) CRYPTO 1992. LNCS, vol. 740, pp. 139–147. Springer, Heidelberg (1993). https://doi.org/10.1007/3-540-48071-4_10

32. Dziembowski, S., Faust, S., Kolmogorov, V., Pietrzak, K.: Proofs of space. In: Gennaro, R., Robshaw, M. (eds.) CRYPTO 2015. LNCS, vol. 9216, pp. 585–605. Springer, Heidelberg (2015). https://doi.org/10.1007/978-3-662-48000-7_29

33. Mathieu, É.: Mémoire sur l'étude des fonctions de plusieurs quantités sur la manière de les former et sur les substitutions qui les laissent invariables. J. Math. Pures Appl. **6**(2), 241–323 (1861)

34. Garay, J., Kiayias, A., Leonardos, N.: The Bitcoin backbone protocol: analysis and applications. Cryptology ePrint Archive # 2014/765 (2014)

35. Gennaro, R., Gentry, C., Parno, B., Raykova, M.: Quadratic span programs and succinct NIZKs without PCPs. In: Johansson, T., Nguyen, P.Q. (eds.) EUROCRYPT 2013. LNCS, vol. 7881, pp. 626–645. Springer, Heidelberg (2013). https://doi.org/10.1007/978-3-642-38348-9_37

36. Goldschlag, D.M., Stubblebine, S.G.: Publicly verifiable lotteries: applications of delaying functions. In: Hirchfeld, R. (ed.) FC 1998. LNCS, vol. 1465, pp. 214–226. Springer, Heidelberg (1998). https://doi.org/10.1007/BFb0055485

37. Guralnick, R.M., Müller, P.: Exceptional polynomials of affine type. J. Algebra **194**(2), 429–454 (1997)

38. Hou, X.-d.: Permutation polynomials over finite fieldsa survey of recent advances. Finite Fields Appl. **32**, 82–119 (2015)

39. Jerschow, Y.I., Mauve, M.: Non-parallelizable and non-interactive client puzzles from modular square roots. In: Availability, Reliability and Security (ARES) (2011)

40. Alwen, J., Blocki, J., Pietrzak, K.: Depth-robust graphs and their cumulative memory complexity. In: Coron, J.-S., Nielsen, J.B. (eds.) EUROCRYPT 2017. LNCS, vol. 10212, pp. 3–32. Springer, Cham (2017). https://doi.org/10.1007/978-3-319-56617-7_1

41. Juels, A., Kaliski Jr., B.S.: PORs: proofs of retrievability for large files. In: Proceedings of the 14th ACM Conference on Computer and Communications Security, pp. 584–597. ACM (2007)

42. Jules, A., Brainard, J.: Client-puzzles: a cryptographic defense against connection depletion. In: Proceedings of Network and Distributed System Security Symposium (NDSS 1999), pp. 151–165 (1999)
43. Kiayias, A., Russell, A., David, B., Oliynykov, R.: Ouroboros: a provably secure proof-of-stake blockchain protocol. In: Katz, J., Shacham, H. (eds.) CRYPTO 2017. LNCS, vol. 10401, pp. 357–388. Springer, Cham (2017). https://doi.org/10.1007/978-3-319-63688-7_12
44. King, S., Nadal, S.: Peercoin–secure & sustainable cryptocoin, August 2012. https://peercoin.net/whitepaper
45. Kogan, D., Manohar, N., Boneh, D.: T/key: second-factor authentication from secure hash chains. In: ACM Conference on Computer and Communications Security (2017)
46. Lenstra, A.K., Wesolowski, B.: A random zoo: sloth, unicorn, and trx. IACR Cryptology ePrint Archive, 2015 (2015)
47. Lidl, R., Mullen, G.L., Turnwald, G.: Dickson Polynomials, vol. 65. Chapman & Hall/CRC, Boca Raton (1993)
48. Lipmaa, H.: Secure accumulators from euclidean rings without trusted setup. In: Bao, F., Samarati, P., Zhou, J. (eds.) ACNS 2012. LNCS, vol. 7341, pp. 224–240. Springer, Heidelberg (2012). https://doi.org/10.1007/978-3-642-31284-7_14
49. Mahmoody, M., Moran, T., Vadhan, S.: Publicly verifiable proofs of sequential work. In: Proceedings of the 4th Conference on Innovations in Theoretical Computer Science. ACM (2013)
50. Maurer, U., Renner, R., Holenstein, C.: Indifferentiability, impossibility results on reductions, and applications to the random oracle methodology. In: Naor, M. (ed.) TCC 2004. LNCS, vol. 2951, pp. 21–39. Springer, Heidelberg (2004). https://doi.org/10.1007/978-3-540-24638-1_2
51. Micali, S.: CS proofs. In: 1994 Proceedings of the 35th Annual Symposium on Foundations of Computer Science, pp. 436–453. IEEE (1994)
52. Micali, S.: Algorand: the efficient and democratic ledger. arXiv preprint arXiv:1607.01341 (2016)
53. Miller, A., Juels, A., Shi, E., Parno, B., Katz, J.: Permacoin: repurposing bitcoin work for data preservation. In: 2014 IEEE Symposium on Security and Privacy (SP), pp. 475–490. IEEE (2014)
54. Moran, T., Naor, M., Segev, G.: An optimally fair coin toss. In: Reingold, O. (ed.) TCC 2009. LNCS, vol. 5444, pp. 1–18. Springer, Heidelberg (2009). https://doi.org/10.1007/978-3-642-00457-5_1
55. Morris, R., Thompson, K.: Password security: a case history. Commun. ACM 22(11), 594–597 (1979)
56. Müller, P.: A weil-bound free proof of Schur's conjecture. Finite Fields Appl. 3(1), 25–32 (1997)
57. Nakamoto, S.: Bitcoin: a peer-to-peer electronic cash system (2008)
58. Park, S., Pietrzak, K., Kwon, A., Alwen, J., Fuchsbauer, G., Gai, P.: SpaceMint: a cryptocurrency based on proofs of space. Cryptology ePrint Archive, Report 2015/528 (2015). http://eprint.iacr.org/2015/528
59. Parno, B., Howell, J., Gentry, C., Raykova, M.: Pinocchio: nearly practical verifiable computation. In: IEEE Security and Privacy (2013)
60. Pierrot, C., Wesolowski, B.: Malleability of the blockchains entropy. Cryptogr. Commun. 10, 211–233 (2016)
61. Pietrzak, K.: Unique proofs of sequential work from time-lock puzzles (2018). Manuscript

62. Rabin, M.O.: Transaction protection by beacons. J. Comput. Syst. Sci. **27**, 256–267 (1983)
63. Rivest, R.L., Shamir, A.: PayWord and MicroMint: two simple micropayment schemes. In: Lomas, M. (ed.) Security Protocols 1996. LNCS, vol. 1189, pp. 69–87. Springer, Heidelberg (1997). https://doi.org/10.1007/3-540-62494-5_6
64. Rivest, R.L., Shamir, A., Wagner, D.A.: Time-lock puzzles and timed-release crypto (1996)
65. Syta, E., et al.: Scalable bias-resistant distributed randomness. In: 2017 IEEE Symposium on Security and Privacy (SP), pp. 444–460. IEEE (2017)
66. Valiant, P.: Incrementally verifiable computation or proofs of knowledge imply time/space efficiency. In: Canetti, R. (ed.) TCC 2008. LNCS, vol. 4948, pp. 1–18. Springer, Heidelberg (2008). https://doi.org/10.1007/978-3-540-78524-8_1
67. Van Oorschot,P.C., Wiener, M.J.: Parallel collision search with application to hash functions and discrete logarithms. In: ACM Conference on Computer and Communications Security (1994)
68. Wahby, R.S., Setty, S.T., Ren, Z., Blumberg, A.J., Walfish, M.: Efficient RAM and control flow in verifiable outsourced computation. In: NDSS (2015)

Proofs of Work From Worst-Case Assumptions

Marshall Ball[1], Alon Rosen[2], Manuel Sabin[3(✉)],
and Prashant Nalini Vasudevan[4]

[1] Columbia University, New York, USA
marshall@cs.columbia.edu
[2] Efi Arazi School of Computer Science, IDC Herzliya, Herzliya, Israel
alon.rosen@idc.ac.il
[3] UC Berkeley, Berkeley, USA
msabin@berkeley.edu
[4] MIT, Cambridge, USA
prashvas@mit.edu

Abstract. We give Proofs of Work (PoWs) whose hardness is based on well-studied *worst-case* assumptions from fine-grained complexity theory. This extends the work of (Ball et al., STOC '17), that presents PoWs that are based on the Orthogonal Vectors, 3SUM, and All-Pairs Shortest Path problems. These, however, were presented as a 'proof of concept' of provably secure PoWs and did not fully meet the requirements of a conventional PoW: namely, it was not shown that multiple proofs could not be generated faster than generating each individually. We use the considerable *algebraic structure* of these PoWs to prove that this non-amortizability of multiple proofs does in fact hold and further show that the PoWs' structure can be exploited in ways previous heuristic PoWs could not.

This creates full PoWs that are provably hard from worst-case assumptions (previously, PoWs were either only based on heuristic assumptions or on much stronger cryptographic assumptions (Bitansky et al., ITCS '16)) while still retaining significant structure to enable extra properties of our PoWs. Namely, we show that the PoWs of (Ball et al., STOC '17) can be modified to have much faster verification time, can be proved in zero knowledge, and more.

Finally, as our PoWs are based on evaluating low-degree polynomials originating from average-case fine-grained complexity, we prove an *average-case direct sum theorem* for the problem of evaluating these polynomials, which may be of independent interest. For our context, this implies the required non-amortizability of our PoWs.

1 Introduction

Proofs of Work (PoWs), introduced in [DN92], have shown themselves to be an invaluable cryptographic primitive. Originally introduced to combat Denial of Service attacks and email spam, their key notion now serves as the heart of most

© International Association for Cryptologic Research 2018
H. Shacham and A. Boldyreva (Eds.): CRYPTO 2018, LNCS 10991, pp. 789–819, 2018.
https://doi.org/10.1007/978-3-319-96884-1_26

modern cryptocurrencies (when combined with additional desired properties for this application).

By quickly generating easily verifiable challenges that require some quantifiable amount of work, PoWs ensure that adversaries attempting to swarm a system must have a large amount of computational power to do so. Practical uses aside, PoWs at their core ask a foundational question of the nature of hardness: Can you prove that a certain amount of work t was completed? In the context of complexity theory for this theoretical question, it suffices to obtain a computational problem whose (moderately) hard instances are easy to sample such that solutions are quickly verifiable.

Unfortunately, implementations of PoWs in practice stray from this theoretical question and, as a consequence, have two main drawbacks. First, they are often based on *heuristic* assumptions that have no quantifiable guarantees. One commonly used PoW is the problem of simply finding a value s so that hashing it together with the given challenge (e.g. with SHA-256) maps to anything with a certain amount of leading 0's. This is based on the *heuristic* belief that SHA-256 seems to behave unpredictably with no provable guarantees.

Secondly, since these PoWs are not provably secure, their heuristic sense of security stems from, say, SHA-256 not having much discernible *structure* to exploit. This lack of structure, while hopefully giving the PoW its heuristic security, limits the ability to use the PoW in richer ways. That is, heuristic PoWs do not seem to come with a structure to support any useful properties beyond the basic definition of PoWs.

This work, building on the techniques and the proof of concept of our results in [BRSV17a], addresses both of these problems by constructing PoWs that are based on *worst-case complexity theoretic assumptions* in a provable way while also having considerable *algebraic structure*. This simultaneously moves PoWs in the direction of modern cryptography by basing our primitives on well-studied worst-case problems and expands the usability of PoWs by exploiting our algebraic structure to create, for example, PoWs that can be proved in Zero Knowledge or that can be distributed across many workers in a way that is robust to Byzantine failures. Our biggest use of our problems' structure is in proving a direct sum theorem to show that our proofs are non-amortizable across many challenges; this was the missing piece of [BRSV17a] in achieving PoWs according to their usual definition [DN92].

1.1   On Security From Worst-Case Assumptions

We make a point here that if SHA-256 is secure then it can be made into the aforementioned PoW whereas, if it is not, then SHA-256 is broken. While tautological, we point out that this is a Win-Lose situation. That is, either we have a PoW, or a specific instantiation of a heuristic cryptographic hash function is broken and no new knowledge is gained.

This is in contrast to our provably secure PoWs, in which we either have a PoW, or we have a breakthrough in complexity theory. For example, if we base a PoW on the Orthogonal Vectors problem which we define in Sect. 1.2, then either we have a PoW or the Orthogonal Vectors problem can be solved in sub-quadratic time which has been shown [Wil05] to be sufficient to break the Strong Exponential Time Hypothesis (SETH), giving a faster-than-brute-force algorithm for CNF-SAT formulas and thus a major insight to the P vs NP problem.

By basing our PoWs on well-studied complexity theoretic problems, we position our conditional results to be in the desirable position for cryptography and complexity theory: a Win-Win. Orthogonal Vectors, 3SUM, and All-Pairs Shortest Path are the central problems of fine-grained complexity theory precisely because of their many quantitative connections to many other computational problems and so breaking any of their associated conjectures would give considerable insight into computation. Heuristic PoWs like SHA-256, however, aren't even known to have natural generalizations or asymptotics much less connections to other computational problems and so a break would simply say that that specific design for that specific input size happened to not be as secure as we thought.

1.2 Our Results

In this paper we introduce PoWs based on the Orthogonal Vectors (OV), 3SUM, and All-Pairs Shortest Path problems, which comprise the central problems of the field of fine-grained complexity theory. Similar PoWs were introduced in [BRSV17a], although these failed to prove non-amortizability of these PoWs – that many challenges take proportionally more work, as is required by the definition of PoWs [DN92, BGJ+16]. We show here that the PoWs of [BRSV17a] can be extended to exploit their considerable algebraic structure to show non-amortizability via a direct sum theorem and, thus, that they are genuine PoWs according to the conventional definition. Further, we show that this structure to can be used to allow for much quicker verification and zero-knowledge PoWs. We also note that our structure plugs into the framework of [BK16b] to obtain distributed PoWs robust to Byzantine failure.

While all of our results and techniques will be analogous for 3SUM and APSP, we will use OV as our running example for our proofs and results statements. Namely, OV (defined in Sect. 2.2) is a well-studied problem that is conjectured to require $n^{2-o(1)}$ time in the *worst-case* [Wil15]. Roughly, we show the following.

Informal Theorem. *Suppose* OV *takes* $n^{2-o(1)}$ *time to decide for sufficiently large n. A challenge* c *can be generated in* $\widetilde{O}(n)$ *time such that:*

- *A valid proof* π *to* c *can be computed in* $\widetilde{O}(n^2)$ *time.*
- *The validity of a candidate proof to* c *can be verified in* $\widetilde{O}(n)$ *time.*
- *Any valid proof to* c *requires* $n^{2-o(1)}$ *time to compute.*

This can be scaled to $n^{k-o(1)}$ hardness for all $k \in \mathbb{N}$ by a natural generalization of the OV problem to the k-OV problem, whose hardness is also supported by SETH. Thus fine-grained complexity theory props up PoWs of any complexity that is desired.

Further, we show that the verification can still be done in $\widetilde{O}(n)$ time for all of our $n^{k-o(1)}$ hard PoWs, allowing us to *tune* hardness. The corresponding PoW for this is interactive but we show how to remove this interaction in the Random Oracle model in Sect. 5.

We also note that a straightforward application of [BK16b] allows our PoWs to be distributed amongst many workers in a way that is robust to byzantine failure or errors and can detect malicious party members. Namely, that a challenge can be broken up amongst a group of provers so that partial work can be error-corrected into a full proof.

Further, our PoWs admit zero knowledge proofs such that the proofs can be simulated in *very low* complexity – i.e. in time comparable to the verification time. While heuristic PoWs can be proved in zero knowledge as they are NP statements, the exact polynomial time complexities matter in this regime. We are able to use the algebraic structure of our problem to attain a notion of zero knowledge that makes sense in the fine-grained world.

A main lemma which may be of independent interest is a direct sum theorem on evaluating a specific low-degree polynomial $f\mathsf{OV}^k$.

Informal Theorem. *Suppose k-OV takes $n^{k-o(1)}$ time to decide. Then, for any polynomial ℓ, any algorithm that computes $f\mathsf{OV}^k(x_i)$'s correctly on ℓ uniformly random x_i's with probability $1/n^{O(1)}$ takes time $\ell(n) \cdot n^{k-o(1)}$.*

1.3 Related Work

As mentioned earlier, PoWs were introduced by Dwork and Naor [DN92]. Definitions similar to ours were studied by Jakobsson and Juels [JJ99], Bitansky et al. [BGJ+16], and (under the name Strong Client Puzzles) Stebila et al. [SKR+11] (also see the last paper for some candidate constructions and further references).

We note that, while PoWs are often used in cryptocurrencies, the literature studying them in that context have more properties than the standard notion of a PoW (e.g. [BK16a]) that are desirable for their specific use within cryptocurrency and blockchain frameworks. We do not consider these and instead focus on the foundational cryptographic primitive that is a PoW.

In this paper we build on the work of [BRSV17a], which introduced PoWs whose hardness is based on the same worst-case assumptions we consider here. While [BRSV17a] introduced the PoWs as a proof-of-concept that PoWs can be based on well-studied worst-case assumptions, they did not fully satisfy the definition of a PoW in that the PoWs were not shown to be non-amortizable. That is, it was not proven that many challenges could not be batch-evaluated faster than solving each of them individually. We show here that these PoWs are in fact non-amortizable by proving a direct sum theorem in Sect. 4. Further, the k-OV-based PoWs of [BRSV17a] have verification times of $\widetilde{O}(n^{k/2})$ whereas we show how to achieve verification in time $\widetilde{O}(n)$, which makes the PoWs much more realistic for use. These are both properties that are *expected* of a PoW that were not included in [BRSV17a]. Beyond that, we show that our PoWs can be proved in zero knowledge and note that our PoWs can be distributed across many worker in way that is robust to Byzantine error, both of which are properties seemingly *not achievable* from the current 'structureless' heuristic PoWs that are used.

Provably secure PoWs have been considered before in [BGJ+16] where PoWs are achieved from *cryptographic assumptions* (even stronger than an average-case assumption). Namely, they show that if there is a worst-case hard problem that is non-amortizable *and* succinct randomized encodings exist, then PoWs are achievable. In contrast, our PoWs are based on solely on *worst-case* assumptions on well-studied problems from fine-grained complexity theory.

Subsequent to our work, Goldreich and Rothblum [GR18] have constructed (implicitly) a PoW protocol based on the worst-case hardness of the problem of counting t-cliques in a graph (for some constant t); they show a worst-case to average-case reduction for this problem, a doubly efficient interactive proof, and that the average-case problem is somewhat non-amortizable, which are the properties needed to go from worst-case hardness to PoWs.

A previous version of this paper appeared under the title Proofs of Useful Work [BRSV17b], where we had presented the same protocol as in this paper as a PoW scheme where the prover's work could be made "useful" by using it to perform independently useful computation. However, it was pointed out to us (by anonymous reviewers) that a naive construction satisfied our definition of a "Useful PoW."

2   Proofs of Work from Worst-Case Assumptions

In this section, we first define Proof of Work (PoW) schemes, and then present our construction of such a scheme based on the hardness of Orthogonal Vectors (OV) and related problems. In Sect. 2.1, we define PoWs; in Sect. 2.2, we introduce OV and related problems; in Sect. 2.3, we describe an interactive proof for these problems that is used in our eventual construction, which is presented in Sect. 2.4. Our PoWs, while similar, will differ from those of [BRSV17a] in that we allow interaction to significantly speed the verification time by exploiting the PoWs' algebraic structure. We will show how to remove interaction in the Random Oracle model in Sect. 5.

2.1   Definition

Syntactically, a Proof of Work scheme involves three algorithms:

- $\mathsf{Gen}(1^n)$ produces a *challenge* c.
- $\mathsf{Solve}(c)$ solves the challenge c, producing a *proof* π.
- $\mathsf{Verify}(c, \pi)$ verifies the proof π to the challenge c.

Taken together, these algorithms should result in an efficient proof system whose proofs are hard to find. This is formalized as follows.

Definition 2.1 (Proof of Work). *A $(t(n), \delta(n))$-Proof of Work (PoW) consists of three algorithms (Gen, Solve, Verify). These algorithms must satisfy the following properties for large enough n:*

- **Efficiency:**
  - Gen(1^n) *runs in time* $\widetilde{O}(n)$.
  - *For any* $c \leftarrow$ Gen(1^n), Solve(c) *runs in time* $\widetilde{O}(t(n))$.
  - *For any* $c \leftarrow$ Gen(1^n) *and any* π, Verify(c, π) *runs in time* $\widetilde{O}(n)$.
- **Completeness:** *For any* $c \leftarrow$ Gen(1^n) *and any* $\pi \leftarrow$ Solve(c),

$$\Pr[\text{Verify}(c, \pi) = accept] = 1$$

where the probability is taken over Verify*'s randomness.*

- **Hardness:** *For any polynomial* ℓ, *any constant* $\epsilon > 0$, *and any algorithm* Solve$_\ell^*$ *that runs in time* $\ell(n) \cdot t(n)^{1-\epsilon}$ *when given* $\ell(n)$ *challenges of size n as input,*

$$\Pr\left[\forall i : \text{Verify}(c_i, \pi_i) = acc \,\middle|\, \begin{array}{l} (c_i \leftarrow \text{Gen}(1^n))_{i \in [\ell(n)]} \\ \pi \leftarrow \text{Solve}_\ell^*(c_1, \ldots, c_{\ell(n)}) : \\ \pi = (\pi_1, \ldots, \pi_{\ell(n)}) \end{array}\right] < \delta(n)$$

where the probability is taken over Gen *and* Verify*'s randomness.*

The efficiency requirement above guarantees that the verifier in the Proof of Work scheme runs in nearly linear time. Together with the completeness requirement, it also ensures that a prover who actually spends roughly $t(n)$ time can convince the verifier that it has done so. The hardness requirement says that any attempt to convince the verifier without actually spending the prescribed amount of work has only a small probability of succeeding, and that this remains true even when amortized over several instances. That is, even a prover who gets to see several independent challenges and respond to them together will be unable to reuse any work across the challenges, and is effectively forced to spend the sum of the prescribed amount of work on all of them.

In some of the PoWs we construct, Solve and Verify are not algorithms, but are instead parties in an interactive protocol. The requirements of such interactive PoWs are the natural generalizations of those in the definition above, with Verify deciding whether to accept after interacting with Solve. And the hardness requirement applies to the numerous interactive protocols being run in any form of composition – serial, parallel, or otherwise. We will, however, show how to remove interaction in Sect. 5.

Heuristic constructions of PoWs, such as those based on SHA-256, easily satisfy efficiency and completeness (although not formally, given their lack of asymptotics), yet their hardness guarantees are based on nothing but the heuristic assumption that the PoW itself is a valid PoW. We will now reduce the hardness of our PoW to the hardness of well-studied worst-case problems in fine-grained complexity theory.

2.2   Orthogonal Vectors

We now formally define the problems – Orthogonal Vectors (OV) and its generalization k-OV – whose hardness we use to construct our PoW scheme. The properties possessed by OV that enable this construction are also shared by other well-studied problems mentioned earlier, including 3SUM and APSP as noted in [BRSV17a], and an array of other problems [BK16b, GR17, Wil16]. Consequently, while we focus on OV, PoWs based on the hardness of these other problems can be constructed along the lines of the one here. Further, the security of these constructions would also follow from the hardness of other problems that reduce to OV, 3SUM, etc. in a fine-grained manner with little, if any, degradation of security. Of particular interest, deciding graph properties that are statable in first-order logic all reduce to (moderate-dimensional) OV [GI16], and so we can obtain PoWs if *any* problem statable as a first-order graph property is hard.

All the algorithms we consider henceforth – reductions, adversaries, etc. – are *non-uniform Word-RAM algorithms* (with words of size $O(\log n)$ where n will be clear from context) unless stated otherwise, both in our hardness assumptions and our constructions. Security against such adversaries is necessary for PoWs to remain hard in the presence of pre-processing, which is typical in the case of cyrptocurrencies, for instance, where specialized hardware is often used. In the case of reductions, this non-uniformity is solely used to ensure that specific parameters determined completely by instance size (such as the prime $p(n)$ in Definition 2.5) are known to the reductions.

Remark 2.2. All of our reductions, algorithms, and assumptions can easily be made uniform by having an extra Setup procedure that is allowed to run in $t(n)^{1-\epsilon}$ for some $\epsilon > 0$ for a $(t(n), \delta(n))$-PoW. In our setting, this will just be used to find a prime on which to base a field extension for the rest of the PoW to satisfy the rest of its conditions. This makes sense for a PoW scheme to do and, for all the problems we consider, this can be done be done so that all the conjectures can be made uniformly. We leave everything non-uniform, however, for exposition's sake.

Definition 2.3 (Orthogonal Vectors). *The* OV *problem on vectors of dimension d (denoted* OV_d*) is to determine, given two sets U, V of n vectors from $\{0, 1\}^{d(n)}$ each, whether there exist $u \in U$ and $v \in V$ such that $\langle u, v \rangle = 0$ (over $\mathbb{Z}$). If left unspecified, d is to be taken to be $\lceil \log^2 n \rceil$.*

OV is commonly conjectured to require $n^{2-o(1)}$ time to decide, for which many conditional fine-grained hardness results are based on [Wil15], and has been shown to be true if the Strong Exponential Time Hypothesis (SETH) holds [Wil05]. This hardness and the hardness of its generalization to k-OV of requiring $n^{k-o(1)}$ time (which also holds under SETH) are what we base the hardness of our PoWs on. We now define k-OV.

Definition 2.4 (k-Orthogonal Vectors). *For an integer $k \geq 2$, the k-OV problem on vectors of dimension d is to determine, given k sets $(U_1, \ldots, U_k)$ of* *n vectors from $\{0,1\}^{d(n)}$ each, whether there exist $u^s \in U_s$ for each $s \in [k]$ such that over $\mathbb{Z}$,*

$$\sum_{\ell \in [d(n)]} u_\ell^1 \cdots u_\ell^k = 0$$

We say that such a set of vectors is k-orthogonal. If left unspecified, d is to be taken to be $\lceil \log^2 n \rceil$.

While these problems are conjectured worst-case hard, there are currently no widely-held beliefs for distributions that it may be average-case hard over. [BRSV17a], however, defines a related problem that is shown to be average-case hard when assuming the worst-case hardness of k-OV. This problem is that of evaluating the following polynomial:

For any prime number p, we define the polynomial $f\mathsf{OV}_{n,d,p}^k : \mathbb{F}_p^{knd} \to \mathbb{F}_p$ as follows. Its inputs are parsed in the manner that those of k-OV are: below, for any $s \in [k]$ and $i \in [n]$, u_i^s represents the ith vector in U_s, and for $\ell \in [d]$, $u_{i\ell}^s$ represents its ℓth coordinate.

$$f\mathsf{OV}_{n,d,p}^k(U_1,\ldots,U_k) = \sum_{i_1,\ldots,i_k \in [n]} \prod_{\ell \in [d]} \left(1 - u_{i_1\ell}^1 \cdots u_{i_k\ell}^k\right)$$

When given an instance of k-OV (from $\{0,1\}^{knd}$) as input, $f\mathsf{OV}_{n,d,p}^k$ counts the number of tuples of k-orthogonal vectors (modulo p). Note that the degree of this polynomial is kd; for small d (e.g. $d = \lceil \log^2 n \rceil$), this is a fairly low-degree polynomial. The following definition gives the family of such polynomials parameterized by input size.

Definition 2.5 ($\mathcal{F}\mathsf{OV}^k$). *Consider an integer $k \geq 2$. Let $p(n)$ be the smallest prime number larger than $n^{\log n}$, and $d(n) = \lceil \log^2 n \rceil$. $\mathcal{F}\mathsf{OV}^k$ is the family of functions $\left\{f\mathsf{OV}_{n,d(n),p(n)}^k\right\}$.*

Remark 2.6. We note that most of our results would hold for a much smaller choice of $p(n)$ above – anything larger than n^k would do. The reason we choose p to be this large is to achieve negligible soundness error in interactive protocols we shall be designing for this family of functions (see Protocol 1.1). Another way to achieve this is to use large enough extension fields of $\mathbb{F}_p$ for smaller p's; this is actually preferable, as the value of $p(n)$ as defined now is much harder to compute for uniform algorithms.

2.3   Preliminaries

Our final protocol and its security consists, essentially, of two components – the hardness of evaluating $f\mathsf{OV}^k$ on random inputs, and the the ability to certify the correct evaluation of $f\mathsf{OV}^k$ in an efficiently verifiable manner. We explain the former in the next subsection; here, we describe the protocol for the latter

(Protocol 1.1), which we will use as a sub-routine in our final PoW protocol. This protocol is a $(k-1)$-round interactive proof that, given $U_1, \ldots, U_k \in \mathbb{F}_p^{nd}$ and $y \in \mathbb{F}_p$, proves that $f\mathsf{OV}_{n,d,p}^k(U_1, \ldots, U_k) = y$.

In the special case of $k = 2$, a non-interactive (MA) protocol for OV was shown in [Wil16] and this MA protocol was used to construct a PoW scheme based on OV, 3SUM, and APSP in [BRSV17a], albeit one that only satisfies a weaker hardness requirement (i.e. non-batchability was not considered or proved). We introduce interaction to greatly improve the verifier's efficiency and show how interaction can be removed in Sect. 5. The following interactive proof is essentially the sum-check protocol, but in our case we need to pay close attention to the complexity of the prover and the verifier and so use ideas from [Wil16].

We will set up the following definitions before describing the protocol. For each $s \in [k]$, consider the univariate polynomials $\phi_1^s, \ldots, \phi_d^s : \mathbb{F}_p \to \mathbb{F}_p$, where ϕ_ℓ^s represents the ℓth column of U_s – that is, for $i \in [n]$, $\phi_\ell^s(i) = u_{i\ell}^s$. Each ϕ_ℓ^s has degree at most $(n-1)$. $f\mathsf{OV}_{n,d,p}^k$ can now be written as:

$$f\mathsf{OV}_{n,d,p}^k(U_1, \ldots, U_k) = \sum_{i_1,\ldots,i_k \in [n]} \prod_{\ell \in [d]} \left(1 - u_{i_1\ell}^1 \cdots u_{i_k\ell}^k\right)$$

$$= \sum_{i_1,\ldots,i_k \in [n]} \prod_{\ell \in [d]} \left(1 - \phi_\ell^1(i_1) \cdots \phi_\ell^k(i_k)\right)$$

$$= \sum_{i_1,\ldots,i_k \in [n]} q(i_1, \ldots, i_k)$$

where q is defined for convenience as:

$$q(i_1, \ldots, i_k) = \prod_{\ell \in [d]} \left(1 - \phi_\ell^1(i_1) \cdots \phi_\ell^k(i_k)\right)$$

The degree of q is at most $D = k(n-1)d$. Note that q can be evaluated at any point in $\mathbb{F}_p^k$ in time $\widetilde{O}(knd \log p)$, by evaluating all the $\phi_\ell^s(i_s)$'s (these polynomials can be found using fast interpolation techniques for univariate polynomials [Hor72]), computing each term in the above product and then multiplying them.

For any $s \in [k]$ and $\alpha_1, \ldots, \alpha_{s-1} \in \mathbb{F}_p$, define the following univariate polynomial:

$$q_{s,\alpha_1,\ldots,\alpha_{s-1}}(x) = \sum_{i_{s+1},\ldots,i_k \in [n]} q(\alpha_1, \ldots, \alpha_{s-1}, x, i_{s+1}, \ldots, i_k)$$

Every such q_s has degree at most $(n-1)d$ – this can be seen by inspecting the definition of q. With these definitions, the interactive proof is described as Protocol 1.1 below. The completeness and soundness of this interactive proof is then asserted by Theorem 2.7, which is proven in Sect. 3.

Theorem 2.7. *For any $k \geq 2$, let d and p be as in Definition 2.5. Protocol 1.1 is a $(k-1)$-round interactive proof for proving that $y = \mathcal{F}\mathsf{OV}^k(x)$. This protocol has perfect completeness and soundness error at most $\left(\frac{knd}{p}\right)$. The prover runs in time $\widetilde{O}(n^k d \log p)$, and the verifier in time $\widetilde{O}(knd^2 \log p)$.*

Interactive Proof for $\mathcal{F}\mathsf{OV}^k$:

The inputs to the protocol are $(U_1, \ldots, U_k) \in \mathbb{F}_p^{knd}$ (a valid input to $f\mathsf{OV}_{n,d,p}^k$), and a field element $y \in \mathbb{F}_p$. The polynomials q are defined as in the text.

- The prover sends the coefficients of a univariate polynomial q_1^* of degree at most $(n-1)d$.
- The verifier checks that $\sum_{i_1 \in [n]} q_1^*(i_1) = y$. If not, it rejects.
- For s from 1 up to $k-2$:
  - The verifier sends a random $\alpha_s \leftarrow \mathbb{F}_p$.
  - The prover sends the coefficients of a polynomial $q_{s+1,\alpha_1,\ldots,\alpha_s}^*$ of degree at most $(n-1)d$.
  - The verifier checks that $\sum_{i_{s+1} \in [n]} q_{s+1,\alpha_1,\ldots,\alpha_s}^*(i_{s+1}) = q_{s,\alpha_1,\ldots,\alpha_{s-1}}^*(\alpha_s)$. If not, it rejects.
- The verifier picks $\alpha_{k-1} \leftarrow \mathbb{F}_p$ and checks that $q_{k-1,\alpha_1,\ldots,\alpha_{k-2}}^*(\alpha_{k-1}) = q_{k-1,\alpha_1,\ldots,\alpha_{k-2}}(\alpha_{k-1})$, computed using the fact that $q_{k-1,\alpha_1,\ldots,\alpha_{k-2}}(\alpha_{k-1}) = \sum_{i_k \in [n]} q_{k,\alpha_1,\ldots,\alpha_{k-1}}(i_k)$. If not, it rejects.
- If the verifier hasn't rejected yet, it accepts.

Protocol 1.1: Interactive Proof for $\mathcal{F}\mathsf{OV}^k$.

As observed earlier, Protocol 1.1 is non-interactive when $k = 2$. We then get the following corollary for $\mathcal{F}\mathsf{OV}$.

Corollary 2.8. *For $k = 2$, let d and p be as in Definition 2.5. Protocol 1.1 is an MA proof for proving that $y = \mathcal{F}\mathsf{OV}(x)$. This protocol has perfect completeness and soundness error at most $\left(\frac{2nd}{p}\right)$. The prover runs in time $\widetilde{O}(n^2)$, and the verifier in time $\widetilde{O}(n)$.*

2.4 The PoW Protocol

We now present Protocol 1.2, which we show to be a Proof of Work scheme assuming the hardness of k-OV.

Theorem 2.9. *For some $k \geq 2$, suppose k-OV takes $n^{k-o(1)}$ time to decide for all but finitely many input lengths for any $d = \omega(\log n)$. Then, Protocol 1.2 is an (n^k, δ)-Proof of Work scheme for any function $\delta(n) > 1/n^{o(1)}$.*

Remark 2.10. As is, this will be an interactive Proof of Work protocol. In the special case of $k = 2$, Corollary 2.8 gives us a non-interactive PoW. If we want to remove interaction for general k-OV, however, we could use the MA proof in [Wil16] at the cost of verification taking time $\widetilde{O}(n^{k/2})$ as was done in [BRSV17a]. To keep verification time at $\widetilde{O}(n)$, we instead show how to remove interaction in the Random Oracle model in Sect. 5. This will allow us to *tune* the gap between the parties – we can choose k and thus the amount of work, $n^{k-o(1)}$, that must be done by the prover while always only needing $\widetilde{O}(n)$ time for verification.

Proof of Work based on hardness of k-OV:

- $\mathsf{Gen}(1^n)$:
    - Output a random $c \in \mathbb{F}_p^{knd}$.
- $(\mathsf{Solve}, \mathsf{Verify})$ work as follows given c:
    - Solve computes $z = f\mathsf{OV}_{n,d,p}^k(c)$ and outputs it.
    - Solve and Verify run Protocol 1.1 with input (c, z), Solve as prover, and Verify as verifier.
    - Verify accepts iff the verifier in the above instance of Protocol 1.1 accepts.

Protocol 1.2: Proof of Work based on the hardness of k-OV.

Remark 2.11. We can also exploit this PoW's algebraic structure on the Prover's side. Using techniques from [BK16b], the Prover's work can be distributed amongst a group of provers. While, cumulatively, they must complete the work required of the PoW, they can each only do a portion of it. Further, this can be done in a way robust to Byzantine errors amongst the group. See Remark 3.4 for further details.

We will use Theorem 2.7 to argue for the completeness and soundness of Protocol 1.2. In order to prove the hardness, we will need lower bounds on how well the problem that Solve is required to solve can be batched. We first define what it means for a function to be non-batchable in the average-case in a manner compatible with the hardness requirement. Note that this requirement is stronger than being non-batchable in the worst-case.

Definition 2.12. *Consider a function family $\mathcal{F} = \{f_n : \mathcal{X}_n \to \mathcal{Y}_n\}$, and a family of distributions $\mathcal{D} = \{D_n\}$, where D_n is over $\mathcal{X}_n$. $\mathcal{F}$ is not (ℓ, t, δ)-batchable on average over $\mathcal{D}$ if, for any algorithm Batch that runs in time $\ell(n)t(n)$ when run on $\ell(n)$ inputs from $\mathcal{X}_n$, when it is given as input $\ell(n)$ independent samples from D_n, the following is true for all large enough n:*

$$\Pr_{x_i \leftarrow D_n} \left[\mathsf{Batch}(x_1, \ldots, x_{\ell(n)}) = (f_n(x_1), \ldots, f_n(x_{\ell(n)})) \right] < \delta(n)$$

We will be concerned with the case where the batched time $t(n)$ is less than the time it takes to compute f_n on a single instance. This sort of statement is what a direct sum theorem for $\mathcal{F}$'s hardness would guarantee. Theorem 2.13, then, claims that we achieve this non-batchability for $\mathcal{F}\mathsf{OV}^k$ and, as $\mathcal{F}\mathsf{OV}^k$ is one of the things that Solve is required to evaluate, we will be able to show the desired hardness of Protocol 1.2. We prove Theorem 2.13 via a direct sum theorem in Appendix A, and prove a weaker version for illustrative purposes in Sect. 4.

Theorem 2.13. *For some $k \geq 2$, suppose k-OV takes $n^{k-o(1)}$ time to decide for all but finitely many input lengths for any $d = \omega(\log n)$. Then, for any constants*

$c, \epsilon > 0$ and $\delta < \epsilon/2$, $\mathcal{F}OV^k$ is not $(n^c, n^{k-\epsilon}, 1/n^\delta)$-batchable on average over the uniform distribution over its inputs.

We now put all the above together to prove Theorem 2.9 as follows.

Proof of Theorem 2.9. We prove that Protocol 1.2 satisfies the various requirements demanded of a Proof of Work scheme assuming the hardness of k-OV.

Efficiency:

- Gen(1^n) simply samples knd uniformly random elements of $\mathbb{F}_p$. As $d = \log^2 n$ and $p \leq 2n^{\log n}$ (by Bertrand-Chebyshev's Theorem), this takes $\widetilde{O}(n)$ time.
- Solve computes $f OV^k_{n,d,p}(c)$, which can be done in $\widetilde{O}(n^k)$ time. It then runs the prover in an instance of Protocol 1.1, which can be done in $\widetilde{O}(n^k)$ time by Theorem 2.7. So in all it takes takes $\widetilde{O}(n^k)$ time.
- Verify runs the verifier in an instance of Protocol 1.1, taking $\widetilde{O}(n)$ time, again by Theorem 2.7.

Completeness: This follows immediately from the completeness of Protocol 1.1 as an interactive proof for $\mathcal{F}OV^k$, as stated in Theorem 2.7, as this is the protocol that Solve and Verify engage in.

Hardness: We proceed by contradiction. Suppose there is a polynomial ℓ, an (interactive) algorithm Solve*, and a constant $\epsilon > 0$ such that Solve* runs in time $\ell(n)n^{k-\epsilon}$ and makes Verify accept on $\ell(n)$ independent challenges generated by Gen(1^n) with probability at least $\delta(n) > 1/n^{o(1)}$ for infinitely many input lengths n.

For each of these input lengths, let the set of challenges (which are $f OV$ inputs) produced by Gen(1^n) be $\{c_1, \ldots, c_{\ell(n)}\}$, and the corresponding set of solutions output by Solve* be $\{z_1, \ldots, z_{\ell(n)}\}$. So Solve* succeeds as a prover in Protocol 1.1 for *all* the instances $\{(c_i, z_i)\}$ with probability at least $\delta(n)$.

By the negligible soundness error of Protocol 1.1 guaranteed by Theorem 2.7, in order to do this, Solve* has to use the correct values $f OV^k_{n,d,p}(c_i)$ for all the z_i's with probability negligibly close to $\delta(n)$ and definitely more than, say, $\delta(n)/2$. In particular, with this probability, it has to explicitly compute $f OV^k_{n,d,p}$ at $c_1, \ldots, c_{\ell(n)}$, all of which are independent uniform points in $\mathbb{F}_p^{knd}$ for all of these infinitely many input lengths n. But this is exactly what Theorem 2.13 says is impossible under our assumptions. So such a Solve* cannot exist, and this proves the hardness of Protocol 1.2.

We have thus proven all the properties necessary and hence Protocol 1.2 is indeed an (n^k, δ)-Proof of Work under the hypothesised hardness of k-OV for any $\delta(n) > 1/n^{o(1)}$. $\qquad\square$

3   Verifying $\mathcal{F}OV^k$

In this section, we prove Theorem 2.7 (stated in Sect. 2), which is about Protocol 1.1 being a valid interactive proof for proving evaluations of $\mathcal{F}OV^k$. We use here terminology from the theorem statement and protocol description. Recall the the input to the protocol is $U_1, \ldots, U_k \in \mathbb{F}_p^{nd}$ and $y \in \mathbb{F}_p$, and the prover wishes to prove that $y = f\mathsf{OV}_{n,d,p}^k(U_1, \ldots, U_k)$.

Completeness. If indeed $y = f\mathsf{OV}_{n,d,p}^k(U_1, \ldots, U_k)$, the prover can make the verifier in the protocol accept by using the polynomials $(q_1, q_{2,\alpha_1}, \ldots, q_{k,\alpha_1,\ldots,\alpha_k})$ in place of $(q_1^*, q_{2,\alpha_1}^*, \ldots, q_{k,\alpha_1,\ldots,\alpha_k}^*)$. Perfect completeness is then seen to follow from the definitions of these polynomials and their relation to q and hence $f\mathsf{OV}_{n,d,p}^k$.

Soundness. Suppose $y \neq f\mathsf{OV}_{n,d,p}^k(U_1, \ldots, U_k)$. We now analyze the probability with which a cheating prover could make the verifier accept.

To start with, note that the prover's q_1^* has to be different from q_1, as otherwise the check in the second step would fail. Further, as the degree of these polynomials is less than nd, the probability that the verifier will then choose an α_1 such that $q_1^*(\alpha_1) = q_1(\alpha_1)$ is less than $\frac{nd}{p}$.

If this event does not happen, then the prover has to again send a q_{2,α_1}^* that is different from q_{2,α_1}, which again agree on α_2 with probability less than $\frac{nd}{p}$. This goes on for $(k-1)$ rounds, at the end of which the verifier checks whether $q_{k-1}^*(\alpha_{k-1})$ is equal to $q_{k-1}(\alpha_{k-1})$, which it computes by itself. If at least one of these accidental equalities at a random point has not occurred throughout the protocol, the verifier will reject. The probability that no violations occur over the $(k-1)$ rounds is, by the union bound, less than $\frac{knd}{p}$.

Efficiency. Next we discuss details of how the honest prover and the verifier are implemented, and analyze their complexities. To this end, we will need the following algorithmic results about computations involving *univariate* polynomials over finite fields.

Lemma 3.1 (Fast Multi-point Evaluation [Fid72]). *Given the coefficients of a univariate polynomial* $q : \mathbb{F}_p \to \mathbb{F}_p$ *of degree at most* N, *and* N *points* $x_1, \ldots, x_N \in \mathbb{F}_p$, *the set of evaluations* $(q(x_1), \ldots, q(x_N))$ *can be computed in time* $O(N \log^3 N \log p)$.

Lemma 3.2 (Fast Interpolation [Hor72]). *Given* $N+1$ *evaluations of a univariate polynomial* $q : \mathbb{F}_p \to \mathbb{F}_p$ *of degree at most* N, *the coefficients of* q *can be computed in time* $O(N \log^3 N \log p)$.

To start with, both the prover and verifier compute the coefficients of all the ϕ_ℓ^s's. Note that, by definition, they know the evaluation of each ϕ_ℓ^s on n points, given by $\{(i, u_{i\ell}^s)\}_{i \in [n]}$. This can be used to compute the coefficients of each ϕ_ℓ^s in time $\widetilde{O}(n \log p)$ by Lemma 3.2. The total time taken is hence $\widetilde{O}(knd \log p)$.

The proof of the following proposition specifies further details of the prover's workings.

Proposition 3.3. *The coefficients of the polynomial* $q_{s,\alpha_1,\ldots,\alpha_{s-1}}$ *can be computed in time* $\widetilde{O}((n^{k-s+1}d + nd^2) \log p)$ *given the above preprocessing.*

Proof. The procedure to do the above is as follows:

1. Fix some value of $s, \alpha_1, \ldots, \alpha_{s-1}$.
2. For each $\ell \in [d]$, compute the evaluation of ϕ_ℓ^s on nd points, say $\{1, \ldots, nd\}$.
   - Since its coefficients are known, the evaluations of each ϕ_ℓ^s on these nd points can be computed in time $\widetilde{O}(nd \log p)$ by Lemma 3.1, for a total of $\widetilde{O}(nd^2 \log p)$ for all the ϕ_ℓ^s's.
3. For each setting of $i_{s+1}, \ldots, i_k$, compute the evaluations of the polynomial $\rho_{i_{s+1}, \ldots, i_k}(x) = q(\alpha_1, \ldots, \alpha_{s-1}, x, i_{s+1}, \ldots, i_k)$, on the points $\{1, \ldots, nd\}$.
   - First substitute the constants $\alpha_1, \ldots, \alpha_{s-1}, i_{s+1}, \ldots, i_k$ into the definition of q.
   - This requires computing, for each $\ell \in [d]$ and $s' \in [k] \setminus \{s\}$, either $\phi_\ell^{s'}(\alpha_s)$ or $\phi_\ell^{s'}(i_s)$. All of this can be done in time $\widetilde{O}(knd \log p)$ by direct polynomial evaluations since the coefficients of the $\phi_\ell^{s'}$'s are known.
   - This reduces q to a product of d univariate polynomials of degree less than n, whose evaluations on the nd points can now be computed in time $\widetilde{O}(knd \log p)$ by multiplying the constants computed in the above step with the evaluations of $\phi_\ell^{s'}$ on these points, and subtracting from 1.
   - The product of the evaluations can now be computed in time $\widetilde{O}(nd^2 \log p)$ to get what we need.
4. Add up the evaluations of $\rho_{i_{s+1}, \ldots, i_k}$ pointwise over all settings of $(i_{s+1}, \ldots, i_k)$.
   - There are n^{k-s} possible settings of $(i_{s+1}, \ldots, i_k)$, and for each of these we have nd evaluations. All the additions hence take $\widetilde{O}(n^{k-s+1}d \log p)$ time.
5. This gives us nd evaluations of $q_{s, \alpha_1, \ldots, \alpha_{s-1}}$, which is a univariate polynomial of degree at most $(n-1)d$. So its coefficients can be computed in time $\widetilde{O}(nd \log p)$ by Lemma 3.2.

It can be verified from the intermediate complexity computations above that all these operations together take $\widetilde{O}((n^{k-s+1}d + nd^2) \log p)$ time. This proves the proposition. $\qquad\square$

Recall that what the honest prover has to do is compute $q_1, q_{2,\alpha_1}, \ldots,$ $q_{k, \alpha_1, \ldots, \alpha_{k-1}}$ for the α_s's specified by the verifier. By the above proposition, along with the preprocessing, the total time the prover takes is:

$$\widetilde{O}(knd \log p + (n^k d + nd^2) \log p) = \widetilde{O}(n^k d \log p)$$

The verifier's checks in steps (2) and (3) can each be done in $\widetilde{O}(n \log p)$ time using Lemma 3.1. Step (4), finally, can be done by using the above proposition with $s = k$ in time $\widetilde{O}(nd^2 \log p)$. Even along with the preprocessing, this leads to a total time of $\widetilde{O}(knd^2 \log p)$.

Remark 3.4. Note the Prover's work of finding coefficients of polynomials is mainly done by evaluating the polynomial on many points and interpolating. Similarly to [BK16b], this opens the door to distributing the Prover's work.

Namely, the individual evaluations can be split amongst a group of workers which can then be recombined to find the final coefficients. Further, since the evaluations of a polynomial is a Reed-Solomon code, this allows for error correction in the case that the group of provers make errors or have some malicious members. Thus, the Prover's work can be distributed in a way that is robust to Byzantine errors and can identify misbehaving members.

4   A Direct Sum Theorem for $\mathcal{F}$OV

A direct sum theorem for a problem roughly states that solving m independent instances of a problem takes m times as long as a single instance. The converse of this is attaining a non-trivial speed-up when given a *batch* of instances. In this section we prove a direct sum theorem for the problem of evaluating $\mathcal{F}$OV and thus its non-batchability.

Direct sum are typically elusive in complexity theory and so our results, which we prove for generic problems with a certain set of properties, may be of independent interest to the study of hardness amplification. That our results show that batch-evaluating our multivariate low-degree polynomials is *hard* may be particularly surprising since batch-evaluation for *univariate* low-degree polynomials is known to be *easy* [Fid72, Hor72] and, further, [BK16b, GR17, Wil16] show that batch-evaluating multivariate low-degree polynomials (including our own) is *easy to delegate*. For more rigorous definitions of direct sum and direct product theorems, see [She12].

We now prove the following weaker version of Theorem 2.13 on $\mathcal{F}$OV's non-batchability (Theorem 2.13 is proven in Appendix A using an extension of the techniques employed here). The notion of non-batchability used below is defined in Definition 2.12 in Sect. 2.

Theorem 4.1. *For some $k \geq 2$, suppose k-OV takes $n^{k-o(1)}$ time to decide for all but finitely many input lengths for any $d = \omega(\log n)$. Then, for any constants $c, \epsilon > 0$, $\mathcal{F}OV^k$ is not $(n^c, n^{k-\epsilon}, 7/8)$-batchable on average over the uniform distribution over its inputs.*

Throughout this section, $\mathcal{F}$, $\mathcal{F}'$ and $\mathcal{G}$ are families of functions $\{f_n : \mathcal{X}_n \to \mathcal{Y}_n\}$, $\{f'_n : \mathcal{X}'_n \to \mathcal{Y}'_n\}$ and $\left\{g_n : \hat{\mathcal{X}}_n \to \hat{\mathcal{Y}}_n\right\}$, and $\mathcal{D} = \{D_n\}$ is a family of distributions where D_n is over $\hat{\mathcal{X}}_n$.

Theorem 4.1 is the result of two properties possessed by $\mathcal{F}OV^k$. We define these properties below, prove a more general lemma about functions that have these properties, and use it to prove this theorem.

Definition 4.2. *$\mathcal{F}$ is said to be (s, ℓ)-downward reducible to $\mathcal{F}'$ in time t if there is a pair of algorithms* (Split, Merge) *satisfying:*

- *For all large enough n, $s(n) < n$.*
- Split *on input an $x \in \mathcal{X}_n$ outputs $\ell(n)$ instances from $\mathcal{X}'_{s(n)}$.*

$$\mathsf{Split}(x) = (x_1, \ldots, x_{\ell(n)})$$

- *Given the value of $\mathcal{F}'$ at these $\ell(n)$ instances,* Merge *can reconstruct the value of $\mathcal{F}$ at x.*

$$\text{Merge}(x, f'_{s(n)}(x_1), \ldots, f'_{s(n)}(x_{\ell(n)})) = f_n(x)$$

- Split *and* Merge *together run in time at most $t(n)$.*

If $\mathcal{F}'$ is the same as $\mathcal{F}$, then $\mathcal{F}$ is said to be downward self-reducible.

Definition 4.3. *$\mathcal{F}$ is said to be ℓ-robustly reducible to $\mathcal{G}$ in time t if there is a pair of algorithms* (Split, Merge) *satisfying:*

- Split *on input an $x \in \mathcal{X}_n$ (and randomness r) outputs $\ell(n)$ instances from $\mathcal{X}_n$.*

$$\text{Split}(x; r) = (x_1, \ldots, x_{\ell(n)})$$

- *For such a tuple $(x_i)_{i \in [\ell(n)]}$ and any function g^* such that $g^*(x_i) = g_n(x_i)$ for at least 2/3 of the x_i's,* Merge *can reconstruct the function value at x as:*

$$\text{Merge}(x, r, g^*(x_1), \ldots, g^*(x_{\ell(n)})) = f_n(x)$$

- Split *and* Merge *together run in time at most $t(n)$.*
- *Each x_i is distributed according to D_n, and the x_i's are pairwise independent.*

The above is a more stringent notion than the related non-adaptive random self-reducibility as defined in [FF93]. We remark that to prove what we need, it can be shown that it would have been sufficient if the reconstruction above had only worked for *most* r's.

Lemma 4.4. *Suppose $\mathcal{F}$, $\mathcal{F}'$ and $\mathcal{G}$ have the following properties:*

- *$\mathcal{F}$ is (s_d, ℓ_d)-downward reducible to $\mathcal{F}'$ in time t_d.*
- *$\mathcal{F}'$ is ℓ_r-robustly reducible to $\mathcal{G}$ over D in time t_r.*
- *$\mathcal{G}$ is $(\ell_a, t_a, 7/8)$-batchable on average over D, and $\ell_a(s_d(n)) = \ell_d(n)$.*

Then $\mathcal{F}$ can be computed in the worst-case in time:

$$t_d(n) + \ell_d(n)t_r(s_d(n)) + \ell_r(s_d(n))\ell_d(n)t_a(s_d(n))$$

We note, that the condition $\ell_a(s_d(n)) = \ell_d(n)$ above can be relaxed to $\ell_a(s_d(n)) \leq \ell_d(n)$ at the expense of a factor of 2 in the worst-case running time obtained for $\mathcal{F}$. We now show how to prove Theorem 4.1 using Lemma 4.4, and then prove the lemma itself.

Proof of Theorem 4.1. Fix any $k \geq 2$. Suppose, towards a contradiction, that for some $c, \epsilon > 0$, $\mathcal{F}OV^k$ is $(n^c, n^{k-\epsilon}, 7/8)$-batchable on average over the uniform distribution. In our arguments we will refer to the following function families:

- $\mathcal{F}$ is k-OV with vectors of dimension $d = \left(\frac{k}{k+c}\right)^2 \log^2 n$.

- $\mathcal{F}'$ is k-OV with vectors of dimension $\log^2 n$.
- $\mathcal{G}$ is fOV^k (over $\mathbb{F}_p^{knd}$ for some p that definitely satisfies $p > n$).

Let $m = n^{k/(k+c)}$. Note the following two properties :

- $\frac{n}{m^{c/k}} = m$
- $d = \left(\frac{k}{k+c}\right)^2 \log^2 n = \log^2 m$

We now establish the following relationships among the above function families.

Proposition 4.5. $\mathcal{F}$ is (m, m^c)-downward reducible to $\mathcal{F}'$ in time $\widetilde{O}(m^{c+1})$.

Split$_d$, when given an instance $(U_1, \dots, U_k) \in \{0,1\}^{k(n \times d)}$, first divides each U_i into $m^{c/k}$ partitions $U_{i1}, \dots, U_{im^{c/k}} \in \{0,1\}^{m \times d}$. It then outputs the set of tuples $\{(U_{1j_1}, \dots, U_{kj_k}) \mid j_i \in [m^{c/k}]\}$. Each U_{ij} is in $\{0,1\}^{m \times d}$ and, as noted earlier, $d = \log^2 m$. So each tuple in the set is indeed an instance of $\mathcal{F}'$ of size m. Further, there are $(m^{c/k})^c = m^c$ of these.

Note that the original instance has a set of k-orthogonal vectors if and only if at least one of the m^c smaller instances produced does. So Merge$_d$ simply computes the disjunction of the $\mathcal{F}'$ outputs to these instances.

Both of these can be done in time $O(m^c \cdot k \cdot md + m^c) = \widetilde{O}(m^{c+1})$.

Proposition 4.6. $\mathcal{F}'$ is $12kd$-robustly reducible to $\mathcal{G}$ over the uniform distribution in time $\widetilde{O}(m)$.

Notice that for any $U_1, \dots, U_k \in \{0,1\}^{m \times d}$, we have that k-OV$(U_1, \dots, U_k) = fOV^k_m(U_1, \dots, U_k)$. So it is sufficient to show such a robust reduction from $\mathcal{G}$ to itself. We do this now.

Given input $\boldsymbol{x} \in \mathbb{F}_p^{knd}$, Split$_r$ picks two uniformly random $\boldsymbol{x}_1, \boldsymbol{x}_2 \in \mathbb{F}_p^{knd}$ and outputs the set of vectors $\{\boldsymbol{x} + t\boldsymbol{x}_1 + t^2\boldsymbol{x}_2 \mid t \in \{1, \dots, 12kd\}\}$. Recall that our choice of p is much larger than $12kd$ and hence this is possible. The distribution of each of these vectors is uniform over $\mathbb{F}_p^{knd}$, and they are also pairwise independent as they are points on a random quadratic curve through $\boldsymbol{x}$.

Define the univariate polynomial $g_{\boldsymbol{x},\boldsymbol{x}_1,\boldsymbol{x}_2}(t) = fOV^k_m(\boldsymbol{x} + t\boldsymbol{x}_1 + t^2\boldsymbol{x}_2)$. Note that its degree is at most $2kd$. When Merge$_r$ is given $(y_1, \dots, y_{12kd})$ that are purported to be the evaluations of fOV^k_m on the points produced by Split, these can be seen as purported evaluations of $g_{\boldsymbol{x},\boldsymbol{x}_1,\boldsymbol{x}_2}$ on $\{1, \dots, 12kd\}$. This can, in turn, be treated as a corrupt codeword of a Reed-Solomon code, which under these parameters has distance $10kd$.

The Berlekamp-Welch algorithm can be used to decode any codeword that has at most $5kd$ corruptions, and if at least $2/3$ of the evaluations are correct, then at most $4kd$ evaluations are wrong. Hence Merge$_r$ uses the Berlekamp-Welch algorithm to recover $g_{\boldsymbol{x},\boldsymbol{x}_1,\boldsymbol{x}_2}$, which can be evaluated at 0 to obtain fOV$^k_n(\boldsymbol{x})$.

Thus, Split$_r$ takes $\widetilde{O}(12kd \cdot kmd) = \widetilde{O}(m)$ time to compute all the vectors it outputs. Merge$_r$ takes $\widetilde{O}((12kd)^3)$ time to run Berlekamp-Welch, and $\widetilde{O}(12kd)$ time to evaluate the resulting polynomial at 0. So in all both algorithms take $\widetilde{O}(m)$ time.

By our assumption at the beginning, $\mathcal{G}$ is $(n^c, n^{k-\epsilon}, 7/8)$-batchable on average over the uniform distribution. Together with the above propositions, this satisfies all the requirements in the hypothesis of Lemma 4.4, which now tells us that $\mathcal{F}$ can be computed in the worst-case in time:

$$\tilde{O}(m^{c+1} + m^c \cdot m + 12kd \cdot m^c \cdot m^{k-\epsilon}) = \tilde{O}(m^{c+1} + m^{c+k-\epsilon})$$
$$= \tilde{O}(n^{k(c+1)/(k+c)} + n^{k(k+c-\epsilon)/(k+c)})$$
$$= \tilde{O}(n^{k-\epsilon'})$$

for some $\epsilon' > 0$. But this is what the hypothesis of the theorem says is not possible. So $\mathcal{F}OV^k$ cannot be $(n^c, n^{k-\epsilon}, 7/8)$-batchable on average, and this argument applies for any $c, \epsilon > 0$. $\square$

Proof of Lemma 4.4. Given the hypothesised downward reduction (Split$_d$, Merge$_d$), robust reduction (Split$_r$, Merge$_r$) and batch-evaluation algorithm Batch for $\mathcal{F}$, f_n can be computed as follows (for large enough n) on an input $x \in \mathcal{X}_n$:

- Run Split$_d(x)$ to get $x_1, \ldots, x_{\ell_d(n)} \in \mathcal{X}'_{s_d(n)}$.
- For each $i \in [\ell_d(n)]$, run Split$_r(x_i; r_i)$ to get $x_{i1}, \ldots, x_{i\ell_r(s_d(n))} \in \hat{\mathcal{X}}_{s_d(n)}$.
- For each $j \in [\ell_r(s_d(n))]$, run Batch$(x_{1j}, \ldots, x_{\ell_d(n)j})$ to get the outputs $y_{1j}, \ldots, y_{\ell_d(n)j} \in \hat{\mathcal{Y}}_{s_d(n)}$.
- For each $i \in [\ell_d(n)]$, run Merge$_r(x_i, r_i, y_{i1}, \ldots, y_{i\ell_r(s_d(n))})$ to get $y_i \in \mathcal{Y}'_{s_d(n)}$.
- Run Merge$_d(x, y_1, \ldots, y_{\ell_d(n)})$ to get $y \in \mathcal{Y}_n$, and output y as the alleged $f_n(x)$.

We will prove that with high probability, after the calls to Batch, enough of the y_{ij}'s produced will be equal to the respective $g_{s_d(n)}(x_{ij})$'s to be able to correctly recover all the $f'_{s_d(n)}(x_i)$'s and hence $f_n(x)$.

For each $j \in [\ell_r(s_d(n))]$, define I_j to be the indicator variable that is 1 if Batch$(x_{1j}, \ldots, x_{\ell_d(n)j})$ is correct and 0 otherwise. Note that by the properties of the robust reduction of $\mathcal{F}'$ to $\mathcal{G}$, for a fixed j each of the x_{ij}'s is independently distributed according to $D_{s_d(n)}$ and further, for any two distinct j, j', the tuples (x_{ij}) and $(x_{ij'})$ are independent.

Let $I = \sum_j I_j$ and $m = \ell_r(s_d(n))$. By the aforementioned properties and the correctness of Batch, we have the following:

$$E[I] \geq \frac{7}{8}m$$
$$\mathrm{Var}[I] \leq \frac{7}{64}m$$

Note that as long as Batch is correct on more than a 2/3 fraction of the j's, Merge$_r$ will get all of the y_i's correct, and hence Merge$_d$ will correctly compute $f_n(x)$. The probability that this does not happen is bounded using Chebyshev's inequality as:

$$\Pr\left[I \le \frac{2}{3}m\right] \le \Pr\left[|I - \mathrm{E}[I]| \ge \left(\frac{7}{8} - \frac{2}{3}\right)m\right]$$

$$\le \frac{\mathrm{Var}[I]}{(5m/24)^2}$$

$$\le \frac{63}{25 \cdot m} < \frac{3}{m}$$

As long as $m > 9$, this probability of failure is less than $1/3$, and hence $f_n(x)$ is computed correctly in the worst-case with probability at least $2/3$. If it is the case that $\ell_r(s_d(n)) = m$ happens to be less than 9, then instead of using Merge_r directly in the above algorithm, we would use Merge_r' that runs Merge_r several times so as to get more than 9 samples in total and takes the majority answer from all these runs.

The time taken is $t_d(n)$ for the downward reduction, $t_r(s_d(n))$ for each of the $\ell_d(n)$ robust reductions on instances of size $s_d(n)$, and $\ell_d(n)t_a(s_d(n))$ for each of the $\ell_r(s_d(n))$ calls to Batch on sets of $\ell_d(n) = \ell_a(s_d(n))$ instances, summing up to the total time stated in the lemma. $\qquad\square$

5   Removing Interaction

In this section we show how to remove the interaction in Protocol 1.2 via the Fiat-Shamir heuristic and thus prove security of our non-interactive PoW in the Random Oracle model.

Remark 5.1. Recent papers have constructed hash functions for which provably allow the Fiat-Shamir heuristic to go through [KRR17,CCRR18]. Both of these constructions require a variety of somewhat non-standard sub-exponential security assumptions: [KRR17] uses sub-exponentially secure indistinguishability obfuscation, sub-exponentially secure input-hiding point function obfuscation, and sub-exponentially secure one-way functions; while [CCRR18] needs symmetric encryption schemes with strong guarantees against key recovery attacks (they specifically propose two instantiating assumptions that are variants on the discrete-log assumption and the learning with errors assumption). While for simplicity we present our work in the context of the random oracle model, [KRR17,CCRR18] give evidence that our scheme can be made non-interactive in the plain model.

We also note that our use of a Random Oracle here is quite different from its possible direct use in a Proof of Work similar to those currently used, for instance, in the cryptocurrency blockchains. There, the task is to find a pre-image to H such that its image starts (or ends) with at least a certain number of 0's. In order to make this only moderately hard for PoWs, the security parameter of the chosen instantiation of the Random Oracle (which is typically a hash function like SHA-256) is necessarily not too high. In our case, however, there is no such need for such a task to be feasible, and this security parameter can be set very high, so as to be secure even against attacks that could break the above kind of PoW.

It is worth noting that because of this use of the RO and the soundness properties of the interactive protocol, the resulting proof of work is effectively unique in the sense that it is computationally infeasible to find two accepting proofs. This is markedly different from proof of work described above, where random guessing for the same amount of time is likely to yield an alternate proof.

In what follows, we take H to be a random oracle that outputs an element of $\mathbb{F}_p$, where p is as in Definition 2.5 and n will be clear from context. Informally, as per the Fiat-Shamir heuristic, we will replace all of the verifier's random challenges in the interactive proof (Protocol 1.1) with values output by H so that secure challenges can be gotten without interaction. Using the definitions of the polynomials $q(i_1, \ldots, i_k)$ and $q_{s,\alpha_1,\ldots,\alpha_{s-1}}(x)$ from Sect. 2, the non-interactive proof scheme for $\mathcal{F}\mathsf{OV}^k$ is described as Protocol 1.3.

Non-Interactive Proof for $\mathcal{F}\mathsf{OV}^k$:

The inputs to the protocol are $\boldsymbol{x} = (U_1, \ldots, U_k) \in \mathbb{F}_p^{knd}$ (a valid input to $f\mathsf{OV}_{n,d,p}^k$), and a field element $y \in \mathbb{F}_p$. The polynomials q are defined as in the text.

Prover($\boldsymbol{x}, y$):

- Compute coefficients of q_1. Let $\tau_1 = (q_1)$.
- For s from 1 to $k - 2$:
  - Compute $\alpha_s = H(\boldsymbol{x}, y, \tau_s)$.
  - Compute coefficients of $q_{s+1} = q_{s+1,\alpha_1,\ldots,\alpha_s}$, with respect to $\boldsymbol{x}$.
  - Set $\tau_{s+1} = (\tau_s, \alpha_s, q_{s+1})$.
- Output τ_{k-1}

Verifier($\boldsymbol{x}, y, \tau^*$):

Given $\tau^* = (q_1, \alpha_1, q_2, \ldots, \alpha_{k-2}, q_{k-1})$, do the following:

- Check $\sum_{i_1 \in [n]} q_1(i_1) = y$. If check fails, reject.
- For s from 1 up to $k - 2$:
  - Check that $\alpha_s = H(\boldsymbol{x}, y, q_1, \alpha_1, \ldots, \alpha_{s-1}, q_s)$.
  - Check that $\sum_{i_{s+1} \in [n]} q_{s+1}(i_{s+1}) = q_s(\alpha_s)$. If check fails, reject.
- Pick $\alpha_{k-1} \leftarrow \mathbb{F}_p$.
- Check that $q_{k-1}(\alpha_{k-1}) = \sum_{i_k \in [n]} q_{k,\alpha_1,\ldots,\alpha_{k-1}}(i_k)$. If check fails, reject.

If verifier has yet to reject, accept.

Protocol 1.3: A Non-Interactive Proof for $\mathcal{F}\mathsf{OV}^k$

Overloading the definition, we now consider Protocol 1.2 as our PoW as before except that we now use the *non-interactive* Protocol 1.3 as the the basis of our Solve and Verify algorithms. The following theorem states that this substitution gives us a *non-interactive* PoW in the Random Oracle model.

Theorem 5.2. *For some $k \geq 2$, suppose k-OV takes $n^{k-o(1)}$ time to decide for all but finitely many input lengths for any $d = \omega(\log n)$. Then, Protocol 1.2, when* *using Protocol 1.3 in place of Protocol 1.1, is a* non-interactive (n^k, δ)-*Proof of Work for k-*OV *in the Random Oracle model for any function* $\delta(n) > 1/n^{o(1)}$.

Efficiency and completeness of our now non-interactive Protocol 1.2 are easily seen to follow identically as in the proof of Theorem 2.9 in Sect. 2. Hardness also follow identically to the proof of Theorem 2.9's hardness except that the proof there required the *soundness* of Protocol 1.1, the interactive proof of $\mathcal{F}OV^k$ that was previously used to implement Solve and Verify. To complete the proof of Theorem 5.2, then, we prove the following lemma that Protocol 1.3 is also sound.

Lemma 5.3. *For any* $k \geq 2$, *if Protocol 1.1 is sound as an interactive proof, then Protocol 1.3 is sound as a non-interactive proof system in the Random Oracle model.*

Proof Sketch. Let P be a cheating prover for the non-interactive proof (Protocol 1.3) that breaks soundness with non-negligible probability $\varepsilon(n)$. We will construct a prover, P', that then also breaks soundness in the *interactive* proof (Protocol 1.1) with non-negligible probability.

Suppose P makes at most $m = \text{poly}(n)$ queries to the random oracle, H; call them $\rho_1, \ldots, \rho_m$, and call the respective oracle answers $\beta_1, \ldots, \beta_m$.

For each $s \in [k-2]$, in order for the check on α_s to pass with non-negligible probability, the prover P must have queried the point $(\boldsymbol{x}, y, q_1, \alpha_1, \ldots, q_s)$. Hence, when P is able to make the verifier accept, except with negligible probability, there are $j_1, \ldots, j_{k-2} \in [m]$ such that the query ρ_{j_s} is actually $(\boldsymbol{x}, y, q_1, \alpha_1, \ldots, q_s)$, and β_{j_s} is α_s.

Further, for any $s < s'$, note that α_s is part of the query whose answer is $\alpha_{s'}$. So again, when P is able to make the verifier accept, except with negligible probability, $j_1 < j_2 < \cdots < j_{k-2}$. The interactive prover P' now works as follows:

- Select $(k-1)$ of the m query indices, and guess these to be the values of $j_1 < \cdots < j_{k-1}$.
- Run P until it makes the j_1^{th} query. To all other queries, respond uniformly at random as an actual random oracle would.
- If ρ_{j_1} is not of the form $(\boldsymbol{x}, y, q_1)$, abort. Else, sent q_1 to the verifier.
- Set the response to this query β_{j_1} to be the message α_1 sent by the verifier.
- Resume execution of P until it makes the j_2^{th} query from which q_2 can be obtained, and so on, proceeding in the above manner for each of the $(k-1)$ rounds of the interactive proof.

As the verifier's messages $\alpha_1, \ldots, \alpha_{k-2}$ are chosen completely at random, the oracle that P' is simulating for P is identical to the actual random oracle. So P would still be producing accepting proofs with probability $\varepsilon(n)$. By the earlier arguments, with probability nearly $\varepsilon(n)$, there are $(k-1)$ oracle queries of P that contain all the q_s's that make up the proof that it eventually produces. Whenever this is the case, if P' guesses the positions of these oracle queries correctly, the transcript of the interactive proof that it produces is the same as the proof produced by P, and is hence an accepting transcript.

Hence, when all of the above events happen, P succeeds in fooling the verifier. The probability of this happening is $\Omega(\varepsilon(n)/m^{k-1})$, which is still non-negligible as k is a constant. This contradicts the soundness of the interactive proof, proving our lemma. $\square$

6   Zero-Knowledge Proofs of Work

In this section we show that the algebraic structure of the protocols can easily be exploited with mainstream cryptographic techniques to yield new protocols with desirable properties. In particular, we show that our Proof of Work scheme can be combined with ElGamal encryption and a zero-knowledge proof of discrete logarithm equality to get an non-repudiatable, non-transferable proof of work from the Decisional Diffie-Hellman assumption on Schnorr groups.

It should be noted that while general transformations are known for zero-knowledge protocols, many such transformations involve generic reductions with (relatively) high overhead. In the proof of work regime, we are chiefly concerned with the *exact* complexity of the prover and verifier. Even efficient transformations that go through circuit satisfiability must be adapted to this setting where no efficient deterministic verification circuit is known. That all said, the chief aim of this section is to exhibit the ease with which known cryptographic techniques used in conjunction the algebraic structure of the aforementioned protocols.

For simplicity of presentation, we demonstrate a protocol for $\mathcal{F}OV^2$, however the techniques can easily be adapted to the protocol for general $\mathcal{F}OV^k$.

Preliminaries.   We begin by introducing a notion of honest verifier zero-knowledge scaled down to our setting. As the protocols under consideration have polynomial time provers, they are, in traditional sense, trivially zero-knowledge. However, this is not a meaningful notion of zero-knowledge in this setting, because we are concerned with the exact complexity of the verifier. In order to achieve a meaningful notion of zero-knowledge, we must restrict ourselves to considering simulators of comparable complexity to the verifier (in this case, running in quasi-linear time). Similar notions are found in [Pas03,BDSKM17] and perhaps elsewhere.

Definition 6.1. *An interactive protocol, $\Pi = \langle P, V \rangle$, for a function family, $\mathcal{F} = \{f_n\}$, is T(n)-simulatable, if for any $f_n \in \mathcal{F}$ there exists a simulator, $\mathcal{S}$, such that any x in the domain of f_n the following distributions are computationally indistinguishable,*

$$View_{P,V}(x) \qquad \mathcal{S}(x),$$

where $View_{P,V}(x)$ denotes the distribution interactions between (honest) P and V on input x and $\mathcal{S}$ is randomized algorithm running in time $O(T(n))$.

Given the exposition above it would be meaningful to consider such a definition where we instead simply require the distributions to be indistinguishable with respect to distinguishers running in time $O(T(n))$. However, given that our protocol satisfies the stronger, standard notion of computational indistinguishability, we will stick with that.

Recall that *El Gamal encryption* consists of the following three algorithms for a group G of order p_λ with generator g.

$\mathsf{Gen}(\lambda; y) = (\mathsf{sk} = y, \mathsf{pk} = (g, g^y))$.
$\mathsf{Enc}(m, (a, b); r) = (a^r, mb^r)$.
$\mathsf{Dec}((c, d), y) = dc^{-y}$

El Gamal is a semantically secure cryptosystem (encryptions of different messages are computationally indistinguishable) if the *Decisional Diffie-Hellman assumption (DHH)* holds for the group G. Recall that DDH on G with generator g states that the following two distributions are compuationally indistinguishable:

- (g^a, g^b, g^{ab}) where a, b are chosen uniformly,
- (g^a, g^b, g^c) where a, b, c are chosen uniformly.

Protocol. Let $\mathbb{Z}_p$ be a Schnorr group such of size $p = qm + 1$ such that DDH holds with generator g.

Let (E, D) denote an ElGamal encryption system on G.

In what follows, we will take $R_{U,V}$ (or R^* for the honest prover) to be q (or q_1) as defined in Sect. 2.3

- Challenge is issued as before: $(U, V) \leftarrow \mathbb{Z}_q^{2nd}$.
- Prover generates a secret key $x \leftarrow \mathbb{Z}_{p-1}$, and sends encryptions of the coefficients of the challenge response over the subgroup size q to Verifier with the public key $(g, h = g^x)$:

$$\mathsf{E}(R^*(\cdot); S(\cdot)) = \mathsf{E}(mr_0^*; s_0), \ldots, \mathsf{E}(mr_{nd-1}^*; s_{nd-1})$$
$$= (g^{s_0}, g^{r_0^*} h^{xs_0}), \ldots, (g^{s_{nd-1}}, g^{mr_{nd-1}^*} h^{xs_{nd-1}}).$$

Prover additionally draws $t \leftarrow \mathbb{Z}_{p-1}$ and sends $a_1 = g^t, a_2 = h^t$.
- Verifier draws random $z \leftarrow \mathbb{Z}_q$ and challenge $c \leftarrow \mathbb{Z}_p^*$ and sends to Prover.
- Prover sends $w = t + cS(z)$ to verifier.
- Verifier evaluates $y = f\mathsf{OV}_V(\phi_1(z), \ldots, \phi_d(z))$ to get g^{my}. Then, homomorphically evaluates $\mathsf{E}(R^*; S)$ on z so that $\mathsf{E}(R^*(z); S(z))$ equals

$$\left((g^{s_0})(g^{s_1})^z \cdots (g^{s_{nd-1}})^{z^d}, (g^{r_0^*} h^{s_0})(g^{mr_1^*} h^{s_1})^z \cdots (g^{mr_{nd-1}^*} h^{s_{nd-1}})^{z^d} \right)$$
$$= (u_1, u_2)$$

Then, Verifier accepts if and only if

$$g^w = a_1(u_1)^c \quad \& \quad h^w = a_2(u_2/g^{my})^c.$$

Recall that the success probability of a subquadratic prover (in the non-zero-knowledge case) does not have negligible success probability.

Remark 6.2. Note that the above protocol is public coin. Therefore, we can apply the Fiat-Shamir heuristic, and use a random oracle on partial transcripts to make the protocol non-interactive.

More explicitly, let H be a random oracle. Then:

- Prover computes

$$(g, h),$$
$$\mathsf{E}(R^*; S),$$
$$a_1 = g^t, a_2 = h^t,$$
$$z = H(U, V, g, h, \mathsf{E}(R^*; S), a_1, a_2),$$
$$c = H(U, V, g, h, \mathsf{E}(R^*; S), a_1, a_2, z),$$
$$w = t + cS(z)$$

and sends $(g, h, \mathsf{E}(R^*; S), a_1, a_2, w)$.
- Verifier calls random oracle twice to get

$$z = H(U, V, g, h, \mathsf{E}(R^*; S), a_1, a_2), c = H(U, V, g, h, \mathsf{E}(R^*; S), a_1, a_2, z).$$

Then, the verifier homomorphically evaluates $\mathsf{E}(R^*; S)(z) = (u_1, u_2)$, it then computes the value $y = f\mathsf{OV}_V(\phi_1(z), \ldots, \phi_d(z))$. Finally, accepts if and only if

$$g^w = a_1(u_1)^c \quad \& \quad h^w = a_2(u_2/g^{my})^c.$$

Theorem 6.3. *Suppose* OV *takes* n^2 *time to decide for all but finitely many input lengths for any* $d = \omega(\log n)$ *and the DDH the holds in Schnorr groups, then the above protocol is a* $\tilde{O}(n)$-*simulatable* (n^2, δ)-*interactive Proof of Work scheme for any function* $\delta(n) > 1/n^{o(1)}$.

Proof. Completeness. From before, if $R^* \equiv R_{U,V}$ as is the case for an honest prover, then for any $z \in \mathbb{Z}_q$ we have $R^*(z) = R_{U,V}(z) = f\mathsf{OV}_V(\phi_1(z), \ldots, \phi_d(z))$. Moreover

$$g^w = g^{t+cS(z)} = g^t(g^{S(z)})^c = a_1\left((g^{s_0})(g^{s_1})^z \cdots (g^{s_{nd-1}})^{z^d}\right)^c,$$

and

$$h^w = h^{t+cS(z)}$$
$$= h^t(g^0 h^{S(z)})^c$$
$$= a_2\left((g^{r_0^*}h^{s_0})(g^{mr_1^*}h^{s_1})^z \cdots (g^{mr_{nd-1}^*}h^{s_{nd-1}})^{z^d} g^{-f\mathsf{OV}_V(\phi_1(z),\ldots,\phi_d(z))}\right)^c.$$

Hardness. Suppose a cheating prover runs in subquadratic time, then by the hardness of Protocol 1.2 with high probability $R^* \not\equiv R_{U,V}$, and so for random z, $R^*(z) \neq f\mathsf{OV}_V(\phi_1(z), \ldots, \phi_d(z))$ with overwhelming probability. Suppose this is the case in what follows, namely: $R^*(z) = y^* \neq y = f\mathsf{OV}_V(\phi_1(z), \ldots, \phi_d(z))$. In particular,

$$\log_g u_1 \neq \log_h u_2/g^{f\mathsf{OV}_V(\phi_1(z),\ldots,\phi_d(z))}.$$

Note that $u_1, u_2/g^{f\text{OV}_V(\phi_1(z),...,\phi_d(z))}$ can be calculated from the Prover's first message.

As is standard, we will fix the prover's first message and (assuming $y \neq y^*$) rewind any two accepting transcripts with distinct challenges to show that $\log_g u_1 = \log_h u_2/g^y$. Fix a_1, a_2 as above and let $(c, w), (c', w')$ be the two transcripts. Recall that if a transcript is accepted, $g^w = a_1 u_1^c$ and $h^w = a_2(u_2/g^y)^c$. Then,

$$g^{w-w'} = u_1^{c-c'} \Rightarrow \log_g u_1 = \frac{w - w'}{c - c'} = \log_h u_2/g^y \Leftarrow h^{w-w'} = (u_2/g^y)^{c-c'}.$$

Therefore, because $u_1 \neq u_2/g^y$ there can be at most one c for which a Prover can convince the verifier. Such a c is chosen with negligible probability.

$\tilde{O}(nd)$-*simulation.* Given the verifier's challenge z, c, (which can simply be sampled uniformly, as above) we can efficiently simulate the transcript with respect to an honest prover as follows:

- Draw public key (g, h).
- Compute the ElGamal Encryption $\mathsf{E}_{g,h}(R'; S)$ where R' is the polynomial with constant term $f\text{OV}_V(\phi_1(z), \ldots, \phi_d(z))$ and zeros elsewhere.
- Draw random w.
- Compute $a_1 = \frac{g^w}{g^{cS(z)}}$ and $a_w = \frac{h^w}{h^{cS(z)}}$.
- Output $((g, h), a_1, a_2, z, c, w)$.

Notice that do to the semantic security of ElGamal, the transcript output is computationally indistinguishable from that of an honest Prover. Moreover, the simulator runs in $\tilde{O}(nd)$ time, the time to compute R', encrypt, evaluate S and exponentiate. Thus, the protocol is $\tilde{O}(nd)$-simulatable.

Efficiency. The honest prover runs in time $\tilde{O}(n^2)$, because the nd encryptions can be performed in time polylog(n) each. The verifier takes $\tilde{O}(nd)$ time as well. Note that the homomorphic evaluation requires $O(d \log z^d) = O(d^2 \log z) = $ polylog(d) exponentiations and $d = $ polylog(n) multiplications. $\qquad\square$

Acknowledgements. We are grateful to Oded Goldreich and Guy Rothblum for clarifying definitions of direct sum theorems, and for the suggestion of using interaction to increase the gap between solution and verification in our PoWs. We would also like to thank Tal Moran and Vinod Vaikuntanathan for several useful discussions. We also thank the anonymous reviewers for comments and references.

The bulk of this work was performed while the authors were at IDC Herzliya's FACT center and supported by NSF-BSF Cyber Security and Privacy grant #2014/632, ISF grant #1255/12, and by the ERC under the EU's Seventh Framework Programme (FP/2007-2013) ERC Grant Agreement #07952. Marshall Ball is supported in part by the Defense Advanced Research Project Agency (DARPA) and Army Research Office (ARO) under Contract #W911NF-15-C-0236, NSF grants #CNS-1445424 and #CCF-1423306, the Leona M. & Harry B. Helmsley Charitable Trust, ISF grant no. 1790/13, and the Check Point Institute for Information Security. Alon Rosen is also supported by ISF grant no. 1399/17. Manuel Sabin is also supported by the National Science Foundation Graduate Research Fellowship under Grant #DGE-1106400. Prashant Nalini Vasudevan is also supported by the IBM Thomas J. Watson Research Center (Agreement #4915012803), by NSF Grants CNS-1350619 and CNS-1414119, and by the Defense Advanced Research Projects Agency (DARPA) and the U.S. Army Research Office under contracts W911NF-15-C-0226 and W911NF-15-C-0236.

References

BDSKM17. Ball, M., Dachman-Soled, D., Kulkarni, M., Malkin, T.: Non-malleable codes from average-case hardness: AC0, decision trees, and streaming space-bounded tampering. Cryptology ePrint Archive, Report 2017/1061 (2017). https://eprint.iacr.org/2017/1061

BGJ+16. Bitansky, N., Goldwasser, S., Jain, A., Paneth, O., Vaikuntanathan, V., Waters, B.: Time-lock puzzles from randomized encodings. In: Sudan, M. (ed.) Proceedings of the 2016 ACM Conference on Innovations in Theoretical Computer Science, Cambridge, MA, USA, 14–16 January 2016, pp. 345–356. ACM (2016)

BK16a. Biryukov, A., Khovratovich, D.: Egalitarian computing. In: Holz, T., Savage, S. (eds.) 25th USENIX Security Symposium, USENIX Security 16, Austin, TX, USA, 10–12 August 2016, pp. 315–326. USENIX Association (2016)

BK16b. Björklund, A., Kaski, P.: How proofs are prepared at Camelot. In: Proceedings of the 2016 ACM Symposium on Principles of Distributed Computing, pp. 391–400. ACM (2016)

BRSV17a. Ball, M., Rosen, A., Sabin, M., Vasudevan, P.N.: Average-case fine-grained hardness. In: Hatami, H., McKenzie, P., King, V. (eds.) Proceedings of the 49th Annual ACM SIGACT Symposium on Theory of Computing, STOC 2017, Montreal, QC, Canada, 19–23 June 2017, pp. 483–496. ACM (2017)

BRSV17b. Ball, M., Rosen, A., Sabin, M., Vasudevan, P.N.: Proofs of useful work. IACR Cryptology ePrint Archive 2017:203 (2017)

CCRR18. Canetti, R., Chen, Y., Reyzin, L., Rothblum, R.D.: Fiat-Shamir and correlation intractability from strong KDM-secure encryption. In: Nielsen, J.B., Rijmen, V. (eds.) EUROCRYPT 2018. LNCS, vol. 10820, pp. 91–122. Springer, Cham (2018). https://doi.org/10.1007/978-3-319-78381-9_4

CPS99. Cai, J., Pavan, A., Sivakumar, D.: On the hardness of permanent. In: Meinel, C., Tison, S. (eds.) STACS 1999. LNCS, vol. 1563, pp. 90–99. Springer, Heidelberg (1999). https://doi.org/10.1007/3-540-49116-3_8

DN92. Dwork, C., Naor, M.: Pricing via processing or combatting junk mail. In: Brickell, E.F. (ed.) CRYPTO 1992. LNCS, vol. 740, pp. 139–147. Springer, Heidelberg (1993). https://doi.org/10.1007/3-540-48071-4_10

FF93. Feigenbaum, J., Fortnow, L.: Random-self-reducibility of complete sets. SIAM J. Comput. 22(5), 994–1005 (1993)

Fid72. Fiduccia, C.M.: Polynomial evaluation via the division algorithm: the fast Fourier transform revisited. In: Fischer, P.C., Zeiger, H.P., Ullman, J.D., Rosenberg, A.L. (eds.) Proceedings of the 4th Annual ACM Symposium on Theory of Computing, 1–3 May 1972, Denver, Colorado, USA, pp. 88–93. ACM (1972)

GI16. Gao, J., Impagliazzo, R.: Orthogonal vectors is hard for first-order properties on sparse graphs. In: Electronic Colloquium on Computational Complexity (ECCC), vol. 23, p. 53 (2016)

GR17. Goldreich, O., Rothblum, G.: Simple doubly-efficient interactive proof systems for locally-characterizable sets. Electronic Colloquium on Computational Complexity Report TR17-018, February 2017

GR18. Goldreich, O., Rothblum, G.N.: Counting t-cliques: worst-case to average-case reductions and direct interactive proof systems. In: Electronic Colloquium on Computational Complexity (ECCC), vol. 25, p. 46 (2018)

Hor72. Horowitz, E.: A fast method for interpolation using preconditioning. Inf. Process. Lett. **1**(4), 157–163 (1972)

JJ99. Jakobsson, M., Juels, A.: Proofs of work and bread pudding protocols (extended abstract). In: Preneel, B. (ed.) Secure Information Networks. ITIFIP, vol. 23, pp. 258–272. Springer, Boston (1999). https://doi.org/10.1007/978-0-387-35568-9_18

KRR17. Kalai, Y.T., Rothblum, G.N., Rothblum, R.D.: From obfuscation to the security of Fiat-Shamir for proofs. In: Katz, J., Shacham, H. (eds.) CRYPTO 2017, Part II. LNCS, vol. 10402, pp. 224–251. Springer, Cham (2017). https://doi.org/10.1007/978-3-319-63715-0_8

Pas03. Pass, R.: Simulation in quasi-polynomial time, and its application to protocol composition. In: Biham, E. (ed.) EUROCRYPT 2003. LNCS, vol. 2656, pp. 160–176. Springer, Heidelberg (2003). https://doi.org/10.1007/3-540-39200-9_10

RR00. Roth, R.M., Ruckenstein, G.: Efficient decoding of Reed-Solomon codes beyond half the minimum distance. IEEE Trans. Inf. Theory **46**(1), 246–257 (2000)

She12. Sherstov, A.A.: Strong direct product theorems for quantum communication and query complexity. SIAM J. Comput. **41**(5), 1122–1165 (2012)

SKR+11. Stebila, D., Kuppusamy, L., Rangasamy, J., Boyd, C., Gonzalez Nieto, J.: Stronger difficulty notions for client puzzles and denial-of-service-resistant protocols. In: Kiayias, A. (ed.) CT-RSA 2011. LNCS, vol. 6558, pp. 284–301. Springer, Heidelberg (2011). https://doi.org/10.1007/978-3-642-19074-2_19

Wil05. Williams, R.: A new algorithm for optimal 2-constraint satisfaction and its implications. Theor. Comput. Sci. **348**(2–3), 357–365 (2005)

Wil15. Williams, V.V.: Hardness of easy problems: basing hardness on popular conjectures such as the strong exponential time hypothesis. In: Proceedings of International Symposium on Parameterized and Exact Computation, pp. 16–28 (2015)

Wil16. Williams, R.R.: Strong ETH breaks with Merlin and Arthur: short non-interactive proofs of batch evaluation. In: 31st Conference on Computational Complexity, CCC 2016, 29 May to 1 June 2016, Tokyo, Japan, pp. 2:1–2:17 (2016)

A   A Stronger Direct Sum Theorem for $\mathcal{F}OV$

In this section, we prove a stronger direct sum theorem (and, thus, non-batchable evaluation) for $\mathcal{F}OV^k$. That is, we prove Theorem 2.13.

In particular, it is sufficient to define a notion of batchability for parametrized families of functions with a monotonicity constraint. In our case, monotonicity will essentially say "adding more vectors of the same dimension and field size does not make the problem easier." This is a natural property of most algorithms. Namely, it is the case if for any fixed d, p, $\mathcal{F}OV_{n,d,p}^k$ is (n, t, δ) − batchable.

Instead, we generalize batchability in a parametrized fashion for $\mathcal{F}OV_{n,d,p}^k$.

Definition A.1. *A parametrized class, $\mathcal{F}_\rho$, is not (ℓ, t, δ)-batchable on average over $\mathcal{D}_\rho$, a parametrized family of distributions if, for any fixed parameter ρ and algorithm* Batch_ρ *that runs in time $\ell(\rho)t(\rho)$ when it is given as input $\ell(\rho)$ independent samples from D_ρ, the following is true for all large enough n:*

$$\Pr_{x_i \leftarrow D_\rho} \left[\mathsf{Batch}(x_1, \ldots, x_{\ell(\rho)}) = (f_\rho(x_1), \ldots, f_\rho(x_{\ell(\rho)})) \right] < \delta(\rho).$$

Remark A.2. We use a more generic parameterization of $\mathcal{F}_\rho$ by ρ rather than just n since we need the batch evaluation procedure to have the property that it should still run quickly as n shrinks, as we use downward self-reducibility of $\mathcal{F}OV_{n,d,p}^k$, even when p and d remain the same.

We now show how a generalization of the list decoding reduction of [BRSV17a] yields strong batch evaluation bounds. Before we begin, we will present a few Lemmas from the literature to make certain bounds explicit.

First, we present an inclusion-exclusion bound from [CPS99] on the polynomials consistent with a fraction of m input-output pairs, $(x_1, y_1), \ldots, (x_m, y_m)$. We include a laconic proof here with the given notation for convenience.

Lemma A.3. ([CPS99]). *Let q be a polynomial over $\mathbb{F}_p$, and define $\mathrm{Graph}(q) := \{(i, q(i)) \mid i \in [p]\}$. Let $c > 2$, $\delta/2 \in (0, 1)$, and $m \le p$ such that $m > \frac{c^2(d-1)}{\delta^2(c-2)}$ for some d. Finally, let $I \subseteq [p]$ such that $|I| = m$. Then, for any set $S = \{(i, y_i) \mid i \in I\}$, there are less than $\lceil c/\delta \rceil$ polynomials q of degree at most d that satisfy $|\mathrm{Graph}(q) \cap S| \ge m\delta/2$.*

Corollary A.4. *Let S be as in Lemma A.3 with $I = \{m+1, \ldots, p\}$, for any $m < p$. Then for $m > 9d/\delta^2$, there are at most $3/\delta$ polynomials, q, of degree at most d such that $|\mathrm{Graph}(q) \cap S| \ge m\delta/2$.*

Proof. Reproduced from [CPS99] for convenience; see original for exposition.

Suppose there exist at least $\lceil c/\delta \rceil$ such polynomials. Consider a subset of exactly $N = \lceil c/\delta \rceil$ such polynomials, $\mathcal{F}$. Define $\mathsf{S}_f := \{(i, f(i)) \in \mathrm{Graph}(f) \cap S\}$, for each $f \in \mathcal{F}$.

$$m \geq \left| \bigcup_{f \in \mathcal{F}} S_f \right| \geq \sum_{f \in \mathcal{F}} |S_f| - \sum_{f, f' \in \mathcal{F}: f \neq f'} |S_f \cap S_{f'}|$$

$$\geq N \frac{m\delta}{2} - \frac{N(N-1)(d-1)}{2} > \frac{N}{2} \left(m\delta - \frac{c(d-1)}{\delta} \right)$$

$$\geq \frac{c}{2\delta} \left(m\delta - \frac{c(d-1)}{\delta} \right) = \frac{cm}{2} - \frac{c^2(d-1)}{2\delta^2}$$

$$= m + \frac{1}{2} \left((c-2)m - \frac{c^2(d-1)}{\delta^2} \right) > m.$$

$\square$

Now, we give a theorem based on an efficient list-decoding algorithm, related to Sudan's, from Roth and Ruckenstein [RR00].

Lemma A.5. ([RR00]). *List decoding for $[n, k]$ Reed-Solomon (RS) codes over $\mathbb{F}_p$ given a code word with almost $n - \sqrt{2kn}$ errors (for $k > 5$), can be performed in*

$$O\left(n^{3/2} k^{-1/2} \log^2 n + (n-k)^2 \sqrt{n/k} + (\sqrt{nk} + \log q) n \log^2(n/k) \right)$$

operations over $\mathbb{F}_q$.

Plugging in specific parameters and using efficient list decoding, we get the following corollary which will be useful below.

Corollary A.6. *For parameters $n \in \mathbb{N}$ and $\delta \in (0, 1)$, list decoding for $[m, k]$ RS over $\mathbb{F}_p$ where $m = \Theta(d \log n / \delta^2)$, $k = \Theta(d)$, $p = O(n^2)$, and $d = \Omega(\log n)$ can be performed in time*

$$O\left(\frac{d^2 \log^{5/2} n \, \text{Arith}(n)}{\delta^5} \right),$$

where $\text{Arith}(n)$ is a time bound on arithmetic operations over prime fields size $O(n)$.

Theorem A.7. *For some $k \geq 2$, suppose k-OV takes $n^{k-o(1)}$ time to decide for all but finitely many input lengths for any $d = \omega(\log n)$. Then, for any positive constants $c, \epsilon > 0$ and $0 < \delta < \varepsilon/2$, $\mathcal{F}OV^k$ is not*

$$(n^c \text{poly}(d, \log(p)), n^{k-\epsilon} \text{poly}(d, \log(p)), n^{-\delta} \text{poly}(d, \log(p)))$$

-batchable on average over the uniform distribution over its inputs.

Proof. Let $k = 2c' + c$ and $p > n^k$. Suppose for the sake of contradiction that $\mathcal{F}OV_{n,d,p}$ is $(n^c \text{poly}(d, \log(p)), n^{2c'+c-\epsilon} \text{poly}(d, \log(p)), n^{-c'} \text{poly}(d, \log(p)))$-batchable on average over the uniform distribution.

Let $m = n^{k/(k+c)}$, as before. By Proposition 4.5, k-OV with vectors of dimension $d = (\frac{k}{k+c}))^2 \log^2 n$ is (m, m^c)-downward reducible to k-OV with vectors of dimension $\log^2(n)$, in time $\tilde{O}(m^{c+1})$.

For each $j \in [m^c]$ $X_j = (U^{j1}, \ldots, U^{jk}) \in \{0,1\}^{kmd}$ is the instance of boolean-valued orthogonal vectors from the above reduction. Now, consider splitting these lists in half, $U^{ji} = (U_0^{ji}, U_1^{ji})$ $(i \in [k])$, such that $(U_{a_1}^{j1}, \ldots, U_{a_k}^{jk}) \in \{0,1\}^{kmd/2}$ for $a \in \{0,1\}^k$. Interpret a as binary number in $\{0, \ldots, 2^k - 1\}$. Then, define the following 2^k sub-problems:

$$A^a = ((U_{a_1}^{j1}, \ldots, U_{a_k}^{jk})), \forall a \in \{0, \ldots, 2^k - 1\}$$

Notice that given solutions to $f\mathsf{OV}_d^k$ on $\{A^a\}_{a \in \{0,1\}^k}$ we can trivially construct a solution to OV_d^k on X_j.

Now, draw random $B_j, C_j \in \mathbb{F}_p^{kmd/2}$ and consider the following degree 2^k polynomial in x:

$$D_j(x) = \sum_{i=1}^{2^k} \delta_i(x) A^{i-1} + (B_j + xC_j) \prod_{i=1}^{2^k} (x - i),$$

where δ_i is the unique degree $2^k - 1$ polynomial over $\mathbb{F}_p$ that takes value 1 at $i \in [2^k]$ and 0 on all other values in $[2^k]$. Notice that $D_j(i) = A^{i-1}$ for $i \in [2^k]$.

Let $r > 2^{k+1}d/\delta^2 \log m$. $D_j(2^k + 1), D_j(6), \ldots, D_j(r + 2^k)$. By the properties of Batch and because the $D_j(\cdot)$'s are independent, $D_1(i), \ldots, D_{m^c}(i)$ are independent for any fixed i. Thus,

$$\mathsf{Batch}(D_1(i), \ldots, D_{m^c}(i)) = f\mathsf{OV}^k(D_1(i)), \ldots, f\mathsf{OV}^k(D_{m^c}(i))$$

for $\delta r/2$ i's with probability at least $1 - \frac{4}{\delta r} = 1 - 1/\mathrm{polylog}(m)$, by Chebyshev.

Now, because $\delta r/2 > \sqrt{16dr}$, we can run the list decoding algorithm of Roth and Ruckenstein, [RR00], to get a list of all polynomials with degree $\leq 2^{k+1}d$ that agree with at least $\delta r/2$ of the values. By Corollary A.4, there are at most $L = 3/\delta$ such polynomials.

By a counting argument, there can be at most $2^k d\binom{L}{2} = O(dL^2)$ points in $\mathbb{F}_p$ on which any two of the L polynomials agree. Because $p > n^k > 2^k d\binom{L}{2}$, we can find such a point, ℓ, by brute-force in $O(L \cdot dL^2 \log^3(dL^2) \log p)$ time, via batch *univariate* evaluation [Fid72]. Now, to identify the correct polynomials $f\mathsf{OV}^k(D_j(\cdot))$, one only needs to determine the value $f\mathsf{OV}^k(D_j(\ell))$. To do so, we can recursively apply the above reduction to all the $D_j(\ell)$s until the number of vectors, m, is constant and $f\mathsf{OV}^k$ can be evaluated in time $O(d \log p)$.

Because each recursive iteration cuts m in half, the depth of recursion is $\log(m)$. Additionally, because each iteration has error probability $< 4/(\delta r)$, taking a union bound over the $\log(m)$ recursive steps yields an error probability that is $\varepsilon < 4 \log m/(\delta r)$.

We can find the prime p via $O(\log m)$ random guesses in $\{m^k + 1, \ldots, 2m^k\}$ with overwhelming probability. By Corollary A.6, taking $r = 8d \log m/\delta^2$, Roth and Ruckenstein's algorithm takes time $O(d^2/\delta^5 \log^{5/2} m \operatorname{Arith}(m^k))$ in each recursive call. The brute force procedure takes time $O(d/\delta^3 \log^3(d/\delta^2) \log m)$, which is dominated by list decoding time. Reconstruction takes time $O(\log m)$ in each round, and is also dominated. Thus the total run time is

$$T = O(m^c(m^{k-\varepsilon}d \log^2 m/\delta^2 + d^2/\delta^5 \log^{7/2} m \operatorname{Arith}(m^k))),$$

with error probability $\varepsilon < 4 \log m\delta/d$.   □

Author Index